FINELY DISPERSED PARTICLES

Micro-, Nano-, and Atto-Engineering

SURFACTANT SCIENCE SERIES

FINELY DISPERSED PARTICLES

PARTICLES

Micro-, Nano-, and Atto-Engineering

Edited by

Aleksandar M. Spasic
Institute for Technology of Nuclear and Other Mineral Raw Materials
Belgrade, Serbia, Serbia & Montenegro

Jyh-Ping Hsu
National Taiwan University
Taipei, Taiwan

CRC Press
Taylor & Francis Group
Boca Raton London New York

CRC Press is an imprint of the
Taylor & Francis Group, an **informa** business
A TAYLOR & FRANCIS BOOK

CRC Press
Taylor & Francis Group
6000 Broken Sound Parkway NW, Suite 300
Boca Raton, FL 33487-2742

First issued in paperback 2019

ISBN-13: 978-1-57444-463-6 (hbk)
ISBN-13: 978-0-367-39197-3 (pbk)

Library of Congress Card Number 2005048543

Library of Congress Cataloging-in-Publication Data

Finely dispersed particles : micro-, nano-, and atto-engineering / [edited by] Aleksander M.
  Spasic & Jyh-Ping Hsu.
      p. cm. – (Surfucant science series ; v. 130)
  Includes bibliographical references and index.
  ISBN 1-57444-463-8
  1, Colloids – electric properties 2. Colloids – Transport properties. 3. Colloids – Industrial Applications.
  4. Nanoscience. 5. Transport Theory. I. Spasic, Aleksander M. II. Hsu, Jyh-Ping, 1955- III. Series

QD549.F395 2005
620'.5—dc22   2005048543

Visit the Taylor & Francis Web site at
http://www.taylorandfrancis.com

and the CRC Press Web site at
http://www.crcpress.com

Preface

This book describes recent developments in basic and applied science and engineering of finely dispersed particles and related systems. Written by a team of outstanding scientists, this book takes an interdisciplinary approach to the elucidation of the heat, mass, and momentum transfer phenomena as well as the electron transfer phenomenon, at well-characterized interfaces. The considered scales are milli-, micro-, nano-, and atto-, using both coherence and decoherence theoretical approaches. Milli- and microscales may cover more or less classical chemical engineering insight, while nano- and attoscales focus on modern molecular and atomic engineering. In this context, "atomic engineering" recalls the ancient idea of interplay of particles that are small, indivisible, and integer (Greek "ατομοζ"). In the recent scientific literature, terms such as nanoscience and nanotechnology, functional artificial nanoarchitectures, nanosystems and molecular machinery, once considered merely futuristic, have become focuses of attention. The aim of this book is to provide the readers with recent concepts in the physics and chemistry of well-studied interfaces of rigid and deformable particles in homo- and hetero-aggregate dispersed systems. As many such systems are non-Newtonian, apart from classical momentum, heat, and mass transfer phenomena, the electron transfer phenomenon is also introduced into their description. Examples of such systems are: emulsions, dispersoids, suspensions, nanopowders, foams, fluosols, polymer membranes, biocolloids, and plasmas. Thus, the central themes of this book are the hydrodynamic, electrodynamic, and thermodynamic instabilities that occur at interfaces and the rheological properties of the interfacial layers responsible for the existence of droplets, particles, and droplet–particle–film structures in finely dispersed systems.

Part I, Introduction, written by Spasic, Mitrovic, and Krstic, gives a brief overview of the finely dispersed systems through their classification considering surface and line continua and point discontinua, states of aggregation, homo and hetero, and their shape, rigid or deformable.

In Part II, General, several overviews are presented, beginning with Ohshima's chapter on electrokinetic behavior of charged particles and droplets, then Delgado and González-Caballero present a chapter on electrokinetic phenomena in suspensions, followed by Schramm and Stasiuk's overview of emulsions, and finally Saboni and Alexandrova close this section with heat and mass transfer in finely dispersed systems.

Part II continues with a section on various approaches and transitions. Chapter 6 covers polymer networks and transitions from nano- to macroscale by Plavsic. The following chapter is on the atomic scale imaging of oscillation and chemical waves at catalytic surface reactions by Elokhin and Gorodetskii. Then next chapter relates the characterization of catalysts by means of an oscillatory reaction written by Kolar-Anić, Anić, and Čupić. Then Dugić, Raković, and Plavsic address polymer conformational stability and transitions based on a quantum decoherence theory approach. Chapter 10 of this section, by Jarić and Kuzmanović, presents a perspective of the physics of interfaces from a standpoint of continuum physics.

Finally, Part II ends with a section on tools. The first chapter, written by Petkovska, discusses nonlinear dynamics methods for estimation of equilibrium and kinetics in heterogeneous solid–fluid systems. Then Oldshue discusses current principles of mixing related to the scale up and scale down. This section ends with Jovanić's chapter on quantification of visual information.

Part III deals with homo-aggregate finely dispersed systems and contains chapters about emulsions, dispersoids, and liquid–liquid dispersions. Oldshue presents a brief chapter on non-Newtonian aspects of emulsification. The following chapter by Spasic, Lazarevic, and Krstic discusses a new theory of electroviscoelasticity using different mathematical tools. Then, a review of experimental results on the production of mono-dispersed emulsions using Shirasu membranes is presented by Vladisavljevic, Shimizu, Nakashima, Schubert, and Nakajima. The *Dispersoids* section contains a single chapter by Zdujić who gives a short account of the main aspects of the mechanical treatment of inorganic solids. Part III concludes with a chapter by Bart on *Liquid–Liquid Dispersions* introducing reactive extraction in electric fields.

Part IV covers hetero-aggregate finely dispersed systems and includes four chapters. The first section *Foams* contains a chapter written by Creux, Lachaise, and Graciaa on gas bubbles within electric fields. In the chapter on section *Fluosols*, Jokanovic presents nano-designing of structures and substructures in spray pyrolysis processes. Further on, Alexandrova, Amang, Garcia, Rollet, and Saboni address transfer phenomena through polymer membranes. The chapter on *Multiphase Dispersed Systems*, written by Duduković and Nikačević, is concerned with gas-flowing solids-fixed bed contactors. The following chapter discusses reaction and capillary condensation in dispersed porous particles by Ostrovskii and Wood. This section ends with a chapter by Skala and Orlovic on particle production using supercritical fluids.

The book closes with Part V "Hetero-Aggregate Finely Dispersed Systems of Biological Interest" and contains eight chapters under the one section head of *Biocolloids*. Kuo and Hsu begin this section by discussing the effects of electrical field on the behavior of biological cells. Then Dzwinel, Boryczko, and Yuen present methods, algorithms, and results of modeling meso-scopic fluids with discrete particles. That is followed by a chapter discussing nonlinear dynamics of a DNA chain presented by Zdravković. The next chapter, written by Partch, Powell, Lee, Varshney, Shah, Baney, Lee, Dennis, Morey, and Flint, discusses surface modification of dispersed phases designed for *in vivo* removal of overdosed toxins. Following on, Pasqualini and López present their chapter on carbon nanocapsules and their nuclear application. In chapter 30, Markvicheva presents methods of bioencapsulation in polymer micro- and nanocarriers and their application in biomedical fields. The penultimate chapter by Bugarski, Obradovic, Nedovic, and Goosen describes a method of electrostatic droplet generation for cell immobiliz-ation. The final chapter of the book, written by Mojovic and Jovanovic, is dedicated to a micro-biosensor based on immobilized cells.

The intended audience of this book includes: chemical engineers — researchers in fundamen-tals of finely dispersed particles — separation, sorption, membrane processes, nanoscience and nanotechnology; physical chemists — researchers in colloid, biocolloid and interface science; theoretical and applied mechanicians — rheologists; biologists and medicine researchers — hematology, genetics and electroneurophysiology; researchers in food, pharmaceutical, petro-chemical, and metallurgical science.

Applications and implications of the material presented in the book are supposed to contribute to the advanced fundamentals of interfacial and colloidal phenomena. Related subject examples are:

- Entrainment problems in solvent extraction
- Colloid and interface science
- Chemical and biochemical sensors
- Electroanalytical methods
- Biology and biomedicine (hematology, genetics, electroneurophysiology)
- Interface surface, line, point and overall barriers-symmetries (surface — bilipid membrane cells, free bubbles of surfactants, Langmuir Blodgett films; line — genes, liquid crystals, microtubules; point — fullerenes, micro-emulsions; overall — dry foams, polymer elastic and rigid foams)

Editors

Aleksandar M. Spasic is a research fellow at Institute for Technology of Nuclear and Other Mineral Raw Materials, Department of Chemical Engineering, Belgrade, Serbia. After he received an IAEA fellowship realized in the Laboratory of Ultra-Refractory Materials, CNRS, Odeillo Font-Romeu, France, his research activities were related to the finely dispersed systems and, in particular, to the electroviscoelastic phenomena at rigid and deformable liquid–liquid interfaces.

Jyh-Ping Hsu is the dean of the College of Engineering, National Ilan University (on leave from the Department of Chemical Engineering, National Taiwan University). Among his research interests are flocculation, adsorption, and electrokinetic phenomena.

Acknowledgment

We would like to thank Professor Arthur T. Hubbard for suggestions and supporting us as the editors of this volume, and also to thank the contributors for their efforts in writing their chapters. Finally, we are pleased to acknowledge the efficiency and care of the people in charge of this project, at Marcel Dekker, Inc., and last but not least, to thank the executive staff of CRC Press and Taylor & Francis.

Aleksandar M. Spasic
Institute for Technology of Nuclear and
Other Mineral Raw Materials

Jyh-Ping Hsu
National Taiwan University

Contributors

Silvia Alexandrova
Department of Chemical Engineering
University of Caen
Caen, France

Dieudonné N. Amang
Department of Chemical Engineering
University of Caen
Caen, France

Slobodan Anić
Department of Physical Chemistry
University of Belgrade
Belgrade, Serbia and Montenegro

R. Baney
Department of Materials Science and
    Engineering
University of Florida
Gainesville, Florida, USA

Hans-Jörg Bart
Department of Mechanical and
    Process Engineering
Institute of Chemical Engineering
Technical University of
    Kaiserslautern
Kaiserslautern, Germany

Krzysztof Boryczko
Institute of Computer Science
AGH University of Science and
    Technology
Kraków, Poland

Branko M. Bugarski
Department of Chemical Engineering
University of Belgrade,
Belgrade, Serbia and Montenegro

Patrice Creux
Laboratory for Complex Fluids
University of Pau
Pau, France

Željko Čupić
Institute of Chemistry,
Technology and Metallurgy
Belgrade, Serbia and Montenegro

Ángel V. Delgado
Department of Applied Physics
University of Granada
Granada, Spain

D. Dennis
Department of Anesthesiology,
    Pharmacology and Experimental
    Therapeutics
University of Florida
Gainesville, Florida, USA

Aleksandar P. Duduković
Department of Chemical
    Engineering
University of Belgrade
Belgrade, Serbia and Montenegro

Miroljub Dugić
Department of Physics
University of Kragujevac
Kragujevac, Serbia and Montenegro

Witold Dzwinel
Institute of Computer Science
AGH University of Science and
    Technology
Kraków, Poland

Vladimir I. Elokhin
Boreskov Institute of Catalysis
Novosibirsk, Russia

J. Flint
Department of Anesthesiology,
    Pharmacology and Experimental
    Therapeutics
University of Florida
Gainesville, Florida, USA

François Garcia
Department of Chemical Engineering
University of Caen
Caen, France

F. González-Caballero
Department of Applied Physics
University of Granada
Granada, Spain

Mattheus F. A. Goosen
Department of Chemical Engineering
University of Puerto Rico
San Juan, Puerto Rico

Vladimir V. Gorodetskii
Boreskov Institute of Catalysis
Novosibirsk, Russia

Alain Graciaa
Laboratory for Complex Fluids
University of Pau
Pau, France

Jyh-Ping Hsu
Department of Chemical Engineering
National Taiwan University
Taipei, Taiwan

Jovo P. Jarić
Department of Mathematics
University of Belgrade
Belgrade, Serbia and Montenegro

Vukoman Jokanovic
Institute of Technical Sciences
Serbian Academy of Sciences and Arts
Belgrade, Serbia and Montenegro

Predrag B. Jovanić
Department of Chemical
  Engineering
Institute for Technology for Nuclear and
  Other Mineral Raw Materials
Belgrade, Serbia and Montenegro

Goran N. Jovanovic
Department of Chemical Engineering
Oregon State University
Corvallis, Oregon, USA

Ljiljana Kolar-Anić
Department of Physical Chemistry
University of Belgrade
Belgrade, Serbia and Montenegro

Dimitrije N. Krstic
Department of Metallurgical Engineering
University of Belgrade
Belgrade, Serbia and Montenegro

Yung-Chih Kuo
Department of Chemical Engineering
National Chung Cheng University
Chia-Yi, Taiwan

Dragoslav S. Kuzmanović
Department of Transport and Traffic
  Engineering
University of Belgrade
Belgrade, Serbia and Montenegro

Jean Lachaise
Laboratory for Complex Fluids
University of Pau
Pau, France

Mihailo P. Lazarevic
Department of Mechanical Engineering
University of Belgrade
Belgrade, Serbia and Montenegro

D-W. Lee
Department of Materials Science and
  Engineering
University of Florida
Gainesville, Florida, USA

Y-H. Lee
Department of Chemistry
Kyungwon University
Sungnam City, Korea

Marisol López
Department of Nuclear Fuels
National Commission of Atomic Energy
Buenos Aires, Argentina

Elena Markvicheva
Shemyakin and Ovchinnikov Institute of
  Bioorganic Chemistry
Moscow, Russia

Milan Mitrovic
Department of Chemical Engineering
University of Belgrade
Belgrade, Serbia and Montenegro

Ljiljana Mojovic
Department of Biochemical
   Engineering
University of Belgrade
Belgrade, Serbia and Montenegro

T. Morey
Department of Anesthesiology,
   Pharmacology and Experimental
   Therapeutics
University of Florida
Gainesville, Florida, USA

Mitsutoshi Nakajima
National Food Research Institute
Tsukuba, Ibaraki, Japan

Tadao Nakashima
Miyazaki Prefectural Industrial Support
   Foundation
Sadowara, Miyazaki, Japan

Viktor A. Nedovic
Department of Food Technology and
   Biochemistry
Zemum, University of Belgrade
Belgrade, Serbia and Montenegro

Nikola M. Nikačević
Department of Chemical Engineering
University of Belgrade
Belgrade, Serbia and Montenegro

Bojana Obradovic
Department of Chemical Engineering
University of Belgrade
Belgrade, Serbia and Montenegro

Hiroyuki Ohshima
Science University of Tokyo
Tokyo, Japan

James Y. Oldshue
Oldshue Technologies International
Sarasota, Florida, USA

Aleksandar Orlovic
Department of Organic Chemical Technology
University of Belgrade
Belgrade, Serbia and Montenegro

Nickolay M. Ostrovskii
Boreskov Institute of Catalysis
Omsk, Russia

Richard Partch
University of Florida
Gainesville, Florida and
Clarkson University
Potsdam, New York, USA

Enrique E. Pasqualini
Department of Nuclear Fuels
National Commission of Atomic Energy
Buenos Aires, Argentina

Menka Petkovska
Department of Chemical Engineering
University of Belgrade
Belgrade, Serbia and Montenegro

Milenko Plavsic
Department of Organic Chemical
   Technology
University of Belgrade
Belgrade, Serbia and Montenegro

E. Powell
Department of Chemistry
Clarkson University
Potsdam, New York, USA

Dejan Raković
Department of Electrical Engineering
University of Belgrade
Belgrade, Serbia and Montenegro

Véronique Rollet
Department of Chemical Engineering
University of Caen
Caen, France

Abdellah Saboni
Department of Chemical
   Processes and Hazards
University of Rouen
Rouen, France

Laurier L. Schramm
Saskatchewan Research Council
Saskatoon and University of Calgary
Calgary, Canada

Helmar Schubert
University of Karlsruhe
Karlsruhe, Germany

D. Shah
Department of Chemical Engineering
University of Florida
Gainesville, Florida, USA

Masataka Shimizu
Miyazaki Prefectural Industrial
    Support Foundation
Sadowara, Miyazaki, Japan

Dejan Skala
Department of Organic
    Chemical Technology
University of Belgrade
Belgrade, Serbia and Montenegro

Aleksandar M. Spasic
Department of Chemical Engineering
Institute for Technology of Nuclear and
    Other Mineral Raw Materials
Belgrade, Serbia and Montenegro

Elaine N. Stasiuk
University of Calgary
Calgary, Canada

M. Varshney
Department of Chemistry
Hamdara University
New Delhi, India

Goran T. Vladisavljevic
Department of Food Technology and
    Biochemistry
University of Belgrade
Belgrade-Zemun, Serbia and
    Montenegro

Joseph Wood
Center of Formulation Engineering
University of Birmingham
Birmingham, UK

David A. Yuen
Minnesota Supercomputing Institute
University of Minnesota
Minneapolis, Minnesota, USA

Slobodan Zdravković
Department of Electrical Engineering
University of Priština
Kosovska Mitrovica
Serbia and Montenegro

Miodrag Zdujić
Institute of Technical Sciences
Serbian Academy of Sciences and Arts
Belgrade, Serbia and Montenegro

Table of Contents

Part I

Introduction

1 Classification of Finely Dispersed Systems

Aleksandar M. Spasic
Institute for Technology of Nuclear and Other Mineral Raw Materials,
Belgrade, Serbia and Montenegro

Milan Mitrovic and Dimitrije N. Krstic
University of Belgrade, Belgrade, Serbia and Montenegro

CONTENTS

I. CHEMICAL ENGINEERING AND SCALES — MACRO, MICRO, NANO, AND ATTO

A. MACRO- AND MICRO-SCALE

Classical chemical engineering has been intensively developed during the last century. Theoretical backgrounds of momentum, mass, energy balances, and equilibrium states are commonly used as well as chemical thermodynamics and kinetics. Physical and mathematical formalisms are related to heat, mass, and momentum transfer phenomena as well as to homogeneous and heterogeneous catalyses. Entire object models, continuum models, and constrained continuum models are frequently used for the description of the events, and for equipment designing. Usual, principal, equipments are reactors, tanks, and columns. Output is, generally, demonstrated as conventional products, precision products, chemicals (solutions), and biochemicals.

B. NANO-SCALE

Molecular engineering still suffers substantial development. Besides heat, mass, and momentum transfer phenomena, commonly used in classical chemical engineering, it is also necessary to introduce the electron transfer phenomenon. Description of the events is based on molecular mechanics, molecular orbits, and electrodynamics. Principal tools and equipment are: micro-reactors, membrane systems, micro-analytical sensors, and micro-electronic devices. Output is, generally, demonstrated as molecules, chemicals (solutions), and biochemicals.

C. ATTO-SCALE

Atto-engineering for more than a whole century is in permanent and almost infinite development. Theoretical background is related to the surface physics and chemistry, quantum and wave mechanics, and quantum electrodynamics. Discrete and constrained discrete models are convenient for describing related events. Tools and equipment are: nano- and atto-dispersions and beams (demons, ions, phonons, infons, photons, electrons), ultra-thin films and membranes, fullerenes and bucky tubules, Langmuir–Blodgett systems, molecular machines, nano-electronic devices, and various beam generators. Output is, generally, demonstrated as finely dispersed particles (plasma, fluosol–fog, fluosol–smoke, foam, emulsion, suspension, metal, vesicle, dispersoid).

II. NANO CONTINUA, DISCONTINUA, AND SPACES OF INTERACTIONS

A. GEOMETRY AND FORCES OF INTERACTIONS

The hypothetical continuous material may be called *continuous medium* or *continuum*. The adjective *continuous* refers to the concept when molecular structure of matter is disregarded and being without gaps or empty spaces. Also, it is supposed that mathematical functions entering the theory are continuous functions, except possibly at a finite number of interior surfaces separating regions of continuity [1]. Boundaries of condensed homogeneous phases (liquids and some solids) in the nano-scale also, present, continua, but only with two dimensions; in the third dimension their characteristics change. Such boundaries may be defined as bonded surface continua. Besides a surface tension force, all characteristics of nanolayers at boundaries of condensed phases are different from their bulk properties (e.g., density, viscosity, heat, and mass transfer rates) in the surface toward bulk direction. Further, in the nanospace outside a condensed phase, another layer with interacting forces appears. Molecules and ions inside this nanospace are attracted or repulsed, and attracted molecules or ions behave either as two-dimensional gasses or as another condensed nanophase. In both cases, characteristics (e.g., polarity, reactivity, stereo position) of these adsorbed molecules or ions are different from the properties of "free" molecules or ions, and may be defined as another two-dimensional nano continua. These continua are also bonded and asymmetric (Figure 1.1).

Generally, bonded two-dimensional (surface) continua exist at the contact of two immiscible homogeneous phases. At the contact of three immiscible homogeneous phases, only line nanoelements exist. The examples are contact lines between two liquids and one gaseous phase (Figure 1.2). Line elements are defined as one-dimensional bonded nano continua, with line tension forces, molecules, or ions attraction or repulsion forces in the nanospace over a line element and a complex asymmetric structure. Gasses adsorbed at a line element have only one degree of freedom (one-dimensional gasses can also be a condensed matter). Molecules or ions at line elements also change their characteristics.

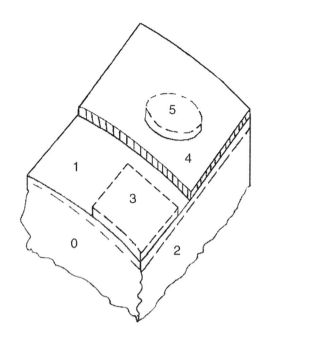

FIGURE 1.1 Bonded surface nano continua: 0 — bulk material (volume continuum), 1 — interface, 2 — bonded surface continuum, 3 — external space of interaction, 4 — condensed adsorbed layer, 5 — new external space.

Bonded point elements exist at contacts of four phases, analogous to line elements at contacts of two liquids, one gaseous phase, and one solid phase. These elements are discontinuous in all dimensions. In the nanospace over point elements, the adsorbed molecules or ions have zero degree of freedom. Point elements are presented in Figure 1.3.

Solid crystals have surface-, line-, and point-bonded elements as a part of their morphology (Figure 1.4). Surfaces of solid adsorbents and catalysts are fractional micro to nano compositions of surface-, line-, and point-bonded elements (Figure 1.5).

Free nano continua or discontinua can be defined, also, as surface, line, and point elements, and as their combinations. Free surface elements are ideally composed of two nanolayers placed symmetrically each other (Figure 1.6). Such structures are "ideal membranes" and can be composed of solid materials, liquids, and even gasses. Ideal membranes exist in nature, e.g., bilipide cell membranes and "black" surfactant bubbles (Figure 1.7a and b). It is visible that both membranes are composed of only two molecular layers of asymmetric surface-active molecules, with hydrophilic ends placed outward (cell membranes) or inward (surfactant bubbles). Ideal membranes, composed

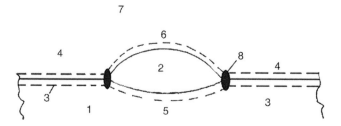

FIGURE 1.2 Line continuum at the contact of three phases: 1 — bulk liquid A, 2 — bulk liquid B, 3 — bonded surface continuum A, 4 — nanospace A, 5 — bonded surface continuum B, 6 — nanospace B, 7 — gaseous phase, 8 — circular line continuum.

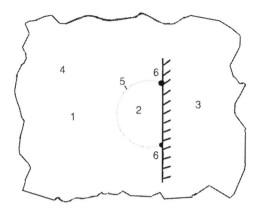

FIGURE 1.3 Point element at the contact of four phases: 1 — liquid A, 2 — liquid B, 3 — solid C, 4 — gaseous phase (over A and B), 5 — line continuum, 6 — point elements.

of two or more monomolecular layers, placed on porous or nonporous carriers, can be produced by Langmuir–Blodgett technique [2], using surface-active polymers (Figure 1.8).

Ideal gas or vapor membranes can be composed of two or oligo molecular gas or vapor layers, in a gap between two solid or liquid surfaces, which "borrow" an external nanospace to the gas (Figure 1.9). Similar effects can be obtained with a gas or vapor inside nanopores of solid hydrophobic membranes.

Free line nanoelements are liquid crystals, nanotubules [3], and linear fullerenes [4]. Tubular organization of carbon atoms and some organic molecules is a very interesting example of a self-organization in the nanoscale. Besides, double helix structures, carriers of genes in chromosomes and plasmids, regardless of their molecular organization, are line nanoelements. All free surface elements (two-dimensional continua) and free line elements (one-dimensional continua) possess some kind of symmetry. Gasses or vapors inside nanotubules are, also, gaseous line nano continua and ought to have different properties than free gasses. Some line nanoelements are presented in Figure 1.10.

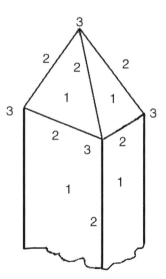

FIGURE 1.4 Bonded surface, line, and point elements at solid crystals: 1 — surface continuums, 2 — line continuums, 3 — point elements.

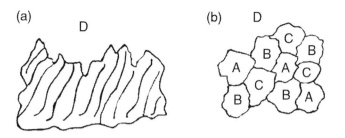

FIGURE 1.5 Bonded fractal surface, line, and point elements of an adsorbent or a catalyst: (a) rough surface of an adsorbent or a catalyst; (b) flat surface of a solid composed of three phases (A–C), with gaseous phase D (over).

Ideal point free nanoelements are fullerenes C_{60} or C_{70} (Figure 1.11) [4]; such structures could be defined as zero-dimensional continua. These elements are symmetric and nearly independent of all external influences. Some nanostructures in liquids, due to their forms and electrical charges, are also close to ideal point nanoelements. Clusters of water molecules or atoms of metals, are nonideal point nanoelements.

Complex free nano continua are foams. Foams can be divided into two groups: cream foams, in which free gas bubbles are submerged in a liquid matrix, and dry foams, in which nearly complete liquid phase is bonded inside foam structures. The first system is a simple two-phase liquid–gas mixture. The second system is a three-dimensional network of surface, line, and point elements. Dry foam is presented in Figure 1.12. Dry foam is composed of three elements: (i) flat surface elements (membranes), composed of two opposite monomolecular liquid surfactant layers, with a nonorganized liquid between the layers; (ii) line elements, which are limiting surface elements, of the tubular shape, with nonattached liquid phase inside; and (iii) point elements, surfactant hollow spheres, connecting line elements, also, filled with a nonattached liquid phase. Sizes of all dry foam elements are large, compared to thicknesses of outside nanolayers, due to the non-attached liquid phase; so, when a foam material is polymerized into solid, elastic or rigid structures, then the mechanical characteristics of solid foams depend upon strengths of all elements and the geometry of a foam and relative strengths of the ordered material at surface of the elements and inside the elements.

B. Processes in Nano Continua, Discontinua, and Spaces of Interactions

Basically, there are four kinds of physical or chemical processes in nano continua, discontinua, and spaces of interactions: (1) processes in the nanospace over surface, line, or point elements; (2) processes in the nanospaces inside surface, line, and point elements; (3) barrier processes — mass and energy transfer processes, with or without physical and chemical transformations, through internal and external nanospaces; (4) membrane processes — mass and energy transport processes through membranes (double gas, liquid, and solid surface, line and point elements). Membrane processes are presented in Figure 1.13.

Furthermore, processes in nanosystems could be defined as follows.

FIGURE 1.6 Free surface elements (continua).

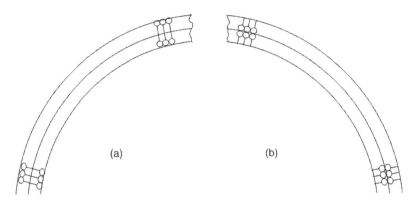

FIGURE 1.7 "Black" bilipide cell membrane: (a) and surfactant bubble (b).

1. Diffusion

Besides unrestricted diffusion in three-dimensional continua and restricted (Knudsens) diffusion, five other types of diffusion processes may be distinguished:

1. Surface diffusion — unrestricted diffusion along nanospaces over or under surfaces in bonded surface continua, over membranes (at both sides), and inside a membrane (Figure 1.14a).
2. Diffusion through surface continua — diffusion through external and internal nanospaces for bonded surface continua, or through two external nanospaces and one internal nano-space for membranes (Figure 1.14b).
3. Diffusion along line continua (through external and internal nanospaces in entities as nanotubules, Figure 1.14c).
4. Diffusion across line continua (from an external surface nanospace to another surface nanospace over a line continua, Figure 1.14d).
5. Point diffusion — diffusion from one point nanoelement to another is a process common for complex nanosystems such as sorbents and catalysts. Such "jumping diffusion" is a process with energy of activation. Point diffusion occurs in nanospaces over complex sur-faces, inside the pores of porous media, with micro- and sub-micropores, and through crystal lattices, from one "point hole" to another (e.g., Na ions through β-Al_2O_3, in batteries S–Na. All these types of point diffusion are presented in Figure 1.15).

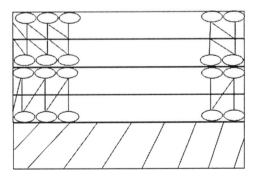

FIGURE 1.8 Langmuir–Blodgett membrane (four molecular layers).

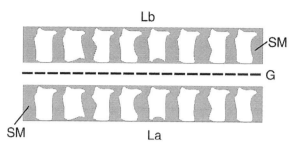

FIGURE 1.9 Ideal gas or vapor membrane: L_a — liquid phase A, L_b — liquid phase B, SM — solid microporous membrane (hydrophilic), G — gaseous phase.

2. Sorption Processes

Interactions in nanospaces (sorption processes) may be classified into following groups:

1. *Adsorption in external nanospaces.* Basically, there are three kinds of interactions in external nanospaces of solids or liquids — adsorption at surface elements, adsorption at line elements, and adsorption at point elements. All these processes present attraction or repulsion of molecules, ions, or particles, influenced by physical forces. Positive adsorption (attraction) is a reversible process and, in the case of noncondensed gasses, only monomolecular layer is adsorbed. This is happening probably because attraction forces over adsorbed molecules are negligible.

2. *Adsorption with condensation (external nanospaces).* At vapor pressures, for adsorbed gasses over corresponding condensation pressures (due to a surface tension, at lower pressures in narrow capillaries), adsorbed gasses convert into a liquid phase. Regardless of whether the primary adsorption occurred at surfaces, lines, or points, a condensed phase generates its own nanospace; so an adsorption with the condensation, generally, produces multi-molecular adsorbed layers.

3. *Chemisorption.* Chemisorption is a surface, line, or point process, where chemical bonds exist between a sorbent and a sorbate. The process occurs both in external and internal nanospaces of a chemisorbent. It, generally, produces multi-molecular layers, because the process proceeds into internal nanolayers of sorbents. Chemisorption, in a number of cases, is irreversible. A chemisorbed film produces its own nanospace, in many cases with completely different characteristics from a primary nanospace.

4. *Adsorption in internal nanospaces.* In nanospaces under surfaces of liquids or gels, or inside membranes, both positive and negative adsorption processes occur (increase or decrease of concentrations of molecules or ions). Tenzides are adsorbed in the nanospace inside water surfaces, and one method of removing and concentrating surface-active agents from aqueous solutions is the foam flotation (high increasing of surface internal nanospace areas). Contrary to this, mineral anions and cations are negatively adsorbed in the same nanospaces, especially if water molecules are activated; in gel structures, inside

Linear fullerene   Carbon nanotube

FIGURE 1.10 Line nanoelements (line continua).

FIGURE 1.11 Fullerene (ideal free point nanoelement).

hydrophilic polymers, in zeolite cages, or in nanopores of hydrophilic minerals. The process of water desalination, by reverse osmosis, can be explained by the exclusions of ions and other solutes, from bonded water inside reverse osmosis (RO) membranes and the consecutive transport of water molecules, through the membrane, by pressure differences. Molecules or ions, adsorbed in internal nanospaces of nanotubules, and linear and point fullerenes, can be stored. Some new electrochemical systems are designed using this phenomenon.

5. *Adhesion.* Adhesion can be defined as a bulk, surface, or line adsorption, or an attraction of surface nanospaces. For a solid–solid system, adhesion forces are well defined only for

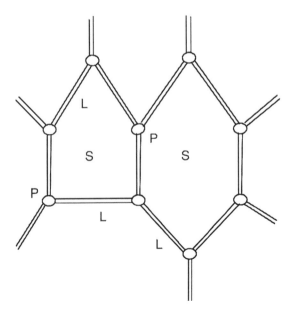

FIGURE 1.12 Model of a dry foam: S — surface elements, L — line elements, P — point elements.

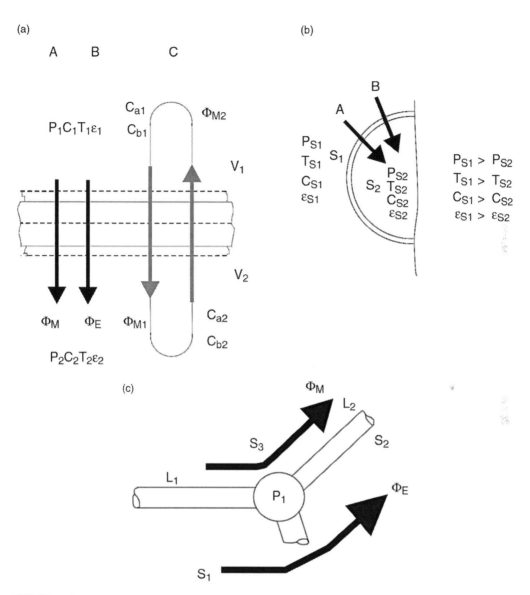

FIGURE 1.13 Mass and energy transport processes (surface, line, and point membranes): (a) processes through surface membrane: A — mass transport, B — energy transport, C — coupled mass transport, $V_{1,2}$ — bulk fluids, P_1, C_1, T_1, ε_1, C_{a2}, $C_{b2} > P_2$, C_2, T_2, ε_2, C_{a1}, C_{b1}; (b) processes over line membrane (two-dimensional gasses): Line membrane, S_1 and S_2 — liquid surfaces (liquids 1.2); (c) processes over point membrane (one-dimensional gasses), S_1, S_2, S_3 — surface membranes, L_1, L_2, L_3 — line membranes, P_1 — point membrane (foam), P_{L1}, C_{L1}, T_{L1}, $\varepsilon_{L1} > P_{L2}$, C_{L2}, T_{L2}, ε_{L2}.

highly polished surfaces, without any adsorbed films. For liquid–solid interactions, adhesive forces are expressed by interphase surface tensions and can be measured by angles of wetting (Figure 1.16). Adhesives are defined as liquid agents, which wet solid surfaces and upon drying, cooling or polymerization glue surfaces together after the solidification. The mechanical adhesive force to separate glued surfaces expresses the quality of an adhesive. Mechanical or chemical roughening of solid surfaces produces three effects: cleaning, increasing specific surface areas, and introducing line and point nanoelements with high specific energies.

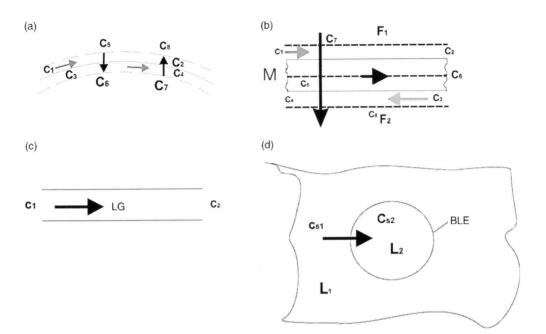

FIGURE 1.14 Types of diffusion processes at a bonded continuum or at a membrane: (a) surface diffusion, (b) diffusion through surface continua, (c) diffusion along line continuums, (d) diffusion across line continua; C_x — bulk or nanospace concentrations, G — gaseous phases, L — liquid phases, F — fluid phases, BLE — bonded line element, LG — line gas.

6. *Adhesion of gas bubbles to solid particles.* In the hydrometallurgical process, flotation, gas bubbles are attached to ore particles and carry them to the surface of the liquid phase. The attraction force is adhesion of circular line elements at the contact of a nonwetted solid phase (ore), liquid phase (aqueous solutions of flotation agents), and gas phase (air or hydrogen bubble). This system is presented in Figure 1.17.

By different flotation agents that are adsorbed at surfaces of components, in a complex ore mixture, the flotation process can be made to be selective, with bubbles attached only to a specified ore particle.

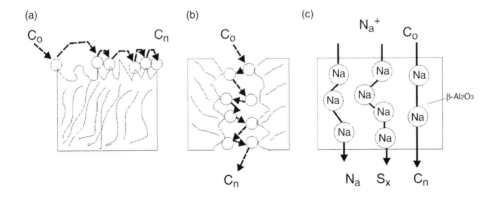

FIGURE 1.15 Types of point diffusion processes.

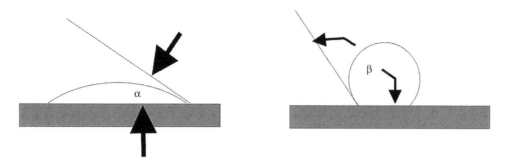

FIGURE 1.16 Contact angles as a measure of surface interactions: lipophilic α and lipophobic β.

3. Heterogeneous Catalysis

Heterogeneous catalysis is an activation process to carry out specific chemical reactions, occurring in different nanospaces. Concerning types of nanospaces, following catalytic systems exist:

1. *Catalysis in internal nanospaces of liquid catalysis.* A catalysis of this sort probably occurs in molten salts carried in pores of inert porous carriers, to increase gas–liquid contact surface areas — vanadium pentoxide catalysts for the oxidation of SO_2 to SO_3 [5]. Reactants diffuse into the nanolayer below the catalysts surface and the product countercurrently diffuses out.
2. *Catalytic reactions in the surface external nanospace, at the surface of impermeable liquids.* Example of this catalytic system is Viladsen's catalyst [6,7], with microspheres of molten indium inside the "cages" of a porous inert ceramics (Figure 1.18). Since reaction rates are proportional to the catalysts area, this arrangement ensures very high accessible specific surface areas.
3. *Catalytic reactions in external nanospaces over solid, line, and point elements.* A very high percentage of all heterogeneous catalytic systems is of this type. Combinations of catalysts, promoters, and carriers obtain line and point active centers, with specified characteristics. Besides a catalytic activity and selectivity, good catalytic system ought

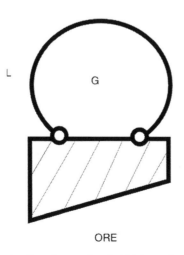

ORE

FIGURE 1.17 Adhesion of the circular line element (gas bubble) to a hydrophobic ore particle.

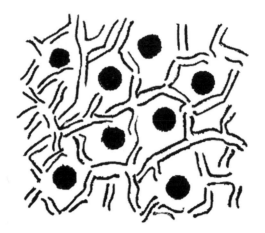

FIGURE 1.18 Viladsen's liquid-phase catalyst.

  to have high mass transfer rates, in the case of a diffusion-controlled process, and also high
  heat transfer rates, for strongly exothermal or endothermic processes.
  4. *Stereo-specific catalytic reactions.* External nanospaces for stereo-specific catalytic synth-
     eses are "nanomachines" with such arrangement of active centers that only molecules
     of defined three-dimensional shapes can be synthesized. An example is crown ethers,
     with hollows in which active atomic groups are distributed in such a way that only one
     stereo-specific molecule can be produced.

C. Membrane Processes

Owing to its importance in nature, science and technology of membrane processes have the special
place in nanospace processes. Generally, membrane processes can be classified into three interconnected groups: mass transport processes, without or with chemical transformations; energy
transport processes, without or with energy transformations; and information transport and
processing. In this section, only mass transport processes interconnected with corresponding
energy transport and transformations will be discussed.

  Transport of ions, molecules, and molecule aggregations through different membranes can be
achieved by following mechanisms: viscous flows through porous membranes; having with pores
much larger than 100 nm; diffusion (Knudsens, line or point), for pores of about 10 nm; diffusion
with adsorptions and desorptions (point diffusion); point diffusion from one hole to other (ions);
line diffusion through line ionic conductors; and diffusion through internal nanospaces of nanoporous membranes. Driving forces are gravity, gradients of pressures, temperatures, and concentrations, electrical and magnetic fields, and chemical interactions (reversible or irreversible
reactions, countercurrent transport of ions).

  By aims membrane mass transport processes can be classified as controlled release, separation,
separation and enrichment, chemical transformation (catalytic or noncatalytic), and biochemical
transformation processes.

1. Controlled Release Processes

Controlled release is used mostly to make different liquid or gaseous mixtures and to distribute
drugs or pesticides. Process is controlled by diffusion rates (point diffusion or diffusion in internal
nanospaces) and the driving force is a concentration gradient.

2. Separation Processes

Separation processes may be classified by sizes of entities to be separated, mechanisms of separation, types of membranes, and driving forces. Classification by mechanisms is as follows:

1. *Sieving*. Sieving presents a separation of particles considering their sizes, e.g., the separations of microorganisms, macromolecules, molecules, and atoms. In sub-micro- and nanoscale, sieving is seldom the only separation mechanism. Mostly it is combined with interactions of membrane elements and systems to be separated, as an adsorption or repulsion of molecules, ions, and particles, on membrane sub-micro- or nanoelements, and electric charges of a membrane and components to be separated. Driving force is mostly a pressure difference, and membranes are porous. Rates of transport are, generally, governed by viscous flow rates.

2. *Diffusion*. In most cases, diffusion, without other combined mechanisms (solubility, adsorption, exclusion), produce low separation factors. High separation factors may be obtained only if diffusion is highly selective, e.g., point diffusion in "holes" of β-Al_2O_3, for Na ions.

3. *Barrier membrane process*. Processes in which one component, in a mixture, interacting with internal or external nanospaces of a membrane, present a barrier for another components, so that only the interacting component passes through the membrane. In the simplest case, a surface-active micro-filtration membrane, if saturated with a wetting liquid, is the barrier for a gas or a nonwetting liquid. Using lipophilic and lipophobic micro-filtration membranes, this phenomenon is applied in multiphase membrane contactors [7] and in the production of fine emulsions.

    Another type of a barrier makes use of adsorbed molecules or ions, inside membrane sub-micro- or nanopores, which prevent non-adsorbed components to pass through a porous membrane (Sourirajans model for RO water desalination). In this case, a membrane is composed of an adsorbed component.

    Third type of barrier separation membranes work with the exclusion of components nonpassing through the membrane, by changed solubilities of nonpassing components in a passing component, due to interaction with the internal nanospace of the membrane. Such a mechanism explains the water separation, by reverse osmosis, from dissolved salts and organics, using hydrophilic polymer or gel-type membranes.

    Fourth type of barrier membrane process works on the principle of exclusion, but by an external power field, was applied in O_2/N_2 separation, using porous membranes under a strong magnetic field (paramagnetic O_2 was excluded from the field and the membrane).

4. *Solvent-type separation membranes*. Membranes used mostly for separations of gasses or vapors, which dissolve and pass through one component, or a group of components, from a mixture. It is not clear whether membranes composed of only two symmetric nanolayers have advantages over thicker membranes, besides shorter diffusion paths, but one theory claims that for some ultra-thin (nano) membranes, very high separation factors can be obtained, for pairs of gasses, of which molecules of one can "jump" through the membrane, while the other gas have to pass by the slow diffusion process.

5. *Other interactions in a membrane internal nanospace*. The first is evaporation through a nonporous membrane (pervaporation). Interaction of separating components with a polymer in a membrane internal nanospace, in some cases, moves liquid–vapor equilibrium and removes azeotropic points.

    Separation and enrichment of metal ions, using reversible chemical reactions in liquid or gel-type membranes: Using selective extragents, it is possible to obtain high separation and enrichment factors for metal ions (e.g., Cu, In, U, Pu), with a simultaneous extraction and stripping (pertraction). Driving force for this process is the countercurrent diffusion of

H ions. It is not clear whether molecular forces in the nanospace, inside a membrane, activate these processes. In our experiments on the simultaneous pertraction of two metals (Cu and In), from a common aqueous solution, Cu : In separation factors of 5000 : 1 and enrichment factors for both metals up to 1000 : 1 were obtained, and so it would be worthwhile to further investigate such processes.

6. *Membrane catalytic reactors.* Catalytic elements, in the form of membranes: thin two-sided layers of a permeable liquid catalyst, or a porous solid catalyst, has the main advantage in its two-sided configurations, so it is possible to design more complex and efficient catalytic reactors, as some, already investigated, or only considered systems:

Membrane catalytic reactors with separated reactants:

Suitable for fast and potentially dangerous reactions, or for reactions which could produce unwanted products, without a catalyst. When the reaction rate is much higher than diffusion rates of both reactants, the reaction zone, inside a catalytic membrane, is a thin layer, moving in the membrane according to stoichiometric ratios. A product is moving out of the catalyst countercurrently to one of reactants (adjusted by pressure differences).

Membrane catalytic reactors with the separation of products from reactants:

In the case when products could react with reactants, or when equilibria are far from complete conversions, a combination of the catalytic membrane reactor with a membrane, which separates products from reactants, would make possible such processes as ammonia synthesis, at low pressures and with high yields.

D. Stability of Nanosystems

By definition, surface nano continua, either bonded or free, ought to be infinitely stable entities, and so their shapes are closed spheres or ellipsoids. Destabilizing factors are: power fields, e.g., a large liquid drops, on a nonwetting surface becomes unstable and breaks into smaller drops (Figure 1.19); shear stresses, e.g., liquid drops or gas bubbles moving in a fluid with velocities over critical *Re* numbers change their shapes and break into smaller entities; changes in chemical compositions, e.g., evaporation of water from soap bubbles; Marangoni instabilities at liquid internal surface nanospaces; and mechanical or electromechanical oscillations of surface elements.

Surface continua can be limited and stabilized by line elements (Figure 1.20), but line-bonded or free continua are stable only if they are unlimited structures (rings). It is an interesting observation that stable genetic elements are plasmids — closed ring genetic elements inside the cells protoplasm — while open chains of genes, in chromosomes, are inherently unstable and are temporarily stabilized by "telomeres," stacks of point elements at the ends of a gene double helix. Open line elements in a dry foam are limited and stabilized by point elements (Figure 1.12). Linear fullerenes, are also, stabilized by half of the point fullerene. Ring nanotubules or fullerenes though not synthesized till now, it could be expected that such nanoentities do exist.

Point elements are stable, if they have a structure of a closed symmetric sphere. Asymmetric point structures as different clusters are unstable and easily agglomerate or break.

Complex nanostructures such as sols or gels, with networks of fractal surface, line, and point elements, are highly unstable. Sols are jellified by removing a part of the solvent (sol–gel transformations to obtain ultra-thin ceramic membranes), by temperature changes, or by changing pH. Gels are broken by heating, by mechanical forces, or by ultra-high electromagnetic waves. Drying of hydrogels, to produce sorbents with nanopores, generates large forces inside submicro- and nanopores, due to water meniscus in the pores. With such processes, it is impossible to obtain sufficiently large surface areas. Supercritical drying — heating gels

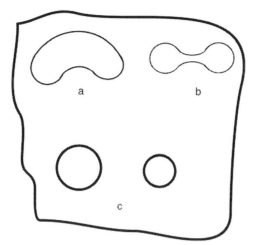

FIGURE 1.19 Unstable liquid drops at a lipophobic surface.

over the critical temperature for water, at pressures over the critical pressure, and gradually lowering the pressure — ensure the drying of gels without destabilizing pores.

Instabilities in beds of porous catalysts, for strongly exothermic processes, occur when heat transfer rates from the bed are not adequate, so the temperature rises, increasing reaction rates, and thus producing heat energy. This instability produces "hot spots" inside the bed and melts or inactivates the catalyst. Besides other solutions, our microcapillary catalytic reactor elements, using very thin porous layers of a catalyst (6–10 μm), attached to the heat-transferring body (glassy carbon), ensure stable exothermal processes, as the ammonia synthesis, or CO oxidation to CO_2.

E. Conclusions

The concept of surfaces and line-bonded or free nano continua and point discontinua, besides extending mechanics of continua to two- and one-dimensional entities, make possible an unifying approach to processes such as diffusion, adsorption, and heterogeneous catalysis. Practically, it would enable R&D of new micro- or nanochemical reactors, separation units, and integrated reactor-separator systems. With the introduction of electric, magnetic, and optic characteristics of nano continua and influences of corresponding power fields, the concept could be extended to electrochemistry, nanoelectronics, and optics. Also, studies of mechanical characteristics of surface, line, and point nanostructures would make possible design and construction of new nanomachines.

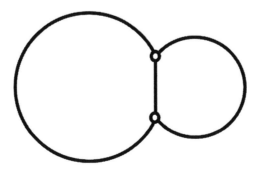

FIGURE 1.20 Stabilizing of surface continua with a circular line continuum.

III. GENERAL COMMENTS RELATED TO THE NEW CLASSIFICATION OF FINELY DISPERSED SYSTEMS

A. RESEARCH PHILOSOPHY

One brief answer to the questions What? and Why? (Philosophy) could be schematically presented using the proposition: "Along the Path Question Mark toward Exclamation Mark" (Figure 1.21).

E — Engineering — takes the behavior as given and study how to build the object that will behave — Engineering enable the things to work, e.g., the process, operation, and equipment B can fulfill the objective C; proof is the demonstration!

AS — Applied Science — theories related to particular classes of things, just different of the Science that studies the general or universal laws; some processes, operations, and equipment B can fulfill the objective C; when the demonstration become possible then the applied science disappears!

S — Science — takes the object as given and studies its behavior; sciences discover and explain how the things work!

B. RESEARCH STRATEGY AND METHODOLOGY

Using the same proposition "Along the Path Question Mark toward Exclamation Mark" one brief answer to the question How? (strategy) could be schematically presented following methodological weight hierarchy (capability or feasibility), as it is shown in Figure 1.22, although, researcher is not always aware of the strategic path or the step he is using at the time!

C. CHARACTERISTICS, APPROXIMATION, AND ABSTRACTION LEVELS

A new idea, using deterministic approach, has been applied for the elucidation of the electron and momentum transfer phenomena at both the rigid and deformable interfaces in finely (micro, nano, atto) dispersed systems. Since the events at the interfaces of finely dispersed systems have to be considered at the molecular, atomic, and entities level, it is inevitable to introduce the electron transfer besides the classical heat, mass, and momentum transfers commonly used in chemical engineering [8]. Therefore, an entity can be defined as the smallest indivisible element of matter that is related to the particular transfer phenomenon. Hence, the entity can be either a differential element of mass or demon, or an ion, or a phonon as quantum of acoustic energy, or an infon as quantum of information, or a photon, or an electron [9,10].

A possible approach is proposed to the general formulation of the links between the basic characteristics, levels of approximation, and levels of abstraction related to the existence of finely dispersed systems (DS) [11]. At first, for simplicity and easy physical and mathematical modeling, it is convenient to introduce the terms: homo-aggregate (phases in the same state of aggregation [HOA]) and hetero-aggregate (phases in a more than one state of aggregation [HEA]). Now the

FIGURE 1.21 Research Philosophy.

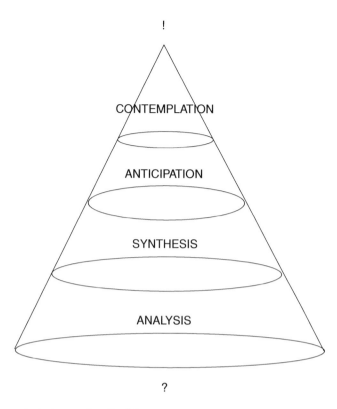

FIGURE 1.22 Research strategy and methodology.

matrix presentation of all finely dispersed systems is given by

$$[(DS)^{ij} = (HOA)\delta^{ij} + (HEA)\tau^{ij}] \tag{1.1}$$

where i and j refer to the particular finely dispersed system position; i.e., when $i = j$ then diagonal positions correspond to the homo-aggregate finely dispersed systems (plasmas, emulsions, and dispersoids, respectively), and when $i \neq j$ then tangential positions correspond to the hetero-aggregate systems (as already mentioned fluosols–fog, fluosols–smoke, foam, suspension, metal, and vesicle, respectively). Furthermore, the interfaces may be deformable D and rigid R that is presented in Table 1.1.

Now, related to the levels of abstraction and approximation it is possible to distinguish continuum models (the phases considered as a continuum, i.e., without discontinuities inside entire phase, homogeneous, and isotropic) and discrete models (the phases considered according to the Born–Oppenheimer approximation: entities and nucleus/center of total energy (CTE), i.e. their motions are considered separately). Continuum models are convenient for microscale description (entire object models), e.g., conventional products, precision products, chemical solutions, biochemicals while discrete models are convenient for either nanoscale description (molecular mechanics, molecular orbits), e.g., chemical solutions, biochemicals, molecular engineering, and attoscale description (quantum electrodynamics), e.g., molecular engineering, attoengineering. Since the interfaces in finely dispersed systems are much developed, it is sensible to consider the discrete models approach for description of a related events [12]. For easier understanding it is convenient to consult, among others, e.g., Refs. [13–23].

TABLE 1.1
A New Classification of Finely Dispersed Systems

	Dispersed Medium		
Dispersed Phase	**Gas**	**Liquid**	**Solid**
Gas	PLASMA *D*	FOAM *D*	METAL *R*
Liquid	FLUOSOL/fog *D*	EMULSION *D*	VESICLE *D*
Solid	FLUOSOL/smoke *R*	SUSPENSION *R*	DISPERSOIDE *R*

D. HIERARCHY OF ENTITIES

Figure 1.23a shows a stereographic projection or mapping from Riemann sphere, whereas Figure 1.23b shows a "hierarchy" of entities, which have to be understood as a limit value of the ratio u_0/Z (withdrawn from magnetic Reynolds criteria $[Re_m = 4\pi lGu_0/c^2]$, where the conductivity G is expressed as a reciprocal of viscosity/impedance Z ($G = l/Z$), l the path length that an entity "override," u_0 the characteristic velocity, and c the velocity of light). In general S corresponds to the slow system or superfluid and F corresponds to the fast system or superconductor; now, it is possible to propose that all real dynamic systems are situated between these limits. Also, it seems sensible to think about the further structure of entities, namely, demon, phonon, infon, photon, and electrons; for example, electron, the basic entity can be understood as an energetic ellipsoid shown in Figure 1.23c (based on the model of electron following Maxwell–Dirac isomorphism (MDI). Electron is an entity which is dual in nature, one and the same time, it is quantum–mechanical or microscopic ($N = -2$) and electrodynamics or macroscopic ($N = 3$).

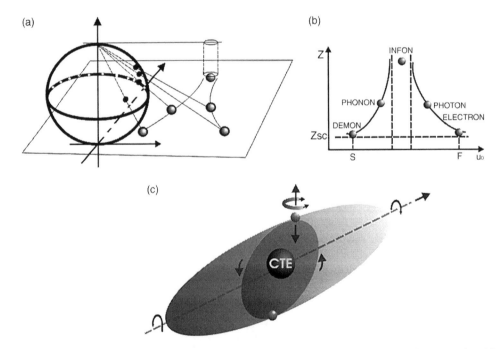

FIGURE 1.23 (a) A stereographic projection or mapping from Riemann sphere; (b) hierarchy of entities, correlation viscosity and impedance-characteristic velocity, S — slow/demon (superfluid) and F — fast/electron (superconductor); (c) entity as an energetic ellipsoid (at the same time macroscopic and microscopic), CTE — center of total energy, motions (translation, rotation, vibration, precession, angle rotation).

Now, spatio-temporal, five-dimensional existence of an entity, e.g., electron may be presented by equation $[(x - a)^2 + (y - b)^2 + (z - d)^2 - (ct - e)^2 - (^kN_0\omega_N - f)^2 = 0]$, where $^kN_0$ is the factor of spatio-temporal synergy (cm sec), and ω_N the isotropic angle rotation (sec^{-1}) [4].

ACKNOWLEDGMENT

This work was supported by Ministry of Science and Environmental Protection of Republic of Serbia as a part of fundamental project "Multiphase Dispersed Systems" (No. 101822).

REFERENCES

1. Malvern, E.L., *Introduction to the Mechanics of a Continuous Medium*, Prentice-Hall, Inc., Englewood Cliffs, NJ, 1971, 1 pp.
2. Stroeve, P. and Frances, E., *Molecular Engineering of Ultra-Thin Polymeric Films*, Elsevier, London and New York, 1987.
3. Dresselhaus, M., Dresselhaus, G., and Avouris, Ph., *Carbon Nanotubes: Synthesis, Structure, Properties and Applications*, Springer-Verlag, Berlin, 2001.
4. Koruga, Dj., *Fulerenes Nanotubes Rev.*, 1, 75, 1997.
5. Dojcinovic, M., Susic, M., and Mentus, S., An investigation of a catalytically active V_2O_5–$K_2S_2O_7$–K_2SO_4 melt, *J. Mol. Catal.*, 11, 276, 1981.
6. Viladsen, J. et al., Catalyst for dehydrogenating organic compounds, in particular amines, thiols, and alcohols, and for its preparation. US Patent 4,224,190. September 23, 1980.
7. Mitrović, M., Integrated dual and multiple separation systems applying combined separation techniques, in: *Precision Process Technology*, Weijnen, M.P.C. and Drinkenburg, A.A.H., Eds., Kluwer Academic Publishers, Dordrecht, 1993, 393 pp.
8. Spasic, A.M., Electroviscoelasticity of liquid–liquid interfaces, in: *Interfacial Electrokinetics and Electrophoresis*, Delgado, A.V., Ed., Marcel Dekker, New York, 2002, 837–867.
9. Spasic, A.M., Babic, M.D., Marinko, M.M., Djokovic, N.N., Mitrovic, M., and Krstic, D.N., A new classification of finely dispersed systems, Abstract of Papers, Part 5, in: *Proceedings of the Fourth European Congress of Chemical Engineering*, Granada, Spain, September 21–25, 2003, p.5.2.39.
10. Lazarevic, M.P. and Spasic, A.M., Electroviscoelasticity of liquid–liquid interfaces: fractional approach, Abstract of Papers, Part 5, in: *Proceedings of the Fourth European Congress of Chemical Engineering*, Granada, Spain, September 21–25, 2003, p. 5.2.33.
11. Spasic, A.M., Babic, M.D., Marinko, M.M., Djokovic, N.N., Mitrovic, M., and Krstic, D.N., Classification of rigid and deformable interfaces in finely dispersed systems: micro-, nano-, and atto-engineering, CD of Papers and Abstract of Papers, Part 2, in: *Proceedings of the 16th International Congress of Chemical and Process Engineering*, Prague, Czech Republic, August 22–26, 2004, p. D5.1.
12. Spasic, A.M. and Lazarevic, M.P., Theory of electroviscoelasticity: fractional approach, CD of Papers and Abstract of Papers, Part 2, in: *Proceedings of the 16th International Congress of Chemical and Process Engineering*, Prague, Czech Republic, August 22–26, 2004, p. D5.4.
13. Probstein, R.F., *Physicochemical Hydrodynamics*, Wiley, New York, 1994, pp. 352–369.
14. Kruyt, H.R., *Colloid Science*, Vol. I, Elsevier, Amsterdam, Houston, New York, London, 1952, pp.302–341.
15. Spasic, A.M. and Krstic, D.N., Structure and stability of electrified liquid–liquid interfaces in finely dispersed systems, in: *Chemical and Biological Sensors and Analytical Electrochemical Methods*, Rico, A.J., Butler, M.A., Vanisek, P., Horval, G., and Silva, A.E., Eds., ECS, Pennington, NJ, 1997, pp. 415–426.
16. Reid, C.R., Prausnitz, J.M., and Pauling, B.E., *The Properties of Gasses and Liquids*, McGraw-Hill, New York, 1989, pp. 632–655.
17. Condon, E.U. and Odishaw, H., Eds., *Handbook of Physics*, McGraw-Hill, New York, 1958, 4/13 pp.
18. Pilling, M.J., *Reaction Kinetics*, Clarendon Press, Oxford, 1975, pp. 37–48.
19. Yeremin, E.N., *The Foundation of Chemical Kinetics*, Mir Publishers, Moscow, 1979, pp. 103–149.

20. Leonard I.S., *Quantum Mechanics*, McGraw-Hill, New York, 1955, pp. 7–90.
21. Krall, A.N. and Trivelpiece, W.A., *Principles of Plasma Physics*, McGraw-Hill, New York, 1973, pp. 98–128.
22. Gasser, R.P.H. and Richards, W.G., *Entropy and Energy Levels*, Clarendon Press, Oxford, 1974, pp. 27–38.
23. Hirchfelder, J.O., Curtiss, Ch. F., and Bird, R.B., *Molecular Theory of Gases and Liquids*, Wiley, New York, 1954, 139 pp.

Part II

General

Overview

2 Charged Particles and Droplets – Overview

Hiroyuki Ohshima
Science University of Tokyo, Tokyo, Japan

CONTENTS

I. INTRODUCTION

In this chapter, we discuss electrokinetic behaviors of particles and drops. The motion of charged colloidal particles in a liquid under a steady external electric field, which is called electrophoresis, depends on the thickness of the electrical diffuse double layer formed around the charged particles and the zeta potential ζ [1–9]. The zeta-potential ζ is defined as the potential at the plane where the liquid velocity relative to the particle is zero. We assume that this plane, which is called the slipping plane or shear plane, is located at the particle surface so that the zeta-potential ζ is equal to the particle surface potential ψ_0. We also assume that the magnitude of the applied electric field E is not very large so that the velocity U of the particles, which is called electrophoretic velocity, is proportional to E in magnitude. The ratio of the magnitude of U to that of E is called electrophoretic mobility μ, which is given by $\mu = U/E$ (where $U = |U|$ and $E = |E|$). In this chapter, we derive equations relating μ to ζ or electric charges of various types of colloidal particles.

II. GENERAL MOBILITY EXPRESSION

A charged particle immersed in a liquid containing an electrolyte is surrounded by the electrical diffuse double layer. The thickness of the electrical double layer is given by the Debye length $1/\kappa$ (κ = Debye–Hückel parameter). For a general electrolyte composed of N ionic mobile species of valence z_i and bulk concentration (number density) n_i^∞, κ is defined by

$$\kappa = \left(\frac{1}{\varepsilon_r \varepsilon_0 kT} \sum_{i=1}^{N} z_i^2 e^2 n_i^\infty \right)^{1/2} \tag{2.1}$$

where ε_r is the relative permittivity of the electrolyte solution, ε_0 the permittivity of a vacuum, e the elementary electric charge, k is Boltzmann's constant, and T the absolute temperature. The velocity U of the particle under an applied electric field E strongly depends on the relative magnitudes of $1/\kappa$ and the particle size.

The first attempt to derive the relation between μ and ζ was made by Von Smoluchowski [10] and Hückel [11], and later by Henry [12]. Full electrokinetic equations determining electrophoretic mobility μ of spherical particles with arbitrary values of κa and ζ were derived independently by Overbeek [13] and Booth [14]. Wiersema et al. [15] solved the equations numerically. The computer calculation of the electrophoretic mobility was considerably improved by O'Brien and White [16]. Approximate analytic mobility expressions have been proposed by several authors [17–19].

Consider a spherical hard particle of radius a and zeta potential ζ moving with a velocity U in a liquid containing a general electrolyte composed of N ionic species with valence z_i and bulk concentration (number density) n_i^∞, and drag coefficient λ_i ($i = 1, 2, \dots, N$). The origin of the spherical polar coordinate system (r, θ, ϕ) is held fixed at the center of the particle. Ohshima et al. [19] derived the following general mobility expression:

$$\mu = \frac{a^2}{9} \int_a^\infty \left(1 - \frac{3r^2}{a^2} + \frac{2r^3}{a^3} \right) G(r)\, dr \tag{2.2}$$

with

$$G(r) = -\frac{e}{\eta r} \frac{dy}{dr} \sum_{i=1}^{N} z_i^2 n_i^\infty \exp(-z_i y) \phi_i \tag{2.3}$$

$$y = \frac{e\psi}{kT} \tag{2.4}$$

where ψ is the equilibrium potential, y the corresponding scaled potential, η the viscosity, and the function ϕ_i, which is related to the deviation $\delta\mu_i(r)$ of the electrochemical potential of the ith ionic species at position r due to the applied electric field, is defined by

$$\delta\mu_i(r) = -z_i e \phi_i(r) E \cos\theta \tag{2.5}$$

III. MOBILITY EXPRESSION CORRECT TO ORDER ζ (HENRY'S FORMULA)

When ζ is low, the electrical double layer around a spherical particle maintains its spherical symmetry during electrophoresis. In this situation, Equation (2.2) gives

$$\mu = \frac{\varepsilon_r \varepsilon_0}{\eta} \zeta f(\kappa a) \tag{2.6}$$

with

$$f(\kappa a) = 1 - e^{\kappa a} \{ 5E_7(\kappa a) - 2E_5(\kappa a) \} \tag{2.7}$$

where $E_n(\kappa a)$ is the exponential integral of order n. Equation (2.6) was first derived by Henry [12] and $f(\kappa a)$ is called Henry's function. As $\kappa a \to \infty$, $f(\kappa a) \to 1$ and Equation (2.6) tends to Smoluchowski's formula applicable for large κa [10], viz.,

$$\mu = \frac{\varepsilon_r \varepsilon_0}{\eta} \zeta \qquad (2.8)$$

whereas if $\kappa a \to 0$, then $f(\kappa a) \to 2/3$ and Equation (2.6) becomes Hückel's formula applicable for small κa [11], viz.,

$$\mu = \frac{2\varepsilon_r \varepsilon_0}{3\eta} \zeta \qquad (2.9)$$

Ohshima [20] has derived the following simple approximate formula for Henry's function $f(\kappa a)$ with relative errors less than 1%:

$$f(\kappa a) = \frac{2}{3}\left[1 + \frac{1}{2(1 + 2.5/\kappa a\{1 + 2\ \exp(-\kappa a)\})^3} \right] \qquad (2.10)$$

For the case of a cylindrical particle, the electrophoretic mobility depends on the orientation of the particle with respect to the applied electric field. When the cylinder is oriented parallel to the applied electric field, its electrophoretic mobility $\mu_\parallel$ is given by Smoluchowski's equation (2.8), that is,

$$\mu_\parallel = \frac{\varepsilon_r \varepsilon_0}{\eta} \zeta \qquad (2.11)$$

If the cylinder is oriented perpendicular to the applied field, then the mobility depends on ζ and κa and is given by (Equation (3) in Ref. [21])

$$\mu_\perp = \frac{\varepsilon_r \varepsilon_0}{\eta} \zeta f(\kappa a) \qquad (2.12)$$

with

$$f(\kappa a) = 1 - \frac{4(\kappa a)^4}{K_0(\kappa a)} \int_{\kappa a}^{\infty} \frac{K_0(t)}{t^5} dt + \frac{(\kappa a)^2}{K_0(\kappa a)} \int_{\kappa a}^{\infty} \frac{K_0(t)}{t^3} dt \qquad (2.13)$$

where $K_0(x)$ is the zero-order modified Bessel function of the second kind. Ohshima [23] obtained an approximate formula for Henry's function for a cylinder,

$$f(\kappa a) = \frac{1}{2}\left[1 + \frac{1}{(1 + 2.55/\kappa a\{1 + \exp(-\kappa a)\})^2} \right] \qquad (2.14)$$

the relative error being less than 1%. As $\kappa a \to \infty$, $f(\kappa a) \to 1$ and Equation (2.12) gives Smoluchowski's equation (2.8), while if $\kappa a \to 0$, then $f(\kappa a) \to 1/2$. For a cylindrical particle oriented at an arbitrary angle between its axis and the applied electric field, its electrophoretic mobility averaged over a random distribution of orientation is given by [22,24]

$$\mu_{av} = \frac{1}{3}\mu_{\parallel} + \frac{2}{3}\mu_{\perp} \tag{2.15}$$

IV. MOBILITY EXPRESSION CORRECT TO ORDER $1/\kappa a$

Henry's equation (2.6) assumes that ζ is low, in which case the double layer remains spherically symmetrical during electrophoresis. For high zeta potentials, the double layer is no longer spherically symmetrical. This effect is called the relaxation effect. Henry's equation (2.6) does not take into account the relaxation effect, and thus this equation is correct to the first order of ζ. Ohshima et al. [19] derived an accurate analytic mobility expression correct to order $1/\kappa a$ in a symmetrical electrolyte of valence z and bulk concentration (number density) n with the relative error less than 1% for $10 \leq \kappa a < \infty$, which is

$$E_m = \text{sgn}(\zeta)\Bigg[\frac{3}{2}\tilde{\zeta} - \frac{3F}{1+F}H + \frac{1}{\kappa a}\bigg\{-18\Big(t + \frac{t^3}{9}\Big)K + \frac{15F}{1+F}\Big(t + \frac{7t^2}{20} + \frac{t^3}{9}\Big)$$

$$-6(1+3\tilde{m})(1 - e^{-\tilde{\zeta}/2})G + \frac{12F}{(1+F)^2}H + \frac{9\tilde{\zeta}}{1+F}(\tilde{m}G + mH)$$

$$-\frac{36F}{1+F}\Big(\tilde{m}G^2 + \frac{m}{1+F}H^2\Big)\bigg\}\Bigg] \tag{2.16}$$

with

$$E_m = \frac{3\eta ze}{2\varepsilon_r\varepsilon_0 kT}\mu \tag{2.17}$$

$$\tilde{\zeta} = \frac{ze|\zeta|}{kT} \tag{2.18}$$

$$F = \frac{2}{\kappa a}(1+3m)(e^{\tilde{\zeta}/2}-1) \tag{2.19}$$

$$G = \ln\left(\frac{1+e^{-\tilde{\zeta}/2}}{2}\right) \tag{2.20}$$

$$H = \ln\left(\frac{1+e^{\tilde{\zeta}/2}}{2}\right) \tag{2.21}$$

$$K = 1 - \frac{25}{3(\kappa a + 10)}\exp\left[-\frac{\kappa a}{6(\kappa a - 6)}\tilde{\zeta}\right] \tag{2.22}$$

$$t = \tanh\left(\frac{\tilde{\zeta}}{4}\right) \tag{2.23}$$

$$m = \frac{2\varepsilon_r\varepsilon_0 kT}{3\eta z^2 e^2}\lambda \tag{2.24}$$

$$\tilde{m} = \frac{2\varepsilon_r\varepsilon_0 kT}{3\eta z^2 e^2}\tilde{\lambda} \tag{2.25}$$

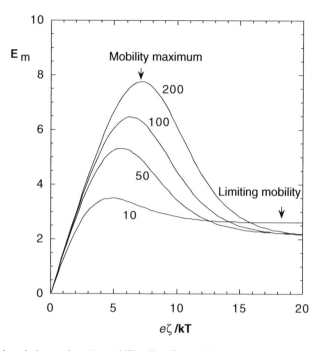

FIGURE 2.1 The reduced electrophoretic mobility E_m of a positively charged spherical colloidal particle of radius a in a KCl solution at 25°C as a function of reduced zeta-potential $e\zeta/kT$ for various values of κa. E_m is defined by Equation (2.17). Calculated via Equation (2.16).

where E_m is the scaled electrophoretic mobility, $\text{sgn}(\zeta) = +1$ if $\zeta > 0$ and -1 if $\zeta < 0$, $\tilde{\zeta}$ the magnitude of the scaled zeta potential are λ and $\tilde{\lambda}$ respectively, corresponding to the ionic drag coefficients of counterions and co-ions; m and $\tilde{m}$ are the corresponding scaled quantities.

In Figure 2.1, we plot the mobility zeta-potential relationship for a positively charged spherical particle in KCl ($\tilde{m} = 0.176$ for K and $m = 0.169$ for Cl) for several values of κa. It is seen that there is a mobility maximum due to the relaxation effect.

V. MOBILITY EXPRESSION CORRECT TO ORDER ζ^3

Equation (2.16) is applicable for all ζ for $10 \leq \kappa a < \infty$. To obtain an approximate mobility expression applicable for $\kappa a < 10$, it is convenient to express the mobility in powers of ζ and make corrections to higher powers of ζ in Henry's mobility equation (2.6), which is correct to the first power of ζ. Ohshima [25] derived a mobility formula for a spherical particle of radius a in a symmetrical electrolyte solution of valence z and bulk (number) concentration n under an applied electric field. The drag coefficient of cations, λ_+, and that of anions, λ_-, may be different. The result is

$$
\mu = \frac{2\varepsilon_r\varepsilon_0\zeta}{3\eta}\left(1 + \frac{1}{2[1 + 2.5/\{\kappa a(1 + 2e^{-\kappa a})\}]^3}\right)
$$

$$
- \frac{2\varepsilon_r\varepsilon_0\zeta}{3\eta}\left(\frac{ze\zeta}{kT}\right)^2\left[\frac{\kappa a\{\kappa a + 1.3\ \exp(-0.18\kappa a) + 2.5\}}{2\{\kappa a + 1.2\ \exp(-7.4\kappa a) + 4.8\}^3}\right.
$$

$$
\left. + \left(\frac{m_+ + m_-}{2}\right)\frac{9\kappa a\{\kappa a + 5.2\ \exp(-3.9\kappa a) + 5.6\}}{8\{\kappa a - 1.55\ \exp(-0.32\kappa a) + 6.02\}^3}\right]
$$

(2.26)

where m_+ and m_- are dimensionless ionic drag coefficients, defined by

$$m_\pm = \frac{2\varepsilon_r\varepsilon_0 kT}{3\eta z^2 e^2}\lambda_\pm \qquad (2.27)$$

The first term on the right-hand side of Equation (2.26) corresponds to an approximate Equation (2.10) for Henry's function (Equation (2.6)). Equation (2.26) excellently agreed with exact numerical results [15] especially for small κa, in which region no simple analytic mobility formula is available other than Henry's equation (2.6). Thus Equation (2.26) is a considerable improvement of Henry's equation (2.8). For example, the relative error is less than 1% for $|\zeta| \leq 7$ at $\kappa a = 0.1$ and for $|\zeta| \leq 3$ at $\kappa a = 1$.

VI. LIMITING MOBILITY OF HIGHLY CHARGED PARTICLES

It is seen from Figure 2.1 and also from Equation (2.16) that μ tends to a constant limiting value μ^∞, which is independent of ζ, as $\zeta \to \infty$. This limiting mobility μ^∞ is given by

$$\mu^\infty = \frac{\varepsilon_r\varepsilon_0}{\eta}\left(\frac{kT}{ze}\right)2 \ln 2 + O\left(\frac{1}{\kappa a}\right) \qquad (2.28)$$

Equation (2.28) implies that a highly charged particle behaves as if its effective zeta-potential ζ_{eff} were $(kT/ze)2 \ln 2$ independent of the real zeta-potential ζ. This is due to the fact that for highly charged particles, counterions are accumulated near the particle surface and these counterions do not contribute to the mobility. Recently, Ohshima [26,27] derived an expression for the limiting mobility of a positively charged particle in a general electrolyte solution, viz.,

$$\mu^\infty = \frac{\varepsilon_r\varepsilon_0}{\eta}\zeta_{\text{eff}} + O\left(\frac{1}{\kappa a}\right) \qquad \text{as } \zeta \to \infty \qquad (2.29)$$

with

$$\zeta_{\text{eff}} = -\frac{kT}{2e}\sum{}' z_i n_i \int_0^y \left\{\int_0^y \frac{\{\exp(-z_i y') - 1\}}{\sqrt{\sum_{i=1}^N n_i\{\exp(-z_i y') - 1\}}}dy'\right\} \frac{dy}{\sqrt{\sum_{i=1}^N n_i\{\exp(-z_i y) - 1\}}} \qquad (2.30)$$

where the summation $\sum'$ is taken over only co-ions ($z_i > 0$ for positively charged particles). Equation (2.29) gives

$$\zeta_{\text{eff}} = \left(\frac{kT}{e}\right)\ln 6 \quad \text{and} \quad \mu^\infty = \frac{\varepsilon_r\varepsilon_0}{\eta}\left(\frac{kT}{e}\right)\ln 6 \qquad (2.31)$$

for the case of 2−1 electrolyte and

$$\zeta_{\text{eff}} = \left(\frac{kT}{e}\right)2 \ln\left(\frac{2}{1 + 1/\sqrt{3}}\right) \quad \text{and} \quad \mu^\infty = \frac{\varepsilon_r\varepsilon_0}{\eta}\left(\frac{kT}{e}\right)2 \ln\left(\frac{2}{1 + 1/\sqrt{3}}\right) \qquad (2.32)$$

for 1−2 electrolyte. For the case of a positively charged particle in an aqueous electrolyte solution at 25°C, ζ_{eff} is 35.6 mV for 1−1 electrolytes, 46.0 mV for 2−1 electrolytes, and 12.2 mV for 1−2 electrolytes. It is also found that μ^∞ increases as the valence of co-ions increases, whereas μ^∞ decreases as the valence of counterions increases.

VII. LIQUID DROPS

The electrophoretic mobility of liquid drops is quite different from that of rigid particles since the flow velocity of the surrounding liquid is conveyed into the drop interior [28–32]. The electrophoretic mobility of a drop thus depends on the viscosity η_d of the drop as well as on the liquid viscosity η. Here we treat the case of mercury drops, in which case the drop surface is always equipotential. The general mobility expression for a mercury drop having a zeta-potential ζ is derived by Ohshima et al. [30],

$$\mu = \frac{a^2}{9} \int_a^\infty \left[\left(\frac{3\eta_d}{3\eta_d + 2\eta} \right) - \frac{3r^2}{a^2} + \left\{ \frac{6(\eta_d + \eta)}{3\eta_d + 2\eta} \right\} \frac{r^3}{a^3} \right] G(r)\, dr \tag{2.33}$$

where $G(r)$ is given in Equation (2.3). For low ζ, Equation (2.33) leads to

$$\mu = \frac{\varepsilon_r \varepsilon_0}{\eta} \zeta \left[\frac{\eta}{3\eta_d + 2\eta} \kappa a + \frac{3\eta_d + \eta}{3\eta_d + 2\eta} + 2e^{\kappa a} E_5(\kappa a) - \frac{15\eta_d}{3\eta_d + 2\eta} e^{\kappa a} E_7(\kappa a) \right] \tag{2.34}$$

For $\eta_d \to \infty$, Equation (2.34) reduces to Henry's equation (2.6) for a rigid sphere. Ohshima [33] has shown that Equation (2.34) is further approximated well by

$$\mu = \frac{2\varepsilon_r \varepsilon_0}{3\eta} \zeta \left[1 + \frac{1}{2(1 + \frac{1.86}{\kappa a})^3} \right] \left[\frac{\eta}{3\eta_d + 2\eta} \kappa a + \frac{3(\eta_d + \eta)}{3\eta_d + 2\eta} \right] \tag{2.35}$$

An approximate analytic mobility equation applicable for arbitrary values of ζ was derived by Levich [28], and Ohshima et al. [30] derived a more accurate mobility expression correct to order $1/\kappa a$ for the case of symmetrical electrolytes of valence z. The leading term of their expression is given by [30]

$$\mu = \frac{2\varepsilon_r \varepsilon_0 kT}{3\eta ze} \frac{\text{sgn}(\zeta)}{(\frac{\eta_d}{\eta} + \frac{2}{3} + D)} \left(\frac{\kappa a}{2} \right) (e^{\tilde{\zeta}/2} - e^{-\tilde{\zeta}/2}) \tag{2.36}$$

with

$$D = m(1 - e^{\tilde{\zeta}/2})^2 + \tilde{m}(1 - e^{-\tilde{\zeta}/2})^2 \tag{2.37}$$

where $\tilde{\zeta}$, F, m, and $\tilde{m}$ are already given in Equations (2.18), (2.19), (2.24), and (2.25), respectively. We find that in the limit of large κa,

$$\frac{\mu(\text{mercury})}{\mu(\text{rigid})} = O(\kappa a) \tag{2.38}$$

That is, in this limit, the mobility of mercury drops is much larger than that of rigid particles by the order of κa. It is also to be noted that mercury drops behave like rigid particles at very high zeta potentials (solidification effect) [28,30] and exhibit the same limiting mobility given by Equation (2.28).

VIII. GENERAL MOBILITY EXPRESSION FOR SOFT PARTICLES

Now we consider the electrokinetic behavior of soft particles, i.e., colloidal particles covered with a polymer layer (Figure 2.2). A number of theoretical studies have been made [34–46] on the basis of the model of Debye and Bueche [47], which assumes that the polymer segments are regarded as resistance centers distributed in the polymer layer, exerting frictional forces γu on the liquid flowing in the polymer layer, where u the liquid flow velocity and γ a frictional coefficient. The Navier–Stokes equation for the liquid flow inside the polymer layer is thus given by

$$\eta \nabla \times \nabla \times u + \gamma u + \nabla p + \rho_{cl} \nabla \psi = 0 \tag{2.39}$$

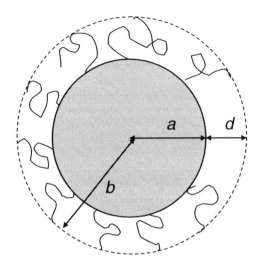

FIGURE 2.2 A soft particle consisting of a hard particle core of radius a covered with a polymer layer of thickness d; $b = a + d$.

Ohshima [42–45] presented a theory for electrophoresis of a soft particle. The general mobility expression of a soft particle that consists of the hard particle core of radius a covered with a layer of polyelectrolytes of thickness d ($= b - a$) and moves in an electrolyte solution of viscosity η is given by [45]

$$
\mu = \frac{b^2}{9} \int_b^\infty \left[3\left(1 - \frac{r^2}{b^2}\right) - \frac{2L_2}{L_1}\left(1 - \frac{r^3}{b^3}\right) \right] G(r)\,dr + \frac{2L_3}{3\lambda^2 L_1} \int_a^\infty \left(1 + \frac{r^3}{2b^3}\right) G(r)\,dr
$$

$$
- \frac{2}{3\lambda^2} \int_a^b \left[1 - \frac{3a}{2\lambda^2 b^3 L_1} \{(L_3 + L_4\lambda r)\cosh[\lambda(r-a)] - (L_4 + L_3\lambda r)\sinh[\lambda(r-a)]\} \right] G(r)\,dr
$$

$$(2.40)$$

where

$$
L_1 = \left(1 + \frac{a^3}{2b^3} + \frac{3a}{2\lambda^2 b^3} - \frac{3a^2}{2\lambda^2 b^4}\right)\cosh[\lambda(b-a)] - \left(1 - \frac{3a^2}{2b^2} + \frac{a^3}{2b^3} + \frac{3a}{2\lambda^2 b^3}\right)\frac{\sinh[\lambda(b-a)]}{\lambda b}
$$

$$(2.41)$$

$$
L_2 = \left(1 + \frac{a^3}{2b^3} + \frac{3a}{2\lambda^2 b^3}\right)\cosh[\lambda(b-a)] + \frac{3a^2}{2b^2}\frac{\sinh[\lambda(b-a)]}{\lambda b} - \frac{3a}{2\lambda^2 b^3}
$$

$$(2.42)$$

$$
L_3 = \cosh[\lambda(b-a)] - \frac{\sinh[\lambda(b-a)]}{\lambda b} - \frac{a}{b}
$$

$$(2.43)$$

$$
L_4 = \sinh[\lambda(b-a)] - \frac{\cosh[\lambda(b-a)]}{\lambda b} + \frac{\lambda a^2}{3b} + \frac{2\lambda b^2}{3a} + \frac{1}{\lambda b}
$$

$$(2.44)$$

and

$$\lambda = \left(\frac{\gamma}{\eta}\right)^{1/2} \tag{2.45}$$

IX. CHARGED-POLYMER-COATED PARTICLE

Consider a particle coated with a charged polymer layer (i.e., a polyelectrolyte layer). For the case where dissociated groups of valence Z are distributed with a uniform density N in the polyelectrolyte layer and the electrolyte is symmetrical with valence z and bulk concentration (number density) n, Ohshima [42–45] obtained

$$\mu = \frac{\varepsilon_r \varepsilon_0}{\eta} \frac{\psi_0/\kappa_m + \psi_{DON}/\lambda}{1/\kappa_m + 1/\lambda} f\left(\frac{d}{a}\right) + \frac{ZeN}{\eta\lambda^2} \tag{2.46}$$

with

$$f\left(\frac{d}{a}\right) = \frac{2}{3}\left[1 + \frac{1}{2(1+d/a)^3}\right] \tag{2.47}$$

$$\psi_0 = \frac{kT}{ze}\left(\ln\left[\frac{ZN}{2zn} + \left\{\left(\frac{ZN}{2zn}\right)^2 + 1\right\}^{1/2}\right] + \frac{2zn}{ZN}\left[1 - \left\{\left(\frac{ZN}{2zn}\right)^2 + 1\right\}^{1/2}\right]\right) \tag{2.48}$$

$$\psi_{DON} = \frac{kT}{ze}\ln\left[\frac{ZN}{2zn} + \left\{\left(\frac{ZN}{2zn}\right)^2 + 1\right\}^{1/2}\right] \tag{2.49}$$

$$\kappa_m = \kappa\left[1 + \left(\frac{ZN}{2zn}\right)^2\right]^{1/4} \tag{2.50}$$

where ψ_0 is the surface potential (i.e., the potential at the boundary of the polyelectrolyte layer and the surrounding solution), ψ_{DON} the Donnan potential of the polyelectrolyte layer, and κ_m the effective Debye–Hückel parameter of the polyelectrolyte layer that involves the contribution of the fixed charges ZeN.

Equation (2.46) is applicable for the case where the relaxation effect is negligible and $\lambda a \gg 1$, $\kappa a \gg 1$, $\lambda d \gg 1$, and $\kappa d \gg 1$. For $d \ll a$ ($f(d/a) \approx 1$), the polyelectrolyte layer can be regarded as planar and Equation (2.46) becomes

$$\mu = \frac{\varepsilon_r \varepsilon_0}{\eta} \frac{\psi_0/\kappa_m + \psi_{DON}/\lambda}{1/\kappa_m + 1/\lambda} + \frac{ZeN}{\eta\lambda^2} \tag{2.51}$$

whereas for $d \gg a$ ($f(d/a) \approx 2/3$), the soft particle behaves like a spherical polyelectrolyte with no particle core. In the limit $a \to 0$, the particle core vanishes and the particle becomes a spherical polyelectrolyte (a porous charged sphere). For low potentials, in particular, Equation (2.46) tends to

$$\mu = \frac{ZeN}{\eta\lambda^2}\left[1 + \frac{2}{3}\left(\frac{\lambda}{\kappa}\right)^2 \frac{1 + \lambda/2\kappa}{1 + \lambda/\kappa}\right] \tag{2.52}$$

which agrees with Hermans–Fujita's equation for spherical polyelectrolytes [48].

For cylindrical soft particles, Ohshima [49,50] derived the following mobility expressions:

$$\mu_\parallel = \frac{\varepsilon_r \varepsilon_0}{\eta} \frac{\psi_0/\kappa_m + \psi_{DON}/\lambda}{1/\kappa_m + 1/\lambda} + \frac{ZeN}{\eta\lambda^2} \tag{2.53}$$

and

$$\mu_\perp = \frac{\varepsilon_r \varepsilon_0}{\eta} \frac{\psi_0/\kappa_m + \psi_{DON}/\lambda}{1/\kappa_m + 1/\lambda} f\left(\frac{d}{a}\right) + \frac{ZeN}{\eta\lambda^2} \tag{2.54}$$

with

$$f\left(\frac{d}{a}\right) = \frac{1}{2}\left[1 + \frac{1}{2(1+d/a)^2}\right] \tag{2.55}$$

Ohshima [51] also derived an expression for the electroosmotic velocity in an array of parallel soft cylinders.

For the case where the fixed charges are not uniformly distributed in the polyelectrolyte layer and that the relative permittivity in the polyelectrolyte layer does not take the same value as that in the bulk solution phase, the above theory must be modified as discussed by Ohshima and Kondo [52] and Hsu et al. [54]. Tseng et al. [55] considered the effects of charge regulation on the mobility in the polyelectrolyte layer. The case where the polyelectrolyte layer is not fully ion-penetrable is considered in Ref. [56]. Varoqui [57] considered the case where electrically neutral polymers are adsorbed with an exponential segment density distribution onto the particle surface with a charge density. Ohshima [58] extended Varoqui's theory [57] to the case where adsorbed polymers are charged. Saville [59] and Hill et al. [60] considered the relaxation effects of soft particles in electrophoresis.

X. UNCHARGED-POLYMER-COATED PARTICLE

The general mobility expression (2.76) is also applicable for the case where a hard particle of radius a carrying with zeta-potential ζ is covered with an uncharged polymer layer. Ohshima [61] derived the following mobility expression applicable for arbitrary values of κa but for low zeta potentials ζ:

$$\mu = \frac{\varepsilon_r \varepsilon_0 \zeta}{\eta} f_m(\kappa a, \kappa b, \lambda a, \lambda b) \tag{2.56}$$

with

$$
\begin{aligned}
f_m(\kappa a, \kappa b, \lambda a, \lambda b) =& \frac{\kappa^2 a^4}{6} \int_b^\infty \frac{e^{-\kappa(r-a)}}{r^3} dr - \frac{\kappa^2 a^4 b^2}{2} \left(1 - \frac{2L_2}{3L_1} + \frac{2L_3}{\lambda^2 b^2 L_1}\right) \int_b^\infty \frac{e^{-\kappa(r-a)}}{r^5} dr \\
&+ \frac{3\kappa^2 a^5 (L_3 + L_4)}{4\lambda^3 b^3 L_1} \left[\frac{\kappa}{3} \int_a^b \frac{e^{-(\kappa+\lambda)(r-a)}}{r^3} dr - \frac{1}{\lambda} \int_a^b \frac{e^{-(\kappa+\lambda)(r-a)}}{r^5} dr\right] \\
&- \frac{3\kappa^2 a^5 (L_3 - L_4)}{4\lambda^3 b^3 L_1} \left[\frac{\kappa}{3} \int_a^b \frac{e^{-(\kappa-\lambda)(r-a)}}{r^3} dr + \frac{1}{\lambda} \int_a^b \frac{e^{-(\kappa-\lambda)(r-a)}}{r^5} dr\right] \\
&+ \frac{\kappa^2 a^4}{\lambda^2} \left(1 - \frac{L_3}{L_1}\right) \int_a^b \frac{e^{-\kappa(r-a)}}{r^5} dr \\
&- \frac{\kappa^2 a^2 (L_3 + L_4)}{2\lambda^3 b^3 L_1} \left[\left\{\frac{1}{\lambda b}\left(1 + \frac{a^3}{2b^3}\right) + \frac{\kappa}{\kappa+\lambda}\right\} e^{-(\kappa+\lambda)(b-a)} - \frac{3}{2\lambda a} - \frac{\kappa}{\kappa+\lambda}\right] \\
&- \frac{\kappa^2 a^2 (L_3 - L_4)}{2\lambda^3 b^3 L_1} \left[\left\{\frac{1}{\lambda b}\left(1 + \frac{a^3}{2b^3}\right) - \frac{\kappa}{\kappa-\lambda}\right\} e^{-(\kappa-\lambda)(b-a)} - \frac{3}{2\lambda a} + \frac{\kappa}{\kappa-\lambda}\right] \\
&+ \frac{\kappa^2}{\lambda^2}\left[\frac{L_3}{L_1} - 1 + \frac{a^3 L_3}{2b^3 L_1}\left(1 + \frac{2}{\kappa a} + \frac{2}{\kappa^2 a^2}\right)\right] \\
&- \frac{2\kappa a}{3}\left[1 - \frac{L_2}{L_1}\left(1 + \frac{1}{\kappa b}\right) - \frac{\kappa}{\lambda^2 b}\left(1 + \frac{a^3}{2b^3}\right)\right] e^{-\kappa(b-a)} \tag{2.57}
\end{aligned}
$$

where $f_m(\kappa a, \kappa b, \lambda a, \lambda b)$ is a modified Henry function for polymer-coated spherical particles, L_1–L_4 are given by Equations (2.41)–(2.44), and λ is given by Equation (2.45). Ohshima [62] also derived a mobility formula applicable for arbitrary ζ but large κa in a z–z symmetrical electrolyte, viz.,

$$\mu = \frac{3a(L_3 + \lambda a L_4)\varepsilon_r \varepsilon_0}{2\eta \lambda^2 b^3 L_1} \left\{ \zeta - \frac{2F}{1+F}\left(\frac{kT}{ze}\right) \ln\left[\frac{1 + \exp(ze\zeta/2kT)}{2}\right] \right\} \tag{2.58}$$

where F is given by Equation (2.19).

XI. ELECTROPHORETIC MOBILITY IN SALT-FREE MEDIA

So far we have discussed electrophoresis of particles and drops in electrolyte solutions, i.e., salt-containing media. For those in salt-free media containing only counterions (e.g., nonaqueous media [63]), special consideration is needed. To treat a suspension of spherical particles in a salt-free medium, one usually employs a free volume model, in which each sphere of radius a is surrounded by a spherical free volume of radius R within which counterions are distributed so that electrical neutrality as a whole is satisfied. The particle volume fraction ϕ is given by $\phi = (a/R)^3$. We treat the case of dilute suspensions, viz., $\phi \ll 1$ or $a/R \ll 1$. Let the concentration (number density) and valence of the counterions be n and $-z$, respectively. For a spherical particle carrying surface charge density σ or total surface charge $Q = 4\pi a^2 \sigma$, it follows from the electroneutrality condition that

$$Q = 4\pi a^2 \sigma = \frac{4}{3}\pi(R^3 - a^3)zen \tag{2.59}$$

The Debye–Hückel parameter thus depends on the particle charge, viz.,

$$\kappa = \left(\frac{z^2 e^2 n}{\varepsilon_r \varepsilon_0 kT}\right)^{1/2} = \left[\frac{ze}{\varepsilon_r \varepsilon_0 kT}\frac{3a^2\sigma}{(R^3 - a^3)}\right]^{1/2} = \left[\frac{3zeQ}{4\pi\varepsilon_r \varepsilon_0 kT(R^3 - a^3)}\right]^{1/2} \tag{2.60}$$

Consider first the equilibrium potential distribution around a spherical particle in a salt-free medium. The Poisson–Boltzmann equation for electric potential $\psi(r)$ (or its scaled form $y = ze\psi/kT$) at a distance r from the origin of one particle is given by

$$\frac{d^2 y}{dr^2} + \frac{2}{r}\frac{dy}{dr} = \kappa^2 e^y \tag{2.61}$$

Here $\psi(r)$ is set equal to zero at points where the concentration of counterions equals its average value n. An approximate solution to Equation (2.61) for the case of dilute suspensions has been obtained by Imai and Oosawa [64–66] and later by Ohshima [67]. They showed that there are two distinct cases separated by a certain critical value of the particle charge Q, that is, (case 1) the low-charge case and (case 2) the high-charge case. In the latter case, counterions are condensed near the particle surface (counterion condensation). For the dilute case ($\phi \ll 1$), approximate values of $\psi(a)$ and $\psi(R)$ together with the particle surface potential ψ_s defined by $\psi_s = \psi(a) - \psi(R)$ are given below.

The low-charge case: If

$$Q^* \leq \ln\left(\frac{1}{\phi}\right) \tag{2.62}$$

is satisfied, then

$$\psi(a) = \frac{Q}{4\pi\varepsilon_r\varepsilon_0 a} \tag{2.63}$$

$$\psi(R) \approx 0 \tag{2.64}$$

$$\psi_s = \frac{Q}{4\pi\varepsilon_r\varepsilon_0 a} \tag{2.65}$$

where

$$Q^* = \frac{Q}{4\pi\varepsilon_r\varepsilon_0 a}\frac{ze}{kT} \tag{2.66}$$

is the scaled particle charge.
   The high-charge case: If

$$Q^* > \ln\left(\frac{1}{\phi}\right) \tag{2.67}$$

is satisfied, then

$$\psi(a) = \frac{kT}{ze}\ln\left(\frac{Q^*}{6\phi}\right) \tag{2.68}$$

$$\psi(R) = -\frac{kT}{ze}\ln\left[\frac{Q^*}{\ln(1/\phi)}\right] \tag{2.69}$$

$$\psi_s = \frac{kT}{ze}\ln\left[\frac{Q^{*2}}{6\phi\,\ln(1/\phi)}\right] \tag{2.70}$$

In the low-charge case, ψ_s is proportional to Q (Equation (2.65)). In the high-charge case, however, because of the counterion condensation effect, the increase of ψ_s with Q is suppressed (Equation (2.70)). An approximate expression for the critical value Q_{cr}^* of Q^* separating the low-charge case and the high-charge case for dilute suspensions is given by

$$Q_{cr}^* = \ln\left(\frac{1}{\phi}\right) \tag{2.71}$$

Ohshima [68,69] derived approximate expressions for the electrophoretic mobility of a spherical particle in a salt-free medium. The results are:

$$\mu = \frac{Q}{6\pi\eta a} \quad \text{or} \quad \mu = \frac{2\varepsilon_r\varepsilon_0\zeta}{3\eta} \quad \text{(for the low-charge case)} \tag{2.72}$$

$$\mu = \frac{2\varepsilon_r\varepsilon_0}{3\eta}\frac{kT}{ze}\ln\left(\frac{1}{\phi}\right) \quad \text{(for the high-charge case)} \tag{2.73}$$

That is, in the low-charge case ($Q^* \leq \ln(1/\phi)$), μ is given by Hückel's formula (Equation (2.9)) and in the high-charge case ($Q^* > \ln(1/\phi)$), because of the effect of counterion condensation, μ becomes constant and is the same as if the zeta potential were always equal to $(kT/ze)\ln(1/\phi)$, independent of Q.

The above theory can be extended to electrophoresis of liquid drops [70] and soft particles [71a,b] in salt-free media. The result for liquid drops of viscosity η_d is as follows [70]:

$$\mu = \frac{Q(\eta_d + \eta)}{2\pi\eta a(3\eta_d + 2\eta)} \quad \text{or} \quad \mu = \frac{2\varepsilon_r\varepsilon_0\zeta(\eta_d + \eta)}{\eta(3\eta_d + 2\eta)} \quad \text{(for the low-charge case)} \tag{2.74}$$

$$\mu = \frac{2\varepsilon_r\varepsilon_0\zeta(\eta_d + \eta)}{\eta(3\eta_d + 2\eta)} \frac{kT}{ze}\ln\left(\frac{1}{\phi}\right) \quad \text{(for the high-charge case)} \tag{2.75}$$

For soft particles [71a,b],

$$\mu = \frac{Q}{D_H} \quad \text{(for the low-charge case)} \tag{2.76}$$

$$\mu = \frac{4\pi\varepsilon_r\varepsilon_0 b}{D_H} \frac{kT}{ze}\ln\left(\frac{1}{\phi}\right) \quad \text{(for the high-charge case)} \tag{2.77}$$

with

$$D_H = 6\pi\eta b\left(\frac{L_2}{L_1} + \frac{3L_3}{2\lambda^2 b^2 L_1}\right)^{-1} \tag{2.78}$$

where D_H is the drag coefficient of a soft particle, L_1–L_3 are defined by Equations (2.41)–(2.43), and λ is given by Equation (2.45).

REFERENCES

1. Dukhin, S.S. and Derjaguin, B.V., Electrokinetic phenomena, in: *Surface and Colloid Science*, Matievic, E., Ed., Wiley, New York, 1974, Vol. 7.
2. Hunter, R.J., *Zeta Potential in Colloid Science*, Academic Press, New York, 1981.
3. van de Ven, T.G.M., *Colloid Hydrodynamics*, Academic Press, New York, 1989.
4. Hunter, J., Ed., *Foundations of Colloid Science*, Clarendon Press University Press, Oxford, 1989, Vol. 2, Chapter 13.
5. Dukhin, S.S., Non-equilibrium electric surface phenomena, *Adv. Colloid Interface Sci.*, 44, 1, 1993.
6. Lyklema, J., *Fundamentals of Interface and Colloid Science, Solid–Liquid Interfaces*, Academic Press, New York, 1995, Vol. 2.
7. Ohshima, H. and Furusawa, K., Eds., *Electrical Phenomena at Interfaces: Fundamentals, Measurements, and Applications*, 2nd ed., revised and expanded, Dekker, New York, 1998.
8. Delgado, A.V., Ed., *Electrokinetics and Electrophoresis*, Dekker, New York, 2000.
9. Ohshima, H., Electrokinetic behavior of particles: theory, in: *Encyclopedia of Surface and Colloid Science*, Somasundaran, P., Ed., Dekker, New York, 2000.
10. von Smoluchowski, M., Electrische Endosmose und Strömungsströme, in: *Handbuch der Elektrizität und des Magnetismus*, Greatz, L., Ed., Barth, Leipzig, 1921, Vol. 2, p. 366.
11. Hückel, E., *Phys. Z. Die Kataphoresese der Kugel.*, 25, 204, 1924.
12. Henry, D.C., The cataphoresis of suspended particles, *Proc. R. Soc. London, Ser. A*, 133, 106, 1931.
13. Overbeek, J.Th.G., Theorie der Elektrophorese. Der Relaxationseffekt, *Kolloid-Beihefte*, 54, 287, 1943.

14. Booth, F., The cataphoresis of spherical, solid non-conducting particles in a symmetrical electrolyte, *Proc. R. Soc. London, Ser. A*, 203, 514, 1950.

15. Wiersema, P.H., Loeb, A.L., and Overbeek, J.Th.G., Calculation of the electrophoretic mobility of a spherical colloid particle, *J. Colloid Interface Sci.*, 22, 78, 1966.

16. O'Brien, R.W. and White, L.R., Electrophoretic mobility of a spherical colloidal particle, *J. Chem. Soc., Faraday Trans. 2*, 74, 1607, 1978.

17. Dukhin, S.S. and Semenikhin, N.M., Theory of double-layer polarization and its influence on the electrokinetic and electrooptical phenomena and the dielectric permeability of disperse systems: calculation of the electrophoretic and diffusiophoretic mobility of solid spherical particles, *Kolloid Zh.*, 32, 360, 1970.

18. O'Brien, R.W. and Hunter, R.J., The electrophoretic mobility of large colloidal particles, *Can. J. Chem.*, 59, 1878, 1981.

19. Ohshima, H., Healy, T.W., and White, L.R., Approximate analytic expressions for the electrophoretic mobility of spherical colloidal particles and the conductivity of their dilute suspensions, *J. Chem. Soc., Faraday Trans. 2*, 79, 1613, 1983.

20. Ohshima, H., A simple expression for Henry's function for the retardation effect in electrophoresis of spherical colloidal particles, *J. Colloid Interface Sci.*, 168, 269, 1994.

21. van der Drift, W.P.J.T., Keizer, A. de, and Overbeek, J.Th.G., Electrophoretic mobility of a cylinder with high surface charge density, *J. Colloid Interface Sci.*, 71, 67, 1979.

22. Stigter, D., Electrophoresis of highly charged colloidal cylinders in univalent salt solutions. 2. Random orientation in external field and application to polyelectrolytes, *J. Phys. Chem.*, 82, 1424, 1978.

23. Ohshima, H., Henry's function for electrophoresis of a cylindrical colloidal particle, *J. Colloid Interface Sci.*, 180, 299, 1996.

24. Keizer, A. de, van der Drift, W.P.J.T., and Overbeek, J.Th.G., Electrophoresis of randomly oriented cylindrical particles, *Biophys. Chem.*, 3, 107, 1975.

25. Ohshima, H., Approximate analytic expression for the electrophoretic mobility of a spherical colloidal particle, *J. Colloid Interface Sci.*, 239, 587, 2001.

26. Ohshima, H., On the limiting electrophoretic mobility of a highly charged colloidal particle in an electrolyte solution, *J. Colloid Interface Sci.*, 263, 337, 2003.

27. Ohshima, H., Electrophoretic mobility of a highly charged colloidal particle in a solution of general electrolytes, *J. Colloid Interface Sci.*, 275, 665, 2004.

28. Levich, V.G., *Physicochemical Hydrodynamics*, Prentice-Hall, Englewood Cliffs, NJ, 1962.

29. Levine, S. and O'Brien, R.N., A theory of electrophoresis of charged mercury drops in aqueous electrolyte solution, *J. Colloid Interface Sci.*, 43, 616, 1973.

30. Ohshima, H., Healy, T.W., and White, L.R., Electrokinetic phenomena in a dilute suspension of charged mercury drops, *J. Chem. Soc., Faraday Trans. 2*, 80, 1643, 1984.

31. Baygents, J.C. and Saville, D.A., Electrophoresis of drops and bubbles, *J. Chem. Soc., Faraday Trans.*, 87, 1883, 1991.

32. Baygents, J.C. and Saville, D.A., Electrophoresis of small particles and fluid globules in weak electrolytes, *J. Colloid Interface Sci.*, 146, 9, 1991.

33. Ohshima, H., A simple expression for the electrophoretic mobility of charged mercury drops, *J. Colloid Interface Sci.*, 189, 376, 1997.

34. Donath, E. and Pastuschenko, V., Electrophoretical study of cell surface properties: the influence of the surface coat on the electric potential distribution and on general electrokinetic properties of animal cells, *Bioelectrochem. Bioenerg.*, 6, 543, 1979.

35. Jones, I.S., A theory of electrophoresis of large colloidal particles with adsorbed polyelectrolyte, *J. Colloid Interface Sci.*, 68, 451, 1979.

36. Wunderlich, R.W., The effect of surface structure on the electrophoretic mobilities of large particles, *J. Colloid Interface Sci.*, 88, 385, 1982.

37. Levine, S., Levine, M., Sharp, K.A., and Brooks, D.E., Theory of the electrokinetic behavior of human erythrocytes, *Biophys. J.*, 42, 127, 1983.

38. Scharp, K.A. and Brooks, D.E., Calculation of the electrophoretic mobility of a particle bearing bound polyelectrolyte using the nonlinear Poisson–Boltzmann equation, *Biophys. J.*, 47, 563, 1985.

39. Ohshima, H. and Kondo, T., Electrophoresis of large colloidal particles with surface charge layers: position of the slipping plane and surface charge layer thickness, *Colloid Polym. Sci.*, 264, 1080, 1986.

40. Ohshima, H. and Kondo, T., Electrophoretic mobility and Donnan potential of a large colloidal particle with a surface charge layers, *J. Colloid Interface Sci.*, 116, 305, 1987.

41. Ohshima, H. and Kondo, T., Approximate analytic expression for the electrophoretic mobility of colloidal particles with surface charge layers, *J. Colloid Interface Sci.*, 130, 281, 1989.

42. Ohshima, H., Electrophoretic mobility of soft particles, *J. Colloid Interface Sci.*, 163, 474, 1994.

43. Ohshima, H., Electrophoretic mobility of soft particles, *Adv. Colloid Interface Sci.*, 62, 443, 1995.

44. Ohshima, H., Electrophoretic mobility of soft particles, *Colloids Surf. A: Physicochem. Eng. Aspects*, 103, 249, 1995.

45. Ohshima, H., On the general expression for the electrophoretic mobility of a soft particle, *J. Colloid Interface Sci.*, 228, 190, 2000.

46. Ohshima, H. and Ohki, S., Donnan potential and surface potential of a charged membrane, *Biophys. J.*, 47, 673, 1985.

47. Debye, P. and Bueche, A.M., Intrinsic viscosity, diffusion, and sedimentation rate of polymers in solution, *J. Chem. Phys.*, 16, 573, 1948.

48. Hermans, J.J. and Fujita, H., Electrophoresis of charged polymer molecules with partial free drainage, *Koninkl. Ned. Akad. Wetenschap. Proc.*, B58, 182, 1955.

49. Ohshima, H., Electrophoretic mobility of cylindrical soft particles, *Colloid Polym. Sci.*, 275, 480, 1997.

50. Ohshima, H., On the electrophoretic mobility of a cylindrical soft particle, *Colloid Polym. Sci.*, 279, 88, 2001.

51. Ohshima, H., Electroosmotic velocity in an array of parallel soft cylinders, *Colloids Surf. A*, 192, 227, 2001.

52. Ohshima, H. and Kondo, T., On the electrophoretic mobility of biological cells, *Biophys. Chem.*, 39, 191, 1991.

53. Hsu, J.P., Hsu, W.C., and Chang, Y.I., Effects of fixed-charge distribution and pH on the electrophoretic mobility of biological cells, *Colloid Polym. Sci.*, 265, 911, 1987.

54. Hsu, J.P. and Fan, Y.P., Electrophoretic mobility of a particle coated with a charged membrane: effects of fixed charge and dielectric constant distributions, *J. Colloid Interface Sci.*, 172, 230, 1995.

55. Tseng, S., Sung-Hwa Lin, S.-H., and Jyh-Ping Hsu, J.-P., Effect of pH on the electrophoretic mobility of a particle with a charge-regulated membrane in a general electrolyte solution, *Colloids Surf. B*, 13, 277, 1999.

56. Ohshima, H. and Makino, K., Electrophoretic mobility of colloidal particles covered with a partially ion-penetrable surface layer, *Colloids Surf. A*, 13, 277, 1999.

57. Varoqui, R., Effect of polymer adsorption on the electrophoretic mobility of colloids, *Nouv. J. Chim.*, 6, 187, 1982.

58. Ohshima, H., Electrophoretic mobility of polyelectrolyte-adsorbed colloidal particle: effect of segment distribution, *J. Colloid Interface Sci.*, 185, 269, 1997.

59. Saville, D.A., Electrokinetic properties of fuzzy colloidal particles, *J. Colloid Interface Sci.*, 185, 269, 2000.

60. Hill, R.J., Saville, D.A., and Russel, W.B., Electrophoresis of spherical polymer-coated colloidal particles, *J. Colloid Interface Sci.*, 259, 56, 2003.

61. Ohshima, H., Modified Henry function for the electrophoretic mobility of a charged spherical colloidal particle covered with an ion-penetrable uncharged polymer layer, *J. Colloid Interface Sci.*, 119, 252, 2002.

62. Ohshima, H., Approximate expression for the electrophoretic mobility of a spherical colloidal particle covered with an ion-penetrable uncharged polymer layer, *Colloid Polym. Sci.*, 283, 819, 2005.

63. Kitahara, A., Nonaqueous media, in: *Electrical Phenomena at Interfaces: Fundamentals, Measurements, and Applications*, Ohshima, H. and Furusawa, K., Eds., 2nd ed., revised and expanded, Dekker, New York, 1998, Chapter 7.

64. Imai, N. and Oosawa, F., Counterion distribution around a highly charged spherical polyelectrolyte, *Busseiron Kenkyu, I*, 52, 42, 1952 (in Japanese).

65. Imai, N. and Oosawa, F., Counterion distribution around a highly charged spherical polyelectrolyte, *Busseiron Kenkyu, II*, 59, 99, 1953 (in Japanese).

66. Oosawa, F., *Polyelectrolytes*, Dekker, New York, 1971.

67. Ohshima, H., Surface charge density/surface potential relationship for a spherical colloidal particle in a salt-free medium, *J. Colloid Interface Sci.*, 225, 233, 2000.

68. Ohshima, H., Electrophoretic mobility of a spherical colloidal particle in a salt-free medium, *J. Colloid Interface Sci.*, 248, 499, 2002.

69. Ohshima, H., Numerical calculation of the electrophoretic mobility of a spherical particle in a salt-free medium, *J. Colloid Interface Sci.*, 262, 294, 2003.

70. Ohshima, H., Electrophoretic mobility of a liquid drop in a salt-free medium, *J. Colloid Interface Sci.*, 263, 333, 2003.

71a. Ohshima, H., Electrophoretic mobility of a soft particle in a salt-free medium, *J. Colloid Interface Sci.*, 269, 255, 2004.

71b. Ohshima, H., Electrophoretic mobility of a soft particle in a salt-free medium, *J. Colloid Interface Sci.*, 272, 503, 2004.

3 Electrokinetic Phenomena in Suspensions

Ángel V. Delgado and F. González-Caballero
Universidad de Granada, Granada, Spain

CONTENTS

I. INTRODUCTION

Everybody involved in the use, design, or investigation of colloidal particles dispersed in aqueous solution is aware of the importance of the surface charge acquired by the particles. Many properties of the suspension are in fact dramatically controlled by the large surface-to-volume ratio of the particles and by the additional phenomenon of charging of the interface. With respect to the first issue, recall that while a spherical particle 1 cm in diameter has a surface-to-volume ratio of 600 m^{-1}, the figure increases by as much as a factor of 10^4 if the same particle was subdivided into 1-μm particles. But it is the second issue mentioned (the existence of the charged interface) that is the core of this chapter. Figure 3.1 shows approximately the stability ratio of charged spheres 100 nm in diameter suspended in 1 mmol/l KCl solutions. This quantity is the ratio of the number of collisions between particles to the number of collisions resulting in coagulation, and it measures how effective is the surface charge to avoid aggregation in the suspension. Note the dramatic effect of the surface charge on the colloidal stability of the system: it increases by orders of magnitude even for the relatively modest charges accessible to the colloidal particles.

Not in all cases are those charges directly accessible by experiment (for instance, by conductimetric or potentiometric titration, see Refs. [1–3]). It is the rule rather than the exception that the

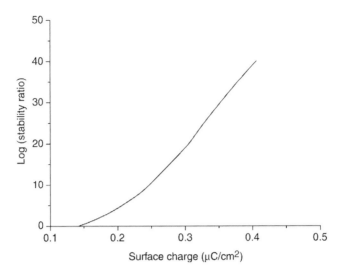

FIGURE 3.1 Stability ratio (ratio between the number of collisions and the number of collisions leading to coagulation) of a suspension of 100-nm spheres in 1 mmol/l KCl solution, for different surface charge densities.

surface charge must be inferred or indirectly determined, and a number of methods available for such a determination are grouped under the common name of electrokinetic techniques, based on the occurrence of *electrokinetic phenomena.* In order to understand them, it is first necessary to consider that electroneutrality conditions impose that the liquid adjacent to the suspended particle has a net electric charge, opposite to that on the surface: part of the ions in that liquid will likely be strongly attached to the surface by short-range attractive forces, and can be considered immobile, and the same will be admitted with respect to the liquid in that region. On the contrary, both ions and liquid outside it can be moved by an external field, for instance, an electric one: in fact the electric force will act on the ions (mainly, counterions, that is, ions with charge opposite in sign to that of the surface) and they will drag liquid in their motion. The electric potential existing at the boundary between the mobile and immobile phases is known as electrokinetic or zeta (ζ) potential, and the relative motion between the solid and the liquid referred to above is in fact the very essence of electrokinetic phenomena.

Briefly, one could say that they involve the generation of a liquid flow (or a particle motion) when an external field is applied to the solid–liquid interface, or, conversely, the generation of an electric potential when a relative solid–liquid motion is provoked by, for instance, a pressure gradient or a gravitational field. In this context, electrokinetic phenomena and the techniques associated with them demonstrate their importance. They are manifestations of the electrical properties of the interface, and hence deserve attention by themselves. But, furthermore, they are a valuable (unique, in many cases) source of information on those electrical properties, because of the possibility of being experimentally determined [4–6].

From these comments, it appears reasonable to admit that, prior to a thorough description of the electrokinetic effects in suspensions, some sketch of the structure of the ionic atmosphere surrounding the particle should be given. This is the aim of the following section.

II. THE ELECTRICAL DOUBLE LAYER AROUND A COLLOID PARTICLE

It is an experimental fact that most solids acquire an electric surface charge when dispersed in a polar solvent, in particular, in an aqueous electrolyte solution. The origins of this charge are diverse [4–8], and include:

1. *Preferential adsorption of ions in solution.* This is the case of ionic surfactant adsorption. The charged entities must have a high affinity for the surface to avoid electrostatic repulsion by already adsorbed ions.
2. *Adsorption–desorption of lattice ions.* Silver iodide particles in Ag^+ or I^- solutions are the typical example: the crystal lattice ions can easily find their way into crystal sites and become part of the surface. They are called potential-determining ions (p.d.i.).
3. *Direct dissociation or ionization of surface groups.* This is the mechanism through which most polymer latexes get their charge. Thus, the dissociation of sulfate and carboxyl groups is responsible for the negative charge of many anionic polymer lattices. In the case of oxides, zwitterionic MOH (M is the metal) surface groups can generate either positive or negative charge, depending on pH. H^+ and OH^- would hence be p.d.i. for oxides.
4. *Charge-defective lattices: isomorphous substitution.* This is a mechanism typical of clay minerals [9]: some of the Si^{4+} and Al^{3+} cations of ideal structure are substituted by others with lower charge and almost the same size. As a consequence, the crystal would be negatively charged, although this structural charge is compensated by surface cations, easily exchangeable in solution.

The net surface charge must be compensated for by ions around the particle to maintain the electroneutrality of the system. Both the surface charge and its compensating countercharge in solution form the so-called electrical double layer (EDL). In spite of the traditional use of the word "double," its structure can be very complex, not fully resolved in many instances, and it may contain three or more layers, extending over varying distances from the solid surface (Figure 3.2).

It is normally admitted that the charges responsible for the surface charge σ_0, the so-called *titratable charge*, are located on the surface itself. If titration is possible, this is the charge that will be measured. The corresponding potential is the surface potential Ψ_0. Ions capable of undergoing specific adsorption might be located at a distance from the wall that will be of the order of an ionic radius, because it is assumed that they have lost their hydration shell, at least in the direction of the solid surface. We will call σ_i the surface charge and Ψ_i the potential at such a plane of atoms, located at a distance β from the solid (see Figure 3.2), and often called *inner Helmholtz*

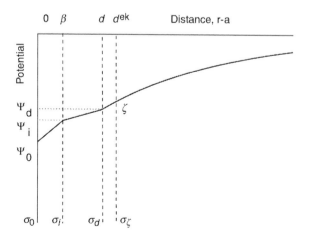

FIGURE 3.2 Schematic representation of the potential distribution around a negatively charged spherical particle of radius a.

plane (IHP). If r is the radial coordinate of a coordinate system centered in the spherical particle of radius a, then the region between $r = a$ and $r = a + \beta$, or *Stern layer* (also known as *inner part of the double layer*, or *compact part of the double layer*), is free of charge. Let us note that ions responsible for Ψ_i will have not only electrostatic interactions with the surface: in fact they often overcome electric repulsions, and are capable, for instance, of increasing the positive charge of an already positive surface. It is usual to say that the missing interactions are of a chemical or *specific* nature.

At a larger distance from the surface, $r = a + d$ and beyond, only ions undergoing electrostatic interactions with the surface and subject to thermal collisions with the molecules of solvent are located, and they are in fact distributed over a certain distance to the solid. This third layer in the ionic distribution can be characterized by a volume charge density $\rho(r)$, although it is of practical use to introduce a diffuse charge density σ_d, located at d, according to [5,8]:

$$\sigma_d = \frac{1}{(a+d)^2} \int_d^\infty r^2 \rho(r)\, dr \tag{3.1}$$

for a spherical interface. The potential at d is called diffuse or Stern potential, Ψ_d, and the volume ionic distribution extending from $r = a + d$ is called the *diffuse layer*, or *diffuse part of the double layer*. The ideal surface located at d is the *outer Helmholtz plane* (OHP), and it identifies the beginning of the diffuse layer.

The diffuse layer can be described mathematically in a rather simple way. The distribution of ions in space is given by the Boltzmann distribution:

$$n_i(\mathbf{r}) = n_i^0(\infty)\exp\left[\frac{-ez_i\Psi(\mathbf{r})}{k_B T}\right], \quad i = 1,\ldots,N \tag{3.2}$$

where $n_i(\mathbf{r})$ is the number concentration of type-i ions (it is assumed that there are N different ionic species, with charge $z_i e$) in position $\mathbf{r}$, $n_i^0(\infty)$ the equilibrium number concentration of those ions, far from the particle, k_B the Boltzmann constant, and T the absolute temperature. Finally, the Poisson equation will give us the relationship between the potential Ψ at position $\mathbf{r}$ and ionic concentrations in the same position:

$$\nabla^2 \Psi(\mathbf{r}) = -\frac{1}{\varepsilon_m}\rho(\mathbf{r}) = -\frac{1}{\varepsilon_m}\sum_{i=1}^N ez_i n_i^0(\infty)\exp\left[-\frac{ez_i\Psi(\mathbf{r})}{k_B T}\right] \tag{3.3}$$

ε_m being the electric permittivity of the dispersion medium. Equation (3.4) (the *Poisson–Boltzmann equation*) is the starting point of the Gouy–Chapman description of the diffuse layer.

It will be clear that there is no general solution to this partial differential equation but in certain cases [10,11]. We detail some of them as follows.

1. *A flat interface, with low diffuse potential.* In this case:

$$\Psi = \Psi_d e^{-\kappa x} \tag{3.4}$$

the so-called Debye approximation. κ^{-1} is the Debye length and it is clearly a measure of the diffuse layer thickness. Its value is

$$\kappa^{-1} = \left\{\frac{\varepsilon_m k_B T}{\sum_{i=1}^N e^2 z_i^2 n_i^0(\infty)}\right\}^{1/2} \tag{3.5}$$

The following is a practical formula for the calculation of κ^{-1} in an $1-1$ electrolyte ($N = 2$, $z_1 = 1$, $z_2 = -1$) dissolved in water at 25°C:

$$\kappa^{-1} = 0.308c^{-1/2} \text{ nm}$$

if c is the molar concentration of electrolyte ($n_1^0(\infty) = n_2^0(\infty) = 10^3 N_A c$, with N_A being the Avogadro number).

2. *A flat interface, in a symmetrical z-valent electrolyte ($z_1 = -z_2 = z$) for arbitrary Ψ_d:*

$$y(x) = 2 \ln \left[\frac{1 + e^{-\kappa x} \tanh(y_d/4)}{1 - e^{-\kappa x} \tanh(y_d/4)} \right] \tag{3.6}$$

where y is the dimensionless potential and given as

$$y = \frac{ze\Psi}{k_B T} \tag{3.7}$$

and a similar expression can be given for y_d.

3. *A spherical interface (radius a) at low potential:*

$$\Psi(r) = \Psi_d \left(\frac{a}{r}\right) e^{-\kappa(r-a)} \tag{3.8}$$

Note that numerical solutions or approximate analytical expressions have to be applied in other cases. It must also be pointed out that if the complex structure of the double layer is simplified to a model considering that the diffuse layer starts right on the solid surface (no Stern layer is present), Ψ_0 can be used instead of Ψ_d in Equations (3.4), (3.7), and (3.8).

Summarizing, the main hypotheses and simplifications of the Gouy–Chapman–Stern approach are [5]:

- Ion sizes are neglected: they are considered as point charges.
- The electric permittivity of the medium is constant at any point of the ionic atmosphere, with a value identical to that in the bulk solution.
- The incomplete dissociation of the solute is not considered.
- The solvent is considered as a continuum not altered or polarized by the charged surface.
- The surface charge is considered homogeneous, and the surface itself is supposed to be molecularly flat.

This is not the proper place to discuss these issues. The reader is referred to Ref. [5] for a detailed discussion. By way of example, we will only consider the ion size problem in some detail. The picture of the diffuse layer in Figure 3.2 must be changed: let us assume that there are only two types of ions in solution, cations with radius r_1, and anions, with radius r_2. Several authors have contributed to the evaluation of the potential and ion concentration profiles in this case: early works were carried out by Valleau and Torrie [12] and Bhuyan et al. [13]. Figure 3.3 [12] is an example illustrating the distributions of electric potential and ionic concentrations assuming finite ion sizes and a flat interface. This figure corresponds to the case in which the surface charge density is zero: the difference in ionic radii gives rise to a charge separation and, correspondingly, to a potential drop that would be absent for point ions. It must be mentioned that at high charged densities counterions govern the whole distribution and there is no difference between the several treatments.

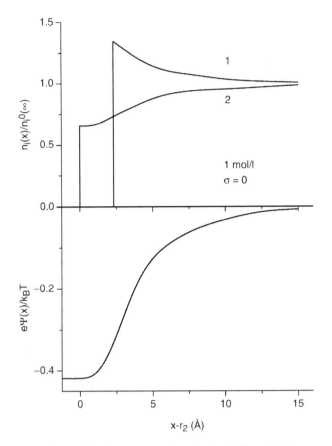

FIGURE 3.3 Position dependence of the ion concentrations (top) and dimensionless potential (bottom) for an uncharged interface with ion sizes r_2 (anions) and $(r_2 + 2.6 \text{ Å})$ (cations) with 1 mol/l concentration in the bulk. (From Valleau, J.P. and Torrie, G.M., *J. Chem. Phys.*, 76, 4623, 1982. With permission of the American Institute of Physics.)

Another assumption of the Gouy–Chapman theory that might not be fulfilled in practical situations is that the solid–liquid interface is rigid. There are a number of situations in which the surface contains dissociable groups distributed in a surface layer of certain thickness, giving rise, upon dissociation, to the formation of a volume distribution of charges. This is the case of many biological cells or artificial membranes as well as polyelectrolyte-covered surfaces (*soft particles*). The electrolyte solution in this fixed-charge region may also have a viscosity different from that in the bulk, due to the additional drag force provoked by polymer chains extending in the region. Contributions by Ohshima and co-workers have been especially noticeable in this field [14–16]. Let us also mention that Kuo and Hsu [17] first considered a general case in which both soft interfaces and ions of generally different charges and radii were investigated.

III. ELECTROKINETIC PHENOMENA AND THE ZETA POTENTIAL

If an external electric field is applied to the interface in Figure 3.2, it will exert some force on the ions (and hence on the liquid) in the diffuse part of the double layer, whereas both ions and liquid closer to the wall (the so-called *stagnant* layer) can be considered immobile for the moment. As a consequence, a relative motion between the phases in contact will occur. The exact location of the boundary between the mobile and immobile phases, the so-called slip or shear-plane, is not precisely known. It corresponds to the distance d^{ek} in Figure 3.2 at such a distance to the surface, the potential equals the *electrokinetic* or *zeta* potential, ζ. Note that the existence of that boundary, and even of the zeta potential itself, is strictly an abstraction, as both notions are based on the assumption that the viscosity of the liquid medium jumps discontinuously from infinite in the Stern layer to a finite value in the diffuse atmosphere. However, as all treatments on electrokinetic phenomena rely in fact on the existence of the zeta potential, we will not pursue this question any more, and admit the model of viscosity jump as one that works reasonably well.

Having discussed the notion of the electrokinetic potential, we will now focus on the different electrokinetic phenomena, which can be distinguished by the mobile–immobile phases, the nature of the applied field, and the quantity that must be experimentally determined. A brief list of definitions follows:

1. *Electrophoresis.* It is the translation of a colloidal particle under the action of an externally applied field, $\mathbf{E}$, constant in time and position-independent. For not very large applied fields, a linear relationship exists between the steady electrophoretic velocity, $\mathbf{v}_e$ (attained by the particle typically a few microseconds after application of the field) and the applied field,

$$\mathbf{v}_e = u_e \mathbf{E} \tag{3.9}$$

where u_e is the quantity of interest known as electrophoretic mobility.

2. *Sedimentation potential.* It is the potential difference, U_{sed}, sensed by two electrodes placed at a known vertical distance in the suspension, subjected to a gravitational (or, equivalently, centrifugal) field, $\mathbf{g}$. If the density of the particles is lower than that of the supporting fluid, we can also speak of *flotation potential.*

3. *Electrorotation.* It is the rotational motion of colloidal particles. It is provoked by an applied rotating field, and the quantity of interest is the angular velocity of the particle, Ω. It depends on the frequency of the field, ω, and can be zero, positive (i.e., co-field rotation), or negative (counter-field rotation).

4. *Dielectrophoresis.* An alternating, nonhomogeneous (spatially varying) electric field is applied in this case to the particles. They undergo a translational motion toward or away from the high-field region, depending on the difference between the electric permittivities of particle and liquid. This translation is known as dielectrophoresis.

5. *Diffusiophoresis.* It is the motion of the suspended particles under the action of an externally applied concentration gradient of the electrolyte solution (or a gradient of solvent composition) that constitutes the dispersion medium. The presence of this macroscopic concentration gradient induces a local gradient of electric potential in the vicinity of the particle thus provoking its motion. The reciprocal phenomenon of diffusiophoresis is termed *capillary osmosis*: the concentration-gradient-induced electric field sets the liquid in the vicinity of the double layer into motion.

6. *Electroosmosis.* It is the motion of the liquid adjacent to a charged surface due to an externally imposed electric field. The phenomenon may occur in, for example, flat or cylindrical capillaries, membranes, porous plugs, etc.

7. *Streaming potential and streaming current.* In these phenomena, the motion of the liquid is forced by an applied pressure gradient. The displacement of the charged liquid gives rise to an electric current (streaming current) if there is a return path for the charges, or an electric potential (streaming potential) if the sensing electrodes are connected to a high input-impedance voltmeter (open circuit).

8. *Dielectric dispersion.* It is the change with the frequency of an applied alternating current (AC) field of the dielectric permittivity of a suspension of colloidal particles. The phenomenon

is also dependent on the concentration of particles, their zeta potential, the ionic composition of the medium, and appears to be very sensitive to most of these quantities.

9. *Electroacoustic phenomena.* They are electrokinetic phenomena that have recently gained interest, both experimentally and theoretically. In the ESA (electrokinetic sonic amplitude) technique, an alternating electric field is applied to the suspension and the sound wave produced in the system is detected and analyzed. The *colloid vibration potential* (CVP) or *colloid vibration current* (CVI) is the reciprocal of the former: a mechanical (ultrasonic) wave is forced to propagate through the system, and the resulting alternating potential difference (or current) is measured.

The list could be made longer, taking the idea of electrokinetics in a wide sense (response of the colloidal system to an external field that affects differently to particles and liquid). Thus, we could include: *electroviscous effects* (the presence of the EDL alters the viscosity of a suspension in the Newtonian range); *suspension conductivity* (the effect of the solid–liquid interface on the direct current (DC) conductivity of the suspension); *particle electroorientation* (the torque exerted by an external field on anisotropic particles will provoke their orientation; this affects the refractive index of the suspension, and its variation, if it is alternating, is related to the double-layer characteristics).

Not all these phenomena have the same significance for the investigation of colloidal suspensions, as only some of them occur in dispersed systems, and not all have received the same interest because of their different case of application to the determination of interface properties in practical situations. A wider account than that given here can be found in, for instance, Refs. [5,6,8]. In the following sections, we describe in some detail the physics underlying a selection of those that can be considered more interesting from the authors' point of view.

IV. ELECTROPHORESIS

A. Physical Principles

We must solve the problem of finding the steady velocity acquired by a colloidal sphere of radius a and total surface charge Q upon application of an electric field that, far from the particle, equals $\mathbf{E}$. The particle is considered nonconducting and with an electric permittivity much smaller than that of the dispersion medium. For the moment, we will also assume that the electrolyte concentration is very low and that a is small enough for the following inequality to hold between the double-layer thickness (Equation (3.5)) and the radius:

$$\kappa^{-1} \gg a \quad \text{or} \quad \kappa a \ll 1 \tag{3.10}$$

that is, we are in the thick double-layer (or Hückel) approximation. Because the ionic atmosphere extends over such long distances, the volume charge density in the surrounding liquid will be very low, and the applied field will hence not provoke any liquid motion around the particle. As a consequence, the only forces acting on the latter are Stokes drag ($\mathbf{F}_S$) and electrostatic ($\mathbf{F}_E$). As the particle moves with constant velocity (the electrophoretic velocity $\mathbf{v}_e$), the net force must vanish:

$$\mathbf{F}_S = -6\pi\eta a \mathbf{v}_e, \qquad \mathbf{F}_E = Q\mathbf{E}, \qquad \mathbf{F}_S + \mathbf{F}_E = \mathbf{0} \tag{3.11}$$

From these equations,

$$\mathbf{v}_e = \frac{Q}{6\pi\eta a}\mathbf{E} \tag{3.12}$$

It is admitted in this and the following that the surface, diffuse, and zeta potentials coincide, so that the potential on the surface of the charged sphere is in fact the zeta potential. Its value will be

$$\zeta = \frac{1}{4\pi\varepsilon_m}\frac{Q}{a}$$
(3.13)

that, upon substitution in Equation (3.13), leads to

$$\mathbf{v}_e = \frac{2}{3}\frac{\varepsilon_m}{\eta}\zeta\mathbf{E}$$
(3.14)

or the electrophoretic mobility:

$$u_e = \frac{2}{3}\frac{\varepsilon_m}{\eta}\zeta$$
(3.15)

which is the so-called Hückel's formula for the electrophoretic mobility of a sphere; its usefulness is limited, as it is only valid for $\kappa a \ll 1$.

Let us now consider the opposite situation, for which analytical solution exists, that is, the large κa approximation:

$$\kappa^{-1} \ll a \quad \text{or} \quad \kappa a \gg 1$$
(3.16)

In this case, the surface charge is screened by double-layer ions in a comparatively thin region, in which electroneutrality is lost. The field will hence provoke a liquid motion that affects the particle motion itself, and that is essential in fact to understand the phenomenon of electrophoresis.

As before, we assume that the particle is spherical with a constant potential, ζ, along the surface; another important assumption is that the equilibrium potential distribution is not altered by the presence of the external field. The electric potential distribution will simply be the superposition of the equilibrium potential (such as given by Equations (3.4), (3.6), and (3.8) in their ranges of validity) plus a potential due to the presence of the external field. Assuming that the zeta potential is low (this may be an important restriction for the solution that we will find), the particle plus its double layer can be considered as insulating (the electrical conductivity of the particle, K_p, is almost zero) and the electric permittivity of the particle will equal ε_p. The electrolyte solution is a leaky dielectric, because of its finite conductivity ($K_m \neq 0$). The problem is better solved if a reference system is used that is centered in the sphere (recall that it is moving at constant velocity, so it is an inertial reference system): the particle can thus be considered as an obstacle at rest to the flow of charged liquid. Because the liquid does not move far from the particle in the laboratory system, the use of the coordinate system fixed to the particle will yield a liquid velocity equal to $-\mathbf{v}_e$ at long distances. In other words, calculating the velocity of liquid far from the obstacle will give us the electrophoretic velocity by simply changing the sign to our result.

It is frequent [11,18] to solve the problem by considering it as the superposition of two situations. The steady electrophoretic velocity is just the result of the balance between two forces: one is the force that would be necessary to keep the particle at rest in the presence of an external field that sets the charged liquid in the double layer into motion; the second force is the one that would be necessary to maintain the particle fixed under a uniform liquid flow of velocity $-\mathbf{v}_e$.

We will consider the first case. If E is the modulus of the applied field, assumed to be directed along the z-axis of the reference system, the electric potential perturbation provoked by the field,

$\delta \Psi(\mathbf{r})$, is given by

$$\delta \Psi(\mathbf{r}) = -Er \cos \theta + C_0 \frac{Ea^3}{r^2} \cos \theta \tag{3.17}$$

where r, θ are, respectively, the radial and angular polar coordinates and C_0 is known as the *induced dipole coefficient*. Its value depends on the conductivity and permittivity ratios between the particles and the medium. We will return to it later. Now we are interested in evaluating the fluid velocity profile, $\mathbf{v}(\mathbf{r})$, provoked by this potential perturbation on the charged diffuse layer. This requires solving the Navier–Stokes equations for a problem with spherical symmetry, in which the body force term is given by $-\rho(\mathbf{r})\nabla[\delta \Psi(\mathbf{r})]$, in an incompressible fluid in steady motion at low Reynolds number. The equations read [18–20]:

$$\eta \nabla \times \nabla \times \mathbf{v}(\mathbf{r}) + \nabla p(\mathbf{r}) + \rho(\mathbf{r})\nabla \delta \Psi(\mathbf{r}) = 0, \qquad \nabla \cdot \mathbf{v}(\mathbf{r}) = 0 \tag{3.18}$$

where $p(\mathbf{r})$ is the pressure distribution. As we are not interested in this quantity, it is useful to take the curl operator in the first Equation (3.18) and make use of the second one to write

$$\eta \nabla^2 (\nabla \times \mathbf{v}) + \nabla \times [\rho(\mathbf{r})\nabla \delta \Psi(\mathbf{r})] = 0 \tag{3.19}$$

Taking into account Poisson equation relating the volume charge density $\rho(\mathbf{r})$ to the *equilibrium* potential distribution, $\Psi(\mathbf{r})$ (recall that the geometry of the double layer is assumed to be the same as in the absence of applied field; hence, Ψ only depends on the radial coordinate r), Equation (3.19) transforms into

$$\eta \nabla^2 (\nabla \times \mathbf{v}) + \varepsilon_m \left(\frac{\mathrm{d}^3 \Psi}{\mathrm{d}r^3} \hat{\mathbf{e}}_r \right) \times \nabla \delta \Psi = 0 \tag{3.20}$$

where $\hat{\mathbf{e}}_r$ is the radial unit vector of a spherical coordinate system.

We will not go into the full details of the calculation (see Refs. [11,19]), but some crucial steps will be given. First, it can be shown that derivatives of the velocity with respect to the angular spherical coordinate θ are negligible when compared with radial derivatives, because the double layer is so thin compared with the particle radius. Then

$$\nabla^2 (\nabla \times \mathbf{v}) \simeq \frac{\partial^3 v_\theta}{\partial r^3} \hat{\mathbf{e}}_\phi \tag{3.21}$$

where v_θ is the velocity component in the tangential direction and $\hat{\mathbf{e}}_\phi$ the unit vector in the direction of the spherical coordinate ϕ. Finally, using Equations (3.20) and (3.21), we find the partial differential equation linking the fluid velocity and potential distributions in the double layer:

$$\eta \frac{\partial^3 v_\theta}{\partial r^3} = -\varepsilon_m \frac{\partial \delta \Psi}{\partial \theta} \frac{\partial^3 \Psi}{\partial r^3} = -\varepsilon_m (1 - C_0)E \sin \theta \frac{\partial^3 \Psi}{\partial r^3} \tag{3.22}$$

after making use of Equation (3.17). From Equation (3.22),

$$v_\theta(r) = \frac{\varepsilon_m}{\eta}(1 - C_0)E \sin \theta [\zeta - \Psi(r)] \tag{3.23}$$

Outside the double layer (at distances $r \approx a + \kappa^{-1}$), the electroneutrality of the liquid leads to a zero volume force in Navier–Stokes equation, and the velocity field is given by the simple form

$$\nabla^2 (\nabla \times \mathbf{v}) = 0 \tag{3.24}$$

Following Landau and Lifshitz [19] and O'Brien and White [18], symmetry arguments suggest that the velocity can be expressed in terms of an unknown function f depending on the radial coordinate r as follows:

$$\mathbf{v}(\mathbf{r}) = \nabla \times \nabla \times [f(r)\mathbf{E}] \tag{3.25}$$

and considering that $\mathbf{E}$ is a constant vector,

$$\nabla \times \mathbf{v} = -\nabla^2(\nabla f \times \mathbf{E}) \tag{3.26}$$

and

$$\nabla^2(\nabla \times \mathbf{v}) = 0 = (\nabla^2 \nabla^2 \nabla f) \times \mathbf{E} \tag{3.27}$$

The solution of Equation (3.27), making again use of the constancy of $\mathbf{E}$, is [19]:

$$f(r) = Ar + \frac{B}{r} \tag{3.28}$$

where A and B are constants to be evaluated. According to Russel et al. [11], this can be done by matching at the double-layer limit the velocities calculated outside (3.25) and inside (3.23). First, we evaluate the radial, v_r, and tangential, v_θ, components of the outside velocity:

$$v_r = 2E \cos \theta \left(-\frac{A}{r} + \frac{B}{r^3} \right), \qquad v_\theta = E \sin \theta \left(\frac{A}{r} + \frac{B}{r^3} \right) \tag{3.29}$$

The matching is carried out by equating v_θ and dv_θ/dr in (3.29) with the corresponding expressions from (3.23) assuming that the double layer is so thin that the double-layer limit can be taken at $r \approx a$. This yields

$$A = \frac{3}{2}\frac{\varepsilon_m}{\eta}a(1 - C_0)\zeta, \qquad B = -\frac{1}{2}\frac{\varepsilon_m}{\eta}a^3(1 - C_0)\zeta \tag{3.30}$$

We are now in conditions of calculating the total force on the sphere for the first problem: a constant applied field. For symmetry reasons, it will be directed in the direction of the applied field, that is, along the z-axis. As the electrical force is negligible at the limit of the double layer and beyond, we will need to calculate only the pressure and hydrodynamic forces on the particle's surface. The z-component of the resultant is [19]:

$$F'_z = \oiint_{r=a} (\sigma_{rr} \cos \theta - \sigma_{r\theta} \sin \theta) \, dS \tag{3.31}$$

where σ_{ij} are the components of the stress tensor:

$$\sigma_{rr} = -p + 2\eta \frac{\partial v_r}{\partial r}, \qquad \sigma_{r\theta} = \eta \left(\frac{1}{r}\frac{\partial v_r}{\partial \theta} + \frac{\partial v_\theta}{\partial r} - \frac{v_\theta}{r} \right) \tag{3.32}$$

and the pressure distribution can be calculated from Navier–Stokes equation in the absence of volume forces Equation (3.18). The procedure is described in Ref. [19], and it yields

$$p = p_0 - 2\eta \frac{a}{r^2}E \cos \theta \tag{3.33}$$

where p_0 is the (isotropic) pressure in the absence of the field. The result of the integration can be easily found:

$$F'_z = \frac{8}{3}\pi\eta A E \tag{3.34}$$

Concerning the second problem, the force on the particle is simply the Stokes drag. For a liquid moving with velocity $-v_e\hat{z}$ far from the particle:

$$F''_z = 6\pi\eta a(-v_e) \tag{3.35}$$

As the particle is moving at a constant velocity, the net force must be zero, $F'_z + F''_z = 0$, that is:

$$v_e = \frac{4}{9}\frac{A}{a} = \frac{2}{3}\frac{\varepsilon_m}{\eta}(1 - C_0)\zeta E \tag{3.36}$$

and the electrophoretic mobility:

$$u_e = \frac{2}{3}\frac{\varepsilon_m}{\eta}(1 - C_0)\zeta \tag{3.37}$$

The value of the dipole coefficient for different situations will lead to various expressions for the mobility. A discussion on this issue can be found in, for example, Refs. [20–22], and a general expression for a constant (DC) field is

$$C_0 = \frac{2Du - 1}{2Du + 2} \tag{3.38}$$

where Du, the Dukhin number, relates the conductivity of the double layer (the conductivity K_p of the particle is assumed to be negligible) or surface conductivity, K^σ, to that of the medium, K_m, by means of

$$Du = \frac{K^\sigma}{K_m a} \tag{3.39}$$

For a given conductivity of the electrolyte solution, the Dukhin number is mainly dependent on the zeta potential. A simple expression, valid for $\kappa a \gg 1$, was first found by Bikerman (see Ref. [5]) for a symmetrical electrolyte of valence z:

$$K^\sigma = \frac{2e^2 z^2 c}{k_B T \kappa}\left[D^+(e^{-ze\zeta/2k_B T} - 1)\left(1 + \frac{3m^+}{z^2}\right) + D^-(e^{ze\zeta/2k_B T} - 1)\left(1 + \frac{3m^-}{z^2}\right)\right] \tag{3.40}$$

where D^+ (D^-) is the diffusion coefficient of cations (anions) and m^+ (m^-) the dimensionless mobility of cations (anions):

$$m^\pm = \frac{2}{3}\frac{\varepsilon_m}{\eta}\left(\frac{k_B T}{e}\right)^2 \frac{1}{D^\pm} \tag{3.41}$$

Note that if the Dukhin number is low (absence of surface conductivity), the dipole coefficient equals $-1/2$, and the electrophoretic mobility will be

$$u_e = \frac{\varepsilon_m}{\eta}\zeta \tag{3.42}$$

which is the well-known Helmholtz–Smoluchowski (H–S) equation. It is useful because of its simplicity and, additionally, because it remains valid whatever the shape of the particle with the condition that the double-layer thickness be much smaller than the curvature radius everywhere on the particle surface [23,24].

As the zeta potential increases, Du also increases and the dipole coefficient changes from its originally negative value to zero or positive, and so for high ζ, the mobility will be lower than predicted by the H–S formula. Figure 3.4 is a representation of the variation of C_0 and the electrophoretic mobility with the zeta potential. Note that our expression (3.37) fails for very high electrokinetic potentials, due to the fact that in these conditions the double layer gets polarized, that is, it loses its equilibrium configuration. We will return to this point later.

Henry [25] was the first author who solved the problem for spheres of any radius (also for infinite cylinders), that is, of any κa value, although for small zeta potentials. Restricting ourselves to the case of spheres, Henry's equation for nonconducting particles reads

$$u_e = \frac{2}{3}\frac{\varepsilon_m}{\eta}\zeta f(\kappa a) \tag{3.43}$$

where

$$f(\kappa a) = 1 + \frac{(\kappa a)^2}{16} - 5\frac{(\kappa a)^3}{48} + \cdots \tag{3.44}$$

and the following approximate formula has been given by Ohshima [10]:

$$f(\kappa a) = \left[1 + \frac{1}{2(1 + 2.5/\kappa a[1 + 2\exp(-\kappa a)])^3}\right] \tag{3.45}$$

Key contributions to the understanding and evaluation of the electrophoretic mobility, and, in general, of the physical basis of electrokinetic phenomena is due to Overbeek [26], and also to Booth [27], who produced theories that followed similar lines, for spheres in both cases. These

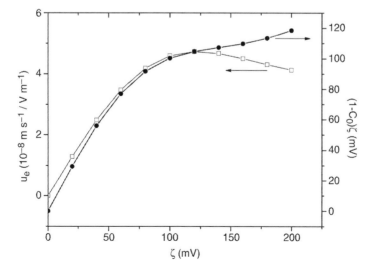

FIGURE 3.4 Comparison of the zeta-potential (ζ) dependencies of the electrophoretic mobility u_e (left axis) and the product $(1 - C_0)\zeta$ (Equation 3.37) for spherical particles of radius $a = 100$ nm in 1 mM KCl solutions.

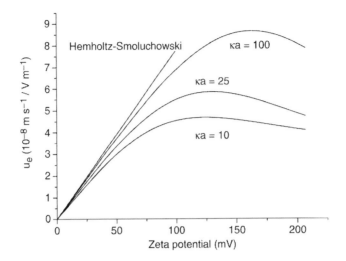

FIGURE 3.5 Full calculation of the electrophoretic mobility of spheres as a function of the zeta potential for the κa values indicated, compared with the Helmholtz–Smoluchowski formula.

authors first considered that during the electrophoretic migration the double layer loses its original symmetry, and gets polarized: the nonequilibrium potential distribution is no longer the simple addition of those of the applied field on an uncharged particle, and of the equilibrium EDL. The mathematical problem is much more involved in this case, and it can only be fully solved numerically. The first (numerical) treatments of the problem, valid for arbitrary values of the radius, the zeta potential, or the ionic concentrations, were elaborated by Wiersema et al. [28] and O'Brien and White [18]. Some numerical results are displayed in Figure 3.5. The validity of the Helmholtz–Smoluchowski formula for large κa and low-to-moderate zeta potentials is clearly observed.

B. Experimental Determination

1. Microelectrophoresis

It is probably the most widespread method: it is based on the direct observation, with a suitable magnifying optics, of individual particles in their electrophoretic motion. In fact, it is not the particle what is seen but its scattering pattern when illuminated in a dark background field. It allows direct observation of particles in their medium and the observer can in principle select a range of sizes to be tracked in case of polydispersed suspensions [29,30]. As the observations are possible only if the suspensions are dilute enough, even moderately unstable systems can be measured, as aggregation times are expectedly large for such dilute systems.

The main drawback of these methods is related to the bias and subjectivity of the observer, that can easily select only a narrow range of velocities, which might be little representative of the true average value of the suspension. Hence, some manufacturers (see a few Web sites in Ref. [31]) have modified their designs to include automatic tracking by digital image processing: the observer's eye is substituted by a video camera and a computer.

Another source of error is the location of the so-called *stationary level*: if the electrophoresis channel is cylindrical, electroosmotic flow in the channel walls will provoke a velocity distribution in the cylinder [8] given by

$$v_L = v_{ea}\left[2\frac{r^2}{R^2} - 1\right]$$

(3.46)

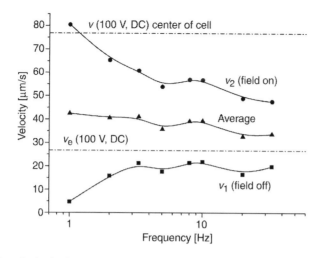

FIGURE 3.6 Particle velocity in the center of a cylindrical electrophoresis cell as a function of the frequency of an applied pulsed potential. Data obtained during field on and field off are shown together with their average. The dashed-dotted horizontal line corresponds to the electrophoretic velocity measured in a DC field at the stationary level (labeled v_e) and at the center of the channel (labeled v).

where v_{eo} is the electroosmotic liquid velocity close to the wall, R the capillary radius, and r the radial distance from the cylinder axis. From Equation (3.46) it is clear that $v_L = 0$ if $r = R/\sqrt{2}$, so that the true electrophoretic velocity will be displayed only by particles moving in a cylindrical shell placed at $0.292R$ from the channel wall.

Minor et al. [32] have analyzed the time dependence of both the electroosmotic flow and electrophoretic mobility in an electrophoresis cell. They concluded that, for most experimental conditions, the colloidal particle reaches its steady motion after the application of an external field in a much shorter time than electroosmotic flow does. Hence, if electrophoresis measurements are performed in an alternating field with a frequency much larger than the reciprocal of the characteristic time for steady electroosmosis ($\tau \sim 10^0$ sec), but smaller than that of steady electrophoresis ($\tau \sim 10^{-7}$ sec), the electroosmotic flow cannot develop. In such conditions, electroosmosis is suppressed, and the velocity of the particle is independent of the position in the cell. Figure 3.6 is an example: we measured the velocity of polystyrene particles in the center of a cylindrical cell using a pulsed field with the frequency indicated: when the frequency is above 10 Hz, the velocity (average between the field-on and field-off values) tends to the true electrophoretic velocity measured at the stationary level. Another way to overcome the electroosmosis problem is to place both electrodes providing the external field, inside the cell, completely surrounded by electroneutral solution; as no net external field acts on the charged layer close to the cell walls, the associated electroosmotic flow will not exist [33].

2. Electrophoretic Light Scattering

Visual microelectrophoresis techniques are increasingly replaced by automatic methods based on the analysis of the (laser) light scattered by particles during their electrophoresis [34–36]. These are known as electrophoretic light scattering (ELS) methods [37] and have different principles of operation. In general, due to the Doppler shift of the scattered light, a beat pattern is produced in the detection photomultiplier, the frequency of which, ω_e, can be related to the electrophoretic velocity. Such a frequency can be measured by means of a spectrum analyzer or by analysis of the correlation function of the scattered light. Knowledge of ω_e leads immediately to that of the velocity as

$$v_e = \frac{\omega_e}{Ek \cos \alpha} \tag{3.47}$$

where k is the modulus of the scattering vector (difference between the wavevectors of scattered and incident light, meaured in the dispersion medium) and α the angle between $\mathbf{k}$ and the direction of electrophoretic migration. Another technique, phase analysis light scattering or PALS [38,39], is capable of determining the electrophoretic mobility with high accuracy, especially in the case of particles moving with very low electrophoretic velocities. The method is capable of detecting electrophoretic mobilities as low as 10^{-12} m^2 V^{-1} s^{-1}, that is, 10^{-4} μm s^{-1}/V cm^{-1} in practical mobility units.

V. SEDIMENTATION POTENTIAL (DORN EFFECT)

When a colloidal particle has a density different from that of the surrounding liquid, sedimentation (or buoyancy) will take place. The presence of the double layer gives rise additionally to the generation of an electric field that, summed over all the particles (if their average separation is larger than their size), generates the sedimentation (or flotation) potential. This is the Dorn effect, a simplified theory of which will be described below [40–42].

When the particle is falling, the flow of liquid around it will alter the spatial distribution of double-layer charges (Figure 3.7, see Ref. [42]): the normal fluxes of counterions (cations) and coions (anions) carried upward by the liquid are roughly identical, as they are due to the convective motion of electroneutral solution. However, the double layer is enriched in cations (we assume a negative surface charge), and hence their tangential flux will be much larger than that of anions, so the lower pole of the particle will be enriched in anions that are scarce in the double layer, while the upper region will undergo accumulation of cations. This process originates a dipole oriented against the gravitational field $\mathbf{g}$ (in the same direction as $\mathbf{g}$ for a positive particle). If the dipole moment is $\mathbf{d}$, the electric field it will generate is

$$\delta \Psi = -\frac{1}{4\pi\varepsilon_m} \frac{\mathbf{d} \cdot \mathbf{r}}{r^3} \tag{3.48}$$

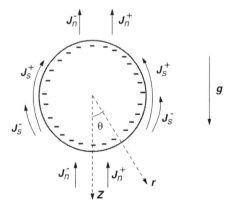

FIGURE 3.7 Diagrammatic illustration of the charge fluxes (normal, $\vec{j}_n^{\pm}$, and tangential $\vec{j}_s^{\pm}$) around a sedimenting particle. (From Delgado, A.V. and Shilov, V.N., in *Encyclopedia of Surface and Colloid Science*, Marcel Dekker, New York, 2002, pp. 1920–1950. With permission from Marcel Dekker, Inc.)

Calling n to the number of (identical) particles per unit volume in the suspension, the total field, $\mathbf{E}_{\text{sed}}$ can be calculated as [24,41]:

$$\vec{E}_{\text{sed}} = -\frac{n\mathbf{d}}{\varepsilon_{\text{m}}} \tag{3.49}$$

The following expression was given by Dukhin and Derjaguin [24] for the induced dipole moment

$$\mathbf{d} = 6\pi \frac{\varepsilon_{\text{m}}^2 \zeta a}{K_{\text{m}}} \mathbf{U}_{\text{sed}} \tag{3.50}$$

where $\mathbf{U}_{\text{sed}}$ is the limiting sedimentation velocity of an uncharged sphere given by

$$\mathbf{U}_{\text{sed}} = \frac{2}{9}\frac{a^2(\rho_{\text{p}} - \rho_{\text{m}})}{\eta}\mathbf{g} \tag{3.51}$$

and $\rho_{\text{p}}(\rho_{\text{m}})$ is the density of the particles (medium). Using Equations (3.50) and (3.51):

$$\mathbf{d} = \frac{4}{3}\frac{\pi\varepsilon_{\text{m}}^2 \zeta a^3}{K^{\infty}\eta}(\rho_{\text{p}} - \rho_{\text{m}})\mathbf{g} \tag{3.52}$$

Combining (3.49) and (3.52), we obtain

$$\mathbf{E}_{\text{sed}} = -\frac{\varepsilon_{\text{m}}\zeta}{K_{\text{m}}\eta}(\rho_{\text{p}} - \rho_{\text{m}})\phi\mathbf{g} \tag{3.53}$$

that is the Smoluchowski equation for the sedimentation potential. In Equation (3.54), use has been made of the relationship between volume fraction ϕ and particle radius and concentration: $\phi = 4\pi a^3 n/3$.

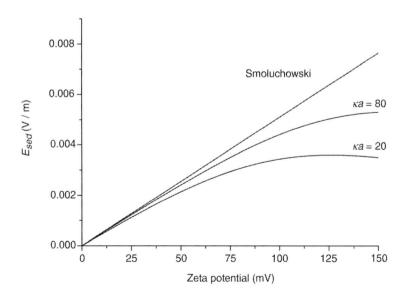

FIGURE 3.8 Full calculation of the sedimentation field (E_{sed}) of spherical particles of 100 nm radius as compared with Smoluchowski formula, as a function of zeta potential for the κa values indicated.

Like in the case of electrophoresis, the Smoluchowski equation is only valid for particles with thin double layers and negligible surface conductance (low zeta potentials). The theory was later generalized to arbitrary κa values by Booth [43] for low zeta potentials, and was developed for arbitrary ζ by Stigter [44]. Considering the fact that rather concentrated suspensions are often used in sedimentation potential determinations, theories have also been elaborated to include these situations [45–47].

Figure 3.8 shows some examples in which full numerical results are compared with Smoluchowski equation. Note that, as mentioned, the Smoluchowski approach is less valid, the higher the zeta potential and the thicker the double layer.

Because of the experimental difficulties involved in the determination of the Dorn effect, sedimentation potentials are seldom used in routine determinations of zeta potential. The reader is referred to Ref. [42] for the latest methods of sedimentation and flotation potential determinations.

VI. DIELECTROPHORESIS

Observation of this electrokinetic phenomenon involves the application of a spatially nonhomogeneous electric field to the suspension: this provoking the migration (or *dielectrophoresis*) of the polarized particles [48,49]. The particles will move toward the high-field region if they are more polarizable than the dispersion medium, or will be repelled toward the low-field region otherwise.

A great deal of information on the electrical properties of the interface can be obtained if the measurements are performed in alternating fields. In such a case, an effect of the frequency of the sinusoidal external field on the velocity of motion (the dielectrophoretic velocity) can be observed. The relaxations observed can be interpreted in terms of the different contributions to the polarization of the particle and its double layer.

In general, because of the differences between the permittivities and conductivities of the dispersed phase and the medium, the induced dipole moment of the particle in an electrolyte solution consists of two contributions. One is due to the polarization (orientation) of the molecular dipoles of both phases, and the other is due to the process of accumulation of charges of different signs on opposite poles of the particle. Thus, at very high frequencies (in practice, several MHz) ionic motions in the electrolyte solution and in the double layer toward and around the particle are too rapid for charge accumulation to proceed. Hence, only orientation of dipoles in both the particle and the liquid medium can participate in the dipole.

When the frequency is of the order of the so-called Maxwell–Wagner relaxation frequency, ω_{MW},

$$\omega_{MW} \cong \frac{K_m}{\varepsilon_m} \tag{3.54}$$

ionic migrations in both the solution and the double layer can participate in the polarization of the latter. At this frequency and below, the zeta potential of the particles starts to play a role: if the Dukhin number $Du \ll 1$ (low ζ), normal fluxes brought about by the field (in the case of a negative particle, accumulation of cations at the left pole and depletion at the right, when the field points from left to right) cannot be compensated for by the double-layer conductance, whereas for $Du \gg 1$ (high zeta potential), the cations brought normally to the particle are transported at a faster rate tangentially to it (high surface conductivity), and they accumulate at the right, being depleted at the left.

The last fundamental frequency scale is related to the fact that on the right-hand side of the particle, normal outward fluxes of cations from the double layer find inward fluxes of anions brought normally from the bulk by the field. As a consequence, an increase in neutral electrolyte concentration is produced, and, for the same reasoning, a decrease will occur at the left side. A gradient of neutral electrolyte concentration (*concentration polarization*) is thus produced, with a characteristic frequency

$$\omega_\alpha \cong \frac{2D_{\text{eff}}}{a^2} \tag{3.55}$$

where D_{eff} is the effective diffusion coefficient given by

$$D_{\text{eff}} = \frac{2D^+ D^-}{D^+ + D^-} \tag{3.56}$$

This concentration gradient gives rise to diffusion fluxes that oppose the tangential fluxes provoked by the field. If the time dependence of the applied field is of the form $\mathbf{E} = \mathbf{E}_0\, e^{-i\omega t}$, the strength and phase of the induced dipole moment in each of the situations described can be interpreted by considering that the latter is a complex quantity, $\mathbf{d}^*$, and so is the dipole coefficient (complex quantities will be denoted by asterisks):

$$\mathbf{d}^* = 4\pi\varepsilon_m a^3 C_0^* \mathbf{E} \tag{3.57}$$

As the dielectrophoretic force acting on the particles depends on the real part of the induced dipole coefficient [50,51] as follows:

$$\mathbf{F}_{\text{DP}} = \varepsilon_m \operatorname{Re}(C_0^*)\nabla E_{\text{rms}}^2 \tag{3.58}$$

where E_{rms} is the root-mean-square amplitude of the applied field, it will be clear that the change of the velocity of motion with the frequency of the external alternating field must be due to the frequency dependence (dispersion) of the induced dipole moment [52,53]. Figure 3.9 illustrates the behavior of $\operatorname{Re}(C_0^*)$ as a function of frequency for a spherical particle with $\zeta = -100$ mV and $a = 100$ nm in a 1-mM KCl solution. In view of the large variations that can be attained by

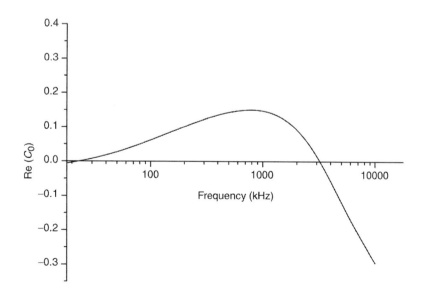

FIGURE 3.9 Frequency dependence of the real part of the induced dipole coefficient for a spherical particle 100 nm in radius with a 100 mV zeta potential in a 1-mmol/l KCl solution.

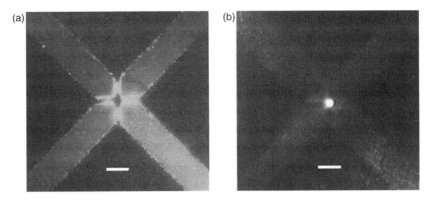

FIGURE 3.10 Fluorescence micrographs (bar length = 20 μm) showing the positive (a) and negative (b) dielectrophoresis of latex spheres. In (a) the frequency of the applied field is 1 MHz and the particles are attracted toward the region of high field (between adjacent electrodes); in (b) the frequency is 1 MHz, and the spheres are repelled to the region of low field (central area between the electrodes). (Taken from Hughes, M.P. and Green, N.G., *J. Colloid Interface Sci.*, 250, 266, 2002. With permission from Elsevier Science.)

the dipole coefficient, it might happen that its successive relaxations would cause that the polarizability of the particle changes from being larger to being smaller than that of the medium. At the frequency at which this takes place (the *crossover* frequency), the particle would invert the direction of its dielectrophoretic motion, as shown in Figure 3.10. The analysis of the behavior of the crossover frequency can therefore yield important information on the dielectric properties of the particles, and it has been particularly applied to submicrometer particles [54].

Let us mention that dielectrophoresis has also found wide application in manipulation and sorting of particles and biological cells. Together with standard electrophoresis, it is perhaps the most often used electrokinetic phenomenon with practical applications in mind. Even particle separation can be achieved by using microelectrode arrays [55]. Based on the dielectrophoresis phenomenon, a new technique has recently become available for particle or cell separation, namely the dielectrophoresis/gravitational field-flow fractionation (DEP/G-FFF). In DEP/G-FFF, the relative positions and velocities of unequal particles or cells are controlled by the dielectric properties of the colloid and the frequency of the applied field. The method has been applied to model polystyrene beads, but, most interestingly, to suspensions of different biological cells [56].

Experimental determinations of the dielectrophoretic motion in conventional dielectrophoresis have often been based on the changes in the light scattered by the suspensions because of the dielectrophoretic aggregation of the particles. Gimsa and co-workers [50] have set up a precise optical technique (based on PALS, and thus called DPALS by the authors): the nonhomogeneous electric field is generated in the measuring cell by two peaked electrodes, and the scattered light is phase-analyzed in a very similar way to that described for electrophoresis and electrorotation. As observed in Figure 3.11 [50], the particle velocity can be measured with an accuracy of the order of 1 μm/s, and the effect of the frequency of the applied field (above ∼1 kHz, to avoid unwanted electrode effects) on the translational velocity is clearly appreciable.

VII. LOW-FREQUENCY DIELECTRIC DISPERSION OF SUSPENSIONS

A. PHYSICAL BASES FOR THE DIELECTRIC RELAXATION IN SUSPENSIONS OF COLLOIDAL PARTICLES

The dipole moment induced in a colloidal particle by an external field can be probed through its contribution to a *collective* behavior of the suspension, namely the dielectric constant or the

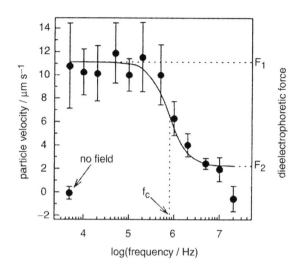

FIGURE 3.11 Dielectrophoretic velocity of latex particles in an external conductivity of 3.4 mS/cm, as a function of the frequency of the external field. (From Eppmann, P., Prüger, B., and Gimsa, J., *Colloids Surf. A*, 149, 443, 1999. With permission from Elsevier Science.)

conductivity of the system. We will now briefly describe how the latter two macroscopic quantities can be expressed in terms of the microscopic individual dipole moments of the spherical particles (permittivity ε_p) dispersed in a medium with permittivity ε_m. Let us point out that, because of the complex character of the induced dipole moment, the suspension will also be characterized by a complex permittivity, which will be denoted by $\varepsilon^* = \varepsilon' + i\varepsilon''$.

A rich source of information is the variation, with the frequency of the externally applied AC field, of the complex dielectric constant or the complex conductivity. Similar to dielectrophoresis, the relaxations observed can be interpreted in terms of the different contributions to the polarization of the particle and its double layer. This can be seen by first evaluating the current density, **J**, through a suspension in the presence of a time-varying field:

$$\mathbf{J} = \mathbf{J}_{DC} - i\omega\varepsilon^*\mathbf{E} = K_{DC}\mathbf{E} - i\omega\varepsilon^*\mathbf{E} \qquad (3.59)$$

where $\mathbf{J}_{DC}$ is the zero-frequency (conduction) current. From this, the complex conductivity K^* relating **J** and **E** will be

$$K^* = K_{DC} - i\omega\varepsilon^* \qquad (3.60)$$

K_{DC} being the constant field (DC) conductivity. For the low values of volume fraction ϕ that will be assumed, a linear dependence between suspension quantities and ϕ:

$$K^* = K_s^* + \phi\Delta K^* \qquad (3.61)$$

$$K_{DC} = K_m + \phi\Delta K_{DC} \qquad (3.62)$$

$$\varepsilon^* = \varepsilon_m + \phi\Delta\varepsilon^* \qquad (3.63)$$

where K_s^* is the complex conductivity of the solution given by

$$K_s^* = K_m + i\omega\varepsilon_m \qquad (3.64)$$

and the so-called increments of complex conductivity, DC conductivity, and complex permittivity, ΔK^*, ΔK_{DC}, and $\Delta \varepsilon^*$, respectively, represent the role of the particles and their double layers on the overall conductive and dielectric properties of the suspension. In terms of their real and imaginary parts,

$$\Delta \varepsilon = \Delta \varepsilon' + i\Delta \varepsilon'' \tag{3.65}$$

$$\Delta K^* = \Delta K_{DC} - i\omega \Delta \varepsilon^* = \Delta K_{DC} + \omega \Delta \varepsilon'' - i\omega \Delta \varepsilon' \tag{3.66}$$

Note that, according to Equation (3.66), any significant portion of the current out of phase with respect to the field can be interpreted macroscopically as a large real part of the dielectric constant of the suspension. From our earlier qualitative description, the slowest processes are the diffusion fluxes originated by concentration polarization. At low frequencies, a high dielectric constant is thus expected for the suspension. As the frequency increases, these slow processes cannot follow the field: as a consequence, they are frozen and the dielectric constant decreases. This is the α- (or volume diffusion-) relaxation of the suspension, which will occur for $\omega \sim \omega_\alpha$. At still higher frequencies, the Maxwell–Wagner relaxation ($\omega \sim \omega_{MW}$) will be observable: for those frequencies, ions cannot rearrange back and forth around the particle as fast as required by the field.

The quantitative relationship between the conductivity of the suspension, K^*, and the complex dipole coefficient reads [24,57–60]:

$$K^* = K_s^*(1 + 3\phi C_0^*) \tag{3.67}$$

and from this, the main experimentally accessible quantities, ε' and K (real parts of the dielectric constant and conductivity of the suspension, respectively) are given, with $C_0^* = c_1 + ic_2$, by

$$\varepsilon' = \varepsilon_m + 3\phi\varepsilon_m \left[c_1 - \frac{K_m}{\omega\varepsilon_m} c_2 \right], \qquad K = K_m + 3\phi K_m \left[c_1 + \frac{\omega\varepsilon_m}{K_m} c_2 \right] \tag{3.68}$$

The last equation demonstrates that the starting point for the solution of the problem is the calculation of $c_1(\omega)$ and $c_2(\omega)$. There is no simple expression for these quantities that are extremely sensitive to the structure and dynamics of the double layer (this makes low-frequency dielectric dispersion [LFDD] measurements a most valuable electrokinetic technique). Probably, the first theoretical treatment is the one due to Schwarz [61], who considered only surface diffusion of counterions (it is the so-called *surface diffusion model*). In fact, the model is inconsistent with any explanation of dielectric dispersion based on double-layer polarization. The generalization of the theory of diffuse atmosphere polarization to the case of alternating external fields and its application to the explanation of LFDD were first achieved by Dukhin and Shilov [20]. A full numerical approach to the LFDD in suspensions is due to DeLacey and White [60], and comparison with this numerical model allowed to show that the thin double-layer approximations [20,62,63] worked reasonably well in a wider than expected range of values of both ζ and κa [64]. Figure 3.12 is an example of the calculation of $\Delta \varepsilon'$. From this it will be clear that: (i) at low frequencies $\Delta \varepsilon'$ can be very high; and (ii) the relaxation of the dielectric constant takes place in the few-kHz frequency range, in accordance with Equations (3.56) and (3.57).

B. THE MEASUREMENT OF THE DIELECTRIC CONSTANT OF SUSPENSIONS

Although measurements have been reported in the time domain [65], the most usual technique for measuring the dielectric constant (or, recall its equivalence, the conductivity) of suspensions as a function of the frequency of the applied field is based on the use of a conductivity cell connected

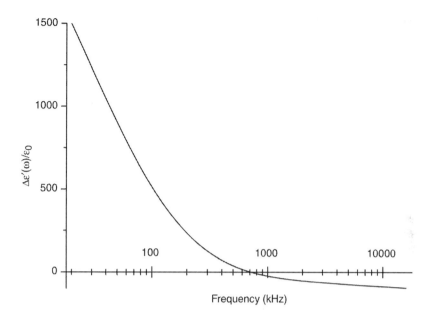

FIGURE 3.12 Real part of the dielectric increment (relative to the vacuum permittivity ε_0) of a suspension of spherical particles 100 nm in diameter with a 100-mV zeta potential. Dispersion medium: 1 mmol/l KCl solution.

to an impedance meter. In most cases, either the distance between the electrodes can be changed (see, e.g., Refs. [66–72]) or a four-electrode cell [73–76] is used.

These experimental requirements are designed to overcome or, at least, minimize the problem of electrode polarization at low frequencies. It is important to stress that the electrode impedance dominates over that of the sample at sufficiently low frequencies. Grosse and Tirado [77] have recently introduced a method (the *quadrupole method*), in which the correction for electrode polarization is optimized by following a suitable measurement routine. Finally, it has been shown that the so-called *logarithmic derivative* method can help in separating the effect of the electrodes from the true relaxation of the suspension permittivity [78]. Figure 3.13 allows the comparison between the accuracies achieved with the different procedures.

VIII. ELECTROACOUSTIC PHENOMENA

As mentioned during the description of the sedimentation potential, when a gravitational or centrifugal field produces the deformation (polarization) of the EDL, a dipole is generated in each particle of the suspension. If, instead of a constant field, it is a harmonic sound wave that passes through the suspension, the relative motion between the particles and the surrounding liquid (that are assumed to have different densities) provokes again a distortion of the ionic atmosphere, and as a consequence, an alternating electric field. This simple idea is the core of the group of electrokinetic phenomena known as *electroacoustic phenomena* [79–82]. The one described above, in which a pressure wave produces an AC field, is called colloid vibration potential (CVP) or current (CVI) if it is the current that is measured instead of the potential; the reciprocal phenomenon also occurs: a sound wave is produced by application of an AC electric field to the suspension. The latter electrokinetic phenomenon and technique are called electrokinetic sonic amplitude (ESA).

O'Brien [81,82] introduced the concept of *dynamic electrophoretic mobility* of the particles, u_d^*, as a basic quantity for the electroacoustic description of dispersed systems. The macroscopic (average) current, $\mathbf{J}$, and particle velocity, $\mathbf{v}_p$, are related to the pressure gradient ∇p and external

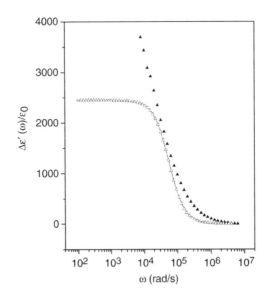

FIGURE 3.13 Real part of the dielectric increment (relative to the vacuum permittivity ε_0) of ethylcellulose (Aquacoat®) latex particles as a function of frequency, in 1 mmol/l sodium salicylate: ($\triangle$) logarithmic derivative technique; ($\blacktriangle$) standard electrode-separation procedure.

field $\mathbf{E}$ by

$$\mathbf{J} = \alpha\nabla p + K^*\mathbf{E}, \qquad \mathbf{v}_p = \beta\nabla p + u_d^*\mathbf{E} \qquad (3.69)$$

where α, β, K^*, u_d^* are the transport coefficients, and because the applied field or pressure gradient are alternating, they all have a time dependence of the type $\exp(-i\omega t)$. Note that β is a measure of the electric field generated by a sound wave, while u_d^* contains information about the acoustic wave generated by an applied AC field. The evaluation of β and u_d^* requires the solution of the electrokinetic equations for zero field and zero pressure gradient, respectively. However, it is not necessary to calculate both coefficients. The generalization of Onsager's reciprocity relationship leads to

$$\alpha \propto \phi\frac{\rho_p - \rho_m}{\rho_m}u_d^* \qquad (3.70)$$

where ϕ is the volume fraction of solids, and ρ_p and ρ_m are the densities of the particles and the liquid medium, respectively. As the CVP is measured in open-circuit conditions ($\mathbf{J} = 0$ in Equation (3.69)), the potential gradient provoked by the pressure one will be

$$\nabla V = \frac{\alpha}{K^*}\nabla p \qquad (3.71)$$

and from this, the CVP (voltage drop per unit pressure drop) will be

$$U_{CV} = \frac{\alpha}{K^*} \propto \phi\frac{\rho_p - \rho_m}{\rho_m}\frac{u_d^*}{K^*} \qquad (3.72)$$

If, as is presently the case with commercial devices, it is the current and not the potential that is measured, the CVI (per unit pressure gradient) will be given by

$$I_{CV} \propto \phi\frac{\rho_p - \rho_m}{\rho_m}u_d^* \qquad (3.73)$$

with the advantage that the dependence with the complex conductivity disappears from the measured quantity. A formally identical expression [82] can be given for the amplitude of the sound induced by the applied field per unit field strength (A_{ESA}):

$$A_{ESA} \propto \phi \frac{\rho_p - \rho_m}{\rho_m} u_d^* \tag{3.74}$$

Assuming (as it is reasonable) that for conditions in which the approximation $\kappa a \gg 1$ is valid, the dynamic mobility also contains the $(1 - C_0^*)$ dependence displayed by the static mobility (Equation (3.37)), one can expect a qualitative dependence of the dynamic mobility on the frequency of the field as shown in Figure 3.14. The first relaxation (the one at lowest frequency) in the modulus of u_d^* can be expected at the α-relaxation frequency (Equation (3.55)): as the dipole coefficient increases at such frequency, the mobility should decrease. If the frequency is increased, one finds the Maxwell–Wagner relaxation (Equation (3.54)), where the situation is reversed: $Re(C_0^*)$ decreases and the mobility increases. In addition, it can be shown [19,82] that at frequencies of the order of $(\eta/a^2\rho_p)$ the inertia of the particle hinders its motion, and the mobility decreases in a monotonic fashion. Depending on the particle size and the conductivity of the medium, the two latter relaxations might superimpose on each other and be impossible to distinguish.

Calculations performed by O'Brien [81] have shown that the expression for the mobility in the case of dilute suspensions with double-layer thickness smaller than the particle radius is

$$u_d^* = \frac{2}{3}\frac{\varepsilon_m \zeta}{\eta} G\left(\frac{\omega a^2}{v}\right)[1 + f] \tag{3.75}$$

where $v = \eta/\rho_m$ is the kinematic viscosity of the liquid, and the function G carries the information of the inertia of the particle, which is given by

$$G(x) = \frac{1 + (1+i)\sqrt{x/2}(3 + 2\Delta\rho/\rho_m)}{1 + (1+i)\sqrt{x/2} + i(x/9)(3 + 2\Delta\rho/\rho_m)} \tag{3.76}$$

with $\Delta\rho \equiv \rho_p - \rho_m$. The function f contains the information on the effect of Dukhin number and frequency on the polarizability of the particle (the zeta potential cannot be high, as the

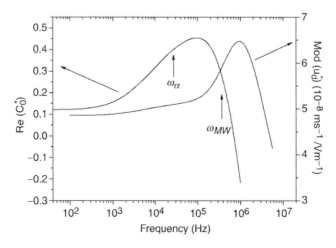

FIGURE 3.14 Modulus of the dynamic mobility and real part of the induced dipole coefficient as a function of frequency for particles 500 nm in radius in 0.1 mmol/l KCl solution. Zeta potential: -150 mV.

concentration polarization is not considered in this approach):

$$f = \frac{1 - i\omega' - (2Du - i\omega'\varepsilon_p/\varepsilon_m)}{2(1 - i\omega') + (2Du - i\omega'\varepsilon_p/\varepsilon_{rs})} \tag{3.77}$$

where $\omega' \equiv \omega/\omega_{MW}$. Figure 3.15 illustrates the typical dependences of the modulus and phase of u_d^* on the frequency.

Dukhin et al. [83–85] have performed the direct calculation of the CVI in the situation of concentrated systems. In fact, it must be mentioned here that one of the most promising potential applicabilities of these methods is their usefulness with concentrated systems (high volume fractions of solids, ϕ) because the effect to be measured is also in this case a collective one. The first generalizations of the dynamic mobility theory to concentrated suspensions made use of the Levine and Neale cell model [86,87] to account for particle–particle interactions. An alternative method estimated the first-order volume fraction corrections to the mobility by detailed consideration of pair interactions between particles at all possible different orientations [88–90]. A comparison between these approaches and calculations based on the cell model of Zharkikh and Shilov [91] has been carried out in Refs. [92,93].

Experimental determinations with both electrokinetic techniques have increased during the last 10 years mainly because there are commercial instruments devised for their determination [94]. In the basic experimental procedures, an AC voltage is applied to a transducer which produces a sound wave that after traveling through a delay line passes into the suspension. This acoustic excitation causes a very small periodic displacement of the fluid in turn giving rise to a small oscillating dipole. Like in the Dorn effect, the electric field detected by the receiving transducer is the result of the addition of the dipolar fields of each individual particle. The voltage difference, which then appears between the electrodes (measured at zero current flow), is proportional to U_{CV}. The current, measured under short-circuit conditions, is a measure of I_{CV}. Alternatively, an AC electric field can be applied across the electrodes. A sound wave is thereby generated near the electrodes in the suspension, and this wave travels along the delay line to the transducer. The transducer output is then a measure of the ESA effect. For both techniques, measurements can be performed for a set of frequencies typically in the range of 1–100 MHz.

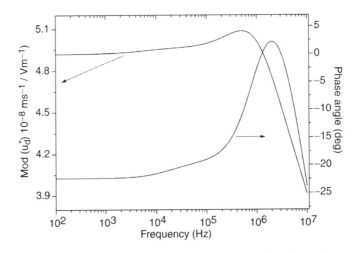

FIGURE 3.15 Modulus and phase angle of the dynamic mobility as a function of frequency for particles 170 nm in radius in a 0.2 mmol/l KCl solution. Zeta potential: −90 mV.

IX. ELECTROKINETICS OF NONRIGID PARTICLES

So far we have limited ourselves to solid, rigid particles, for which the interface and, in particular, the electrokinetic potential and the slip plane are well defined. We will briefly discuss two situations in which this hypothesis fails. The first one is the electrophoretic mobility of charged liquid drops in an emulsion. Clearly, we cannot assume that the fluid velocity relative to the particle is zero at the separation between the drop surface and the liquid medium itself, as the motion of the latter will be transmitted to the drop, the viscosity of which is η_p.

A careful account of the problem can be found in Ref. [95]. Ohshima et al. [96] first found a numerical solution of the problem, valid for arbitrary values of the zeta potential or the product κa. In the same paper, they dealt with the problem of finding the sedimentation potential and the DC conductivity of a suspension of mercury drops. The problems are solved following the lines of the electrophoresis theory of rigid particles previously derived by O'Brien and White [18]. The liquid drop is assumed to behave as an ideal conductor, so that electric fields and currents inside the drop are zero, and its surface is equipotential. The main difference between the treatment of the electrophoresis of rigid particles and that of drops is that there is a velocity distribution of the fluid inside the drop, $\mathbf{v_I}$, governed by the Navier–Stokes equation with zero body force (in the case of electrophoresis), and related to the velocity outside the drop, $\mathbf{v}$, by the boundary conditions:

a) $\mathbf{v}$ and $\mathbf{v_I}$ are equal on the drop surface:

$$\mathbf{v}\Big|_{r\to a^+} = \mathbf{v_I}\Big|_{r\to a^-} \tag{3.78}$$

b) The normal components of both velocities must vanish on the surface:

$$\mathbf{v}\cdot\hat{\mathbf{r}}\Big|_{r\to a^+} = \mathbf{v_I}\cdot\hat{\mathbf{r}}\Big|_{r\to a^-} = 0 \tag{3.79}$$

c) The tangential component of the hydrodynamic stress tensor $\tilde{\sigma}$ is continuous:

$$(\tilde{\sigma}\cdot\hat{\mathbf{r}})\times\hat{\mathbf{r}}\Big|_{r\to a^+} = (\tilde{\sigma}\cdot\hat{\mathbf{r}})\times\hat{\mathbf{r}}\Big|_{r\to a^-} = 0 \tag{3.80}$$

As in the case of rigid particles, the general solution of the problem must be obtained numerically, but in the same paper [96] the authors obtained approximate analytical solutions correct to order ζ or ζ^2, or valid for large κa. For instance, their expression valid for low zeta potential and arbitrary κa reads

$$u_e = \frac{\varepsilon_m\zeta}{\eta}\left[\frac{\eta}{3\eta_p+2\eta}\kappa a + \frac{3\eta_p+\eta}{3\eta_p+2\eta} + 2e^{\kappa a}E_5(\kappa a) - \frac{15\eta_p}{3\eta_p+2\eta}e^{\kappa a}E_7(\kappa a)\right] \tag{3.81}$$

where $E_n(z)$ is the exponential integral of order n:

$$E_n(z) = \int_1^\infty \frac{e^{-zt}}{t^n}\,dt$$

In Equation (3.81) the first term, linearly dependent on κa, coincides with the expression previously found by Levine and O'Brien [97].

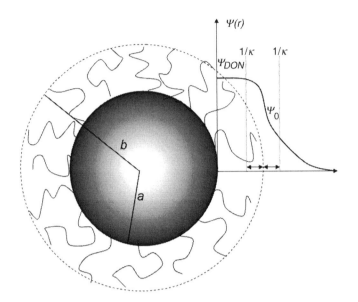

FIGURE 3.16 Schematic representation of the interface structure for a solid particle covered with a polyelectrolyte layer in an electrolyte solution. The Donnan and diffuse potentials are marked.

The second problem of interest is that of the electrophoresis of soft (polyelectrolyte-covered) particles. This problem is not merely of academic interest: polyelectrolytes are often used to control the surface charge of the particles and hence the stability and other properties of their suspensions. In the presence of such polyelectrolytes, the particles get covered by a layer of the polymer in an amount controlled by its interactions with the solid and by the quality of the solvent for the polyelectrolyte used.

The essential point is that such an adsorption process markedly influences the hydrodynamic and electrical properties of the interface. Figure 3.16 is a schematic representation of the structure formed: polymer chains with fixed charges extend out of the solid surface to an average distance $d = b - a$; in this region, the fluid can move, although with an increased viscosity because of the hydrodynamic resistance of the polyelectrolyte layer (also called hydrogel layer). To take this into account, a friction term $-\gamma \mathbf{v}$ is included in the Navier–Stokes equation.

In addition, the polymer layer is assumed to be characterized by a constant volume concentration of fixed charges (N per unit volume) of valence Z, and it is permeable to the ionic species in the solution. This contains a z-valent symmetrical electrolyte with concentration n (number of each kind of ions per unit volume). Hence, the electric potential at the solid surface, $r = a$, is the Donnan potential, Ψ_{DON}, given by [14–16]:

$$\Psi_{\mathrm{DON}} = \frac{k_B T}{ze} \ln \left[\frac{ZN}{2zn} + \left\{ \left(\frac{ZN}{2zn}\right)^2 + 1 \right\}^{1/2} \right] \tag{3.82}$$

The limit of the charged layer, $r = b$, coincides with the beginning of the diffuse atmosphere, with potential Ψ_0. Note that this potential has no relationship with ζ, in fact an undefined quantity for this type of particles, as there is no true slip plane. The Ψ_0 potential is

$$\Psi_0 = \frac{k_B T}{ze} \left(\ln \left[\frac{ZN}{2zn} + \left\{ \left(\frac{ZN}{2zn}\right)^2 + 1 \right\}^{1/2} \right] + \frac{2zn}{ZN} \left[1 - \left\{ \left(\frac{ZN}{2zn}\right)^2 + 1 \right\}^{1/2} \right] \right) \tag{3.83}$$

Both potentials govern the evaluation of the electrophoretic mobility that can be expressed as

$$u_e = \frac{\varepsilon_m}{\eta} \frac{\Psi_d/\kappa_m + \Psi_{DON}/\lambda}{1/\kappa_m + 1/\lambda} f\left(\frac{d}{a}\right) + \frac{ZeN}{\eta\lambda^2} \qquad (3.84)$$

where

$$\lambda = \left(\frac{\gamma}{\eta}\right)^{1/2} \qquad (3.85)$$

is a dimensionless friction parameter and

$$\kappa_m = \kappa\left[1 + \left(\frac{ZN}{2zn}\right)^2\right]^{1/4} \qquad (3.86)$$

is an effective Debye length. The function $f(d/a)$ changes from 1 to $2/3$ as d/a increases; if $d \ll a$ the hydrogel layer is almost planar. In the opposite case, one is effectively describing the behavior of a sphere of polyelectrolyte.

A very significant result of this theory, widely confirmed by experiments, is that at the limit of high electrolyte concentration ($n \gg N$) the mobility does not tend to zero as it does in the case of rigid particles (where the zeta potential is much reduced by high ionic strength because of double-layer compression). Instead, a finite mobility is always found, precisely related to the characteristics of the hydrogel layer [10]:

$$u_e(n \gg N) = \frac{ZeN}{\eta\lambda^2} \qquad (3.87)$$

As an example, Figure 3.17 [98] shows the effect of $NaNO_3$ concentration and pH on the electrophoretic mobility of magnetite particles covered by a layer of the negatively charged polymer Carbopol 941® (a registered trade mark of Noveon, USA): note that the mobility can be very

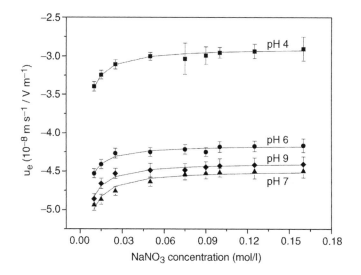

FIGURE 3.17 Effect of $NaNO_3$ concentration on the electrophoretic mobility of magnetite particles covered with a layer of a negatively charged polyelectrolyte (Carbopol 941®) for different pH values. (From Viota, J.L., de Vicente, J., Durán, J.D.G., and Delgado, A.V., *J. Colloid Interface Sci.*, 284, 527, 2005. Copyright (2005), with permission from Elsevier Science).

high (close to $-5\ \mu\mathrm{m\ s^{-1}/V\ cm^{-1}}$ at pH 7, when the charge/viscosity of the polymer is optimum) even for 0.16 mol/l concentration. The lines are the best fit to Ohshima's theory, which is clearly capable to explain the results.

Let us finally mention that other approaches have been proposed that numerically solve the problem for arbitrary values of the parameters of interest, specifically, zeta potential and ionic strength. Details can be found in Refs. [99,100]. The problem of the dielectric spectroscopy of suspensions of this type of particles has been analyzed in Refs. [101,102].

X. DYNAMIC STERN LAYER AND ELECTROKINETIC PHENOMENA

In our previous discussion of the different electrokinetic phenomena, we have relied on the notion of the electrokinetic or zeta potential for their explanation. It should also be clear that there is no possibility of directly measuring ζ, and all our estimations on it must be based on a theoretical model. According to this so-called *classical* or *standard electrokinetic model*, any ions located in the stagnant layer of liquid immediately below the slip plane are completely immobile, and do not participate in any transport processes taking place in the system.

However, the significant differences found between theory and experiment for a number of electrokinetic phenomena prompted a number of authors to reconsider the validity of this hypothesis. Differences of this kind have in fact been found by comparing the zeta potentials obtained from two or more electrokinetic techniques on the very same systems. Comparisons have been carried out between, for instance, electrophoretic mobility and DC conductivity [103,104], electrophoretic mobility and dielectric dispersion [72,75,105,106]. As an example, Figure 3.18 shows the significant differences that can be found between experimental data of dielectric dispersion of latex particles and those calculated by means of the classical DeLacey and White theory [60] using the zeta potentials obtained from electrophoretic mobility. It has long been recognized that a possible explanation for these divergences must lie on the fact that ionic transport in the inner part of the double layer is possible, that is, a *dynamic Stern layer* (DSL) model must be used rather than the standard one in order to reconcile the results from different electrokinetic techniques.

Rosen et al. [107], Simonova and Shilov [108], and Zukoski and Saville [109,110] were among the first in modifying, in a quantitative manner, the classical electrokinetic theories to account for

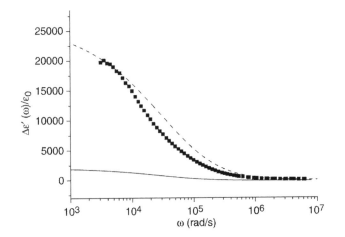

FIGURE 3.18 Frequency dependence of the relative dielectric increment of 265-nm radius polystyrene spheres in 0.1 mmol/l KCl solution. Symbols: experimental data; solid line: classical calculation; dashed line: DSL calculation. In both calculations, the zeta potential used was the one best-fitting simultaneously electrophoretic mobility and dielectric dispersion data.

the possibility of lateral transport of ions in the inner part of the double layer, and the improvements achieved were checked against DC conductivity and electrophoretic mobility data. Later, the theory was also elaborated for the analysis of the dielectric constant of suspensions in the DSL scenario [107,111]. Mangelsdorf and White [112–114] performed a slightly different study of the problem, and the ability of this theory to reach a close similarity between theory and experiment has also been demonstrated [71,72]. Figure 3.18 also includes calculations based on this theory and its closeness to experimental data is evident. The DSL theory has also been applied to other electrokinetic phenomena like sedimentation potential [47]. Let us mention that Kijlstra et al. [73–75] were also successful in modifying the large-κa model of dielectric dispersion to include ionic motions in the stagnant (immobile fluid) layer adjacent to the solid.

ACKNOWLEDGMENTS

Financial support from CICYT, Spain, under Project No. MAT2004-00866 is gratefully acknowledged. The collaboration of Dr. F.J. Arroyo, University of Jaén, Spain, and Dr. S. Ahualli, University of Granada, Spain, is also acknowledged.

REFERENCES

1. Van den Hul, H.J. and Vanderhoff, J.W., The characterization of latex particles by ion exchange and conductimetric titration, *J. Electroanal. Chem.*, 37, 161, 1972.
2. Yates, D.E. and Healy, T.W., Titanium-dioxide electrolyte interface. 2. Surface-charge (titration) studies, *J. Chem. Soc., Faraday Trans. 1*, 76, 9, 1980.
3. Plaza, R.C., Delgado, A.V., and González-Caballero, F., Electrical surface charge and potential of hematite/yttrium oxide core-shell colloidal particles, *Colloid Polym. Sci.*, 179, 1206, 2001.
4. Hunter, R.J., *Foundations of Colloid Science*, Oxford University Press, Oxford, 2001.
5. Lyklema, J., *Fundamentals of Interface and Colloid Science*, Vol. II, *Solid–Liquid Interfaces*, Academic Press, New York, 1995, chap. 3.
6. Delgado, A.V. and Arroyo, F.J., Electrokinetic phenomena and their experimental determination: an overview, in *Interfacial Electrokinetics and Electrophoresis*, Delgado, A.V., Ed., Surfactant Science Series, Marcel Dekker, New York, 2002, Vol. 106, chap. 1.
7. Lyklema, J., The structure of the solid/liquid interface and the electrical double layer, in *Solid/Liquid Dispersions*, Tadros, Th. F., Ed., Academic Press, London, 1987, pp. 63–90.
8. Hunter, R.J., *Zeta Potential in Colloid Science*, Academic Press, New York, 1981.
9. van Olphen, H.J., *An Introduction to Clay Colloid Chemistry*, Wiley, New York, 1977, chap. 7.
10. Ohshima, H., Interfacial electrokinetic phenomena, in *Electrical Phenomena at Interfaces*, Ohshima, H. and Furusawa, K., Eds., Surfactant Science Series, Vol. 76, Marcel Dekker, New York, 1998, pp. 19–55.
11. Russel, W.B., Saville, D.A., and Schowalter, W.R., *Colloidal Dispersions*, Cambridge University Press, Cambridge, 1989, chaps. 4 and 7.
12. Valleau, J.P. and Torrie, G.M., The electrical double-layer. 3. Modified Gouy–Chapman theory with unequal ion sizes, *J. Chem. Phys.*, 76, 4623, 1982.
13. Bhuyan, L.B., Blum, L., and Henderson, D., The application of the modified Gouy–Chapman theory to an electrical double layer containing asymmetric ions, *J. Chem. Phys.*, 78, 442, 1983.
14. Ohshima, H. and Makino, K., Electrophoretic mobility of a particle covered with a partially ion-penetrable polyelectrolyte layer, *Colloids Surf. A*, 109, 71, 1996.
15. Ohshima, H., On the general expression for the electrophoretic mobility of a soft particle, *J. Colloid Interface Sci.*, 228, 190, 2000.
16. Ohshima, H., Dynamic electrophoretic mobility of a soft particle, *J. Colloid Interface Sci.*, 233, 142, 2001.
17. Kuo, Y.C. and Hsu, J.P., Double-layer properties of an ion-penetrable charged membrane: effect of sizes of charged species, *J. Phys. Chem. B*, 103, 9743, 1999.

18. O'Brien, R.W. and White, L.R., Electrophoretic mobility of a spherical colloidal particle, *J. Chem. Soc., Faraday Trans. 2*, 74, 1607, 1978.

19. Landau, L.D. and Lifshitz, E.M., *Fluid Mechanics*, Butterworth-Heinemann, Oxford, 2000.

20. Dukhin, S. and Shilov, V.N., *Dielectric Phenomena and the Double Layer in Dispersed Systems and Polyelectrolytes*, Wiley, New York, 1974.

21. Shilov, V.N., Delgado, A.V., González-Caballero F., Horno J., López-García, J.J., and Grosse, C.J., *Colloid Interface Sci.*, 232, 141, 2000.

22. Grosse, C., Relaxation mechanisms of homogeneous particles and cells suspended in aqueous electrolyte solutions, in *Interfacial Electrokinetics and Electrophoresis*, Delgado, A.V., Ed., Surfactant Science Series, Marcel Dekker, New York, 2002, Vol. 106, chap. 11.

23. Morrison, F.A., Jr., Electrophoresis of a particle of arbitrary shape, *J. Colloid Interface Sci.*, 34, 210, 1970.

24. Dukhin, S.S. and Derjaguin B.V., Equilibrium double layer and electrokinetic phenomena, in *Surface and Colloid Science*, Matijević, E., Ed., Wiley, New York, 1974, Vol. 7, chap. 2.

25. Henry, D.C., The cataphoresis of suspended particles. I. The equation of cataphoresis, *Proc. Roy. Soc.*, 133A, 106, 1931.

26. Overbeek, J.Th.G., Theorie der elektrophorese. Der relaxationeffekt, *Kolloid Beih.*, 54, 287, 1943.

27. Booth, F., The cataphoresis of spherical, solid non-conducting particles in a symmetrical electrolyte, *Proc. Roy. Soc.*, 203A, 514, 1950.

28. Wiersema, L.H., Loeb, A.L., and Overbeek, J.Th.G., Calculation of the electrophoretic mobility of a spherical colloidal particle, *J. Colloid Interface Sci.*, 22, 78, 1966.

29. James, A.M., Electrophoresis of particles in suspension, in *Surface and Colloid Science*, Good, R.J. and Stromberg, R.S., Eds., Plenum Press, New York, 1979, Vol. 9, chap. 4.

30. Smith, A.L., Electrical phenomena associated with the solid/liquid interface, in *Dispersions of Powders in Liquids*, Parfitt, G.D., Ed., Applied Science, London, 1981, chap. 2.

31. Web sites: www.cad-inst.com, www.lavallab.com, www.zeta-meter.com, www.webzero.co.uk.

32. Minor, M., van der Linde, A.J., Leeuwen, H.P., and Lyklema, J., Dynamic aspects of electrophoresis and electroosmosis: a new fast method for measuring electrophoretic mobilities, *J. Colloid Interface Sci.*, 189, 370, 1997.

33. Uzgiris, E.E., Laser Doppler spectrometer for study of electrophoretic phenomena, *Rev. Sci. Instrum.*, 45, 74, 1974.

34. Uzgiris, E.E., Electrophoresis of particles and biological cells measured by the Doppler shift of scattered light, *Opt. Commun.*, 6, 55, 1972.

35. Wave, R. and Flyare, W.H., Invention and development of electrophoretic light scattering, *J. Colloid Interface Sci.*, 39, 670, 1972.

36. Malher, E., Martin, D., Duvivier, C., Volochine, B., and Stoltz, J. F., New device for determination of cell electrophoretic mobility using Doppler velocimetry, *Biorheology*, 19, 647, 1982.

37. Web sites: www.beckman.com/coulter, www.bic.com, www.malvern.co.uk.

38. Miller, J.F., Schätzel, K., and Vincent, B., The determination of very small electrophoretic mobilities in polar and non-polar colloidal dispersions using phase analysis light scattering, *J. Colloid Interface Sci.*, 143, 532, 1991.

39. Manerwatson, F., Tscharnuter, W., and Miller, J., A new instrument for the measurement of very small electrophoretic mobilities using phase analysis light scattering (PALS), *Colloids Surf. A*, 140, 53, 1988.

40. Dukhin, S.S., Development of notions as to the mechanisms of electrokinetic phenomena and the structure of the colloid micelle, in *Surface and Colloid Science*, Matijević, E., Ed., Wiley, New York, 1974, Vol. 7, chap. 1.

41. Ozaki, M. and Sasaki, H., Sedimentation potential and flotation potential, in *Electrical Phenomena at Interfaces*, Ohshima, H. and Furusawa, K., Eds., Surfactant Science Series, Marcel Dekker, New York, 1998, Vol. 76, chap. 10.

42. Delgado, A.V. and Shilov, V.N., Electrokinetics of suspended solid colloid particles, in *Encyclopedia of Surface and Colloid Science*, Hubbard, H., Ed., Marcel Dekker, New York, 2002, pp. 1920–1950.

43. Booth, F., Sedimentation potential and velocity of solid spherical particles, *Proc. Roy. Soc. Lond., Ser. A*, 203, 514, 1950.

44. Stigter, D., Sedimentation of highly charged colloidal spheres, *J. Phys. Chem.*, 84, 2758, 1980.

45. Levine, S., Neale, G., and Epstein, N., The prediction of electrokinetic phenomena within multiparticle systems. II. Sedimentation potential, *J. Colloid Interface Sci.*, 57, 424, 1976.

46. Ohshima, H., Sedimentation potential in a concentrated suspension of spherical colloidal particles, *J. Colloid Interface Sci.*, 208, 295, 1998.

47. Carrique, F., Arroyo, F.J., and Delgado, A.V., Effect of a dynamic Stern layer on the sedimentation velocity and potential in a dilute suspension of colloidal particles, *J. Colloid Interface Sci.*, 227, 212, 2000.

48. Gimsa, J., Particle characterization by AC electrokinetic phenomena. 1. A short introduction to dielectrophoresis (DP) and electrorotation (ER), *Colloids Surf. A*, 149, 451, 1999.

49. Pohl, H.A., *Dielectrophoresis*, Cambridge University Press, Cambridge, 1978.

50. Eppmann, P., Prüger, B., and Gimsa, J., Particle characterization by AC electrokinetic phenomena. 2. Dielectrophoresis of latex particles measured by dielectrophoretic phase analysis of light scattering (DPALS), *Colloids Surf. A*, 149, 443, 1999.

51. Zimmerman, V., Grosse, C., and Shilov, V.N., Contribution of electroosmosis to electrorotation spectra in the frequency-range of the beta dispersion, *Colloids Surf. A*, 159, 299, 1999.

52. Baygents, J.C., Electrokinetic effects on the dielectric response of colloidal particles: dielectrophoresis and electrorotation, *Colloids Surf. A*, 92, 67, 1994.

53. Shramko, O., Shilov, V.N., and Simonova, T., Polarization interaction of disperse particles with not thin Debye atmosphere, *Colloids Surf. A*, 140, 385, 1998.

54. Hughes, M.P. and Green, N.G., The influence of Stern layer conductance on the dielectrophoretic behavior of nanospheres, *J. Colloid Interface Sci.*, 250, 266, 2002.

55. Suehiro, J. and Pethig, R., The dielectrophoretic movement and positioning of a biological cell using a 3-dimensional grid electrode system, *J. Phys. D: Appl. Phys.*, 31, 3298, 1998.

56. Yang, J., Huang, Y., Wuang, X., Becker, F.F., and Gascoyne, P.R.C., Cell separation on microfabricated electrodes using dielectrophoretic/gravitational field-flow fractionation, *Anal. Chem.*, 71, 911, 1999.

57. Dukhin, S.S., Non-equilibrium electric surface phenomena, *Adv. Colloid Interface Sci.*, 44, 1, 1993.

58. O'Brien, R.W., The electrical conductivity of a dilute suspension of charged particles, *J. Colloid Interface Sci.*, 81, 234, 1980.

59. O'Brien, R.W., The response of a colloidal suspension to an alternating electric field, *Adv. Colloid Interface Sci.*, 16, 281, 1982.

60. DeLacey, E.H.B. and White, L.R., Dielectric response and conductivity of dilute suspensions of colloidal particles, *J. Chem. Soc., Faraday Trans. 2*, 77, 2007, 1981.

61. Schwarz, G., A theory of the low-frequency dielectric dispersion of colloidal particles in electrolyte solution, *J. Phys. Chem.*, 66, 2636, 1962.

62. Fixman, M., Charged macromolecules in external fields. I. The sphere, *J. Chem. Phys.*, 72, 5177, 1980.

63. Fixman, M., Thin double-layer approximation for electrophoresis and dielectric response, *J. Chem. Phys.*, 78, 1483, 1983.

64. Carrique, F., Zurita, L., and Delgado, A.V., Thin double-layer approximation and exact prediction for the dielectric response of a colloidal suspension, *J. Colloid Interface Sci.*, 170, 176, 1995.

65. Hayashi, Y., Livshits, L., Caduff, A., and Feldman, Y., Dielectric spectroscopy study of specific glucose influence on human erythrocyte membranes, *J. Phys. D: Appl. Phys.*, 36, 369, 2003.

66. Tirado, M.C., Arroyo, F.J., Delgado, A.V., and Grosse, C., Measurement of the low-frequency dielectric properties of colloidal suspensions: comparison between different methods, *J. Colloid Interface Sci.*, 227, 141, 2000.

67. Springer, M.M., *Dielectric relaxation of dilute polystyrene lattices*, Ph.D. thesis, University of Wageningen, The Netherlands, 1979.

68. Lim, K. and Frances, E.I., Electrical properties of aqueous dispersions of polymer microspheres, *J. Colloid Interface Sci.*, 110, 201, 1986.

69. Grosse, C., Hill, A.J., and Foster, K.R., Permittivity of suspensions of metal particles in electrolyte solution, *J. Colloid Interface Sci.*, 127, 167, 1989.

70. Carrique, F., Zurita, L., and Delgado, A.V., Some experimental and theoretical data on the dielectric relaxation in dilute polystyrene suspensions, *Acta Polymerica*, 45, 115, 1994.

71. Arroyo, F.J., Carrique, F., Bellini, T., and Delgado, A.V., Dielectric dispersion of colloidal suspensions in the presence of Stern layer conductance: particle size effects, *J. Colloid Interface Sci.*, 210, 194, 1999.

72. Arroyo, F.J., Carrique, F., and Delgado, A.V., Effects of temperature and polydispersity on the dielectric relaxation of dilute ethylcellulose suspensions, *J. Colloid Interface Sci.*, 217, 411, 1999.

73. Kijlstra, J., Ph.D. thesis, University of Wageningen, The Netherlands, 1992.

74. Kijlstra, J., van Leeuwen, H.P., and Lyklema, J., Low-frequency dielectric relaxation of hematite and silica sols, *Langmuir*, 9, 1625, 1993.

75. Kijlstra, J., van Leeuwen, H.P., and Lyklema, J., Effect of surface conduction on the electrokinetic properties of colloids, *J. Chem. Soc., Faraday Trans.*, 88, 3441, 1992.

76. Myers, D.F. and Saville, D.A., Effect of surface conduction on the electrokinetic properties of colloids, *J. Colloid Interface Sci.*, 131, 448, 1989.

77. Grosse, C. and Tirado, M.C., Measurement of the dielectric properties of polystyrene particles in electrolyte solutions, *Mater. Res. Soc. Symp. Proc.*, 430, 287, 1996.

78. Jiménez, M.L., Arroyo, F.J., van Turnhout, J., and Delgado, A.V., Analysis of the dielectric permittivity of suspensions by means of the logarithmic derivative of its real part, *J. Colloid Interface Sci.*, 249, 327, 2002.

79. Takeda, S., Tobori, N., Sugawara, H., and Furusawa, K., Dynamic electrophoresis, in *Electrical Phenomena at Interfaces*, Ohshima, H. and Furusawa, K., Eds., Surfactant Science Series, Marcel Dekker, New York, 1998, Vol. 76, chap. 13.

80. Kissa, E., *Dispersions: Characterization, Testing and Measurement*, Surfactant Science Series, Marcel Dekker, New York, 1999, Vol. 84, chap. 13.

81. O'Brien, R.W., Electroacoustic effects in a dilute suspension of spherical particles, *J. Fluid Mech.*, 190, 71, 1988.

82. O'Brien, R.W., The electroacoustic equations for a colloidal suspension, *J. Fluid Mech.*, 212, 81, 1990.

83. Dukhin, A.S. and Goetz, P.J., Electroacoustic phenomena in concentrated dispersions: new theory and CVI experiments, *Langmuir*, 12, 4987, 1996.

84. Dukhin, A.S., Shilov, V.N., Ohshima, H., and Goetz, P.J., Dynamic electrophoretic mobility in concentrated disperse systems: cell model, *Langmuir*, 15, 3445, 1999.

85. Dukhin, A.S., Shilov, V.N., Ohshima, H., and Goetz, P.J., Electroacoustic phenomena in concentrated dispersions: effect of surface conductance, *Langmuir*, 15, 6692, 1999.

86. Levine, S. and Neale, G.H., The prediction of electrokinetic phenomena within multiparticle systems. I. Electrophoresis and electroosmosis, *J. Colloid Interface Sci.*, 47, 520, 1974.

87. Levine, S. and Neale, G.H., Electrophoretic mobility of multiparticle systems, *J. Colloid Interface Sci.*, 49, 330, 1974.

88. O'Brien, R.W., Midmore, B.R., Lamb, B., and Hunter, R.J., Electroacoustic studies of moderately concentrated colloidal suspensions, *Faraday Discuss., Chem. Soc.*, 90, 301, 1990.

89. Rider, P. and O'Brien, R.W., The dynamic mobility in non-dilute suspensions, *J. Fluid Mech.*, 257, 607, 1993.

90. O'Brien, R.W., Jones, A., and Rowlands, W.N., A new formula for the dynamic mobility in a concentrated colloid, *Colloids Surf. A*, 218, 89, 2003.

91. Zharkikh, N.I. and Shilov, V.N., Theory of collective electrophoresis of spherical particles in the Henry approximation, *Colloid J.*, 43, 865, 1981.

92. Dukhin, A.S., Goetz, P.J., Shilov, V.N., and Ohshima, H., Electroacoustic phenomena in concentrated dispersions: theory, experiment, applications, in *Interfacial Electrokinetics and Electrophoresis*, Delgado, A.V., Ed., Surfactant Science Series, Marcel Dekker, New York, 2002, Vol. 106, chap. 17.

93. Arroyo, F.J., Carrique, F., Ahualli, S., and Delgado, A.V., Dynamic mobility of concentrated suspensions: comparison between different calculations, *Phys. Chem. Chem. Phys.*, 6, 1446, 2004.

94. Web sites: www.colloidal-dynamics.com, www.dispersion.com, www.matec.com.

95. Ohshima, H., Electrophoresis of charged particles and drops, in *Interfacial Electrokinetics and Electrophoresis*, Delgado, A.V., Ed., Surfactant Science Series, Marcel Dekker, New York, 2002, Vol. 106, chap. 5.

96. Ohshima, H., Healy, T.W., and White, L.R., Electrokinetic phenomena in a dilute suspension of charged mercury drops, *J. Chem. Soc., Faraday Trans. II*, 80, 1643, 1984.

97. Levine, S. and O'Brien, R.N., Theory of electrophoresis of charged mercury drops in aqueous electrolyte solution, *J. Colloid Interface Sci.*, 43, 716, 1973.

98. Viota, J.L., de Vicente, J., Durán, J.D.G., and Delgado, A.V., Stabilization of magnetorheological suspensions by polyacrylic acid polymers, *J. Colloid Interface Sci.*, 284, 527, 2005.

99. Hill, R.J., Saville, D.A., and Russel. W.B., Electrophoresis of spherical polymer-coated colloidal particles, *J. Colloid Interface Sci.*, 258, 56, 2003.

100. López-García, J.J., Grosse, C., and Horno, J., Numerical study of colloidal suspensions of soft spherical particles using the network method. 1. DC electrophoretic mobility, *J. Colloid Interface Sci.*, 265, 327, 2003.

101. Hill, R.J., Saville, D.A., and Russel, W.B., Polarizability and complex conductivity of dilute suspensions of spherical colloidal particles with charged (polyelectrolyte) coatings, *J. Colloid Interface Sci.*, 263, 478, 2003.

102. López-García, J.J., Grosse, C., and Horno, J., Numerical study of colloidal suspensions of soft spherical particles using the network method. 2. AC electrokinetic and dielectric properties, *J. Colloid Interface Sci.*, 265, 341, 2003.

103. Zukoski IV, C.F. and Saville, D.A., An experimental test of electrokinetic theory using measurements of electrophoretic mobility and electrical conductivity, *J. Colloid Interface Sci.*, 107, 322, 1985.

104. van der Put, A.G. and Bijsterbosch, B.J., Electrical conductivity of dilute and concentrated aqueous dispersions of monodisperse polystyrene particles: influence of surface conductance and double-layer polarization, *J. Colloid Interface Sci.*, 75, 512, 1980.

105. Myers, D.F. and Saville, D.A., Dielectric spectroscopy of colloidal suspensions. II. Comparison between experiments and theory, *J. Colloid Interface Sci.*, 131, 461, 1989.

106. Rosen, L.A. and Saville, D.A., Dielectric spectroscopy of colloidal dispersions: comparison between experiment and theory, *Langmuir*, 7, 36, 1991.

107. Rosen, L.A., Baygents, J.C., and Saville, D.A., The interpretation of dielectric response measurements on colloidal dispersions using the dynamic Stern layer model, *J. Chem. Phys.*, 98, 4183, 1993.

108. Simonova, T.S. and Shilov, V.N., Influence of the mobilities of the ions in the compact part of the electric double layer of spherical particles on the electrophoresis and electric conductivity of disperse systems, *Colloid J.*, 48, 370, 1986.

109. Zukoski IV, C.F. and Saville, D.A., The interpretation of electrokinetic measurements using a dynamic model of the Stern layer. I. The dynamic model, *J. Colloid Interface Sci.*, 114, 32, 1986.

110. Zukoski IV, C.F. and Saville, D.A., The interpretation of electrokinetic measurements using a dynamic model of the Stern layer. II. Comparisons between theory and experiment, *J. Colloid Interface Sci.*, 114, 45, 1986.

111. Rosen, L.A. and Saville, D.A., The dielectric response of polystyrene latexes: effects of alterations in the structure of the particle surface, *J. Colloid Interface Sci.*, 140, 82, 1990.

112. Mangelsdorf, C.S. and White, L.R., Effects of Stern-layer conductance on electrokinetic transport properties of colloidal particles, *J. Chem. Soc., Faraday Trans.*, 86, 2859, 1990.

113. Mangelsdorf, C.S. and White, L.R., The dynamic double layer. 1. Theory of a mobile Stern layer, *J. Chem. Soc., Faraday Trans.*, 94, 2441, 1998.

114. Mangelsdorf, C.S. and White, L.R., The dynamic double layer. 2. Effects of Stern-layer conduction on the high-frequency electrokinetic transport properties, *J. Chem. Soc., Faraday Trans.*, 94, 2583, 1998.

4 Emulsions: Overview

Laurier L. Schramm
Saskatchewan Research Council, Saskatoon, Canada
University of Calgary, Calgary, Canada

Elaine N. Stasiuk
University of Calgary, Calgary, Canada

CONTENTS

I. INTRODUCTION

This chapter provides an introduction to the occurrence, properties, and importance of practical emulsions. Spanning a wide array of bulk physical properties and stabilities, a common starting point for understanding emulsions is provided by the fundamental principles of colloid science. These principles may be applied to emulsions in different ways to achieve quite different results. A desirable emulsion, which must be carefully stabilized to assist one stage of an industrial process, may be undesirable in another stage and may necessitate a demulsification strategy. With an emphasis on the definition of important terms, the importance of interfacial properties to emulsion making, stability, and breaking is demonstrated.

A. IMPORTANCE OF EMULSIONS

If two immiscible liquids are mixed together in a container and shaken, one of the two phases will become a collection of droplets that are dispersed in the other phase. Thus, an emulsion has been formed. Some important kinds of familiar emulsions include those occurring in foods (milk, mayonnaise, etc.), cosmetics (creams and lotions), pharmaceuticals (soluble vitamin and hormone products), agricultural products (insecticide and herbicide emulsion formulations), and the petroleum recovery and processing industry (drilling fluid, production, process plant, and transportation emulsions). Emulsions have long been of great industrial interest, they may be applied or encountered at all stages in the processing industries. In addition to their wide occurrence, emulsions have important properties that may be desirable, in a natural or formulated product, or undesirable, like an unwanted emulsion in an industrial process. This chapter is intended to provide an introduction to the basic principles involved in the occurrence, making, and breaking of practical emulsions.

Examples in Table 4.1 show that real-world emulsions may be desirable or undesirable. In the petroleum industry for example, one kind of oil well drilling fluid (or "mud") is emulsion based. Here, a stable emulsion (usually oil dispersed in water) is used to lubricate the cutting bit, and to carry cuttings up to the surface. This is obviously a desirable emulsion and great care goes into its proper preparation. On the other hand crude oil, when spilled on the ocean, tends to become emulsified in the form of "chocolate mousse" emulsions, so named for their color and semi-solid consistency. These high water content, water-in-oil emulsions tend to be quite stable due to the presence of strong stabilizing films, increase the quantity of pollutant and are usually very much more viscous than the oil itself. This kind of environmentally damaging emulsion is highly undesirable.

All practical emulsion applications or problems tend to have in common the same basic principles of colloid science that govern the nature, stability, and properties of emulsions. The widespread importance of emulsions in general, and scientific interest in their formation, stability and properties, have precipitated a wealth of published literature on the subject. This chapter provides an introduction and overview. A number of books provide very useful introductions to the properties, importance, and treatment of emulsions [1–5] in industry. Most good colloid chemistry texts contain introductory chapters on emulsions [6–9], while for much more detailed treatment of advances in specific emulsion areas the reader is referred to some of the chapters available in specialist monographs and encyclopedias [10–12].

TABLE 4.1
Some Occurrences of Emulsions

Field	Emulsion Examples
Environment and meteorology	Water or sewage treatment emulsions, oil spill mousse emulsions
Foods	Milk, butter, mayonnaise, cream liqueurs, ice cream, creams, soft drink syrups, mayonnaise, sauces, margarine, spreads, salad dressings, cheeses
Geology, agriculture, and soil science	Insecticides and herbicides, sulfidic melt in magma
Manufacturing and materials science	Polishes, asphalt (paving) emulsion
Biology and medicine	Soluble vitamin and hormone products, biological membranes, blood
Petroleum production and mineral processing	Emulsion drilling fluids; emulsion fracturing, stimulation, acidizing fluids; enhanced oil recovery (EOR) *in situ* emulsions; produced (well-head) emulsions; bituminous oil sand process and froth emulsions; heavy oil pipeline emulsions; fuel oil and tanker emulsions
Home and personal care products	Hair and skin creams and lotions

B. DEFINITION AND CLASSIFICATION OF EMULSIONS

Emulsions are a special kind of colloidal dispersion — one in which a liquid is dispersed in a continuous liquid phase of different composition. The dispersed phase is sometimes referred to as the internal (disperse) phase. The continuous phase is sometimes referred to as the external phase. Emulsions form one of the several kinds of colloidal systems, being so designated when their droplets have diameters between about 1 and 1000 nm [13]. In practical emulsions, the droplet sizes may well exceed the size limits given above (e.g., the fat droplets in milk) and extend into the hundreds of micrometers. In most emulsions, one of the liquids is aqueous while the other is an oil of some kind. Several types of emulsion are readily distinguished in principle, depending upon which kind of liquid forms the continuous phase:

- Oil-in-water (O/W) for oil droplets dispersed in water
- Water-in-oil (W/O) for water droplets dispersed in oil
- Oil-in-water-in-oil (O/W/O) double emulsion
- Water-in-oil-in-water (W/O/W) double emulsion

Double emulsions have applications in, for example, cosmetics, agriculture, food, photography, leather, petroleum, and drug-delivery [5,14]. There also exist even more complex types of emulsions, like O/W/O/W, etc. [5]. Figure 4.1 (from Ref. [15]) shows an example of a W/O/W/O emulsion.

The type of emulsion that is formed depends upon a number of factors. If the ratio of phase volumes is very large or very small then the phase having the smaller volume is frequently the dispersed phase. If the ratio is closer to 1 then other factors determine the outcome. Table 4.1 lists some simple examples of industrial emulsion types.

Lyophobic colloids, which include all emulsions other than the microemulsions, are not formed spontaneously on contact of the phases, because they are thermodynamically unstable when compared with the separated states. These dispersions can be formed by other means, however. Most emulsions that will be encountered in practice contain oil, water, and an emulsifying agent. The emulsifier may comprise one or more of the following: simple inorganic electrolytes, surfactants,

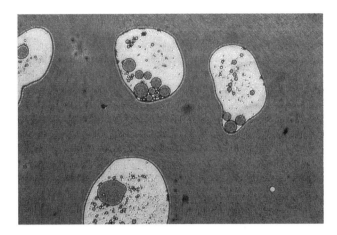

FIGURE 4.1 Example of a water-in-oil-in-water-in-oil (W/O/W/O) emulsion. (From Schramm and Kutay [15]. Reprinted with the permission of Cambridge University Press.)

polymers, or finely divided solids. The emulsifying agent may be needed to reduce interfacial tension and aid in the formation of the increased interfacial area with a minimum of mechanical energy input, or it may be needed to form a protective film, at the droplet surfaces, that acts to prevent coalescence with other droplets. Although, most emulsions are not thermodynamically stable, in practice they can be quite stable (lasting as long as months to years in some cases) and may resist explicit demulsification treatments.

In some systems the addition of a fourth component, a cosurfactant, to an oil/water/surfactant system can cause the interfacial tension to drop to near-zero values, on the order of 10^{-3} to 10^{-4} mN/m, allowing spontaneous or nearly spontaneous emulsification to very small drop sizes, ca. 10 nm or smaller. The droplets can be so small that they scatter little light, thus the emulsions can appear to be transparent and they do not break on standing or centrifuging. Unlike coarse emulsions these microemulsions are usually thought to be thermodynamically stable [16,17]. The thermodynamic stability is frequently attributed to transient negative interfacial tensions, but this remains an area of ongoing research [18,19]. The ultralow interfacial tensions, high solubilization capacity, and long-term stability of microemulsion systems have application in areas such as enhanced oil recovery (EOR), soil and aquifer decontamination and remediation, foods, pharmaceuticals (drug delivery systems), cosmetics, and pesticides [1,5,17,20–22].

II. PHYSICAL CHARACTERISTICS OF EMULSIONS

A. Appearance

A tremendous range of emulsion appearances are possible, depending upon the droplet sizes and the difference in refractive indices between the phases. An emulsion can be transparent if either the refractive index of each phase is the same, or alternatively, if the dispersed phase is made up of droplets that are sufficiently small compared with the wavelength of the illuminating light. Thus, an O/W microemulsion of even a crude oil in water may be transparent. If the droplets are of the order of 1 μm diameter a dilute O/W emulsion will take on a somewhat milky-blue cast; if the droplets are very much larger then the oil phase will become quite distinguishable and apparent.

Physically the nature of the simple emulsion types can be determined by methods such as [23]:

Texture. The texture of an emulsion frequently reflects that of the external phase. Thus O/W emulsions usually feel "watery or creamy" while W/O emulsions feel "oily or greasy." This distinction becomes less evident as the emulsion viscosity increases, so that a highly viscous O/W emulsion may feel oily.

Mixing. An emulsion readily mixes with a liquid that is miscible with the continuous phase. Thus, milk (O/W) can be diluted with water while mayonnaise (W/O) can be diluted with oil. Usually, an emulsion that retains a uniform and milky appearance when greatly diluted is more stable than the one that aggregates upon dilution [10].

Dyeing. Emulsions are most readily and consistently colored by dyes soluble in the continuous phase (e.g., methylene blue for water or fuschin for oil [24]).

Conductance. O/W emulsions usually have a very high specific conductance, like that of the aqueous phase itself, while W/O emulsions have a very low specific conductance. A simple test apparatus is described in Ref. [10].

Inversion. If an emulsion is very concentrated, it will probably invert when diluted with additional internal phase.

Fluorescence. If the oil phase fluoresces then fluorescence microscopy can be used to determine the emulsion type as long as the drop sizes are larger than the microscope's limit of resolution (>0.5 μm) (see Ref. [24]).

Magnetic resonance imaging (MRI). MRI is sensitive to any NMR-active nuclei, like protons. This allows one to distinguish different chemical environments in which these nuclei find themselves, including oil versus water (see Refs. [15,25]).

Emulsions do not always occur in the idealized form of drops of one phase dispersed in another. The occurrence of double emulsions, of the types O/W/O and W/O/W, and more complex multiple emulsions, such as O/W/O/W and W/O/W/O, have already been mentioned. In addition, emulsions may contain oil and water, solid particles, and even gas [5]. Emulsions may also occur within another type of colloidal dispersion like inside foams [26]. Figure 4.2 (from Ref. [15]) shows an example of an aqueous foam with oil droplets residing in its Plateau borders. In these cases, the above techniques can sometimes be used to determine the continuous

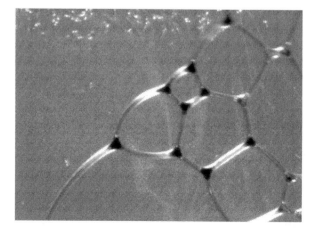

FIGURE 4.2 Example of a foam containing emulsified oil droplets. (From Schramm and Kutay [15]. Reprinted with the permission of Cambridge University Press.)

phase, but not always. For example, the dye test may produce misleading results when applied to a multiple emulsion, or to a bicontinuous emulsion. The characterization of multiple emulsions is usually best accomplished by techniques such as optical and electron microscopy that permit direct observation of the dispersed phases.

B. RHEOLOGY

The rheological properties of an emulsion are very important. High viscosity may be the reason that an emulsion is troublesome, a resistance to flow that must be dealt with, or a desirable property for which an emulsion is formulated.

A summary of the various rheological classifications can be found in Refs. [27,28]. Emulsions are frequently pseudoplastic, as shear rate increases viscosity decreases, this is also termed as shear-thinning. An emulsion may also exhibit a yield stress, that is, the shear rate (flow) remains zero until a threshold shear stress is reached — the yield stress (τ_Y) — then pseudoplastic or Newtonian flow begins. The pseudoplastic flow may also be time dependent, or thixotropic, in which case viscosity decreases under constant applied shear rate.

Some very useful descriptions of experimental techniques have been given by Whorlow and others [27,29,30]. Very often measurements are made with an emulsion sample placed in the annulus between two concentric cylinders. The shear stress is calculated from the measured torque required to maintain a given rotational velocity of one cylinder with respect to the other. Knowing the geometry, the effective shear rate can be calculated from the rotational velocity. One reason for the relative lack of rheological data for emulsions, compared with that for other colloidal systems, is the difficulty associated with performing the measurements in these systems. A practical O/W emulsion sample may contain suspended particles in addition to the oil drops. In attempting to conduct a measurement, a number of changes may occur in the sample chamber making the measurements irreproducible and not representative of the original emulsion [31]. Physical changes such as creaming/sedimentation, centrifugal separation, and shear-induced coalescence can cause non-uniform distributions, altered droplet size distributions, or even phase separations within the measuring chamber [23].

It is frequently desirable to be able to describe emulsion viscosity in terms of the viscosity of the continuous phase (η_0) and the amount of emulsified material. A very large number of equations have been advanced for estimating suspension (or emulsion, etc.) viscosities. Most of these are empirical extensions of Einstein's equation for a dilute suspension of spheres:

$$\eta = \eta_0(1 + 2.5\phi)$$

where η_0 is the medium viscosity and ϕ the dispersed phase volume fraction, $\phi < 1$.

For example, the empirical equations typically have the form [1]:

$$\eta = \eta_0(1 + \alpha_0\phi + \alpha_1\phi^2 + \alpha_2\phi^3 + \cdots)$$

and may include other terms, as in the Thomas equation,

$$\eta = \eta_0(1 + 2.5\phi + 10.5\phi^2 + 0.00273\exp[16.6\phi]).$$

These equations assume Newtonian behavior, or at least apply to the Newtonian region of a flow curve, and they usually apply if the droplets are not too large, and if there are no strong electrostatic interactions. Additional relationships are compiled and discussed in more detail elsewhere [5,13,19,26].

Emulsions can show varying rheological or viscosity behaviors. Sometimes these properties are due to the emulsifier or other agents in the emulsion. However, if the internal phase has a sufficiently high volume fraction (typically anywhere from 10% to 50%) the emulsion viscosity increases due to droplet "crowding," or structural viscosity, and becomes non-Newtonian. The maximum volume fraction possible for an internal phase made up of uniform, incompressible spheres is

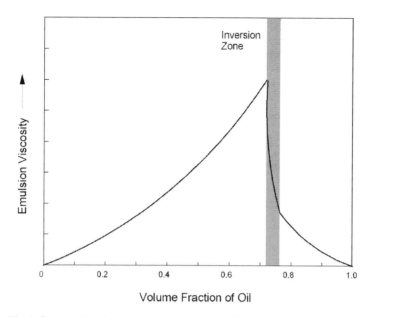

Volume Fraction of Oil

FIGURE 4.3 The influence of volume fraction on the emulsion type and viscosity of a model emulsion. (Adapted from data in Becher [1].)

74%, although emulsions with an internal volume fraction of 99% have been made [10]. Figure 4.3 shows how emulsion viscosity tends to vary with volume fraction; the drop in viscosity at $\phi = 0.74$ signifies inversion. At this point, the dispersed phase volume fraction becomes 0.26, in this example, and the lower value of ϕ is reflected by a much lower viscosity. If inversion does not occur, then the viscosity continues to increase. This is true for both W/O and O/W types.

A graphic and important example is furnished by the oil spill "chocolate mousse" emulsions formed when crude oil spills into seawater. These W/O emulsions have high water contents which may exceed 74% and reach $\phi = 0.80$ or more without inverting. As their common name implies, these mousse emulsions not only have viscosities that are much higher than the original crude oil, but can also become a semisolid. With increasing time after a spill, these emulsions weather (the oil becomes depleted in its lower boiling fractions), apparently making the emulsions more stable, solid-like, and considerably more difficult to handle and break.

In addition to bulk viscosity properties, a closely related and very important property is the interfacial viscosity, which can be thought of as the two-dimensional equivalent of bulk viscosity, operative in the oil–water interfacial region. As droplets in an emulsion approach each other the thinning of the films between the drops, and their resistance to rupture, are thought to be of great importance to the ultimate stability of the emulsion. Thus, a high interfacial viscosity can promote emulsion stability by retarding the rate of droplet coalescence, as discussed in later sections. Further details on the principles, measurement, and applications to emulsion stability of interfacial viscosity are reviewed by Malhotra and Wasan [32].

III. MAKING AND STABILIZING EMULSIONS

A. MAKING EMULSIONS

An emulsion can be made by simply mixing oil into water with sufficient mechanical shear. Emulsions of any significant stability, however, contain oil, water, and at least one emulsifying agent. The emulsifying agent may lower interfacial tension thus making it easier to create small droplets. Another emulsifying agent may be needed to stabilize the small droplets so that they do not coalesce to form larger drops, or even separate out as a bulk phase. In the classical method of emulsion preparation, the emulsifying agent is dissolved into the phase where it is most soluble, after which the second phase is added, and the whole mixture is vigorously agitated. The agitation must be turbulent and is crucial to produce sufficiently small droplets [33]. Frequently, after an initial mixing, a second mixing with very high applied mechanical shear forces is required.

1. Mechanical Aspects

High-shear mixing can be provided by a propeller-style mixer or turbine agitator, but more commonly a colloid mill or ultrasound generator is employed. A wide range of techniques are now available including, for example, atomizers and nebulizers of various designs used to produce emulsions having relatively narrow size distributions. Food emulsions are usually made using colloid mills if very small droplet sizes are not required, such as for mayonnaise or salad cream [34]. When very small droplet sizes are needed then a coarse emulsion is usually passed through a high-pressure homogenizer, which can produce droplets smaller than 2 μm diameter [34]. When making high-viscosity emulsions, agitators capable of scraping the walls of the container are used [33,35].

Phase inversion can also be used. For example, butter results from the creaming, breaking and inversion of emulsified fat droplets in milk. More generally, if ultimately a W/O emulsion is desired, then a coarse O/W emulsion can be first prepared with mechanical energy addition, and then the oil content is progressively increased. At some volume fraction above 60–70% the emulsion will suddenly invert, producing a W/O emulsion of much smaller water droplet sizes than were the oil drops in the original O/W emulsion. A phase inversion method requiring much less mechanical energy uses a controlled temperature change to cause an emulsion to suddenly change form, from a coarse O/W emulsion, through a microemulsion phase, and into a fine W/O emulsion [36]. This is known as the phase inversion temperature (PIT) method (Figure 4.4).

2. Surface Chemical Aspects

In simple two-phase colloidal systems a thin intermediate region or boundary, known as the interface, lies between the dispersed and dispersing phases. Interfacial properties are very important because emulsified droplets have a large interfacial area, and even a modest interfacial energy

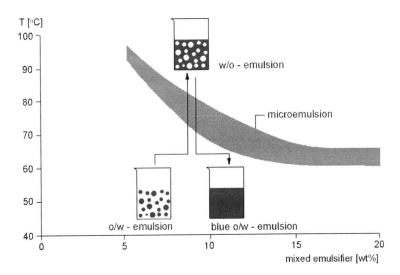

FIGURE 4.4 Illustration of the preparation of a finely divided O/W emulsion from a coarse W/O emulsion using a heating–cooling cycle and the phase inversion temperature (PIT) principle. (From Förster [36]. Reprinted with permission of Blackwell Publishing Ltd.)

per unit area can become a considerable total interfacial energy to be reckoned with. For example, when emulsifying an oil into water every subdivision step in which the average droplet radius is halved increases the interfacial area by a factor of 2, with a commensurate increase in total free surface energy needed. In practice, the energy requirement is even greater than estimated by a surface area calculation due to the need for droplets and bubbles to deform before smaller drops can pinch-off [37,38].

The exact result is influenced by the viscosities of the two liquid phases, the amount of mechanical shear produced, and the interfacial tension, among others. One representation is given by the critical Weber number [37–39]. The Weber number, We, is given by:

$$We = \frac{\eta_1 \dot{\gamma} R}{\gamma_{12}} \qquad (4.1)$$

where η_1 is the viscosity of the continuous phase, $\dot{\gamma}$ the shear rate, R the droplet radius produced, and γ_{12} the interfacial tension (the subscript 2 represents the dispersed phase). The critical Weber value has to be exceeded in order for a droplet to be disrupted and burst. In laminar flow We can range from 0.5 to ∞. For turbulent flow the conditions are somewhat different [38,39]. Figure 4.5 shows the critical Weber number as a function of viscosity ratio, η_2/η_1, between the dispersed (η_2) and continuous (η_1) phases [39]. This shows that high velocity gradients are needed to deform droplets when the continuous phase viscosity is low.

The type of emulsion that will be formed is also influenced by the critical Weber number [36–38,40]. Figure 4.6 shows that for a given viscosity ratio, η_2/η_1, between the dispersed (η_2) and continuous (η_1) phases, reducing the interfacial tension increases the Weber number, and lowers the energy needed to cause droplet breakup (see [37,40]). For a given flowing system involving a viscous oil, the viscosity ratio will be smaller, and an emulsion is easier to form, if it is a W/O emulsion rather than an O/W emulsion.

To reduce the energy requirement for emulsification, surfactants are usually added to lower the interfacial free energy or interfacial tension. A small quantity of surfactant addition can lower the amount of mechanical energy needed for emulsification by several orders of magnitude (an illustration is provided in [23]). Surfactants adsorb at interfaces, frequently concentrating in one molecular layer at the interface. These interfacial films often provide the stabilizing influence in emulsions since they can both lower interfacial tension and increase the interfacial viscosity. The

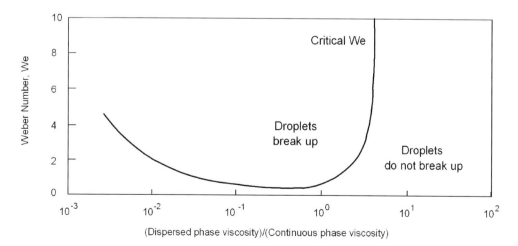

FIGURE 4.5 The critical Weber number for disruption of droplets in simple shear flow (solid curve), and for the resulting average droplet size in a colloid mill (hatched area) as a function of the viscosity ratio for disperse to continuous phases. (From data in Walstra [39].)

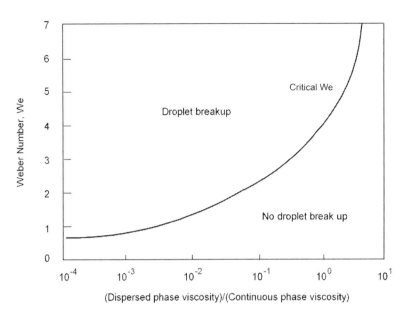

FIGURE 4.6 Droplet breakup as a function of viscosity ratio. The solid line represents the critical Weber number value above which droplet breakup will occur. (From Isaacs and Chow [37]. Copyright 1992 American Chemical Society. With permission.)

former makes emulsions easier to form while the latter provides a mechanical resistance to coalescence. In-depth discussions of surfactant structure and chemistry can be found in Refs. [41–43].

There are a number of "rules of thumb" for predicting the emulsion type that will form from the specific chemical systems. For example,

- Bancroft's rule, is based on the concept that if an emulsifying agent is preferentially wetted by one of the phases then the greatest amount of the emulsifying agent can be accommodated at the interface if it is convex toward the preferentially wetting phase (i.e., if that phase is the continuous phase). Simply stating that the liquid in which the surfactant is most soluble tends to become the continuous phase.
- The oriented-wedge theory is based on the concept that, for anionic surfactant emulsifiers, polyvalent metal ions will tend to form each coordinate to the polar groups of two surfactant molecules, forcing the hydrocarbon tails into a wedge-like orientation. In this case, the hydrocarbon tails in a close-packed interfacial layer are most easily accommodated if the oil phase is the continuous phase. As a result, anionic surfactants associated with monovalent metal cations should tend to produce O/W emulsions, while those of polyvalent metal cations should tend to produce W/O emulsions. This works best for carboxylate surfactants.
- By analogy with the Bancroft and oriented-wedge rules, the liquid that preferentially wets solid particles should tend to form the continuous phase in a solids-stabilized emulsion. Thus if there is a low contact angle (measured through the water phase) then an O/W emulsion should form.

It is emphasized that there are exceptions to each of these rules, and sometimes one will work where the others do not. They do, however, remain useful for making initial predictions.

The hydrophile–lipophile balance (HLB) concept is probably the most useful approach to predict the type of emulsion that will be stabilized by a given surfactant or surfactant formulation.

The HLB concept was introduced [44,45] as an empirical scale that could be used to describe the balance of the size and strength of the hydrophilic and lipophilic groups on an emulsifier molecule. Originally used to classify Imperial Chemical Industries' nonionic surfactant series of "Spans" and "Tweens" the HLB system has now been applied to many other surfactants, including ionics and amphoterics.

This dimensionless scale ranges from 0 to 20; a low HLB (<9) refers to a lipophilic surfactant (oil soluble) and a high HLB (>11) to a hydrophilic (water soluble) surfactant. In general, W/O emulsifiers exhibit HLB values in the range 3–8 while O/W emulsifiers have HLB values of about 8–18. There exist empirical tables of HLB values required to make emulsions out of various materials [44] and tables and equations to determine emulsifier HLB values [10,44,46]. If the value is not known, then a series of lab emulsification tests are required, using a series of emulsifying agents of known HLB values.

Experimentally, bottle tests can be used to determine the HLB value for a prospective emulsifier by mixing it with an emulsifier of known HLB, and an oil for which the HLB required for emulsification is known [23,47].

A limitation of the HLB system is that other factors are important as well. Also, the HLB is an indicator of the emulsifying characteristics of an emulsifier, but not the efficiency of an emulsifier. Thus, while all emulsifiers having a high HLB tend to promote O/W emulsions, there will be a considerable variation in the efficiency with which those emulsifiers act for any given system. For example, usually mixtures of surfactants work better than pure compounds of the same HLB.

There is some evidence to suggest that, depending upon the phase volume ratios employed, the emulsification technique used can be of greater importance in determining the final emulsion type than the HLB values of the surfactants themselves [47]. As an empirical scale the HLB values are determined by a standardized test procedure. However, the HLB classification for oil phases in terms of the required HLB values is apparently greatly dependent on the emulsification conditions and process for some phase volume ratios. When an emulsification procedure involves high shear, or when a 50/50 phase volume ratio is used, interpretations based on the classical HLB system appear to remain valid. However, at other phase volume ratios and especially under low shear emulsification conditions, inverted, concentrated emulsions may form at unexpected HLB values [47]. This is illustrated in Figure 4.7a and b.

Just as solubilities of emulsifying agents vary with temperature, so does the HLB, especially for the nonionic surfactants. A surfactant may stabilize O/W emulsions at low temperature, but W/O emulsions at some higher temperature. The transition temperature, at which the surfactant changes from stabilizing O/W to W/O emulsions, is known as the PIT, or sometimes as the HLB temperature. At the PIT, the hydrophilic and oleophilic natures of the surfactant are essentially the same. As a practical matter, emulsifying agents are chosen so that their PIT is far from the expected storage and use temperatures of the desired emulsions. In one method [48], an emulsifier with a PIT of about 50°C higher than the storage or use temperature is selected. The emulsion is then prepared at the PIT where very small droplet sizes are most easily created. Next, the emulsion is rapidly cooled to the desired use temperature, where now the coalescence rate will be slow, and a stable emulsion results.

Microemulsions are stable emulsions of hydrocarbons and water in the presence of surfactants and co-surfactants. They are characterized by spontaneous formation, ultra-low interfacial tension, and thermodynamic stability.

The optimum surfactant formulation for a microemulsion system is dependent on many variables (i.e., pH, salinity, temperature, etc.). References [17,49] list some of the components in a typical formulation. The surfactants and co-surfactants must be available in large amounts at a reasonable cost. In addition, they should also be chemically stable, brine soluble, and compatible with the other formulation components. Common surfactants used are petroleum sulfonates and ethoxylated alcohol sulfates [50,51]. The degree of interfacial tension lowering depends on the

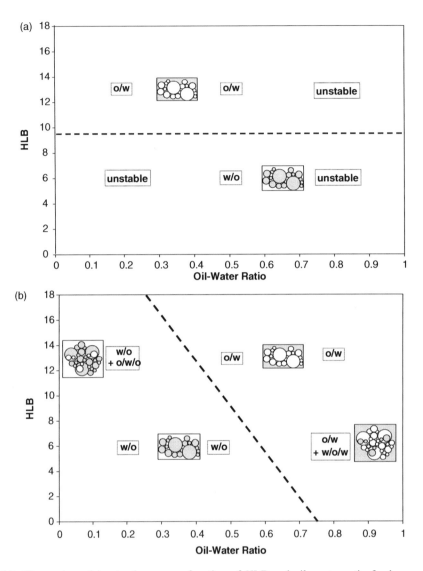

FIGURE 4.7 Observed emulsion tendency as a function of HLB and oil–water ratio for kerosene/water emulsions prepared under high shear (a) and low shear (b). (From Vander Kloet and Schramm [47]. Reproduced with permission of American Oil Chemists' Society.)

phase behavior of the oil or brine or surfactant mixture. Surfactants are generally used at concentrations much higher than their critical micelle concentration (cmc). Phase behavior will depend on the surfactant partition coefficient between the oil and brine.

The widespread interest in microemulsions and its use in industrial applications are based mainly on their high solubilization capacity for both hydrophilic and lipophilic compounds, their large interfacial areas, and on the ultralow interfacial tensions achieved when they coexist with excess aqueous and oil phases. The properties of microemulsions have been extensively reviewed elsewhere [18,52–57]. The ultralow interfacial tension achieved in microemulsion systems has application in several phenomena involved in oil recovery and in other extraction processes (i.e., soil decontamination and detergency).

B. STABILIZING EMULSIONS

A consequence of the small drop size and presence of an interfacial film on the droplets in emulsions is that quite stable dispersions of these species can be made. That is, the suspended droplets do not settle out or float rapidly, and also the drops do not coalesce quickly.

Emulsion droplets can come together in very different ways, so emulsion stability has to be considered in terms of three different processes: creaming (sedimentation), aggregation, and coalescence. Creaming is the opposite of sedimentation and results from a density difference between the two liquid phases. In aggregation two or more droplets clump together — touching only at certain points, and with virtually no change in total surface area. Aggregation is sometimes referred to as flocculation or coagulation. In coalescence two or more droplets fuse together to form a single larger unit, reducing the total surface area.

In aggregation, the species retain their identity, but lose their kinetic independence since the aggregate moves as a single unit. Aggregation of droplets may lead to coalescence and the formation of larger droplets until the phases become separated. In coalescence, on the other hand, the original species lose their identity and become part of a new species. Kinetic stability can thus have different meanings. An emulsion can be kinetically stable with respect to coalescence, but unstable with respect to aggregation. Or, a system could be kinetically stable with respect to aggregation, but unstable with respect to sedimentation or flotation. This means that kinetic stability has to be understood in terms of a clearly defined process, degree of change, and time scale.

Encounters between particles in dispersion can occur frequently due to any of the Brownian motion, sedimentation, or stirring. The stability of the dispersion depends upon how the particles interact when this happens. Table 4.2 lists some factors involved in determining the stability of emulsions. The main causes of repulsive or barrier forces may be electrostatic, steric, or mechanical. The main attractive forces are the van der Waals forces between objects.

1. Charged Interfaces and the Electric Double Layer

Most substances acquire a surface electric charge when brought into contact with a polar medium like water. For emulsions, the origin of the charge can be due to ionization, as when surface acid functionalities ionize when oil droplets are dispersed into an aqueous solution, or it can be due to adsorption, as when surfactant ions or charged particles adsorb onto an oil droplet surface. For solid particles there can be additional mechanisms of charging. One is the unequal dissolution of cations and anions that make up the crystal structure (e.g., in the salt-type minerals). Another is the diffusion of counterions away from the surface of a solid whose internal crystal structure carries an opposite charge due to isomorphic substitution (e.g., in clays). The surface charge influences the distribution of nearby ions in the polar medium. Ions of opposite charge (counterions) are attracted

---

TABLE 4.2
Some Factors Involved in Determining the Stability of Emulsions

Low interfacial tension (makes it easier to form and maintain large interfacial area)
Electric double layer (EDL) repulsion (reduces the rates of aggregation and coalescence)
Surface viscosity (retards coalescence)
Steric repulsion (reduces the rates of aggregation and coalescence)
Small droplet size (may reduce the rate of aggregation)
Small volume of dispersed phase (reduces the rate of aggregation)
Bulk viscosity (reduces the rates of creaming and aggregation)
Small density difference between phases (reduces the rates of creaming and aggregation)
Dispersion force attraction (increases the rates of aggregation and coalescence)

---

to the surface while those of like charge (coions) are repelled. An electric double layer (EDL), which is diffuse because of mixing caused by thermal motion, is thus formed.

In emulsions where the oil phase contains partitioned organic acids, like fatty acids, the degree of surface charging is more complicated than this. In this case, the oil–water interface becomes negatively charged in alkaline aqueous solutions due to the ionization of the acid groups at the interface. Adsorption of surfactants at the interface also influences the overall charge. Therefore, the degree of negative charge at the interface depends on the pH and ionic strength of the solution [58,59], and also on the concentration of natural surfactant monomers present in the aqueous phase [60,61]. Examples can be found in heavy oil [17,49, 51,62] and bitumen-containing systems [58–61,63,64].

The EDL consists of the charged surface and excess of a neutralizing counterions over coions, distributed near the surface. The EDL can be viewed as being composed of two layers:

1. An inner layer that may include adsorbed ions
2. A diffuse layer where ions are distributed according to the influence of electrical forces and thermal motion.

In a simple model for the diffuse double layer, such as that of Göuy and Chapman, the electrical potential (ψ) decays exponentially with distance (x) into solution:

$$\psi = \psi^\circ \exp(-\kappa x) \tag{4.2}$$

where ψ° is the surface potential and $1/\kappa$ is the double layer thickness.

The variation of potential out into solution is actually more complex. For example, a layer of specifically adsorbed ions bounded by a plane, the Stern plane, can be added to the model represented by Equation (4.1), in which case the potential changes from ψ° at the surface, to $\psi(\delta)$ at the Stern plane, to $\psi = 0$ in bulk solution. Additional details are provided elsewhere [65–67].

In the simplest example of colloid stability, emulsion droplets would be stabilized entirely by the repulsive forces created when two charged surfaces approach each other and their EDLs overlap. The repulsive energy V_R for spherical droplets is given approximately as,

$$V_R = \frac{B\varepsilon k^2 T^2 a \gamma^2}{z^2} \exp[-\kappa H] \tag{4.3}$$

where

$$\gamma = \frac{\exp[ze\psi(\delta)/2kT] - 1}{\exp[ze\psi(\delta)/2kT] + 1}$$

Here the spheres have radius a and are separated by distance H, B is a constant, and z the counterion charge number.

2. Attractive Forces

van der Waals postulated that neutral molecules exert forces of attraction on each other, which are caused by electrical interactions between three types of dipolar configurations. The attraction results from the orientation of dipoles which may be (1) two permanent dipoles, (2) dipole–induced dipole, or (3) induced dipole–induced dipole. The third kind of forces between nonpolar molecules are also called London dispersion forces. Except for quite polar materials, the London dispersion forces are the most significant of the three.

For molecules the dispersion force varies inversely with the sixth power of the intermolecular distance. For dispersed emulsion droplets, the dispersion forces can be approximated by adding up the attractions between all interdroplet pairs of molecules. When added this way the dispersion force between two droplets decays less rapidly as a function of separation distance than is the case for individual molecules. For two spheres of radius a separated by distance H, the attractive energy V_A can be approximated by,

$$V_A = -\frac{Aa}{12H} \tag{4.4}$$

for $H < 10$–20 nm and $H \ll a$. The constant A is known as the Hamaker constant and depends on the density and polarizability of atoms in the droplets. The Hamaker constant is usually not well known and must be approximated [68,69].

3. DLVO Theory

Derjaguin and Landau, and independently Verwey and Overbeek [70], developed a quantitative theory for the stability of lyophobic colloids, now known as the DLVO theory. It was developed in an attempt to account for the observation that colloids coagulate quickly at high electrolyte concentrations, slowly at low concentrations, and with a very narrow electrolyte concentration range over which the transition from one to the other occurs. The latter defines the critical coagulation concentration (CCC). The DLVO theory accounts for the energy changes that take place when two droplets (or particles) approach each other, and involves estimating the energy of attraction (London–van der Waals) versus interparticle distance, and the energy of repulsion (electrostatic) versus distance. These, V_A and V_R, are then added together to yield the total interaction energy V. There is a third important force at very small separation distances where the atomic electron clouds overlap, causing a strong repulsion, called Born repulsion. The theory has been developed for several special cases, including the interaction between two spheres, and refinements are constantly being made.

V_R decreases exponentially with increasing separation distance, and has a range about equal to κ^{-1}, while V_A decreases inversely with increasing separation distance. Figure 4.8 shows a single

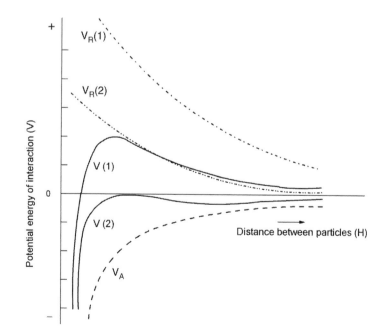

FIGURE 4.8 Two interaction energy curves, $V(1)$ and $V(2)$, resulting from the summation of an attraction curve (V_A) with two different repulsive energy curves $V_R(1)$ and $V_R(2)$. (From Shaw [6]. Copyright 1981, with permission from Elsevier Ltd.)

attractive energy curve and two different repulsive energy curves, representing two very different levels of electrolyte concentration. The figure also shows the total interaction energy curves that result in each case. It can be seen that either the attractive van der Waals forces or the repulsive EDL forces can predominate at different interdroplet distances.

Where there is a positive potential energy maximum, a dispersion should be stable if $V \gg kT$, that is, if the energy is large compared to the thermal energy of the particles (15 kT is considered unsurmountable). In this case colliding droplets should rebound without contact, and the emulsion should be stable to aggregation. If, on the other hand, the potential energy maximum is not very great, $V \approx kT$, then slow aggregation should occur. The height of the energy barrier depends on the surface potential, $\psi(\delta)$ and on the range of the repulsive forces, κ^{-1}. The figure shows that an energy minimum can occur at larger interparticle distances. If this is reasonably deep compared to kT then a loose, easily reversible aggregation should occur.

4. Practical Guidelines for Electrostatic Stabilization

The DLVO calculations can become quite involved, requiring considerable knowledge about the systems of interest. Also, regardless of the equations used, there are some problems. For example, there will be some distortion of the spherical emulsion droplets as they approach each other and begin to seriously interact, causing a flattening. Also, views of the validity of the theory have been changing as more becomes known about the influence of additional forces such as those due to surface hydration. The DLVO theory nevertheless forms a very useful starting point in attempting to understand complex colloidal systems like petroleum emulsions. There are empirical "rules of thumb" that can be used to give a first estimate of the degree of colloidal stability that a system is likely to have if the zeta potentials of the droplets are known.

Many types of colloids tend to adopt a negative surface charge when dispersed in aqueous solutions having ionic concentrations and pH typical of natural waters. For such systems one rule of thumb stems from observations that the colloidal particles are quite stable when the zeta potential is about -30 mV or more negative, and quite unstable due to agglomeration when the zeta potential is between $+5$ and -5 mV. An expanded set of guidelines, developed for particle suspensions, is given in Ref. [71]. Such criteria are frequently used to determine optimal dosages of polyvalent metal electrolytes, for example, alum, used to effect coagulation in water-treatment plants.

Any of maximizing, minimizing, or reversing the charge on surfaces can be important to optimize the efficiency of industrial separation processes, including flotation processes [69,72,73]. Here again, batch and on-line measurements of the emulsified droplet zeta potentials can be used to monitor and control such processes [69,73,74].

In general, where electrical surface charge is an important determinant of stability it tends to be easier to formulate a stable O/W emulsion than a W/O emulsion because the EDL thickness is much greater in water than in oil. W/O emulsions tend to be stabilized by other mechanisms, including steric and mechanical film stabilization.

5. Steric Stabilization

Protective agents can act in several ways. They can increase double layer repulsion if they have ionizable groups. The adsorbed layers can lower the effective Hamaker constant. An adsorbed film may necessitate desorption before particles can approach closely enough for van der Waals forces to cause attraction, or approaching particles may simply cause adsorbed molecules to become restricted in their freedom of motion (volume restriction). The use of natural and synthetic polymers to stabilize aqueous colloidal dispersions is technologically important, with research in this area being focused on adsorption and steric stabilization [75–80].

With sufficiently high surface loading, long-chain surfactants and high molecular-weight polymers can become adsorbed at the surfaces of emulsion droplets such that a significant amount of adsorbate extends out from the surfaces. In this situation, an entropy decrease can accompany the approach of a droplet, providing a short-range, volume-restriction, stabilization mechanism referred to as protection, or steric stabilization. The term protection has been used because this effect can cause significant salt tolerance on the part of an O/W emulsion. In most cases there will be adsorption layers on each approaching droplet surface so there will also be an osmotic pressure contribution, again at close approach, due to overlap of the adsorption layers. The osmotic contribution may result in increased repulsion, or not, depending on the favorability of mixing the protruding chains of the adsorbed species. Damodaran [81] provides an equation for the net steric repulsive energy potential between protein-stabilized emulsion droplets that illustrate the relative contributions of an osmotic repulsive contribution that favors stretching the chains and elastic energy of the chains that opposes the stretching.

In steric stabilization, adsorbed polymer molecules must extend outward from the droplet surface, yet be strongly enough attached that they remain adsorbed in the presence of applied shear. It is also possible to have particles stabilized by both electrostatic and steric stabilization; these are said to be electrosterically stabilized.

For effective stabilization the droplet surfaces should be fully covered by the adsorbed surfactant or polymer, otherwise uncovered regions of adjacent particles or droplets may come into contact with each other, or bridging flocculation between them may occur. Further, the stabilizing surfactant or polymer should be strongly adsorbed (firmly anchored) to the surfaces. Molecular structure and solvation, adsorption layer thickness and hydrodynamic volume, and temperature also determines the effectiveness of steric stabilization [75–79]. One way to predict whether steric stabilization is likely for a given dispersion is to estimate the protrusion distance of the surfactant or polymer chains [80].

6. Mechanical Film Stabilization

Some kinds of emulsions can be stabilized by the presence of a protective film around the dispersed droplets. Oilfield W/O emulsions may be stabilized by the presence of a protective film formed from asphaltenes, resins, and possibly solid particles. When drops approach each other during the process of aggregation, the rate of oil-film drainage will be determined initially by the bulk oil viscosity, but within a certain distance of approach, the interfacial viscosity becomes important. A high interfacial viscosity will significantly retard the final stage of film drainage and promote kinetic emulsion stability. These films can be viscoelastic, possibly rigid, and provide a mechanical barrier to coalescence, yielding a high degree of emulsion stability. More detailed descriptions are given in [32,82,83].

Emulsions can also be stabilized solely by an interfacial layer of adsorbed fine particles. Such emulsions are termed Pickering emulsions. A film of close-packed, interacting solid particles can have considerable mechanical strength, which provides the mechanism for stabilization. The most stable Pickering emulsions occur when the contact angle is close to $90°$, so that the particles will collect at the interface. If the contact angle for such a fine particles is $\theta < 90°$ then most of the particle will reside in the aqueous phase, and an O/W emulsion is indicated. Conversely, if $\theta > 90°$ then the particle will be mostly in the oil phase, and W/O emulsion would be predicted. This represents another application of the oriented wedge theory which, stated for solids, is: if the particles are preferentially wetted by one of the phases then many of them can be accommodated at the interface if that interface is convex toward that phase — that is, if that phase is the continuous phase. There will be exceptions to this rule, but it remains useful for making initial predictions.

In Pickering emulsions the particles can be quite close-packed and the stabilizing film between droplets can be quite rigid, providing a strong mechanical barrier to coalescence. This can lead to emulsions having stabilities on the order of 1 yr [84]. For this to occur the solid particles need to be well anchored at the oil–water interface and to have significant lateral attractive interactions with each other [84]. An example of the kind of solid particles that can be used to make Pickering emulsions is provided by hydrophilic silica particles that have been partially hydrophobized by treatment with organosilane compounds. Such particles can be bi-wetting and have sufficient mutual attraction to be at least weakly aggregated in bulk suspension, and therefore to exhibit the necessary lateral–mutual attraction when adsorbed at the O/W interface. Arditty et al. [88] described the preparation of O/W, W/O, and multiple emulsions using only solid particles as the stabilizing agents.

IV. EMULSION SEPARATION AND BREAKING

A. CREAMING AND SEDIMENTATION

Droplets in an emulsion will have some tendency to rise or settle according to Stokes' law. An uncharged spherical droplet in a fluid will sediment out if its density is greater than that of the fluid. The driving force is gravity, the resisting force is viscosity and is approximately proportional to the droplet velocity. After a short period of time the particle reaches terminal (constant) velocity dx/dt when the two forces are matched. Thus,

$$\frac{dx}{dt} = \frac{2a^2(\rho_2 - \rho_1)g}{9\eta} \tag{4.5}$$

where a is the particle radius, ρ_2 the droplet density, ρ_1 the external fluid density, g the gravitational constant, and η the bulk viscosity. If the droplet has a lower density than the external phase then it rises instead (negative sedimentation). Since emulsion droplets are not rigid spheres, they may deform in shear flow. Also, with the presence of emulsifying agents at the interface, the drops will not be noninteracting, as is assumed in the theory. Thus, Stokes' law will not strictly apply and may underestimate or even overestimate the real terminal velocity.

The process in which emulsion droplets rise or settle without significant coalescence is called creaming (Figure 4.9). This is not emulsion breaking, but creaming produces two separate layers of emulsion that have different droplet concentrations, which are usually distinguishable from each other by color or opacity. The term comes from the familiar separation of cream from raw milk. It can be seen from Stokes' law that creaming will occur faster when there is a larger density difference and when the droplets are larger. The rate of separation can be enhanced by replacing the gravitational driving force by a centrifugal field. Centrifugal force, like gravity, is proportional to the mass, but the proportionality constant is not g but $\omega^2 x$, where ω is the angular velocity ($=2\pi x$ rev. per sec) and x the distance of the particle from the axis of rotation. The driving force for sedimentation becomes $(\rho_2 - \rho_1) \omega^2 x$. Since $\omega^2 x$ is substituted for g, one speaks of multiples of "g" in a centrifuge. The centrifugal acceleration in a centrifuge is not really a constant throughout the system, but varies with x. Since the actual distance from top to bottom of a sedimenting column is usually small compared to the distance from the center of revolution, the average acceleration is used.

The terminal velocity then becomes:

$$\frac{dx}{dt} = \frac{2a^2(\rho_2 - \rho_1)\omega^2 x}{9\eta} \tag{4.6}$$

In general, for given liquid densities, creaming will occur more slowly, the greater the electrical charge on the droplets the higher the emulsion viscosity [85]. Although, a distinct process, creaming does promote coalescence by increasing the droplet crowding and hence the probability of droplet–droplet collisions.

B. COALESCENCE

Up to this point, stability to aggregation has been considered. However, once aggregation has taken place in an emulsion, there remains the question of stability to coalescence. Usually emulsions made

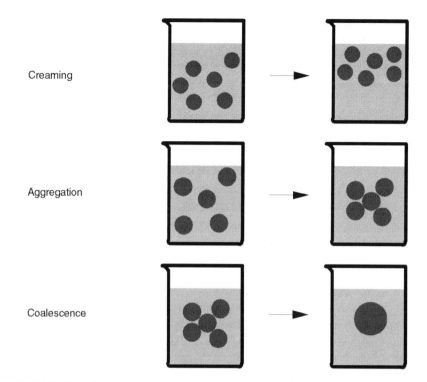

FIGURE 4.9 Creaming, aggregation, and coalescence in an O/W emulsion. (Reprinted from Schramm, L.L., in Emulsions, Fundamentals and Applications in the Petroleum Industry, Schramm, L.L., Ed., American Chemical Society, Washington, and Oxford University Press, New York, 1992, pp. 1–49. Copyright 1992 American Chemical Society. With permission.)

by mixing together two pure liquids are not very stable. To increase the stability an additional component is usually needed, which forms a film around the dispersed droplets providing a barrier to both aggregation and coalescence. Although, there are numerous effective agents and mechanisms, the additional component is frequently a surfactant. Stability to coalescence involves the mechanical properties of the interfacial films. This will be considered further in the next section.

Considering stability to both aggregation and coalescence, the factors favoring emulsion stability can be summarized as:

Low interfacial tension — low interfacial free energy makes it easier to maintain large interfacial area.

Mechanically strong film — this acts as a barrier to coalescence and may be enhanced by adsorption of fine solids, or of close-packed surfactant molecules.

EDL repulsion — this repulsion acts to prevent collisions and aggregation.

Small volume of dispersed phase — this reduces the frequency of collisions and aggregation. Higher volumes are possible, for close-packed spheres the dispersed phase volume fraction would be 0.74, but in practice the fraction can even be higher.

Small droplet size — if the droplets are electrostatically or sterically interacting.

High viscosity — this slows the rates of creaming and coalescence.

An assessment of emulsion stability involves the determination of the time variation of some emulsion property such as those described in the section, Physical Characteristics of Emulsions above. The classical methods are well described in Ref. [4]. Some newer approaches include the use of pulse nuclear magnetic resonance or differential scanning calorimetry [86].

C. INVERSION

Inversion refers to the process in which an emulsion suddenly changes form, from O/W to W/O, or vice versa. The maximum volume fraction possible for an internal phase made up of uniform, incompressible spheres is 74%. Although, emulsions with a higher internal volume fraction do occur (such as W/O oil spill mousse and O/W cosmetic "night cream" emulsions), it is more usual that inversion occurs when the internal volume fraction exceeds some value reasonably close to $\phi = 0.74$. Other factors have a bearing as well, of course, including the nature and concentration of emulsifiers and physical influences such as temperature or the application of mechanical shear.

The exact mechanism of inversion remains unclear, although, obviously some processes of coalescence and dispersion are involved. In the region of the inversion point multiple emulsions may be encountered. The process is also not always exactly reversible. That is, hysteresis may occur if the inversion point is approached from different sides of the composition scale (an example is given in Ref. [23]).

D. EMULSION BREAKING

The stability of an emulsion is often a problem. Demulsification involves two steps. First, there must be agglomeration or coagulation of droplets. Then, the agglomerated droplets must coalesce. Only after these two steps can complete phase separation occur. In a typical W/O emulsion some of the water may quite readily settle out (the "free water"); the rest will require some kind of specific emulsion treatment. In processing industries, chemical demulsification is commonly used to separate water from oil in order to produce a fluid suitable for further processing. The specific kind of emulsion treatment required can be highly variable, even within the same industry.

The first step in systematic emulsion breaking would be to characterize the emulsion to be broken in terms of its nature (O/W, W/O, or multiple emulsion), the number and nature of immiscible phases, the presence of a protective interfacial film around the droplets, and the sensitivity of the emulsifiers [32,82,83]. Based on an emulsion characterization, a chemical addition can be prescribed to neutralize the effect of the emulsifier, followed by mechanical means to complete the phase separation.

It follows directly from the previous considerations of emulsion stability that if an emulsion is stabilized by electrical repulsive forces, then demulsification could be brought about by overcoming or reducing these forces. In this context the addition of electrolyte to an O/W emulsion could be used to achieve the CCC, in accord with the Schulze–Hardy rule. Similarly, demulsifying agents, designed to reduce emulsion stability by displacing or destroying the effectiveness of protective agents, can be applied. An example is antagonistic action — addition of an O/W promoter to break a W/O emulsion. These considerations are discussed in further detail in [1,87,88].

A wide range of chemical demulsifiers are available in order to effect this separation [89–93]. Selecting the best demulsifier however, is complicated by the wide range of factors that can affect demulsifier performance, including oil type, the presence and wettability of solids, oil viscosity, and the size distribution of the dispersed water phase.

Demulsifiers are frequently surfactants; the relationship between demulsifier structure and performance has been studied for over 50 years [94]. Mikula and Munoz [88] trace the historical evolution of demulsifier chemistry and effective concentration range, and several reviews of demulsifier chemistry and properties are also available [95–99]. A demulsifier must displace or counteract the emulsifying agent stabilizing the emulsion, and promote aggregation and coalescence of the dispersed phase into large droplets that can be separated [92,93].

Examples of the primary active agents in commercial demulsifiers include ethoxylated (crosslinked or uncross-linked) propylene oxide or ethylene oxide polymers or alkylphenol resins. There are, of course, many others. These products are formulated to provide specific properties including

HLB, solubility, rate of diffusion into the interface, and effectiveness at destabilizing the interface [4,87]. A wide range of properties and parameters are used to characterize surfactant demulsifiers and predict their performance, from physical properties to compositional and structural analyses [88,99–102], although in practice the degree of characterization needed to tailor a demulsifier based on first principles is prohibitive.

Apart from the above noted chemical treatments, a variety of physical methods are used in emulsion breaking. These are all designed to accelerate coagulation and coalescence. For example, oilfield W/O emulsions may be treated by some or all of settling, heating, electrical dehydration, chemical treatment, centrifugation, and filtration. The mechanical methods, such as centrifuging or filtering rely on increasing the collision rate of droplets, and applying an additional force driving coalescence. An increase in temperature will increase thermal motions to enhance the collision rate, and also reduce viscosities (including interfacial viscosity), thus increasing the likelihood of coalescence. In the extremes, very high temperatures will cause dehydration due to evaporation, while freeze–thaw cycles will break some emulsions. Electrical methods may involve electrophoresis of oil droplets, causing them to collide, to break O/W emulsions. With W/O emulsions, the mechanism involves deformation of water droplets, since these are essentially nonconducting emulsions. Here the electric field causes an increase in the droplet area, disrupting the interfacial film. Increased droplet contacts increase the coalescence rate, breaking the emulsion. More details on the application of these methods in large-scale continuous processes are given elsewhere [87,103,104].

V. APPLICATIONS

Emulsions are commonly used in many industries. Emulsions are important in foods, pharmaceuticals, insecticides and herbicides, polishes, drugs, biological systems, cosmetics and other personal care creams and lotions, paints, lacquers, varnishes, and electrically and thermally insulating materials.

In addition to desirable emulsions, a recurring feature in many process industries is the rag layer, a gel-like emulsion that forms and accumulates at the oil–water interface in the separation vessels of many industrial processes. Rag layers tend to concentrate a range of emulsion stabilizing components. Once formed, it can also trap additional components that would otherwise have creamed or settled out of the way. Rag layers can interfere with level-sensing monitors, short-out electrostatic grids, promote channeling flows, and, of course, prevent oil or water separation [87,88].

A. MANUFACTURING

Metalworking fluids provide lubricity and cooling during the various metal-working and metal-cutting operations. Some modern synthetic metalworking oils are actually O/W microemulsions [105]. Such microemulsions switch readily to O/W macroemulsions when diluted with water at the time of application. Once applied, the surfactants involved adsorb onto metal surfaces with their hydrophobic groups oriented away from the surfaces thus reducing friction and ensuring wetting of the metal by hydrocarbons present in the metalworking emulsion [105].

Asphalt emulsions are used in road paving for the production of a smooth, water-repellant surface. First, an asphalt O/W emulsion is formulated, which has sufficiently low viscosity to be easy to handle and apply, and which has sufficient stability to survive transportation, brief periods of storage, and the application process itself. The emulsion needs to be able to shear thin during application, and then break quickly when it contacts the aggregate. The asphalt emulsions are usually 40–70% bitumen and stabilized either by natural naphthenic surfactants released by treatment with alkali (for a somewhat similar situation involving bitumen processing see also [64]), or else by the addition of anionic or cationic surfactant [106]. Emulsified asphalt can be applied to gravel or rock even when wet [1,20,105,106].

B. Environment

Marine oil spills can cause significant environmental damage. Following a spill, wind and wave energy advection and turbulence can cause an O/W emulsion to be formed, the droplets of which typically become weathered and settle out, which helps to disperse oil into the water column and away from sensitive shorelines [107]. Otherwise, the oil may pick up water to form a W/O "mousse" emulsion, probably stabilized by mechanically strong films [108,109] comprised of asphaltenes and natural surfactants [108,110–112]. These high water content mousse emulsions tend to be quite stable, increase the quantity of pollutant, and are usually very much more viscous than the oil itself. With weathering they can become semi-solid and considerably more difficult to handle, very much like the rag layer emulsions referred earlier. The presence of mechanically strong films makes it hard to get demulsifiers into these emulsions, so they are hard to break [113]. These factors, together with weather and sea conditions, complicate the use of dispersants and demulsifiers [107,108,113–116].

The contamination of groundwater by nonaqueous phase liquids (NAPLs) is a common environmental concern. Typical NAPLs such as tetrachloroethylene (PCE), trichloroethylene (TCE), and 1,1,1-trichloroethane (TCA) can easily invade the subsurface, are difficult to remove, are of concern with respect to drinking water standards, and have low biodegradability [117,118]. The techniques used to achieve the displacement, solubilization, and flushing of the NAPLs are adapted from surfactant-based EOR technology [119–121], including microemulsion-flooding [122,123]. With a good surfactant formulation based upon good phase behavior, up to 99.9% of the NAPL can be recovered from a soil column in as little as 1.0–2.0 pore volumes of surfactant-flooding [121]. A major constraint for such processes is that surfactants, if left behind must not impose an environmental threat [124].

Industrial processing may lead to emulsions being discharged into tailings ponds, such as in the tailings ponds created by surface processing of mined oil sands [64,125].

C. Foods

1. Food Emulsion Stabilization

The principles of colloid stability, particularly including DLVO theory and steric stabilization, can be applied to many food emulsions [80,81]. The applicability of DLVO theory is restricted, however, partly because the primary potential-energy minima are somewhat shallow and partly due to the tendency of adsorbed proteins extend outward from surfaces so far that steric stabilization becomes more important [34,126]. The presence of protein in an adsorption layer can also contribute a viscoelastic restriction to coalescence. Finally, the oils in food colloids are usually triglycerides (of either animal or vegetable origin). These oils may exist in liquid or crystalline states at room temperature; frequently both simultaneously. The existence of the crystal form at interfaces contributes yet another stabilizing component [34].

Most of the surfactants used to stabilize food emulsions and foams fall into one of two categories [81,127,128]:

1. Low molar mass species, such as lipids, phospholipids (lecithin), mono- and di-glycerides, sorbitan monostearate, polyoxyethylenesorbitan monostearate, and
2. High molar mass species, such as proteins and gums.

The low molar mass species usually have long-chain fatty acid groups which provide the hydrophobic character. The polar groups may be, for example, glycerol (in mono- and di-glycerides) or substituted phosphoglyceryl species (in phospholipids) [34]. These tend to be used in the preparation of, for example, W/O emulsions.

The high molar mass species reside mostly in the aqueous phase with a number of peptide groups residing in the oil–water interface [81]. Although, these surfactants are less effective at reducing interfacial tension, they can form a viscoelastic membrane-like film around oil droplets and tend to be used in the preparation of, for example, O/W emulsions.

These distinctions and trends are by no means exclusive, mixtures are the norm and competitive adsorption is prevalent. Caseinate, one of the most commonly used surfactants in the food industry, is itself a mixture of interacting proteins of varying surface activity [128]. The phospholipids can also interact with proteins and lecithins to form independent vesicles [34], thus creating an additional dispersed phase.

2. Protein-Stabilized Emulsions

Some food emulsions, including milk, cream, ice cream, and coffee whiteners and toppings, are stabilized by proteins like casein that form a coating around the fat globules [34,129]. These products also need some of the fat (oil) to be partly crystalline to link the droplets in a network structure and prevent complete coalescence of the oil droplets, especially under shear [34]. In milk, homogenization involves breaking the stabilizing films surrounding both fat droplets and casein micelles, allowing interaction between them [126]. The newly created, smaller, fat droplets become stabilized by casein micelles and fragments of casein micelles [126]. Products such as ice cream (see next section), proteins, additional emulsifiers (such as monoglycerides), and gum hydrocolloids may all take part in forming a stabilizing film around the oil droplets. In all these cases emulsion stabilization is thought to be mostly due to steric rather than electrostatic forces. Stability may also arise from reduction of density contrast between the droplets and external phase [85].

Casein or egg-yolk proteins are used as emulsifiers in another category of O/W food emulsions [34,126]. A key difference here is that in these caseinate-stabilized oil emulsions, the casein forms essentially monolayers and there are no casein micelles or any calcium phosphate. Such emulsions are thought to be stabilized more by electrostatic repulsive forces and less by steric stabilization [126]. Similarly, mayonnaise, hollandaise, and béarnaise sauces, for example, are O/W emulsions mainly stabilized by egg-yolk protein [34,129]. Here, the protein-covered oil (fat) droplets are stabilized by a combination of electrostatic and steric stabilization [129]. Perram et al. [130] described the application of DLVO theory to emulsion stability in sauce béarnaise.

Ice cream is interesting in that it is a partially frozen, foamed emulsion [131]. The first step in formulating ice cream is to create an emulsion that is, essentially, homogenized milk. Here, proteins and other emulsifiers (such as egg yolk) stabilize the fat droplets against coalescence [132,133]. The second stage in ice cream production is foaming and emulsion destabilization. Air is incorporated by whipping or by air injection and egg white acts as a foaming agent. The added shear causes partial coalescence, causing air bubbles to be trapped in clumped fat globules. When whipping and freezing occur simultaneously, good fat destabilization is achieved and a complex internal structure is achieved that has fat globules both adsorbed onto air bubbles and aggregated to each other [131,133]. Here also, the DLVO theory has been applied to describe the stability of this complex kind of emulsion [133].

Cream liqueurs are emulsions that have a cream-like appearance and need to remain stable for years despite having a high alcohol content (which makes the aqueous phase a poorer solvent for proteins) [128]. As is the case in making ice cream, the homogenization process ruptures the protective membrane surrounding oil (milk fat) droplets leaving droplets that are coated with casein protein, except that in this case smaller droplets (<0.8 μm diameter) are produced.

3. Nonprotein-Stabilized Emulsions

Some products, like butter and margarine are stabilized by fat crystals. Margarine comes from a hot W/O emulsion, stabilized by fat crystals that is formulated when quickly set by rapid chilling.

Lecithin, a typical ingredient in margarine, enhances the solubility of monoglycerides in the oil blend, and monoglycerides reduce the interfacial tension between the oil and water phases [134]. Mayonnaise, hollandaise, and béarnaise, are examples of O/W emulsions in which the emulsifier is egg yolk (a source of phospholipids) [129,134].

Salad dressings and mayonnaise can be stabilized by ionic surfactants, which provide some electrostatic stabilization as described by DLVO theory, or by nonionic surfactants which provide a viscoelastic surface coating. The protein-covered oil (fat) droplets tend to be mostly stabilized by steric stabilization (rather than electrostatic stabilization) [34,126,129], particularly at very high levels of surface protein adsorption, in which case the adsorption layer can include not just protein molecules but structured protein globules (aggregates). In some cases, lipid liquid crystal layers surround and stabilize the oil droplets, such as the stabilization of O/W droplets by egg-yolk lecithins in salad dressing [34,135].

Carbonated soft drinks are frequently prepared by diluting and carbonating "bottler's" soft drink syrups, which are O/W emulsions of flavoring oils (about 10 vol.%) in aqueous solutions of sugars, coloring and preservatives [128]. These emulsions must be stable enough to survive shipping and storage prior to bottling, plus the dilution and storage prior to sale and use. In this case the emulsifiers tend to be of the low molar mass type like polysaccharides (not proteins) [128].

D. BIOLOGY AND MEDICINE

One of the several shapes that micelles can take is laminar. Since the ends of such micelles have their lyophobic portions exposed to the surrounding solvent, they can curve upward to form spherical structures called vesicles. Vesicles comprise one or more bilayers surrounding a pocket of liquid, and are formed by molecules such as phospholipids. Multilamellar vesicles have concentric spheres of unilamellar vesicles, each separated from one another by a layer of solvent [105]. Figure 4.10 provides an illustration. Vesicles can be about the same size as living cells, but have a much simpler structure [136–139]. Vesicles can be used as drug delivery vehicles by solubilizing pharmacologically active species in the hydrocarbon core of the bilayers [140,141]. This approach has been used in the treatment of tumors and rheumatic arthritis [142].

Other kinds of microencapsulation also provide means for transporting and controlling the release of chemical species in a variety of applications, from pharmaceuticals to cosmetics to agrochemicals [142]. The microencapsulating surface, or barrier, can be:

1. A polymer film, in which case chemical release is controlled by wall thickness and the nature and properties of pores in the film, if any, or
2. A liquid membrane (as in a multiple emulsion), in which case chemical release is controlled by diffusion and dissolution.

Emulsions are involved when interfacial polymerization is used to microencapsulate a lipophilic drug. In this case the drug is placed in an oil phase together with a hydrophobic monomer, an added surfactant and suitable mechanical shear allows these to be incorporated into the dispersed oil droplets of an O/W emulsion. A hydrophilic monomer is then dissolved in the aqueous phase and the two monomers interact at the oil/aqueous interface to create polymer films that form the capsule walls. For a hydrophilic drug all of the above could simply be reversed.

A multiple emulsion can also be used for microencapsulation by, for example, first making a stable "primary" O/W emulsion using a high HLB surfactant and a mechanical stirrer. The primary emulsion is then itself emulsified into an oil phase that contains a low HLB surfactant, this time employing gentle, low shear agitation. The result is an O/W/O multiple emulsion. The first surfactant (high HLB) should provide a strong interfacial film in order to protect the droplets during the subsequent preparation steps, and also later, during storage. The second surfactant (low HLB) should provide a barrier to aggregation and coalescence. A polymer coating on the exterior

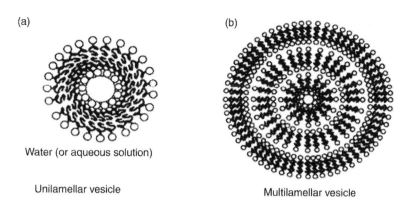

FIGURE 4.10 (a) Unilamellar and (b) multilamellar vesicles. (From Rosen and Dahanayake [105]. Reproduced with permission of American Chemical Society.)

surfaces of the O/W/O droplets can provide additional stability [142]. The multiple emulsion also allows for slow release of the delivered drug and the time release mechanism can be varied by adjusting the emulsion stability. Again, for a W/O/W emulsion, all of the above can be reversed. W/O/W emulsions have been used for transporting hydrophilic drugs such as vaccines, vitamins, enzymes, and hormones [143].

E. PERSONAL CARE PRODUCTS

1. Cosmetic Skin Care Products

Cosmetic skin care products contain many ingredients, have both functional and aesthetic requirements, and are frequently formulated as emulsions. These emulsions can provide an "otherwise impractical combination of ingredients into a single, stable, formulation and enable regulation of rheological properties without significantly affecting the efficacy of active ingredients" [144–146]. This kind of product may be required to serve many functions, including [145] moisturizing and blocking harmful UV radiation, smoothness, velvety feel, pleasant smell, and appearance. The aqueous phase typically contains humectants to prevent water loss, co-solvents to solubilize fragrances or preservatives, water-soluble surfactants, viscosity builders, proteins, vitamins, and minerals [145]. The oleic phase typically contains oils and waxes, dyes and perfumes, and oil-soluble surfactants [145].

The nature of the emulsion may be W/O, O/W, multiple, or microemulsion. Both PIT emulsions and microemulsions are used, characterized by fine droplet sizes and good stability [147]. Selection of the actual emulsifiers for cosmetic skin care products is usually based on HLB while being careful about their possible toxicity and irritability. Good cosmetic formulations also yield good skin–product interactions and therefore good penetration of active ingredients into the skin layers. A hand cream, for instance, may be an O/W macroemulsion with a 10–25% oil phase, or a W/O emulsion, which has a greasier feel and leaves a longer-lasting residue [147]. As the product dries on the skin, phase behavior is an important factor in producing an effective product [144].

The rheological properties of cosmetic creams and lotions are an important aspect of consumer approval. Cosmetic emulsions that are intended for moisturizing, cleansing, and protecting skin require rheological properties that permit rapid application and the deposition of continuous, protective oleic films onto skin surfaces [33]. The rheological properties also determine the final thickness of the oleic layers that are deposited onto the skin surface, which in turn determines the effectiveness of the product since skin moisturizing occurs by the formation of an occlusive film over the skin surface, which slows down moisture loss [33,145]. Cosmetic creams and lotions are generally nonNewtonian fluids, exhibiting shear thinning behavior and a yield stress [148]. They may also be viscoelastic due to the incorporation of hydrocolloid thickeners that contribute a gel network.

Although, forming a protective barrier on the skin is important, some cosmetic products also contain physiologically active ingredients that will improve skin conditions only if they penetrate the skin [35]. Having the product formulated as an emulsion has additional benefits when it comes to the delivery of active ingredients which, for example, are released more slowly from multiple emulsions than from pure solutions containing the same substances [149].

2. Other Personal Care Products

A number of personal care foams are produced from what are commonly known as aerosol products.[1] These include cosmetic foams like hair styling mousse and shaving foam. There are also some similar nonpersonal care products such as aerosol foods (i.e., canned whipped cream), household aerosol foams (like glass and carpet cleaning foams) and household insulating foams (polyurethane foam). Originally these tended to use chlorofluoro carbons (CFCs) as the pressurized propellant phase (which in application becomes the dispersed gas phase). With increasing environmental awareness and concerns, the CFCs have largely been replaced by propane–butane blends. These products are formulated as emulsions in the pressurized containers and therefore contain a number of components that are needed to, among other things, stabilize the emulsion in the can, and then stabilize the foam that is produced during the use of the aerosol product [150].

F. PETROLEUM PRODUCTION

Emulsions occur or are created throughout the full range of processes in the petroleum producing industry, including drilling and completion, fracturing and stimulation, reservoir recovery, surface treating, transportation, oil spill and tailings treating, refining and upgrading. These emulsions are desirable in some process contexts and undesirable in others.

1. Near-Well Emulsions

Emulsions have been used as alternatives to suspensions (muds) in drilling fluid formulation. Two kinds of oilwell drilling fluid (or "drilling mud") are emulsion based: water-continuous and oil-continuous (invert) emulsion drilling fluids. Here a stable emulsion (usually oil dispersed in water) is used to maintain hydrostatic pressure in the hole. This is obviously a desirable kind of emulsion. However, just as with classical suspension drilling muds, careful formulation is needed in order to minimize fluid loss into the formation, cool and lubricate the cutting bit, and to carry drilled rock cuttings up to the surface [151]. The oils used to make the emulsions were originally crude oil or diesel oil, but are now more commonly refined mineral oils [151]. Oil-continuous, or invert, emulsion fluids are typically stabilized by long-chain carboxylate or branched polyamide surfactants. Borchardt [152] lists a number of other emulsion stabilizers that have been used. Organophilic clays have also been used as stabilizing agents. Invert emulsion fluids provide good rheological and fluid-loss properties, are particularly useful for high-temperature applications, and can be used to minimize clay-hydration problems in shale formations [151].

Other desirable near-wellbore emulsions are used to increase the injectivity or productivity of wells by fracturing or acidizing. In either case, the goal is to increase flow capacity in the near-well region of a reservoir. Either water- or oil-based fracturing emulsions are injected at high pressure and velocity, through a wellbore, and into a formation at greater than its parting pressure. As a

---

[1]This common use of the term aerosol refers to formulated products that are packaged under pressure, and released through a fine orifice to produce a foam. This is different from the meaning of the term aerosol in colloid and interface science, which refers to a dispersion of either liquid droplets or solid particles in a gas (liquid aerosols or solid aerosols, respectively).

result, fractures (cracks) are created and propagated. Emulsified fracturing fluids are typically O/W emulsions that may consist of 60–70% liquid hydrocarbon dispersed in 30–40% of a viscous polymer solution or gel. They can provide excellent fluid loss control, possess good transport properties, and can be less damaging to the reservoir than other fluids. However, emulsions are more difficult to prepare and can be more expensive. In acidizing emulsions, the acids used to increase the productivity of reservoirs by dissolving fine particles in flow channels are usually hydrochloric acid (carbonate reservoirs) or hydrofluoric acid (sandstone reservoirs) [153], or blends of these [152].

2. Reservoir Occurrences of Emulsions

Emulsions are commonly produced at the wellhead during primary (natural pressure driven) and secondary (water-flood driven) oil production. For these processes the emulsification has not usually been attributed to formation in reservoirs, but rather to formation in, or at the face of, the wellbore itself [154]. However, at least in the case of heavy oil production, laboratory [162] and field [156,157] results suggest that W/O emulsions can be formed in the reservoir itself during water and steam-flooding. Macroemulsions, as opposed to microemulsions, can be injected or produced *in situ* in order to either for blocking and diverting [158,159], or for improved mobility control [160].

After the primary and secondary cycles of oil recovery, chemicals may be injected to drive out additional oil in an EOR process, which may involve creating *in situ* emulsions in the reservoir. In a water-wet reservoir the water would have been imbibed most strongly into the smallest radius pores, while the largest pores will retain high oil contents. Chemical-flooding involves the injection of a surfactant solution, which can cause the oil or aqueous interfacial tension to drop from about 30 mN/m to near-zero values, on the order of 10^{-3} to 10^{-4} mN/m, allowing spontaneous or nearly spontaneous emulsification and displacement of the oil [50,161,162]. The exact kind of emulsion formed can be quite variable, ranging from fine macroemulsions, as in alkali or surfactant or polymer-flooding [51], to microemulsions [18,163]. Microvisualization studies suggest that with such low interfacial tensions, multiple emulsions may form, even under the low flow rates that would be produced in a reservoir [15]. Microemulsions can produce a more efficient oil displacement than alkali or surfactant or polymer-flooding, but microemulsion-flooding has developed slowly because of its complex technology and higher costs [49,121,164–167].

3. Emulsions in Surface Operations

An emulsion that was beneficial in a reservoir may be an undesirable type of emulsion (usually W/O) when produced at a wellhead. Such emulsions have to be broken and separated because pipeline and refinery specifications place severe limitations on the water, solids, and salt contents of oil they will accept (in order to avoid corrosion, catalyst poisoning, and process upset problems). A variety of physical methods are used in emulsion breaking, all of which are designed to accelerate coagulation and coalescence. Examples include settling, heating, electrical dehydration, chemical treatment, centrifugation, and filtration. The mechanical methods, such as centrifuging or filtering, rely on increasing the collision rate of droplets and applying an additional force driving coalescence. Thermal methods increase droplet collision rates and also reduce viscosity. Electrical methods may involve electrophoresis-induced droplet collisions (O/W emulsions) or deformation of water droplets causing disruption of interfacial films (W/O emulsions). More details on the application of these methods in large-scale continuous processes are given elsewhere [103,104]. In addition to physical methods, with suitable emulsion characterization, chemical treatments can be applied to displace or destroy the effectiveness of the original stabilizing agents at the interfaces [4,87], thus making an emulsion easier to break.

Some emulsions are made to reduce viscosity so that an oil can be made to flow, such as when heavy oils are formulated as O/W emulsions to reduce their viscosity for economic pipeline transportation over large distances [168,169]. Here the emulsions need to be stable enough for handling, transportation, and able to survive brief flow stoppages, but they also need to be amenable to breaking at the end of the pipeline. Asphalt emulsions, for example, should shear thin upon application and break up to form a suitable water-repelling roadway coating material.

The large Canadian oil sands surface-mining and processing operations involve a number of kinds of emulsions in a variety of process steps. Here, bitumen is separated from the sand matrix, in large tumblers, and forms an O/W emulsion containing not just oil and water, but also dispersed solids and gas. The emulsified oil is further separated from the solids by a flotation process which produces an oleic foam termed bituminous froth, which may be either gas dispersed in the oil (primary or secondary flotation) or the reverse, gas dispersed in water (secondary or tertiary flotation). The bituminous froths contain not just oil and gas, but also emulsified water and some dispersed solids. The froth has to be broken in order to permit pumping and subsequent removal of entrained water and solids before the bitumen can be upgraded to synthetic crude oil. The diluted froth contains multiple emulsion types including tenacious multiple emulsions [23], which complicate the downstream separation processes. These aspects are reviewed elsewhere [25,64,125].

Finally, at an oil upgrader or refinery, any emulsified water will have to be broken and separated out in order to avoid operating problems [87,170]. As with other crude oil emulsions, the presence of solid particles and film-forming components from the crude oil can make this very difficult.

ACKNOWLEDGMENT

The authors thankfully acknowledge the long-standing financial support of their research by the Natural Sciences and Engineering Research Council of Canada.

REFERENCES

1. Becher, P., *Emulsions, Theory and Practice*, Reinhold, New York, 1965.
2. Sumner, C.G., *Clayton's the Theory of Emulsions and their Technical Treatment*, 5th ed., Blakiston Co. Inc., New York, 1954.
3. Becher, P., Ed., *Encyclopedia of Emulsion Technology*, vols. 1–3, Dekker, New York, 1983–1988.
4. Lissant, K.J., *Demulsification, Industrial Applications*, Dekker, New York, 1983.
5. Schramm, L.L., Ed., *Emulsions, Fundamentals and Applications in the Petroleum Industry*, American Chemical Society, Washington, and Oxford University Press, New York, 1992.
6. Shaw, D.J., *Introduction to Colloid and Surface Chemistry*, 3rd ed., Butterworth's, London, 1981.
7. Osipow, L.I., *Surface Chemistry Theory and Industrial Applications*, Reinhold, New York, 1962.
8. Ross, S. and Morrison, I.D., *Colloidal Systems and Interfaces*, Wiley, New York, 1988.
9. Kruyt, H.R., Ed., *Colloid Science*, vol. 1, Elsevier, Amsterdam, 1952.
10. Griffin, W.C., Emulsions. in *Kirk-Othmer Encyclopedia of Chemical Technology*, Vol. 8, 2nd ed., Interscience, New York, 1965, pp. 117–154.
11. Birdi, K.S., Ed., *Handbook of Surface and Colloid Chemistry*, CRC Press, Boca Raton, 1997.
12. Hubbard, A.T., Ed., *Encyclopedia of Surface and Colloid Science*, Marcel Dekker, New York, 2002.
13. Schramm, L.L., *Dictionary of Colloid and Interface Science*, John Wiley & Sons, New York, 2001.
14. Garti, N. and Bisperink, C., Double emulsions: progress and applications *Curr. Opin. Coll. Interf. Sci.*, 3, 657–667, 1988.
15. Schramm, L.L. and Kutay, S., Emulsions and foams in the petroleum industry, in *Surfactants: Fundamentals and Applications in the Petroleum Industry*, Schramm, L.L., Ed., Cambridge University Press, Cambridge, UK, 2000, pp. 79–117.
16. Mittal, K.L., Ed., *Micellization, Solubilization and Microemulsions* vols. 1–2, Plenum, New York, 1977.
17. Shah, D.O., Ed., *Macro- and Microemulsions*, American Chemical Society, Washington, 1985.
18. Prince, L.M., Ed., *Microemulsions Theory and Practice*, Academic Press, New York, 1977.

19. Tadros, Th.F., Surface chemistry in agriculture, in *Handbook of Applied Surface and Colloid Chemistry*, Vol. 1, Holmberg, K., Ed., Wiley, New York, 2001, pp. 73–83.

20. L.L. Schramm, Ed., *Surfactants: Fundamentals and Applications in the Petroleum Industry*, Cambridge University Press, Cambridge, UK, 2000.

21. Sabatini, D.A., Knoz, R.C., and Harwell, J.H., Eds., *Surfactant-Enhanced Subsurface Remidiation Emerging Technologies*, American Chemical Society, Washington, 1995.

22. El-Nokaly, M. and Cornell, D., Eds., *Microemulsions and Emulsions in Foods*, American Chemical Society, Washington, 1991.

23. Schramm, L.L., Petroleum emulsions, basic principles, in *Emulsions, Fundamentals and Applications in the Petroleum Industry*, Schramm, L.L., Ed., American Chemical Society, Washington, and Oxford University Press, New York, 1992, pp. 1–49.

24. Mikula, R.J., Emulsion characterization, in *Emulsions, Fundamentals and Applications in the Petroleum Industry*, Schramm, L.L., Ed., American Chemical Society, Washington and Oxford University Press, New York, 1992, pp. 79–129.

25. Shaw, R.C., Czarnecki, J., Schramm, L.L., and Axelson, D., Bituminous froths in the hot-water flotation process, in *Foams Fundamentals and Applications in the Petroleum Industry*, Schramm, L.L., Ed., American Chemical Society, Washington, and Oxford University Press, New York, 1994, pp. 423–459.

26. Schramm, L.L., Ed., *Foams, Fundamentals and Applications in the Petroleum Industry*, American Chemical Society, Washington, and Oxford University Press, New York, 1994.

27. Whorlow, R.W., *Rheological Techniques*, Ellis Horwood, Chichester, England, 1980.

28. Schramm, G., *Optimization of Rotovisco Tests* Gebrüder Haake GmbH, Dieselstrasse, Germany, 1981.

29. van Wazer, J.R., Lyons, J.W., Kim, K.Y., and Colwell, R.E., *Viscosity and Flow Measurement*, Wiley, New York, 1963.

30. Fredrickson, A.G., *Principles and Applications of Rheology*, Prentice-Hall, Englewood Cliffs, NJ, 1964.

31. Schramm, L.L., The influence of suspension viscosity on bitumen rise velocity and potential recovery in the hot water flotation process for oil sands, *J. Can. Petrol. Technol.*, 28, 73–80, 1989.

32. Malhotra, A.K. and Wasan, D.T., Interfacial rheological properties of adsorbed surfactant films with applications to emulsion and foam stability, in *Thin Liquid Films*, Ivanov, I.B., Ed., Dekker, New York, 1988, pp. 829–890.

33. Breuer, M.M., Cosmetic emulsions, in *Encyclopedia of Emulsion Technology*, vol. 3, Becher, P., Ed., Marcel Dekker, New York, 1985.

34. Dalgleish, D.G., Food emulsions, in *Emulsions and Emulsion Stability*, Sjoblom, J., Ed., Marcel Dekker, New York, 1996, pp. 287–325.

35. Förster, Th., Jackwerth, B., Pittermann, W., von Tybinskin, W., and Schmitt, M., Properties of emulsions, *Cosmetics Toiletries*, 112, 73–82, 1997.

36. Förster, T., Principles of emulsion formation, in *Surfactants in Cosmetics*, Rieger, M.M. and Rhein, L.D., Eds., Marcel Dekker, New York, 1997, pp. 105–125.

37. Isaacs, E.E. and Chow, R.S., Practical aspects of emulsion stability, in *Emulsions, Fundamentals and Applications in the Petroleum Industry*, Schramm, L.L., Ed., *American Chemical Society*, Washington and Oxford University Press, New York, 1992, pp. 51–77.

38. Walstra, P. and Smulders I., Making emulsions and foams: an overview, in *Food Colloids: Proteins, Lipids and Polysaccharides*, Royal Society Chemistry, London, 1997, pp. 367–381.

39. Walstra, P., Principles of emulsion formation, *Chem. Eng. Sci.* 48, 333–349, 1993.

40. Groeneweg, F., van Dieren, F., and Agterof, W.G.M., Droplet break-up in a stirred water-in-oil emulsion in the presence of emulsifiers, *Coll. Surf. A*, 91, 207–214, 1994.

41. Rosen, M.J., *Surfactants and Interfacial Phenomena*, 2nd ed., Wiley, New York, 1989.

42. Karsa, D.R., Ed., *Industrial Applications of Surfactants*, Royal Society Chemistry, London, 1987.

43. Lucassen-Reynders, E.H., Ed., *Anionic Surfactants Physical Chemistry of Surfactant Action*, Marcel Dekker, New York, 1981.

44. *The HLB System*, ICI Americas Inc., Wilmington, 1976.

45. Griffin, W.C., Classification of surface-active agents by "HLB", *J. Soc. Cosmetic Chem.* 1949, 1, 311–326.

46. *McCutcheon's Emulsifiers and Detergents*, Vol. 1, MC Publishing Co., Glen Rock, NJ, 1990.

47. Vander Kloet, J. and Schramm, L.L., The effect of shear and oil–water ratio on the required hydrophile–lipophile balance for emulsification, *J. Surfactants Deterg.*, 5, 19–24, 2002.

48. Shinoda, K. and Saito, H., The stability of O/W type emulsions as functions of temperature and the HLB of emulsifiers: the emulsification by PIT-method, *J. Coll. Interf. Sci.* 30, 258–263, 1969.

49. Austad, T. and Milter, J., Surfactant flooding in enhanced oil recovery, in *Surfactants, Fundamentals and Applications in the Petroleum Industry*, Schramm, L.L., Ed., Cambridge University Press, Cambridge, 2000, pp. 203–249.

50. Sharma, M.K., Surfactants in enhanced petroleum recovery processes: a review, in *Particle Technology and Surface Phenomena in Minerals and Petroleum*, Sharma, M.K. and Sharma, G.D., Eds., Plenum, New York, 1991, pp. 199–222.

51. Taylor, K. and Hawkins, B., Emulsions in enhanced oil recovery, in *Emulsions, Fundamentals and Applications in the Petroleum Industry*, Schramm, L.L., Ed., American Chemical Society, Washington, and Oxford University Press, New York, 1992, pp. 263–293.

52. Robb, I.D., Ed., *Microemulsions*, Plenum, New York, 1977.

53. Overbeek, J.T.G., de Bruy, P.L., and Verhoeckx, F., Microemulsions, in *Surfactants*, Tadros, Th.F., Ed., Academic Press, New York, 1984, pp. 111–131.

54. Tadros, Th.F., *Surfactants in Solution*, Mittal, K.L. and Lindman, B., Eds., Plenum, New York, 1984, pp. 1501–1532.

55. Robinson, B.H., Applications of microemulsions, *Nature*, 320, 309, 1986.

56. Friberg, S.E. and Bothorel, P., Eds., *Microemulsions: Structure and Dynamics*, CRC Press, Boca Raton, FL, 1987.

57. Leung, R., Jeng Hou, M., and Shah, D.O., *Surfactants in Chemical/Process Engineering*, Wasan, D.T., Ginn, M.E., and Shah, D.O., Eds., Marcel Dekker, New York, 1988, pp. 315–367.

58. Takamura, K., Microscopic structure of Athabasca oil sand, *Can. J. Chem. Eng.* 60, 538–545, 1982.

59. Takamura, K. and Chow, R.S., A mechanism for initiation of bitumen displacement from oil sand, *J. Can. Petrol. Technol.,* 22, 22–30, 1983.

60. Schramm, L.L., Smith, R.G., and Stone, J.A., The influence of natural surfactant concentration on the hot water process for recovering bitumen from the Athabasca oil sands, *AOSTRA J. Res.* 1, 5–14, 1984.

61. Schramm, L.L. and Smith, R.G., The influence of natural surfactants on interfacial charges in the hot water process for recovering bitumen from the Athabasca oil sands, *Coll. Surf.*, 14, 67–85, 1985.

62. Acevedo, S., Gutierrez, X., and Rivas, H., Bitumen-in-water emulsions stabilized with natural surfactants, *J. Coll. Interf. Sci.*, 242, 230–238, 2001.

63. Schramm, L.L. and Smith, R.G., Two classes of anionic surfactants and their significance in hot water processing of oil sands, *Can. J. Chem. Eng.* 65, 799–811, 1987.

64. Schramm, L.L., Stasiuk, E.N., and MacKinnon, M., Surfactants in Athabasca oil sands slurry conditioning, flotation recovery, and tailings processes, in *Surfactants, Fundamentals and Applications in the Petroleum Industry*, Schramm, L.L., Ed., Cambridge University Press, Cambridge, UK, 2000, pp. 365–430.

65. Hiemenz, P. and Rajagopalan, R., *Principles of Colloid and Surface Chemistry*, 3rd ed., Marcel Dekker, New York, 1997.

66. Hunter, R.J., *Zeta Potential in Colloid Science*, Academic Press, New York, 1981.

67. Overbeek, J.Th.G., Electrokinetic phenomena, in *Colloid Science*, vol. 1, Kruyt, H.R., Ed., Elsevier, Amsterdam, 1952, pp. 194–244.

68. Overbeek, J.Th.G., in *Colloidal Dispersions*, Goodwin, J.W., Ed., Royal Society Chemistry, London, 1982, pp. 1–21.

69. Nguyen, A.V. and Schulze, H.J., *Colloidal Science of Flotation*, Marcel Dekker, New York, 2004, 850 pp.

70. Verwey, E.J.W. and Overbeek, J.Th.G., *Theory of the Stability of Lyophobic Colloids*, Elsevier, New York, 1948.

71. Riddick, T.M., *Control of Stability Through Zeta Potential*, Zeta Meter Inc., New York, 1968.

72. Fuerstenau, M.C. and Han, K.N., Metal-surfactant precipitation and adsorption in froth flotation, *J. Coll. Interf. Sci.*, 256, 175–182, 2002.

73. Leja, J., *Surface Chemistry of Froth Flotation*, Plenum Press, New York, 1982.

74. Schramm, L.L. and Smith, R.G., Monitoring electrophoretic mobility of bitumen in hot water extraction process plant water to maximize primary froth recovery, Canadian Patent 1,265,463 (February 6, 1990).

75. Sato, T. and Ruch, R., *Stabilization of Colloidal Dispersions by Polymer Adsorption*, Marcel Dekker, New York, 1980.

76. Tadros, Th.F., Ed., *The Effect of Polymers on Dispersion Properties*, Academic Press, London, 1982.

77. Lipatov, Y.S. and Sergeeva, L.M., *Adsorption of Polymers*, Halsted, New York, 1974.

78. Finch, C.A., Ed., *Chemistry and Technology of Water-Soluble Polymers*, Plenum Press, New York, 1983.

79. Vincent, B., The stability of solid/liquid dispersions in the presence of polymers, in *Solid/Liquid Dispersions*, Tadros, Th.F., Ed., Academic Press, New York, 1987, pp. 147–162.

80. Walstra, P., Introduction to aggregation phenomena in food colloids, in *Food Colloids and Polymers: Stability and Mechanical Properties*, Royal Society Chemistry, London, 1993, pp. 3–15.

81. Damodaran, S., Protein-stabilized foams and emulsions, *Food Sci. Technol.* 80, 57–110, 1997.

82. Cairns, R.J.R., Grist, D.M., and Neustadter, E.L., The effects of crude oil-water interfacial properties on water–crude oil emulsion stability, in *Theory and Practice of Emulsion Technology*, Smith, A.L., Ed., Academic Press, New York, 1976, pp. 135–151.

83. Jones, T.J., Neustadter, E.L., and Whittingham, K.P., Water-in-crude oil emulsion stability and emulsion destabilization by chemical demulsifiers, *J. Can. Petrol. Technol.*, 17, 100–108, 1978.

84. Arditty, S., Schmitt, V., Giermanska-Kahn, J., and Leal-Calderon, F., Materials based on solid-stabilized emulsions, *Coll. Surf. A*, 275, 659–664, 2004.

85. Robins, M.M., Emulsions: creaming phenomena, *Curr. Opin. Coll. Interf. Sci.*, 5, 265–272, 2000.

86. Cavallo, J.L. and Chang, D.L., Emulsion preparation and stability, *Chem. Eng. Progr.*, 86, 54–59, 1990.

87. Grace, R., Commercial emulsion breaking, in *Emulsions, Fundamentals and Applications in the Petroleum Industry*, Schramm, L.L., Ed., American Chemical Society, Washington, and Oxford University Press, New York, 1992, pp. 313–339.

88. Mikula, R.J. and Munoz, V.A., Characterization of demulsifiers. in *Surfactants, Fundamentals and Applications in the Petroleum Industry*, Schramm, L.L., Ed., Cambridge University Press, Cambridge, UK, 2000, pp. 51–78.

89. Tambe, D., Paulis, J., and Sharma, M.M., Factors controlling the stability of colloid-stabilized emulsions. IV. Evaluating the effectiveness of demulsifiers, *J. Coll. Interf. Sci.*, 171, 463–469, 1995.

90. Fuestel, M., *Oil Gas Eur. Mag.*, 21, 42, 1995.

91. Taylor, S.E., Resolving crude oil emulsions, *Chem. Ind.*, 20, 770–775, 1992.

92. Bessler, D.U., *Demulsification of Enhanced Oil Recovery Produced Fluids*, Petrolite Corporation, St. Louis, 1983.

93. Mukherjee, S. and Kushnick, A.P., Effect of demulsifiers on interfacial properties governing crude oil demulsification, in *Oil-Field Chemistry, Enhanced Recovery and Production Stimulation*, Borchardt, J.K. Yen, T.F., Eds., American Chemical Society, Washington, 1989, pp. 364–374.

94. Zaki, N. and Al-Sabagh, L., De-emulsifiers for water-in-crude oil-emulsions. *Tenside Surf. Deter.*, 34, 12–17, 1997.

95. Smith, V.H. and Arnold, K.E., Crude oil emulsions, in *Petroleum Engineering Handbook*, Bradley, H.B. Ed., Society of Petroleum Engineers, Richardson, TX, 1992, pp. 19-1 to 19-34.

96. Staiss, F., Bohm, R., and Kupfer, R., Improved demulsifier chemistry: a novel approach in the dehydration of crude oil, *SPE Prod. Eng.*, 6, 334–338, 1991.

97. Berger, P.D., Hsu, C., and Arendell, J.P., Designing and selecting demulsifiers for optimum field performance on the basis of production fluid characteristics *SPE Prod. Eng.*, 2, 522–529, 1987.

98. Monson, L.T. and Stenzel, R.W., The technology of resolving petroleum emulsions, in *Colloid Chemistry*, Vol. VI, Alexander, J., Ed., Reinhold, New York, 1946, pp. 535–552.

99. van Os, N.M., Haak, J.R., and Rupert, L.A.M., *Physico-Chemical Properties of Selected Anionic, Cationic and Nonionic Surfactants*, Elsevier, New York, 1993.

100. Zana, R., *Surfactant Solutions. New Methods of Investigations*, Marcel Dekker, New York, 1986.

101. Sonntag, H. and Strenge, K., *Coagulation and Stability of Disperse Systems*, Halsted Press, New York, 1972.

102. Gennaro, A.R., Ed., *Remington's Pharmaceutical Sciences*, 17th ed., Mack Publishing Co., 1985.

103. Scoular, R.J., Breaking wellhead emulsions, in *Supplementary Notes for Petroleum Emulsions and Applied Emulsion Technology*, Schramm, L.L., Ed., Petroleum Recovery Institute, Calgary, AB, 1992.

104. Leopold, G., in *Emulsions, Fundamentals and Applications in the Petroleum Industry*; Schramm, L.L., Ed., American Chemical Society: Washington, and Oxford University Press: New York, 1992, pp. 342–383.

105. Rosen, M.J. and Dahanayake, M., *Industrial Utilization of Surfactants*, AOCS Press, Champaign, Illinois, 2000.

106. *The Asphalt Handbook*, The Asphalt Institute, College Park, MD, 1965.

107. National Research Council, *Using Oil Spill Dispersants on the Sea*, National Academy Press, Washington, DC, 1989.

108. Mackay, D., *Formation and Stability of Water-in-Oil Emulsions*, Manuscript Report EE-93, Environment Canada, Ottawa, 1987.

109. Urdahl, O. and Sjöblom, J., Water-in-crude oil emulsions from the Norwegian continental shelf, *J. Dispers. Sci. Technol.*, 16, 557–574, 1995.

110. Bobra, M.A., Water-in-oil emulsification a physicochemical study, *Proceedings of the International Oil Spill Conference*, American Petroleum Institute, Washington, 1991, pp. 483–488.

111. Bobra, M.A., *A Study of Water-in-Oil Emulsification*, Manuscript Report EE-132, Environment Canada, Ottawa, 1992.

112. Canevari, G.P., *Proceedings of the Oil Spill Conference*, API Publication 4452, American Petroleum Institute, Washington, DC, 1987, pp. 293–296.

113. Cormack, D., Lynch, W.J., and Dowsett, B.D., Evaluation of dispersant effectiveness, *Oil Chem. Pollut.*, 3, 87–103, 1986/87.

114. Fingas, M., Use of surfactants for environmental applications, in *Surfactants, Fundamentals and Applications in the Petroleum Industry*, Schramm, L.L., Ed., Cambridge University Press, Cambridge, UK, 2000, pp. 461–539.

115. Fingas, M.F., *Spill Technology Newsletter*, 19, 1, 1994.

116. Bocard, C. and Gatellier, C., *Proceedings of the Oil Spill Conference*, API Publication 4452, American Petroleum Institute, Washington, DC, 1981, pp. 601–607.

117. Pankow, J.F. and Cherry, J.A., *Dense Chlorinated Solvents*, Waterloo Press, Portland, OR, 1996.

118. Mackay, D.M. and Cherry, J.A., Groundwater contamination: pump-and-treat remediation. *Environ. Sci. Technol.*, 23, 630–636, 1989.

119. Fountain, J.C., Klimek, A., Beikirch, M.G., and Middleton, T.M., The use of surfactants for in situ extraction of organic pollutants from a contaminated aquifer, *J. Hazard. Mater.*, 28, 295–311, 1991.

120. Mulligan, C.N., Yong, R.N., and Gibbs, B.F., Surfactant-enhanced remediation of contaminated soil, *Eng. Geol.*, 60, 371–380, 2001.

121. Dwarakanath, V. and Pope, G.A., Surfactant enhanced aquifer remediation, in *Surfactants, Fundamentals and Applications in the Petroleum Industry*, Schramm, L.L., Ed., Cambridge University Press, Cambridge, UK, 2000, pp. 433–460.

122. Baran, J.R., Pope, G.A., Wade, W.H., and Weerasooriya, V., Phase behavior of water/perchloroethylene/anionic surfactant systems, *Langmuir*, 10, 1146–1150, 1994.

123. Weerasooriya, V., Yeh, S.L., Pope, G.A. Integrated demonstration of surfactant-enhanced aquifer remediation (SEAR) with surfactant regeneration and reuse, in *Proceedings of the Symposium on Recent Advances in ACS Surfactant-based Separations*, Dallas, Texas, March 29–30, 1998.

124. West, C.C. and Harwell, J.H., Surfactants and subsurface remediation, *Environ. Sci. Technol.*, 26, 2324–2330, 1992.

125. Shaw, R.C., Schramm, L.L., and Czarnecki, J., Suspensions in the hot water flotation process for Canadian oil sands, in *Suspensions, Fundamentals and Applications in the Petroleum Industry*; Schramm, L.L., Ed., American Chemical Society, Washington, and Oxford University Press, New York, 1996, pp. 639–675.

126. Dalgleish, D.G., Aspects of stability in milk and milk products, in *Food Colloids*, Bee, R.D., Richmond, P., and Mingins, J., Eds., Royal Society Chemistry, Cambridge, 1990, pp. 295–305.

127. St. Angelo, A.J., A Brief Introduction to Food Emulsions and Emulsifiers, in *Food Emulsifiers: Chemistry, Technology, Functional Properties and Applications*, Charalambous, G. and Doxastakis, G., Eds., Elsevier, Amsterdam, 1989, pp. 1–8.

128. Stainsby, G. and Dickinson, E., Food emulsions and foams, *Actual. Chim.*, 3, 35–39, 1988.

129. Dickinson, E., Food colloids — An overview, *Coll. Surf.*, 42, 191–204, 1989.

130. Perram, C.M., Nicolau, C., and Perram, J.W., Interparticle forces in multiphase colloid systems: the resurrection of coagulated sauce bearnaise, *Nature*, 270, 572–573, 1977.

131. Goff, H.D., Colloidal aspects of ice cream — A review, *Int. Dairy J.*, 7, 363–373, 1997.

132. Krog, N., The use of emulsifiers in ice cream, in *Ice Cream*, Buchheim, W., Ed., Int. Dairy Fed., Brussels, 1998, pp. 37–43.

133. Campbell, I.J. and Pelan, B.M.C., The influence of emulsion stability on the properties of ice cream, in *Ice Cream*, Buchheim, W., Ed., Int. Dairy Fed., Brussels, 1998, pp. 25–36.

134. Dziezak, J.D., Microencapsulation and encapsulated ingredients, *Food Technol.*, 172, 136–148, 1988.

135. Krog, N., Barfod, N.M., and Sanchez, R.M., Interfacial phenomena in food emulsions. *J. Dispers. Sci. Technol.*, 10, 483–504, 1989.

136. Prost, J. and Rondelez, F., Structures in colloidal physical chemistry, *Nature*, 350, 11, 1991.

137. Israelachvili, J.N., *Intermolecular and Surface Forces*, 2nd ed., Academic Press, New York, 1992.

138. Cevc, G. and Marsh, D., *Phospholipid Bilayers*, Wiley, New York, 1987; Marsh, D., *Handbook of Lipid Bilayers*, CRC Press, Boca Raton, FL, 1990.

139. Miller, R. and Kretzschmar, G., Adsorption kinetics of surfactants at fluid interfaces, *Adv. Colloid Interf. Sci.*, 37, 97–121, 1991.

140. Yang, L., Alexandridis, P. Physicochemical aspects of drug delivery and release from polymer-based colloids. *Curr. Opin. Colloid. Interf. Sci.*, 5, 132, 2000.

141. Barenholz, Y., Liposome application: problems and prospects, *Curr. Opin. Colloid Interf. Sci.*, 6, 66–77, 2001.

142. Tadros, Th.F., Industrial applications of dispersions, *Adv. Colloid Interf. Sci.*, 46, 1–47, 1993.

143. Bibette, J., Leal Calderon, F., and Poulin, P., Emulsions, basic principles, *Rep. Prog. Phys.*, 62, 969–1033, 1999.

144. Umbach, W., The importance of colloid chemistry in industrial practice, *Progr. Colloid Polym. Sci.*, 111, 9–16, 1998.

145. Bhargava, H.N., The present status of formulation of cosmetic emulsions, *Drug Devel. Indus. Pharm.*, 13, 2363–2387, 1987.

146. Binks, B.P., Emulsions, *Ann. Rep. Roy. Soc. Chem.*, 92-C, 97–133, 1995.

147. Schueller, R. and Romanowski, P., The science of reactive hair-care products, *Cosmetics Toiletries*, 113, 39–44, 1998.

148. Miner, P.E., Emulsion rheology: creams and lotions, in *Rheological Properties of Cosmetics and Toileteries*, Laba, D., Ed., Marcel Dekker, New York, 1993, pp. 313–370.

149. Seiller, M., Puisieux, F., and Grossiard, J. L., Multiple emulsions in cosmetics, in *Surfactants in Cosmetics*, 2nd ed., Reiger, M.M. and Rhein, L.D., Eds., Marcel Dekker, New York, 1997, pp. 139–154.

150. Hoffbauer, B., Foam aerosols, *Aerosol Spray Rep.*, 35, 508–515, 1996.

151. Jones, T.G.J. and Hughes, T.L., Drilling fluid suspensions, in *Suspensions, Fundamentals and Applications in the Petroleum Industry*, Schramm, L.L., Ed., American Chemical Society, Washington, and Oxford University Press, New York, 1996, pp. 463–564.

152. Borchardt, J.K., Chemicals used in oil-field operations, in *Oil-Field Chemistry*, Borchardt, J.K. and Yen, T.F., Eds., American Chemical Society, Washington, DC, 1989, pp. 3–54.

153. Gdanski, R. and Behenna, R., Design considerations for foam diversion of acid stimulation treatments, in *Proceedings, Field Application of Foams for Oil Production Symposium*, Olsen, D.K. and Sarathi, P.S., Eds., U.S. Dept. of Energy, Bartlesville, OK, 1993, Paper FS7, pp. 163–172.

154. Dow, D.B., *Oil-Field Emulsions*, Bulletin 250, US Bureau of Mines, Washington, 1926.

155. Chung, K.H. and Butler, R.M., In situ emulsification by the condensation of steam in contact with bitumen, *J. Can. Petrol. Technol.*, 28, 48–55, 1989.

156. Vittoratos, E., Flow regimes during cyclic steam stimulation at Cold Lake, *J. Can. Petrol. Technol.*, 30, 82–86, 1991.

157. Vittoratos, E., In-situ emulsification: interpreting production data, in *Supplementary Notes for Petroleum Emulsions and Applied Emulsion Technology*, Schramm, L.L., Ed., Petroleum Recovery Institute, Calgary, AB, 1992.

158. French, T.R., Broz, J.S., Lorenz, P.B., and Bertus, K.M., Use of emulsions for mobility control during steamflooding, in *Proceedings of the, 56th Califoria Regional Meeting*, Society of Petroleum Engineers, Richardson, Texas, 1986, Paper SPE 15052.

159. Broz, J.S., French, T.R., and Corroll, H.B., Blocking of high permeability zones in steam flooding by emulsions, in *Proceedings of the, 3rd UNITAR/UNDP International Conference Heavy Crude and Tar Sands*, United Nations Institute for Training and Research, New York, 1985, pp. 444–451.

160. Woo, R., Jackson, C., and Maini, B.B., unpublished results, Petroleum Recovery Institute, Calgary, AB, Canada, 1992.

161. Poettmann, F.H., Microemulsion flooding, in *Improved Oil Recovery*, Interstate Compact Commission, Oklahoma City, OK, 1983, pp. 173–250.

162. Lake, L.W., *Enhanced Oil Recovery*, Prentice Hall, Englewood Cliffs, NJ, 1989.

163. Neogi, P., Oil recovery and microemulsions, in *Microemulsions: Structure and Dynamics*, Friberg, S.E. and Bothorel, P., Eds., CRC Press, Boca Raton, FL, 1987, pp. 197–212.

164. Thomas, S. and Farouq Ali, S.M., Micellar-polymer flooding: status and recent advances. *J. Can. Petrol. Technol.*, 31, 53–60, 1992.

165. Moritis, G., EOR dips in U.S. but remains a significant factor, *Oil Gas J.*, 92, 51–71, 1992.

166. Chapotin, D., Lomer, J.F., and Putz, A., The Chateaurenard (France) industrial microemulsion pilot design and performance, in *Proceedings of the, SPE/DOE 5th Symposium on EOR*, Society of Petroleum Engineers, Richardson, Texas, 1986, Paper SPE/DOE 14955.

167. Reppert, T.R., Bragg, J.R., Wilkinson, J.R., Snow, T.M., Maer, N.K., Jr., and Gale, W.W., Second Ripley surfactant flood pilot test, in *Proceedings of the, 7th Symposium on Enhanced Oil Recovery*, Society of Petroleum Engineers, Richardson, Texas, 1990, Paper SPE/DOE 20219.

168. Rimmer, D.P., Gregoli, A.A., Hamshar, J.A., and Yildirim, E., Pipeline emulsion transportation for heavy oils, in *Emulsions, Fundamentals and Applications in the Petroleum Industry*, Schramm, L.L., Ed., American Chemical Society, Washington, and Oxford University Press, New York, 1992, pp. 295–312.

169. Plegue, T.H., Frank, S.G., Fruman, D.H., and Zakin, J.L., Studies of water-continuous emulsions of heavy crude oils prepared by alkali treatment, *SPE Prod. Eng.*, 3, 181–188, 1989.

170. Lewis, V.E. and Minyard, W.F., Antifoaming and defoaming in refineries, in *Foams, Fundamentals and Applications in the Petroleum Industry*, Schramm, L.L., Ed., American Chemical Society, Washington, and Oxford University Press, New York, 1994, pp. 461–483.

5 Hydrodynamics and Heat or Mass Transfer in Finely Dispersed Systems

Abdellah Saboni
University of Rouen, Rouen, France

Silvia Alexandrova
University of Caen, Caen, France

CONTENTS

I. INTRODUCTION

The knowledge of the basic phenomena involved between a fluid sphere and an external flow is important for the comprehension of liquid–liquid, solid–liquid, and gas–liquid systems in the fields of chemical, petrochemical, or environmental engineering. The system considered here is a rigid or a fluid sphere of radius a moving with a constant velocity U_∞ in another immiscible fluid of infinite extent, which is at a different temperature. Some of the early works on the motion and heat or mass transfer from a fluid sphere, as well as recent advances, are exposed. Particular emphasis is placed on drag coefficients and Nusselt and Sherwood numbers. This information should be useful for the study of the dispersed systems (liquid–liquid, gas–liquid, or solid–liquid), particularly for the design of liquid–liquid contactors (in the field of chemical engineering) like liquid–liquid extraction equipment or direct contact exchangers. This review deals with transport at low (creeping flow) and high Reynolds numbers ($0 < Re < 400$).

II. GOVERNING EQUATIONS

We consider Reynolds numbers not exceeding 400; this is the higher limit for which the flow is axisymmetric in the case of the rigid sphere and for which a bubble of air in water remains quasi-spherical. As the flow is considered axisymmetric, the Navier–Stokes equations can be written in terms of stream function and vorticity (ψ and ω) in spherical coordinates r and θ [1, 2]:

$$E^2 \Psi_d = \omega_d r \sin \theta \tag{5.1}$$

$$\frac{\mu_c}{\mu_d} \frac{\rho_d}{\rho_c} \frac{Re}{2} \left[\frac{\partial \Psi_d}{\partial r} \frac{\partial}{\partial \theta} \left(\frac{\omega_d}{r \sin \theta} \right) - \frac{\partial \Psi_d}{\partial \theta} \frac{\partial}{\partial r} \left(\frac{\omega_d}{r \sin \theta} \right) \right] \sin \theta = E^2 (\omega_d r \sin \theta) \tag{5.2}$$

where

$$E^2 = \frac{\partial^2}{\partial r^2} + \frac{\sin \theta}{r^2} \frac{\partial}{\partial \theta} \left(\frac{1}{\sin \theta} \frac{\partial}{\partial \theta} \right)$$

Outside the fluid sphere the above equations are still valid, but for numerical reasons the radial coordinate r is transformed via $r = e^z$, where z is the logarithmic radial coordinate. The results are as follows:

$$E^2 \Psi_c = \omega_c r \sin \theta \tag{5.3}$$

$$\frac{Re}{2} \left[\frac{\partial \Psi_c}{\partial z} \frac{\partial}{\partial \theta} \left(\frac{\omega_c}{e^z \sin \theta} \right) - \frac{\partial \Psi_c}{\partial \theta} \frac{\partial}{\partial z} \left(\frac{\omega_c}{e^z \sin \theta} \right) \right] e^z \sin \theta = e^{2z} E^2 (\omega_c e^z \sin \theta) \tag{5.4}$$

Introducing the following dimensionless quantities normalizes all variables: $r = r'/a$; $\omega = \omega' a / U_\infty$; $\psi = \psi'/(U_\infty a^2)$; $Re = 2aU_\infty/v_c$, where a is the radius of the sphere, Re the Reynolds number, U_∞ the terminal velocity, and v the kinematic viscosity. The primes denote the dimensional quantities and subscripts d and c refer to dispersed and continuous phase, respectively.

In terms of dimensionless stream function ψ, the dimensionless radial and tangential velocities are given by

$$u = -\frac{1}{r^2 \sin \theta} \frac{\partial \Psi}{\partial \theta}, \qquad v = \frac{1}{r \sin \theta} \frac{\partial \Psi}{\partial r}$$

The boundary conditions to be satisfied are:

1. Far from the fluid sphere ($z = z_\infty$), undisturbed parallel flow is assumed: $\omega_c = 0$; $\psi_c = 0.5 e^{2z} \sin^2 \theta$
2. Along the axis of symmetry ($\theta = 0, \pi$): $\psi_c = 0$, $\omega_c = 0$, $\psi_d = 0$, $\omega_d = 0$
3. Across the interface ($z = 0$ or $r = 1$), the following relations take into account, respectively: negligible material transfer, continuity of tangential velocity, and continuity of tangential stress:

$$\psi_c = 0, \quad \psi_d = 0, \quad \frac{\partial \Psi_c}{\partial z} = \frac{\partial \Psi_d}{\partial r}, \quad \frac{\mu_c}{\mu_d} \left(\frac{\partial^2 \Psi_c}{\partial z^2} - 3 \frac{\partial \Psi_c}{\partial z} \right) = \left(\frac{\partial^2 \Psi_d}{\partial r^2} - 2 \frac{\partial \Psi_d}{\partial r} \right)$$

with μ being the dynamic viscosity.

The drag coefficient C_D and surface pressure P are obtained by integrating the appropriate component and are given as follows.

Stagnation pressure:
$$P_0 = \frac{8}{Re} \int_1^{r_\infty} \left(\frac{\partial \omega}{\partial \theta} \right)_{\theta=0^\circ} \frac{dr}{r} \tag{5.5}$$

Surface pressure distribution:
$$P_\theta = P_0 + \frac{4}{Re} \int_0^\theta \left(\frac{\partial \omega}{\partial r} + \omega \right)_{r=1} d\theta - (u^2)_{r=1} \tag{5.6}$$

Friction drag coefficient:
$$C_{Df} = \frac{4}{Re} \int_0^\pi \left(\frac{\partial \omega}{\partial \theta} + \omega \cotg \theta \right)_{r=1} \sin 2\theta \, d\theta \tag{5.7}$$

Pressure drag coefficient:
$$C_{Dp} = \int_0^\pi P_\theta \sin 2\theta \, d\theta \tag{5.8}$$

Total drag coefficient:
$$C_D = C_{Dp} + C_{Df} \tag{5.9}$$

The unsteady convective heat transfer from a fluid sphere is governed by the following dimensionless energy equation:

$$\frac{\partial T}{\partial t} + \frac{Pe}{2e^z} \left(u_c \frac{\partial T}{\partial z} + v_c \frac{\partial T}{\partial \theta} \right) = \frac{1}{e^{2z}} \left(\frac{\partial^2 T}{\partial z^2} + \frac{\partial T}{\partial z} + \frac{\partial^2 T}{\partial \theta^2} + \cotg \theta \frac{\partial T}{\partial \theta} \right) \tag{5.10}$$

where Pe is the Peclet number ($Pe = 2aU_\infty/\alpha$), α the thermal diffusivity, t the time ($t = \alpha t'/a^2$), and T the dimensionless temperature based on the initial temperature difference: $T = (T' - T'_\infty)/(T'_{d,0} - T'_\infty)$.

The boundary conditions to be satisfied are:

Far from the fluid sphere ($r \to \infty$): $T = 0$
Along the axis of symmetry ($\theta = 0$ and π): $\partial T/\partial \theta = 0$
Across the interface ($r = 1$): $T = 1$

The Nusselt number is computed from the flux of heat transfer from the surface of the sphere:

$$Nu = - \int_0^\pi \frac{\partial T}{\partial r} \bigg|_{r=1} \sin \theta \, d\theta \tag{5.11}$$

For constant properties, in the absence of dissipation and for dilute solutions, the equations governing mass transfer and heat transfer are identical. The results in this chapter are given in terms of Nusselt number $Nu = f(Re, Pe)$ in which $Pe = Re \, Pr$. The equivalent results for mass transfer can be found by simply replacing Nu by Sherwood number $Sh = f(Re, Pe)$ in which $Pe = Re \, Sc$.

III. TRANSPORT AT LOW REYNOLDS NUMBERS

A. FLUID MECHANICS

Rigid sphere: When $Re \to 0$, the nonlinear inertia terms in the Navier–Stokes equations vanish and an analytical solution exists. The steady creeping flow past a rigid sphere was first determined by

Stokes [3]. The stream function representing the motion is given by

$$\psi = U_\infty r^2 \sin^2 \theta \left(\frac{3}{4} \frac{a}{r} - \frac{1}{4} \frac{a^3}{r^3} \right)$$ (5.12)

The drag coefficient is given by "Stokes's law":

$$C_D = \frac{24}{Re}$$ (5.13)

The surface vorticity is given by

$$\omega_s = \frac{3}{2a} U_\infty \sin \theta$$ (5.14)

Fluid sphere: Studies by Hadamard [4] and Rybczynski [5] have addressed the problems of steady creeping flow past a fluid sphere analytically. The stream functions representing the motion are given by

$$\psi_c = \frac{U_\infty r^2 \sin^2 \theta}{2} \left(1 - \frac{a(2 + 3\kappa)}{2r(1 + \kappa)} + \frac{\kappa a^3}{2r^3(1 + \kappa)} \right)$$ (5.15)

$$\psi_d = \frac{U_\infty r^2 \sin^2 \theta}{4(1 + \kappa)} \left(1 - \frac{r^2}{a^2} \right)$$ (5.16)

where $\kappa = \mu_d/\mu_c$ is the viscosity ratio.

This solution yields the following expression for the drag coefficient of a viscous fluid sphere:

$$C_D = \frac{8}{Re} \left(\frac{2 + 3\kappa}{1 + \kappa} \right)$$ (5.17)

The vorticity at the interface is given by

$$\omega_i = \frac{U_\infty \sin \theta}{2a} \left(\frac{2 + 3\kappa}{1 + \kappa} \right)$$ (5.18)

B. HEAT OR MASS TRANSFER

For a stagnant drop, in the absence of any motion in the continuous phase ($Re = 0$), conduction is in the radial direction. The steady-state conduction–convection equation is reduced to

$$\frac{\partial}{\partial r} \left(r^2 \frac{\partial T}{\partial r} \right) = 0$$ (5.19)

This equation gives the Nusselt number:

$$Nu = 2$$

which represent the lower limit for heat or mass transfer from a sphere.

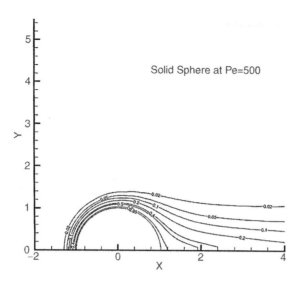

FIGURE 5.1 Concentration contours for a solid sphere in creeping flow. (From Feng, Z.G., *Powder Technol.*, 112, 63–69, 2000. With permission.)

For creeping flow ($0 < Re < 1$), the solutions of the conduction–convection equation with flow field of the Hadamard–Rybczynski or Stokes are given by numerical integration [1]. The numerical results show that the concentration contours are not symmetrical (Figure 5.1 and Figure 5.2) and that the flow inside and outside the sphere largely influences heat or mass transfer. In the case of a sphere with weak viscosity ratio, the heat or mass transfer is facilitated.

The Feng and Michaelides [6] results show a higher interface mass-transfer rate for the gas bubble than that of a solid sphere.

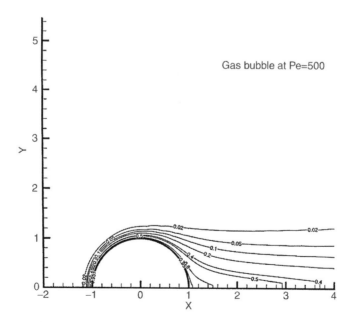

FIGURE 5.2 Concentration contours for a gas bubble in creeping flow. (From Feng, Z.G., *Powder Technol.*, 112, 63–69, 2000. With permission.)

For fluid spheres in creeping flow ($Re < 1$) with $\kappa = 0$, the conduction–convection equation with the Hadamard–Rybczynski flow field has been solved numerically by several authors. Clift et al. [1] have correlated the available numerical data in the form:

$$Nu = 1 + (1 + 0.564Pe^{2/3})^{3/4} \tag{5.20}$$

For rigid spheres in Stokes's regime ($Re < 1$), Clift et al. [1] proposed an empirical correlation on the basis of results from numerical solutions to conduction–convection equation, which follows:

$$Nu = 1 + (1 + Pe)^{1/3} \tag{5.21}$$

When the Peclet number is very large ($Pe \to \infty$), the temperature or the concentration varies only in a very thin layer adjacent to the surface of the sphere. Using the Stokes flow field and the thin boundary layer approximation, Friedlander [8] and Lochiel and Calderbank [9] derive the following expression for rigid sphere:

$$Nu = 0.991Pe^{1/3} \tag{5.22}$$

For a fluid sphere with $Pe \to \infty$, the thin boundary layer approximation and the flow field of Hadamard–Rybczynski give the following equation [1]:

$$Nu = 0.65\sqrt{\frac{Pe}{1 + \kappa}} \tag{5.23}$$

More recently Feng, and Michaelides [7] have studied the effect of the viscosity ratio on the heat or mass transfer from a fluid sphere. For fluid spheres in creeping flow ($Re < 1$) and viscosity ratio ranging from 0 to ∞, they correlated their numerical data by the following equation:

$$Nu = \left(\frac{0.651}{1 + 0.95\kappa}Pe^{1/2} + \frac{0.991\kappa}{1 + \kappa}Pe^{1/3}\right)\left(1 + \frac{0.61Re}{Re + 21} + 0.032\right)$$
$$+ \left(\frac{1.65(1 - 0.61Re/(Re + 21) - 0.032)}{1 + 0.95\kappa} + \frac{\kappa}{1 + \kappa}\right) \tag{5.24}$$

TABLE 5.1
Nusselt Number at Low Reynolds Number ($Re \to 0$)

		Pe			
		10	**100**	**500**	**1000**
Equation (5.21)		3.62	7.91	15.86	21.85
Equation (5.24)	$\kappa = 0$	3.72 (3.68)	8.33 (8.20)	16.66 (16.29)	22.90 (22.37)
	$\kappa = 1$	3.51 (3.46)	7.15 (6.97)	13.11 (12.93)	17.37 (17.27)
	$\kappa = 10$	3.27 (3.28)	6.03 (5.90)	9.90 (9.78)	12.41 (12.38)
	$\kappa = 100$	3.22 (3.25)	5.79 (5.69)	9.22 (9.10)	11.38 (11.30)
	$\kappa = 1000$	3.21 (3.24)	5.76 (5.64)	9.15 (8.97)	11.27 (11.09)
Equation (5.20)		3.22	5.66	8.94	11.00

Values of Nusselt number from Equation (5.24) are reported in Table 5.1 for Peclet numbers ranging between 10 and 1000 and for viscosity ratio ranging between 0 and 1000. Nusselt number from the numerical results of Feng and Michaelides [7] is given in parantheses. In the same table, we incorporate the results from Equations (5.20) and (5.21) valid for $\kappa = 0$ and $\kappa = \infty$. From Table 5.1, one can see that the results from Equation (5.24) agree well with those carried out by Clift et al. [1] (Equation (5.20) and Equation (5.21)) for the bubble and the rigid sphere).

IV. TRANSPORT AT HIGHER REYNOLDS NUMBERS

A. FLUID MECHANICS

For higher Reynolds numbers, analytical solutions do not exist, so the numerical solutions must be considered. When $\kappa \to \infty$, this problem corresponds to the viscous flow around a rigid particle and was studied by several authors [10–15]. When $\kappa = 0$, this problem corresponds to the viscous flow around a spherical bubble and was also studied by several authors [15–18]. The significant phenomena are very well explained in the books of Clift et al. [1] and Sadhal et al. [2]. Values of drag coefficients from numerical solutions for bubbles and rigid spheres are presented in Table 5.2, which shows a good agreement between the different studies.

For intermediate viscosities ratios, the resolution of the Navier–Stokes equations is more difficult because of the coupled flows inside and outside the fluid sphere. In this case there are only few works. Abdel-Alim and Hamielec [19] used a finite-difference method to calculate the steady motion for $Re \leq 50$ and viscosity ratio $\kappa \leq 1.4$. This work was extended to higher Reynolds number (up to 200) by Rivkind and Ryskin [20] and Rivkind et al. [21]. Oliver and Chung [22] used a different method (series truncation method with a cubic finite element method) for moderate Reynolds numbers $Re \leq 50$. Feng and Michaelides [7], Saboni and Alexandrova [23], Saboni et al. [15] used a finite-difference method to calculate the flow field inside and outside the fluid sphere. The results provide information on the two-flow field and values for drag coefficients of viscous sphere over the entire range of the viscosity ratio.

At low viscosity ratio, from the numerical results [15,20–22], it appears that internal circulation is sufficiently rapid to prevent flow separation and the formation of trailing vortex, and a small

TABLE 5.2
Drag Coefficients for Bubbles and Rigid Spheres

1	10	20	30	Re 50	100	200	300	400	Reference
					Bubbles				
—	—	—	—	0.664	0.369	0.200	0.138	0.104	Blanco and Magnaudet [13]
17.59	2.35	1.362	—	—	—	0.197	—	—	Brabston and Keller [10]
17.44	2.411	1.322	—	—	0.369	0.200	0.138	—	Magnaudet et al. [12]
—	—	1.380	—	0.661	0.364	—	—	—	Raymond [14]
17.50	2.430	1.410	—	0.670	0.380	0.220	—	—	Ryskin and Leal [11]
17.58	2.490	1.429	1.035	0.690	0.391	0.206	0.143	0.108	Saboni et al. [15] ($\kappa = 0$)
					Rigid spheres				
26.97	—	2.682	—	—	1.095	—	—	—	Feng and Michaelides [18]
27.37	4.337	2.736	2.126	—	1.096	0.772	0.632	0.552	Leclair et al. [17]
27.54	4.317	2.707	—	—	1.092	0.765	0.645	—	Magnaudet et al. [12]
—	4.398	—	—	—	1.014	0.727	0.610	—	Rimon and Cheng [16]
27.55	4.424	2.768	2.148	1.589	1.084	0.776	0.629	0.534	Saboni et al. [15] ($\kappa = \infty$)

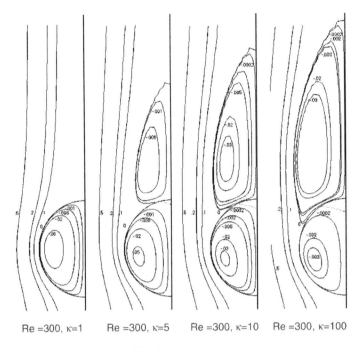

Re =300, κ=1   Re =300, κ=5   Re =300, κ=10   Re =300, κ=100

FIGURE 5.3 Streamlines for flow past a fluid sphere.

asymmetry exists between upstream and downstream regions near the sphere. While for high viscosity ratio, the results show a recirculation region followed by a wake at the rear of the sphere. Figure 5.3 shows the streamlines contours inside and outside a fluid sphere for a high Reynolds number ($Re = 300$) and different viscosity ratio ($\kappa = 1$, 5, 10, and 100). From this figure, one can notice that the eddy length and the angle of flow separation increase with viscosity ratio.

The surface vorticity and tangential velocity for different viscosity ratio are plotted as function of the angular coordinate in Figure 5.4 and Figure 5.5. From these figures, it arises that except for

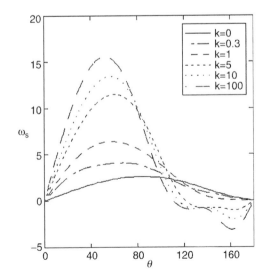

FIGURE 5.4 Vorticity distribution at surface of fluid sphere for $Re = 300$ [15].

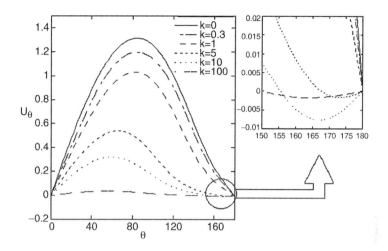

FIGURE 5.5 Velocity distribution at surface of fluid sphere for $Re = 300$ [15].

the area close to the stagnation point at the rear of the sphere, the vorticity and tangential velocity strongly depend on the Reynolds number and on the viscosity ratio. Indeed, for a fixed viscosity ratio, the vorticity and velocity raise with the Reynolds number, whereas the decrease in κ is accompanied for a fixed Re by a vorticity diminution and a simultaneous increase of the tangential surface velocity.

The main vortex inside the fluid sphere is driven by the tangential stress exerted by the external flow downstream over a large part of the sphere. For high Reynolds numbers, the stress exerted by the circulation in the trailing vortex is sufficient to create a small secondary vortex inside the fluid sphere with a circulation in an opposite direction to that of the main vortex. For example, at $Re = 300$ and for $\kappa = 5$, 10, and 100 (Figure 5.5), the presence of negative values for the tangential velocity at the rear of the sphere indicates the presence of such a secondary circulation inside the fluid sphere.

The values of the drag coefficient resulting from Saboni et al. [15] calculations are given in Figure 5.6. It is observed that the drag coefficient decreases as the Reynolds number increases

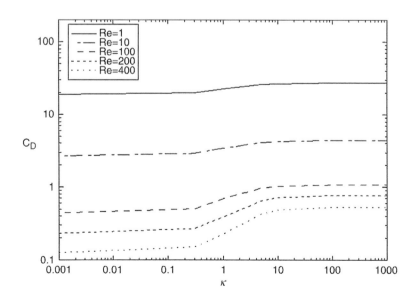

FIGURE 5.6 Variation of drag coefficient with viscosity ratio.

for a fixed viscosity ratio. In addition, for a fixed Reynolds number, the drag coefficient increases as the viscosity ratio increases and reaches a limit value corresponding to the drag coefficient for a rigid spherical particle.

From the numerical results, correlations, which give the variation of the drag coefficient as function of Reynolds number, and viscosity ratio were proposed. For $Re \leq 50$ and viscosity ratio $\kappa \leq 1.4$, the Abdel-Alim and Hamielec [19] results were fitted in an empirical equation:

$$C_D = \frac{26.5}{Re^{0.78}} \left(\frac{(1.3 + \kappa)^2 - 0.5}{(1.3 + \kappa)(2 + \kappa)} \right) \tag{5.25}$$

The work of Abdel-Alim and Hamielec [19] was extended to higher Reynolds number (up to 200) by Rivkind and Ryskin [20] and Rivkind et al. [21]. A correlation based on the best fitting of these numerical results was proposed [20]:

$$C_D = \frac{1}{1 + \kappa} \left(\kappa \left(\frac{24}{Re} + 4 Re^{-1/3} \right) + 14.9 \ Re^{-0.78} \right) \tag{5.26}$$

For moderate Reynolds numbers, Oliver and Chung [22] proposed an equation which best fit their own numerical results for low Reynolds ($0 \leq Re \leq 2$):

$$C_D = C_{D_{cf}} + 0.4 \left(\frac{3\kappa + 2}{1 + \kappa} \right)^2 \tag{5.27}$$

where $C_{D_{cf}}$ is the drag coefficient for creeping flow [4,5] which is given by

$$C_{D_{cf}} = \frac{8}{Re} \frac{3\kappa + 2}{1 + \kappa} \tag{5.28}$$

Correlations based on the best fitting of the numerical results of Feng and Michaelides [7] were proposed. Feng and Michaelides [7] remark that the series of the results for $0 < \kappa < 2$ exhibit a different behavior from the results for $2 < \kappa < \infty$. The resulting correlations are as follows.

For $0 < \kappa < 2$ and $5 < Re < 1000$:

$$C_D = \left(\frac{2 - \kappa}{2} \right) C_{D0} + 4 \left(\frac{\kappa}{6 + \kappa} \right) C_{D2} \tag{5.29}$$

For $2 < \kappa < \infty$ and $5 < Re < 1000$:

$$C_D = \left(\frac{4}{2 + \kappa} \right) C_{D2} + 4 \left(\frac{\kappa - 2}{\kappa + 2} \right) C_{D\infty} \tag{5.30}$$

For $Re < 5$ and $0 < \kappa < \infty$:

$$C_D = \frac{8}{Re} \frac{3\kappa + 2}{\kappa + 1} \left(1 + 0.05 \left(\frac{3\kappa + 2}{\kappa + 1} \right) Re \right) - 0.01 \left(\frac{3\kappa + 2}{\kappa + 1} \right) Re \ \ln (Re) \tag{5.31}$$

where C_{D0}, C_{D2}, and $C_{D\infty}$ are given by the following expressions:

$$C_{D0} = \frac{48}{Re}\left(1 - \frac{2.21}{\sqrt{Re}} + \frac{2.14}{Re}\right) \tag{5.32}$$

$$C_{D2} = 17\ Re^{-2/3} \tag{5.33}$$

$$C_{D\infty} = \frac{24}{Re}\left(1 + \frac{1}{6}Re^{2/3}\right) \tag{5.34}$$

In a more recent study, which is an extension of the previous works, Saboni et al. [15] proposed a predictive equation for drag coefficients covering Reynolds numbers in the range $0.01 \leq Re \leq 400$ and viscosity ratio from 0 to 1000. This correlation, which is reduced to the solution of Hadamard [4] and Rybczynski [5] for $Re \to 0$, is as follows:

$$C_D = \frac{(\kappa(24/Re + 4/Re^{0.36}) + 15/Re^{0.82} - 0.02(\kappa\ Re^{0.5}/(1+\kappa)))Re^2 + 40(3\kappa+2)/Re + 15\kappa + 10}{(1+\kappa)(5 + 0.95\ Re^2)}$$

$$\tag{5.35}$$

In Saboni et al. [15] study, values of the drag coefficient resulting from Equation (5.35) were compared with the numerical results [15,20] and with the correlations of Rivkind and Ryskin [20], Oliver and Chung [22] and of Hadamard–Rybczynski [4,5]. The comparison shows that this correlation gives values of the drag coefficient which coincide with those calculated numerically with an error not exceeding 7% for all the range $Re < 400$ and $0 < \kappa < 1000$. It also comes out from this study that the results for small Reynolds numbers are quite in conformity with those obtained by the correlations of Hadamard–Rybczynski and Oliver and Chung (Equation (5.17) and Equation (5.27)). This is not true for the correlation of Rivkind and Ryskin (Equation (5.26)), which significantly underestimates the values of the drag coefficient for small Reynolds numbers.

Table 5.3 and Table 5.4 give a comparison between predicted (Equation (5.25), Equation (5.29), Equation (5.31) and Equation (5.35)) and experimental drag coefficients obtained by Abdel-Alim and Hamielec [19] for two systems (cyclohexanol–water and n-butyl lactate–water) at low to moderate Reynolds number ($Re < 50$) and different viscosity ratio.

These tables show good agreement between the Abdel-Alim and Hamielec's experimental results and data from Saboni et al. [15] (Equation (5.35)) and Feng and Michaelides correlations

TABLE 5.3
Predicted and Experimental Drag Coefficients for the System Cyclohexanol–Water

κ Re	0.0995 1	0.301 1	0.554 1	0.0995 5	0.301 5	0.554 5	0.0995 10	0.301 10	0.554 10	Reference
Experiments	18.00	19.00	20.00	4.60	5.00	5.20	2.80	3.10	3.30	(Abdel-Alim and Hamielec [19])
Equation (5.25)	13.15	14.84	16.43	3.99	4.51	4.99	2.39	2.70	2.99	(Abdel-Alim and Hamielec [19])
Equation (5.31)	18.47	19.84	21.07	—	—	—	—	—	—	(Feng and Michaelides [7])
Equation (5.29)	—	—	—	4.39	4.69	5.02	2.59	2.80	3.03	(Feng and Michaelides [7])
Equation (5.35)	18.52	19.90	21.12	4.66	5.09	5.48	2.62	2.89	3.13	(Saboni et al. [15])

TABLE 5.4
Predicted and Experimental Drag Coefficients for the System *n*-Butyl Lactate–Water

κ	0.266	0.708	1.40	0.266	0.708	1.40	0.266	0.708	1.40	Reference
Re	5	5	5	25	25	25	50	50	50	
Experiments	4.50	4.90	5.00	1.45	1.75	1.80	0.90	1.20	1.25	Abdel-Alim and Hamielec [19]
Equation (5.25)	4.43	5.23	5.95	1.35	1.59	1.81	0.81	0.95	1.08	Abdel-Alim and Hamielec [19]
Equation (5.29)	4.64	5.18	5.66	1.41	1.64	1.87	0.82	0.98	1.16	Feng and Michaelides [7]
Equation (5.35)	5.03	5.66	6.18	1.38	1.61	1.82	0.81	0.98	1.13	Saboni et al. [15]

[7] (Equation (5.29) and Equation (5.31)). The predictive equation proposed by Abdel-Alim and Hamielec [19] is in poor agreement with their own experimental results.

In Table 5.5, values of the drag coefficient resulting from Equation (5.35) are compared with those given by Equation (5.29) and Equation (5.30) and with the experimental data by Winnikow and Chao [24] obtained for several systems (chlorobenzene droplet in water [$\kappa = 0.8$], bromobenzene droplet in water [$\kappa = 1.18$], nitrobenzene droplet in water [$\kappa = 2.01$], *m*-nitrotoluene droplet in water [$\kappa = 2.39$] at higher Reynolds number ($Re > 100$). Here also, it is seen that the agreement between the Winnikow and Chao's experimental results and data from Saboni et al. [15] and Feng and Michaelides correlations [7] is very good.

The correlations of Feng and Michaelides and Saboni et al. give comparable results that are in agreement with the various experimental results. As it is simpler, compact, and thus easier to use, we recommend the correlation of Saboni et al. [15] as a first approximation of the more precise results from the fully numerical simulations.

B. Heat or Mass Transfer

As the flow field depends on Reynolds number, the rate of transfer will depend on both the Reynolds and the Peclet numbers. Thus it is necessary to solve the Navier–Stokes equations, to obtain the velocity field, and to use the latter for the resolution of the conduction–convection equation.

For finite Reynolds number and very large Peclet number ($Pe \to \infty$), Clift et al. [1] used the boundary layer approximation and the surface velocities of Abdel-Alim and Hamielec to obtain Sherwood numbers at intermediate values of κ and Re. Their numerical results for $0 \leq \kappa \leq 2$

TABLE 5.5
Predicted and Experimental Drag Coefficients for Systems: Chlorobenzene Droplet in Water ($\kappa = 0.8$), Bromobenzene Droplet in Water ($\kappa = 1.18$), Nitrobenzene Droplet in Water ($\kappa = 2.01$), *m*-Nitrotoluene Droplet in Water ($\kappa = 2.39$)

κ	0.80	0.80	1.18	1.18	2.01	2.01	2.01	2.39	2.39	Reference
Re	138	318	245	299	246	288	360	187	319	
Experiments	0.45	0.27	0.35	0.29	0.42	0.35	0.28	0.64	0.42	Winnikow and Chao [24]
Equation (5.29)	0.47	0.25	0.36	0.31	—	—	—	—	—	Feng and Michaelides [7]
Equation (5.30)	—	—	—	—	0.43	0.39	0.34	0.55	0.39	Feng and Michaelides [7]
Equation (5.35)	0.49	0.26	0.37	0.32	0.44	0.40	0.34	0.55	0.39	Saboni et al. [15]

were correlated by

$$Nu = \frac{2}{\sqrt{\pi}}\sqrt{1 - \frac{(2 + 3\kappa)/3(1 + \kappa)}{\{1 + (2 + 3\kappa)\sqrt{Re}/[(1 + \kappa)(8.67 + 6.54\kappa^{0.64})]^n\}^{1/n}}} \tag{5.36}$$

in which $n = 4/3 + 3\kappa$.

In a similar way, Oliver and DeWitt [25] used the boundary layer approximation and their own surface velocities to obtain Sherwood numbers for the dimensionless parameters in the range $0 < Re < 100$, $0 < \kappa < 50$ and $Pe \rightarrow \infty$. Their numerical results for were correlated by

$$Nu = \frac{0.65}{(1 + \kappa)^{3/2}}\left(1 + \kappa(1 + 0.1\sqrt{Re}) + 0.11 \ln (Re + 1)\right)\sqrt{Pe} \tag{5.37}$$

Feng and Michaelides [7] resolved energy equation in the continuous phase by using a finite-difference method with their own numerical velocity field. In this work, steady-state numerical solutions were obtained for Reynolds number, viscosity ratio, and Peclet number ranges of $0 < Re \le 500$, $0 \le \kappa \le \infty$ and $0 \le Pe \le 1000$, respectively.

In the work of Saboni and Alexandrova [26], the transient heat transfer from a fluid sphere was investigated. However for steady state, the Saboni and Alexandrova [26] results are in agreement with those of Feng and Michaelides [7].

Variations of the Nusselt number with the Peclet number from Saboni and Alexandrova [26] are shown in Figure 5.7 and Figure 5.8 for two values of Reynolds numbers ($Re = 10$ and 100) and a fixed viscosity ratio ($\kappa = 1$).

The figures show that the Nusselt number is quite large initially and conduction dominates as the heat transfer mode. The Nusselt number decreases with time and approaches an asymptotic value depending on the predominance of the convective transfer (the asymptotic Nusselt number increases with increasing Peclet number). From Figure 5.7 and Figure 5.8, we can see the influence of the circulation on the heat transfer: the instantaneous Nusselt numbers and their asymptotic values are greater for $Re = 100$ than those for $Re = 10$ regime.

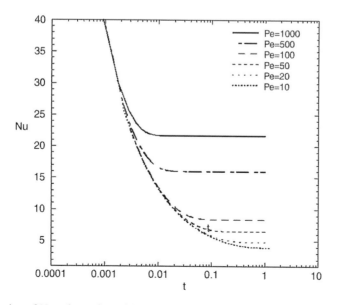

FIGURE 5.7 Variation of Nusselt number with dimensionless time for $Re = 10$ ($\kappa = 1$).

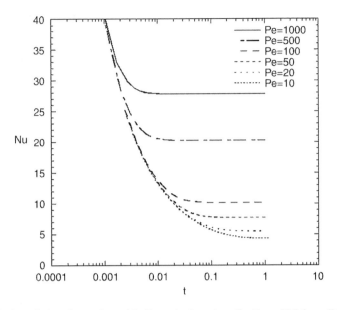

FIGURE 5.8 Variation of Nusselt number with dimensionless time for $Re = 100$ ($\kappa = 1$).

Feng and Michaelides [7] correlated their numerical data by the following equations:

$$Nu = \frac{2 - \kappa}{2}Nu_0 + \frac{4\kappa}{6 + \kappa}Nu_2 \quad \text{for } 0 \le \kappa \le 2 \text{ and } 10 \le Pe \le 1000 \tag{5.38}$$

$$Nu = \frac{4}{\kappa + 2}Nu_2 + \frac{\kappa - 2}{\kappa + 2}Nu_\infty \quad \text{for } 2 \le \kappa \le \infty \text{ and } 10 \le Pe \le 1000 \tag{5.39}$$

in which

$$Nu_0 = 0.651\sqrt{Pe}\left(1.032 + \frac{0.61Re}{Re + 21}\right) + \left(1.60 - \frac{0.61Re}{Re + 21}\right) \tag{5.40}$$

$$Nu_\infty = 0.852\,Pe^{1/3}(1 + 0.233\,Re^{0.287}) + 1.3 - 0.182\,Re^{0.355} \tag{5.41}$$

$$Nu_2 = 0.64\,Pe^{0.43}(1 + 0.233\,Re^{0.287}) + 1.41 - 0.15\,Re^{0.287} \tag{5.42}$$

In Figure 5.9, the Nusselt number from the Feng and Michaelides [7] numerical solution is compared with results from Equations (5.36) and (5.37) (based on the boundary layer approximation) and with results from Equations (5.38) to (5.42) for $Re = 50$ and $Pe = 1000$. For very low viscosity ratio, the different equations agree with the numerical solution. At larger viscosity ratio, Equations (5.36) and (5.37) depart increasingly from the numerical results. As the same observations are true for other Reynolds numbers, it emerges that the correlation of Feng and Michaelides [7] (Equations (5.38)–(5.42)) is more precise and more general than the other correlations suggested before.

V. CONCLUSION

In this chapter, the fluid dynamics and heat or mass transfer associated with a moving fluid sphere were examined. Although realistic situations require the consideration of different sizes, evaporation

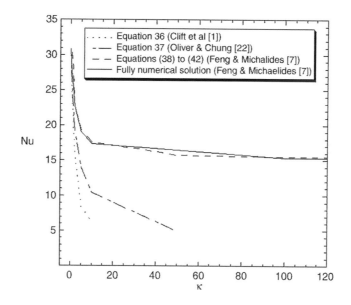

FIGURE 5.9 Variation of Nusselt number with viscosity ratio for $Re = 50$ and $Pe = 1000$.

or condensation, break-up, and coalescence phenomena, it is still important to understand the single fluid sphere problem before attempting a more general analysis.

In this study, we were interested in the hydrodynamics on the one hand and its influence on the heat or mass transfer on the other hand. Recent numerical results based on the resolution of the Navier–Stokes equations and the heat equation were used to confront various correlations of the literature for the drag coefficient and the Nusselt number (or Sherwood number).

Concerning the drag coefficient, we retained the correlation of Saboni et al. [15] (Equation (5.35)), which is valid for Reynolds numbers in the range $0.01 \leq Re \leq 400$ and viscosity ratio from 0 to 1000. For the Nusselt number, the correlation of Feng and Michaelides [7] (Equation (5.24)) is adapted to the heat transfer for small Reynolds number (creeping flow) and Peclet number ranging between 10 and 1000. For higher Reynolds number, other correlations (Equations (5.38) to (5.42)) suggested by Feng and Michaelides [7] can be used according to the value of the viscosity ratio.

REFERENCES

1. Clift, R., Grace, J.R., and Weber, M.E., *Bubbles, Drops and Particles*, Academic Press, New York, 1978.
2. Sadhal, S.S., Ayyaswamy, P.S., and Chung, J.N.C., *Transport Phenomena with Drops and Bubbles*, Springer, Berlin, 1996.
3. Stokes, G.G., *Trans. Camb. Philos. Soc.*, 9, 8–27, 1851.
4. Hadamard, J., Mouvement permanent lent d'une sphere liquide et visqueuse dans un liquide visqueux, *C. R. Acad. Sci. Paris*, 152, 1735–1738, 1911.
5. Rybczynski, W., Über die fortschreitende Bewegung einer flüssigen Kugel in einem zähen medium, *Bull. Acad. Sci. de Cracovie A*, 1, 40–46, 1911.
6. Feng, Z.G. and Michaelides, E.E., Mass and heat transfer from fluid spheres at low Reynolds numbers, *Powder Technol.*, 112, 63–69, 2000.
7. Feng, Z.G. and Michaelides, E.E., Heat and mass transfer coefficients of viscous spheres, *Int. J. Heat Mass Transfer*, 44 (23), 4445–4454, 2001.
8. Friedlander, S.K., Mass and heat transfer to single spheres and cylinders at low Reynolds numbers, *AIChE J.*, 3, 43–48, 1957.

9. Lochiel, A.C. and Calderbank, P.H., Mass transfer in the continuous phase around axisymmetric bodies of revolution, *Chem. Eng. Sci.*, 19, 471–484, 1964.

10. Brabston, D.C. and Keller, H.B., Viscous flows past spherical gas bubbles, *J. Fluid Mech.*, 69, 179–189, 1975.

11. Ryskin, G. and Leal, L.G., Numerical solution of free-boundary problems in fluid mechanics. Part 1. The finite-difference technique, *J. Fluid Mech.*, 148, 1–17, 1984.

12. Magnaudet, J., Rivero, M., and Fabre, J., Accelerated flows past a rigid sphere or a spherical bubble. Part I. Steady straining flow, *J. Fluid Mech.*, 284, 97–135, 1995.

13. Blanco, A. and Magnaudet, J., The structure of the axisymmetric high-Reynolds number flow around an ellipsoidal bubble of fixed shape, *Phys. Fluids*, 7 (6), 1265–1274, 1995.

14. Raymond, F., Thèse de doctorat de l'Ecole Centrale de Nantes, ED82-129, 1995.

15. Saboni, A., Alexandrova, S., and Gourdon, C., Détermination de la traînée engendrée par une sphère fluide en translation, *Chem. Eng. J.*, 98 (1–2), 175–182, 2004.

16. Rimon, Y. and Cheng, I., Numerical solution of a uniform flow over a sphere at intermediate Reynolds number, *Phys. Fluids*, 12, 949–965, 1969.

17. Leclair, B.P., Hamielec, A.E., and Pruppacher, H.R., A numerical study of the drag on a sphere at low and intermediate Reynolds numbers, *J. Atmos. Sci.*, 27, 308–315, 1970.

18. Feng, Z.G. and Michaelides, E.E., A numerical study on the transient heat transfer from a sphere at high Reynolds and Peclet numbers, *Int. J. Heat Mass Transfer*, 43 (2), 219–229, 2000.

19. Abdel-Alim, A.H. and Hamielec, A.E., A theoretical and experimental investigation of the effect of internal circulation on the drag of spherical droplets falling at terminal velocity in liquid media, *Ind. Eng. Chem. Fundam.*, 14, 308–312, 1975.

20. Rivkind, V.Y. and Ryskin, G.M., Flow structure in motion of a spherical drop in a fluid medium at intermediate Reynolds numbers, *Fluid Dyn.*, 11, 5–12, 1976.

21. Rivkind, V.Y., Ryskin, G.M., and Fishbein, G.A., Flow around a spherical drop in a fluid medium at intermediate Reynolds numbers, *Appl. Math. Mech.*, 40, 687–691, 1976.

22. Oliver, D.L.R. and Chung, J.N., Steady flows inside and around a fluid sphere at low Reynolds numbers, *J. Fluid Mech.*, 154, 215–230, 1985.

23. Saboni, A. and Alexandrova, S., Numerical study of the drag on a fluid sphere, *AIChE J.*, 48 (12), 2992–2994, 2002.

24. Winnikow, S. and Chao, B.T., Droplet motion in purified systems, *Phy. Fluids*, 9 (1), 50–61, 1966.

25. Oliver, D.L.R. and DeWitt, K.J., Mass transfer from fluid spheres at moderate Reynolds numbers: a boundary layer analysis, Paper No. AIAA-92-0105, in: *Proceedings of the AIAA 30th Aerospace Sciences Meeting*, January 1992, Reno, Nevada, 1992.

26. Saboni, A. and Alexandrova, S., Transfert de chaleur entre une inclusion sphérique et un milieu visqueux: régime transitoire, rapport d'avancement, LTP2002-5, 2002.

Various Approaches and Transitions

6 Interactions of Nanostructured Fillers with Polymer Networks — Transition from Nano- to Macroscale

Milenko Plavsic
University of Belgrade, Belgrade, Serbia and Montenegro

CONTENTS

I. INTRODUCTION

Polymer networks, in general sense, are of the unique molecular type, extending in size (and structure) from micro- to macroworld [1–8]. In very simplified terms, characteristic dimensions of a polymer network, following the same constitutive relations, can be extended, from nano- to kilometers, that is, for a dozen of magnitudes in space scale. In principle, one can expect that using such properties can bridge over, and connect in synergetic way, properties of some entities from nano- to macroscale. But, in the common approach to polymers (and polymer-type carbon networks) this way of thinking is still very rare. On the other side, some issues obviously related to structure parameters on both, nano- and macroscale, stay unsolved in spite of the great efforts and investments, extensive research, and large pool of accumulated data, for a very long time. Such a problem is the reinforcement of rubbers with active fillers. Although based on some obvious phenomena, following from the fundamental principles (as the interactions due to uncompensated forces at a surface of nanoparticles, making the architecture of active fillers), quantification of that influence on properties of materials, on macroscale, is not yet possible [2–4].

It is exactly a 100 years since carbon black (CB) has started to be added into rubber compounds [5], and soon polymer engineers will celebrate 200 years of rubber networking [6]. Such a network made possible the layout of the first telegraph cable between Dover and Calle. It (at least in principle) means that if "Maxwell demon" would move down the covalent bonds, it is possible that it finds a continuous chain of covalent bonds of 200 km. But, again in principle (and with very pragmatic consequences), cannot be rationalized in quantitative and even not in qualitative way, how nanoparticles of fillers (which can be organic as CB, or inorganic as silica, in constitution) make possible a hyperelastic deformations of such a chain network. Moreover, similar elastic behavior in connection to small entities can be observed on microscale of some new photonic nanosensors, biopolymer tissues, and organelles [9,10].

It comes out, that polymer network is a very specific ensemble of entities [11] in terms of classification in the introductory chapter of this book, it has been possible to transform some nanoeffects to macroscale. At the same time, as we shall see, it is expected to be liable to statistical, mechanical, and even quantum statistics treatment. Recent experiments with networks of nanotube type, as "quantum wires" capable of "ballistic" transfer of charge [12,13], can be reported as the extension of a such approach to carbon–carbon network.

An example of the inverse physical approach to nanoarchitecture is the application of carbon–carbon networks, as templates for building of nanostructures of nonpolymer materials [14,15]. It can be realized by the orientation of copolymer chains in thin films to control interfacial interactions. Relatively simple way of achieving this control is anchoring a random copolymer to the interface, which consists of monomer units identical to those of the block copolymer. Through the variation of particular monomer units in the random copolymer, interfacial interactions can be balanced and the diblock copolymer applied will orient normal to the surface. Such a structure provides possibilities for many further applications. In the case of cylindrical micro-domain morphology, the thin film is essentially an array of nanocylinders, oriented normal to the surface, with an area density that is dictated by the size of copolymer molecular weight. With some of lithographic procedures, the blocks of one type are cross-linked and of other type are degraded. It produces an array of nanopores in the network. That can be used as a template for transfer of patterns to the substrate, or as porous separation media, after removal from the substrate. By chemical or electrochemical methods, the pores can be filled with metal, glass, or other materials, and this can produce, for example, arrays of nanowires with exceptionally high aspect ratios. For example, with such a method, it is possible to obtain area densities in excess of 9×10^{11} copolymer cylinders per cm^2 [14].

It is not clear whether the filler material fine particles are stabilized by polymer network, or the metastable states of the network chains are stabilized with nanostructured fillers, or the highly sensitive biomaterials are protected by polymer network vesicles transferred (e.g., DNA in gene therapy, or in the case of some other target drag delivery application [10]). Whatever the case be, we face the fascinating issue of controlling the coupling between adjacent phases, when they are driven far from an equilibrated state. The issues of interface interactions and metastable states are related to the broad spectrum of material properties such as mechanical strength and durability of hybrid materials, and their optical, electronic, catalytic, medical, and other applications. Interest for them is highly grown especially on topics of shared commonalities with nanotechnologies. Numerous new, very sophisticated experimental techniques and devices for characterization of interfacial interactions have been developed. Still, a number of fundamental questions stay open, as we have seen. Those issues, in particular connected to interactions of large-size networks with fillers of nanosize architecture, leading to unique hyperelasticity phenomena, constitute the subject of this chapter.

II. THE SCALING THEORIES AND INTERFACIAL INTERACTIONS

A change of some properties with change of size of the system provides us with the opportunity to recognize the entities of its structure. The great theoretical support has been provided in polymer science with excluded volume theory of Flory [1]. It provides the following law that describes scaling of a polymer molecules radius of gyration R_G with number of its backbone bonds N_b:

$$R_G = \kappa N_b^v \tag{6.1}$$

According to Flory, $v = 3/5$ for all kinds of flexible chains in good solvents. It represents a global property of macromolecules in terms of de Gennes scaling concept in physics [16]. Global properties are of general validity for polymer molecules (or for some class of them) describing some essential features of the structure while the prefactor κ represents local properties, characteristic of each polymer type. Starting with Equation (6.1), de Gennes has tried to find other laws of the same form, for polymer systems. An understanding the reasons, which dictate the exponents would help us to extract the essence of the structure characterizing the considered systems. Once such global parameters are recognized, it is much easier to handle numerous experimental data describing details of particular system behavior. But, scaling of properties at the phase boundaries is of main interest here that is quite specific in nature.

Different approaches to the problem are possible. Generally speaking, theoretical considerations treating the problem can be sorted, according to the following main approaches:

- Macro or micro (scale of structure) theories
- Static or dynamic (state formulation) theories

The first formulation of the problem was given by Laplace [17] for two adjacent fluid phases:

$$p_a - p_b = \gamma_{ab}(K_1 + K_2) \tag{6.2}$$

where p_a and p_b are hydrostatic pressures in phases a and b, respectively, and K_1 and K_2 are the principal curvatures at the considered locus, and γ_{ab} the interfacial tension. In modern terms, it is merely a macroscopic manifestation of anisotropy of the pressure tensor within the interfacial layer. Trying to generalize this phenomenological presentation, one faces two possibilities. The first is, understanding what really happens between two phases in contact at the molecular level. The second is a replacement of curvatures K_i with real geometry, especially when solid phases are involved.

As we know, thermodynamics of surface phenomena was first formulated by Gibbs [18] in a very elegant way. But for phenomena of interest here, one must be very careful in its application. In short, the classical thermodynamics was devised for handling locally homogeneous systems, that is, the change of any intensive variable is completely negligible over a differential volume of the continuum, which is much larger than molecules dimensions. But, this is clearly not so within the transition layer, the density of witch may drop, for example, by a factor of 1000 over 1–2 nm, when passing from the liquid to the gas phase. Gibbs superseded the difficulty by introducing the concept of dividing surface. It is passing through "points which are similarly situated with respect to the condition of the adjacent matter" [18]. The actual system is then replaced by two homogeneous bulk phases, extending on each side of chosen dividing surface. This surface itself is treated as a homogeneous two-dimensional phase and provided with appropriate mechanical and physical–chemical properties. For example, free energy G is split into three parts

$$G = G_a + G_b + G_s \tag{6.3}$$

for phases a, b and surfaces s. In the same way, bulk concentrations per unit volume V of each phase are defined by $^{a}C_i = {}^{a}N_i/V_a, {}^{b}C_i = {}^{b}N_i/V_b$; it is convenient to introduce the adsorption $\Gamma_i = {}^{s}N_i/A_S$, as the mean number of molecules of species i $\{N_i\}$ adsorbed on a unit area A_S of the interface. Then

$$d\gamma = -S_s \, dT - \sum \Gamma_i \, d\mu_i \tag{6.4}$$

where S_s is the surface entropy per unit area and μ_i the chemical potential of i.

The connections of defined macro-parameters with molecular structure system provide statistical mechanics. The route proceeds from the partition (or better to say grand partition function) Z of the system going to the free energy G and finally to γ by means of the general relationship:

$$G = -k_B T \ln Z \tag{6.5}$$

and

$$\gamma = \left(\frac{\partial G}{\partial A_S}\right)_{T,V,\{N_i\}} \tag{6.6}$$

The first statistical mechanical theory describing the interface of liquid–gas far from critical point was formulated by Prigogine and Saraga [19]. They used a cell model with each molecule of the liquid locked in a cell, with u being potential energy at the cage center and z the partition function of a molecule in its cell. They obtained

$$\gamma = \frac{1}{a^2}\left[\frac{1}{2}(u_{Sr} - u_{Lq}) - kT \ln\left(\frac{z_{Sr}}{z_{Lq}}\right)\right] \tag{6.7}$$

where a^2 is the area of a surface cell at the interface and indexes Lq and Sr correspond to "liquid" and "surface" cells, respectively. One, who wants to extend that model to polymer systems, immediately faces two possibilities. The first possibility is that one molecule occupies more than one cell. The second is application of more realistic interaction potentials. Prigogine used such an approach for studying short chains [20,21]. He sliced the system into monolayers and then used Flory–Huggins theory developed for bulk phases [1] to analyze each layer by the mean field method. But the calculations become very complicated for long chains. A number of authors later proposed theories, trying to supersede those difficulties: Silberberg [22] and Hoeve et al. [23]. Flory [24,25] himself, returned to this problem several times. In theories on molecular organization in micelles, in vesicles, and in lipid bilayer membranes [26,27], and in the theory on interface structure of lamellar semicrystalline polymers, Flory et al. [28] developed methods for accounting long polymer transitions between layers and space-filling constrains due to geometry differences.

With colossal development of computer facilities, the new method based on "numerical experiments," that is, models based on computer simulation of interphase interactions, is developed. We are interested in the model of Theodoru [29] which is an extension of the Suter and Theodorou model. This model treats ensemble of undeformed and deformed model structures of glassy amorphous vinyl polymer, to probe the local structure and to explore the microscopic mechanisms of small-strain deformations. It was later applied to polymer-film model systems. Ultrathin glassy polymer films with high fraction of interphase layer molecules are very active areas of research with high technological importance, as well. Theodorou extended the probing method with molecular dynamics models; so it belongs to the finestructure dynamics group in terms of different approaches, mentioned at the start of this section. The results indicate that "dynamical interfacial thickness" is twice as large as the thickness over which the mass-material's density varies. The center of mass motion of chains located as far as $2-3$ times the unperturbed radius of gyration R_{G0} from the edge of the free surface is enhanced and becomes highly anisotropic.

The third method in statistical mechanics which uses exact enumeration techniques for interphase problems. In short, it is a kind of a midway between the mean field and computer simulation methods. Again, the grand partition function of the system can be formulated, considering distribution of segments and other substance molecules in a regular lattice. The number of possible chain conformations decreases significantly in the interface layers according to DiMarzio and Rubin [30,31]. A combination of the mean field method with exact enumeration technique has been used very successfully in some more general theories considering polymer bulk properties and equilibrium between phases of rigid and semi-rigid chains, for example, in theories of Flory with Abe, Ronca, and Matheson [32,33].

The fourth method is analytical. For changes of surface tension with solution concentration, the necessary expansion can be obtained in the form of an exact series. But, analytical theories, in general, are very difficult to develop for long chains. The edges effects for two-dimensional ensemble are obtained only for dimers and trimers. But even for the more general problem of chain configurational statistics influenced by external fields solutions are possible [34–36]. This is achieved by a combination with other techniques such as graph methods or exact enumeration supported with appropriate experimental information (starting from dimers and trimers up to long chains).

Another approach has been suggested by de Gennes [37]. Any analytical function (as mentioned above) describing $\gamma(x)$, for polymer solutions with strong forces, driving chain adsorption on the solid wall, and rather weak forces between that chains with polymer molecules in the bulk of the system, can be split into two parts:

$$\gamma = \gamma_d(c_s) + I_b(c_s, c_b) \tag{6.8}$$

where c_s is the concentration profile, $c(x)$ is the concentration profile toward the wall and c_b is the bulk concentration. For $x = a$, c_s corresponds to the first layer, a being the cell length and x the coordinate normal to the wall. Then γ_d describes interactions in the proximal range and I_b effects of the concentration profile at larger distances. The free energy per chain g_{ch} can be estimated from that model in the form

$$(k_B T)^{-1} g_{ch} = \left(\frac{R_G}{D_l}\right)^{5/3} - |\gamma_1| a^2 \left(\frac{a}{D_b}\right) N_b \tag{6.9}$$

where R_G is the Flory radius from Equation (6.1), N_b the number of bonds, and D_l the thickness of layer where adsorbed chains have loops, extending from the wall. γ_1 is the part of γ_d describing interaction energy per unit surface. Instead of looking for more precise analytical expression for γ, three regions of concentration scaling in the space are established in the model: (1) proximal ($x \sim a < D_l$), (2) optimal ($D_l < x < \xi_b$), and (3) distal ($x > \xi_b$), where parameter ξ_b can be approximated for semidilute solutions with R_G. Comparisons of this scaling pattern with possible scaling in other type systems such as melts, thin films, and networks, are of fundamental importance.

Besides the above factors we should also point to yet another fundamental factor: as the flexible chain (connected with some segments to solid surface), makes more or less large loops in space, the solid surface cannot be considered flat at that scale. Mandelbrot [38] and later many authors advocated methods for handling that difficulty (already mentioned for interfaces in connection with K_i values in Equation (6.2) [38–40]. They called attention to the particular geometrical properties of some objects such as shoreline of continents, some cluster surfaces, or volume of clouds. Mandelbrot coined the name "fractal" for these complex shapes to express that they can be characterized by a noninteger (fractal) dimensionality. A broad class of growing patterns (from some snowflake crystals up to tree branches) characterized by open branching structure can be described in terms of fractal geometry. In the present cases, this means that the growing structures are self-similar in a statistical sense, and the volume $V(R)$ of the region bounded by interface, scales with the increasing linear size R of the object in a nontrivial way:

$$V(R) \sim R^{D_f} \qquad (6.10)$$

where $D_f < d_E$ is typically a noninteger number called the fractal dimension and d_E the Euclidian dimension of the space in which the fractal is embedded in. For a real object, the above scaling holds, of course, only for length scales between a lower and upper cut-off. The volume of a growing fractal $V(R)$ can be measured with number of balls (of unit size) needed to cover the object completely: $N_V(L)$ where L is the linear size of the whole structure. Then

$$D_f = \lim_{L \to \infty} \frac{\ln N_V(L)}{\ln (L)} \qquad (6.11)$$

Obviously, the previous definitions can easily be related to the general relation describing scaling in Equation (6.1). Indeed, a number of authors considered a polymer chain as a fractal object and the scaling exponent as fractal dimension. But, one should keep in mind that fractal models are concerned with geometrical features of the object, and Flory formula describes scaling that results from complex interactions between chain elements. It provides the solution for some multi-body problems, what is a very tough subject in physics. Besides geometry aspects, it incorporates a number of other relations. It indicates connections of Flory exponent with dynamical properties and fraction dimension of Alexander and Orbach [41], according to some experiments with biopolymers. Now, some recent experiments providing a new level of insight into polymer interphase interactions, molecular dynamics scaling, and scaling of structure should also be pointed.

A central question in such experiments is: how molecular motions are arrested in the glassy state to result in very slow dynamics. A number of researches attempted, using sophisticated techniques, to determine the film thickness at which glass transition temperature T_g differs from the bulk value. For example, a threshold thickness of 40 nm was found for polystyrene (PS) on Si−H substrate in experiments with PS films using spectroscopic ellipsometry [42]. A similar decrease in T_g for PS on SiO_x using nulling ellipsometry was obtained [42]. Positron annihilation spectroscopy measurements of PS on Si−H near the PS−vacuum interphase gave results consistent with ellipsometry [43]. Measurements by quartz crystal microbalance for PS on gold substrate suggest that T_g decreases with film thickness. Results from scanning viscoelasticity microscopy indicate that local T_g decreases even in relatively thick films of 200 nm [42]. Also, x-ray refractivity measurements of PS on Si−H substrate indicate departure from bulk value, but in opposite direction. According to these measurements T_g increases. Similar results were obtained from diffusion of deuterated PS into hydrogenated PS: the diffusion was slower in thin films than in thicker. The authors suggested that an "effective" T_g of PS increases by 25°C [44]. Slower PS chain diffusion relative to the bulk in films thinner than 150 nm was found by measuring fluorescence recovery (after patented photo-blanching) but the results cannot be explained solely on the basis of a decrease in the T_g [45].

Obviously the phenomenon is not simple and needs extensive theoretical treatment as will be discussed in the following sections. But, for our further discussion it is of special interest to consider the univocal conclusion on always-present transitions in chain dynamics and scaling of structure in interphase layers. The scaling transition problems will be considered in the next section.

III. TRANSITIONS AND SELF-CONSISTENT FIELDS PHILOSOPHY

Transition from local chain parameters to system properties is realized in self-consistent field approach. This is evidenced in classical works such as the Flory–Huggins model [1] implicitly and later explicitly in a number of papers of de Gennes and coworkers [16]. Considering possible configurations of chains in a mixture (described by lattice model of cell dimension a), one can add a unit link to the chain which is subject to an average potential of its environment. The probability of chain growth can be estimated from the balance of a certain concentration profile $c(r)$ and average repulsive potential $U(r)$, and is proportional to the local concentration. The change of that probability for a chain of N_s links at temperature T can be represented as

$$W_{N+1}(\mathbf{r}', \mathbf{r}) - W_N(\mathbf{r}', \mathbf{r}) \approx \frac{\partial W}{\partial N}(\mathbf{r}', \mathbf{r}) = -\frac{U(\mathbf{r})}{T} W_N(\mathbf{r}', \mathbf{r}) + \frac{a}{b}\nabla^2 W_N(\mathbf{r}', \mathbf{r}) + \cdots \quad (6.12)$$

Equation (6.12) can be rearranged into

$$-\frac{\partial W}{\partial N} = -\frac{a^2}{b}\nabla^2 W + \frac{U(\mathbf{r})}{T} W \quad (6.13)$$

which is remarkably similar to the Schrödinger equation for a nonrelativistic particle of wave function $\psi(\mathbf{r}, t)$:

$$-i\hbar \frac{\partial \psi}{\partial t} = -\frac{\hbar^2}{2m}\nabla^2 \psi + V(\mathbf{r})\psi \quad (6.14)$$

In that analogy N corresponds to time and can be understood as the number of steps of a walker on the lattice, moving with constant speed. One particular chain conformation corresponds to one particular path for the particle and the wave function appears as a coherent superposition of amplitudes for different paths. For so-defined states of the system, one can estimate relevant properties, and if necessary change the value for local concentration to $c'(\mathbf{r})$ and repeat the procedure with new potential till the sequence of approximations $c(\mathbf{r}) \rightarrow c'(\mathbf{r}) \rightarrow \cdots c_{\text{fin}}(\mathbf{r})$ converges to a stable solution. The potential U_{final} is then self-consistent.

Another type of problem appears in systems, where scaling transitions are also present between different chains. The local concentration profiles and local potentials then have to be accounted for total transition. The evolution of such systems can be analyzed using Fokker–Planck equation:

$$\frac{\partial W}{\partial t} = -v_d \frac{\partial W}{\partial x} + D_L \frac{\partial^2 W}{\partial x^2} \quad (6.15)$$

where W is the probability density of configurations, v_d the "drift" velocity, D_L the "diffusion" coefficient, and x and t the space and time coordinates, respectively. For example, x can be interpreted here as a measure of the extent of transition, v_d as the rate of the transition, and D_L as a measure of dissipative character of the transition.

For fractal systems (as solid interfaces of interest here), additional complexity appears. The progress of a transition is space-dependent. It can be illustrated with a relatively simple example, originally proposed by Le Mehaute, of electrical deposition of mass on fractal surface of an electrode [39,40]. If Φ is the density of flow per unit length L_e of ideally flat electrode, the transfer rate will be $J = L_e\Phi$. If the flow is not stationary, we can write the time dependence equation as $J(t) = L_e\Phi(t)$. But if the surface is flexible, L_e also changes with time and additional nonstationary process occurs. The reaction occurs with a delay caused also by a longer route taken by the particles in traveling to the surface and the shape of the particle is also changing. During this delay, the length again changes, and we cannot use above expression to describe the process. In such a situation, previous changes in the system should also be accounted for. It can be done by applying nonintegrity derivative (fraction derivative) in Liouville–Riemann form.

For some function $f(t)$ of time t, it is

$$_{t_0}D_t^q f(t) = \frac{1}{\Gamma(1-q)} \frac{d}{dt} \int_0^t \frac{f(\tau)}{(t-\tau)^q} d\tau \qquad (6.16)$$

Now for process with fractal electrode, having fractal dimension $D_f = D_2 + 1$ we can write

$$D_t^{(1/D_2)-1} J(t) \sim \Phi(t) \qquad (6.17)$$

In other terms, deposition of mass on a fractal surface is a noninteger integral of density of flow. Considering the driving force of the process as Dirac δ-function potential $\Delta\mu(t)$, two main cases are of interest here. The first is $\Phi(t) \sim \Delta\mu(t)$, then we can write

$$D_t^{(1/D_2)-1} J(t) \sim \Delta\mu(t) \qquad (6.18)$$

The second case is a more complex process in fractal media. For instance, when a $(D_{ff} - 1)$ fractal is embedded in D_f fractal, the hyperscaling behavior occurs [37]. Then, Equation (6.18) has to be extended to the density of flow but now through the relation:

$$D_t^{(1/D_{ff})-1}[\Phi(t)] \sim \Delta\mu(t) \qquad (6.19)$$

Without going more into details of the hypothetical experiment described, here, I point that $\Delta\mu(t)$ can be understood generally as a thermodynamic force related to departure from thermodynamic equilibrium or more generally, from the stationary state. Then it can be written in the usual form as:

$$\Delta\mu(t) \sim [\mu^*(t) - \mu(t)] \qquad (6.20)$$

where $\mu(t)$ represents the actual thermodynamic potential at the flexible interface and $\mu^*(t)$ its thermodynamic asymptotic value at long time t.

IV. HYPERELASTICITY PHENOMENOLOGY AND REINFORCEMENT

There are only a few material groups such as elastomers, soft foams, and some biological tissues, which can undergo large deformations without permanent set; that is, exhibit large nonlinear elastic behavior and we call them hyperelastic materials [5–8,46–58]. At the first glance they have not much in common of their structure, but the same unique properties exposition, incite both theoretical and experimental research of their possible hidden relations, indicating more fundamental reasons for such behavior. Especially, there is increased interest for such effects in different domains of modern material science, medicine, and engineering in general. This is because some new materials such as hyperswelling elastomer hydrogels, some biomaterials, etc., use hyperelastic deformations to transform small changes in a system to different signals or mechanical responses, opening in that way a broad field for new sensors and nanodevices [9,55]. Incorporation of a hyperelastic component in different composite materials also provides new areas for its application. Further, the composite properties, especially for elastomer systems, provide a new quality of information for understanding hyperelastic behavior, as well.

Addition of fillers can dramatically change mechanical properties of elastomer materials. For example, a pure gum vulcanizate of general purpose styrene–butadiene rubber (SBR) has a tensile strength of no more then 2.2 MPa but, by mixing in 50 parts per hundred weight parts of rubber (p.p.h.r) of a active CB, this value rises more than 10 times to 25 MPa. How CB, being fine powder of practically no mechanical strength, can make reinforcement in rubbers, similar to a metal armature in composite materials? Extensive research has been done in the area of reinforcement with "active" fillers in the last 100 years (it is almost exactly 100 years, since first CB has been applied to rubber compound in 1904) but, complete answer is not yet obtained [2–8,57–59].

Generally, high surface area per unit weight of filler (i.e., very fine particles) is necessary for reinforcement, but there is no direct proportionality between change of reinforcement activity and difference in surface area for CB of different grades. Moreover, according to some authors moduli at high strain for elastomers reinforced by silica filers decrease with the increase of surface area [59]. This discrepancy can be defined in terms of de Gennes's "scaling concepts in polymer physics" as difference in global and local properties of active fillers. To obtain reinforcement activity, all fillers must achieve some level of surface area. But, differences in activity between grades of the same filler follow from local properties, that is, finestructure of the filler [2–4]. Here, the answer has been found in the interpretation of surface activity as interaction of polymer with some active centers on the filler surface. Without enough large surface area, it is not possible to obtain enough active centers, but the same area does not automatically guarantee the same concentration and accessibility of active centers. But, this raises the difficult question about finestructure of filler surface, the estimation of number of active centers, and the nature of their interaction with polymer matrix. Fortunately, application of new experimental techniques meanwhile developed for characterization of materials such as atomic force microscopy, x-ray scattering, neutron scattering, infinite dilution inverse gas chromatography, etc., provides a new level of insight to the fine structure of fillers at nano- and sub-nanolevel [57–59].

The direct observations of atomic organization of the CB particle surface with new techniques made possible proposition of new model of the elementary particle with surface covered by platen-like layers with zigzag edges similar to scales. The zigzag edges of scales can provide almost perfect fit to some carbon chains in *trans–trans* conformation absorbed on the surface. Such chains, perfectly compatible with filler and also incorporated in the matrix, will be subject to all strains and constrains imposed to filled compounds, which will influence their conformational mobility, intramolecular interactions, and chain flexibility, as well. But, it raises immediately following three issues. The first: if such a mechanism is causing reinforcement interactions, what are the sources of reinforcement by silica fillers, with completely different surface structure? The second: if all active CB particles have the same surface structure, what are the reasons for high difference in reinforcement activity of different CB grades? The third: how significant is polymer chain and network structure, for reinforcement? Common approach to such multi-variable problems is partial variation of one-by-one class of factors, while all other system characteristics stay same. As factors of the highest influence and technological importance have been studded, some investigator have dealt with the properties without involving the volume fraction of added filler. This approach is in distinct contrast with the theories of rubber elasticity, show that increase of degree of networking increases the modulus of gum networks (i.e., the networks without filler) but also that modulus decreases with extension up to some level. In the first molecular theory of rubber elasticity, based on contributions of Kuhn, Wall, and Flory, the relation between tensile force per unit of unstrained cross-section σ and strain ratio $\lambda = l/l_0$ is formulated as

$$\sigma = M(\lambda - \lambda^{-2}) \tag{6.21}$$

what is saved in most other theories (differing in formulation of modulus M). Strain ratio λ is the ratio of sample dimensions in strained (l) and unstrained (l_0) state. The decrease of M, characteristics in the range of small and medium λ, is well described with semi-phenomenological theory of Mooney and Rivlin as

$$\sigma = \left(\frac{C_1 + C_2}{\lambda}\right)(\lambda - \lambda^{-2}) \tag{6.22}$$

where C_1 and C_2 are constants for solid elastomer materials. The increase of modulus M at some level of strain is explained in Kuhn–Wall–Flory theory by transition of network chain statistics from Gaussian to Langevin type, due to finite extensibility of polymer chain. In the modern rubber elasticity theories of Flory, Erman, Mark, and others [32,63,64], these are the well-understood phenomena. Filled elastomer materials follow the same behavior pattern, but at much higher M level, and not in strict linear way, as described in Equation (6.22). Also some regular chain structure polymers such as natural rubber (NR) crystallize partially at high extension, producing additional reinforcement. This well-known phenomenon is however, not to be considered here. Herein, we will focus our attention on the first two issues cited earlier (or better to say: interactions of active filler of different constitution and space architecture, with polymer networks). But, we should also take note of some additional aspects of the problem. The reinforcement by fillers is defined at present as "improvement of modulus and *failure properties* of the final vulcanizate" [2–4]. Already in 1920 Wiegand proposed resilient energy as the measure of reinforcement [6]. In general, rubber reinforcement represents one of the two main types of material reinforcement. The other type is reinforcement by armature of a material with much higher modulus (and stiffness) than the matrix. Again in general terms, the rigid armature extends end-to-end of the product. With active filler hyperelasticity reinforcement, the situation is just the opposite: the flexibility and elasticity are saved (in parallel with the (relative) increase of moduli and improvement of failure properties).

V. THEORIES OF RUBBER REINFORCEMENT — A BRIEF REVIEW OF THE PREVIOUS WORK

The issue of no-armature reinforcement of materials and also the high interest of growing industry for them has naturally attracted a lot of work to the field of elastomer reinforcement. The theories, and some qualitative models (or just lucid proposals), trying to explain the phenomenon can be sorted in several classes [2–8]. Moreover, some classes can be divided into groups of theories. In brief, we can describe it with three classes:

1. Continuum theories
2. Rheological model theories
3. Structural theories

The predictions of the elastic moduli in the class of continuum theories are based on phenomenological laws, variation methods, and self-consistent shims, but without any insight into the structure of hyperelastic materials. In the class of rheological model theories, some crude insight into the structure is given through different types of coupling of elastic and viscous elements that can be related to coupling of phases. But it cannot explain the behavior of fine dispersed particle systems.

The structural class can be divided into 10 groups, based on particular approach to the problem as, for example, hydrodynamic theories. They belong models of Guth and Gold and Smallwood, but some authors include also the occluded volume theory of Medalia and theories of filler and bonded rubber compact spheres of Brennan and Jermyn and some others. The groups can be described as:

1. Hydrodynamic theories (Guth and Gold, Smallwood, Medalia, Brennan and Jermyn, Raos and Allegra).
2. Strain amplification in the rubber phase due to filler (Mullins and Toubin).
3. Nonaffine deformation of network junctions causing stress relaxation (Mullins and Toubin).
4. Stress softening caused by weak and strong linkages (Blanchard and Parkinson).

5. Stress hardening due to alignment and tightening of highly stretched chains (Blanchard).
6. Tightening of short chains between filer particles (Blanchard and Hess).
7. Breaking of filler-to-network-bonds and molecular rearrangement by slippage on filler-to-network-Bonds surface (Bueche, Aleksandrov and Lazurkin, Houvink, Dannenberg, Rigbi, Fujiwara and Fujimora, Maier and Goritz).
8. Networking of filler particles (Bouche, Payne, Dogatkin, Furukawa, Goritz, Wang, Lin and Lee).
9. Influence of filler surface energy and morphology (Kraus, Donnet, Wagner, Wolf, Gespacher, Lablanc, Niedermeier, Wang).
10. Cluster–cluster aggregation (Kraus, Payne, Gespacher, Rubinstain, Vilgis, Heinrich, Kluppel).

Obviously, rubber is very extensive material but a number of explanations proposed can be already seen from the list. Here we will describe the main ideas following from the listed contributions, while details of the theories can be found in the literature [2–8,56–69].

(a) The first hydrodynamic approach follows the general theory of energy dissipation in dispersions [6–8]. In 1906, Einstein explained the increase of the viscosity of low-concentrated suspension. When a particle is suspended in a flowing fluid, the velocity profiles near the particle are perturbed due to longer distance that the fluid must travel (to keep the continuity of the flow), compared with those which it would travel if the particle were not present. A consumption of energy for that process produces energy dissipation in the system, causing viscosity effects. When a second particle is placed (according to self-consistent field method!) in the same flow field, it is subjected to a flow field which is different from that which would occur if the first particle were not present. But, for low concentrations of particles total effects can be presented simply by the sum of individual particle contributions as

$$\eta_f = \eta_u(1 + 2.5c) \tag{6.23}$$

where η_f is the viscosity of the suspension, c the volume concentration of the particles, and η_u, the viscosity of the pure solvent. In 1938, Guth and Gold [2–4] used the same formula, considering reinforced rubber compound as a dispersion of fillers in polymer media. But, to use Equation (6.23) the particles should be spheres, ideally wettable by the fluid, of uniform size and should not interact. These theoretical requirements are not met by filler particles. To correct for the interactions, Guth et al. added a quadratic term for concentration in Equation (6.23) and later, to correct for the shape, they introduced f_{sh} as the shape factor (equal to the ratio of the longest to the shortest diameter of the particle), and proposed the following equation:

$$\eta_u = \eta_u(1 + 0.67fc + 1.62f^2c^2) \tag{6.24}$$

In 1944, Smallwood showed [2–4] that Young's modulus E of several filled vulcanizates at low concentration fit the relation

$$E = E_0(1 + 2.5c) \tag{6.25}$$

where c is the volume concentration of the filler. It is the same relation as Equation (6.23) but now describing elasticity instead of viscosity η, which permitted substitution of E instead of η in Equation (6.24). The new formulation of Equations (6.24) and (6.25) fitted the data for the fine thermal CB particles, up to a volume fraction of 0.3. With different choices of the value of f_{sh} Equation (6.24) gives fair agreement with the modulus–concentration data for many system, but the physical meaning of f_{sh} is lost. Instead of employing shape factors to obtain reasonable correlation, Medalia proposed in 1973, "occluded rubber volume" as the actual filler volume in Equation (6.24), for elasticity, as a procedure to improve the agreement between theory and facts [2–8]. Occluded rubber was defined as the elastomeric matrix which penetrated the void space of the individual carbon particle aggregates, partially shielded from deformation. Occluded volume was estimated from aggregate morphology and void-volume measurements where dibutylphtalate absorption (DBPA) was found as a dominant parameter.

(b) Theories on breaking filler–rubber network interactions and on molecular slippage on filler surface belong to the second approach. Bueche proposed in 1960, a mechanism to explain stress softening, based on the simple concept of the breakage of network chains, or their attachments to filler particles in the course of extension [2–4,6–8]. It is similar to Blanchard's idea to explain Mullins effect by formation of strong and weak linkages between network segments and particle surface sites. The interparticle chains usually have a distribution of lengths and the shorter chains will rupture first, at small elongations. As the stress on a chain before rupture is large, it contributes greatly to the stiffness or modulus. Chains broken on the first stretch will not be able to affect stiffness on the second stretch and softening will occur. Bueche attributed stress recovery to the return to a random distribution of interparticle chains by the replacement of the original broken chains by other similar chains but that all will consume a lot of energy and degrade the material very soon. Dannenberg elaborated in 1966 (but a number of others have proposed similar mechanisms in the period 1944–2000) reinforcement by chain slippage on filler surface, which attempts to explain reinforcement of filled rubber against break [2–8]. According to the mechanism, the surface-adsorbed network segments move relative to the surface, accommodating the imposed stress and preventing molecular rupture. As a consequence of the slippage, the stress is redistributed to neighboring molecules. Stress redistribution results in molecular alignment and increased strength. This process first absorbs strain energy and then dissipates it by slippage as frictional heat; in this manner it acts as a major source of hysteresis. Energy requirements for the process are thus increased by dissipation of strain energy as heat. Although molecular mechanisms proposed in theories of Bueche and Dannenberg are quite different, the main explanation of phenomena as recombination processes at the filler surface are very similar.

(c) The third approach is filler–filler network theories and cluster–cluster aggregation [2–8]. During 1963–1970 Payne found sigmoidal decline of real part of dynamic modulus, from a limiting zero amplitude value to some higher amplitude, as a typical behavior of reinforced rubbers. He explained it as a result of breakage of physical (van der Waals) bonds between filler particles. Kraus elaborated it by a simple model and a numerous authors extended the model since 1970s up to now as the breakdown of secondary aggregate network, formed by fillers through the whole material. In earlier studies, the filler network has been interpreted by percolation theory developed for polymer network formation and as a fractal cluster structure [16,39].

(d) The fourth approach represents formulation of models on the basis of data from detailed investigation of filler structure and properties. The data obtained experimentally for fillers are correlated with reinforcement parameters, for example, moduli, obtained from typical compounds with most important rubbers. It has been a pragmatic way to find the best additives for industry, hoping at the same time that such correlations will point dominant factors in reinforcement mechanism. In 1930s, when the first x-ray diffraction measurements provided data for formulation of the first CB-primary particle model, a huge pool of data from different kinds of measurements were presented in more than thousand papers [2–4,56–58]. In short, the main sources of information in that approach are data from x-ray diffraction, electron microscopy, surface chemistry, inverse gas chromatography, gas adsorption calorimetry, electrical conductivity, nuclear magnetic resonance (NMR), electron spin resonance spectra (ESR), secondary ion mass spectroscopy, scanning tunneling microscopy, atomic force microscopy, and even neutron scattering. Besides the practical importance of polymer industry consuming fillers and development of new filler grades and even new production technologies, a univocal conclusion about reinforcement mechanism is yet to be obtained.

VI. THE UNSOLVED PROBLEMS AND SOME PARADOXES

Many of the models described in a previous section include ideas implicitly or explicitly on filler particle contribution to reinforcement as additional junctions to the chemical cross-linkages of covalent network. The main idea is that bonding of chains to filler surface increases the total density of a network, influencing the increase of modulus but, at the same time with much less decrease of dynamical degrees of freedom of polymer chains. It provides possibilities for stress relaxation, for example by "slippage" (according to Dannenberg) for chains or "saltating" on the filler surface and prevents the formation of cracks. The crack formation, according to the general theories of Born and Griffith, initialize fracture of the material. At the same time, friction due to slippage produces dissipation of energy in agreement with hydrodynamic theory. Although filler particles are covered with adsorbed "bond rubber," breaking and recombination of filler–filler particle contacts (as proposed in filler-network theories) is possible, due to the two levels of their structure: the primary particles and their aggregates. The latter are present in the rubber matrix. Reversible interactions between filler particles, explaining Payne's experiments, can be rationalized as interactions between primary particles inside aggregates (or between aggregates inside agglomerates). The breakage of chain-to-filler contacts at high strains (from Bueche model) can be added to the breakage mechanism. This mechanism predominates at high extensions, leading to the final fracture. All these conclusions lead to consistent total mechanism of the process and provide a background for formulation of a general theory of rubber reinforcement.

Unfortunately, experimental results on interface structure and dynamics, obtained by modern techniques, give the evidence for effects opposite to that proposed in previous theories. As already mentioned in Section II, chain dynamics abruptly changes at the interface layer. Analyses based on NMR, ESR, and dynamical measurements indicate that effective T_g of adsorbed rubber chains increase even up to 150°C. It follows that at room conditions, hard and immobile layer forms as an armor at active filler surface. Logically, it makes background for further growth of bond rubber layer around the filler particles that are impossible to remove, even with good solvents. Thus the models of slippage (and saltating) on the filler surface become irrelevant at once. Indeed, according to Einstein model, viscosity is a measure of energy dissipation but rubber elasticity modulus is, according to de Gennes and percolation theory, a parameter with the opposite meaning [16]. That is, a measure of connectivity of the system. So, hydrodynamic interpretation at once becomes a paradox. Moreover, according to results of Leblanc, the connectivity around filler particles increases with time. He found a significant development of bond rubber within a month after the preparation of different rubber compounds [65]. It is a completely overlooked or ignored effect in all previous theories. Of course, some rationalization of the new results in the frame of previous theories can be found by looking for specific interactions in the course of the glassy layer formation. Following that logic, two main sources can be expected to produce differences in reinforcement due to interactions at the surface of fillers: different number of active centers on the surface and its different roughness. A number of early investigations have considered these issues in detail. But, the results are again surprising. The hypothesis about differences in surface-active center population between CB grades is rather old. A decrease of bond rubber and reinforcement ability after graphitization of some active fillers is strong support to it. Donnet, Wolff, Wang, and collaborators have shown by inverse gas chromatography at infinite dilution that activity of CB increases parallelly with the surface extent. However, the active-site species, which are evidenced by this technique, are similarly present on the surface of all CB grades [56–58]. They formulated the intrinsic surface activity factor as the ratio of adsorption free energy (see Equation (6.3)) of a given probe ΔG_S^0 to that of an alkane (real or hypothetical) whose surface area is identical with that of the given adsorbent $(\Delta G_S^0)_{alk}$: $S_f = \Delta G_S^0/(\Delta G_S^0)_{alk}$. They demonstrated very clearly that S_f is almost constant for all industrial CBs. It provides quite a new view to numerous papers and some of the previously listed theories based on CB surface chemistry for reinforcement explanation. Of course for systems such as silicon rubber and silica filler, chemical bonding of chains provides a good fit with the simple model of united network, but this is more an exception than the rule. The numerous kind of groups found on the surface of CBs obtained before furnace process has been developed (e.g., quinone, lactone, carboxylic) with, for example, 3–8% oxygen for channel black, are not valid for models describing present CB systems, with total oxygen content around 0.1%. Now the total amount of groups on the surface of furnace black is found to be only 10–40 milliequivalent per 100 g of CB. This contributes a very small amount (~2%) of surface coverage by chemical groups. The same can be said for microporosity, which seems to be absent from the surface of furnace-type CB [2–8].

Very interesting measurements of CB surface properties have been obtained by Raman spectroscopy, scanning tunneling microscopy (STM), and neutron scattering. The last two methods measured in fact pertain to the roughness of the surface. Gerspacher et al. [66] have shown for a large series of CBs via small-angle neutron scattering (SANS) that CB fillers exhibit a fractal surface on length scales from 6 to 17 nm. But, they found that all investigated blacks from the N 100- up to the N 700-series have the same fractal dimensions $D_f = 2.4$ independent of the CB grade and the size of primary particles. Kluppel et al. compared surface roughness of furnace blacks to that of graphitized furnace blacks by means of gas adsorption techniques [60]. They found for all untreated furnace blocks only one linear range with the fractal dimension $D_f = 2.6$. According to their results, the graphitization diminished only for the roughness of the surface on length scales below 1 nm, while on large scales D_f remains unchanged at $D_f = 2.6$. Instead of obtaining a much clearer situation with improved experimental techniques, we face two paradoxes. In spite of the obvious and technologically very important differences in reinforcing activity of different CB grades, no difference is obtained in their surface activity and roughness, as well. Even with NMR measurement there are some disagreements. Early studies by NMR and dynamic testing of Fujiwara et al. on NR at 60 MHz and Smit for SBR in forced shear vibration of 2.4% amplitude at 8.5 Hz indicated immobilized interphase layer thickness of 5.0 and 2.0 nm, respectively. But Kraus found at the same time "no significant layer of immobilized rubber" in both dynamic and NMR studies [2–8]. McBrierty et al. found later that T_2-relaxation NMR data are consistent with existence of two layers: a tightly bond layer in the immediate vicinity of the filler particle, in addition to a loosely bond component [59,68]. The inner layer accounts for about one-fifth of the total bond layer thickness but interpretation of results is model-dependent. Obviously, for further progress in rubber reinforcement a new model and theoretical interpretation is necessary.

VII. THE THEORY OF RUBBER REINFORCEMENT — A NOVEL APPROACH

A. GENERAL PHENOMENOLOGICAL THEORY

1. Qualitative Description of the Model

Trying to explain the mechanism of rubber reinforcement, we must consider three main issues that follow from previous analyses:

A. What are the parameters of filler structure we have to account as responsible for reinforcing activity of different types of fillers?

B. If fillers act as additional fixed cross-links of the rubber network, why reinforcing effects are not alike to gum network with additional cross-links?

C. Is it the essence of rubber reinforcement in (some kind of) superposition of polymer matrix modulus with moduli of some much stronger and stiffer substance (but with preserved rubber-like deformability)?

In the list, issues are defined in pragmatic engineering terms, but other aspects will be considered soon. Let as start from the third, the most general problem. Practically all previous theories, explicitly or as a silent aspect (to be accepted per se, implicitly), include condition that filler modulus must be higher than the polymer one. For example, in the most recent review of "the-state-of-the-art" in rubber reinforcement theories, it can be found that: "A necessary condition for rubber reinforcement by filler clusters is the rigidity condition $G_A \gg G_R$, where G_A is the elastic modulus of filler cluster and G_R that of the rubber. This is obvious because the structure that is weaker than the rubber cannot contribute to stiffening of the polymer matrix" [60]. Following this logic we have to look at how the strength of the reinforcing material can be added onto the mechanical strength of a polymer matrix or some composite matrix in general. We can propose three solutions. Two of them are well known:

I. The existence of a filler–filler particle network that contributes in parallel with covalent polymer network to total modulus.
II. The existence of a united polymer network of chains and filler particles with higher density.
III. We shall propose now a third solution, possible for finely structured fillers. It is a temporary structure arising as the result of force balances in the system interlayer. Its character is more in changes of dynamics than in space organization of the system, and its main actors are polymer chains, not the filler particles.

Let us consider the case III. In the case of percolation, that is, if transition layers in a polymer matrix, around embedded active filler aggregates, connect (or touch) each other, the effect is significantly increased, extending itself through the whole macrosystem and resulting in reinforcement effects. So, polymer chains not filler particles, are the main actors of reinforcement. The connections between layers are strain-dependent, so different kinds of reinforcement effects arise.

After the general presentation, let us consider these effects one by one. Naturally, hyperelasticity as a dominant dynamic property is saved in our system. The model can easily be extended to some biopolymers and biological systems with polymers, but this will not be discussed here. We have seen that the morphology of filler aggregates, in many cases, can be correlated to reinforcement parameters, for example, to the increase of moduli. Free space between primary nanoparticles embedded into the aggregates is the locus of strong fields of uncompensated forces, coming from nanoparticle structure. Let us call this space as "macropores" to simplify the terminology. But, it should not be misinterpreted with classical pore effects in physical chemistry. In Section VI, we have seen that for some active fillers aggregates micropores do not exist at all. Fields of uncompensated forces in nanosystems described earlier, can vary much in density in such macropores, depending on aggregate architecture. The polymer chains entering the macropores before network formation will significantly change their dynamics, but saving much random space organization (which will be discussed later). The closest chains of the network will adsorb at the filler surface under strong influence of surface forces, but more distant layers will freeze their dynamics, in some extent and become denser. In such a state, they will be networked during vulcanization process. Their state is much more similar to an amorphous networked segments in some plastics than to the bulk of their own network. It is well known that such plastics have much higher modulus than rubber materials. But in macropores this is not the equilibrium state at low temperature. In macropores these exists a gradient of structures produced by superposition of filler forces with forces influenced by local network structure and forces from connections with bulk network matrix. If the network matrix deforms under extending stress, some of chains in that gradient material will change their conformation. Those chains will extend much more in outer layers than close to the bottom of the pore. The whole polymer content of the macropore cannot exit from the macropore, and come back to the bulk of the network. It has an almost solid, rigid surface and soft core connected with bulk of the network. This presentation nanolevel structure of rubber network–filler gives the explanation of several phenomena well known for reinforced rubbers:

1. It is well known that for some filler–polymer pairs, bond rubber content is more than 20% of total polymer [2–8,65]. On the other side, from NMR (as we have seen in Section VI) the hard

"glassy" layer is only $\sim$2-nm thick. The gradient layer field of interactions around filler particles can immediately help to explain so high an extent of bond rubber in the cases described. Moreover, its conformational dynamic character provides logical explanation for time evolution of bond rubber content up to the equilibrium.

2. The new model can immediately rationalize the differences in reinforcing activity exhibited by some very similar types of CB, discussed earlier. Especially because of the recent results, indicating no significant difference in surface activity and fractal character of their structure. Modern, very sophisticated methods for estimations of both types of surface parameters describe the surface at sub-nanolevel. So one cannot see primary particle space configurations (responsible for differences) in relief of aggregates (e.g., macropores). In simple terms, if the macropores are of appropriate shape and size, for example, very shallow with flat bottom, concentration in the local internal field domain will be much lower. The external forces transferred by the network continuity from the bulk can in this case much easily draw out the dominant amount of chains from macropore; connections decrease in the local layer.

3. Our model immediately provides explanation of the Payne's effect. It is so much elaborated in the literature by the model of filler–filler interactions. As described earlier (Section V, issue c), many authors have followed an opinion that so high modulus of filled rubber at low deformations can only be described by interaction between filler particles having much higher modulus than polymer. It suddenly decreases at some critical strain amplitude (we shall call it the first threshold). Some of them, go so far to believe that filler particles make parallel network with polymer covalent network, throughout all the final product (as we have seen in Sections V and VI). But, this is in complete contradiction with the existence of bond rubber layer around each particle. Even if there is some hidden way that filler particles are connected to each other, in spite of bond rubber layer around aggregates (as discussed in previous sections), another contradictory appears. The chain of particles, which is a part of such a filler–filler network (making a kind of rigid armature for the reinforced material), must be completely broken at high deformations and reinforcement should disappear. But, just at very high deformations filled rubbers expose the highest degree of reinforcement. So, the concept of filler–filler network, parallel to the polymer network, must be rejected. As mentioned in Sections V and VI, some authors believe that Payne's effect describes separation (and again association) of primary particles (or their sub-aggregates) inside filler aggregates. Now, the outer surface of the aggregates makes with bond rubber a kind of additional junctions in combined polymer–filler network, and in that way reinforces the material. Even if such a "soft" disconnection is possible (in spite of armor of hard glassy polymer layer around aggregates, and obvious rigidity of aggregates), already some simple calculations point two substantial contradictions in this model. Both the force intensity and strain amplitudes at the first threshold (as it is described, by the filler–filler interaction model) are not in agreement with calculations. First, the forces which would be expected to arise from inter-aggregate interactions are too low to explain the magnitude of the modulus at low strain. Even more problematic than the actual value of modulus is the strain at which the change in dynamic modulus occurs (the first threshold). Typically, the change in modulus for filled rubbers occurs near 6–8% double-strain amplitude. The change in aggregate separation must be large enough to accommodate this strain, and for primary particles substantially larger than the range for effective surface forces between primary particles. The dynamic properties of CB in oil have often been quoted as verifying the inter-aggregate model. However, for these blacks in oil systems, the change in modulus occurs near 0.01% double-strain amplitude. The value for the strain at the first threshold for the carbon-black-in-oil system occurs at a strain, consonant with the expected change in surface force with separation. The difference in strain at the threshold is a critical difference in strains between the two systems. The rubber data are not consistent with the expectations of the inter-aggregate model, while the carbon-black-in-oil system is completely consistent with the model. Obviously the model is not wrong physically but it is not applicable for rubber reinforcement process.

Let us now get back at our model and apply it to conditions corresponding to Payne's experiments. Under low deformations, only the outer part of the gradient reinforcing layer around the filler aggregates will be active, because it is directly exposed to large and inert bulk network, transferring stress. The chains in the outer part will be under influence of two kinds of forces: outer deformation forces and forces from interactions in the gradient layer, which are dominated with the field caused by fillers. Because of this the chains in the outer part will be highly extended, producing very high average modulus at low deformations corresponding to the upper plateau of Payne's diagram. The mechanism is similar to self-reinforcement in plastics, by local yielding, where in front of the crack tip highly extended, load-bearing fibrils span the gap. It is also in agreement with predictions from Born and latter Griffith theories about decay of materials. But, here bulk part of the polymer is much softer and starts to deform on that level of stress. It takes then over the large part of stress. The same is valid for the soft core inside the macropores. In fact that core will be first activated, mainly due to direct connections to the outer part of the layer and the lowest inertness. That activation of the soft core is abrupt. In simple terms, in most cases, the soft core will go out of the macropores. This corresponds to a sudden decrease of modulus in Payne's experiments in the 6–8% strain range. This range fits well with the deformations of polymer layer but not of hard filler aggregates. It is the interpretation of the first threshold. In addition two facts support our mechanism. The same effect of surmounting the energy barrier, as realized by outer action of mechanical forces at characteristic 6–8% strain at room temperature, is possible to realize at 0.1% strain at 90–100°C. This can be easily understood from the changes in the gradient layer resulting in appropriate dynamical transition with temperature. Activation energy of the process obtained from log G_M' versus $1/T$ plot (where G_M' is real part of dynamics modulus and T the temperature) corresponds to the level of van der Waals interactions of network polymers [57–60]. After this transition, the bulk network will extend easier than the more compact and rigid gradient layer, which is under influence of the filler.

4. Let us now consider reinforcement in the middle range of strain. The percolated gradient layer with much lower chain conformational dynamics and higher compactness will serve here as a kind of armature in the rest of bulk rubber network. This will shift the modulus of the total material to higher values. But, the elastic behavior of the gradient layer is much the same as the rest of the system (although it is slower and stiffer). Material flexibility and rubber-like elasticity of the material are thus saved. On the other hand, for gum networks with increase of cross-linking density, fluctuations of network segments at medium deformations will decrease significantly. It leads to a different behavior (as indicated earlier). The dynamic character of this armature based on chain conformational and orientational adjustment, conforms the important fact of significant loss of reinforcement, if the material is biaxially deformed. At the same time due to similarity of gradient layer and bulk network behavior at medium deformations, more and more chains at the layer border will escape from the filler's influence and change their dynamics to the pattern of bulk network behavior. In this way gradient layer will decay with deformation, as described by Equation (6.22). At the minimum of Mooney–Rivlin curve the abrupt change arises. This is the second threshold.

5. At the second threshold, the inner, hard part of the gradient layer is directly activated in reinforcement. It starts to carry significant part of the load. But it is glassy and strongly bonded to aggregate surface. Because of this the role of aggregates is changed. The filler particles behave now as additional junctions of united network. Of course there is no chain slippage or saltating on filler surface. But, the network is now highly extended and fluctuations of particles are possible. It enables particle orientation in space and stress relaxation. Large particles are a good barrier to crack propagation as well. It is also in agreement with theories of Flory, Erman, Ronka, Mark, and others [7,32,63,64] on transition from affine to fantom behavior of the gum network. The second threshold appears for almost all filled rubbers in the range of 300% strain modulus (in terminology used in industry). Interpretation of this parameter for reinforcement is however confusing, especially when the modulus at 300% is compared with lower strain moduli for CB of different grades.

6. At very high deformations, the main contribution to reinforcement comes from the change of network chain statistics due to finite chain extensibility, but supported here with filler particle as

additional junctions in united network. The change of chain conformational statistics due to finite extensibility of chains is a well-known phenomenon, already elaborated in the theory of Kuhn–Wall–Flory. But here, the change disconnection from the filler surface is possible, due to high stress. That leads (together with chain breakage, similar to Bueche mechanism) to total failure of the material, what is the third threshold of reinforcement.

7. In the model proposed, answers to all the three issues (listed at the start of this section) are offered. The model describes the change of behavior of the gradient layer between filler particles and bulk network with deformation, as a continual process. In fact, it is the change of chain conformational statistics under the influence of two force fields: outer field causing deformations, and the field of uncompensated forces from the filler nanoparticle surface. In the quantitative formulation of the model, in Part B of this section, statistical mechanical model of the system will be proposed. Here it should be noted that the concept of necessity to use a material with much higher modulus for reinforcement of composites is not indispensable in physics. A new concept of network conformational reinforcement by (dynamical) self-assembled chains into gradient layer is proposed. The quantitative formulation of that process will be proposed also in the Part B of this chapter. But, here it should be pointed to one experimental confirmation of that statement, that is, significant reinforcement of rubber networks is possible. With polymer particles of appropriate nano-architecture, this is possible [67]. However, the influence of nano-architecture of filler particles deserves to be elaborated in more detail.

2. Fractal Geometry, Thermodynamics, and Conformational Statistics

Active filler aggregates are fractal objects, and the fractal dimension of their surface has been determined by different techniques, as described in Section VI. However, many aggregates are mass fractals, due to the architecture of nanoparticles embedded in them. On comparing the (already numerous) results of different microscopy methods, applied to both CB and silica active fillers, we can see that the particles can be sorted in several types similar in shape:

- Fumed silica shape
- CB N 220-like shape
- Precipitated silica shape

More careful analyses of data show [69] that the main difference between that types can be described using resolution-dependent volume (see Equation (6.10)) of an aggregate described by

$$V(l) \cong V(0) + \zeta \, l^{D_{f\delta}} \tag{6.26}$$

where ζ is a constant and $D_{f\delta}$ an exponent quantifying fractal properties, which can be calculated from

$$\zeta = \lim_{l \to 0} \frac{\ln[N_V(l)l^d - V(0)]}{\ln l} \tag{6.27}$$

where $N_V(l)$ is the number of d-dimensional balls needed to cover the structure. In other terms active filler aggregates belong to types of "thin" (as fumed silica) and "fat" (as precipitated silica) fractals. The main difference in grades of active filler comes from the position on the scale between those two types. According to preliminary results of the present author, it can be quantified using Equation (6.26) and Equation (6.27) and appropriate combination of data from BET method, DBPA method, and SAXS measurements [59]. But it was overlooked in all previous analyses and theories.

In more rigorous terms of fractal geometry, earlier formulations about filler aggregates can be understood as multi-fractal object definitions and prefactor ζ as lacunarity of corresponding fractal part of the system [38,39]. For thin fractal types $V(0) \to 0$ and

$$V(l) \sim l^{d-D_f} \tag{6.28}$$

where $D_f < d$ (see Equation (6.11)). The prefactor ζ can be understood as a measurement of local field scaling in accordance with Equation (6.1). Both filler aggregates and polymer network can be understood as fractal objects. As described in Part 1 of this section, during formation of the network (below percolation threshold of covalent network) a set of sub-networks is formed. In the same way, we can present formation of the gradient layer around filler aggregates, for example, at different concentrations of filler or at different deformation levels. That sub-networks can be treated as fractal aggregates that percolate under appropriate conditions to continual dynamic armature, reinforcing the system. There is some analogy of such subsystems (of metastable gradient chain interactions) with polymer blend networks. Both systems can be presented as dispersions of fractal clusters, with special interactions on the border [59,69]. But, here we have combination of filler fractal objects embedded to gradient conformation fractal layer, that is, a hyperscaling in terms of Section III and Equation (6.1), as well. It is a quite complex organization of the system. But, with combination of the geometry of the system and scaling properties, as we learned from Flory, let us turn to thermodynamic interpretation.

As already mentioned in Section IV, the relevant parameter for analyses of reinforcement can be the elastic energy accumulated by the system during deformation. The part of the network chains belonging to the gradient layer, accumulates (according to that physical presentation) some amount of additional elastic energy by reinforcement of the system. This additional accumulation of elastic energy can be the background for quantification of our model. On the other hand, the system as a whole can be understood as a dynamical structure instead of detailed analyses of its space organization. We use here the term "structure" for both the space configuration of the system and dynamics of it. For presentation of the conformational dynamics in the gradient layer, it is more convenient to combine Fokker–Planck equation with Liouville–Riemann derivative, than to use the hyperscaling method described in Section III. The formulations in Fokker–Planck equation provide better tools for description of filler influence upon the gradient layer. The fractal character of interactions inside macropores due to geometry of aggregate nanoparticle architecture can be described by the dissipative coefficient D_L (see Equation (6.15)) changing distribution of chain conformations. But, the second, independent influence upon this distribution, comes from the fractality of the local polymer network itself. It can be described in several ways, but we shall use in the next part of this section Liouville–Riemann derivative of the conformational distribution function. It is very convenient for description of viscoelastic properties of polymer systems.

Attraction of some chain parts, for example, some chain groups or double bonds, to the active filler surface changes significantly its number of possible conformations and configurational distribution in the layer connected to filler surface. But that influence of filler fractal surface on the polymer network, which is itself a fractal (but of different character), will decay slowly through the gradient layer, like a memory function. Because of this Liouville–Riemann differential is a very convenient approach to changes in conformational distribution.

We have now enough elements for formulation of the quantitative part of the theory. Still one very fundamental question stays open: what is the main source of forces coming from the nanoparticles, to influence chain conformation? It is much more general an issue than the problem of rubber reinforcement and asks for the extra space and extra analyses [69,70]. A brief glance at this attractive issue, will be provided in the next part.

3. Quantum Mechanics and the Basic Statements of Conclusions

Raman spectroscopy confirming indications from the tunneling electron microscopy (see Section VI) showed that the surface of CB nanoparticles is characterized by the existence of crystallites and amorphous domains. In furnace-black production, the reaction time is very short and a thermodynamic equilibrium is never reached. Amorphous carbon regions in that metastable state, of high entropy, can be explained by incomplete dehydrogenation.

In a similar way, chemically induced dimmer configuration prepared on the silicon Si(1 0 0) surface is essentially untitled and differs, both electronically and structurally, from the dynamically tilting dimers normally found on this surface [71]. The dimer units that compose the bare Si(1 0 0) surface tilt back and forth in a low-frequency ($\sim$5 THz) seesaw mode. In contrast, dimers that have reacted with H_2 have their Si—Si dimer bonds elongated and locked in the horizontal plane of the surface. They are more reactive than normal dimers. For molecular hydrogen (H_2) adsorption, the enhancement is even 10^9 at room temperature. In a similar way, boundaries between crystallites and amorphous regions seem to be active sites of chain adsorption on CB surface. CB nanoparticles can be understood as open quantum systems, and the uncompensated forces can be analyzed in terms of quantum decoherence effects [70]. The dynamic approach to reinforcement proposed in this chapter becomes an additional support: in epistemology of it, and with data from sub-nanolevel.

From the proposed mechanism and methods of solution for issues of interactions of nanostructured filler with polymer network we can conclude:

1. Rubber reinforcement can be described as an additional accumulation of elastic energy by the system, due to the contribution of a fraction of network chains, self-assembled to gradient interlayer around active filler particles.
2. This self-organization produces dynamical system changing with deformation (or time, as well).
3. Polymer network chains, not the filler particle chains, are the main actors of reinforcement. The role of filler nanoparticles is passive in reinforcement. They produce field of forces influencing conformational changes of gradient layer chains.
4. Influence of the gradient layer on reinforcement increases essentially at percolation of particle-coating layers (connected by tacking or intersection).
5. The local layers around filler aggregates can be understood as fractal subsystems.
6. The difference in activity of filler types and grades can be quantified by the special fractal classification in the scale between thin and fat fractals.
7. Rubber reinforcement mechanism can be explained as continuum process but with three thresholds. In simple terms the thresholds can be described as:
   - The exit of soft network core from the macropore at 6–8% deformations (Payne's effect).
   - The transition of bulk network from Gaussian to Langevin statistics accompanied with transfer of the part of a load to inner hard polymer layer at the filler surface (in the range of Mooney–Rivlin curve minimum).
   - Decay of filler–carbon network connections at the start of material destruction (for very high deformations or high frequencies).

B. Definition of the Model

1. Definition of the System and its Transitions

We consider a typical covalent polymer network in the rubbery state at room conditions, consisting of ν linear chains whose ends are joined to multifunctional junctions of any functionality higher than 2 (i.e., more than two chains are attached at each junction). The chains that join pairs of junctions are long, very flexible, and of equal size, all of which can be easily related to real situations described in previous section (e.g., for chains of size equal to average size for many real systems of 400 bonds where, with high flexibility of random network, some differences in size are often not important). It follows that for tetrafunctional junctions, the concentration of cross-linkages is in typical order of 3×10^{19} junctions per cm^3. Let positions of the junction j be determined by radius vector $\mathbf{R}_j$ and distances between junctions i and j by chain vectors $\mathbf{r}_{ij} = \mathbf{R}_j - \mathbf{R}_i$. The system contains active filler in the optimal load region and dispersion for reinforcement (e.g., active CB of 0.2 volume fraction), as also described in the previous section. The grand partition function of the filler–polymer network can be described as

$$Z_D = \sum_{N=0}^{\infty} z^N Q_N(V, T) \tag{6.29}$$

where N is the number of network configurations and Q_N the factor for state N, with z as its degenerative term. We can sort the subsystems of the grand canonical ensemble into two classes: those under the influence of the filler and the "bulk polymer" subsystems. For the time being, we shall split states to $N = N_1 + N_2$ where N_1 is the number of chains in one subsystem. Such a subsystem can be interpreted as set of chains. So our treatment considers the probability of cooperative rearrangement in a fixed subsystem (k) as a function of its size N_k. Then

$$Q_N(V, T) \equiv \int \frac{d^{3N} p \, d^{3N} q}{N! h^{3N}} e^{-\beta H(p, q)} \tag{6.30}$$

where Q_N corresponds to the volume in a Γ-space with $6N$ dimensions and impulses p and space positions q as coordinates. H is the Hamiltonian of N state and h factor making Q_N dimensionless, with $\beta = 1/kT$ as well. Our definitions can be easily translated into common terms of general statistical mechanics, interpreting also N_k as a number of a rigid links forming the flexible chains. In the subsystem, they can be understood as the particles. Then, N_k at the same time represents possible configurations of the subsystem (accounting the appropriate degeneration of the states with z). For the probability density of conformations of 1, we can write, in principle, to describe the influence of the rest system on it:

$$p_c(p, q, N_1)_1 = \frac{Q_{N_2}(V_2, T) e^{-\beta H(p, q, N_1)_1}}{Q_N(V, T)} \frac{}{N_1! h^{3N_1}} \tag{6.31}$$

where $N \to \infty$. According to the general principles of statistical mechanics, we can write Equation (6.31) as

$$p_c(p, q, N_1)_1 = \text{Const } e^{-[A(N_1, V_1, T)]/kT} e^{-[A(N-N_1, V-V_1, T) - A(N_1, V_1, T)]/kT} \tag{6.32}$$

where $A(N, V, T)$ is the Helmholtz free energy of the system. For $N \gg N_1$ and $V \gg V_1$ we may use

$$A(N - N_1, V - V_1, T) - A(N, V, T) \approx -N_1 \mu + V_1 P \tag{6.33}$$

where μ and P are the chemical potential and pressure of the part of the system external to the subsystem of the volume V_1, respectively.

The second factor on the right-hand side of Equation (6.32) can be understood as transition probability changing conformational states of 1, due to the influence of other chains of the system. In general

$$P_{tr}(\{X\}, t + \Delta t) = \int P(\{X\}, t + \Delta t | \{Y\}, t) P(\{Y\}, t) \, d\{Y\} \tag{6.34}$$

where $\{X\}$ and $\{Y\}$ are the sets of system coordinates in time t and $(t + \Delta t)$. Needless to explain that in unstrained rubber network, only a restricted number of surrounding chains can influence conformations of the considered chain. So we can write for transition probability:

$$P_{tr} = \text{const}\, e^{-n(\Delta x)\Delta\mu/kT} \tag{6.35}$$

where $n(\Delta x)$ is the number of subsystems making cooperative rearranging region, which upon a sufficient fluctuations in energy initialized by some external influence (e.g., of the filler) can rearrange into another configuration, independent of the rest of the network. This is exactly what we described qualitatively in the phenomenological part of the theory as a change of conformational dynamics of chain layer around the filler particles, causing apparent glass transition in the layer. But, following this logic in formulation of the partition function of the chain, we must extend the model with formulation of probability functions, including networking of the cooperative domain and then, to the influence of the hard domain to elastic behavior of the network as a whole. We can do it following self-consistent field method, the principles of which have been described in Section III.

In the first approximation, elastic free energy of a network can be presented as the sum of elastic free energies of the individual chains [1]. It is in good agreement with experiments for small deformations and we shall apply it for the time being, for the stiffer domains as the subsystems of v_k chains. To the subsystem corresponds configurational probability function $W_n = W(p_c)$. Existence of stiff domains (i.e., subsystems) is induced by chain interaction with filler particles. It follows that distribution of the domains in space corresponds to the space distribution of filler aggregates in the material. So the total elastic effect in the system can be understood as a cooperative response of the bulk polymer network and reinforcing domains. It can be presented as convolution of distribution functions for those two parts of total system:

$$W^*(\Delta X) = \int_V W_n(\delta \bar{X}) * w(\Delta \bar{X})\, d^3 \Delta \bar{X} \tag{6.36}$$

where $\Delta \bar{X}$ and $\delta \bar{X}$ are the vectors of coordinates for balk network domains and stiffer domains, respectively. Isothermal, isobaric elastic deformation of the system will introduce changes in conformation of bulk and stiff domains, but in different ways. The configurational partition function is then

$$Z_D^* = \prod_j \left(\frac{\chi_j \eta}{\eta_j}\right)^{\eta_j} \tag{6.37}$$

where $\chi_j = W_n\, d\delta X$ and $\eta_j/\eta = W^*\, d\Delta X$. For the free elastic energy of the system, in accordance with general principles (see Equation (6.5)) we can write now

$$\Delta A_{el} = \Delta A_{balk} + \Delta A_{fd} \tag{6.38}$$

where ΔA_{fd} correspond to the contribution of stiffer domains and ΔA_{balk} is for balk network.

2. Probability Current and Network Response

The change of conformational distribution function of the network chains in the vicinity of active filler aggregates, described in Section VII.A.1, can be formulated as probability current:

$$J_W = \frac{\partial}{\partial x}\left(\frac{U(x)}{m\xi_{D_f}} + D_L \frac{\partial}{\partial x}\right) W(x, t) \tag{6.39}$$

where $U(x)$ is a potential of attraction field transferred from active filler surface to the gradient layer around filler particles, ξ_{D_f} the friction coefficient describing the resistance of chains to conformational transitions, and m represents average mass of network segments. The fractal character of macropores (see Section VII.A.1) can be described by diffusion coefficient D_L. For the change of probability current due to step transition (as are the three threshold transitions described in Section VII.A.1), we can use additional term r_{TR}:

$$J_W^{TR} = \frac{\partial}{\partial x}\left(\frac{U(x)}{m\xi_{D_f}} + D_L\frac{\partial}{\partial x}\right)W(x,t) + r_{TR} = L_{FP}W + r_{TR} \tag{6.40}$$

where L_{FP} is the Fokker–Planck operator. The change of conformational distribution function of the gradient layer chains, due to fractal character of the gradient network, can be described by $D_t^{1-D_f}W(x,t)$, where $D_t^{1-D_f}$ is fractional derivative (see Equation (6.16)). Then the total change is described by

$$\dot{W}(x,t) = D_t^{1-D_f}L_{FP}W \tag{6.41}$$

This equation can be used for estimation of addition free energy A_{fd} according to Equation (6.38) as contribution of the gradient layer to the reinforcement of a system. Of course time variable t can be interpreted in terms of continuum approach described in Section III as number of steps, etc.

3. An Engineering Approach and Scaling Transitions

We can use the criteria approach commonly used in chemical engineering for flow analyses in reaction system by application of Fokker–Planck equation, also for analyses of conformational distribution function changes under some characteristics conditions. A couple of examples will be given here.

Let us define reduced variables of space and time $[X] = x/L_{sys}$ and $[t] = t/\tau_{ve}$ where L_{sys} is some space dimension characterizing the system as a whole, and in the same manner τ_{ve} is a characteristic relaxation time for the system. From Equation (6.39) and Equation (6.40) we can obtain probability criteria analog of Peclet (Pe) and Damkeller (Da) criteria, $Pe_W = L_{sys}^2/D_L\tau_{ve}$, $Da_W = \tau_{ve}/\tau_{dcy}$. For boundary conditions $d[W]/d[t] = 0$ we obtain from Equation (6.39);

$$[W] - Pe_W^{-1}\frac{d[W]}{d[x]} = 1 \quad \text{for } [x] = 0 \tag{6.42}$$

$$\frac{d[W]}{d[x]} = 0 \quad \text{for } [x] = 1 \tag{6.43}$$

Because of discontinuity for $[x] = 0$ in Equation (6.42) let us use additional boundary conditions:

$$[W] = 1, \quad [x] = 0 \tag{6.44}$$

From Equation (6.43) and Equation (6.44), we obtain immediately the possibilities for analyses of the two characteristics states of our system. If $[x] \to 0$ we are close to second threshold and if $[x] \to 1$ we are close to the first threshold, as described in Section VII.A.1.

C. Particular Solutions

Let us suppose that W is the Gaussian distribution commonly used for rubber networks. The change of the network conformational distribution function in the gradient layer can be obtained by Equation (6.41). In the equilibrium of external forces and filler field forces at some extension state in the range between the first and second thresholds, J_W will be constant. For that stationary state we can write

$$\frac{U(x)W_{st}}{(m\xi_{D_f})} + D_L d(x^{-\delta}W_{st}) = C_0 \tag{6.45}$$

by solution of Equation (6.45) we obtain

$$W_{st}(x) = A_c^{\delta}x^{\omega}\left\{C_0'\exp\left\{-\left(\frac{\int x^{\omega}U(x)dx}{m\xi_{D_f}D_L}\right)\right\} + C_0\left[\frac{m\xi_{D_f}}{A^{\delta}x^{\omega}U(x)} - F_{\Omega}\right]\right\},$$

$$F_{\Omega} = e^{-\int x^{\omega}U(x)dx/m\xi_{D_f}D_L}\int e^{\int x^{\omega}U(x)dx/m\xi_{D_f}D_L}d\left(\frac{m\xi_{D_f}}{A^{\delta}x^{\omega}U(x)}\right) \tag{6.46}$$

For low ω and $C_0 = 0$, Equation (6.46) follows distribution of Boltzmann type:

$$W_{st} \propto \exp\left\{\frac{-U(x)}{k_B T}\right\} \tag{6.47}$$

At the same time diffusion coefficient satisfies Einstein relation $D_L = k_B T/m\xi_{D_f}$ connecting in that way two dissipative effects due to conformational changes and fractal geometry of filler surface producing that change. Our system obeys the generalized fluctuation-dissipation theorem. As a comment on the obtained distribution function we can say that from NMR experiments it is indeed possible to obtain Boltzmann distribution of energy sites (within the accuracy of the data) in the form $I = I_0 \exp(395 \pm 50/T)$.

VIII. EXPERIMENTAL IMPLICATIONS

The purpose of the model presented is to provide a better understanding of the essence of rubber reinforcement mechanism and its entities, not to give correlations for practical use in industry. Of course, once the mechanism is understood, it will be much easier to make a formula for practical correlation, but it is a long-standing problem with large pool of discrepancies accumulated and surprisingly growing number of issues. The simple answer could be that it is much more a fundamental problem than it looks out at the first sight, and to be handled only by an industrial approach. Unfortunately, the large number of accumulated data is just of that kind. Rubber materials are complex systems with many additives (producing different effects) that screen the influences considered in our model. The optimization of material commodity properties is commonly realized as a combination of several factor influences. But, for understanding the nature of those factors, it is necessary to consider them one by one, that is, to separate their effects each. It is just the opposite of the common industrial practice and approach to the optimization by formulation of commodity materials. Of course, when these factors are well understood, it will be much easier to make predictions of their united effects. But for the check of relations described in our model, an appropriate separation of factors of influence is necessary. For example, plasticizers have quite an opposite influence on rubber properties than on that of active fillers. The good balance between fillers and plasticizers is necessary for commodity material production and performances. But, check of the model relations on accumulated filler data on compounds with commodity plasticizers (varying in types, properties, amounts, etc.) is impossible. In more complex analyses, comparison of fractal properties of filler aggregates prior to multifractality classification is not appropriate. In the previous section, some calculations are presented, to check logical consistency more than compare the model with experiments. For this purpose, appropriate systems and measuring methods should be developed, or available data should be selected very carefully [56–69]. Some preliminary results are very encouraging.

IX. CONCLUSIONS

The new theory of rubber reinforcement is proposed as a qualitative explanation of the mechanism and the quantitative formulation of reinforcement dependence on system structure parameters. It connects the long-standing problem of different reinforcing ability between types of fillers and even grades of the same filler type (of practically the same standard structure parameters), with more general approach to the structure of polymer networks and nanostructured filler aggregates (of interest today for a number of other technologies and scientific fields). In parallel, a comparative analysis of accumulated experience and data about CB and silica nanostructure, with new polymer systems of interest, is made. According to the theory, not the filler particle chains but polymer network chain domains are the main actors of reinforcement. It is based on the contribution of self-assembled dynamical structures in the material under the influence of two field forces: forces from nanostructured filler aggregates and elastic forces of polymer network.

REFERENCES

1. Flory, P.J., *Principles of Polymer Chemistry*, Cornell University Press, Ithaca, NY, 1953.
2. Kraus, G., *Reinforcement of Elastomers*, Interscience Publishers, New York, 1965.
3. Bansal, R.C., Donnet, J.B., and Stoeckli, H.F., *Active Carbon*, Marcel Dekker, New York, 1988.
4. Bansal, R.C. and Donnet, J.B., *Carbon Black*, Marcel Dekker, New York, 1993.
5. Zimmerman, B.N., *International Rubber Science Hall of Fame*, Acron Rubber Division, American Chemical Society, Washington, DC, 1989.
6. Blow, C.M., *Rubber Technology and Manufacture*, Butterworth, London, 1982.
7. Plavsic, M.B., *Polymer Materials Science and Engineering* (in Serbian), N.K. Belgrade, 1996.
8. Plavsic, M.B., Popovic, R.S., and Popovic, R.G., *Elastomeric Materials* (in Serbian), N.K. Belgrade, 1995.
9. Holtz, J.H. and Acher, S.A., Polymerized colloidal crystal hydrogel films as intelligent chemical sensing materials, *Nature*, 389, 829–832, 1997.
10. Koltover, I., Salditt, T., Readler, J.V., and Safinya, C.R., An inverted hexagonal phase of cationic liposome-DNA complexes related to DNA release and delivery, *Science*, 281, 78, 1998.
11. Spasic, A.M., Electroviscoelasticity of liquid/liquid interfaces, in *Interfacial Electrokinetics and Electrophoresis*, Delgrado, A.V., Ed., Marcel Dekker, New York, 2002, pp. 837–867.
12. Lieber, C.M., Rueckes, T., Kim, K., Joselevich, E., Tseng, G.Y., and Cheung, C.L., Carbon nanotube-based nonvolatile random access memory for molecular computing, *Science*, 289, 94–97, 2000.
13. Kwiat, P.G., Berglund, A.J., Altepeter, J.B., and White, A.G., Experimental verification of decoherence-free subspaces, *Science*, 290, 498–501, 2000.
14. Thurn-Albrecht, T., Schotter, J., Kastle, G.A., Emley, N., Shibauchi, T., Krusin-Elbaum, L., Guuarini, K., Black, C.T., Tuominen, M.T., and Russell, T.P., Ultrahigh-density nanowire arrays grown in self-assembled diblock copolymer templates, *Science*, 290, 2126–2129, 2000.
15. Fytas, G., Anastasiadis, S.H., Seghrouchini, R., Vlassopoulos, D., Li, J., Factor, B.J., Theobald, W., and Toprakcioglu, C., Probing collective motions of terminally anchored polymers, *Science*, 274, 2041, 1996.
16. de Gennes, P.G., *Scaling Concept in Polymer Physics*, Cornell University Press, Ithaca, NY, 1979.
17. Laplace, P.S., De l'equilibre des fluides, Oevres Completes de Laplace, publiees de l' Academie des Sciences, Tome premier, No. 17, Goutiiez, Paris, 1879.
18. Gibbs, J.W., *Collected Works*, Vol. II., ACS, Washington, DC, 1928.
19. Prigogine, I. and Saraga, L. Sur la tension superficie lle et le modele cellulaire de l'etat liquide, *J. Chim. Phys.*, 49, 399–407, 1952.

20. Prigogine, I., Sur la tension superficielle de soltions de molecules de dimensions differentes, *J. Chim. Phys.*, 47, 33–40, 1950.

21. Prigogine, I. and Sarolea, L., Sur la tension superficielle des solutions de molecules de dimensions differentes, *J. Chim. Phys.*, 47, 807–815, 1950.

22. Silberberg, A., Theoretical aspects of the adsorption of macromolecules, *J. Polym. Sci., Part C*, 30, 393–397, 1970.

23. Hoeve, C.A.J., DiMarzio, E.A., and Peyser, P., Adsorption of polymer molecules at low surface coverage, *J. Chem. Phys.*, 42, 2558–2563, 1965.

24. Flory, P.J., Treatment of disordered and ordered systems of polymer chains by lattice methods, *Proc. Natl. Acad. Sci. USA*, 79, 4510–4514, 1982.

25. Flory, P.J., Private communication in the period 1977–1985.

26. Dill, K.A. and Flory, P.J., Interphases of chain molecules: monolayers and lipid bilayer membranes, *Proc. Natl. Acad. Sci. USA*, 77, 3115–3119, 1980.

27. Dill, K.A. and Flory, P.J., Molecular organization in micelles and vesicles, *Proc. Natl. Acad. Sci. USA*, 78, 676–680, 1981.

28. Flory, P.J., Yoon, D.Y., and Dill, K.A., The interphase in lamellar semicrystalline polymers, *Macromolecules*, 17, 862–868, 1984.

29. Mansfield, K.F., and Theodorou, D.N., Molecular dynamics simulation of a glassy polymer surface, *Macromolecules*, 24, 6283–6294, 1991.

30. DiMarzio, E.A. Proper accounting of conformations of a polymer near a surface, *J. Chem. Phys.*, 42, 2101–2106, 1965.

31. DiMarzio, E.A. and Rubin, R.J., Adsorption of a chain polymer between two plates, *J. Chem. Phys.*, 55, 4318–4336, 1971.

32. Flory, P.J., *Collected Papers*, Stanford University, San Francisco, 1985.

33. Matheson, R.R., Jr., and Flory, P.J., Statistical thermodynamics of mixtures of semirigid macromolecules: chains with rod-like sequences at fixed locations, *Macromolecules*, 14, 954–960, 1981.

34. Saiz, E., Hummel, P., Flory, P.J., and Plavsic, M., Direction of the dipole moment in the ester group, *J. Phys. Chem.*, 85, 3211–3215, 1981.

35. Yarim-Agaev, Y., Plavsic, M., and Flory, P.J., Conformational analysis and dipole moments of oligomers of poly(methyl acrylate), *Polym. Prep.*, 24, 233–234, 1983.

36. Flory, P.J. and Plavsic, M., Unpublished.

37. de Gennes, P.G., Polymer solutions near an interface. 1. Adsorption and depletion layers, *Macromolecules*, 14, 1637–1644, 1981.

38. Mandelbrot, B.B., *The Fractal Geometry of Nature*, Freeman and Co., San Francisco, 1982.

39. Avnir, D., Ed., *The Fractal Approach to Heterogeneous Chemistry*, Wiley, New York, 1989.

40. Pietronero, L. and Tosatti, E., Eds., *Fractals in Physics*, North-Holland, Amsterdam, 1985.

41. Alexander, S. and Orbach, R., Density of states on fractals: "fractons." *J. Phys. (Paris) Lett.*, 43, 623–631, 1982.

42. Fryer, D.S., Nealey, P.F., and de Pablo, J., Thermal probe measurements of the glass transition temperature for ultrathin polymer films as a function of thickness, *Macromolecules*, 33, 6439–6447, 2000.

43. DeMagio, G.B., Frieze, W.E., Gidley, D.W., Zhu, M., Hristov, H.A., and Yee, A.F., Interface and surface effects on the glass transition in thin polystyrene films, *Phys. Rev. Lett.*, 78, 1524–1529, 1997.

44. Zheng, X., Rafailovich, M.H., Sokolov, J., Strzhemechny, Y., Schwarz, S.A., Sauer, B.B., and Rubinstein, M. Long-range effects on polymer diffusion induced by a bounding interface, *Phys. Rev. Lett.*, 79, 241–244, 1997.

45. Frank, B., Gast, A.P., Russell, T.P., Brown, H.R., and Hawker, C. Polymer mobility in thin films, *Macromolecules*, 29, 6531–6534, 1996.

46. Plavsic, M.B., Molecular engineering of polymer materials, *Mater. Sci. Forum*, 214, 123–1130, 1996.

47. Plavsic, M.B. and Lazic, N.L., Nanostructure of silica fillers and reinforcement of SBR elastomers—a fractal approach, *Mater. Sci. Forum*, 413, 213–218, 2003.

48. Plavsic, M., Pajic-Lijakovic, I., Cubric, B., Popovic, R.S., Bugarski, B., Cvetkovic, M., and Lazic, N., Chain conformational statistics and mechanical properties of elastomer composites, *Mater. Sci. Forum*, 453–454, 485–490, 2004.

49. Popovic, R.S., Plavsic, M., and Popovic R.G., The resolution of natural rubber (NR)/polydimethylsiloxane (MQ) binary blend properties by stress–strain modeling, *Kautschuk-Gummi-Kunstoffe*, 49, 826–830, 1996.

50. Popovic, R.S., Plavsic, M., Popovic, R.G., and Milosavljevic, M., Mechanical properties, crosslink density and surface morphology of SBR/silicone rubber blends, *Kautschuk-Gummi-Kunstoffe*, 50, 861–867, 1997.

51. Ristic, R., Vrhovac, Lj., and Plavsic, M., The influence of stabilizers on mechanochemical processes in SBR rubbers, *J. Appl. Polym. Sci.*, 72, 835–847, 1999.

52. Ignjatovic, N., Tomic, S., Dakic, M., Miljkovic, M., Plavsic, M., and Uskokovic, D., Synthesis and properties of hydroxyapatite/poly-L-lactide composite biomaterials, *Biomaterials*, 20, 809–817, 1999.

53. Ignjatovic, N., Plavsic, M., Miljkovic, M., Zivkovic, Lj., and Uskokovic, D., Microstructural characteristics of calcium hydroxyapatite/poly-L-lactide based composites, *J. Microsc.*, 196, 243–249, 1999.

54. Ignjatovic, N., Savic, V., Najman, S., Plavsic, M., and Uskokovic, D., A study of HAp/PLLA composite as a substitute for bone powder, using FT-IR spectroscopy, *Biomaterials*, 22, 571–578, 2001.

55. Lee, K. and Ascher, S.A., Photonic crystal chemical sensors: pH and ionic strength, *J. Am. Chem. Soc.*, 122, 9534–9537, 2000.

56. Donnet, J.B., Black and white fillers and tire compound, *Rubb. Chem. Technol.*, 71, 323–341, 1998.

57. Wang, M.J., Effect of polymer-filler and filler-filler interactions on dynamic properties of filled vulcanizates, *Rubb. Chem. Technol.*, 71, 520–589, 1998.

58. Wolff, S., Chemical aspects of rubber reinforcement by fillers, *Rubb. Chem. Technol.*, 69, 325–346, 1996.

59. Plavsic, M., in preparation.

60. Heinrch, G. and Kluppel, M., Recent advances in the Theory of filler networking in elastomers, *Adv. Polym. Sci.*, 160, 1–44, 2002.

61. Lazic, N., Adnadjevic, B., Plavsic, M.B., and Vucelic, D., Interaction of silica with styrene–butadiene rubber, *International Rubber Conference 98*, Proceedings C27-1, 2, Paris, France, May 12–14, 1998.

62. Plavsic, M., Lazic, N., and Paunovic, M., Spectroscopic studies of the influence of silica surface chemistry on rubber networking, in *International Conference on Fundamental and Applied Aspects of Physical Chemistry*, Proceeding 501–503, Belgrade, SCG, September 2000.

63. Mark, J.E. and Erman, B., *Rubber Elasticity — A Molecular Primer*, Wiley, New York, 1988.

64. Mark, J.E., Some recent theory, experiments and simulations on rubberlike elasticity, *J. Phys. Chem. B*, 107, 903–9013, 2003.

65. Lablanc, J.L. and Hardy, P., Evolution of bound rubber during the storage of uncured compounds, *Kautschuk-Gummi-Kunststoffe*, 44, 1119–1124, 1991.

66. Gerspacher, M., O'Farrel, C.P., and Yang, H.H., Carbon black network responds to dynamic strains: methods and results, *Elastomerics*, 123, 35, 1901.

67. Westermann, S., Kraistschmann, M., Pickhot-Hintzen, W., Richter, D., Strabe, E., Farago, B., and Goerigk, G., Matrix chain deformation in reinforced networks: a SANS approach, *Macromolecules*, 32, 5793–5802, 1999.

68. McBrierty, V.J., NMR and ESR of filled elastomers: nature of the elastomer/particle interface, in *International Rubber Conference*, Proceedings CPS 1–6, Paris, France, May 12–14, 1998.

69. Plavsic, M.B., Pajič-Lijakovič, I., Putanov, P., Chain Conformational Statistics and Mechanical Properties of Elastomer Blends, in *New Polymeric Materials*, MacKnight, W.J., Martuscelli E., Korugic-Karasz Lj., Eds., American Chemical Society Washington, in press.

70. Dugic, M., Rakovic, D., and Plavsic, M., The polymer conformational stability and transitions: a quantum decoherence theory approach, in *Finely Dispersed Particles: Micro-, Nano- and Atto-Engineering*, Spasic, A.M., Hsu, J.P., Eds., Marcel Dekker/CRC Press/Taylor & Francis, Chap. 9.

71. Buehler, E.J. and Boland, J.J., Dimer preparation that mimics the transition state for the adsorption of H_2 on the Si(1 0 0)-2×1 surface, *Science*, 290, 506–509, 2000.

7 Atomic Scale Imaging of Oscillation and Chemical Waves at Catalytic Surface Reactions: Experimental and Statistical Lattice Models

Vladimir I. Elokhin and Vladimir V. Gorodetskii
Boreskov Institute of Catalysis, Novosibirsk, Russia

CONTENTS

I. INTRODUCTION

The catalytic oxidation of CO over platinum group metals is relatively simple and also important from the ecological viewpoint. In addition, this reaction exhibits a rich kinetic behavior, including regimes with sustained kinetic oscillations (for reviews, see [1–12]). Great interest in self-oscillatory phenomena in catalytic reaction over metal surfaces is for a large part caused by the possibility to perform more effectively the catalytic processes using the unsteady-state operation. The CO and H_2 oxidation on metals (Pt, Pd) is a nonlinear system, in which temporal and spatial organization becomes possible [1,2]. In the oscillatory regime, the reaction mixture periodically affects the properties of metal surfaces. As there is a synergy between the concentrations of the adsorbed species and the structure of the surface throughout those oscillations, and as the different products often display different oscillation cycles, and are also objected by changes in surface phases, valuable information can be extracted about the mechanism of such reactions from kinetic and characterization studies on the surface species. In the last decades, CO oxidation reaction has became a model for testing the newest physical methods for studying the structure and composition of catalysts. Specifically, it has been reported that the mechanisms of oscillatory oxidation reactions are connected with a periodic change of surface structure (from a reconstructed hexagonal phase to the unreconstructed surface in surface structure on Pt(1 0 0)), with subsurface oxygen formation (at least on Pd(1 1 0)), and with the "explosive" nature of interactions between adsorbed species [1–3]. A common feature in all these mechanisms is the spontaneous periodical transitions of the metal from inactive to highly active states. Since the first discovery of a relationship between reconstruction and kinetic oscillations in CO oxidation on Pt(1 0 0) by Ertl [1], this has become one of the most extensively investigated oscillatory systems in heterogeneous catalysis. The use of spatially resolved ($\sim$1 μm) photoelectron emission microscopy made it then possible to discover the formation of chemical waves on the surfaces of Pt and Pd single crystals [1]. Real metal and support catalysts usually consist of nanosized metal particles on which different crystal planes are exposed. The important question is, can a small supported particle be compared with macroscopic single-crystal surfaces that are normally used in surface science studies. Recent experimental work has shown that field electron microscopy (FEM), which has a sharp tip with a lateral resolution of $\sim$20 Å, can also serve as an *in situ* catalytic flow reactor for the study of these oscillations [3,13].

In the literature, there is much information about the adsorption of small molecules on Pt, Rh, and Pd (see, e.g., [3,13]) on such samples as single-crystal surfaces and supported metal catalysts. The FEM enables us to bridge the gap between these two extremes, because it allows a very high resolution look at sharp metal tips ($\sim$1000 Å), that are in many cases only about one order of magnitude larger than in a supported catalyst. This surface science approach, for example, permits the study of the interaction of adjacent planes on the reactivity of one another. Many of the oscillatory reactions seen on field emitters *in situ* are examples of such interplay of the different nanosized surfaces present [11,14]. This interaction can obviously not be studied with large single crystals and is lost in the *black box* techniques of the macroscopic world of the supported catalysts.

The aims of this contribution are: (1) the investigation of $CO + O_2$ catalytic reaction on Pd and Pt metals varying from single-crystal surfaces up to model sharp-tip catalysts ($\sim$10^3 Å) on an atomic and molecular scale; (2) the elucidation of fundamental problems like the formation of chemical wave, the transformation of the regular oscillations to the complicated and chaotic ones, and the synchronization of local oscillators on various levels of a catalytic system; (3) the study of oscillatory behavior as a function of the dimension of metal surfaces and as a result of differences in the underlying mechanisms of the formation of spatio-temporal structures (oscillations, chemical waves) like oxide formation, phase transitions; (4) the development of realistic mathematical models, describing the oscillatory behavior, coupling mechanisms, and the formation of chemical waves. The ultimate objective of this study is to understand on the molecular level the various kinds of nonlinear processes during catalytic reactions on different levels of catalytic systems: to bridge the gap between single crystals, sharp tips, and nanosized supported metal particles. The purpose of this chapter is to give the experimental and theoretical studies of spatio-temporal self-organization on oscillatory CO oxidation reaction over Pd and Pt nanosized surfaces.

II. CATALYTIC CO OXIDATION

A. REACTION MECHANISM

For the description of the CO oxidation reaction over platinum metals (Pt, Pd, Ir, Ru, Rh), the following elementary steps were used over a long period of time:

$$(1) \quad CO + * \Longleftrightarrow CO_{ads}$$
$$(2) \quad O_2 + 2* \longrightarrow 2O_{ads}$$
$$(3) \quad CO_{ads} + O_{ads} \longrightarrow CO_2 + 2*$$
$$(4) \quad CO + O_{ads} \longrightarrow CO_2 + *$$

where CO_{ads}, O_{ads}, and $*$ are the adsorbed CO, oxygen, and unoccupied surface sites, respectively. The foregoing mechanism includes two routes: "impact" [steps (1) and (4)] and "adsorption" [steps (1)–(3)]. The "impact" mechanism is often called the Eley–Rideal (E–R) mechanism and the "adsorption" one is referred to as the Langmuir–Hinshelwood (L–H) mechanism. Strictly speaking, however, this is incorrect because both these mechanisms date back to Langmuir [15].

The solution of the dilemma as to whether it is an "impact" or an "adsorption" mechanism was the framework within which many catalytic reactions were studied. Evolution in the interpretation of CO oxidation reaction over noble metals (in high-vacuum experiments) can be characterized in a rather simplified form by the three periods: I — from Langmuir studies until the 1970s, the impact (E–R) mechanism; II — during the first half of the 1970s, a combination of the impact (E–R) and adsorption (L–H) mechanisms [16–18]; III — the present time starting in the second part of the 1970s. In our opinion, this period is characterized by two major viewpoints: (i) CO oxidation follows the adsorption mechanism (this viewpoint was reported in the pioneering studies using the molecular beam technique [19,20]) whose kinetic characteristics (reaction rate coefficients) depend significantly on the surface composition (ii) alongside the adsorption mechanism, there is a contribution from the interaction of adsorbed oxygen with CO in the weakly physical adsorption state ("precursor state"). This mechanism can be treated as either a modified impact (for Pd, e.g., [21]) or any version of the adsorption mechanism and can be valid for the reaction at atmospheric pressure. One can consider now the adsorption (L–H) mechanism for CO oxidation reaction over platinum metals as unambiguously proven [22]. This reaction characterized by "volcano-shaped" dependence of the CO_2 formation rate versus bond energy of oxygen with the metal surface [23]. The activity range has been established for the platinum group metals: Pt > Pd > Ir > Rh > Ru.

The first step (1) of the adsorption mechanism describes the monomolecular adsorption of CO. On all the platinum metals except Ir, adsorption of CO proceeds through the preadsorbed ("precursor") state. The activation energy is practically zero and the initial sticking coefficient is high (0.5–1.0). Desorption kinetic curves are of the first order. The activation energy for desorption is in the $100-160 \text{ kJ mol}^{-1}$ range and depends on the surface coverages of CO (primarily) and O_2. As a rule, oxygen adsorption over Pt metals is dissociative with practically zero activation energy (step (2)). The sticking coefficient varied significantly depending on particular type of catalyst. Oxygen desorption proceeds usually at $T > 600$ K, therefore in the temperature range of CO oxidation (300–600 K), step (2) could be considered as irreversible. The final step (3) of interaction between O_{ads} and CO_{ads} is accompanied by immediate desorption of CO_2 into gas phase with the formation of two adsorption centers.

It is well known that the catalytic CO oxidation reaction proceeds with a high heat release. Studies of this reaction over Pt and Pd reveal an unusual phenomenon: carryover of the part of the reaction energy by resulting products — CO_2 molecules [23–25]. The formation of the excited CO_2^* molecules is connected with the step (3) — interaction between O_{ads} and CO_{ads}. The possibility of the local accumulation of free reaction energy in the states of the excited intermediates or particular active centers on the catalyst surface has been discussed by Boreskov [26] by the examination of "chain theory of catalysis." A conclusion was drawn that such excited states could appear only as a "side" effect in the course of high exothermal reactions and, according to Boreskov, it could not serve as a basis for the mechanism of heterogeneous catalysis action. Beginning from the middle of 1970s, the direct experimental evidences confirming Boreskov's conception about the possibility of the nonequilibrium distribution of energy in various steps of heterogeneously catalyzed reaction have been appearing. It has been shown that the CO_2 molecules, formed in CO oxidation reaction on Pt, leave the surface with the excited vibrational, rotational, and translation degrees of freedom [27,28]. For example, the translational energy of CO_2 molecules at $T = 880$ K was found to be ~ 28 kJ/mol corresponding to the molecule's temperature ~ 3650 K [27]. Hence the energy carryover by CO_2 molecules make up to $\sim 10\%$ of reaction heat. The high temperature of CO_2 translation energy (~ 2000 K) has been revealed by titration reaction $CO + O_{ads}$ over Pd(1 1 0) in the temperature range $T \sim 170$–300 K [29]. A plain model of the activated complex in $CO + O_{ads} \rightarrow CO_2$ over Pd(1 1 0) has been proposed in Ref. [30] that reveals the partial carryover of energy by CO_2 molecules formed in the course of the reaction. It is assumed when a CO_2 molecule has been formed it has no time to go down to the potential well to achieve the equilibrium state. Desorption of the CO_2 molecule proceeds from the top of the potential barrier; therefore it should be characterized by high excess of translation and vibrational energies.

B. EXPERIMENTAL: CO AND O_2 ADSORPTION

1. Experimental

The high-resolution electron energy loss spectroscopy (HREELS), temperature programmed reaction spectroscopy (TPR), and FEM experiments were performed in an ultrahigh-vacuum (UHV) chamber with a base pressure below 10^{-10} mbar. The energy loss spectra were obtained at the specular direction by using a VG ADES 400 electron spectrometer, an electron energy of ~ 2.5 eV, and an incident angle of $\sim 35°$ with respect to the surface normal. The resolution of the elastically reflected beam was about 9–11 meV (~ 70–90 cm^{-1}). The TPR spectra were measured with a quadrupole mass spectrometer by using a heating rate of 6–10 K sec^{-1}. The experimental setup has been described in detail elsewhere [31]. In terms of the FEM experiments, the surface analysis of the field tip emitter is based on the changes in local work function ($\Delta \Phi$) with adsorption of CO and oxygen, which can be correlated with the total field electron currents, as described earlier [32]. Electrostatic field effects at the ~ 0.4 V/Å fields used during these investigations are low enough not to affect the chemistry studied [33]. The FEM experimental device, based on the use of sharp field emitter tips, has been described earlier [32,34]. The molecular beam work was carried out in a separate UHV chamber [35,36]. The crystal was exposed to mixed $CO + O_2$ molecular beams by using a capillary array dozer. The gas flow was controlled by mass spectrometry, and calibrated by comparison of the areas under flash desorption curves obtained by adsorbing CO using the beam setup versus isotropically by backfilling of the system. The cleaning procedure of the Pt(1 0 0), Pd(1 1 1), and Pd(1 1 0) surfaces included Ar$^+$ etching and annealing cycles in oxygen and in vacuum. The structures of the clean single-crystal surfaces were confirmed by low-energy electron diffraction (LEED). The temperature of the single crystals was measured by means of chromel and alumel thermocouples spot-welded to the sample, and could be controlled to within 1 K by using a heating power supply with a feedback loop. The reaction gasses, CO and O_2, were of the highest purity available, and were always checked by a mass spectrometry before use.

2. Adsorption of CO

The fact that the Pt(1 0 0) surface shows two surface structures that exhibit dramatically different adsorption and catalytic properties has made it a popular system for surface science investigations. The quasistable unreconstructed clean Pt(1 0 0)-1×1 surface is reconstructed under UHV condition and demonstrates the hexagonal (hex) surface structure [37–40]. The temperature of the (hex) $\leftrightarrow$ (1 × 1) back-phase transition depends strongly on the coverage and on the composition of the adsorption layers. The reconstruction of the Pt(1 0 0) is supposed to be responsible for oscillations of the reaction rate of CO oxidation by O_2, which are accompanied by the propagation of surface concentration waves [41].

The adsorption of CO on the Pt(1 0 0) surface has been studied in detail with several surface science techniques. The following scheme of CO adsorption on Pt(1 0 0)-(hex) surface at 300 K can be formulated on the basis of data from the literature. According to Refs. [42–45], CO adsorption on the Pt(1 0 0)-(hex) surface occurs without lifting the (hex) → (1 × 1) back-reconstruction of the topmost layer of metal. CO molecules on the (hex) surface are coordinated in on-top configuration. This state is denoted as CO_{hex}. The back-reconstruction starts as soon the CO coverage on the (hex) surface exceeds the critical values of approximately 0.05–0.1 ML (one monolayer [ML] is equal to the number of platinum atoms of the topmost layer of the unreconstructed Pt(1 0 0)-1 × 1 surface [43,44,46]). As a result, the formation of adsorbed islands of CO with (1 × 1) structure is observed. The local coverage inside the (1 × 1) islands is assumed to be 0.5 ML during growth of the islands [44]. The (hex) phase surrounds the islands and is nearly free of CO molecules.

The atomic density of the (1 × 1) surface is 1.28×10^{15} Pt atoms cm^{-2} [39]. This is approximately 25% lower than the atomic density of the (hex) phase (i.e., 1.61×10^{15} Pt atoms cm^{-2} [47]). This density difference results in significant mass transport of platinum atoms during the back-reconstruction. Scanning tunneling microscopy (STM) data demonstrate [47–49] that expelled platinum atoms form clusters with size of $\sim 15–25$ Å. The clusters are randomly distributed within the CO-islands boundaries. Infrared absorption spectroscopy (IRAS) [42,43] and HREELS [31,44,45] have shown the existence of two molecularly adsorbed CO species on the surface of the (1 × 1) islands: in bridge (CO_{br}) state and in the on-top (CO_{top}) state. The saturation coverage of CO on the Pt(1 0 0) surface at 300 K, estimated by a nuclear microanalysis, is 0.75 ML [39].

Measurements of the heat of adsorption made by Yeo et al. [50] showed that the (hex) reconstruction is lifted at very low CO coverage. Hopkinson et al. [51,52] reported that the growth rate of the CO_{ads}/(1 × 1) islands depends nonlinearly on the local coverage of CO on the (hex) phase. This results in strong power-law dependence of the CO sticking probability on CO partial pressure. According to LEED and STM data [47,53], the formation of the CO_{ads}/(1 × 1) islands proceeds through a nuclear and trapping mechanism in two stages. The first stage is the generation of the centers of nucleation of the islands on the (hex) phase. The second stage is the growth of the islands with increasing total coverage.

It is now generally accepted that CO is adsorbed on platinum group metal surfaces via carbon atom with its axis parallel to the surface normal. Experimental evidence comes from photoelectron spectroscopy and vibration spectroscopy. The CO_{ads} bond is formed by electron transfer from the 5σ-orbital of CO to unoccupied metal orbitals, implying a donation of electrons to the metal, accompanied by back-donation of electrons from occupied d-orbitals into the unoccupied 2π-orbitals of CO. For the first time, the electronic structure of a semiinfinite Pt(1 1 1) crystal was calculated applying the linear muffin-tin orbitals-tight-binding (LMTO-TB) approximation to aid interpretation of the angle-resolved photoemission spectra (ARUPS) by Tapilin et al. [54]. The experimental photoemitted electron spectra were recorded in the reflected angle of $\sim 60°$ and an incident angle of photon flux of $\sim 80°$ (He I, $h\nu = 22.2$ eV) with respect to the surface normal. Both the experiment and calculation reveal a surface state (S) near the Fermi level in the neighborhood of the $\bar{K}$ point of the surface Brillouin zone. To examine the surface localization of the S peak, the effect of CO adsorption on its intensity has been studied. It is seen that CO adsorption causes peak S to vanish [54]. By comparing the experimental and calculated data, a consistent picture is obtained for the contribution of back-donation from surface state in the bonding of CO to Pt(1 1 1) surface.

In more traditional methods, the initial heat of adsorption of CO on Pt, ranges from 151 kJ mol^{-1} on Pt(2 1 0) to 109 kJ mol^{-1} on Pt(1 1 0) [55]. The advent of single-crystal adsorption calorimetry (SCAC) has led to calorimetric measurements of heats of adsorption, ΔH_{cal}, with the added advantage that irreversible processes can also be studied [56]. On Pt(1 1 0), ΔH_{cal} decreases from 193 kJ mol^{-1} at low coverage to 140 kJ mol^{-1} at coverage of 0.7 [57]. On Pt(1 1 1), ΔH_{cal} decreases from 187 kJ mol^{-1} at low coverage to 60 kJ mol^{-1} at coverage of 0.5 [58]. More specifically on Pd, the initial heats of adsorption range from 142 to 167 kJ mol^{-1} on Pd(1 1 1) and Pd(1 1 0), respectively [59]. So far, only Pd(1 0 0) has been measured with SCAC [60]. Starting at 163 kJ mol^{-1}, ΔH_{cal} drops to 70 kJ mol^{-1} at coverage of 0.5. On polycrystalline-evaporated Pd films, q_{CO} continuously decreased with increasing coverage from 167 kJ mol^{-1} [61]. The same uninterrupted decrease in q_{CO} was found for different Pd surfaces and particles in later works [62–64]. In the case of Pd(1 1 0), q_{CO} remained constant at 167 kJ mol^{-1} until the coverage rose above ~0.2 [59]. For both small Pd particles supported on SiO$_2$/Si(1 0 0) and Pd(1 1 1) showed, within experimental error, the same changes in q_{CO}. Particles of Pd, with 54 Å average size, had a q_{CO} of 148 $\pm$ 5 kJ mol^{-1} at low coverage to 120 kJ mol^{-1} at a coverage of 0.5 [64].

The isosteric heat of adsorption of CO on Pd-tip surface was found to be ~154 kJ mol^{-1} in the limit of zero coverage, decreasing to ~133 kJ mol^{-1} at coverage of 0.5 [32]. Heats of adsorption measured with FEM tend to be close to values for open surfaces such as (2 1 0). The only measured value that exists for Pd(2 1 0) is 147 kJ mol^{-1} [59].

At 300 K, a saturated layer of CO on Pd-tip surface caused a work function (WF) increase of 0.98 eV above that of the clean surface. At 300 K, the WF increases rapidly with increasing CO exposure up to a maximum of 0.98 eV above that of the clean surface [32]. This value is already reached at an exposure of ~2.3 × 10^{-6} mbar sec, suggesting a sticking probability near unity. Measurements on different Pd single-crystal surfaces give a maximum increase in WF in the range from 0.75 to 1.27 eV for the various planes, with Pd(1 1 1) showing a maximum increase of 0.98 eV [56].

For reference, CO adsorption at 50 L exposure (1 L = 1.3 × 10^{-6} mbar s) at 100 K on the clean Pd(1 1 0) surface leads to several molecular CO$_{ads}$ states with desorption peak temperatures at 230 K (α_1), 280 K (α_2), and 330 K (α_3), and higher temperature features at 410 K (β_1) and 470 K (β_2) [65,66]. The CO desorption spectrum for Pd(1 1 1) after an exposure to $\theta = 0.6$ ML at 200 K is similar to those on Pd(1 1 0), and also shows four peaks at 250, 330, 380 (α), and 470 K (β).

3. Adsorption of O$_2$

Based on the HREELS data, an adsorption of O$_2$ on Pt(1 1 1) at 105 K results in formation of two molecular species: (i) peroxide O$_{2ads}^{2-}$ with frequency v(O—O) band at 870 cm^{-1} and v(Pt—O$_2$) band at 380 cm^{-1}; (ii) superoxide O$_2$ with v(O—O) band at 1240 cm^{-1}. Adsorption is accompanied by formation of atomic O$_{ads}$ with a frequency of v(Pt—O) band at 490 cm^{-1}. Heating up to 250 K of a layer adsorbed at 105 K causes the transition of molecular oxygen to an atomic state: a sharp increase of the intensity band is seen at 460 cm^{-1} (Figure 7.1) [67]. Based on the TDS data, a partial desorption of O$_{2ads}$ to a gas phase as O$_2$ peak with T_{des} ~ 160 K proceeds simultaneously with dissociation.

According to HREELS data, at 90 K, O$_2$ adsorption on reconstructed Pt(1 0 0)-hex surface results in formation of molecular peroxide O$_{2ads}^{2-}$ state with bond axis parallel to the surface and characterized by v(OO) band at 900 cm^{-1}. Molecular oxygen is stable below 120 K and is entirely desorbed at 140 K. Opposite to Pt(1 0 0)-hex, heating the molecular oxygen layer on Pt(1 0 0)-1 × 1 surface from 90 to 200 K is accompanied by dissociation of O$_{2ads}^{2-}$ state with formation of oxygen adatoms O$_{ads}$, characterized by v(Pt—O) band at 500 cm^{-1}. According to TDS data, a partial O$_2$ desorption occurs simultaneously with dissociation at 160 K from the Pt(1 0 0) in accordance with Pt(1 1 1) results.

Figure 7.1b presents a set of HREELS spectra of an atomic oxygen layer formed after 3 L NO exposure on the nonreconstructed Pt(1 0 0)-(1 × 1) surface at 300 K with the temperature increasing up to 500 K for NO$_{ads}$ desorption and after titration reaction O$_2$ + H$_{ads}$/Pt(1 0 0)-(hex) at 220 K. The band at 540 cm^{-1} represents v(PtO) stretching of adsorbed atomic oxygen as a result

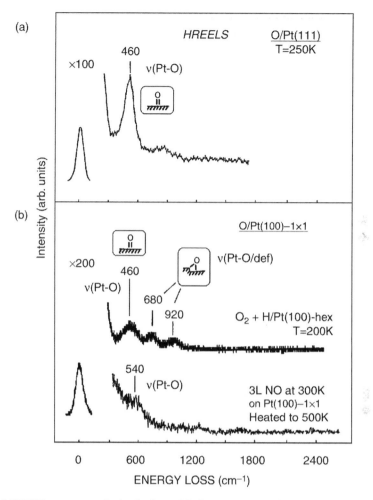

FIGURE 7.1 (a) HREEL spectrum obtained after a 9 L O_2 exposure on the Pt(1 1 1) surface at 105 K and subsequent (O_{2ads}) heating in vacuum up to 250 K. (b) HREEL spectra (i) for the atomic oxygen appearing in the course of 220 K H_{ads} titration by oxygen (30 L) on the Pt(1 0 0)-(hex) surface; (ii) for the atomic oxygen produced by NO dissociation on the Pt(1 0 0)-(1 × 1) surface at 300 K. (Reprinted from Gorodetskii, V.V., Matveev, A.V., Cobden, P.D., and Nieuwenhuys, B.E., J. Mol. Catal. A: Chem., 158, 155–160, 2000. With permission from Elsevier.)

of NO dissociation on defect sites. For comparison, on the flat Pt(1 1 1) surface a single $\nu(PtO)$ band is observed at 460 cm^{-1} after O_2 adsorption at 250 K (see Figure 7.1a). Figure 7.1b shows that interaction of an atomic hydrogen layer H/Pt(1 0 0)-(hex) with oxygen molecules (220 K, 30 L) is accompanied by the formation of an oxygen adatom layer (O_{ads}) with Pt—O bond vibration frequencies of 460, 720, and 920 cm^{-1} [67]. According to TDS data, oxygen desorption occurs with a maximum around 710 K. Comparable TDS results were found after exposure to O_2 at high temperature (575 K, 3×10^5 L) on the Pt(1 0 0)-(hex): two desorption peaks at 660 and 710 K are observed. Hence, in agreement with O_2/Pt(3 2 1) [68] the bands at 720 and 920 cm^{-1} are intrinsic to atomic states of oxygen on the structural defects (presumably like steps or kinks) induced by the O + H reaction and can serve as a good spectral indicator of the strong structural transformation of the Pt(1 0 0)-(hex) to the Pt(1 0 0)-(1 × 1) surface.

The oxygen adsorption on Pt is known to proceed through the sequence of different adsorbed states: $O_2 \rightarrow O_2^- \rightarrow O_2^{2-} \rightarrow O^- \rightarrow O^{2-}$.

Figure 7.2 displays a series of O_2 TPR spectra from palladium single-crystal surface. After O_2 adsorption at 78 K on the Pd(1 1 1) surface, several molecular (O_{2ads}) states are detected, at 100 K (α_1), 130 K (α_2), and 155 K (α_3). In addition, a peak at 775 K is also seen due to recombination of O_{ads} surface atoms (Figure 7.2a). Figure 7.2b shows O_2 TPR spectra from Pd(1 1 0) after different oxygen exposures at 100 K. Above a 0.3 L exposure, there are three peaks in the spectrum, around 125 K (α-O_2 molecular state), 740 K (β_1-O atomic state), and 815 K (β_2-O atomic state). The β_1-O state has been attributed to subsurface oxygen ($O_{ads} + {}^*v \rightarrow {}^*O_{sub}$), as indirectly indicated by WF measurements [69].

C. DETECTION OF SURFACE INTERMEDIATES

The CO oxidation that occurs during the phase transition from the Pt(1 0 0)-(hex) reconstructed plane to the (1 × 1) upon oxygen and CO coadsorption has been studied by HREELS. Exposures of a clean Pt(1 0 0) surface to 1 L of CO at 300 K results in the appearance of three main loss peaks

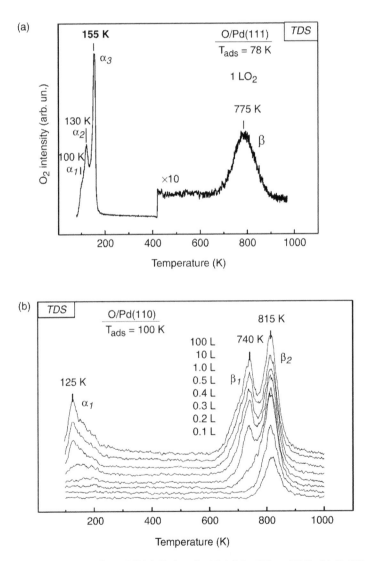

FIGURE 7.2 (a) O_2 TPR spectrum from Pd(1 1 1) dosed with 1.0 L of O_2 at 78 K. (b) O_2 TPR spectrum from Pd(1 1 0) dosed with different exposures of O_2 at 100 K. (Reprinted from Jones, I.Z., Bennett, R.A., and Bowker, M., Surf. Sci., 439, 235–248, 1999. With kind permission of Springer Science and Business Media.)

respectively at 480, 1880, and 2105 cm^{-1}, attributed to the ν(Pt—CO) and to the ν(CO) stretching modes of CO_{ads} molecules in the bridge and on-top states, respectively [31]. It has been reported that at this temperature adsorbed islands of CO with (1×1) surface structure are formed while the (hex) phase surrounding the islands remains nearly free of CO_{ads} molecules [44]. According to the HREELS data, subsequent O_2 adsorption on the reconstructed Pt(1 0 0)-(hex) surface precovered with these $CO_{1\times1}$ islands at 90 K results in the formation of molecular peroxide, O_{2ads}^{2-}, with its bond axis parallel to the surface and a characteristic ν(OO) band at 920 cm^{-1} (Figure 7.3a) [70]. The CO vibrational bands are not significantly affected by the O_2 coadsorption (Figure 7.3a). CO_2, CO, and O_2 TPR obtained from the resulting mixed $CO_{1\times1} + O_{2ads}$ adlayers indicate that all molecular oxygen and CO desorbs molecularly at 150 and 510 K, respectively, without any CO_2 formation (Figure 7.3b). This observation supports the idea that the $CO_{1\times1}$ islands have a high local coverage (0.5 ML) [44], and therefore do not allow for the dissociation of the molecular peroxide, O_{2ads}^{2-}, surface species.

Molecular CO adsorption on the hexagonal Pt(1 0 0) at 90 K occurs in on-top sites without lifting of the reconstruction, and is characterized by ν(Pt—CO) and ν(CO) bands at 450 and 2120 cm^{-1}, respectively. Exposure of this CO_{hex} precovered Pt(1 0 0)-(hex) surface to 10 L of O_2 again results in the formation of molecular peroxide O_{2ads}^{2-}, in this case characterized by the

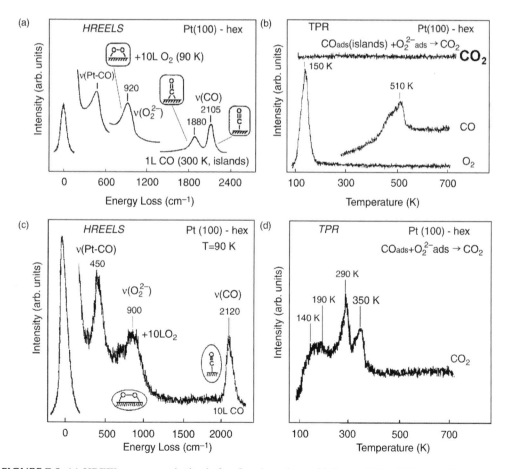

FIGURE 7.3 (a) HREEL spectrum obtained after O_2 adsorption at 90 K on a Pt(1 0 0)-(hex) surface covered by $CO_{ads}/(1 \times 1)$ islands generated by a 1.0 L CO exposure at 300 K. (b) TPR spectra from the coadsorbed molecular oxygen and carbon monoxide $CO_{1\times1}$ prepared in (a). (c) HREEL spectrum obtained after CO and O_2 coadsorption at 90 K on the Pt(1 0 0)-(hex) surface. (d) CO_2 TPR spectrum from the surface prepared in (c). (Reprinted from Gorodetskii, V.V. and Drachsel, W., Appl. Catal. A: Gen., 188, 267–275, 1999. With permission from Elsevier.)

$\nu(OO)$ band at 900 cm^{-1} (Figure 7.3c). Contrary to the case in Figure 7.1b, however, heating of this mixed $CO_{hex} + O_{2ads}^{2-}$ adlayer leads to the desorption of significant amount of CO_2 in four peaks respectively at 140, 190, 290, and 350 K (Figure 7.3d). It is known that the hexagonal reconstruction can be removed at low temperatures by CO adsorption [43]. The packing density of the reconstructed Pt(1 0 0)-(hex) surface is known to be higher by 20% than that of the nonreconstructed Pt(1 0 0)-(1 × 1) surface, and, according to STM data [49], the Pt(1 0 0)-(hex) → (1 × 1) phase transition induced by CO molecular adsorption causes the excess Pt atoms to be ejected over the upper layer of the metal. After this, O_{2ads}^{2-} can dissociate on the active centers of the (1 × 1) phase, and the $CO_{ads} + O_{ads} \rightarrow CO_{2,gas}$ reaction can occur.

D. Effect of Subsurface Oxygen

The effect of subsurface oxygen on the rate of the $CO + O_{ads} \rightarrow CO_2$ reaction was also studied by TPR experiments [71]. It has become clear that the differentiation between adsorbed (O_{ads}) and subsurface (O_{sub}) oxygen is difficult. An attempt was made in these studies to isolate the chemistry of the subsurface oxygen toward CO by using ^{18}O labeling. Figure 7.4 shows TPR results for cases where a clean Pd(1 1 0) crystal was first exposed to 0.2 L of $^{18}O_2$ at 100 K and then preheated to 500 K to produce a $^{18}O/Pd(1\ 1\ 0)$-(1 × 2) reconstructed surface. This was followed by adsorption of 0.8 L of $^{16}O_2$ at 100 K and preheating to up to 200 K (to desorb any molecular $^{16}O_{2ads}$). This procedure was designed to prepare a Pd(1 1 0) surface with a $^{16}O_{ads}$ layer on top of $^{18}O_{sub}$ subsurface oxygen. The reconstruction of the surface from its (1 × 1) phase to a (1 × 2) structure facilitates the formation of the O_{sub} species [72], because on the (1 × 2) reconstructed surface approximately 50% of the surface sites are located below the uppermost Pd atoms, and that makes the diffusion of oxygen (^{18}O) into the subsurface region easier. TPR results for this mixed isotope system show that the interaction of CO_{ads} with subsurface $^{18}O_{sub}$ produces a single $C^{18}O^{16}O$ desorption peak at 420 K (Figure 7.4b). In contrast, the TPR spectra for the reaction of the coadsorbed oxygen (atomic β-$^{16}O_{ads}$ and β-$^{18}O_{sub}$ states) with CO_{ads} on the Pd(1 1 0)-(1 × 2) surface, shown in Figure 7.4a, indicates low-temperature CO_2 formation, starting at CO exposures above 0.2 L.

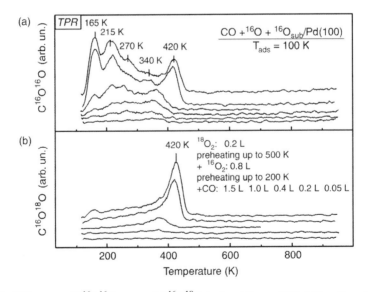

FIGURE 7.4 CO_2 TPR spectra ($C^{16}O^{16}O$, top, and $C^{16}O^{18}O$, bottom) from a Pd(1 1 0) surface, first pre-treated with $^{18}O_2$ (0.2 L, 500 K) and $^{16}O_2$ (0.8 L, 200 K), and then dosed with varying amounts of CO at 100 K. The $C^{16}O^{16}O$ and $C^{16}O^{18}O$ traces represent products from reactions with adsorbed and subsurface oxygen, respectively. (Reprinted from Gorodetskii, V.V., Matveev, A.V., Podgornov, E.A., and Zaera, F., Topics Catal., 32, 17–28, 2005. With kind permission of Springer Science and Business Media.)

Three distinct $C^{16}O^{16}O$ desorption peaks are observed respectively at 165, 215, and 270 K. These experiments directly confirm that the low-temperature (<200 K) CO_2 evolution corresponds to the reaction between CO_{ads} and the active O_{ads} atoms generated by O_{2ads}^{2-} dissociation on the $^*O_{sub}$/ Pd(1 1 0)-(1 × 2) surface. Therefore, the formation of subsurface oxygen ($^*O_{sub}$) promotes the appearance of weakly bound oxygen atoms highly active toward reactions with CO. A similar remarkable reactivity of weakly adsorbed oxygen atoms with CO on Au(1 1 0)-(1 × 2) surfaces has been reported recently [73]. In that case, TPR spectra showed three separate CO_2 formation peaks respectively at 67, 105, and 175 K from oxygen-precovered gold surfaces.

E. STEADY-STATE REACTION

As mentioned in Section I, the oxidation of CO on platinum group metals is often explained by using a Langmuir–Hinshelwood mechanism. To understand the occurrence of kinetic oscillations, though, an additional feedback mechanism is required. It is believed that the oscillations on Pt(1 0 0) are connected with the Pt(1 0 0)-(hex) ↔ (1 × 1) surface reconstruction [1]. In contrast, the clean Pd(1 1 0) surface does not change its surface structure upon catalytic CO oxidation, and the oscillations in reaction rate are associated with changes in oxygen adsorption probabilities (S_0) induced by depletion of subsurface oxygen: O_{ads} ↔ O_{sub} [1,2]. The HREELS and TPR results show two different ways by which CO_2 can be produced at low temperatures: *via* the reaction of CO_{ads} with active weakly bound oxygen adatom generated by O_{2ads}^{2-} dissociation during the Pt(1 0 0)-(hex) → (1 × 1) phase transition, and by a conversion of weakly bound O_{ads} atoms in mixed $O_{ads} + O_{sub} + CO$ layers on the Pd(1 1 0) surface.

Figure 7.5a shows molecular beam results on the temperature dependence of the steady-state rate of CO oxidation over a Pd(1 1 0) surface. At 300 K the reaction is limited by oxygen adsorption because the surface is covered with CO_{ads}. In the temperature interval between 370 and 650 K, however, the rate for CO_2 production increases rapidly, presumably because of desorption of some of the CO, which reduce the CO_{ads} coverage on the surface. A bistability is seen in this temperature region, as indicated by the hysteresis in CO_2 formation rate seen between experiments with increase and decrease of temperature (Figure 7.5a). As the temperature is increased, the transfer from the CO layer to the O_{ads} layer is delayed, while when the temperature is decreased, the reverse is true. Local single oscillations are also seen for the CO_2 rate at 372 and 382 K in the

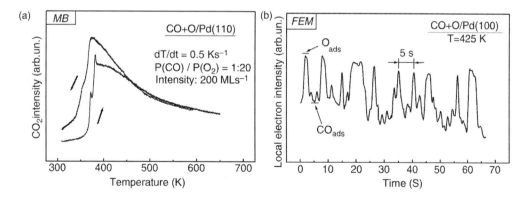

FIGURE 7.5 (a) Steady-state CO_2 formation rate on Pd(1 1 0) during a heating–cooling cycle under exposure to a 1:20 CO + O_2 molecular beam. A hysteresis is seen in the behavior with rise and decrease of the surface temperature. (b) Variations in local emission current from a Pd(1 0 0)$_{step}$ plane as a function of time during an oscillatory reaction between CO and O_2 on a Pd-tip under constant reaction conditions, $T = 425$ K, $P(O_2) = 2.6 \times 10^{-3}$ mbar, and $P(CO) = 1.3 \times 10^{-4}$ mbar. Low current levels reflect a Pd(1 0 0) nanoplane covered by CO_{ads}. Field value: 0.4 V/Å. (Reprinted from Gorodetskii, V.V., Matveev, A.V., Kalinkin, A.V., and Nieuwenhuys, B.E., Chem. Sustain. Dev., 11, 67–74, 2003. With kind permission of Springer Science and Business Media.)

rise of temperature part of this experiment. A typical example of the oscillations in CO_2 production obtained when the Pd(1 1 0) single-crystal surface is exposed at 390 K to a gas mixture of $P(O_2) = 5 \times 10^{-2}$ mbar and $P(CO) = 5 \times 10^{-5}$ mbar was reported recently [66].

F. CONCLUSIONS

The low-temperature adsorption and reaction between CO and O_2 on Pd(1 1 1), Pd(1 1 0), Pt(1 0 0) surfaces was investigated using different macroscopic (MB, TPR, HREELS) and on Pd- and Pt-tip surfaces microscopic (FEM) analytical tools to learn about the details of the reaction dynamics at those catalyst surfaces. The key results from these studies are summarized below.

Both Pt and Pd surfaces are catalytically active for the CO + O_2 reaction due to their ability to dissociate O_2 molecules. Results from ^{18}O labeling experiments directly support the conclusion that adsorbed (weakly bound) atomic oxygen is the active form of oxygen that reacts with carbon monoxide at low (160–200 K) temperatures to form CO_2. A path leading to the formation of these active O atoms on the Pt(1 0 0) and Pd(1 1 0) surfaces was isolated. It was determined that the local concentration of CO_{ads} on the Pt(1 0 0)-hex prevents oxygen atoms from occupying the hollow position, and apparently leads to the formation of weakly bound active O_{ads} atoms. The subsurface oxygen O_{sub} that can be formed on the Pd(1 1 0) surface may be an important species for the formation of these new weakly bound O atoms. The adsorbed O_{ads} is highly active compared with the O_{sub} species, and rapidly reacts with adsorbed carbon monoxide, CO_{ads}, to produce CO_2. Finally, comparison of the kinetics for CO oxidation in molecular beam experiments with and without preadsorbed atomic oxygen shows that the surface reaction between adsorbed CO and adsorbed atomic oxygen is rate limiting in instances where the dissociation of molecular oxygen does not occur [71].

III. OSCILLATIONS AND CHEMICAL WAVES

The reaction kinetics on the supported metal catalyst might be quite different when compared with that on the single-crystal surfaces, as a result of interplay between different nanoplanes present on small particles. These surfaces, with a crystallite size of 100–300 Å, are mainly formed by the most dense (1 1 1), (1 0 0), and (1 1 0) planes which differ dramatically in adsorption and oscillation behavior. The FEM is an experimental tool for performing *in situ* investigations of real dynamic surface processes in which different crystallographic nanofaces of emitter-tip are simultaneously exposed to the reacting gas. In this work, the mechanism of movable waves generation in the oscillating CO + O_2 reaction has been studied on a Pt-tip, $\sim$700 Å in radius, by an FEM with a lateral resolution of $\sim$20 Å. Mobile reaction zone structure was analyzed by mass-to-charge resolved field-ion microscopy.

A. EXPERIMENTAL TECHNIQUES ON AN ATOMIC SCALE

Since Taylor [74] coined the concept of active centers in heterogeneous catalysis, a variety of methods have been developed to characterize their properties. A long-standing aim has been the direct microscopic imaging of catalytically reacting surface sites. Three methods exist for imaging a metal surface in real space on an atomic scale: the electron microscope developed by Ruska in 1933 [75], the field ion microscope (FIM) invented by Müller and Tsong in 1951 [76], and the scanning tunneling microscope established by Binning and Rohrer in 1983 [77]. Among these methods, only the FIM provides the possibility of performing *in situ* investigation of a real dynamic surface processes in which different crystallographic nanoplanes of a tip-emitter are simultaneously exposed to the reacting gas. This advantage was accompanied, however, so far by the disadvantage of the high electrostatic fields ($\sim$3.5 V/Å) required for surface imaging with noble gases (He or Ne) on an atomic scale (Figure 7.6b). For the first time the oscillating CO oxidation is investigated on a Pt-tip by FIM that uses the catalytically reacting molecules as imaging gas (O_2) [78–80] or reaction products (H_2O) [81] at rather low electric fields $\sim$1.0–1.5 V/Å.

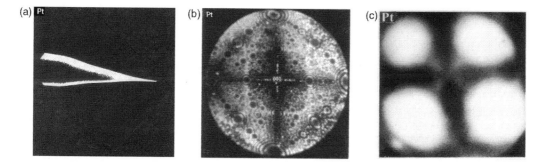

FIGURE 7.6 (a) Transmission electron image of Pt-tip and (b) Field Ion Microscope image of a Pt-tip surface used for investigations of the oscillating CO oxidation. Pt-tip imaged with Ne at 78 K, $P_{Ne} = 2 \times 10^{-4}$ mbar, $F = 3.5$ V/Å, $r = 680$ Å. (c) Field Emission Microscope image of the same Pt-tip at $F = 0.3$ V/Å.

In the present communication, we review an FIM experimental approach that uses the negative electrostatic field strength for imaging the surface at F ~ 0.3 V/Å (Figure 7.6c). All experiments were performed with a conventional field-ion/field-electron microscope.

B. Phase Transitions: Pt-tip Experiments

As shown in earlier work by Lim et al. [82], the existence of monostable and bistable regions as well as of the oscillating regime can be evaluated with the help of FEM, and displayed in a kinetic phase diagram. Such a diagram has been established by us for the Pt(1 0 0) nanoplanes (~200 Å dimension) on a Pt-tip at 340 K (Figure 7.7a) [34]. This diagram compares well with that for a single-crystal Pt(1 0 0) surface over a much higher (440–520 K) temperature range [83].

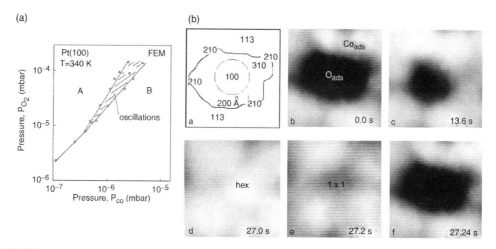

FIGURE 7.7 (a) Phase diagram for CO oxidation on the (1 0 0) nanoplane of a Pt-tip, taken from the wave propagation front at 340 K. The catalytically active O_{ads} layer, region A, and inactive CO_{ads} layer, region B, are separated by the indicated lines. Oscillations occur in the dashed region of partial pressures of reactants. (b) Magnified view of a sequence of FEM images obtained along the (1 0 0) plane of the tip during a $CO + O_2$ oscillating cycle at 340 K and constant $P(CO) = 2.3 \times 10^{-6}$ mbar and $P(O_2) = 4 \times 10^{-4}$ mbar pressures. (a) Stereographic projection of the Pt[1 0 0] oriented tip; (b) $\tau = 0$, when the (1 0 0) plane is covered by O_{ads}. (c) After 13.6 sec. The O_{ads} layer shrinks due to a slow reaction. (d) After 27.0 sec. The very bright image indicates a clean and reconstructed (1 0 0)-hex surface. (e) $\tau = 27.20$ sec. A small CO_{ads} coverage remains, and the reconstruction is lifted; and (f) Renewal of the O_{ads} layer and return to the same condition as (b). (Reprinted from Gorodetskii, V.V. and Drachsel, W., Appl. Catal. A: Gen., 188, 267–275, 1999. With permission from Elsevier.)

Figure 7.7b represents a typical sequence of FEM images for (1 0 0) nanoplanes obtained when a Pt-tip with [1 0 0]-orientation is exposed to a gas mixture of $P(O_2) = 4 \times 10^{-4}$ mbar and $P(CO) = 2.3 \times 10^{-6}$ mbar at 340 K [34]. Oscillations are seen with amplitudes ranging from O_{ads} (low current) to CO_{ads} (high current) layers and a periodicity of 27 sec. The "clean-off" reaction mechanism that take place on the (1 0 0) nanoplanes at 340 K is illustrated in Figure 7.7b. At the beginning of the cycle (panel (b)), the unreconstructed Pt(1 0 0)-(1 × 1) surface is completely covered with a O_{ads} island (shown as a dark patch). That island slowly reacts with CO_{ads}, and results in a gradual shrinkage of the O_{ads} layer (panel (c)), until a very fast catalytic surface process (the "clean-off" reaction) removes the CO_{ads} layer. At that point the clean unstable (1 × 1) surface suddenly reconstructs to the (hex) structure [84], creating an unoccupied clean (1 0 0) surface (panel (d)), with high local electron emission current (hence the bright picture). The small sticking coefficient for oxygen on this surface, 10^{-3}, keeps this reconstructed plane free from O_{ads} until, at a certain local CO coverage, the reconstruction is lifted again, and the (1 × 1) structure is returned (panel (e)). This is accompanied by the increase in oxygen sticking coefficient, from $\approx 10^{-3}$ (hex) to $\approx 10^{-1}$ (1 × 1) that induces a transition from a catalytically inactive state to an active surface for CO oxidation. Subsequently, in a very fast reaction (within one video frame, $\Delta t \approx 40$ ms), the initial state of the O_{ads}-covered surface is regained, as seen in the transition from panels (e) to (f).

The details of the oscillatory behavior depend sensitively on the external control parameters, that is, the selected temperature and partial pressures. Figure 7.8 represents a typical series of oscillations (FEM) when the Pt-tip with (1 0 0)-orientation is exposed at 365 K to a gas mixture of $P(O_2) = 5 \times 10^{-4}$ mbar and $P(CO) = 8 \times 10^{-6}$ mbar. The oscillation amplitude ranges from the O_{ads} layer (low current) to the CO_{ads} layer (high current) with a periodicity of 120 sec. A difference of work functions between O_{ads} and CO_{ads} of ~ 0.4 eV is connected with a change in the electron current. For $T = 478$ K, for instance, the reaction/diffusion fronts propagate with a speed of 5000 Å s^{-1} [85]. Rates within this order of magnitude (~ 3 μm s^{-1}) have been found previously on Pt(1 1 0) surfaces at 485 K in PEEM experiments [86]. Qualitatively similar results were also found on Pt(1 0 0) [87].

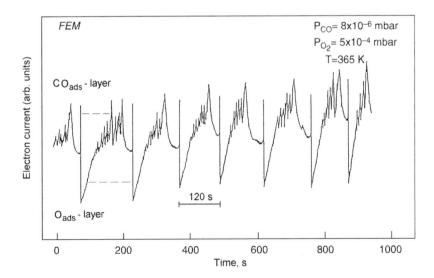

FIGURE 7.8 Oscillatory behavior of the total field electron current for the Pt field emitter at constant control parameters under conditions of pronounced surface selectivity. Low current levels reflect O_{ads}-covered planes. $F \approx 0.4$ V/Å. (Reprinted from Gorodetskii, V.V. and Drachsel, W., Appl. Catal. A: Gen., 188, 267–275, 1999. With permission from Elsevier.)

The FEM images have surprisingly shown very sharp reaction fronts, therefore the nature of the *local reaction rates* are of great chemical interest. To obtain the information about the chemical identity of the field emitted ions during the reaction, a mass analysis by the atom-probe hole technique has been applied (Figure 7.9) [88]. For low partial pressure conditions ($P(O_2) = 1.5 \times 10^{-4}$ mbar, $P(CO) = 1 \times 10^{-5}$ mbar) and the reaction temperature of 450 K, the yield of CO_2 molecules (CO_2^+ ions) has shown the maximum intensity when the traveling reaction zone, ~40 Å in width between O_{ads} and CO_{ads} layers (oxygen wave front), crosses the hole of atom-probe. This CO_2^+ yield shows low intensity on the CO_{ads}-side of the reaction and high intensity on the O_{ads}-side of the reaction; the highest intensity is obtained when the reaction zone front passes. Possible mechanisms describing such a sharp reaction zone formation are based on a local transformation of the Pt(1 0 0)-(hex) $\leftrightarrow$ (1 × 1) and the appearance of empty sites.

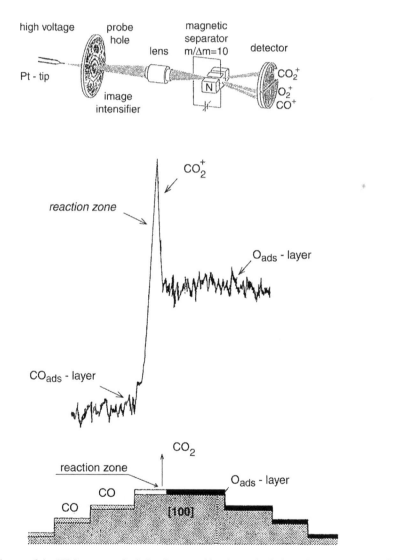

FIGURE 7.9 Scheme of the FIM mass-analysis implemented by the probe hole technique. Reaction/diffusion front monitored during $CO + O_2$ reaction on the [1 0 0]-oriented Pt field emitter. The preadsorbed CO_{ads} layer in the presence of $P(CO) = 1 \times 10^{-5}$ mbar reacts with oxygen ($P(O_2) = 1.5 \times 10^{-4}$ mbar), producing a sharp O_{ads} wave front crossing the hole in the period of time ~100 ms. $F \approx 2$ V/Å. (Reprinted from Gorodetskii, V.V. and Drachsel, W., Appl. Catal. A: Gen., 188, 267–275, 1999. With permission from Elsevier.)

C. SUBSURFACE OXYGEN: Pd-TIP EXPERIMENTS

Isothermal, nonlinear dynamic processes for the $CO + O_2$ reaction on Pd-tips and the formation of face-specific adsorption islands and the mobility of reaction–diffusion fronts were studied by FEM [66]. The initiating role of Pd {1 1 0} nanoplanes for the generation of local waves on the Pd-tip surface was established. Figure 7.5b reports the series of FEM oscillations detected when the Pd-tip was exposed to a $CO + O_2$ reaction mixture at 425 K. The oscillation amplitude obtained when going from CO_{ads} (low current) to O_{ads} (high current) layers has a periodicity of approximately 5 sec. A difference in WF between the CO_{ads} and O_{ads} covered surfaces of ~ 0.6 eV is connected with these electron current intensity changes. Analysis of Pd-tip surfaces with a local resolution of ~ 20 Å shows the availability of a sharp boundary between the mobile CO_{ads} and O_{ads} fronts (Figure 7.10). It was found that subsurface oxygen formation also acts as a source in the generation of chemical waves over the Pd-tip surface. Two more observations derive from these experiments: (i) maximum initial rate was observed on the {1 1 0} plane and (ii) two spatially separated adlayers are formed on the surface of the tip [66]. Oxygen layers form only on the {1 1 1}, {1 1 0}, {3 2 0}, and {2 1 0} planes, whereas CO_{ads} layers form on the {1 0 0} and {1 0 0}$_{step}$ planes. The O_{ads} and CO_{ads} layers interact via a sequence of reaction steps, including a reversible $O_{ads} \leftrightarrow O_{sub}$ transition, which acts as the feedback step during the oscillations, to cause the chemical waves.

D. CONCLUSIONS

The low-temperature reaction between CO and O_2 on Pd(1 1 1), Pd(1 1 0), and Pt(1 0 0) surfaces and the mechanism of surface wave generation in the oscillating regime of this reaction on Pd- and Pt-tip surfaces were investigated using different macroscopic (MB, TPR, HREELS) and microscopic (FEM) analytical tools to learn about the details of the reaction dynamics at those catalyst surfaces. The key results from these studies are summarized below.

(i) Both Pt and Pd surfaces are catalytically active for the $CO + O_2$ reaction due to their ability to dissociate O_2 molecules. Results from [18]O labeling experiments directly support the conclusion that adsorbed (weakly bound) atomic oxygen is the active form of oxygen that reacts with carbon monoxide at low (160–200 K) temperatures to form CO_2. It was determined that the local concentration of CO_{ads} on the Pt(1 0 0)-hex prevents oxygen atoms from occupying the hollow position, and apparently leads to the formation of weakly bound active O_{ads} atoms. The subsurface oxygen O_{sub} that can be formed on the Pd(1 1 0) surface may be an important species for the formation of this new weakly bound O atoms. The adsorbed O_{ads} is highly active compared with the O_{sub} species, and rapidly reacts with adsorbed carbon monoxide, CO_{ads}, to produce CO_2. Finally, comparison of

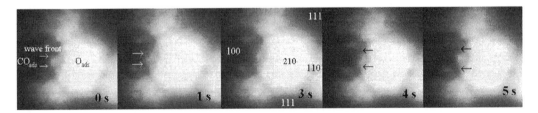

FIGURE 7.10 Magnified view of a sequence of FEM-images over (1 1 0)-oriented Pd-tip obtained during the $CO + O_2$ oscillating cycle at $T = 425$ K, $P(O_2) = 2.6 \times 10^{-3}$ mbar, $P(CO) = 1.3 \times 10^{-4}$ mbar. At $\tau = 0 \div 3$ s formation of a large CO_{ads}-island (dark area over Pd(1 0 0) plane) and the (2 1 0) plane temporarily covered by oxygen O_{ads} layer appear bright – moving reaction front during regular oscillation, starting from (1 0 0) toward (2 1 0) plane. At 3 s the CO_{ads} layer forms a final island size. After 4.0 sec a reverse reaction front starts from (2 1 0) toward (1 0 0) plane. (Reprinted from Gorodetskii, V.V., Matveev, A.V., Kalinkin, A.V., and Nieuwenhuys, B.E., Chem. Sustain. Dev., 11, 67–74, 2003. With permission from Publishing house of Siberian Branch of Russian Academy of Sciences (SB RAS).)

the kinetics for CO oxidation in molecular beam experiments with and without preadsorbed atomic oxygen shows that the surface reaction between adsorbed CO and adsorbed atomic oxygen is rate limiting in instances where the dissociation of molecular oxygen does not occur.

(ii) FEM has been developed to investigate the dynamics of the surface phenomena associated with CO oxidation on a nano-scale level. Sharp tips of Pt and Pd, up to several hundreds of Angstroem units in size, have been used to perform investigations of real dynamic surface processes simultaneously on different crystallographic nanoplanes of the emitter-tip. These tips were proven to be excellent models for metal-supported catalyst to study $CO + O_2$ oscillations *in situ*.

IV. MATHEMATICAL MODELING

A. General Background

Detailed studies of the coadsorption of oxygen and carbon monoxide, hysteresis phenomena, and oscillatory reaction of CO oxidation on Pt(1 0 0) and Pd(1 1 0) single crystals, Pt- and Pd-tip surfaces have been carried out with the MB, FEM, TPR, XPS, and HREELS techniques. It has been found that the Pt(1 0 0) nanoplane under self-oscillation conditions passes reversibly from a catalytically inactive state (hex) into a highly active state (1 × 1). The occurrence of kinetic oscillations over Pd nanosurfaces is associated with periodic formation and depletion of subsurface oxygen (O_{sub}). Transient kinetic experiments show that CO does not react chemically with subsurface oxygen to form CO_2 below 300 K. It has been found that CO reacts with an atomic O_{ads}/O_{sub} state beginning at temperature ~150 K. Analysis of Pd- and Pt-tip surfaces with a local resolution of ~20 Å shows the availability of a sharp boundary between the mobile CO_{ads} and O_{ads} fronts. The study of CO oxidation on Pt(1 0 0) and Pd(1 1 0) *nanosurfaces* by FEM has shown that the surface "phase transition" and oxygen penetration into the subsurface can lead to critical phenomena such as hysteresis, self-oscillations, and chemical waves.

Below we intend to compare the specific features of the statistical lattice models (based on the Monte Carlo technique) for imitating the oscillatory and autowave dynamics in the adsorbed layer during carbon monoxide oxidation over Pt(1 0 0) and Pd(1 1 0) single crystals differing by the structural properties of catalytic surfaces. The statistical lattice model constructed for $(CO + O_2)/Pd$ reaction takes into account the change of surface properties due to the penetration of the adsorbed oxygen into subsurface layer. Autowave processes on the model palladium surface accompany the oscillations of the rate of CO_2 formation and the concentrations of the adsorbed species. The existence of the reaction zone between the moving adsorbate islands has been shown. The statistical lattice model has been constructed for the $(CO + O_2)/Pt(1 0 0)$ reaction which takes into account the change of surface properties due to the adsorbate-induced reversible surface transformation hex ↔ 1 × 1. The model reproduces qualitatively the hysteresis and the oscillations of reaction rate, O_{ads}, CO_{ads} coverages, hex, and 1 × 1 surface phases under the conditions close to the experimental ones. Autowave processes accompany self-oscillations of the reaction rate. The existence of the reaction zone between the moving adsorbate islands characterized by the elevated concentration of the free active centers has been shown.

In both cases (Pt(1 0 0) and Pd(1 1 0)), the synchronous oscillations of the reaction rate and surface coverages are exhibited within the range of the suggested model parameters under the conditions very close to the experimental observations [89,90]. These oscillations are accompanied by the autowave behavior of surface phases and adsorbates coverages. The intensity of CO_2 formation in the CO_{ads} layer is low, inside oxygen island it is intermediate and the highest intensity of CO_2 formation is related to a narrow zone between the growing O_{ads} island and surrounding CO_{ads} layer ("reaction zone"). The presence of the narrow reaction zone was found experimentally by means of the field ion probe-hole microscopy technique with ~5 Å resolution (Figure 7.9) [34]. The boundaries of oscillatory behavior and hysteresis effects have been revealed. The possibility for the appearance of the turbulent patterns, a spiral and elliptic wave on the surface, in the cases under study has been shown.

The mechanism of local oscillator synchronization is one of the fundamental problems arising when studying the oscillatory behavior of heterogeneous catalytic reaction [81]. The isothermal kinetic oscillations in different oxidation reactions are observed as a rule [81,91] on the supported metals and on metal tips considered as a superposition of the interrelated single-crystal nanoplanes. The following factors should play the dominant role in synchronizing a catalytic system consisting of separate oscillators with different properties, especially in the case of high-pressure experiments: (i) the global coupling through the gas phase and (ii) in the case of the support with high thermal conductivity, the coupling *via* heat transfer [92 and references therein]. In the case of single crystals and metal-tips studies under UHV conditions, the surface diffusion of the adsorbed species can be responsible in general for the synchronization of local oscillators. Furthermore, according to Boudart [93] considering the kinetic features of CO oxidation reaction over Pd/Al_2O_3 it is necessary to take into account the contribution of CO_{ads} diffusion over the support onto the active metal particles surface (spillover). In our contribution, we shall consider the possible consequences of the several catalytic surface sections coupling (in our case that is the surface of the Pd-tip with four Pd(1 1 0) faces) exhibiting the surface wave behavior with some time shift in the period of oscillations. The analysis would be provided by means of statistical lattice modeling by the example of the CO oxidation reaction over Pd(1 1 0).

The current knowledge of the $CO + O_2$ reaction mechanism makes possible to state rather justified theoretical models giving insight to the features of spatio-temporal dynamics of reaction on the platinum surface. Carbon monoxide oxidation over Pt(1 0 0) single crystal has been studied comprehensively. It was shown that under certain conditions (partial pressures of reactants and temperature), the adsorbate coverages and the reaction rate undergo self-oscillations attended by the spatio-temporal pattern of CO_{ads} and O_{ads} formation on the surface [1,2,94]. The observed phenomena are associated with the reversible adsorbate-induced surface phase transition hex $\leftrightarrow$ 1 × 1. The platinum state in unreconstructed 1 × 1 phase is catalytically active due to the ease of oxygen molecules dissociation: $S^{1 \times 1}(O_2) \approx 0.3-0.4 \gg S^{hex}(O_2) \approx 10^{-3}$. The CO adsorption on the reconstructed hex surface is described by the nucleation and trapping mechanism. STM data evidence that upon attaining some critical coverages $\approx 0.05-0.1$ ML, the hex to 1 × 1 surface phase transition proceeds with formation of $CO_{ads}/1 \times 1$ islands [1,2]. This phase transition is accompanied by the formation of structural defects since hex phase of Pt(1 0 0) is more than 20% dense than the 1 × 1 phase. When the CO_{ads} coverage falls below a critical value, than the reverse surface phase transition (1 × 1) $\rightarrow$ hex is initiated. The (1 × 1) phase of Pt(1 0 0) is unstable and at $T > 400$ K transforms quickly into the nonactive hex phase.

In the early theoretical models of $(CO + O_2)/Pt(1 0 0)$ reaction, the rate of hex $\rightarrow$ (1 × 1) transition was assumed to be linearly dependent on the local CO_{ads} coverage over the hex phase (e.g., [95]). However, it has been shown recently by molecular beam studies that the (1 × 1)-CO island growth rate, and therefore, the (1 × 1) phase, is governed by a strongly nonlinear power law $\dot{\Theta}_{1 \times 1} \sim (\Theta_{CO}^{hex})^n$, where $\Theta_{1 \times 1}$ is a part of the surface transformed into (1 × 1) phase, Θ_{CO}^{hex} the CO coverage on the hex phase, and $n \approx 4$ [52]. Subsequently, this was included in the new model accounting for the oscillatory behavior of $CO + O_2/Pt(1 0 0)$ reaction [96].

B. Pd: MODELING OF OSCILLATIONS AND WAVE PATTERNS

The catalytic oxidation of CO over palladium surfaces exhibits temporal oscillations under a certain range of reaction parameters, T and P_i. A feedback mechanism for these oscillations is associated with the changes in the sticking probability of oxygen (S_0) induced by depletion of subsurface oxygen (Pd(1 1 0): $O_{ads} \leftrightarrow O_{sub}$). The "oxide" model [2] assumes that the O_{sub} layer simultaneously blocks oxygen adsorption and helps the growth of CO_{ads} layers, leading to surface reaction poisoning (low rate of CO_2 formation). Nevertheless, a slow CO_{ads} reaction with O_{sub} removes the subsurface oxygen, after which O_2 adsorption is again possible (high rate of CO_2 formation). Then, the subsurface oxygen layer forms again and the cycle is restored. This subsurface oxygen is also likely to induce significant changes in the adsorption energies of CO_{ads} and O_{ads} [97]. Based on our TPR, MB, and FEM data regarding CO oxidation over Pd surfaces, some elementary steps have been added to the classic LH scheme, namely

Scheme I

Atomic oxygen route

(1) $O_{2(gas)} + 2* \longrightarrow 2O_{ads}$
(2) $CO_{(gas)} + * \longleftrightarrow CO_{ads}$
(3) $CO_{ads} + O_{ads} \longrightarrow CO_{2(gas)} + 2*$

Subsurface oxygen route

(4) $O_{ads} + *_v \longrightarrow *O_{sub}$
(5) $CO_{ads} + O_{sub} \longrightarrow CO_{2(gas)} + 2* + *_v$
(6) $CO_{gas} + *O_{sub} \longleftrightarrow CO_{ads}O_{sub}$
(7) $CO_{ads}O_{sub} \longrightarrow CO_{2(gas)} + * + *_v$

Here $*$ and $*_v$ are the active sites on the surface and subsurface layers, respectively. The first step describes irreversible oxygen adsorption; the second step is the adsorption and desorption of CO. The third step corresponds to the reaction between CO_{ads} and oxygen atoms O_{ads}. Formation of the subsurface oxygen proceeds according to step (4). The reaction between the nearest-neighbor CO_{ads} molecules and subsurface oxygen is described by step (5). The formation of a complex of CO_{ads} molecules on dissolved oxygen in the form $CO_{ads}O_{sub}$ occurs both *via* the direct adsorption of CO from the gas phase (step (6)), and *via* CO diffusion along the surface. The decomposition of the $CO_{ads}O_{sub}$ complex is accompanied by the formation of CO_2 molecules and freeing adsorption sites $*$ and $*_v$ (step (7)). We suppose that the heat of CO adsorption on the oxidized centers [O_{sub}] is less than that on the initial $*$ one, that is, the probability of $CO_{ads}O_{sub}$ desorption (step (6)) is greater than that of CO_{ads} (step (2)) one. The adsorbed CO_{ads} species can diffuse over the surface according to the following rules: (i) $CO_{ads} + * \leftrightarrow * + CO_{ads}$; (ii) $CO_{ads} + *O_{sub} \leftrightarrow * + CO_{ads}O_{sub}$; (iii) $CO_{ads}O_{sub} + *O_{sub} \leftrightarrow *O_{sub} + CO_{ads}O_{sub}$. The sequence of steps (1)–(5) is often used for modeling of oscillations in oxidation catalytic reactions including the Monte Carlo models. In our study, in addition to steps (1)–(5), the possible process of $CO_{ads}O_{sub}$ complex formation has been considered both due to CO adsorption (step (6)), and to the CO_{ads} diffusion over the surface. Step (4) is supposed to be irreversible.

The following sequence of an oscillatory cycle has been proposed: (i) O_{sub} formation takes place only on the O_{ads}-covered palladium surface, accompanied by a decrease of the sticking coefficient for the oxygen adsorption S_{O_2}; (ii) the formation of CO_{ads} layer is a result of the fast reaction $CO_{gas} + O_{ads}$ with the formation of CO_2 molecules and their desorption; (iii) the elevated concentration of the empty active sites appears either due to reverse diffusion process $O_{sub} \rightarrow O_{ads}$ with subsequent removal of O_{ads} in the reaction with CO_{ads}, or due to slow reaction O_{sub} with CO_{ads} to form CO_2; (iv) the transition to the initial oxygen layer proceeds from $S(O_2)$ increase due to the decrease of O_{sub} concentration.

In our modeling, we used a square $N \times N$ lattice ($N = 400-1600$) with periodical boundary conditions. The states of square cells were set according to the rules determined by the detailed mechanism of the reaction (e.g., in the case of Pd(1 1 0) each lattice cell can exist in one of five states: $*$, CO_{ads}, O_{ads}, [$*O_{sub}$], [$CO_{ads} *O_{sub}$]). The time was measured in terms of the so-called Monte Carlo steps (MC step) consisting of $N \times N$ trials to choose and realize the main elementary processes. For an MC step, each cell was called once in the average. The probability of each step for the processes of adsorption, desorption, and reaction was determined by the ratio of the rate constant of a given step to the sum of the rate constants of all steps.

After each choice of one of the processes and an attempt to realize it, the program considered the internal diffusion cycle that involved M diffusion attempts for CO_{ads} molecules (usually $M = 50-100$). The reaction rate of CO oxidation and the surface coverages with reactants were calculated after each MC step as a ratio of the amount of CO_2 molecules formed (or the number of lattice cells in the corresponding state) to the overall amount of cells N^2 (the procedure was described in detail in Refs. [89,90,98,99]).

In the whole range of parameters at which the reaction rate oscillated, the dependencies of a change in the adsorbate concentration and the reaction rate have a similar nature (Figure 7.11). A drastic increase in the reaction rate occurs simultaneously with the removal of the CO_{ads} layer and filling the surface with the oxygen layer (Figure 7.12a–d). This moment of time on the curve of the concentration of subsurface oxygen corresponds to the minimal value for a given period. When the rate approaches its maximal value, the concentration of adsorbed oxygen redistributes: $O_{ads} \rightarrow O_{sub}$. The position of the maximum on the curve of the concentration of subsurface oxygen determines the moment of a decrease in the reaction rate. Simultaneously, the surface accumulates CO_{ads}, and this process is accompanied by the complete O_{ads} removal. The minimal value of the reaction rate is caused by the interaction of CO_{ads} molecules with subsurface oxygen. A decrease in the concentration (O_{sub}) to a certain critical value again creates favorable conditions for the reaction, and this completes the oscillation cycle. Changes in the coverages occur via the propagation of the mobile waves, whose front is characterized by the high concentration of catalytically active sites responsible for the maximal rate of CO_2 molecule formation. A decrease in the parameter M, which determines the rate of CO_{ads} diffusion, leads to the chaotic nature of oscillations and to the appearance of complex spatio-temporal structures on the surface.

It is known that CO adsorption or the dissociative adsorption of oxygen leads to the added or missing row reconstruction of the Pd(1 1 0) plane $(1 \times 1) \rightarrow (1 \times 2)$ (Figure 7.13). As a result, the diffusion of CO_{ads} molecules along the rows of metal atoms occurs more rapidly than across the rows. We found that, if this effect is taken into account, such integral characteristics as the reaction rate and the values $\theta_{(CO)}$, $\theta_{(O)}$, and $\theta_{(O_{sub})}$ do not change. However, the propagation of a wave on the Pd(1 1 0)-(1×2) surface becomes noticeably anisotropic. Figure 7.14 shows the moments when coverages are changed. They correspond to approximately the same values of $\theta(O_{ads})$ and different

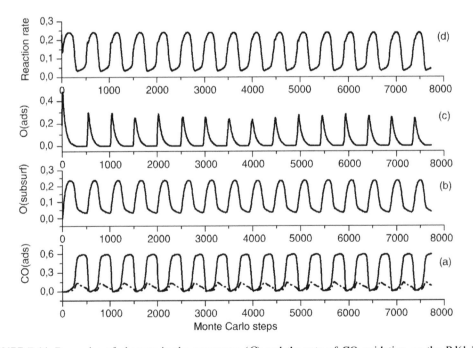

FIGURE 7.11 Dynamics of changes in the coverages (Θ) and the rate of CO oxidation on the Pd(1 1 0) surface: (a) $[CO_{ads}*O_{sub}]$ (dot-and-dash line), CO_{ads} (solid line), (b) $[*O_{sub}]$, (c) O_{ads}, and (d) the reaction rate; $N = 768$; $M = 100$. The values of the rate constants of steps (s^{-1}) (Scheme I): $k_1 = 1$, $k_2 = 1$, $k_{-2} = 0.2$, $k_3 = \infty$, $k_4 = 0.03$, $k_5 = 0.01$, $k_6 = 1$, $k_{-6} = 0.5$, and $k_7 = 0.02$. The partial pressures of reagents (CO and O_2) and the concentration of active sites on the palladium surface are included in the rate constants of adsorption k_1, k_2, and k_7. (Reprinted from Latkin, E.I., Elokhin, V.I., Matveev, A.V., and Gorodetskii, V.V., J. Mol. Catal. A: Chem., 158, 161–166, 2000. With permission from Elsevier.)

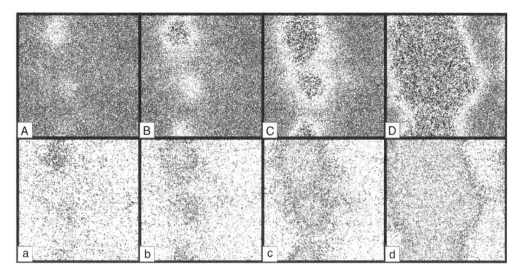

FIGURE 7.12 The distribution of adsorbates over the surface (A–D) and the intensity of CO_2 formation (a–d) at the moment when the coverages change on the Pd(1 1 0) surface. (Reprinted from Latkin, E.I., Elokhin, V.I., Matveev, A.V., and Gorodetskii, V.V., J. Mol. Catal. A: Chem., 158, 161–166, 2000. With permission from Elsevier.)

M_x/M_y ratios (x and y are the directions of axes; the direction of the x-axis coincides with the direction [1 $\bar{1}$ 0]). It is seen that, with an increase in M_x/M_y the mobile wave extends along the direction [1 $\bar{1}$ 0].

The effect of anisotropy of CO_{ads} molecule diffusion becomes more apparent if one monitors the spatio-temporal structures, such as spiral waves, which are permanently present on the surface (detailed modeling of such structures was described in Ref. [98]). Figure 7.15 shows the evolution of modeled spiral waves after the transition from the isotropic regime ($M_x/M_y = 50/50$) to the anisotropic one ($M_x/M_y = 80/20$). Figure 7.15a shows the form of the spiral wave at the moment when the anisotropy of diffusion starts to work. Figure 7.15b and c shows how spiral waves extend after passing through the first and second turns of the spiral. Such an asymmetric behavior of the spiral agrees very well with experimental data obtained using the photoelectron emission microscopy (PEEM) [100]. Figure 7.16 shows images of the adsorbed layer in the case of the spiral wave in the reaction of CO oxidation on Pd(1 1 0). It is clearly seen that the spiral wave observed in the experiment [1 0 0] reflects the anisotropy of the single-crystal Pd(1 1 0) surface in the direction [1 $\bar{1}$ 0].

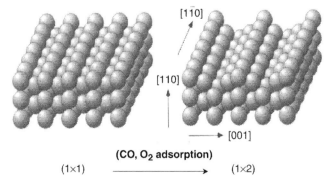

(CO, O$_2$ adsorption)

(1×1) ⟶ (1×2)

FIGURE 7.13 The restructuring of the Pd(1 1 0)-(1 × 1) into (1 × 2) with missing/added rows in the course of O_2 or CO adsorption. (Reprinted from Matveev, A.V., Latkin, E.I., Elokhin, V.I., and Gorodetskii, V.V., Chem. Sustain. Dev., 11, 173–180, 2003. With permission from Publishing house of SB RAS.)

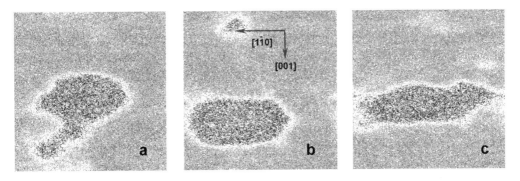

FIGURE 7.14 The distribution of adsorbates over the surface in the course of the oxygen wave propagation ($N = 768$, $(O_{ads}) \approx 0.05$): (a) 2979 MC step, $M_x/M_y = 75/25$; (b) 3951 MC step, $M_x/M_y = 80/20$; (c) 7344 MC step, $M_x/M_y = 85/15$. (Reprinted from Matveev, A.V., Latkin, E.I., Elokhin, V.I., and Gorodetskii, V.V., Chem. Sustain. Dev., 11, 173–180, 2003. With permission from Publishing house of SB RAS.)

C. Pt: MODELING OF OSCILLATIONS AND WAVE PATTERNS

Our FEM results have strongly suggested that the CO-induced (hex) $\leftrightarrow$ (1×1) phase transition observed on Pt(1 0 0) single crystals is the driving force for the isothermal oscillations seen on that surface, in the same way as on the (1 0 0) planes of microscopic Pt grains. The reactive phase diagram for the (1 0 0) nanoplanes of a Pt-tip surface does correlate with that of large Pt single crystals [82,83]. The oscillation cycles of the CO oxidation on macroscopic Pt(1 0 0) single crystals has been described by Cox et al. [101] and Imbihl et al. [95]. According to the Ertl model [102], it is possible to account for these reactions with a mechanism having the following key ingredients:

(i) On a starting (hex) surface (CO side of reaction), the negligible value of the oxygen sticking coefficient ($S_0 \approx 10^{-3}$) results in both preferential adsorption of CO ($S_0 \approx 0.8$) from the gas mixture and surface diffusion from {1 1 3} planes (Figure 7.7b(d)). As the local CO_{ads} coverage approaches an average coverage of $CO_{hex} > 0.08$, the (hex) $\rightarrow$ (1×1) phase transition is initiated (Figure 7.7b(e)), and islands with $CO_{1 \times 1}$ structure grow at the expense of the (hex) surface area.

(ii) According to HREELS data, both the local $CO_{1 \times 1}$ (0.5 ML) layers and the (hex) surface areas inhibit O_2 dissociative adsorption and the $CO_{1 \times 1} + O_{2\,ads}^{2-}$ reaction (Figure 7.3). On the other hand, the production of empty sites as a result of the $CO_{hex} \rightarrow CO_{1 \times 1}$ phase transition [103] permits O_{ads} formation, and O_2 molecular dissociation is accompanied by active O adatoms

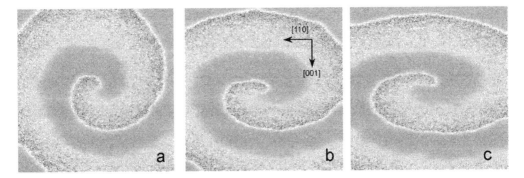

FIGURE 7.15 Changes in the form of the spiral wave in the reaction of CO oxidation on the Pd(1 1 0) surface after the transition from the regime with isotropic diffusion ($M_x/M_y = 50/50$) to anisotropic ($M_x/M_y = 80/20$), $N = 1536$: (a) isotropic diffusion; (b) after the first turn of the spiral; and (c) after the second turn of the spiral. (Reprinted from Matveev, A.V., Latkin, E.I., Elokhin, V.I., and Gorodetskii, V.V., Chem. Sustain. Dev., 11, 173–180, 2003. With permission from Publishing house of SB RAS.)

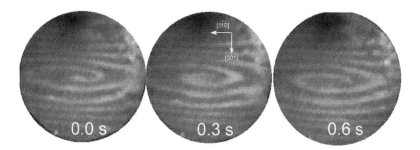

FIGURE 7.16 PEEM images of the surface in the case of existence of a spiral wave in CO oxidation on Pd(1 1 0) [1 0 0]: $P(O_2) = 4 \times 10^{-3}$ Torr, $P(CO) = 1.6 \times 10^{-5}$ Torr, $T = 349$ K. Dark regions show CO_{ads}, light regions show O_{ads}; they corresponds to different values of work function. (Reprinted from Matveev, A.V., Latkin, E.I., Elokhin, V.I., and Gorodetskii, V.V., Chem. Sustain. Dev., 11, 173–180, 2003. With permission from Publishing house of SB RAS.)

generation (Figure 7.3d). A fast reaction between these O_{ads} atoms and CO_{ads}, and the creation of vacant surface sites for fast oxygen adsorption (O_{2ads}^{2-} + empty sites → $O_{ads} + O_{ads, active}$), leads to an autocatalytic CO_2 formation step. The $CO_{1 \times 1}$ coverage then decreases, and an O_{ads} layer on the (1 0 0) plane is formed (Figure 7.7b(f)).

(iii) The areas, where CO molecules are consumed, are replenished either by CO molecules from the gas phase or by surface diffusion from the adjacent areas where CO_{ads} layers are still present (Figure 7.7b(b)). The reduced local oxygen coverage destabilizes the (1 × 1) phase and leads to the transformation of the (hex) phase (Figure 7.7b(c,d)), thus cutting down further oxygen adsorption.

(iv) CO molecules adsorbed on the (hex) patches diffuse into the (1 × 1) areas and react with O_{ads}. The (hex) phase grows at the expense of the (1 × 1) patches until the initial state (i) is reached again. Possible mechanisms describing a sharp reaction zone formation are based on: (a) local observations of the Pt(1 0 0)-(hex) ↔ (1 × 1) transformation, and the appearance of empty sites and (b) formation of the weakly bound O_{ads} atoms, highly active toward oxidation of the adsorbed CO molecules. The detailed mechanism of this CO oxidation process is as follows [89]:

Scheme II

(1) $CO + * \longrightarrow CO_{ads}$

(2) $CO_{ads}^{hex} \longrightarrow CO + *_{hex}$

(3) $CO_{ads}^{1 \times 1} \longrightarrow CO + *_{1 \times 1}$

(4) (hex) → (1 × 1): $4CO_{ads} \longrightarrow 4CO_{ads}^{1 \times 1}$

(5) (1 × 1) → (hex): $*_{1 \times 1} \longrightarrow *_{hex}$

(6) $O_2 + 2*_{1 \times 1} \longrightarrow 2O_{ads}^{1 \times 1}$

(7) $O_{ads}^{1 \times 1} + CO_{ads} \longrightarrow CO_2 + *_{1 \times 1} + *$

(8) $CO_{ads} + * \longrightarrow * + CO_{ads}$

(1) CO adsorption: the absence of indices near the center * implies that CO, in contrast to oxygen, is considered to have equal sticking probability on both $*_{hex}$ and $*_{1 \times 1}$. (2 and 3) CO desorption: the rate coefficients for CO desorption on hex and 1 × 1 phases differ widely (approximately by three to four orders of magnitude). (4) Structural phase transformation (hex) → (1 × 1): in accordance with Ref. [52], let us assume that the adsorption of four CO molecules on the 2 × 2 neighboring centers of the lattice would transform these centers (with some probability) into the (1 × 1) structure. (5) Back-structural phase transition (1 × 1) → (hex). (6) Oxygen adsorption: oxygen adsorbs dissociatively only on the two neighboring (1 × 1) centers. (7) CO_2 formation: the surface reaction proceeds via the Langmuir–Hinshelwood mechanism conserving the type of the active centers. Adsorbed oxygen interacts equally with both CO_{ads}^{hex} and $CO_{ads}^{1 \times 1}$. (8) CO_{ads} diffusion: adsorbed carbon monoxide can diffuse via hopping from their sites to vacant nearest-neighbor site and the type of active centers remains the same. Along with the stage (7), this process offers an additional source of empty active centers $*_{1 \times 1}$ required for dissociative oxygen adsorption. Monte Carlo simulations could again reproduce both the oscillations seen for the rate of CO_2 formation and the appearance of surface waves on Pt(1 0 0) using this mechanism [89].

Figure 7.17 shows autooscillations of the rate of CO_2 formation, the surface coverage with O_{ads} and CO_{ads}, and the portion of the vacant surface in the (hex) and (1×1) structures. At the initial moment, the platinum surface is in the (hex) state and only CO adsorption is allowed. Despite the low rate of (hex) $\rightarrow$ (1×1) restructuring, the portion of the (1×1) rapidly grows. As a result of CO_{ads} diffusion, sites for oxygen adsorption are formed. However, the value of $O_{(1 \times 1)}$ coverage is very low due to the fast reaction with neighboring CO_{ads} molecules. When the maximal value of $\theta(CO_{ads}) \sim 0.8$ is attained, the adsorbed layer largely consists of $CO_{(hex)}$ and $CO_{(1 \times 1)}$ (Figure 7.18A), which is accompanied by a sharp decrease in the reaction rate. Figure 7.18a shows that the rate of CO_2 molecule formation in the CO_{ads} layer is low, but regions with an elevated concentration of vacant sites are available on the surface. In these regions, the $O_{(1 \times 1)}$ islands nucleate and propagate along the metal surface (Figures 7.18B and C, b and c). Figure 7.18b and c shows that the maximal intensity of CO_2 molecule formation is observed in the narrow zone at the boundary of growing O_{ads} phase and the CO_{ads} layer. The appearance of such a narrow reaction zone was experimentally observed by the field ion probe-hole microscopy technique with a resolution of ~ 5 Å (Figure 7.9). Inside the oxygen island, the rate of CO_2 formation has an intermediate value; the lower value is in the CO_{ads} layer. The highest value of the reaction rate along the oscillation period corresponds to the moment when the perimeter of the reaction zone is at maximum (Figure 7.18c). At the final stage of the oscillation cycle, the $\theta(CO_{ads})$ coverage increases due to CO molecule adsorption on the vacant sites (both (hex) and (1×1)) with the further transition of the (1×1) phase into (hex) (Figures 7.18D and d).

D. SPATIO-TEMPORAL CHAOS

A decrease in the parameter M from 100 to 50 does not affect the regular nature and uniformity of oscillations but leads to a small decrease in the period and amplitude of oscillations. However, with

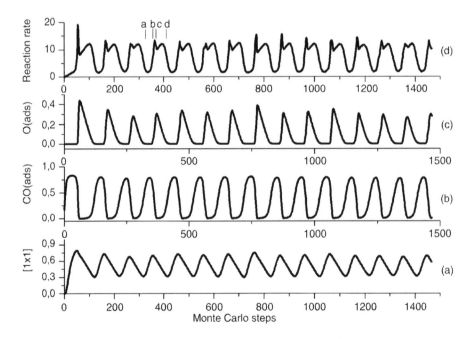

FIGURE 7.17 Oscillations of the reaction rate and the surface coverage (Θ) of Pt(1 0 0) for the lattice size $N = 384$ and the parameter $M = 100$. The values of the rate constants of elementary steps (Scheme II): $k_1 = 2.94 \times 10^5$ ML sec^{-1} Torr^{-1} (1 ML $= 9.4 \times 10^{14}$ atom/cm^2), $P_{CO} = 5 \times 10^{-5}$ Torr, $k_2 = 4c^{-1}$, $k_3 = 0.03c^{-1}$, $k_4 = 3c^{-1}$, $k_5 = 2c^{-1}$, $k_6 = 5.6 \times 10^5$ ML s^{-1} Torr^{-1}, $P_{O_2} = 10^{-4}$ Torr, $k_7 = \infty$. (Reprinted from Latkin, E.I., Elokhin, V.I., and Gorodetskii, V.V., J. Mol. Catal. A: Chem., 166, 23–30, 2001. With permission from Elsevier.)

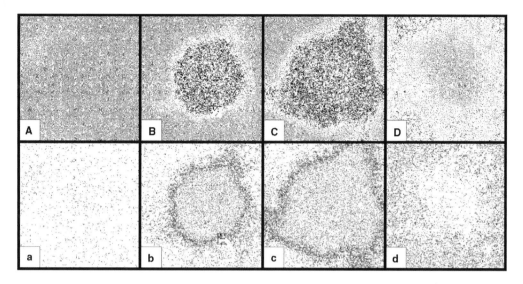

FIGURE 7.18 The distribution of adsorbates over the surface (A–D) and the intensity of CO_2 formation (a–d) at the moment when the surface coverage changes on Pt(1 0 0). The concentration of O_{ads} is shown in black, CO_{ads} is in gray, and free surface is in white. Parts A(a)–D(d) correspond to the moments 332, 360, 364, and 408 MC steps shown in Figure 7.17, curve d — reaction rate. (Reprinted from Latkin, E.I., Elokhin, V.I., and Gorodetskii, V.V., J. Mol. Catal. A: Chem., 166, 23–30, 2001. With permission from Elsevier.)

a decrease in M to 30, the period and amplitude of oscillations become irregular (Figure 7.19). In this case, O_{ads} atoms are always present on the surface and mobile islands in the form of cellular structures, spiral fragments, etc., are formed (Figure 7.20). Similar turbulent spatio-temporal structures were experimentally observed in the reaction of CO oxidation on Pt(1 0 0) using the method of ellipsomicroscopy for surface imaging (EMSI) [91].

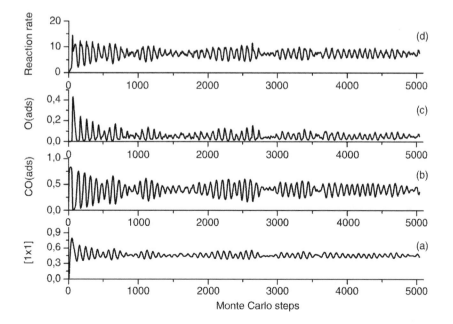

FIGURE 7.19 Oscillations of the reaction rate and the surface coverage of Pt(1 0 0) for the lattice size $N = 384$ and the parameter $M = 30$. (Reprinted from Latkin, E.I., Elokhin, V.I., and Gorodetskii, V.V., J. Mol. Catal. A: Chem., 166, 23–30, 2001. With permission from Elsevier.)

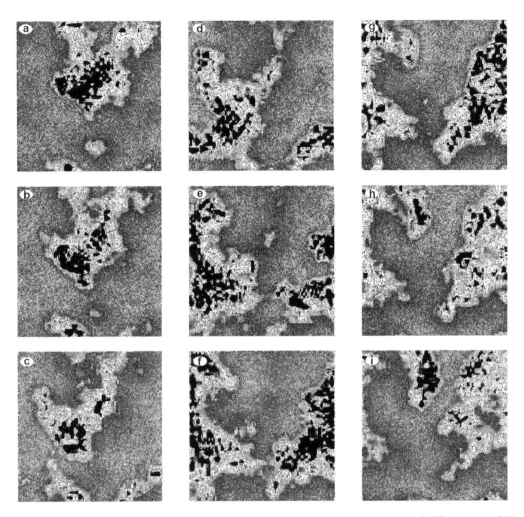

FIGURE 7.20 The distribution of adsorbates over the surface in the case of the low rate of diffusion ($M = 30$). The concentration of O_{ads} is shown in black, CO_{ads} is in gray, and the free surface is shown in white. (Reprinted from Latkin, E.I., Elokhin, V.I., and Gorodetskii, V.V., J. Mol. Catal. A: Chem., 166, 23–30, 2001. With permission from Elsevier.)

V. SUMMARY AND CONCLUSIONS

The reaction between CO and O_2 on Pd(1 1 0) and Pt(1 0 0) surfaces and the mechanism of surface wave generation in the oscillating regime of this reaction on Pd- and Pt-tip surfaces was investigated using different macroscopic (MB, TPR, HREELS, WF) and microscopic (FEM, FIM) analytical tools to learn about the details of the reaction dynamics at those catalyst surfaces. The key results from these studies are summarized below.

The principal result of this work is that the nonlinear reaction kinetics is not restricted to macroscopic planes as: (i) the planes $\sim$200 Å in diameter show the same nonlinear kinetics; (ii) the regular waves appear under the reaction rate oscillations; (iii) the propagation of reaction — diffusion waves includes the participation of the different crystal nanoplanes and indicates an effective coupling of adjacent planes.

Chemical wave patterns corresponding to moving surface concentration patches are the result of coupling of surface diffusion, surface reconstruction, and surface reaction. Depending on the reaction condition, such spatio-temporal phenomena can also lead to an oscillatory behavior of the overall reaction rate. It becomes possible to study the catalysis on an atomic level which is necessary for understanding the mechanism of the action of the high-dispersion-supported metal catalysts having the metal microcrystallites $\sim100-300$ Å in size as an active part of the catalyst. This result opens new fields for the development of theoretical concepts of heterogeneous catalysis.

Thus, we constructed and studied statistical lattice models that describe the oscillation and wave dynamics in the adsorbed layer for the reaction of CO oxidation over the Pt(1 0 0) and Pd(1 1 0) single-crystal surfaces. These models differ in the mechanisms of formation of oscillations: a mechanism involving the phase transition of planes (Pt(1 0 0): (hex)$\longleftrightarrow$(1 $\times$ 1)) and a mechanism via formation of subsurface oxygen (Pd(1 1 0)). The models demonstrate the oscillations of the rate of CO_2 formation and the concentrations of adsorbed reactants. These oscillations are accompanied by various wave processes on the lattice that models single crystalline surfaces. The effects of the size of the model lattice and the intensity of CO_{ads} diffusion on the synchronization and the form of oscillations and surface waves are studied. We showed the presence of a narrow zone of the reaction when the wave front propagates along the metal surface. This is supported by the results obtained by the methods of FEM and FIM. We found that the inclusion of CO_{ads} diffusion anisotropy, which reflects the real symmetry of the single-crystal Pd(1 1 0) surface does not affect the dynamics of oscillations of the integral characteristics of the reaction (the rate and the surface coverage), but it leads to the formation of ellipsoid spatio-temporal structures on the surface, which were observed experimentally with modern physical methods for surface science studies. It was shown that it is possible to obtain a wide spectrum of chemical waves (cellular and turbulent structures, and spiral and ellipsoid waves) using the lattice models developed. These waves have been observed in experimental studies of oscillatory dynamics of catalytic reactions.

Finally, it may be said that the character of the CO + O_2 oscillating reaction on Pd differs remarkably from that on Pt because: (a) different subsurface oxygen (Pd) and (hex) $\leftrightarrow$ (1 $\times$ 1) phase transition (Pt) mechanisms apply and (b) the oxygen front in CO + O_2 waves travel in reverse directions: on Pd it goes from (1 1 0) to (1 0 0) surface, on Pt it travels in the opposite direction.

ACKNOWLEDGMENT

This work was supported by RFBR Grant # 05-03-32971 and NWO Grant # 047.015.002.

REFERENCES

1. Ertl, G., Oscillatory catalytic reactions at single-crystal surfaces, *Adv. Catal.*, 37, 213–277, 1990.
2. Imbihl, R., Oscillatory reactions on single-crystal surfaces, *Progr. Surf. Sci.*, 44, 185–343, 1993.
3. Nieuwenhuys, B.E., The surface science approach toward understanding automotive exhaust conversion catalysis at the atomic level, *Adv. Catal.*, 44, 259–328, 1999.
4. Sheintuch, M. and Schmitz, R.A., Oscillations in catalytic reactions, *Catal. Rev.-Sci. Eng.*, 15, 107–172, 1977.
5. Slinko, M.G. and Slinko, M.M., Self-oscillations of heterogeneous catalytic reaction rates, *Catal. Rev.-Sci. Eng.*, 17, 119–153, 1978.
6. Razon, L.F. and Schmitz, R.A., Intrinsically unstable behavior during the oxidation of carbon monoxide on platinum, *Catal. Rev.-Sci. Eng.*, 28, 89–164, 1986.
7. Schüth, F., Henry, B.E., and Schmidt, L.D., Oscillatory reactions in heterogeneous catalysis, *Adv. Catal.*, 39, 51–127, 1993.
8. Eiswirth, M. and Ertl, G., Pattern formation on catalytic surfaces, in: *Chemical Waves and Patterns*, Understanding Chemical Reactivity, Kapral, R., Showalter, K., Eds., Kluwer Academic Publishers, Dordrecht, 1994, Vol. 10.
9. Slinko, M.M. and Jaeger, N.I., Oscillating heterogeneous catalytic systems, *Studies in Surface Science and Catalysis*, Delmon, B., Yates, J.T., Eds., Elsevier, Amsterdam, 1994, Vol. 86, 393 pp.

10. Eiswirth, M., Chaos in surface-catalysed reactions, in: *Chaos in Chemistry and Biochemistry*, Fields, R.J., Györgi, L., Eds., World Scientific, Singapore, 1993, Chap. 6, 141–174.

11. Imbihl, R. and Ertl, G., Oscillatory kinetics in heterogeneous catalysis, *Chem. Rev.*, 95, 697–733, 1995.

12. Ertl, G., Dynamics of reactions at surfaces, *Adv. Catal.*, 45, 1–69, 2000.

13. Cobden, P.D., Janssen, N.M.H., van Breugel Y., and Nieuwenhuys, B.E., Non-linear behaviour in the $NO-H_2$ reaction over single crystals and field emitters of some Pt-group metals, *Faraday Discuss*, 105, 57–72, 1996.

14. Janssen, N.M.H., Cobden P.D., and Nieuwenhuys, B.E., Non-linear behaviour of nitric oxide reduction reactions over metal surfaces, *J. Phys. Condens. Matter*, 9, 1889–1917, 1997.

15. Langmuir, I., I. Chemical reaction on surfaces. II. The mechanism of the catalytic action of platinum in the reaction $2CO + O_2 = 2CO_2$ and $2H_2 + O_2 = 2H_2O$, *Trans. Faraday Soc.*, 17, 607–654, 1922.

16. Matsushima, T., The mechanism of the CO_2 formation on Pt(1 1 1) and polycrystalline surfaces at low temperatures, *Surf. Sci.*, 127, 403–423, 1983.

17. Ertl, G. and Rau, P., Chemisorption und Reaktion von Sauerstoff und Kohlenmonoxid an einer Palladium (1 1 0)-Oberfläche, *Surf. Sci.*, 15, 443–465, 1969.

18. Bonzel, H.P. and Ku, R., Mechanisms of the catalytic carbon monoxide oxidation on Pt(1 1 0), *Surf. Sci.*, 33, 91–106, 1972.

19. Palmer, P.L. and Smith, J.N., Molecular beam study of CO oxidation on a (1 1 1) platinum surface, *J. Chem. Phys.*, 60, 1453–1463, 1974.

20. Engel, T. and Ertl, G., A molecular beam investigation of the catalytic oxidation of CO on Pd(1 1 1), *J. Chem. Phys.*, 69, 1267–1281, 1978.

21. Matsushima, T. and White, J.M., Kinetics of the reaction of oxygen with carbon monoxide adsorbed on palladium, *Surf. Sci.*, 67, 122–138, 1977.

22. Engel, T. and Ertl, G., Elementary steps in the catalytic oxidation of carbon monoxide on platinum metals, *Adv. Catal.*, 28, 1–77, 1979.

23. Nieuwenhuys, B.E., Adsorption and reactions of CO, NO, H_2, and O_2 on group VIII metal surfaces, *Surf. Sci.*, 126, 307–336, 1983.

24. Matsushima, T., Ohno, Y., and Rar, A., Reaction sites for carbon monoxide oxidation on a stepped platinum (1 1 2) surface: a spatial distribution study a product desorption, *Surf. Sci.*, 293, 145–151, 1993.

25. Matsushima, T., Angle-resolved measurements of product desorption and reaction dynamics on individual sites, *Surf. Sci. Rep.*, 52, 1–62, 2003.

26. Boreskov, G.K., Development of ideas on the nature of heterogeneous catalysis. *Kinet. Katal.*, 18, 1111–1121, 1977 (in Russian).

27. Becker, C.A., Cowin, J.P., Wharton, I., and Auerbach, D.J., CO_2 product velocity distributions for CO oxidation on platinum, *J. Chem. Phys.*, 67, 3394–3395, 1977.

28. Watanabe, K., Uetsuka, H., Ohnuma, H., and Kunimori, K., The dynamics of CO oxidation on Pt(1 1 0) studied by infrared chemiluminescence of the product CO_2: effect of CO coverage, *Catal. Lett.*, 47, 17–20, 1997.

29. Matsushima, T., Anisotropic velocity distribution of desorbing product in carbon monoxide oxidation on palladium (1 1 0), *J. Chem. Phys.*, 97, 2783–2789, 1992.

30. Matsushima, T., The crystal azimuth dependence of the angular distribution of the desorption flux of carbon dioxide produced on palladium (1 1 0) surfaces, *J. Chem. Phys.*, 91, 5722–5730, 1987.

31. Smirnov, M.Yu., Zemlyanov, D., Gorodetskii. V.V., and Vovk, E.I., Formation of the mixed $(NO + CO)/(1 \times 1)$ islands on the Pt(1 0 0)-(hex) surface. *Surf. Sci.*, 414, 409–422, 1998.

32. Cobden, P.D., Nieuwenhuys, B.E., and Gorodetskii, V.V., Adsorption of some small molecules on a Pd field emitter, *Appl. Catal. A: Gen.*, 188, 69–77, 1999.

33. Suchorski, Yu., Imbihl, R., and Medvedev, V.K., Compatibility of field emitter studies of oscillating surface reactions with single-crystal measurements: catalytic CO oxidation on Pt, *Surf. Sci.*, 401, 392–399, 1998.

34. Gorodetskii, V.V. and Drachsel, W., Kinetic oscillations and surface waves in catalytic $CO + O_2$ reaction on Pt surface: field electron microscope, field ion microscope and high resolution electron energy loss spectroscope studies, *Appl. Catal. A: Gen.*, 188, 267–275, 1999.

35. Liu, J., Xu, M., Nordmeyer, T., and Zaera, F., Sticking probabilities for CO adsorption on Pt(1 1 1) surfaces revisited, *J. Phys. Chem.*, 99, 6167–6175, 1995.

36. Gopinath, C.S. and Zaera, F., NO + CO + O_2 reaction kinetics on Rh(1 1 1): a molecular beam study, *J. Catal.*, 200, 270–287, 2001.

37. McCarroll, J.J., Surface physics and catalysis, *Surf. Sci.*, 53, 297–316, 1975.

38. Heilmann, P., Heinz, K., and Müller, K., The superstructures of the clean Pt(1 0 0) and Ir(1 0 0) surfaces, *Surf. Sci.*, 83, 487–497, 1979.

39. Norton, P.R., Davies, J.A., Creber, D.K., Sitter, C.W., and Jackman, T.E., The Pt(1 0 0) (5 × 20) ⇔ (1 × 1) phase transition: a study by Rutherford back-scattering, nuclear microanalysis, LEED and thermal desorption spectroscopy, *Surf. Sci.*, 108, 205–234, 1981.

40. Bonzel, H.P., Broden, G., and Pirug, G., Structure sensitivity of NO adsorption on a smooth and stepped Pt(1 0 0) surface, *J. Catal.*, 53, 96–105, 1978.

41. Rotermund, H.H., Self-organized reactions on surfaces, *Physica Scripta*, T49, 549–553, 1993.

42. Gardner, P., Martin, R., Tushaus, M., and Bradshaw, A.M., A vibrational spectroscopic investigation of NO and CO co-adsorption on Pt{1 0 0}, *J. Electron Spectrosc. Relat. Phenom.*, 54/55, 619–628, 1990.

43. Martin, R., Gardner, P., and Bradshaw, A.M., The adsorbate-induced removal of the Pt{1 1 0} surface reconstruction. Part II: CO, *Surf. Sci.*, 342, 69–84, 1995.

44. Behm, R.J., Thiel, P.A., Norton, P.R., and Ertl, G., The interaction of CO and Pt(1 0 0). I. Mechanism of adsorption and Pt phase transition, *J. Chem. Phys.*, 78, 7437–7447, 1983.

45. Thiel, P.A., Behm, R.J., Norton, P.R., and Ertl, G., The interaction of CO and Pt(1 0 0). II. Energetic and kinetic parameters, *J. Chem. Phys.*, 78, 7448–7458, 1983.

46. Jackman, T.E., Griffiths, K., Davies, J.A., and Norton, P.R., Absolute coverages and hysteresis phenomena associated with the CO induced Pt(1 0 0) hex ⇔ 1 × 1 phase transition, *J. Chem. Phys.*, 79, 3529–3533, 1983.

47. Ritter, E., Behm, R.J., Potschke, G., and Wintterlin, J., Direct observation of a nucleation and growth process on an atomic scale, *Surf. Sci.*, 181, 403–411, 1987.

48. Hösler, W., Ritter, E., and Behm, R.J., Topological aspects of the (1 × 1) ⇔ "hexagonal" phase transition on Pt(1 0 0), *Ber. Bunsenges. Phys. Chem.*, 90, 205–208, 1986.

49. Borg, A., Hilmen, A.M., and Bergene, E., STM studies of clean, CO- and O_2-exposed Pt(1 0 0)-hex-RO.7°, *Surf. Sci.*, 306, 10–20, 1994.

50. Yeo, Y.Y., Vattuone, L., and King, D.A., Energetics and kinetics of CO and NO adsorption on Pt(1 0 0) — restructuring and lateral interaction, *J. Chem. Phys.*, 104, 3810–3821, 1996.

51. Hopkinson, A., Bradley, J.M., Guo, X.-C., and King, D.A., Nonlinear island growth dynamics in adsorbate-induced restructuring of quasihexagonal reconstructed Pt{1 0 0} by CO, *Phys. Rev. Lett.*, 71, 1597–1600, 1993.

52. Hopkinson, A., Guo, X.-C., Bradley, J.M., and King, D.A., A molecular beam study of the CO induced surface phase transition on Pt(1 0 0). *J. Chem. Phys.*, 99, 8262–8269, 1993.

53. Thiel, P.A., Behm, R.J., Norton, P.R., and Ertl, G., Mechanism of an adsorbate-induced surface phase transformation: CO on Pt(1 0 0). *Surf. Sci.*, 121, L553–L558, 1982.

54. Tapilin, V.M., Zemlyanov, D.Y., Smirnov, M. Yu., and Gorodetskii, V.V., Angle-resolved photo-emission study and calculation of the electronic structure of Pt(1 1 1) surface, *Surf. Sci.*, 310, 155–162, 1994.

55. Nieuwenhuys, B.E., Correlation between work function changes and degree of electron back-donation in the adsorption of carbon monoxide and nitrogen on group VIII metals, *Surf. Sci.*, 105, 505–521, 1981.

56. Brown, W.A., Kose, R., and King, D.A., Femtomole adsorption calorimetry on single-crystal surfaces, *Chem. Rev.*, 98, 797–832, 1998 and references therein.

57. Wartnaby, C.E., Stuck, A., Yeo, Y.Y., and King, D.A., Microcalorimetric heats of adsorption for CO, NO, and oxygen on Pt{1 1 0}, *J. Phys. Chem.*, 100, 12483–12488, 1996.

58. Yeo, Y.Y., Vattuone, L., and King, D.A., Calorimetric heats for CO and oxygen adsorption and for the catalytic CO oxidation reaction on Pt(1 1 1). *J. Chem. Phys.*, 106, 392–401, 1997.

59. Conrad, H., Ertl, G., Koch J., and Latta, E.E., Adsorption of CO on Pd single-crystal surfaces, *Surf. Sci.*, 43, 462–480, 1974.

60. Yeo, Y.Y., Vattuone, L., and King, D.A., Calorimetric investigation of NO and CO adsorption on Pd(1 0 0) and influence of preadsorbed carbon monoxide, *J. Chem. Phys.*, 106, 1990–1999, 1997.

61. Brennan, D. and Hayes, F.H., The adsorption of carbon monoxide on evaporated metal films, *Phil. Trans. Roy. Soc. (London)*, A258, 347–386, 1965.

62. Kuhn, W.K., Szanyi, J., and Goodman, D.W., CO adsorption on Pd(1 1 1): the effect of temperature and pressure, *Surf. Sci. Lett.*, 274, L611–L618, 1992.

63. Kuhn, W.K., Szanyi, J., and Goodman, D.W., Adsorption isobars for CO on Pd/Ta(1 1 0) at elevated pressures and temperatures using infrared reflection–adsorption spectroscopy, *Surf. Sci.*, 303, 377–385, 1994.

64. Voogt, E.H., Coulier, L., Gijzeman, O.L.J., and Geus, J.W., Adsorption of carbon monoxide on Pd(1 1 1) and palladium model catalysts, *J. Catal.*, 169, 359–364, 1997.

65. Ehsasi, M., Seidel, C., Ruppender, H., Drachsel, W., Block, J.H., and Christmann, K., Kinetic oscillations in the rate of CO oxidation on Pd(1 1 0), *Surf. Sci.*, 210, L198–L208, 1989.

66. Gorodetskii, V.V., Matveev, A.V., Kalinkin, A.V., and Nieuwenhuys, B.E., Mechanism for CO oxidation and oscillatory reactions on Pd tip and Pd(1 1 0) surfaces, *Chem. Sustain. Dev.*, 11, 67–74, 2003.

67. Gorodetskii, V.V., Matveev, A.V., Cobden, P.D., and Nieuwenhuys, B.E., Study of H_2, O_2, CO adsorption and $CO + O_2$ reaction on Pt(1 0 0), Pd(1 1 0) monocrystal surfaces, *J. Mol. Catal. A: Chem.*, 158, 155–160, 2000.

68. Gland, J.L., McClellan, M.R., and McFeely, F.R., Carbon monoxide oxidation on the kinked Pt(3 2 1) surface, *J. Chem. Phys.*, 79, 6349–6356, 1983.

69. He, J.-W. and Norton, P.R., Thermal desorption of oxygen from a Pd(1 1 0) surface, *Surf. Sci.*, 204, 26–34, 1988.

70. Schmidt, J., Stuhlmann, Ch., and Ibach, H., Oxygen adsorption on the Pt(1 1 0)(1 × 2) surfaces studied with EELS, *Surf. Sci.*, 284, 121–128, 1993.

71. Gorodetskii, V.V., Matveev, A.V., Podgornov, E.A., and Zaera, F., Study of the low-temperature reaction between CO and O_2 over Pd and Pt surfaces, *Topics Catal.*, 32, 17–28, 2005.

72. Jones, I.Z., Bennett, R.A., and Bowker, M., CO oxidation on Pd(1 1 0): a high-resolution XPS and molecular beam study, *Surf. Sci.*, 439, 235–248, 1999.

73. Gottfried, J.M., Schmidt, K.J., Schroeder, S.L.M., and Christmann, K., Oxygen chemisorption on Au(1 1 0)-(1 × 2). II. Spectroscopic and reactive thermal desorption measurements, *Surf. Sci.*, 525, 197–206, 2003.

74. Taylor, H.S., A theory of the catalytic surface, *Proc. R. Soc. London (A)*, 108, 105–111, 1925.

75. Ruska, E., *The Early Development of Electron Lenses and Electron Microscopy*, S. Hirzel Verlag, Stuttgart, 1980.

76. Müller, E.W. and Tsong, T.T., *Field Ion Microscopy: Principles and Applications*, American Elsevier Publishing Company, Inc., New York, 1969, 354 pp.

77. Binning, G., Rohrer, H., Gerber, Ch., and Weibel, E., 7×7 reconstruction on Si(1 1 1) resolved in real space, *Phys. Rev. Lett.*, 50, 120–123, 1983.

78. Gorodetskii, V., Drachsel, W., and Block, J.H., Imaging the oscillation CO-oxidation on Pt-surfaces with field ion microscopy, *Catal. Lett.*, 19, 223–231, 1993.

79. Gorodetskii, V., Drachsel, W., and Block, J.H., Field ion microscopic studies of the CO oxidation on platinum: field ion imaging and titration reactions, *J. Chem. Phys.*, 100, 6907–6914, 1994.

80. Gorodetskii, V., Drachsel, W., Ehsasi, M., and Block, J.H., Field ion microscopic studies of the CO oxidation on platinum: bistability and oscillations, *J. Chem. Phys.*, 100, 6915–6922, 1994.

81. Gorodetskii, V., Lauterbach, J., Rotermund, H.-H., Block, J.H., and Ertl, G., Coupling between adjacent crystal planes in heterogeneous catalysis by propagating reaction diffusion waves, *Nature*, 370, 276–279, 1994.

82. Lim, Y.-S., Berdau, M., Naschitzki, M., Ehsasi, M., and Block, J.H., Oscillating catalytic CO oxidation on a platinum field emitter tip: determination of a reactive phase diagram by field electron microscopy, *J. Catal.*, 149, 292–299, 1994.

83. Eiswirth, M., Schwankner, R., and Ertl, G., Conditions for the occurrence of kinetic oscillations in the catalytic oxidation of CO on a Pt(1 0 0) surfaces, *Z. Phys. Chem.*, 144, 59–67, 1985.

84. Ertl, G., Norton, P.R., and Rüstig, J., Kinetic oscillation in the platinum-catalyzed oxidation of CO, *Phys. Rev. Lett.*, 42, 177–182, 1982.

85. Gorodetskii, V., Block, J.H., Drachsel, W., and Ehsasi, M., Oscillations in the carbon monoxide oxidation on platinum surfaces observed by field electron microscopy, *Appl. Surf. Sci.*, 67, 198–205, 1993.

86. Rotermund, H.H., Jakubith, S., von Oertzen, A., and Ertl, G., Solitons in a surface reaction, *Phys. Rev. Lett.*, 66, 3083–3086, 1991.

87. Lauterbach, J., and Rotermund, H.H., Spatio-temporal pattern formation during the catalytic CO-oxidation on Pt(1 0 0), *Surf. Sci.*, 311, 231–246, 1994.

88. Drachsel, W., Wesseling, C., and Gorodetskii, V., Field desorption pathways of water during the H_2 oxidation on a Pt field emitter, *J. Phys. IV*, 6, 31–36, 1996.

89. Latkin, E.I., Elokhin, V.I., and Gorodetskii, V.V., Monte Carlo model of oscillatory CO oxidation having regard to the catalytic properties due to the adsorbate-induced Pt(1 0 0) structural transformation, *J. Mol. Catal. A: Chem.*, 166, 23–30, 2001.

90. Latkin, E.I., Elokhin, V.I., Matveev, A.V., and Gorodetskii, V.V., The role of subsurface oxygen in oscillatory behaviour of $CO + O_2$ reaction over Pd metal catalysts: Monte Carlo model, *J. Mol. Catal. A: Chem.*, 158, 161–166, 2000.

91. Lauterbach, J., Bonilla, G., and Pletcher, T.D., Non-linear phenomena during CO oxidation in the mbar pressure range: a comparison between Pt/SiO_2 and Pt(1 0 0), *Chem. Eng. Sci.*, 54, 4501–4512, 1999.

92. Slinko, M.M., Ukharskii, A.A., and Jaeger, N.I., Global and non-local coupling in oscillating heterogeneous catalytic reactions: the oxidation of CO on zeolite-supported palladium, *Phys. Chem. Chem. Phys.*, 3, 1015–1021, 2001.

93. Boudart, M., Model catalysts: reductionism for understanding, *Topics Catal.*, 13, 147–149, 2000.

94. Cox, M.P., Ertl G., Imbihl R., and Rustig, J., Non-equilibrium surface phase transitions during the catalytic oxidation of CO on a Pt(1 0 0), *Surf. Sci.*, 134, L517–L523, 1983.

95. Imbihl, R., Cox, M.P., Ertl, G., Müller, H., and Brenig, W., Kinetic oscillation in the catalytic CO oxidation on Pt(1 0 0): theory, *J. Chem. Phys.*, 83, 1578–1587, 1985.

96. Gruyters, M., Ali, T., and King, D.A., Theoretical inquiry into the microscopic origins of the oscillatory CO oxidation reaction on Pt(1 0 0), *J. Phys. Chem.*, 100, 14417–14423, 1996.

97. Ladas, S., Imbihl, R., and Ertl, G., Kinetic oscillations and faceting during the catalytic CO oxidation on Pt(1 1 0). *Surf. Sci.*, 280, 42–68, 1993.

98. Latkin, E.I., Elokhin, V.I., and Gorodetskii, V.V., Spiral concentration waves in the Monte Carlo model of CO oxidation over Pd(1 1 0) caused by synchronisation via CO_{ads} diffusion between separate parts of catalytic surface, *Chem. Eng. J.*, 91, 123–131, 2003.

99. Matveev, A.V., Latkin, E.I., Elokhin, V.I., and Gorodetskii, V.V., Manifestation of the adsorbed CO diffusion anisotropy caused by the structure properties of the Pd(1 1 0)-(1 × 2) surface on the oscillatory behavior during CO oxidation reaction — Monte-Carlo model, *Chem. Sustain. Dev.*, 11, 173–180, 2003.

100. Block, J.H., Ehsasi, M., Gorodetskii, V., Karpowicz, A., and Berdau, M., Direct observation of surface mobility with microscopic techniques: photoemission electron and field electron microscopy, in: *New Aspects of Spillover Effect in Catalysis: Studies in Surface Science and Catalysis*, Inui T., et al., Eds., Elsevier, Amsterdam, 1993, Vol. 77, pp. 189–194.

101. Cox, M.P., Ertl, G., and Imbihl, R., Spatial self-organization of surface structure during an oscillating catalytic reaction, *Phys. Rev. Lett.*, 54, 1725–1728, 1985.

102. Imbihl, R., Cox, M.P., and Ertl, G., Kinetic oscillations in the catalytic CO oxidation on Pt(1 0 0): experiments, *J. Chem. Phys.*, 84, 3519–3534, 1986.

103. Kim, M., Sim, W.S., and King, D.A., CO-induced removal of the Pt{1 0 0}-hex reconstruction studied by RAIRS, *J. Chem. Soc., Faraday Trans.*, 92, 4781–4785, 1996.

8 Characterization of the Catalysts by Means of an Oscillatory Reaction

Ljiljana Kolar-Anić and Slobodan Anić
University of Belgrade, Belgrade, Serbia and Montenegro

Željko Čupić
Institute of Chemistry, Technology and Metallurgy, Belgrade, Serbia and Montenegro

CONTENTS

I. INTRODUCTION

One of the most known physical chemists Wilhelm Ostwald defined catalyst as a substance that participates in a particular chemical reaction and thereby increases its rate but without a net change in the amount of that substance in the system [1–3]. Hereinafter catalyst will referred as the common name for both, catalyst and inhibitor, where inhibitor has an opposite role decreasing the rate of chemical reaction. Having such important function in chemical kinetics and different applications, catalysts are the permanent subjects of scientific investigations. These investigations contain discovering of new catalysts, determining their physicochemical characteristics, and examinations of their influence on particular reactions. By the interaction between catalyst and reaction system, the control of considered process, as well as the characteristics of catalyst, can be analyzed in parallel. If our aim is the investigation of catalyst characteristics, the selection of reaction system as a matrix for its examination is of great importance. Therefore, we discussed in the following the different reaction systems including their main characteristics important for mentioned investigations (Sections I.A and I.B) with particular attention on oscillatory reactions in general (Section II) and Bray–Liebhafsky [4,5] in particular (Section III). In Section IV, the characterization of the catalysts by means of the Bray–Liebhafsky oscillatory reaction as the matrix system will be presented.

A. MATRIX REACTION SYSTEMS

All reactions from simple linear to complex nonlinear ones, sensitive on the presence of considered catalyst, could be used for its characterization. The manipulation with simpler reactions is easier whereas the number of information that can be obtained by complex reactions is richer. Although, our aim here is to examine catalysts by a complex oscillatory reaction, we shall begin the explanations with relatively simple reaction of the homogeneous hydrogen peroxide decomposition in the aqueous solution [6] given by the following reaction scheme:

$$2H_2O_2 \xrightarrow{k} 2H_2O + O_2 \tag{8.1}$$

This reaction, as almost all decompositions, is the pseudo-first order with respect to hydrogen peroxide

$$\frac{d[H_2O_2]}{dt} = -k[H_2O_2] \tag{8.2}$$

The obtained activation energy for this reaction is $E_a = 75\,kJ/mol$. If the same reaction is catalyzed by iodide ion, by the following reaction scheme

$$H_2O_2 + I^- \xrightarrow{k_1} H_2O + IO^-\,(slow) \tag{8.3a}$$

$$H_2O_2 + IO^- \xrightarrow{k_2} H_2O + O_2 + I^-\,(fast) \tag{8.3b}$$

where hydrogen peroxide is either oxidant (8.3a) or reductant (8.3b), the net reaction is the same as in the previous case (8.1)

$$2H_2O_2 \xrightarrow{I^-} 2H_2O + O_2 \tag{8.3c}$$

whereas the activation energy is lower, $E_a = 56\,kJ/mol$ [6]. The explanation is in the role of iodide or hypoiodite ions as the catalytic couple. Obviously both, the hypoiodite, the product in reaction (8.3a) and the reactant in reaction (8.3b), and iodide, the product in reaction (8.3b) and the reactant in reaction (8.3a), are intermediates that do not exist in the net reaction (8.3c). As the stoichiometry of reactions (8.1) and (8.3c) are the same, the enthalpy of reactions would be equal. However, rates of reactions are different. In reaction (8.3c) where the reaction rate is higher, the activation energy is lower. More precisely, as the rate of the reaction (8.3a) is lower than (8.3b), it is the one that determine overall rate of hydrogen peroxide decomposition, and approximately the activation energy of overall process. Thus, the effect of a catalyst in increasing the rate of a reaction is to provide an alternate pathway with lower activation energy. Although the catalytic pathway (8.3c) is more convenient for hydrogen peroxide decomposition than (8.1), in the reaction system described by reaction (8.3c), the noncatalyzed hydrogen peroxide decomposition is also present. Beside others, the ratio of rates between catalyzed and uncatalyzed reaction pathways is important for characterization of catalyst.

The rate of reaction (8.3c) can be analyzed when the reaction system is in the steady state. Then, the concentrations of intermediates are approximately constant. It means that

$$\frac{d\,[IO^-]}{dt} = k_1[H_2O_2][I^-] - k_2[H_2O_2][IO^-] \tag{8.4}$$

where k_1 and k_2 denote the corresponding rate constants of reactions (8.3a) and (8.3b), is equal to zero. The corresponding concentrations $[IO^-]_{ss}$ and $[I^-]_{ss}$ are

$$[IO^-]_{ss} = \frac{k_1}{k_2}[I^-]_{ss} \tag{8.5}$$

With this assumption, the rate of overall reaction (8.3c) is given by the expression

$$\frac{d\,[H_2O_2]}{dt} = -k_1\,[H_2O_2][I^-] - k_2\,[H_2O_2][IO^-] \tag{8.6}$$

In the steady state

$$\frac{d\,[H_2O_2]}{dt} = -2k_1\,[H_2O_2][I^-]_{ss} \tag{8.7}$$

As the rate of hydrogen peroxide decomposition is higher in reaction (8.3c) than in reaction (8.1), it means that $2k_1\,[I^-]_{ss} > k$.

Thus, the presence of catalyst changes the reaction mechanism. The manner in which a catalyst modifies the reaction pathways depends on both considered catalyst and selected matrix reaction system. One catalyst in different reaction systems can exhibit different properties. Also, if the reaction system is more complex, the number of characteristic properties that can be found is often larger. If, for example, the reaction system is more complex than the previous one (8.3c), the considered decomposition can perform through several reaction pathways. Then, the domination of reaction pathways by changing the ratio between catalyst and reactant, or other external conditions such as temperature and pressure, can be of great importance for characterization of possible structure and reactivity of considered catalyst. Therefore, instead of analyzing the catalyst in the simple reaction system, we decided to do this in the complex oscillatory one.

B. REACTION SYSTEM WITH SEVERAL STEADY STATES

Some complex reaction systems can be in several steady states. During time evolution, they can even change their steady states with characteristic concentrations of the species therein. In general if we, for example, consider the process having two reactants (A and B), two products (P_1, P_2) and one intermediate (X), described by the following scheme

$$A \xrightarrow{k_1} X \tag{8.8a}$$

$$B + X \xrightarrow{k_2} P_1 \tag{8.8b}$$

$$X \xrightarrow{k_3} P_2 \tag{8.8c}$$

and if the rate of the first reaction is much slower than that of the second one whereas the rate of the second one is higher than that of third one ($k_1 \ll k_3 < k_2[B]_0$; $[B]_0$ denotes the initial concentration of [B]), together with the condition that the concentration of species A is much lower than the concentration of species B, we shall have one steady state until the end of reaction, in other words, during the reactant A exist in the system. (Detailed analysis of the model (8.8a)–(8.8c) may be found in Ref. [7].) The reaction (8.8c), which is parallel with reaction (8.8b), will be present, but unimportant for overall process. If the rates of reactions (8.8b) and (8.8c) are approximately equal, the net reaction of the overall process is

$$2A + B \longrightarrow P_1 + P_2 \tag{8.8d}$$

The concentration of the intermediate species X in the steady state is given by the expression

$$[X]_{ss} = \frac{k_1[A]}{k_2[B] + k_3} \tag{8.9}$$

However, if the concentration of species A is much larger than that of species B, we shall have at the beginning the steady state defined by the reactions (8.8a) and (8.8b), whereas later, when the concentration of reactant B will be close to zero, the steady state defined by the reactions (8.8a) and (8.8c). In that case, the net stoichiometric equation will also change from $A + B \to P_1$ to $A \to P_2$. The pseudo-steady-state concentration of intermediate X change from the initial value of about $k_1[A]/k_2[B]$ to the end value of about $k_1[A]/k_3$.

The net stoichiometric equation can be the same for several pathways in the other more complex system. Such kind of reaction mechanism possesses the model of the mechanism for the Bray–Liebhafsky oscillatory reaction, which will be considered later.

II. OSCILLATORY CHEMICAL REACTION

The reactions consisting of reactant intermediates and products with concentrations that vary periodically in time are known as the oscillatory ones. Strictly speaking, the oscillatory time evolution is the characteristic of intermediates only, whereas reactants and products have a stepwise either decreasing or increasing evolution. Variety of different oscillatory dynamic states appears in a narrow range of initial conditions. Consequently, such a system is in the oscillatory state or its vicinity is extremely sensitive to external conditions and any external perturbation [8–22].

The oscillatory chemical reactions can only be realized in some nonlinear system having feedback in the form of autocatalysis or autoinhibition [7,23–29].

A. LINEAR AND NONLINEAR REACTION SYSTEMS

Formally, if in the differential equation for the rate of a particular chemical reaction there are only linear terms with respect to sum of exponentials over concentrations in them, it is a linear process. All other cases are nonlinear. It means that there are more nonlinear than linear chemical reactions [7,23–29].

Linear reactions are all first-order reactions such as $A \to P, A \leftrightarrows P, A \to X \to P, A \leftrightarrows X \leftrightarrows P$, etc., if there are no thermal effects in them. We can see the physical meaning of the linearity at the following example:

$$A \underset{k_{-1}}{\overset{k_1}{\rightleftharpoons}} X \underset{k_{-2}}{\overset{k_2}{\rightleftharpoons}} P \tag{8.10}$$

Writing the rate equation with respect to intermediate X in the form

$$\frac{d[X]}{dt} = k_1[A] + k_{-2}[P] - (k_{-1} + k_2)[X] \tag{8.11}$$

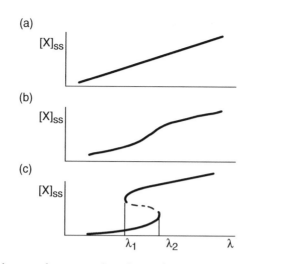

FIGURE 8.1 The effect of the control parameter λ on the steady-state concentration of the intermediate $[X]_{ss}$. (a) Linear law; (b and c) nonlinear law with monostability and multistability, respectively. The bifurcation points are denoted by λ_1 and λ_2.

we can see the linear terms with respect to exponents on the concentrations in all three terms at the right-hand side. The last equation can be rewritten in the form

$$\frac{d[X]}{dt} = \lambda - k[X] \tag{8.12}$$

where $k = k_{-1} + k_2$ is the constant, whereas $\lambda = k_1[A] + k_{-2}[P]$ the parameter that varies in time as a function of time-dependent concentrations of species A and P. Such parameter that defines the state of the system is control parameter. The nonequilibrium stationary state (steady state) is satisfied always when $d[X]/dt$ is equal to zero. Consequently, the stationary-state concentration of intermediate X, as the only function of state of considered reaction system,

$$[X]_{ss} = \frac{\lambda}{k} \tag{8.13}$$

is a linear function of λ (Figure 8.1a).

Analyzing little more complex system

$$A + 2X \underset{k_{-1}}{\overset{k_1}{\rightleftharpoons}} 3X \tag{8.14a}$$

$$X \underset{k_{-2}}{\overset{k_2}{\rightleftharpoons}} P \tag{8.14b}$$

with net reaction $A \rightleftharpoons P$, equal to the one in the previous example, we can see that rate equation with respect to intermediate X is nonlinear one

$$\frac{d[X]}{dt} = k_1[A][X]^2 - k_{-1}[X]^3 - k_2[X] + k_{-2}[P] \tag{8.15}$$

The stationary-state solution is satisfied when Equation (8.15) is equal to zero

$$k_{-1}[X]_{ss}^3 - k_1[A][X]_{ss}^2 + k_2[X]_{ss} - k_{-2}[P] = 0 \tag{8.16}$$

As any cubic equation, it can be rewritten in the form

$$Y_{ss}^3 - \mu Y_{ss} - \lambda = 0 \tag{8.17}$$

where

$$Y_{ss} = [X]_{ss} - \frac{k_1[A]}{3k_{-1}} \tag{8.18}$$

The control parameters μ and λ are given by the following expressions:

$$\mu = \frac{k_2}{k_{-1}} - \frac{1}{3}\left(\frac{k_1[A]}{k_{-1}}\right)^2 \tag{8.19}$$

$$\lambda = -\frac{2}{27}\left(\frac{k_1[A]}{k_{-1}}\right)^3 + \frac{k_1 k_2[A]}{3k_{-1}^2} - \frac{k_{-2}[P]}{k_{-1}} \tag{8.20}$$

The three solutions of Equation (8.18) can be either all real or one real and two conjugate complexes, depending on the parameters μ and λ. In other words, this nonlinear system has one (Figure 8.1b) or three (Figure 8.1c) steady states for same external conditions. They can be stable or unstable. A steady state is stable if a small perturbation of the system tends to decay. It is unstable if the perturbation tends to grow, displacing the system in another state. The unstable steady state is always surrounded by the stable steady states (Figure 8.1c, $\lambda_1 < \lambda < \lambda_2$).

B. MULTISTABILITY AND FEEDBACK

We have multistability when more than one stable steady state exist for a given set of the values of the parameters. These steady states depend on the values of the parameters (control parameters) [7,23–31]. A bifurcation occurs when their number or stability changes as the value of a parameter changes (Figure 8.2).

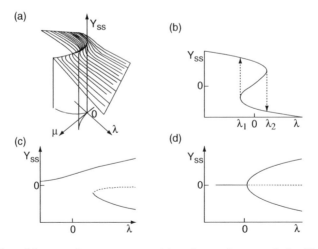

FIGURE 8.2 The effect of the control parameters μ and λ on the steady-state solution Y_{ss} (Equation (8.17)) in the vicinity of the bifurcation point. (a) The folded surface $F(\mu, \lambda, Y_{ss}) = 0$, together with the region in the (μ, λ) plane of existence of the tree real solutions. (b) The intersection with a plane $\mu = \text{const} > 0$. (c) The intersection with a plane $\lambda = \text{const} \neq 0$. (d) The intersection with a plane $\lambda = 0$ [27].

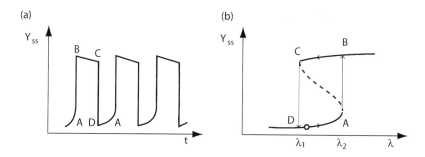

FIGURE 8.3 The possible movements of the bistable reaction system through the phase space ($[Y]_{ss}$, λ) in the vicinity of the bifurcation points and the corresponding time evolution $[Y]_{ss}$. The empty circle denotes the starting state of the system [32].

In Figure 8.2b, we see two bifurcation points λ_1 and λ_2 with respect to the parameter λ. Obviously, when λ is close to zero, there are three steady states that are either stable or unstable. The states lying on the upper and lower branches of the S-shaped curve are stable, whereas the ones lying on its intermediate part are unstable. If the initial steady state of the system belongs to the upper branch and λ increases, it will first follow this upper branch. At the point λ_2 it will jump onto the lower branch. Then, if λ decreases, the system will stay on the low branch until the point λ_1 is reached, where after it will jump onto the upper branch. Hence, in the vicinity of $\lambda = 0$, the states of the system depend on its history and we have multistability. Since the system can stay longer in two of these three states (two stable ones), we shall say that it possesses bistability. The bifurcation diagrams when $\lambda = \text{const} \neq 0$, and $\lambda = 0$, are presented at Figure 8.2c and d, respectively.

If the considered system possesses the feedback in a form of autocatalysis or autoinhibition, as it is in the case (8.14), we can expect its oscillatory evolution in time under particular values of external conditions as the control parameters. In that case the system can alternate between at least two steady states as long as external parameters have the values necessary to hold a system in the region of multistability. In the isothermal well-stirred closed reactor, the number of oscillations is regulated with reactants' concentrations. In the open reactor, the particular dynamic state, independent of whether it is the oscillatory or nonoscillatory one, is sustained by permanent inflow of feed substances, and therefore can be maintained, as it is necessary for investigations. Such situation is illustrated in Figure 8.3a. The corresponding phase space diagram denoting the time interdependence between two intermediate species in their concentration space is given in Figure 8.3b.

III. BRAY–LIEBHAFSKY OSCILLATORY REACTION

The Bray–Liebhafsky reaction is the decomposition of hydrogen peroxide into the water and oxygen in the presence of iodate and hydrogen ions:

$$2H_2O_2 \xrightarrow{\;IO_3^-,\; H^+\;} 2H_2O + O_2 \tag{D}$$

This apparently simple reaction comprises a complex homogeneous catalytic oscillatory process involving numerous iodine intermediates [4,5,7–9,25,27,29,30,32–92]. The global reaction (D) is the result of the reduction (R) of iodate to iodine and the oxidation (O) of iodine to iodate by the following complex reaction scheme:

$$2IO_3^- + 2H^+ + 5H_2O_2 \longrightarrow I_2 + 5O_2 + 6H_2O \tag{R}$$
$$I_2 + 5H_2O_2 \longrightarrow 2IO_3^- + 2H^+ + 4H_2O \tag{O}$$

The sum of reactions (R) and (O) gives the overall decomposition (D). Their rates tend to become equal and we usually observe only the smooth decomposition. In a narrow range of concentrations, however, it is also possible that processes (R) and (O) are alternately dominant, the iodine concentration increases and decreases alternately, and the reaction is periodic (Figure 8.4).

Even in these narrow external conditions, the dynamic states could be very different [4,5,33,34,42,44,47,51,62,77,83,84]. We can note the simple and complex oscillations with different amplitudes, periods ($\Delta\tau$), the number of oscillations (n), the preoscillatory period (τ_1), the duration from the beginning of the reaction to the end of the oscillatory state (τ_{end}), and the duration of oscillatory state ($\tau_{end} - \tau_1$). Consequently, the kinetics and the activation energies of overall reaction and particular pathways vary [33,49,51,54,66,69,70,86]. They are also a function of the reactor, either closed or open, in which reaction is generated. The sensitivity of the system on the control parameters is illustrated by two examples, one with respect to initial hydrogen peroxide concentration in a closed reactor and the other with respect to temperature in the open reactor (Figure 8.5 and Figure 8.6).

Obviously small differences in initial concentrations of hydrogen peroxide (Figure 8.5) and temperatures (Figure 8.6) can perturb the previously established dynamic state significantly. The situation is very similar if control parameters are the concentrations of hydrogen ion, iodate, iodine, iodide, or some other species that interact with the reaction system although they are not intrinsic ones [4,8,17–22,33,34,44,45,51,53,54,70,77].

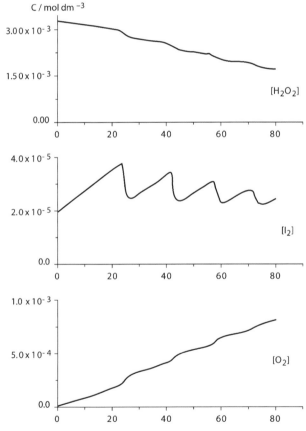

FIGURE 8.4 The time evolution of hydrogen peroxide, iodine, and iodide in the Bray–Liebhafsky reaction system generated in the well-stirred closed isothermal reactor ($T = 60°C$). $[HClO_4]_0 = 6.5 \times 10^{-2}$ mol dm^{-3}, $[NaIO_3]_0 = 9.5 \times 10^{-2}$ mol dm^{-3}, $[H_2O_2]_0 = 1.00 \times 10^{-1}$ mol dm^{-3} [50].

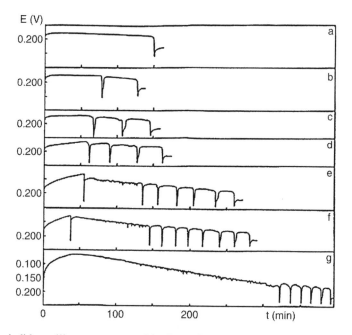

FIGURE 8.5 The iodide oscillograms generated in the well-stirred isothermal closed reactor in the order of increasing the initial concentrations of hydrogen peroxide (in $mol\,dm^{-3}$): (a) 1.36×10^{-3}, (b) 3.30×10^{-3}, (c) 3.60×10^{-3}, (d) 4.98×10^{-3}, (e) 1.98×10^{-2}, (f) 2.99×10^{-2}, and (g) 3.28×10^{-1}. $T = 62°C$, $[H_2SO_4]_0 = 2.45 \times 10^{-2}\,mol\,dm^{-3}$, $[KIO_3]_0 = 7.35 \times 10^{-2}\,mol\,dm^{-3}$ [48].

A. MECHANISM OF THE BRAY–LIEBHAFSKY REACTION

The decomposition of hydrogen peroxide in the presence of hydrogen and iodate ions is a complex process catalyzed by iodate and iodine as catalytic couple. All three reactions (D), (R), and (O) that we mentioned so far to explain possible mechanism are the net ones. Besides, it is well known [4] that the rate of oxygen production during domination of reaction (O) is several times higher than during domination of reaction (R), although oxygen is not a product of reaction (O). Hence, some reaction steps ought to be included in the model of mechanism, having (D), (R), and (O) as the net reactions of corresponding pathways. At present, there is one such model with its different variants [30,50,61,63,64,78,79]. Although all variants, including the core of a model consisted of the first six reactions proposed by Schmitz [50], are able to describe the oscillatory evolution, the version with the following eight reactions has been found as the most successful in simulating the lot of experimentally observed phenomena in a closed and open reactor [30,63,64]:

$$IO_3^- + I^- + 2H^+ \rightleftarrows HIO + HIO_2 \qquad \text{(R1), (R-1)}$$

$$HIO_2 + I^- + H^+ \longrightarrow I_2O + H_2O \qquad \text{(R2)}$$

$$I_2O + H_2O \rightleftarrows 2HIO \qquad \text{(R3), (R-3)}$$

$$HIO + I^- + H^+ \rightleftarrows I_2 + H_2O \qquad \text{(R4), (R-4)}$$

$$HIO + H_2O_2 \longrightarrow I^- + H^+ + O_2 + H_2O \qquad \text{(R5)}$$

$$I_2O + H_2O_2 \longrightarrow HIO + HIO_2 \qquad \text{(R6)}$$

$$HIO_2 + H_2O_2 \longrightarrow IO_3^- + H^+ + H_2O \qquad \text{(R7)}$$

$$IO_3^- + H^+ + H_2O_2 \longrightarrow HIO_2 + O_2 + H_2O \qquad \text{(R8)}$$

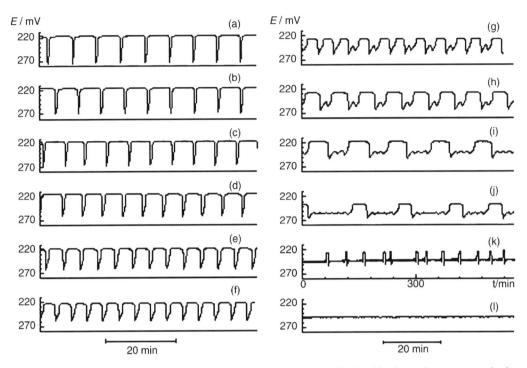

FIGURE 8.6 The sequences of iodide oscillations generated in the well-stirred isothermal open reactor in the order of decreasing temperature (in mol dm^{-3}): (a) 60.0°C, large amplitude relaxation oscillations, 1^0; (b) 58.8°C, 1^0; (c) 57.5°C, 1^0; (d) 55.6°C, 1^0; (e) 54.4°C, 1^0; (f) 52.8°C, 1^0; (g) 50.3°C, chaos; (h) 49.8°C, mixed mode oscillations, 1^1, (i) 49.3°C, chaos; (j) 48.8°C, chaos; (k) 47.8°C, chaos, (l) 47.6°C, stable steady state [84].

The proposed model is based on the liquid-phase reactions; the rates of escape of volatile species and gaseous O_2 and I_2 from the system are not considered. There is no direct autocatalytic or autoinhibition step of the form $A + xB \rightarrow (x \pm 1)B$. The feedback is an intrinsic part of the model as a result of mutual combinations between reactions. Moreover, this model has all characteristics necessary to explain the considered catalytic hydrogen peroxide decomposition as one complex nonlinear process having a region of multistability wherein the different dynamic states from simple oscillatory to complex ones and chaos are found.

By the stoichiometric network analysis of the proposed model, it was shown that the overall process of hydrogen peroxide decomposition into the water and oxygen (reaction (D)) could be realized by four reaction pathways. They are given in Table 8.1.

Each net processes (R) and (O) can be realized by three reaction pathways (Table 8.1). By such model having several different pathways for reactions (R), (O), and (D), different dynamic states as a function of external conditions can be explained.

B. PHENOMENOLOGICAL KINETICS

The phenomenological kinetic analysis of the oscillatory reactions used for the examination of a catalyst is of essential importance for its characterization. However, the phenomenological kinetics of oscillatory reactions has specific properties based on their specific features. Thus, the amplitude of oscillations, the periods between them ($\Delta \tau$), their number (n), the preoscillatory period (τ_1), the duration from the beginning of the reaction to the end of the oscillatory state (τ_{end}), and the duration of oscillatory state ($\tau_{osc} = \tau_{end} - \tau_1$) are all the kinetic parameters specific for kinetic and dynamic states of the system [47,48,51,54,66,69,70].

TABLE 8.1
The Reaction Pathways

Reaction Pathway	Net Reaction
(R2) + (R5) + (R6)	$2H_2O_2 \rightarrow 2H_2O + O_2$
(R1) + (R5) + (R7)	
(R-1) + (R2) + (R6) + (R8)	
(R7) + (R8)	
$2 \times$ (R1) + $2 \times$ (R2) + $2 \times$ (R3) + (R4) + $5 \times$ (R5)	$2IO_3^- + 2H^+ + 5H_2O_2 \rightarrow I_2 + 5O_2 + 6H_2O$
$3 \times$ (R-1) + $2 \times$ (R2) + $2 \times$ (R3) + (R4) + $5 \times$ (R8)	
$2 \times$ (R2) + $2 \times$ (R3) + (R4) + $3 \times$ (R5) + $2 \times$ (R8)	
$2 \times$ (R-1) + $3 \times$ (R2) + $2 \times$ (R-3) + (R-4) + $5 \times$ (R6)	$I_2 + 5H_2O_2 \rightarrow 2IO_3^- + 2H^+ + 4H^+ + 4H_2O$
(R1) + $2 \times$ (R-3) + (R-4) + $2 \times$ (R6) + $3 \times$ (R7)	
(R2) + $2 \times$ (R-3) + (R-4) + $3 \times$ (R6) + $2 \times$ (R7)	

In the case of Bray–Liebhafsky oscillatory reaction, it is found that the overall reaction, independent of its complexity, is the pseudo-first order with respect to hydrogen peroxide:

$$-\frac{d[H_2O_2]}{dt} = k_D[H_2O_2] \tag{8.21}$$

In the particular system consisting of hydrogen peroxide, potassium iodate, and sulfuric acid, the corresponding rate constant k_D is a function of potassium iodate and sulfuric acid given in the form

$$k_D = k[H_2SO_4]^q[KIO_3] \tag{8.22}$$

The influence of sulfuric acid on the overall process is complex, between first and second order depending on the range of values of $[H_2SO_4]_0$ [51].

It is also found that there is a correspondence between the rate constants of the reactions (D), (R), and (O), and the characteristic kinetic parameters such as $\Delta\tau$, n, τ_1, τ_{end}, and τ_{osc}, presented in Figure 8.7 [47,51,69,70].

It is shown that

$$k_D = C_1\frac{1}{\tau_{end}} \tag{8.23}$$

$$k_D = \frac{1}{C_2 n} \tag{8.24}$$

$$k_R = C_3\frac{1}{\tau_1} \tag{8.25}$$

where C_1, C_2, and C_3 are the constants characteristic for the considered system.

The maximum concentrations of iodide ions (lower potentials of iodide-sensitive electrode) correspond to the pseudo-steady states when reaction (R) dominates, whereas its minimum concentrations (higher potentials of iodide-sensitive electrode) correspond to the pseudo-steady states when reaction (O) dominates (Figure 8.7). These states depend on the iodide ion concentration,

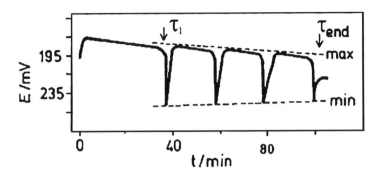

FIGURE 8.7 The iodide oscillogram generated in the well-stirred isothermal closed reactor with denoted kinetic parameters; τ_1 — the preoscillatory period, τ_{end} — the duration from the beginning of the reaction to the end of the oscillatory state. By max and min, the pseudo-steady states characterized by the maximum and minimum iodide concentrations are denoted, respectively. $[H_2O_2]_0 = 4.00 \times 10^{-3}$ mol dm^{-3}, $[H_2SO_4]_0 = 4.90 \times 10^{-2}$ mol dm^{-3}, $[KIO_3]_0 = 7.35 \times 10^{-2}$ mol dm^{-3}, $T = 62.0°C$ [70].

which is one of a function of the states of the system. The rate law, which explains the evolution of the maximum iodide concentration, is the first order with respect to the iodide. The corresponding rate constant is proportional to the rate constant of the process (R), k_R. In a similar manner the rate law, which explains the evolution of the minimum iodide concentration, is the first order with respect to the iodide and the rate constant is proportional to the rate constant k_O. We have found that both processes (R) and (O), under considered conditions, are the first order with respect to iodide ion concentration.

Obviously, any perturbation of the system can shift all introduced parameters in a different manner. Following their changes, one can obtain a variety of different informations about the perturber.

IV. EXAMINATIONS OF CATALYSTS USING THE BRAY–LIEBHAFSKY REACTION AS THE MATRIX SYSTEM

After reading previous sections, we can conclude that the Bray–Liebhafsky reaction system, as any oscillatory one, is extremely sensitive to various perturbations, and therefore suitable for analytical applications. They can be used as the matrix for analyzing properties of the substances that already exist in the system [17,18,42,56], but also the ones that only interact with it [10–17,19–22]. Beside others, such substances can be catalysts [93,94].

However, before application we must know that results here, more than in the simple reaction system, depend on the experimental conditions and applied perturbations. Generally, because an oscillatory reaction system can be in essentially different dynamic states far from equilibrium, their selection is of important interest for desired application.

The perturbations can be performed in different manners depending on the perturber and the aim of examination. They also depend on the reaction setup and aggregate state of perturber. Thus, if the reaction is performed in a closed reactor, the matrix system can be perturbed by adding the perturber in the reaction vessel in the selected moment of reaction evolution [9], but also at the beginning of reaction with or without previous preparation with other reactants [32]. In the case of heterogeneous catalysis, the catalyst must be in the reaction vessel before any perturbation. In the open reactor, the solid catalyst must also be in the system before perturbations are performed. During last investigations, beside the normal pulse perturbations with selected analyte, the control parameter can be regarded as a perturber [42,73,74]. For desired application, several perturbers can be used in different manners depending on the knowledge and scientific intuition of the analyst. Thus, amongst these, the methods based on the excitability and phase-response behavior are defined [95–97]. We applied these methods on the iodide perturbation of the Bray–Liebhafsky reaction system being in the oscillatory state and successfully simulated by the proposed model [18].

By perturbations of the Bray–Liebhafsky system being in the stable steady state in the vicinity of a bifurcation point, we found the kinetic method for quantitative determination of several substances such as Cl^-, Br^-, I^-, Mn^{2+}, malonic acid, quercetin, paracetamol, rutin, ascorbic acid, and others [17–22,98] with very low detection limit in comparison with other methods.

The similar quantitative method for sodium thiosulfate, gallic acid, glutathione, resorcinol, paracetamol, vitamin B_6, ascorbic acid, and vanillin was also found, but applied on the H_2O_2–NaSCN–$CuSO_4$ reaction system being in the oscillatory state [10,11,13–16].

With aim to examine catalysts by the Bray–Liebhafsky reaction system as the matrix, two examples, one in the closed and the other in the open reactor, will be given in the following.

A. Examinations in the Closed Reactor

Two different catalysts for hydrogen peroxide decomposition, the enzyme peroxidase (isolated from the horseradish root, HRP), and polymer-supported catalyst (acid form of poly-4-vinylpyridine functionalized by ferric sulfate, apFe) [99,100], are examined with an aim to compare their activity. The active center in the peroxidases is the ferric ion in protoporphyrin IX. Besides the complex made of ferric ion and protoporphyrin IX, that is ferricprotoporphyrin IX, also known as ferric heme or hemin, peroxidase possesses a long chain of proteins [101,102]. On the other hand, the macroporous acid form of polyvinyl pyridine functionalized by ferricsulfate is obtained from cross-linked polyvinyl pyridine in macroporous bead form [103]. Pyridine enables it to form coordination complexes or quaternary salts with different metal ions such as iron (III) [104]. An active center on the polymeric matrix functionalized by iron, as metallic catalyst immobilized on polymer by pyridine, has similar microenvironment conditions as active center in an enzyme [105].

With the aim to compare the activities of the two mentioned catalysts (HRP and apFe) we had to examine their kinetic properties. This is performed in the Bray–Liebhafsky reaction being in the oscillatory state. The reaction was conducted in the closed well-stirred reactor.

The experiments are performed in two different manners. In one series, the polymer-supported catalyst and hydrogen peroxide was injected into the aqueous solution of potassium iodate and sulfuric acid simultaneously, denoting the beginning of reaction. In the other series, the polymer-supported catalyst and hydrogen peroxide was mixed before adding to the solution of potassium iodate and sulfuric acid. The period of their mutual reacting Q was different in different experiments.

1. The Influence of the Amount of Catalysts on the Hydrogen Peroxide Decomposition

The Bray–Liebhafsky system without an additional catalyst, the same system with peroxidase, and the same system with apFe are considered. The initial conditions for all experiments were identical. The initial amounts of catalysts were used in such a way to ensure equal concentrations of iron in the corresponding reaction system possessing one of the two examined catalysts [93]. The results are presented in Figure 8.8.

For any added amount of examined catalyst, all mentioned kinetic variables are changed. In Figure 8.9 and Figure 8.10, τ_1 and τ_{end} are presented as a function of added amount of catalysts, respectively. Every point is obtained from several experiments repeated under same conditions; the deviations from average values are under $\pm 5\%$.

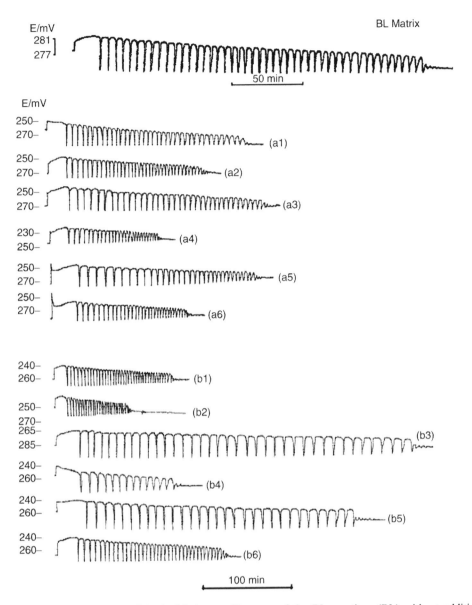

FIGURE 8.8 A typical series of the iodide-ion oscillograms of the BL reaction: (BL) without additional catalysts; (a) with various amounts of an enzyme catalyst and (b) with various amounts of polymer-supported catalyst [93].

2. The Influence of the Interaction between Catalysts and Hydrogen Peroxide before Initiation of Hydrogen Peroxide Decomposition in the Bray–Liebhafsky System

In this series, the catalyst and hydrogen peroxide were mixed before adding to the solution of potassium iodate and sulfuric acid, where, in analogy with other investigations, the moment when two mentioned mixtures were connected was taken as the beginning of the Bray–Liebhafsky reaction. The period of their mutual reaction Q before adding to the solution of potassium iodate and sulfuric acid was different in different experiments. All other parameters were kept constant ($[H_2SO_4]_0 = 4.8 \times 10^{-2}$ mol dm^{-3}, $[KIO_3]_0 = 7.2 \times 10^{-2}$ mol dm^{-3}, $[H_2O_2]_0 = 4.6 \times 10^{-2}$ mol dm^{-3}, $T = 332$ K).

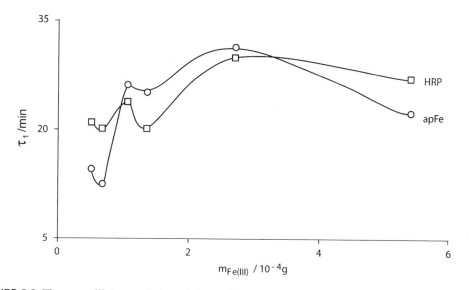

FIGURE 8.9 The preoscillatory period τ_1 of the oscillograms presented in Figure 8.8 as a function of the amount of the examined catalysts $m_{Fe(III)}$ [93].

Under influence of mixing period Q, all kinetic variables were changed. The oscillograms of the BL system perturbed by the enzyme are presented in Figure 8.11a, and perturbed by the polymer-supported catalyst in Figure 8.11b [106].

The kinetic variables, n, τ_1, and τ_{end} for the BL system perturbed with both catalysts, are given as a function of the mixing period Q, in Figure 8.12–Figure 8.14, respectively.

By the presence of either HRP or apFe, the Bray–Liebhafsky reaction is changed in a similar manner. Some amounts of the mentioned catalysts influence decrease, whereas the other amounts influence increase of the characteristic periods τ_1 and τ_{end}. In other words, some amounts of mentioned catalysts cause the acceleration of the reactions (R), (O), and (D), whereas the other amounts cause their inhibition. Anyhow, by the presence of either HRP or apFe in the BL reaction, the new reaction system for hydrogen peroxide decomposition is formed.

Moreover, as the kinetic parameters such as n, τ_1, and τ_{end} depend on the added amount of both considered catalysts in a complex but analogous manner (Figure 8.9, Figure 8.10, Figure 8.12–Figure 8.14),

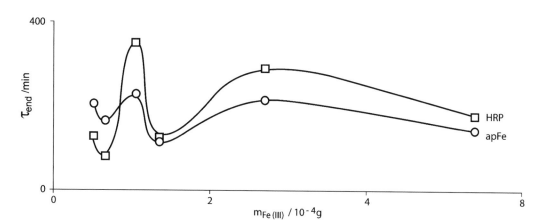

FIGURE 8.10 Duration from the beginning of the reaction to the end of the oscillatory state τ_{end} of the oscillograms presented in Figure 8.8 as a function of the amount of the examined catalysts $m_{Fe(III)}$ [106].

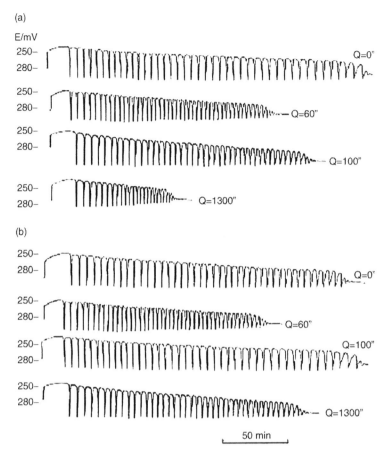

FIGURE 8.11 A typical series of the iodide-ion oscillograms of the BL reaction for different mixing period Q: (a) with an enzyme catalyst and (b) with polymer-supported catalyst [106].

a similar reaction mechanism can be expected in both cases. This can be explained only by similar activity of the Fe(III) centers in both catalysts.

B. Examinations in the Open Reactor

In analogy with kinetic determinations of numerous substances in open reactor where the Bray–Liebhafsky reaction was used as the matrix system, we decided to analyze activity of

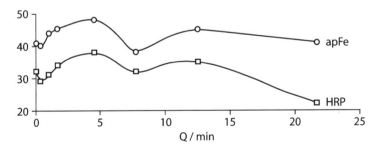

FIGURE 8.12 The number of the oscillations n of the oscillograms presented in Figure 8.11 as a function of the mixing period Q [106].

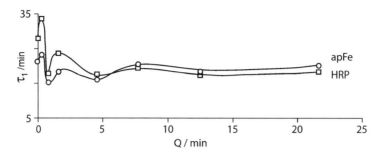

FIGURE 8.13 The preoscillatory period τ_1 of the oscillograms presented in Figure 8.11 as a function of the mixing period Q [106].

polymer-supported catalyst having different granulation. For this purpose, the bifurcation analysis with temperature as the control or bifurcation parameter was selected.

The examined polymer-supported catalyst having two different granulations (bead size) A and B was the macroporous cross-linked copolymer of 4-vinylpyridine and 25% (4/1) divinylbenzene–styrene copolymer (commercial product named Reillex 425) [103]. Their average bead diameter (d), skeletal density (ρ), surface area (S_{BET}), pore volume (V_p), and relative swelling ratio (SR) are given in Table 8.2.

The experiments were performed in the light-protected, thermostated ($\pm 0.1°C$), and well-stirred ($r = 900$ rpm) reaction vessel filled up by the three separate inflows of the feed substances, $[KIO_3] = 5.9 \times 10^{-2}$ mol l^{-1}, $[H_2SO_4] = 5.5 \times 10^{-2}$ mol l^{-1} and $[H_2O_2] = 2.0 \times 10^{-1}$ mol l^{-1}. The excess of the reaction mixture was sucked out through the U-shaped glass tube having a sintered glass inside, to reach the actual reaction mixture volume, $V = 22.2 \pm 0.2$ ml [20].

Three experimental series where temperature was the control parameter were performed under equal initial conditions. In the first series, the pure Bray–Liebhafsky reaction system was analyzed, whereas in the other two series the same system together with either catalyst A or catalyst B was considered. In the first series, the temperature was varied between 45.4 and 55.3°C, in the second series (BL system with catalyst A) between 48.2 and 58.1°C, and in the third series (BL system with catalyst B) between 45.9 and 60.0°C.

The obtained time sequences are presented in Figure 8.15.1–Figure 8.17.1. They are generally similar. For the lowest temperatures only stable steady state was noted (a). With increasing temperature, chaotic dynamics with various complexities was obtained. First the trains of burst-like oscillations emerge chaotically from an irregular procession of small-amplitude oscillations

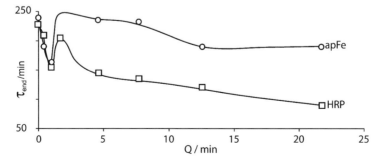

FIGURE 8.14 Duration from the beginning of the reaction to the end of the oscillatory state τ_{end} of the oscillograms presented in Figure 8.11 as a function of the mixing period Q [106].

TABLE 8.2
The Catalyst Properties [94]

	ρ (g cm^{-3})	S_{BET} (m^2 g^{-1})	V_p (cm^3 g^{-1})	d (mm)	S.R.
A	1.17	42.60	0.189	0.6	2.0
B	1.18	58.45	0.234	0.3	1.8

(b and c). Before transition to simple periodic oscillations, the dynamic state with an irregular mixture of 1^0 and 1^1 oscillations, where the first number denotes the number of large-amplitude oscillations and the second, the number of small ones, are found (d). When temperature increases further, only regular oscillations having large amplitudes are noted (e and f). Their amplitudes and the periods between them increase with increasing temperature.

With decreasing temperature, whole scenario was repeated. The hysteresis was not found. The bifurcation diagrams showing the envelope of the oscillations and the locus of the stable non-equilibrium stationary states (stable steady states) are given in Figure 8.15.2–Figure 8.17.2. Stars denote the average values of small amplitudes, whereas empty circles denote the average values of large amplitudes. The stable steady states are denoted by solid circles. The plots of the squares of the oscillations amplitudes as a function of the temperature T are presented in

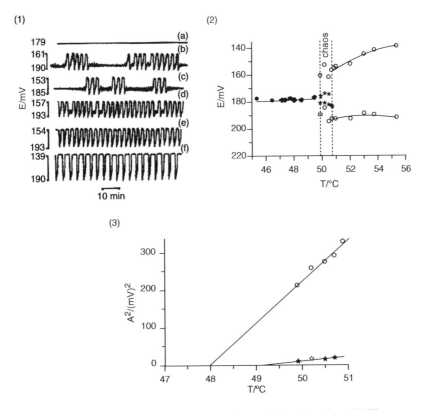

FIGURE 8.15 The Bray–Liebhafsky reaction system without additional catalyst. (1) Time sequences with respect to temperature in °C (a) 49.5, (b) 49.9, (c) 50.2, (d) 50.7, (e) 51.0, (f) 55.3. (2) Bifurcation diagram with stable steady states (solid circles) and envelop of large-amplitude oscillations (empty circles) and small amplitude oscillations (stars). (3) Square of oscillation amplitudes versus temperature [94].

Figure 8.15.3–Figure 8.17.3. The oscillations having small and the ones having large amplitudes are analyzed separately. In the vicinity of the bifurcation point denoting the transition between the stable and unstable steady states, that is, when the chaotic behavior emerges, the approximately constant periods between oscillations and linear response of the squares of amplitudes with respect to control parameter (temperature) can be noted for both type of oscillations. Thus, two intersections between mentioned straight lines and abscissa can be determined for every experimental series (T_c-large and T_c-small). These intersections could be considered as the bifurcation points. The linear response of the squares of amplitudes with respect to control parameter (temperature) was found, but these intersections cannot be simple Hopf bifurcation points since both type of oscillations have the intersections with abscissa at a temperature where the stable steady state is found. They cannot correspond to subcritical Hopf bifurcation point since hysteresis is not obtained. Moreover, the bifurcation point here is a complex one with two kinds of oscillations that emerge from it.

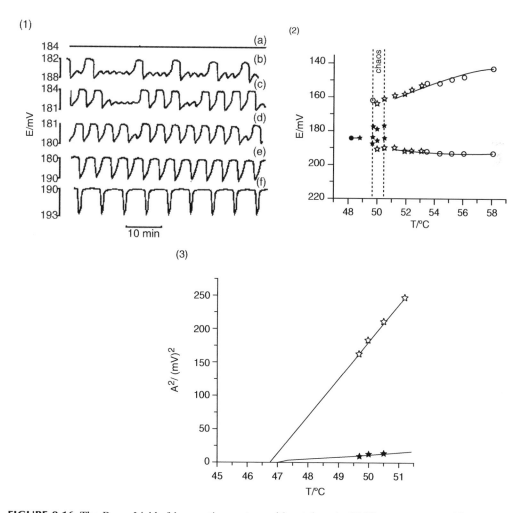

FIGURE 8.16 The Bray–Liebhafsky reaction system with catalyst A. (1) Time sequences with respect to temperature in °C (a) 48.8, (b) 49.7, (c) 50.0, (d) 50.5, (e) 51.2, (f) 55.2. (2) Bifurcation diagram with stable steady states (solid circles) and envelop of large-amplitude oscillations (empty circles) and small-amplitude oscillations (stars). (3) Square of oscillation amplitudes versus temperature [94].

Anyhow, we found that defined intersections can be used as the measure of the interaction between the particular catalyst and the matrix system, that is, as the measure of its activity. For this purpose, only large-amplitude oscillations are applied since the amplitudes of small oscillations are very irregular. In Figure 8.18, the temperatures T_c-large are given as a function of the S_{BET} of the catalysts having different granulations. The same value obtained in a matrix system without catalyst is presented in Figure 8.18, as the point where $S_{BET} = 0$. Thus we have obtained that the BL system is sensitive to the granulations of considered catalyst, and that the bifurcation analysis of their activities gives the parameters that can be used for recognizing the granulations.

We can also note that the slope of linear function between square of large-oscillation amplitudes and temperature depends on granulation. Particularly, the ratio between slopes $(\text{tg } \alpha(B)/\text{tg } \alpha(A))$ of the catalysts B and A is 1.4. The ratio between surfaces $(S_{BET}(B)/S_{BET}(A))$

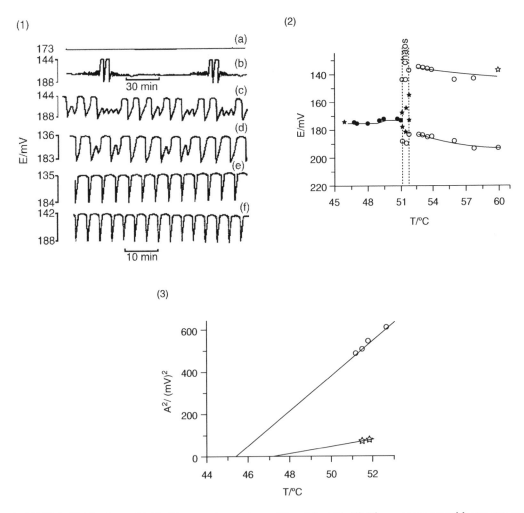

FIGURE 8.17 The Bray–Liebhafsky reaction system with catalyst B. (1) Time sequences with respect to temperature in °C (a) 51.0, (b) 51.2, (c) 51.5, (d) 51.8, (e) 53.5, (f) 56.0. (2) Bifurcation diagram with stable steady states (solid circles) and envelop of large-amplitude oscillations (empty circles) and small-amplitude oscillations (empty stars). (3) Square of oscillation amplitudes versus. temperature. The point denoting the square of the amplitudes of small oscillations for 51.2°C could not be determined [94].

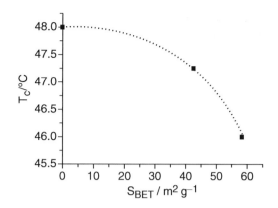

FIGURE 8.18 T_c of large-amplitude oscillations versus. S_{BET} [94].

is also 1.4. It means that the slope, which is the measure of the amplitude increase with control parameter in the vicinity of the bifurcation point, depends on the polymer surface.

Because polyvinyl pyridine reacts with iodine, it consequently influences on the iodine hydrolysis, one of the most important reactions in the Bray-Liebhafsky system [100]. As this reaction is crucial for the ratio between reduction and oxidation pathways, it is responsible for the value of the oscillations amplitudes. Thus, observed correspondence between $(tg\ \alpha(B)/tg\ \alpha(A))$ and $(S_{BET}(B)/S_{BET}(A))$ ratios is once more confirmation of the interaction of polyvinyl pyridine with iodine from the BL system.

V. CONCLUSION

Analyzing different catalysts by means of an oscillatory reaction conducted in open and closed reactors as a matrix, it was shown that their characterization under mentioned conditions is, generally, possible and useful. Thus, by comparison with respect to dynamical effects of several catalysts in the matrix reaction system, the structure of active centers should be discussed. Particularly, analyzing two catalysts for hydrogen peroxide decomposition, the natural enzyme peroxidase and synthetic polymer-supported catalyst, the similarity in their catalytic activity is found. Hence, we can note that the evolution of the matrix oscillatory reaction can be used for determination of the enzyme activity. Moreover, one can see that the analysis of the granulation and active surface may also be performed by the oscillatory reaction.

ACKNOWLEDGMENTS

We are grateful to our colleague N. Pejić for unpublished data presented here. The present investigations are partially supported by Ministry of Sciences, Technologies and Development of Serbia, grants No. 1448 and 1807.

REFERENCES

1. Veljković, S., *Hemijska kinetika*, Gradevinska knjiga, Beograd, 1969.
2. Yeremin, E.N., *The Foundations of the Chemical Kinetics*, MIR, Moscow, 1979.
3. Pilling, M.J., *Reaction Kinetics*, Oxford University, Oxford, 1997.
4. Bray, W.C., A periodic reaction in homogeneous solution and its relation to catalysis, *J. Am. Chem. Soc.*, 43, 1262–1267, 1921.

5. Liebhafsky, H.A., Furuichi, R., and Roe, G.M., Reaction involving hydrogen peroxide, iodine, and iodate ion. 7. The smooth catalytic decomposition of hydrogen peroxide, mainly at 50°C, *J. Am. Chem. Soc.*, 103, 51–56, 1981, and references therein.

6. Miller, M., *Chemistry — Structure and Dynamics*, McGraw-Hill, New York, 1985, pp. 417, 419.

7. Schmitz, G., *Stationnarite et reactions periodiques*, Thesis, Universite libre de Bruxelles, Bruxelles, 1983.

8. Schmitz, G., Transient behaviours in the Bray–Liebhafsky reaction, in: *Spatial Inhomogeneities and Transient Behaviour in Chemical Kinetics*, Gray, P., Nicolis, G., Baras, F., Borkmans, P., Scott, S.K., Eds., Manchester University Press, Manchester, 1990, pp. 666–668.

9. Vukojević, V., Sorensen, P.G., and Hynne, F., Quenching analysis of the Briggs–Rauscher reaction, *J. Phys. Chem.*, 97, 4091–4100, 1993.

10. Jiménez-Prieto, R., Silva, M., and Pérez-Bendito, D., Analyte pulse perturbation technique: a tool for analytical determinations in far-from-equilibrium dynamic systems, *Anal. Chem.*, 67, 729–734, 1995.

11. Jiménez-Prieto, R., Silva, M., and Pérez-Bendito, D., Determination of gallic acid by an oscillating chemical reaction using the analyte pulse perturbation technique, *Anal. Chim. Acta.*, 321, 53–60, 1996.

12. Vukojević, V., Sorensen, P.G., and Hynne, F., Predictive value of a model of the Briggs–Rauscher reaction fitted to quenching experiments, *J. Phys. Chem.*, 100, 17175–17185, 1996.

13. Jiménez-Prieto, R., Silva, M., and Pérez-Bendito, D., Simultaneous determination of gallic acid and resorcinol based on an oscillating chemical reaction by the analyte pulse perturbation technique, *Anal. Chim. Acta.*, 334, 323–330, 1996.

14. Jiménez-Prieto, R., Silva, M., and Pérez-Bendito, D., Determination of trace amounts of reduced glutathione by a chemical oscillating reaction, *Analyst*, 121, 563–566, 1996.

15. Jiménez-Prieto, R., Silva, M., and Pérez-Bendito, D., Application of oscillating reaction-based determinations to the analysis of real samples, *Analyst*, 122, 287–292, 1997.

16. Strizhak, P.E., Determination haos v khimii, Akademperiodika, Kiev, 2002.

17. Vukojević, V., Pejić, N., Stanisavljev, D., Anić, S., and Kolar-Anić, Lj., Determination of Cl^-, Br^-, I^-, Mn^{2+}, malonic acid and quercetin by perturbation of a non-equilibrium stationary state in the Bray–Liebhafsky reaction, *Analyst*, 124, 147–152, 1999.

18. Vukojević, V., Anić, S., and Kolar-Anić, Lj., Investigation of dynamic behavior of the Bray–Liebhafsky reaction in the CSTR: properties of the system examined by pulsed perturbations with I^-, *Phys. Chem. Chem. Phys.*, 4, 1276–1283, 2002.

19. Vukojević, V., Pejić, N., Stanisavljev, D., Anić, S., and Kolar-Anić, Lj., Micro-quantitative determination of the quercetin by perturbation of the non-equilibrium stationary state in the Bray–Liebhafsky reaction system, *Die Pharmazie*, 56, 897–898, 2001.

20. Pejić, N., Anić, S., Kuntić, V., Vukojević, V., and Kolar-Anić, Lj., Kinetic determination of micro-quantities of rutin by perturbation of the Bray–Liebhafsky oscillatory reaction in an open reactor, *Microchim. Acta.*, 143, 261–267, 2003.

21. Pejić, N., Anić, S., Mijatović, M., Milenković, S., Ćirić, J., and Grozdić, T., Doprinos razvoju nove mikrozapreminske/mikrokoncentracione kvantitativne analize: odredjivanje morfina, *Nauka Tehnika Bezbednost*, 1, 67–74, 2003.

22. Pejić, N. and Blagojević, S., Odredjivanje sub-mikrograma hesperidina metodom perturbacije Bray–Liebhafsky oscilatorne reakcije realizovane u otvorenom reaktoru, *Nauka Tehnika Bezbednost*, 2, 81–88, 2003.

23. Gray, P. and Scott, S.K., *Chemical Oscillations and Instabilities — Non-linear Chemical Kinetics*, Clarendon Press, Oxford, 1990.

24. Scott, S.K., *Chemical Chaos*, Clarendon Press, Oxford, 1991.

25. Field, R.J. and Burger, M., Eds., *Oscillations and Traveling Waves in Chemical Systems*, Wiley, New York, 1985.

26. Nicolis, G., *Introduction to Nonlinear Science*, University Press, Cambridge, 1995.

27. Kolar-Anić, Lj., Čupić, Ž., and Anić, S., Multistabilnost i nelinearni dinamički sistemi, *Hem. Ind.*, 52, 337–342, 1998.

28. Kolar-Anić, Lj., Anić, S., and Vukojević, V., Dinamika nelinearnih procesa-Od monotone do oscilatorne evolucije, Fakultet za fizičku hemiju Univerziteta u Beogradu, Beograd, 2004.

29. Kolar-Anić, Lj. and Anić, S., Autokataliza i autoinhibicija. Oscilatorne reakcije, in *Novi izazovi u katalizi*, Putanov, P., Ed., SANU, Novi Sad, 1977, pp. 139–162.

30. Kolar-Anić, Lj., Čupić, Ž., Anić, S., and Schmitz, G., Pseudo-steady states in the model of the Bray–Liebhafsky oscillatory reaction, *J. Chem. Soc., Faraday Trans.*, 93(12), 2147–2152, 1997.

31. Schmitz, G., Kolar-Anić, Lj., Anić, S., and Čupić, Ž., The illustration of multistability, *J. Chem. Educ.*, 77, 1502–1505, 2000.

32. Anić, S., Kolar-Anić, Lj., Čupić, Ž., Pejić, N., and Vukojević, V., Oscilatorna hemijska reakcija kao modelni sistem za karaterizaciju katalizatora, *Svet polimera*, 4, 55–66, 2001.

33. Peard, M.G. and Cullis, C.F., A periodic chemical reaction: the reaction between hydrogen peroxide and iodic acid, *Trans. Faraday Soc.*, 47, 616–630, 1951.

34. Degn, H., Evidence of a branched chain reaction in the oscillating reaction hydrogen peroxide, iodine and iodate, *Acta Chem. Scand.*. 21, 1057–1066, 1967.

35. Woodson, J.H. and Liebhafsky, H.A., Iodide-selective electrodes in reacting and in equilibrium systems, *Anal. Chem.*, 41, 1894–1897, 1969.

36. Matsuzaki, I., Woodson, J.H., and Liebhafsky, H.A., pH and temperature pulses during the periodic decomposition of hydrogen peroxide, *Bull. Chem. Soc. Jpn.*, 43, 3317, 1970.

37. Vavalin, V.A., Zhabotinskii, A.M., and Zakain, A.N., Avtokolebanya koncentracii iodid-iona v hode reakcii razlozheniya perekisi vodoroda, kataliziruemoi iodatom, *Russ. J. Phys. Chem.*, 44, 1345–1346, 1970.

38. Matsuzaki, I., Alexander, R.B., and Liebhafsky, H.A., Rate measurements of highly variable gas evolution with a mass flourometer, *Anal. Chem.*, 42, 1690–1693, 1970.

39. Matsuzaki, I., Simić, R., and Liebhafsky, H.A., The mechanism of decomposition of hydrogen peroxide by iodine in acid solution: the rates of associated reactions, *Bull. Chem. Soc. Jpn.*, 45, 3367–3371, 1972.

40. Liebhafsky, H.A. and Wu, L.S., Reaction involving hydrogen peroxide, iodine, and iodate ion. V. Introduction to the oscillatory decomposition of hydrogen peroxide, *J. Am. Chem. Soc.*, 96, 7180–7187, 1974.

41. Furuichi, R. and Liebhafsky, H.A., Rate of the Dushman reaction in iodic acid at low iodide concentration: complexity of ionic acid, *Bull. Chem. Soc. Jpn.*, 48, 745–750, 1975.

42. Charma, K.R. and Noyes, R.M., Oscillations in chemical systems. VIII. Effect of light and oxygen on the Bray–Lebhafsky reaction, *J. Am. Chem. Soc.*, 97, 202–204, 1975.

43. Charma, K.R. and Noyes, R.M., Oscillations in chemical systems. 13. A detailed molecular mechanism for the Bray–Lebhafsky reaction of iodate and hydrogen peroxide, *J. Am. Chem. Soc.*, 98, 4345–4360, 1976.

44. Chopen-Dumas, J., Diagramme d'etat de la reaction de Bray, *J. C. R. Acad. Sci. Ser. C*, 287, 533–556, 1978.

45. Edelson, D. and Noyes, R.M., Detailed calculations modeling the oscillatory Bray–Lebhafsky reaction, *J. Phys. Chem.*, 83, 212–220, 1979.

46. Anić, S., Mitić, D., and Kolar-Anić, Lj., The Bray–Liebhafsky reaction. I. Controlled development of oscillations, *J. Serb. Chem. Soc.*, 50, 53–59, 1985.

47. Anić, S. and Kolar-Anić, Lj., Some new detail in the kinetic considerations of the oscillatory decomposition of hydrogen peroxide, *Ber. Bunsenges. Phys. Chem.*, 90, 539–542, 1986.

48. Anić, S. and Kolar-Anić, Lj., The oscillatory decomposition of H_2O_2 monitored by the potentiometric method with Pt and Ag^+/S^{2-} indicator electrode, *Ber. Bunsenges. Phys. Chem.*, 90, 1084–1086, 1986.

49. Anić, S. and Kolar-Anić, Lj., The influence of potassium iodate on hydrogen peroxide decomposition in Bray–Liebhafsky reaction, *Ber. Bunsenges. Phys. Chem.*, 91, 1010–1013, 1987.

50. Schmitz, G., Cinetic de la reaction Bray, *J. Chim.Phys.*, 84, 957–965, 1987.

51. Anić, S. and Kolar-Anić, Lj., Kinetic aspects of the Bray–Liebhafsky oscillatory reaction, *J. Chem. Soc. Faraday Trans. I.*, 84, 3413–3421, 1988.

52. Anić, S. and Mitić, D., The Bray–Liebhafsky reaction. IV. New results in the studies of hydrogen peroxide oscillatory decomposition of high acidity, *J. Serb. Chem. Soc.*, 53, 371–376, 1988.

53. Buchholtz, F.G. and Broecher, S., Oscillations of the Bray–Liebhafsky reaction of low flow rates in a continuous flow stirred-tank reactor, *J. Phys. Chem. A*, 102, 1556–1559, 1988.

54. Anić, S., Stanisavljev, D., Krnajski Belovljev, G., and Kolar-Anić, Lj., Examination of the temperature variations on the Bray–Liebhafsky oscillatory reaction, *Ber. Bunsenges. Phys. Chem.*, 93, 488–491, 1989.

55. Noyes, R.M., Mechanism of some chemical oscillators, *J. Phys. Chem.*, 94, 4404–4412, 1990.

56. Kolar-Anić, Lj., Mišljenović, Đ.M., Stanisavljev, D.R., and Anić, S.R., Applicability of Schmitz's model to dilution-reinitiated oscillations in the Bray–Liebhafsky reaction, *J. Phys. Chem.*, 94, 8144–8146, 1990.

57. Anić, S. and Kolar-Anić, Lj., Deterministic aspects of the Bray–Liebhafsky oscillatory reaction, in *Spatial Inhomogeneities and Transient Behaviour in Chemical Kinetics*, Gray, P., Nicolis, G., Baras, F., Borkmans, P., Scott, S.K., Eds., Manchester University Press, Manchester, 1990, pp. 664–665.

58. Schmitz, G., Etude du Brayalator per la methode de Clarke, *J. Chim. Phys.*, 88, 15–26, 1991.

59. Kolar-Anić, Lj., Stanisavljev, D., Krnajski Belovljev, G., Peeters, Ph., and Anić, S., The first maximum of the iodide concentrations in the Bray–Liebhafsky reaction, *Comput. Chem.*, 14, 345–347, 1990.

60. Anić, S., Kolar-Anić, Lj., Stanisavljev, D., Begović, N., and Mitić, D. Dilution reinitiated oscillations in the Bray–Liebhafsky system, *React. Kinet. Catal. Lett.*, 43, 155–162, 1991.

61. Kolar-Anić, Lj. and Schmitz, G., Mechanism of the Bray–Liebhafsky reaction: effect of the oxidation of iodius acid by hydrogen peroxide, *J. Chem. Faraday Soc.*, 88, 2343–2349, 1992.

62. Treindl, L. and Noyes, R.M., A new explanation of the oscillations in the Bray–Liebhafsky reaction, *J. Phys. Chem.*, 287, 533–556, 1993.

63. Kolar-Anić, Lj., Mišljenović, Đ., Anić, S., and Nicolis, G., Influence of the reaction of iodate by hydrogen peroxide on the model of the Bray–Liebhafsky reaction, *React. Kinet. Catal. Lett.*, 54, 35–41, 1995.

64. Kolar-Anić, Lj., Vukelić, N., Mišljenović, Đ., and Anić, S., On the instability domains of some models for the Bray–Liebhafsky oscillatory reaction, *J. Serb. Chem. Soc.*, 60, 1005–1013, 1995.

65. Kolar-Anić, Lj., Mišljenović, Đ., and Anić, S., Kinetic model for the Bray–Liebhafsky process with out reaction $IO_3^- + I^- + 2H^+ \rightarrow HIO + HIO_2$, *React. Kinet. Catal. Lett.*, 57, 37–42, 1996.

66. Anić, S. and Stanisavljev, D., Bray–Liebhafsky oscillatory reaction. V. New kinetic data on low-acidity reaction systems, *J. Serb. Chem. Soc.*, 61, 125–127, 1996.

67. Anić, S. and Kolar-Anić, Lj., The Bray–Liebhafsky reaction. VI. Kinetics in iodide oscillations, *J. Serb. Chem. Soc.*, 61, 885–891, 1996.

68. Čupić, Ž., Anić, S., and Mišljenović, Đ., The Bray–Liebhafsky reaction. VII. Concentration of the external species H^+ and IO_3^-, *J. Serb. Chem. Soc.*, 61, 893–902, 1996.

69. Anić, S., Relation between the number of oscillations and activation energy of an oscillatory process, *J. Serb. Chem. Soc.*, 62, 65–69, 1997.

70. Anić, S., Kolar-Anić, Lj., and Koros, E., Methods to determine activation energies for two kinetic states of the oscillatory Bray–Liebhafsky reaction, *React. Kinet. Catal. Lett.*, 61, 111–116, 1997.

71. Stanisavljev, D., Consideration of the thermodynamic stability of iodine species in the Bray–Liebhafsky reaction, *Ber. Bunsenges. Phys. Chem.*, 101, 1036–1039, 1997.

72. Radenković, M., Schmitz, G., and Kolar-Anić, Lj., Simulation of iodine oxidation by hydrogen peroxide in acid media, *J. Serb. Chem. Soc.*, 62, 367–369, 1997.

73. Ševčik, P. and Adamčikova L. Effect of a pressure decrease and stirring on oscillatory Bray–Liebhafsky reaction, *Chem. Phys. Lett.*, 267, 307–312, 1997.

74. Ševčik, P. and Adamčikova L. Effect of gas bubbling and stirring on the oscillatory Bray–Liebhafsky reaction, *J. Phys. Chem. A.*, 102, 1288–1291, 1998.

75. Valent, I., Adamčikova, L., and Ševčik, P. Simulations of the iodine interphase transport effect on the oscillatory Bray–Liebhafsky reaction, *J. Phys. Chem. A.*, 102, 7576–7579, 1998.

76. Čupić, Ž., Modeliranje mehanizma oscilatornih kinetičkih procesa sa primenom na razlaganje vodonikperoksida, Thesis, Fakultet za fizičku hemiju Univreziteta u Beogradu, 1998.

77. Anić, S., Stanisavljev, D., Čupić, Ž., Radenković, M., Vukojević, V., and Kolar-Anić, Lj., The self-organization phenomena during catalytic decomposition of hydrogen peroxide, *Sci. Sintering*, 30, 49–57, 1998.

78. Schmits, G., Models for the oscillating reactions nullclines and steady states, in *Physical Chemistry '98*, Ribnikar, S., Anić, S., Eds., Soc. Phys. Chemists of Serbia, Belgrade, 1998, pp. 173–179.

79. Čupić, Ž. and Kolar-Anić, Lj., Contration of the model for the Bray–Liebhafsky oscillatory reaction by eliminating intermediate I_2O, *J. Chem. Phys.*, 110, 3951–3954, 1999.

80. Čupić, Ž. and Kolar-Anić, Lj., Contration of the complex model by the stoichiometric network analysis, in: *Advanced Sciences and Technology of Sintering*, Stojanović, B.D., Shorokhod, V.V., Nikolić, M.V., Eds., Kluwer Acad. Planum Publ., New York, 1999, pp. 75–80.

81. Schmitz, G., Kinetics and mechanism of the iodate-iodide reaction ant other related reactions, *Phys. Chem. Chem. Phys.*, 1, 1909–1914, 1999.

82. Schmitz, G., Kinetics of Dushman reaction of low I^- concentrations, *Phys. Chem. Chem. Phys.* 2, 4041–4044, 2000.

83. Vukojević, V., Anić, S., and Kolar-Anić, Lj., Investigation of dynamic behaviour of the Bray–Liebhafsky reaction in the CSTR. Determination of bifurcations points, *J. Phys. Chem. A.*, 104, 10731–10739, 2000.

84. Vukojević, V., *Bifurkaciona i perturbaciona analiza Bray–Liebhafsky reakcije*, Thesis, Faculty of Physical Chemistry University of Belgrade, Belgrade, 2000.

85. Schmitz, G., Kinetics of the halates–halides–halogens reactions: apparent differences and fundamental similarities, in *Physical Chemistry 2000*, Ribnikar, S., Anić, S., Eds., Soc. Phys. Chemists of Serbia, Belgrade, 2000, pp. 129–140.

86. Ćirić, J., Anić, S., Čupić, Ž., Kolar-Anić, Lj., The Bray–Liebhafsky oscillatory reaction. Kinetic investigation in reduction and oxidation pathways based on hydrogen peroxide concentration monitoring, *Sci. Sintering*, 32, 187–196, 2000.

87. Ševčik, P., Kicsiminova, K., and Adamčikova L., Oxygen production in the oscillatory Bray–Liebhafsky reaction, *J. Phys. Chem.*, 104, 3958–3963, 2000.

88. Schmitz, G., The oxidation of iodine to iodate by hydrogen peroxide, *Phys. Chem. Chem. Phys.*, 3, 4741–4746, 2001.

89. Kicsiminova, K., Valent, I., Adamčikova L., and Ševčik, P., Numerical simulations of the oxygen production in the oscillatory Bray–Liebhafsky reaction, *Chem. Phys. Lett.*, 341, 345–350, 2001.

90. Anić, S., Kolar-Anić, Lj., Vukojević, V., Čupić, Ž., Stanisavljev, D., and Radenković, M., Brej–Liebhafski oscilatorna reancija, in *Profesoru Draganu Veselinoviću*, Anić, S., Marković, D., Eds., Soc. Phys. Chemists of Serbia and Faculty of Physical Chemistry, Belgrade, 2001, pp. 175–192.

91. Anić, S., Kolar-Anić, Lj., Vukojević, V., Ćirić, J., Mehanizam oscilatorne hemijske reakcije: Bray–Liebhafsky reakcija, *Nauka Tehnika Bezbednost*, 2, 47–64, 2001.

92. Schmitz, G., Thermodynamics and kinetics of some inorganic reactions of iodine, in: *Physical Chemistry 2002*, Anić, S., Ed., Soc. Phys. Chemists of Serbia, Belgrade, 2002, pp. 137–144.

93. Pejić, N., Čupić, Ž., Anić, S., Vukojević, V., and Kolar-Anić, Lj., The oscillatory Bray–Liebhafsky reaction as a matrix for analyzing enzyme and polymeric catalysts for hydrogen peroxide, *Sci. Sintering*, 33, 107–115, 2001.

94. Milošević, M., Pejić, N., Čupić, Ž., Anić, S., and Kolar-Anić, Lj., Examinations of cross-linked polyvinylpyridine in open reactor, *Mater. Sci. Forum*, 494, 369–374, 2005.

95. Ruof, P. and Noyes, R.M., Phase response behaviors of different oscillatory states in the Belousov–Zhabotinsky reaction, *J. Chem. Phys.*, 89, 6247–6254, 1988.

96. Ruof, P., Försterling, H.D., Györgyi, L., and Noyes, R.M., Bromous acid perturbations in the Belousov–Zhabotinskii reaction: experiments and model calculations of phase response curves, *J. Phys. Chem.*, 95, 9314–9320, 1991.

97. Stemwedel, J.D., Ross, J., and Schreiber, I., Formulation of oscillatory reaction mechanisms by deduction from experiments, *Adv. Chem. Phys.*, 89, 327–388, 1995.

98. Pejić, N., Blagojević, S., Anić, S., Vukojević, V., and Kolar-Anić, Lj., Microquantitative determination of hesperidin by pulse perturbation of the oscillatory reaction system, *J. Anal. Bioanal. Chem.*, 381, 775–780, 2005.

99. Čupić, Ž., Anić, S., Terlecki-Baričević, A., and Kolar-Anić, Lj., Bray–Liebhafsky Reaction: the influence of some polymers based on poly (4-vinylpyridine), *React. Kinet. Catal. Lett.*, 54, 43–49, 1995.

100. Terlecki-Baričević, A., Čupić, Ž., Anić, S., Kolar-Anić, Lj., Mitrovski, S., and Ivanović, S., Polyvinylpyridine supported iron (III) catalyst in hydrogen peroxide decomposition, *J. Serb. Chem. Soc.*, 60, 969–979, 1995.

101. Loew, G. and Dupuis, M., Structure of a model transient peroxide intermediate of peroxidases by ab initio methods, *J. Am. Chem. Soc.*, 118, 10584–10587, 1996.

102. Savenkova, M.I., Kuo, J.M., and Ortiz de Montellano, P.R., Improvement of peroxygenase activity by relocation of a catalytic histidine within the active site of horseradish peroxidase, *Biochemistry*, 37, 10828–10836, 1998.

103. Goe, G.L., Marston, C.R., Scriven, E.F.V., and Sowers, E.E., Applications of pyridine-containing polymers in organic chemistry, in *Catalysis of Organic Reactions*, Blackburn, D.W., Ed., Marcel Dekker, New York, 1990, pp. 275–286.

104. REILLEX Report 3, Catalysts, Reilly Tar and Chemical Corporation, Indianapolis, 1985.

105. Ekerdt, J.G., The role of substrate transport in catalyst activity, in *Polymeric Reagents and Catalysts*, Ford, W.T., Ed., ACS, Washington DC, 1985, Vol. 4, pp. 68–83.

106. Pejić, N., Private communication.

9 The Polymer Conformational Stability and Transitions: A Quantum Decoherence Theory Approach

Miroljub Dugić
University of Kragujevac, Kragujevac, Serbia and Montenegro

Dejan Raković and Milenko Plavsic
University of Belgrade, Belgrade, Serbia and Montenegro

CONTENTS

I. INTRODUCTION

In this chapter, we describe the problem of polymer conformational stability and transitions in the framework of the so-called quantum decoherence theory. We propose a rather qualitative scenario yet bearing generality in the context of the quantum decoherence theory, enabling us to reproduce both, existence and stability of the polymers conformations, *and* the short time scales for the quantum-mechanical processes resulting effectively in the conformational transitions. The proposed model is qualitative yet providing us with the possibility to *overcome the main obstacle* in resolving the problem of (semi-)classically unreasonably long time necessary for the change of conformation of the polymers in a solution.

The long-standing problem of the polymer conformational transitions is an open issue of the cross-disciplinary research work and interest [1–4]. It is usually referred to as the Levinthal's paradox with an emphasis on the substantial discrepancy between the phenomenological data and the theoretical background of the issue [5]. The original Levinthal's analysis [5] has led to extensive search for the "preferred pathways (trajectories)" in the conformation space of a macro-molecule. Recently, the problem is sharpened by the new approach that calls for the funnel-like form of the conformation space [6], probably offering the possibility to overcome the Levinthal's paradox.

In this chapter, we offer *a new approach* to the problem. Actually, we show that the fully *quantum-mechanical* approach within the decoherence theory [7] offers both, existence and stability of the molecules conformations, *and* the rather fast decoherence-like transition between the different conformations. Within our approach, the Levinthal's paradox completely disappears.

II. THE PROBLEM

In this section, we precisely outline the problem we are interested in.

A. THE BORN–OPPENHEIMER ADIABATIC APPROXIMATION

It is well known that the Born–Oppenheimer adiabatic approximation establishes geometrical shape of a molecule. The atoms (atomic groups) constituting a molecule are imagined to be placed in the vertices of certain three-dimensional (3D) shape as illustrated in Figure 9.1.

At the zeroth approximation, the atoms (more precisely atomic groups) sitting in the vertices are frozen — their oscillations around the equilibrium positions being neglected. In reality, the atoms are rather quickly vibrating thus giving rise to the fast changes of the molecule's shape, which, on average, is presented by the zeroth approximation shape. Yet, a word of caution is worth saying in this regard.

The geometrical forms of the molecules should not be too literally understood and interpreted. Even in the zeroth approximation, the molecule's shape is subject to the Heisenberg (position versus momentum) uncertainty relation. However, the *relative positions* of the atoms are still well-defined variables. These variables' quantum-mechanical *averages* justify the zeroth approxi-mation as defined earlier. Fortunately enough, these relative positions can be "measured" by the low-energy particles. Bearing these subtleties in mind, one may consider the molecules *effectively* to bear the "definite" geometrical shapes — as it is *generally assumed* in chemistry.

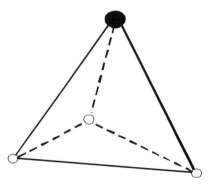

FIGURE 9.1 The ammonia molecule geometric (pyramidal) shape or form.

B. THE MOLECULES CONFORMATIONS

The larger molecules are always found in oriented states, which assume the definite geometrical form in the sense of the preceding section. In the simplified terms, the larger molecules may be viewed as the 3D (semi-)classical clusters as presented in Figure 9.2.

Actually, the explanation of the molecules' optical activity originates from the assumption that the different geometrical forms of the molecules, being energetically different, necessarily give rise to different responses upon the external optical stimulus. More precisely: the relative distances as well as the valence angles (cf. Figure 9.2) between the adjacent atoms (atomic groups) are *well-defined* variables of the molecules so much as they can be referred as to the *molecules' conformations.* Furthermore, the different conformations of a molecule are mutually related by *conformational transitions* — the geometrical transformations keeping the mutual distances of the atomic groups as well as the valence angles (ϑ_i) as depicted earlier — allowing (semi-)classically for a succession of local rotations of adjacent molecular segments over the preceding valence bonds. While keeping the primary structure of the molecules, the conformational transformations effect in the different (bio)chemical 3D-structurally dependent activity of the molecule subjected to the transformation.

The experimental evidence [1–4] in this regard can be summarized as follows: The molecules *dissolved in a liquid* are found both to bear as well as to maintain their geometrical shapes or forms — which refers to every single molecule in the liquid. Therefore, in a liquid, the molecules' forms can be described by a *statistical ensemble of shapes* (*conformations*) generally depending on the parameters of the composite system "molecules + liquid," such as the composition, temperature, viscosity, etc. For the fixed parameters of the system, the above-mentioned ensemble maintains its definition — which we here refer to as the *stationary state*: in general, the concentration of the molecules bearing a given shape (from the set of the possible conformations) is constant. In other words: If left intact, the ensemble will maintain its state. However, certain external actions can give rise to the change of the conformations of the molecules in the liquid [8]. Such external action can be described as the *nonstationary process*, which finally gives rise to the relaxation process eventually leading to the *new stationary state* of the system, with different concentrations of the conformations, including (possibly) appearance of the new ones.

Therefore, in simplified terms, the evidence about the conformations can be described as follows:

(i) A stationary state is initially defined by the molecules' conformations (statistical) distribution and concentration, which remains intact as long as the stationary state is conserved.

(ii) Certain external actions can destroy the stationary state, and can be characterized by the change of the composite system's parameters.

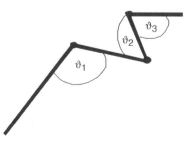

FIGURE 9.2 A 3D cluster (semi-)classically representing a molecule. The vertices are occupied by the atoms or atomic groups, the straight lines representing the chemical valence bonds.

(iii) The external action eventually gives rise to the relaxation process, which, in turn, gives rise to another stationary state, which is characterized by another conformation distribution or concentration.

C. Levinthal's Paradox: A Survey

Owing to the influence of the environment, the large molecules may *change their conformations.* According to Figure 9.2, these changes can be viewed and interpreted as the externally induced deformations of the 3D molecular cluster as presented in Figure 9.3. For instance, the *successive*, *local* rotations can effect in change of the lattice shape. These conformational transformations keep both the relative positions of the vertices and the angles characterizing the initial conformation.

Physically, the different conformations of a molecule are described by the different (conformational) energies $V(k)$ — minima as shown in Figure 9.4 of the molecule [1–4]. Actually, even for the "frozen" molecule (neglected vibrational degrees of freedom), the different conformations are ascribed to the different energies. Denoting the conformation, K, as the molecules' variable, the molecular electronic energy dependence on K can be qualitatively presented by Figure 9.4.

The horizontal axis refers (for simplicity) to the "position" of one-dimensional "particle" K — the configuration space of the system. The vertical axis refers to the conformational molecular electronic energy, $V(k)$, as a potential energy for the adiabatically decoupled (vibrational and) one-dimensional conformational system K, k representing a value of the variable K. The local minima represent the (meta)stable conformations of a molecule with the following characteristic: the (one-dimensional) particle sitting in the vicinity of a local minimum is attracted toward the minimum, finally centering around the bottom of the minimum (of a stable conformation). As a consequence, not every geometrically possible conformation may be taken by a molecule. Rather, only the conformations referring to the energetically preferable shapes are allowed — as defined by the local minima in Figure 9.4. Certainly, the *continuous* change of shape of the molecule follows the $V(k)$ plot in K-space. Once centered around the bottom of a local minimum,

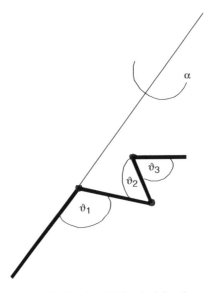

FIGURE 9.3 Conformational transition by the (semi-)classical local rotation for the angle α in the 3D molecular cluster. Here, only the segment defined by the valence angles $\vartheta_{2,3}$ rotates over the preceding chemical valence bond, coinciding with the axis of rotation.

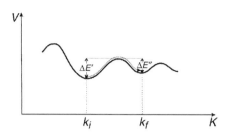

FIGURE 9.4 The molecular electronic energy as a potential energy for the adiabatically decoupled (vibrational and) one-dimensional conformational system K.

the particle does not have energy enough to change its position — *unless* it is *externally forced* to do so.

Therefore, the conformational change of a molecule can be (semi-)classically viewed as the continuous change of its geometrical shape originating from some initial k_i to the final conformation k_f (in two-conformational example of Figure 9.4). Being in the vicinity of a local minimum, the one-dimensional particle presented in Figure 9.4 will tend to reach the minimum. This (semi-)classical model gives a background for the experimentally verified findings about the large molecules conformations. Unfortunately, it immediately gives rise to the problem — the so-called Levinthal's paradox [5].

Actually, as Levinthal emphasizes, the conformational transitions can be realized through a *sequence* of local rotations (cf. Figure 9.3) eventually giving rise to another (energetically preferable) conformation. The *core of the Levinthal's paradox* can be presented as follows:

> For certain large molecules, the number of the local rotations necessary for effecting the conformational change (e.g. $k_i \rightarrow k_f$ in Figure 9.4) may be so large that the effective time necessary for completing the transformation becomes unreasonably long, thus making the whole procedure physically unrealistic a process.

More precisely: assuming $2n$ torsional angles of an n-residue protein, each having three stable rotational states; this yields $3^{2n} \approx 10^n$ possible conformations for the chain (even with rather gross underestimating). If a protein can explore new conformations in a *random way*, at the rate that single bond can rotate, it can find approximately 10^{13} conformations per seconds, which is here an overestimating. We can then calculate the time t (sec) required for a protein to explore all the conformations available to it: $t = 10^n/10^{13}$. For a rather small protein of $n = 100$ residues, one obtains $t = 10^{87}$ sec, which is immensely more than the apparent age of the universe ("Levinthal paradox"). Yet, according to some experiments, proteins can fold to their native conformation in less than a few seconds [8]. It follows that conformational changes of proteins in solution, due to compositional, thermal, and other influences of the environment, do not occur in a random way (as e.g., movements of gas particles) — but fold to their native conformation in some sort of *ordered set of pathways* in which the approach to the native state is accompanied by sharp increasing conformational stability — this being one of the most crucial questions in all life sciences.

The Levinthal's paradox raised the search for the preferred (ordered set of) pathways in K-space. The core of the research work in this regard refers to this task essentially pointed out by Levinthal. However, some recent approaches shed some new light in this concern.

D. LEVINTHAL'S PARADOX: REVISITED

The new approach [6] (and references therein) goes beyond the concept of the preferred pathways. Actually, it refers to the details in the single molecule's conformational transitions yet searching for the fast transitions.

The approach calls for the specific *funnel* structure of the more realistic multidimensional $V(k)$-hypersurface, which might provide the basis for fast conformational transitions. Actually, existence of the local *funnel-like* regions in $V(k)$-hypersurface is assumed, with the following main idea: The funnel-like shape *restricts* the set of the allowed trajectories (pathways) for the conformational change. The different trajectories should be stochastically taken by the different (single) molecules in the ensemble of molecules from the set of the possible trajectories in the *restricted K*-space.

Essentially, this proposal relaxes the original idea about the preferred pathways, yet in an elaborate fashion. Unfortunately, this is still a *qualitative* (semi-)classical model requiring much research work yet to be done; it is therefore hard to predict the success of this approach. For example, one may notice that this proposal does not substantially go beyond the standard "pathways" (semi-)classical approach to the issue, especially in its *kinematical* context.[1]

III. NEW APPROACH: THE CLUE

The Levinthal's paradox emerges from the (semi-)classical picture of the molecules conformations defined in the K-space of the one-dimensional model of Figure 9.4. Within this strategy, the particle bears a definite position k_1 in every instant of time. Thus every conformational change can be represented by a "trajectory" (path) in K-space, following the shape of $V(k)$.

Bearing this in mind, we speculate about the possible solution of the problem in the context of the following idea:

> To find the theoretical background allowing us to abandon the concept of the trajectory (the path) in *K*-space, in order to avoid the problems of the *kinematical nature*.

Fortunately enough, there is a theory justifying this idea — quantum mechanics. Actually, in the full quantum-mechanical treatment, the one-dimensional particle K might *be allowed* the linear superpositions of the different "positions," thus making the concept of the position (and consequently of "trajectory") physically meaningless.[2]

At first sight, this approach may seem unreasonable, because it should simultaneously provide both existence and maintenance of the (stable) conformations in the stationary state of the system, *and* the model for fast conformational transitions. Fortunately enough, there is a quantum-mechanical theory meeting these criteria and requirements — the so-called *decoherence theory* [7]. Section V justifies this claim. In Section IV, we outline the fundamentals of the decoherence theory.

Finally to this section, we answer the following question:

> Why should one believe in quantum-mechanical behavior of the large molecules? After all, the (semi-) classical approach seems perfectly to work for most purposes in chemistry.

Our answer can be given in few steps.

---

[1]The funnel approach restricts the number of the possible pathways. Kinematically, it means that for n local rotations — no matter which path down the funnel has been taken by the molecule — there appears the constraint of small n. This constraint comes from the spectroscopic data on the poorly dimensionally sensitive dispersion laws of the internal quasiparticle excitations [9,10], which stem the same order of magnitude for the two time intervals, for the molecular conformational transition (τ), as well as for the (average) time of the local segmental rotations (τ_r), while bearing $\tau = n\tau_i$ in mind. Certainly, this might be a serious restriction, in principle, for the large molecules conformational transitions in the still (semi-)classical funnel approach.

[2]To this end, one may use an analogy. For example, in the interference experiments in optics, the concept of trajectory (of a particle traversing a slit in the diffraction grating) becomes meaningless — existence of trajectories wipes out the interference (diffraction) pattern on the screen. Interference is analogous to the coherent (quantum-mechanical) superpositions of the different "positions" in K-space.

First, the molecules are ultimately quantum-mechanical systems. It is therefore *per se* interesting to investigate this approach to the issue. Second, the "borderline" between the "quantum" and "classical" is a matter of extensive research science-wide [7]. Particularly, it means that experience with the (semi-)classical behavior does not necessarily remove the possibility of purely quantum-mechanical behavior of the large molecules. Third, the recent experiments on the macromolecules (spatial) interferometry [11,12] directly address the following question: Under *which circumstances*, one may expect the quantum-mechanical behavior of the *mesoscopic* systems, with the view to even much larger ("macroscopic") systems? To this end, the lesson is rather simple: Rejecting the quantum-mechanical behavior of large molecules is just a matter of *stipulation*, not yet a scientific truth — as much as we know to date. Fourth, there is a quantum-mechanical basis for the macromolecules *individuality* in a liquid (solution), thus giving rise to the (seemingly) (semi-)classical basis of the kinetic theory[3] [13].

Therefore, we conclude:

The macromolecules dynamics in a solution is an interplay between the different (even mutually competing) processes that are only poorly known to date, some of them bearing quantum-mechanical origin.

IV. THE FUNDAMENTALS OF THE DECOHERENCE THEORY

A. TERMINOLOGY

A word of caution is worth saying. Sometimes, some physically different processes or effects are misidentified or misinterpreted as "decoherence." Below, we give a precise definition of decoherence, that is, of the *decoherence-induced superselection rules effect* [14], which we refer to as "decoherence."

B. OPEN QUANTUM SYSTEMS

Standard quantum-mechanical theory deals with the *isolated* quantum systems, whose dynamics is governed by Schrodinger equation. For such systems, the environment effectively plays the role of the external potential, which is an additive term of the system's Hamiltonian. Needless to say, such dynamics is exactly reversible. On the other side, if the environment nontrivially acts on the system, such a system is referred to as the *open* quantum system: the interaction between the (open) system (S) and its environment (E) changes also the state of the environment, thus effecting irreversibly the dynamics of the system S.

The open system's dynamics is crucially determined by the interaction in the composite system S + E. Usually, for the composite system S + E, one adopts validity of the Schrodinger equation. Then, the task is to properly describe the open system's dynamics. Probably the best-known examples of open systems are: the object of quantum measurement, a quantum particle in Wilson chamber, and detection of a quantum particle on the screen.

From the mathematical point of view, the distinction between the isolated and the open systems can be described as follows:

(A) For the isolated systems, the effect of its environment is encapsulated by the "external potential" (potential energy) of the particle, $V(\widehat{x}_{Si}, \widehat{p}_{Si}, a_{Sj}, A_{Ek})$, where $\widehat{x}_{Si}$ and $\widehat{p}_{Si}$ denote the system's degrees of freedom and their conjugate momenta, respectively, while a_{Sj} and A_{Ek} represent the system's and the environment's parameters (such as the mass, electric charge, etc.), respectively. Needless to say, V is the "one-particle" observable changing the states of the open system, not yet of the environment.

---

[3]The individuality refers to distinguishing the particles in a liquid not yet compromising their quantum-mechanical behavior.

(B) The interaction Hamiltonian $\widehat{H}_{S+E} \equiv \widehat{H}_{int}$ is a two-body observable, coupling the observables of both the system S and environment E. The composite system's Hamiltonian reads:

$$\widehat{H} = \widehat{H}_S \otimes \widehat{I}_E + \widehat{I}_S \otimes \widehat{H}_E + \widehat{H}_{int} \tag{9.1}$$

where the third term denotes the interaction energy, which is a "two-system" observable changing, in general, the states of both S and E

C. The Task

The central issue of quantum mechanics of open systems is calculating the open system's dynamics. Let us by $\widehat{\rho}_S(t = 0)$ denote the open system's initial state, while by $\widehat{\sigma}_E(t = 0)$ we denote the initial state of the environment. Then, by definition, the initial state of the composite systems $S + E$ reads

$$\widehat{\rho}_S(t = 0) \otimes \widehat{\sigma}_E(t = 0) \tag{9.2}$$

Then (9.1) gives uniquely rise to the unitary evolution operator $\widehat{U}$ for the composite system, that is, to the unitary dynamics of the composite system:

$$\widehat{\rho}_{S+E}(t) = \widehat{U}\,\widehat{\rho}_S(t = 0) \otimes \widehat{\sigma}_E(t = 0)\,\widehat{U}^* \tag{9.3}$$

Now, according to the general rules of quantum mechanics, the subsystem's state is defined as

$$\widehat{\rho}_S(t) = tr_E \widehat{\rho}_{S+E}(t) \tag{9.4}$$

with tr_E denoting the "tracing out" (integrating over) the environmental degrees of freedom.

It is apparent that the interaction term $\widehat{H}_{int}$ is central to the open system's dynamics. For example, if one may write (in accordance with Section IV.B(A)):

$$\widehat{H}_{int} = \widehat{V}_S \otimes \widehat{I}_E + \widehat{I}_S \otimes \widehat{V}'_E \tag{9.5}$$

then the two subsystems evolve mutually independently in time:

$$\widehat{\rho}_{S+E}(t) = \widehat{\rho}_S(t) \otimes \widehat{\sigma}_E(t) \tag{9.6}$$

However, for the nontrivial coupling of the observables of the two subsystems, one obtains (cf. Section IV.B(B)) the correlations of states of the subsystems.

As to the decoherence theory, the most interesting is the situation in which the initial state of the system S is a "pure" quantum state (an element of the system's Hilbert state space), $|\psi\rangle_S$:

$$\widehat{\rho}_S(t = 0) = |\psi\rangle_S\langle\psi| \tag{9.7}$$

Writing

$$|\psi\rangle_S = \sum_i C_i |\varphi\rangle_{Si} \tag{9.8}$$

one may directly obtain

$$\widehat{\rho}_S(t=0) = \sum_{i,j} C_i C_j^* |\varphi_i\rangle_S \langle\varphi_j| \tag{9.9}$$

with the nonzero off-diagonal ($i \neq j$) terms of $\widehat{\rho}_S$.

One of the central findings of the decoherence theory is the observation that for certain special states (orthonormalized basis) $\{|\varphi_i\rangle_S\}$, the evolution in time effects in the loss of the off-diagonal terms of $\widehat{\rho}_S$. That is, one may write for the off-diagonal terms of $\widehat{\rho}_S$:

$$\lim_{t\to\infty} \rho_{Sij}(t) = 0, \quad i \neq j \tag{9.10}$$

for the rather *short time intervals* of the order of τ_D — the decoherence time.

More precisely: The matrix representation of $\widehat{\rho}_S$ *in the basis* $\{|\varphi_i\rangle_S\}$ — the "pointer basis" — is of the quasi-diagonal form, thus giving rise to the *effective superselection rules* [14,15] for the open system that is described by the orthogonal decomposition of the system's Hilbert state:

$$H_S = \sum_n {}^\oplus H_n \tag{9.11}$$

Physically, the environment influences the loss of coherence in the open system's state, H, for the states belonging to the different subspaces H_n; $\langle\varphi_i|\varphi_j\rangle = 0$, if $|\varphi_i\rangle \in H_n$, $|\varphi_j\rangle \in H_{n'}$, $n \neq n'$.

D. THE ENVIRONMENT-INDUCED DECOHERENCE

The loss of the initial coherence can be presented as follows:

$$\sum_{i,j} C_i C_j^* |\varphi_i\rangle_S \langle\varphi_j| \xrightarrow{\tau_D} \sum_i |C_i|^2 |\varphi_i\rangle_S \langle\varphi_i| \tag{9.12}$$

meaning that the coherence between the states belonging to the different subspaces, H_n, is effectively forbidden after the (rather *very short* [7,15]) time interval of the order of the decoherence time.

Alternatively, the decoherence effect can be described by the *existence* of the "pointer observable", $\widehat{\Lambda}_S$, whose spectral form reads [14,15]:

$$\widehat{\Lambda}_S = \sum_n \lambda_n \widehat{P}_{Sn} \tag{9.13}$$

where the projectors $\widehat{P}_{Sn}$ are in one-to-one correspondence with the subspaces H_n (appearing in (9.11)). This observable is the center of algebra of the observables of the open system S, while fulfilling the commutator relation [14–16]:

$$[\widehat{H}_{int}, \widehat{\Lambda}_S \otimes \widehat{I}_E] = 0 \tag{9.14}$$

E. THE PHYSICAL CONTENTS

The decoherence effect is a striking effect: effectively, there is the loss of coherence in the open system's state space. This restricts both the states and the observables of the open system that can be observed by an independent observer [7].

The lack of coherence is exactly what is expected from a *macroscopic* (a classical) system, which is subject to the classical *determinism* and *reality*. Actually, for the macroscopic systems, one may say that coherent superpositions of the type (9.8) have never been observed. Paradigmatic in this concern are the macroscopic "center-of-mass" coordinates of a macroscopic body or object. This is the reason why the decoherence effect is sometimes considered as the main candidate for finding out the solution to the problem of the transition from "quantum to classical" [7,15]. To this end, the "pointer basis" and the "pointer observable" are considered to bear the macroscopic characteristics of an open quantum system.

Also crucial is the following observation: *relative to* $\widehat{H}_{\mathrm{int}}$, the elements of a "pointer basis" (which is also an eigenbasis of $\widehat{\Lambda}_S$) are *robust* [15,16]. Physically, it means that, once effected, the decoherence will keep the states of a "pointer basis" effectively intact in the course of the unitary evolution of the composite system S + E. This robustness of certain system's states is crucial for the macroscopic context of the decoherence theory. Particularly, it means that the decoherence effect gives rise to both, *existence* and *maintenance* of states of a "pointer basis" — that is, the relevance of the superselection rules — of an open system in the course of the unitary evolution of the combined system S + E. In other words, the decoherence effect tends to freeze the open system's dynamics as defined by the decomposition (9.11). The decoherence time τ_D is usually very short, including the *mesoscopic systems* such as certain macromolecules [12]. It is therefore not for surprise that the decoherence effect has been observed in the quite controlled circumstances only recently [12,17,18].

Finally, in principle, certain *external actions* on the composite system can effect in breaking the superselection rules, that is, of the decoherence effect. Actually, for certain interactions with another external system E′, if one may write:

$$[\widehat{H}'_{\mathrm{int}}, \widehat{\Lambda}_S \otimes \widehat{I}_E] \neq 0 \qquad (9.15)$$

there might appear the coherent superpositions of states from the pointer basis. For example, if $\widehat{H}'_{\mathrm{int}}$ is such that: (i) it dominates in the system and (ii) it defines another decomposition in contradistinction with (9.11), then one may obtain another "pointer basis," $\{|\chi_i\rangle_S\}$, such that

$$|\chi_j\rangle_S = \sum_i d_{ij}|\varphi_i\rangle_S \qquad (9.16)$$

In the macroscopic considerations, the possibility of such an effect is generally neglected. However, this need not be the case for the *mesoscopic systems* such as the macromolecules.

V. NEW APPROACH

Formally, we deal with the one-dimensional system (S), $(\widehat{K}_S, \widehat{P}_S)$, where $\widehat{K}_S$ stands for the "coordinate" *conformation*, while the momentum $\widehat{P}_S$ satisfies

$$\left[\widehat{K}_S, \widehat{P}_S\right] = \frac{ih}{2\pi} \qquad (9.17)$$

However, the system S is an *open system* — as distinguished in Section II. It inevitably interacts with its environment, which physically consists of the (liquid's) molecules. Therefore, the system of interest is the composite system "conformation + liquid (S + E)."

In this section, we address the following tasks:

(A) To establish existence *and* maintenance of an ensemble of conformations in the *stationary state* of the composite system, where the conformations $|k_i\rangle_S$ satisfy

$$\widehat{K}_S|k_i\rangle_S = k_i|k_i\rangle_S \tag{9.18}$$

(B) To model the conformation change, $|k_i\rangle_S \to |k_f\rangle_S$, from the set of the allowed conformations.

Physically, the task (b) refers to the *nonstationary state* of the composite system that is induced by the external action (point (ii) of Section II.B). At the first sight, these two tasks might seem formidable. Fortunately enough, there is a quantum-mechanical theory fulfilling these tasks — the decoherence theory.

A. THE CONFORMATION STABILITY

The composite system $S + E$ is formally defined by the Hamiltonian of the form (9.1). For simplicity, we shall further deal exclusively with the interaction term, $\widehat{H}_{S+E} \equiv \widehat{H}_{int}$.

As we learned from Section IV, a proper choice (model) of H_{int} might provide both existence and stability of an ensemble of conformations in the stationary state of the composite system. To this end, it is crucial to recognize the conformation K_S as the "pointer observable" of the system. In an ensemble of conformations, in general, every conformation, $|k_i\rangle_S$, should be ascribed a probability $p_i; \sum_i p_i = 1$. Then, the initial stationary state (point (i) of Section II.B) of the ensemble is described by the statistical operator:

$$\widehat{\rho}_S = \sum_i p_i|k_i\rangle_S\langle k_i| \tag{9.19}$$

More precisely, to obtain existence (appearance *due to decoherence*) and maintenance (stability *due to decoherence*) of the conformations in the initial state (9.19), it is necessary (and for certain simple models, it is sufficient) to have satisfied the condition or requirement [16]:

$$[\widehat{H}_{int}, \widehat{K}_S \otimes \widehat{I}_E] = 0 \tag{9.20}$$

Fortunately enough, this gives rise to applicability of a rather wide class (and types) of the interaction Hamiltonians.

Actually, one may write (for the time-independent interaction) for the unitary operator of the composite system:

$$\widehat{U} \cong \widehat{U}_{int} = \exp\left\{\frac{-ith\widehat{H}_{int}}{2\pi}\right\} \tag{9.21}$$

which for the initial state (*before* the "initial stationary state" (9.19))

$$|\psi(t=0)\rangle = \sum_i C_i|k_i\rangle_S \otimes |0\rangle_E \tag{9.22}$$

gives for the state in an instant t:

$$|\psi(t)\rangle = \sum_i C_i |k_i\rangle_S \otimes |\chi_i(t)\rangle_E \tag{9.23}$$

satisfying the condition

$$\lim_{t\to\infty} |\langle\chi_i(t)|\chi_j(t)\rangle| = 0, \quad i \neq j \tag{9.24}$$

Needless to say, then one obtains satisfiability of the condition (9.10) — existence (appearance) of the preferred states of the open system S, the set of preferred conformations. Because (9.20) guarantees stability (robustness) of the conformations, relative to H_{int}, the requirement of maintenance of the conformations is also satisfied, where $p_i = |C_i|^2$. Physically: The proper model (9.20) gives rise to appearance and stability of conformation for every single molecule in the liquid in the initial stationary state (9.19).

B. Nonstationary State

The external action on the composite system (point (ii) of Section II.B) gives rise to the change of the system's parameters. It is the fact: Such external actions might induce the change of the stationary state as well as of the conformational transitions for every single molecule. In effect, there appears another ensemble of the possible conformations.

Modeling the nonstationary state in the general terms is rather simple. Actually, it seems quite natural to assume that the external action *changes the interaction* in the composite system. Furthermore, it is certainly the case that the new (effective) environment, E', appears. Therefore, one should assume the new interaction term, $H_{S+E'} \equiv H'_{int}$. Setting (cf. (9.15)):

$$[\widehat{H}'_{int}, \, \widehat{K}_S \otimes \widehat{I}_{E'}] \neq 0 \tag{9.25}$$

and consequently $[\widehat{H}_{int}, \, \widehat{H}'_{int}] \neq 0$, one may obtain the change of conformations.

Assuming that the external action is so strong that H'_{int} *dominates* in the system, then the spectral form (9.19) necessarily changes. Actually, then one may write

$$[\widehat{\rho}_S, \, \widehat{\rho}'_S] \neq 0 \tag{9.26}$$

where the state $\widehat{\rho}'_S$ refers to the nonstationary state, and is defined as

$$\widehat{\rho}'_S(t) = \text{tr}_E \widehat{U}'_{int} \widehat{\rho}_{S+E'}(t') \widehat{U}'^*_{int}, \quad t > t' \tag{9.27}$$

where $\widehat{U}'_{int} = \exp\{-ith\widehat{H}'_{int}/2\pi\}$.

These general considerations do not restrict significantly the possible set of states $\widehat{\rho}'_S(t)$. Furthermore, due to (9.26), one may be free to write

$$\widehat{\rho}'_S(t) = \sum_i \pi_i(t) |\chi_{it}\rangle_S \langle\chi_{it}| \tag{9.28}$$

where, in general, there *appears coherence of the different conformations*:

$$|\chi_{jt}\rangle = \sum_i \alpha_{ijt} |k_i\rangle_S \tag{9.29}$$

Physically, the external action breaks the initial loss of coherence, which is caused by decoherence giving rise to the initial stationary state (9.19). This observation is virtually totally independent on the assumptions about the new interaction (in the system S + E′) [16].

The nonstationary state is expected to be terminated by the relaxation process.

C. THE RELAXATION PROCESS

Following the evidence (point (iii) of Section II.B), we assume that the external action terminates, eventually giving rise to the relaxation process. Actually, we assume that the relaxation process gives rise to *re-establishing* of the stationary state, which, in turn, should be determined by the *same kind* of interaction — selecting the conformational states as the "pointer basis." Actually, we assume that the new composite system, S + E′, is defined by the interaction of the *kind*[4] (9.20). With this *natural assumption*, we easily reproduce reappearance of conformations, as the final states for every single molecule (cf. Section IV).

Namely, validity of (9.20) in the final stationary state gives rise to the possible *occurrence of decoherence* in the composite system S + E′. This is exactly what we have required in Section II. Essentially, the stationary states (initial and final ones) should qualitatively coincide — being characterized by the appearance and stability of conformations.

Formally, if we denote the final state (after the relaxation process) by $\widehat{\rho}''$, then the relaxation process gives rise to the transition

$$\widehat{\rho}'_S \xrightarrow{\text{relaxation}} \widehat{\rho}''_S \tag{9.30}$$

while the subsequent decoherence gives rise to

$$\widehat{\rho}''_S \xrightarrow{\text{decoherence in the new stationary state}} \widehat{\rho}'''_S \tag{9.31}$$

The point strongly to be emphasized is that $\widehat{\rho}'''_S$ now reads, in general, as

$$\widehat{\rho}'''_S = \sum_j q_j |k'_j\rangle_S \langle k'_j| \tag{9.32}$$

where, in general, appear both the different conformations,[5] $|k'_j\rangle_S$, as well as the *different probabilities* q_j (of the possible conformations in the final ensemble) than in the initial stationary state (9.19).

Needless to say, physically, this means that the *net effect* is the conformational change:

$$\widehat{\rho}_S = \sum_i p_i |k_i\rangle_S \langle k_i| \quad \longrightarrow \quad \widehat{\rho}'''_S = \sum_j q_j |k'_j\rangle_S \langle k'_j| \tag{9.33}$$

as it is experimentally observed.

D. THE MODEL

We keep in mind the requirements of Section II. Then, with respect to the experience with the decoherence theory (Section IV), we call for the rather general, hopefully *realistic assumptions* about the composite system, that is, about the interactions in the composite systems — (9.20) and (9.25).

Particularly, we assume:

(A) Every *stationary state* (initial, intermediate — if such exist — states, as well as the final one) is characterized by interaction in the composite system that is of the *same*

---

[4]Not necessarily of the same type (of the same form).

[5]An extreme case is $|k_i\rangle = |k'_j\rangle$, while $p_i \neq q_j$. This is just the change of concentrations of the initial conformations.

kind — being able to give rise to the occurrence of decoherence with the conformations as the "pointer basis states."

(B) Nonstationary state is characterized by the *change in the character of interaction* in the composite system.

The net effect takes the following "phases," each having its own characteristic time:

(a) External action (producing the nonstationary state), taking time T_{ext}.

(b) Relaxation process (establishing the new, final stationary state), taking time T_{relax}.

(c) Decoherence process (*in* the final stationary state), taking time of the "decoherence time," τ_D.

Therefore, to summarize, the conformation transition (9.33) *takes time*:

$$T_{\text{ext}} + T_{\text{relax}} + \tau_D \tag{9.34}$$

which (cf. Section IV) gives (plausibly) rise to

$$T_{\text{ext}} + T_{\text{relax}} + \tau_D \approx T_{\text{ext}} + T_{\text{relax}} \tag{9.35}$$

Therefore, we conclude, "In our model, the Levinthal's paradox completely disappears."

VI. DISCUSSION

We essentially make a couple of plausible assumptions or interpretations of the phenomenological data which allow the natural accounting for the decoherence effect in the composite system "conformation + environment." These assumptions are worth repeating. First, we assume that every stationary state of the composite system — that is characterized by the constant values of the system's parameters — is characterized by the same *kind* of interaction in the composite system (cf. (9.20)). Second, we assume that the external action — eventually giving rise to the conformational transitions — substantially change the kind of interaction in the (new) composite system (cf. (9.25)). It is a matter of the general decoherence theory straightforwardly to prove the final result (9.33), as well as (9.35) [7,14–16,19].

Needless to say, the system S (the "conformation") is (likewise in the (semi-)classical approach) a characteristic of a (single) molecule as a *whole*. That is, as usual, we do not take into account the local details of the conformational rotations themselves, which essentially take into account the electron-state transitions. As much as we can see, these are of the secondary importance to our model, which abandons the concept of the transitions in K-space. Abandoning the K-space is key to the possible success of our model. It is a decoherence-like process that breaks stability of conformations (in the nonstationary state), eventually giving rise to the possibility of rather fast conformational transitions.

There is the following prediction from our model: Even for the single (unique) initial conformation in the composite system ($p_0 = 1$, $p_i = 0$, $\forall i \neq 0$ — cf. (9.19)), our model *predicts* appearance of a *set* (nonunique) of the *final conformations* ($q_j \neq 1$, $\forall j$ — cf. (9.32)). Distinguishing experimentally between this prediction and the opposite possibility — for example, one-to-one conformational transitions — might sharpen the role of our proposal in the context of the conformational transitions problem.

Bearing in mind the foundations of the decoherence process, it should also be stressed: A definition of an open system goes *simultaneously* with defining the system's environment [20,21]. A strong, local interaction with a part of the environment may redefine the open system, simultaneously defining the rest of the environment as the new environment for the new open system.

This way, even the larger "pieces" of a living cell may be allowed the quantum-mechanical behavior — which, we believe, might be of interest in the biomolecular recognition process [22,23].

VII. CONCLUSION

The Levinthal's paradox is an open problem still. To avoid the *core* of the problem — it's *kinematical* aspect — we propose a new approach in this regard. Actually, we treat the macromolecules conformations as the quantum-mechanical observable. Bearing in mind the foundations of the decoherence theory, we are able to model both, existence and maintenance of the conformations *as well as* the conformational transitions in the *rather short* time intervals. Our model is rather qualitative yet a general one — while completely removing the Levinthal's paradox — in contradistinction with the (semi-)classical approach to the issue.

REFERENCES

1. Volkenstein, J.V., *Configurational Statistics of Polymer Chains*, Interscience-Wiley, New York, 1963 (transl. from Russian ed., 1959).
2. a. Flory, P.J., *Principles of Polymer Chemistry*, Cornell University Press, Ithaca, 1953.
   b. Flory, P., *Statistics of Chain Molecules*, Interscience-Wiley, New York, 1969.
3. De Gennes, P.-G., *Scaling Concepts in Polymer Physics*, Cornell University Press, Ithaca, 1979.
4. Plavšić, M., *Polymer Materials Science and Engineering*, Naučna Knjiga, Belgrade, 1996 (in Serbian).
5. Levinthal, C., Are there pathways for protein folding. *J. Chem. Phys.*, 65, 44–45, 1968.
6. Dill, K. and Chau, H.S., From Levinthal to pathways to funnels, *Nat. Struct. Biol.*, 4 (1), 10–19, 1997.
7. Giulini, D., Joos, E., Kiefer, C., Kupsch, J., Stamatescu, I.-O., and Zeh, H.-D., *Decoherence and the Appearance of a Classical World in Quantum Theory*, Springer, Berlin, 1996.
8. Anfinsen, C.B., Principles that govern the folding of protein chains, *Science*, 181, 223–230, 1973, and references therein.
9. Raković, D., *Physical Bases and Characteristics of Electrical Materials*, Faculty of Electrical Engineering, Belgrade, 1995 (in Serbian).
10. Gribov, L.A., *Theory of Infrared Spectra of Polymers*, Nauka, Moscow, 1977 (in Russian).
11. Arndt, M., et al., Wave–particle duality of C_{60} molecules, *Nature*, 401, 680–682, 1999.
12. Hackermüller, L., et al., Decoherence by the emission of thermal radiation, *Nature*, 427, 711–714, 2004.
13. Dugić, M., Quantum entanglement suppression, *Europhys. Lett.*, 60 (1), 7–13, 2002.
14. Zurek, W.H., Environment-induced superselection rules, *Phys. Rev. D*, 26 (8), 1862–1880, 1982.
15. Zurek, W.H., Preferred states, predictability, classicality and the environment-induced decoherence, *Prog. Theor. Phys.*, 89 (2), 281–302, 1993.
16. Dugić, M., On diagonalization of a composite-system observable, *Phys. Scripta*, 56, 560–565, 1997.
17. Brune, M., et al., Observing the progressive decoherence of the meter in a quantum measurement, *Phys. Rev. Lett.*, 77, 4887–4890, 1996.
18. Ammann, H., et al., Quantum delta-kicked rotor: experimental observation of decoherence, *Phys. Rev. Lett.*, 80 (19), 4111–4115, 1998.
19. Dugić, M., On diagonalization of a composite-system observable, *J. Res. Phys.*, 27 (2), 141–153, 1998.
20. Dugić, M., A contribution to the foundations of decoherence theory in nonrelativistic quantum mechanics, Ph.D. Thesis in Physics, University of Kragujevac, 1997 (in Serbian).
21. Dugić, M., Ćirković, M.M., and Raković, D., On a possible physical theory of consciousness, *Open Syst. Information Dyn.*, 9 (2), 153–166, 2002.
22. Ćosić, I., *The Resonant Recognition Model of Macromolecular Bioactivity: Theory and Applications*, Birkhauser Verlag, Basel, 1997, and references therein.
23. Raković, D., Dugić, M., and Plavšić, M., Biopolymer chain folding and biomolecular recognition: a quantum decoherence theory approach, *Material Science Forum*, 494, 513–518, 2005.

10 Reality and Compatibility of Physical and Mathematical Formalisms

Jovo P. Jarić and Dragoslav S. Kuzmanović
University of Belgrade, Belgrade, Serbia & Montenegro

CONTENTS

I. INTRODUCTION

The objective of this chapter is to present perspective of physics of interfaces from standpoint of continuum physics. This approach develops the foundations more carefully than the traditional approach where there is a tendency to hurry on to the applications, and moreover, provides a background for advanced study in modern nonlinear continuum physics. Our ultimate intension was to present the subject in a sound manner as clear as possible. We hope that the text provides enough insights for understanding of terminology used in scientific state-of-the-art papers and to find the "right and straightforward path" in the scientific world of material surface phenomena. In what follows a few words of general meaning are worthwhile.

While at one time certain theoretical statements were regarded as "laws" of physics, nowadays many theories prefer to regard each theory as a mathematical model of some aspect of nature. But, any mathematical theory of physics must have idealized nature. Then, every theory is only "approximate" in respect to nature itself. Particularly, it stands for continuous distribution of matter which is the principal assumption of continuum physics.

In physical theory, mathematical rigor is of the essence. Then, a theory is tested by experiment. In this sense, a given theory is "good," if a range of application is greater than another's, it is "better" of the two. This holds as well for a continuum physics approach.

Having these in mind, we organize the text as follows. The text consists of six sections: Section I gives the introduction. Section II — Interface in problems of continuum physics — is designed to cover the essential features of interfaces problems in continuum physics. For those who have not been exposed to necessary mathematics we have included Section III — Basic notion of geometry and kinematics of surface — because the proper understanding of the subject requires knowledge of tensor calculus. In Section IV — Material displacement derivative — we concentrated on notion of displacement derivative, the quantity which is of most importance in problems of interfaces. In Section V — The theory of singular surfaces — the basic techniques for study of propagation of interfaces is derived. In Section VI — Balances laws of bulk material and interface — we start with the formulation of a general equation of balance and proceed by listing special cases that are of particular interest in continuum mechanics and surface of discontinuity.

Finally, in writing this part of the monograph, it has been our hope to make available to the physical and material scientists and engineers some of the more sophisticated mathematical techniques.

II. INTERFACE IN PROBLEMS OF CONTINUUM PHYSICS

Continuum physics is concerned with the description of physical phenomena as observed at the macroscopic level, with no reference to the underlying microstructure of the matter constituting the medium in which the phenomena occur. The medium itself is regarded as a continuous distribution of matter and is referred to as a continuous medium (or simply continuum). Physical quantities (such as mass or velocity) are distributed through the medium, and in mathematical terms are treated as fields. These fields are subject to a number of physical laws which express general principles common to all forms of matter.

The balance laws are formulated as integral equations governing fields defined on regions of space occupied by a material body in motion. A disturbance in the continuity of a phenomenon or physical field is termed a singularity. The aim of our contribution is to present a unified view of the theory of nonrelativistic thermodynamics incorporating phenomena with singularities. These singularities will be presented as discontinuous functions or their derivatives, and in the form of the discontinuity in respect to the Lebesgue measure of physical quantities. The examples of the first type of singularity are shock and acceleration waves. The second type is usually associated with surface concentrations of physical quantities. Discontinuities in fields may be caused by discontinuities in material properties or by some discontinuous behavior of the source which gives rise to the fields. In most problems, discontinuities in the source function propagate through the medium, and if the source function is prescribed at the boundary, that is on some initial surface, the carrier of the discontinuity is a moving surface in the medium, which in chemical physics is called interface.

Our aim is to draw attention to the fact, which so far has rarely absorbed theoreticians studying continuous media, that a moving surface may carry not only disturbances, but also physical properties different from those of the surrounding media. As an example, we consider the direct interaction of two different phases of a material. The phases are usually defined so that we can imagine an interfacial region between the phases. We can model a situation of this kind by the movement of a surface separating two well-behaved material media, while attributing to the surface the physical properties of a phase change. Thus, the term interphase mass transfer simply means the transfer of a component between two or more phases in contact with each other. The component being transferred can undergo reactions in one or both of the phases, or it can be conservative (i.e., nonreactive). Other examples may be provided by phenomena such as the motion of surface dislocations, or the propagation of cracks. In fluid mechanics, the surface tension of drops provides an example.

Since the pioneering paper by Gibbs [1], phase transformation phenomena in three-dimensional (3D) continua have also been described by introducing into the body a movable singular surface separating two different material phases in thermodynamic equilibrium state. The phase transformation phenomenon manifests itself best in thin layers of matter: films, membranes, plates, and shells. For example, thin films made of shape-memory alloys like NiTi, NiMnGa, NiTiCu, or NiAl can considerably alter their shapes under appropriate stress and temperature changes. Full analysis of the phenomenon in such thin-walled structures is often infeasible if one wants to apply the 3D continuum model. The mechanical description of behavior of such structures can conveniently be based on various 2D models consisting of a base surface endowed with various fields modeling an additional microstructure. Then the notion of a movable surface curve separating 2D regions with different material phases in an appropriate and convenient tool for modeling the phase interface in thin-walled shell structures.

A special focus exists on the fracture at interfaces. Current topics include the role of thermal residual or processing-induced stresses, the detailed role of plasticity, and geometric effects on interface crack driving forces. Of particular note in a few of the papers is the focus on multiscale modeling, a critical link for complete material behavior descriptions. The research described is fundamental by nature, but has engineering relevance in the following areas: thin films, multilayers and assemblies in the semiconductor industry, thermal barrier coatings, and structural engineering composites. It is clear that many research opportunities exist in this field, and it is expected that new contributions will provide direction for this future work [2–4].

Great effort has been devoted in recent years to determining the bending rigidity K. Conceptually, two different approaches can be distinguished. In the mechanical approach, the response of the membrane to an applied force is measured, from which the bending rigidity is deduced. The extreme softness of these systems is exploited by the second type of method where the bending rigidity is derived from the thermally excited membrane fluctuations. One example of the mechanical approach [5,6] is provided by studies of tether formation from giant vesicles which are aspirated with a micro-pipette [7]. Bending elasticity or, in its mathematical formulation, curvature energy not only generates a large variety of shapes, but also leads to different fluctuation or excitation spectra of these shapes and different dynamics distinct from that shown by simple liquid interfaces. These phenomena require different mathematical tools for their description such as conformal transformations in three dimensions.

A prime motivation to investigate membranes arises from biology in our 3D world [8]. The lipid bilayer is the most elementary and indispensable structural component of biological membranes, which form the boundary of all cells and cell organelles [9]. In biological membranes, the bilayer consists of many different lipids and other amphiphiles. Biomembranes are "decorated" with embedded membrane proteins, which ensure the essential functional properties of a bio-membrane such as ion pumping, conversion from light energy to chemical energy, and specific recognition. Often a polymeric network is anchored to the membrane endowing it with further structural stability. This stability is particularly spectacular in red blood cells which can squeeze through tiny capillaries and still recover their rest shape countless times in a life cycle.

Many properties of polymeric systems are not controlled directly by the bulk of the materials but its surface and by internal interfaces. This has been recognized very early and the study of polymers at interfaces has become a major area of polymer science.

The study of polymer solutions at interfaces has most often been motivated by the effects of polymers on the stability of colloidal suspensions.

Interfacial effects are also very important in polymer melts. The variation of the surface tension of a polymer melt with molecular weight is associated to subtle interactions between the end points and the interface and only starts to be understood.

Some of the major issues for applications such as adhesion or friction are related to the mechanical properties of interfaces.

The interfacial rheology of polymers at a solid interface dominates the friction properties and strongly depends on the degree of slip of polymers on a surface.

In addition to the possible existence of interfacial-tension gradients at surfactant-adsorbed fluid interfaces, other interfacial rheological stresses of a viscous nature may arise, such as those relating to interfacial shear and dilatational viscosities (see [10,11]).

Understanding the physical properties of the bilayer through the study of vesicles should provide valuable insight into the physical mechanisms that also govern the more complex bio-membranes for which, from this perspective, the artificial vesicle is a model system. Striking phenomenological similarities between the budding and exocytosis, where small vesicles bud off the cell membrane, encourage a thorough analysis of these artificial membranes. Referring to the biological motivation, a distinction has been emphasized between classical "biophysics" and a field which acquired the somewhat fancy notion of "biologically inspired physics" [12]. While the former field is concerned with the detailed modeling of real biological processes often at the cost of many parameters in a theoretical description, the latter approach takes the biological material as inspiration for asking questions biologists often may not even find (yet).

It is also well known that certain chemical components of liquid and gaseous phases will accumulate at interfaces. For example, surfactants in detergents accumulate at the interface of soils and water, thereby allowing dirt to be removed from soiled clothing. Some organic molecules with hydrophobic characteristics can accumulate at the air–water interface to such an extent that very few solvent molecules are present at the interface. These surface films have been studied extensively by physical chemists, and in some cases the mechanics of the film has yielded information on molecular dimensions. Nitrogen gas is also known to accumulate at solid–gas interfaces, and this property allows us to use nitrogen to determine the surface area of very fine or porous surfaces or adsorbents. Likewise, some molecules or ions may be depleted (negatively adsorbed) at interfaces. In either case, something about the interface is either liked or disliked by the molecules in question. Accumulation of molecules at the interface at these three examples suggests that this configuration somehow minimizes the Gibbs function for these systems.

When we are dealing with the propagation of interfaces, we are facing with two basic issues:

- The problem of interface morphology (planar or curved interface, cellular structure; unstationary shape, chaotic, turbulent) as a function of the control parameters.
- The problem of propagation velocity, or growth velocity as a function dendritic of the same control parameters.

From mathematical point of view, interface motion is equivalent to the solution of a free boundary problem. The question is to determine a solution for a scalar field (pressure, temperature, concentration, and charge of an impurity) or a vector field (such as a fluid velocity field) satisfying a partial derivatives equation (diffusion equation, Euler, or Navier–Stokes equation) with boundary conditions applied on the interface. Saffman–Taylor interface dynamics and dendritic growth appear to be prototypes for the understanding of the dynamics of curved fronts.

III. BASIC NOTION OF GEOMETRY AND KINEMATICS OF SURFACE

For a proper understanding of interfacial (transport) processes, one needs to be familiar with the basic geometrical description of a surface, as shown in the following section. For a terminology or notation, the reader may wish to consult any of the standard textbooks of tensor analysis and differential geometry (see, for instance, [13–16] and the bibliography in them).

Here we discuss those elements of geometry and kinematics of surface S_t which are important for our further presentation. As we confine to the real physical problems we consider E_3 — three-dimensional Euclidean space — as a space of physical events. Thus a (material) surface S_t is a subspace of E_3, i.e., $S_t \in E_3$.

The analytical expression for surface S_t, for each t in an open real interval I, is given by

$$S_t : f(\mathbf{x}, t) = 0 \tag{10.1}$$

where by $\mathbf{x}$ and t we denote a point in E_3 and time, respectively. It is assumed that function f is of class $r \geq 1$ in $(\mathbf{x}, t)$.

Taking into account that S_t is two-dimensional manifold, position of any of its point can be defined in relation to allowable coordinate system $u^\alpha (\alpha = 1, 2)$ of S_t. In that case, the point $\mathbf{x}(x^k), (k = 1, 2, 3)$, as the point of the surface S_t, is defined by the relation $\mathbf{x} = \mathbf{x}(\mathbf{u}, t)$, i.e.,

$$x^k = x^k(u^\alpha, t), \quad \mathrm{rank}\left(\frac{\partial x^i}{\partial u^\alpha}\right) = 2 \tag{10.2}$$

for every $x \in S_t$ (or every $\mathbf{u}(u_\alpha) \in D$, where D is a domain of R^2). The second expression in (10.2) represents the condition under which the surface S_t is a regular one.

The coordinates u^α are called *Gaussian parameters* of surface S, and they are intrinsic to the surface. This way of presentation is called *parameterization of surface*. Relations (10.2) represent one of many possible parameterizations of surface S_t, unlike its representation (10.1), which is unique.

Taking into consideration (10.2), the position vector $\mathbf{p}$, with respect to a frame in E_3, of the point $\mathbf{x} \in S_t$ is given by the expression

$$\mathbf{p}(\mathbf{x}) = \mathbf{p}\left[x^i(u^\alpha, t)\right] \tag{10.3}$$

or

$$\mathbf{p}(\mathbf{u}) = \mathbf{p}(u^\alpha, t) \tag{10.4}$$

In order to be able to apply differential calculus to problems we are going to investigate, we must require the existence of a certain number of partial derivatives of $\mathbf{p}$ (or $\mathbf{x}$) with respect to u^α and t. Further, we note that $\mathbf{p}(u^\alpha, t)$ is of class $r \geq 1$.

A. THE GEOMETRY OF SURFACES

When considering the geometry of surface, the value of parameter t is fixed.

We denote by

$$\mathbf{g}_i \overset{\text{def}}{=} \frac{\partial \mathbf{p}}{\partial x^i} \tag{10.5}$$

the *spatial covariant base vectors* or *natural basis of the curvilinear system* x^k. Also, by $\mathbf{g}^i$ we denote *contravariant base vectors* (*dual basis* or *reciprocal basis of natural basis* $\mathbf{g}_i$).

By definition

$$\mathbf{g}_i \cdot \mathbf{g}^j = \delta_i^j \tag{10.6}$$

Here and further, δ-systems are generally referred to as the Kronecker deltas.

The metric tensor of Euclidean space E_3 is the identity tensor $\mathbf{I}$. Its componental representation is: g_{ij}, g^{ij}, or δ_i^j, depending on basis, that is,

$$g_{kl} = \mathbf{g}_k \cdot \mathbf{g}_l, \qquad g^{kl} = \mathbf{g}^k \cdot \mathbf{g}^l \tag{10.7}$$

It is easy to show, taking into account (10.6) and (10.7), that[1]

$$\mathbf{g}^k = g^{kl}\mathbf{g}_l, \qquad \mathbf{g}_k = g_{kl}\mathbf{g}^l \tag{10.8}$$

Then (see [17])

$$\mathbf{g}_{i,j} = \frac{\partial \mathbf{g}_i}{\partial x^j} - \Gamma_{ij}^k \mathbf{g}_k = 0 \tag{10.9}$$

where

$$\Gamma_{ij}^k = \frac{\partial \mathbf{g}_i}{\partial x^j} \cdot \mathbf{g}^k$$

is *Cristoffel's symbol of the second kind*.

In the same way, from (10.3) and (10.5), it follows that

$$\mathbf{a}_\alpha \equiv \frac{\partial \mathbf{p}}{\partial u^\alpha} = x_{,\alpha}^i \mathbf{g}_i \tag{10.10}$$

$\mathbf{a}_\alpha$ are called *covariant base vectors* (or *natural basis*) of curvilinear system u^α on surface S and $x_{,\alpha}^i = \partial x^i / \partial u^\alpha$.

From (10.6) and (10.10), it follows that

$$x_{,\alpha}^i = \mathbf{g}^i \cdot \mathbf{a}_\alpha \tag{10.11}$$

Reciprocal base vectors $\mathbf{a}^\alpha$ of the base vectors $\mathbf{a}^\alpha$ are defined, as in (10.6), by

$$\mathbf{a}_\alpha \cdot \mathbf{a}^\beta = \delta_\alpha^\beta \tag{10.12}$$

We write $a_{\alpha\beta}, a^{\alpha\beta}$, and δ_α^β for the components of *metric tensor* $\mathbf{1}$ of the surface S. Thus, in particular,

$$a_{\alpha\beta} \equiv \mathbf{a}_\alpha \cdot \mathbf{a}_\beta = g_{ij} x_{,\alpha}^i x_{,\beta}^j \tag{10.13}$$

---

[1] Here and further, we adopt the Einstein summation convention: if an index appears twice in the same term, once as a subscript and once as a superscript, the sign $\sum$ will be omitted.

Further, it is easy to show, using (10.12), that

$$\mathbf{a}_\alpha \equiv a_{\alpha\beta}\mathbf{a}^\beta, \qquad \mathbf{a}^\alpha \equiv a^{\alpha\beta}\mathbf{a}_\beta \tag{10.14}$$

where

$$a^{\alpha\beta} = \mathbf{a}^\alpha \cdot \mathbf{a}^\beta \tag{10.15}$$

Also,

$$a_{\alpha\gamma}a^{\gamma\beta} = \delta_\alpha^\beta \tag{10.16}$$

which directly follows from (10.12), (10.14), and (10.15).

The unit normal vector $\mathbf{n}$ to the surface is given by

$$\mathbf{n} = \frac{\text{grad } f}{|\text{grad } f|}, \qquad \mathbf{n} \cdot \mathbf{n} = 1 \tag{10.17}$$

Relative to bases $\mathbf{g}_i$ and $\mathbf{g}^i$, components of the vector $\mathbf{n}$ are, respectively,

$$n^i = \mathbf{n} \cdot \mathbf{g}^i = \frac{g^{ij}f_{,j}}{|\text{grad } f|}, \qquad n_i = \mathbf{n} \cdot \mathbf{g}_i = \frac{f_{,i}}{|\text{grad } f|} \tag{10.18}$$

We note that the direction of $\mathbf{n}$ is such that the *space orientation* of $(\mathbf{a}_1, \mathbf{a}_2, \mathbf{n})$ is positive. This set of vectors makes a vector basis of E_3 on S.

The orthogonality of $\mathbf{n}$ on S implies the following relations:

$$\mathbf{a}_\alpha \cdot \mathbf{n} = 0 \quad \text{or} \quad \mathbf{a}^\alpha \cdot \mathbf{n} = 0 \tag{10.19}$$

or equivalently,

$$n_j x^j_{,\alpha} = g_{ij}n^i x^j_{,\alpha} = 0 \tag{10.20}$$

because of (10.10) and (10.18).

In many cases, we shall make use of the decomposition of basic vectors $\mathbf{g}^i$ with respect to the bases $(\mathbf{a}^1, \mathbf{a}^2, \mathbf{n})$. Then, in view of (10.11), (10.17)–(10.19)

$$\mathbf{g}^i = n^i\mathbf{n} + x^i_{,\alpha}\mathbf{a}^\alpha \tag{10.21}$$

Also, from $\mathbf{n} \cdot \mathbf{n} = 1$ we obtain

$$\frac{\partial \mathbf{n}}{\partial u^\alpha} \cdot \mathbf{n} = 0$$

so that

$$\mathbf{n}_{,\alpha} \equiv \frac{\partial \mathbf{n}}{\partial u^\alpha} = -b_\alpha^\beta\mathbf{a}_\beta \tag{10.22}$$

The symmetric tensor

$$b_{\alpha\beta} = \mathbf{n} \cdot \frac{\partial \mathbf{a}_\alpha}{\partial u^\beta} = -\mathbf{a}_\alpha \cdot \frac{\partial \mathbf{n}}{\partial u^\beta}$$

$$b_\alpha^\beta = a^{\beta\gamma} b_{\alpha\gamma}$$

(10.23)

is known as the *fundamental quantity of the second order* of surface $\mathcal{S}$.
Then

$$\frac{\partial \mathbf{a}_\alpha}{\partial u^\beta} = b_{\alpha\beta}\mathbf{n} + \Gamma_{\alpha\beta}^\gamma \mathbf{a}_\gamma$$

(10.24)

where

$$\Gamma_{\alpha\beta}^\gamma = \mathbf{a}^\gamma \cdot \frac{\partial \mathbf{a}_\alpha}{\partial u^\beta}$$

(10.25)

is *Cristoffel's symbol of the second kind* defined on $\mathcal{S}_t$.
In view of (10.24), we write

$$\mathbf{a}_{\alpha,\beta} = \frac{\partial \mathbf{a}_\alpha}{\partial u^\beta} - \Gamma_{\alpha\beta}^\gamma \mathbf{a}_\gamma = b_{\alpha\beta}\mathbf{n}$$

(10.26)

Furthermore, making use of

$$x_{;\alpha\beta\gamma}^i - x_{;\alpha\gamma\beta}^i = x_{;\delta}^i R_{.\alpha\beta\gamma}^\delta$$

(10.27)

(see [18]), we have

$$\mathbf{a}_{\alpha,\beta\gamma} - \mathbf{a}_{\alpha,\gamma\beta} = \mathbf{a}_\delta R_{.\alpha\beta\gamma}^\delta$$

(10.28)

where $R_{.\alpha\beta\gamma}^\delta$ is the Riemann–Christoffel tensor of a surface. On the other hand, by means of (10.22) and (10.26), it follows that

$$\mathbf{a}_{\alpha,\beta\gamma} - \mathbf{a}_{\alpha,\gamma\beta} = (b_{\alpha\beta,\gamma} - b_{\alpha\gamma,\beta})\,\mathbf{n} + (b_{\alpha\gamma}b_\beta^\delta - b_{\alpha\beta}b_\gamma^\delta)\,\mathbf{a}_\delta$$

(10.29)

Then, from (10.28) and (10.29) we have obtained the following equations.

Mainardi–Codazzi equations: $b_{\alpha\beta,\gamma} = b_{\alpha\gamma,\beta}$

(10.30)

and

Gauss equations: $R_{.\alpha\beta\gamma}^\delta = b_{\alpha\gamma}b_\beta^\delta - b_{\alpha\beta}b_\gamma^\delta$

(10.31)

Furthermore by

$$K_M = \frac{1}{2}b_\alpha^\alpha$$

(10.32)

$$K_G = \det(b_\beta^\alpha)$$

(10.33)

we denote the *mean curvature* and *Gaussian curvature* of $\mathcal{S}$, respectively.

Further, we need to know the rate of change (with respect to time) of some geometrical and physical quantities defined on $\mathcal{S}_t$.

B. Kinematics of Surface

Let us continually observe a point $\mathbf{x}$, defined by (10.3), as it moves. Now, if we differentiate the equation $f = 0$ with respect to time, then

$$\frac{\partial f}{\partial t} + \operatorname{grad} f \cdot \frac{\partial \mathbf{p}}{\partial t} = 0 \tag{10.34}$$

where $\partial \mathbf{p}/\partial t$ is called the *velocity of a point* $\mathbf{x}$ (or equivalently, the point $\mathbf{u}$) of surface S.

The velocity of point $\mathbf{x}$ is given by

$$\frac{\partial \mathbf{p}}{\partial t} = \underset{n}{u}\,\mathbf{n} + v^{\alpha}\mathbf{a}_{\alpha} \quad \text{or} \quad \frac{\partial x^i}{\partial t} = \underset{n}{u}\,n^i + v^{\alpha}x^i_{,\alpha} \tag{10.35}$$

where v_{α} is the tangential velocity of the point $\mathbf{x}$.

The normal velocity of the surface, $\underset{n}{u}$, or the *speed of displacement*, is given by

$$\underset{n}{u} \overset{\text{def}}{=} \frac{\partial \mathbf{p}}{\partial t} \cdot \mathbf{n} \tag{10.36}$$

or, by means of (10.17) and (10.34),

$$\underset{n}{u} = -\frac{\partial f/\partial t}{|\operatorname{grad} f|} \tag{10.37}$$

As the right-hand side of (10.37) is determined by the spatial equation (10.1) alone, it is independent of the choice of the parameterization (10.2) or (10.3). Clearly, the velocity $\partial \mathbf{p}/\partial t$ depends on a particular choice of the surface coordinates.

1. Orthogonal Parameterization

When the parameterization u^{α} of the surface S is such that velocities of the surface points are always orthogonal on it, that is, when $v^{\alpha} = 0$, then (10.35) becomes

$$\left.\frac{\partial \mathbf{p}}{\partial t}\right|_{\mathbf{u}} = \underset{n}{u}\,\mathbf{n} \quad \text{or} \quad \left.\frac{\partial x^i}{\partial t}\right|_{u^{\alpha}} = \underset{n}{u}\,n^i \tag{10.38}$$

Such parameterization of the surface S is called *orthogonal* (see Figure 10.1).

With respect to such parameterization, for observer who is outside the surface, position of the surface points is fixed, because there is no tangential component of the velocity to the surface. Therefore, such coordinates $u^{\alpha} \equiv \zeta^{\alpha}$ have advantage over other allowable coordinate systems of the surface because, with respect to them, expressions are considerably simplified.

We point out that the simplest forms of parameterization of surface S are not always its orthogonal parameterization. Also, in general, orthogonal parameterization has local character.

IV. MATERIAL DISPLACEMENT DERIVATIVE

A. Componental Formulation and Consideration

Here, we concentrate on notion of a *displacement derivative*, the quantity which is of most importance in dynamical problems of (material) interfaces.

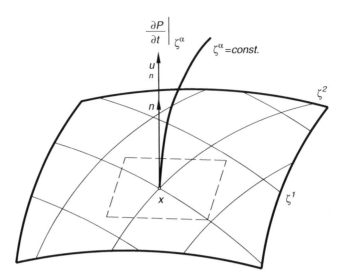

FIGURE 10.1

We emphasize that we will here deal, in general, with double tensor fields, which are defined with respect to E_3 and surface $S(t)$ embedded in E_3. Thus, we will deal with the quantities which obey the transformation law for a tensor under the following groups of transformations:

$$\bar{x}^i = \bar{x}^i(x^j), \qquad \bar{u}^\alpha = \bar{u}^\alpha(u^\beta, t)$$

In other words, we are interested in the rate of change with respect to time of these quantities when their tensor character is preserved.

Following Thomas [19], we have introduced the notion of *displacement derivative* (or δ-time derivative) for an arbitrary time-dependent field f in $\mathcal{D}$ confined to the surface $S(t)$, defined as the time derivative of f along the normal trajectory. In order to improve this definition, Truesdell and Toupin [20] suggested a generalization of Thomas's derivative to two-point tensor fields. Their derivative reduces to Thomas's derivative in the case of one-point fields.

However, as pointed out by Bowen and Wang [21] (see also [22]), the generalization of the displacement derivation given by Truesdell and Toupin has to be modified: the value of the Truesdell–Toupin derivative of a given geometrical object depends on the basis in which that object is represented, that is, whether it is the spatial basis in E_3 or the basis of surface vectors. To rectify this error, Bowen and Wang introduced what they called the *total displacement derivative* of a function, this being the partial time derivative of the function defined on a surface given in *convected parameterization*.[2] Because, in a convected parameterization ζ^α [21], the geometrical locus of the surface point $\zeta^\alpha = $ const is the normal trajectory of the surface, the displacement derivative is $\delta \tilde{f}/\delta t$, where $\tilde{f}(\zeta^\alpha, t) = f(\mathbf{x}(\zeta^\alpha, \mathbf{t}), \mathbf{t})$. Thus, it is defined as the time derivative of $\tilde{f}$ if the moving surface $S(t)$ is given in the convected parameterization $\mathbf{x} = (\mathbf{x}(\zeta^\alpha, \mathbf{t}), \mathbf{t})$. Jarić and Milanović-Lazarević [23] have extended the definition of the displacement derivative to any (not necessarily convected) surface coordinate system. Cohen and Wang [17] call it *transverse displacement derivative*.

We have applied this notion to the time derivative of tensor quantities along the trajectory of a material points of the surface $S(t)$, that is, when $U^\Delta = $ const. We call the time derivative of this kind *material displacement derivative* in order to emphasize its physical meaning. Because of its

---

[2]Concept of convected coordinates, in both mathematics and continuum mechanics, has wider meaning (see [13,20]).

importance, we give its derivation in detail. As before, $S(t)$ is given with respect to the orthogonal parameterization u^α.

Let

$$\Psi^{i\cdots j\alpha\cdots\beta}_{k\cdots l\kappa\cdots\lambda}(\mathbf{x}, \mathbf{u}, t)$$

be a double tensor field on $S(t)$, under the group of transformations $\bar{\mathbf{x}} = \bar{\mathbf{x}}(\mathbf{x}), \bar{\mathbf{u}} = \bar{\mathbf{u}}(\mathbf{u}, t)$. Then

$$\bar{\Psi}^{i\cdots j\alpha\cdots\beta}_{k\cdots l\kappa\cdots\lambda} = \Psi^{p\cdots q\pi\cdots\theta}_{r\cdots s\varrho\cdots\sigma} \frac{\partial\bar{x}^i}{\partial x^p} \cdots \frac{\partial\bar{x}^j}{\partial x^q} \frac{\partial\bar{u}^\alpha}{\partial u^\pi} \cdots \frac{\partial\bar{u}^\beta}{\partial u^\theta} \frac{\partial x^r}{\partial\bar{x}^k} \cdots \frac{\partial x^s}{\partial\bar{x}^l} \frac{\partial u^\varrho}{\partial\bar{u}^\kappa} \cdots \frac{\partial u^\sigma}{\partial\bar{u}^\lambda}$$

As, $\mathbf{x} = \mathbf{x}(\mathbf{u}, t)$ and $\mathbf{u} = \mathbf{u}(\mathbf{U}, t)$, where $\mathbf{U}$ is material parameterization, then

$$\bar{\mathbf{u}} = \bar{\mathbf{u}}(\mathbf{u}(\mathbf{U}, t), t) = \bar{\mathbf{u}}(\mathbf{U}, t), \qquad \bar{\mathbf{x}} = \bar{\mathbf{x}}[\mathbf{x}(\mathbf{u}, t)] = \bar{\mathbf{x}}\{\mathbf{x}[\mathbf{u}(\mathbf{U}, t), t]\} = \bar{\mathbf{x}}(\mathbf{U}, t)$$

Thus, when $U^\Delta = \text{const}$

$$
\begin{aligned}
\frac{\mathrm{d}\bar{\Psi}^{i\cdots j\alpha\cdots\beta}_{k\cdots l\kappa\cdots\lambda}}{\mathrm{d}t} = &\frac{\mathrm{d}\Psi^{p\cdots q\pi\cdots\theta}_{r\cdots s\varrho\cdots\sigma}}{\mathrm{d}t} \frac{\partial\bar{x}^i}{\partial x^p} \cdots \frac{\partial u^\sigma}{\partial\bar{u}^\lambda} \\
&+ \Psi^{p\cdots q\pi\cdots\theta}_{r\cdots s\varrho\cdots\sigma} \frac{\mathrm{d}}{\mathrm{d}t}\left(\frac{\partial\bar{x}^i}{\partial x^p}\right) \cdots \frac{\partial u^\sigma}{\partial\bar{u}^\lambda} + \cdots \\
&+ \Psi^{p\cdots q\pi\cdots\theta}_{r\cdots s\varrho\cdots\sigma} \frac{\partial\bar{x}^i}{\partial x^p} \cdots \frac{\mathrm{d}}{\mathrm{d}t}\left(\frac{\partial u^\sigma}{\partial\bar{u}^\lambda}\right)
\end{aligned}
\tag{10.39}
$$

Now, we have to calculate the terms which are derivatives of second order, like

$$\frac{\mathrm{d}}{\mathrm{d}t}\left(\frac{\partial\bar{x}^i}{\partial x^p}\right), \ldots, \frac{\mathrm{d}}{\mathrm{d}t}\left(\frac{\partial u^\sigma}{\partial\bar{u}^\lambda}\right)$$

We start with

$$\frac{\mathrm{d}\bar{x}^i}{\mathrm{d}t} = \frac{\partial\bar{x}^i}{\partial x^a}\frac{\partial x^a}{\partial t}, \qquad \frac{\mathrm{d}}{\mathrm{d}t}\left(\frac{\partial\bar{x}^i}{\partial x^p}\right) = \frac{\partial^2\bar{x}^i}{\partial x^p\partial x^a}\frac{\mathrm{d}x^a}{\mathrm{d}t}$$

But,

$$\frac{\partial^2\bar{x}^i}{\partial x^p\partial x^a} = \Gamma^b_{pa}\frac{\partial\bar{x}^i}{\partial x^b} - \bar{\Gamma}^i_{cd}\frac{\partial\bar{x}^c}{\partial x^p}\frac{\partial\bar{x}^d}{\partial x^a}$$

so that

$$\frac{\mathrm{d}}{\mathrm{d}t}\left(\frac{\partial\bar{x}^i}{\partial x^p}\right) = \Gamma^b_{pa}\frac{\mathrm{d}x^a}{\mathrm{d}t}\frac{\partial\bar{x}^i}{\partial x^b} - \bar{\Gamma}^i_{cd}\frac{\mathrm{d}\bar{x}^d}{\mathrm{d}t}\frac{\partial\bar{x}^c}{\partial x^p}
\tag{10.40}$$

On the other hand,

$$\frac{\mathrm{d}}{\mathrm{d}t}\frac{\partial\bar{u}^\alpha}{\partial u^\pi} = \frac{\partial}{\partial t}\left(\frac{\partial\bar{u}^\alpha}{\partial u^\pi}\right) + \frac{\partial^2\bar{u}^\alpha}{\partial u^\pi\partial u^\nu}\frac{\mathrm{d}u^\nu}{\mathrm{d}t}
\tag{10.41}$$

Also,

$$\frac{d\bar{u}^\alpha}{dt} = \frac{\partial\bar{u}^\alpha}{\partial t} + \frac{\partial\bar{u}^\alpha}{\partial u^\nu}\frac{du^\nu}{dt}$$

from which we get

$$\frac{\partial}{\partial u^\pi}\frac{d\bar{u}^\alpha}{dt} = \frac{\partial}{\partial u^\pi}\frac{\partial\bar{u}^\alpha}{\partial t} + \frac{\partial^2\bar{u}^\alpha}{\partial u^\pi\partial u^\nu}\frac{du^\nu}{dt} + \frac{\partial\bar{u}^\alpha}{\partial u^\pi}\frac{\partial}{\partial u^\nu}\frac{du^\pi}{dt} \tag{10.42}$$

From (10.41) and (10.42), we have

$$\frac{d}{dt}\frac{\partial\bar{u}^\alpha}{\partial u^\pi} = \frac{\partial}{\partial u^\pi}\frac{d\bar{u}^\alpha}{dt} - \frac{\partial\bar{u}^\alpha}{\partial u^\pi}\frac{\partial}{\partial u^\nu}\frac{du^\pi}{dt} \tag{10.43}$$

The term $(d/dt)\partial u^\alpha/\partial\bar{u}^\pi$ follows directly from (10.43) when we strictly interchange the role of u^α and $\bar{u}^\alpha$. Then, substituting (10.40) and (10.41) into (10.39), after long calculation and rearranging the terms, we get

$$\frac{\delta_m\bar{\Psi}^{i\cdots j\alpha\cdots\beta}_{k\cdots l\kappa\cdots\lambda}}{\delta t} = \frac{\delta_m\Psi^{p\cdots q\pi\cdots\theta}_{r\cdots s\varrho\cdots\sigma}}{\delta t}\frac{\partial\bar{x}^i}{\partial x^p}\cdots\frac{\partial x^s}{\partial\bar{x}^l}\frac{\partial u^\varrho}{\partial\bar{u}^\kappa}\cdots\frac{\partial u^\sigma}{\partial\bar{u}^\lambda} \tag{10.44}$$

where by $\delta_m/\delta t$ we denote *material displacement derivative*, that is,

$$\frac{\delta_m\Psi^{i\cdots j\alpha\cdots\beta}_{k\cdots l\kappa\cdots\lambda}}{\delta t} \stackrel{\text{def}}{=} \left.\frac{d\Psi^{i\cdots j\alpha\cdots\beta}_{k\cdots l\kappa\cdots\lambda}}{dt}\right|_U + \underset{\dot{u}}{\pounds}\,\Psi^{i\cdots j\alpha\cdots\beta}_{k\cdots l\kappa\cdots\lambda}$$

$$= \left.\frac{\partial\Psi^{i\cdots j\alpha\cdots\beta}_{k\cdots l\kappa\cdots\lambda}}{\partial t}\right|_{\mathbf{x},\mathbf{u}} + \Psi^{i\cdots j\alpha\cdots\beta}_{k\cdots l\kappa\cdots\lambda,m}\frac{dx^m}{dt} + \underset{\dot{u}}{\pounds}\,\Psi^{i\cdots j\alpha\cdots\beta}_{k\cdots l\kappa\cdots\lambda} \tag{10.45}$$

But, in view of (10.38),

$$\frac{\delta_m\mathbf{x}}{\delta t} = \left.\frac{\partial\mathbf{x}}{\partial t}\right|_{\mathbf{u}} + \dot{u}^\alpha\mathbf{a}_\alpha \quad\text{or}\quad \frac{dx^i}{dt} = \left.\frac{\partial x^i}{\partial t}\right|_{\mathbf{u}} + \dot{u}^\alpha x^i_{,\alpha} = u_n n^i + \dot{u}^\alpha x^i_{,\alpha} \tag{10.46}$$

so that

$$\frac{\delta_m\Psi^{i\cdots j\alpha\cdots\beta}_{k\cdots l\kappa\cdots\lambda}}{\delta t} \stackrel{\text{def}}{=} \left.\frac{d\Psi^{i\cdots j\alpha\cdots\beta}_{k\cdots l\kappa\cdots\lambda}}{dt}\right|_{\mathbf{x},\mathbf{u}}$$

$$+ \Psi^{i\cdots j\alpha\cdots\beta}_{k\cdots l\kappa\cdots\lambda,m}\underset{n}{u}n^m + \Psi^{i\cdots j\alpha\cdots\beta}_{k\cdots l\kappa\cdots\lambda,m}x^m_{;\gamma}\dot{u}^\gamma + \underset{\dot{u}}{\pounds}\,\Psi^{i\cdots j\alpha\cdots\beta}_{k\cdots l\kappa\cdots\lambda} \tag{10.47}$$

where

$$
\begin{aligned}
\pounds_{\dot{u}}\, \Psi^{i\cdots j\alpha\cdots\beta}_{k\cdots l\kappa\cdots\lambda} &\overset{\text{def}}{=} \dot{u}^{\nu}\frac{\delta\Psi^{i\cdots j\alpha\cdots\beta}_{k\cdots l\kappa\cdots\lambda}}{\partial u^{\nu}} + \Psi^{i\cdots j\alpha\cdots\beta}_{k\cdots l\nu\cdots\lambda}\frac{\partial \dot{u}^{\nu}}{\partial u^{\kappa}} + \cdots + \Psi^{i\cdots j\alpha\cdots\beta}_{k\cdots l\kappa\cdots\nu}\frac{\partial \dot{u}^{\nu}}{\partial u^{\lambda}} \\
&\quad - \Psi^{i\cdots j\nu\cdots\beta}_{k\cdots l\kappa\cdots\lambda}\frac{\partial \dot{u}^{\alpha}}{\partial u^{\nu}} - \cdots - \Psi^{i\cdots j\alpha\cdots\nu}_{k\cdots l\kappa\cdots\lambda}\frac{\partial \dot{u}^{\beta}}{\partial u^{\nu}} \\
&= \dot{u}^{\nu}\Psi^{i\cdots j\alpha\cdots\beta}_{k\cdots l\kappa\cdots\lambda,\nu} + \Psi^{i\cdots j\alpha\cdots\beta}_{k\cdots l\nu\cdots\lambda}\dot{u}^{\nu}_{,\kappa} + \cdots + \Psi^{i\cdots j\alpha\cdots\beta}_{k\cdots l\kappa\cdots\nu}\dot{u}^{\nu}_{,\lambda} \\
&\quad - \Psi^{i\cdots j\nu\cdots\beta}_{k\cdots l\kappa\cdots\lambda}\dot{u}^{\alpha}_{,\nu} - \cdots - \Psi^{i\cdots j\alpha\cdots\nu}_{k\cdots l\kappa\cdots\lambda}\dot{u}^{\beta}_{,\nu}
\end{aligned}
\tag{10.48}
$$

is the *Lie derivative* of the field $\Psi^{i\cdots j\alpha\cdots\beta}_{k\cdots l\kappa\cdots\lambda}$ with respect to the velocity field $\dot{\mathbf{u}} = \dot{u}^{\alpha}\mathbf{a}_{\alpha}$. We point out that, according to this definition,

$$
\dot{u}^{\alpha} \overset{\text{def}}{=} \frac{\delta_m u^{\alpha}}{\delta t}\bigg|_{U^{\Delta}=\text{const}} = \frac{du^{\alpha}}{dt}
$$

It is easy to show that the material displacement derivative has the following properties:

1. The material displacement derivative of a sum of tensor fields equals the sum of the material displacement derivative of the summands.
2. The rule of Leibnitz holds for:
    (a) Tensor product of tensor quantities
    (b) Product of tensor fields.

Note: The expression (10.47) for material displacement derivative differs both by form and by content from displacement derivative defined by Truesdell and Toupin [20, Equation (179.5)]

$$
\frac{\delta_d\Psi^{i\cdots j\alpha\cdots\beta}_{k\cdots l\kappa\cdots\lambda}}{\delta t} \overset{\text{def}}{=} \frac{\partial\Psi^{i\cdots j\alpha\cdots\beta}_{k\cdots l\kappa\cdots\lambda}}{\delta t}\bigg|_{\mathbf{x},\mathbf{u}} + \Psi^{i\cdots j\alpha\cdots\beta}_{k\cdots l\kappa\cdots\lambda,m}\,un^{m}_{\ n} + \pounds_{\dot{u}}\,\Psi^{i\cdots j\alpha\cdots\beta}_{k\cdots l\kappa\cdots\lambda}
\tag{10.49}
$$

This is obvious if (10.47) is written in the following form:

$$
\frac{\delta_m\Psi^{i\cdots j\alpha\cdots\beta}_{k\cdots l\kappa\cdots\lambda}}{\delta t} = \frac{\delta_d\Psi^{i\cdots j\alpha\cdots\beta}_{k\cdots l\kappa\cdots\lambda}}{\delta t} + \Psi^{i\cdots j\alpha\cdots\beta}_{k\cdots l\kappa\cdots\lambda,m}x^{m}_{,\gamma}\dot{u}^{\gamma}
\tag{10.50}
$$

Thus $\delta_m/\delta t$ and $\delta_d/\delta t$ are the same only when $u^{\alpha} = \delta^{\alpha}_{\Delta}U^{\Delta}$.

B. Material Displacement Derivative of Basic Surface Quantities

a. Surface Base Vectors

Material displacement derivative can be applied to all systems, independent of their nature, which satisfy the law of transformation of tensor quantities. Particularly, for the base vectors $\mathbf{g}_i$ it is easy to show that

$$
\frac{\delta_m\mathbf{g}_k}{\delta t} = 0
\tag{10.51}
$$

However, it is not the case when the material displacement derivative of surface base vectors $\mathbf{a}_\alpha$ are in question. Then $\mathbf{a}_\alpha$ depends explicitly on time so that, in view of (10.45),

$$\frac{\delta_m \mathbf{a}_\alpha}{\delta t} = \frac{\partial \mathbf{a}_\alpha}{\partial t} + \mathbf{a}_{\alpha,k}\frac{dx^k}{dt} + \underset{\dot{u}}{\pounds}\,\mathbf{a}_\alpha$$

But, according to (10.22) and (10.38),

$$\frac{\partial \mathbf{a}_\alpha}{\partial t} = \frac{\partial}{\partial t}\frac{\partial \mathbf{p}}{\partial u^\alpha} = \frac{\partial}{\partial u^\alpha}\frac{\partial \mathbf{p}}{\partial t} = \frac{\partial}{\partial u^\alpha}(\underset{n}{u}\,\mathbf{n}) = \underset{n}{u}_{,\alpha}\mathbf{n} - \underset{n}{u}\,b_\alpha^\beta \mathbf{a}_\beta$$

$\mathbf{a}_{\alpha,k} = 0$ and

$$\underset{\dot{u}}{\pounds}\,\mathbf{a}_\alpha = \dot{u}^\beta \mathbf{a}_{\beta,\alpha} + \mathbf{a}_\beta \dot{u}^\beta_{,\alpha} = \dot{u}^\beta b_{\alpha\beta}\mathbf{n} + \mathbf{a}_\beta \dot{u}^\beta_{,\alpha}$$

so that

$$\frac{\delta_m \mathbf{a}_\alpha}{\delta t} = \left(\underset{n}{u}_{,\alpha} + b_{\alpha\beta}\dot{u}^\beta\right)\mathbf{n} + \left(\dot{u}^\beta_{,\alpha} - \underset{n}{u}\,b_\alpha^\beta\right)\mathbf{a}_\beta \tag{10.52}$$

Than making use of $\mathbf{a}_\alpha = x^i_{;\alpha}\mathbf{g}_i$, (10.51) and (10.52), we get

$$\frac{\delta_m x^i_{;\alpha}}{\delta t} = \left(\underset{n}{u}_{,\alpha} + b_{\alpha\beta}\dot{u}^\beta\right)n^i + \left(\dot{u}^\beta_{,\alpha} - \underset{n}{u}b_\alpha^\beta\right)x^i_{,\beta} \tag{10.53}$$

b. Unit Normal Vector to a Surface
In the same way, it is easy to show that

$$\frac{\delta_m \mathbf{n}}{\delta t} = \frac{d\mathbf{n}}{dt} = -\left(\underset{n}{u}_{,\alpha} + b_{\alpha\beta}\dot{u}^\beta\right)\mathbf{a}^\alpha \tag{10.54}$$

c. Metric Tensor (First Fundamental Tensor) of a Surface
Trivially, from (10.7) and (10.51), we have

$$\frac{\delta_m g_{ij}}{\delta t} = 0, \qquad \frac{\delta_m g^{ij}}{\delta t} = 0 \tag{10.55}$$

Then, from (10.13), (10.53), and (10.55), we get

$$\frac{\delta_m a_{\alpha\beta}}{\delta t} = 2\left[\dot{u}_{(\alpha,\beta)} - \underset{n}{u}b_{\alpha\beta}\right] \tag{10.56}$$

d. Determinant of Metric Tensor of a Surface
Let

$$a = \det(a_{\alpha\beta})$$

Then

$$\frac{\delta_m a}{\delta t} = \frac{\partial a}{\partial a_{\alpha\beta}}\frac{\delta_m a_{\alpha\beta}}{\delta t} = a a^{\alpha\beta}\frac{\delta_m a_{\alpha\beta}}{\delta t}$$

But, $a_{,\alpha} = 0$ so that

$$\frac{\delta_m a}{\delta t} = 2a \left(\dot{u}^\alpha_{,\alpha} - u b^\alpha_\alpha \right) \tag{10.57}$$

where we made use of (10.56).

For further discussion, the following relation is required:

$$\frac{\delta_m \sqrt{a}}{\delta t} = \sqrt{a} \left(\dot{u}^\alpha_{,\alpha} - u b^\alpha_\alpha \right) \tag{10.58}$$

which follows directly from (10.57).

e. Second Fundamental Tensor of a Surface

To calculate the material displacement derivative $b_{\alpha\beta}$, we start with (10.26). Then,

$$\frac{\delta_m b_{\alpha\beta}}{\delta t} = \frac{\delta_m \mathbf{a}_{\alpha\beta}}{\delta t} \cdot \mathbf{n} \tag{10.59}$$

$$= \left(\frac{\partial \mathbf{a}_{\alpha,\beta}}{\partial t} + \underset{u}{\mathcal{L}} \mathbf{a}_{\alpha,\beta} \right) \cdot \mathbf{n} = \dot{u}_{,\alpha\beta} - u b_{\alpha\gamma} b^\gamma_\beta + b_{\alpha\beta,\gamma} \dot{u}^\gamma + b_{\alpha\gamma} \dot{u}^\gamma_{,\beta} + b_{\beta\gamma} \dot{u}^\gamma_{,\alpha}$$

f. Contravariant and Mixed Representation of Material Derivative

From (10.55) we see that the tensors g_{ij}, g^{ij}, and δ^i_j behave as though they were constants in material differentiation with respect to t. However, this is not the case with surface metric tensor $a_{\alpha\beta}$ as can be seen from (10.56), and it is the consequence of its explicit dependence of time. Then the operation of raising and lowering of indices of tensor fields with respect to $a_{\alpha\beta}$ is not, generally, commutative with material time derivative. Particularly, this is true for $a^{\alpha\beta}$. Indeed, it is easy to show that

$$\frac{\delta_m a^{\alpha\beta}}{\delta t} = -a^{\alpha\gamma} a^{\beta\delta} \frac{\delta_m a_{\gamma\delta}}{\delta t} = -2 \left[\dot{u}^{(\alpha,\beta)} - u b^{\alpha\beta} \right] \tag{10.60}$$

where we have used (10.56).[3]

Another quantity of importance for further investigation is b^β_α. Then, from relation

$$b^\beta_\alpha = b_{\alpha\gamma} a^{\gamma\beta}$$

after some calculation, we get

$$\frac{\delta_m b^\gamma_\alpha}{\delta t} = \dot{u}^\gamma_{,\alpha.} + u b_{\alpha\beta} b^{\beta\gamma} + b^\gamma_\beta \dot{u}^\beta_{,\alpha} + b^\gamma_{\alpha,\beta} \dot{u}^\beta - b^\beta_\alpha \dot{u}^\gamma_{,\beta} \tag{10.61}$$

---

[3]It is important to notice that

$$\frac{\delta_m \delta^\alpha_\beta}{\delta t} = 0$$

since, by definition

$$\frac{\delta_m \delta^\alpha_\beta}{\delta t} = \frac{\partial \delta^\alpha_\beta}{\partial t} + \underset{u}{\mathcal{L}} \delta^\alpha_\beta = \underset{u}{\mathcal{L}} \delta^\alpha_\beta \equiv 0$$

g. Mean Curvature of the Surface

In the same way, we can derive material derivatives of the other quantities as $b = \det b_{\alpha\beta}$, K_M, K_G, etc. Thus, from (10.61), contraction with respect to indices α and β leads to

$$2\frac{\delta_m K_M}{\delta t} = u_{n,\alpha.}^{\alpha} + u b_{\alpha}^{\gamma} b_{\gamma}^{\alpha} + 2K_{M,\gamma}\dot{u}^{\gamma} \tag{10.62}$$

This relation can be reduced, by using the Cayley–Hamilton's theorem,

$$\mathbf{B}^2 - I_B \mathbf{B} + \mathcal{I} \det \mathbf{B} = 0 \tag{10.63}$$

where $\mathbf{B} = \|b_{\beta}^{\alpha}\|$ and $\mathcal{I} = \|\delta_{\beta}^{\alpha}\|$. But, $I_B = \operatorname{tr}\mathbf{B} = 2K_M$ and $\det \mathbf{B} = K_G$, which follows from (10.32) and (10.33). Also,

$$\operatorname{tr}\mathbf{B}^2 = b_{\alpha}^{\gamma} b_{\gamma}^{\alpha} = 2(2K_M^2 - K_G)$$

Then (10.62) can be written in the following form:

$$\frac{\delta_m K_M}{\delta t} = K_{M,\alpha}\dot{u}^{\alpha} + u_n(2K_M^2 - K_G) + \frac{1}{2}u_{n,\alpha.}^{\alpha} \tag{10.64}$$

or more concisely

$$\frac{\delta_m K_M}{\delta t} = \frac{1}{2}\Delta_s u_n + u_n(2K_M^2 - K_G) + \dot{\mathbf{u}} \cdot \nabla_s K_M \tag{10.65}$$

where surface gradient and surface Laplace operator on S_t are denoted by ∇_s and Δ_s, respectively.

h. Gaussian Curvature of a Surface

In order to determine $\delta_m K_G / \delta t$ we use the following identities:

$$\frac{\delta_m K_M}{\delta t} \equiv \frac{1}{2}\frac{\delta_m(\operatorname{tr}\mathbf{B})}{\delta t} = \frac{1}{2}\operatorname{tr}\frac{\delta_m \mathbf{B}}{\delta t}, \qquad \frac{\delta_m K_G}{\delta t} \equiv 4\frac{\delta_m K_M}{\delta t}K_M - \operatorname{tr}\left(\frac{\delta_m \mathbf{B}}{\delta t}\mathbf{B}\right)$$

where last identity follows from (10.63). From (10.59), after some calculation, we get

$$\operatorname{tr}\left(\frac{\delta_m \mathbf{B}}{\delta t}\mathbf{B}\right) = u_{n,\alpha.}^{\beta} b^{\alpha\beta} + 2u_n K_M(4K_M^2 - 3K_G) + b_{\beta}^{\alpha} b_{\alpha,\gamma}^{\beta}\dot{u}^{\gamma}$$

Then

$$\frac{\delta_m K_G}{\delta t} = u_{n,\alpha.}^{\beta}(2K_M\delta_{\beta}^{\alpha} - b_{\beta}^{\alpha}) + 2u_n K_M K_G + (4K_M K_{M,\gamma} - b_{\beta}^{\alpha} b_{\alpha,\gamma}^{\beta})\dot{u}^{\gamma}$$

or finally

$$\frac{\delta_m K_G}{\delta t} = u_{n,\alpha.}^{\beta}(2K_M\delta_{\beta}^{\alpha} - b_{\beta}^{\alpha}) + 2u_n K_M K_G + \dot{\mathbf{u}} \cdot \nabla_s K_G \tag{10.66}$$

1. The List of Basic Results

For later reference, we record the following formulas:

$$\frac{\delta_m \mathbf{a}_\alpha}{\delta t} = \left(u_{,\alpha} + b_{\alpha\beta}\dot{u}^\beta\right)\mathbf{n} + \left(\dot{u}^\beta_{,\alpha} - u b^\beta_\alpha\right)\mathbf{a}_\beta$$

$$\frac{\delta_m \mathbf{n}}{\delta t} = -\left(u_{,\alpha} + b_{\alpha\beta}\dot{u}^\beta\right)\mathbf{a}^\alpha$$

$$\frac{\delta_m a_{\alpha\beta}}{\delta t} = 2\left[\dot{u}_{(\alpha,\beta)} - u b_{\alpha\beta}\right]$$

$$\frac{\delta_m a^{\alpha\beta}}{\delta t} = 2\left[u b^{\alpha\beta} - \dot{u}^{(\alpha,\beta)}\right]$$

$$\frac{\delta_m \sqrt{a}}{\delta t} = \sqrt{a}\left(\dot{u}^\alpha_{,\alpha} - u b^\alpha_\alpha\right) = \sqrt{a}\left(\nabla_s \cdot \dot{\mathbf{u}} - 2u K_M\right) \qquad (10.67)$$

$$\frac{\delta_m b_{\alpha\beta}}{\delta t} = u^\beta_{,\alpha.} - u b_{\alpha\gamma}b^\gamma_\beta + b_{\alpha\gamma}\dot{u}^\gamma_{,\beta} + b_{\beta\gamma}\dot{u}^\gamma_{,\alpha} + b_{\alpha\beta,\gamma}\dot{u}^\gamma$$

$$\frac{\delta_m b^\beta_\alpha}{\delta t} = u_{,\alpha}\cdot^\beta + u b_{\alpha\gamma}b^{\beta\gamma} + b^\beta_\gamma \dot{u}^\gamma_{,\alpha} + b^\beta_{\alpha,\gamma}\dot{u}^\gamma - b^\gamma_\alpha\dot{u}^\beta_{,\gamma}$$

$$\frac{\delta_m K_M}{\delta t} = \frac{1}{2}\Delta_s u + u\left(2K_M^2 - K_G\right) + \dot{\mathbf{u}} \cdot \nabla_s K_M$$

$$\frac{\delta_m K_G}{\delta t} = u^\beta_{,\alpha.}\left(2K_M \delta^\alpha_\beta - b^\alpha_\beta\right) + 2u K_M K_G + \dot{\mathbf{u}} \cdot \nabla_s K_G$$

Remark: It would be appropriate here to emphasize the difference between material surface derivatives $\delta_m/\delta t$ and D_m/D defined by Truesdell and Toupin [20]. Namely, the concept of $\delta_m/\delta t$ is more general than D_m/D and it reduces to D_m/D in the case when the surface is stationary, that is, when $u = 0$.

When

$$U^\Delta = \delta^\Delta_\alpha u^\alpha$$

then $U^\Delta = $ const is the orthogonal trajectory of material particles of surface. From mathematical point of view, it means that we are talking about orthogonal parameterization as the referent one. Then

$$\frac{\delta_m \mathbf{a}_\alpha}{\delta t} = u_{,\alpha}\mathbf{n} - u b^\beta_\alpha \mathbf{a}_\beta$$

$$\frac{\delta_m \mathbf{n}}{\delta t} = -u_{,\alpha}\mathbf{a}^\alpha$$

$$\frac{\delta_m a_{\alpha\beta}}{\delta t} = -2u b_{\alpha\beta}$$

$$\frac{\delta_m a^{\alpha\beta}}{\delta t} = 2u b^{\alpha\beta} \qquad (10.68)$$

$$\frac{\delta_m \sqrt{a}}{\delta t} = -u \sqrt{a} b^\alpha_\alpha = -2u \sqrt{a}K_M$$

$$\frac{\delta_m b_{\alpha\beta}}{\delta t} = u_{,\alpha\beta} - u b_{\alpha\gamma}b^\gamma_\beta$$

$$\frac{\delta_{\mathrm{m}} b_{\alpha}^{\beta}}{\delta t} = \underset{n}{u}_{,\alpha}^{\beta} \cdot + \underset{n}{u} \, b_{\alpha\gamma} b^{\beta\gamma}$$

$$\frac{\delta_{\mathrm{m}} K_{\mathrm{M}}}{\delta t} = \frac{1}{2} \Delta_{\mathrm{s}} \underset{n}{u} + \underset{n}{u} (2K_{\mathrm{M}}^2 - K_{\mathrm{G}})$$

$$\frac{\delta_{\mathrm{m}} K_{\mathrm{G}}}{\delta t} = \underset{n}{u}_{,\alpha}^{\beta} \cdot (2K_{\mathrm{M}} \delta_{\beta}^{\alpha} - b_{\beta}^{\alpha}) + 2\underset{n}{u} \, K_{\mathrm{M}} K_{\mathrm{G}}$$

These relations are identical with corresponding expressions of Truesdell and Toupin [20] and then $\delta_{\mathrm{m}}/\delta t \equiv \delta_{\mathrm{d}}/\delta t$ (see also Refs. [22,24,25]).

2. General Consideration

So far, we have discussed the material displacement derivative of tensor quantities which are defined by their components with respect to an arbitrary admissible coordinate systems x^i in E_3 and u^α in S_t, for example, for $\Psi_{k \cdots l \kappa \cdots \lambda}^{i \cdots j \alpha \cdots \beta}(\mathbf{x}, \mathbf{u}, t)$. However, generally we need the expression of material displacement derivative for

$$\Psi = \Psi_{p \cdots q \Lambda \cdots \Sigma}^{k \cdots m \Gamma \cdots \Delta} \mathbf{g}_k \cdots \otimes \mathbf{g}_m \otimes \mathbf{g}^p \otimes \cdots \mathbf{g}^q \otimes \mathbf{a}_\Gamma \cdots \otimes \mathbf{a}_\Delta \otimes \mathbf{a}^\Lambda \cdots \otimes \mathbf{a}^\Sigma \tag{10.69}$$

given in U^Δ.

Then, in view of its properties and (10.51), we have

$$\begin{aligned}
\frac{\delta_{\mathrm{m}}}{\delta t} \Psi &= \left(\frac{\delta_{\mathrm{m}}}{\delta t} \Psi_{p \cdots q \Lambda \cdots \Sigma}^{k \cdots m \Gamma \cdots \Delta} \right) \mathbf{g}_k \cdots \otimes \mathbf{g}_m \otimes \mathbf{g}^p \otimes \cdots \mathbf{g}^q \otimes \mathbf{a}_\Gamma \cdots \otimes \mathbf{a}_\Delta \otimes \mathbf{a}^\Lambda \cdots \otimes \mathbf{a}^\Sigma \\
&+ \Psi_{p \cdots q \Lambda \cdots \Sigma}^{k \cdots m \Gamma \cdots \Delta} \mathbf{g}_k \cdots \otimes \mathbf{g}_m \otimes \mathbf{g}^p \otimes \cdots \mathbf{g}^q \otimes \frac{\delta_{\mathrm{m}}}{\delta t} \mathbf{a}_\Gamma \cdots \otimes \mathbf{a}_\Delta \otimes \mathbf{a}^\Lambda \cdots \otimes \mathbf{a}^\Sigma \\
&\quad\quad\quad\quad\quad\quad\quad\quad \vdots \\
&+ \Psi_{p \cdots q \Lambda \cdots \Sigma}^{k \cdots m \Gamma \cdots \Delta} \mathbf{g}_k \cdots \otimes \mathbf{g}_m \otimes \mathbf{g}^p \otimes \cdots \mathbf{g}^q \otimes \mathbf{a}_\Gamma \cdots \otimes \mathbf{a}_\Delta \otimes \mathbf{a}^\Lambda \cdots \otimes \frac{\delta_{\mathrm{m}}}{\delta t} \mathbf{a}^\Sigma
\end{aligned} \tag{10.70}$$

Particularly, in the case of orthogonal parameterization we obtained

$$\begin{aligned}
\frac{\delta_{\mathrm{d}}}{\delta t} \Psi &= \left(\frac{\delta_{\mathrm{d}}}{\delta t} \Psi_{p \cdots q \Lambda \cdots \Sigma}^{k \cdots m \Gamma \cdots \Delta} \right) \mathbf{g}_k \cdots \otimes \mathbf{g}_m \otimes \mathbf{g}^p \cdots \otimes \mathbf{g}^q \otimes \mathbf{a}_\Gamma \cdots \otimes \mathbf{a}_\Delta \otimes \mathbf{a}^\Lambda \cdots \otimes \mathbf{a}^\Sigma \\
&+ \Psi_{p \cdots q \Lambda \cdots \Sigma}^{k \cdots m \Gamma \cdots \Delta} \mathbf{g}_k \cdots \otimes \mathbf{g}_m \otimes \mathbf{g}^p \otimes \cdots \mathbf{g}^q \otimes \frac{\delta_{\mathrm{d}}}{\delta t} \mathbf{a}_\Gamma \cdots \otimes \mathbf{a}_\Delta \otimes \mathbf{a}^\Lambda \cdots \otimes \mathbf{a}^\Sigma \\
&\quad\quad\quad\quad\quad\quad\quad\quad \vdots \\
&+ \Psi_{p \cdots q \Lambda \cdots \Sigma}^{k \cdots m \Gamma \cdots \Delta} \mathbf{g}_k \cdots \otimes \mathbf{g}_m \otimes \mathbf{g}^p \otimes \cdots \mathbf{g}^q \otimes \mathbf{a}_\Gamma \cdots \otimes \mathbf{a}_\Delta \otimes \mathbf{a}^\Lambda \cdots \otimes \frac{\delta_{\mathrm{d}}}{\delta t} \mathbf{a}^\Sigma
\end{aligned} \tag{10.71}$$

Note that in both cases, (10.70) and (10.71), it can be seen that the change of base surface vectors affects material derivatives of the quantity Ψ.

For example, displacement derivative (10.49), defined by Truesdell and Toupin [20], depends on base vectors used to present the object, that is, whether the base is of E_3 (spatial) or surface S.

a. Material Displacement Derivative of Higher Order

The material displacement derivative of higher order, that is, $\delta_m^k/\delta t^k \Psi$, $k = 2, 3, \ldots$, can be obtained from (10.70). Obviously, their expressions are, generally, very long and complicated. As such they are not of much use. The simplest case appears when we calculate $\delta_m/\delta t$ along $u^\alpha = \text{const}$, that is, when $\dot{u}^\alpha = 0$. From mathematical point of view, it means that we are talking about orthogonal parameterization as the referent one. In any case, for their derivation we need the material displacement derivatives of $\delta_m^k \mathbf{a}^\alpha/\delta t^k$, $\delta_m^k \mathbf{n}/\delta t^k$, $k = 2, 3, \ldots$.[4] Fortunately, in practice, we need material displacement derivatives of second and, eventually, of third order.

Then in addition to (10.68), we need

$$\frac{\delta \mathbf{a}^\alpha}{\delta t} = \underset{n}{u^{\cdot\alpha}} \mathbf{n} + \underset{n}{u}\, b_\beta^\alpha \mathbf{a}^\beta \tag{10.72}$$

and

$$\begin{aligned}
\frac{\delta^2 \mathbf{a}^\alpha}{\delta t^2} &= \frac{\delta \underset{n}{u^{\cdot\alpha}}}{\delta t}\mathbf{n} + \underset{n}{u^{\cdot\alpha}}\frac{\delta \mathbf{n}}{\delta t} + \frac{\delta(\underset{n}{u}\, b_\beta^\alpha)}{\delta t}\mathbf{a}^\beta + \underset{n}{u}\, b_\beta^\alpha \frac{\delta \mathbf{a}^\beta}{\delta t} \\
&= \frac{\delta \underset{n}{u^{\cdot\alpha}}}{\delta t}\mathbf{n} - \underset{n}{u^{\cdot\alpha}}\, \underset{n}{u}\, u_{,\beta}\mathbf{a}^\beta + \frac{\delta(\underset{n}{u}\, b_\beta^\alpha)}{\delta t}\mathbf{a}^\beta + \underset{n}{u}\, b_\beta^\alpha \left[\underset{n}{u}^{,\beta}\mathbf{n} + \underset{n}{u}\, b_\beta^\alpha \mathbf{a}^\beta \right] \Rightarrow \\
&= \left[\frac{\delta \underset{n}{u^{\cdot\alpha}}}{\delta t} + \underset{n}{u}\, b_\beta^\alpha \underset{n}{u}^{\cdot\beta} \right]\mathbf{n} + \left[-\underset{n}{u^{\cdot\alpha}}\, \underset{n}{u}_{,\beta} + \frac{\delta(\underset{n}{u}\, b_\beta^\alpha)}{\delta t} + \underset{n}{u}^2\, b_\gamma^\alpha b_\beta^\gamma \right]\mathbf{a}^\beta \tag{10.73}
\end{aligned}$$

Further, from (10.68)$_2$ and (10.72), we obtain

$$\frac{\delta^2 \mathbf{n}}{\delta t^2} = -\frac{\delta \underset{n}{u}_{,\alpha}}{\delta t}\mathbf{a}^\alpha - u_{,\alpha}\frac{\delta \mathbf{a}^\alpha}{\delta t} = \underset{n}{u^{\cdot\alpha}}\, \underset{n}{u}_{,\alpha}\mathbf{n} + \left[\frac{\delta \underset{n}{u}_{,\alpha}}{\delta t} - \underset{n}{u}\, \underset{n}{u}_{,\beta}b_\alpha^\beta \right]\mathbf{a}^\alpha \tag{10.74}$$

Under the above assumption from (10.46), we obtain

$$\frac{\delta \mathbf{x}}{\delta t} = \underset{n}{u}\, \mathbf{n} \tag{10.75}$$

Then, in view of (10.68)$_2$,

$$\frac{\delta^2 \mathbf{x}}{\delta t^2} = \frac{\delta \underset{n}{u}}{\delta t}\mathbf{n} + \underset{n}{u}\frac{\delta \mathbf{n}}{\delta t} = \frac{\delta \underset{n}{u}}{\delta t}\mathbf{n} - \underset{n}{u}\, \underset{n}{u}_{,\alpha}\mathbf{a}^\alpha \tag{10.76}$$

3. Decomposition of a General Tensor Field

Generally, the decomposition of tensor field defined on surface $\sigma(t)$ gives better insight on the geometrical structures and physical nature of the field. For instance, from (10.35) and (10.37) we see that the normal and tangential components of velocity of geometrical point $\mathbf{x}$ of $\sigma(t)$ are of different nature. Thus, $\underset{n}{u}$ is independent of the parameterization contrary to v^α which depends on parameterization $\sigma(t)$.

In what follows, we shall see the other advantages of such presentation of tensor fields. Because of that, we proceed keeping the argument on the highest level of generality, in order to present the

---

[4]Here, to simplify the notation, instead of $\delta_m^k/\delta t^k$ we write $\delta^k/\delta t^k$.

theory which can be applied to different field quantities of importance in material sciences, instead of giving the final formulas.

Particularly, applying this procedure to the problem of surface of singularity, we call this approach a direct one contrary to the iterative approach given by Truesdell and Toupin [20], as this decomposition consists in representing a tensor field, defined on the singular surface, with respect to the natural bases, which consists of the tangent vectors $\mathbf{a}_\alpha$, $\alpha = 1, 2$, and unit normal vector of the surface.

Let

$$\mathbf{T}(\mathbf{x}, t) = T_{i_1 \cdots i_k} \mathbf{g}^{i_1} \otimes \cdots \otimes \mathbf{g}^{i_k} \tag{10.77}$$

be a tensor field in E_3. In general case $\mathbf{T}$ has 3^k independent components. At the points of $\sigma(t)$ tensor $\mathbf{T}$, according to (10.21), is given by

$$\mathbf{T}(\mathbf{u}, t) = \mathbf{T}(\mathbf{x}(\mathbf{u}, t), t) = T_{i_1 \cdots i_k} \left(n^{i_1} \mathbf{n} + x^{i_1}_{,\alpha_1} \mathbf{a}^{\alpha_1} \right) \otimes \cdots \otimes \left(n^{i_k} \mathbf{n} + x^{i_k}_{,\alpha_k} \mathbf{a}^{\alpha_k} \right) \tag{10.78}$$

To further simplify the calculation, we make use of the following quantities:

$$\mathbf{A}^i = n^i \mathbf{n}, \qquad \mathbf{B}^i = x^i_{,\alpha} \mathbf{a}^\alpha \tag{10.79}$$

Then

$$\mathbf{T} = T_{i_1 \cdots i_k} (\mathbf{A}^{i_1} + \mathbf{B}^{i_1}) \otimes \cdots \otimes (\mathbf{A}^{i_k} + \mathbf{B}^{i_k}) \tag{10.80}$$

Obviously,

$$(\mathbf{A}^{i_1} + \mathbf{B}^{i_1}) \otimes \cdots \otimes (\mathbf{A}^{i_k} + \mathbf{B}^{i_k})$$

has 2^k addends. The explicit form of (10.80) can be writhen by collecting the addends which contain the same numbers of terms $\mathbf{A}$ or $\mathbf{B}$.

First term of representative addend is

$$\mathbf{A}^{i_1} \otimes \cdots \otimes \mathbf{A}^{i_\lambda} \otimes \mathbf{B}^{i_{\lambda+1}} \otimes \cdots \otimes \mathbf{B}^{i_k} \tag{10.81}$$

All other terms of this addend have the same order of indices

$$i_1, \ldots, i_\lambda, i_{\lambda+1}, \ldots, i_k$$

It is convenient to write them in the form of a table. For example, for terms of the form $\mathbf{A}^i \otimes \mathbf{A}^j \otimes \mathbf{A}^k \otimes \mathbf{B}^l \otimes \mathbf{B}^m$, Table 10.1 is appropriate.

This reduces to the combination without repetition of k elements of λth class. The total numbers of them is $\binom{k}{\lambda}$. Thus, for all possible classes $\lambda = 0, 1, \ldots, k$, we will have

$$\sum_{\lambda=0}^{k} \binom{k}{\lambda} = 2^k \tag{10.82}$$

elements.

TABLE 10.1

i	j	k	l	m
A	A	A	B	B
A	A	B	A	B
A	A	B	B	A
A	B	A	A	B
A	B	A	B	A
A	B	B	A	A
B	A	A	A	B
B	A	A	B	A
B	A	B	A	A
B	B	A	A	A

From (10.79), (10.80), and (10.81) it follows that the first representative term of decomposition (10.80) is

$$T_{i_1 \cdots i_\lambda i_{\lambda+1} \cdots i_k} n^{i_1} \cdots n^{i_\lambda} x^{i_{\lambda+1}}_{,\alpha_1} \cdots x^{i_k}_{,\alpha_{k-\lambda}} \mathbf{n} \otimes \cdots \otimes \mathbf{n} \otimes \mathbf{a}^{\alpha_1} \otimes \cdots \otimes \mathbf{a}^{\alpha_{k-\lambda}}, \quad 0 \le \lambda \le k \tag{10.83}$$

Particularly, for $\lambda = k$

$$T_{i_1 \cdots i_k} n^{i_1} \cdots n^{i_k} \mathbf{n} \otimes \cdots \otimes \mathbf{n} \tag{10.84}$$

and for $\lambda = 0$

$$T_{i_1 \cdots i_k} x^{i_1}_{,\alpha_1} \cdots x^{i_k}_{,\alpha_k} \mathbf{a}^{\alpha_1} \otimes \cdots \otimes \mathbf{a}^{\alpha_k} \tag{10.85}$$

In the case when $\mathbf{T}$ is symmetric,

$$T_{i_1 \cdots i_\lambda i_{\lambda+1} \cdots i_k} n^{i_1} \cdots n^{i_\lambda} x^{i_{\lambda+1}}_{,\alpha_1} \cdots x^{i_k}_{,\alpha_{k-\lambda}} \tag{10.86}$$

will be common for all terms which can be derived from

$$\mathbf{n} \otimes \cdots \otimes \mathbf{n} \otimes \mathbf{a}^{\alpha_1} \otimes \cdots \otimes \mathbf{a}^{\alpha_{k-\lambda}} \tag{10.87}$$

as the elements of the combination without repetition of k elements of λth class.

This is the case of kth gradient of tensor $\mathbf{T}$, that is, of tensor

$$\nabla^{(k)} \mathbf{T} = \mathbf{T}_{,i_1 \cdots i_k} \otimes \mathbf{g}^{i_1} \otimes \cdots \otimes \mathbf{g}^{i_k} \tag{10.88}$$

where ∇ denotes gradient. Obviously, this decomposition is very complicated and the expressions are very large. In writing them we need to express

$$\mathbf{T}_{,i_1 \cdots i_\lambda i_{\lambda+1} \cdots i_k} n^{i_1} \cdots n^{i_\lambda} x^{i_{\lambda+1}}_{,\alpha_1} \cdots x^{i_k}_{,\alpha_{k-\lambda}} \tag{10.89}$$

in final form over all indices α.

We shall illustrate this decomposition for gradients of $\mathbf{T}$ in E_3 up to order 3:

(a)

$$\nabla \mathbf{T} = \mathbf{T}_{,i} n^i \otimes \mathbf{n} + \mathbf{T}_{,i} x^i_{,\alpha} \otimes \mathbf{a}^\alpha$$

$$= \partial_\mathbf{n} \mathbf{T} \otimes \mathbf{n} + \mathbf{T}_{,\alpha} \otimes \mathbf{a}^\alpha \tag{10.90}$$

where $\partial_\mathbf{n} \mathbf{T}$ denotes the normal derivative of tensor $\mathbf{T}$.

(b) Next,

$$\nabla^{(2)} \mathbf{T} = \mathbf{T}_{,ij} n^i n^j \otimes \mathbf{n} \otimes \mathbf{n}$$

$$+ \mathbf{T}_{,ij} n^i x^j_{,\alpha} \otimes (\mathbf{n} \otimes \mathbf{a}^\alpha + \mathbf{a}^\alpha \otimes \mathbf{n})$$

$$+ \mathbf{T}_{,ij} x^i_{,\alpha} x^j_{,\beta} \otimes \mathbf{a}^\alpha \otimes \mathbf{a}^\beta$$

But,

$$\mathbf{T}_{,ij} x^i_{,\alpha} x^j_{,\beta} = \mathbf{T}_{,\alpha\beta} - b_{\alpha\beta} \partial_\mathbf{n} \mathbf{T}, \quad \mathbf{T}_{,ij} n^i x^j_{,\alpha} = (\partial_\mathbf{n} \mathbf{T})_{,\alpha} + b^\beta_\alpha \mathbf{T}_{,\beta} \tag{10.91}$$

or finally

$$\nabla^{(2)} \mathbf{T} = \partial_\mathbf{n}^{(2)} \mathbf{T} \otimes \mathbf{n} \otimes \mathbf{n}$$

$$+ [(\partial_\mathbf{n} \mathbf{T})_{,\alpha} + b^\beta_\alpha \mathbf{T}_{,\beta} \otimes (\mathbf{n} \otimes \mathbf{a}^\alpha + \mathbf{a}^\alpha \otimes \mathbf{n})$$

$$+ (\mathbf{T}_{,\alpha\beta} - b_{\alpha\beta} \partial_\mathbf{n} \mathbf{T}) \otimes \mathbf{a}^\alpha \otimes \mathbf{a}^\beta \tag{10.92}$$

In the same method, one can derive the expression of $\delta^{(q)} \mathbf{T}$, $q \geq 3$. So obtained results as well as the procedure can be compared with the results given by Podio-Guidugli [26].

4. The Decomposition of Displacement (Material) Derivative

Let $\mathbf{T} = \mathbf{T}(\mathbf{x}, t)$, $\mathbf{x} = (x^1, \ldots, x^n)$. Obviously, $\mathbf{T} = \mathbf{x}$ is particular case. Therefore, the expression for displacement (material) derivative of $\mathbf{T}(\mathbf{x}, t)$ can be used for this particular case.

Then on $\sigma(t)$ tensor field $\mathbf{T}$ is, according to (10.78), a function of u^α and t. The displacement derivative of $\mathbf{T}$ on $\sigma(t)$ is the quantity defined by

$$\frac{\delta \mathbf{T}}{\delta t} = \frac{\partial \mathbf{T}}{\partial t} + \frac{\delta \mathbf{x}}{\delta t} \cdot \nabla \mathbf{T} \tag{10.93}$$

(see $(10.45)_2$ and (10.70)). By means of (10.75), this can be written as

$$\frac{\delta \mathbf{T}}{\delta t} = \frac{\partial \mathbf{T}}{\partial t} + u_n \, \mathbf{n} \cdot \nabla \mathbf{T} = \frac{\partial \mathbf{T}}{\partial t} + u_n \, \partial_\mathbf{n} \mathbf{T} \tag{10.94}$$

or equivalently,

$$\frac{\partial \mathbf{T}}{\partial t} = -u_n \, \partial_\mathbf{n} \mathbf{T} + \frac{\delta \mathbf{T}}{\delta t} \tag{10.95}$$

The same process yields the following expression for $\delta \nabla \mathbf{T} / \delta t$. Thus,

$$\frac{\delta \nabla \mathbf{T}}{\delta t} = \frac{\partial \nabla \mathbf{T}}{\partial t} + \frac{\delta \mathbf{x}}{\delta t} \cdot \nabla^{(2)} \mathbf{T} = \frac{\partial \nabla \mathbf{T}}{\partial t} + \underset{n}{u} \, \mathbf{n} \cdot \nabla^{(2)} \mathbf{T} \tag{10.96}$$

But, because of (10.92), it follows that

$$\mathbf{n} \cdot \nabla^{(2)} \mathbf{T} = \partial_{\mathbf{n}}^{(2)} \mathbf{T} \otimes \mathbf{n} + [(\partial_{\mathbf{n}} \mathbf{T})_{,\alpha} + b_{\alpha}^{\beta} \mathbf{T}_{,\beta}] \otimes \mathbf{a}^{\alpha} \tag{10.97}$$

so that (10.96) becomes

$$\frac{\delta \nabla \mathbf{T}}{\delta t} = \frac{\partial \nabla \mathbf{T}}{\partial t} + \underset{n}{u} \, \{ \partial_{\mathbf{n}}^{(2)} \mathbf{T} \otimes \mathbf{n} + [(\partial_{\mathbf{n}} \mathbf{T})_{,\alpha} + b_{\alpha}^{\beta} \mathbf{T}_{,\beta}] \otimes \mathbf{a}^{\alpha} \} \tag{10.98}$$

On the other hand

$$\frac{\delta \nabla \mathbf{T}}{\delta t} = \mathbf{C} \otimes \mathbf{n} + \mathbf{C}_{\alpha} \otimes \mathbf{a}^{\alpha} \tag{10.99}$$

But, from (10.99), we obtain

$$\mathbf{C} = \mathbf{n} \cdot \frac{\delta \nabla \mathbf{T}}{\delta t} = \frac{\delta \mathbf{n} \cdot \nabla \mathbf{T}}{\delta t} - \frac{\delta \mathbf{n}}{\delta t} \cdot \nabla \mathbf{T} = \frac{\delta \partial_{\mathbf{n}} \mathbf{T}}{\delta t} + \underset{n}{u^{\alpha}} \mathbf{a}_{\alpha} \cdot \nabla \mathbf{T}$$

$$= \frac{\delta \partial_{\mathbf{n}} \mathbf{T}}{\delta t} + \underset{n}{u^{\alpha}} \mathbf{T}_{,\alpha}$$

where we made use of (10.90).

Furthermore, it is easily seen from (10.99) and (10.68)$_1$ that

$$\mathbf{C}_{\alpha} = \mathbf{a}_{\alpha} \cdot \frac{\delta \nabla \mathbf{T}}{\delta t} = \frac{\delta \mathbf{a}_{\alpha} \cdot \nabla \mathbf{T}}{\delta t} - \frac{\delta \mathbf{a}_{\alpha}}{\delta t} \cdot \nabla \mathbf{T}$$

$$= \frac{\delta \mathbf{T}_{,\alpha}}{\delta t} - (\underset{n,\alpha}{u} \, \mathbf{n} - \underset{n}{u} \, b_{\alpha}^{\beta} \mathbf{a}_{\beta}) \cdot \nabla \mathbf{T} = \frac{\delta \mathbf{T}_{,\alpha}}{\delta t} - \underset{n,\alpha}{u} \, \partial_{\mathbf{n}} \mathbf{T} + \underset{n}{u} \, b_{\alpha}^{\beta} \mathbf{T}_{,\beta}$$

The substitution of these two last expressions into (10.99) yields

$$\frac{\delta \nabla \mathbf{T}}{\delta t} = \left(\frac{\delta \partial_{\mathbf{n}} \mathbf{T}}{\delta t} + \underset{n}{u^{\alpha}} \mathbf{T}_{,\alpha} \right) \otimes \mathbf{n} + \left(\frac{\delta \mathbf{T}_{,\alpha}}{\delta t} - \underset{n,\alpha}{u} \partial_{\mathbf{n}} \mathbf{T} + \underset{n}{u} \, b_{\alpha}^{\beta} \mathbf{T}_{,\beta} \right) \otimes \mathbf{a}^{\alpha} \tag{10.100}$$

Finally, from (10.98) and (10.100), it therefore follows that

$$\frac{\partial \nabla \mathbf{T}}{\partial t} = \left(\frac{\delta \partial_{\mathbf{n}} \mathbf{T}}{\delta t} + \underset{n}{u^{\alpha}} \mathbf{T}_{,\alpha} + \underset{n}{u} \partial_{\mathbf{n}}^{(2)} \mathbf{T} \right) \otimes \mathbf{n} + \left[\frac{\delta \mathbf{T}_{,\alpha}}{\delta t} - (\underset{n}{u} \partial_{\mathbf{n}} \mathbf{T})_{,\alpha} \right] \otimes \mathbf{a}^{\alpha} \tag{10.101}$$

We now wish determine the expression for $\delta^2 \mathbf{T}/\delta t^2$. Proceeding in the same manner as earlier, the following important formula is derived:

$$\frac{\delta^2 \mathbf{T}}{\delta t^2} = \frac{\delta}{\delta t}\left(\frac{\partial \mathbf{T}}{\partial t} + \frac{\delta \mathbf{x}}{\delta t} \cdot \nabla \mathbf{T}\right) = \frac{\delta}{\delta t}\frac{\partial \mathbf{T}}{\partial t} + \frac{\delta^2 \mathbf{x}}{\delta t_2} \cdot \nabla(\mathbf{T}) + \frac{\delta \mathbf{x}}{\delta t}\frac{\delta}{\delta t}\nabla \mathbf{T}$$

$$= \frac{\partial^2 \mathbf{T}}{\partial t^2} + 2\frac{\delta \mathbf{x}}{\delta t} \cdot \frac{\partial \nabla \mathbf{T}}{\partial t} + \mathrm{tr}\left(\frac{\delta \mathbf{x}}{\delta t} \otimes \frac{\delta \mathbf{x}}{\delta t}\right)\nabla^{(2)}\mathbf{T} + \frac{\delta^2 \mathbf{x}}{\delta t^2} \cdot \nabla \mathbf{T} \qquad (10.102)$$

By virtue of (10.75) and (10.76), after some manipulation, we obtain

$$\frac{\delta^2 \mathbf{T}}{\delta t^2} = \frac{\partial^2 \mathbf{T}}{\partial t^2} + u_n^2 \partial_{\mathbf{n}}^{(2)}\mathbf{T} + 2u_n\frac{\delta \partial_{\mathbf{n}} \mathbf{T}}{\delta t} + \frac{\delta u_n}{\delta t}\partial_{\mathbf{n}}\mathbf{T} + u_n u_n^\alpha \mathbf{T}_{,\alpha} \qquad (10.103)$$

or equivalently,

$$\frac{\partial^2 \mathbf{T}}{\partial t^2} = -u_n^2 \partial_{\mathbf{n}}\mathbf{T} - 2u_n\frac{\delta \partial_{\mathbf{n}} \mathbf{T}}{\delta t} - \frac{\delta u_n}{\delta t}\partial_{\mathbf{n}}\mathbf{T} - u_n u_n^\alpha \mathbf{T}_{,\alpha} - \frac{\delta^2 \mathbf{T}}{\delta t^2} \qquad (10.104)$$

V. THE THEORY OF SINGULAR SURFACES

Following the classical approach, the *phases* are described by a *field* φ. In this theory, an *interface* is not a surface but, rather, a transition layer across which φ varies smoothly. The thickness of such layers is constitutively determined. We can consider a version of the phase-field theory that, due to a special choice of *constitutive equations* and a special scaling, allows us to control the thickness of transition layers. We may then investigate the ramification of shrinking that thickness. The phase-field theory allows for two approaches to deriving sharp-interface equations. We refer to these approaches as "direct" and "indirect." While these yield the same analytical results, the insights that they afford are very different.

The direct approach, which involves, for instance, the *configurational force balance* of the phase-field theory, yields more insight.

This is the main reason why, in this text, we have adopted this widely accepted device of representing a phase interface by a *singular surface*, rather than as a three-dimensional region of some thickness (see [10,27]). Like in continuum mechanics, this should be regarded as a model for reality. Our understanding of the phase interface is by no means complete, but there is good experimental evidence that indicates density may be a continuous function of position through the interfacial regions. Perhaps all the intensive variables we are concerned with, including velocity, should more accurately be regarded as continuous functions of position in going from one phase to another.

The phase interface in general is not material. We observe mass moving across a phase interface when an ice cube melts. Here the *speed of displacement* of the phase interface is controlled by the rate of heat transfer to the system. Sometimes the speed of displacement of the phase interface might be specified by the rate of a chemical reaction. In general, the speed of displacement is given in the problem statement, or it is one of the unknowns which must be determined.

Generally, from physical point of view, field φ suffer discontinuity at the interface. Then such interface is called as *surface of discontinuity*. More precisely, a disturbance in the continuity of a phenomenon or physical field is termed as *singularity*. The singularity will present as discontinuous functions or their derivatives. The examples of the first type are *shock* and *acceleration waves*. (A surface that is singular with respect to some quantity and that has a nonzero speed of propagation is said to be a propagating singular surface or *wave*.)

The second type is usually associated with the surface concentrations of physical quantities. Discontinuity in fields may be caused by some discontinuous behavior of the source which gives rise to the fields. In most problems, discontinuities in the source function propagate through the medium. If the source function is prescribed at the boundary, that is on some initial surface, the carrier of the discontinuity is a moving surface in the medium. In the rest of the chapter, we are concerned with the problem of surface singularities, that is, with the derivation of compatibility relations for functions suffering jump discontinuities across a surface.

Compatibility conditions are representation formulas for the *jumps* of partial derivatives of tensor fields in general in terms of the jumps of the tangential, the normal, and the displacement derivatives of the tensor field at its singular surface.

To find the formulas of compatibility conditions, we proceed from very general point of view.

This study provides a natural generalization and unification of the classical treatments of compatibility conditions for moving surfaces and curves as submanifolds of E_3. The motivation for such a generalization is twofold.

First, it is desirable to exhibit the compatibility conditions in a single unified set of formulas expressed in terms of standard quantities from *differential geometry* and explicitly displaying the features that are common to all submanifolds regardless of their dimensions.

Second, it may be of some benefit to the science of continuum physics to have a general theory which treats the phenomena connected with continua of diverse dimensions on an equal footing.

Here we present the essential ideas of the theory and gather the results which are necessary for the following sections. The interested reader is referred to the article by Truesdell and Toupin [20, Chapters 172–194], perhaps, to the best single reference in connection with this topic.

Consider the surface $\sigma(t)$ which is the common boundary of two regions $\mathfrak{R}^+$ and $\mathfrak{R}^-$ in E_3. The unit normal $\mathbf{n}$ of $\sigma(t)$ is directed toward the region $\mathfrak{R}^+$. Let $\varphi(\mathbf{x}, \mathbf{u}, t)$ be a scalar-valued, vector-valued, or tensor-valued function such that $\varphi(\cdot, \cdot, t)$ is continuous within each of the regions $\mathfrak{R}^+$ and $\mathfrak{R}^-$, and let $\varphi(\cdot, \cdot, t)$ have definite limits φ^+ and φ^- as $\mathbf{x}$ approaches a point on the surface $\sigma(t)$ from the paths entirely within the regions $\mathfrak{R}^+$ and $\mathfrak{R}^-$, respectively.

Definition 10.1: The jump of $\varphi(\cdot, \cdot, t)$ across $\sigma(t)$ is defined by

$$[\![\varphi]\!] = \varphi^+ - \varphi^- \tag{10.105}$$

Clearly, for each time t, the jump $[\![\varphi]\!]$ of $\varphi(\cdot, \cdot, t)$ can be a function of the position on $\sigma(t)$. Therefore, it is expressible in surface coordinates and time only.

Definition 10.2: The surface $\sigma(t)$ is said to be the singular surface with respect to $\varphi(\cdot, \cdot, t)$ if

$$[\![\varphi]\!] \neq 0 \tag{10.106}$$

This definition may be extended to include the spatial and temporal derivatives of φ.

Definition 10.3: If the jump $[\![\varphi]\!]$ of tensor field φ is normal to $\sigma(t)$, the discontinuity of φ is said to be longitudinal; if tangent to $\sigma(t)$, transversal.

In a metric space, the jump of any tensor may be resolved unequally into longitudinal and transversal components.

The entire differential theory of singular surfaces grows from the application of modified Hadamard's lemma to

$$\varphi_{;\alpha} = \varphi_{,\alpha} + \varphi_{,k} x^k_{,\alpha} \tag{10.107}$$

so that

$$\llbracket \varphi \rrbracket_{;\alpha} = \llbracket \varphi \rrbracket_{,\alpha} + \llbracket \varphi_{,k} \rrbracket x_{,\alpha}^{k} \tag{10.108}$$

which asserts that the jump of a total tangential derivative is total tangential derivative of the jump.
    Particularly, for $\varphi(\mathbf{x}, t)$

$$\llbracket \varphi \rrbracket_{,\alpha} = \llbracket \varphi_{,\alpha} \rrbracket = \llbracket \varphi_{,k} x_{,\alpha}^{k} \rrbracket = \llbracket \varphi_{,k} \rrbracket x_{,\alpha}^{k} \tag{10.109}$$

that is, the jump of a tangential derivative is the tangential derivative of the jump.
    As the values of $\varphi(\mathbf{x}, \mathbf{u}, t)$ in $\mathfrak{R}^{+}$ and $\mathfrak{R}^{-}$ are in general entirely unrelated to one another, the limiting values of the normal derivatives of $\varphi(\cdot, \cdot, t)$ on two sides of the singular surface $\sigma(t)$

$$\left\llbracket \frac{\partial \varphi}{\partial n} \right\rrbracket \quad \text{are unrestricted} \tag{10.110}$$

Assuming also that the limiting values φ^{+} and φ^{-} are continuously differentiable functions of t in $\mathfrak{R}^{+}$ and $\mathfrak{R}^{-}$, respectively, we derive a condition that the discontinuity in $\varphi(\cdot, \cdot, t)$ persists in time rather than appearing and disappearing at some particular instant. In a metric space, however, the existence of a definite speed of displacement u_{n} for the moving surface enable one to write the *kinematical condition of compatibility* for a spatial tensor field

$$\left\llbracket \frac{\delta_{d} \varphi}{\delta t} \right\rrbracket = \left\llbracket \frac{\partial \varphi}{\partial t} \right\rrbracket + u_{n} \llbracket \varphi_{,k} \rrbracket n^{k} \tag{10.111}$$

where $\delta_{d} \varphi / \delta t$ is the displacement derivative defined by Truesdell and Toupin [20]. We shall come to this formula latter. The formulas (10.108)–(10.111) are essential in the theory of singular surfaces. Henceforth, we shall make very often use of them.
    Note that

$$\left\llbracket \frac{\delta_{d} \varphi}{\delta t} \right\rrbracket = \left(\frac{\delta_{d} \varphi}{\delta t} \right)^{+} - \left(\frac{\delta_{d} \varphi}{\delta t} \right)^{-} = \frac{\delta_{d} \varphi^{+}}{\delta t} - \frac{\delta_{d} \varphi^{-}}{\delta t} = \frac{\delta_{d}}{\delta t} \llbracket \varphi \rrbracket \tag{10.112}$$

    To simplify the calculation, we consider the notion of the jump defined by (10.105) as the application of the operator $\llbracket$ applied to the tensor field φ. Then we state the following properties of the operator $\llbracket \, \rrbracket$:

(i) $\llbracket \, \rrbracket$ is linear operator.
    Indeed, from the definition (10.105), we have

$$\llbracket a\varphi + b\psi \rrbracket = a\llbracket \varphi \rrbracket + b\llbracket \psi \rrbracket \tag{10.113}$$

for any $a, b \in R$ and any tensor fields φ, ψ of the same order and type. R is the set of real numbers.
(ii)

$$\llbracket \varphi\psi \rrbracket = \langle \varphi \rangle \llbracket \psi \rrbracket + \langle \psi \rangle \llbracket \varphi \rrbracket$$

where $\langle \varphi \rangle$ and $\langle \psi \rangle$ are the mean values of φ and ψ, respectively, that is,

$$\langle \varphi \rangle = \frac{1}{2}(\varphi^+ + \varphi^-), \qquad \langle \psi \rangle = \frac{1}{2}(\psi^+ + \psi^-)$$

We already state the Hadamard's lemma by (10.108).

From (ii) and (10.105) we conclude that if φ is continuous, that is, if $\varphi_1 = \varphi_2 = \varphi$ then

$$\llbracket \varphi \psi \rrbracket = \varphi \llbracket \psi \rrbracket \tag{10.114}$$

A. Singular Surfaces Associated with a Motion

So far we have introduced the basic concepts of the theory of moving singular surfaces. However, there are certain conditions, *the geometrical conditions of compatibility and kinematical conditions of compatibility*, which must be satisfied across the singular surfaces. *The geometrical conditions of compatibility* relate the jump in the derivatives of $\varphi^{\cdots}(\,\cdot\,,t)$ to the jump of the normal derivatives of $\varphi^{\cdots}(\,\cdot\,,t)$, the tangential derivatives of the jump of $\varphi^{\cdots}(\,\cdot\,,t)$, and the geometrical properties of the singular surfaces. Usually these conditions are iterated to yield higher-order conditions of compatibility relating the jumps of the higher-order derivatives $\varphi^{\cdots}(\,\cdot\,,t)$ and their derivatives. The derivations of these conditions of compatibility are quite lengthy though rather straightforward. The interested reader should consult the work of Thomas [19] and Truesdell and Toupin [20] in which detailed derivations of these conditions are presented.

Definition 10.4:   The order of a singular surface is (usually defined as) the lower order $k + l$ of the derivative $(\partial^l / \partial t^l) \varphi^{\cdots}_{,i_1 \cdots i_k}$ which suffers a finite jump across the surface.
    Therefore, the zeroth-order singular surface is such that the tensor field $\varphi^{\cdots}(\,\cdot\,,t)$ itself suffers a discontinuity across it.
    Here and in what follows, we assume that in regions $\mathfrak{R}^+$ and $\mathfrak{R}^-$ on each side of the singular surface $\sigma(t)$ the function $\varphi(\mathbf{x}, \mathbf{u}, t)$ and all its derivatives up to the highest order considered exist and are continuously differentiable functions of $\mathbf{x}, \mathbf{u}$, and t, while on $\sigma(t)$ they approach definite limits which are continuously differentiable functions of position.
    There is no compelling reason to allow only discontinuities of this special type. Jump discontinuities upon surfaces are not the only ones that occur in physical problems. Boundaries, slip surfaces, dislocations, and tears are excluded as not being defined by sufficiently smooth jump discontinuities in function of the material variables. Singularities at isolated lines or points are common. In the case of jump discontinuities on surfaces, there is no a priori ground to expect that the limit values on each side of the surface be continuously differentiable on the surface, as we have assumed. The reasons for considering here only singularities of this kind are: (a) for more general singularities other than those analyzed earlier scarcely any definite results are known except in very particular cases and (b) singular surfaces of the above types are frequently found useful in special theories of materials.
    This definition of order of the singular surfaces is independent of the motion of any material medium. We now suppose that a medium consisting of particles $\mathbf{X}$ is in motion through the space of places $\mathbf{x}$ according to

$$\mathbf{x} = \mathbf{x}(\mathbf{X}, t), \qquad \mathbf{X} = \mathbf{X}(\mathbf{x}, t) \tag{10.115}$$

We assume that these functions are single-valued and continuous. Modifications appropriate to motion suffering discontinuity will be given later. We consider a surface $\sigma(t)$ given by a representation of the form (10.1), and set

$$F(\mathbf{X},t) \equiv f(\mathbf{x}(\mathbf{X},t),t), \quad \text{so that } f(\mathbf{x},t) \equiv F(\mathbf{x}(\mathbf{X},t),\ t) \tag{10.116}$$

identically in $\mathbf{x}$, $\mathbf{X}$, and t. Alternative representations of the moving surface are thus

$$\sigma(t) : f(\mathbf{x},t) = 0, \qquad \Sigma(t) : F(\mathbf{X},t) = 0 \tag{10.117}$$

The two representatives are the duals of one another. The other dual quantities, we are going to use frequently here, are the outward unit normal vectors $\mathbf{n}$ and $\mathbf{N}$ of σ and Σ, respectively; also, grad $\mathbf{x} \equiv \mathbf{F}$ and grad $\mathbf{X} \equiv \mathbf{F}^{-1}$ denote material and space gradients of motion.

In the special case when $f(\mathbf{x}) = 0$, we say that the surface σ is stationary; when $F(\mathbf{X}) = 0$, the surface Σ is material. In the former case, the surface σ consists always of the same places; in the latter, Σ, of the same particles.

Although $\sigma(t)$ and $\Sigma(t)$ are but different means of representing the same phenomenon, the two surfaces so defined are, in general, entirely different from one another geometrically. The surface $f(\mathbf{x},t) = 0$ is a surface in the space of places, while the surface $F(\mathbf{X},t) = 0$ is the locus, in the space of particles, of the *initial* positions of the particles $\mathbf{X}$ that are situate upon the surface $f(\mathbf{x},t) = 0$ at time t.

The dual of the speed of displacement, u_n, is the *speed of propagation*

$$U_N = -\frac{\partial F/\partial t}{|\text{grad } F|} \tag{10.118}$$

Many of the singularities of greatest interest are included in the case when

$$\varphi = \mathbf{x}(\mathbf{X},t) \tag{10.119}$$

that is, are surfaces across which the motion itself, or one of its derivatives, is discontinuous. By the order of a singular surface henceforth we shall mean, unless some other quantity is mentioned explicitly, that we are taking $\varphi = \mathbf{x}$. Then at a singular surface of order 0, the motion $\mathbf{x} = \mathbf{x}(\mathbf{X},t)$ suffers a jump discontinuity. This must be interpreted as starting that the particles $\mathbf{X}$ upon the singular surface at time t are simultaneously occupying two places $\mathbf{x}^+$ and $\mathbf{x}^-$ or jump instantaneously from $\mathbf{x}^+$ to $\mathbf{x}^-$. Such discontinuities have been found in field theory of fracture mechanics.

Because fracture of the body is excluded from our consideration, the motion on a surface is assumed to be continuous. Therefore, a singular surface of order zero is assumed not to exist, and on every singular surface the relation $[\![\mathbf{x}]\!] = 0$ is supposed to hold.

On a singular surface of order 1, the deformation gradient and the velocity of the medium may suffer finite discontinuity. Such a propagating singular surface will be called a *shock wave*.

On a singular surface of order 2, the deformation gradients and the velocity of the medium will be continuous, while the second gradients of motion and the acceleration of material particles may suffer jumps. Such a propagating singular surface will be called an *acceleration wave*.

Higher-order singular surfaces are similarly defined.

Clearly, the definition of the order of a singular surface may be expressed alternatively in terms of the covariant derivatives with respect to material variables $\mathbf{X}$, that is, $\varphi_{,K_1 \cdots K_q}$. No modification in the results is needed to allow us to substitute double tensors of the type $\varphi_{p\cdots q\gamma\cdots\delta}^{k\cdots m\alpha\cdots\beta}$ in the various jump conditions. For example, in the case of a surface which is singular with respect to φ and also a singular surface of order 2 or greater with respect to the motion itself, the principle of duality when applied to (10.111) yields

$$\left[\!\!\left[\frac{\delta_d \varphi}{\delta t}\right]\!\!\right] = \left[\!\!\left[\frac{\partial \varphi}{\partial t}\right]\!\!\right] + U_N [\![\varphi_{,K}]\!] N^K \tag{10.120}$$

where the displacement derivative $\delta_d/\delta t$ is defined in terms of the motion of the material diagram $F(\mathbf{X}, t) = 0$. This result follows at once because, corresponding to any selected initial state, there is a unique speed of propagation $\underset{N}{U}$.

B. CONDITIONS OF COMPATIBILITY

Now, we are ready to write the compatibility conditions making use of the results of Sections IV.B.3 and IV.B.4. In this way, we demonstrate the advantage of this procedure over iterative procedure (see Refs. [20, Sections 176 and 181]). Particularly, we confine ourselves to the expressions: (10.90), (10.92), (10.95), (10.101), and (10.104). Then, making use of (10.109), (10.110), (10.112), and (10.114) we write the following.

1. Geometrical Conditions of Compatibility

$$[\![\nabla \mathbf{T}]\!] = [\![\partial_\mathbf{n} \mathbf{T}]\!] \otimes \mathbf{n} + [\![\mathbf{T}]\!]_{,\alpha} \otimes \mathbf{a}_\alpha \tag{10.121}$$

$$[\![\nabla^{(2)} \mathbf{T}]\!] = [\![\partial_\mathbf{n}^{(2)} \mathbf{T}]\!] \otimes \mathbf{n} \otimes \mathbf{n}$$
$$+ [\![\partial_\mathbf{n} \mathbf{T}]\!]_{,\alpha} + b_\alpha^\beta [\![\mathbf{T}]\!]_{,\beta} \otimes (\mathbf{n} \otimes \mathbf{a}^\alpha + \mathbf{a}^\alpha \otimes \mathbf{n}) \tag{10.122}$$
$$+ ([\![\mathbf{T}]\!]_{,\alpha\beta} - b_{\alpha\beta} [\![\partial_\mathbf{n} \mathbf{T}]\!] \otimes \mathbf{a}^\alpha \otimes \mathbf{a}^\beta$$

2. Kinematical Conditions of Compatibility

$$\left[\!\!\left[\frac{\partial \mathbf{T}}{\partial t}\right]\!\!\right] = -\underset{n}{u} [\![\partial_\mathbf{n} \mathbf{T}]\!] + \frac{\delta [\![\mathbf{T}]\!]}{\delta t} \tag{10.123}$$

$$\left[\!\!\left[\frac{\partial \nabla \mathbf{T}}{\partial t}\right]\!\!\right] = \left(\frac{\delta [\![\partial_\mathbf{n} \mathbf{T}]\!]}{\delta t} + \underset{n}{u^\alpha} [\![\mathbf{T}]\!]_{,\alpha} - \underset{n}{u} [\![\partial_\mathbf{n}^{(2)} \mathbf{T}]\!]\right) \otimes \mathbf{n}$$
$$+ \left[\frac{\delta [\![\mathbf{T}]\!]_{,\alpha}}{\delta t} - (\underset{n}{u} [\![\partial_\mathbf{n} \mathbf{T}]\!])_{,\alpha}\right] \otimes \mathbf{a}_\alpha \tag{10.124}$$

$$\left[\!\!\left[\frac{\partial^2 \mathbf{T}}{\partial t^2}\right]\!\!\right] = -\underset{n}{u^2} [\![\partial_\mathbf{n}^{(2)} \mathbf{T}]\!] - 2\underset{n}{u} \frac{\delta [\![\partial_\mathbf{n} \mathbf{T}]\!]}{\delta t} - \frac{\delta \underset{n}{u}}{\delta t} [\![\partial_\mathbf{n} \mathbf{T}]\!]$$
$$- \underset{n}{u} \underset{n}{u^\alpha} [\![\mathbf{T}]\!]_{,\alpha} - \frac{\delta_2}{\delta t_2} [\![\mathbf{T}]\!] \tag{10.125}$$

There are several special cases of importance in continuum physics.

(a) If $\mathbf{T}$ is continuous, that is, if $[\![\mathbf{T}]\!] = 0$, then

$$[\![\nabla \mathbf{T}]\!] = [\![\partial_\mathbf{n} \mathbf{T}]\!] \otimes \mathbf{n} \tag{10.126}$$

$$[\![\nabla^{(2)} \mathbf{T}]\!] = [\![\partial_\mathbf{n}^{(2)} \mathbf{T}]\!] \otimes \mathbf{n} \otimes \mathbf{n} + [\![\partial_\mathbf{n} \mathbf{T}]\!]_{,\alpha} \otimes (\mathbf{n} \otimes \mathbf{a}^\alpha + \mathbf{a}^\alpha \otimes \mathbf{n})$$
$$- b_{\alpha\beta} [\![\partial_\mathbf{n} \mathbf{T}]\!] \otimes \mathbf{a}^\alpha \otimes \mathbf{a}^\beta \tag{10.127}$$

$$\left[\!\!\left[\frac{\partial \mathbf{T}}{\partial t}\right]\!\!\right] = -\underset{n}{u} [\![\partial_\mathbf{n} \mathbf{T}]\!] \tag{10.128}$$

$$\left[\!\!\left[\frac{\partial\nabla\mathbf{T}}{\partial t}\right]\!\!\right] = \left(\frac{\delta[\![\partial_\mathbf{n}\mathbf{T}]\!]}{\delta t} - \underset{n}{u}[\![\partial_\mathbf{n}^{(2)}\mathbf{T}]\!]\right)\otimes\mathbf{n} - (\underset{n}{u}[\![\partial_\mathbf{n}\mathbf{T}]\!])_{,\alpha}\otimes\mathbf{a}^\alpha \tag{10.129}$$

$$\left[\!\!\left[\frac{\partial^2\mathbf{T}}{\partial t^2}\right]\!\!\right] = -\underset{n}{u}^2\,[\![\partial_\mathbf{n}^{(2)}\mathbf{T}]\!] - 2\underset{n}{u}\frac{\delta[\![\partial_\mathbf{n}\mathbf{T}]\!]}{\delta t} - \frac{\delta\underset{n}{u}}{\delta t}[\![\partial_\mathbf{n}\mathbf{T}]\!] \tag{10.130}$$

(b) If in addition to $\mathbf{T}$, $\nabla\mathbf{T}$ is continuous, that is, if $[\![\mathbf{T}]\!] = [\![\partial_\mathbf{n}\mathbf{T}]\!] = 0$, then $\partial\mathbf{T}/\partial t$ is also continuous. But,

$$[\![\nabla^{(2)}\mathbf{T}]\!] = [\![\partial_\mathbf{n}^{(2)}\mathbf{T}]\!]\otimes\mathbf{n}\otimes\mathbf{n} \tag{10.131}$$

$$\left[\!\!\left[\frac{\partial\nabla\mathbf{T}}{\partial t}\right]\!\!\right] = -\underset{n}{u}[\![\partial_\mathbf{n}^{(2)}\mathbf{T}]\!]\otimes\mathbf{n} \tag{10.132}$$

$$\left[\!\!\left[\frac{\partial^2\mathbf{T}}{\partial t^2}\right]\!\!\right] = -\underset{n}{u}^2[\![\partial_\mathbf{n}^{(2)}\mathbf{T}]\!] \tag{10.133}$$

Let $\mathbf{T} = \mathbf{x}$ and $[\![\mathbf{x}]\!] = 0$. Then, in accordance with the definition of the order of singular surfaces, we have the following.

(c) For a singular surface of order 1,

$$[\![\text{grad }\mathbf{x}]\!] = \mathbf{a}\otimes\mathbf{N} \tag{10.134}$$

$$[\![\dot{\mathbf{X}}]\!] = -\underset{N}{U}\,\mathbf{a} \tag{10.135}$$

where

$$\mathbf{a} = [\![\text{grad }\mathbf{x}\cdot\mathbf{N}]\!]$$

In componental form these read

$$[\![x^k_{;K}]\!] = a^k N_K \tag{10.136}$$

$$[\![\dot{x}^k]\!] = -\underset{n}{u}\,a_k \tag{10.137}$$

where

$$a^k = [\![x^k_{;K}N^K]\!]$$

The vector $\mathbf{a}$ is the singularity vector; while (10.135) shows it to be parallel to the jump of velocity, its magnitude varies with the choice of the initial state and thus does not furnish a measure of the strength of the singularity. It is convenient to divide singular surfaces of order 1 into two classes:

1. Material singularities, which affect only the deformation gradients.
2. Waves, including both shock waves and propagating vortex sheets. For the former, the choice of the initial state is of prime importance.

For the latter, it is not, and the nature of the waves is best specified in terms of the jump of velocity itself, $[\![\dot{\mathbf{x}}]\!]$, which may be arbitrary both in direction and in magnitude. Indeed, if we adopt a strictly spatial standpoint, we may say the only geometrical and kinematical requirement is that discontinuities in velocity be propagated, both the amount of the discontinuity and the speed of propagation being arbitrary. Moreover, it follows that a jump in velocity is impossible unless it is accompanied by jumps in the deformation gradients.

(d) For a singular surface of order 2

$$\llbracket \text{grad}^2 \, \mathbf{x} \rrbracket = \mathbf{b} \otimes \mathbf{N} \otimes \mathbf{N} \tag{10.138}$$

$$\llbracket \text{grad} \, \dot{\mathbf{x}} \rrbracket = -\underset{N}{U} \mathbf{b} \otimes \mathbf{N} \tag{10.139}$$

$$\llbracket \ddot{\mathbf{x}} \rrbracket = -\underset{N}{U}^2 \mathbf{b} \tag{10.140}$$

where

$$\mathbf{b} = \llbracket \mathbf{N} \cdot (\text{grad}^2 \, \mathbf{x}) \mathbf{N} \rrbracket$$

In componental form these read

$$\llbracket x^k_{;KL} \rrbracket = b^k N_K N_L \tag{10.141}$$

$$\llbracket \dot{x}^k_{;K} \rrbracket = -\underset{N}{U} b^k N_K \tag{10.142}$$

$$\llbracket \ddot{x}_k \rrbracket = -\underset{N}{U}^2 b^k \tag{10.143}$$

where

$$b^k = [[x^k_{;KL} N^K N^L]]$$

These formulae show that a singular surface of order 2 is completely determined by a vector $\mathbf{b}$ and the speed of propagation, $\underset{N}{U}$. In particular, material discontinuities of second order affect only the derivatives x^k_{KL}, while discontinuities in the acceleration and in the velocity gradient are necessarily propagated, and conversely, every wave of second order carries jumps in the velocity gradient and the acceleration. Waves of second order are therefore called *acceleration waves*.

3. Dynamical Conditions of Compatibility

When a singular surface involves field variables that are affected by the motion and deformation of the medium, the geometrical and kinematical compatibility conditions should be supplemented by restrictions originating from the local balance equations. These conditions are called the *dynamical conditions of compatibility*. The dynamical conditions of compatibility are due to the local con-servation of mass, balance of linear and angular momenta, balance of energy, and the local Clausius–Duhem inequality on $\sigma(t)$.

These conditions are of fundamental importance in the investigation of many theoretical and practical problems, such as a wave propagation, in continuum physics. Their investigation will not be considered here.

VI. BALANCE LAWS OF BULK MATERIAL AND INTERFACE

As the primary objective of continuum physics is to determine the fields of density, motion, and temperature, field equations are needed. It is customary to base such field equations upon the equations of balance of mechanics, thermodynamics, and electrodynamics. These are the equations of balance of mass, momentum, moment of momentum, and energy or, in other terms, the continuity equation, Newton's laws of motion and the first, as well as the second law of thermodynamics for both: bulk material and interface.

In case of electromagnetism, the Maxwell equations must be taken into consideration. They are the set of four fundamental equations governing electromagnetism (i.e., the behavior of electric and magnetic fields).

Particularly, we consider materials surfaces such as *permeable, semipermeable* as well as *impermeable*.

Rather than considering the individual transport processes separately governing the balance of mass, momentum, species, etc., we focus now on a single, abstract, generic conservation law, known under the name *general balance law*, governing the transport of all extensive physical properties, in continuous three-dimensional media, and then in discontinuous media. Ultimately, the generic balance equations will be applied in later chapters to specific physical circumstances.

A. TRANSPORT THEOREM

A kinematic theorem that proves useful in the derivation of balance laws in continuum physics is the Reynolds theorem [28], in the literature known as *transport theorem*. We give it in a generalized form in order that it might apply to a material region through which a phase interface is moving.

Once more, instead of a phase interface, we say that the material region is divided by a surface which is discontinuous (singular) with respect to a quantity Ψ.

In the analysis of phenomena involving singular surfaces two cases should be distinguished:

(a) σ is a material surface
(b) σ is not a material surface, but a surface passing through the medium

In case (a), the same material particles remain on the surface during its motion, and the analysis concerns a film or layer, while in case (b) the analysis is applicable mainly to wave propagation problems and phase transition phenomena. Moreover, in a number of free boundary problems surfaces may be applied in modeling as well as in analysis.

We state here standard forms of transport theorems: for a volume which contains discontinuity surfaces and for a surface in E_3 [24].

1. Transport Theorem for a Volume which Contains Discontinuity Surfaces

We consider the material volume v of the body B which is divided by singular surface σ into two parts v^+ and v^- (Figure 10.2).

The outward unit normal to ∂v, the boundary of v, is denoted by $\mathbf{n}$. The velocity of particle of the body is denoted by $\dot{\xi}$.

The singular surface, assumed smooth, may be in motion with any speed of displacement u. Is also assumed that $\sigma(t)$ is a persistent singular surface with respect to a quantity Ψ and possibly also with respect to $\overset{n}{\dot{\xi}}$, the velocity of particle of the body denoted by $\dot{\xi}$.[5]

---

[5]It should be noticed that in the previous considerations we write $\mathbf{x}$ and $\dot{\mathbf{x}}$ for the placement of the material particle and its velocity independent of the dimension of the body, that is, whether the body is three-dimensional. This kind of notation will be used later as well except when one-dimensional or two-dimensional continuum is observed. It is the case, for instance, when the material surface is contained in three-dimensional body. In that case $\mathbf{x}$ and $\dot{\mathbf{x}}$ are quantities which are related to two-dimensional body. With $\xi = \xi(\Xi, t)$ we denote the position of the material particle ξ of three-dimensional material body. Also with

$$\dot{\xi} = \frac{D_m \xi}{Dt} = \frac{d\xi}{dt}\bigg|_{\xi=\text{const}}$$

we denote the velocity of that particle [20].

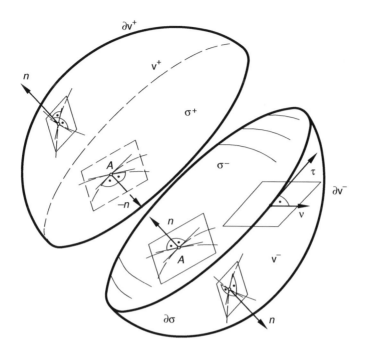

FIGURE 10.2

Further the outward normal of σ, pointing to v^+ is denoted also by **n**. The outward normal of $\partial\sigma$, the intersection of σ and $\partial\sigma$, is denoted by v, which is tangent vector field to σ defined at the points of $\partial\sigma$.

Then, for any additive quantities, Ψ associated to the body B the following transport theorem holds:

$$\frac{D_m}{Dt}\int_{v-\sigma}\Psi\,dv = \int_{v-\sigma}\left(\dot{\Psi} + \Psi\,div\dot{\xi}\right)dv + \int_\sigma [\![\Psi(\dot{\xi} - \dot{x})]\!]\cdot\mathbf{n}\,da \tag{A}$$

where $[\![\Psi]\!] = \Psi^+ - \Psi^-$ indicates the jump of Ψ across σ.

2. Transport Theorem for a Surface

Let φ be any additive quantity defined on surface $s(t)$ (see Figure 10.3). Then the following transport theorem is valid:

$$\frac{\delta_m}{\delta t}\int_s \varphi\,da = \int_s\left[\frac{\delta_m\varphi}{\delta t} + \varphi\left(\nabla_s\cdot\dot{\mathbf{u}} - 2u K_M\right)\right]da \tag{B}$$

where

$$\frac{\delta_m}{\delta t}da = \left(\nabla_s\cdot\dot{\mathbf{u}} - 2u K_M\right)da$$

Thus follows from $da = \sqrt{a}\,du^1\,du^2$ and $(10.67)_5$.

For more detailed analysis of the transport theorem for a surface, which contains discontinuity line 10 (see [29]).

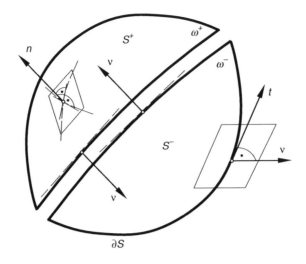

FIGURE 10.3

B. Balance Laws for a Single Body

Generally an equation of balance can be written for all additive quantities, irrespective of their physical nature. Therefore, this chapter starts with the formulation of a general equation of balance and it proceeds by listing special cases that are of particular interest to continuum mechanics. To start in a fairly general manner, we shall consider a material volume of a body which is separated into two parts v^+ and v^- by a singular surface σ (see Figure 10.2).

Let Ψ be an additive quantity associated with the body so that its amount in v may be written as

$$\Psi = \int_{v+\cup v^-} \psi_v \, dv + \int_\sigma \psi_\sigma \, da$$

where ψ_v and ψ_σ are the volume and surface densities, of Ψ, respectively.

The existence of the second integral is evidence of the occurrence of surface effects associated with a concentration of the quantity Ψ on a singular surface σ.

In the analysis of phenomena involving singular surfaces, two cases should be distinguished: (a) σ is a material surface and (b) σ is not a material surface, but a surface passing through the medium. In case (a), the same material particles remain on the surface during its motion, and the analysis concerns a film or layer, while in case (b) the analysis is applicable mainly to wave propagation problems and phase transition phenomena. Moreover, in a number of free boundary problems surfaces may be applied in modeling as well as in analysis.

Then the balance of the rate of change of Ψ is given by (see [22,24,30])

$$\frac{D_m}{Dt} \int_{v+\cup v^-} \psi_v \, dv + \frac{\delta_m}{\delta t} \int_\sigma \psi_\sigma \, da = \int_{\partial v-\sigma} \Phi_v \mathbf{n} \, da + \int_{v-\sigma} p_v \, dv + \int_{\partial \sigma} \Phi_\sigma v \, d\ell + \int_\sigma p_\sigma \, da \quad (10.144)$$

where Φ_v and Φ_σ are flux densities of ψ, p_v and p_σ are supply (production) densities in the volume and on the singular surface, respectively.

The use of transport theorems (A) and (B), as well as the divergence theorems

$$\int_{v-\sigma} \operatorname{div} \mathbf{w} \, dv + \int_{\sigma} [\![\mathbf{w}]\!] \cdot \mathbf{n} \, da = \int_{\partial v-\sigma} \mathbf{w} \cdot \mathbf{n} \, da,$$

$$\int_{S-\omega} \operatorname{div}_S \mathbf{t} \, da + \int_{\omega} [\![\mathbf{t}]\!] \cdot v \, ds = \int_{\partial S-\omega} \mathbf{t} \cdot v \, ds$$

for all vectors $\mathbf{w}$ in E_3, and tangential vectors field $\mathbf{t}$, will provide more explicit expressions for (10.144), that is,

$$\int_v \left(\frac{D_m \psi_v}{Dt} + \psi_v \nabla \cdot \dot{\boldsymbol{\xi}} - \nabla \cdot \Phi_v - p_v \right) dv$$

$$+ \int_\sigma \left[\frac{\delta_m \psi_\sigma}{\delta t} + \psi_\sigma \left(\nabla_\sigma \cdot \dot{\mathbf{u}} - 2u K_M \atop n \right) - \nabla_\sigma \cdot \Phi_\sigma - p_\sigma \right] da$$

$$+ \int_\sigma [\![\psi_v(\dot{\boldsymbol{\xi}} - \dot{\mathbf{x}}) - \Phi_v]\!] \cdot \mathbf{n} \, da = 0 \tag{10.145}$$

The localization of (10.145) now gives the local balance laws

$$\frac{D_m \psi_v}{Dt} + \psi_v \nabla \cdot \dot{\boldsymbol{\xi}} - \nabla \cdot \Phi_v - p_v = \hat{p}_v, \quad \text{in } v - \sigma \tag{10.146}$$

$$\frac{\delta_m \psi_\sigma}{\delta t} + \psi_\sigma (\nabla_\sigma \cdot \dot{\mathbf{u}} - 2u \atop n K_M) - \nabla_\sigma \cdot \Phi_\sigma - p_\sigma$$

$$+ [\![\psi_v(\dot{\boldsymbol{\xi}} - \dot{\mathbf{x}}) - \Phi_v]\!] \cdot \mathbf{n} = \hat{p}_\sigma, \quad \text{on } \sigma \tag{10.147}$$

where

$$\int_{v-\sigma} \hat{p}_v \, dv + \int_\sigma \hat{p}_\sigma \, da = 0 \tag{10.148}$$

The quantities $\hat{p}_v$ and $\hat{p}_\sigma$ are called *nonlocal volume and surface effects* (or *residuals*), respectively (see [31]).

Taking into account the influence of the quantities $\hat{p}_v$ and $\hat{p}_\sigma$ when describing the behavior of continuum takes us out from the local into nonlocal continuum theory. In the local continuum theory, the absence of such quantities is a priori assumed; then $\hat{p}_v = 0$ and $\hat{p}_\sigma = 0$. Restriction (10.148) to nonlocal residuals is generally valid. In most cases it is assumed that

$$\int_{v-\sigma} \hat{p}_v \, dv = 0, \qquad \int_\sigma \hat{p}_\sigma \, da = 0 \tag{10.149}$$

This assumption is physically justified in case when surface and volume residuals $\hat{p}_v$ and $\hat{p}_\sigma$ are independent or when their interaction is poor.

The relations (10.146) and (10.147) constitute the generic, volumetric, and surface balance equations for continuous three-dimensional media at each point of the continuum with a surface of discontinuity.

It is important to notice that adequate quantities in general balance law are defined over volume, surface, and length units, respectively. Because volume, surface, and line are geometrical concepts, it is more appropriate, from the physical point of view, to define physical quantities over mass unit whenever it is possible.

Having in mind that $dm = \varrho\,dV$ and $dm_\sigma = \gamma\,da$, where ϱ and γ are mass density of three- and two-dimensional bodies, respectively, we write

$$\psi_v \longrightarrow \varrho\psi_v, \qquad p_v \longrightarrow \varrho p_v, \qquad \hat{p}_v \longrightarrow \varrho\hat{p}_v \tag{a}$$

$$\psi_\sigma \longrightarrow \gamma\psi_\sigma, \qquad p_\sigma \longrightarrow \gamma p_\sigma \tag{b}$$

We call the reader's attention to the fact that so-defined new quantities ψ_v and ψ_σ do not change their physical dimensions. Now local balance laws (10.146) and (10.147) become

$$\frac{D_m \varrho\psi_v}{Dt} + \varrho\psi_v \nabla \cdot \dot{\boldsymbol{\xi}} - \nabla\Phi_v - \varrho p_v = \hat{p}_v \tag{10.150}$$

$$\frac{\delta_m \gamma\psi_\sigma}{\delta t} + \gamma\psi_\sigma(\nabla_\sigma \cdot \dot{\mathbf{u}} \cdot 2\underset{n}{u}\,K_M) - \nabla_\sigma \cdot \Phi_\sigma - \gamma p_\sigma$$

$$+ \left[\!\left[\varrho\psi_v(\dot{\boldsymbol{\xi}} - \dot{\mathbf{x}}) - \Phi_v \right]\!\right] \cdot \mathbf{n} = \hat{p}_\sigma \tag{10.151}$$

Particularly, for $\psi_v = 1$, $\Phi_v = 0$, $\varrho p_v = 0$, $\varrho\hat{p}_v = \hat{\varrho}$, from (10.150), we obtain the following.

(a) Balance of mass

$$\frac{D_m \varrho}{Dt} + \varrho\nabla \cdot \dot{\boldsymbol{\xi}} = \hat{\varrho} \tag{10.152}$$

or equivalently

$$\frac{\partial\varrho}{\partial t} + \nabla \cdot (\varrho\dot{\boldsymbol{\xi}}) = \hat{\varrho} \tag{10.153}$$

By substituting (10.152) into (10.150) we get, for bulk material,[6] the following.

(b) Local balance law of ψ_v

$$\varrho\frac{D_m \psi_v}{Dt} - \nabla \cdot \Phi_v - \varrho p_v = \varrho\hat{p}_v - \hat{\varrho}\psi_v \tag{10.154}$$

Here we confine our investigation mainly to the physical phenomena of nonpolar[7] nonlinear continuous bodies which are exposed to the thermodynamical (mechanical) effects. Other effects, like chemical, electrical, electromagnetic, etc., can be treated in the same way.

Now, the basic fields of thermodynamics in the bulk materials on the interface, other than mass density, are: motion and temperature. Then the field equations are based upon the equations of balance (10.154), with Ψ chosen as: momentum, moment of momentum, energy, and entropy. The other quantities are identified in accordance with their physical meaning in continuum mechanics. They are concisely given in Table 10.2 (see Refs. [24,30,31]).

Physical meaning of the quantities given in the Table 10.2 are: $\mathbf{T} = T^{kl}\mathbf{g}_l \otimes \mathbf{g}_k$ is the stress tensor,[8] ε the internal energy, $\mathbf{q} = q^k\mathbf{g}_k$ the heat flux, η the entropy density, $\mathbf{s} = s^k\mathbf{g}_k$ the

---

[6]The material occupying the region $v - \sigma$ is called bulk material.

[7]When material bodies are referred to as nonpolar, it means that all torques acting on the material are the results of forces.

[8]This way of representing tensors $\mathbf{T} = T^{kl}\mathbf{g}_l \otimes \mathbf{g}_k$ differs from the representation $\mathbf{T} = T^{kl}\mathbf{g}_k \otimes \mathbf{g}_l$ only when $\mathbf{T}$ is not symmetric. In continuum mechanics, this difference comes from the representation of stress vector $\mathbf{t}_{(n)} = \mathbf{t}^k n_k$, where $\mathbf{n} = n^k\mathbf{g}_k$, $\mathbf{t}_k$ — stress vectors acting on coordinate surface. Then, if we write $\mathbf{t}_k = t^{kl}\mathbf{g}_l$ we arrive to the representation $\mathbf{T} = T^{kl}\mathbf{g}_l \otimes \mathbf{g}_k$ (see, for instance, Refs. [31,32], etc.). But, if we write $\mathbf{t}_k = t^{lk}\mathbf{g}_l$ we arrive to the representation $\mathbf{T} = T^{kl}\mathbf{g}_k \otimes \mathbf{g}_l$ (see, for instance, Refs. [20,33,34], etc.).

TABLE 10.2

ψ	ψ_v	Φ_v	p_v	$\hat{p}_v$
Momentum	$\dot{\xi}$	$\mathbf{T}$	$\mathbf{f}$	$\hat{f}$
Moment of momentum	$\mathbf{p} \times \dot{\xi}$	$\mathbf{p} \times \mathbf{T}$	$\mathbf{p} \times \mathbf{f}$	$\mathbf{p} \times \hat{f}$
Energy	$\frac{1}{2}\dot{\xi} \cdot \dot{\xi} + \varepsilon$	$\mathbf{T}^T \dot{\xi} + \mathbf{q}$	$\mathbf{f} \cdot \dot{\xi} + h$	$\hat{f} \cdot \dot{\xi} + \hat{h}$
Entropy	η	$\mathbf{s}$	h/θ	$\hat{h}/\theta$

entropy flux, $\mathbf{f} = f^k \mathbf{g}_k$ the body force per unit mass, h the supply of energy per unit mass, θ the absolute temperature, and h/θ the entropy production.

Further, all quantities with superscript are nonlocal residuals of corresponding quantities. For definiteness, they are called nonlocal volume and surface effects (or residuals).

By "$\cdot\cdot$" we denote the summation convention over two pair of successive indices. Thus,

$$\epsilon \cdot\cdot \mathbf{T} = \varepsilon_{ijk} T^{jk} \mathbf{g}^i$$

where ϵ is the Ricci alternation tensor. Also $\mathbf{p} \times \mathbf{T} = \epsilon_{ijk} \xi^i T^{lj} \mathbf{g}^k \otimes \mathbf{g}_l$.[9]

With the notation introduced in the table, the equations of balance of microinertia, momentum, moment of momentum, energy, and entropy read as follows.

(c) Momentum

$$\nabla \cdot \mathbf{T} + \varrho(\mathbf{f} - \dot{\xi}) = \hat{\varrho}\dot{\xi} - \varrho\hat{f} \qquad (10.155)$$

(d) Moment of momentum

$$\epsilon \cdot\cdot \mathbf{T} = \varrho \mathbf{p} \times \hat{f} \qquad (10.156)$$

(e) Energy (first law of thermodynamics)

$$-\varrho\dot{\varepsilon} + \text{tr}\,\mathbf{T}(\nabla\dot{\xi})^T + \nabla \cdot \mathbf{q} + \varrho h = \hat{\varrho}\left(\varepsilon - \frac{1}{2}\dot{\xi} \cdot \dot{\xi}\right) + \varrho\hat{f} \cdot \dot{\xi} - \varrho\hat{h} \qquad (10.157)$$

(f) Entropy inequality (second law of thermodynamics)

$$\varrho\dot{\eta} - \nabla \cdot \mathbf{s} - \varrho\frac{h}{\theta} \geq \varrho\frac{\hat{h}}{\theta} - \hat{\varrho}\eta \qquad (10.158)$$

In classical, unlike rational, thermodynamics entropy flux is postulated in the form

$$\mathbf{s} = \frac{\mathbf{q}}{\theta} \qquad (10.159)$$

---

[9]The cross product of a vector and a tensor is a tensor. It is a consequence of the product $\mathbf{g}_i \times (\mathbf{g}_j \otimes \mathbf{g}_k) = (\mathbf{g}_i \times \mathbf{g}_j) \otimes \mathbf{g}k = \epsilon_{ijl}\mathbf{g}^l \otimes \mathbf{g}_k$. But, $(\mathbf{g}_i \otimes (\mathbf{g}_j) \otimes \mathbf{g}_k) = \mathbf{g}_i \otimes (\mathbf{g}_j \times \mathbf{g}_k) = \epsilon_{jkl}\mathbf{g}_i \otimes \mathbf{g}^l$. Then $\mathbf{a} \times \mathbf{A} = \epsilon_{ijk}a^i A^{jl}\mathbf{g}^k \otimes \mathbf{g}_l$, $\mathbf{A} \times \mathbf{a} = \epsilon_{ijk}A^{li}a^j\mathbf{g}_l \otimes \mathbf{g}^k$, where $\mathbf{a}$ and $\mathbf{A}$ are any vector and second-order tensor (see [35]). Then $\mathbf{p} \times \mathbf{An} = (\mathbf{p} \times \mathbf{A})\mathbf{n}$. Also $\mathbf{n}\mathbf{A} \otimes \mathbf{a} = -\mathbf{a} \times (\mathbf{n}\mathbf{A}) = -\mathbf{a} \times \mathbf{A}^T\mathbf{n} = (-\mathbf{a} \times \mathbf{A}^T)\mathbf{n}$. These relations are useful for the calculation of the surface integrals (see Ref. [36]).

Then from (10.157) and (10.158) it follows that

$$-\varrho(\dot{\varepsilon} - \theta\dot{\eta}) + \operatorname{tr}\mathbf{T}(\nabla\dot{\boldsymbol{\xi}})^{\mathsf{T}} + \mathbf{q}\cdot\nabla(\ln\theta) \geq \hat{\varrho}\left(\varepsilon - \theta\eta - \frac{1}{2}\dot{\boldsymbol{\xi}}\cdot\dot{\boldsymbol{\xi}}\right) + \varrho\hat{\mathbf{f}}\cdot\dot{\boldsymbol{\xi}} \qquad (10.160)$$

In solving specific problems, it is necessary to express balance laws in the componental form. It is useful to write them in general coordinate system. But, the choice of a particular coordinate system depends on the problem which has to be solved. In general, from mathematical point of view, we have to deal with the system of partial differential equations.

Thus, the balance laws (10.152) and (10.155)–(10.158), read as follows.

(g) Balance of mass

$$\dot{\varrho} + \varrho\dot{\xi}^k_{,k} = \hat{\varrho} \qquad (10.161)$$

(h) Balance of momentum

$$T^{kl}_{,k} + \varrho(f^l - \ddot{\xi}^l) = \hat{\varrho}\dot{\xi}^l - \varrho\hat{f}^l \qquad (10.162)$$

(i) Balance of moment of momentum

$$\epsilon^{lmn}T_{mn} = \varrho\epsilon^{lmn}p_m\hat{f}_n \qquad (10.163)$$

(j) Balance of energy (first law of thermodynamics)

$$-\varrho\dot{\varepsilon} + T^{kl}\dot{\xi}_{l,k} + q^k_{,k} + \varrho h = \hat{\varrho}\left(\varepsilon - \frac{1}{2}\dot{\xi}^k\dot{\xi}_k\right) + \varrho\hat{f}^k\dot{\xi}_k - \varrho\hat{h} \qquad (10.164)$$

(k) Entropy inequality (second law of thermodynamics)

$$-\varrho(\dot{\varepsilon} - \theta\dot{\eta}) + T^{kl}\dot{\xi}_{l,k} + q^k(\ln\theta)_{,k} \geq \hat{\varrho}\left(\varepsilon - \theta\eta - \frac{1}{2}\dot{\xi}^k\dot{\xi}_k\right) + \varrho\hat{f}^k\dot{\xi}_k \qquad (10.165)$$

It is important to note that $(10.149)_1$ holds for the set $(\hat{\varrho}, \varrho\hat{f}^l, \varrho\hat{h})$ of the residuals over their domains of definitions, that is:

$$\int_{v-\sigma} (\hat{\varrho}, \varrho\hat{f}^l, \varrho\hat{h})\,dv = 0 \qquad (10.166)$$

In many cases, it is more convenient to use the material representation of these balance laws:

$$T^{Kl}_{,K} + \varrho_0(f^l - \ddot{\xi}^l) = \hat{\varrho}_0\dot{\xi}^l - \varrho_0\hat{f}^l \qquad (10.167)$$

$$-\varrho_0\dot{\varepsilon} + T^K_l\dot{\xi}^k_{,K} + Q^K_{,K} = \hat{\varrho}_0\left(\varepsilon - \frac{1}{2}\dot{\xi}^k\dot{\xi}_k\right) + \varrho_0\hat{f}^k\dot{\xi}_k - \varrho_0(\hat{h} + h) \qquad (10.168)$$

and entropy inequality

$$-\varrho_0(\dot{\varepsilon} - \theta\dot{\eta}) + T_k^K \dot{\xi}_{;K}^{\,k} + Q^K(\ln\theta)_{,K} \geq \hat{\varrho}_0\left(\varepsilon \cdot \theta\eta - \frac{1}{2}\dot{\xi}^k\dot{\xi}_k\right) + \varrho_0\hat{f}^k\dot{\xi}_k \qquad (10.169)$$

where[10]

$$\varrho_o = \varrho J, \quad J = \det(\xi_{;K}^k), \qquad \hat{\varrho}_0 = \hat{\varrho}J, \qquad T^{Kl} \equiv JX_{;k}^K t^{kl}, \qquad Q^K \equiv JX_{;k}^K q^k$$

C. Nonmaterial Interface: Boundary Conditions

In the case when discontinuity surface is not a material (e.g., the surface propagates as a wave), $\gamma = 0$ by definition. Then, all the quantities which are related to the surface are equal to zero, except surface nonlocal effect $\hat{p}_\sigma$ (see (b)): ψ_σ, Φ_σ, and p_σ. Then (10.151) becomes

$$[\![\varrho\psi_\nu(\dot{\xi} - \dot{x}) - \Phi_\nu]\!] \cdot \mathbf{n} = \hat{p}_\sigma \qquad (10.170)$$

Formally, (10.170) states the general boundary condition corresponding to the equation by which a balance law of the quantity ψ_ν is expressed. Specially, (10.170) defines the boundary condition for ϱ

$$[\![\varrho(\dot{\xi} - \dot{x})]\!] \cdot \mathbf{n} = \hat{\gamma} \qquad (10.171)$$

as then $\psi_\nu = 1$, $\Phi_\nu = 0$, and $\hat{p}_\sigma = \hat{\gamma}$.

The other explicit form of boundary conditions for the specific physical quantity ψ_ν is obtained by using data from Table 10.1:

$$[\![\varrho\dot{\xi} \otimes (\dot{\xi} - \dot{x}) - \mathbf{T}]\!]\mathbf{n} = \hat{\mathbf{f}}_\sigma \qquad (10.172)$$

$$0 = \hat{\ell}_\sigma - \mathbf{p} \times \hat{\mathbf{f}}_\sigma \qquad (10.173)$$

$$\left[\!\left[\varrho\left(\varepsilon + \frac{1}{2}\dot{\xi} \cdot \dot{\xi}\right)(\dot{\xi} - \dot{x}) - \dot{\xi}\mathbf{T} - \mathbf{q}\right]\!\right]\mathbf{n} = \hat{\varepsilon}_\sigma \qquad (10.174)$$

$$[\![\varrho\eta(\dot{\xi} - \dot{x}) - \mathbf{s}]\!] \cdot \mathbf{n} = \hat{n}_\sigma \qquad (10.175)$$

which present boundary conditions for balance of momentum (10.155), balance of moment of momentum (10.156), balance of energy (10.157), and balance of entropy (10.158).

1. Material Interface

The singular surface will also be used as a mathematical model for a thin wall or a membrane which separates one part of the body under consideration from another part.

Then the interfacial balance law (10.151) can be used in the way which is completely analogous to the procedure of using balance law (10.150) of three-dimensional body (bulk material). Thus for $\psi_\sigma = 1$, $\Phi_\sigma = 0$, $p_\sigma = 0$, $\hat{p}_\sigma = \hat{\gamma}$, as well as $\psi_\nu = 1$ and $\Phi_\nu = 0$ we obtain the following

(a) Balance of mass of interface

$$\frac{\delta_m\gamma}{\delta t} + \gamma\left(\nabla_\sigma \cdot \dot{\mathbf{u}} - 2uK_M\right) + [\![\varrho(\dot{\xi} - \dot{x})]\!] \cdot \mathbf{n} = \hat{\gamma} \qquad (10.176)$$

---

[10]Note that ϱ_0 is reduced to referent density only in the case when $\hat{\varrho} = 0$.

From (10.151) and (10.176) we get, in the case of general parameterization of interface, the following.

(b) Local balance law of the quantity ψ_σ

$$\gamma\frac{\delta_m\psi_\sigma}{\delta t} - \nabla_\sigma \cdot \Phi_\sigma - \gamma p_\sigma + [\![\varrho(\psi_v - \psi_\sigma)(\dot{\xi} - \dot{x}) - \Phi_v]\!]\mathbf{n} = \hat{p}_\sigma - \hat{\gamma}\psi_\sigma \qquad (10.177)$$

where

$$\frac{\delta_m\psi_\sigma}{\delta t} = \frac{\partial\psi_\sigma}{\partial t} + \underset{u}{\pounds}\,\psi_\sigma$$

It is more usual in the literature to use orthogonal parameterization, because then (10.177) can be written in more simplified form.

From mathematical point of view, problems of two-dimensional bodies are more complex because of the geometry of the bodies. In general case, here we are dealing with the Riemann geometry of surface, which is much more complicated than Euclidean geometry. Having this in mind, mathematical models of two-dimensional bodies are primarily simplified by disregarding the effects which can be physically justified, as some nonlocal influences, such as surface mass residual, microinsertion influence, etc. Then for the general balance law (10.177) we write

$$\gamma\frac{\delta_m\psi_\sigma}{\delta t} - \nabla_\sigma \cdot \Phi_\sigma - \gamma p_\sigma + [\![\varrho(\psi_v - \psi_\sigma)(\dot{\xi} - \dot{x}) - \Phi_v]\!]\mathbf{n} = 0 \qquad (10.178)$$

For such mathematical models, we write balance laws by using general balance law (10.178) and Table 10.3.

Here, $\mathbf{S} = S^{i\alpha}\mathbf{g}_i \otimes \mathbf{a}_\alpha$ is the surface stress, $\mathbf{f}_\sigma$ the external body force per unit mass of material surface, ε_σ the specific internal surface energy, $\mathbf{q}_\sigma = q_\sigma^\alpha\mathbf{a}_\alpha$ the surface heat flux vector, η_σ the surface entropy density, $\mathbf{s}_\sigma = s^\alpha\mathbf{a}_\alpha$ the surface entropy flux vector, and h_σ/θ the surface entropy production.

It is also

$$\mathbf{x} \times \mathbf{S} = \varepsilon_{ijk}x^j S^{k\alpha}\mathbf{g}^i \otimes \mathbf{a}_\alpha$$

Next, by substituting corresponding quantities from Table 10.2 into (10.178) we get the following balance laws for material interface.

(c) Balance of momentum of material interface

$$\gamma\ddot{x} - \nabla_\sigma \cdot \mathbf{S} - \gamma\mathbf{f}_\sigma + [\![\varrho(\dot{\xi} - \dot{x}) \otimes (\dot{\xi} - \dot{x}) - \mathbf{T}]\!]\mathbf{n} = -\hat{\gamma}\dot{x} \qquad (10.179)$$

(d) Balance of moment of momentum of material interface

$$\varepsilon_{ijk}x^i_{,\alpha}S^{j\alpha}\mathbf{g}^k = -\hat{\gamma}\mathbf{x} \times \dot{x} \qquad (10.180)$$

TABLE 10.3

ψ	ψ_v	Φ_v	p_σ
Momentum	$\dot{x}$	$\mathbf{S}$	$\mathbf{f}_\sigma$
Moment of momentum	$\mathbf{x} \times \dot{x}$	$\mathbf{x} \times \mathbf{S}$	$\mathbf{x} \times \mathbf{f}_\sigma$
Energy	$\frac{1}{2}\dot{x} \cdot \dot{x} + \varepsilon_\sigma$	$\mathbf{S}\dot{x} + \mathbf{q}_\sigma$	$\mathbf{f} \cdot \dot{x} + h_\sigma$
Entropy	η_σ	$\mathbf{s}_\sigma s$	h_σ/θ

(e) Balance of energy of material interface

$$\gamma\dot{\varepsilon}_\sigma - \operatorname{tr}\mathbf{S}^{\mathrm{T}}(\nabla_\sigma\dot{\mathbf{x}}) - \nabla_\sigma\mathbf{q}_\sigma - \gamma h_\sigma$$

$$+ \left[\!\!\left[\varrho\left[\frac{1}{2}(\dot{\boldsymbol{\xi}} - \dot{\mathbf{x}})^2 + (\varepsilon_v - \varepsilon_\sigma)\right](\dot{\boldsymbol{\xi}} - \dot{\mathbf{x}}) - \mathbf{T}^{\mathrm{T}}(\dot{\boldsymbol{\xi}} - \dot{\mathbf{x}}) - \mathbf{q} \right]\!\!\right] \cdot \mathbf{n}$$

$$= -\hat{\gamma}\left(\varepsilon_\sigma - \frac{1}{2}\dot{\mathbf{x}}\dot{\mathbf{x}}\right) \tag{10.181}$$

(f) Balance of entropy of material interface

$$\gamma\dot{\eta}_\sigma - \nabla_\sigma \cdot s_\sigma - \gamma\frac{h_\sigma}{\theta} + [\![\varrho(\eta - \eta_\sigma)(\dot{\boldsymbol{\xi}} - \dot{\mathbf{x}}) - \mathbf{s}]\!] \cdot \mathbf{n} = -\hat{\gamma}\eta_\sigma \tag{10.182}$$

Remark: In the special case when the mathematical model is a nonpolar continuum, where the influence of nonlocality is disregarded, that is, when

$$\hat{\gamma} = 0$$

balance laws of material interface (10.176) and (10.179)–(10.182) become (see [30])

$$\frac{\partial\gamma}{\partial t} + \nabla_\sigma(\gamma\mathbf{u}) - 2\gamma u K_{\mathrm{M}} + [\![\varrho(\dot{\boldsymbol{\xi}} - \dot{\mathbf{x}})]\!]\mathbf{n} = 0 \tag{10.183}$$

$$\gamma\ddot{\mathbf{x}} - \nabla_\sigma\mathbf{S} - \gamma\mathbf{f}_\sigma + [\![\varrho(\dot{\boldsymbol{\xi}} - \dot{\mathbf{x}}) \otimes (\dot{\boldsymbol{\xi}} - \dot{\mathbf{x}}) - \mathbf{T}]\!]\mathbf{n} = 0 \tag{10.184}$$

$$\varepsilon_{ijk}x^i_{,\alpha}S^{j\alpha}\mathbf{g}^k = 0 \tag{10.185}$$

$$\gamma\dot{\varepsilon}_\sigma - \operatorname{tr}\mathbf{S}^{\mathrm{T}}(\nabla_\sigma\dot{\mathbf{x}}) - \nabla_\sigma \cdot \mathbf{q}_\sigma - \gamma h_\sigma$$

$$+ \left[\!\!\left[\varrho\left[\frac{1}{2}(\dot{\boldsymbol{\xi}} - \dot{\mathbf{x}})^2 + (\varepsilon_v - \varepsilon_\sigma)\right](\dot{\boldsymbol{\xi}} - \dot{\mathbf{x}}) - \mathbf{T}^{\mathrm{T}}(\dot{\boldsymbol{\xi}} - \dot{\mathbf{x}}) - \mathbf{q} \right]\!\!\right]\mathbf{n} = 0 \tag{10.186}$$

$$\gamma\dot{\eta}_\sigma - \nabla_\sigma S_\sigma - \gamma\frac{h_\sigma}{\theta} + [\![\varrho(\eta - \eta_\sigma)(\dot{\boldsymbol{\xi}} - \dot{\mathbf{x}}) - \mathbf{s}]\!]\mathbf{n} = 0 \tag{10.187}$$

Relation (10.185) is significantly simplified by decomposing the stress tensor on normal and tangential components

$$S^{j\alpha} = S^\alpha n^j + S^{\beta\alpha}x^j_{,\beta} \tag{10.188}$$

Then (10.185) is reduced to

$$S^\alpha = 0, \qquad \varepsilon_{\alpha\beta}S^{\alpha\beta} = 0 \tag{10.189}$$

Thus, surface stress

$$\mathbf{S} = S^{\alpha\beta}\mathbf{a}_\alpha \otimes \mathbf{a}_\beta \tag{10.190}$$

is a symmetrical tensor.

These balance laws are valid for the most general class of material surfaces which allow mass transport of the bulk material through it, that is, when

$$(\dot{\boldsymbol{\xi}} - \dot{\mathbf{x}}) \cdot \mathbf{n} \neq 0 \tag{10.191}$$

Such material surfaces are said to be *permeable*.

In some cases, material surfaces allow transport of just one kind of a bulk material. For that material, which we are going to denote by α, (10.191) is valid in the form:

$$(\dot{\boldsymbol{\xi}}_\alpha - \dot{\mathbf{x}}) \cdot \mathbf{n} \neq 0 \tag{10.192}$$

Then the condition of impermeability for the bulk material β reads

$$(\dot{\boldsymbol{\xi}}_\beta - \dot{\mathbf{x}}) \cdot \mathbf{n} = 0 \tag{10.193}$$

Such material surfaces are said to be *semi-permeable*.

2. Impermeable Material Surface

It is the most restricted class of material surfaces. In that case the particles of the bulk material do not pass through surface. Mathematically, it is equivalent to the condition

$$(\dot{\boldsymbol{\xi}} - \dot{\mathbf{x}}) \cdot \mathbf{n} = 0 \tag{10.194}$$

Then balance *laws of the impermeable material surface* reads

$$\frac{\partial \gamma}{\partial t} + \nabla_\sigma \cdot (\gamma \dot{\mathbf{u}}) - 2 u \gamma K_M = 0 \tag{10.195}$$

$$\gamma \ddot{\mathbf{x}} - \nabla_\sigma \cdot \mathbf{S} - \gamma \mathbf{f}_\sigma - [\![\mathbf{T}]\!]\, \mathbf{n} = 0 \tag{10.196}$$

$$\gamma \dot{\varepsilon}_\sigma - \operatorname{tr} \mathbf{S}^{\mathrm{T}}(\nabla_\sigma \dot{\mathbf{x}}) - \nabla_\sigma \cdot \mathbf{q}_\sigma - \gamma h_\sigma \cdot [\![\mathbf{T}^{\mathrm{T}}(\dot{\boldsymbol{\xi}} - \dot{\mathbf{x}}) + \mathbf{q}]\!] \cdot \mathbf{n} = 0 \tag{10.197}$$

$$\gamma \dot{\eta}_\sigma - \nabla_\sigma \cdot \mathbf{s}_\sigma - \gamma \frac{h_\sigma}{\theta} + [\![s]\!] \cdot \mathbf{n} = 0 \tag{10.198}$$

D. Balance Laws for a Mixture

In single-component systems (or pure substances), which have been considered, up to now, the chemical composition in all phases is the same. But, in many areas within the field of continuum physics it is necessary to use the fact that the material being described may be composed of several different constituents.

Such multicomponent systems are called *mixtures*. In these systems, the chemical composition of a given phase changes in response to pressure and temperature and these compositions are not the same in all phases.

The constituents of mixture, generally, may react with each other to produce new constituents. Such a general material will be called a *heterogeneous reacting continuum* or simply a *reacting continuum*. If the constituents composing the material do not react, then it will be called a *heterogeneous continuum*. An example of a reacting continuum is a dissociating and ionizing gas. Liquid helium II, an electrically conducting plasma, and a suspension of solid particles in a fluid are examples of heterogeneous continua.

Most of the literature on reacting continua and on heterogeneous continua deals with chemically reacting fluids. This literature has been unified and generalized by Truesdell and Toupin [20], who presented the differential balance equations for a mixture of chemically reacting continua. They do not restrict the continuum to be a solid, liquid, or gas.

Thus, the theory of mixture is more complicated than the theory of a single body but not different in kind.

The approach presented here is an extension of the program started by Truesdell [37] to the problem of nonlocal heterogeneous continuum.

Modeling of the behavior of multicomponent systems can be done using several methods and looking at the problem at different spatial scales. Eringen and co-workers [38,39] have developed the micromorphic theory of mixture of several constituents in anticipation of the possible application, for example, to crystal lattices in which the lattice sites are regularly occupied by two or more different ions or molecules, to granular or polycrystalline mixture, to composite materials, or to fluid suspensions.

For derivation of complete theory, we refer the reader to the above papers and literature cited in them. Because of that here we give the basic concepts and expressions which are going to be used in what follows.

In order to treat motion of physical mixtures possibly undergoing chemical changes, Fick [40] and Stefan [41] suggested that each place $\mathbf{x}$ may be regarded as occupied simultaneously by several different particles $\mathbf{X}_\alpha, \alpha = 1, 2, \ldots, k$, one for each constituent α. The mixture is thus represented as a superposition of α continuous media, each of which follows its own individual motion

$$\mathbf{x} = \mathbf{x}_\alpha(\mathbf{X}_\alpha, t) \tag{10.199}$$

Henceforth, media whose motion is described by (10.199) will be called *heterogeneous* if $\alpha > 1$; if $\alpha = 1$, they will be called *simple*. In considering kinematics of heterogenous systems, we follow Truesdell and Toupin [20].

Then the *constituent* (individual) *velocity* $\mathbf{v}_\alpha$ is defined by

$$\mathbf{v}_\alpha \equiv \frac{\partial \mathbf{x}}{\partial t}\bigg|_{\mathbf{X}_\alpha=\text{const}} \quad \text{or} \quad v_a^k \equiv \frac{\partial x^k}{\partial t}\bigg|_{\mathbf{X}_\alpha=\text{const}} \tag{10.200}$$

Further, because each constituent has its individual density ϱ_α, we define the total density ϱ by

$$\varrho = \sum_{\alpha=1}^{k} \varrho_\alpha \tag{10.201}$$

The concentration c_α of the constituent α is defined by

$$c_\alpha = \frac{\varrho_\alpha}{\varrho} \tag{10.202}$$

so that (10.201) is equivalent to

$$\sum_{\alpha=1}^{k} c_\alpha = 1 \tag{10.203}$$

The *mean velocity* $\mathbf{v}$ of the mixture is defined by the requirement that the total mass flow is the sum of the individual mass flows:

$$\varrho\mathbf{v} = \sum_{\alpha=1}^{k} \varrho_\alpha\mathbf{v}_\alpha \quad \text{or} \quad \mathbf{v} = \sum_{\alpha=1}^{k} c_\alpha\mathbf{v}_\alpha \tag{10.204}$$

The *diffusion velocity* or *peculiar velocity* of the constituent α is its velocity relative to the mean velocity:

$$\mathbf{u}_\alpha = \mathbf{v}_a - \mathbf{v} \tag{10.205}$$

From (10.204) it follows

$$\sum_{\alpha=1}^{k} \varrho_\alpha \mathbf{u}_\alpha = 0, \qquad \sum_{\alpha=1}^{k} c_\alpha \mathbf{u}_\alpha = 0 \qquad (10.206)$$

That is to say, the mean velocity has been defined in such a way that the total mass flow of the diffusive motions is zero.

We now introduce two different material derivatives $\dot{\psi}$ and $\acute{\psi}$; the former, which coincides with that used for simple media, follows the mean motion, while the latter follows the individual motion of the constituent α:

$$\dot{\psi} = \frac{\partial \psi}{\partial t} + \mathbf{v} \cdot \operatorname{grad}\psi, \qquad \acute{\psi} = \frac{\partial \psi}{\partial t} + \mathbf{v}_\alpha \cdot \operatorname{grad}\psi \qquad (10.207)$$

Hence

$$\acute{\psi} - \dot{\psi} = \mathbf{u}_\alpha \cdot \operatorname{grad}\psi \qquad (10.208)$$

so that the two derivatives coincide, in the case when ψ is a nonconstant scalar, if and only if (iff) the diffusion velocity of the constituent α is tangent to the surface $\psi = \text{const}$.

Further, we set

$$\varrho\psi \equiv \sum_\alpha \varrho_\alpha \psi_\alpha \qquad (10.209)$$

and then, making use of (10.207), we obtain the following *fundamental identity*:

$$\sum_\alpha \varrho_\alpha \acute{\psi}_\alpha = \varrho\dot{\psi} + \psi\left[\frac{\partial \varrho}{\partial t} + \operatorname{div}(\varrho\mathbf{v})\right]$$
$$+ \sum_\alpha \operatorname{div}(\varrho_\alpha \psi_\alpha \mathbf{u}_\alpha) - \sum_\alpha \psi_\alpha\left[\frac{\partial \varrho_\alpha}{\partial t} + \operatorname{div}(\varrho_\alpha \mathbf{v}_\alpha)\right] \qquad (10.210)$$

or

$$\sum_\alpha \varrho_\alpha \acute{\psi}_\alpha = \varrho\dot{\psi} + \psi\left[\frac{\partial \varrho}{\partial t} + (\varrho v^k)_{,k}\right]$$
$$+ \sum_\alpha (\varrho_\alpha \psi_\alpha u_\alpha^k)_{,k} - \sum_\alpha \psi_\alpha\left[\frac{\partial \varrho_\alpha}{\partial t} + (\varrho_\alpha v_\alpha^k)_{,k}\right] \qquad (10.211)$$

Upon this identity, which relates the material derivative of the mean value (10.209) to the mean value of the material derivatives, all our proofs of equations of balance in a heterogeneous medium are founded.

Then, the general balance laws can be written in the form

$$\left(\int_v \psi_\alpha \, dv\right)^{\cdot} = \int_{\partial v} \phi_\alpha \mathbf{n} \, da + \int_v p_\alpha \, dv \qquad (10.212)$$

As this holds for all v, however small, a classical argument yields the differential form of the general balance:

$$\acute{\psi}_\alpha + \psi_\alpha \operatorname{div} \mathbf{v}_\alpha - \operatorname{div} \phi_\alpha - p_\alpha = \hat{p}_\alpha \qquad (10.213)$$

$$[\![\psi_\alpha(\mathbf{v}_\alpha - \mathbf{u}) - \phi_\alpha]\!]\mathbf{n} = \hat{\hat{p}}_\alpha \qquad (10.214)$$

where

$$\int_{v-\sigma} \hat{p}_\alpha \, dv + \int_\sigma \hat{\hat{p}}_\alpha \, da = 0$$

Remark: It is very important to underline that $\hat{p}_\alpha$ and $\hat{\hat{p}}_\alpha$ contain both influences: nonlocality and chemical reactions of the constituents. A very general theory of mixture for micromorphic material with chemical reactions can be found in Cvetković [42]; herein we follow this approach, which is based on Eringen's paper [43]. Also we make use of their notations. Note that these papers contain only parts of $\hat{p}_\alpha$ and $\hat{\hat{p}}_\alpha$, that is, the influence of a chemical reaction. In other words, they did not take into account the effects of nonlocality. Here we underline that the influence of nonlocality will be taken into account through the constitutive equations.

By using arguments quite similar to those presented in Section IV.B, we derive constituent balance equations for mixture.

1. The **balance of mass**

In this case $\psi_\alpha = \varrho_\alpha$, $\phi_\alpha = 0$, $p_\alpha = 0$, and $\hat{p}_\alpha = \varrho\hat{\beta}_\alpha$. Then from (10.213) and (10.214) we obtain

$$\dot{\varrho}_\alpha + \varrho_\alpha \operatorname{div} \mathbf{v}_\alpha = \varrho\hat{\beta}_\alpha \quad \text{or} \quad \frac{\partial \varrho_\alpha}{\partial t} + \operatorname{div}(\varrho_\alpha \mathbf{v}_\alpha) = \varrho\hat{\beta}_\alpha \tag{10.215}$$

$$[\![\varrho_\alpha(\mathbf{v}_\alpha - \mathbf{u})]\!]\mathbf{n} = 0 \tag{10.216}$$

In order to further simplify the calculation, particularly having in mind (10.210), we need the expression $\partial\varrho/\partial t + \operatorname{div}(\varrho\mathbf{v})$. This can be achieved by summing (10.215) and (10.216) over all constituents. In this way, we obtain the local balance equation of mass of mixture

$$\frac{\partial \varrho}{\partial t} + \operatorname{div}(\varrho\mathbf{v}) = 0, \qquad \sum_\alpha \hat{\beta}_\alpha = 0 \tag{10.217}$$

as a consequence of the assumption that the mass of mixture does not change.

Making use of (10.215) and (10.217) in (10.210), we reduced the fundamental identity to the form

$$\sum_\alpha \varrho_\alpha \dot{\psi}_\alpha = \varrho\dot{\psi} + \sum_\alpha \operatorname{div}(\varrho_\alpha \psi_\alpha \mathbf{u}_\alpha) - \sum_\alpha \varrho\hat{\beta}_\alpha \psi_\alpha \tag{10.218}$$

Also, we write

$$\psi_\alpha \longrightarrow \varrho_\alpha \psi_\alpha, \qquad p_\alpha \longrightarrow \varrho_\alpha p_\alpha, \qquad \hat{p}_\alpha \longrightarrow \varrho\hat{p}_\alpha \tag{10.219}$$

Then, in view of (10.213)–(10.215), and (10.219), we obtain

$$\operatorname{div} \phi_\alpha + \varrho_\alpha (p_\alpha - \psi'_\alpha) = \varrho(\hat{\beta}_\alpha \psi_\alpha - \hat{p}_\alpha) \tag{10.220}$$

$$[\![\varrho_\alpha \psi_\alpha(\mathbf{v}_\alpha - \mathbf{u}) - \phi_\alpha]\!]\mathbf{n} = \hat{\hat{p}}_\alpha \tag{10.221}$$

These relations are fundamental in obtaining the local form of *constituent balance equations* for mixture. In order to derive them, we can use Table 10.2 for a constituent of mixture. In this way, we obtain the following.

2. The *constituent local balance equations for momentum*:

$$t^k_{\alpha,k} + \varrho_\alpha(\mathbf{f}_\alpha - \grave{\mathbf{v}}_\alpha) = \varrho\hat{\beta}_\alpha \mathbf{v}_\alpha \tag{10.222}$$

$$[\![t^k_\alpha - \varrho_\alpha \mathbf{v}_\alpha(v^k_\alpha - u_k)]\!] n_k = 0 \tag{10.223}$$

Further, we need the following.

3. The *constituent local balance equations for moment of momentum*:

$$t^{kl}_{\alpha,k} + t^l_\alpha - \tilde{t}^l_\alpha + \varrho_\alpha \mathbf{f}^l_\alpha = \varrho\hat{\beta}^l_\alpha \mathbf{v}_\alpha \tag{10.224}$$

$$[\![t^{kl}_\alpha]\!] n_k = 0 \tag{10.225}$$

We do not need the *constituent local balance laws of energy and entropy*. Again, we refer the reader to the original literature if needed (see, for instance, Ref. [42]).

The mixture local balance of momentum and moment of momentum and the jump conditions are obtained by summing (10.222)–(10.225) over all constituents. The results are as follows.

(a) The *balance of momentum*

$$t^k_{,k} + \varrho(\mathbf{f} - \dot{\mathbf{v}}) = 0 \tag{10.226}$$

$$[\![t^k - \varrho\mathbf{v}(v^k - u^k)]\!] n^k = 0 \tag{10.227}$$

where

$$\varrho\mathbf{f} = \sum_\alpha \varrho_\alpha \mathbf{f}_\alpha, \qquad t^k = \sum_\alpha (t^k_\alpha - \varrho_\alpha \mathbf{u}_\alpha u^k_\alpha) \tag{10.228}$$

(b) The *balance of moment of momentum*

$$t^{km}_{,k} + t^m - \tilde{t}^m + \varrho\mathbf{f}^m = \varrho\mathbf{v}\sum_\alpha \hat{\beta}^m_\alpha \tag{10.229}$$

under the conditions

$$\sum_\alpha \hat{\beta}^m_\alpha = 0 \tag{10.230}$$

Also, by definition

$$t^{km} = \sum_\alpha t^{km}_\alpha, \qquad t^m - \tilde{t}^m = \sum_\alpha (t^m_\alpha - \tilde{t}^m_\alpha), \qquad \varrho\mathbf{f}^m = \sum_\alpha \mathbf{f}^m_\alpha, \tag{10.231}$$

$$[\![t^{kl}]\!] n_k = 0 \tag{10.232}$$

VII. CONCLUSION

Motion, stress, energy, entropy, and electromagnetism are the concepts upon which field theories are constructed. Laws of conservation or balance are laid down as relating these quantities in all cases. These basic principles, which are in integral form, in regions where the variables change sufficiently smoothly are equivalent to differential field equations; at surfaces of discontinuity, to jump conditions.

The field equations and jump conditions form an undetermined system, insufficient to yield specific answers unless further equations are supplied.

The balance laws of continuum physics make no reference to the constitution of the body. Material bodies of the same mass and geometry respond to the same external effects in different ways. Internal constitution of matter is responsible for these differences. From a continuum point of view, we may develop equations which reflect the nature of the material and the constitution of the body. Such a set of the equations are known as constitutive equations. Thus, the characterization of particular materials is brought within the framework of continuum physics through the formulation of constitutive equations (or equation of state).

From theoretical point of view, constitutive equations define an ideal material. Mathematically the purpose of these relations is to support connections between kinematic, mechanical, thermal, and electromagnetic fields which are compatible with the field equations and which, in conjunction with them, yield a theory capable of providing solutions correctly set problems.

Each field of continuum mechanics deals with certain continuous media including fluids, which are liquids or gases (such as water, oil, air, etc.) and solids (such as rubber, metal, ceramics, wood, living tissue, etc.). If the constitutive equations are valid for physical objects such as fluids we call the field of continuum, mechanics fluid mechanics. Another important field in which constitutive equations are valid for solids is known as solid mechanics.

Physically, constitutive equations represent various forms of idealized material response which serve as model of the behavior of actual substances. The predictive value of models, as assessed experimentally over particular ranges of physical conditions, affords justification for the special continuum mentioned above.

Because the constitutive theory is very broad and is a specific subject, owing to the limited space here we refer the reader to [31–34,44].

In this way the theoretical approach is completed. Then, the particular problems of bulk material and interfaces can be considered and solved.

Thus, the constitutive equations and field equations together, along with the jump conditions and boundary condition, should lead to a definite theory, predicting specific answers to particular problems.

ACKNOWLEDGMENT

This work is a part of the project No. 1793 which is supported by MNTR of Republic Serbia.

REFERENCES

1. Gibbs, J.W., On the equilibrium of heterogeneous substances, in *The Collected Works of J. Willard Gibs*, Longman/Green, New York, 1928, pp. 55–353, reprint.
2. Eshelby, J.D., The elastic energy–momentum tensor, *J. Elast.*, 5, 321–335, 1975.
3. Maugin, G.A., *Material Inhomogeneities in Elasticity*, Chapman & Hall, London, 1993.
4. Gurtin, E.M., Configurational Forces as basic concepts of continuum physics, *Applied Mathematical Sciences*, Springer-Verlag, 2000, Vol. 137.
5. Evans, E., Bending elastic modulus of red blood cell membrane derived from buckling instability in micropipet aspiration tests, *Biophys. J.*, 43, 27–30, 1983.
6. Evans, E. and Needham, D., Physical properties of surfactant bilayer membranes: thermal transitions, elasticity, rigidity, cohesion, and colloidal interactions, *J. Chem. Phys.*, 91, 4219–4228, 1987.
7. Bo, L. and Waugh, R.E., Determination of bilayer membrane bending stiffness by tether formation from giant, thin-walled vesicles, *Biophys. J.*, 55, 509–517, 1989.
8. Seifert, U., Configurations of fluid membranes and vesicles, *Adv. Phys.*, 46 (1), 13–137, 1997.
9. Alberts, B.D., Lewis, J., Raff, M., Roberts, K., and Watson, J.D., *Molecular Biology of the Cell*, 2nd ed., Garland, New York, 1989.

10. Edwards, D.A., Brenner, H., and Darsh, T.W., Interfacial transport processes and rheology, *Series in Chemical Engineering*, Butterworth-Heinemann, 1991.

11. Wasan, D.T. and Mohan, V., Interfacial rheological properties of fluid interfaces containing surfactants, in *Improved Oil Recovery by Surfactant and Polymer Flooding,* Shah, D.O. and Schechter, R.S., Eds., Academic Press, New York, pp. 161–201, 1977.

12. Peliti, L., *Biologically Inspired Physics*, Plenum Press, New York and London, 1991.

13. Aris, R., *Vectors, Tensors, and the Basic Equations of Fluid Mechanics*, Dover Publications, Inc., New York, 1989.

14. do Carmo, P.M., *Differential Geometry of Curves and Surfaces*, Prentice-Hall, 1976.

15. Millman, R.S. and Parker, G.D., *Elements of Differential Geometry*, Prentice-Hall, 1977.

16. Kreyszig, E., *Differential Geometry*, Dover, 1991.

17. Cohen, H. and Wang, C.-C., On compatibility conditions for singular surfaces, *Arch. Rat. Mech. Anal.*, 80, 205–261, 1982.

18. McConnell, A.J., *Applications of Tensor Analysis*, Dover Publications, New York, 1957.

19. Thomas, T.Y., Extended compatibility conditions for the study of surfaces of discontinuity in continuum mechanics, *J. Math. Mech.*, 6, 311–322, 1957.

20. Truesdell, C.A. and Toupin, R., Classical field theories, in *Handbuch der Physik III/1*, Flügge, S., Ed., Springer Verlag, Berlin, 1960.

21. Bowen, R.M. and Wang, C.-C., On displacement derivatives, *Quart. Appl. Math.*, 29 (1), 29–39, 1971.

22. Kosiński, W., *Field Singularites and Wave Analysis in Continuum Mechanics*, Ellis Horwood Limited and PWN-Polish Scientific Publishers, 1981.

23. Jarić, P.J. and Milanović-Lazarević, S., Micropolar theory of interface, *Theor. Appl. Mech.*, 4, 73–81, 1978.

24. Müller, I., *Thermodynamics*, Pitman Advanced Publishing Program, 1985.

25. Jarić, J., Cvetković, P., Golubović, Z., and Kuzmanović, D., Advances in continuum mechanics, Monographical booklets, in *Applied and Computer Mathematics*, Fazekas, F., Ed., Pannonian Applied Mathematical Meetings, Budapest, MB-25/PAMM, 2002.

26. Podio-Guidugli, P., Contact interactions, stress and material symmetry, *Theor. Appl. Mech.*, 28–29, 2002.

27. Slattery, C.J., *Momentum, Energy, and Mass Transfer in Continua*, McGraw-Hill, 1972.

28. Reynolds, O., *The Sub-Mechanics of the Universe*, Collective Papers, 1903, Vol. 3.

29. Jarić, J. and Golubović, Z., The balance laws of interline and bulk material, ZAMM, 12, 518, 1991.

30. Moeckel, G.P., Thermodynamics of an Interface, *Arch. Rat. Mech. Anal.*, 57 (3), 255–280, 1975.

31. Eringen, A.C., Ed., Nonlocal polar field theories, *Continuum Physics*, Academic Press, New York, London, 1976, Vol. 4.

32. Chadwick, P., *Continuum Mechanics*, George Allen & Unwin Ltd., London, 1976.

33. Gurtin, M.E., *An Introduction to Continuum Mechanics*, Academic Press, 1981.

34. Holzapfel, A.G., *Nonlinear Solid Mechanics*, Wiley, 2001.

35. Fredrickson, G.A., *Principles and Applications of Rheology*, Prentice-Hall, New York, 1964.

36. Brand, L., *Vestor and Tensor analysis*, Wiley, 1953.

37. Truesdell, C., Sulle basi della termomeccanica, *Rend. Lincei*, 8 (22), 33–38, 1957.

38. Twiss, J.R. and Eringen, A.C., Theory of mixtures for micromorphic materials — 1, *Int. J. Eng. Sci.*, 9, 1019–1044, 1971.

39. Twiss, J.R. and Eringen, A.C., Theory of mixtures for micromorphic materials — II. Elastic constitutive equations, *Int. J. Eng. Sci.*, 10, 437–465, 1972.

40. Fick, A., Über diffusion, *Ann. Phys.*, 94, 59–86, 1855.

41. Stefan, J., Über das Gleichgewicht und die Bewegung, insbesondere die Diffusion von Gasmengen, *Sitzgsber. Akad. Wiss. Wien.*, 63(2), 63–124, 1871.

42. Cvetković, P., Theory of high order for micromorphic heterogeneous media (in Serbian), Ph.D. thesis, University of Belgrade, 1976.

43. Eringen, A.C., Balance laws of micromorphic mechanics, *Int. J. Eng. Sci.*, 8, 819–828, 1970.

44. Truesdell, C.A. and Noll, W., The Nonlinear field theories of mechanics, in *Handbuch der Physik III/3*, Flügge, S., Ed., Springer-Verlag, Berlin, 1965.

TOOLS

11 Nonlinear Frequency Response Method for Investigation of Equilibria and Kinetics of Adsorption Systems

Menka Petkovska
University of Belgrade, Belgrade, Serbia and Montenegro

CONTENTS

I. INTRODUCTION

Like other heterogeneous solid–fluid systems, adsorption systems generally involve a number of interacting phenomena and processes. For their proper design, the knowledge of both equilibrium and kinetics is essential. Most of the methods for estimation of equilibrium and kinetic parameters described in the literature address either the problem of equilibrium or the one of kinetics, but not both. Also, most of these methods are used for estimation of parameters of *a priori* assumed models.

The method presented in this chapter is based on nonlinear frequency response and the concept of higher-order frequency response functions, which have been proven as very convenient tools for analyzing weakly nonlinear systems. The basics for their application lay in the facts that:

- In addition to the first (basic) harmonic, frequency response (FR) of a nonlinear system also contains a DC component and a number (theoretically indefinite) of higher harmonics.
- A model of a weakly nonlinear system can be replaced by an indefinite sequence of linear models of different orders. In the frequency domain, these linear models are defined as frequency response functions (FRFs) of different orders, which can be estimated from different harmonics of the FR.

Being generally weakly nonlinear, adsorption systems make good candidates for investigation by means of nonlinear FR. As it will be shown further in the text, the nonlinear FR method enables identification of the kinetic mechanism and estimation of both equilibrium and kinetic parameters from the same experimental data.

The nonlinear frequency response (NLFR) method for investigation of adsorption systems has been developed as an extension of the classical FR method, by applying the mathematical tools of Volterra series and the concept of higher-order FRFs. For that reason, after a very brief survey of other methods for adsorption equilibrium and kinetic measurements, we will give a short overview of the application of the classical FR method and a brief description of the concept of higher-order FRFs.

A. BRIEF SURVEY OF METHODS FOR ADSORPTION EQUILIBRIUM AND KINETIC MEASUREMENTS

A large number of methods for measurements of adsorption isotherms and kinetic data (mostly diffusion coefficients) have been developed and published. A number of good reviews in this area can be found in classical adsorption books.

1. Equilibrium Measurements

Reviews of experimental methods for determination of adsorption isotherms are given, for example, by Rouquerol et al. [1], Guiochon et al. [2], and Do [3]. Also, there are a number of good review papers that cover some experimental techniques [4–7]. Details about operational procedures can be found in Ref. [1], while Ref. [7] gives a critical review of standard sorption-measuring instruments.

The most commonly used methods for gas adsorption systems are manometric or volumetric, gravimetric, and different chromatographic methods. Combinations of two methods can be used for measuring adsorption of gas mixtures [8].

For liquid adsorption systems, different variations of chromatographic methods are used most commonly, although static methods have also been reported [2,6].

As mentioned, there are different variations of chromatographic methods. The most well known are [2,4–6]: frontal analysis, frontal analysis by characteristic point, perturbation method, elution method by characteristic point, and inverse chromatography.

2. Kinetic Measurements

Although different kinetic mechanisms are generally influencing the overall rate of adsorption processes [9], some kind of a diffusion process is usually recognized (or assumed) as a limiting step. Accordingly, most of the efforts in this area are focused on developing methods for measuring diffusion coefficients. Good reviews of such methods are given in Refs. [3,10,11]. These methods could generally be divided into two groups:

1. Microscopic, such as NMR spectroscopy, pulsed field gradient NMR (PFG-NMR), quasi-elastic neutron scattering (QENS), or isotope exchange technique
2. Macroscopic, such as the analysis of uptake curves, Wicke-Callanbach methods based on steady-state or transient diffusion cell, time lag method, chromatographic methods, zero length column (ZLC) method, and FR method

Diffusion coefficients determined using different methods can differ substantially [11]. In the last couple of years, several research groups investigating different methods for diffusivity measurements have been performing collaborative research to understand and overcome this problem [12].

B. Methods Based on Linear FR

FR (quasi-stationary response of a system to a periodic input change around a steady-state value) is one of the most commonly used methods for investigation of process dynamics and for model identification. After the pioneer work of Naphtali and Polinski [13], it has also been used for investigation of adsorption systems, mainly of adsorption kinetics. Owing to its potential for dealing with reasonably fast systems and its role as an alternative to other methods, in the last two decades the FR method became attractive to a number of investigators, dealing with both theoretical and experimental aspects of this technique. Several research groups have made considerable contributions in the application of the FR method: at the Toyama University in Japan [14–17], at CNRS-LIMSI, Orsay, France [18–22], at the University of Queensland, Australia [23–29], at the University of Edinburgh, Scotland [30–39], and recently, also at the Hungarian Academy of Science, Budapest [37–39], at the Kyungnam University in Masan, South Korea [40,41], and at the Vanderbilt University, Neshville, USA [42,43]. In the majority of these references, the FR method is used for estimation of kinetic parameters in gas–solid adsorption systems and only single adsorbates are considered. Some exceptions are: Gregorczyk and Carta [44], who used FR for investigation of adsorption from liquid phase on polymeric adsorbents, Boniface and Ruthven [45], who considered FR of a chromatographic column, and Park et al. [40], who considered multicomponent adsorption.

The commonly investigated system consists of a reservoir in which gaseous adsorbate and adsorbent particles are placed together, with periodic change of the reservoir pressure, usually caused by forced periodic change of the reservoir volume (Figure 11.1), although the change of the inlet flow rate in a flow-through adsorber has also been considered [29,40,41]. Most investigators measure only the reservoir pressure response and analyze the so-called "in-phase" and "out-of-phase" characteristic functions, first introduced by Yasuda [15], which are directly related to the real and imaginary parts of the frequency transfer function [28]. The LIMSI group also measures the particle temperature [21], thus extending the method to nonisothermal systems.

The FR method was introduced as a method for macroscopic measurement of diffusion coefficients, but it can also be used for measurement of equilibrium data. The diffusion coefficient is obtained from the locus of the maximum of the "out-of-phase" function, while the slope of the adsorption isotherm can be obtained from the low-frequency asymptote of the "in-phase" function [15]. Some applications of the FR methods to investigation of heterogeneous reaction systems have also been reported [46–51].

In all these investigations, very small input amplitudes were used to justify the use of a linear technique for investigation of generally nonlinear adsorption systems. One of the main drawbacks of the classical, linear FR method for investigation of adsorption kinetics is that, in a number of cases, same shapes of linear FR characteristic functions are obtained for different kinetic mechanisms, and the method is reduced to estimation of kinetic parameters of an assumed model. One example of such behavior is the same shape of characteristic curves obtained for adsorption governed by micropore, pore, or pore–surface diffusion mechanisms [28]. Another very characteristic example is the case of bimodal characteristic functions (with two maxima of the "out-of-phase" and two inflection points of the "in-phase" characteristic functions [31,33,35]). These results were shown to fit equally well to three different kinetic models [19,33,35]. As it will be shown in Section III, the nonlinear FR method can overcome this problem.

C. NLFR AND THE CONCEPT OF HIGHER-ORDER FRFs

NLFR is a quasi-stationary response of a nonlinear system to a periodic (sinusoidal or cosinusoidal) input, around a steady state. One of the most convenient tools for treating nonlinear FRs is the concept of higher-order FRFs [52], which is based on Volterra series and generalized Fourier transform. This concept will be briefly presented below.

Let us consider a stable system with a single input x and single output y. Dynamic response of a linear system to an arbitrary input $x(t)$ can be defined using a convolution integral

$$y(t) = \int_{-\infty}^{\infty} g(\tau)\, x(t - \tau)\, d\tau \qquad (11.1)$$

where $g(\tau)$ is the so-called impulse-response function of the system, or its kernel.

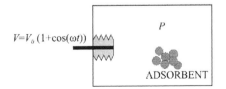

FIGURE 11.1 Schematic representation of a batch ideally mixed adsorber with volume variation.

On the other hand, the response of a weakly nonlinear system, for which the system nonlinearity has a polynomial form (or can be developed in a Taylor series) can be represented in the form of a Volterra series:

$$y(t) = \sum_{n=1}^{\infty} y_n(t) \tag{11.2}$$

with the nth element of the series defined as

$$y_n(t) = \int_{-\infty}^{\infty} \cdots \int_{-\infty}^{\infty} g_n(\tau_1, \ldots, \tau_n) \, x(t - \tau_1) \cdots x(t - \tau_n) \, d\tau_1 \cdots d\tau_n \tag{11.3}$$

$g_n(\tau_1, \ldots, \tau_n)$ is the nth order Volterra kernel, or the generalized impulse-response function of order n.

The first element of the Volterra series y_1 corresponds to the linearized model, while $y_2, y_3, \ldots$ are the "correction" functions of the 1st, 2nd, ... orders.

Similar to Taylor series expansion, a Volterra series of indefinite length is needed for exact representation of a nonlinear system, but for practical applications, finite series can be used.

Frequency response functions. By applying the Fourier transform to the function $g(\tau)$, the FRF, or frequency transfer function is obtained:

$$G(\omega) = \int_{-\infty}^{\infty} g(\tau) \, e^{-j\omega\tau} \, d\tau \tag{11.4}$$

This function is directly related to the amplitude and phase of the quasi-stationary response to a single harmonic input:

$$x = A\cos(\omega t) \implies t \to \infty: \quad y(t) = A|G(\omega)| \cos(\omega t + \arg(G(\omega))) \tag{11.5}$$

On the other hand, by applying multidimensional Fourier transform on the function $g_n(\tau_1, \ldots, \tau_n)$, the nth-order FRF or the nth-order generalized transfer function is obtained:

$$G_n(\omega_1, \ldots, \omega_n) = \int_{-\infty}^{\infty} \cdots \int_{-\infty}^{\infty} g_n(\tau_1, \ldots, \tau_n) \, e^{j(\omega_1 \tau_1 + \cdots + \omega_n \tau_n)} \, d\tau_1 \cdots d\tau_n \tag{11.6}$$

If the input is a periodic function of the general form

$$x(t) = \sum_{k=1}^{N} A_k e^{j\omega_k t} \tag{11.7}$$

the nth element of the Volterra series defined in Equation (11.3) is

$$y_n(t) = \sum_{k_1=1}^{N} \sum_{k_2=1}^{N} \cdots \sum_{k_n=1}^{N} A_{k_1} A_{k_2} \cdots A_{k_n} \, G_n(\omega_{k_1}, \omega_{k_2}, \ldots, \omega_{k_n}) \, e^{j(\omega_{k_1} + \omega_{k_2} + \cdots + \omega_{k_n})t} \tag{11.8}$$

For a single harmonic input

$$x(t) = A\cos(\omega t) = \frac{A}{2} e^{j\omega t} + \frac{A}{2} e^{-j\omega t} \tag{11.9}$$

the first three elements of the Volterra series become

$$y_1(t) = G_1(\omega)\frac{A}{2}e^{j\omega t} + G_1(-\omega)\frac{A}{2}e^{-j\omega t} \tag{11.10}$$

$$y_2(t) = G_2(\omega, \omega)\left(\frac{A}{2}\right)^2 e^{2j\omega t} + 2G_2(\omega, -\omega)\left(\frac{A}{2}\right)^2 e^0 + G_2(-\omega, -\omega)\left(\frac{A}{2}\right)^2 e^{-2j\omega t} \tag{11.11}$$

$$y_3(t) = G_3(\omega, \omega, \omega)\left(\frac{A}{2}\right)^3 e^{3j\omega t} + 3G_3(\omega, \omega, -\omega)\left(\frac{A}{2}\right)^3 e^{j\omega t}$$

$$+ 3G_3(-\omega, -\omega, \omega)\left(\frac{A}{2}\right)^3 e^{-j\omega t} + G_3(-\omega, -\omega, -\omega)\left(\frac{A}{2}\right)^3 e^{-3j\omega t} \tag{11.12}$$

In this way, the nonlinear model of the system is replaced by an indefinite sequence of functions of the 1st, 2nd, 3rd, ... orders, and the output is obtained as a sum of the basic harmonic (of the same frequency as the input), a DC (nonperiodic) term and indefinite number of higher harmonics. This is schematically shown in Figure 11.2.

The DC component of the output is obtained by collecting the nonperiodic terms:

$$y_{DC} = 2\left(\frac{A}{2}\right)^2 G_2(\omega, -\omega) + 6\left(\frac{A}{2}\right)^4 G_4(\omega, \omega, -\omega, -\omega) + \cdots \tag{11.13}$$

the first harmonic by collecting the terms of frequency ω:

$$y_1 = B_1 \cos(\omega t + \varphi_1) = \left\{\left(\frac{A}{2}\right)G_1(\omega) + 3\left(\frac{A}{2}\right)^3 G_3(\omega, \omega, -\omega) + \cdots\right\} e^{j\omega t}$$

$$+ \left\{\left(\frac{A}{2}\right)G_1(-\omega) + 3\left(\frac{A}{2}\right)^3 G_3(\omega, -\omega, -\omega) + \cdots\right\} e^{-j\omega t} \tag{11.14}$$

the second harmonic by collecting the terms of frequency 2ω:

$$y_{II} = B_{II} \cos(2\omega t + \varphi_{II}) = \left\{\left(\frac{A}{2}\right)^2 G_2(\omega, \omega) + 4\left(\frac{A}{2}\right)^4 G_4(\omega, \omega, \omega, -\omega) + \cdots\right\} e^{2j\omega t}$$

$$+ \left\{\left(\frac{A}{2}\right)^2 G_2(-\omega, -\omega)\right.$$

$$\left. + 4\left(\frac{A}{2}\right)^4 G_4(\omega, -\omega, -\omega, -\omega) + \cdots\right\} e^{-2j\omega t} \tag{11.15}$$

$$x = A\cos(\omega t) \longrightarrow \boxed{\begin{array}{c} \textbf{WEAKLY NONLINEAR SYSTEM} \\ \mathbf{G} \equiv G_1(\omega), G_2(\omega_1, \omega_2), G_3(\omega_1, \omega_2, \omega_3), \ldots \end{array}} \xrightarrow{\; y = y_{DC} + y_I + y_{II} + y_{III} + \cdots \;}$$

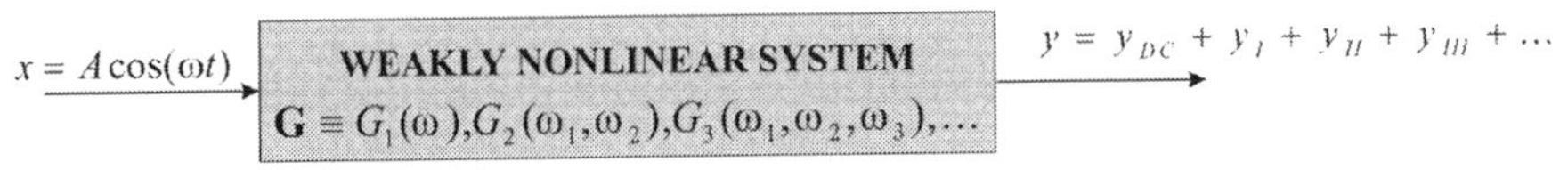

FIGURE 11.2 Schematic representation of FR of a nonlinear system.

the third by collecting the terms of frequency 3ω:

$$
Y_{\mathrm{III}} = B_{\mathrm{III}}\cos(3\omega t + \varphi_{\mathrm{III}}) = \left\{ \left(\frac{A}{2}\right)^3 G_3(\omega, \omega, \omega) \right.
$$

$$
\left. + 5\left(\frac{A}{2}\right)^5 G_5(\omega, \omega, \omega, \omega, -\omega) + \cdots \right\} e^{3j\omega t}
$$

$$
+ \left\{ \left(\frac{A}{2}\right)^3 G_3(-\omega, -\omega, -\omega) \right.
$$

$$
\left. + 5\left(\frac{A}{2}\right)^5 G_5(\omega, \omega, -\omega, -\omega, -\omega) + \cdots \right\} e^{-3j\omega t} \qquad (11.16)
$$

and so on.

For weakly nonlinear systems, the contributions of the higher harmonics and higher FRFs decrease with the increase of their order. Different harmonics of the output can be estimated directly by harmonic analysis of the output signal. On the other hand, as can be seen from Equation (11.13) to Equation (11.16), the first-order function $G_1(\omega)$ corresponds to the dominant term of the first harmonic, the second-order functions $G_2(\omega, \omega)$ and $G_2(\omega, -\omega)$ to the dominant terms of the second harmonic and the DC component, respectively, the third-order function $G_3(\omega, \omega, \omega)$ to the dominant term of the third harmonic, etc. This fact enables estimation of different FRFs from the harmonics of the output obtained for different input amplitudes, as shown by Lee [53].

II. NLFR OF AN IDEALLY MIXED ADSORBER

The NLFR method was first applied to the reservoir-type adsorber, such as the one presented in Figure 11.1. It is assumed that the gas phase in the reservoir is ideally mixed and all particles are equally exposed to the gas phase. The method is not restricted to batch adsorbers with forced periodic modulation of the reservoir volume, but can be used for other configurations as well (e.g., for continuous flow adsorbers with periodic modulation of the inlet molar flow rate). The system is considered to be in concentration and temperature equilibrium in the initial state (before the start of the input modulation).

A. DEFINITION OF TRANSFER FUNCTIONS

The usual way of using the FR technique for investigation of adsorption kinetics is to analyze the response of the adsorber (the periodic change of the pressure in the adsorber reservoir). Nevertheless, the final aim of the FR investigation is to reveal the kinetic mechanism and to obtain the equilibrium and kinetic parameters of the adsorption process. In practice, this aim is reduced to identification of the best mathematical model of the adsorbent particle. Accordingly, we recognize the particle as a subsystem of the adsorber (Figure 11.3) and we define two sets of FRFs (generalized transfer functions), one representing the model on the adsorber and the other on the particle scale [54,55]. The particle FRFs depend only on the kinetic mechanism, while the adsorber ones depend on the adsorber type as well.

1. Adsorber Scale FRFs

When a chosen adsorber input x is modulated periodically in the quasi-stationary state, all the output variables of the adsorber become periodic functions of time as well. In the general non-isothermal case, four adsorber outputs can be defined: pressure P and temperature T_g of the gas

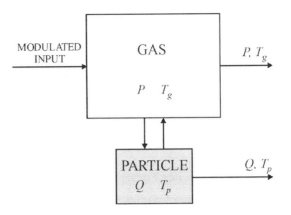

FIGURE 11.3 Schematic representation of an ideally mixed adsorber. (From Petkovska, M., *Nonlinear Dyn.*, 26, 351–370, 2001. With permission.)

phase in the reservoir, and the concentration in the solid phase (loading) Q and the particle temperature T_p (Figure 11.3). For the general nonlinear case, each output is related to the input change via an indefinite sequence of FRFs, as shown in the schematic block diagram given in Figure 11.4.

2. Particle Scale FRFs

The mathematical model of the particle relates the sorbate concentration in the particle Q and the particle temperature T_p (the particle outputs) to the pressure P and the temperature T_g of the gas surrounding the particle (the particle inputs). A general block diagram of a particle is presented in Figure 11.5. Six sets of FRFs are needed to define the particle model, four of them relating each output to each input and two series of cross-functions relating each output to both inputs. We use F to denote the FRFs corresponding to the output Q, and H for those corresponding to T_p. The subscript represents the input variable (P for pressure and T for the gas temperature). The particle FRFs depend only on the kinetic mechanism and not on the adsorber type, so they can be used for identification of the kinetic model and estimation of its parameters.

The definitions of the adsorber and particle transfer functions can be applied to various kinetic mechanisms and different types of adsorbers. They are valid for the general nonlinear,

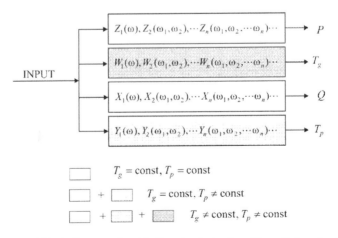

FIGURE 11.4 A general block diagram of an ideally mixed adsorber. (From Petkovska, M., *Nonlinear Dyn.*, 26, 351–370, 2001. With permission.)

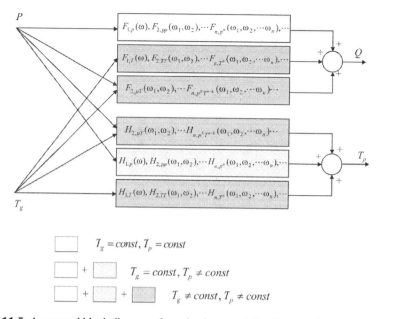

FIGURE 11.5 A genearal block diagram of an adsorbent particle. (From Petkovska, M., *Nonlinear Dyn.*, 26, 351–370, 2001. With permission.)

nonisothermal case, as well as for special cases. For example, for a special linear case, each set of FRFs shown in Figure 11.4 and Figure 11.5 reduces to a single (the first order) FRF. Regarding the isothermality, we consider three cases: the general nonisothermal case for which both the gas and particle temperatures are variable, a special nonisothermal case with constant gas temperature and variable particle temperature, and the isothermal case for which both the gas and particle temperatures are constant. In these two later cases, the model is significantly simplified: for $T_p \neq$ const, $T_g =$ const the complete model reduces to three series of adsorber FRFs and two series of particle FRFs (only the white and light shadowed boxes in Figure 11.3 and Figure 11.4), while for $T_p =$ const, $T_g =$ const it reduces to two series of adsorber and one series of particle FRFs (only the white blocks in Figure 11.3 and Figure 11.4).

B. Procedure for Practical Application of the NLFR Method

In this section we present a procedure for practical application of the general concept of FR investigation of adsorption kinetics, presented in the previous section. The procedure consists of several steps.

Step 1. Experimental FR measurements. The chosen adsorber input is modulated in a periodic way and, after a quasi-stationary behavior is reached, all directly measurable adsorber outputs are measured and recorded. The experiments are performed for a number of different frequencies and for several different values of the input amplitude.

Step 2. Harmonic analysis of the output signals. The recorded output signals are analyzed using fast Fourier transform. As a result, the DC components and the amplitudes and phases of the first, second, third, . . . , harmonics of the measured outputs are obtained.

Step 3. Estimation of the adsorber FRFs. From the results of Step 2, obtained for different input amplitudes, the FRFs corresponding to different adsorber outputs are estimated, using Equations (11.1)–(11.3) and the procedure given by Lee [53]. If some of the adsorber outputs cannot be measured directly (usually that is the case with the loading Q, and sometimes with the particle temperature T_p) the FRFs corresponding to the unmeasured outputs are calculated using the adsorber model equations [55]. As a result of this step, all FRFs defined in Figure 11.3 (the Z, W, X, and Y functions) are known.

Step 4. Calculation of the particle FRFs from the adsorber ones. Starting from the adsorber FRFs, obtained in Step 3, the particle FRFs defined in Figure 11.4 (the F and the H functions) are calculated. The key equations for this calculation are obtained by expressing Q and T_p once using the model on the adsorber scale (using Figure 11.3) and the other time using the model on the particle scale (Figure 11.4). As a result, the following expressions are obtained [55].

For the first-order FRFs:

$$X_1(\omega) = Z_1(\omega)F_{1,p}(\omega) + W_1(\omega)F_{1,T}(\omega) \tag{11.17}$$

$$Y_1(\omega) = Z_1(\omega)H_{1,p}(\omega) + W_1(\omega)H_{1,T}(\omega) \tag{11.18}$$

For the second-order FRFs:

$$
\begin{aligned}
X_2(\omega_1, \omega_2) = {} & Z_2(\omega_1, \omega_2)F_{1,p}(\omega_1 + \omega_2) + W_2(\omega_1, \omega_2)F_{1,T}(\omega_1 + \omega_2) \\
& + Z_1(\omega_1)Z_1(\omega_2)F_{2,pp}(\omega_1, \omega_2) + W_1(\omega_1)W_1(\omega_2)F_{2,TT}(\omega_1, \omega_2) \\
& + \frac{Z_1(\omega_1)W_1(\omega_2)F_{2,pT}(\omega_1, \omega_2) + Z_1(\omega_2)W_1(\omega_1)F_{2,pT}(\omega_2, \omega_1)}{2}
\end{aligned}
\tag{11.19}
$$

$$
\begin{aligned}
Y_2(\omega_1, \omega_2) = {} & Z_2(\omega_1, \omega_2)H_{1,p}(\omega_1 + \omega_2) + W_2(\omega_1, \omega_2)H_{1,T}(\omega_1 + \omega_2) \\
& + Z_1(\omega_1)Z_1(\omega_2)H_{2,pp}(\omega_1, \omega_2) + W_1(\omega_1)W_1(\omega_2)H_{2,TT}(\omega_1, \omega_2) \\
& + \frac{Z_1(\omega_1)W_1(\omega_2)H_{2,pT}(\omega_1, \omega_2) + Z_1(\omega_2)W_1(\omega_1)H_{2,pT}(\omega_2, \omega_1)}{2}
\end{aligned}
\tag{11.20}
$$

and so on.

From this set of equations, the functions $F_{1,p}(\omega)$, $F_{1,T}(\omega)$, $H_{1,p}(\omega)$, $H_{1,T}(\omega)$, $F_{2,pp}(\omega_1, \omega_2)$, $F_{2,TT}(\omega_1, \omega_2)$, $F_{2,pT}(\omega_1, \omega_2)$, $H_{2,pp}(\omega_1, \omega_2)$, $H_{2,TT}(\omega_1, \omega_2)$, $H_{2,pT}(\omega_1, \omega_2)$ etc., are determined, knowing the functions $Z_1(\omega)$, $W_1(\omega)$, $X_1(\omega)$, $Y_1(\omega)$, $Z_2(\omega_1, \omega_2)$, $W_2(\omega_1, \omega_2)$, $X_2(\omega_1, \omega_2)$, $Y_2(\omega_1, \omega_2)$, and so on. This procedure is not straightforward for the general nonisothermal case, but becomes quite simple for some special cases. The details can be found in Ref. [55].

Step 5. Identification of the kinetic model. The particle FRFs calculated in Step 4 are compared with theoretically derived sets of particle FRFs corresponding to different mechanisms. Recognizing the significant patterns of the FRFs, the most probable model or models are chosen. This is possible owing to the fact that the second- and higher-order particle FRFs corresponding to different kinetic models have different shapes [56]. This step assumes that a library of theoretical sets of particle FRFs corresponding to different kinetic mechanisms has been previously formed. The procedure for theoretical derivation of particle FRFs is given in Section III.A, and some examples in Sections III.B–III.E.

Step 6. Parameter estimation. Using the model chosen in Step 5 and the particle FRFs estimated from the experimental FR measurements, obtained in Step 4, the model parameters are estimated. Estimation of the model parameters for some particular cases is discussed in Sections IV and V.

III. LIBRARY OF THEORETICAL PARTICLE FRFs

One of the important assumptions for application of the procedure for identification of adsorption kinetics from NLFR is that a certain library of sets of higher-order FRFs corresponding to different kinetic models is available. The number of cases which could be included in such a library is very big. Here, we will first describe the procedure for theoretical derivation of the particle FRFs, and then list the FRFs for some specific simple isothermal and nonisothermal adsorption mechanism, and for some more complex ones. All these cases correspond to adsorption of pure gases.

A. Procedure for Derivation of Particle FRFs

The procedure for theoretical derivation of the particle FRFs for a general nonisothermal, nonlinear case is shortly given below.

Step 1. Setting up the model equations on the particle scale. These equations are generally nonlinear partial differential equations (PDEs). For analysis in the frequency domain, it is most convenient to use nondimensional concentrations and temperatures, defined as relative deviations from their steady-state values.

Step 2. Definition of the inputs. The particle inputs (P and T_g) are defined as harmonic functions of the general form:

$$P = A_1 e^{ju_1 t} + A_2 e^{ju_2 t} + A_3 e^{ju_3 t} + \cdots \tag{11.21}$$

$$T_g = B_1 e^{jw_1 t} + B_2 e^{jw_2 t} + B_3 e^{jw_3 t} + \cdots \tag{11.22}$$

Step 3. Definition of the outputs. The particle outputs (Q and T_p) are represented in the form of Volterra series [52] of the general form:

$$
\begin{aligned}
y(t) = &\sum_{i=1}^{\infty} A_i G_{1,p}(u_i)\, e^{ju_i t} + \sum_{i=1}^{\infty} B_i G_{1,T}(w_i)\, e^{jw_i t} \\
&+ \sum_{i=1}^{\infty}\sum_{j=1}^{\infty} A_i A_j G_{2,pp}(u_i, u_j)\, e^{j(u_i + u_j)t} + \sum_{i=1}^{\infty}\sum_{j=1}^{\infty} B_i B_j G_{2,TT}(w_i, w_j)\, e^{j(w_i + w_j)t} \\
&+ \sum_{i=1}^{\infty}\sum_{j=1}^{\infty} A_i B_j G_{2,pT}(u_i, w_j)\, e^{j(u_i + w_j)t} + \sum_{i=1}^{\infty}\sum_{j=1}^{\infty} A_j B_i G_{2,pT}(w_i, u_j)\, e^{j(w_i + u_j)t} + \cdots
\end{aligned}
\tag{11.23}
$$

Using the same notation as in Figure 11.4, $G \equiv F$ if the considered output is $y = Q$, and $G \equiv H$ for $y = T_p$. In most cases the concentration and temperature within the particle are not uniform and the outputs are defined as their mean values.

Step 4. Substitution of the inputs and outputs into the model equations. The substitution itself is trivial, but the resulting equations are generally very cumbersome.

Step 5. Application of the method of harmonic probing. This method is performed by collecting the terms with $A_1 e^{ju_1 t}, B_1 e^{jw_1 t}, A_1 A_2 e^{j(u_1 + u_2)t}, B_1 B_2 e^{j(w_1 + w_2)t}, A_1 B_1 e^{j(u_1 + w_1)t}$, and so on, in the equations obtained in Step 4, and equating them to zero. In the resulting sets of equations, time as independent variable is replaced by frequency, while the dependent variables are replaced by the corresponding sets of FRFs. These sets of equations define the first, second, third, ..., FRFs.

Step 6. Solving the equations obtained in Step 5. The solution procedure is recursive, that is, the equations defining the first-order FRFs are solved first, next the ones defining the second-order FRFs, and so on.

Practical application of this procedure for different cases can be found in Refs. [56–59]. We recommend to interested readers to look at the detailed derivation of the first- and second-order FRFs for a simple nonisothermal adsorption mechanism in Ref. [57].

B. Particle FRFs for Simple Isothermal Mechanisms

Our library contains FRFs for four simple isothermal mechanisms: Langmuir kinetics, film resistance model, micropore diffusion, and pore–surface model. For each mechanism, a short description with the model equations is given, together with the expressions for the first-order FRF $F_{1,p}(\omega)$, and two second-order FRFs, $F_{2,pp}(\omega, \omega)$ and $F_{2,pp}(\omega, -\omega)$. No details about the FRF derivation will be given here. They can be found in Refs. [56,60].

All concentrations in the considered models are defined as nondimensional relative deviation from their steady-state values. In that way, the FRFs correlate the relative deviation of the concentration in the adsorbent particle to the relative deviation of the pressure or bulk concentration of the gas around the particles.

Standard Bodé-plots (amplitudes versus frequency in log–log and phases versus frequency in semi-log diagrams) will be used for graphical representation of the FRFs.

1. Surface Barrier Model (Langmuir Kinetics Model)

This is the simplest kinetic case in which the overall adsorption process is governed by the rates of adsorption and desorption of the solute molecules onto and from the surface. For this case, the adsorption kinetics is usually described by the well-known Langmuir kinetic equation:

$$\frac{dq}{dt} = k_a p(q_0 - q) - k_d q \tag{11.24}$$

In this equation, t is the time, p the nondimensional pressure (for isothermal case equal to the nondimensional bulk concentration in the gas phase c), and q the nondimensional concentration in the solid phase. k_a and k_d are the modified adsorption and desorption rate constants, and q_0 the non-dimensional concentration in the solid phase corresponding to maximal coverage (all these modified parameters depend on the steady state around which the system is perturbed and their definitions can be found in Refs. [56,60]).

Here are the expressions for the *first- and second-order FRFs* obtained for this model [56]:

$$F_{1,p}(\omega) = \frac{K_L}{\tau_L \omega j + 1}, \quad K_L = \frac{k_a}{k_d} q_0, \quad \tau_L = \frac{1}{k_d} \tag{11.25}$$

$$F_{2,pp}(\omega, \omega) = -\frac{K_L^2}{q_0} \left(\frac{1}{2\tau_L \omega j + 1} \right) \left(\frac{1}{\tau_L \omega j + 1} \right) \tag{11.26}$$

$$F_{2,pp}(\omega, -\omega) = -\frac{K_L^2}{q_0} \frac{1}{1 + \tau_L^2 \omega^2} \tag{11.27}$$

The graphical representation of these functions is given in Figure 11.6.

2. Film Resistance Model

If the overall mass transfer resistance is lumped in the fluid film surrounding the particle or in the thin skin at the particle surface, a simple, lumped parameter, film resistance model can be used:

$$\frac{dq}{dt} = k_m(p - \Gamma(q)) = k_m \left(p - \left(\left. \frac{\partial p}{\partial q} \right|_s q + \frac{1}{2} \left. \frac{\partial^2 p}{\partial q^2} \right|_s q^2 + \cdots \right) \right)$$

$$= k_m(p - a_q q - b_{qq} q^2 - \cdots) \tag{11.28}$$

k_m is the modified mass transfer coefficient and $\Gamma(q)$ the equilibrium relation (generally nonlinear), which has been replaced by its Taylor series expansion (the coefficients of this expansion, a_q, b_{qq}, ... depend on the steady-state concentration Q_s and pressure P_s).

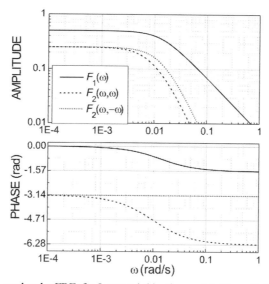

FIGURE 11.6 First- and second-order FRFs for Langmuir kinetics model. (From Petkovska, M. and Do, D.D., *Nonlinear Dyn.*, 21, 353–376, 2000. With permission.)

The first- and second-order FRFs for this case are [56]:

$$F_{1,p}(\omega) = \frac{K_F}{\tau_F \omega j + 1}, \qquad K_F = \frac{1}{a_q}, \qquad \tau_F = \frac{1}{a_q k_m} \tag{11.29}$$

$$F_{2,pp}(\omega, \omega) = -\frac{b_{qq}}{a_q^3} \left(\frac{1}{2\tau_F \omega j + 1}\right) \left(\frac{1}{\tau_F \omega j + 1}\right)^2 \tag{11.30}$$

$$F_{2,pp}(\omega, -\omega) = -\frac{b_{qq}}{a_q^3} \frac{1}{1 + \tau_F^2 \omega^2} \tag{11.31}$$

Their graphical representation is given in Figure 11.7. This figure corresponds to a favorable isotherm ($b_{qq} > 0$).

3. Micropore Diffusion Model

In a number of adsorbents, the adsorbent particle is composed of a large number of microporous microparticles, with larger pores between them. If the dominant mass transfer resistance is within the microparticles, the adsorption process is controlled by the rate of micropore diffusion and the model is defined by the material balance on the microparticle level. For one-dimensional Fickian diffusion, it can be described by the following equation:

$$\frac{\partial q_\mu}{\partial t} = \frac{1}{r_\mu^{\sigma_\mu}} \frac{\partial}{\partial r_\mu} \left(r_\mu^{\sigma_\mu} D_\mu \frac{\partial q_\mu}{\partial r_\mu} \right) \tag{11.32}$$

with the boundary conditions:

$$r_\mu = 0: \quad \frac{\partial q_\mu}{\partial r_\mu} = 0 \tag{11.33}$$

$$r_\mu = R_\mu: \quad q_\mu = \Theta(p) = \frac{\partial q_\mu}{\partial p}\bigg|_s p + \frac{1}{2} \frac{\partial^2 q_\mu}{\partial p^2}\bigg|_s p^2 + \cdots = a_p p + b_{pp} p^2 + \cdots \tag{11.34}$$

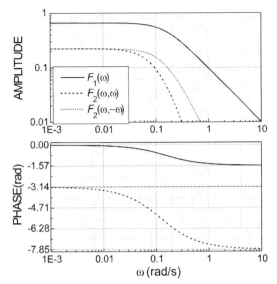

FIGURE 11.7 First- and second-order FRFs for film resistance model. (From Petkovska, M. and Do, D.D., *Nonlinear Dyn.*, 21, 353–376, 2000. With permission.)

In these equations, r_μ is the microparticle space coordinate and R_μ its half-dimension, q_μ is the non-dimensional concentration of the adsorbate in the micropores, D_μ the micropore diffusion coefficient and σ_μ the microparticle shape factor ($\sigma_\mu = 0$ for plane, $\sigma_\mu = 1$ for cylindrical, and $\sigma_\mu = 2$ for spherical microparticle geometry). Θ is the adsorption isotherm relation (generally non-linear), which is again replaced by its Taylor series expansion (the coefficients of which, $a_p, b_{pp}, \ldots$ depend on the steady-state pressure and concentration). The meaning of the boundary condition (11.33) is that the concentration profile in the microparticle is symmetrical, and of the boundary condition (11.34) that adsorption equilibrium is established at the micropore mouth.

The concentration in the micropores depends on the position in the microparticle. Therefore, the mean concentration defined as

$$\langle q_\mu \rangle = \frac{\sigma_\mu + 1}{R_\mu^{\sigma_\mu+1}} \int_0^{R_\mu} r_\mu^{\sigma_\mu} q_\mu(r_\mu)\, dr_\mu \tag{11.35}$$

is used as the particle output.

For the case of constant micropore diffusion coefficient, derivation of the first- and higher-order FRFs is relatively simple, for all three geometries. The first- and second-order FRFs are [56]:

$$F_{1,p}(\omega) = a_p \Phi(\omega) \tag{11.36}$$
$$F_{2,pp}(\omega, \omega) = b_{pp} \Phi(2\omega) \tag{11.37}$$
$$F_{2,pp}(\omega, -\omega) = b_{pp} \tag{11.38}$$

with

$$\Phi(\omega) = \begin{cases} \dfrac{\tanh(\alpha\sqrt{\omega}R_\mu)}{\alpha\sqrt{\omega}R_\mu}, & \sigma_\mu = 0 \\[2ex] 2\dfrac{I_1(\alpha\sqrt{\omega}R_\mu)}{\alpha\sqrt{\omega}R_\mu I_0(\alpha\sqrt{\omega}R_\mu)}, & \sigma_\mu = 1 \quad \text{and} \quad \alpha = \sqrt{\dfrac{j}{D_\mu}} \\[2ex] 3\dfrac{\alpha\sqrt{\omega}R_\mu \coth(\alpha\sqrt{\omega}R_\mu) - 1}{\alpha^2 \omega R_\mu^2}, & \sigma_\mu = 2 \end{cases} \tag{11.39}$$

One of the interesting results is that, for the micropore diffusion model with constant diffusivity, the functions $F_{1,p}(\omega)$ and $F_{2,pp}(\omega, \omega)$ have the same form. This conclusion can be extended to higher-order functions $(F_{3,ppp}(\omega, \omega, \omega),\ F_{4,pppp}(\omega, \omega, \omega, \omega),$ etc.). The asymmetrical second-order FRF $F_{2,pp}(\omega, -\omega)$ is constant (as well as the FRFs $F_{4,pppp}(\omega, \omega, -\omega, -\omega),\ F_{6,pppp}(\omega, \omega, \omega, -\omega,$ $-\omega, -\omega)$, etc.).

An example of the first- and second-order FRFs, corresponding to slab geometry $(\sigma_\mu = 0)$ and favorable isotherm $(b_{pp} < 0)$ is shown in Figure 11.8. The characteristic behavior of the FRFs for this case is that the amplitudes and phases of $F_{1,p}(\omega)$ and $F_{2,pp}(\omega, \omega)$ have identical shapes (the curves could be overlapped by horizontal translation), while the amplitude and phase of the function $F_{2,pp}(\omega, -\omega)$ are constant.

4. Pore–Surface Diffusion Model

In most activated carbon adsorbents, the adsorption kinetic is governed by particle diffusion which generally takes place by two parallel mechanisms: pore diffusion (diffusion of the solute molecules within the particle pores) and surface diffusion (diffusion of the adsorbate molecules adsorbed on the pore walls). For this case, the model is defined by the particle material balance:

$$(1 - \varepsilon_p)\frac{\partial q_i}{\partial t} + \varepsilon_p \frac{C_{is}}{Q_{is}} \frac{\partial c_i}{\partial t} = (1 - \varepsilon_p)\frac{1}{r^\sigma} \frac{\partial}{\partial r}\left(D_s r^\sigma \frac{\partial q_i}{\partial r}\right) + \varepsilon_p \frac{C_{is}}{Q_{is}} \frac{1}{r^\sigma} \frac{\partial}{\partial r}\left(D_p r^\sigma \frac{\partial c_i}{\partial r}\right) \qquad (11.40)$$

with the following boundary conditions:

$$r = 0: \quad \frac{\partial c_i}{\partial r} = \frac{\partial q_i}{\partial r} = 0 \qquad (11.41)$$

$$r = R: \quad c_i = p \qquad (11.42)$$

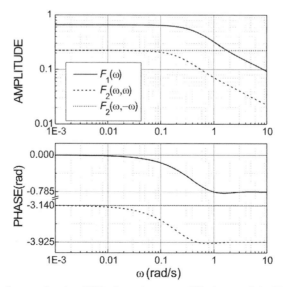

FIGURE 11.8 First- and second-order FRFs for micropore diffusion model. (From Petkovska, M. and Do, D.D., *Nonlinear Dyn.*, 21, 353–376, 2000. With permission.)

Local equilibrium within the particle pores is usually assumed:

$$\forall r: \quad q_i = \Theta(c_i) = \left.\frac{\partial q_i}{\partial c_i}\right|_s c_i + \frac{1}{2}\left.\frac{\partial^2 q_i}{\partial c_i^2}\right|_s c_i^2 + \cdots = a_p c_i + b_{pp} c_i^2 + \cdots \tag{11.43}$$

In these equations, r is the particle spatial coordinate and R its half dimension, ε_p the particle porosity, q_i and c_i are the nondimensional concentrations in the solid phase (adsorbed) and in the gas phase within the pores, respectively, and Q_{is} and C_{is} the corresponding dimensional concentrations in steady state, which are in equilibrium, D_p and D_s are the pore and surface diffusion coefficients, σ the particle shape factor, and Θ the adsorption isotherm relation, which is again replaced by its Taylor series expansion around the steady state. The boundary conditions are based on the assumptions of concentration profiles symmetry (Equation (11.41)) and no mass transfer resistance at the particle surface, that is, equal concentrations at the pore mouth and in the bulk gas (Equation (11.42)).

The total concentration of the adsorbate in the particle is a weighted sum of the concentrations in the solid and in the gas phase within the pores. As it changes with the position in the particle, the mean concentration is used again. The nondimensional mean concentration is

$$\langle q \rangle = \frac{1}{(1-\varepsilon_p)Q_{is} + \varepsilon_p C_{is}}\frac{\sigma+1}{R^{\sigma+1}}\int_0^R r^\sigma((1-\varepsilon_p)Q_{is}q_i(r) + \varepsilon_p C_{is}c_i(r))\,dr \tag{11.44}$$

The FRFs for the pore–surface diffusion model were derived for the case of constant pore and surface diffusion coefficients. The first-order FRF can be derived analytically for all three particle geometries (the solution is analogous to the one obtained for the micropore diffusion model). On the other hand, the second- and higher-order FRFs can be derived analytically only for the slab particle geometry. These are the expressions for the first- and second-order FRFs, for $D_p = $ const, $D_s = $ const, and $\sigma = 0$ [56,58]:

$$F_{1,p}(\omega) = \frac{(1-\varepsilon_p)a_p Q_{is} + \varepsilon_p C_{is}}{(1-\varepsilon_p)Q_{is} + \varepsilon_p C_{is}}\,\Phi(\omega) \tag{11.45}$$

$$F_{2,pp}(\omega,\omega) = \frac{(1-\varepsilon_p)b_{pp}Q_{is}}{(1-\varepsilon_p)Q_{is} + \varepsilon_p C_{is}}$$
$$\times [(1+j\alpha^2 D_s)\Phi(\omega) - \alpha^2 R^2 \omega(1+j\alpha^2 D_s)\Phi^2(\omega)\Phi(2\omega) - j\alpha^2 D_s \Phi(2\omega)] \tag{11.46}$$

$$F_{2,pp}(\omega,-\omega) = \frac{(1-\varepsilon_p)b_{pp}Q_{is}}{(1-\varepsilon_p)Q_{is} + \varepsilon_p C_{is}}\left[\frac{1+j\alpha^2 D_s}{2}(\Phi(\omega) + \Phi(-\omega)) - j\alpha^2 D_s\right] \tag{11.47}$$

with

$$\Phi(\omega) = \frac{\tanh(\alpha\sqrt{\omega}R)}{\alpha\sqrt{\omega}R}, \quad \alpha = \sqrt{j\frac{(1-\varepsilon_p)a_p Q_{is} + \varepsilon_p C_{is}}{(1-\varepsilon_p)a_p Q_{is}D_s + \varepsilon_p C_{is}D_p}} = \sqrt{\frac{j}{D_{eff}}} \tag{11.48}$$

D_{eff} is the effective or apparent diffusion coefficient, which is often defined for the pore–surface diffusion model.

An example of the first- and second-order FRFs for the pore–surface diffusion model, corresponding to a favorable adsorption isotherm ($b_{pp} < 0$), is shown in Figure 11.9.

Unlike the FRFs corresponding to micropore diffusion model, for this case the shapes of $F_{1,p}(\omega)$ and $F_{2,pp}(\omega,\omega)$ are different (the amplitude of $F_{2,pp}(\omega,\omega)$ changes the slope several times and the phase has a distinct minimum), and the amplitude of $F_{2,pp}(\omega,-\omega)$ is not constant, but a descending

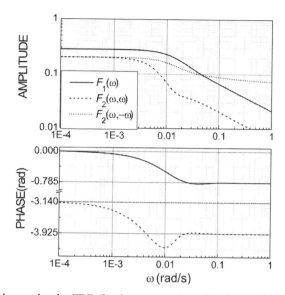

FIGURE 11.9 First- and second-order FRFs for the pore–surface diffusion model. (From Petkovska, M. and Do, D.D., *Nonlinear Dyn.*, 21, 353–376, 2000. With permission.)

curve with horizontal asymptotes for low and high frequencies. These characteristics can be used for clear discrimination between the pore–surface diffusion and the micropore diffusion mechanism.

Figure 11.10 shows the first- and second-order FRFs for a special case of the pore–surface diffusion model, for $D_s = 0$ (pure pore diffusion). For this case, the characteristic behavior of the second-order FRFs become even more significant: the amplitude of $F_{2,pp}(\omega, \omega)$ has a local minimum and its phase a very distinctive minimum, while the amplitude of $F_{2,pp}(\omega, -\omega)$ approaches 0 at high frequencies and not to some finite value. For this case, the effective diffusion coefficient D_{eff}, defined in Equation (11.48), reduces to the pore diffusion coefficient D_p.

5. Influence of the Isotherm Shape

The FRFs shown in Figure 11.7–Figure 11.10 all correspond to favorable adsorption isotherms. As an example, in Figure 11.11 we show the FRFs corresponding to the micropore diffusion model and unfavorable isotherm ($b_{pp} > 0$). Comparison with Figure 11.8 shows that the amplitudes remain unchanged, while the phases of the second-order FRFs are shifted by $+\pi$ (start at 0, for $\omega = 0$, instead at $-\pi$, which is the case for favorable isotherms). Analogous results were obtained for the other kinetic mechanisms under investigation [56]. The value of the phase of $F_{2,pp}(\omega, \omega)$ (and the low-frequency asymptotic value of the phase of $F_{2,pp}(\omega, \omega)$) is directly related to the sign of the second derivative of the adsorption isotherm b_{pp} and, consecutively, to the isotherm shape: it is 0 for $b_{pp} > 0$ (unfavorable isotherm) and $-\pi$ for $b_{pp} < 0$ (favorable isotherm). The change of the phase of $F_{2,pp}(\omega, -\omega)$ from one steady state to another is a signal that an inflection point exists between those two steady states.

6. Summary of the FRFs Characteristics for Simple Isothermal Mechanisms

The characteristic features of the first- and second-order FRFs corresponding to the four simple isothermal mechanisms are summarized in Table 11.1. The slopes of the high-frequency asymptotes of the amplitudes and the high-frequency asymptotic values of the phases are listed in this table. The phases are given for the case of a favorable isotherms and the values corresponding to unfavorable

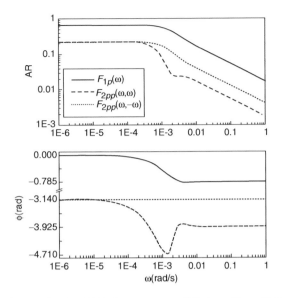

FIGURE 11.10 First- and second-order FRFs for pure pore diffusion. (From Petkovska, M., *Bull. Chem. Technol. Macedonia*, 18, 149–160, 1999. With permission.)

isotherms are given in parentheses. These features can be used for identification of the kinetic mechanism governing the adsorption process.

C. Particle FRFs for Complex Isothermal Mechanisms: Bidispersed Sorbents

In the case of bidispersed adsorbents, diffusion on both macroparticle (pellet) and microparticle scale usually influence the overall adsorption rate [9]. Often, surface barrier (finite adsorption or

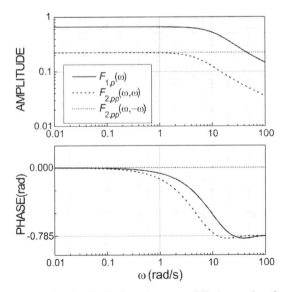

FIGURE 11.11 First- and second-order FRFs for micropore diffusion and unfavorable isotherm. (From Petkovska, M. and Do, D.D., *Nonlinear Dyn.*, 21, 353–376, 2000. With permission.)

TABLE 11.1

Summary of the high-frequency behavior of the first- and second-order FRFs for different simple isothermal kinetic mechanisms

Mechanisms	Slope of the amplitude for $\omega \to \infty$			Phase for $\omega \to \infty$		Specific features
	$F_{1,p}(\omega)$	$F_{2,pp}(\omega, \omega)$	$F_{2,pp}(\omega, \omega)$	$F_{1,p}(\omega)$	$F_{2,pp}(\omega, \omega)$	
Langmuir kinetics	-1.0	-2.0	-2.0	$-\pi/2$	-2π	
Film resistance	-1.0	-3.0	-2.0	$-\pi/2$	$-5\pi/2\ (-3\pi/2)$	
Micropore diffusion	-0.5	-0.5	0	$-\pi/4$	$-5\pi/4\ (-\pi/4)$	$\|F_{2,pp}(\omega, -\omega)\| = \text{const}$ $F_{1,p}(\omega)$ and $F_{2,pp}(\omega, \omega)$ have identical shapes
Pore–surface diffusion	-0.5	-0.5	0	$-\pi/4$	$-5\pi/4\ (-\pi/4)$	$\|F_{2,pp}(\omega, -\omega)\| \downarrow$ $F_{1,p}(\omega)$ and $F_{2,pp}(\omega, \omega)$ have different shapes
Pure pore diffusion	-0.5	-0.5	-0.5	$-\pi/4$	$-5\pi/4\ (-\pi/4)$	$F_{2,pp}(\omega, \omega)$: local minimum of the amplitude and distinct minimum of the phase

desorption rate at the micropore mouth) and film mass transfer resistance at the macroparticle surface also have to be taken into account. The general isothermal model, in which the kinetics of diffusion on both scales, as well as surface barrier and film mass transfer resistance are taken in account, for Fickian diffusion processes, Langmuiran adsorption kinetics at the micropore mouth, and linear driving force in the stagnant film at the particle surface, is obtained as the following set of equations.

The material balance for the microparticle:

$$\frac{\partial q_\mu}{\partial t} = \frac{1}{r_\mu} \frac{\partial}{\partial r_\mu} \left(r_\mu^{\sigma_\mu} D_\mu \frac{\partial q_\mu}{\partial r_\mu} \right) \tag{11.49}$$

with the boundary conditions:

$$r_\mu = 0: \quad \frac{\partial q_\mu}{\partial r_\mu} = 0 \tag{11.50}$$

$$r_\mu = R_\mu: \quad D_\mu \frac{\partial q_\mu}{\partial r_\mu} = k_a c_i (q_0 - q_\mu) - k_d q_\mu \tag{11.51}$$

The material balance for the macroparticle:

$$(1 - \varepsilon) \frac{\partial q_i}{\partial t} + \varepsilon \frac{C_{is}}{Q_{is}} \frac{\partial c_i}{\partial t} = \varepsilon \frac{C_{is}}{Q_{is}} \frac{1}{r^\sigma} \frac{\partial}{\partial r} \left(D_p r^\sigma \frac{\partial c_i}{\partial r} \right) \tag{11.52}$$

with

$$q_i = \frac{\sigma_\mu + 1}{R_\mu^{\sigma_\mu + 1}} \int_0^{R_\mu} r_\mu^{\sigma_\mu} q_\mu(r_\mu) \, dr_\mu \tag{11.53}$$

and the boundary conditions:

$$r_\mu = 0: \quad \frac{\partial q_i}{\partial r} = \frac{\partial c_i}{\partial r} = 0 \tag{11.54}$$

$$r = R: \quad D_p \frac{\partial c_i}{\partial r} = k_m (p - c_i) \tag{11.55}$$

The mean value of the concentration in the particle:

$$\langle q \rangle = \frac{1}{(1-\varepsilon_p)Q_{is} + \varepsilon_p C_{is}} \frac{\sigma+1}{R^{\sigma+1}} \int_0^R r^\sigma (1-\varepsilon_p)Q_{is}q_i(r) + \varepsilon_p C_{is}c_i(r)\, dr \qquad (11.56)$$

The notations used in Equations (11.49)–(11.56) are the same as those used in Section III.B. In case of fast adsorption and desorption kinetics, when equilibrium is practically established at the micropore mouth, the boundary condition (11.51) is transformed in the following way:

$$r_\mu = R_\mu: \quad q_\mu = \Theta(c_i) = \left.\frac{\partial q_\mu}{\partial c_i}\right|_s c_i + \frac{1}{2}\left.\frac{\partial^2 q_\mu}{\partial c_i^2}\right|_s c_i^2 + \cdots = a_p c_i + b_{pp} c_i^2 + \cdots \qquad (11.57)$$

On the other hand, if the mass transfer resistance of the film at the particle surface is negligible, the boundary condition (11.55) becomes

$$r = R: \quad c_i = p \qquad (11.58)$$

The first- and second-order FRFs for the general model (micropore + macropore diffusion + surface barrier + film resistance), defining the relation between $\langle q \rangle$ and p, were derived analytically for the case of constant diffusion coefficients and slab geometry. The following expressions were obtained [61]:

$$F_{1,p}(\omega) = \frac{\varepsilon C_{is} + (1-\varepsilon)k_a/k_d q_0 Q_{is}}{\varepsilon C_{is} + (1-\varepsilon)Q_{is}} \gamma(\omega)\Phi(\alpha(\omega)\sqrt{\omega}R) \qquad (11.59)$$

$$F_{2,pp}(\omega,\omega) = \Lambda(\omega)\left(\xi(\omega)G_1(2\omega) - \frac{\alpha^2(\omega)}{2\alpha^2(\omega) - \alpha^2(2\omega)}\Phi(\alpha(\omega)\sqrt{\omega}R) \right.$$
$$\left. - \frac{1-j}{2}\frac{1}{\cosh^2(\alpha(\omega)\sqrt{\omega}R)} \right) \qquad (11.60)$$

$$F_{2,pp}(\omega,-\omega) = -\frac{(1-\varepsilon)(k_a/k_d)^2 q_0 Q_{is}}{\varepsilon C_{is} + (1-\varepsilon)Q_{is}} \mathrm{Re}(\varphi(\omega))|\gamma(\omega)|^2$$
$$\times \frac{\alpha^2(\omega)\Phi(\alpha(\omega)\sqrt{\omega}R) + \alpha^2(-\omega)\Phi(j\alpha(-\omega)\sqrt{\omega}R)}{\alpha^2(\omega) + \alpha^2(-\omega)} \qquad (11.61)$$

In Equations (11.59)–(11.61) the following groups have been defined:

$$\beta = \sqrt{\frac{j}{D_\mu}}$$

$$\Phi(X) = \frac{\tanh(X)}{X}$$

$$\varphi(\omega) = \left(\frac{R_\mu}{k_d}\Phi(\beta\sqrt{\omega}R_\mu)\omega j + 1 \right)^{-1}$$

$$\alpha(\omega) = \sqrt{\frac{j}{\varepsilon D_p}\left(\varepsilon + (1-\varepsilon)\frac{k_a}{k_d}\frac{q_0 Q_{is}}{C_{is}}\varphi(\omega)\Phi(\beta\sqrt{\omega}R_\mu)\right)}$$

$$\gamma(\omega) = \left(\left(\varepsilon + (1-\varepsilon)\frac{k_a}{k_d}\frac{q_0 Q_{is}}{C_{is}}\varphi(\omega)\Phi(\beta\sqrt{\omega}R_\mu)\right)\frac{R}{k_m}\Phi(\alpha(\omega)\sqrt{\omega}R)\omega j + 1\right)^{-1}$$

$$\Lambda(\omega) = -\frac{(1-\varepsilon)(k_a/k_d)^2 q_0 Q_{is}}{\varepsilon C_{is} + (1-\varepsilon)Q_{is}}\varphi(\omega)\Phi(\alpha(\omega)\sqrt{\omega}R)\varphi(2\omega)\Phi(\alpha(2\omega)\sqrt{2\omega}R)\gamma^2(\omega)$$

$$\xi(\omega) = \frac{Q_{is}}{\varepsilon D_p C_{is}}\left\{\left(\frac{j}{2\alpha^2(\omega) - \alpha^2(2\omega)} - \frac{1}{\alpha^2(2\omega)}\right)\frac{1}{\cosh^2(\alpha(\omega)\sqrt{\omega}R)}\right.$$
$$\left. + \frac{\gamma(\omega) - 2}{\gamma(\omega)}\frac{j}{2\alpha^2(\omega) - \alpha^2(2\omega)}\right\}$$

For special cases, that is, for simpler models, these expressions simplify, in the following way:

1. For negligible film mass transfer resistance: $\gamma(\omega) \to 1$
2. For fast adsorption–desorption and equilibrium at the micropore mouth: $\varphi(\omega) \to 1$, $q_0 k_a/k_d \to a_p$, $-q_0(k_a/k_d)^2 \to b_{pp}$
3. For fast diffusion on the microparticle level (high D_μ/R_μ^2): $\Phi(\beta\sqrt{\omega}R_\mu) \to 1$
4. For fast diffusion on the macroparticle level (high D_p/R^2): $\Phi(\alpha\sqrt{\omega}R) \to 1$

A number of models of different complexity can be derived from the general model defined by Equations (11.49)–(11.56). We will limit our analysis to four models, with two, three, or four mass transfer mechanisms influencing the overall adsorption rate, all of them based on the existence of two major mass transfer resistances in the adsorption particle, on the micro- and on the macroparticle level:

1. Model 1 — the micropore–macropore model. This model is obtained from the general one for negligible mass transfer resistance in the stagnant film and at the micropore mouth.
2. Model 2 — the micropore–macropore-adsorption model. This model takes into consideration the finite adsorption or desorption rate at the micropore mouth, but neglects the film mass transfer resistance.
3. Model 3 — the micropore–macropore-film model. This model takes into account the film mass transfer resistance, but assumes equilibrium at the micropore mouth.
4. Model 4 — the micropore–macropore-adsorption-film model. This is the general model taking into account all four mechanisms of mass transfer.

The simulated first- and second-order FRFs for these four cases are given in Figure 11.12–Figure 11.15.

An overview of the characteristic behavior of the first- and second-order FRFs for the complex mechanisms shown in these figures is given in Table 11.2.

From Figure 11.12–Figure 11.15 and Table 11.2, it can be seen that different high-frequency behavior of the second-order FRFs is obtained for different mechanism combinations so that they can be used for model identification even for complex kinetic mechanisms.

D. PARTICLE FRFs FOR NONISOTHERMAL MICROPORE DIFFUSION MECHANISM WITH VARIABLE DIFFUSIVITY

In a number of cases, the adsorption process is not isothermal and the heat effects have to be taken into account for its proper description. As stated in Section II.A, in that case six series of FRFs are

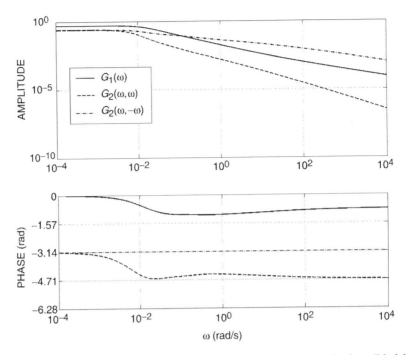

FIGURE 11.12 First- and second-order FRFs for micropore–macropore mechanism (Model 1). (From Petkovska, M., *Adsorption*, 11, 497–502, 2005. With permission.)

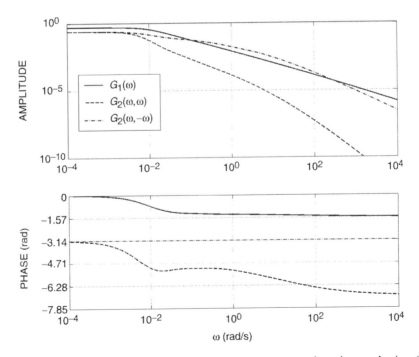

FIGURE 11.13 First- and second-order FRFs for micropore–macropore-adsorption mechanism (Model 2). (From Petkovska, M., *Adsorption*, 11, 497–502, 2005. With permission.)

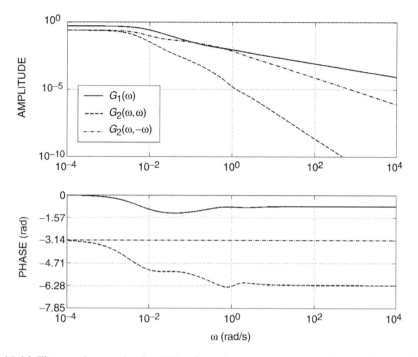

FIGURE 11.14 First- and second-order FRFs for micropore–macropore-film mechanism (Model 3). (From Petkovska, M., *Adsorption*, 11, 497–502, 2005.

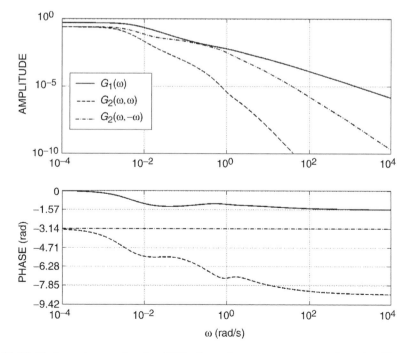

FIGURE 11.15 First- and second-order FRFs for micropore–macropore-adsorption-film mechanism (Model 4). (From Petkovska, M., *Adsorption*, 11, 497–502, 2005.

TABLE 11.2

Summary of the high-frequency behavior of the first- and second-order FRFs for different complex isothermal kinetic mechanisms

	Slope of the amplitude for $\omega \to \infty$			Phase for $\omega \to \infty$	
	$F_{1,p}(\omega)$	$F_{2,pp}(\omega, \omega)$	$F_{2,pp}(\omega, -\omega)$	$F_{1,p}(\omega)$	$F_{2,pp}(\omega, \omega)$
Model 1	-0.5	-1.0	-0.5	$-\pi/4$	$-3\pi/2$
Model 2	-0.5	-2.0	-1.0	$-\pi/4$	-2π
Model 3	-1.0	-2.5	-1.5	$-\pi/2$	$-9\pi/4$
Model 4	-1.0	-3.5	-2.0	$-\pi/2$	$-11\pi/4$

From Ref. [61].

needed to describe the process on the particle level. In our library of particle FRFs, we included one simple nonisothermal case in which the limiting mass transfer mechanism is micropore diffusion and the limiting heat transfer mechanism convection (so that the microparticle temperature can be considered as uniform). As the micropore diffusion coefficient is generally not constant, its concentration and temperature dependence was also considered: $D_\mu = \Xi(q_\mu, \theta_p)$.

Taking this into account, as well as the fact that the equilibrium concentration in the solid phase is a nonlinear function of both pressure and the particle temperature, the microparticle material balance and its boundary conditions can be written in the following form:

$$\frac{\partial q_\mu}{\partial t} = \frac{1}{r_\mu^{\sigma_\mu}} \frac{\partial}{\partial r_\mu} \left(D_{\mu s}(1 + D_q^{(1)} q_\mu + D_T^{(1)} \theta_p + D_{qq}^{(2)} q_\mu^2 + D_{TT}^{(2)} \theta_p + D_{qT}^{(2)} q_\mu \theta_p + \cdots) r_\mu^{\sigma_\mu} \frac{\partial q_\mu}{\partial r_\mu} \right) \quad (11.62)$$

$$r_\mu = 0: \quad \frac{\partial q_\mu}{\partial r_\mu} = 0 \quad (11.63)$$

$$r_\mu = R_\mu: \quad q = \Theta(p, \theta_p) = a_p p + a_T \theta_p + b_{pp} p^2 + b_{TT} \theta_p^2 + b_{pT} p \theta_p + \cdots \quad (11.64)$$

where the nonlinear functions Ξ and Θ have been replaced by their Taylor series expansions.

Using the assumption about lumped heat transfer resistance, the heat balance for the microparticle can be written in the following way:

$$\frac{d\theta_p}{dt} = \xi \frac{d\langle q_\mu \rangle}{dt} + \varsigma(\theta_g - \theta_p) \quad (11.65)$$

In these equations, θ_p and θ_g are the nondimensional particle and gas temperatures, respectively (defined as relative deviations from their steady-state values $T_{ps} = T_{gs} = T_s$). The other variables are defined in the same way as in Section III.B.3. The coefficients in the equations are: $D_{\mu s}$ — the steady-state value of the micropore diffusivity, $D_q^{(1)}$, $D_T^{(1)}$, $D_{qq}^{(2)}$, etc., — the coefficients of the Taylor series expansion of the function Ξ, and a_p, a_T, b_{pp}, etc., — the ones of the function Θ, while ξ and ς represent the modified heat of adsorption and the modified heat transfer coefficient, respectively [57].

For the presented model, analytical derivation of the first- and second-order FRFs is possible for slab microparticle geometry ($\sigma_\mu = 0$). The following expressions were obtained [57].

The first order functions:

$$F_{1,p}(\omega) = \frac{a_p \Phi(\omega)}{1 - a_T \Lambda(\omega)\Phi(\omega)} \tag{11.66}$$

$$H_{1,p}(\omega) = F_{1,p}(\omega)\Lambda(\omega) = \frac{a_p \Lambda(\omega)\Phi(\omega)}{1 - a_T \Lambda(\omega)\Phi(\omega)} \tag{11.67}$$

$$F_{1,T}(\omega) = \frac{a_T \Omega(\omega)\Phi(\omega)}{1 - a_T \Lambda(\omega)\Phi(\omega)} \tag{11.68}$$

$$H_{1,T}(\omega) = F_{1,T}(\omega)\Lambda(\omega) + \Omega(\omega) = \frac{\Omega(\omega)}{1 - a_T \Lambda(\omega)\Phi(\omega)} \tag{11.69}$$

where

$$\Phi(\omega) = \frac{\tanh(R_\mu \alpha \sqrt{\omega})}{R_\mu \alpha \sqrt{\omega}}, \qquad \Lambda(\omega) = \frac{\xi j\omega}{j\omega + \varsigma}, \qquad \Omega(\omega) = \frac{\varsigma}{j\omega + \varsigma}, \qquad \alpha = \sqrt{\frac{j}{D_{\mu s}}}$$

The symmetrical second-order FRFs (corresponding to the second harmonics):

$$F_{2,pp}(\omega, \omega) = \frac{\varphi_1(\omega, \omega)\Phi(2\omega) + 2\varphi_2(\omega, \omega)\Phi(\omega)}{1 - a_T \Lambda(2\omega)\Phi(2\omega)} \tag{11.70}$$

$$H_{2,pp}(\omega, \omega) = F_{2,pp}(\omega, \omega)\Lambda(2\omega) \tag{11.71}$$

$$F_{2,TT}(\omega, \omega) = \frac{\psi_1(\omega, \omega)\Phi(2\omega) + 2\psi_2(\omega, \omega)\Phi(\omega)}{1 - a_T \Lambda(2\omega)\Phi(2\omega)} \tag{11.72}$$

$$H_{2,TT}(\omega, \omega) = F_{2,TT}(\omega, \omega)\Lambda(2\omega) \tag{11.73}$$

$$F_{2,pT}(\omega, \omega) = \frac{\eta_1(\omega, \omega)\Phi(2\omega) + (\eta_2(\omega, \omega) + \eta_3(\omega, \omega))\Phi(\omega)}{1 - a_T \Lambda(2\omega)\Phi(2\omega)} \tag{11.74}$$

$$H_{2,pT}(\omega, \omega) = \Lambda(2\omega)\Phi(2\omega) \tag{11.75}$$

with

$$\varphi_1(\omega, \omega) = b_{pp} + b_{pT}H_{1,p}(\omega) + b_{TT}H_{1,p}^2(\omega)$$
$$+ \frac{D_q^{(1)}}{2}(a_p + a_T H_{1,p}(\omega))^2 (1 + \tanh^2(\alpha\sqrt{\omega}R_\mu))$$
$$+ D_T^{(1)}(a_p + a_T H_{1,p}(\omega))H_{1,p}(\omega)$$

$$\varphi_2(\omega, \omega) = \frac{a_p + a_T H_{1,p}(\omega)}{2}\left[D_T^{(1)}H_{1,p}(\omega) - \frac{D_q^{(1)}}{2}(a_p + a_T H_{1,p}(\omega))\right]$$

$$\psi_1(\omega, \omega) = H_{1,T}^2(\omega)\left[b_{TT} + \frac{D_q^{(1)}a_T^2}{2}(1 + \tanh^2(\alpha\sqrt{\omega}R_\mu)) - D_T^{(1)}a_T\right]$$

$$\psi_2(\omega, \omega) = H_{1,T}^2(\omega)\left[\frac{D_T^{(1)}a_T}{2} - \frac{D_q^{(1)}a_T^2}{4}\right]$$

$$\eta_1(\omega, \omega) = b_{pT}H_{1,T}(\omega) + b_{TT}H_{1,p}(\omega)H_{1,T}(\omega) + D_q^{(1)}a_T(a_p + a_TH_{1,p}(\omega))H_{1,T}(\omega)$$

$$\times (1 + \tanh^2(\alpha\sqrt{\omega}R_\mu)) - D_T^{(1)}(a_p + a_T + a_TH_{1,p}(\omega))H_{1,T}(\omega)$$

$$\eta_2(\omega, \omega) + \eta_3(\omega, \omega) = -D_q^{(1)}a_T(a_p + a_TH_{1,p}(\omega))H_{1,T}(\omega)$$

$$- D_T^{(1)}(a_p + a_T + a_TH_{1,p}(\omega))H_{1,T}(\omega)$$

The asymmetrical second-order functions (corresponding to the DC components):

$$F_{2,pp}(\omega, -\omega) = \varphi_1(\omega, -\omega) + 2\text{Re}(\varphi_2(\omega, -\omega)\Phi(\omega)) \tag{11.76}$$

$$H_{2,pp}(\omega, -\omega) = 0 \tag{11.77}$$

$$F_{2,TT}(\omega, -\omega) = \psi_1(\omega, -\omega) + 2\text{Re}(\psi_2(\omega, -\omega)\Phi(\omega)) \tag{11.78}$$

$$H_{2,TT}(\omega, -\omega) = 0 \tag{11.79}$$

$$F_{2,pT}(\omega, -\omega) = \eta_1(\omega, -\omega) + \eta_2(\omega, -\omega)\Phi(\omega) + \eta_3(\omega, -\omega)\Phi(-\omega) \tag{11.80}$$

$$H_{2,pT}(\omega, -\omega) = 0 \tag{11.81}$$

with

$$\varphi_1(\omega, -\omega) = b_{pp} + b_{pT}\text{Re}(H_{1,p}(\omega)) + b_{TT}|H_{1,p}(\omega_1)|^2$$

$$+ \frac{D_q^{(1)}}{2}(|a_p + a_TH_{1,p}(\omega)|)^2 - D_T^{(1)}\text{Re}[(a_p + a_TH_{1,p}(\omega))H_{1,p}(-\omega)]$$

$$\varphi_2(\omega, -\omega) = -\frac{a_p + a_TH_{1,p}(\omega)}{2}\left[D_T^{(1)}H_{1,p}(-\omega) + \frac{D_q^{(1)}}{2}(a_p + a_TH_{1,p}(-\omega))\right]$$

$$\psi_1(\omega, -\omega) = |H_{1,T}(\omega)|^2\left[b_{TT} + \frac{D_q^{(1)}a_T^2}{2} + D_T^{(1)}a_T\right]$$

$$\psi_2(\omega, -\omega) = -|H_{1,T}(\omega)|^2\left[\frac{D_T^{(1)}a_T}{2} + \frac{D_q^{(1)}a_T^2}{4}\right]$$

$$\eta_1(\omega, -\omega) = b_{pT}H_{1,T}(-\omega) + b_{TT}H_{1,p}(\omega)H_{1,T}(-\omega)$$

$$+ D_q^{(1)}a_T(a_p + a_TH_{1,p}(\omega))H_{1,T}(-\omega) - D_T^{(1)}(a_p + a_T + a_TH_{1,p}(\omega))H_{1,T}(-\omega)$$

$$\eta_2(\omega, -\omega) = -\left(D_T^{(1)} + \frac{D_q^{(1)}a_T}{2}\right)(a_p + a_TH_{1,p}(\omega))H_{1,T}(-\omega)$$

$$\eta_3(\omega, -\omega) = -\left(D_T^{(1)} + \frac{D_q^{(1)}(a_p + a_TH_{1,p}(\omega_1))}{2}\right)a_TH_{1,T}(-\omega)$$

Some simulation results of the first- and second-order FRFs for the nonisothermal micropore diffusion model with variable diffusion coefficient are given in Figure 11.16. They correspond to literature data for adsorption of CO_2 on silicalite-1 [34], $P_s = 10$ kPa and $T_s = 298$ K, and to moderate heat transfer resistances [57]. The functions $H_{2,pp}(\omega, -\omega)$, $H_{2,TT}(\omega, -\omega)$, and $H_{2,pT}(\omega, -\omega)$, which are identically equal to zero, are not shown. In Figure 11.16a we also give the FRFs corresponding to isothermal case (the parameter ζ very large). Notice that for that case the F_p set of FRFs describes the system completely.

Comparison of the F_p functions corresponding to nonisothermal and isothermal cases shows distinctive differences in the shapes of the amplitudes of the second-order functions. The differences in the first-order functions also exist, but they become more obvious if they are shown in the form of real and imaginary parts: the imaginary part of $F_{1,p}(\omega)$ for isothermal case has one minimum, while for the nonisothermal case a curve with two minima is obtained [54,59]. Also, for the case of variable micropore diffusivity, the low- and high-frequency asymptotic values of $F_{2,pp}(\omega, -\omega)$ are not the same as in the case of constant diffusivity.

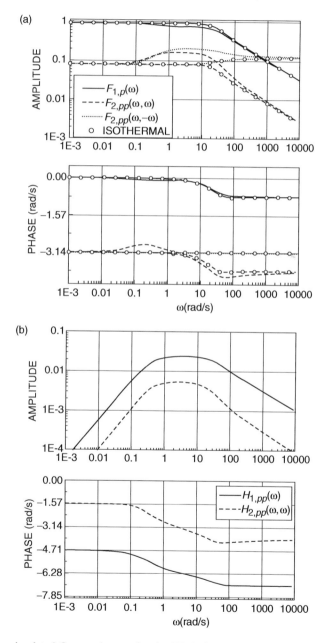

FIGURE 11.16 The simulated first- and second-order FRFs for adsorption of CO_2 on silicalite-1 [34] at 10 kPa and 298 K: (a) F_p (q vs. p) functions + F_p functions for isothermal case; (b) H_p (θ_p vs. p) functions; (c) F_T (q vs. θ_g) and F_{pT} (q vs. p and θ_g) functions; (d) H_T (θ_p vs. θ_g) and H_{pT} (θ_p vs. p and θ_g) functions. (From Petkovska, M., *J. Serbian Chem. Soc.*, 65, 939–961, 2000. With permission.)

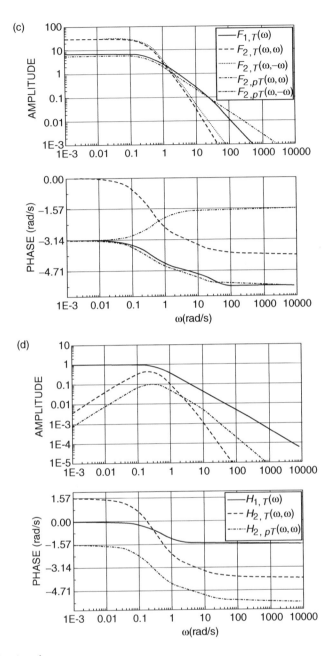

FIGURE 11.16 *Continued.*

E. PARTICLE FRFs FOR MODELS USED TO DESCRIBE BIMODAL CHARACTERISTIC CURVES

As mentioned in Section I, a very characteristic example of the inability of the linear FR method to identify the kinetic model is the case of bimodal characteristic functions, with two maxima of the imaginary and two inflection points of the real part. An example of bimodal characteristic curves is shown in Figure 11.17. Such results were obtained experimentally by Rees and co-workers [31,33,35], for adsorption of some substances on silicalite-1. Their experimental results obtained by the linear FR method were shown to fit equally well to three different mechanisms: [19,33,35]. A short

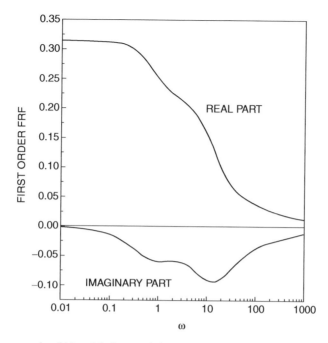

FIGURE 11.17 An example of bimodal characteristic curves.

description of the models, with the model equations, and the first- and second-order FRFs corresponding to them is given subsequently.

1. Model 1

This model assumes an isothermal mechanism of *two parallel, independent diffusion processes* in two different types of micropores. Each diffusion process is defined in the identical way as in Section III.B.3. If the diffusion coefficients differ enough, bimodal characteristic curves, such as those shown in Figure 11.17, are obtained. For constant diffusivities, the material balance equations for the two types of micropores are:

$$\frac{\partial q_{\mu_1}}{\partial t} = D_{\mu_1} \nabla^2 q_{\mu_1}, \qquad \frac{\partial q_{\mu_2}}{\partial t} = D_{\mu_2} \nabla^2 q_{\mu_2} \tag{11.82}$$

with the boundary conditions:

$$r_\mu = 0: \quad \frac{\partial q_{\mu_1}}{\partial r_\mu} = \frac{\partial q_{\mu_2}}{\partial r_\mu} = 0 \tag{11.83}$$

$$r_\mu = R_\mu: \quad q_{\mu_1} = \Theta_1(p) = a_{p_1} p + b_{pp_1} p^2 + \cdots,$$

$$q_{\mu_2} r = \Theta_2(p) = a_{p_2} p + b_{pp_2} p^2 + \cdots \tag{11.84}$$

and the mean sorbate concentration in the microparticle is

$$\langle q_\mu \rangle = \chi \frac{\sigma_\mu + 1}{R_\mu^{\sigma_\mu}} \int_0^{R_\mu} r_\mu^{\sigma_\mu} q_{\mu_1}(r_\mu) \, dr_\mu + (1 - \chi) \frac{\sigma_\mu + 1}{R_\mu^{\sigma_\mu}} \int_0^{R_\mu} r_\mu^{\sigma_\mu} q_{\mu_2}(r_\mu) \, dr_\mu \tag{11.85}$$

In these equations, the subscripts 1 and 2 correspond to the pores of the first and second type, respectively, and χ is the fraction of the sorbate corresponding to the micropores of the first type, at equilibrium.

Based on these equations, the following expressions for the first- and second-order FRFs were obtained for slab microparticle geometry [59]:

$$F_{1,p}(\omega) = \chi a_{p_1} \Phi(\alpha_1(\omega)) + (1 - \chi)a_{p_2} \Phi(\alpha_2(\omega)) \tag{11.86}$$

$$F_{2,pp}(\omega, \omega) = \chi b_{pp_1} \Phi(2\alpha_1(\omega)) + (1 - \chi)b_{pp_2} \Phi(2\alpha_2(\omega)) \tag{11.87}$$

$$F_{2,pp}(\omega, -\omega) = \chi b_{pp_1} + (1 - \chi)b_{pp_2} \tag{11.88}$$

with

$$\Phi(\alpha(\omega)) = \frac{\tanh(\alpha)}{\alpha(\omega)}, \quad \alpha_1(\omega) = \sqrt{\frac{j\omega}{D_{\mu_1}}}R_\mu, \quad \alpha_2(\omega) = \sqrt{\frac{j\omega}{D_{\mu_2}}}R_\mu$$

2. Model 2

The *isothermal diffusion-rearrangement* model [23] also corresponds to adsorbents with two types of micropores, but it assumes that diffusion takes place only in one type (transport pores), while the other one serves only for storage of the sorbed molecules (storage pores). This mechanism can also result with bimodal characteristic curves. The model equations for this case are:

• Mass balances:

$$\frac{\partial q_{\mu_1}}{\partial t} + \delta\frac{\partial q_{\mu_2}}{\partial t} = D_{\mu_1}\nabla^2 q_{\mu_1} \tag{11.89}$$

$$\frac{\partial q_{\mu_2}}{\partial t} = k_1(q_{\mu_1} + 1)(q_{0_2} - q_{\mu_2}) - k_2(q_{\mu_2} + 1)(q_{0_1} - q_{\mu_1})$$

$$= K_1 q_{\mu_1} - K_2 q_{\mu_2} + K_3 q_{\mu_1} q_{\mu_2} \tag{11.90}$$

• Boundary conditions:

$$r_\mu = 0: \quad \frac{\partial q_{\mu_1}}{\partial r_\mu} = 0 \tag{11.91}$$

$$r_\mu = R_\mu: \quad q_{\mu_1} = \Theta_1(p) = a_{p_1}p + b_{pp_1}p^2 + \cdots \tag{11.92}$$

Mean sorbate concentration in the particle is defined in the same way as for Model 1.

In these equations, k_1 and k_2 are the rate constants defining the reversible mass transfer between the transport and the storage pores, χ the fraction of the sorbate corresponding to the transport pores, at equilibrium, and δ the ratio of the equilibrium sorbate concentrations in the storage and in the transport pores. The subscripts 1 and 2 correspond to the transport and storage pores, respectively.

For slab microparticle geometry, the following expressions for the first- and second-order FRFs were obtained [59]:

$$F_{1,p}(\omega) = (\chi + (1 - \chi)\beta(\omega))a_{p_1} \Phi(\alpha(\omega)) \tag{11.93}$$

$$F_{2,pp}(\omega, \omega) = (\chi + (1 - \chi)\beta(2\omega))\{C^* \Phi(\alpha(2\omega)) + D_1^* \Phi(\alpha(\omega)) + D_2^* \cosh^{-2}(\alpha(\omega))\}$$

$$+ 0.5(1 - \chi)K_1^{-1}K_3 a_{p_1}^2 \beta(\omega)\beta(2\omega)\{\Phi(\alpha(\omega)) + \cosh^{-2}(\alpha(\omega))\} \tag{11.94}$$

$$F_{2,pp}(\omega, -\omega) = (\chi + (1 - \chi)K_1 K_2^{-1})b_{pp_1}$$

$$+ K_2^{-1}K_3 a_{p_1}^2 \text{Re}(\beta(\omega))\{\alpha^2(\omega) - \bar{\alpha}^2(\omega)\}^{-1}\{\alpha(\omega)\Phi(\alpha(\omega)) - \bar{\alpha}(\omega)\Phi(\bar{\alpha}(\omega))\} \quad (11.95)$$

with

$$\beta(\omega) = \frac{K_1}{K_2 + j\omega}, \quad \alpha(\omega) = \sqrt{\frac{j\omega}{D_{\mu_1}}(1 + \delta\beta(\omega))R_\mu}, \quad \bar{\alpha}(\omega) = \text{conj}(\alpha(\omega))$$

$$D_1^* = \frac{K_3 \delta a_{p_1}^2 \beta(\omega)\beta(2\omega)\omega j}{K_1 D_{\mu_1}(4\alpha^2(\omega) - \alpha^2(2\omega))}, \quad D_2^* = -\frac{K_3 \delta a_{p_1}^2 \beta(\omega)\beta(2\omega)\omega j}{K_1 D_{\mu_1}\alpha^2(2\omega)}$$

$$C^* = b_{pp_1} - 2D_1^* + (D_1^* - D_2^*)\cosh^{-2}(2\alpha(\omega))$$

$\Phi(\alpha(\omega))$ is defined in the same way as for Model 1.

3. Model 3

As mentioned in Section III.D, the *nonisothermal micropore diffusion model* can also result with bimodal characteristic curves [19]. As the model equations and the derived FRFs have already been listed in Section III.D, they will not be given here again.

The analysis of the second-order FRFs corresponding to the three models show that they have different shapes, so they could be used for model discrimination [59]. The asymmetrical functions $F_{2,pp}(\omega, -\omega)$ which look especially useful for that purpose are shown in Figure 11.18.

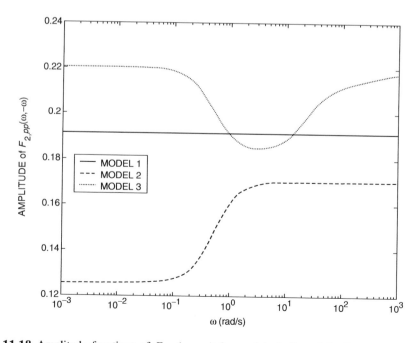

FIGURE 11.18 Amplitude functions of $F_{2,pp}(\omega, -\omega)$ for models 1, 2, and 3. (From Petkovska, M. and Petkovska, L.T., *Adsorption*, 9, 133–142, 2003. With permission.)

F. IDENTIFICATION OF THE KINETIC MECHANISM FROM PARTICLE FRFS OF THE FIRST AND SECOND ORDER

For all cases that have been analyzed in Sections III.B–III.E, the second-order FRFs had different patterns of their amplitudes and phases. Based on this fact, the correct adsorption mechanism could be reliably identified, and the most appropriate model for the particular case chosen, by comparing the particle FRFs estimated from experimental NLFR measurements, with the theoretical ones collected in the library.

The fact that different second-order FRFs were obtained for all investigated mechanisms, together with the facts that the significance of different terms in the output decreases, whereas the complexity of the involved algebra increases considerably, with the increase of the FRF order, was the reason that we limited our analysis only to the FRFs of the first and second order. Nevertheless, the FRFs of higher orders can be derived following the same procedure described in Section III.A, if needed.

After the identification of the correct mathematical model, the next step is to determine the model parameters. As stated in Section I, the NLFR method enables estimation of both equilibrium and kinetic parameters of the model.

IV. ESTIMATION OF EQUILIBRIUM PARAMETERS FROM THE PARTICLE FRFS

The equilibrium parameters are obtained easily from the low-frequency asymptotes of the particle FRFs.

A. LANGMUIR KINETIC MODEL

From Equations (11.25)–(11.27) the following low-frequency limiting values are obtained:

$$\lim_{\omega \to 0} F_{1,p}(\omega) = K_L = \frac{k_a}{k_d} q_0 = \frac{(K_a/K_d)C_s}{1 + (K_a/K_d)C_s} \frac{Q_0 - Q_s}{Q_s} \tag{11.96}$$

$$\lim_{\omega \to 0} F_{2,pp}(\omega, \omega) = \lim_{\omega \to 0} F_{2,pp}(\omega, \omega)$$

$$= -\frac{1}{q_0} K_L^2 = -\left(\frac{k_a}{k_d}\right)^2 q_0 = -\left(\frac{(K_a/K_d)C_s}{1 + (K_a/K_d)C_s}\right)^2 \frac{Q_0 - Q_s}{Q_s} \tag{11.97}$$

In these equations, K_a, K_d, and Q_0 are the adsorption rate, desorption rate, and concentration at maximal coverage in the original model with dimensional concentrations [56] ($k_a = K_a C_s$, $k_d = K_d + K_a C_s$, $q_0 = (Q_0 - Q_s)/Q_s$).

The ratio of the limiting values of the second- and first-order FRFs

$$\frac{\lim_{\omega \to 0} F_{2,pp}(\omega, \omega)}{\lim_{\omega \to 0} F_{1,p}(\omega)} = -\left(\frac{(K_a/K_d)C_s}{1 + (K_a/K_d)C_s}\right) \tag{11.98}$$

enables estimation of the ratio K_a/K_d, defining the equilibrium constant. The concentration at maximal coverage Q_0 can be estimated next, from either of the limits defined by Equation (11.96) and Equation (11.97).

B. Film Resistance Model

Based on the expressions for the first- and second-order FRFs given in Equations (11.29)–(11.31), the following results are obtained:

$$\lim_{\omega \to 0} F_{1,p}(\omega) = K_F = \frac{1}{a_q} = a_p = \frac{\partial Q}{\partial P}\Big|_s \frac{P_s}{Q_s} \tag{11.99}$$

$$\lim_{\omega \to 0} F_{2,pp}(\omega, \omega) = \lim_{\omega \to 0} F_{2,pp}(\omega, -\omega) = -\frac{b_{qq}}{a_q^3} = b_{pp} = \frac{1}{2} \frac{\partial^2 Q}{\partial P^2}\Big|_s \frac{P_s^2}{Q_s} \tag{11.100}$$

In these equations, we use the fact that a_p, b_{pp}, ... and a_q, b_{qq}, ... are essentially derivatives of two inverse functions (Θ and Γ). $(\partial Q/\partial P)_s$ and $(\partial^2 Q/\partial P^2)_s$ are the first and second derivative of the adsorption isotherm written in the dimensional form [56] at steady state defined by P_s and Q_s. It can be shown that the low-frequency asymptotic values of the third- and higher-order functions are proportional to the third- and higher-order derivatives of the adsorption isotherm

C. Micropore Diffusion Model

Similar to the previous case, the derivatives of the adsorption isotherm with respect to pressure are directly related to the low-frequency limiting values of the corresponding F_p functions:

$$\lim_{\omega \to 0} F_{1,p}(\omega) = a_p = \frac{\partial Q_\mu}{\partial P}\Big|_s \frac{P_s}{Q_{\mu s}} \tag{11.101}$$

$$\lim_{\omega \to 0} F_{2,pp}(\omega, \omega) = \lim_{\omega \to 0} F_{2,pp}(\omega, -\omega) = b_{pp} = \frac{1}{2} \frac{\partial^2 Q_\mu}{\partial P^2}\Big|_s \frac{P_s^2}{Q_{\mu s}} \tag{11.102}$$

and so on.

For the case of nonisothermal adsorption governed by micropore diffusion, treated in Section III.D, the isotherm derivatives with respect to temperature can also be estimated from the low-frequency asymptotes of the F_T functions:

$$\lim_{\omega \to 0} F_{1,T}(\omega) = a_T = \frac{\partial Q_\mu}{\partial T_p}\Big|_s \frac{T_{ps}}{Q_{\mu s}} \tag{11.103}$$

$$\lim_{\omega \to 0} F_{2,TT}(\omega, \omega) = \lim_{\omega \to 0} F_{2,TT}(\omega, -\omega) = b_{TT} = \frac{1}{2} \frac{\partial^2 Q_\mu}{\partial T_P^2}\Big|_s \frac{T_{Ps}^2}{Q_{\mu s}} \tag{11.104}$$

and so on, and the mixed derivatives from the low-frequency asymptotes of the F_{pT} functions:

$$\lim_{\omega \to 0} F_{2,pT}(\omega, \omega) = \lim_{\omega \to 0} F_{2,pT}(\omega, -\omega) = b_{pT} = \frac{\partial^2 Q_\mu}{\partial P \partial T_p}\Big|_s \frac{P_s T_{ps}}{Q_{\mu s}} \tag{11.105}$$

The derivatives of the third- and higher-order could be obtained from the low-frequency asymptotes of the third- and higher-order FRFs.

D. PORE–SURFACE DIFFUSION MODEL

The low-frequency asymptotic values of first- and second-order FRFs for the pore–surface diffusion model are:

$$\lim_{\omega \to 0} F_{1,p}(\omega) = \frac{\varepsilon_p C_{is} + (1 - \varepsilon_p) a_p Q_{is}}{\varepsilon_p C_{is} + (1 - \varepsilon_p) Q_{is}} = \frac{\varepsilon_p C_{is} + (1 - \varepsilon_p)\left.\frac{\partial Q_i}{\partial P}\right|_s P_s}{\varepsilon_p C_{is} + (1 - \varepsilon_p) Q_{is}} \tag{11.106}$$

$$\lim_{\omega \to 0} F_{2,pp}(\omega, \omega) = \lim_{\omega \to 0} F_{2,pp}(\omega, -\omega) = \frac{(1 - \varepsilon_p) b_{pp} Q_{is}}{\varepsilon_p C_{is} + (1 - \varepsilon_p) Q_{is}} = \frac{(1 - \varepsilon_p)\left.\frac{\partial^2 Q_i}{\partial P^2}\right|_s P_s^2}{\varepsilon_p C_{is} + (1 - \varepsilon_p) Q_{is}} \tag{11.107}$$

They are also directly related to the first- and second-order derivatives of the adsorption isotherm. Analogous relations exist between the isotherm derivatives and FRFs of the third and higher orders, as well.

E. COMPLEX KINETIC MODELS

For the general complex isothermal kinetic model treated in Section III.C (taking into account micropore and macropore diffusion, as well as surface barrier and film resistance mechanisms), the low-frequency asymptotes of the first- and second-order FRFs are:

$$\lim_{\omega \to 0} F_{1,p}(\omega) = \frac{\varepsilon_p C_{is} + (1 - \varepsilon_p)(k_a/k_d) q_0 Q_{is}}{\varepsilon_p C_{is} + (1 - \varepsilon_p) Q_{is}}$$

$$= \frac{\varepsilon_p C_{is} + (1 - \varepsilon_p)\dfrac{(K_a/K_d) C_{is}(Q_0 - Q_{\mu s})}{1 + (K_a/K_d) C_{is}}}{\varepsilon_p C_{is} + (1 - \varepsilon_p) Q_{is}} \tag{11.108}$$

$$\lim_{\omega \to 0} F_{2,pp}(\omega, \omega) = \lim_{\omega \to 0} F_{2,pp}(\omega, -\omega) = -\frac{(1 - \varepsilon_p)(k_a/k_d)^2 q_0 Q_{is}}{\varepsilon_p C_{is} + (1 - \varepsilon_p) Q_{is}}$$

$$= -\frac{(1 - \varepsilon_p)\left(\dfrac{(K_a/K_d) C_{is}}{1 + (K_a/K_d) C_{is}}\right)^2 (Q_0 - Q_{\mu s})}{\varepsilon_p C_{is} + (1 - \varepsilon_p) Q_{is}} \tag{11.109}$$

For the case of local equilibrium at the micropore mouth, these expressions reduce to the ones corresponding to the pore–surface diffusion model (Equation (11.106) and Equation (11.107)).

V. ESTIMATION OF KINETIC PARAMETERS FOR SIMPLE KINETIC MODELS

A. SIMPLE ISOTHERMAL MODELS

The problem of estimation of the kinetic parameters from linear FR characteristic functions, has been solved long ago, for simple isothermal kinetic models [15]. The process time constant can be estimated from the extremum of the so-called "out-of-phase" function [15], which is identical to the negative imaginary part of the first-order particle FRF $F_{1,p}(\omega)$ [28].

1. Isothermal Langmuir Kinetics and Film Resistance Models

Figure 11.19 shows the shape of the negative imaginary part of the first-order FRFs corresponding to the Langmuir kinetics and the film resistance models. If the product of the frequency and the characteristic time constant is used on the abscissa, the maximum is obtained for $\omega\tau = 1$, that is, for $\omega = 1/\tau$. In this way, the time constant can be estimated directly from the position of the maximum of $-\text{Imag}(F_{1,p}(\omega))$.

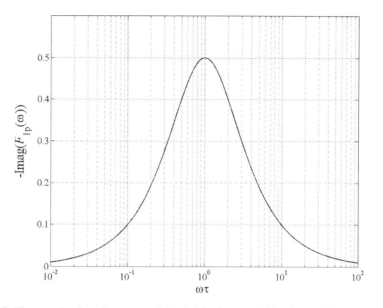

FIGURE 11.19 The negative imaginary part of $F_{1,p}(\omega)$ for Langmuir kinetics and film resistance models.

The time constant for the Langmuir kinetic model was defined in Section III.B.1:

$$\tau_L = \frac{1}{k_d} = \frac{1}{K_d + K_a C_s} \tag{11.110}$$

Together with the value of the ratio K_a/K_d, which can be estimated from the low-frequency asymptotes of the first- and second-order FRFs, this enables determination of the adsorption and desorption rate constants K_a and K_d separately.

On the other hand, the time constant for the film resistance model was defined as (Equation (11.29)):

$$\tau_F = \frac{1}{a_q k_m} = \frac{a_p}{k_m} \tag{11.111}$$

The mass transfer coefficient can be calculated as the ratio of the estimated values of a_p and τ_F.

2. Isothermal Micropore and Pore–Surface Diffusion Models

For the adsorption process governed by a single Fickian diffusion process, the time constant is defined as the ratio $\tau = L^2/D$, where L is the characteristic half-dimension and D the diffusion coefficient. Accordingly, the time constant for the micropore diffusion model would be

$$\tau_M = \frac{R_\mu^2}{D_\mu} \tag{11.112}$$

and for the pore–surface diffusion model:

$$\tau_{PS} = \frac{R^2}{D_{eff}} \tag{11.113}$$

The negative imaginary parts of the $F_{1,p}(\omega)$ functions, corresponding to a single diffusion mechanism and to three different geometries (plane, cylindrical, and spherical), are shown in Figure 11.20. The maxima of these curves correspond to: $\omega\tau = 2.5492$ for plane geometry; $\omega\tau = 6.3504$ for cylindrical geometry; and $\omega\tau = 11.5630$ for spherical geometry.

Using these results, the time constant of the diffusion process is obtained from the locus of the maximum of the $-\mathrm{Imag}(F_{1,p}(\omega))$ curve and the knowledge of the microparticle or macroparticle geometry. The corresponding diffusion coefficient is obtained as the ratio of the square of the micro- or macroparticle half-dimension and the time constant: $D = L^2/\tau$.

The expressions for the FRFs for the isothermal micropore and pore–surface diffusion models were obtained for constant diffusion coefficients. If this assumption is not met, that is, if the concentration dependence of the diffusion coefficient has to be taken into account, the value estimated from the maximum of the $-\mathrm{Imag}(F_{1,p}(\omega))$ curve is the diffusion coefficient corresponding to the steady-state concentration.

For the pore–surface diffusion model, only the effective diffusion coefficient

$$D_{\mathrm{eff}} = \frac{(1-\varepsilon_p)D_s a_p Q_{is} + \varepsilon_p D_p C_{is}}{(1-\varepsilon_p)a_p Q_{is} + \varepsilon_p C_{is}} \tag{11.114}$$

can be estimated from the first-order function and not the pore and surface diffusion coefficients separately. Nevertheless, the high-frequency asymptotic value of the second-order FRF $F_{2,pp}(\omega, -\omega)$ is

$$\lim_{\omega\to\infty} F_{2,pp}(\omega,-\omega) = \frac{(1-\varepsilon_p)b_{pp}Q_{is}}{(1-\varepsilon_p)Q_{is} + \varepsilon_p C_{is}}\frac{D_s}{D_{\mathrm{eff}}} \tag{11.115}$$

This, together with the estimated value of D_{eff} and the values of the equilibrium parameters a_p and b_{pp}, enables calculation of the separate values of the pore and surface diffusion coefficients D_p and D_s, which is not possible by standard methods.

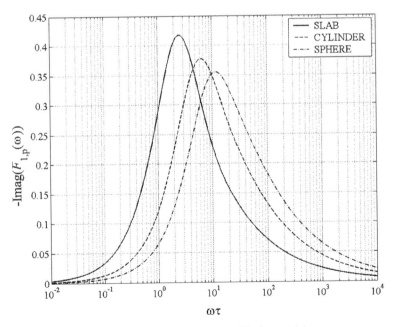

FIGURE 11.20 The negative imaginary part of $F_{1,p}(\omega)$ for diffusion models.

B. Nonisothermal Micropore Diffusion Model

1. Estimation of the Micropore Diffusion Coefficient

For the nonisothermal case, the position of the maximum of $-\mathrm{Imag}(F_{1,p}(\omega))$ is different, compared with the isothermal micropore diffusion mechanism. Actually, in most cases this function has two maximums, as can be seen from Figure 11.21 (from Ref. [57]). Nevertheless, the analysis of the FRFs corresponding to the nonisothermal micropore model shows that the ratio

$$\frac{F_{1,T}(\omega)}{H_{1,T}(\omega)} = a_T \Phi(\omega) \tag{11.116}$$

has the same shape and the same position of the maximum as the isothermal $F_{1,p}(\omega)$ function (see Figure 11.21). Accordingly, this ratio and its imaginary part can be used for estimation of the micropore diffusion coefficient D_μ.

2. Estimation of Other Model Parameters

The particle FRFs can be used for estimation of other model parameters, such as the modified heat of adsorption ξ and modified heat transfer coefficient ζ. The parameter ξ can be estimated from the ratio [57]:

$$\frac{H_{1,p}(\omega)}{F_{1,p}(\omega)} = \Lambda(\omega) = \frac{\xi j\omega}{\varsigma + j\omega} \tag{11.117}$$

that is, from its high-frequency asymptote

$$\lim_{\omega \to \infty} \Lambda(\omega) = \xi \tag{11.118}$$

On the other hand, the ratio

$$\frac{H_{1,p}(\omega)}{F_{1,T}(\omega)} = \frac{a_p}{a_T} \frac{\Lambda(\omega)}{\Omega(\omega)} = \frac{a_p}{a_T} \frac{\xi}{\varsigma} j\omega \tag{11.119}$$

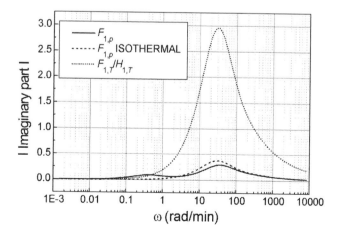

FIGURE 11.21 Imaginary parts of $F_{1,p}$ for nonisothermal and isothermal case and $F_{1,T}/H_{1,T}$. (From Petkovska, M. *J. Serbian. Chem. Soc.*, 65, 939–961, 2000. With permission.)

is a linear function of frequency, with the slope proportional to the ratio of ξ/ζ. Having the value of ξ and the equilibrium constants a_p and a_T, the value of ζ can be obtained from this slope. The heat of adsorption can be obtained from the value of ξ and the value of heat transfer coefficient from the value of ζ [57].

3. Estimation of Parameters for the Case of Variable Micropore Diffusion Coefficient

For the case of variable diffusivity, the micropore diffusion coefficient obtained from the position of the extremum of the imaginary part of $F_{1,p}(\omega)$ (for isothermal case) or of the ratio $F_{1,T}(\omega)/H_{1,T}(\omega)$ (for nonisothermal case) corresponds to the steady-state concentration $Q_{\mu s}$ and temperature T_s.

The first-order concentration coefficient of the micropore diffusion coefficient $D_q^{(1)}$ can be estimated from the high-frequency asymptote of second-order FRF $F_{2,pp}(\omega, -\omega)$ [57]:

$$\lim_{\omega \to \infty} F_{2,pp}(\omega, -\omega) = b_{pp} + \frac{D_q^{(1)} a_p^2}{2} \tag{11.120}$$

assuming that the equilibrium parameters a_p and b_{pp} have been estimated as shown in Section IV.

On the other hand, the high-frequency asymptote of the ratio

$$\lim_{\omega \to \infty} \frac{F_{2,TT}(\omega, -\omega)}{|H_{1,T}(\omega)|^2} = b_{TT} + \left(\frac{D_q^{(1)} a_T^2}{2} + D_T^{(1)} a_T \right) \tag{11.121}$$

can be used for estimation of the first-order temperature coefficient of the micropore diffusion coefficient $D_T^{(1)}$, for known $D_q^{(1)}$ and the equilibrium coefficients b_{TT} and a_T. The higher-order coefficients $(D_{qq}^{(2)}, D_{TT}^{(2)}, D_{qT}^{(2)}, \ldots)$ could be estimated from the third- and higher-order FRFs.

Estimation of the kinetic parameters for complex kinetic mechanisms is a more difficult problem. At the moment, we are working on its solution.

VI. NLFR OF A CHROMATOGRAPHIC COLUMN

The assumption of perfect mixing in a reservoir-type adsorber is usually not acceptable for adsorption from liquid phase. That was a reason for developing a method based on NLFR of a chromatographic column, for which the fluid flow is much better defined. The FRFs of a chromatographic column were defined in such a way to relate the nondimensional outlet concentration from the column c_o to the periodic change of the nondimensional inlet concentration c_i, for constant flow rate of the fluid phase [62]. The nondimensional concentrations were defined as relative deviations from the steady-state concentration C_s. For the time being, only an isothermal system with a single adsorbing component has been considered.

The FRFs of a chromatographic column were derived starting from the commonly used equilibrium-dispersion model [2], written in its nondimensional form:

$$\frac{\partial c}{\partial \tau} + \frac{1-\varepsilon}{\varepsilon} \frac{Q_s}{C_s} \frac{\partial q}{\partial \tau} + \frac{\partial c}{\partial x} = \frac{1}{2N} \frac{\partial^2 c}{\partial x^2}, \quad \forall \tau, \forall x: \; q = \Theta(c) = \tilde{a}c + \tilde{b}c^2 + \tilde{c}c^3 + \cdots \tag{11.122}$$

with the following initial and boundary conditions:

$$x = 0: \quad c(0, \tau) = c_i(\tau) + \frac{1}{2N} \frac{\partial c}{\partial x}\bigg|_{x=0}, \qquad x = 1: \quad \frac{\partial c}{\partial x}\bigg|_{x=1} = 0 \tag{11.123}$$

$$\tau \le 0: \quad c(x) = c_i(x) = q(x) = 0 \tag{11.124}$$

In Equations (11.122)–(11.124) τ and x are the nondimensional time and axial coordinate, respectively, c and q the nondimensional concentrations in the fluid and in the solid phase, respectively, defined as relative deviations from their steady-state values C_s and Q_s, ε is the bed porosity, and N the number of theoretical plates. The nonlinear equilibrium relation (adsorption isotherm) Θ is again represented in the Taylor series form (the coefficients $\tilde{a}$, $\tilde{b}$, $\tilde{c}$, ... depend on the steady-state concentration).

The derivation procedure of the FRFs is similar to the one presented in Section III.A [62]. The derived FRFs are rather cumbersome and will not be given here. Instead, only their main characteristics will be discussed.

As illustration, one example of the simulated first-, second-, and third-order FRFs of a chromatographic column is presented in Figure 11.22. These functions correspond to a favorable isotherm and to $N = 1000$.

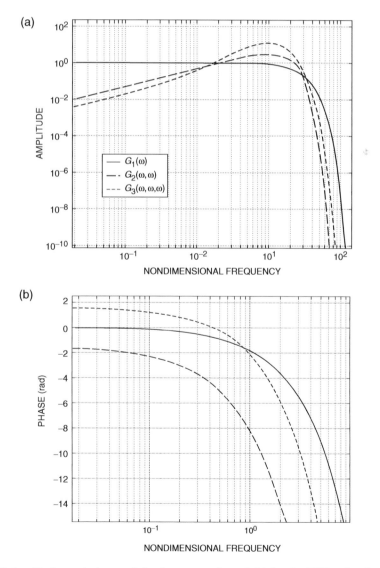

FIGURE 11.22 Amplitudes and phases of the first-, second-, and third-order FRFs of a chromatographic column for favorable isotherm.

The main characteristics of the FRFs of a chromatographic column are the following:

- $G_1(\omega)$: amplitude tends to 1 for $\omega \to 0$ and to 0 for $\omega \to \infty$ and phase tends to 0 for $\omega \to 0$ and to $-\infty$ for $\omega \to \infty$.
- $G_2(\omega, \omega)$ and $G_3(\omega, \omega, \omega)$: amplitudes tend to 0 for both $\omega \to 0$ and $\omega \to \infty$ and phases tend to either $-\pi/2$ or $+\pi/2$ (depending on the sign of the coefficient $\tilde{b}$ or $\tilde{c}$, that is, on the isotherm shape) for $\omega \to 0$ and to $-\infty$ for $\omega \to \infty$.
- $G_2(\omega, -\omega) \equiv 0$.
- The FRFs depend both on the number of theoretical plates N and on the isotherm coefficients $\tilde{a}, \tilde{b}, \tilde{c}, \ldots$, but their low-frequency asymptotes depend only on the isotherm coefficients (for $G_1(\omega)$ only on $\tilde{a}$, for $G_2(\omega, \omega)$ only on $\tilde{b}$, and for $G_3(\omega, \omega, \omega)$ only on $\tilde{c}$).

The following results were obtained for the low-frequency characteristics of the column FRFs [62] (in these equations the frequency is defined as a nondimensional variable):

$$\lim_{\omega \to 0} \left| \frac{dG_1(\omega)}{d\omega} \right| = 1 + \tilde{a} \frac{1-\varepsilon}{\varepsilon} \frac{Q_s}{C_s} = 1 + \frac{dQ}{dC}\bigg|_s \frac{1-\varepsilon}{\varepsilon} \tag{11.125}$$

$$\lim_{\omega \to 0} \left| \frac{dG_2(\omega, \omega)}{d\omega} \right| = 2|\tilde{b}| \frac{1-\varepsilon}{\varepsilon} \frac{Q_s}{C_s} = \left| \frac{d^2Q}{dC^2} \right|_s \frac{1-\varepsilon}{\varepsilon},$$

$$\text{sign}(\tilde{b}) = -\text{sign}\left(\lim_{\omega \to 0} (\arg(G_2(\omega, \omega))) \right) \tag{11.126}$$

$$\lim_{\omega \to 0} \left| \frac{dG_3(\omega, \omega, \omega)}{d\omega} \right| = 3|\tilde{c}| \frac{1-\varepsilon}{\varepsilon} \frac{Q_s}{C_s} = \frac{1}{2} \left| \frac{d^3Q}{dC^3} \right|_s \frac{1-\varepsilon}{\varepsilon},$$

$$\text{sign}(\tilde{c}) = -\text{sign}\left(\lim_{\omega \to 0} (\arg(G_3(\omega, \omega, \omega))) \right) \tag{11.127}$$

These results can be used for estimation of the equilibrium parameters from the low-frequency asymptotes of the FRFs of a chromatographic column. Their applicability has been checked experimentally [63].

The results shown in Equations (11.125)–(11.127) are valid, not only for the equilibrium-dispersion model, but also for a more general case of finite mass transfer rate between the fluid and the gas phase in the column [62].

The number of theoretical plates can also be estimated from the minimum of a function defined as $d|G_1(\omega)|/d\omega$ (Figure 11.23). The frequency $\omega_{\min}$ for which this minimum is obtained is proportional to the square root of the number of theoretical plates N:

$$\omega_{\min} = \frac{\sqrt{N}}{1 + (\partial Q/\partial C)|_s (1-\varepsilon)/\varepsilon} \tag{11.128}$$

VII. CONCLUSIONS

A. SIGNIFICANCE, ADVANTAGES, AND LIMITATIONS OF THE NLFR METHOD

The significance of the NLFR method for investigation of adsorption equilibria and kinetics, presented in this chapter, can be summarized in several points:

1. A set of FRFs of different orders is obtained as a result of NLFR experiments. Owing to the fact that the second- and higher-order FRFs corresponding to different adsorption mechanisms have different shapes, this method enables identification of the actual

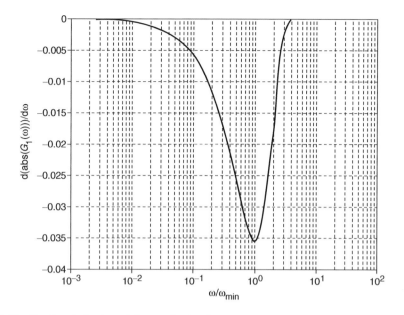

FIGURE 11.23 The first derivative of the amplitude of $G_1(\omega)$.

mechanism by comparing and matching the patterns of the FRFs estimated from experimental NLFR data with theoretically derived sets corresponding to various mechanisms, collected in a library of FRFs. In this way, our method is free of any assumptions regarding the adsorption mechanisms and is not limited to any specific ones.

2. Both equilibrium and kinetic parameters can be estimated from the same set of experimentally obtained FRFs. As the experiments are performed by periodic modulation around a steady state, these parameters correspond to the steady state values of the system variables.

3. The equilibrium parameters are obtained from the low-frequency asymptotes of the FRFs. The first, second, third, etc., derivatives of the adsorption isotherm, corresponding to the defined steady state, are obtained (for nonisothermal cases, the derivatives with respect to temperature are obtained, along with the derivatives with respect to pressure or concentration). This result is significant for several reasons:

   (a) By using several isotherm derivatives instead of only the first one, it is possible to reconstruct the adsorption isotherm in a wide range of concentrations from fewer steady-state points. In Ref. [62] we have shown that a complete isotherm can be reconstructed from only six steady-state points, even for a complex isotherm with an inflection point.

   (b) The sign of the second isotherm derivative is directly related to the isotherm shape in the particular steady-state point. A change of this sign from one steady-state point to another indicates existence of an inflection point between those two points. In this way, the NLFR method is especially convenient for investigation of adsorption systems with complex isotherms (with one or more inflection points).

   (c) No assumption about the isotherm model is needed for estimation of the adsorption isotherm by the NLFR method. On the contrary, the isotherm model can be identified by analyzing the set of different isotherm derivatives. For example, it can easily be shown that for the popular Langmuir isotherm the following is valid [63]:

$$\frac{(\partial Q/\partial P)/(\partial^2 Q/\partial P^2)}{(\partial^2 Q/\partial P^2)/(\partial^3 Q/\partial P^3)} = 1.5$$

4. The kinetic data are estimated for a model that has previously been identified based on the second-order FRFs, and not for an assumed model. In addition:
    (a) Although the estimated kinetic parameters correspond to the used steady state, their concentration and temperature coefficients of different order can be obtained from the second- and higher-order FRFs, so the parameters can be used in a relatively wide range. No assumption regarding the form of the concentration and temperature dependence of the kinetic parameters is needed.
    (b) More complete kinetic data are obtained, compared with classical, linear models. For example, for the pore–surface diffusion mechanism, the effective pore and surface diffusion coefficients are obtained from the same experimental data.

All this makes the NLFR method attractive for investigation of adsorption systems. Nevertheless, we can also see some of its limitations:

1. The mathematics involved looks rather complex at first sight, which discourages many of the researchers working in the area of adsorption.
2. The experimental work is rather long and tedious. For each chosen steady state, a number of experiments have to be performed: for different frequencies and for at least two or three different input amplitudes. The proper choice of the frequencies and amplitudes is important.
3. In the investigation of adsorption systems, the input change is performed by using mechanical means (e.g., a bellow combined with a piston for the change of the reservoir volume, a valve for the change of the inlet flow rate, pumps for adjusting the inlet concentration into a chromatographic column, etc.). The upper limit of the frequencies obtained with these mechanical means is relatively low: in practice a maximum of 10 Hz could be obtained. There are no limits at the low-frequency end (used for estimation of the equilibrium data), but the experiments with low frequencies last very long. Also, producing sinusoidal input changes with mechanical means is rather difficult, especially for higher frequencies. On the other hand, use of other periodic input changes, for example, a square-wave, which can be produces easier results with experimental data which are much more difficult for analysis [63].
4. Reliable measurements of the periodically changing outputs are not always available, as they are highly dependent on the sensor dynamic characteristics.

B. Potentials for Application of the NLFR Method to Other Heterogeneous Systems

In this chapter, we limited our analysis to application of the NLFR method to investigation of adsorption systems. Nevertheless, the method can also be applied to other heterogeneous systems. It is important to clearly define the possible inputs that can be varied periodically, the measurable variables, and the generalized transfer functions (sets of FRFs) relating them, that can be used for description of the system. Also, it is important to recognize the potential mechanisms for the particular system and build a library of theoretical FRFs which can be used for comparison with experimental FRFs in order to identify the real model. Some work has already been done for membrane systems. We expect that the method could give good results for investigation of heterogeneous reaction systems. We especially see a great potential for application on investigation of electrochemical systems, as the inputs and outputs in such systems are usually electrical currents and voltages, which, on one hand, can be modulated, an on the other, can be measured, rather easily.

REFERENCES

1. Rouquerol, F., Rouquerol, J., and Sing, K., *Adsorption by Powders and Porous Solids*, Academic Press, London, 1999.
2. Guiochon, G., Shirazi, S.G., and Katti, A.M., *Fundamentals of Preparative and Nonlinear Chromatography*, Academic Press, London, 1994.
3. Do, D.D., *Adsorption Analysis: Equilibria and Kinetics*, Imperial College Press, London, 1998.
4. Katsanos, N.A., Thede R., and Roubani-Kalantzopolou, F., Diffusion, adsorption and catalytic studies by gas chromatography, *J. Chromatogr. A*, 795, 133–184, 1998.
5. Thielmann F. Introduction into the characterization of porous materials by gas chromatography, *J. Chromatogr. A*, 1037, 115–123, 2004.
6. Seidel-Morgenstern, A., Experimental determination of single solute and competitive adsorption isotherms, *J. Chromatogr. A*, 1037, 255–272, 2004.
7. Keller, J.U. and Robens, E., A note on sorption measuring instruments, *J. Therm. Anal. Calorimetry*, 71, 37–45, 2003.
8. Keller, J.U., Dreisbach, F., Rave, H., Staudt, R., and Tomalla, M., Measurement of gas mixture adsorption equilibria of natural gas compounds on microporous sorbents, *Adsorption*, 5, 199–214, 1999.
9. Do, D.D., Hierarchy of rate models for adsorption and desorption in bidispersed structured sorbents, *Chem. Eng. Sci.*, 45, 1373–1381, 1990.
10. Karger, J. and Ruthven, D.M., *Diffusion in Zeolites and Other Microporous Solids*, Wiley, New York, 1992.
11. Karger, J., Measurement of diffusion in zeolites — never ending challenge, *Adsorption*, 9, 29–35 2003.
12. Jobic, H., Karger, J., Krause, C., Brandani, S., Gunadi, A., Methivier, A., and Ehlers, G., Diffusivities of *n*-alkanes in 5 Å zeolite measured by neutron spin echo, pulsed-field gradient NMR and zero-length column techniques, *Adsorption*, 11, 403–407, 2005.
13. Naphtali, L.M. and Polinski, L.M., A novel technique for characterisation of adsorption rates on heterogeneous surfaces, *J. Phys. Chem.*, 67, 369–375, 1963.
14. Yasuda, Y. and Saeki, M., Kinetic details of a gas-surface systems by the frequency response method, *J. Phys. Chem.*, 82, 74–80, 1978.
15. Yasuda, Y., Determination of vapor diffusion coefficients in zeolite by the frequency response method, *J. Phys. Chem.*, 86, 1913–1917, 1982.
16. Yasuda, Y. and Sugasawa, G., A frequency response technique to study zeolitic diffusion of gases, *J. Catal.* 88, 530–534, 1984.
17. Yasuda, Y., Frequency response method for investigation of gas/surface dynamic phenomena, *Heterogeneous Chem. Rev.*, 1, 103–124, 1994.
18. Sun, L.M., Meunier, F., and Karger, J., On the heat effect in measurements of sorption kinetics by the frequency response method, *Chem. Eng. Sci.*, 48, 715–722, 1993.
19. Sun, L.M. and Bourdin, V., Measurement of intracrystalline diffusion by the frequency response method: analysis and interpretation of bimodal response curves, *Chem. Eng. Sci.*, 48, 3783–3793, 1993.
20. Sun, L.M., Meunier, F., Grenier, Ph., and Ruthven, D.M., Frequency response for nonisothermal adsorption in biporous pellets, *Chem. Eng. Sci.*, 49, 373–381, 1994.
21. Grenier, Ph., Malka-Edery, A., and Bourdin, V., A temperature frequency response method for adsorption kinetic measurements, *Adsorption*, 5, 135–143, 1999.
22. Malka-Edery, A., Abdullah, K., Grenier Ph., and Meunier F., Influence of traces of water on adsorption and diffusion of hydrocarbons in NaX zeolite, *Adsorption*, 7, 17–25, 2001.
23. Jordi, R.G. and Do, D.D., Frequency response analysis of sorption in zeolite crystals with finite intra-crystal reversible mass exchange, *J. Chem. Soc., Faraday Trans.*, 88, 2411–2419, 1992.
24. Jordi, R.G. and Do, D.D., Analysis of the frequency response method for sorption kinetics in bidispersed structured sorbents, *Chem. Eng. Sci.*, 48, 1103–1130, 1993.
25. Jordi, R.G. and Do, D.D., Analysis of the frequency response method applied to non-isothermal sorption studies, *Chem. Eng. Sci.*, 49, 957–979, 1994.
26. Sun, L.M. and Do, D.D., Dynamic study of a closed diffusion cell by using a frequency response method: single resonator, *J. Chem. Soc., Faraday Trans.*, 91, 1695–1705, 1995.
27. Sun, L.M. and Do, D.D., Frequency response analysis of a closed diffusion cell with two resonators, *Adsorption*, 2, 265–277, 1996.

28. Park, I.S., Petkovska, M., and Do, D.D., Frequency response of an adsorber with the modulation of the inlet molar flow-rate: Part I. A semi-batch adsorber, *Chem. Eng. Sci.*, 53, 819–832, 1998.

29. Park, I.S., Petkovska, M., and Do, D.D., Frequency response of an adsorber with the modulation of the inlet molar flow-rate: Part II. A continuous flow adsorber, *Chem. Eng. Sci.*, 53, 833–843, 1998.

30. Rees, L.V.C. and Shen, D., Frequency-response measurements of diffusion of xenon in silicalite-1. *J. Chem. Soc., Faraday Trans.*, 86, 3687–3692, 1990.

31. Shen, D.M. and Rees, L.V.C., Adsorption and diffusion of *n*-butane and 2-butane in silicalite-1, *Zeolites*, 11, 666–684, 1991.

32. Shen, D. and Rees, L.V.C., Frequency response study of single-file diffusion in Theta-1, *J. Chem. Soc., Faraday Trans.*, 90, 3017–3022, 1994.

33. Shen, D. and Rees, L.V.C., Analysis of bimodal frequency-response behaviour of *p*-xylene diffusion in silicalite-1, *J. Chem. Soc., Faraday Trans.*, 91, 2027–2033, 1995.

34. Shen, D. and Rees, L.V.C., Study of carbon dioxide diffusion in zeolites with one- and three-dimensional channel networks by MD simulations and FR methods, *J. Chem. Soc., Faraday Trans.*, 92, 487–491, 1996.

35. Song, L. and Rees, L.V.C., Adsorption and transport of *n*-hexane in silicalite-1 by the frequency response technique, *J. Chem. Soc., Faraday Trans.*, 93, 649–657, 1997.

36. Song, L.J., Sun, Z.L., and Rees, L.V.C., Experimental and molecular simulation studies of adsorption and diffusion of cyclic hydrocarbons in silicalite-1, *Microp. Mesop. Mater.*, 55, 21–49, 2002.

37. Valyon, J., Onyestyak, G., and Rees, L.V.C., A frequency-response study of the diffusion and sorption dynamics of ammonia in zeolites, *Langmuir*, 16, 1331–1336, 2000.

38. Onyestytak, G., Valyon, J., Bota, A., and Rees, L.V.C., Frequency-response evidence for parallel diffusion processes in the bimodal micropore system of an activated carbon, *Helvetica Chim. Acta*, 85, 2463–2468, 2002.

39. Onyestyak, G., Valyon, J., Hernadi, K., Kiricsi, I., and Rees, L.V.C., Equilibrium and dynamics of acetylene sorption in multiwalled carbon nanotubes, *Carbon*, 41, 1241–1248, 2003.

40. Park, I.S., Kwak C., and Hwang Y.G., Frequency response of continuous-flow adsorber for multicomponent system, *Kor. J. Chem. Eng.*, 17, 704–711, 2000.

41. Park, I.S., Kwak C., and Hwang Y.G., Frequency response of adsorption of a gas onto bidisperse pore-structured solid with modulation of inlet molar flow rate, *Kor. J. Chem. Eng.*, 18, 330–335, 2001.

42. Wang, Y., Sward, B.K., and LeVan, M.D., New frequency response method for measuring adsorption rates via pressure modulation: application to oxygen and nitrogen in a carbon molecular sieve, *Ind. Eng. Chem. Res.*, 42, 4213–4222, 2003.

43. Sward, B.K. and LeVan, M.D., Frequency response method for measuring mass transfer rates in adsorbents via pressure perturbation, *Adsorption*, 9, 37–54, 2003.

44. Gregorczyk, D.S. and Carta, G., Frequency response of liquid-phase adsorption on polymeric adsorbents, *Chem. Eng. Sci.*, 52, 1589–1608, 1997.

45. Boniface, H.A. and Ruthven, D.M., Chromatographic adsorption with sinusoidal input, *Chem. Eng. Sci.*, 40, 2053–2061, 1985.

46. Yasuda, Y., Frequency response method for study of a heterogeneous catalytic reaction of gases, *J. Phys. Chem.*, 93, 7185–7190, 1989.

47. Yasuda, Y., Frequency response method for study of kinetic details of a heterogeneous catalytic reaction of gases. 1. Theoretical treatment, *J. Phys. Chem.*, 97, 3314–3318, 1993.

48. Yasuda, Y. and Nomura, K., Frequency response method for study of kinetic details of a heterogeneous catalytic reaction of gases. 2. A methanol conversion to olefins, *J. Phys. Chem.*, 97, 3319–3323, 1993.

49. Yasuda, Y., Mizusawa, H., and Kamimura, T., Frequency response method for investigation of kinetic details of a heterogeneous catalyzed reaction of gases, *J. Phys. Chem., B*, 106, 6706–6712, 2002.

50. Ortelli, E.E., Wambach, J., and Wokaun, A., Methanol synthesis reactions over a CuZr-based catalyst investigated using periodic variations of reactant concentrations, *Appl. Catal. A Gen.*, 216, 227–241, 2001.

51. Garayhi, A.R. and Keil, F.J., Determination of kinetic expressions from the frequency response of a catalytic reactor — theoretical and experimental investigations, *Chem. Eng. Sci.*, 56, 1317–1325, 2001.

52. Weiner, D.D. and Spina, J.F., *Sinusoidal Analysis and Modeling of Weakly Nonlinear Circuits*, Van Nostrand Reinhold Company, New York, 1980.

53. Lee, G.M., Estimation of nonlinear system parameters using higher-order frequency response functions, *Mech. Syst. Signal Process.*, 11, 219–228, 1997.
54. Petkovska, M. and Do, D.D., Nonlinear frequency response of adsorption systems: general approach and special cases, in *Fundamentals of Adsorption*, Meunier, F., Ed., Elsevier, Paris, 1998, pp. 1189–1194.
55. Petkovska, M., Nonlinear frequency response of nonisothermal adsorption systems, *Nonlinear Dyn.*, 26, 351–370, 2001.
56. Petkovska, M. and Do, D.D., Use of higher order FRFs for identification of nonlinear adsorption kinetics: single mechanisms under isothermal conditions, *Nonlinear Dyn.*, 21, 353–376, 2000.
57. Petkovska, M., Non-linear frequency response of non-isothermal adsorption controlled by micropore diffusion with variable diffusivity, *J. Serbian Chem. Soc.*, 65, 939–961, 2000.
58. Petkovska, M., Nonlinear frequency response of isothermal adsorption controlled by pore–surface diffusion, *Bull. Chem. Technol. Macedonia*, 18, 149–160, 1999.
59. Petkovska, M. and Petkovska, L.T., Use of nonlinear frequency response for discriminating adsorption kinetics mechanisms resulting with bimodal characteristic functions, *Adsorption*, 9, 133–142, 2003.
60. Petkovska, M. and Do, D.D., Nonlinear frequency response of adsorption systems: isothermal batch and continuous flow adsorber, *Chem. Eng. Sci.*, 53, 3081–3097, 1998.
61. Petkovska, M., Application of nonlinear frequency response to adsorption systems with complex kinetic mechanisms, *Adsorption*, 11, 497–502, 2000.
62. Petkovska, M. and Seidel-Morgenstern, A., Nonlinear frequency response of a chromatographic column. Part I: Application to estimation of adsorption isotherms with inflection points, *Chem. Eng. Commun.*, 192, 1300–1333, 2005.
63. Petkovska, M., Živković, V., Kaufmann, J., and Seidel-Morgenstern, A., Estimation of adsorption isotherms using nonlinear frequency response of a chromatographic column — experimental study, The 2003 AIChE Annual Meeting, San Francisco, USA, November 16–21, 2003, Proceedings on CD.

12 Current Principles of Mixing

James Y. Oldshue
Oldshue Technologies International, Sarasota, Florida, USA

CONTENTS

I. INTRODUCTION

The fluid mixing process involves three different areas of viscosity which affects flow patterns and scale up, and two different scales within the fluid itself, macro-scale and micro-scale. Design questions come up when looking at the design and performance of mixing processes in a given volume. Considerations must be given to proper impeller and tank geometry as well as the proper speed and power for the impeller. Similar considerations come up when it is desired to scale up or down and this involves another set of mixing considerations.

If the fluid discharge from an impeller is measured with a device that has a high frequency response, one can track the velocity of the fluid as a function of time. The velocity at a given point of time can then be expressed as an average velocity v plus fluctuating component v'. Average velocities can be integrated across the discharge of the impeller and the pumping capacity normal to an arbitrary discharge plane can be calculated. This arbitrary discharge plane is often defined as the plane bounded by the boundaries of the impeller blade diameter and height. Because there is no casing, however, an additional 10–20% of flow typically can be considered as the primary flow of an impeller.

The velocity gradients between the average velocities operate only on larger particles. Typically, these larger size particles are greater than 1000 μm. This is not a proven definition, but it does give a feeling for the magnitudes involved. This defines macro-scale mixing. In the turbulent

329

region, these macro-scale fluctuations can also arise from the finite number of impeller blades passing a finite number of baffles. These set-up velocity fluctuations can also operate on the macro-scale.

Smaller particles primarily see only the fluctuating velocity component. When the particle size is much less than 100 μm, the turbulent properties of the fluid become important. This is the definition of the boundary size for micro-scale mixing.

All of the power applied by a mixer to a fluid through the impeller appears as heat. The conversion of power to heat is through viscous shear and is approximately 2500 Btu/hr/hp. Viscous shear is a major component of the phenomenon of micro-scale mixing. At 1-μm level, in fact, it does not matter what specific impeller design is used to apply the power.

Numerous experiments show that power per unit volume in the zone of the impeller (which is about 5% of the total tank volume) is about 100 times higher than the power per unit volume in the rest of the vessel. Making some reasonable assumptions about the fluid mechanics parameter, the root-mean-square (RMS) velocity fluctuation in the zone of the impeller appears to be 5–10 times higher than in the rest of the vessel. This conclusion has been verified by experimental measurements.

The ratio of the RMS velocity fluctuation to the average velocity in the impeller zone is about 50% with many open impellers. If the RMS velocity fluctuation is divided by the average velocity in the rest of the vessel, however, the ratio is in the order of 5–10%. This is also the level of RMS velocity fluctuation to the mean velocity in the pipeline flow. There are phenomena in micro-scale mixing that can occur in mixing tanks that do not occur in pipeline reactors. Whether this is good or bad depends upon the process requirements.

Figure 12.1 shows velocity versus time for three different impellers. The differences between the impellers are quite significant and can be important for mixing processes.

All three impellers are calculated for the same impeller flow, Q and same diameter. The A310 (Figure 12.2) draws the least power, and has the least velocity fluctuations. This gives the lowest micro-scale turbulence and shear rate.

The A200 (Figure 12.3) shows increased velocity fluctuations and draws more power.

The R100 (Figure 12.4) draws the most power and has the highest micro-scale shear rate.

The proper impeller should be used for each individual process requirement.

The velocity spectra in the axial direction for an axial flow impeller A200 is shown in Figure 12.5. A decibel correlation has been used in Figure 12.5 because of its well-known applicability in mathematical modeling as well as the practicality of putting many orders of magnitude of data on a reasonably sized chart. Other spectra of importance are the power spectra (the square of

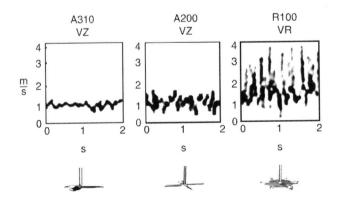

FIGURE 12.1 Typical velocity as a function of time for three different impellers, all at the same time total pumping capacity.

FIGURE 12.2 FluidFoil impeller (A310).

the velocity) and the Reynolds stress, (the product of the R and Z components) which is a measure of the momentum at a point.

The ultimate question is this: How do all of these phenomena apply to process design in mixing vessels? No one today is specifying mixers for industrial processes based on meeting criteria of this

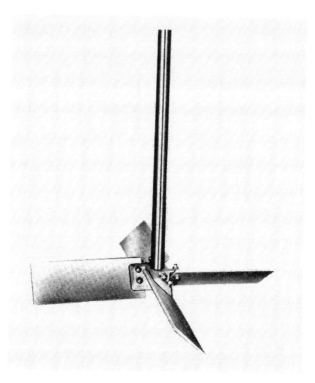

FIGURE 12.3 Typical axial flow turbine (A200).

FIGURE 12.4 Radial flow Rushton turbine (R100).

type. This is largely because processes are so complex that it is not possible to define the process requirements in terms of these fluid mechanics parameters. If the process results could be defined in terms of these parameters, sufficient information probably exists to permit the calculation of an approximate mixer design. It is important to continue studying fluid mechanics parameters in both mixing and pipeline reactors to establish what is required by different processes in fundamental terms.

Recently, one of the most practical results of these studies has been the ability to design pilot plant experiments (and, in many cases, plant-scale experiments) that can establish the sensitivity of process to macro-scale mixing variables (as a function of power, pumping capacity, impeller diameter, impeller tip speeds, and macro-scale shear rates) in contrast to micro-scale mixing variables (which are relative to power per unit volume, RMS velocity fluctuations, and some estimation of the size of the micro-scale eddies).

Another useful and interesting concept is the size of the eddies at which the power of an impeller is eventually dissipated. This concept utilizes the principles of isotropic turbulence developed by Kolmogorov [1]. The calculations assume some reasonable approach to the degree of isotropic turbulence, and the estimate do give some idea as to how far down in the micro-scale size the power

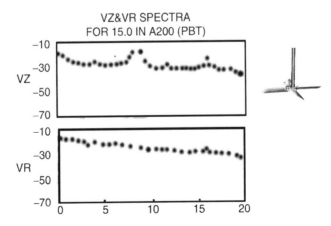

FIGURE 12.5 Typical velocity spectrum as a function of fluctuation frequency.

per unit volume can effectively reach.

$$L = \left(\frac{\nu^3}{\varepsilon}\right)^{\frac{1}{4}}$$

(12.1)

II. SCALE UP/SCALE DOWN

Two aspects of scale up frequently arise. One is building a model based on pilot plant studies that develop an understanding of the process variables for an existing full-scale mixing installation. The other is taking a new process and studying it in the pilot plant in such a way that pertinent scale up variables are worked out for a new mixing installation.

There are a few principles of scale up that can indicate what approach to take in either case. Using geometric similarity, the macro-scale variables can be summarized as follows [2]:

- Blend and circulation times in the large tank will be much longer than in the small tank [3].
- Maximum impeller zone shear rate will be higher in the larger tank, but the average impeller zone shear rate will be lower; therefore, there will be a much greater variation in shear rates in a full-scale tank than in a pilot unit.
- Reynolds numbers in the large tank will be higher, typically in the order of 5–25 times higher than those in a small tank.
- Large tanks tend to develop a recirculation pattern from the impeller through the tank back to the impeller. This results in a behavior similar to that exhibited by a number of tanks in a series. The net result is that the mean circulation time is increased over what would be predicted from the impeller pumping capacity. This also increases the standard deviation of the circulation times around the mean.
- Heat transfer is normally much more demanding on a large scale. The introduction of helical coils, vertical tubes, or other heat transfer devices causes an increased tendency for areas of low recirculation to exist.
- In gas–liquid systems, the tendency for an increase in the gas superficial velocity upon scale up can further increase the overall circulation time.

What about the micro-scale phenomena? These are dependent primarily on the energy dissipation per unit volume; however, one must also be concerned about energy spectra. In general, the energy dissipation per unit volume around the impeller is approximately 100 times higher than in the rest of the tank. This results in an RMS velocity fluctuation ratio to the average velocity in the order of 10:1 between the impeller zone and the rest of the tank.

Because each year there are thousands of specific processes that involve mixing, there will be at least hundreds of different situations requiring a somewhat different pilot plant approach. Unfortunately, no set of rules states how to carry out studies for any specific program, but here are a few guidelines that can help one carry out a pilot plant program.

- For any given process, take a qualitative look at the possible role of fluid shear stresses. Try to consider path-ways related to fluid shear stress that may affect the process. If there are none, then this extremely complex phenomena can be dismissed and the process design can be based on such things as uniformity, circulation time, blend time, or velocity specifications. This is often the case in the blending of miscible fluids and the suspension of solids.
- If fluid shear stresses are likely to be involved in obtaining a process result, then one must qualitatively look at the scale at which the shear stresses influence the result. If the particles, bubbles, droplets, or fluid clumps are in the order of 1000 μm or larger, the variables are macro-scale and average velocities at a point are the predominant variable.

When macro-scale variables are involved, every geometric design variable can affect the role of shear stresses. They can include such items as power, impeller speed, impeller diameter, impeller blade shape, impeller blade width or height, thickness of the material used to make the impeller, number of blades, impeller location, baffle location, and number of impellers.

Micro-scale variables are involved when the particles, droplets, baffles, or fluid clumps are in the order of 100 μm or less. In this case, the critical parameters usually are power per unit volume, distribution of power per unit volume between the impeller and the rest of the tank, RMS velocity fluctuation, energy spectra, dissipation length, the smallest micro-scale eddy size for the particular power level, and viscosity of the fluid.

- The overall circulating pattern, including the circulation time and the deviation of the circulation times, can never be neglected. No matter what else a mixer does, it must be able to circulate fluid throughout an entire vessel appropriately. If it cannot, then that mixer is not suitable for the tank being considered.
- Qualitative and, hopefully, quantitative estimates of how the process results will be measured in advance. The evaluations must allow one to establish the importance of the different steps in a process, such as gas–liquid mass transfer, chemical reaction rate, or heat transfer.
- It is seldom possible, either economically or time-wise, to study every potential mixing variable or to compare the performance of many impeller types. In many cases, a process needs a specific fluid regime that is relatively independent of the impeller type used to generate it. Because different impellers may require different geometries to achieve an optimum process combination, a random choice of only one diameter of each of two or more impeller types may not tell what is appropriate for the fluid regime ultimately required.
- Often, a pilot plant will operate in the viscous region while the commercial unit will operate in the transition region or alternatively, the pilot plant may be in the transition region and the commercial unit in the turbulent region. Some experience is required to estimate the difference in performance to be expected upon scale up.
- In general, it is not necessary to model Z/T ratios between pilot and commercial units.
- In order to make the pilot unit more like a commercial unit in macro-scale characteristics, the pilot unit impeller must be designed to lengthen the blend time and to increase the low maximum impeller zone shear rate. This will result in a greater range of shear rates than is normally found in a pilot unit.

III. EFFECT OF CIRCULATION TIME AND SPECTRUM OF SHEAR RATES ON 10 MIXING TECHNOLOGIES

A. GAS–LIQUID DISPERSIONS

The macro-scale shear rate change effects the bubble size distribution in various size tanks. As processes are scaled up, the linear, superficial gas velocity tends to be higher in the larger tank. This is the major contributor to the energy input of the gas stream. If the power per unit volume put in by the mixer remains relatively constant, then small tanks have a different ratio of mixing to gas expansion energy which affects the flow pattern and a variety of other fluid mechanics parameters. The large tank will tend to have a larger variation of the size distribution of bubbles than will the small tank.

This entire phenomenon is affected by the fact that the surface tension and viscosity vary all the way from that of a relatively pure liquid phase through all types of situations with dissolved chemicals, either electrolytes or non-electrolytes and other types of surface-active agents.

B. GAS–LIQUID MASS TRANSFER

If we are concerned with only the total volumetric mass transfer, then we can achieve very similar kGa values in large tanks and in small tanks [4,5].

Blend time enters into the picture primarily for other process steps immediately preceding or following the gas–liquid mass transfer step. Blending can play an important role in other steps in the total process in which gas–liquid mass transfer is only one component.

C. SOLIDS SUSPENSIONS AND DISPERSION

Solids suspension in general is not usually effected by blend time or shear rate changes in the relatively low to medium solids concentration in the range of 0–40% by weight. However, as solids become more concentrated, the effect of solids concentration on power required has a change in criteria from the settling velocity of the individual particles in the mixture to the apparent viscosity of the more concentrated slurry. This means that we enter into an area where the blending of non-Newtonian fluid regions the shear rates and circulation patterns to play a marked role (Figure 12.6).

The suspension of a single solid particle should depend primarily on the upward velocity at a given point, and also should be affected by the uniformity of this velocity profile across the entire tank cross-section. There are upward velocities in the tank and there must also be corresponding downward velocities.

Using a draft tube in the tank for solids suspension introduces another different set of variables. There are other relationships that are very much affected by scale up in this type of process, as shown in Figure 12.6. Different scale up problems exist whether the impeller is pumping up or down within the draft tube.

If the process involves the dispersion of solids in a liquid, then we may either be involved with breaking up agglomerates or possibly, physically breaking or shattering particles than have a low cohesive force between their components. Normally, we do not think of breaking up ionic bonds with the shear rates available in mixing machinery.

If we know the shear stress required to break up a particle, we can then determine the shear rate required from the machinery by various viscosities with the equation.

$$\text{Shear stress} = \text{viscosity (shear rate)} \tag{12.2}$$

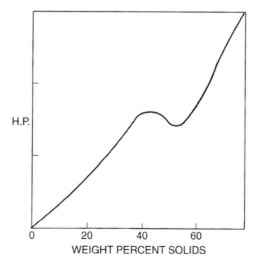

FIGURE 12.6 Effect of percent solids by weight and power required for uniformity and fluid motion.

The shear rate available from various types of mixing and dispersion devices is known approximately and also the range of viscosities in which they can operate. This makes the selection of the mixing equipment subject to calculation of the shear stress required for the viscosity to be used.

In the previous equation, it is assumed that there is 100% transmission of the shear rate in the shear stress. However, with the slurry viscosity determined essentially by the properties of the slurry, at high concentrations of slurries there is a slippage factor. Internal motion of particles in the fluids over and around each other can reduce the effective transmission of viscosity efficiencies from 100% down as low as 30%.

Animal cells in biotechnology do not normally have a tough skin like that of fungal cells and are very sensitive to mixing effects. Many approaches have been and are being tried to minimize the effect of increased shear rates on scale up, and these include encapsulating the organism in or on micro-particles and/or conditioning cells selectively to shear rates. In addition, traditional fermentation processes have maximum shear rate requirements in which cells become progressively more and more damaged until they become motile.

D. Solid–Liquid Mass Transfer

There is potentially a major effect of both shear rate and circulation time in these processes. The solids can either be inorganic, in which case we are looking at the slip velocity of the particle, and also whether we can break up agglomerates of particles which may enhance the mass transfer. When the particles become small enough, they tend to follow the flow pattern, so the slip velocity necessary to affect the mass transfer becomes less and less available.

What this shows is that, from the definition of off-bottom motion to complete uniformity, the effect of mixer power is much less, from going to on-bottom motion to off-bottom suspension. The initial increase in power causes more and more solids to become active in communication with the liquid and has a much greater mass transfer rate than that occurring above the power level for off-bottom suspension in which slip velocity between the particles of fluid is the major contributor.

Since, there may well be chemical or biological reactions happening on or in the solid phase, depending on the size of the process participants, macro- or micro-scale effects may or may not be appropriate to consider.

In the case of living organisms, their access to dissolved oxygen throughout the tank is of great concern. Large tanks in the fermentation industry often have a Z/T ratio of 2:1 to 4:1, and thus, top to bottom blending can be a major factor. Some biological particles are facultative and can adapt and re-establish their metabolism at different dissolved oxygen levels. Other organisms are irreversibly destroyed by sufficient exposure to low dissolved oxygen levels.

E. Liquid–Liquid Emulsions

Almost every shear rate parameter we have effects liquid–liquid emulsion formation. Some of the effects are dependent upon whether the emulsion is both dispersing and coallesing in the tank, or whether there are sufficient stabilizers present to maintain the smallest droplet size produced for long periods of time. Blend time and the standard deviation of circulation times effects the length of time it takes for a particle to be exposed to the various levels of shear work and thus the time it takes to achieve the ultimate small particle size desired.

When a large liquid droplet is broken up by shear stress, it tends to initially elongate into a "dumbbell" type of shape which determines the particle size of the two large droplets formed. Then, the neck in the center between the "dumbbell" may explode or shatter. This would give debris of particle sizes which can be quite different from the two major particles produced.

F. LIQUID–LIQUID EXTRACTION

If our main interest is in the total volumetric mass transfer between the liquids, the role of shear rate and blend time is relatively minor. However, if we are interested in the bubble-size distribution, and we often are because that affects the settling time of an emulsion in a multi-stage co-current or counter-current extraction process, then the change in macro- and micro-rates on scale up is a major factor. Blend time and circulation time are usually not a major factor on scale up.

G. BLENDING

If the blending process is between two or more fluids with relatively low viscosity such that the blending is not effected by fluid shear rates, then the difference in blend time and circulation between small and large tanks is the only factor involved. However, if the blending involves wide disparities in the density of viscosity and surface tension between the various phases, then a certain level of shear rate may be required before blending can proceed to its ultimate degree of uniformity.

The role of viscosity is a major factor in going from the turbulent regime, through the transition region, into the viscous regime and the change in the role of energy dissipation discussed previously. The role of non-Newtonian viscosities comes into the picture very strongly since that tends to markedly change the type of influence of impellers and determines the appropriate geometry that is involved.

Another factor here is that the relative increase in Reynolds number on scale up. This means that we could have pilot plants running in the turbulent region as well as the plant. We could have the pilot plant running in the transition region and the plant in the turbulent, and the pilot plant could be in the viscous region while the plant is in the transition region. There is no apparent way to prevent this Reynolds number change upon scale up in marked way. In reviewing the qualitative flow pattern in a pilot scale system, it should be realized that the flow pattern in the large tank will be at apparently much lower viscosity and therefore, at a much higher Reynolds number than is being observed in the pilot plant. This means that the role of tank shape, D/T ratio, baffles, and impeller locations can be based on different criteria in the plant size unit than in the pilot size unit under observation.

H. CHEMICAL REACTIONS

Chemical reactions are influenced by the uniformity of concentration both at the feed point and in the rest of the tank and can be markedly affected by the change in overall blend time and circulation time as well as the micro-scale environment. It is possible to keep the ratio between the power per unit volume at the impeller and in the rest of the tank relatively similar on scale up, but much detail needs to be considered when talking about the reaction conditions, particularly where it involves selectivity. This means reactions that can take different paths depending upon chemistry and fluid mechanics, and is a major consideration in what should be examined. The method of introducing the reagent stream can be projected in several different ways depending upon the geometry of the impeller and the feed system.

I. FLUID MOTION

Sometimes the specification is purely in terms of pumping capacity. Obviously, the change in volume and velocity relationships depends upon the size of the two- and three-dimensional area or volume involved. The impeller flow is treated in a head/flow concept and the head required for various types of mixing systems can be calculated or estimated.

J. Heat Transfer

In general, the fluid mechanics of the film on the mixer side of the heat transfer surface is a function of what happens at that surface rather than the fluid mechanics going on around the impeller zone. The impeller provides largely flow across and adjacent to the heat transfer surface and that is the major consideration of the heat transfer result obtained. Many of the correlations are in terms of traditional dimensionless groups in heat transfer, while the impeller performance is often expressed as the impeller Reynolds number.

IV. COMPUTATIONAL FLUID DYNAMICS

There are several software programs that are available to model flow patterns of mixing tanks. They allow the prediction of flow patterns based on certain boundary conditions. The most reliable models use accurate fluid mechanics data generated for the impellers in question and a reasonable number of modeling cells to give the overall tank flow pattern. These flow patterns can give velocities, streamlines, and localized changes in mixing variables based on doing certain things to the mixing process. These programs can model velocity, shear rates, and kinetic energy, but probably cannot adapt to the actual chemistry of diffusion or mass transfer kinetics of actual industrial process at the present time.

Relatively uncomplicated transparent tank studies with tracer fluids or particles can give a similar feel for the overall flow pattern. It is important that a careful balance be made between the time and expense of calculating these flow patterns with computational fluid dynamics compared to their applicability to an actual industrial process. The future of computational fluid dynamics appears very encouraging and a reasonable amount of time and effort placed in this regard can yield immediate results as well as potential for future process evaluation.

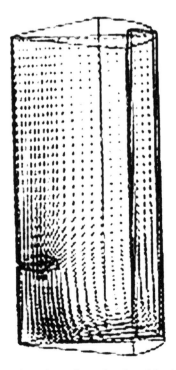

FIGURE 12.7 Typical velocity pattern for a three-dimensional model using computational fluid dynamics for an axial flow impeller (A310).

FIGURE 12.8 Typical contours of kinetic energy of turbulence using a three-dimensional model with computational fluid dynamics of an axial flow impeller (A310).

FIGURE 12.9 Typical particle trajectory using an axial flow impeller (A310) with 100 μm particles using computational fluid dynamics.

Figures 12.7–12.9 show some approaches. Figure 12.7 shows velocity vectors for an A310 impeller. Figure 12.8 shows contours of kinetic energy of turbulence. Figure 12.9 uses a particle trajectory approach with neutral buoyancy particles.

Numerical fluid mechanics can define many of the fluid mechanics parameters for an overall reactor system. Many of the models break up the mixing tank into small microcells. Suitable material and mass transfer balances between these cells throughout the reactor are then made. This can involve long and massive computational requirements. Programs are available that can give reasonably acceptable model of experimental data taken in mixing vessels. Modeling the three-dimensional aspect of a flow pattern in a mixing tank can require a large amount of computing power.

NOMENCLATURE

N	Impeller speed
D	Impeller diameter
T	Tank diameter
Z	Liquid level
P/V	Power per unit volume
SR	Solidity ratio obtained by dividing the projected area of the impeller blades by the area of a disk circumscribing the impeller blades
N_p	Power number
N_q	Flow number
H	Velocity head ($v^2/2G$)
P	Power
L	Length scale
v	Fluid velocity
v'	Fluid velocity fluctuation
$K_g a$	Gas–liquid mass transfer
K, a	Liquid–liquid mass transfer coefficient
k_s	Liquid–solid mass transfer coefficient
$\in$	Energy dissipation rate
μ	Dynamic viscosity
$\sim v$	Average fluid velocity
ν	Kinematic viscosity

REFERENCES

1. Levich, V., *Physio-Chemical Hydrodynamics*, Prentice-Hall, New Jersey, 1962.
2. Oldshue, J.Y., *Mixing'89, Chem. Eng. Progr.*, Vol. 5, 1989, pp. 33–42.
3. Middleton, J.C., Proceedings of the Third European Conference on Mixing, Vol. 4, 1989, BHRA, pp. 15–36.
4. Oldshue, J.Y., Post, T.A., and Weetman, R.J., Comparison of mass transfer characteristics of radial and axial flow impellers, BHRA *Proceedings of the Sixth European Conference on Mixing*, Vol. 5, 1988.
5. Neinow, A.W., Buckland, B., and Weetman, R.J., *Proceedings of the Mixing XII Research Conference*, Potosi, MO, Vol. 8, 1989.

13 Quantification of Visual Information

Predrag B. Jovanić
Institute for Technology of Nuclear and Other Mineral Raw Materials,
Belgrade, Serbia, Serbia and Montenegro

CONTENTS

I. INTRODUCTION

Contemporary society demands a culture based on the images. Endless quantities of information must be collected in purpose to exploit the natural resources, increase the productivity, keep the order, lead the war, or give jobs to the birocracy. Double properties of visual information, to make reality subjective or objective, are ideal for these purposes. Sets of visual information, called images, define reality in two very important ways of existence of the contemporary society: as spectacle and as the subject of control (process, technology, phenomenon, or system). Image production creates leading ideology. Real changes are substituted with changes of the images. Freedom to consume the huge amounts of the information, mostly visual, may be equalized with the "freedom". Decreasing the freedom of choice to free economical consumption demands, endless production, and consumption of visual information replaces the freedom. Since we produce images, we use them by creating the need for more information and this goes toward eternity. Sets of visual information are not a kind of constant treasure, they exist everywhere and the only thing we have to do is to memorize them using a suitable system. An image could induce certain stimulations in a person in a manner similar to that of a desire, and such stimulations are not clearly definable. Since the sets of the visual information are endless, each project from this field swallows itself. Author's attempts to fix the worn-out meaning of reality, just makes the very attempt worn-out. Our bitter feeling of permanent motions and instabilities related to the real world sharpened from the moment the means for fixing fluctuations in visual information were available to us. We spare the images on ever-increasing rate. As Balzac suspected that cameras spend the body parts, the images spend reality. Sets of visual information are at the same time antidotes and the illness, means to acquire the reality and to overcome it. Power of visual information blurred our understanding of reality, in that way we do not think about our experience through differences between images or differences between copy and original. Platoon compared images to the shadows, transitional entities with minimum information, immaterial and weak followers of real things that make them. The power of visual informations comes from the fact that they are themselves the material reality and a treasury of information. They remain as the consequences of entities' emission, powerful enough to turn-over the reality. This is the way an image becomes the shadow of reality. If there is a better way to acquire the entities of visual information, then we will need not only the ecology of real things but also the ecology of visual information.

Information archived as the visual information represents the way to shorten the reality or exclude complicated analog to digital conversions and vice versa [1,2]. Man could not have the reality, he could only possess the existent moment, but still he could possess the past known as a memory. Image means immediate access to the reality, and at the same time, creation of reality distance. Possessing the world in the form of images, means, once again experiencing the virtual and distant reality. The present-day concept of "faith" is not related to reality but to its visual perception — image or illusion takes the place of reality. Foyerbah's vision of the twentieth century becomes the widely accepted diagnosis of how one society becomes modern, when one of its main activities becomes production and consumption of images, which are glorious replacements for experience. If we would be able to tell everything using words, then we will not need the systems for visual information acquisition and storage. Nowadays, everything could be image. The era of visual perception is taking place, more and more with inevitable tendency to replace the era of linguistical confusion [2].

What does quantification of visual information mean? The main fact lies in the difference between the art and the science. The purpose of the science is to find the quantitative relations between the input and the output, from the analyzed entity. Art does not need any relations. When we are talking about artistic images or paintings, we have binary relation to them, whether we like them or not. Most of us once in a lifetime said "I don't know what art is but I like it (or not)." The only objective measure of quality of artistic images is time, or better to say, beauty and innocence have the only one enemy — time. The dark side of the technical disciplines and research is that they always must create quantitative measures about all the entities that they analyze in real time and in the shortest possible period of time, even if we are talking about some complex set of visual information as the artistic painting.

For the moment, take a look at the author's favorite painting, the famous "Mane's Breakfast on the grass" which is exposed at "Jeu de Pomme" in Paris (Figure 13.1).

At the moment of creation, almost one century ago, this artwork was so revolutionary that it created a scandal. At that time, police would have stopped this kind of picnic and the spectators assumed that Mane presented a real event. Many years later, one of the researchers found the origin of figures presented at the painting. Models for figures on Mane's painting were figures of classical gods presented on Rafael's relief and relief of river gods [1] (Figure 13.2 and Figure 13.3).

When the attention was focused, the connection was obvious, but we miss it because Mane was not copied or presented Rafael's composition. He simply borrowed main lines from the characters and transferred them in his time. In the present analysis, we used the oldest method of the quantification of visual information — comparison of original with the set of other reliefs or paintings and standards, which we defined as the source, permanent or persistent. Standards and the analyzed paintings are not the same; only some forms are present, which exists on the standards. Basic forms of older Rafael's figures are upgraded in the new environment. If we go further in analysis of the paintings, we will find that even Rafael's reliefs have source in the older Roman art or even further in the so-called relief of Roman Gods. So, we can make the chain of images Mane, Rafael, and Roman Gods; a connection which starts somewhere in the blurred past; thinking on Breakfast will influence some other contemporary artistic works too. This analysis shows that

FIGURE 13.1 Breakfast on the grass.

the different presentations have almost the same basic source, or basic cell, which takes us to the story of fractals [3].

If the wisdom that "man is not the island" is true, the same is valid for the sets of visual information. All the chains of images make a carpet in which each set of visual information has a specific position, which in the art we call "tradition," and in the science we call "genesis." In the art, we use quantification of visual information to measure and define originality of piece, and in the science, we use quantification of visual information to allow reproductivity of the best created entity. But, wherever we use, the methods of quantification are always the same. We must have

FIGURE 13.2 Rafael's relief.

FIGURE 13.3 Relief of River Gods.

the system of objective and precious measure for elements or structural entities. All is done in purpose to obtain information that could be used for comparison and reproduction.

II. WHAT DO WE SEE?

The human eye is a good optical device, with theoretical resolution about 10^{-4} rad or 60 arcseconds. Most of the information comes as the visual one when compared with the other sensual information such as taste, touch, smell, or hearing. It is estimated that over the 95% of all acquired information is visual.

We will concentrate our attention on the visual information that is obtained from the experiments or from the scientific observations of entities or objects. Visual part of electromagnetic spectrum is almost infinitesimally narrow (400–700 nm) when compared with the whole radiation spectrum. Hence, for visualization of all scientifically observable universes, we have a small part of spectrum including one property that is named color. There are several ways to describe the sensitivity of the human visual system. To begin, let us assume that a homogeneous region in an image has an intensity as a function of wavelength (color) given by $I(\lambda)$; further on, let us assume that $I(\lambda) = I_0$, where "I_0" is fix value. Wavelength sensitivity of human observer or the perceived intensity as a function of λ. The spectral sensitivity, for the "typical observer" is shown in Figure 13.4.

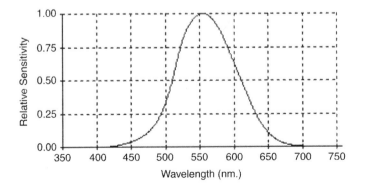

FIGURE 13.4 Spectral sensitivity of the typical human observer.

If the intensity (brightness) I_0 is allowed to vary, then, to a good approximation, the visual response, R, is proportional to the logarithm of the intensity, which is called stimulus sensitivity. This is known as the Weber–Fechner law:

$$R = \log(I_0)$$

The implications of this statement are easy to illustrate. Equal *perceptual* steps in brightness, $\Delta R = k$, require that the physical brightness (the stimulus) increases exponentially. This is illustrated in Figure 13.5.

The *Mach band effect* is visible in Figure 13.5. Although the physical brightness is constant across each vertical stripe, the human observer perceives an "undershoot" and "overshoot" at a physically explainable step edge in brightness. Thus, just before the step, we see a slight undershoot in brightness when compared with the true physical value. After the step, we see a slight overshoot in brightness when compared with the true physical value. The total effect is one of an increased, local, *percept* contrast at a step edge in brightness. Human color perception is an extremely

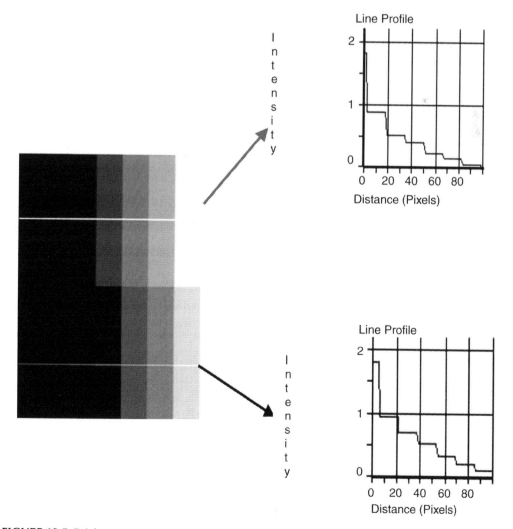

FIGURE 13.5 Brightness step actual brightnesses plus interpolated values.

complex topic. As such, we can only present a brief introduction here. The physical perception of color is based upon three color pigments in the retina [4].

A. COLOR

On the basis of psychophysical measurements, standard curves have been adopted by the Commission Internationale de l'Eclairage (CIE) as the sensitivity curves for the typical or standard observer for the three pigments, $\bar{x}(\lambda)$, $\bar{y}(\lambda)$, and $\bar{z}(\lambda)$. These are not the *actual* pigment absorption characteristics found in the "standard" human retina but rather sensitivity curves derived from actual data. For an arbitrary homogeneous region in an image that has intensity as a function of wavelength (color) given by $I(\lambda)$, the three responses are called the *tristimulus values*:

$$X = \int_0^\infty I(\lambda)\bar{x}(\lambda)\,d\lambda \quad Y = \int_0^\infty I(\lambda)\bar{y}(\lambda)\,d\lambda \quad Z = \int_0^\infty I(\lambda)\bar{z}(\lambda)\,d\lambda$$

The *chromaticity coordinates*, which describe the perceived color information, are defined as

$$x = \frac{X}{X+Y+Z} \quad y = \frac{Y}{X+Y+Z} \quad z = 1 - (x+y)$$

The red chromaticity coordinate is given by x, and the green chromaticity coordinate is given by y. The tristimulus values are linear in $I(\lambda)$ and thus the absolute intensity information has been lost in the calculation of the chromaticity coordinates $\{x, y\}$. All color distributions, $I(\lambda)$, that appear to an observer as having the same color will have the same chromaticity coordinates.

If we use a tunable source of pure color (such as a dye laser), then the intensity can be modeled as $I(\lambda) = d(\lambda - \lambda_0)$ with $d(*)$ as the impulse function. The collection of chromaticity coordinates $\{x, y\}$ that will be generated by varying λ_0 gives the *CIE chromaticity triangle* as shown in Figure 13.6.

The description of color on the basis of chromaticity coordinates not only permits an analysis of color but also provides a synthesis technique. Using a mixture of two color sources, it is possible to

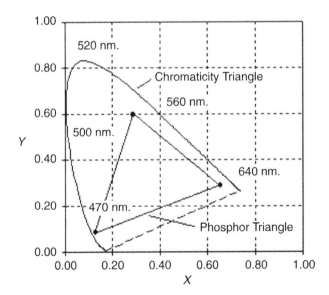

FIGURE 13.6 Chromaticity diagram containing the CIE chromaticity triangle associated with pure spectral colors.

generate any of the colors along the line connecting their respective chromaticity coordinates. Since we cannot have a negative number of photons, the mixing coefficients must be positive.

The formulas for converting from the tristimulus values (X, Y, Z) to the well-known colors (R, G, B) are widely used for various applications. As long as the position of a desired color (X, Y, Z) is inside the phosphor triangle as shown in Figure 13.6, the values of R, G, and B will be positive as computed. It is incorrect to assume that a small displacement anywhere in the chromaticity diagram (Figure 13.6) will produce a proportionally small change in the *perceived* color.

B. VISUALIZATION

If we look at the simplified schematic diagram presented in Figure 13.7, which presents the whole electromagnetic spectrum, we will notice that direct visual observable universe is very narrow.

The electromagnetic part of the radiation spectrum, except visible, must be transformed into a suitable form, to be presented to our eyes. So, we came to the first step of visualization, devices for transformation of electromagnetic radiation to form, which could be processed in purpose to obtain an image or a visual information. This is the wide field of detectors, which is directly connected with technological developments in the field of materials and electronics. First detectors and the information storage devices were the photographic plates or films, which were totally dependent on the chemical reactions, between incoming radiation and sensitive layer. After that, the main detectors, over the years, were photomultiplication tubes, based on the photoelectric effect, transforming the incoming radiation to the electric signal, which could be further processed.

The breakthrough came with the solid-state detectors (CCD and CID), which could convert two-dimensional distribution of incoming radiation to the two-dimensional array of electric signals, for processing. As we see, most of the today's detectors convert incoming radiation into the electric signal, which could be, with additional conversions, presented in the visual form. One property of detectors which must be known is the relation between incoming radiation and the output electric signal. Preferably, it must be linear. There are many other properties of detectors and processing devices which play an important role in the signal processing, but they are beyond the scope of this chapter.

We can now speak about two visualization processes. First, direct visualization received from strong or perceptive fields, which could be seen through our unaided eyes or converted into the other signals using the suitable detector. Information obtained from perceptive fields could be directly stored into the image using the chemical conversion of incoming radiation.

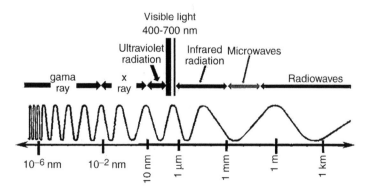

FIGURE 13.7 Electro-magnetic radiation spectra.

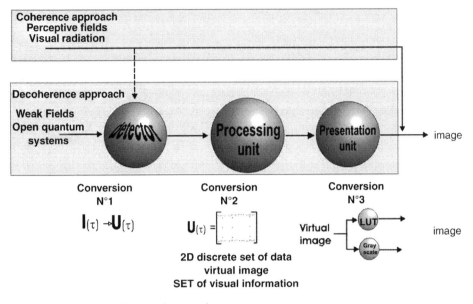

FIGURE 13.8 Generation of the visual sets — images.

Second, indirect visualization received from weak fields or all other radiation from the spectrum that must undergo several conversions to be presented in a visual form. Conversions are presented schematically in Figure 13.8.

Processing unit assumes two-dimensional discrete sets of quantified detector signal intensities, which we can call virtual image, or sets of visual information. At this point, we still do not see information. This is where the data manipulation algorithms are applied in purpose to improve the signal-to-noise ratio of the obtained data. Human vision can extract several different kinds of information in the form of images and much of the processing that takes place has been optimized by evaluation and experience. Some of them are transformations of obtained data sets in some other spaces, where original data never exist, but they help us in our goal to extract the valid data sets for analysis. We always know the return route through the Alice mirror. After this conversion, called image processing, we can apply two other conversions or processes. One is the visual presentation of the obtained and processed data, and the other is their quantification.

Visualization is the process of presenting the obtained data in the visual part of spectrum or in the visual form. One more conversion of original data must be done. Contemporary visualization algorithms are based on the look-up tables (LUT) or gray scale methodology. No matter which of them we apply, the principle is the same. Range of obtained data is related either to the part or complete visible spectrum if we apply the LUT or the part or whole gray scale methodology.

In the first case, we obtain the color image, and in the second, we have the black and white image of the obtained data. In both cases, two-dimensional intensity distribution of incoming radiation to the detector is presented as image with the related distribution of colors, valers, or levels of gray. We see the image of something that is far from the image in common sense. It is just the visual presentation of incoming entity (radiation or interaction) to the detector and its response in the form which is recognizable by the processing unit.

The best example is the visual presentation of signals obtained by the radio telescope observations. The whole instrument is the detector. Rare people, like Jody Foster in Carl Sagans

(a)   (b)

FIGURE 13.9 (a) Listening signals from the very large area of telescopes in New Mexico and (b) the visual presentation, "image," of radio signals. (From Cosmos Warner Bross. and "NRAO/AUI/NSF." With permission.)

Contact, are listening to cosmic radio signals (Figure 13.9a). Even in that case, we have the same principle of the output generation, while in this case, we spread received radio spectrum to the audio spectrum to "hear" the signals. In most other cases, then Hollywood, researchers made visual presentations of received radio signals (Figure 13.9b) by connecting radio signals with the adequate LUT table for this purpose. The results are seen in Figure 13.9b, presenting the radio intensities distribution from quasars, captured by the VLA radio telescopes. So, we see the radio waves, their intensity presentation is in the selected range of gray scale.

At this point, it is better to speak about sets of visual information rather than about the images. First of all, these sets have some random elements and subjectivity because the researcher is choosing the LUT table or range of gray scale, by a subjective set of criteria. So, why do we want to present visually all of the obtained data by experiments or observations, when visual presentations could be different even for the same data? The proverb, "a picture is worth a thousand words," is always used as the illustration of apparently rich information content of images and it is wrong in many ways. Typically, nowadays, visual presentation or image of any kind occupies the space of several millions of words. As the communication tool for information exchange is very inefficient. There is a little reason to believe that two researchers will extract the same information from the same image without some additional information, which creates context for interpretation. We always must have the key how to read visual presentation to obtain the closest information as its author. There is a chance that some of the other kinds of information are either ignored or suppressed and are not normally observed. Seeing, but not observing, is the common mistake of many research tasks. Contemporary research devices allow us to see practically any object or process. To observe them, we must quantify the obtained data in any form. Observing means that we have to see visual information on different ways, from the different points of view, and extract the useful data for our goal. Set of visual information is the frizzed moment of the existence of an entity. Observing such information means that researcher must find the path of the process that creates the set. At this point, we could make an analogy with the holograms. We presume that entropy is just the measure of total information capacity of some system, so all information connected to the analyzed phenomenon could be stored into the two-dimensional boundary regions, such as the holographic image. In reverse process, when the laser light passes through the holographic image, the image of the object which creates the hologram is formed.

The scientific sets of visual information are also two-dimensional visual presentations of the obtained data. Acquired data have all the information about the process that creates actual set.

The problem is the reverse process, how to extract them and put in some logical sense, called models or theories. Modern digital technology has made it possible to manipulate multidimensional signals with systems that range from simple digital circuits to advanced parallel computers. When talking about sets of visual information, the process of their quantification for further analysis or observation can be set as follows:

Data acquisition   Radiation in $\rightarrow$ processable signal out
Data processing and presentation   Signal in $\rightarrow$ image out
Image processing   Image in $\rightarrow$ enhanced image out
Quantification of visual information   Image in $\rightarrow$ measured data sets out
Analysis   Data sets in $\rightarrow$ high-level observations out

III. DATA ACQUISITION

In the days of Internet, analog-to-digital (A/D) converters, and all other miracles of nanotechnology, the problems of data acquisition are not focused by the researchers. Sometimes, the fact that data will be acquired by some devices is accepted like an axiom.

Schematic scale and the corresponding instruments or imaging devices, which create the visual information, are presented in Figure 13.10. This could be named as the quest for the resolution. Only the small fraction of the scale is directly observable by the human eye. In all other cases, the effects or the interactions are used to create the visual information. Even the optical microscope use photon interactions with material to create the image. Electron microscopes, transmission and scanning, use different kinds of electron interactions with material to obtain information.

For the dimensions lower than 10^{-9} m, tunneling effect is used to obtain the information. Atomic force microscope uses the measure of force to obtain information about the material structure. Only optical microscope without the cameras gives real visual image of the specimen. All other instruments gives the visual presentation of the specific interaction, which takes place in the specimen under investigation. When the information have to be collected from the scales lower than 10^{-12} m, only the induced "visible" macroeffects can be used for visualization of such structures. Accelerators, the ultimate microscopes, combine high-energy particle beams and

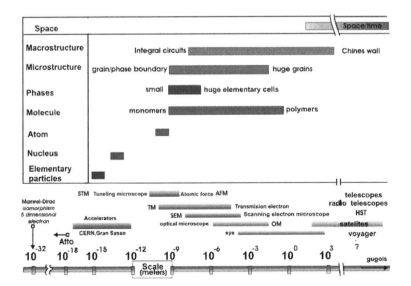

FIGURE 13.10 Scale and imaging devices.

their interactions with target materials to see the subatomic particles. The particles are not seen by themselves. What we see is the particle–detector medium interaction in the suitable detector. It is interesting to note that the resolution demands of the contemporary research request bigger and more sophisticated observation devices, whether we observe the subatomic particles or the universe. The size does matter.

A. QUANTIFICATION OF VISUAL INFORMATION

The goal of human vision is recognition. Whatever we are searching for, the first thing that attracts our attention in an image is something familiar. To be recognized, an object or objects must have a name, some label that our consciousness can assign. The label is mental model of the object, which can be expressed either in words, images, or other forms, which captures the important characteristics of the object [4]. The basic technique that is in the root of human vision is comparison. Nothing in images is measured by the eye and mind. There are no rulers and standards in our head. Machine comparison works at the same way, but it takes time. If the stored object is known, then corresponding model consists of a set of object properties, which may be compared. Scale or comparison, the set of rules must be known in advance (Figure 13.11).

The basic process of quantification of visual information is the process of the object extraction or object identification in the visual set. The set of visual information could contain the color, patterns, and objects, which have to be quantified. Objects in the visual set could be defined as the set of data with the same properties or with the properties which fall in the defined range of values. Patterns could be defined as the traces of the structures frozen in time. If the LUT methodology for the data presentation is used, the obtained visual presentation already has defined objects by the selected LUT table. Standard processes for the object or pattern extraction from the set of visual information are thresholding and segmentation and their cosmeceutical derivations. In the analysis of the objects and patterns in sets of visual information, it is essential that we can

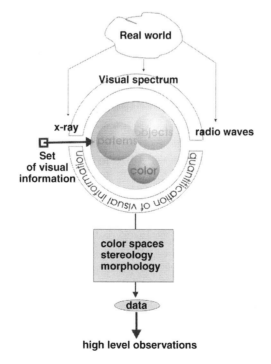

FIGURE 13.11 Schematic diagram of quantification of visual information.

distinguish the objects of interest and the rest or background that is not of interest for the current analysis. It is important to understand that there is no universally applicable segmentation technique that will work for all images; none of segmentation techniques is perfect.

The output is the label "object" or "background" which, due to its dichotomous nature, can be represented as a Boolean variable "1" or "0". In principle, the test condition could be based upon any visual property of data set. The central question in thresholding then becomes "How do we choose the threshold?" While there is no universal procedure for threshold selection that will work on all images, there are a variety of alternatives. A variety of techniques have been devised to automatically choose a threshold.

B. MORPHOLOGICAL AND STEREOLOGICAL PRINCIPLES

The generic term *object* or *pattern* that is used to describe the physical elements, which make up any structure comes from the concept of a *phase* borrowed from classical thermodynamics. Each phase in a structure is a set of three-dimensional objects. To belong to the same phase, the objects must have the same internal physical properties, which are presented with the same properties in the set of visual information. Commonly, this means that the objects of one phase have the same chemical makeup and the same atomic, molecular, crystal, or biological structure. The collection of parts of such structure that belong to the same object is one example of an *object set*. Structures are space filling, nonregular, nonrandom arrangements of the object sets of objects in three-dimensional space. A structure may be a single-phase tessellation consisting of four object sets, polyhedral cells that have faces, edges, and vertices arranged to fill the three-dimensional space. Alternatively, a structure may consist of two phases, two distinguishable object sets, either or both of which may be cell structures that fit together with precision to fill up the space occupied by the structure. Structures frequently consist of several distinguishable phases, each contributing to the total collection of object sets in the system. In some cases, voids or porosity may be present and is also treated as a measurable phase.

In the description of a structure, it may be useful to focus on the properties of a particular object such as the phase and its boundaries. Each member of the object has its own collection of *geometric* properties. Properties of the whole collection of objects in the set are called *global* properties. Examples include the total volume, boundary surface area, or number of particles in the set. A more complete description of the structure might incorporate information about the spatial distribution of the phase in the context of other object sets in the structure.

To facilitate the organization structure characterization, three levels of structure characterization could be introduced: the *qualitative* structural state, the *quantitative* structural state, and the *topographic* structural state.

The first of these levels of description is a list of the classes of object that exist in the structure. The second level makes the description quantitative by assigning numerical values to geometric properties that are appropriate to each object set. The third level of description deals with no uniformities in the spatial distribution, which may exist in the structure.

1. Qualitative Microstructural State

Structures consist not only of the three-dimensional objects that make up the objects and patterns in the structure but also of two-dimensional surfaces, one-dimensional lines, and zero-dimensional points associated with these objects. The qualitative microstructural state is a list of all of the classes of object sets that are found in the structure. Surfaces, edges, and points arise from the incidence of three-dimensional particles or cells. For example, the incidence of two cells in space forms a surface. The kind of surface is made explicit by reporting the class of the two cells that form it. The qualitative microstructural state can be assessed or inferred by inspection of a sufficient number of fields to represent the structure. In many cases, one field will be enough for this qualitative purpose. In making this assessment, keep in mind that the process of sectioning the structure reduces the dimensions of the objects by one. Sections through three-dimensional cells of volume objects appear as two-dimensional areas on the section. Sections through two-dimensional surfaces appear as one-dimensional lines or curves on the section. Lineal objects in space appear only as points of intersection with the sectioning plane; triple lines in a cell structure intersect as triple points (Figure 13.12).

Characterization of any given microstructure should begin with an explicit list of the objects it contains [4].

2. Quantitative Microstructural State

To each of the classes of objects listed in the previous section, one or more geometric properties may be associated. Use of stereology to estimate values for these properties constitutes specification of the quantitative structural state. Properties that are stereologically accessible have the useful attribute and have unambiguous meaning for object sets of arbitrary shape or complexity. These geometric properties may be associated with individual objects in the visual set or as global properties of the whole objects in the set.

Stereological measures are generally related to these global properties of the full set of objects. They are usually reported in normalized units as the value of the property per unit volume of structure. The volume fraction occupied by an object set may be quantitatively estimated through the point count, the most used relationship in stereology (Figure 13.13).

Surfaces that exist as two-dimensional object sets in the three-dimensional microstructure possess the property area. Each object in a surface or interfacial object set in the structure has a value of its area. The object set as a whole has a global value of its surface area. The concept of the area of a surface has unambiguous meaning for object sets of arbitrary size, shape, size distribution, or complexity. The surface that separates particles from the gas phase in a stack of powder has an area. The normalized global property measured stereologically is the surface area per unit volume, sometimes called the surface area density of the two-dimensional object set. Lines or space curves that exist in the three-dimensional microstructure possess a length. Individual line segments, such as edges in a cell network, have a length. The full object set has a global value of its length, the length per unit volume in stereological measurements. Many objects, such as fibers in

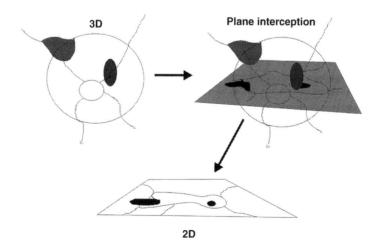

FIGURE 13.12 The dimension of each object is reduced by one when a three-dimensional structure is sectioned by a plane. (Russ, C.J. and DeHoff, R.T., *Practical Stereology*, 2nd ed., Plenum Press, New York, 2002, pp. 29. With permission.)

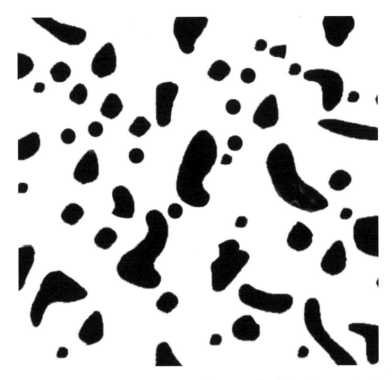

FIGURE 13.13 A simple two-phase microstructure provides a sample of the volumes and surfaces that exist in the three-dimensional structure. (Russ, C.J. and DeHoff, R.T., *Practical Stereology*, 2nd ed., Plenum Press, New York, 2002, pp. 31. With permission.)

composite materials, axons in neurons, capillary blood vessels, or plant roots, are approximate lineal objects [4].

3. Topological Properties

Line length, surface area, and volume are called metric properties because they depend explicitly on the dimensions of the objects under examination. Geometric properties that do not depend upon shape, size, or size distribution are called the topological properties. Most familiar of these properties is the number of disconnected parts of an object. The surfaces bounding particles also may be counted. In this case, the number of disconnected parts in the surfaces is the same as the number of particles in the three-dimensional objects. For example, this will not be true if the particles are hollow spheres. Then, each particle is bounded by two surfaces and the number of surfaces is twice the number of particles. The normalized value of this generic topological property is the number density. A less familiar topological property is the connectivity of objects. Connectivity reports the number of extra connections that objects in the structure have with themselves. To visualize the connectivity of a three-dimensional object, determine how many times the object could be sliced with a knife without dividing it into two parts. Objects that may be deformed into a sphere without tearing or making new joints are "topologically equivalent to a sphere." The objects in the second row in Figure 13.14 can all be cut once without separating the object into two parts. These objects all have connectivity equal to one and are topologically equivalent to a torus. Connectivity of the remaining objects is given in Figure 13.14. A microstructure that consists of a network, such as a powder stack or the capillaries in an organ, may have a very large value of connectivity.

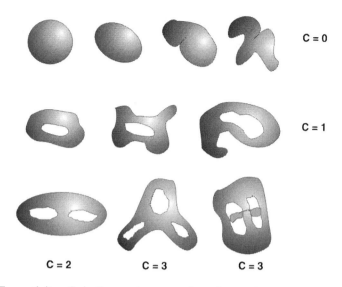

C = 0

C = 1

C = 2   C = 3   C = 3

FIGURE 13.14 Connectivity, C, is the maximum number of cuts that can be made through a three-dimensional object without separating it into two parts. (Russ, C.J. and DeHoff, R.T., *Practical stereology*, 2nd ed., Plenum Press, New York, 2002, pp. 33. With permission.)

At the end, we can conclude that there are three levels of characterization in the description of the geometric state of a structure. The qualitative structural state is simply a list of the three-, two-, one-, and zero-dimensional objects that exist in the structure. Each of the objects in the list has geometric properties, which have unambiguous meaning for objects of arbitrary complexity. These properties may be visualized for individual objects in the structure or as global properties for the whole objects [4].

Evaluation of one or more of these geometric properties constitutes a step in the assessment of the quantitative microstructural state. Real structures exhibit variations of some properties with macroscopic position in the structure. Some objects and patterns may display variation with orientation in the macroscopic specimen. Proper sample design may provide appropriate averages of global properties. Alternatively, these gradients and anisotropies may be assessed quantitatively. Comparison of appropriate combinations of these properties with predictions from random or uniform models for the structure provides measures of tendencies for objects to be positively or negatively associated with each other.

IV. CHAOS AND FRACTALS

The world of mathematics has been confined to the linear world for centuries. That is to say, mathematicians and physicists have overlooked dynamical systems as random and unpredictable. The only systems that could be understood in the past were those that were believed to be linear and systems that follow predictable patterns and arrangements. Linear equations, linear functions, linear algebra, linear programming, and linear accelerators are all areas that have been understood and mastered by the human race. However, the problem arises that we humans do not live in an even remotely linear world; in fact, our world should indeed be categorized as nonlinear; hence, proportion and linearity are scarce. How one may go about pursuing and understanding a nonlinear system in a world that is confined to the easy, logical linearity of everything? This is the question that scientists and mathematicians became burdened within the nineteenth century. The very name "chaos theory" seems to contradict reason; in fact, it seems somewhat of an oxymoron [5]. The name *chaos theory* leads the reader to believe that mathematicians have discovered some new and definitive knowledge about utterly random and incomprehensible phenomena. The acceptable definition of chaos theory is: "chaos theory is the qualitative study of unstable aperiodic behavior in deterministic nonlinear dynamical systems." A dynamical system may be defined to be a simplified model for the time-varying behavior of an actual system, and aperiodic behavior is simply the behavior that occurs when no variable describing the state of the system undergoes a regular repetition of values. Aperiodic behavior never repeats and it continues to manifest the effects of any small perturbation; hence, any prediction of a future state in a given system that is aperiodic is impossible.

What is so incredible about chaos theory is that unstable aperiodic behavior can be found in mathematically simple systems. These simple mathematical systems display behavior so complex and unpredictable that it is acceptable to merit their descriptions as random. An interesting question arises concerning why chaos has just recently been noticed. If chaotic systems are so mandatory to our everyday life, how come mathematicians have not studied chaos theory earlier? The answer can be given in one word: computers. The calculations involved in studying chaos are repetitive, boring and number in the millions, and computers have always been used for endless repetition.

Before advancing into the more precocious and advanced areas of chaos, it is necessary to mention the basic principle that adequately describes chaos theory, the butterfly effect. Small butterfly wing waving in the one part of the world induces the storm in the other. When applied to the system, small variations in initial conditions result in huge, dramatic effects in following events. The graphs of what seem to be identical, dynamic systems appear to diverge as time goes on until all resemblance disappears [5]. Perhaps the most identifiable symbol linked with the butterfly effect is the famous Lorenz attractor. Edward Lorenz, meteorologist, was looking for a way to model the action of the chaotic behavior of a gaseous system. Hence, he took a few equations from the physics field of fluid dynamics, simplified them and plotted the three differential equations on a three-dimensional plane, with the help of a computer of course, no geometric structure or even complex curve would appear; instead, a weaving object known as the Lorenz attractor appeared. This is because the system never exactly repeats itself, the trajectory never intersects itself, instead it loops around forever.

The attractor will continue weaving back and forth between the two wings, its motion seemingly random, and its very action mirroring the chaos, which drives the process. Lorenz proved that complex, dynamical systems show order, but they never repeat. Since our world is classified as a dynamical, complex system, our lives, our weather, and our experiences will never repeat, they should form patterns [6] (Figure 13.15).

Chaos and randomness are no longer ideas of a hypothetical world, they are quite realistic. A basis for chaos is established in the butterfly effect, the Lorenz attractor, and there must be an immense world of chaos beyond the rudimentary fundamentals. This new form mentioned is highly complex, repetitive, and replete with intrigue.

The extending and folding of chaotic systems give strange attractors, such as the Lorenz attractor, the distinguishing characteristic of a nonintegral dimension. This nonintegral dimension is most commonly referred to as a *fractal* dimension. Fractals appear to be more popular for their aesthetic nature than for their mathematics [7]. Everyone who has seen a fractal has admired the beauty of a colorful, fascinating image, but what is the formula that makes up this handsome image? The classical Euclidean geometry is quite different from the fractal geometry mainly because fractal geometry concerns nonlinear, nonintegral systems, while Euclidean geometry is mainly concerns linear, integral systems. Euclidean geometry is a description of regular geometrical figures.

Fractal geometry is a description of algorithms. There are two basic properties that constitute a fractal. First, it is self-similarity, which is to say that most magnified images of fractals are essentially indistinguishable from the unmagnified version. A visual form of fractal will look almost, or even exactly, the same no matter what size it is viewed at. This repetitive pattern gives fractals their aesthetic nature [5,7].

Second, fractals have noninteger dimensions. This means that they are entirely different from the graphs of lines and conic sections that we have in fundamental Euclidean geometry. The

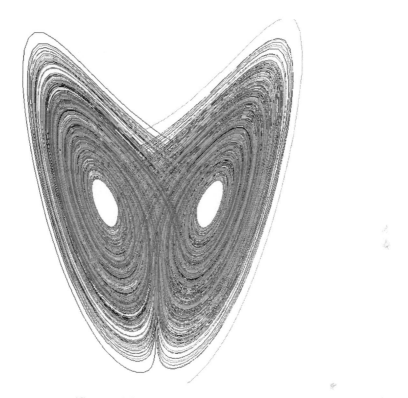

FIGURE 13.15 Lorentz attractor. (Generated by the ManpWin v2.8. http://www.deleeuv.com.au. With permission.)

Sierpenski triangle and the Koch snowflake are the best known representatives of fractal community (Figure 13.16). The iterations are repeated an infinite number of times and eventually a very simple fractal arises.

Gaston Maurice Julia masterpiece entitled "Memoire sur l'iteration des fonctions" dealt with the iteration of a rational function. The following fractals belong to the Julia set (Figure 13.17).

The work of Julia was reviewed and popularized by Benoit Mandelbrot, and his paper entitled "*How long is the coast of Britain*" becomes the starting point of the whole new Universe of fractals [3]. Fractals are in close relation to the "fragments," which should also mean "irregular." With the

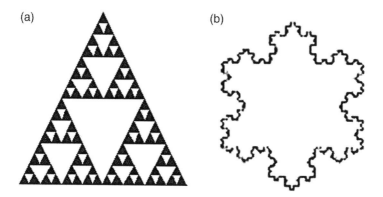

FIGURE 13.16 Two simple fractals. (Mandelbrot, B.B., The Fractal Geometry of Nature, W.H. Freeman and Company, San Francisco, 1977, pp. 49, 142. With permission.)

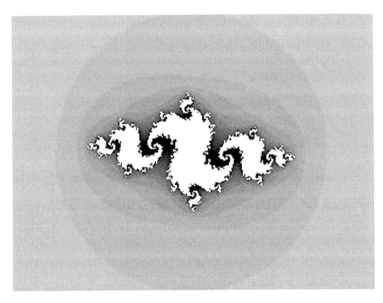

FIGURE 13.17 Visual presentations of the Julia sets. (Generated by the ManpWin v2.8. http://www.deleeuv.com.au. With permission.)

aid of computers, today, anyone is able to show how iteration is the source of some of the most beautiful fractals known today. The simple equation that is the basis of the Mandelbrot set is included as follows (Figure 13.18) [3].

$$\text{Variable Number} + \text{Fixed Number} = \text{Result}$$

It is now established that fractals are quite real and beautiful, but what do these patterns have to do with real life? Is there a real process behind these fascinating images? Fractals patterns could be found in the biological world. Pulmonary alveolar structure, regional myocardial blood flow heterogeneity, clouds, arteries, veins, nerves, parotid gland ducts, and the bronchial tree, all show some sort of fractal pattern organization. Fractals are one of the most interesting visual presentation of chaos theory, and they are beginning to become important tool in biology, medicine, and technology.

Why fractals stand in the quantification of visual information? First of all, the fractals could be only analyzed visually. They are created on the computer screen and present the patterns of iteration processes. First, we see them, then we observe them. It is interesting that mathematical games and visual presentations, which, in the most cases, are the beautiful images of them, are the facts that consume our attention to them. The connection with the real world comes later. This is opposite to what was presented earlier. In the research, we have a set of visual information acquired by the some system and the analysis is focused on finding the process which creates such forms. In the case of chaos and fractals, we have the visual presentation of the model and we know how it was generated (iterations). The only thing that we have to do is to recognize the real world presentation of obtained fractals. One of the best cases for that is the Cantor set.

What would happen when an infinite number of line segments were removed from an initial line interval. Cantor devised an example, which portrayed classical fractal made by iteratively taking away something. This operation created a "dust" of points-Cantor dust. The operation is shown in Figure 13.19 [3].

The Cantor set is simply the dust of points that remain. The number of these points is infinite, but their total length is zero. Mandelbrot recognized the Cantor set as a model for the occurrences of

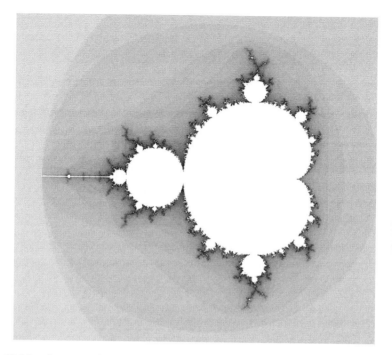

FIGURE 13.18 Visual presentation of the Mandelbrot set. (Generated by the ManpWin v2.8. http://www.deleeuv.com.au. With permission.)

errors in an electronic transmission line [3,5]. Engineers saw periods of errorless transmission, mixed with periods when errors would come in packages. When these packages of errors were analyzed, it was found that they contained error-free periods within them. As the transmissions were analyzed with higher and higher resolution, it was found that such dusts, as in the case of Cantor dust, were indispensable in modeling intermittency; the proof that fractals exist in the real world [3,5].

The fractals are visual patterns nice to look at; they are present in the real world, but what breakthroughs can be made in terms of discovery? Is chaos theory anything more than looking at the phenomena and processes from the different point of view? The future of chaos theory is

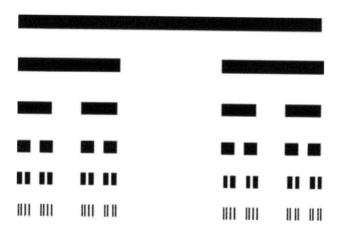

FIGURE 13.19 Visual presentation of the Cantor set. (Mandelbrot, B.B., The Fractal Geometry of Nature, W.H. Freeman and Company, San Francisco, 1977, pp. 80. With permission.)

unpredictable, but if a breakthrough is made, it will be huge. Minor discoveries have been made in the field of chaos within the past century, and as expected, they had the butterfly effect influence on the science.

A. PROCESS VISUALIZATION

One of the most fascinating problems is generation of patterns in real systems. According to that, reaction–diffusion systems can be treated as the minimal models for various patterns observed in the nature. To present the possible patterns, a model of excitable reaction, diffusion systems have been elaborated. The model consists of two coupled catalytic (enzymatic) reactions. One of them is allosterically inhibited by an excess of its reactant and product. The other is the usual catalytic reaction, which proceeds in its saturation regime. It is assumed that reactions occur in the open systems, with boundaries impermeable to the reagents. Patterns generated by the model analysis are present in Figure 13.20, which are exactly the old-Hebrew alphabet [8].

Not all letters have elegant forms. Some of them are similar to scribble, but they are readable, especially when used in sets (words). More elegant form of the letters can be obtained, if the reactors with smooth boundaries are used instead of the rectangular polygons. It is noteworthy that the reaction–diffusion model is structurally stable, which means that small changes in its parameters do not change the shapes of the asymptotic solutions. In addition, small changes in sizes of the reactors and positions of initial excitations do not change the qualitative properties of the asymptotic solutions. All letters have been obtained as the asymptotic solutions of the deterministic problem with well-defined inhomogeneities as the initial distributions of reagents. In real systems, inhomogeneities can appear due to internal, local fluctuations. Therefore, there is a probability greater than zero that the patterns can appear spontaneously in real systems [8].

It is noteworthy that the model is not exceptional one. The identical patterns can be generated in many reaction–diffusion systems, provided that they have the similar qualitative properties to the presented model. Moreover, it is worth to stress that the model contains two variables only, and therefore, it is simple one. One can expect more rich patterns in systems with three, four, and more variables.

The other process of patterns generated is the analysis of magnesium influence on the wettability properties of ceramic substrates. Data obtained with experiments were analyzed in two ways. Kinetic models available in literature were tested, which proved that they could not be used for real system modeling. Results directed to the use of methodology of dynamic systems analysis for modeling of wetting process. Two parameter models were used for simulation of the said wetting process [9].

Assumptions were that final layer of metal is formed by the "n" passes of metal over substrate points. In the "nth" pass, the first connection is formed, as the first contact. Isolated connection are bridged to each other via various patterns depending on the applied temperature to the final metal

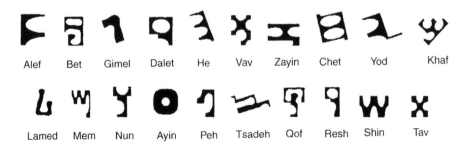

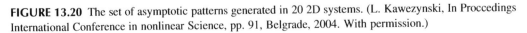

FIGURE 13.20 The set of asymptotic patterns generated in 20 2D systems. (L. Kawezynski, In Proccedings International Conference in nonlinear Science, pp. 91, Belgrade, 2004. With permission.)

layer [9]. For modeling, the iteration method was used as the main modeling tool. Generally, it could be assumed that the each dynamic, nonlinear, process can be described as

$$\frac{dI_m}{d\tau} = F_\mu(I_m(\tau))$$

This general model describes the evolution of the system in time. I_m is the set of functions of all possible parameters, which have influence on the system.

a. When $\lambda = 1.3578$, number of iterations $= 120$
b. When $\lambda = 1.31456$, number of iterations $= 567$
c. When $\lambda = 1.31466$, number of iterations $= 498$

Two-parameter model was used for simulation of the patterns formation during the wetting process. General form of the model is

$$x = \lambda x(1 - X).$$

Small changes of parameter λ directed the process from slow to very fast changes as could be seen from Figure 13.21. The forms generated by iteration simulation are in close agreement with the observed forms in real system. All of the simulated forms could be seen in the real process, as shown in Figure 13.22.

The system is dynamic and nonlinear, so basically, it is sensitive to a lot of parameters which could lead to the dramatic changes in behavior and, at the end, on the layer quality.

Intermediate, unstable forms are critical for the layer quality, because they have influence on the bonding quality of the metal–substrate connection. A system can be random in a short term and deterministic in a long term.

The applications of chaos theory are infinite. Random systems produce patterns of spooky understandable irregularity. From the Mandelbrot set to turbulence, feedback, and strange

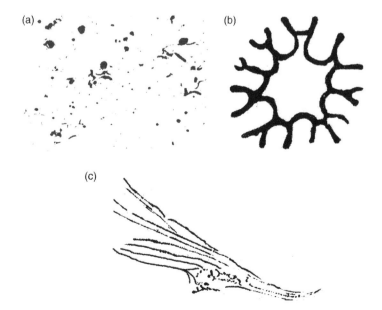

FIGURE 13.21 Patterns generated with the simulation model during the formation of MMC composite.

FIGURE 13.22 Real forms generated during MMC formation.

attractors, chaos appears to be everywhere. Breakthroughs have been made in the area of chaos theory, and, to achieve any more colossal accomplishments in the future, they must continue to be made.

V. CONCLUSION

Endless quantities of the information which are collected in everyday life are presented to the mankind in the visual form, using various presentation devices. "Seeing is a believing" is the motto of the unidentified flying object (UFO) observers, which could be used for all other areas of the high-level observations. Contemporary research is based on the detection of variations in the electromagnetic spectrum or interactions between fields or particles in the open systems, which are then presented in the visual form. Figure 13.23 presents the time–space scale and the visualization devices which are used to obtain the information.

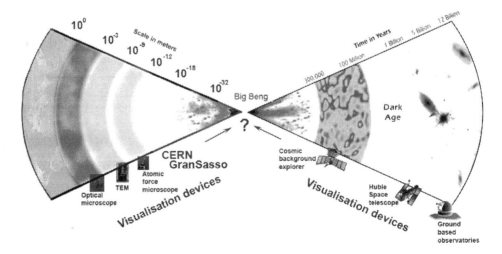

FIGURE 13.23 Visualization devices for the time–space scale.

Theories and models of the material structure and cosmos were developed long time ago, but the proof and general acceptance come with the visual evidence of the theoretical predictions. All the mathematical formalisms of models and theories could rise or fall if some experimental structure pattern, does or does not appear on the visualization device. It is interesting to note that as we want to see further back in time or to see the smaller and smaller basic material ingredients, we must use more and more energy. From 30 eV in the SEM, to 209 GeV in ultimate CERN's microscope, or from Hawaii Keck's telescope to Cosmic background explorer or the next generation space telescopes, energy demands for the unit visual information rise. Besides, the energy demands for extraction of the unit information, the complexity of the visual devices and algorithms for the data processing also rise. What then the needs for visualization of the neutrinos, mass bozon particles, p-brains, or the first atto seconds of creation will be? There is no doubt that the missing links in the material structure theories will be found and there is no doubt that it will be the visualization of interaction with the other particles. Quantification of observable interaction, probably by measure of path angle or path length, will give the desired proof of particle existence. The same principle is valid for the space research, where all the obtained data will be presented in the visual form and then quantified, wheather they are uniform or some irregularity exist to find the ultimate answer how matter was created.

Contemporary scientists have new toy, for the visualization and quantification, called supercomputers. Fast, reliable, energy consumable, and far from the simple devices, they are the new universal laboratories. This is another proof for the research philosophy of chaos or fractals, first see patterns, then analyze them. Super computers or the ultimate tele-microscope could be used to simulate all of the possible cases of theory or model and to visualize the results. Inputs are collected information, and outputs are transformed information, which can be compared with the information obtained from the experiments in the real world.

We exchange, transform and compare information in purpose to determine a relevant solution. Fingerprints of the subatomic particles or cosmic microwave radiation patterns are sets of quantified information obtained in the research processes. Visual presentations of the events, interactions, and patterns information obtained from the real or virtual experiments, and their quantification becomes the milestone tool in the endless search of the matter origin and the moment of creation.

ACKNOWLEDGMENTS

This project was financially supported by Ministry of Science and Environmental Protection of Serbia, Entitled "Multiphase Dispersed Systems" 101822.

REFERENCES

1. Jonson, W.H., *Introduction, History of art*, Harry.N Abrams Inc, New York 1989.
2. Spasic, A.M. *Multiphase Dispersed Systems*, ITNMS, Belgrade, 1997, pp. 5.
3. Mandelbrot, B.B., *The Fractal Geometry of Nature*, W.H. Freeman and Company, San Francisco, 1977.
4. Russ, C.J. and DeHoff, R.T., *Practical Stereology*, 2nd ed., Plenum Press, New York, 2002.
5. Donahue M.J. III, *An Introduction to Mathematical Chaos Theory and Fractal Geometry*. http://www.duke.edu/~mjd/chaos/chaos.html (retrieved October 2004).
6. Jürgens, H., O Peitgen, H., and Saupe, D., *Chaos and Fractals*, New Frontiers of Science, Springer-Verlag, New York, 1992.
7. A. Dewdney, A.K., *Computer Recreations*, Scientific American, 1989, pp.108–112.
8. Kawezynski, L., Model of reaction diffusion system as generator of old Hebrew Alphabet, in *Selforganization in Nonequilibrium Systems*, Proceedings of International Conference in Nonlinear Science, Belgrade, Serbia and Montenegro, September, pp. 89–91, Belgrade, 2004.
9. Jovanić, B.P. and Stanković, G.D., Modeling of the SiC wetting process using morphological analysis of the SEM images, in *Composite Formation Analysis*, Proceedings of XIV SPIM Meeting, Vol. II, Singapore, September 1–10, 1999, Nanyang University, Singapore, 1999.

Part III

Homo-Aggregate Finely
Dispersed Systems

Emulsions

14 Non-Newtonian Effects on Particle Size in Mixing Systems

James Y. Oldshue
Oldshue Technologies International, Sarasota, Florida, USA

Countercurrent-type fluid mixers are used in industry to make a variety of two-phase dispersions. A common application is in liquid–liquid systems, in which there is a disperse phase and a continuous phase. There are also systems where there may be three liquid phases, and the complexity of predicting bubble size is of course correspondingly greater. In this chapter, we will talk about two-phase dispersions and relate these to the common practice in macroscale, microscale, and nanoscale. The so-called macroscale is more properly addressed as the milliscale and occurs in bubbles ranging from 1 mm to several millimeter in size.

Referring to Chapter 12 on Current Principles of Mixing, a vessel with a rotating impeller has a spectrum of shear rates, the highest being at the impeller zone and the lowest near the tank walls and in the liquid top or bottom surface. The fact that the shear stress is what actually does the dispersion work is sometimes overlooked; the shear stress is the product of viscosity and shear rate. In non-Newtonian behavior, there is a spectrum of fluid velocities throughout the impeller zone and tank zone, which give a spectrum of shear rates. The viscosity of the fluid at these various shear rates gives another spectrum of the shear stress throughout the tank. Thus, it is common for there to be a bubble-size distribution throughout the tank. It is also common for liquid bubbles to get smaller and smaller in size. Equilibrium may take several hours. Sometimes, as the mixer speed is increased, these bubbles reach a minimum bubble size in which they almost appear as solid particles. Thus, the typical Gaussian distribution of bubble sizes that might appear at low power levels may become extremely biased at high power levels where there is a minimum size that cannot be broken easily.

The industrial complexity of liquid–liquid dispersions and emulsions is complicated by the fact that small amounts of impurities make marked differences in the bubble size produced. There is a proliferation of data on bubble sizes and droplet sizes in industrial systems and also in academic studies where the presence of trace chemicals can often be eliminated. However, there is a multitude of equations giving the bubble size for various kinds of two-phase well-purified systems, and no equation has been set up to give bubble-size distributions more than on an estimated or predictive basis, particularly where there is a statistical distribution of drop sizes. Many of the papers in the literature are very useful in ratio form. It means that we experimentally determine a drop-size distribution and then use the effect of other liquid properties, the effect of geometry, tank size, and the effect of baffles, etc. in a relative sense to predict what would happen to bubble size when we scale up or scale down.

Obviously, in industrial production of polymers and emulsions, there has been a lot of work done by manufacturers on their products. Usually, after a trial-and-error procedure with parameters such as speed and diameter of the impeller, there emerges a point at which the combination of all the effects gives a drop-size distribution, yielding a product that is controllable and is useful as an industrial process. Going down to the microsize, one finds guess involved in the limit of shear-rate properties that occur at the minimum scale, predicted by the Komolgarof equation. This equation is given as:

$$L = \left(\frac{v^3}{\varepsilon}\right)^{\frac{1}{4}}$$

and incorporation of chemical values for higher power levels and other factors in the equation gives a minimum droplet size of about 5 to 0.5 μm diameter.

Down at the nanoscale, it is not practical to make impellers much less than 15–20 mm diameter. Sometimes, all that can be done is to machine out blades and other cavities or projections in the shaft. Once we get down to the nanosize, it is usually necessary to consider what kind of jet action of fluids is impinging on each other, depending on the energy in the two impinging phases. Very high power levels can be generated when two phases impinge on each other, and there are several devices recorded in literature for nanoscale shear rates to minimize bubble size. By putting a process through a microscale reactor, there is a limit to the diameter of the impeller to about 10–15 mm. Therefore, down at the nanoscale, we have to depend primarily on options in putting mechanical shapes in the flow stream to generate shear rates at that level. Typically, a lot of operations are studied at this level of shear-rate size and the role of fluid mechanics in mixing is predominant. Several papers in the literature look at these kinds of high-intensity reactors, but there are no equations for predicting the performance of these units. There are many kinds of high density–high shear mixers, which are made using high-power and high-speed operations. These mixers sometimes have a separate high-shear type of impeller or may be a rotor–stator. There has not been a cohesive body of literature in research on these devises, since they have all tended to be developed individually by separate suppliers. There is a consortium at the University of Maryland, which is planned to give a technical basis for the dispersion characteristics of these devices. Sometimes, these devices are in a batch tank of almost any volume or they may be in continuous flow in a pipeline-type system in which the high-shear mixer is comprised in a compartment in the pipeline.

The use of impinging jets or sonic or ultrasonic or ultra types of vibration characteristics and the dispersion in the nano- and attotechnology rates are a popular and important topic in current research. There is not a comprehensive technical background for the design of these devices. Since the requirements of many types of dispersion are not quantitatively available before trial, considerable work on small-scale operations is required. However, the atto- and nanotechnology area is already a small-scale one; the device that will be the ultimate choice in the plant will be used in an actual experiment on the actual scale, sometimes.

The use of laser Doppler-velocity meters or particle-image-tracking devices, or modeling with computational fluid mechanics, techniques are very important in the overall understanding of the mechanism involved. However, one thing to bear in mind is that many of these devices give an average rate or velocity patterns and do not give a real-time output for the actual high-frequency fluctuations going on in a fluid-mechanics system.

In research with these instruments, the computing time required to analyze the real-time effect can sometimes be overwhelming and the use of parallel or series computing and recording devices must be carefully considered in terms of the cost and time required for the study and duration of the proper flow patterns.

For the mechanical high-shear devices, there is quite a bit of empirical and proprietary data on how these devices disperse solids, liquids, and gasses in a continuous medium. It turns out that many types of these devices have become inherent tools in a particular industry and that the performance is empirically studied and correlated to give the resulting final product.

The resolution of performance down on the nano- and attoscale requires looking at the resolutions involved with visible light, and these might be resolved with x-rays or electron microscope concepts. This whole area is involved, of course, with many living-cell processes, since bacteria are in the milli- and macroscale, while viruses are in the nanoscale range, and individual cells may be in the nano- or atto-area. This is becoming a fertile field for various types of impeller or nonimpeller devices to give dispersion down on the nano- and atto-scale to study their effect on vaccine, medicine, or actual living cell response to these shear rates.

A different kind of mixing reactor is a flow reactor, which uses flow chemistry and is carried out in glass microreactors, which has different sizes in the order of a few tenths of a millimeter (Syrris Royston, UK).

15 Theory of Electroviscoelasticity

Aleksandar M. Spasic
Institute for Technology of Nuclear and Other Mineral Raw Materials, Belgrade, Serbia, Serbia and Montenegro

Mihailo P. Lazarevic and Dimitrije N. Krstic
University of Belgrade, Belgrade, Serbia, Serbia and Montenegro

CONTENTS

I. INTRODUCTION

A. ELECTROVISCOSITY–ELECTROVISCOELASTICITY OF LIQUID–LIQUID INTERFACES

Electroviscosity and electroviscoelasticity are terms that may be dealing with fluid flow effects on physical, chemical, and biochemical processes. The hydrodynamic or electrodynamic motion is considered in the presence of both potential (elastic forces) and nonpotential (resistance forces) fields. The elastic forces are gravitational, buoyancy, and electrostatic or electrodynamic (Lorentz), and the resistance forces are continuum resistance or viscosity and electrical resistance or impedance.

According to the classical deterministic approach, the phases that constitute the multiphase dispersed systems are assumed to be a continuum, that is, without discontinuities inside the one entire phase, homogeneous and isotropic. The principles of conservation of momentum, energy, mass, and charge are used to define the state of a real fluid system quantitatively. In addition to the conservation equations, which are insufficient to uniquely define the system, statements on the material behavior are required. These statements are termed constitutive relations, for example, Newton's law, Fourier's law, Fick's law, and Ohm's law. In general, the constitutive equations are defined empirically, although the coefficients in these equations (e.g., viscosity coefficient, heat conduction coefficient, and complex resistance coefficient or impedance) may be determined at the molecular level. Often, these coefficients are determined empirically from related phenomena; therefore, such a description of the fluid state is termed a phenomenological description or model. Sometimes, particular modifications are needed when dealing with fine-dispersed systems. An example is Einstein's modification of the Newtonian viscosity coefficient in dilute colloidal suspensions [1]; further on, Smoluchowski's modification of Einstein's relation for particles carrying electric double layers [2] and recent more profound elaboration of the entropic effects [3].

According to the approach introduced here, an interrelation between three forms of "instabilities" is postulated: rigid, elastic, and plastic. Figure 15.1 shows that the events are understood as interactions between internal–immanent and external–incident periodical physical fields.

Since both electric or electromagnetic and mechanical physical fields are present in a droplet, they are considered as internal, whereas ultrasonic, temperature, or any other applied periodical physical fields are considered as external. Hence, the rigid form of instability has the possibility of two-way disturbance spreading or dynamic equilibrium. This form of instability, when all forces involved (electrostatic, van der Waals, solvation, and steric) are in equilibrium, permits a two-way disturbance spreading (propagation or transfer) of entities either by tunneling (low energy dissipation and occurrence probability) or by induction (medium or high energy dissipation and occurrence probability). The elastic form of instability has the possibility of reversible disturbance spreading with or without hysteresis. Finally, the plastic form of instability has the possibility of irreversible disturbance spreading with a low or high intensity of influence between two entities. Now, a disperse system consists of two phases, "continuous" and "dispersed." The continuous phase is modeled as an infinitely large number of harmonic electromechanical oscillators with low strength interactions among them. Furthermore, the dispersed phase is a macrocollective consisting of a finite number of microcollectives or harmonic electromechanical oscillators (clusters) with strong interactions between them. The cluster can be defined as the smallest repetitive unit that has a character of integrity. Clusters appear in a micro- and nanodispersed systems, while

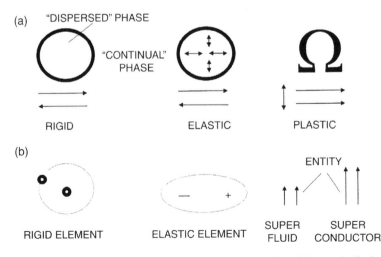

FIGURE 15.1 Interactions between internal–immanent and external–incident periodical physical fields: (a) three forms of instabilities and (b) the constructive elements of phases. (Taken from Spasic, A.M. et al., *J Colloid Interf. Sci.*, 185, 434–446, 1997. With permission from Academic Press.)

atto-clusters or entities appear in atto-dispersed systems. The microcollective consists of the following elements: the rigid elements (atoms or molecules), the elastic elements (dipoles or ions that may be recombined), and the entities (as the smallest elements).

B. Previous Work

In normal viscous fluids, only the rate of deformation is of interest. In the absence of external and body forces, no stresses are developed and there is no means of distinguishing between a natural state and a deformed state [4]. It is rather disturbing to think of the very large overall deformations obtained in the flow of fluids being associated in any way with substances that have elasticity. The rationalization lies in realizing that for the substances considered here, the behavior is essentially that of a fluid; although much translation and rotation may occur, the "elastic" distortion of the elementary volumes around any point is generally small. This "elastic" distortion or material's strain is nevertheless present and is a feature that cannot be neglected. It is responsible for recovery of reverse flow after the removal of applied forces and for all the other non-Newtonian effects [4]. These distortions or strains are determined by the stress history of the fluid and cannot be specified kinematically in terms of the large overall movement of the fluid. Another way of looking at the situation is to say that the natural state of the fluid changes constantly in flow and tries to catch up with the instantaneous state or the deformed state. It never does quite succeed in doing so, and the lag is a measure of the memory or the elasticity. In elastic solids, the natural state does not change and there is a perfect memory [4].

The entropy of elasticity of a droplet is a measure of the increase in the available volume in configuration space. This increase occurs with a transition from rigid, regular structure to an ensemble of states that include many different structures. If the potential wells in the liquid state were as narrow as those in the solid state and if each of those potential wells was equally populated and corresponded to a stable amorphous structure (and vice versa), then the entropy of elasticity would be a direct measure of the increase in number of wells or a direct measure of the number of available structures [3,5].

In the last two centuries, a lot of attempts and discussion have been made on the elucidation and development of the various constitutive models of liquids. Some of the theoretical models that can be mentioned here are: Boltzmann, Maxwell (UCM, LCM, COM, IPM), Voight or Kelvin, Jeffrey, Reiner-Rivelin, Newton, Oldroyd, Giesekus, graded fluids, composite fluids, retarded fluids with a strong backbone and fading memory, and so on. Further and deeper knowledge related to the physical and mathematical consequences of the structural models of liquids and of the elasticity of liquids can be found in Ref. [6].

For the studies and modeling of the structure and dynamics of finely dispersed systems, various concepts may be applied, for example, relativistic theories — potential energy surfaces (PES) (Dirac 1931 and Feynman, Schwinger, and Tomonaga 1940 — all based on the Born–Oppenheimer assumption); nonrelativistic theories — quantum electrodynamics (QED; Schrodinger 1926); molecular orbital methods (*ab initio* and semiempirical), molecular mechanics — MM2, MM3, and CSC (Allinger, Lii, Lippincot, Rigby — ball and stick); and transition state theories — TST (Wigner, etc.).

II. THEORY: STRUCTURE

A. Electrified Interfaces: A New Constitutive Model of Liquids

The secondary liquid–liquid droplet or droplet–film structure is considered as a macroscopic system with internal structure determined by the way the molecules (ions) are tuned (structured) into the primary components of a cluster configuration. How the tuning or structuring occurs depends on the physical fields involved, both potential (elastic forces) and nonpotential (resistance forces). All these microelements of the primary structure can be considered as electromechanical oscillators assembled into groups, so that an excitation by an external physical field may cause oscillations at the resonant or characteristic frequency of the system itself (coupling at the characteristic frequency) [5,7,8].

Figure 15.2 shows a series of graphical sequences that are supposed to facilitate the understanding of the proposed structural model of electroviscoelastic liquids. The electrical analog (Figure 15.2a) consists of passive elements (R, L, and C) and an active element (emitter-coupled oscillator W). Further on, the emitter-coupled oscillator is represented by the equivalent circuit as shown in Figure 15.2b. Figure 15.2c shows the electrical (oscillators j) and mechanical (structural volumes V_j) analogs when they are coupled to each other, for example, in the droplet. Now, the droplet consists of a finite number of structural volumes or spaces or electromechanical oscillators (clusters) V_j, a finite number of excluded surface volumes or interspaces V_s, and a finite

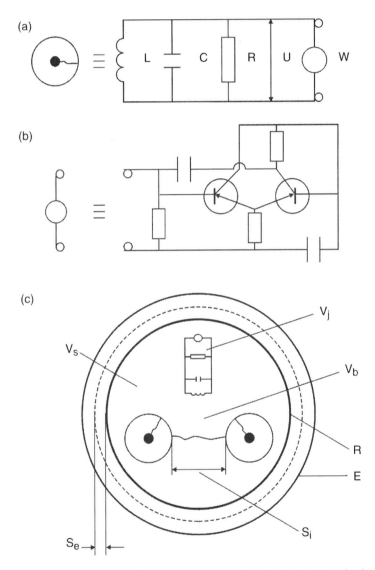

FIGURE 15.2 Graphical interpretation of the structural model: (a) electrical and mechanical analog of the microcollective or cluster; (b) equivalent circuit for the emitter-coupled oscillator; (c) the macrocollective: a schematic cross-section of the droplet and its characteristics (V_j, structural volumes or clusters; V_s, excluded surface volumes or interspaces; V_b, excluded bulk volumes or interspaces; S_i, internal separation; S_e, external separation; R, rigidity droplet boundary; E, elasticity droplet boundary).

number of excluded bulk volumes or interspaces V_b. Furthermore, the interoscillator or cluster distance or internal separation S_i represents the equilibrium of all forces involved (electrostatic, solvation, van der Waals, and steric) [9]. The external separation S_e is introduced as a permitted distance when the droplet is in interaction with any external periodical physical field. The rigidity droplet boundary, R, presents a form of droplet instability when all forces involved are in equilibrium. Nevertheless, the two-way disturbance spreading (propagation or transfer) of entities occur, either by tunneling (low energy dissipation) or by induction (medium or high energy dissipation). The elasticity droplet boundary, E, represents a form of droplet instability when equilibrium of all forces involved is disturbed by the action of any external periodical physical field, but the droplet still exists as a dispersed phase. In the region between the rigidity and elasticity droplet boundaries, a reversible disturbance spreading occurs. After the elasticity droplet boundary, the plasticity as a form of droplet instability takes place; then the electromechanical oscillators or clusters do not exist any more and the beams of entities or atto clusters appear. Atto clusters are the entities that appear in the atto-dispersed systems. In this region, one-way propagation of entities occurs.

Considering all presented arguments and comments, the probability density function (PDF) in general form can be expressed as

$$F_d(V) = F_d(V_j) + [F_d(V_s) + F_d(V_b)] \tag{15.1}$$

where the first term on the right-hand side of the equation is due to energy effects and the second term (consisting of two subterms) is due to entropic effects; subscript j is related to the structural volumes or energies, and subscripts s and b are related respectively to the excluded surface and bulk volumes or energies. Consider an uncertain physical property and a corresponding space describing the range of values that the property can have (e.g., the configuration of a thermally excited N-particle system and the corresponding $3N$-dimensional configuration space). The PDF associated with a property is defined over the corresponding space; its value at a particular point is the probability per unit volume that the property has a value in an infinitesimal region around that point [5].

An alternative expression of this PDF, considering Figure 15.2c may be written as

$$\Delta_{1,2}(V) = \sum_{i=3}^{n} (V_j)_i + \Delta_{1,2} \left[\sum_{i=1}^{n-3} (V_b)_i + \sum_{i=3}^{n} (V_s)_i \right] \tag{15.2}$$

therefore, the number i of clusters V_j remains constant while the droplet passes through the rigid (e.g., the state related to the subscript 1) and elastic (e.g., the state related to the subscript 2) form of instabilities. The integer (i) is the summation integer that takes the values in the interval $[3, n]$ for the number of clusters V_j and for the number of excluded surface volumes V_s and in the interval $[1, n - 3]$ for the number of excluded bulk volumes, while integer n takes the values in the interval $[0, \infty]$. The simplest case, used only to explain, is the droplet that contains three oscillators V_j $[(i) = 3; n = 0]$, one excluded bulk volume V_b $[(i) = 1; n = 3]$, and three excluded surface volumes V_s $[(i) = 3; n = 0]$. Differences $\Delta_{1,2}$ in volumes or energies V_b and V_s occur only in the entropic part, that is, the internal separation S_i (Figure 15.2c) changes (increases or decreases) during the transition of the droplet from rigid to elastic or vice versa. Consequently, the external separation S_e decreases or increases depending on the direction of transition.

1. Classical Assumptions for Interfacial Tension Structure and for Partition Function

(1) The droplet is considered as a unique thermodynamic system, which can be described by a characteristic free energy function expressed as

$$\Delta G = (\sigma_i + T\Delta S) = \left[\sigma_i - T \left(\frac{d\sigma_i}{dT} \right)_{X_i} \right] = -kT \ln Z_p \tag{15.3}$$

where χ_i correspond to the constant chemical potential, σ_i the interfacial tension, T the temperature, S the entropy, k the Boltzmann constant, and Z_p the partition function.

(2) According to quantum mechanical principles, the droplet possesses vacancies or "free volumes" and the relation for interfacial tension can be written as

$$\sigma_i = \frac{G^s - G^b}{\kappa^0} = \frac{1}{\kappa^0}\left[(\Phi^s - \Phi^b) + N^s kT \ln\left(\frac{V_f^b}{V_f^s}\right)\right] \tag{15.4}$$

where G^s and G^b are free energies of the surface and bulk and Φ^s and Φ^b are the overall energies of N heavy-phase molecules in their ideal positions on the surface and in the bulk, respectively. N^s is the number of molecules on the surface, V_f^s and V_f^b the "free volumes" of the molecules on the surface and in the bulk, and κ^0 is the surface of "free surface" [5,7,10–36]. This means, the droplet is a macrosystem with physicochemical properties, which may be described with the help of different thermodynamic parameters.

The phenomenological meaning of the given interfacial tension structure is in agreement with the "free volume" fluid model. Hence, a fluid is a system with ideal or ordered neighborhood elements distribution and with discontinuities of the package density (boundaries of the subsystems or microcollectives or clusters as some particular physical systems) [5,7].

Furthermore, the partition function, that is, its physical meaning states that how molecules are distributed among the available energy levels. It is possible to separate various contributions (the sum of the translational, rotational, vibrational, and electronic energy terms) to the partition function [16].

(3) Using the equivalency of the mean energies W, at the instant of equilibrium, a characteristic free energy function is expressed as

$$\Delta G = \overline{w} = -kT \ln Z_p \tag{15.5}$$

where the partition function for this particular system is derived from

$$W = \frac{h\omega}{2\pi}\frac{\partial \ln Z_p}{\partial \theta} \tag{15.6}$$

were h is Planck's constant, further on,

$$Z_p = \frac{Q_N}{\lambda^{3N}} = \sum_{j=0}^{\infty} \exp\left[-\left(j+\frac{1}{2}\right)\Theta\right] = \frac{1}{2}\cosh\frac{\Theta}{2} \tag{15.7}$$

where Q_N is a configuration integral and j is a number of identical oscillators Θ, where each is given by

$$\Theta = \frac{h\omega}{2\pi kT} \tag{15.8}$$

and λ is a free path between two collisions and it is expressed as

$$\lambda = \frac{h^2}{2\pi mkT} \tag{15.9}$$

Further, more detailed discussion and derivation of the partition function can be found in Refs. [5,7,16].

2. Postulated Assumptions for an Electrical Analog

1. The droplet is a macrosystem (collective of particles) consisting of structural elements that may be considered as electromechanical oscillators.
2. Droplets as microcollectives undergo tuning or coupling processes and so build the droplet as a macrocollective.
3. The external physical fields (temperature, ultrasonic, electromagnetic, or any other periodic) cause the excitation of a macrosystem through the excitation of a microsystems at the resonant or characteristic frequency, where elastic or plastic deformations may occur.

Hence, the study of the electromechanical oscillators is based on electromechanical and electrodynamic principles. At first, during the droplet formation, it is possible that the serial analog circuits are more probable, but later, as a consequence of tuning and coupling processes, the parallel circuitry becomes dominant. In addition, since the transfer of entities by tunneling (although with low energy dissipation) is much less probable, it is sensible to consider the transfer of entities by induction (medium or high energy dissipation).

A nonlinear integral–differential equation of the van der Pol type represents the initial electromagnetic oscillation

$$C\frac{dU}{dt} + \left(\frac{U}{R} - \alpha U\right) + \gamma U^3 + \frac{1}{L}\int U\,dt = 0 \tag{15.10}$$

where U is the overall potential difference at the junction point of the spherical capacitor C and the plate, L the inductance caused by potential difference, R the ohmic resistance (resistance of the energy transformation, electromagnetic into the mechanical or damping resistance), and t the time. The α and γ are constants determining the linear and nonlinear parts of the characteristic current and potential curves. U_0, the primary steady-state solution of this equation, is a sinusoid of frequency close to $\omega_0 = 1/(LC)^{0.5}$ and amplitude $A_0 = [(\alpha - 1)/R/3\gamma/4]^{0.5}$.

The noise in this system, due to linear amplification of the source noise (the electromagnetic force assumed to be the incident external force, which initiates the mechanical disturbance), causes the oscillations of the "continuum" particle (molecule surrounding the droplet or droplet–film structure), which can be represented by the particular integral

$$C\frac{dU}{dt} + \left(\frac{1}{R} - \alpha\right)U + \gamma U^3 + \frac{1}{L}\int U dt = -2A_n \cos \omega t \tag{15.11}$$

where ω is the frequency of the incident oscillations.

Finally, considering the droplet or droplet–film structure formation, "breathing," or destruction processes and taking into account all the noise frequency components, which are included in the driving force, the corresponding equation is given by

$$C\frac{dU}{dt} + \left(\frac{1}{R} - \alpha\right)U + \frac{1}{L}\int U dt + \gamma U^3 = i(t) = \frac{1}{2\pi}\int_{-\infty}^{\infty} \exp(j\omega t)A_n(\omega)\,d\omega \tag{15.12}$$

where $i(t)$ is the noise current and $A_n(\omega)$ is the spectral distribution of the noise current as a function of frequency.

In the case of nonlinear oscillators, however, the problem of determining of the noise output is complicated by the fact that the output is fed back into the system, thus modifying, in a complicated manner, the effective noise input [3,8,19]. The noise output appears as an induced anisotropic effect.

III. THEORY OF ELECTROVISCOELASTICITY: DYNAMICS

A. Tension Tensor Model

Now, using the presented propositions and electromechanical analogies, an approach to non-Newtonian behaviors and to electroviscoelasticity is to be introduced. If Equation (15.13) is applied to the droplet when it is stopped, for example, as a result of an interaction with some periodical physical field, the term on the left-hand side becomes equal to zero.

$$\rho \frac{D\tilde{u}}{Dt} = \sum_i \tilde{F}_i (dx\, dy\, dz) + d\tilde{F}_s \tag{15.13}$$

Furthermore, if the droplet is in the state of "forced" levitation and the volume forces balance each other, then the volume force term is also equal to zero [7,8,20]. It is assumed that the surface forces are, for the general case that includes the electroviscoelastic fluids, composed of interaction terms expressed as

$$d\tilde{F}_s = \tilde{T}^{ij}\, d\tilde{A} \tag{15.14}$$

where the tensor T^{ij} is given by

$$T^{ij} = -\alpha_0 \delta^{ij} + \alpha_1 \delta^{ij} + \alpha_2 \zeta^{ij} + \alpha_3 \zeta_k^i \zeta^{kj} \tag{15.15}$$

where T^{ij} is composed of four tensors, δ^{ij} the Kronecker symbol, ζ^{ij} the tension tensor, and $\zeta_k^i \zeta^{kj}$ the tension coupling tensor. In the first isotropic tensor, the potentiostatic pressure $\alpha_0 = \alpha_0(\rho, U)$ is dominant and the contribution of the other elements is neglected. Here, U represents hydrostatic or electrostatic potential. In the second isotropic tensor, the resistance $\alpha_1 = \alpha_1(\rho, U)$ is dominant and the contribution of the other elements is neglected. In the third tension tensor, its normal elements $\alpha_2 \sigma$ are due to the interfacial tensions and the tangential elements $\alpha_2 \tau$ are presumed to be of the same origin as the dominant physical field involved. In the fourth tension coupling tensor, there are normal, $\alpha_3 \sigma_k^i$ and $\alpha_3 \sigma^{kj}$ elements, and tangential, $\alpha_3 \tau_k^i$ and $\alpha_3 \tau^{kj}$, elements, which are attributed to the first two dominant periodical physical fields involved. Now, the general equilibrium condition for the dispersed system with two periodical physical fields involved may be derived from Equation (15.15) and may be expressed as

$$\tau_d = \frac{-\alpha_0 + \alpha_1 + \alpha_2(\sigma/d) + \alpha_3(\sigma/d)}{2(\alpha_2 + \alpha_3)} \tag{15.16}$$

where τ_d are the tangential elements of the same origin as those of the dominant periodical physical field involved. Figure 15.3 shows the schematic equilibrium of surface forces at any point of a stopped droplet–film structure while in interaction with some periodical physical field [5]. Note that for dispersed systems consisting of, or behaving as Newtonian fluids, $\alpha_3 = \alpha_3(\rho, U)$ is equal to zero.

The processes of formation or destruction of the droplet or droplet–film structure are nonlinear. Therefore, the viscosity coefficients μ_i ($i = 0, 1, 2$), where each consists of bulk, shear, and tensile

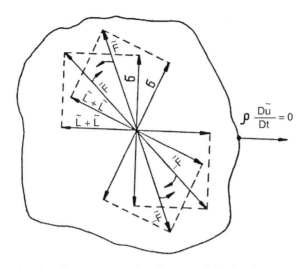

FIGURE 15.3 Balance of surface forces at any point of a stopped droplet–film structure while in interaction with some periodical physical field. A two-dimensional projection. F represents the projection of the resultant surface forces, vector in 3N-dimensional configuration space. T stands for the tangential and σ for the normal components. (Taken from Spasic, A.M. et al., *J Colloid Interf. Sci.*, 185, 434–446, 1997. With permission from Academic Press.)

components, when correlated with the tangential tensions of mechanical origin, τ_v can be written as

$$\tau_v = \mu_0 \frac{du}{dx} + \mu_1 \frac{d^2u}{dx^2} + \mu_2 \left(\frac{du}{dx}\right)^2 \tag{15.17}$$

where u is the velocity and x is the one of the space coordinates.

Using the electrical analog, the impedance coefficients Z_i ($i = 0, 1, 2$), where each consists of ohmic, capacitive, and inductive components, will be correlated with the tangential tensions of electrical origin τ_e as follows:

$$\tau_e = Z_0 \frac{d\phi_e}{dt} + Z_1 \frac{d^2\phi_e}{dt^2} + Z_2 \left(\frac{d\phi_e}{dt}\right)^2 \tag{15.18}$$

where ϕ_e is the electron flux density and t is the time coordinate.

More detailed discussion about derivation of these equations can be found in Refs. [3,7,8,21].

B. VAN DER POL DERIVATIVE MODEL

Figure 15.4 shows a series of graphical sequences that may help in understanding of the proposed theory of electroviscoelasticity. This theory describes the behavior of electrified liquid–liquid interfaces in fine-dispersed systems and is based on a new constitutive model of liquids [3,7,8,20,22]. If an incident periodical physical field (Figure 15.4b), for example, electromagnetic, is applied to the rigid droplet (Figure 15.4a), then the resultant, equivalent electrical circuit can be represented as shown in Figure 15.4c.

The equivalent electrical circuit, rearranged under the influence of an applied physical field, is considered as a parallel resonant circuit coupled to another circuit such as an antenna output circuit. Thus, in Figure 15.4c, W_d, C_d, L_d, and R_d correspond to the circuit elements each; W_d represents active emitter-coupled oscillator; and C_d, L_d, and R_d, represent passive capacitive, inductive, and resistive elements respectively. The subscript d is related to the particular droplet diameter, that is, the droplet under consideration. Now, again the initial electromagnetic oscillation is represented by

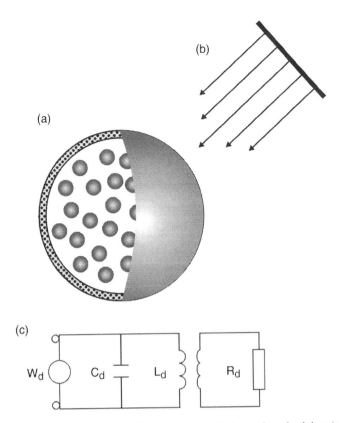

FIGURE 15.4 Definition sketch for understanding the theory of electroviscoelasticity: (a) rigid droplet; (b) incident physical field, for example, electromagnetic; (c) equivalent electrical circuit–antenna output circuit. W_d represents the emitter-coupled oscillator and C_d, L_d, and R_d are capacitive, inductive, and resistive elements of the equivalent electrical circuit, respectively. Subscript d is related to the particular diameter of the droplet under consideration. (Courtesy of Marcel Dekker, Inc.) Spasic, A.M. Ref. 3., p. 854.

the integral–differential equations, Equation (15.11) and Equation (15.12), and when the nonlinear terms are omitted or cancelled, the following linear equation is obtained:

$$C\frac{dU}{dt} + \left(\frac{1}{R} - \alpha\right)U + \frac{1}{L}\int U dt = -2A_n \cos \omega t \qquad (15.19)$$

with a particular solution resulting in the following expression for the amplitude:

$$A = \frac{2\omega C A_n}{\left[4(\omega_0 - \omega)^2 + \left(\frac{1}{R} - \alpha\right)^2\right]^{0.5}} \qquad (15.20)$$

and for all the noise frequency components, the linear equation is given by

$$C\frac{dU}{dt} + \left(\frac{1}{R} - \alpha\right)U + \frac{1}{L}\int U\, dt = i(t) = \frac{1}{2\pi}\int_{-\infty}^{\infty} \exp(j\omega t)A_n(\omega)\, d\omega \qquad (15.21)$$

with the particular solution expressed as

$$U_n = \frac{j\omega A_n \exp(j\omega t)}{C(\omega_0^2 - \omega^2) + j((1/R) - \alpha)\omega} - \frac{j\omega A_n \exp(-j\omega t)}{C(\omega_0^2 - \omega^2) + j((1/R) - \alpha)\omega} \tag{15.22}$$

Furthermore, the electrical energy density w_e inside the capacitor is given by

$$w_e = \frac{1}{8\pi}\varepsilon E^2 \tag{15.23}$$

where ε is the dielectric constant and E is the electric field, and the magnetic energy density w_m inside the capacitor is given by

$$w_m = \frac{1}{8\pi}\mu_e H^2 \tag{15.24}$$

where μ_e is the magnetic permeability constant and H is the magnetic field; hence the overall mean energy may be written as

$$\overline{w} = \frac{1}{8\pi}(\varepsilon E^2 + \mu_e H^2) \tag{15.25}$$

The electromagnetic oscillation causes the tuning or structuring of the molecules (ions) in the "electric double layers." The structuring is realized by complex motions over the various degrees of freedom, whose energy contributions depend on the positions of the individual molecules in and around the stopped droplet–film structure under the action of some periodical physical field.

The hydrodynamic motion is considered to be the motion in the potential (elastic forces) and nonpotential (resistance forces) fields. There are several possible approaches to correlate the electromagnetic and mechanical oscillations; for example, this motion may be represented by the differential equation for the forced oscillation:

$$\frac{d^2\xi}{dt^2} + 2\beta\frac{d\xi}{dt} + \omega_0^2\xi = A\cos\omega t \tag{15.26}$$

with the solution which represents the mechanical oscillation of the ordered group of molecules expressed as

$$\xi = \xi_0 \exp(-\beta t)\cos(\omega t + \zeta) \tag{15.27}$$

Now, if the electromagnetic force is assumed to be the incident (external) force, which initiates the mechanical disturbance, then the oscillation of the "continuum" particles (molecules surrounding the droplet–film structure) is described by the differential equation (15.19), where ω is the frequency of the incident oscillations. After a certain time, the oscillations of the free oscillators ω_0 (molecules surrounding the droplet–film structure) tune with the incident oscillator frequency Ω. This process of tuning between free oscillations of the environmental oscillators and the incident oscillations of the electromechanical oscillator can be expressed as

$$\Omega = (\omega_0^2 - 2\beta^2)^{0.5} \tag{15.28}$$

Thereafter, there are two possibilities, depending on the energy appearing during the tuning process; the first leads toward the constant energy or rigid sphere and the second leads toward the increasing energy or elastic sphere. For example, if a wave with high enough amplitude appears, then the rupture of the droplet–film structure occurs [7].

During the interaction of the droplet or droplet–film structure with an incident periodical physical field at the instant of equilibrium, the mean electric, electromagnetic, and mechanical energies will be equal:

$$\overline{w} = \frac{1}{8\pi}(\varepsilon E^2 + \mu H^2) = \frac{1}{2}\rho \xi_0^2 \omega^2 \tag{15.29}$$

and hence the frequency of the incident wave can be expressed as

$$\omega = \left[\frac{1}{4\pi\rho\xi^2}(\varepsilon E^2 + \mu H^2)\right]^{0.5} \tag{15.30}$$

Figure 15.5 shows the behavior of the circuit depicted in Figure 15.4c, using the correlation impedance–frequency–arbitrary droplet diameter. If the electromagnetic oscillation causes the tuning or structuring of the molecules (ions) in the electric double layer, then the structuring is realized by complex motions over the various degrees of freedom.

Since all events occur at the resonant or characteristic frequency, depending on the amount of coupling, the shape of the impedance–frequency curve is judged using the factor of merit or Q-factor [23]. The Q-factor primarily determines the sharpness of resonance of a tuned circuit and may be represented as the ratio of the reactance to the resistance, as follows:

$$Q = \frac{2\pi f L}{R} = \frac{\omega L}{R} \tag{15.31}$$

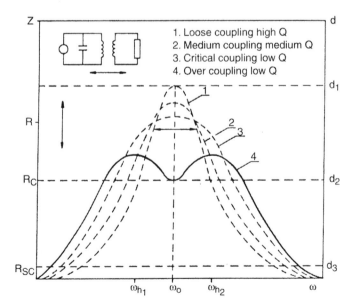

FIGURE 15.5 Impedance of the equivalent electric circuit versus its frequency. (Taken from Spasic, A.M. et al., *J Colloid Interf. Sci.*, 185, 434–446, 1997. With permission from Academic Press.)

Furthermore, the impedance Z can be related to the factor of merit Q as given by

$$Z = \frac{(2\pi fR)^2}{R} = \frac{(\omega L)^2}{R} \tag{15.32}$$

and

$$Z = \omega Q \tag{15.33}$$

From these equations and Figure 15.5, it can be seen that the impedance of a circuit is directly proportional to its effective Q at resonance. In addition, at the resonant frequency ω_0, the impedance Z is equal to the resistance R; R_c and R_{sc} represent critical and supercritical, respectively. These resistances and $Z-\omega$ curves correspond to the various levels of coupling (1 — loose coupling or high Q; 2 — medium coupling or medium Q; 3 — critical coupling or low Q; 4 — overcoupling or low Q). The ω_{h1} and ω_{h2} represent the hump frequencies that appear during the overcoupling or curve 4. On the right axes of Figure 15.5, the corresponding critical diameters d_1, d_2, and d_3 are arbitraryly plotted.

The presented theory has been applied to the representative experimental system as previously described [21–25]. Validation of the theoretical predictions was corroborated experimentally by means of electrical interfacial potential measurements and by means of nuclear magnetic resonance spectroscopy [21–25].

C. VAN DER POL DERIVATIVE MODEL: FRACTIONAL APPROACH

Fractional derivatives provide an excellent instrument for the description of memory and hereditary properties of various materials and processes [37–50]. This is the main advantage of fractional derivatives when compared with the classical integer-order models, in which such effects are in fact neglected. The mathematical modeling and simulation of systems and processes, based on the description of their properties in terms of fractional derivatives, naturally leads to differential equations of fractional order and to the necessity for solving such equations.

1. Fundamentals of Fractional Calculus

The fractional integral–differential operators (fractional calculus) present a generalization of integration and derivation to noninteger order (fractional) operators. First, one can generalize the differential and integral operators into one fundamental D_t^p operator t, which is known as fractional calculus:

$$_aD_t^p = \begin{cases} \dfrac{d^p}{dt^p} & \Re(p) > 0 \\[2mm] 1 & \Re(p) = 0 \\[2mm] \displaystyle\int_a^t (d\tau)^{-p} & \Re(p) < 0 \end{cases} \tag{15.34}$$

The two definitions generally used for the fractional differential–integral are the Grunwald definition and the Riemann–Liouville (RL) definition. The Grunwald definition is given by

$$_aD_t^p f(t) = \lim_{h \to 0} \frac{1}{h^p} \sum_{j=0}^{[(t-a)/h]} (-1)^j \binom{p}{j} f(t - jh) \tag{15.35}$$

where a and t are the limits of operator and $[x]$ means the integer part of x. The RL definition of fractional derivative is given by

$$_aD_t^p f(t) = \frac{1}{\Gamma(n-p)} \frac{d^n}{dt^n} \int_a^t \frac{f(\tau)}{(t-\tau)^{p-n+1}} d\tau \tag{15.36}$$

for $(n - 1 < p < n)$ and for the case of $(0 < p < 1)$, the fractional integral is defined as

$$_0D_t^{-p}f(t) = \frac{1}{\Gamma(p)} \int_0^t \frac{f(\tau)}{(t - \tau)^{1-p}} d\tau \tag{15.37}$$

where $\Gamma(\cdot)$ is the well-known Euler's gamma function written as

$$\Gamma(z) = \int_0^\infty e^{-t} t^{z-1} dt, \quad z = x + iy \quad \Gamma(z + 1) = \Gamma(z) \tag{15.38}$$

One of the basic properties of the gamma function is that it satisfies the functional equation

$$\Gamma(z + 1) = z\Gamma(z) \implies \Gamma(n + 1) = n(n - 1)! = n! \tag{15.39}$$

Another important property of the gamma function is that it has simple poles at the points $z = -n$, $(n = 0, 1, 2, \ldots)$. For convenience, the Laplace domain is usually used to describe the fractional integral–differential operation for solving engineering problems. The formula for the Laplace transform of the RL fractional derivative has the form [37]:

$$\int_0^\infty e^{-st} {}_0D_t^p f(t) \, dt = s^p F(s) - \sum_{k=0}^{n-1} s^k {}_0D_t^{p-k-1} f(t)_{|t=0} \tag{15.40}$$

One may note that the fractional differential operator is not a local operator, that is, the derivative is not only dependent on the value at the point but also on the value of the function on the whole interval.

For numerical calculation of fractional-order derivation, one can use relation derived from the Grunwald definition. This relation has the following form:

$$_{(t-L)}D_t^p f(t) \approx h^{-p} \sum_{j=0}^{N(t)} b_j f(t - jh) \tag{15.41}$$

where L is the "memory length," h the step-size of the calculation

$$N(t) = \min\left\{\left[\frac{t}{h}\right], \left[\frac{L}{h}\right]\right\} \tag{15.42}$$

$[x]$ the integer part of x, and b_j the binomial coefficient given by

$$b_0 = 1, \quad b_j = \left(1 - \frac{1 + \alpha}{j}\right) b_{j-1} \tag{15.43}$$

For the solution of the fractional-order differential equations, most effective and easy analytic methods were developed based on the formula of the Laplace transform method of the Mittag–Leffler function in two parameters; for further details see Refs. [19,38].

2. Example of Analog Realization of a Fractional Element

Constructing an analog realization of a fractional-order element may be much easier than the discreet ladder circuits proposed so far. If a single material exhibited fractance characteristics, then a single component would replace the entire network. In fact, an ideal capacitor does not exist. An ideal dielectric in a capacitor having an impedance of the form $1/j\omega C$ would violate causality. It seems sensible to look for dielectric materials exhibiting the more realistic fractional behavior $1/(j\omega C)^\alpha$, where $p \approx 0.5$. Such a component would display "fractance" attributes and could be termed a "fractor," as opposed to a resistor or capacitor. In this case, a single component would do the job of an entire network of "ideal" components. Moreover, the capacitor's impedance is described by the transfer function

$$Z(s) = 1/Cs^p, \quad 0 < p < 1 \tag{15.44}$$

In addition, such circuits can be obtained if generalized models of resistors, capacitors, and induction coils are taken. For example, authors [27] investigated lithium hydrazinium sulfate ($LiN_2H_5SO_4$) and found that "over a large range of temperature and frequency, the real and imaginary parts of the susceptibility are very large (up to $\varepsilon' \approx \varepsilon'' = 10^6$, $\varepsilon = \varepsilon' + j\varepsilon''$) and vary with frequency somewhat as $f^{-1/2''}$, one can get $\varepsilon = \varepsilon_r \sqrt{2}(j\omega)^{-1/2}$. Using the definition of the relationship between the dielectric function and the impedance

$$Z = \frac{1}{j\omega C_C \varepsilon_r \sqrt{2}(j\omega)^{-1/2}} \tag{15.45}$$

where C_C is the empty cell capacitance, one get the impedance of the "factor"

$$Z_F = \frac{K}{s^{1/2}} \tag{15.46}$$

converting to the Laplace notation $j\omega \to s$, where K represents the lumped constants of the impedance.

3. Solution of the Representative Linear Model

For the sake of clarity, introducing van der Pol derivative model, and looking back in Figure 15.2 and Figure 15.4, governing linear integral–differential equation is given by

$$C\frac{dU}{dt} + \left(\frac{1}{R} - \alpha\right)U + \frac{1}{L}\int U dt = i(t) = \frac{1}{2\pi}\int_{-\infty}^{\infty} \exp(j\omega t)A_n(\omega)\,d\omega \tag{15.21}$$

where $i(t)$ is the noise current and $A_n(\omega)$ is the spectral distribution of the noise current as a function of frequency. Particular solution of Equation (15.21) is expressed as

$$U_n = \frac{j\omega A_n \exp(j\omega t)}{C(\omega_0^2 - \omega^2) + j((1/R) - \alpha)\omega} - \frac{j\omega A_n \exp(-j\omega t)}{C(\omega_0^2 - \omega^2) + j((1/R) - \alpha)\omega} \tag{15.22}$$

Here, the capacitive and inductive elements, using fractional order $p \in (0, 1)$, enable formation of the fractional differential equation, that is, more flexible or general model of liquid–liquid interfaces behavior. Now, using again, for example, RL definition of fractional derivative and integral is given by

$$_0D_t^p[U(t)] = \frac{d^p U}{dt^p} = \frac{1}{\Gamma(1-p)}\frac{d}{dt}\int_0^t \frac{U(\tau)}{(t-\tau)^p}d\tau \quad 0 < p < 1$$

$$_0D_t^{-p}[U(t)] = \frac{1}{\Gamma(p)}\int_0^t \frac{U(\tau)}{(t-\tau)^{1-p}}d\tau \quad p > 0$$

(15.47)

where $\Gamma(\cdot)$ denotes Euler's gamma function

$$\Gamma(z) = \int_0^\infty e^{-t}t^{z-1}dt, \quad z = x + iy \quad \Gamma(z+1) = \Gamma(z)$$

(15.48)

So, in that way, one can obtain linear fractional differential equation with zeros as initial conditions as follows:

$$C_0D_t^p[U(t)] + \left(\frac{1}{R} - \alpha\right)U + \frac{1}{L}{}_0D_t^{-p}[U(t)] = i(t)$$

(15.49)

Using the Laplace transform of Equation (15.49) leads to

$$G(s) = \frac{U(s)}{i(s)} = \frac{1}{Cs^p + 1/Ls^{-p} + ((1/R) - \alpha)} = \frac{s^p}{Cs^{2p} + ((1/R) - \alpha)s^p + 1/L}$$

(15.50)

or

$$G(s) = s^p G_3(s), \quad G_3(s) = \frac{1}{as^{2p} + bs^p + c} \quad a = C, \ b = (1/R - \alpha), \ c = 1/L$$

(15.51)

The term-by-term inversion, based on the general expansion theorem for the Laplace transform [37], produces

$$G_3(t) = \frac{1}{a}\sum_{k=0}^\infty \frac{(-1)^k}{k!}\left(\frac{c}{a}\right)^k t^{2p(k+1)-1} E_{2p-p,\,2p+pk}^{(k)}\left(-\frac{b}{a}t^{2p-p}\right),$$

(15.52)

where $E_{\lambda,\mu}(z)$ is the Mittag−Leffler function in two parameters

$$E_{\lambda,\mu}^{(k)}(t) = \frac{d^k}{dt^k}E_{\lambda,\mu}(t) = \sum_{j=0}^\infty \frac{(j+k)!t^j}{j!\Gamma(\lambda j + \lambda k + \mu)} \quad k = 0, \ 1, \ 2, \ldots$$

(15.53)

Inverse Laplace transform of $G(s)$ is fractional Greens function:

$$G(t) = D^p G_3(t)$$

(15.54)

where the fractional derivative of $G_3(t)$, Equation (15.54), are evaluated with the help of Equation (15.47) and Equation (15.48). Finally, an explicit representation of the solution is

$$U(t) = \int_0^t G(t-u)i(u)\,du$$

(15.55)

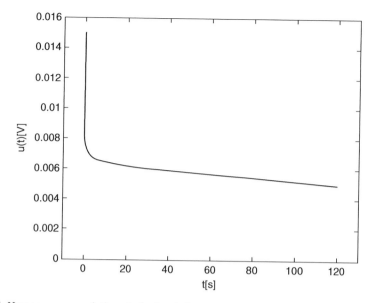

FIGURE 15.6 Homogeneous solution (calculated from Equation (15.10)): $\alpha = 0.9995$; $U_0 = 15$ mV; $p = 0.95$; $T = 0.001$ sec.

Now, again the initial electromagnetic oscillation is represented by the differential equation (15.10), and when the nonlinear terms are omitted or canceled the first step, homogeneous solution may be obtained using numerical calculation derived from the Grunwald definition, Equations (15.41)–(15.43), as shown in Figure 15.6.

The calculation has been done for the following parameters: $\alpha = 0.9995$; $U_0 = 15$ mV; $p = 0.95$; $T = 0.001$ sec.

Further on, considering Equation (15.11) and again using Equations (15.41)–(15.43) nonhomogeneous solution is obtained and presented in Figure 15.7.

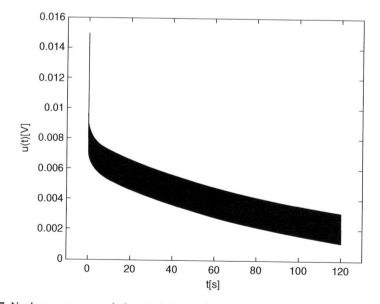

FIGURE 15.7 Nonhomogeneous solution (calculated from Equation (15.11)): $\alpha = 0.95$; $U_0 = 15$ mV; $p = 0.95$; $T = 0.01$ sec; $A_n = 0.05$ nm; $\cos(\omega_0 t) = \mathrm{d}^r/\mathrm{d}t^r$ (sin $\omega_0 t$).

The obtained result appears as a band because the input (cos) is of the fractional order and the output is in a damped oscillatory mode of high frequencies.

The calculation has been done for the parameters [where the derivative of $\cos(\omega_0 t)$ is also fractional of the order r]: $\alpha = 0.95$; $U_0 = 15\,\text{mV}$; $p = 0.95$; $T = 0.01\,\text{sec}$; $A_n = 0.05\,\text{nm}$; $\cos(\omega_0 t) = \mathrm{d}^r/\mathrm{d}t^r\,(\sin \omega_0 t)$.

Here, the presented model is derived only for the linearized van der Pol equation, while the more realistic nonlinear case as well as calculated and experimentally corroborated data follow Refs. [37–50].

4. Solution of the Representative Nonlinear Model

Nonlinear fractional differential equations have received rather less attention in the literature, partly because many of the model equations proposed have been linear. Here, a nonlinear integral–differential equation of the van der Pol type will be considered. This equation represents the droplet or droplet–film structure formation, "breathing," or destruction processes and taking into account the particular frequency component, which is included in the driving force and given by Equation (15.12):

$$C\frac{\mathrm{d}U}{\mathrm{d}t} + \left(\frac{1}{R} - \alpha\right)U + \frac{1}{L}\int U\,\mathrm{d}t + \gamma U^3 = i(t) \tag{15.12}$$

In addition, one can obtain equivalent nonlinear problem by applying differentiation of Equation (15.12) such as:

$$C\frac{\mathrm{d}^2 U}{\mathrm{d}t^2} + \left(\frac{1}{R} - \alpha + 3\gamma U^2\right)\frac{\mathrm{d}U}{\mathrm{d}t} + \frac{1}{L}U = \frac{\mathrm{d}i(t)}{\mathrm{d}t} = i^\circ(t) \tag{15.56}$$

Taking into account expression (15.47), using the same reasoning as in the solution of linear case, it yields:

$$C_0 D_t^{2p} U(t) + \left(\frac{1}{R} - \alpha + 3\gamma U(t)^2\right){}_0 D_t^p U(t) + \frac{1}{L}U(t) = {}_0 D_t^p i(t) = i^\circ(t) \tag{15.57}$$

or

$${}_0 D_t^{2p} U(t) = -\frac{1}{C}\left(\frac{1}{R} - \alpha\right){}_0 D_t^p U(t) - \frac{3\gamma}{C}U^2(t){}_0 D_t^p U(t) - \frac{1}{CL}U(t) + \frac{1}{C}i^\circ(t) \tag{15.58}$$

Now, one may convert previous equation with commensurate multiple fractional derivatives into equivalent system of equations of lower order. Let

$$x_1(t) = U(t), \quad x_2(t) = {}_0 D_t^p U(t), \quad p \in \mathbb{Q} \tag{15.59}$$

In that way, one can get

$${}_0 D_t^p x_1(t) = {}_0 D_t^p U(t) = x_2(t)$$

$${}_0 D_t^p x_2(t) = {}_0 D_t^{2p} U(t) = -\frac{1}{C}\left(\frac{1}{R} - \alpha\right)x_2(t) - \frac{3\gamma}{C}x_1^2(t)x_2(t) - \frac{1}{CL}x_1(t) + \frac{1}{C}i^\circ(t) \tag{15.60}$$

In condensed form, introducing vector $x(t) = (x_1, x_2)^{\mathrm{T}}$, is presented as follows:

$$
_0D_t^p x(t) = \begin{bmatrix} 0 & 1 \\ \dfrac{-1}{LC} & -\dfrac{(1/R - \alpha)}{C} \end{bmatrix} \begin{Bmatrix} x_1(t) \\ x_2(t) \end{Bmatrix} + \begin{bmatrix} 0 & 0 \\ 0 & \dfrac{-3\gamma x_1^2(t)}{C} \end{bmatrix} \begin{Bmatrix} x_1(t) \\ x_2(t) \end{Bmatrix} + \begin{bmatrix} 0 \\ \dfrac{1}{C} \end{bmatrix} i^\circ(t) \quad (15.61)
$$

or

$$
_0D_t^p x(t) = Ax(t) + B(x_1(t))x(t) + Fi^\circ(t) \tag{15.62}
$$

It is easily observed that previous case is the one of the general case for this nonlinear problem, which can be obtained in the form

$$
_0D_t^p x(t) = f(t,\, x(t)) \tag{15.63}
$$

subject to the initial conditions

$$
\frac{d^{p-k}}{dt^{p-k}} x(t)_{|t=0^+} = c_k, \quad k = 0, 1, \ldots, n = [p+1] \tag{15.64}
$$

with given values c_k, where n is integer defined by $n - 1 < p < n$, Equation (15.15); in our case, $n = 1$.

In practical applications, the initial conditions are frequently not available, and it may not even be clear what their physical meaning is [48]. Therefore, to avoid this conflict, Caputo [45] has proposed that one should incorporate the integer order (classical) derivative of function x, as they are commonly used in initial value problems with integer-order equations. In that way, one can use the derivatives of the Caputo type such as:

$$
{}_0^C D_t^p [U(t)] = \frac{d^p U}{dt^p} = \frac{1}{\Gamma(1-p)} \int_0^t \frac{U^{(1)}(\tau)}{(t-\tau)^p}\, d\tau, \quad 0 < p < 1, \quad U^{(1)}(\tau) = \frac{dU}{d\tau} \tag{15.65}
$$

From the definition of RL and Caputo derivatives, one may observe that the relation between the two fractional derivatives is [49]:

$$
{}_0^C D_t^p [U(t)] = {}_0D_t^p[(U - T_{n-1}[U])(t)] \tag{15.66}
$$

where $T_{n-1}[U]$ is the Taylor polynomial of order $(n-1)$ for U, centered at 0. So, one can specify the initial conditions in the classical form

$$
U^{(k)}(0) = U_0^{(k)}, \quad k = 0, 1, \ldots, n-1 \tag{15.67}
$$

The RL and Caputo formulation coincide when the initial conditions are zero. Now, nonlinear equation (15.62) is given by

$$
{}_0^C D_t^p x(t) = f(t,\, x(t)) \tag{15.68}
$$

subject to the initial conditions

$$
x^{(k)}(0) = x_0^{(k)}, \quad k = 0, 1, \ldots, [p] \tag{15.69}
$$

Here, the problem of finding a unique continuous solution of Equation (15.68) and Equation (15.69) is considered. Using some classical results from the fractional calculus, that is, applying the following lemma, one can obtain a solution of initial value problem of Equation (15.68) and Equation (15.69) [50].

Lemma 1: *Let the function f be continuous. The function $x(t)$ is a solution of Cauchy problem (15.68) and (15.69) if and only if*

$$x(t) = \sum_{j=0}^{n-1} \frac{t^j}{j!} x^{(j)}(0) + \frac{1}{\Gamma(p)} \int_0^t f(s, x(s))(t-s)^{p-1} ds, \quad n-1 < p \le n \qquad (15.70)$$

This equation is weakly singular if $0 < p < 1$ and regular for $p \ge 1$. In the former case, we must give explicit proofs for existence and uniqueness of the solution. Hence, results that are very similar to the corresponding classical theorems of existence and uniqueness, known in the scalar case of first-order equations, are discussed subsequently [49].

Theorem 1 (Existence): *Assume that D:$[0, T^\bullet] \times [x_0^{(0)} - c, x_0^{(0)} + c]$ with some $T^\bullet > 0$ and some $c > 0$ and let the function $f : D \to \mathbb{R}$ be continuous. Furthermore, define $T = \min\{T^\bullet, (c\Gamma(p+1)/\|f\|_\infty)^{1/p}\}$. Then, there exists a function $x : [0, T] \to \mathbb{R}$ solving the Cauchy problem (15.68) and (15.69) — scalar case.*

Theorem 2 (Uniqueness): *Assume that D:$[0, T^\bullet] \times [x_0^{(0)} - c, x_0^{(0)} + c]$ with some $T^\bullet > 0$ and some $c > 0$. In addition, let the function $f : D \to \mathbb{R}$ be bounded on D and fulfill a Lipschitz condition with respect to the second variable, that is*

$$|f(t, \tilde{x}) - f(t, \bar{x})| \le \lambda |\tilde{x} - \bar{x}| \qquad (15.71)$$

with some constant $\lambda > 0$ independent of t, $\bar{x}$, $\tilde{x}$. Then, there exists at most one function $x : [0, T] \to \mathbb{R}$ solving the Cauchy problem (15.68) and (15.69) — scalar case.

The generalization of previous theorems to vector-valued functions x is immediate. The proof of the uniqueness theorem will be based on the generalization of Banach's fixed point theorem [50].

Remark 1: *Without the Lipschitz assumption on f*, the solution need not to be unique.

a. Numerical Methods for Nonlinear Equations

In some cases, for numerical calculation of nonlinear equations, one can use a fact that fractional derivative is based on a convolution integral, the number of weights used in the numerical approximation to evaluate fractional derivatives. In addition, one can apply predictor–corrector formula for the solution of systems of nonlinear equations of lower order. This approach is based on rewriting the initial value problem (15.68) and (15.69) as an equivalent fractional integral equation (Volterra integral equation of the second kind)

$$x(t) = x_0 + \frac{1}{\Gamma(p)} \int_0^t f(s, x(s))(t-s)^{p-1} ds \qquad (15.72)$$

Then, one may introduce uniformly distributed grid points $t_k = kh$, $h = T/\omega$, and $\omega \in \mathbb{N}$. Next step is to find an approximation $x_k = x(t_k)$ by predictor–corrector approach. For predictor, one can use fractional Euler's method, and for the corrector, one can use the product trapezoidal formula [49]. In addition, we desire numerical schemes that are *convergent*, *consistent*, and *stable*, but these properties are not treated here.

IV. CONCLUSIONS

So far, three possible mathematical formalisms have been discussed related to the developed theory of electroviscoelasticity. The first is tension tensor model where the normal and tangential forces are considered regardless, only from mathematical point of view, of their origin (mechanical or electrical). The second is van der Pol integral-derivative model. Finally, the third model presents an effort to generalize the van der Pol integral–differential equation; the ordinary time derivative and integral are now replaced with corresponding fractional-order time derivative and integral of order $0 < p < 1$.

Each of these mathematical formalisms, although related to the same physical formalism, facilitate a better understanding of different aspects of a droplet existence (formation, life, and destruction).

Tension tensor model discusses the force equilibrium at the interfaces, either deformable or rigid, but its solution is difficult because the tensor contain nonlinear and complex elements.

van der Pol derivative model is convenient for discussion of the antenna output circuit, the resulting equivalent electrical circuit; but in the case of nonlinear oscillators, that is here the realistic one, the problem of determining the noise output is complicated by the fact that the output is fed back into the system, thus modifying the effective noise input in a complicated manner. The noise output appears as an induced anisotropic effect.

The theory of electroviscoelasticity using fractional approach constitutes a new interdisciplinary approach to colloid and interface science. Hence, (1) more degrees of freedom are in the model, (2) memory storage considerations and hereditary properties are included in the model, and (3) history or impact to the present and future is in the game.

ACKNOWLEDGMENTS

This project was financially supported by Ministry of Science and Environmental Protection of Serbia, entitled "Multiphase Dispersed Systems" 101822.

REFERENCES

1. Osipov, I.L., *Surface Chemistry*, Reinhold, New York, 1964, p. 295.
2. Spasic, A.M., Jokanovic, V. and Krstic, D.N., A theory of electroviscoelasticity: a new approach for quantifying the behavior of liquid–liquid interfaces under applied fields., *J. Colloid Interf. Sci.*, 185, 434–446, 1997.
3. Spasic, A.M., Electroviscoelasticity of liquid–liquid interfaces, in *Interfacial Electrokinetics and Electrophoresis*, Delgado, A.V., Ed., 106 Surfactany Science Series, Marcel Dekker, Inc., New York, 2002, pp. 837–868.
4. Pao, Y.H., Hydrodynamic theory for the flow of viscoelastic fluid, *J. Appl. Phys.*, 28, 591–598, 1957.
5. Drexler, E.K., *Nanosystems: Molecular Machinery, Manufacturing and Computation*, Wiley, New York, 1992, 161–189.
6. Joseph, D.D., *Fluid Dynamics of Viscoelastic Liquids*, Springer-Verlag, New York, 1990, 539 pp.
7. Spasic, A.M. and Jokanovic, V., Stability of the secondary droplet–film structure in polydispersed systems, *J. Colloid Interf. Sci.*, 170, 229–240, 1995.

8. Spasic, A.M. and Krstic, D.N., Structure and stability of electrified liquid–liquid interfaces, in *Chemical and Biological Sensors and Analytical Electrochemical Methods*, Rico, A.J., Butler, M.A., Vanisek, P., Horval, G. and Silva, A.E. Eds., ISE & ECS Penington, N.J., 1997, pp. 415–426.

9. Spasic, A.M. and Krstic, D.N., A new constitutive model of liquids, in *Thirteenth International Congress of Chemical and Process Engineering*, CHISA, Praha, 1998, CD 7.

10. Condon, E.U. and Odishaw, H., Eds., *Handbook of Physics*, McGraw Hill, New York, 1958, p. 4/13.

11. Prigogine, I., *The Molecular Theory of Solutions*, North Holland, New York, 1957.

12. Garstens, M.A., Noise in nonlinear oscillators, *J. Appl. Phys.*, 28, 352–356, 1957.

13. Davis, S.H., Problems in Fluid Mechanics, *Trans ASME Ser. E., J. Appl. Mech.*, 50, 977–982, 1983.

14. Ericksen, J.L., *Isledovanie po mehanike splosnih sred*, Mir, Moscow, 1977.

15. Andjelic, P.T., *Tenzorski racun*, Beograd, N.K., 1967, p. 263.

16. Friberg, S.E. and Bothorel, R., Eds., *Microemulsions: Structure and Dynamics*, CRC Press, Boca Raton, FL, 1987, 173 pp.

17. Godfrey, J., Hanson, C., Slater, M.J. and Tharmalingam, Sh., *AIChE Symp. Ser.*, 74, 1978.

18. Godfrey, J., *The Formation of Liquid–Liquid Dispersions — Chemical and Engineering Aspects — Flow Phenomena of Liquid–Liquid Dispersions in Process Equipment*, Institute of Chemical Engineering London, 1984.

19. Lazarević, M.P. and Spasić, A.M., Electroviscoelasticity of liquid–liquid interfaces: fractional approach, in *Fourth European Congress of Chemical Engineering*, Part 5, Granada, Spain, September 21–25, 2003, 5.2. 33, Abstract of Papers.

20. Spasic, A.M., Djokovic, N.N., Babic, M.D. and Marinko, M.M., A new approach to the existence of micro, nano, and atto dispersed systems, in *Thirteenth International Congress of Chemical and Process Engineering*, CHISA, Praha, 1998, CD 7.

21. Spasic, A.M., A new topics in fine dispersed systems, in *Twelfth International Congress of Chemical and Process Engineering*, CHISA, Praha, 1996, C1.6.

22. Spasic, A.M., Mechanism of the secondary liquid–liquid droplet–film rupture on inclined plate, *Chem. Eng. Sci.* 47 (15/16), 3949–3957, 1992.

23. Orr, W.I., *Radio Handbook*, 17th ed., Editors and Engineers, Ltd, New Augusta, Indiana, 1967, p. 60.

24. Spasic, A.M., Djokovic, N.N., Babic, M.D., Marinko, M.M. and Jovanovic, G.N., Performance of demulsions: entrainment problems in solvent extraction, *Chem. Eng. Sci.*, 52 (5), 657–675, 1997.

25. Jaric, J., *Mehanika kontinuuma*, Beograd, G.K., Ed., 1988, 117 pp.

26. Schramm, L.L., Ed., *Emulsions: Fundamentals and Application in Petroleum Industry*, American Chemical Society, Washington DC, 1992, 131 pp.

27. Schmidt, V.H., Drumheller. Dielectric properties of lithium hydrazinium sulfate. *Phys. Rev. B.*, 4, 4582–4597, 1971.

28. Mitrovic, M. and Jaric, J., Lecture Mathematical Faculty, University of Belgrade, Belgrade, 1997.

29. Tadros, Th.F., *Interfacial Aspects of Emulsification*, Institute of Chemical Engineering, London, 1984.

30. Hartland, S. and Wood, S.M., *AIChE J.*, 19, 810, 1983.

31. Spisak, W., Toroidal bubbles, *Nature*, 23, 349, 1991.

32. Oldshue, J.J., *Fluid Mixing Technology*, McGraw-Hill Pub. Co, New York, 1983, p. 125.

33. Grinfeld, M.A. and Norris, A.N., Hamiltonian and onsageristic approaches in the nonlinear theory of fluid-permeable elastic continua, *Int. J. Eng. Sci.*, 35, 75–87, 1997.

34. Pugh, R.J., Foaming, foam films, antifoaming and defoaming, *Adv. Colloid Interf. Sci.*, 64, 67–142, 1996.

35. Adamson, A., *Physical Chemistry of Surfaces*, John Wiley, New York, 1967, p. 505.

36. Lo, T.C., Baird, M. and Hanson, C., Eds., *Handbook of Solvent Extraction*, Wiley, New York, 1983, p. 275.

37. Oldham, K.B. and Spanier, J., The fractional calculus, *Mathematics in Science and Engineering*, Vol. III, Academic Press, New York, 1974.

38. Podlubny, I., *Fractional Differential Equations*, Academic Press, San Diego, 1999.

39. Mainardi, F., Fractional calculus: some basic problems in continuum and statistical mechanics, in *Fractals and Fractional Calculus in Continuum mechanics*, Carpinteri, A. and Mainardi, F., Eds., Springer, Wien, 1997, pp. 291–348.

40. Sakakibara, S., Properties of vibration with fractional derivative damping of order 1/2, *JSME Int. J.*, C40, 393–399, 1997.

41. Bagley, R.L. and Calico, R.A., Fractional order state equations for the control of viscoelasticity damped structures, *J. Guidance.*, 14 (2), 1412–1417, 1991.
42. Oustaloup, A., *La Commande CRONE*, Hermes, Paris, 1991.
43. Matignon, D., Stability results for fractional differential equations with application to control processing, in *Computation Engineering in Systems Applications*, IMACS, IEEE-SMC Lille, France, July, 1996.
44. Lazarević, P.M., D^α Type iterative learning control for fractional LTI system, in *Proceedings of ICCC2003*, Tatranska Lomnica, Slovak Republic, May, 26–29, 2003, p. 869.
45. Caputo, M., *Elasticita e Dissipazione*, Zanichelli, Bologna, Italy, 1969.
46. Babenko, Yu., *Heat and Mass Transfer*, Chimia, Leningrad, 1986.
47. Torvik, P.J. and Bagley, R.L., A theoretical basis for the application of fractional calculus to viscoelasticity, *J. Rheol.*, 27, 201–210, 1983.
48. Giannantoni, C., The problem of the initial conditions and their physical meaning in linear differential equations of fractional order, *Appl. Math. Comput.*, XXX, 1–15, 2003.
49. Diethelm, K. and Ford, N.J., Analysis of fractional differential equations, *J. Math. Anal. Appl.*, 265, 401–418, 2002.
50. El-Sayed, A., Multivalued fractional differential equation, *J. Appl. Math. Comput.*, 68 (1), 15–25, 1995.

16 Production of Monodispersed Emulsions Using Shirasu Porous Glass Membranes

Goran T. Vladisavljevic
University of Belgrade, Belgrade-Zemun, Serbia, Serbia and Montenegro

Masataka Shimizu and Tadao Nakashima
Miyazaki Prefectural Industrial Support Foundation, Sadowara, Miyazaki, Japan

Helmar Schubert
University of Karlsruhe (T.H.), Karlsruhe, Germany

Mitsutoshi Nakajima
National Food Research Institute, Tsukuba, Ibaraki, Japan

CONTENTS

I. INTRODUCTION

Membrane emulsification (ME) is a new technology for making monodisperse emulsions over a wide spectrum of mean droplet sizes, ranging from ca. 0.5 μm to several tens of micrometer. ME involves production of droplets individually (drop-by-drop) by extrusion of pure disperse phase through a porous membrane into a moving continuous phase (direct ME) or by the passage of a previously prepared coarse emulsion (premix) through the membrane (premix ME). The membrane must not be wetted with the continuous phase; For instance, oil-in-water (O/W) emulsions are prepared using a hydrophilic membrane and water-in-oil (W/O) emulsions are prepared using a hydrophobic membrane. If the membrane wall is wetted with the continuous phase, droplet disruption in coarsely emulsified feeds can be followed by phase inversion, that is, a fine W/O emulsion can be produced from the O/W coarse emulsion and vice versa. To force a pure disperse phase or pre-emulsified feed through the membrane, a pressure gradient across the membrane is applied, usually by pressurizing the disperse phase with a compressed gas in a pressure vessel.

The most suitable membrane for emulsification is the so-called Shirasu porous glass (SPG) membrane. This membrane can be fabricated with a wide range of mean pore sizes (0.05–20 μm), with a uniform pore size distribution (PSD), and with a wall porosity between 50 and 60%. Hydrophobic modification of SPG membrane can be easily carried out by surface coating with silicone resin.

II. SPG MEMBRANE

SPG is a special kind of porous glass, obtained by phase separation of a primary CaO—Al_2O_3—B_2O_3—SiO_2 type glass, made of Shirasu (volcanic ash from the southern part of Kyushu island), calcium carbonate, and boric acid [1]. Shirasu is added as a source of SiO_2 and Al_2O_3, but since it is a natural material, it also contains small amounts of other components such as Na_2O, K_2O, Fe_2O_3, etc. [2]. After the primary glass is formed into tube, it is subjected to heat treatment at a temperature of 650–750°C for the periods ranging from several hours to tens of hours [2,3]. This treatment causes the homogeneous primary glass to transform into a heterogeneous two-phase glass consisting of Al_2O_3—SiO_2 and CaO—B_2O_3 phases. Since the CaO—B_2O_3 phase is readily soluble in acid, a porous-glass membrane is obtained by immersing the phase-separated glass into a dilute solution of hydrochloric acid. The mean pore size of SPG membrane can be controlled by adjusting the heat treatment conditions (time and temperature).

SPG finds many applications as a packing material for HPLC columns, a carrier of enzymes and tissue cultures, an injection needle for blood transfusion and dialysis, a high-functional dispersion medium for preparing uniform droplets and microbubbles [3–5], and a separation medium for microfiltration of emulsions and suspensions [6–8]. The microstructure of SPG is of primary importance in any application. In ME, the droplet size distribution of prepared emulsions depends critically on the PSD and wetting characteristics of SPG membrane used. The geometry of the pore network plays a very important role in providing high mechanical strength and high thermal resistance shock of SPG when compared with the porous Vycor glass and a porous alumina of the same porosity [9].

A. Pore Structure of SPG Membrane

Figure 16.1 and Figure 16.2 are scanning electron micrographs (SEM) of two SPG membranes largely differing in mean pore size, but with virtually the same porosity (56.65 ± 0.05%). It is

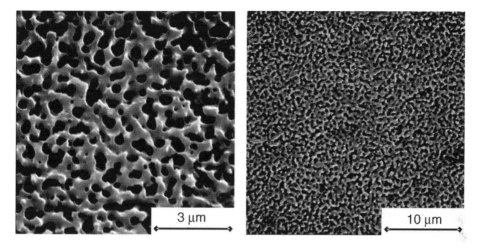

FIGURE 16.1 SEM of the hydrophilic SPG membrane with a mean pore size of 0.262 μm. (From Vladisavljević, G.T., Shimizu, M., and Nakashima, T., *J. Membr. Sci.*, 250 (1–2), 69–77, 2005. With permission.)

clear that the pore structure remains unchanged when the mean pore size increases from 0.262–14.8 μm. In both cases, the membrane contains cylindrical tortuous pores forming a three-dimensional interconnected network. All of the pore cross-sections are not circular, because the pores do not always intersect with the external surfaces at a right angle, and the pore joints also appear in the micrographs. A noncircular cross-section of the pores plays an important role in providing spontaneous detachment of disperse phase from the pore outlets in direct emulsification [11].

Figure 16.3 is a SEM picture of the same membrane as shown in Figure 16.1, but after coating with a silicone resin (KP-18C, a product of Shin-Etsu Chem. Ind. Co., Ltd, Japan). The coating was carried out by immersion of the membrane tube in a diluted resin solution (1:100) at room temperature for 24 h, followed by drying in an oven at 100–120°C for several hours [12]. Although the presence of coated layer is visible in the micrographs, it can be seen that the pores are not blocked by the resin.

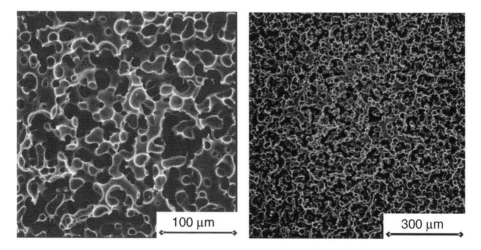

FIGURE 16.2 SEM of the hydrophilic SPG membrane with a mean pore size of 14.8 μm. (From Vladisavljević, G.T., Shimizu, M., and Nakashima, T., *J. Membr. Sci.*, 250 (1–2), 69–77, 2005. With permission.)

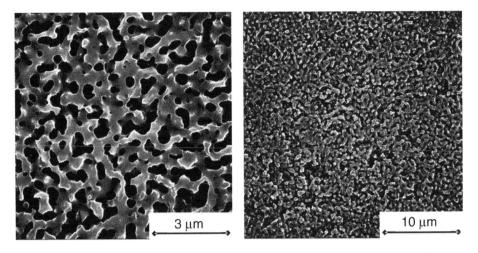

FIGURE 16.3 SEM of the hydrophobic SPG membrane with a mean pore size of 0.262 μm. (From Vladisavljević, G.T., Shimizu, M., and Nakashima, T., *J. Membr. Sci.*, 250 (1–2), 69–77, 2005. With permission.)

The PSD of the SPG membranes given in Figure 16.1 and Figure 16.2 is shown in Figure 16.4. Obviously, the PSD curves of the two membranes are of similar shape and width, in spite of the fact that their mean pore sizes largely differ. The PSD curves of nine different SPG membranes over the range of mean pore sizes of 0.26–20.3 μm are shown in Figure 16.5. The span of PSD was calculated from the experimental data obtained by mercury porosimetry, using the equation: span $= (d_{90} - d_{10})/d_{50}$, where d_X is the pore diameter corresponding to X vol% on a cumulative pore volume curve. The span values were in the range of 0.29–0.68 (Table 16.1), which is comparable with the typical spans of the particle size distribution of the emulsions prepared using SPG membranes [13–15]. The least uniform pores (span $= 0.68$) were obtained at the largest mean pore size of 20.3 μm.

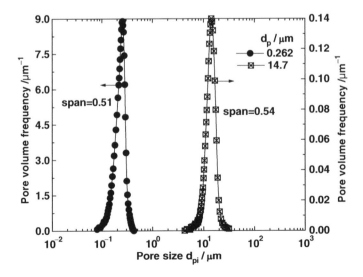

FIGURE 16.4 PSD on a frequency basis of SPG membranes shown in Figure 16.1 and Figure 16.2. (From Vladisavljević, G.T., Shimizu, M., and Nakashima, T., *J. Membr. Sci.*, 250 (1–2), 69–77, 2005. With permission.)

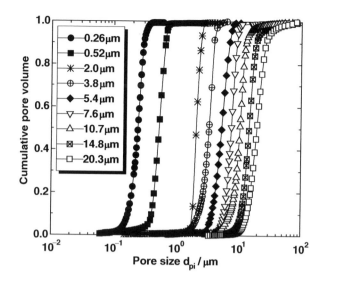

FIGURE 16.5 PSD curves on a cumulative basis of SPG membranes with different mean pore sizes. (From Vladisavljević, G.T., Shimizu, M., and Nakashima, T., *J. Membr. Sci.*, 250 (1–2), 69–77, 2005. With permission.)

B. Permeability of SPG Membrane to Pure Liquids

The permeation of wettable liquids through SPG membrane can be analyzed using the nonuniform cylindrical capillary model. Consider a piece of SPG membrane (tube or disc) with a cross-sectional area of A_m, a thickness of δ_m, and a dry weight of m_m containing nonuniform cylindrical pores. The number of pores per unit membrane weight in the ith range of pore sizes, the mean diameter of which is d_{pi}, can be expressed as:

$$\frac{N_i}{m_m} = \frac{4V_{pi}}{\pi d_{pi}^2 L_i m_m} = \frac{4V_{pi}}{\pi d_{pi}^2 \delta_m \xi_i m_m}$$

(16.1)

where V_{pi} is the pore volume in the ith range of pore sizes and L_i and ξ_i are the mean length and mean tortuosity factor of these pores, respectively. The apparent density of the membrane

TABLE 16.1
Characteristics of Pore Structure of SPG Membranes as a Function of the Mean Pore Size, d_p

d_p (μm)	Span of PSD	ε (%)	ρ_a (kg/dm³)	ρ (kg/dm³)	V_p/m_m (dm³/kg)	A_p/m_m (dm²/g)	δ_m (μm)
0.262	0.51	56.7	1.08	2.28	0.527	95.6	860
0.525	0.46	56.2	1.04	2.37	0.541	46.7	845
1.96	0.29	55.3	1.08	2.43	0.510	11.3	860
3.76	0.49	57.4	0.971	2.28	0.591	8.67	655
5.39	0.60	58.1	0.954	2.28	0.609	6.21	895
7.63	0.44	50.4	0.992	2.00	0.508	2.78	710
10.7	0.55	55.2	0.980	2.19	0.563	3.35	795
14.8	0.54	56.6	1.00	2.31	0.564	2.76	750
20.3	0.68	54.3	0.971	2.13	0.560	1.46	790

Source: From Vladisavljević, G.T., Shimizu, M., and Nakashima, T., *J. Membr. Sci.*, 250 (1–2), 69–77, 2005. With permission.

is given by

$$\rho_a = \frac{m_m}{V_m} = \frac{m_m}{A_m \delta_m} \qquad (16.2)$$

By substituting Equation (16.2) in Equation (16.1) and assuming that the mean tortuosity factor of the pores is independent on the pore size ($\xi_i = \xi = \text{const}$), one obtains the expression for the number of pores per unit membrane area in the ith range of pore sizes:

$$\frac{N_i}{A_m} = \frac{4\rho_a V_{pi}}{\pi d_{pi}^2 \xi m_m} \qquad (16.3)$$

The total number of pores per unit membrane area is represented by

$$\frac{N}{A_m} = \frac{4\rho_a}{\pi \xi} \sum_{i=1}^{k_s} \frac{V_{pi}/m_m}{d_{pi}^2} \qquad (16.4)$$

where k_s is the number of pore size ranges, that is, the number of size channels provided by a pore size analyzer. V_{pi}/m_m vs. d_{pi} data can be easily obtained by mercury porosimetry and thus N/A_m can be calculated using Equation (16.4), if ξ and ρ_a are known.

According to the Hagen–Poiseuille law, the volumetric rate of flow of a Newtonian liquid flowing through the pores with a diameter of d_{pi} under the transmembrane pressure of Δp_{tm} is given by

$$Q_i = \frac{\Delta p_{tm} \pi d_{pi}^4 N_i}{128 \eta \delta_m \xi} \qquad (16.5)$$

where η is the viscosity of permeating liquid. Equation (16.5) holds only if the permeating liquid wets the membrane wall, but the membrane may be either hydrophilic or hydrophobic. Using Equation (16.3) and Equation (16.5), one obtains the total liquid rate of flow through the membrane

$$Q = \sum_{i=1}^{k_s} Q_i = \frac{\Delta p_{tm} A_m \rho_a}{32 \eta \delta_m \xi^2} \sum_{i=1}^{k_s} \frac{V_{pi}}{m_m} d_{pi}^2 \qquad (16.6)$$

The transmembrane flux of permeating liquid can be expressed as follows:

$$J = \frac{Q}{A_m} = \frac{\Delta p_{tm} \rho_a}{32 \eta \delta_m \xi^2} \sum_{i=1}^{k_s} \frac{V_{pi}}{m_m} d_{pi}^2 \qquad (16.7)$$

However,

$$\sum_{i=1}^{k_s} \frac{V_{pi}}{m_m} d_{pi}^2 = \frac{\varepsilon}{\rho_a} \sum_{i=1}^{k_s} r_i d_{pi}^2 = \frac{\varepsilon d_p^2}{\rho_a} \qquad (16.8)$$

where ε is the porosity of membrane wall, r_i the volume fraction of the pores in the ith range of pore sizes, that is, the volume of the pores in the ith range divided by the total pore volume, and d_p the volume-weighted mean pore diameter given by

$$d_p = \sqrt{\sum_{i=1}^{k_s} r_i d_{pi}^2} = \sqrt{\frac{\rho_a}{\varepsilon} \sum_{i=1}^{k_s} \frac{V_{pi}}{m_m} d_{pi}^2} = \sqrt{\frac{4\varepsilon A_m}{\pi \xi N}} \qquad (16.9)$$

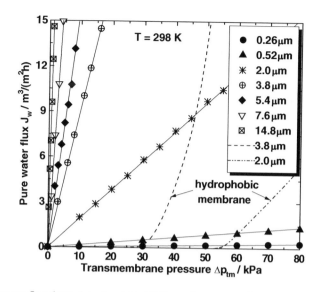

FIGURE 16.6 Pure water flux through hydrophilic SPG membranes as a function of transmembrane pressure at different mean pore sizes (solid lines). Pure water flux through hydrophobic SPG membranes is represented by the dashed lines. (From Vladisavljević, G.T., Shimizu, M., and Nakashima, T., *J. Membr. Sci.*, 250 (1–2), 69–77, 2005. With permission.)

By substituting Equation (16.8) in Equation (16.7) one obtains

$$J = \frac{\Delta p_{tm}}{\eta} \frac{\varepsilon d_p^2}{32\delta_m \xi^2} \qquad (16.10)$$

A typical plot of pure water flux vs. pressure difference for hydrophilic SPG membranes is shown in Figure 16.6 (solid lines). Under these conditions, J_w linearly increases with increasing the transmembrane pressure, which is in correspondence with Equation (16.10) for laminar flow regime in the pores. As predicted from Equation (16.10), the water flux was higher for the larger mean pore size. If the SPG membrane was hydrophobic, the water flux was not observed below a critical pressure (dotted lines in Figure 16.6). After reaching the critical pressure, the water flux sharply increases and approaches a J_w value for the same hydrophilic membrane.

C. HYDRODYNAMIC RESISTANCE OF SPG MEMBRANE

Equation (16.10) can be rewritten as

$$J = \frac{\Delta p_{tm}}{\eta} \frac{1}{R_m} \qquad (16.11)$$

where R_m is the hydrodynamic membrane resistance, which is given by

$$R_m = \frac{32\delta_m \xi^2}{\varepsilon d_p^2} \qquad (16.12)$$

The R_m values for hydrophilic SPG membranes were calculated from the slope of the J_w vs. Δp_{tm} lines using the equation: $R_m = (1/\eta_w)(\partial J_w/\partial \Delta p_{tm})^{-1}$. The R_m values for hydrophobic SPG membranes were determined using 99.5% ethanol as a permeating liquid. As shown in Figure 16.7, R_m was unaffected by the surface treatment with silicone resin. Therefore, no pore

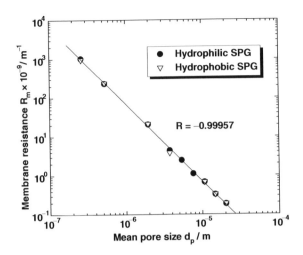

FIGURE 16.7 Hydrodynamic resistance of hydrophilic and hydrophobic SPG membrane as a function of mean pore size. (From Vladisavljević, G.T., Shimizu, M., and Nakashima, T., *J. Membr. Sci.*, 250 (1–2), 69–77, 2005. With permission.)

plugging by the resin occurred at the given conditions, even in the submicron range of mean pore sizes. The linear fit of $\log(R_m)$ vs. $\log(d_p)$ data gave the following equation:

$$R_m = 0.0691 \, d_p^{-1.985} \tag{16.13}$$

where R_m and d_p are in m^{-1} and m, respectively. The exponent on d_p of -1.985 is very close to -2, predicted by Equation (16.12) for laminar flow. Therefore, liquid permeation through any wettable SPG membrane (hydrophilic or hydrophobic) obeys the Hagen–Poiseuille law over the whole range of mean pore sizes. This means that the effects of flow divergence and confluence on liquid permeation are negligibly small. The linear fit of data in Figure 16.7 using a fixed slope of -2 gave the following equation:

$$R_m = 0.0704 d_p^{-2} \tag{16.14}$$

D. Pore Tortuosity and Number of Pores per Unit Cross-Sectional Area of SPG Membrane

In Figure 16.8, the ratio of membrane thickness and membrane resistance, δ_m/R_m, was plotted against $d_p^2 \varepsilon$ using logarithmic coordinates. It is clear that the data for hydrophilic and hydrophobic membranes can be correlated by a single line with a slope of unity, whose equation is

$$\frac{\delta_m}{R_m} = 0.0191 d_p^2 \varepsilon \tag{16.15}$$

The mean tortuosity factor ξ of the pores can be calculated from Equations (16.15) and (16.12) as follows:

$$\xi = \frac{1}{\sqrt{32 \times 0.0191}} = 1.28 \tag{16.16}$$

As a comparison, the mean pore tortuosity of polypropylene hollow fiber membranes ranges between 1.9 and 2.8 [16,17]. Equation (16.16) suggests that in spite of quite different conditions

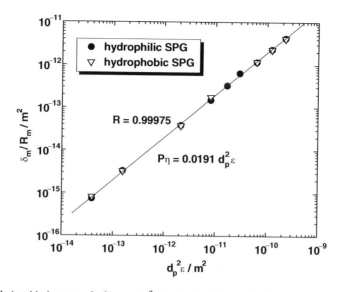

FIGURE 16.8 Relationship between δ_m/R_m and $d_p^2\varepsilon$ for hydrophilic and hydrophobic SPG membranes. (From Vladisavljević, G.T., Shimizu, M., and Nakashima, T., *J. Membr. Sci.*, 250 (1–2), 69–77, 2005. With permission.)

during phase separation in the production of porous glass, the prepared SPG membranes are characterized by the similar morphology of the pore network.

In Figure 16.9, the number of pores per unit membrane area, N/A_m, was plotted against the mean pore size in logarithmic coordinates. The N/A_m values for each membrane were calculated from the equation

$$\frac{N}{A_m} = \frac{4\varepsilon}{\pi \xi d_p^2} \tag{16.17}$$

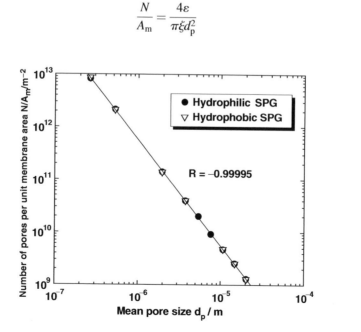

FIGURE 16.9 Number of pores per unit cross-sectional area of membrane as a function of the mean pore size for hydrophilic and hydrophobic SPG membranes. (From Vladisavljević, G.T., Shimizu, M., and Nakashima, T., *J. Membr. Sci.*, 250 (1–2), 69–77, 2005. With permission.)

The ε values were taken from Table 16.1 and the ξ values were calculated for each membrane from the experimental R_m values using Equation (16.12). The number of pores per unit membrane area was inversely proportional to the square of the mean pore size, with the following equation:

$$\frac{N}{A_m} = 0.56 d_p^{-2} \tag{16.18}$$

where N/A_m and d_p are in m^{-2} and m, respectively. Equation (16.18) enables to predict N/A_m for any SPG membrane, irrespective of their mean pore size or surface affinity.

III. MEMBRANE EMULSIFICATION

The first investigation on using membranes for emulsification can be traced back to the later 1980s when Nakashima and Shimizu [1] fabricated the SPG membrane and successfully produced highly uniform kerosene-in-water and water-in-kerosene emulsions [18]. Since this time, the method has continued to attract attention due to its effectiveness in producing narrow droplet size distributions at low energy consumption. To date, in addition to experimentation using SPG membranes, investigations of a broad range of other types of membranes, such as ceramic [19], metallic [20], polymeric [21], and microengineered [11], have been reported.

The development of emulsification methods for production of monodisperse droplets must be rooted in one of two possible manufacturing approaches [22]: (a) reduction of process length scales of the turbulent perturbations and enhancement of their uniformity in the mixing processes that rupture the liquids and (b) manufacture of droplets individually (drop-by-drop) using microporous systems. ME, in which the production of emulsions is achieved by the droplets extruding from the outlet of many individual pores, represents a typical example of the second approach. This chapter aims to introduce the latest development on the utilization of the ME technique following from some of our earlier investigations on using this technique for manufacture of emulsion dispersions.

A. PRINCIPLES AND METHODS OF ME

ME methods reported so far in the literature are depicted schematically in Figure 16.10. In conventional direct ME (Figure 16.10a), fine droplets are formed *in situ* at the membrane–continuous phase interface by pressing a pure disperse phase through the membrane. To ensure a regular droplet detachment from the pore outlets, shear stress is generated at the membrane–continuous phase interface by recirculating the continuous phase using a low-shear pump (Figure 16.11a) [12] or

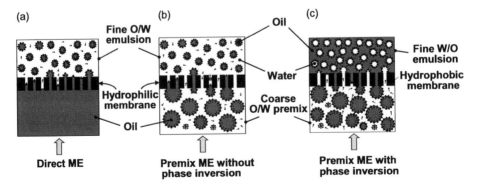

FIGURE 16.10 Schematic diagram of ME methods. (From Vladisavljević, G.T. and Williams, R.A., *Adv. Colloid Interf. Sci.*, 113 (1), 1–20, 2005. With permission.)

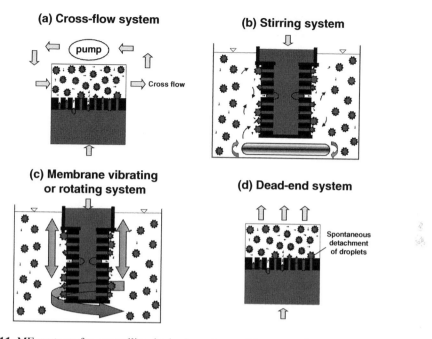

FIGURE 16.11 ME systems for controlling hydrodynamic conditions near the membrane surface. (From Vladisavljević, G.T. and Williams, R.A., *Adv. Colloid Interf. Sci.*, 113 (1), 1–20, 2005. With permission.)

by agitating in a stirring vessel [24] (Figure 16.11b). The rate of mixing should be high enough to provide the required tangential shear on the membrane surface, but not too excessive to induce further droplet break up. Another approach uses systems equipped with a moving membrane, in which the droplet detachment from the pore outlets is stimulated by rotation [25] or vibration [26] of the membrane within a stationary continuous phase (Figure 17.11c). Even in the absence of any tangential shear, droplets can be spontaneously detached from the pore outlets at small disperse phase fluxes (Figure 16.11d), particularly in the presence of fast-adsorbing emulsifiers in the continuous phase and for a pronounced noncircular cross-section of the pores [27]. This type of behavior is similar to the interfacial tension driven-droplet formation process observed in grooved microchannel emulsification [28].

A potential disadvantage of direct ME is the relatively low maximum disperse phase flux through the membrane of typically $0.01–0.1$ $m^3/(m^2$ h), which has to be restrained to avoid the transition from a dripping to jetting flow regime [27] and to avoid the steric hindrance among droplets that may be formed simultaneously at the adjacent pores [30]. The formation of uniform droplets is only possible in the dripping zone, in which the mean droplet size is virtually independent of the disperse phase flux or tangential stress [14]. In ME studies with a microsieve [28] and SPG membranes [29], the disperse phase flux was 2.5 $m^3/(m^2$ h), but polydisperse emulsion droplets were obtained. Owing to low productivity, direct ME is more suitable for the preparation of relatively diluted emulsions with disperse phase contents up to 30 vol%. Nevertheless, this process enables to obtain very narrow droplet size distributions over a wide range of mean droplet sizes. In the SPG emulsification [4,13–15,31–32], the mean droplet size of both O/W and W/O emulsions can range from less than 1 μm to over 60 μm, with the relative span factors under optimal conditions of $0.26–0.45$. For ceramic membranes, the droplet size range can be $0.2–100$ μm with a range of spans [33].

Suzuki et al. [34] first implemented "premix" ME, in which a preliminarily emulsified coarse emulsion (rather than a single pure disperse phase) is forced through the membrane (Figure 16.10b). This is achieved by mixing the two immiscible liquids together first using a conventional stirrer and then passing this preliminarily emulsified emulsion through the membrane. If the disperse phase of coarse emulsion wets the membrane wall and if suitable surfactants are dissolved in both liquid phases, the process may result in a phase inversion, that is, a coarse O/W emulsion may be inverted into a fine W/O emulsion (Figure 16.10c) and vice versa [35]. The main advantage of this method is that a fine emulsion can be easily prepared from a low concentration coarse emulsion at high rates. For a 1 μm poly tetrafluoro ethylene (PTFE) membrane used by Suzuki et al. [35], the maximum volume ratio is of disperse phase in the phase-inverted emulsions were 0.9 and 0.84 for O/W and W/O emulsions, respectively. Flow-induced phase-inversion phenomenon was investigated earlier by Akay [36] using a multiple expansion–contraction static mixer, which is a series of short capillaries with flow dividers, and by Kawashima et al. [37,38] using polycarbonate membranes with mean pore sizes of 3 and 8 μm. Kawashima et al. [37] inverted a W/O/W emulsion consisting of liquid paraffin, Span 80 and Tween-20 into W/O emulsion using hydrophobic polycarbonate membranes. This W/O emulsion was redispersed into an aqueous solution of hydrophilic surfactant to form W/O/W emulsion containing smaller and more uniform internal droplets than the original W/O/W emulsion.

Premix ME holds several advantages over "direct" ME: the optimal transmembrane fluxes with regard to droplet size uniformity are typically above 1 $m^3/(m^2\,h)$, which is one to two orders of magnitude higher than that in direct ME; the mean droplet sizes that can be achieved using the same membrane and phase compositions are smaller than that in direct ME, which can be advantageous; the experimental setup is generally simpler than that in direct ME, for example, no moving parts such as cross-flow pump or stirrer are needed, except for the preparation of pre-emulsion; finally, the premix ME process is easier to control and operate than direct ME, since the driving pressure and emulsifier properties are not so critical for the successful operation as in the direct ME process.

In the first premix ME study [34], a cross-flow system was used, in which the coarse emulsion was diluted by permeation into pure continuous phase or diluted emulsion recirculating at the low-pressure side of the membrane. In the subsequent works (Table 16.2), a dead-end (once-through) system was used, in which the fine emulsion was withdrawn as a product after passing through the membrane, without any recirculation or dilution with the continuous phase. It enables fast preparation of emulsions with a disperse phase content of 50 vol% or more [44]. One of the disadvantages of premix ME is a higher droplet polydispersity when compared with direct ME. To combine the advantages of both techniques, that is, high throughput of premix ME and narrow droplet size distribution of direct ME, a multipass (repeated) premix ME was investigated [40,41,44,45]. In this process, the fine emulsion is repeatedly passed through the same membrane several times to achieve additional droplet size reduction and to enhance droplet size uniformity.

B. Direct ME in Cross-Flow System

Direct emulsification in a cross-flow system is still the most investigated system for ME applications. Figure 16.12 shows a typical laboratory-scale experimental setup for direct ME. The continuous phase or emulsion recirculates inside the membrane tube using a low-shear pump such as the Mohno-pump. The disperse phase is placed in a pressure vessel and introduced in the annular space of the module with compressed gas. The weight of disperse phase permeated through the pores is measured by a balance on which the pressure vessel rests. The balance is interfaced to a computer to continuously collect time vs. weight data. The disperse phase pressure, p_d, is adjusted with a regulating valve and monitored by the manometer P1, which is connected at the entrance of the module. The tube-side pressures at the inlet and outlet of the module, $p_{c,in}$ and $p_{c,out}$, are measured by means of pressure transducers P2 and P3 and used to calculate the transmembrane pressure Δp_{tm} according to the following equation:

$$\Delta p_{tm} = p_d - \frac{p_{c,in} + p_{c,out}}{2} \tag{16.19}$$

TABLE 16.2
Some Premix ME Investigations Reported So Far

Membrane Material	System	Mean Pore Size, $d_p(\mu m)$	Product Emulsion	Mean Droplet Size and Span	Flux $(m^3\,m^{-2}\,h^{-1})$	Ref.
Tubular SPG	Cross-flow	2.7 and 4.2	O/W	$1.4-2.1d_p$, span $= 0.4-0.62$	0.03–3.5	[34]
Flat PTFE	Dead end	1.0	O/W and W/O	$2-4.1d_p$	Up to 9	[39]
	Dead end with phase inversion	1.0	O/W and W/O	$2.8-4.0d_p$	1–5.5	[35]
	Dead end and multipass $(n = 1-3)$	1.0	O/W	$1.2-2.6d_p$ span $= 0.55-0.9$	2–18	[40]
Flat cellulose acetate	Dead end	0.2, 0.45, 0.8, and 3.0	W/O/W	$1.0-3.5d_p$	Not given	[41]
Flat polycarbonate	Dead end and multipass $(n = 1-18)$	0.33, 0.38, 0.44, 0.6, and 1.0	O/W	$\leq 1.6d_p$ for $n > 12$	0.2–0.6	[42]
Tubular SPG	Dead end and multipass $(n = 3)$	1.1	S/O/W	$0.9d_p$	1.6	[43]
	Dead end and multipass $(n = 1-5)$	10.7	W/O/W	$0.41-1.2d_p$ span $= 0.28-0.6$	0.8–37	[44]

Source: From Vladisavljević, G.T. and Williams, R.A., *Adv. Colloid Interf. Sci.*, 113 (1), 1–20, 2005 [23]. With permission.

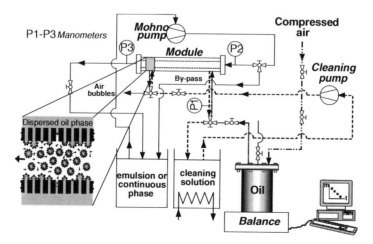

FIGURE 16.12 Schematic diagram of a cross-flow ME system. (From Vladisavljević, G.T., Lambrich, U., Nakajima, M., and Schubert, H., *Colloid Surface A*, 232 (2–3), 199–207, 2004. With permission.)

The minimum transmembrane pressure at which the disperse phase can permeate through the pores is given by the Laplace equation [12]:

$$\Delta p_{tm, min} = p_{cap} = \frac{4\gamma_\infty \cos \theta}{d_p} \tag{16.20}$$

where γ_∞ is the equilibrium interfacial tension between the continuous and disperse phase and θ is the contact angle between the disperse phase and membrane surface in continuous phase.

The shear stress generated at the membrane–continuous phase interface is related to the mean velocity of continuous phase inside the membrane tube, v_t, with the equation [46]:

$$\sigma_w = \frac{\lambda \rho_c v_t^2}{8} \tag{16.21}$$

where ρ_c is the continuous phase density and λ is the Moody's friction factor. In the special case of laminar flow inside the membrane tube ($Re_t < 2300$), the Moody's friction factor is $\lambda = 64/Re_t$ and Equation (16.21) is simplified to: $\sigma_w = 8\eta_c v_t/d_i$, where d_i is the inner diameter of membrane tube and η_c is the continuous phase viscosity. For turbulent flow, in the range of $2300 < Re_t < 10^5$, the friction factor is given by the well-known Blasius equation:

$$\lambda = 0.3164\, Re_t^{-0.25} \tag{16.22}$$

Figure 16.13 shows the relationship between v_t, Re_t, and σ_w if the continuous phase is 2 wt.% aqueous solution of Tween-80 and the internal diameter of membrane tube is $d_i = 8.7$ mm.

To allow production of monodisperse emulsions, the affinities between the membrane surface, disperse and continuous phases, and the electrical charge on the functional groups of emulsifiers must be considered carefully. For a preparation of O/W emulsions, hydrophilic (untreated) SPG membranes must be used. This avoids the wetting and spreading of the disperse phase (oil) on

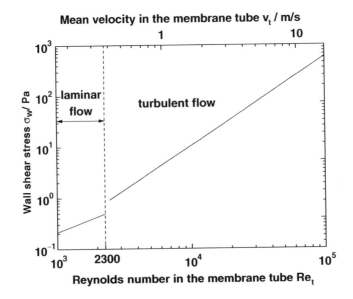

FIGURE 16.13 The variation of wall shear stress with the Reynolds number and the mean velocity of continuous phase (2 wt.% Tween-80 dissolved in demineralized water) in the membrane tube ($d_i = 8.7$ mm).

the membrane surface. Similar principles also apply to the choice of the emulsifiers. The functional groups of the chosen emulsifiers must not carry the charge opposite to that of the membrane surface facilitating the membrane surface to retain its hydrophilicity. For example, an untreated SPG membrane has a negative surface potential of -15 to -35 mV within the pH range of $2-8$, due to the dissociation of acid silanol groups [47]. Hence, for the earlier O/W case, use of cationic emulsifiers, such as cetyltrimethyl ammonium bromide (CTMABr) or tri-n-octylmethyl ammonium chloride (TOMAC), must be avoided.

For production of W/O emulsions, hydrophobic SPG membranes must be used. Hydrophobic treatment of SPG membranes can be successfully carried out by coating the membrane surface with a silicone resin. It was shown in Figure 16.7 that even in the submicron range of mean pore sizes, the membrane resistance was unaffected by this treatment, which means that the pores were not plugged by the resin [10]. However, care has to be taken in this silinization process to avoid blockages.

Production of multiple emulsions involve the preliminary emulsification of two phases (e.g., W/O or O/W), followed by secondary emulsification into a third phase leading to the three-phase mixture such as W/O/W or O/W/O. The primary emulsion is usually prepared under intense homogenization conditions to obtain as fine internal droplets as possible. The secondary emulsification step is carried out under less severe conditions to avoid rupture of the liquid film between the internal disperse phase and continuous phase. If the second step is carried out using conventional emulsification device, the external droplets are, in most cases, highly polydispersed or the entrapment efficiency is very small. ME technique though enables narrow size distribution of the external droplets and maintains a high encapsulation yield of the internal droplets. Production of W/O/W emulsion by ME is shown schematically in Figure 16.14. A fine W/O emulsion is first prepared using a hydrophobic SPG membrane or a conventional emulsification device. This was then followed by secondary emulsification using a hydrophilic membrane and the prepared W/O emulsion as a disperse phase in the second step. Mine et al. [48] revealed that to make a stable W/O/W emulsion, the

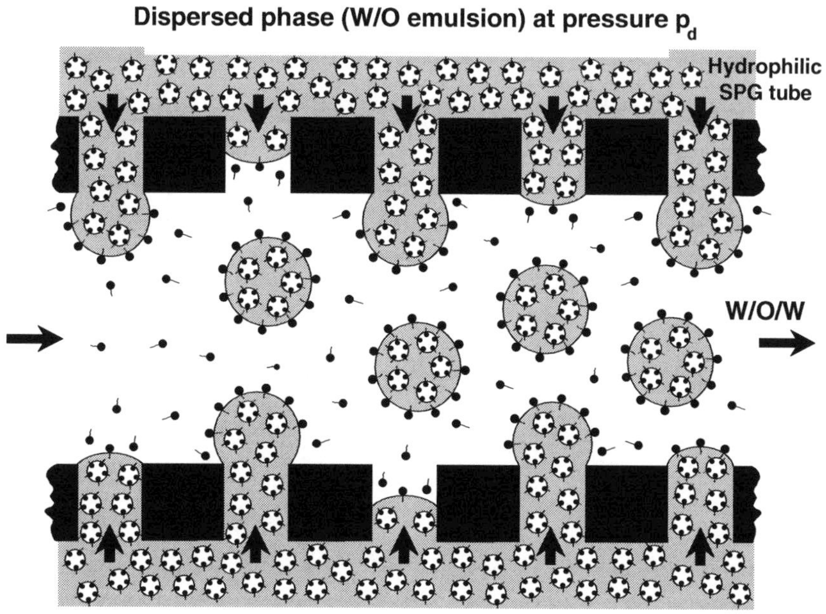

Dispersed phase (W/O emulsion) at pressure p_d

FIGURE 16.14 Schematic diagram of the preparation of multiple W/O/W emulsion using SPG membrane. (From Vladisavljević, G.T. and Schubert, H., *J. Membr. Sci.*, 225 (1–2), 15–23, 2003. With permission.)

concentration of internal aqueous phase droplets must be between 30 and 50 vol% and the membrane pore size for the secondary emulsification must be not less than twice the diameter of the internal droplets.

1. Influence of Operating Parameters on Emulsification Results

a. Influence of Transmembrane Pressure

The influence of $\Delta p_{tm}/p_{cap}$ ratio on the droplet size distribution for a 4.8-μm SPG membrane is shown in Figure 16.15. Even at $\Delta p_{tm}/p_{cap} = 5.7$ corresponding to a disperse phase flux of $771\,m^{-2}\,h^{-1}$ (Figure 16.16), the span of particle size distribution was still rather small (0.52), but nevertheless significantly larger than that at $\Delta p_{tm}/p_{cap} = 2-3$. The formation of larger droplets with diameters between 22 and 34 μm was the main reason for the broadening of the droplet size distribution at the higher $\Delta p_{tm}/p_{cap}$ values.

For the given conditions, the disperse phase flux, J_d, was proportional to $\Delta p_{tm}^{2.3}$, which was due to the higher proportion of simultaneously active pores at the higher pressure. For example, only 1.3–1.4% of the pores was simultaneously active at $\Delta p_{tm}/p_{cap} = 1-2$ and 8.5% at $\Delta p_{tm}/p_{cap} = 5.7$. The fraction of simultaneously active pores at any moment is given by

$$k = \frac{J_d \eta_d R_m}{\Delta p_{tm}} \tag{16.23}$$

A linear relationship between J_d and Δp_{tm} is only possible if the proportion of active pores is independent on the pressure. Abrahamse et al. [29] found that in emulsification with a thin microengineered microsieve possessing highly uniform pores, 16% of the pores was active at $\Delta p_{tm}/p_{cap} = 3$. The fact that all the pores did not become active at the same critical pressure, although they had the same diameter, which was explained by a pressure drop under the membrane as soon as oil phase flows through some pores, preventing other pores to become active [29].

For the given emulsion formulation, the equilibrium interfacial tension γ_∞ between the continuous and disperse phases was 8×10^{-3} N/m and the contact angle θ between the disperse phase and the membrane surface was assumed to be 0. Therefore, from Equation (16.20), one obtains $p_{cap} = 6.7$ kPa for $d_p = 4.8 \times 10^{-6}$ m, which corresponds to 7 kPa deduced from Figure 16.16.

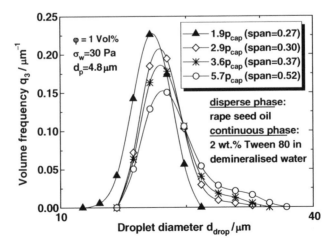

FIGURE 16.15 Influence of transmembrane-to-capillary pressure ratio on droplet size distribution for emulsification using a SPG membrane with the mean pore size of 4.8 μm. (From Vladisavljević, G.T. and Schubert, H., *J. Membr. Sci.*, 225 (1–2), 15–23, 2003. With permission.)

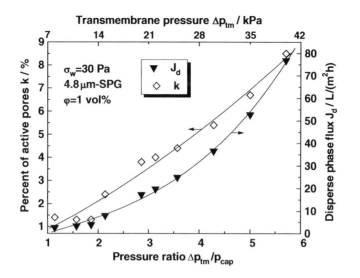

FIGURE 16.16 Influence of transmembrane-pressure-on the disperse phase flux and the percentage of active pores (conditions as in Figure 16.15). (From Vladisavljević, G.T. and Schubert, H., *J. Membr. Sci.*, 225 (1–2), 15–23, 2003. With permission.)

b. Influence of Wall Shear Stress

The influence of wall shear stress on the droplet size distribution for a 4.8-μm SPG membrane is shown in Figure 16.17 and Figure 16.18. With increasing the wall shear stress from 1.3 to 30 Pa, the droplet size distribution curve shifts to smaller droplet diameters and becomes narrower and narrower. For the given pore size and experimental conditions, an emulsion with a narrow droplet size distribution (span = 0.43) was even produced at $\sigma_w = 0.37$ Pa, corresponding to $v_t = 0.3$ m/sec and laminar flow inside the membrane tube (Figure 16.13). Williams et al. [49] obtained a span value of 0.82 at the mean tube velocity of $v_t = 0.6$ m/sec in a semicontinuous

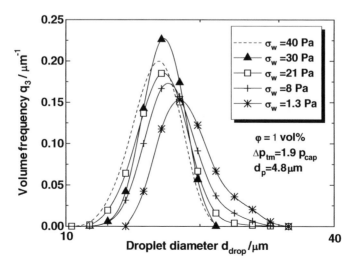

FIGURE 16.17 Influence of wall shear stress on the droplet size distribution (emulsion formulation as in Figure 16.15). (From Vladisavljević, G.T. and Schubert, H., *J. Membr. Sci.*, 225 (1–2), 15–23, 2003. With permission.)

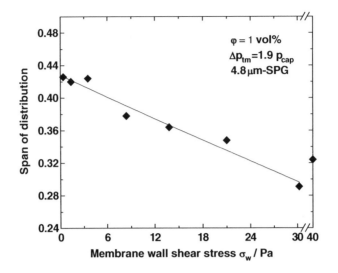

FIGURE 16.18 Influence of wall shear stress on the span of the droplet size distribution (span $= 0$ is an indication of perfectly monodisperse droplets) for the conditions given in Figure 16.17. (From Vladisavljević, G.T. and Schubert, H., *J. Membr. Sci.*, 225 (1–2), 15–23, 2003. With permission.)

preparation of cosmetic model emulsions with alumina membranes. Abrahamse et al. [29] obtained polydisperse O/W emulsions using the microengineered membrane at $\sigma_w = 0.62$ Pa. It was due to droplet–droplet interactions before detachment from the pore tips, caused by a high disperse phase flux of up to 2500 kg m^{-2} h^{-1}. In this work at $\sigma_w = 0.37$ Pa, due to careful control of oil flux, steric hindrance among the droplets was avoided and monodisperse emulsion was produced, although the porosity of SPG membrane (0.5–0.6) was much higher than for the given microsieve (0.15).

The span of droplet size distribution linearly decreases with increasing σ_w up to 30 Pa and then increases with the further increase in σ_w (Figure 16.18). The increase in the span as σ_w increased from 30 to 40 Pa is due to the formation of smaller droplets at 40 Pa, while the maximum droplet size is the same at both σ_w values (Figure 16.17). The broader droplet size distribution at $\sigma_w = 40$ Pa can be explained by (a) partial droplet disruption outside the membrane tube caused by high recirculation rate ($v_c = 3.5$ m/sec) or (b) very intensive droplet deformation before detachment from the pore tips.

Figure 16.19 demonstrates that the mean droplet size, expressed as the mean Sauter diameter $d_{3,2}$, decreases with increasing the wall shear stress, σ_w, for the mean pore sizes of 3.1 and 4.8 μm. The effect of wall shear stress is especially large at $\sigma_w < 10$ Pa and more significant for larger mean pore sizes. As an example, for the 4.8 μm SPG membrane, $d_{3,2}/d_p$ decreases by 15% over a σ_w range of 3.5–40 Pa. Over the same σ_w range and for the similar pressure ratio, $d_{3,2}/d_p$ decreases by 8% for the 3.1 μm membrane. It is in accordance with the results of Schröder and Schubert [19], who found that the effect of wall shear stress on the mean droplet size was much more significant for a 0.8 μm than that for 0.1 μm α-Al$_2$O$_3$ membrane. According to them, the further decrease in mean droplet size at the high wall shear stress is prevented by the rough membrane surface and the forming droplets hindering each other in detaching from the pores. The ratio of mean droplet to mean pore size of 3–4 is typical for SPG membrane and was also found elsewhere [12,15].

c. Influence of Disperse Phase Content

The influence of disperse phase content on droplet size distribution during ME was little investigated in the literature. Figure 16.20 demonstrates that at the small $\Delta p_{tm}/p_{cap}$ ratios, the droplet

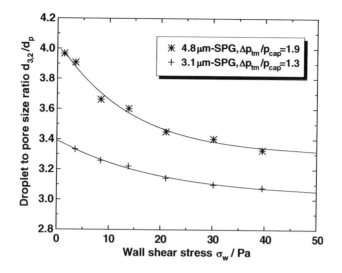

FIGURE 16.19 The variation of mean droplet to mean pore size ratio with the wall shear stress at the disperse phase content of $\varphi \approx 1$ vol% (emulsion formulation as in Figure 16.15). (From Vladisavljević, G.T., Lambrich, U., Nakajima, M., and Schubert, H., *Colloid Surface A*, 232 (2–3), 199–207, 2004. With permission.)

size distribution curve has a similar shape in the range of φ between 1 and 20 vol%. The span of distribution, however, somewhat increases with increasing φ. It is interesting to note that the minimum droplet size in the emulsion decreases with increasing φ (by 17% for the 2.5-μm membrane in the range of 1–16 vol% and by 24% for the 6.6-μm membrane in the range of 1–20 vol%). It could be a consequence of gradual activation of smaller pores.

As shown in Figure 16.21, the oil flux increases with time at the constant transmembrane pressure, reflecting the fact that more and more pores are simultaneously active in droplet formation during the operation. The same type of behavior was found in ME with α-alumina and zirconia membranes [50,51]. It is reasonable to suggest that when $t \to 0$, only the largest pores are partially

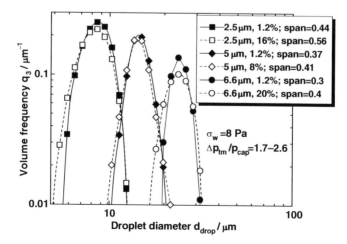

FIGURE 16.20 Influence of disperse phase content, that is, emulsification time on the droplet size distribution. (From Vladisavljević, G.T. and Schubert, H., *J. Membr. Sci.*, 225 (1–2), 15–23, 2003. With permission.)

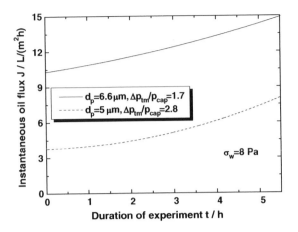

FIGURE 16.21 The variation of disperse phase flux with time for two SPG membranes. (From Vladisavljević, G.T. and Schubert, H., *J. Membr. Sci.*, 225 (1–2), 15–23, 2003. With permission.)

active, while the smallest ones are completely inactive. In the subsequent stage of operation, the smaller pores are gradually activated, and as a result, minimum droplet size decreases. The pore activation till steady state is a slow process at the small $\Delta p_{tm}/p_{cap}$ ratio, so that for the conditions in Figure 16.21, the stationary state was not established even after 5 h of operation.

d. Influence of Mean Pore Size and Comparison of SPG Membranes with Other Membranes and Emulsification Methods

The influence of mean pore size on droplet size distribution of resulting emulsions is shown in Figure 16.22. It can be seen that SPG membrane enables to produce O/W emulsions with a very narrow droplet size distribution over a wide range of mean pore sizes at a small wall shear stress of 8 Pa. The span of droplet size distribution of 0.26–0.45 for emulsification using SPG membranes is much lower than 1.1–2.3 for microfluidization (MF). In contrast, the Microfluidizer® enables to manufacture emulsions with the mean droplet size of less than 0.3 μm, on the condition that the dispersed phase content is up to 20 vol% and that the homogenizing pressure is high enough. Furthermore, using two passes of emulsion through the homogenizing valve at 110 MPa, the mean droplet size of only 0.085–0.087 μm was obtained at the disperse phase content of 1–2.5 vol%. At the same pore size and under the same experimental conditions, the oil droplets produced by utilizing the SPG membrane are more uniform than the droplets produced using α-Al$_2$O$_3$ membrane [14,52].

2. Stability of Prepared O/W Emulsions

The variation of droplet size distribution with time was investigated during stationary storage of samples in a glass cylinder at room temperature (20–25°C). Owing to relatively large mean droplet size, the samples formed a dense creamed layer after only several hours of stationary storage, so that virtually over the whole storage time, oil droplets were highly concentrated on a creamed layer. In spite of that, the droplets were stable and no appreciable change in the mean droplet size or the span of distribution was observed over 3 months, as shown in Figure 16.23. The stability of oil droplets mainly depended on the initial uniformity, while mean droplet size was not an important factor. If the droplets in fresh samples were highly uniform, as in Figure 16.24, the droplet size distribution did not change significantly even after 5 months. The micrographs shown in Figure 16.25 indicate that the oil droplets are still uniform 1–2 months after preparation, irrespective of their mean droplet size.

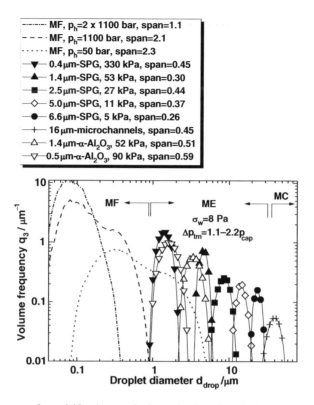

FIGURE 16.22 Influence of emulsification method on droplet size distribution (disperse phase content $\varphi \approx 1$ vol%; emulsion formulation as in Figure 16.15. MF, microfluidization; ME, membrane emulsification; MC, microchannel emulsification. (From Vladisavljević, G.T., Lambrich, U., Nakajima, M., and Schubert, H., *Colloid Surface A*, 232 (2–3), 199–207, 2004. With permission.)

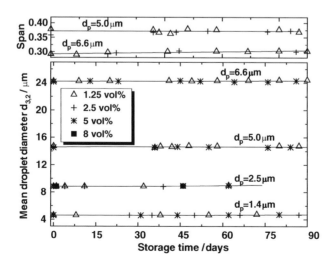

FIGURE 16.23 The variation of the mean droplet size and the span of distribution with time during stationary storage of O/W emulsions prepared with different SPG membranes. (From Vladisavljević, G.T. and Schubert, H., *J. Membr. Sci.*, 225 (1–2), 15–23, 2003. With permission.)

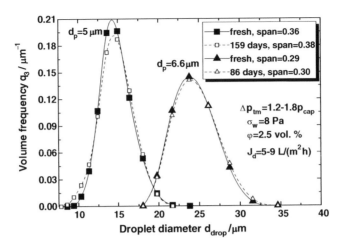

FIGURE 16.24 Influence of storage time on the droplet size distribution of O/W emulsions produced using two different SPG membranes. (From Vladisavljević, G.T. and Schubert, H., *J. Membr. Sci.*, 225 (1–2), 15–23, 2003. With permission.)

3. Direct Microscopic Observation of Droplet Formation at the Surface of SPG Membrane

Yasuno et al. [53] and Vladisavljević et al. [54] succeeded in visualizing SPG ME using a microscope video system, shown in Figure 16.26. A flat SPG membrane was mounted between two transparent plates using rubber spacers to form the upper and lower compartments at both sides of the membrane. The continuous phase flowed through the upper compartment without any recirculation. The disperse phase was pressed through the membrane at a constant flow rate using a syringe pump. The droplet formation process was recorded using a video recorder attached to a CCD camera and an inverted metallographic microscope. The JPEG images were captured from the MPEG video clips by a Canopus V-shot photo grabber.

To get sharp images, as those in Figure 16.27, the membrane surface should be polished with a diamond paste. Figure 16.28 shows the SEM images of a polished and unpolished membrane surface. In the case of the unpolished membrane, the microscopic picture is not clear, due to an irregular reflection of light beam from the membrane surface (Figure 16.29). Therefore, it is necessary to use flat membranes with as smooth surface as possible.

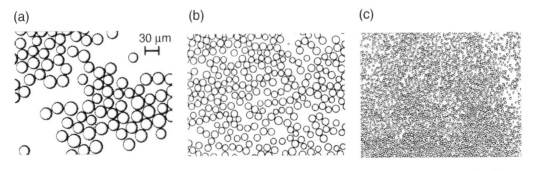

FIGURE 16.25 Photomicrographs of the emulsion droplets prepared by utilizing SPG membranes of different mean pore sizes: (a) $d_p = 6.6$ μm, $d_{3,2} = 24$ μm, 61-day-old emulsion; (b) $d_p = 4.8$ μm, $d_{3,2} = 17.5$ μm, fresh emulsion; (c) $d_p = 1.4$ μm, $d_{3,2} = 4.6$ μm, 48-day-old emulsion. The same magnification was used for all photographs. (From Vladisavljević, G.T. and Schubert, H., *J. Membr. Sci.*, 225 (1–2), 15–23, 2003. With permission.)

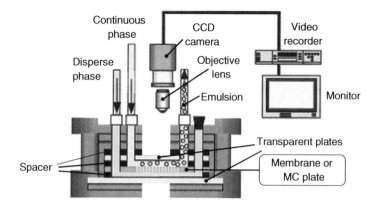

FIGURE 16.26 Experimental setup for microscopic observation of droplet formation in real time.

The droplets were formed in regular time intervals at the same active pore, but this time interval showed large variations over the membrane surface. The mean frequency of droplet formation linearly increased with increasing the disperse phase flux. Even at the frequency of droplet formation of as high as 4 sec^{-1}, the droplets formed at the same pore were highly monodisperse, as shown in Figure 16.30.

At the same disperse phase flux, the mean frequency of droplet formation was substantially smaller for SDS than for Tween-80 emulsifier. It can be explained by the lower interfacial tension at oil–water interfaces in the presence of SDS. In addition, under the same conditions, the mean droplet size was smaller for SDS than for Tween-80. In the case of SDS as an anionic emulsifier, the droplets were detached from the pore openings immediately after formation, due to strong electrostatic repulsions between the charged droplets and identically charged membrane surface. In the case of a nonionic Tween-80 emulsifier, the newly formed droplets were kept at the membrane surface after formation, before being pushed by another droplets forming at the same pore.

C. PREMIX ME

In premix ME, large droplets of a coarse emulsion are disrupted into fine droplets by utilizing the microporous membrane as a special kind of low-pressure homogenizing valve. At the trans-membrane pressures smaller than a critical pressure p_{cap}, the droplets larger than the pores are retained by the membrane and form a concentrated layer at the high-pressure side of the membrane

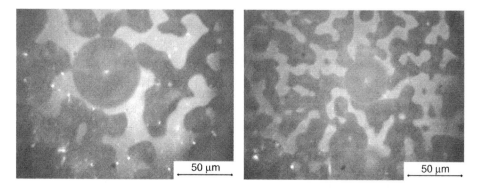

FIGURE 16.27 Droplet formation on the surface of a smooth SPG membrane with $d_p = 15$ μm (left) and 10.2 μm (right). Disperse phase, soybean oil; continuous phase, 0.5 wt% Tween-80 in distilled water.

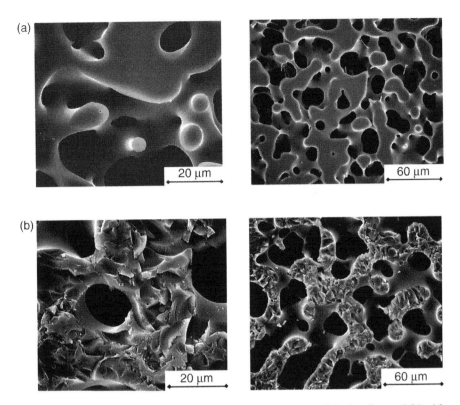

FIGURE 16.28 (a) SEM of a flat 15-μm SPG membrane with a smooth, polished surface and (b) a 16-μm SPG membrane with a rough surface.

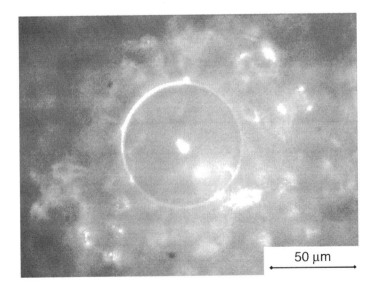

FIGURE 16.29 Droplet formation on the surface of a rough, unpolished SPG membrane with the mean pore size of $d_p = 16.2$ μm. Emulsion formulation as in Figure 16.27.

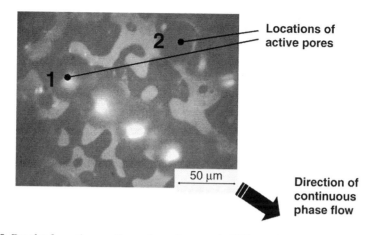

FIGURE 16.30 Droplet formation on the surface of a smooth SPG membrane with $d_p = 15$ μm at the high frequency of droplet formation (about four droplets per second). Four droplets formed successively at the pore 1 and three droplets formed successively at the pore 2 are shown. Emulsion formulation as in Figure 16.27.

(Figure 16.31a). This microfiltration process requires the use of hydrophobic membranes to remove aqueous phase from W/O emulsions [55] or hydrophilic membranes to separate oil phase from O/W emulsions [56].

At the transmembrane pressures above the critical pressure p_{cap}, all droplets pass through the membrane, irrespective of their size. However, at smaller shear stresses inside the pores, the final droplet size, d_2, is larger than the pore size, d_p. In that case, the large droplets of the initial diameter d_1 are deformed at the pore inlets to enter the pores, followed by disruption due to friction between the droplets and the pore walls. The fine droplets are deformed again at the pore outlets to regain a spherical shape (Figure 16.31b). At higher shear stresses, the droplets are more intensively disrupted inside the pores due to collisions between the droplets and collisions with the pore walls, so that the final droplet size can be smaller than the pore size (Figure 16.31c). In that case, the deformation of droplets at the pore outlets does not occur, since the final droplet size is smaller than the pore size.

If the initial droplet size d_1 is not much larger than the pore size d_p, that is, for d_1/d_p ratio close to unity, the critical pressure is given by [42]:

$$p_c = \frac{\gamma_\infty [2 + 2a^6/\sqrt{2a^6 - 1} \times \arccos(1/a^3) - 4a^2]}{a + \sqrt{a^2 - 1}} \qquad (16.24)$$

where $a = d_1/d_p$. If the initial droplet size is much larger than the pore size ($d_1/d_p \gg 1$), the critical pressure is given by Equation (16.20).

The production of multiple $W_1/O/W_2$ emulsions by repeated membrane homogenization of coarsely emulsified feeds was investigated by Vladisavljević et al. [44]. These multiple emulsions are applicable as drug or nutrient delivery systems. The primary W_1/O emulsion prepared by conventional rotor–stator homogenization device was first mixed with the outer aqueous phase W_2 using a stirring bar to prepare a $W_1/O/W_2$ coarse emulsion. This premix was then homogenized by permeation through the SPG membrane using a ME apparatus manufactured by Kiyomoto Iron Works Ltd (Figure 16.32a). The effective membrane length was 12 mm, and the effective cross-sectional membrane area was 3.75 cm². The pressure vessel was filled up with 100 ml of the premix and the required transmembrane pressure was built-up with compressed nitrogen. The fine emulsion which has passed through the membrane was collected into a beaker placed on the balance interfaced to a computer for data acquisition (Figure 16.32b). The feeds were emulsified five to six times through the same membrane to produce fine uniform droplets. The emulsion

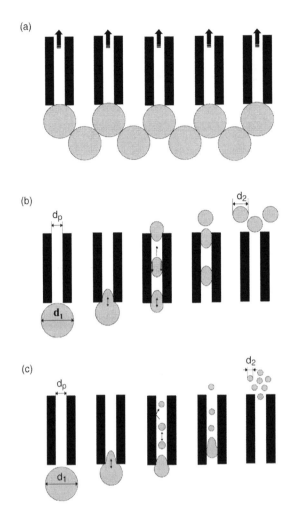

FIGURE 16.31 Droplet retention and breakup as a function of transmembrane pressure: (a) droplet retention below a critical pressure; (b) moderate breakup at moderate shear stresses ($d_m < d_2 < d_1$); (c) intensive breakup at high shear stresses ($d_2 < d_m < d_1$). (From Vladisavljević, G.T., Shimizu, M. and Nakashima, T., *J. Membr. Sci.*, 244 (1–2), 97–106, 2004. With permission.)

formulation is given in Table 16.3. The role of sodium alginate was to increase the viscosity of the outer aqueous phase W_2, thus reducing the creaming tendency of large W_1/O drops prior to membrane homogenization. The role of glucose was to increase the osmotic pressure in both aqueous phases to approximately 0.78 MPa, which is the osmotic pressure of blood and other body liquids and thus to ensure the stability of prepared emulsions in drug delivery systems. In the next paragraph, the effect of operating parameters on transmembrane flux and droplet size distribution is discussed.

1. Influence of Operating Parameters on Emulsification Results

a. *Influence of Transmembrane Pressure*

The influence of transmembrane pressure on the transmembrane flux and the mean size of homogenized droplets is shown in Figure 16.33 and Figure 16.34. It is clear that J increases with increasing Δp_{tm}, which leads to a more intensive droplet break-up and smaller mean particle size at

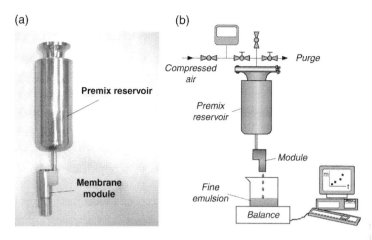

FIGURE 16.32 (a) Apparatus of Kiyomoto Iron Works Ltd. for premix ME and (b) a typical experimental setup for investigation of this process. (From Vladisavljević, G.T., Shimizu, M., and Nakashima, T., *J. Membr. Sci.*, 244 (1–2), 97–106, 2004. With permission.)

higher pressures. The same type of behavior was observed by Suzuki et al. [34,39]. It is in contrast to the experimental results in direct ME [14], in which the mean droplet size at small driving pressures (in the size-stable zone) is independent on driving pressure, and then increases with the further pressure, increase (in the size-expanding zone).

The smaller mean particle sizes at the larger pressures are a consequence of the higher shear stress inside the pores:

$$\sigma_{w,p} = \frac{8\eta_e J \xi}{\varepsilon d_p} \tag{16.25}$$

where η_e is the mean viscosity of emulsion inside the pores. The transmembrane flux J in Figure 16.33 is $1-12 \text{ m}^3/(\text{m}^2 \text{ h})$, which is at three orders of magnitude higher than that in direct emulsification with the same membrane. The ratio of mean droplet to mean pore size of 1.4 to 1.1 obtained after first pass is much smaller than 3–4, found in direct emulsification (Figure 16.19).

Transmembrane pressure is used here to overcome flow resistances in the pores and interfacial tension forces, that is, to deform and disrupt large oil drops into smaller droplets:

$$\Delta p_{tm} = \Delta p_{flow} + \Delta p_{disr} \tag{16.26}$$

TABLE 16.3
The Formulation of $W_1/O/W_2$ Emulsions Prepared by Vladisavljević et al. [44]

Inner aqueous phase, W_1	5 wt% glucose dissolved in distilled water
Oil phase	5 wt% PGPR dissolved in soybean oil
Outer aqueous phase, W_2	0.5 wt% Tween-80, 1 wt% sodium alginate, and 5 wt% glucose dissolved in distilled water
Volume percent of inner aqueous phase in W_1/O emulsion	$\varphi_i = 10-30$ vol%
Volume percent of W_1/O emulsion drops in $W_1/O/W_2$ emulsion	$\varphi_o = 1-60$ vol%
Mean size of inner aqueous phase	$0.37-0.54$ μm
Mean size of homogenized oil droplets	$4.4-13.2$ μm

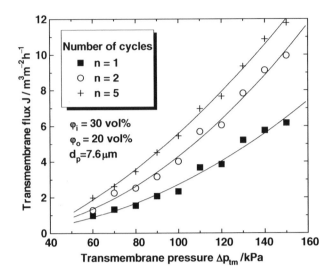

FIGURE 16.33 Effect of transmembrane pressure on the transmembrane flux. Emulsion formulation is given in Table 16.3.

According to Darcy's law, the pressure loss for overcoming flow resistances in the pores, Δp_{flow}, should be proportional to the transmembrane flux, while the expenditure of pressure for droplet disruption, Δp_{disr}, is proportional to the increase in the interfacial area. If $\Delta p_{\text{tm}} = \text{const}$, then

$$\Delta p_{\text{tm}} = \eta_e (R_m + R_{fi}) J_i + C \varphi_o \gamma_\infty \left(\frac{1}{d_i} - \frac{1}{d_{i-1}} \right) = \text{const} \qquad (16.27)$$

The first and second terms in the right-hand side of Equation (16.27) express Δp_{flow} and Δp_{disr}, respectively. Here, C is a parameter independent on the number of cycles, J_i and d_i are the transmembrane flux and the resulting mean particle size corresponding to the ith cycle, and R_{fi} is

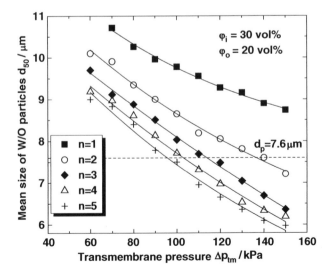

FIGURE 16.34 Effect of transmembrane pressure on the mean size of W_1/O particles.

the overall fouling resistance in the ith cycle. The fouling resistance is a consequence of the accumulation of oil drops on the membrane surface (external fouling) and inside the pores (internal fouling).

b. Influence of Number of Emulsification Cycles and Disperse Phase Content

As shown in Figure 16.35, the flux increases with increasing the number of passes, because the second term in the right-hand side of Equation (16.27) diminishes until it becomes negligible for $d_{i-1} \approx d_i \approx$ const. Figure 16.9 and Figure 16.10 indicate that at constant experimental conditions, the mean particle size tends to a limiting constant value, as the number of cycles increases. The largest flux increase was observed in the second pass, as the largest particle size reduction occurred in the first pass.

The variation of Δp_{flow} and Δp_{disr} with the number of emulsification cycles at $\Delta p_{\text{tm}} = 20$ and 150 kPa is shown in Figure 16.36. The values of Δp_{flow} and Δp_{disr} were calculated from Equation (16.27) by adopting the fluxes J_i from Figure 16.35 and assuming that the limiting flux was established after five passes through the membrane. For the given operating conditions, about 25% of the overall pressure drop was used for droplet disruption in the first pass and the remaining being used for overcoming flow resistances in the pores. Because the content of oil drops was only 10 vol%, it should be expected that at the higher disperse phase contents, the majority of driving pressure in the first pass may be used for overcoming interfacial tension forces. Owing to a substantial dissipation of energy for droplet disruption in the first pass, the mean droplet size was reduced four to nine times. In the subsequent cycles, the mean droplet size continued to decrease but at a decreasing rate. Therefore, the pressure drop due to droplet disruption progressively decreased, until it became negligible after five passes.

Figure 16.37 shows that the transmembrane flux significantly decreases with increasing the content of oil drops, φ_{o}, which is a consequence of the fact that the disruption term in Equation (16.27) is directly proportional to the disperse phase content. For dilute emulsions ($\varphi_{\text{o}} = 1$ vol%), the disruption term can be ignored and the flux is given only by the flow-resistances term. Under these conditions, the maximum flux was observed in the first cycle corresponding to a minimum fouling resistance and emulsion viscosity. Over the φ_{o} range of $5-10$ vol%, the maximum flux

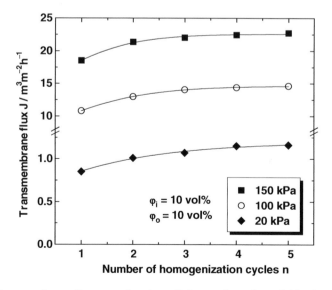

FIGURE 16.35 Transmembrane flux as a function of the number of emulsification cycles at different transmembrane pressures. (From Vladisavljević, G.T., Shimizu, M., and Nakashima, T., *J. Membr. Sci.*, 244 (1–2), 97–106, 2004. With permission.)

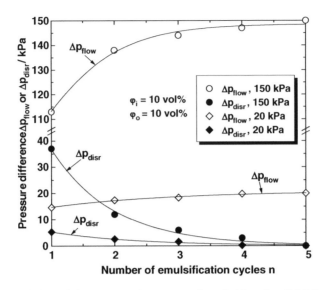

FIGURE 16.36 The variation of the pressure drop terms given in Equation (16.27) with the number of emulsification cycles. (From Vladisavljević, G.T., Shimizu, M., and Nakashima, T., *J. Membr. Sci.*, 244 (1–2), 97–106, 2004. With permission.)

was observed in the third pass. Presumably, for $n < 3$, the flux was more affected by a decrease in Δp_{disr}, then by an increase in the fouling resistance, and consequently, J increased with n. For $n > 3$, a decrease in the disruption term became less significant, then an increase in the fouling resistance and J decreased with further increase in n. For concentrated emulsions ($\varphi_o = 20$–60 vol%), the maximum flux was observed for $n = 5$.

Figure 16.38 demonstrates that at the given operating conditions, the mean droplet size was independent on the content of oil drops over a wide range of 1–60 vol%, in spite of the fact that the flux J was significantly smaller at the higher contents of oil drops (Figure 16.37 and Figure 16.39).

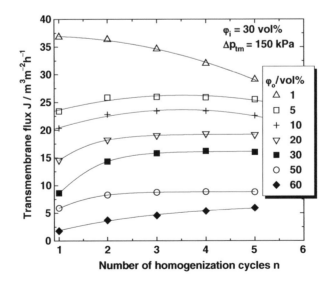

FIGURE 16.37 Transmembrane flux as a function of number of emulsification cycles at different contents of oil drops. (From Vladisavljević, G.T., Shimizu, M., and Nakashima, T., *J. Membr. Sci.*, 244 (1–2), 97–106, 2004. With permission.)

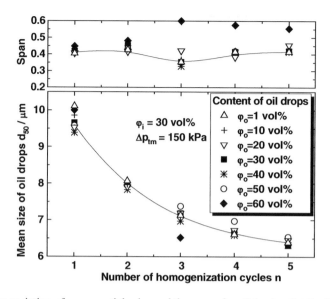

FIGURE 16.38 The variation of mean particle size and the span of particle size distribution with the number of cycles at different contents of oil drops. (From Vladisavljević, G.T., Shimizu, M., and Nakashima, T., *J. Membr. Sci.*, 244 (1–2), 97–106, 2004. With permission.)

It differs from the behavior in high-pressure homogenizers, in which the mean droplet size at a-constant homogenizing pressure is significantly dependent on the disperse phase content, even in the range of 0.05–0.2 vol% [14]. The behavior shown in Figure 16.38 can be explained by the fact that in the absence of coalescence, the mean size of homogenized oil droplets is primarily dictated by the mean pore size and the applied shear stress inside the pores. Equation (16.25) implies that the shear stress inside the pores is proportional to the product $J\eta_e$. At higher disperse phase content, the flux J is smaller, but the emulsion viscosity η_e is higher, so that under certain experimental conditions, the shear stress inside the pores may be independent of the disperse phase

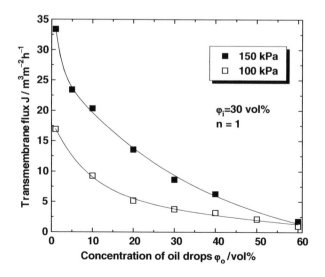

FIGURE 16.39 Effect of the content of oil drops on transmembrane flux at two different transmembrane pressures (single pass through the membrane). (From Vladisavljević, G.T., Shimizu, M., and Nakashima, T., *J. Membr. Sci.*, 244 (1–2), 97–106, 2004. With permission.)

content. Except at $\varphi_o = 60$ vol%, the optimum number of cycles with regard to monodispersity was 3–4 and the minimum span of particle size distribution was 0.28 (Figure 16.38). Altenbach-Rehm et al. [40] found that repeated homogenization using polymeric teflon PTFE membranes also resulted in the smaller mean droplet size and narrower droplet size distribution than a single-stage process. The optimum number of cycles in their investigation was two to three, but the minimum span was 0.55–0.7, which is substantially higher than that in this work.

Figure 16.39 illustrates the influence of operating pressure on the flux vs. disperse phase content curves for $n = 1$. It can be seen that the effect of operating pressure on the flux is more marked at the smaller disperse phase contents. At a content of oil drops of 60 vol%, the transmembrane flux was not influenced by the pressure in the investigated range of 100–150 kPa. However, at $\varphi_o = 1$ vol%, the flux increased by a factor of 2, as the pressure increased from 100 to 150 kPa. It can be explained by the fact that in the region of small contents of oil drops, the pressure difference is predominantly used for overcoming flow resistance forces, while at the high disperse phase contents, the pressure drop is mainly used for overcoming interfacial tension forces, the latter being independent on the transmembrane flux.

Figure 16.40 shows that the optimum number of cycles with regard to droplet uniformity strongly depends on the transmembrane pressure. At the pressure of 100 and 150 kPa, the minimum span of particle size distribution of, respectively, 0.28 and 0.35 was reached after three passes. At the same pressures, a single stage process ($n = 1$) resulted in span = 0.55 and 0.42, respectively. The smallest pressure of 100 kPa was the optimum pressure with regard to droplet size uniformity at $n = 3$, but the least favorable pressure at $n = 1$. It shows that in a single-pass process, the optimum pressure is considerably higher than that in a repeated process.

The micrographs of emulsion particles before and after homogenization taken by optical microscope under the same magnification are shown in Figure 16.41. As can be seen, the particles of coarse emulsion are relatively large and highly polydisperse (some portion of the particles was larger than 300 μm, but 0.7 vol% were smaller than 15 μm). The minimum particle size in the premix was 12 μm, which was above the mean pore size. After first pass through the membrane, 96 vol% of the particles were smaller than 15 μm, but the remaining 4 vol% were still in the

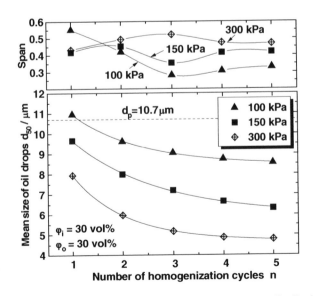

FIGURE 16.40 Variation of the mean particle size and the span of particle size distribution as a function of the number of cycles at different transmembrane pressures. (From Vladisavljević, G.T., Shimizu, M., and Nakashima, T., *J. Membr. Sci.*, 244 (1–2), 97–106, 2004. With permission.)

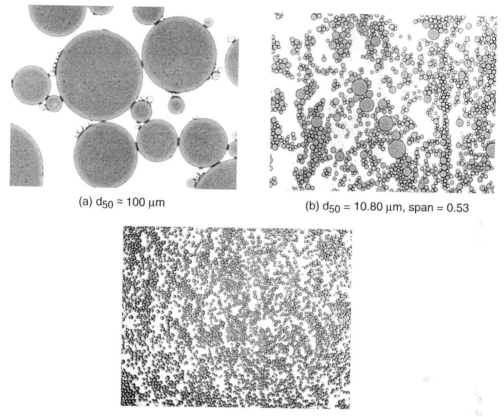

(a) $d_{50} \approx 100 \ \mu m$

(b) $d_{50} = 10.80 \ \mu m$, span = 0.53

(c) $d_{50} = 8.8 \ \mu m$, span = 0.30

FIGURE 16.41 (a) Micrographs of multiple emulsion droplets before homogenization, (b) after first pass, and (c) after six passes through the membrane. Experimental conditions: $d_p = 10.7 \ \mu m$, $\Delta p_{tm} = 100$ kPa, $\varphi_o = 30$ vol%, $\varphi_i = 30$ vol%, emulsion formulation is given in Table 16.3. (From Vladisavljević, G.T., Shimizu, M., and Nakashima, T., *J. Membr. Sci.*, 244 (1–2), 97–106, 2004. With permission.)

range of 15–37 μm (Figure 16.41b). Obviously, the driving pressure of 100 kPa was not large enough to finely break up all particles of coarse emulsion in a single pass. However, when homogenization was repeated six times, all particles were smaller than 10.4 μm and highly uniform, as shown in Figure 16.41c.

ACKNOWLEDGMENTS

The authors wish to thank the Japan Society for the Promotion of Science (JSPS), Tokyo, Japan and the Alexander von Humboldt (AvH) Foundation, Bonn, Germany for the financial support of this work within the scope of the AvH Research Fellowship and the JSPS Postdoctoral Fellowship for foreign researchers.

NOMENCLATURE

a	ratio of initial particle size to pore size (d_1/d_p)
A_m	cross-sectional area of membrane (m^2)
A_p	area of pore walls (m^2)

C	constant in Equation (16.27)
d_p	volume-weighted mean pore diameter (m)
d_i	inner diameter of membrane tube (m)
d_1	initial droplet size (m)
$d_{3,2}$	Sauter mean particle diameter (m)
d_{50}	mean particle diameter (m)
J	transmembrane flux (m s^{-1})
k_s	number of pore size ranges
L	mean length of pores (m)
m_m	dry weight of membrane (kg)
N	number of pores
n	number of homogenization cycles
p	pressure (Pa)
p_{cap}	capillary pressure (Pa)
Δp_{tm}	transmembrane pressure (Pa)
Δp_{flow}	pressure loss for overcoming flow resistances in the pores (Pa)
Δp_{disr}	expenditure of pressure for droplet disruption (Pa)
Q	volumetric rate of flow of permeating liquid (m^3 s^{-1})
q_3	volume frequency of droplets (m^{-1})
R	coefficient of correlation
Re_t	Reynolds number in membrane tube
R_f	fouling resistance (m^{-1})
R_m	hydrodynamic resistance of membrane (m^{-1})
r_i	volume fraction of pores in ith range of pore sizes
span	relative span factor of pore or PSD
t	time (sec)
V	volume of permeating liquid (m^3)
V_m	total volume of membrane (m^3)
V_p	total pore volume (m^3)
v_t	mean velocity of continuous phase in membrane tube (m s^{-1})
δ_m	thickness of membrane wall (m)
ε	porosity of membrane wall (V_p/V_m)
η	viscosity (Pa sec)
λ	Moody friction factor
ρ	true density of membrane (kg m^{-3})
ρ_a	apparent density of membrane (kg m^{-3})
σ_w	shear stress at membrane surface (wall shear stress) (Pa)
$\sigma_{w,p}$	wall shear stress inside pores (Pa)
ξ	mean tortuosity factor of pores
φ	volume proportion of disperse phase in emulsion (vol%)
φ_i	volume proportion of inner droplets of W$_1$ phase in oil drops (vol%)
φ_o	volume proportion of oil drops in outer aqueous phase (vol%)
γ_∞	equilibrium interfacial tension between disperse and continuous phase (N m^{-1})
θ	contact angle between dispersed phase and membrane surface wetted with continuous phase (rad)

Subscripts

| e | refers to emulsion |
| i | refers to ith range of pore sizes or ith emulsification cycle |

w	refers to water
c	refers to continuous phase
d	refers to disperse phase
in	refers to the inlet of membrane module
out	refers to the outlet of membrane module

REFERENCES

1. Nakashima, T. and Shimizu, M., Porous glass from calcium alumino boro-silicate glass, *Ceram. Jap.*, 21 (5), 408–412, 1986.
2. Nakashima, T., *Porous Glass Material and Its Recent Applications*, Society for the Application of SPG Technology, Miyazaki, Japan, 1989, 12 pp.
3. Nakashima, T., Shimizu, M., and Kukizaki, M., Particle control of emulsion by membrane emulsification and its applications, *Adv. Drug Deliv. Rev.*, 45, 47–56, 2000.
4. Vladisavljević, G.T. and Schubert, H., Influence of process parameters on droplet size distribution in SPG membrane emulsification and stability of prepared emulsion droplets, *J. Membr. Sci.*, 225 (1–2), 15–23, 2003.
5. Kandori, K., Applications of microporous glass membranes: membrane emulsification, in *Food Processing: Recent Developments*, Gaonkar, A.G., Ed., 1st ed., Elsevier, Amsterdam, 1995, pp. 113–142.
6. Nakashima, T. and Shimizu, M., Microfiltration of emulsion by porous glass membrane, *Kag. Kog. Ronbunshu*, 15, 645, 1989.
7. Nakashima, T. and Shimizu, M., Effect of oil concentration in microfiltration by porous glass membrane, *Kag. Kog. Ronbunshu*, 20, 468, 1994.
8. Kukizaki, M., Nakashima, T., and Shimizu, M., Preparation of new asymmetric type of porous glass and effect of its structure on microfiltration, *Membrane*, 27 (6), 324–330, 2002.
9. Nakashima, T., Shimizu, M., and Kukizaki, M., Mechanical strength and thermal resistance of porous glass, *J. Cer. Soc. Japan Int.*, 100 (12), 1389–1393, 1992.
10. Vladisavljević, G.T., Shimizu, M., and Nakashima, T., Permeability of hydrohilic and hydrophobic Shirasu-porous-glass (SPG) membranes to pure liquids and its microstructure, *J. Membr. Sci.*, 250 (1–2), 69–77, 2005.
11. Kobayashi, I., Nakajima, M., Chun, K., Kikuchi, Y., and Fujita, H., Silicon array of elongated through-holes for monodisperse emulsion droplets, *AIChE J.*, 48 (8), 1639–1644, 2002.
12. Nakashima, T., Shimizu, M., and Kukizaki, M., *Membrane Emulsification Operation Manual*, 1st ed., Industrial Research Institute of Miyazaki Prefecture, Miyazaki, Japan, 1991, 19 pp.
13. Shimizu, M., Nakashima, T., and Kukizaki, M., Preparation of W/O emulsion by membrane emulsification and optimum conditions for its monodispersion, *Kag. Kog. Ronbunshu*, 28 (3), 310–316, 2002.
14. Vladisavljević, G.T., Lambrich, U., Nakajima, M., and Schubert, H., Production of O/W emulsions using SPG membranes, ceramic α-Al_2O_3 membranes, microfluidizer and a microchannel plate — a comparative study, *Colloid Surface A*, 232 (2–3), 199–207, 2004.
15. Nakashima, T., Shimizu, M., and Kukizaki, M., Effect of surfactant on production of monodispersed O/W emulsion in membrane emulsification, *Kag. Kog. Ronbunshu*, 19 (6), 991–997, 1993.
16. Kiani, A., Bhave, R.R., and Sirkar, K.K., Solvent extraction with immobilized interfaces in a microporous hydrophobic membrane, *J. Membr. Sci.*, 20, 125–145, 1984.
17. Prasad, R. and Sirkar, K., Dispersion-free solvent extraction with microporous hollow-fiber modules, *AIChE J.*, 34 (2), 177–188, 1988.
18. Nakashima, T., Shimizu, M., and Kukizaki, M., Membrane emulsification by microporous glass, Proceedings of the 2nd International Conference on Inorganic Membranes, Montpellier, France, July 1–4, 1991, pp. 513–516.
19. Schröder, V. and Schubert, H., Production of emulsions using microporous, ceramic membranes, *Colloid Surface A*, 152 (1–2), 103–109, 1999.
20. Dowding, P.J., Goodwin, J.W., and Vincent, B., Production of porous suspension polymer beads with a narrow size distribution using a cross-flow membrane and a continuous tubular reactor, *Colloid Surface A*, 180 (3), 301–309, 2001.

21. Vladisavljević, G.T., Brösel, S., and Schubert, H., Preparation of water-in-oil emulsions using microporous polypropylene hollow fibers: influence of some operating parameters on droplet size distribution, *Chem. Eng. Process*, 41 (3), 231–238, 2002.

22. Williams, R.A., Making the perfect particle, *Ingenia*, February (7), 1–7, 2001.

23. Vladisavljević, G.T. and Williams, R.A., Recent developments in manufacturing emulsions and particulate products using membranes, *Adv. Colloid Inter. Sci.*, 113 (1), 1–20, 2005.

24. *Internal Pressure Type Micro Kit, typ MN-20*, Data Sheet, SPG Tecnology Co. Ltd, Sadowara, Japan, 2003.

25. Williams, R.A., Controlled Dispersion Using a Spinning Membrane Reactor, UK Patent Application PCT/GB00/04917, 2001.

26. Hatate, Y., Ohta, H., Uemura, Y., Ijichi, K., and Yoshizawa, H., Preparation of monodispersed polymeric microspheres for toner particles by the Shirasu porous glass membrane emulsification technique, *J. Appl. Polym. Sci.*, 64, 1107–1113, 1997.

27. Kobayashi, I., Nakajima, M., and Mukataka, S., Preparation characteristics of oil-in-water emulsions using differently charged surfactants in straight-through microchannel emulsification, *Colloid Surfaces A*, 229 (1–3), 33–41, 2003.

28. Sugiura, S., Nakajima, M., Kumazawa, N., Iwamoto, S., and Seki, M., Characterization of spontaneous transformation-based droplet formation during microchannel emulsification, *J. Phys. Chem. B*, 106, 9405–9409, 2002.

29. Abrahamse, A.J., van Lierop, R., van der Sman, R.G.M., van der Padt, A., and Boom, R.M., Analysis of droplet formation and interactions during cross-flow membrane emulsification, *J. Membr. Sci.*, 204 (1–2), 125–137, 2002.

30. Katoh, R., Asano, Y., Furuya, A., Sotoyama, K., and Tomita, M., Preparation of food emulsions using a membrane emulsification system, *J. Membr. Sci.*, 113 (1), 131–135, 1996.

31. Shimizu, M., Nakashima, T., and Kukizaki, M., Particle size control of W/O emulsion by means of osmotic pressure as driving force, *Kag. Kog. Ronbunshu*, 28 (3), 304–309, 2002.

32. Vladisavljević, G.T. and Schubert H., Preparation and analysis of oil-in-water emulsions with a narrow droplet size distribution using Shirasu-porous-glass (SPG) membranes, *Desalination*, 144 (1–3), 167–172, 2002.

33. Williams, R.A., Emulsifying with Membranes, Proceedings of the 3rd World Congress on Emulsions, Lyon, France, September 24–27, 2002.

34. Suzuki, K., Shuto, I., and Hagura, Y., Characteristics of the membrane emulsification method combined with preliminary emulsification for preparing corn oil-in-water emulsions, *Food Sci. Technol. Int. Tokyo*, 2 (1), 43–47, 1996.

35. Suzuki, K., Fujiki, I., and Hagura, Y., Preparation of high concentration O/W and W/O emulsions by the membrane phase inversion emulsification using PTFE membranes, *Food Sci. Technol. Int. Tokyo*, 5 (2), 234–238, 1999.

36. Akay, G., Flow-induced phase inversion in the intensive processing of concentrated emulsions, *Chem. Eng. Sci.*, 53 (2), 203–223, 1998.

37. Kawashima, Y., Hino, T., Takeuchi, H., Niwa, T., and Horibe, K., Shear-induced phase inversion and size control of water/oil/water emulsion droplets with porous membrane, *J. Colloid Interf. Sci.*, 145 (2), 512–523, 1991.

38. Hino, T., Kawashima, Y., and Shimabayashi, S., Basic study for stabilization of W/O/W emulsion and its application to transcatheter arterial embolization therapy, *Adv. Drug Deliv. Rev.*, 45 (1), 27–45, 2000.

39. Suzuki, K., Fujiki, I., and Hagura, Y., Preparation of corn oil/water and water/corn oil emulsions using PTFE emulsions, *Food Sci. Technol. Int. Tokyo*, 4 (2), 164–167, 1998.

40. Altenbach-Rehm, J., Suzuki, K., and Schubert H., Production of O/W-Emulsions with Narrow Droplet Size Distribution by Repeated Premix Membrane Emulsification, Proceedings of the third World Congress on Emulsions, Lyon, France, September 24–27, 2002, 6 pp.

41. Shima, M., Kobayashi, Y., Fujii, T., Tanaka, M., Kimura, Y., Adachi, S., and Matsuno, R., Preparation of fine W/O/W emulsion through membrane filtration of coarse W/O/W emulsion and disappearance of the inclusion of outer phase solution, *Food Hydrocolloids*, 18 (1), 61–70, 2004.

42. Park, S.H., Yamaguchi, T., and Nakao, S., Transport mechanism of deformable droplets in microfiltration of emulsions, *Chem. Eng. Sci.*, 56 (11), 3539–3548, 2001.

43. Toorisaka, E., Ono, H., Arimori, K., Kamiya, N., and Goto, M., Hypoglycemic effect of surfactant-coated insulin solubilized in a novel solid-in-oil-in-water (S/O/W) emulsion, *Int. J. Pharm.*, 252 (1–2), 271–274, 2003.

44. Vladisavljević, G.T., Shimizu, M., and Nakashima, T., Preparation of monodisperse multiple emulsions at high production rates by multi-stage premix membrane emulsification, *J. Membr. Sci.*, 244 (1–2), 97–106, 2004.

45. Altenbach-Rehm, J., Schubert, H., and Suzuki, K., Premix-Membranemulgieren mittels hydrophiler und hydrophober PTFE-Membranen zur Herstellung von O/W-Emulsionen mit enger Tropfen-größenverteilung, *Chem. Ing. Tech.*, 74, 587–588, 2002.

46. Vladisavljević, G.T. and Schubert, H., Preparation of emulsions with a narrow particle size distribution using microporous α-alumina membranes, *J. Dispersion Sci. Technol.*, 24 (6), 811–819, 2003.

47. Nakashima, T., Shimizu, M., and Kukizaki, M., Monodisperse Single and Double Emulsions and Method of Producing Same. US Patent 5,326,484, July 5, 1994.

48. Mine, Y., Shimizu, M., and Nakashima, T., Preparation and stabilization of simple and multiple emulsions using a microporous glass membrane, *Colloid Surfaces B*, 6, 261–268, 1996.

49. Williams, R.A., Peng, S.J., Wheeler, D.A., Morley, N.C., Taylor, D., Whalley, M., and Houldsworth, D.W., Controlled production of emulsions using a crossflow membrane, *Chem. Eng. Res. Des.*, 76A (A8), 902–910, 1998.

50. Joscelyne, S.M. and Trägårdh, G., Food emulsions using membrane emulsification: conditions for producing small droplets, *J. Food Eng.*, 39 (1), 59–64, 1999.

51. Schröder, V., *Herstellen von Öl-in-Wasser-Emulsionen mit mikroporösen Membranen*; Ph.D. thesis, University of Karlsruhe (T.H.), Karlsruhe, Germany, 1999, 172 pp.

52. Lambrich, U., Vladisavljević, G.T., Emulgieren mit mikrostrukturierten Systemen (emulsification using microstructured systems), *Chem. Ing. Tech.*, 76 (4), 376–383, 2004.

53. Yasuno, M., Nakajima, M., Iwamoto, S., Maruyama, T., Sugiura, S., Kobayashi, I., Shono, A., and Satoh, K. Visualization and characterization of SPG membrane emulsification, *J. Membr. Sci.*, 210 (1), 29–37, 2002.

54. Vladisavljević, G.T., Shimizu, M., and Nakashima, T., Direct Observation of Droplet Formation in Membrane Emulsification, Proceedings of the DDS Seminar on Development of Lipid Microcarriers and Their Application to Drug Delivery Systems, Sadowara, Japan, August, 2003.

55. Scott, K., Jachuck, R.J., and Hall, D., Crossflow microfiltration of water-in-oil emulsions using corrugated membranes, *Sep. Purif. Technol.*, 22–23, 431–441, 2001.

56. Cheryan, M. and Rajagopalan, N., Membrane processing of oily streams. Wastewater treatment and waste reduction, *J. Membr. Sci.*, 151 (1), 13–28, 1998.

Dispersoids

17 Mechanochemical Treatment of Inorganic Solids: Solid–Solid Fine Dispersions

Miodrag Zdujić
Serbian Academy of Sciences and Arts, Belgrade, Serbia and Montenegro

CONTENTS

I. INTRODUCTION

Milling is an important step in the processing of metallic and ceramic powders. Its primary usage has been for particle size reduction/growth and change of particle shape as well as for homogenization of mixture of two or more powders which will be subsequently compacted, for example, by pressing and sintering. However, milling may lead (and leads) to various physicochemical and chemical changes of the material. In that respect, milling may be described as a *mechanochemical treatment*, the change of reactivity as *mechanical activation* while reaction induced by mechanical energy as *mechanochemical reaction*. It should be forthwith noted that the terms used to denote various processes and phenomena of the powder treatment by milling still are not precisely specified, for instance different terms may be present in literature for the same process. In Appendix A, attempt is made to give provisory definition of the basic terms of mechanochemical treatment by milling.

Although the occurrence of physicochemical changes of the material induced by mechanical energy has been known for more than century (see [1] for the short survey of the history of mechanochemistry), rapid interest for mechanochemical treatment has begun with the finding that milling of mixture of crystalline nickel and niobium powders produces amorphous alloy [2]. Soon, a number of amorphous alloys were prepared by this technique [3–6]. In literature, the synthesis of amorphous alloys by milling is usually named *mechanical alloying* referring to the process, first developed at the end of 1960s of twentieth century, for the production of nickel-based dispersion-strengthened super-alloys [7]. Mechanical alloying may be defined as the mechanochemical treatment of the mixture of two or more powders wherein, at least one is metallic. In the first stage of process, by the repeated particle fracture and coalescence of constitutive powders, composite particles with very fine micro-structure are formed. With prolonged milling, various mechanochemical reactions may take place, which in some cases lead to the formation of amorphous phase.

It was demonstrated that besides the amorphous phase, other metastable materials such as *nano-crystalline alloys* [8] and *supersaturated solid solutions* [9] may also be synthesized by mechanical alloying. Milling is essentially a solid-state process, hence mechanical alloying enables preparation of materials otherwise difficult to obtain from liquid phase, for instance alloys of high-melting elements, for example, $MoSi_2$ [10], alloys of metals with significantly different melting points, for example, Nb–Sn [11] and alloys of mutually immiscible elements, for example, Fe–Cu [12].

Milling of elemental metals, for example, Fe, Zr, Al and Cu produces *nanocrystalline metals* [13,14] characterized by very small grain size, typically 20–25 nm, hence 20–50% of the material consists of grain boundaries. Therefore, since large number of atoms are located in disordered regions between crystals, nanocrystalline materials are expected to posess peculiar mechanical, electrical, magnetic, and catalytic properties.

In contrast, it was found that milling of some intermetallic compounds (YCO_3, $GdCO_3$, $NiTi_2$) leads to *polymorphic transformation of crystalline to amorphous phase* [15–17].

Of course, mechanochemical treatment by milling is not restricted to metallic systems only, but has been already applied to ceramic and polymer materials. Nowadays, mechanochemical treatment has been recognized as a powerful tool for the synthesis of a wide range of materials. Rapid development during last two decades expanded mechanochemistry into two new branches: *mechanically induced self-propagating reaction* and *soft mechanochemistry*. Ball milling can induce *self-sustaining reaction* in many powder mixtures if exothermic heat of reaction is sufficiently high. The process begins with an activation period, referred to as ignition time, during which particle and crystallite size reduction, mixing and defect formation takes place. The mechanically induced self-propagating reaction (MSR) is ignited when the powder reaches a well-defined critical state. Once a reaction front is established, the process becomes a self-propagating high-temperature synthesis (SHS) reaction [18,19]. On the other hand, it has been realized that the use, for instance of hydrated oxides or hydroxides relieves mechanochemical reactions [20,21]. Such a novel approach involving mild mechanochemical synthesis, based on reaction of solid acids, bases, hydrated compounds, basic and acidic salts is known as soft mechanochemistry.

The release of Thiessen's "Grundlagen der Tribochemie" in 1967 [22] was the result of sys-tematic and scientific researches on the phenomena regarding the influence of the mechanical energy on the solid materials that had been started after the Second World War. A monograph by Heinicke published in 1984 [23] is still a good basic source of information about a great number of various mechanochemical reactions, as well as building a framework of the various aspects of mechanochemical phenomena. Even now considerable number of monographs regarding various aspects of mechanochemistry have been published [24–29]. In addition, a number of reviews, to mention some, have appeared in recent years providing a general overview of mechan-ochemical methods [30,31], materials synthesized by mechanical alloying [31–33], the preparation of nanocrystalline [34–36], and amorphous materials [37,38] by milling as well as self-sustaining reactions [39,40] induced by milling.

Mechanochemistry (or *Tribochemistry*) is usually defined as a science dealing with the chemi-cal and physicochemical changes of substances due to the influence of mechanical energy [23] (see Appendix A for other definitions of mechanochemistry). On the other hand, the term *tribology* is commonly used to refer to the science of interacting surfaces in relative motion with respect to friction, lubrication, and wear [41].

Various types of mills are used for mechanochemical treatment, commonly: vibratory, attrition, planetary, and tumbling ball mill. The final product depends on the milling conditions, hence, different types of mill or the alteration of milling parameters may result in diverse reaction paths for mechanochemical reaction. Besides, the milling time necessary to reach desired structure depends on the mill used, sometimes up to order of magnitude. Therefore, because the essence of process, that is, phenomena occurring during milling, is not fully understood, scaling up and/or converting the milling parameters from one to other type of mill is still uncertain.

The mechanochemical treatment by ball milling is a very complex process, wherein a number of phenomena (such as plastic deformation, fracture and coalescence of particles, local heating, phase transformation, and chemical reaction) arise simultaneously influencing each other. The mechanochemical treatment is a non-equilibrium solid-state process whereby, the final product retains a very fine, typically nanocrystalline or amorphous structure. At the moment of ball impact, dissipation of mechanical energy is almost instant. Highly excited state of the short lifetime decays rapidly, hence a "frozen" disordered, metastable structure remains. Quantitative description of the mechanochemical processes is extremely difficult, herewith a mechanochemical reaction still lacks clear interpretations and adequate paradigm.

This review attempts to give a short account of the main aspects of mechanochemical treatment such as: (a) mechanical alloying; (b) commonly used milling units, herewith the influence of milling parameters on the rate and products of mechanochemical treatment; (c) structural changes; (d) mechanochemical reactions; (e) subsequent heat treatment; and (f) powder contamination.

II. MECHANICAL ALLOYING

Since a considerable number of systems have been synthesized by mechanical alloying, we will start our survey with a short description of this process referring to it as a special class of mechanochemical treatment. Mechanical alloying is a milling method for the preparation of composite, macroscopically homogeneous powder with extremely fine microstructure, whereby the starting material is a mixture of constituent powders of a given composition. The essence of the process is the solid-state alloying by repetitive *cold welding* and *fracture* of constitutive powder particles [42]. Its primary industrial application has been for the production of dispersion-strengthened nickel- and iron-based superalloys for working temperatures of 1000°C and higher [43]. Commercial production of these powders is not so large and is about 150 tons per year (according to the data from the year 1994). Investigations in the last two decades, have revealed that the usage of mechanical alloying is not restricted to the materials with second-phase dispersoides. Thus, following prospects may be assigned to mechanical alloying: (i) production of a fine dispersion of second-phase; (ii) decrease of matrix structure to the nanometer sizes; (iii) extension of solution limit; (iv) synthesis of new crystalline phases; and (v) amorphous phase synthesis. The prerequisite for an efficient process is to establish a balance between fracturing and cold welding (coalescence) of constituent particles. It is, therefore, important to choose appropriate milling parameters for a given material system subjected to mechanical alloying. For some powder systems, in order to reduce cold welding and to enhance particle fracturing it is necessary to add some suitable *process-control agent*, that is, a *lubricant* that impedes the clean metal-to-metal contact necessary for cold welding. For this purpose organic compounds such as methanol, ethanol, hexane, toluene, stearic acid, vacuum grease etc., (about 1–5 wt% of the total powder charge) are used. The second approach is to promote particle fracture applying *cryogenic milling*: the mill vial is cooled with liquid nitrogen [44].

Single ball-powder-ball (and ball-powder-vial wall) impact can modify the powder morphology in two ways: by fracturing and by cold welding. When the metal particles are brought in contact, atomically clean surfaces that is, *fresh surfaces* create cold weld forming layered composite particles. Concurrently, work-hardened elemental or composite particles may fracture. Cold welding (along with plastic deformation and agglomeration) and fracturing (particle crush) take place at the same time so that the composite structure persistently becomes refined and homogeneous. Typically, at least for the metallic systems in which one of the powders is ductile, the process of mechanical alloying may be divided in to several stages [44,45]: (i) early — intensive cold welding; (ii) intermediate — pronounced fracturing; and (iii) final — mild cold welding. At the end, a stationary state is reached and further refinement is not possible. Powder particles have an extremely deformed metastable structure.

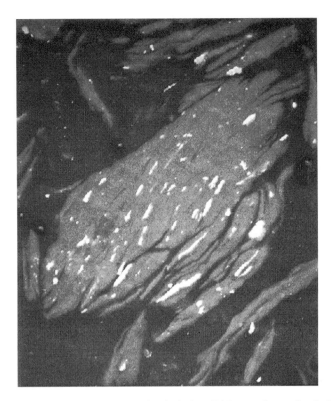

FIGURE 17.1 SEM micrograph of polished sample of Al–3 at.% Mo powder mechanically alloyed for 3 h in tumbler ball mill (bright phase — molybdenum).

Typical lamellar multilayer structure, characteristic of the early stage of milling, is given in Figure 17.1. The majority of dispersoids (bright spots of different sizes) are placed along the weld interfaces. Composite particles significantly differ in sizes from few to several hundred of micrometers. At this stage, fragmented starting powders which are not alloyed also exist. Chemical composition significantly varies from particle to particle as well as within particles. As milling continues, concurrent cold welding and fracturing proceeds leading to progressive structure refinement. Finally, composition of single particle approaches the starting composition of powder mixture. Typical structure evolution with the progress of mechanical alloying is shown in Figure 17.2. As can be seen, with the progress of milling, dispersoid sizes decrease whereby, its distribution becomes more uniform. After 1000 h of milling, Mo dispersoids cannot be resolved (their sizes are under the resolution of scanning electron microscope). Transmission electron microscopy reveals that in the Al–Mo powders mechanically alloyed for 1000 h, Mo dispersoids are in the range of 5–50 nm [46]. In general, in such very fine dispersion systems, formation of supersaturated solution as well as precipitation of a new phase may occur.

III. MILLS AND RELEVANT MILLING PARAMETERS

Milling, as already stated in the introduction, is used for the particle diminution or coarsening, shape change, agglomeration as well as for modification of powdered material such as apparent and tap density and flow rate. Milling is also a necessary step for homogenization of usually complex, multi-component metallic and ceramic materials. Application of milling has been greatly extended to other processes such as mechanical alloying and solid-state reactions. When a mill is intended to carry out mechanochemically that is, *mechanically induced reactions* it may be regarded as *mechanoreactor* [23].

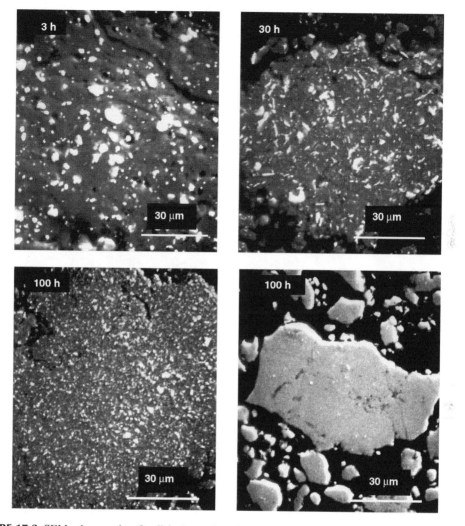

FIGURE 17.2 SEM micrographs of polished samples of Al–10 at.% Mo powders mechanically alloyed for various milling times in tumbler ball mill.

With respect to the objective of milling capacity, mill varies from few milligrams for laboratory mills to several hundreds of kilograms per batch for industrial mills. Regardless of the mill type, the milling process is characterized by the action of milling tool (typically balls) to the powdered material resulting in fragmentation and coalescence of powder particles. During milling, four types of forces act on powder: impact, attrition, shear, and compression [47].

Commonly used mills in practice are: *attrition mills, tumbler ball mills, vibratory ball mills, and planetary ball mills* [31,47]. In *attrition ball mill* (Figure 17.3a), balls and powdered materials are placed in vertical (or horizontal) stationary container that may be cooled with some refrigerant (usually water). Milling is affected by the stirring action of rotating shaft with arms. Angular velocity of a rotating shaft is from $6.3 \ \text{sec}^{-1}$ (60 rpm) for industrial to $31 \ \text{sec}^{-1}$ (300 rpm) for laboratory mills. Typical ball diameter is from 3 to 6 mm while ball velocity is about $0.5 \ \text{msec}^{-1}$. First commercial production of the mechanically alloyed powder was realized in attrition ball mills of the capacity of 34 kg/batch. Subsequently, up to 1 ton of powder was processed in attrition mills of 2 m diameter with more than a million of balls of a total of 10 tons in weight [48]. Analysis of the mill dynamics of this type of mill was given in [49,50].

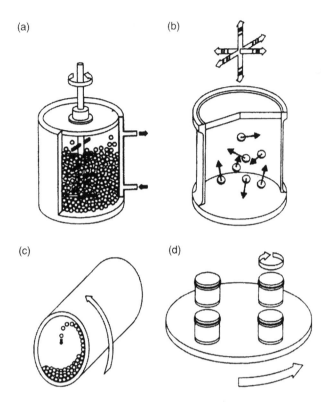

FIGURE 17.3 Schematic view of common mills used for mechanochemical treatment: (a) attrition ball mill; (b) vibratory ball mill; (c) tumbler ball mill; (d) planetary ball mill.

In a *tumbling ball mill* (Figure 17.3c), the angular velocity of the vial should be smaller than critical which attaches balls on vial walls: $\omega = \sqrt{g/(R-r)}$ where ω is the angular velocity of the vial, R, vial radius, r, ball radius and $g = 9.81$ msec^{-1}, gravitational acceleration. Due to simple design, this type of mill is in widespread usage, from laboratory mills with a diameter of 10–40 cm to industrial with a diameter of 1–2 m for the processing of large amounts of powder (135–180 kg/day). Typically, balls of diameter 6–25 mm are used. Ball velocity may be estimated from the relation $v = 2\sqrt{2gR}$, hence for commercial mills it is 4.4–6.3 msec^{-1} [49].

In *vibratory mill* (Figure 17.3b), balls are placed in a vial which oscillates in all three orthogonal direction, whereby the ball oscillatory motion is complicated. The ball and powder motion depends on many factors such as frequency, amplitude, dimension, and curvature of a vial as well as its motion. Required milling time is usually small. Mass of powder treated is in the range of few milligrams to about 4.5 kg. Impact velocity and the frequency of impact are the most important parameters that determine transfer of mechanical energy to the powder charge; consequently, kinetics of mechanochemical treatment. Most frequently used mill for laboratory investigation is SPEX 8000 Mixer/Mill with typical powder charge of 10–20 g, ball mass of 150 g and ball velocity of 1.8–3.3 msec^{-1} [51]. Analysis of vibratory mill dynamics has been reported, for example, in Refs. [49–53].

In *planetary ball mill* (Figure 17.3d), vials filled with balls and powder are placed onto a rotating disc (sun disc), so far as, alike planets, a vial turns around (revolves) and rotates about its own axis. Thenceforth, the name planetary mill. For commercial mills, the ratio of angular velocity of sun disc and vials are fixed. The ball movement in a vial is the result of a superimposition of centrifugal forces. At one moment, the ball is detached from the vial wall and impacts an opposite side.

Milling intensity may be continuously varied by changing the angular velocity of a rotating (supporting) disc. For typical commercial planetary ball mill Fritch Pulverisette 5, maximum angular velocity of a sun disc is about 340 rpm, while the angular velocity of vials is given by the relation $\omega_v = -1.25\, \omega_p$, where ω_v and ω_p are angular velocities of vial and disc, respectively. Typical ball velocity is 2.5–4 msec^{-1}. Considerable efforts have been taken to model mill dynamics of this type of mill [53–61].

The type of mill as well as milling parameters (amplitude and frequency for vibratory mill, angular velocity for planetary mill, ball-to-powder mass ratio, ball diameter, number of balls, etc.) have chief influence on the efficiency and nature of the mechanical energy transfer from the milling tools (typically balls) to powder. As a consequence of the dissipation of mechanical energy to solid, a number of concurrent processes occur: plastic deformation and fracture, consequently formation of fresh surfaces and local temperature rise, that is, *hot spot*. Basic problem in the exploration of mechanochemical processes is the quantitative assessment of mechanical energy transfer and its dissipation [23]. At this moment, because of a great number factors affecting mechanochemical treatment, the influence of milling parameters on the mechanochemical reactions is insufficiently explained. Therefore, these investigations are still predominantly experimental, whereby milling conditions are taken empirically for a concrete system. Therefore, an essential question arises, whether it is possible to prescribe the milling conditions in advance for the attainment of a desired structure or product of mechanochemical treatment?

Energy conveyed to the powdered system during milling has an effect on kinetics as well as on final product of mechanochemical treatment. For example, when Nb_2O_5 was milled in vibrating mill with the amplitude of 20 mm, the starting monoclinic structure transforms to "fully" amorphous phase. With prolonged milling, the amorphous phase transforms to a stable pseudo-hexagonal phase. On the contrary, higher milling amplitude of 50 mm causes direct transformation of monoclinic to pseudo-hexagonal phase, presumably, as a consequence of impact of lower frequency but with higher energy [62]. On the other hand, a number of experimental investigations reveal that "full" amorphization of metallic systems can be attained only for the well-defined milling conditions [56,63–68]. It was postulated that amorphization occurs below some so-called power injected, that is, the product of impact energy and collision frequency and above some minimal energy per impact [64].

According to Butyagin [69] and Cocco, Delogu and coworkers [70,71] relevant milling parameters are: *impact energy, impact frequency, milling intensity* (or power injected), *energy dose,* and *specific dose* (cumulative mechanical energy transferred to the powder during milling time, t), and are defined as follows.

$$\text{Impact energy: } E = \frac{1}{2} m_b\, v_{imp}^2 \text{ (J)} \tag{17.1}$$

$$\text{Impact frequency: } f \text{ (Hz)} \tag{17.2}$$

$$\text{Milling intensity: } I = fE \text{ (W)} \tag{17.3}$$

$$\text{Energy dose: } D = It \text{ (J)} \tag{17.4}$$

$$\text{Specific dose: } D_m = \frac{D}{m_p} \text{ (J/g)} \tag{17.5}$$

Milling intensity (Equation (17.3)) is related to the frequency of collision and the average energy transferred to the powder at each impact. In general, the case of multiple ball milling, f and E depends on their number and mass, so that their quantification requires rather complex and accurate experimental and modeling procedures (see abovementioned reference regarding modeling of mill dynamics). In other words, the foregoing relevant milling parameters should be determined for the respective mill considered. For instance, for the estimation of milling intensity (mill power), I (J/g) of vibratory mill following relation was proposed [72]:

$$I = \frac{n_b m_b (4\pi f a)^2}{2 m_p} \tag{17.6}$$

where I (W/g) is the milling intensity per mass unit (mill power), f (sec^{-1}) frequency, a (m) amplitude, n_b and m_b (g) number and mass of a ball, respectively and m_p (g) powder mass.

For a planetary mill, the intensity (i.e., power consumption or power injected), I (W) may be calculated according to Iasonna and Magini [59]:

$$I = P^* \frac{1}{2} m_b \omega_p^3 r_p^2 n_b \tag{17.7}$$

where P^* is a dimensionless power efficiency factor, m_b (g) the mass of a ball, ω_p (sec^{-1}) and r_p (m) the angular velocity and radius of the planetary mill disc, respectively and n_b the number of balls. The factor P^* depends on the degree of filling, $n_v = N_b/N_{b,tot}$ where N_b is the number of balls used in the experiment and $N_{b,tot}$ is the number of balls necessary to completely fill up the vial.

An illustration on how milling intensity (power injected), Equation (17.7), depends on the number of balls and angular velocity of supporting disc for the Fritch Pulverisette 5, for two different vials charged with balls are given in Figure 17.4. As can be seen, for given angular velocity, intensity increases with ball number up to some maximum values, after which, intensity decreases due to hindering effect.

It seems that the milling intensity (power injected) is the characteristic milling parameter that defines final (steady-state) phase induced by milling, so it has been suggested, that it should be quoted whenever possible in order to compare diverse investigations [73].

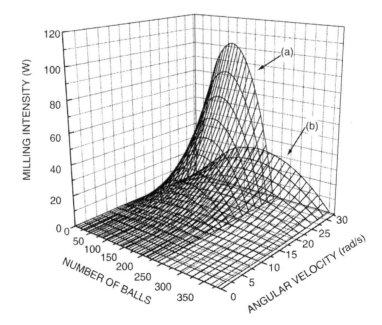

FIGURE 17.4 Milling intensity (power injected) of planetary ball mill as a function of ball number and rotation speed; (a) vial volume 500 ml, balls diameter 13.4 mm; (b) vial volume 220 ml, balls diameter 8 mm.

IV. STRUCTURAL CHANGES DURING MECHANOCHEMICAL TREATMENT

Typically, a mechanochemical treatment leads to the formation of very distorted structures, which in many cases may be described as very fine dispersoids (nanocrystallites) embedded in an amorphous matrix. Structural changes can be quantitatively followed based on detailed analysis of the powders after various milling times. Illustrative example is mechanical alloying of elemental Al and Mo powders (Figure 17.5). Following structural changes were revealed: (i) the crystallite size reduction of both Al and Mo constituents; (ii) the change of Al lattice parameter (the Mo lattice parameter does not change); and (iii) the formation of amorphous phase. The change of Al lattice parameter is a consequence of the gradual solution of Mo into the Al lattice, hence formation of super saturated Al(Mo) solid solution. The equilibrium solid solution of Mo in Al is less than 0.05%, while in the mechanically alloyed Al–17 at.% Mo powder, it reaches 2.4 at.% Mo [46]. Such an *increase of solubility* and the *formation of supersaturated solid solution* is a distinguishing feature for all metallic systems subjected to mechanochemical treatment, also for metals being mutually insoluble neither in liquid- nor in solid-state, for instance Fe–Cu system [12,74].

With reference to Figure 17.5, the structure of Al–17 at.% Mo powder mechanically alloyed for 1000 h may be understood as very fine dispersoids embedded in amorphous matrix. Aluminum and molybdenum crystallites of about 13 and 16 nm in size, respectively dispersed in very disordered, amorphous-like phase amounted about 50% of a total material volume. For powder compositions with nominally higher Mo content, as Mo gradually dissolves, Al crystal lattice being deformed and after certain milling time, Al crystalline phase collapses into an amorphous phase [75]. Similar behavior was also observed for Ni–Mo system [76], though the formation of the amorphous phase is somewhat thermodynamically favorable: heat of mixing is -4.9 and -7.3 kJ mol^{-1} for Al–Mo and Ni–Mo, respectively. In general, heat of mixing is an important

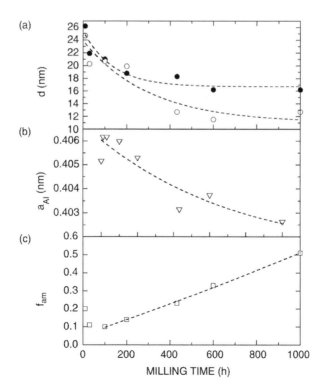

FIGURE 17.5 Structural changes during mechanical alloying of Al–17 at.% Mo powder in tumbler ball mill: (a) crystallite size, d; (b) lattice parameter of aluminum, a_{Al}; (c) amorphous volume fraction, f_{am}.

criterion for the assessment of a metallic system's tendency toward formation of amorphous phase. Thus, for the systems that readily forms an amorphous phase, for instance Cu–Yr, Co–Zr and Ni–Zr, enthalpy of mixing is -29, -42 and -51 kJ mol^{-1}, respectively.

It seems likely that, amorphous phase formation by mechanical alloying of the mixture of elemental metal powders occurs in four stages: (i) formation of very fine composite powder whereby particles my be understood as diffusion couples; (ii) formation of solid solution; (ii) collapse of supersaturated solution to the amorphous phase; and (iv) gradual dissolution of residual crystallites (dispersoids) into the amorphous matrix.

Crystallite size reduction is perhaps the most distinguishing feature of mechanochemical treatment. A number of researchers have observed that average crystallite size of either metal or ceramic powders decreases during milling. Few investigations were devoted exclusively to the preparation of nanocrystalline elemental metallic powders by ball milling [13,14,77–81]. It is a common feature of all materials studied that, the crystallite size decreases with milling time to some minimal value characteristic for the given material. With prolonged milling, the crystallite size remains unchanged, therefore, if nanocrystalline structure does not collapse into amorphous phase, further reduction seems to be difficult. Typical average crystallite size of the nanocrystalline powders prepared by mechanochemical treatment is in the range of 5–20 nm. Crystallite size reduction is always accompanied with an *introduction of atomic level strain*; it was found that the RMS strain varies linearly with reciprocal crystallite size, that is, $(\varepsilon^2)^{1/2} \propto d^{-1}$ [80,81].

Furthermore, in many instances the mechanochemical treatment begins with a particle and crystallite refinement, and the first traces of the reaction product(s) do not appear before the critical crystallite size is reached. Figure 17.6 gives the dependence of α-Fe$_2$O$_3$ crystallite size on milling time [82]. Initial crystallite size of about 223 nm decreases to about 20 nm during the milling period of $\sim$1 h. Then, with prolonged milling, chemical reaction commences. As can be seen, newly formed Fe$_3$O$_4$ phase preserves nanocrystalline structure with the crystallite size varying between 10 and 15 nm.

The development of nanocrystalline structures under the severe plastic deformation introduced by milling is explained by the mechanisms of the generation of a large number of dislocations, that on further deformation forms grain boundaries [14]. It has been suggested [14,77] that a small grain size in itself prevents further plastic deformation via dislocation motion and therefore, further grain

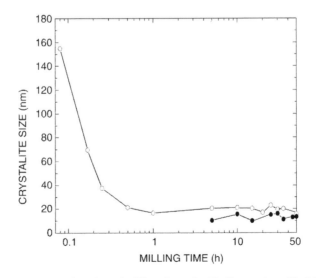

FIGURE 17.6 Crystallite size as a function of milling time of α-Fe$_2$O$_3$ powder milled in planetary ball mill; ($\circ$) α-Fe$_2$O$_3$; ($\bullet$) Fe$_3$O$_4$. (From [82], with permission from Elsevier Science.)

size refinement is not done by milling. However, it seems plausible that recovery (e.g., annihilation of dislocations) and grain growth also occurs during milling, thus limiting the reduction of grain size. Therefore, the final (steady-state) grain size achievable by ball milling is, in general, determined by competition between heavy plastic deformation and the recovery behavior of the materials [77]. However, as already stated, due to the complexity of phenomena occurring during milling, as well as difficulties to follow them *in situ*, the mechanism of crystallite formation has remained uncertain.

V. MECHANOCHEMICAL REACTIONS

During the course of mechanochemical treatment, various solid–solid, solid–liquid and solid–gas chemical reactions may take place. Numerous examples of the mechanochemical reactions, that is, *mechanically induced reactions* have been given in literature [23,83]. Majority of them can be classified into one of the following categories: (i) polymorphous transformation, for example, α-PbO (rhombic) $\leftrightarrow$ β-PbO (tetragonal) [84]; (ii) synthesis reaction, for example, $ZnO + Al_2O_3 \rightarrow ZnAl_2O_4$ [85]; (iii) decomposition reaction, for example, $MCO_3 \rightarrow MO + CO_2$ [83]; and displacement reaction, for example, $CuO + Ni \rightarrow Cu + NiO$ [39].

Mechanochemical reactions are extremely complex, thereby not fully understood. The reason for this probably lies in the fact that the whole numbers of elementary processes throughout the mechanical energy may be dissipated. However, the accumulation of a large number of results published enables, at least tentatively, some generalizations for most mechanochemical reactions to be made: (i) reactions induced by milling takes place in non-equilibrium conditions, whereby the final product retains the non-equilibrium state, that is, the structure is highly disordered, typically nanocrystalline or amorphous; (ii) the kinetics and final products of the mechanically induced reactions depends on the milling conditions; and (iii) in many instances crystallite size reduction preceeded phase transformation or chemical reaction.

Mechanochemical reactions have been believed to display diverse thermodynamic and kinetic characteristics with respect to those thermally induced [23]. Certainly, several phenomena govern the mechanochemical reactions: (i) permanent particle fracture, hence formation of atomically clean ("fresh") surfaces of high reactivity; (ii) permanent particle coalescence which produces very fine composite structure (in the case of mixture of two or more elemental or component powders); (iii) generation of a large amount of structural defects, that is, dislocations, vacancies, interstices etc., and (iv) appearance of highly energetic and localized sites of a short life-time.

Common mechanochemical reaction, is the synthesis reaction, for example: $ZnO + Al_2O_3 \rightarrow ZnAl_2O_4$ [85] or $NiO + Fe_2O_3 \rightarrow NiFe_2O_4$ [86]. Both mechanochemical reactions were realized starting from equimolar mixtures of ZnO and Al_2O_3 or NiO and Fe_2O_3 in corundum or hardened-steel vial and balls, respectively using planetary ball mill. In both cases, the mechanochemical reaction proceeds in a similar manner, however formation of $ZnAl_2O_4$ spinel phase was completed within 4 h of milling, while formation of $NiFe_2O_4$, also with spinel structure, required a milling time of about 35 h. Obviously, different physical properties of the processed systems as well as milling conditions (Retsch PM4 with 17 corundum ball with a diameter of 20 mm and Fritsc Pulverisette 5 with 286 hardened-steel balls with a diameter of 8 mm for $ZnO \cdot Al_2O_3$ and $NiO \cdot Fe_2O_3$ system, respectively) causes different time-scales to achieve complete reaction.

The mechanochemical reaction, $NiO + Fe_2O_3 \rightarrow NiFe_2O_4$ was followed by magnetization measurements (coupled with X-ray powder diffraction analysis) of the samples mechanochemically treated for different milling times. As can be seen, magnetization shows almost the same dependency on milling time for all applied magnetic fields (Figure 17.7). For a milling time of 10 h, magnetization increases slowly, indicating some structural changes causing the transition from weak ferromagnetic state ($NiO + \alpha$-Fe_2O_3) to a ferromagnetic state. For the sample milled for 10 h, the increase in the magnetization is the result of the partial conversion of $NiO + \alpha$-Fe_2O_3 to $NiFe_2O_4$ spinel phase. The change in magnetization is intensive between 20 and 35 h, showing

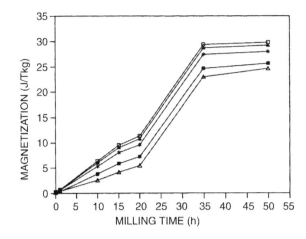

FIGURE 17.7 Magnetization of $NiO \cdot Fe_2O_3$ powder mixture as a function of milling times for various magnetic fields ($\triangle$, 835 Ga; ■, 3290 Ga; *, 6380 Ga; ▲, 8750 Ga; □, 10,540 Ga). (From [86], with permission from Elsevier Science.)

that the mechanochemical reaction mainly takes place in this time interval. After a certain amount of the spinel phase was reached (about 30%, estimated from the magnetization of the sample milled for 20 h), the formation of the spinel phase was pronounced. Thus, about 70% of reactants were converted in the milling period of 20–35 h. With prolonged milling times up to 50, magnetization increases very slightly indicating that reaction is completed in the milling interval up to 35 h.

In general, the mechanochemical synthetic reactions may be divided into at least three distinct stages. The first, in which the formation of the composite powder takes place by repeated fracture and coalescence of the constituent powder particles. Thus, composite intermixing of the constituents occur. The second stage, is characterized by the nucleation and gradual growth of the new phase. After a certain amount of the product phase was reached, the formation of a new phase becomes intensive (third stage). Eventually, prolonged milling is required for very small amount of residual reactants to be consumed completely.

The mechanism of mechanochemical synthesis reaction between two constituents can be roughly explained in the following way. Very intensive milling generates mechanical stresses in the constituent powders particles. Fracture of these particles occurs thereby, creating clean surfaces. Such comminution as well as mixing of powders enable very intimate contact between constituents (e.g., between NiO and α-Fe_2O_3 or ZnO and Al_2O_3). Highly disordered interfaces formed in that way could be suitable nucleation sites for new phase. Once the new phase was formed, reaction continues by permanent particle fracture which removes reaction product and creates new clean (fresh) surfaces. Hence, the reaction is not inhibited and persists until all reactants are consumed. Furthermore, the growth of new phase(s) is promoted by structural defects accumulated during milling as well as by local temperature rises appearing at the moment of ball impact.

Interesting example revealing peculiarity of the mechanochemical reactions is transformation of starting α-Fe_2O_3 powder induced by milling. Owing to the importance of iron oxides, particularly as magnetic materials, the effect of milling on the structural changes of α-Fe_2O_3 has attracted considerable interest, for example, [87–93]. The mechanochemical treatment of α-Fe_2O_3 powder was done concurrently in air and oxygen atmospheres using a conventional planetary ball mill [82,94]. Under appropriate milling conditions, α-Fe_2O_3 completely transforms to Fe_3O_4, and for prolonged milling to the FeO phase, either in air or oxygen atmosphere. Owing to the higher oxygen pressure, the start of the reaction in oxygen is delayed by $\sim$1 h in comparison with the reaction in air (Figure 17.8). The reverse mechanochemical reaction FeO $\rightarrow$ Fe_3O_4 $\rightarrow$ α-Fe_2O_3 takes place under proper oxygen atmosphere. The oxygen partial pressure is the critical parameter

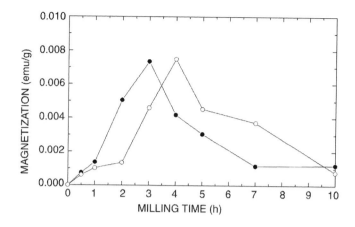

FIGURE 17.8 Magnetization of α-Fe$_2$O$_3$ powder milled for various times in planetary ball mill in ($\bullet$) air and ($\circ$) oxygen (ball-to-powder mass ratio = 40). (From [94], with permission from Elsevier Science.)

responsible for the mechanochemical reactions. The balls-to-powder mass ratio also has a considerable influence on the kinetics of mechanochemical reactions: below the threshold value the reaction does not proceed or proceeds very slowly. Plausibly, three phenomena govern mechanochemical reactions: (i) the generation of highly energetic and localized sites of a short lifetime at the moment of impact; (ii) the adsorption of oxygen at atomically clean surfaces created by particle fracture; and (iii) the change of activities of the constituent phases arising from a very disordered (nanocrystalline) structure. It was postulated that the reaction takes place by the following equation: $6Fe_2O_3 \leftrightarrow 4Fe_3O_4 + O_2 \leftrightarrow 12FeO + 3O_2$. Local modeling of a collision event, coupled with a classical thermodynamic assessment of the Fe$_2$O$_3$–Fe$_3$O$_4$ system, were used to rationalize the experimental results [82]. It was proposed that the mechanochemical reactions proceed at the moment of impact by a process of energization and freezing of highly localized sites of a short lifetime. Excitation on a time scale of $\sim10^{-5}$ sec corresponds to a temperature rise of the order of 10^3 K. Decay of the excited state occurs rapidly at a mean cooling rate higher than 10^6 K sec^{-1}.

VI. KINETICS OF MECHANOCHEMICAL REACTIONS

In recent years, considerable efforts have been taken to follow reactions induced by mechanochemical treatment and to relate them with the milling parameters. Therefrom, attempts have been made toward unification of the influence of the milling parameters on the mechanochemical reactions [70,71,95–100]. Although the mechanisms of mechanochemical that is, mechanically induced reactions is not fully understood the overall kinetics may be derived from suitable measurements (e.g., structural, magnetic) of powder milled for various milling times.

The kinetic evolution is usually represented by a sigmoid-type curve. Such a typical curve is given in Figure 9 for the case of displacement reaction, $Ni + CuO \rightarrow Cu + NiO$ realized in planetary ball mill [58]. Author of this overview analyzed these experimental results by one of the most frequently used kinetic model applied to various solid-state reactions, namely Johnson–Mehl–Avrami equation:

$$X = 1 - \exp(-kt^n) \tag{17.8}$$

where X is the volume fraction transformed at time t (sec), k (sec^{-n}) is the rate constant and n, the Avrami exponent, a parameter depending on the nucleation mechanism and the number of growth dimension. Thus, derived kinetic constants are: $n = 3.57, 2.25, 2.13$, and 2.14 and $k = 7.5 \times 10^{-16}$,

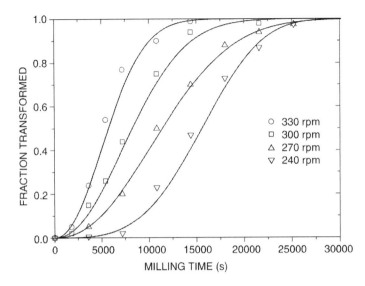

FIGURE 17.9 Fraction transformed for the mechanochemical reaction Ni + CuO → Cu + NiO carried out in planetary ball mill as a function of milling time (experimental points taken from [58]); solid line is Johnson–Mehl–Avrami Equation (17.8).

0.5×10^{-9}, 3.1×10^{-9} and 6.8×10^{-9} sec^{-n} for angular velocity of a supporting disc: 240, 270, 300 and 330 rpm, respectively. However, although Johnson–Mehl–Avrami equation satisfactorily describes overall kinetics, it is hard to give any unambiguous physical interpretations of the derived values of Avrami exponent, n which varies from 2.14 to 3.57.

Delogu and Cocco [70,71,101,102] investigated kinetics of the processes of microstructural refinement taking place during milling, namely amorphization of binary metal mixtures, amorphization of intermetallic compounds, and crystallite size reduction. From the well-defined milling experiments they correlated amorphous phase fraction and crystallite size with milling parameters, that is, impact energy, E, impact frequency, f, milling intensity, I, energy dose, D, and specific dose, D_m, (see Equations (17.1)–(17.5)). Their comprehensive work reveals that the mechanical energy dose, that is, D_m is a characteristic invariant quantity for a given material system subjected to mechanochemical treatment. Thus, it seems that the main factor which controls mechanochemical reaction is the amount of consumed energy that is, mechanical energy dose, as defined by Equation (17.5). For the foregoing reaction Ni + CuO → Cu + NiO, milling intensity can be estimated from the relation, Equation (17.7) (for given milling parameters: $n_b = 8$, $m_b = 8.4$ g, $r_p = 0.122$ m, $P^* = 0.5$, $m_p = 3$ g), hence I is 0.06, 0.08, 0.12, and 0.15 W for angular velocity, $\omega = 25.13$, 28.27, 31.55 and 34.55 sec^{-1}, respectively. The dependence of the fraction transformed on mechanical energy dose, $D = It$ (Equation (17.4)) is given in Figure 17.10. As can be seen, the fraction transformed primarily depends on the total injected energy dose, so the mechanochemical reaction is isokinetic with respect to D (powder mass was constant), in agreement with those found for amorphization of either binary mixtures or intermetallic compounds [71,102]. Therefore, variation of milling intensity (in the case of Ni + CuO → Cu + NiO reaction angular velocity) affects the reaction rate (Figure 17.10), while energy dose (or specific energy dose, Equation (17.5)) should be the invariant quantity characteristic for given material system subjected to mechanochemical treatment by milling. If this important result is confirmed in future by independent studies it would lead to a consistent theory of mechanochemical reactions. However, bearing in mind the complexity of the processes taking place during mechanochemical treatment, other milling parameters as defined in Section III, such as impact energy, collision frequency should also be taken into account.

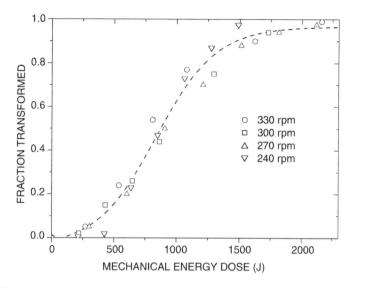

FIGURE 17.10 Fraction transformed for the mechanochemical reaction $Ni + CuO \rightarrow Cu + NiO$ carried out in planetary ball mill as a function of mechanical energy dose.

VII. SUBSEQUENT HEAT TREATMENT OF THE MECHANOCHEMICALLY SYNTHESIZED POWDERS

Mechanochemical treatment, as already emphasized, produces nanocrystalline/amorphous phases. Such pronounced metastable states readily react at elevated temperatures, thereby related equilibrium phases can be obtained relatively easily. For instance, high-temperature intermetallic compound Al_8Mo_3 (melting point: 2123 K), which, due to significantly different melting points of Al and Mo is otherwise difficult to prepare from liquid phase, was obtained by continuous heating up to 1400 K of the mechanically alloyed Al–27 at.% Mo powder [46].

Numerous results of exploration have showed that mechanochemical treatment may serve as a *precursor* technique for the subsequent solid-state reactions, typically thermal reactions. Therefore, subsequent heat treatment is applied for: (i) the reaction completion, for example, Mo–Si powder consisted of Mo and $MoSi_2$ phases after mechanical alloying was transformed to single phase $MoSi_2$ material by continuous heating up to 900°C [10]; (ii) conversion of disordered to well-crystallized phase, for example, $La(NiCoAlMn)_5$ material was prepared by mechanical alloying in amorphous form and subsequently transformed to crystalline form aiming to improve electrochemical properties of metal-hydride batteries, [103]; (iii) thermal reactions, for example, mechanochemically synthesized Fe_3O_4 was converted to γ-Fe_2O_3 by heat treatment in air at 230°C [82] or thermal reaction, $2FeCl_3 + 3CaO \rightarrow Fe_2O_3 + 3CaCl_2$ was realized in vacuum at 200°C from the precursor $FeCl_3$/CaO prepared mechanochemically by milling in vibratory mill of the $FeCl_3$ and CaO mixture in appropriate stoichiometric ratio [104].

Structure attained by mechanochemical treatment influences thermal behavior as well as kinetics of subsequent thermal reactions, for example, [75,105]. Illustrative example is Al–10 at.% Mo powders mechanically alloyed for various milling times and subjected to non-isothermal heating in differential scanning calorimeter (DSC) cell (Figure 17.11). For the powder, milled for 25 h, a very weak exothermic peak at around 808 K can be revealed indicating that some aluminum and molybdenum were in close contact and were able to react giving intermetallic $Al_{12}Mo$ compound. The reaction plausibly took place at the Al–Mo interfaces created by the repeated fracture and cold welding of the constituent powder particles. For the powder, milled for 100 h, the exothermic heat effect is more pronounced and appears at the lower temperature of 750 K. As milling time

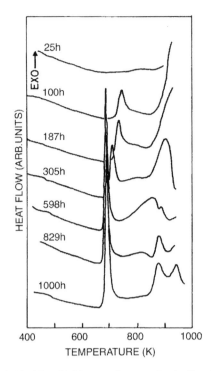

FIGURE 17.11 DSC traces of Al–10 at.% Mo powders mechanically alloyed for various milling times continuously heated at a heating rate of 20 K min^{-1}. (From [75], with permission from Elsevier Science.)

increases, the peak becomes sharper and shifts to lower temperatures (sample milled for 187, 305 and 598 h). At higher temperatures a second broad exothermic heat effect(s) can be assigned to the formation of intermetallic Al_5Mo compound [75]. The dependence of the reaction (peak) temperature and enthalpy of the formation of $Al_{12}Mo$ compound is given in Figure 17.12. The thermal behavior should be closely related to the structural changes occurring during milling. As

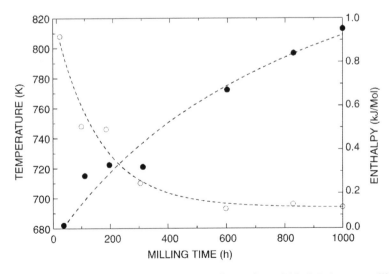

FIGURE 17.12 Temperature (o) and enthalpy (•) of the formation of $Al_{12}Mo$ intermetallic compound in mechanically alloyed Al–10 at.% Mo powders as a function of milling time.

can be seen, although the reaction temperature remains almost constant after about 600 h of milling, the enthalpy continues to increase, suggesting that the total amount of the reaction product, that is, $Al_{12}Mo$ increases. For the temperatures <800 K, diffusion distance of Mo in Al is ~ 1 nm, therefore, such short diffusion distance suggests that formation of $Al_{12}Mo$ plausibly takes place by polymorphous transformation of amorphous phase. Estimated amount of amorphous phase in the powder milled for 1000 h is about 34% [75].

VIII. POWDER CONTAMINATION

In the course of milling, contaminations of powder arising from milling medium (balls and vial walls debris), vial atmosphere and lubricant usually occur. Such "pollutions" may affect properties of the obtained material (perhaps most obvious is magnetic characteristics) as well as mechanochemical reaction(s), consequently final phase(s) attained. To avoid powder oxidation, mechanical alloying, in general, is carried out in inert atmosphere, typically argon. However, oxygen content in mechanically alloyed amorphous Ni–Zr powder was found to be 2–3 at.%, although milling was done in argon atmosphere. Consequently, it was found that kinetics of subsequent thermally-induced crystallization was changed [106]. Lubricant may also be a source of contamination since it incorporates into powder during milling [107,108]. Anyhow, the most serious contamination is debris arising from balls and vial wear and tear. In general, this contamination would depend on the intensity and duration of milling as well on the hardness of powdered material subjected to milling. In the case when vial and balls are made from hardened steel, Fe contamination (as well as other alloying elements such as Cr and Ni) is typically few atomic percents. Milling of various material up to typically 30 h (author's experiments presented in this review) using planetary ball mill with hardened-steel milling medium typically contaminates powders to the amount of 1–2 at.% Fe. In extreme cases, contaminations reaching 60 at.% Fe was reported [31]. An exhaustive survey of the results referring contaminations of different materials mechanochemically treated under various milling conditions is given in literature [31].

Systematic investigation of the accumulation Fe and Cr in the Ni–Zr system during mechanical alloying reveals that, amount of Fe and Cr above some critical concentration "push" system beyond the region of amorphous phase formation and causes the crystallization of amorphous phase [109].

In order to avoid iron contamination, milling should be performed in vials made of tungsten carbide, WC (hard metal), zirconium dioxide, ZrO_2 (zirconia), aluminum trioxide, α-Al_2O_3 (corundum), or silicon dioxide, SiO_2 (agate). However, author's experience with milling in media made of various materials is that corundum and agate is not suitable for high-energy dry milling due to unacceptable debris from such milling tools. For instance, significant wear of corundum balls takes place during mechanochemical treatment of ZnO/Al_2O_3 powder mixtures [85].

It seems that the efficient way to eliminate or at least to reduce iron contamination is to cover balls and vial internal with the appropriate material. In the Mo–Si powders mechanically alloyed in steel vial covered with tungsten carbide using WC balls contamination was not identified [10].

An example of how different milling media affect the final product of milling is provided by α-Bi_2O_3 (bismite) powder in either steel or ZrO_2 milling medium [110]. Thus, when milling was carried out in steel milling medium, virtually pure γ-Bi_2O_3 was obtained after about 10 h of milling; an amount lower than 1.5 wt.% Fe (determined by chemical analysis of the powder milled for 10 h) was sufficient to stabilize γ-Bi_2O_3 phase. On the other hand, milling starting with α-Bi_2O_3 using ZrO_2 instead of steel vials and balls has not resulted in the formation of γ-Bi_2O_3, but a partial transformation to β-Bi_2O_3 (about 36 mass% after 10 h of milling) was observed.

The influence of milling intensity (Equation (17.6)) as well as additives (grinding environment) that is, liquids of various polarity (water, methanol, or toluene) on the contamination of quartz by iron during the course of milling in vibratory ball mill was investigated in detail [72]. Important result derived is that, neither the milling intensity nor milling time influences iron contamination independently, but rather specific milling dose (Equation (17.5)). Thus, this finding supports the importance of the specific milling dose as a relevant milling parameter. Furthermore, Fe contamination was affected by physical properties of additives as well.

Attempt has been made to develop a "clean" mechanical alloying processing for the materials such as those based on titanium or niobium, for which very low contaminations are obligatory in order to obtain good mechanical properties of the compacted material [111,112]. It was demonstrated that by careful manipulation of the powder as well as by performing milling under a high-purity argon atmosphere in sealed vials (interior surface covered with one of the powder alloy constituents or some other compatible material), it is possible to produce powders with contaminations virtually equal to as-received powder [112].

IX. EXAMPLES OF THE MATERIALS SYNTHESIZED BY MECHANOCHEMICAL TREATMENT

In this section, aiming to sustain the foregoing survey, we give some examples of various materials synthesized by milling technique. Table 17.1 provides additional experimental details along with corresponding references.

Example 1: Mechanical alloying of two incompatible elements, that is, metals with significantly different melting point: Mixtures of Al and Mo (melting point of Al and Mo are 660 and 2623°C, respectively) in a broad range of starting compositions (from 3 to 75 at.% Mo) are mechanically alloyed through exceptionally long milling times up to 1000 h of milling. Composite powders having nanocrystalline structures were obtained for all compositions. Equilibrium phases, that is, intermetallic compounds $Al_{12}Mo$, Al_5Mo, Al_4Mo, Al_8Mo_3 and $AlMo_3$ were formed by subsequent heat treatment at significantly lower temperatures than their melting points. Iron contaminations was lower than 2 at.% Fe.

Example 2: Synthesis of high-temperature compounds, that is, carbides: In general, metal carbides have been produced at high temperatures, typically in the range of 1400–2300°C, whereby in some cases simultaneous application of high pressure is required. Milling of Ti, V, Cr, Mn, Fe, Cu, Ni, Zr, Mo, Hf, Ta, W, Re, As, or Si with graphite of appropriate stoichiometric composition in argon or nitrogen (as milling atmosphere) produces carbides TIC, VC, Cr_3C_2, etc. Milling was performed in tungsten carbide or hardened-steel vials up to typically 24 h.

Example 3: Synthesis of multi-component complex compounds: A number of materials such as some lead-containing complex oxides are very difficult to prepare in the form of pure phase. Powder of $Pb(Zn_xMg_{1-x})_{1/3}Nb_{2/3}O_3$ was obtained by soft-mechanochemical procedure, that is, by milling of a stoichiometric mixture of PbO, $Mg(OH)_2$, Nb_2O_5 and $2Zn(OH)_2 \cdot H_2O$ in a multi-ring-type mill up to 3 h. Partial formation of desired phase with perovskite structure already took place during milling. Subsequent heat treatment at 1000°C for 1 h yielded pure perovskite phase.

Example 4: Mechanical activation as a precursor (for subsequent reaction): Owing to its relatively low density of 6.23 g cm^{-3} and corrosion resistance at high temperatures, molybdenum disilicide, $MoSi_2$ (melting point: 2050°C) is an important material for high-temperature structural applications. A precursor, that is, very fine composite powder was obtained by milling of Mo and Si powder mixture. Mechanical activation was achieved by crystallite size reduction of both molybdenum ($d_{Mo} \approx 54$ nm) and silicon ($d_{Si} \approx 34$ nm) phases along with the development of highly developed interfaces of these two reactants. Subsequent self-propagating reaction leads to the production of $MoSi_2$ phase.

Example 5: Polymorphous transformation of crystalline to amorphous phase: Mechanochemical treatment of intermetallic compounds $Ni_{10}Zr_7$, NiZr, $NiZr_2$, Ni_5Zr_2 was carried out in

TABLE 17.1
Materials Prepared by Mechanochemical Treatment

No.	Starting Material	Product	Type of Mill	Atmosphere	Lubricant	References
1	Mixture of Al and 0, 3, 10, 17, 20,27, 50, 75, 100 at.% Mo	Nanocrystalline/amorphous Al–Mo alloy	Tumbling	Argon	Methanol	[46,75]
2	Mixture of Ti (or V, Cr, Mn…) and C	Nanocrystalline TiC (or VC, Cr_3C_2, Mn_3C…)	Planetary or vibratory	Argon or nitrogen	–	[113]
3	Mixture of PbO, $Mg(OH)_2$, Nb_2O_5 and $2Zn(OH)_2 \cdot H_2O$	$Pb(Zn_x Mg_{1-x})_{1/3}Nb_{2/3}O_3$	Multiring-type mill	Air	–	[114]
4	Mixture of Mo and Si	Nanocrystalline composite Mo–Si	Planetary	Argon	–	[115]
5	$Ni_{10}Zr_7$ (or NiZr, $NiZr_2$, Ni_5Zr_2)	Amorphous/nanocrystalline compound $Ni_{10}Zr_7$ (or NiZr, $NiZr_2$, Ni_5Zr_2)	Vibratory	Argon	–	[68]
6	Mixture of Bi_2O_3 and 60 mole% TiO_2 or $Bi_4Ti_3O_{12}$	Amorphous compound $Bi_4Ti_3O_{12}$	Planetary	Air	–	[116]
7	$CuCo_3 \cdot Cu(OH)_2$	CuO, CO_2 and H_2O	Vibratory	Vacuum	–	[117]
8	Mixture of CuO and Ca (or Ni, C, Ti, Al or Fe)	Nanocrystalline Cu and CaO (or NiO, CO_2, TiO_2, Al_2O_3 or Fe_3O_4)	Vibratory	Argon	Dry milling or toluene	[39]
9	Mixture of Ti and BN	TiN and TiB_2	Vibratory	Argon	Stearic acid	[118]
10	Ti (or Zr)	TiN (or ZrN)	Vibratory	Nitrogen	–	[119]
11	Ti (or Zr)	Nanocrystalline TiO_2 (or ZrO_2)	Vibratory	Oxygen	–	[120]

vibratory mill (tungsten carbide vials charged with hardened-steel ball whose mass was varied from 0.1 to 1.0 kg) with various amplitudes and frequencies of oscillations. Vial temperature was also varied from -190 to $250°C$. Fully amorphous steady-state was attained only for well-defined milling conditions, that is, milling intensity. Amorphous phase fraction increases with increasing milling intensity and decreases as temperature increases.

Example 6: Non-equilibrium stationary state as a characteristic of mechanochemical treatment: Concurrent mechanochemical treatment of powder mixture of Bi_2O_3 and TiO_2 in $2:3$ molar ratio and pulverized $Bi_4Ti_3O_{12}$ compound prepared by reactive sintering shows that after some milling time, a steady-state characterized by a very disordered, amorphous-like structure was reached. Thus, the systems evolves toward a non-equilibrium stationary state regardless of different initial thermodynamic states.

Example 7: Decomposition reaction: Milling of $CuCO_3 \cdot Cu(OH)_2$ (malachite) was carried out in vacuum at the pressure of $\sim 10^{-3}$ Pa (the vial was coupled with vacuum pump throughout milling) up to 200 h. Decomposition reaction $CuCO_3 \cdot Cu(OH)_2 \rightarrow 2CuO + CO_2 + H_2O$ was followed by X-ray diffraction and thermogravimetric analysis. In the beginning of milling, crystallite size reduction took place, afterwards material turned out amorphous (sample milled for 150 h). Finally, after 200 h of milling, the powder consisted of about 66% of crystallized CuO phase.

Example 8: Displacement reaction: Displacement reaction between CuO and Ca, Ni, C, Ti, Al or Fe, for example, $2CuO + Ti \rightarrow 2Cu + TiO_2$, which conventionally requires high temperature to be thermally activated, was realized by mechanochemical treatment. When dry milling was carried on, after an activation period, that is, ignition time, milling induces instantaneous self-propagating reaction.

Example 9: Influence of lubricant on mechanochemical reaction: The nanocrystalline TiN/ TiB$_2$ powders were synthesized by milling with the addition of stearic acid as a lubricant (process control agent). Up to 1.5 wt.% of stearic acid, the displacement reaction: $2Ti + 2BN \rightarrow 2TiN + TiB_2$, occurred by mechanically-induced self-propagating reaction (MSR). Over 1.5 wt.% of lubricant, reaction mode changed from MSR to gradual reaction.

Example 10: Nitrogenation reaction: Mechanochemical treatment of either Ti or Zr in mechanoreactor with controlled atmosphere of N_2 and temperature up to $500°C$ induces reaction $2Ti + N_2 \rightarrow 2TiN$ (or $2Zr + N_2 \rightarrow 2ZrN$). After 6 h of milling, conversion degree was about 60%.

Example 11: Oxidation reaction: In the course of mechanochemical treatment of either Ti or Zr in mechanoreactor with controlled atmosphere of O_2, abrupt drop of pressure was detected after several hours of milling as a consequence of instant combustion reaction: $Ti + O_2 \rightarrow TiO_2$ (or $Zr + O_2 \rightarrow ZrO_2$). Pronounced deformation as well as crystallite size reduction preceeded the reaction of oxidation. Prolonged milling following the combustion reaction produced nanocrystalline oxide powder.

X. CONCLUSIONS

Mechanochemistry, is nowadays the object of great interest in the scientific world. Rapid attention for mechanochemistry has begun at the beginning of 1980s and was prompted by the discovery that amorphous and nanocrystalline materials can be synthesized by milling. It seems that a primary reason for the tremendous curiosity of researchers for mechanochemical processes lies in the apparent simplicity of this technique which in turn allows synthesis of novel materials with unusual structures insufficiently examined. Other reason is certainly, the recognition that mechanochemical treatment provides a route to synthesize material which are otherwise difficult to prepare by other methods, for instance high-temperature materials such as carbides and silicides. A huge number of papers published, have demonstrated that mechanochemical processes provide synthesis in solid-state for a wide range of various materials, for instance intermetallic compounds, alloys of mutually insoluble elements and multi-component complex ceramic compounds. Besides, mechanochemical treatment is utilized as a precursor, that is, for the mechanical activation of a material so that a subsequent, typically, thermal treatment is facilitated, that is, thermally induced reactions are promoted at significantly lower temperature than without mechanical activation. Phenomena emerging during the course of mechanochemical treatment are numerous and complex. On that account, mechanism of mechanically-induced reactions, are endowed with an influence of milling parameters on them. However, it is not clearly known as to why the present mechanochemical processes of many useful materials are still not transferred to the commercial large-scale production. Another important drawback of the mechanochemical treatment is powder contaminations, especially for materials demanding high purity as well as exactly defined stoichiometry. Mathematical interpretation of the mechanical energy transfer to the powder charge, design of mills for specified purposes as well as rigorous control of wear and tear of milling medium are perhaps the most important directions for future research.

ACKNOWLEDGEMENT

Support of this work by the Ministry of Science and Ecology of the Republic of Serbia (Grant No. 1822) is gratefully acknowledged.

REFERENCES

1. Boldyrev, V.V., and Tkáčová, K., Mechanochemistry of solids: past, present, and prospects, *J. Mater. Synth. Process*, 8 (2–4), 121–132, 2000.
2. Koch, C.C., Cavin, O.B., McKamey, C.G., and Scarbrough, J.O., Preparation of "amorphous" $Ni_{60}Nb_{40}$ by mechanical alloying, *Appl. Phys. Lett.*, 43 (11), 1017–1019, 1983.
3. Schwarz, R.B., Petrich, R.R. and Saw, C.K., The synthesis of amorphous Ni–Ti alloy powder by mechanical alloying, *J. Non-Cryst. Solids*, 76 (2–3), 281–302, 1985.
4. Politis, C., Amorphous superconducting Nb_3Ge and $Nb_3Ge_{1-x}Al_x$ powders prepared by mechanical alloying, *Physica*, 135B, 286–289, 1985.
5. Hellstern, E., and Schultz, L., Amorphization of transition metal Zr alloys by mechanical alloying, *Appl. Phys. Lett.*, 48 (2), 124–126, 1986.
6. Dolgin, B.P., Vanek, M.A., McGory, T., and Ham, D.J., Mechanical alloying of Ni, Co, and Fe with Ti. Formation of an amorphous phase, *J. Non-Cryst. Solids*, 87 (3), 281–289, 1986.
7. Benjamin, J.S., Dispersion strengthened superalloys by mechanical alloying, *Metall. Trans.*, 1 (10) 2943–2951, 1970.
8. Shingu, P.H., Huang, B., Nishitani, S.R., and Nasu, S., Nano-meter order crystalline structures of Al–Fe alloys produced by mechanical alloying, *Suppl. Trans. JIM*, 29, 3–10, 1988.
9. Shingu, P.H., Huang, B., Kuyama, J., Nishitani, S.R., and Nasu, S., Amorphous and nano-meter order grained structures of Al–Fe and Ag–Fe alloys formed by mechanical alloying, in *Proceedings of the DGM conference on New Materials by Mechanical Alloying Techniques*, Calw-Hirsau, West Germany, October 3–5, 1989, Arzt, E. and Schultz, L., Eds., Deutsche Gesellschaft für Metallkunde, p. 319–326.
10. Schwarz, R.B., Srinivasan, S.R., Petrovic, J.J., and Maggiore, C.J., Synthesis of molybdenum disilicide by mechanical alloying, *Mater. Sci. Eng.*, A155 (1–2), 75–83, 1992.
11. Kim, M.S., and Koch, C.C., Structural development during mechanical alloying of crystalline niobium and tin powders, *J. Appl. Phys.*, 62 (8), 3450–3453, 1987.
12. Uenishi, K., Kobayashi, K.F., Nasu, S., Hatano, H., Ishihara, K.N., and Shingu, P.H., Mechanical alloying in the Fe–Cu system, *Z. Metallkd.*, 83 (2), 132–135, 1992.
13. Fecht, H.J., Hellstern, E., Fu, Z., and Johnson, W.L., Nanocrystalline metals prepared by high-energy ball milling, *Metall. Trans. A*, 21A (9), 2333–2337, 1990.

14. Eckert, J., Holzer, J.C., Krill III, C.E., and Johnson, W.L., Structural and thermodynamic properties of nanocrystalline fcc metals prepared by mechanical attrition, *J. Mater. Res.*, 7 (7), 1751–1761, 1992.

15. Yermakov, A.Ye., Yurchikov, Ye.Ye., and Barinov, V.A., Magnetic properties of amorphous powders of Y–Co alloys produced by grinding, *Fiz. Metl. Metalloved.*, 52 (6), 1184–1193, 1981.

16. Yermakov, A.Ye., Barinov, V.A., and Yurchikov, Ye.Ye., Variation of the magnetic properties of powders of Gd–Co alloys after refinement resulting in amorphism, *Fiz. Metl. Metalloved.*, 54 (5), 935–941, 1982.

17. Schwarz, R.B., and Koch, C.C., Formation of amorphous alloys by the mechanical alloying of crystalline powders of pure metals and powders of intermetallics, *Appl. Phys. Lett.*, 49 (3), 146–148, 1986.

18. Schaffer, G.B., and McCormick, P.G., Combustion synthesis by mechanical alloying, *Scripta Metall.* 23 (6), 835–838, 1989.

19. Schaffer, G.B., and McCormick, P.G., Combustion and resultant powder temperatures during mechanical alloying, *J. Mater. Sci. Lett.*, 9, 1014–1016, 1990.

20. Avvakumov, E.G., Devyatkina, E.T., and Kosova, N.V., Mechanochemical reactions of hydrated oxides, *J. Solid State Chem.*, 113 (2), 379–383, 1994.

21. Senna, M., Incipient chemical interaction between fine particles under mechanical stress — a feasibility of producing advanced materials via mechanochemical routes, *Solid State Ionics*, 63–65 (1), 3–9, 1993.

22. Thiessen, P.A., Mayer, K., and Heinicke, G., *Grundlagen der Tribochemie*, Akademie-Verlag, Berlin, 1967.

23. Heinicke, G., *Tribochemistry*, Akademie-Verlag, Berlin, 1984.

24. Avakumov, E.G., *Mehanicheskie metody himicheskin processov*, Nauka, Novosibirsk, 1986.

25. Tkácová, K., *Mechanical Activation of Minerals*, Elsevier, Amsterdam, 1989.

26. Boldyrev, V.V., *Mehanohimiya i mehanicheskaya aktivaciya tverdyh veshchestv*, SO AN SSSR, Novosibirsk, 1990.

27. Lü, L., and Lai, M.O., *Mechanical Alloying*, Kluwer Academic, Boston, 1997.

28. Gutman, E., *Mechanochemistry of Materials*, Cambridge International Science, Cambridge, 1997.

29. Avvakumov, E.G., Senna, M., and Kosova, N., *Soft Mechanochemical Synthesis*, Springer, Berlin, 2002.

30. Campbell, S.J., and Kaczmarek, W.A., Mössbauer effect studies of materials prepared by mechanochemical methods, in *Mössbauer Spectroscopy Applied to Magnetism and Materials Science*, Long, G.J. and Grandjean, F., Eds., Plenum Press, New York, 1996, Vol. 2, pp. 273–330.

31. Suryanarayana, C., Mechanical alloying and milling, *Prog. Mater. Sci.*, 46 (1–2), 1–184, 2001.

32. Koch, C.C., Materials synthesis by mechanical alloying, *Annu. Rev. Mater. Sci.*, 19, 121–143, 1989.

33. Zhang, D.L., Processing of advanced materials using high-energy mechanical milling, *Prog. Mater. Sci.*, 49 (3–4), 537–560, 2004.

34. Koch, C.C., The synthesis and structure of nanocrystalline materials produced by mechanical attrition: a review, *Nanostruct. Mater.*, 2 (2), 109–129, 1993.

35. Yavari, A.R., Mechanically prepared nanocrystalline materials (overview), *Mater. Trans. JIM*, 36 (2), 228–239, 1995.

36. Gaffet, E., and Caër, G.Le., Mechanical processing for nanomaterials, in *Encyclopedia of Nanoscience and Nanotechnology*, Nalwa, H.S., Ed., American Scientific Publishers, Stevenson Ranch, California, 2004, Vol. 10, pp. 1–39.

37. Schultz, L., Glass formation by mechanical alloying, *J. Less-Common Metals*, 145 (1) 233–249, 1988.

38. Weeber, W., and Bakker, H., Amorphization by ball milling: a review, *Physica B*, 153 (1–3), 93–135, 1988.

39. McCormick, P.G., Application of mechanical alloying to chemical refining (overview), *Mater. Trans. JIM*, 36 (2), 161–169, 1995.

40. Takacs, L., Self-sustaining reaction induced by ball milling, *Prog. Mater. Sci.*, 47 (4) 355–414, 2002.

41. Rigney, D.A., Sliding wear of metals, *Annu. Rev. Mater. Sci.*, 18, 141–163, 1988.

42. Benjamin, J.S., Mechanical alloying, *Sci. Am.*, 234 (5), 40–48, 1976.

43. Fischer, J.J., and Weber, J.H., Mechanical alloying spreads its wings, *Adv. Mater. Process*, 10, 43–50, 1990.

44. Gilman, P.S., and Benjamin, J.S., Mechanical alloying, *Annu. Rev. Mater. Sci.*, 13, 279–300, 1983.

45. Benjamin, J.S., and Volin, T.E., The mechanism of mechanical alloying, *Metall. Trans.*, 5 (8), 1929–1934, 1974.

46. Zdujić, M.V., Kobayashi, K.F., and Shingu, P.H., Structural changes during mechanical alloying of elemental aluminium and molybdenum powders, *J. Mater. Sci.*, 26 (12), 5502–5508, 1991.

47. Kuhn, W.E., Milling of brittle and ductile materials, in *Metals Handbook*, 9th ed., American Society of Metals, Metals Park, Ohio, 1984, Vol. 7, pp. 56–70.

48. Fleetwood, M.J., Mechanical alloying — the development of strong alloys, *Mater. Sci. Technol.*, 2 (12), 1176–1182, 1986.

49. Maurice, D.R., and Courtney, T.H., The physics of mechanical alloying, *Metall. Trans. A*, 21A (2), 289–303, 1990.

50. Rydin, R., Maurice, D., and Courtney, T.H., Milling dynamics: Part I. Attritor dynamics: results of a cinematographic study, *Metall. Trans. A*, 21A (1), 175–185, 1993.

51. Basset, D., Mateazzi, P., and Miani, F., Measuring the impact velocities of balls in high energy mills, *Mater. Sci. Eng.*, A174 (1), 71–74, 1994.

52. Hashimoto, H., and Watanabe, R., Model simulation of energy consumption during vibratory ball milling of metal powder. *Mater. Trans. JIM*, 31 (3), 219–224, 1990.

53. Watanabe, R., Hashimoto, H., and Lee, G.G., Computer simulation of milling ball motion in mechanical alloying (overview), *Mater Trans. JIM*, 36 (2) 102–109, 1995.

54. Burgio, N., Iasonna, A., Magini, M., Martelli, S., and Padella, F., Mechanical alloying of the Fe–Zr system. Correlation between input energy and end products, *Nuovo Cimento*, 13D (4), 459–475, 1991.

55. Brun, P.L., Froyen, L., and Delaey, L., The modeling of the mechanical alloying process in a planetary ball mill: comparison between theory and *in-situ* observations, *Mater. Sci. Eng.*, A161 (1), 75–82, 1993.

56. Abdellaoui, M., and Gaffet, E., The physics of mechanical alloying in a planetary ball mill: mathematical treatment, *Acta Metall. Mater.*, 43 (3), 1087–1098, 1995.

57. Magini, M., and Iasonna, A., Energy transfer in mechanical alloying, *Mater. Trans. JIM*, 36 (2), 123–133, 1995.

58. Dallimore, M.P., and McCormick, P.G., Dynamics of planetary ball milling: a comparison of computer simulated processing parameters with CuO/Ni displacement reaction milling kinetics, *Mater. Trans. JIM*, 37 (5), 1091–1098, 1996.

59. Iasonna, A., and Magini, M., Power measurements during mechanical milling. An experimental way to investigate the energy transfer phenomena, *Acta Mater.*, 44 (3), 1109–1117, 1996.

60. Magini, M., Colella, C., Iasonna, A., and Padella, F., Power measurements during mechanical milling — II. The case of "single path cumulative" solid state reaction, *Acta Mater.*, 46 (8), 2841–2850, 1998.

61. Mio, H., Kano, J., Saito, F., and Kaneko, K., Effects of rotational direction and rotation-to-revolution speed ratio in planetary ball milling, *Mater. Sci. Eng.*, A332 (1–2), 75–80, 2002.

62. Ikeya, T., and Senna, M., Amorphization and phase transformation of niobium pentoxide by fine grinding, *J. Mater. Sci.*, 22, 2497–2502, 1987.

63. Eckert, J., Schultz, L., and Hellstern, E., Glass-forming range in mechanically alloyed Ni–Zr and the influence of the milling intensity, *J. Appl. Phys.*, 64 (6), 3224–3224, 1988.

64. Martin, G., and Gaffet, E., Mechanical alloying: far from equilibrium phase transitions? *J. Phys. (Paris)*, 51 (14), C4-71–C4-77, 1990.

65. Park, Y.H., Yamauchi, K., Hashimoto, H., and Watanabe, R., Some properties of amorphous Ti/Al powder produced by mechanical alloying, *J. Jpn. Soc. Powder Powder Metall*, 38 (7), 914–919, 1991.

66. Padella, F., Paradiso, E., Burgio, N., Magini, M., Martelli, S., and Iasonna, A., Mechanical alloying of the Pd–Si system in controlled conditions of energy transfer, *J. Less-Common Metals*, 175 (1), 79–90, 1991.

67. Magini, M., Burgio, N., Iasonna, A., Martelli, S., Padella, F., and Paradiso, E., Analysis of energy transfer in the mechanical alloying process in the collision regime, *J. Mater. Synth. Process*, 1 (3), 135–144, 1993.

68. Chen, Y., Bibole, M., Hazif, R.L, and Martin, G., Ball-milling induced amorphization in Ni_xZr_y compounds: a parametric study, *Phys. Rev. B*, 48 (1), 14–21, 1993.

69. Butyagin, P.Yu., Problemii i perspektivii razvitiya mekhanokhimii, *Usp. Khim.*, 63 (12), 1031–1043, 1994.
70. Cocco, G., Delogu, F., and Schiffini, L., Toward a quantitative understanding of the mechanical alloying process, *J. Mater. Synth. Process*, 8 (3–4), 167–180, 2000.
71. Delogu, F., Schiffini, L., and Cocco, G., The invariant laws of the amorphization processes by mechanical alloying I. Experimental findings, *Phil. Magn. A*, 81 (8), 1917–1937, 2001.
72. Tkáčová, K., Števulová, N., Lipka, J., and Šepelák, V., Contamination of quartz by iron in energy-intensive grinding in air and liquids of various polarity, *Powder Technol.*, 83 (1), 163–171, 1995.
73. Gaffet, E., Abdellaoui, M., and Malhouroux-Gaffet, N., Formation of nanostructural materials by mechanical processings, *Mater. Trans. JIM*, 36 (2), 198–209, 1995.
74. Eckert, J., Holzer, J.C., Krill III, C.E., and Johnson, W.L., Mechanically driven alloying and grain size changes in nanocrystalline Fe–Cu powders, *J. Appl. Phys.*, 73 (6), 2794–2802, 1993.
75. Zdujić, M., Poleti, D., Karanović, Lj., Kobayashi, K.F., and Shingu, P.H., Intermetallic phases produced by the heat treatment of mechanically alloyed Al–Mo powders, *Mater. Sci. Eng.*, A185 (1–2), 77–86, 1994.
76. Zdujić, M.V., Kobayashi, K.F., and Shingu, P.H., Preparation of amorphous Ni–50 at.%Mo alloy powder by mechanical alloying, *Z. Metallknd.*, 83 (2), 136–139, 1992.
77. Hellstern, E., Fecht, H.J., Fu, Z., and Johnson, W.L. Structural and thermodynamic properties of heavily mechanically deformed Ru an AlRu, *J. Appl. Phys.*, 65 (1), 305–310, 1989.
78. Oleszak, D., and Shingu, P.H., Nanocrystalline metals prepared by low energy ball milling, *J. Appl. Phys.*, 79 (6), 2975–2980, 1996.
79. Goodrich, D.M., and Atzmon, M., Microstructural evolution in ball-milled iron powder, *Mater. Sci. Forum*, 225–227, 223–228, 1996.
80. Tian, H.H., and Atzmon, M., Kinetics of microstructure evolution in nanocrystalline Fe powder during mechanical attrition, *Acta Mater.*, 47 (4), 1255–1261, 1999.
81. Tian, H.H., and Atzmon, M., Comparison of X-ray analysis methods used to determine the grain size and strain in nanocrystalline materials. *Philos. Mag. A*, 79 (4), 1769–1786, 1999.
82. Zdujić, M., Jovalekić, Č., Karanović, Lj., Mitrić, M., Poleti, D., and Skala, D., Mechanochemical treatment of α-Fe$_2$O$_3$ powder in air atmosphere, *Mater. Sci. Eng.*, A245 (1), 109–117, 1998.
83. Lin, I.J., and Nadiv, S., Review of the phase transformation and synthesis of inorganic solids obtained by mechanical treatment (mechanochemical reactions), *Mater. Sci. Eng.*, 39, 193–209, 1979.
84. Senna, M., Problems on the mechanically induced polymorphic transformation, *Cryst. Res. Technol.*, 20 (2), 209–217, 1985.
85. Zdujić, M.V., Milošević, O.B., and Karanović, Lj.Č., Mechanochemical treatment of ZnO and Al$_2$O$_3$ powders by ball milling, *Mater. Lett.*, 13 (2–3), 125–129, 1992.
86. Jovalekić, Č., Zdujić, M., Radaković, A., and Mitrić, M., Mechanochemical synthesis of NiFe$_2$O$_4$ ferrite, *Mater. Lett.*, 24 (6), 365–368, 1995.
87. Kaczmarek, W.A., and Ninham, B.W., Preparation of Fe$_3$O$_4$ and γ-Fe$_2$O$_3$ powders by magneto-mechanical action of hematite, *IEEE Trans. Magn.*, 30 (2), 732–734, 1994.
88. Linderoth, S., Jiang, J.Z., and Mørup, S., Reversible α-Fe$_2$O$_3$ to Fe$_3$O$_4$ transformation during ball milling, *Mater. Sci. Forum*, 235–238, 205–210, 1997.
89. Wu, E., Campbell, S.J., Kaczmarek, W.A., Hofmann, M., Kennedy, S.J., and Studer, A.J., Mechanochemical treatment of haematite — neutron diffraction investigation, *Mater. Sci. Forum*, 312–314, 121–126, 1999.
90. Bid, S., Banerjee, A., Kumar, S., Pradhan, S.K., De, U., and Banarjee, D., Nanophase iron oxides by ball-mill grinding and their Mössbauer characterization, *J. Alloys Comp.*, 326 (1–2), 292–297, 2001.
91. Randrianantoandro, N., Mercier, A.M., Hervieu, M., and Grenèche, J.M., Direct phase transformation from hematite to maghemite during high energy ball milling, *Mater. Lett.*, 47 (3), 150–158, 2001.
92. Janot, R. and Guérard, D., One-step synthesis of maghemite nanometric powders by ball-milling, *J. Alloys Comp.*, 333 (1–2), 302–307, 2002.
93. Hofmann, M., Campbell, S.J., Kaczmarek, W.A., and Welzel, S., Mechanochemical transformation of α-Fe$_2$O$_3$ to Fe$_{3-x}$O$_4$ — microstructural investigation, *J. Alloys Comp.*, 348 (1–2), 278–284, 2003.
94. Zdujić, M., Jovalekić, Č., Karanović, Lj., and Mitrić, M., The ball milling induced transformation of α-Fe$_2$O$_3$ powder in air and oxygen atmosphere, *Mater. Sci. Eng.*, A262 (1–2), 204–213, 1999.

95. Urakaev, F.K., and Boldyrev, V.V., Mechanism and kinetics of mechanochemical processes in comminuting devices. 1. Theory, *Powder Technol.*, 107 (1–2), 93–107, 2000.

96. Urakaev, F.K., and Boldyrev, V.V., *Powder Technol.* 107 (3), 197–206, 2000.

97. Bégin-Colin, S., Girot, T., Caër, G. Le, and Mocellin, A., Kinetics and mechanisms of phase transformations induced by ball-milling in anatase TiO_2, *J. Solid State Chem.*, 149 (1), 41–48, 2000.

98. Bab, M.A., Mendoza-Zélis, L., and Damonte, L.C., Nanocrystalline HfN produced by mechanical milling: kinetic aspects, *Acta Mater.*, 49 (20), 4205–4213, 2001.

99. Vasconcelos, F., and Figueiredo, R.S. de, Transformation kinetics on mechanical alloying, *J. Phys. Chem. B*, 107 (16), 3761–3767, 2003.

100. Delogu, F., Orrù, R., and Cao, G., A novel macrokinetic approach for mechanochemical reactions, *Chem. Eng. Sci.*, 58 (3–6), 815–821, 2003.

101. Delogu, F., and Cocco, G., Relating single-impact events to macrokinetic features in mechanical alloying process, *J. Mater. Synth. Process*, 8 (5–6), 271–277, 2000.

102. Delogu, F., and Cocco, G., Impact-induced disordering of intermetallic phases during mechanical processing, *Mater. Sci. Eng.*, A343 (1–2), 314–317, 2003.

103. Lenain, C., Aymard, L., Salver-Disma, F., Leriche, J.B., Chabre, Y., and Tarascon, J.M., Electrochemical properties of AB_5-type hydride-forming compounds prepared by mechanical alloying, *Solid State Ionics*, 104 (3–4), 237–248, 1997.

104. Ding, J., Tsuzuki, T., and McCormick, P.G., Hematite powders synthesized by mechanochemical processing, *Nanostr. Mater.*, 8 (6), 739–747, 1997.

105. Zdujić, M., Skala, D., Karanović, Lj., Krstanović, I., Kobayashi, K.F., and Shingu, P.H., Thermal behaviour of mechanically alloyed $Ni_{50}Mo_{50}$ powders and associated kinetics of amorphous phase transformation, *Mater. Sci. Eng.*, A161 (1), 237–246, 1993.

106. Brüning, R., Altounian, Z., Strom-Olson, J.O., and Schultz, L., A comparison between the thermal properties of Ni—Zr amorphous alloys obtained by mechanical alloying and melt-spinning, *Mater. Sci. Eng.*, 97, 317–320, 1988.

107. Öveçoglu, M.L., and Nix, W.D., Characterization of rapidly solidified and mechanically alloyed Al–Fe–Ce powders, *Int. J. Powder Metall.*, 22 (1), 17–30, 1986.

108. Schwarz, R.B., Hannigan, J.W., Sheinberg, H., and Tiainen, T., Amorphous powders of Al–Hf prepared by mechanical alloying, in *Proceedings of the Modern Developments in Powder Metallurgy*, Metal Powder Federation, Princeton, 1989, Vol. 21, pp. 415–427.

109. Mizutani, U., and Lee, C.H. Effect of mechanical alloying beyond the completion of glass formation for Ni–Zr alloy powders, *J. Mater. Sci.*, 25, 399–406, 1990.

110. Poleti, D., Karanović, Lj., Zdujić, M., Jovalekić, Č., and Branković, Z., Mechanochemical synthesis of γ-Fe_2O_3, *Solid State Sci.*, 6 (3), 239–245, 2004.

111. Wilkes, D.M.J., Goodwin, P.S., Ward-Close, C.M., Bagnall, K., and Steeds, J., Solid solution of Mg in Ti by mechanical alloying, *Mater. Lett.*, 27 (1–2), 47–52, 1996.

112. Goodwin, P.S., Hinder, T.M.T., Wisbey, A., and Ward-Close, C.M., Recent progress in the clean mechanical alloying of advanced materials, *Mater. Sci. Forum*, 269–272, 53–62, 1998.

113. Matteazzi, P. and Caër, G.Le., Room-temperature mechanosynthesis of carbides by grinding of elemental powders, *J. Am. Ceram. Soc.*, 74 (6), 1382–1390, 1991.

114. Shinohara, S., Baek, J.-G., Isobe, T., and Senna, M., Synthesis of phase-pure $Pb(Zn_xMg_{1-x})_{1/3}Nb_{2/3}O_3$ up to $x = 0.7$ from a single mixture via a soft-mechanochemical route, *J. Am. Ceram. Soc.*, 83 (12), 3208–3210, 2000.

115. Gras, Ch., Gaffet, E., Bernard, F., Charlot, F., and Niepce, J.-C., Mechanically activated self-propagating high temperature synthesis (MASH) applied to the $MoSi_2$ and $FeSi_2$ phase formation, *Mater. Sci. Forum*, 312–314, 281–286, 1999.

116. Zdujić, M., Jovalekić, Č., Mitrić, M., Karanović, Lj., and Poleti, D., Mechanochemical synthesis of some bismuth-containing compounds, in *Proceedings of the Fourth International Conference of the Chemical Societies of the South-East European Countries*, Book of Abstracts, Vol. II, Belgrade, July 18–21, 2004, p. 25.

117. Castricum, H.L., Bakker, H., and Poels, E.K., Mechanochemical reactions on copper-based compounds, *Mater. Sci. Forum*, 312–314, 209–214, 1999.

118. Byun, J.-S., Shim, J.-H., and Cho, Y.W., Influence of stearic acid on mechanochemical reaction between Ti and BN powders, *J. Alloys Comp.*, 365 (1–2), 149–156, 2004.

119. Senna, M., and Okamoto, K., Rapid synthesis of Ti- and Zr-nitrides under tribochemical condition, *Solid State Ionics*, 32–34 (1), 453–460, 1989.
120. Tausimi, K., Botta, W.J., and Yavari, A.R., Explosive formation of titanium and zirconium oxide during milling under O_2 gas, *Mater. Sci. Forum*, 312–314, 73–78, 1999.

APPENDIX A: GLOSSARY OF TERMS USED IN MECHANOCHEMISTRY

The purpose of this glossary is to give basic technical terms used herewith, bearing in mind slight inconsistent and assorted terminology of the mechanochemical treatment by milling. For instance, term *mechanical alloying* is used to denote milling of mixture of metal powders (sometimes, although rarely even ceramic!), *mechanical grinding* for mechanochemical treatment of single metal or component powder, usually, intermetallics and *comminution* for particle diminution. Terms *mechanical activation, mechanosynthesis* and *reactive milling* are frequently in use when milling is intended to induce some structural modification, phase transformation and chemical reactions in powdered material. Plausible hierarchy regarding generality may be as follows: mechanochemical treatment → milling → mechanical activation → mechanochemical reaction.

In addition to the definition of mechanochemistry given in the introduction there are several others that the author of this survey finds through the Internet search. Mechanochemistry is defined as:

1. Chemical conversion of solids induced by mechanical processing usually milling or grinding.
2. Chemistry accomplished by mechanical systems directly controlling reactant molecules; the formation or breaking of chemical bonds under direct mechanical control.
3. Chemistry of processes in which systems operating, with atomic-scale precision either guide, drive, or are driven by chemical transformation. In general usage, the chemistry of processes in which energy is converted from mechanical to chemical form, or vice versa.

Below is the list of terms used in mechanochemical practice:

Amorphous structure — non-crystalline, and without long-range order.

Cold welding — cohesion between two surfaces of metal, generally under the influence of externally applied pressure at room temperature.

Cold working — plastic deformation below recrystallization temperature.

Composite — unified combinations of two (or more) distinct materials.

Crystallite — domain of solid-state matter that has the same structure as single crystal. Domain that coherently diffracts X-ray.

Dispersoid — finely divided particles of relatively insoluble constituents visible in the microstructure of certain alloys.

Excited state — highly energetic state of a short lifetime, such as hot spot (see).

Fresh surface — reactive surface created by particle fracture, in general, atomically clean surface without adsorbed gas.

Grain — individual crystal within a polycrystalline microstructure.

Hot spot — high temperature localized site of a short lifetime, which may appear as a consequence of impact of a milling tool (typically ball) on powder.

Lubricant — see process control agent.

Mechanical alloying — milling of powder mixture in which at least one of the powder is metallic, that after some milling time produce composite particles whose composition is the same or close to the composition of a starting powder mixture.

Mechanical activation — mechanochemical treatment of single powder or powder mixture with the goal, that induced changes enhance or promote subsequent process, typically to decrease temperature of thermal treatment or/and to increase the rate of chemical reaction.

Mechanically induced reaction — see mechanochemical reaction.

Mechanically induced self-propagating reaction — ignition of self-propagating high-temperature synthesis (SHS) (see) by means of mechanochemical treatment, preceded with an activation period, that is, ignition time, during which particle and crystallite size reduction, mixing and defect formation take place.

Mechanochemical reaction — chemical reaction which take places under the influence of mechanical energy, typically between constituents of powder mixture or powder and gas atmosphere.

Mechanochemical treatment — physico-chemical changes of material under the influence of mechanical energy.

Mechanoreactor — apparatus, that is, mill used to carry out chemical reaction under the influence of mechanical energy by the action of milling tools, typically ball(s), sometimes accompanied by heat energy.

Mechanoplasma — hypothetical, highest energetic state in the dissipation of mechanical energy, corresponding to temperatures greater than 10^4 K.

Mill — apparatus for mechanochemical treatment.

Milling — mechanochemical treatment of single or mixture of powder in mill.

Milling medium — vial and tools, typically ball(s), disc(s) or ring(s).

Nanocrystalline structure — consists of crystallites (grains) smaller than 1000 nm.

Particle — a minute portion of matter; particle may consist of one or more crystals.

Powder — an aggregate of discrete particles that are usually in the size range of $1-1000$ μm.

Process control agent — material that is added to the powder mixture during milling to reduce the effect of cold welding.

Reaction milling — see mechanochemical reaction.

Self-propagating high-temperature synthesis (SHS) — the synthesis of compounds (or materials) in a wave of chemical reaction (combustion) that propagates over starting reactive mixture owing to layer-by-layer heat transfer.

Surfactant — see process control agent.

BIBLIOGRAPHY

Heinicke, G., *Tribochemistry*, Academie, Berlin, 1984.

Metals Handbook, Vol. 7, *Powder Metallurgy: Terms and Definitions*, 9th ed., American Society for Metals, Metals Park, Ohio, 1984.

Metals Handbook, Vol. 9, *Metallography and Microstructures: Terms and Definitions*, 9th ed., American Society for Metals, Metals Park, Ohio, 1984.

Parker, S.P., (ed-in-chief), *Dictionary of Scientific and Technical Terms*, 3rd ed., McGraw-Hill, New York, 1984.

Suryanarayana, C., Mechanical alloying and milling, *Prog. Mater. Sci.*, 46 (1–2), 1–184, 2001.

Vlack, L.H. Van, *Elements of Materials Science and Engineering*, 5th ed., Addison-Wesley, Reading, 1985.

http://www.ism.ac.ru/handbook/glossary/term_g.htm (accessed 2004).

http://avogadro.chem.iastate.edu/CHEM571/Lectures/BalemaIII.pdf. (accessed 2004).

http://www.crnano.org/crnglossary.htm (accessed 2004)

http://www.foresight.org/Nanosystems/glossary/glossary_m.html (accessed 2004).

http://www.worldiq.com/definition/Crystallite (accessed 2004).

Liquid–Liquid Dispersions

18 Reactive Extraction in Electric Fields

Hans-Jörg Bart
Technical University of Kaiserslautern, Kaiserslautern, Germany

CONTENTS

I. INTRODUCTION

The first extraction processes were with solid extraction gaining perfumes, waxes, and pharmaceutical active oils in an operation quite similar to a modern Soxhlet apparatus. An extraction pot with an age of about B.C. 3500 was found 250 km north of Bagdad and extraction instructions were documented by a Sumerian text of B.C. 2100 [1]. The next major improvements were in the medieval age with new solvents like ethanol, mineral acids, and amalgams used to extract and purify metals. The first extraction of a metal was reported by Peligot [2] who used diethylether to extract uranyl nitrate which gave a basis to uranium extraction within the "Manhattan" project in the 1940s [3]. Reactive solvent extraction was then a niche for pyrometallurgically difficult to separate metals (Nb/Ta, Zr/Hf) till the 1960s, when there was a breakthrough with copper extraction. Liquid ion exchanger (LIX) chemicals [4] were size-selective extractants for separation of copper from iron, allowing copper recovery from low-grade ores after a sulfate leaching process. Meanwhile, the use of LIXs has expanded to a large variety of ionic species and neutral solutes in hydrometallurgical, environmental, petrochemical, chemical and biochemical industries [5–9].

II. EXTRACTION SYSTEMS

A reactive extraction system usually consists of a LIX diluted in a solvent. The latter is used to adjust rheological or physicochemical properties because most of the ion exchangers are highly viscous or even solid. The solvent should be nonmiscible with water and with a high boiling point (e.g., ~500 K) to avoid losses. If kerosene-like solvents are used sometimes a modifier (e.g., isododecanol) is necessary in order to prevent the formation of a third phase and so to help to solubilize the ion-exchanger–solute complex. The regeneration and reextraction of the extract phase is with a chemical shift as also discussed. This is in contrast to physical extraction where stripping is performed by distillation.

The practical handling and design of a reactive solvent extraction process is given in appropriate handbooks [5–9], but a short review on the principles involved is given here. LIXs are available as either anion or cation or solvating exchangers. An example of an anion exchange is as follows:

$$2\overline{R_4NHCl} + ZnCl_4^{2-} \longleftrightarrow \overline{(R_4NH)_2ZnCl_4} + 2Cl^- \tag{18.1}$$

where the bar denotes organic species.

The quarternary R_4-alkyl-substituted ammonium chlorides are commercially available and can be stripped with a surplus of chloride, hydroxide, etc., and thus the solute is regenerated in the reextraction or stripping step. The quarternary compound has the advantage of being able to be used in alkaline media compared with the frequently used ternary amines. Primary, secondary (both are water soluble, less used), and tertiary amines are only stable in acidic aqueous media because hydroxide destroys the ammonium complex.

$$\overline{R_3NHCl} + OH^- \longleftrightarrow \overline{R_3N} + Cl^- + H_2O \tag{18.2}$$

Volatile anions like acetate, formate, etc. can also be removed and stripped by a temperature swing which yields the free tertiary amine, R_3N, similar to Equation (18.2). The change in counterion concentration and temperature give rise to a reversible extraction process according to Equation (18.1). Generally, the selectivity of anion exchangers is not always good and there are many developments of new host–guest ligands which take advantage of the different sizes of the solutes [10].

The cation exchange mechanism is as follows:

$$2\overline{(HDEHP)} + Ni^{2+} \longleftrightarrow \overline{Ni(DEHP)_2} + 2H^+ \tag{18.3}$$

Here, Ni is extracted with a di(2-ethylhexyl) phosphoric acid in its H-form (HDEHP) and two protons are set free. This causes a pH-shift during extraction, which can be avoided if the ion exchanger is in the Na-form. Typical extraction isotherms are depicted in Figure 18.1. At the

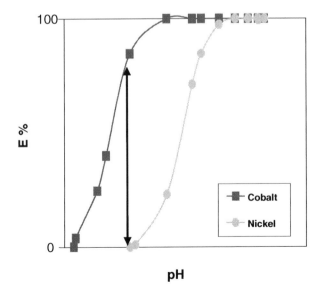

FIGURE 18.1 Typical pH-dependency of Co/Ni extraction isotherms.

indicated pH-value no nickel, but more than 80% of cobalt can be extracted in one equilibrium stage. During cobalt extraction, similar to Equation (18.3), the pH value will shift as seen in Figure 18.1 and the extraction efficiency of cobalt will be reduced. Exchange of Co versus Na will avoid this. However, a problem with Na is the occurrence of microemulsion systems, which will be discussed subsequently. Reextraction is usually with strong mineral acids (preferable H_2SO_4). Besides the alkylated phosphoric compounds there are also phosphonic and phosphinic acids and their thio-forms available. The latter ones are strong extractants and di-thiophosphoric acids are difficult to strip, but can be used at a feed pH-value less than 1. Carboxylic-acid-based ion exchangers are seldom used due to their high water solubility and aryl substituted compounds have also limited applications for steric reasons. Most LIXs have branched alkyl substituents because n-alkyls tend to crystallize and are not liquid. From a chemical point of view, both the solute–anion-exchanger or solute–cation-exchanger complexes are ionic liquids or liquid organic salts.

As mentioned earlier, the breakthrough with reactive solvent systems was with chelating ion exchangers for copper recovery. A size-specific host–guest complexation, in addition to ion exchange, separates copper from impurities such as iron. As seen in Figure 18.2, the nitrogen che-lates the copper ion developing a new six-ring structure only stable with copper as solute. A typical process flowsheet is depicted in Figure 18.3. In the center is the extraction circuit. Here copper is extracted and two protons are set free, which are then consumed in the leaching circuit. The reex-traction is with strong sulfuric acid and the stripping solution is fed to the electrowinning station. Here copper is deposited and on the anode water is transformed to protons and oxygen is set free. Literally, with current new sulfuric acid is produced, which via the extraction cycle is transported to the leaching section, countercurrently to the solute copper. This self-sustaining copper extraction unit (primary needs are only water and current) was a success when built in any remote area or desert. As seen, a copper extraction circuit is a very compact schema with usually three extraction and two reextraction stages. If a separation factor (selectivity) is not so good, a reflux of product or scrubbing section could help to improve that aspect (Figure 18.4). As extra steps to regenerate the ion exchanger (to formulate), a sodium or ammonium salt of the ion exchanger can be incorporated as discussed earlier with the Co/Ni separation. Aqueous phase microdroplets

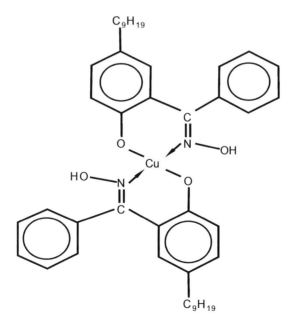

FIGURE 18.2 Copper chelate complex.

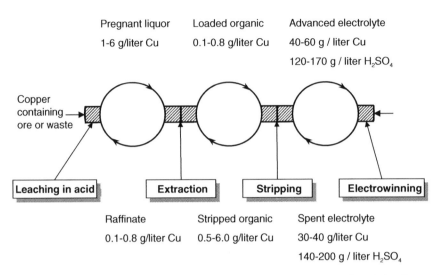

FIGURE 18.3 Hydrometallurgical copper recovery (left, leaching; middle, extraction; right, electrowinning).

entrained in the organic phase can be removed by an extra washing step, especially when there is a switch between different aqueous media (sulfate, nitrate, chloride, etc.) during extraction, regeneration, and strip stages. For more complex separation tasks, for example, the earlier Co/Ni separation, these may be scrub (reflux), followed by extraction, stripping, and regeneration stages with possible one stage water scrub in between, when extraction is from chloride and reextraction from sulfuric acid media.

An example of solvating extraction with solvating agents is shown in Equation (18.4). The chemistry is in the replacement of water molecules by the ion exchanger according to the possible coordination number q.

$$SX_p(H_2O)_q + qL \longleftrightarrow SX_pL_q + qH_2O \qquad (18.4)$$

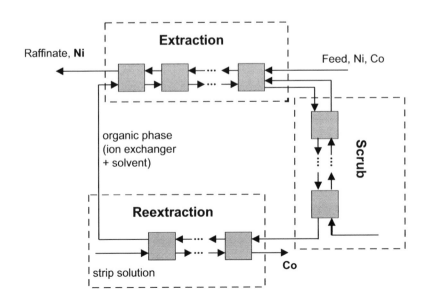

FIGURE 18.4 Flowsheet for Co/Ni separation.

The solute, S, can be a cation (metal ion) or an anion (citrate, nitrate) with an appropriate counterion (X) according to its stoichiometry p, because only neutral compounds (inorganic and organic acids, salts, etc.) can be extracted. Alkyl substituted phosphates, phosphorates, and phosphine oxides (e.g., tri-octyl phosphine oxide, tri-butyl-phosphate (TBP), etc.) are used. From high acidic media, the formation of the complexes $UCl_4 \cdot 3TBP$ and $UCl_4 \cdot 2TBP$ [11] $(UO_2 (NO_3)_2 \cdot 2TBP$ [12]) was reported. As seen, mixed complexes arise from nitric acid media and often water molecules additionally solvate the organic complex [13].

Reextraction is achieved at elevated temperatures and/or low ionic strength (e.g., with pure water). Most of the solvating exchangers can be used undiluted due to appropriate physical properties. Carbon-based compounds (ketones, ethers, etc.) suffer from lower capacity and higher water solubility and the chemistry is thus very similar to physical extraction systems. They can be regenerated by distillation, which is otherwise limited due to high boiling points.

LIXs can be mixed together in order to generate synergistic effects. This means that the effect of a mixture gives a nonlinear improvement in regard to the single systems. Basically, the effect results from an improvement to solvate the new ion-exchanger–solute complex. Such synergistic behavior is also reported with extraction kinetics when Henkel improved their first commercial copper extraction reagent LIX64 with an admixture of a small amount of LIX63. It markedly enhanced the kinetics of the new reagent LIX64N [14], which was a milestone and the commercial start to extract base metals on a large scale. In general, one can find numerous references to synergistic effects reviewed in all the solvent extraction textbooks. However, an equimolar mixture of cation and anion exchanger gives a "mixed" extraction system [15], which can extract salts or acids according to Equation (18.5). Reextraction is then either by a shift of temperature or aqueous ionic strength or acidity/basicity.

$$2\overline{R_3N} + 2\overline{HDEHP} + Zn^{2+} + 2Cl^- \longleftrightarrow \overline{(R_3NH)_2Cl_2 \cdot Zn(DEHP)_2} \tag{18.5}$$

The selection of the right reactive solvent phase is the key to a successful separation process. Some of these are essential for the separation while others are desirable properties, which will improve the separation and make it more economical. The solvent selectivity, recoverability, and a large density difference with the raffinate are essential. Some of the requirements on physical or reactive solvent phases will conflict and a compromise may be necessary.

In general, the requirements for reactive solvent extraction phases are similar to that of physical extraction ones. The viscosity should be lower than 2 mPa sec, boiling range in the region from 420 to 520 K and densities from 750 to 900 kg/m^3. The flash point should be at least 25 K higher than working temperature and a value higher than 330 K is recommended. Aromatic diluents with equivalent molecular weight as aliphatic ones are more polar and thus more water soluble. The higher price and higher toxicity of aromatic diluents lead to a preference of aliphatic diluents in industrial practice. The degradation of the diluent is usually negligible in comparison to that of the ion exchanger. The latter can be chemically, thermally, and radiation-chemically degraded and can also be poisoned by an irreversibly extracted compound. "Crud" (Chalk River Unidentified Deposit; Ritcey, G.M., personal communication, 1998) is the term describing the pollutant phase containing mineral or biological solids that tends to build up at the phase interfaces in the solvent extraction plant. Colloidal and dissolved substances (especially silica) precipitate at high shear rates and humic acids promote this behavior as reported in hydrometallurgical applications [16].

III. REACTIVE EQUILIBRIA

Very often applications with reactive extraction (e.g., metal winning from brines and effluents) are with diluted feed solutions, where ideality can be assumed. There are two attempts to tackle non-identities, which is to introduce activity coefficients or a more complex chemistry. An extended overview in respect to reactive phase equilibria is given in several textbooks and reviews [6,17–20]. The extraction of zinc with the cation exchanger DEHP (RH) may be taken as an example according to the following equation:

$$Zn^{2+} + 1.5\overline{R_2H_2} \longleftrightarrow \overline{ZnR_2(RH)} + 2H^+ \quad (K_{13}) \tag{18.6}$$

The ion exchanger is known to be in dimeric form in aliphalic diluents and the stoichiometry in Equation (18.6) was found with classical slope analysis at low concentrations and confirmed with FTIR-analysis even at high concentrations [19–22]. A compilation of all thermodynamic parameters is available in the URL www.dechema.de/Extraktion/, as this system is a recommended test system for reactive extraction studies by the EFCE (European Federation of Chemical Engineering). The predictability of the model is quite good as is depicted in Figure 18.5, where zinc extraction from chloride media is predicted from data from sulfate media [23]. The nonideality of the aqueous electrolyte solution was tackled with a modified Pitzer model [24], which is applicable up to 6 M ionic strength for weak and strong electrolytes. The regular solution concept of Hildebrand and Scott [25] was used, where one solubility (polarity) parameter is used to characterize the species (ion exchanger, solvent, metal complex) of the system.

An alternative approach is to consider the organic phase as ideal and adopt the deviations from ideality, introducing a more complex chemistry. Wenzel and Maurer [26] assumed the organic phase to be ideal and that there exist a Zn-(DEHP)$_2$ and a Zn-(DEHP)$_4$ complex simultaneously in the organic phase. The complex dissociation constants are estimated according to the following equations:

$$Zn^{2+} + \overline{R_2H_2} \longleftrightarrow \overline{ZnR_2} + 2H^+ \quad (K_{12}) \tag{18.7}$$

$$Zn^{2+} + 2\overline{R_2H_2} \longleftrightarrow \overline{ZnR_2(RH)_2} + 2H^+ \quad (K_{14}) \tag{18.8}$$

The values of the resulting constants are $\ln K_{ij} = -a_{ij}/T + b_{ij}$ with $a_{12} = 1650$, $b_{12} = 0.67$ and $a_{14} = 1447$, $b_{14} = 4.09$ when considering the aqueous nonidealities with the Pitzer equation.

This approach of considering the organic phase as ideal with all the nonidealities in the number of reactions, respectively of the complexes is quite often used in reactive solvent extraction modeling, even neglecting the aqueous phase nonideality. This approach results in system specific

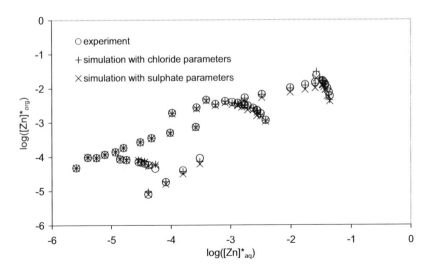

FIGURE 18.5 Zinc equilibria.

parameters, which makes it difficult to transfer the thermodynamic data to slightly different conditions (different aqueous electrolyte matrix or organic diluents). However, the method is simple and fast and of sufficient accuracy as far as moderate concentrations are involved [27]. At very high concentrations, additional reactions have to be taken into account [28–31]. A compilation of equilibrium constants in that respect can be found elsewhere [32–36].

IV. REACTIVE MASS TRANSFER

A. MASS TRANSFER WITHOUT ELECTRICAL FIELDS

The mass transfer at the liquid–liquid interface is composed from diffusional, kinetical, and convective elements. Most of the applications deal with the extraction of ionic species, which are primarily not soluble in the organic phase. After complexation by the LIX, the formed complex is then soluble in organic media. This is why complexation or chemical reaction takes place at the interface. A prerequisite for this is that the ion exchanger is available at the interface. Most of them have an amphiphilic behavior and their adsorption isotherms can be easily estimated when measuring the interfacial tension arising from different bulk concentrations [19]. In general, the assumption is that the adsorbed ion exchanger reacts with the solute and in a second step is replaced by a fresh one while the complex is been released to the bulk. According to a model of Klocker et al. [37] this reads:

$$Zn^{2+} + 2(RH)_{ad} \longleftrightarrow (ZnR_2)_{ad} + 2H^+ \tag{18.9}$$

$$(ZnR_2)_{ad} + \frac{3}{2}\overline{R_2H_2} \longleftrightarrow \overline{ZnR_2(RH)} + 2(RH)_{ad} \tag{18.10}$$

Here ad denotes the adsorbed species and resulting governing kinetics equation is then [19]:

$$-\frac{d[Zn^{2+}]}{dt} = \frac{\kappa_f \cdot [\overline{R_2H_2}]^{1.5} \cdot [Zn^{2+}] - \kappa_r \cdot [H^+]^2 \cdot [\overline{ZnR_2(RH)}]}{[\overline{R_2H_2}]^{1.5} + C_1 \cdot [H^+]^2} \left(\frac{\sqrt{[\overline{R_2H_2}]}}{C_2 + \sqrt{[\overline{R_2H_2}]}}\right)^2 \tag{18.11}$$

The equation has three kinetic parameters, (see URL www.dechema.de/Extraction/), as the forward, κ_f, or the reward reaction constant, κ_r, can be substituted by the equilibrium constant (see Equation (18.6)) according to

$$K_{1,3} = \frac{\kappa_f}{\kappa_r} \tag{18.12}$$

The parameter estimation is in a stirred Lewis-type all (Figure 18.6), where in the plateau region chemical reaction prevails. With increasing stirring speed the initial mass transfer rates will increase since diffusional effects are diminished and a plateau is reached, where mass transfer is independent from diffusion, as discussed elsewhere [19,38]. A further increase of the stirrer speed will disrupture the planar interface, leaving the operating area for determining the kinetics parameters.

The usual mass-transfer geometry for reactive extraction is not planar but spherical due to dispersions produced in mixer-settlers, extraction columns, or centrifuges. Rigid spheres can be found with small droplet diameters, whereas with increasing droplet diameters (usually 1–6 mm) there is an onset of circulations till big oscillating droplets disrupt in smaller new ones. In small droplets nonstationary diffusion from the surface to the centre prevails. In general, with reactive extraction, we usually find bulky organic constituents, which is the ion exchanger and the ion-exchanger–solute complex. Their molecular mass is higher than that of the solute in the aqueous phase.

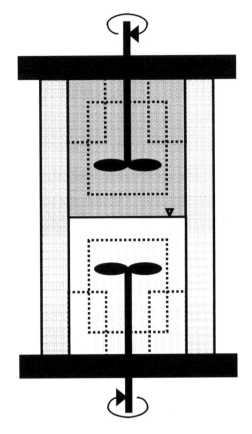

FIGURE 18.6 Stirred Lewis-cell after Nitsch [38].

Additionally, the higher viscosity of the organic phase makes it more likely that the diffusion resistance is predominant in the organic phase in respect to the aqueous one. A correct description is then to model diffusion inside the droplet with the Maxwell–Stefan diffusion law [39], where the chemical potential and not concentration differences represent the driving force. However, a problem up to now is to get information in respect to an exact value of the cross-diffusion coefficients. They represent the frictional coupling of the fluxes of the species involved and can only be neglected in very dilute systems.

Bosse et al. [40] proposed a new model to predict binary Maxwell–Stefan diffusion coefficients D_{ij}, based on Eyring's absolute reaction rate theory [41]. A correlation from Vignes [42], which was shown to be valid only for ideal systems of similar sized molecules without energy interactions [43] was extended with a Gibbs excess energy term

$$\ln\left(\frac{Đ_{12}}{\lambda_m^2}\right) = x_1 \ln\left(\frac{Đ_{12}^{\infty}}{\lambda_1^2}\right) + x_2 \ln\left(\frac{Đ_{21}^{\infty}}{\lambda_2^2}\right) + \frac{g^E}{RT} \tag{18.13}$$

where the mean distance parameter is as

$$\lambda_m = x_1 \lambda_1 + x_2 \lambda_2 \tag{18.14}$$

and the individual λ_i may be estimated from the cubic root of the molecular or the van der Waals volumes. The g^E parameters can be taken from isothermal VLE or LLE data using UNIQUAC,

COSMO-RS, or any other g^E model. The quality of these parameters is then decisive for the model predictions. A bad choice may give up to 30% error compared with 5% in the best case [44].

Modern measurement techniques based on Raman spectroscopy [45] are under way for quick data evaluation with a technical acceptable accuracy. It is clear from theory that in circulating droplets, diffusion no longer prevails as is with rigid ones. This is depicted in Figure 18.7, where a Laser–Lichtschnitt of a 3-mm toluene droplet using laser-induced fluorescence (LIF) is depicted. One can clearly see that the fluorescence signal of the solute in the continuous phase responds to wake and interfacial tension effects affecting mass transfer. This is why it is difficult to recommend mass transfer correlations and direct measurements in rising/falling droplet or Venturi tubes should be preferred. As to this, the Venturi tube (see Figure 18.8) allows arbitrary residence times according to the chemical reaction kinetics involved. In respect to the Zn/HDEHP EFCE test system, a correlation of Steiner [46] for single droplets in combination with the above-derived kinetics model (see Equation (18.11)) gave satisfying results. A better representation of the experimental data was with models having adjustable parameters, like the stagnant cap model of Slater [47] or the model of Henschke and Pfennig [48] but the contribution of the chemical reaction to the overall mass-transfer kinetics could never be neglected, as discussed elsewhere [19,49]. It is well known that amphiphilic substances will adsorb at liquid–liquid interfaces (see Section IV.B.1). Huminic acids, impurities, and even the solute–ion-exchanger complex act in that respect. As a result, during rise of the droplet, the surfactants adsorbed will be swept to the rear of the droplet building up a stagnant cap (Figure 18.9), followed by a wake in the continuous phase. So part of the droplet is then mobile or rigid, where the ratio may be determined experimentally with the model of Slater [47].

B. MASS TRANSFER WITH ELECTRICAL FIELDS

Electric fields influence mass transfer and hydrodynamics in solvent extraction but the chemical equilibria remain untouched. However, high external electrical field strength may influence slightly the dissociation equilibria of ionic species, known as Wien's Second Law and thus their availability

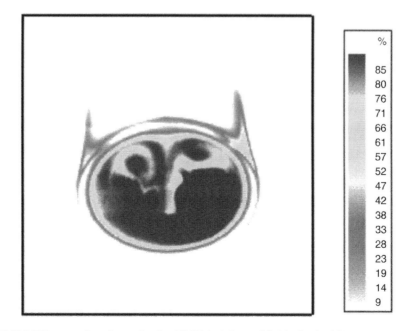

FIGURE 18.7 LIF image of a toluene droplet (100% is 0.1 mmol/l Rhodamin 6G).

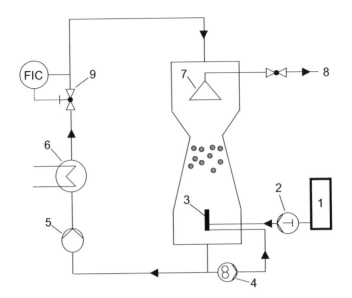

FIGURE 18.8 Venturi tube (1, feed; 2, metering pump; 3, nozzle; 4, rotary gear pump; 5, pump; 6, heat exchanger; 7, funnel; 8, sample valve; 9, valve).

and extrability. The transfer of ionic species may be largely affected by external electrical fields or an ionic Helmholtz double layer near the interface. The latter one is induced by adsorbed ionic surfactants and its influence can be quantified and related to zeta potential (ζ) measurements. This will be discussed in detail with the extraction of organic acids, where negative layers repulse and positive ones attract the anionic solute and thus change the mass transfer velocity. The aspects of electric migration due to external fields on basis of the Nernst–Planck equation will be discussed, too.

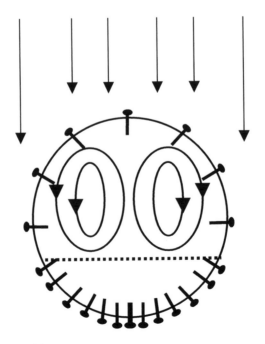

FIGURE 18.9 Stagnant cap model.

1. Surfactants

Surfactants are often found in industrial or natural (humic acids, etc.) feed stocks. Usually they should be avoided because they lower the interfacial tension leading to emulsion formation in an agitated extractor. However, even a metal-loaded ion exchanger is amphiphilic, can adsorb at the interface or aggregate in the bulk phase. This is well known with sodium or other metals [50] and above a critical surfactant concentration (critical micelle concentration, CMC) micellar aggregates are formed. A dimensionless geometric parameter is decisive for the structure of the associates (Figure 18.10); where v is the molecular volume, a_0 the surface and l the length of the surfactant:

$$\frac{v}{a_0 l} \tag{18.15}$$

With $v/a_0 l < 1/3$ one will find spherical micelles, up to a value of $1/2$ are then rod-like micelles followed by lamellar phases. Inverted micelles are then at a value >1. Between 1 and 2 are rod-like structures and above 2 are spherical ones. The phase boundaries for this structure is according to a classification by Winsor [51]. His concept relies on four classes containing surfactant, cosurfactant, oil, and water. The Winsor I system consists of an organic oil phase in equilibrium with an oil-in-water microemulsion. Winsor II is an aqueous phase in equilibrium with a water-in-oil microemulsion. Winsor III consists of three phases, oil, water, and microemulsion. Winsor IV is an isotropic microemulsion phase (Figure 18.11). The microemulsion is translucent because the small micelles (5–50 mm) will not refract light. Biais et al. [52] described this phase equilibria with Gibbs excess energies. Paatero et al. [53] evaluated the phase diagrams for NaOH/Cyanex 272/n-hexan systems. The existence of 1, 2, and 3 phase regions and even of liquid crystalline phases could be found. The same is true with Escaid 120 (Figure 18.12), where we find up to four phases. The liquid phases (L) are translucent after centrifugation but the gel phase (G) remains turbid and has a honey-like to solid consistency. A proper solvent extraction operation is with only one organic phase. As seen, this is only a small region, where about 50% of the ion exchanger is neutralized with 2.4 M NaOH. Figure 18.13 depicts samples of one- to four-phase regions. Quite left is only one phase. In the next sample the interface of the two-phase region is not well visible and indicated by a bar. Next is a sample with 2L and a G phase in between and on the very right side 3L and a G phase are depicted. The water content in all these samples are approximately 40% by mass which has to be considered in process design.

Besides this bulk phase effects surfactants will adsorb at the liquid–liquid interface. Their influence on mass transfer may then be on different mechanism. A blocking effect of adsorption layers in a diffusional transport regime is well known and results in a reduction of mass transfer [54–57] and even Marangoni instabilities [58,59] are found. However, in the kinetical mass-transfer regime, both an enhancement and retartion of mass transfer [59] is with Gibbs surfactant layers. With extracting ionic species, ionic surfactants will induce an electrostatic double layer, which can be related to the ζ-potential. As a result, there exists, in addition to the chemical potentials, an

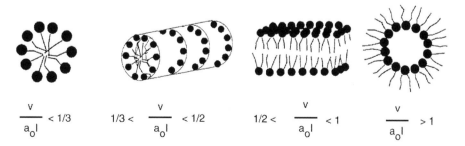

$$\frac{v}{a_0 l} < 1/3 \qquad 1/3 < \frac{v}{a_0 l} < 1/2 \qquad 1/2 < \frac{v}{a_0 l} < 1 \qquad \frac{v}{a_0 l} > 1$$

FIGURE 18.10 Micellar aggregates.

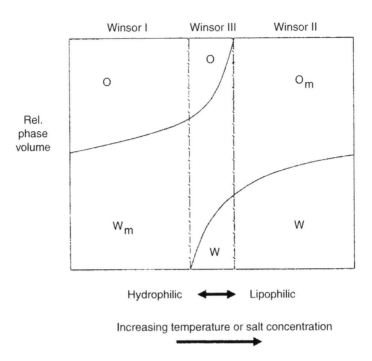

FIGURE 18.11 Microemulsion classes according to Winsor [51].

electrostatical potential difference. In order to quantify these effects, a combination of electrostatic and chemical potential differences as driving force has to be considered [19].

As described in the previous chapter in systems with LIXs, the following transfer regimes may exist [20,60]:

- Chemical reaction regime
- Diffusion controlled regime

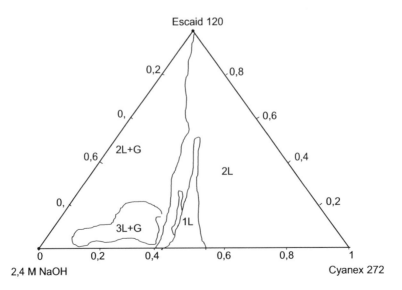

FIGURE 18.12 Phase equilibria of Cyanex 272/NaOH/Escaid 120 (293 K).

FIGURE 18.13 Phase behavior of Ecaid 120/2.4 M NaOH/Cyanex 272 (293 K).

- Mixed regime, with both diffusional and kinetic resistances
- Migrational regime, when electric fields are applied

Again the kinetic parameters can be estimated in the Lewis cell (Figure 18.6) as discussed. The reaction kinetics, here derived for acid (HA) extraction, starts with taking into account the equilibrium equation with the 1:1 complex valid in a dilute system:

$$HA + \overline{TOA} \rightleftharpoons \overline{TOAHA} \quad (K_{EX}) \tag{18.16}$$

It can be splitted into the following molecular steps according to Danesi et al. [61]:

$$H^+ + TOA_{ad} \rightleftharpoons TOAH^+_{ad} \quad (qss: K_1) \tag{18.17}$$

The equation shows protonation of an adsorbed TOA molecule at the interface. This first step is in quasi-stationary equilibrium, which is different from the two following rate determining steps:

$$TOAH^+_{ad} + A^- \rightleftharpoons TOAHA_{ad} \quad (rds: k_2; k_{-2}) \tag{18.18}$$

This reaction represents the formation of an ion-pair molecule which is adsorbed at the interface. This molecule is then replaced by a fresh TOA-molecule from the organic bulk phase in the third and last reaction step:

$$TOAHA_{ad} + \overline{TOA} \rightleftharpoons TOA_{ad} + \overline{TOAHA} \quad (rds: k_3; k_{-3}) \tag{18.19}$$

The interfacially adsorbed anion-exchanger species are balanced according to the Langmuirs' law:

$$[TOA_{ad}] + [TOAH^+_{ad}] + [TOAHA_{ad}] = \frac{\alpha_m[\overline{TOA}]}{\gamma_m + [\overline{TOA}]} \tag{18.20}$$

α_m and γ_m are the adsorption constants which have been determined by interfacial tension measurements to be 1.489×10^{-3} mol/kg for α_m and 8.156×10^{-2} mol/kg for γ_m in isododecane [62]. To guarantee that the model is able to describe the equilibrium state ($t = \infty$), the activities of the species are used instead of their concentrations. Combining the rate determining steps (Equations 18.17–18.19) and Equation (18.20) yields the kinetic rate equation, R':

$$R' = \frac{(a_{H^+} a_{A^-} a_{\overline{TOA}} - 1/K_{EX} a_{\overline{TOAHA}})(\alpha_m a_{\overline{TOA}}/(\gamma_m + a_{\overline{TOA}}))}{C_1 + C_2 a_{H^+} + C_3 a_{H^+} a_{A^-} + C_4 a_{H^+} a_{\overline{TOA}} + C_5 a_{\overline{TOA}} + C_6 a_{\overline{TOAHA}}} \tag{18.21}$$

The constants C_1 to C_6 are functions of the kinetic parameters, where k_{-3} is calculated from K_{EX} ($K_1 = 3.015$ [kg/mol], $k_{-2} = 0.0108$ [sec^{-1}], $k_2 = 83.07$, $k_3 = 1814$, $k_{-3} = 7192$ all in [kg/(mol sec)]):

$$C_1 = \frac{k_{-2}}{k_2 k_3 K_1}, \; C_2 = \frac{k_{-2}}{k_2 k_3}, \; C_3 = \frac{1}{k_3}, \; C_4 = \frac{1}{k_2}, \; C_5 = \frac{1}{k_2 K_1}, \; C_6 = \frac{1}{k_{-2} K_{EX}} \qquad (18.22)$$

At an equal interfacial covering of surfactants the mass transfer of acetate is enhanced by a cationic surfactant (dodecyl trimethylammonium chloride, DTACl), unaffected by a nonionic surfactant (octylpolyethylene glycolether, TX100)$_{10}$ and slowed down by an anionic surfactant (sodium lauryl sulfate, NaLS; sodiumdodecyl sulfate, NaDdS). This can be quantified when calculating the true interfacial concentrations, $m_{PG,i}$, in the electrochemical double layer with the Boltzmann equation:

$$m_{PG,i} = m_{B,i} \, \exp\left(-\frac{ze\Psi_0}{kT}\right) \qquad (18.23)$$

where $m_{B,i}$ is the bulk concentration, z the valence, e the elemental charge, k the Boltzmann constant, and T represents the absolute temperature. The interfacial potential Ψ_0 is calculated in accordance to the electric double layer theory after Stern [63] with the approximation of Ψ is about the zeta potential, ζ [64]:

$$\tanh\left(\frac{ze\Psi}{4kT}\right) = \tanh\left(\frac{ze\Psi_0}{4kT}\right) \exp\left(-\kappa \cdot x\right) \qquad (18.24)$$

Here κ is the Debye–Hückel constant and x the distance to the interface which is estimated to be 3 nm [65] in the discussed system. There is a difference between bulk and interfacial concentrations in magnitudes, for example, bulk pH 3 and the interfacial pH 1.0 after 60 min at 298 K with 7.5×10^{-7} mol/m^2 NaDdS (or pH 6.7 with DTACl) in a system of 1% acetic acid, 10% tri-n-octylamine in isododecane. This is similar for all ionic species involved according to the attraction/repulsion forces.

The model simulations in the dominating kinetic regime are given in Figure 18.14. The nonionic surfactant TX100 has no influence and the cationic (DTACl) shows an enhanced and the anionic

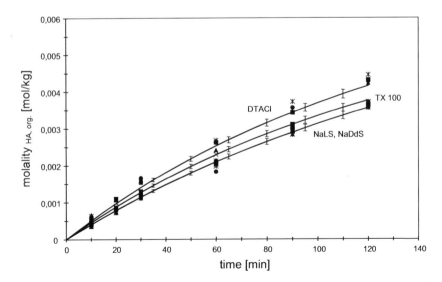

FIGURE 18.14 Kinetics influenced by surfactants (amphiphile interfacial covering of 7.5×10^{-7} mol/m^2, 1% acetic acid, 10% tri-n-octylamine in isododecane, 298 K).

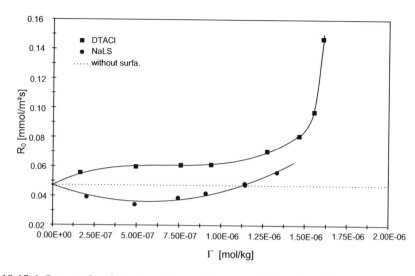

FIGURE 18.15 Influence of surfactants on the reaction rate, variation of amphiphile interfacial covering Γ (1% acetic acid, 10% tri-n-octylamine in isododecane, 298 K, n_{or} = 180 rpm).

surfactants (NaDdS, NaLS) a slow down the kinetics; these effects can quite well be described with the earlier model. However, this interfacial polarization effects will dampen with higher ionic strengths and under system conditions where molecular or eddy diffusion dominates the mass-transfer process in relation to the interfacial chemical reaction-rate resistance.

The influence of the interfacial charge of the anionic surfactant NaDdS and the cationic surfactant DTACl on the transfer rate is shown in Figure 18.15. The upper limit in these experiments is given by the CMC of the components. In the range of lower interfacial charge, the systems behaves as in Figure 18.14, an increase of the surfactant concentrations leads in the same way to a decrease of the initial flux for NaDdS as to an increase for DTACl. But at higher concentrations, the mass transfer for the cationic surfactant increases nonproportional with the interfacial charge and in a similar manner is even an increase instead of a stronger decrease in the anionic case. This can be explained by the constitution of the surfactant film at the interface, because an increase of the surfactant concentration leads to a transformation of the gas-analogous film into a condensed and strictly ordered surfactant-film, which makes penetration and mass transfer easier. This is in accordance with the other systems, for example, water or acetone or toluene in the presence of NaDdS [66].

At surfactant concentrations near the CMC, an enhanced blocking of the surface and change in interfacial rheology, and thus reduction of mass transfer may occur [67]. However, even low surfactant concentrations with species of high adsorption affinity may show the similar effect. As to this, it is essential to know the adsorption behavior of single or mixed adsorption layers to properly predict their impact on interfacial transfer kinetics.

2. External Fields

Similar results, as with surfactants, can be obtained when applying DC fields. The mass transfer of ionic species in an electric field, when there is no current applied, is according to the Nernst–Planck equation [19]:

$$N_i = -D_i \nabla c_i - D_i c_i \nabla \ln \gamma_i + \frac{t_i}{z_i} \sum_{j=1}^{n_I-1} z_j (D_j - D_n) \nabla c_j + \frac{t_i}{z_i} \sum_{j=1}^{n_I} z_j c_j D_j \nabla \ln \gamma_j \qquad (18.25)$$

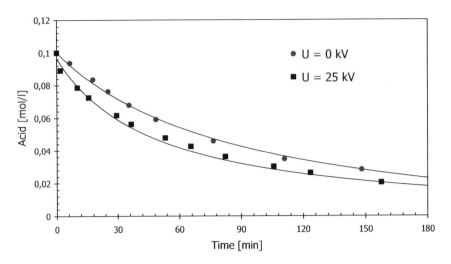

FIGURE 18.16 Mass transfer enhancement of acid extraction by a DC field (0.1 mol/l acid, 5% tri-*n*-ocylamine in toluene, 298 K).

where z is the valence and the transference number of species j is:

$$t_j = \frac{\kappa_j}{\kappa} \qquad (18.26)$$

composed from the equivalent conductivity κ_j of species i and of the mixture κ. In a reaction-controlled regime the first two diffusion-related terms are zero and only the migration term and its thermodynamic correction term remains. A comparison of the model calculations with a field-free system is in Figure 18.16 [68].

As seen, there is a permanent influence in the presence of ionic surfactants on reactive mass transfer due to the induced electrical double layer. External DC fields have only effects, when both phases are in permanent contact with either cathode or anode. This can be achieved with the batch Lewis cell (Figure 18.6) or a continuous Taylor–Couette extractor (Figure 18.17) [69–72], where one can find planar interphases. In the latter one an annular interface will be established between the counter-rotating inner and outer cylinders. With droplets this will work only as far the droplet is on top of a canula during droplet formation and the canula acts as counter electrode. Free moving droplets without a permanent electrode contact will show no effects in contrast to a canula schlieren arrangement [72] shown in Figure 18.18. The experimental set-up consists of a schlieren cell, a high-frequency AC (30 kHz) power supply (FG2-2, Ahlbrandt System GmbH) and a self-constructed DC power supply (Institute of Electrical Engineering, University of Kaiserslautern) were used to apply the desired high-voltage AC or DC signals to the capillary [68,73,74].

Mass transfer from freely moving aqueous droplets was measured in a 50-mm I.D. glass column of length 1 m, depicted in Figure 18.19. The electric field was applied between two vertical parallel

FIGURE 18.17 Taylor–Couette extractor.

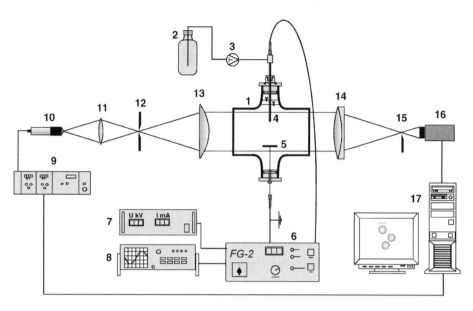

FIGURE 18.18 Schlieren set up. 1, schlieren cell; 2, dispersed phase; 3, metering pump; 4, high-voltage capillary; 5, grounded ring electrode; 6, high-voltage power supply; 7, voltage and current device; 8, oscilloscope; 9, control unit; 10, nanolite; 11, biconvex lens; 12, slit; 13, planoconvex lens; 14, achromat; 15, knife edge; 16, CCD camera; 17, PC.

plate electrodes, 95 cm in length and 2 cm in width, with electrode spacing of 3 cm. The electrodes were coated with a thin layer of PTFE to prevent wetting. A high voltage was applied to one of the plates whereas the other, together with the capillary, was grounded. When high-frequency AC signals were applied, the droplets were deflected in the direction of the high-voltage plate electrode and then fell down keeping a constant distance from the plate. In case of DC voltage, the droplets touched the high-voltage electrode and then moved downward under the action of gravity. For higher DC voltage, if charge transfer takes place at the electrodes, the droplets undergo a sort of zigzag motion between the electrodes. Droplets were formed and droplet size was controlled under applied AC or DC voltage.

With pendant droplets (Figure 18.18) the transfer of acetone from an aqueous droplet (6 wt.%) into a solute-free toluene phase is shown in Figure 18.20. Strong interfacial turbulence and eruptions are visible at 0 kV (left column of images), which decline with time but still are present after 120 sec. After that, mass transfer becomes undisturbed and a diffusional layer around the droplet develops. If AC voltage (2 kV) is applied (middle column of images), no clear change in mass transfer behavior is observed. In comparison to the case without an electric field, AC voltage seems to suppress interfacial turbulence or stabilize the diffusional layer rather than enhance the turbulence at the droplet boundary. In contrast, interfacial turbulence is enhanced with DC voltage (2 kV) and schlieren are radially aligned with the electric field from the droplet to the surrounding phase (right column of images). No diffusional layer appears and mass transfer is nearly completed after 120 sec.

In Figure 18.21 for moving droplets the dimensionless concentration driving force

$$x^+ = \frac{c}{c_0} \tag{18.27}$$

is plotted versus contact time for experiments under AC and DC voltages and without an electric field. The droplet size was kept constant in all cases at 1.5 mm. Acetone, with an initial

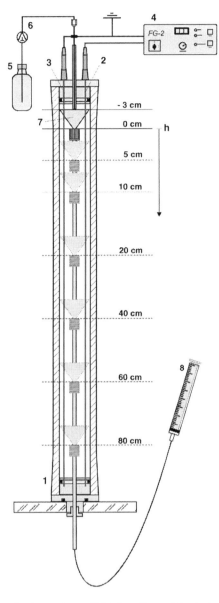

FIGURE 18.19 Droplet column. 1, glass column; 2, high-voltage plate electrode; 3, grounded plate electrode; 4, high-voltage power supply; 5, dispersed phase; 6, metering pump; 7, funnel; 8, syringe.

concentration of $x_0 = 2.5$ wt%, was transferred from aqueous droplets to the surrounding solute-free toluene phase. The experimental data indicates only a slight mass-transfer enhancement under DC voltage during the first 10 sec. For moving droplets the transfer rate is increased by reducing the outer boundary layer and the onset of internal circulation or turbulence in comparison to the pendant droplet. At the same time, electric field effects decline as the droplet detaches from the capillary. Hence, no significant differences in mass transfer behavior for freely moving single droplets were detected in contrast to the results of pendant droplets (Figure 18.20). As expected, the measured concentration driving force for moving droplets decreases much faster than predicted by Newman's model [75] (curve 1). It also decreases faster than predicted by Kronig and Brink's model [76] (curve 2), which assumes laminar internal circulation. Interfacial instabilities

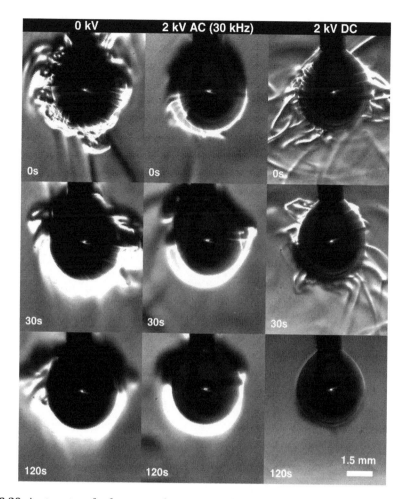

FIGURE 18.20 Acetone transfer from a pendant aqueous droplet (6 wt.%) into a solute-free toluene phase (left column of images: 0 kV; middle column of images: 2 kV AC voltage (30 kHz); right column of images: 2 kV DC voltage.

and turbulence, which are also present, have a promoting effect on mass transfer. The experimental data (0 kV) was correlated with the model of Henschke and Pfennig [48] (curve 3) and a value of the instability constant $C_{IP} = 7370$ was found.

From this it can be concluded that most of the mass-transfer enhancement of free moving droplet phases described in literature are mainly associated with enlargement of interfacial area (smaller droplets are generally formed in electrical fields), enhanced coalescence or other effects, but not due to electric field gradients.

V. LIQUID–LIQUID HYDRODYNAMICS IN ELECTRIC FIELDS

The influence of electrical fields on hydrodynamics and thus on mass transfer has been excellently reviewed by Yamaguchi [77,78]. It is thus possible to produce monodispersed droplet swarms up to extreme viscosities in the nanoscale (Figure 18.22). Here, the force balance on a nozzle leads to a disrupture of drops [79]. Under similar electrical but different geometrical conditions, breakage of an emulsion will occur, when due to polarization droplets form chains and will coalesce to bigger droplets as is technically used in secondary oil recovery, breaking down the water in oil emulsions [80]. Thus, coalescence and droplet formation in the electric field is sensitive to minor

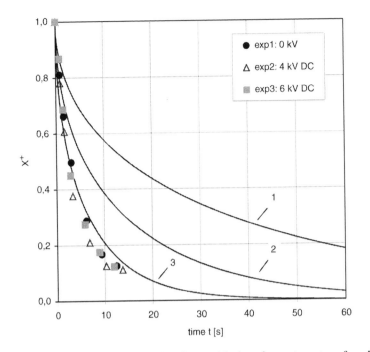

FIGURE 18.21 Decrease of concentration driving force with time for acetone transferred from aqueous droplets into toluene ($x_0 = 2.5$ wt.%, $y_0 = 0$ wt.%, $d = 1.5$ mm) under applied AC and DC signals and without an electric field. 1, Ref. [76]; 2, Ref. [77]; 3, Ref. [48].

changes in geometry or system parameters involved. This may be one of the reasons for the absence of its major application in conventional solvent extraction despite its attractiveness of the use of electrical fields.

However, the formation of monodisperse aqueous or organic drops in liquid–liquid systems is easily possible by the use of electrostatic fields [81]. Here, the applied voltage and thus the electric field at the capillary tip is the most significant parameter. This again is markedly influenced by the electrode and capillary geometry, as also by the electric properties (conductivity, permittivity, and charge relaxation time) of the liquids.

The charging of drops at an electrified capillary results from electrostatic induction. In the case of a conductive disperse phase, the charge carriers are displaced from the bulk of the liquid to the interface. This is due to the fact, that free charges inside a conductor are moved until the surface has reached an equipotential state. The electric force is a vector which acts in the direction of the electric field strength and thus is the highest at the capillary tip. At the interface, the force can be split into a tangential and a normal component. The normal component acts from the conductive to the nonconductive phase and therefore in the case of conductive in nonconductive media from the drop to the surrounding medium. This mechanism was first described by Tsouris et al. [82] and verified by pressure measurements inside the capillary. The resulting drop deformation is depending on the local field strength.

The apparatus comprised of a double casing glass tube, which is sealed with two Teflon plugs (see Figure 18.23). If the lighter organic phase has to be dispersed, the capillary is mounted centrically bottom-up. In contrast to the electrode arrangements described in literature, not only the capillary is electrically insulated (except the capillary tip), but also the grounded electrode. The latter one is an aqueous solution of 0.5 mol/l $NaCl/H_2O$, situated inside the concentric double casing. With this electrode arrangement, the electric field is concentrated at the capillary tip. In our experiments the high voltage power supply (FG2-2, Ahlbrandt System GmbH) was operated between 0 and 4 kV (45 kHz). In this range, the reduction of drop size is the strongest [79]. In all experiments the flow rate was kept constant at 0.05 ml/min and the temperature at 298 K. The conductivity was

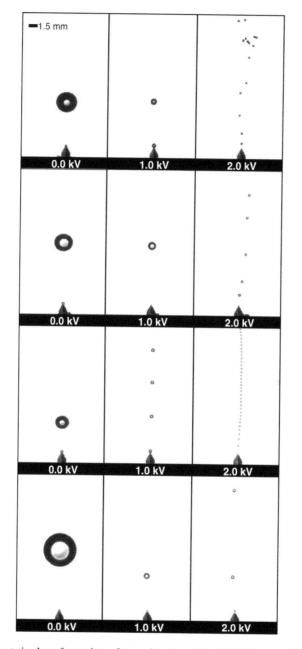

FIGURE 18.22 Electrostatic drop formation of organic solvents in distilled water (top to bottom: toluene $\eta = 0.59$ mPa sec, isododecane $\eta = 1.49$ mPa sec, isotridecanol $\eta = 41.9$ mPa sec, silicone oil DC 200 $\eta = 1000$ mPa sec, 298 K).

adjusted by adding NaCl to the aqueous phase. Two different types of stainless steal capillaries (flat and conical) with different inner and outer diameters were used to investigate the effect of capillary geometry. In all tests the voltage, current, and frequency were measured. The drop sizes were determined using an optical double flash system (Nanolite/Ministrobokin, HSPS) [79].

The drop size decreases with increasing voltage, whereas the drop keeps its spherical shape. Here, the normal component of the electric force is directed inward, which was also verified by pressure measurements inside the capillary [82] and reaches its maximum at the capillary tip.

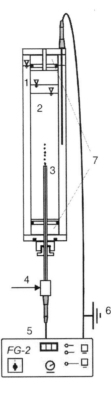

FIGURE 18.23 Experimental setup. 1, double casing glass tube; 2, aqueous phase; 3, capillary; 4, organic phase; 5, high-voltage power supply; 6, grounding; 7, Teflon plugs.

The rising velocity of droplets is unaffected in AC fields. However, DC fields promote the rising velocity with decreasing droplet diameter, which is totally in contrast to field free systems, as depicted in Figure 18.24. The experimental AC data in a distance of 1.5 cm (resp. 3.5 cm) from the capillary tip perfectly matches the correlations for field-free systems from literature. Below 3 mm, all the droplets are rigid and can be well described as a rigid sphere or globule after Vignes [83]. The onset of circulation is best described by the correlation of Klee and Treybal [84]. Also depicted is a limiting curve, where oscillation may start [85]:

$$v_{os} = \sqrt{\frac{2a_{15}\gamma}{\rho_c d_T}} \tag{18.28}$$

with ρ_c as the continuous phase density, γ the interfacial tension, and d_T the droplet diameter. The correlation parameter was set equal unity, guaranteeing an onset of oscillation at a droplet diameter of 5 mm, which is reasonable. As seen, the effect of AC fields on droplet velocity is not significant. In contrast to this, DC fields convert small rigid droplets into small oscillating ones, which have then higher rising velocities and, due to a better Sh-number, also a better mass-transfer efficiency.

VI. SUMMARY

Electric fields in solvent extraction have negligible effects on dissociation equilibria of ionic species (Second Wiensche effect) but influence mass transfer and hydrodynamics.

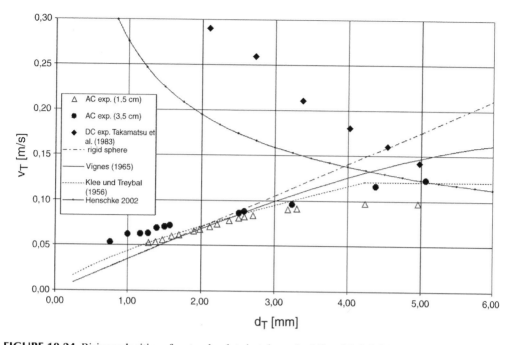

FIGURE 18.24 Rising velocities of water droplets in toluene in AC and DC fields.

The interfacial mass transfer of ionic species is either modified due to polarization or external ionic fields. Polarization occurs when cationic or anionic amphiphiles adsorb at the liquid–liquid interface and thus accelerate or decelerate the solute flux according to electrostatic repulsions. The potential and the resulting concentration gradients can be modeled with double layer theory from literature. If external electrical fields are considered to play a role, the resulting migration effects of ionic species can be described with the Nernst–Planck equation. As to this, there is the necessity to have a permanent DC field between the electrodes with in between the liquid–liquid interface. However, classical extraction equipment, as are extraction columns, mixer-settlers, or centrifuges, produce droplet ensembles or droplet swarms, which cannot be in contact permanently with an electrode. In that respect, a two-fluid Taylor–Couette extractor exhibits an annular interface of two immiscible liquids between corotating cylinders. Both bulk liquid phases can be easily grounded and mass transfer thus be influenced.

The influence of electrical fields on droplet hydrodynamics is noteworthy. During droplet formation on a grounded nozzle, either DC or AC fields can be used to generate droplets uniformly in the microscale size, even at high viscosities. High-frequency AC fields have advantages as frequency and field strength are adaptable. However, under different geometrically constraints coalescence of water-in-oil emulsion can be achieved, technically used with emulsion-breaking in secondary oil recovery. However, the droplet velocity is not influenced by AC fields, which is in contrast to DC fields. As is known from literature, the smaller the droplet, the lower is its rising velocity, which is totally in contrast to systems without a DC field. The result is an improved Sh-number and thus mass-transfer efficiency also for nonionic solutes.

ACKNOWLEDGMENTS

The author thank the National Science Foundations (DFG, German Research Foundation, and AiF, Arbeitsgemeinschaft industrieller Forschungsvereinigungen "Otto von Guericke" e.V.) and BMWA (Bundesministerium für Wirtschaft und Arbeit) for financial support.

REFERENCES

1. Blaß, E., Liebl, T., and Häberl, M., Extraktion — ein historischer Rückblick. *Chem. Ing. Techn.*, 69, 431–437, 1997.
2. Peligot, E., *Ann. Chim. Phys.*, 1842. 5–1.
3. Coleman, C.F. and Leuze, R.E., Some milestone solvent extraction processes at the Oak Ridge National Laboratory, *J. Tennessee Acad. Sci.*, 53 (3), 102–107, 1978.
4. Power, K.L., Operation of the first liquid ion-exchange and electrowinning plant, in *Proceedings of the ISEC 71*, The Hague, Soc. Chem. Ind. London (ed.), 1971, pp. 1409–1415.
5. Ritcey, G.M. and Ashbrook, A.W., *Solvent Extraction Principles and Applications to Process Metallurgy*, Elsevier, New York, 1979.
6. Thornton, J.D. Ed., *Science and Practice of Liquid–Liquid Extraction*. Vol. 1–2, Oxford University Press, Oxford, 1992.
7. Rydberg, J., Musikas, C., and Choppin, G.R., *Principles and Practice of Solvent Extraction*, 2nd edn, Marcel Dekker, New York, 2002.
8. Lo, T.C., Baird, M.H.I., and Hanson, C. Eds., *Handbook of Solvent Extraction*, J. Wiley & Sons, New York, 1983.
9. Hanson, C., *Neuere Fortschritte der Flüssig-Flüssig-Extraktion*, Sauerländer, Aarau, 1971.
10. Gloe, K., Granbaum, H., Wüst, M., Rambusch, T., and Seichter, W., Macrocyclic and open-chain ligands with the redox switchable trithiadiazapentalene unit: synthesis, structures and complexation phenomena, *Coord. Chem. Rev.*, 222, 103–126, 2001.
11. Lopovskii, A.A. and Yakovleva, N.E., Solvation of uranium tetrachloride by tributyl phosphate, *Zh. Neorg. Khim*, 9, 767, 1964.
12. Healy, T. and McKay, H., Extraction of nitrates by tributylphosphate (TBP) II nature of the TBP phase, *Trans. Farad. Soc.*, 52, 633, 1956.
13. Cox, M., Solvent Extraction in Hydrometallurgy, in *Principles and Practice of Solvent Extraction*, Thornton, J.D., Ed., Marcel Dekker, New York, 1992, pp. 357–412.
14. Kordosky, G.A., *The Chemistry of Metals Recovery Using LIX-Reagents*, General Mills, Mines Branch, 1973.
15. Schmuckler, G. and Harel, G., in *Ion Exchange and Solvent Extraction*, Marinsky, J.A. and Marcus, Y., Eds., Vol. 13, Marcel Dekker, New York, 1997, pp. 1–30.
16. Ritcey, G.M., *Solvent Extraction Processing Plants — Problems, Assessment, Solutions, in Value Adding Through Solvent Extraction*, Shallcross, D.C., Paimin, R., and Prvcic, L.M., Eds., The University of Melbourne, Melbourne, 1996, pp. 17–24.
17. Maurer, G., Electrolyte Solutions, *Fluid Phase Equilibria*, 13, 269–296, 1983.
18. Maurer, G., Phase Equilibria in Chemically Reactive Fluid Mixtures, *Fluid Phase Equilibria*, 116, 39–51, 1996.
19. Bart, H.-J., *Reactive Extraction*, Springer, Berlin, 2001.
20. Bart, H.-J. and Stevens, G., Reactive solvent extraction, in *Ion Exchange and Solvent Extraction*, Kertes, M. and Sengupta, A.K., Eds., Vol. 17. Marcel Dekker, New York, 2004, pp. 37–82.
21. Mörters, M., Zum Stoffübergang in Tropfen bei der Reaktivextraktion, Dissertation, TU Kaiserslautern, Kaiserslautern, 2001.
22. Sainz-Diaz, C.I., Klocker, H., Bart, H.-J., and Marr, R., New approach in the modelling of the extraction equilibria of zinc with bis(2-ethylhexyl) phosphoric acid, *Hydrometallurgy*, 42, 1–11, 1996.
23. Mörters, M. and Bart, H.-J., Extraction equilibria of zinc with bis(2-ethyl-hexyl)phosphoric acid, *J. Chem. Eng. Data*, 45 (1), 82–85, 2000.
24. Edwards, T.J., Maurer, G., Newman, J., and Prausnitz, J.M., Vapor liquid equilibria in multicomponent aqueous solutions of volatile weak electrolytes, *AIChE J.*, 24, 966–976, 1978.
25. Hildebrand, J.M. and Scott, R.L., *The Solubility of Nonelectrolytes*, Reinhold, New York, 1950.
26. Wenzel, T., Untersuchung von Phasengleichgewichten bei der Reaktivextraktion von Schwermetallionen mit Di/2-ethylhexyl)phosphorsäure, Dissertation, Universität Kaiserslautern, Kaiserslautern, 2003.
27. Kunzmann, M. and Kolarik, Z., Extraction of zinc(II) with di(2-ethylhexyl)phosphoric acid from perchlorate and sulfate media, *Solvent Extract. Ion Exchange*, 10 (1), 35–49, 1992.

28. Mansur, M.B., Slater, M.J., and Biscaia, E.C., Equilibrium analysis of the reactive liquid–liquid test system $ZnSO_4$/D2EHPA/n-heptane, *Hydrometallurgy*, 63, 117–126, 2002.

29. Kolarik, Z. and Grimm, R., Acidic organophosphorous extractants XXIV: The polymerization behaviour of Cu(II) Cd(II), Zn(II) and Co(II) complexes of D2EHPA in fully loaded organic phases, *J. Inorg. Nucl. Chem.*, 38, 1721–1727, 1976.

30. Morais, B.S. and Mansur, M.B., Characterization of the reactive test system $ZnSO_4$/D2EHPA in n-heptane, *Hydrometallurgy*, 74, 11–18, 2004.

31. Lee, M.-S., Ahn, J.-G., Son, S.-H., Modelling of solvent extraction of zinc from sulfate solutions with D2EHPA, *Materials Transactions*, 42 (12), 2548–2552, 2001.

32. Marcus, Y., *IUPAC Additional Publications Critical Evaluation of Some Equilibrium Constants Involving Organophosphorous Extractants*, Butterworths, London, 1974.

33. Marcus, Y., Kertes, A.S., and Yanir, E., *IUPAC Additional Publications Equilibrium Constants of Liquid–Liquid Distribution Reactions. Introduction and Part I: Organophosphorous Extractants*, Butterworths, London, 1974.

34. Marcus, Y., Kertes, A.S., and Yanir, E., *IUPAC Additional Publications Equilibrium Constants of Liquid–Liquid Distribution Reactions. Part II: Alkylammonium Salt Extractants*, Butterworths, London, 1974.

35. Kertes, A.S., *IUPAC Chemical Data Series Critical Evaluation of Some Equilibrium Constants Involving Alkylammonium Extractants*, Pergamon Press, Oxford, 1977.

36. Marcus, Y., Kertes, A.S., and Yanir, E., *IUPAC Chemical Data Series Equilibrium Constants of Liquid–Liquid Distribution Reactions. Part III: Compound Forming Extractants, Solvating Extractants and Inert Solvents*, Pergamon Press, Oxford, 1977.

37. Klocker, H., Bart, H.-J., Marr, R., and Müller, H., Mass transfer based on chemical potential theory: $ZnSO_4$/H_2SO_4/D2EHPA, *AIChE J.*, 43 (10), 2479–2487, 1997.

38. Nitsch, W., *Transportprozesse und chemische Reaktionen an fluiden Phasengrenzen in Dechema Monographie*, Vol. 114, VCH, Weinheim, 1989, pp. 285–302.

39. Taylor, R. and Krishna, R., *Multicomponent Mass Transfer*, J. Wiley & Sons, New York, 1993.

40. Bosse, D. and Bart, H.-J., *Vorausberechnung von Diffusionskoeffizienten in fluiden Systemen*, GVC-Fachausschuss "Fluidverfahrenstechnik", Leipzig, 2002.

41. Glasstone, S., Laidler, K., and Eyring, H., *The Theory of Rate Processes*, McGraw-Hill, New York, 1941.

42. Vignes, A., Diffusion in binary solutions, *Ind. Eng. Chem. Fundam.*, 5(2), 189–199, 1966.

43. Bearman, R.J., On the molecular basis of some current theories of diffusion, *J. Phys. Chem.*, 65, 1961–1968, 1961.

44. Bosse, D., Diffusion, Viscosity and Thermodynamics In Liquid Systems, PhD thesis, TU Kaiserslautern, Kaiserslautern, 2005.

45. Pfennig, A., Multicomponent diffusion, in *International Workshop on Transport in Fluid Multiphase Systems: From Experimental Data to Mechanistic Models*, Aachen, 2004.

46. Steiner, L., Mass transfer rates from single drops and drop swarms, *Chem. Eng. Sci.*, 41 (8), 1979–1986, 1986.

47. Slater, M.J., A combined model of mass transfer coefficients for contaminated drop liquid–liquid systems, *Can. J. Chem. Eng.*, 73, 462–469, 1995.

48. Henschke, M. and Pfennig, A., Mass transfer enhancement in single-drop extraction experiments, *AIChE J.*, 45 (10), 2079–2086, 1999.

49. Mörters, M. and Bart, H.-J., Mass transfer into droplets in reactive extraction *Chem. Eng. Process.*, 42, 723–731, 2003.

50. Paatero, E. and Sjöblom, J., Phase behaviour in metal extraction systems, *Hydrometallurgy*, 25, 231–256, 1990.

51. Winsor, P.A., Solubilisation with amphiphilic compounds. *Chem. Indus.*, 4, 632–644, 1960.

52. Biais, J., Clin, B., and Lalanne, P., Phase diagrams and pseudophase assumption, in *Microemulsions: Structure and Dynamics*, Friberg, S.E. and Bothorel, P. Eds., CRC Press, Inc., Cleveland, 1987, pp. 1–31.

53. Paatero, E., Ernola, P., Sjöblom, J., and Hummelstedt, L. Formation of microlemulsions in solvent extraction systems containing Cyanex 272. in *Proceedings of the ISEC'88*, Moskau, Nauka, (ed.), 1988, pp. 124–127.

54. Levich, V.G., *Physiochemical Hydrodynamics*, Prentice Halls, Engelwood Cliffs, New York, 1962.

55. Davies, J., *Turbulence Phenomena*, Academic Press, New York, 1972.

56. Nitsch, W. and Weber, G., Overcoming adsorptive inhibition of mass transfer in the range of critical micelle formation concentration, *Chem. Eng. Techn.*, 48, 715, 1976.

57. Lin, K.L. and Osseo-Asare, K., Interfacial charge and mass transfer in the liquid–liquid extraction of metals, *Recent Dev. Separat. Sci.*, 55, 74, 1986.

58. Agbe, D. and Mendes-Tatsis, M.A., The effect of surfactants on interfacial mass transfer in binary liquid–liquid systems, *Int. J. Heat & Mass Transfer*, 43, 1025–1034, 2000.

59. Mohan, V., Padmanabhan, K., and Chandrasekharan, K., Marangoni effects under electric fields, *Adv. Space Res.*, 3 (5), 177–180, 1983.

60. Hancil, V., Slater, M.J., and Yu, W., On the possible use of di-(2-ethylhexyl)phosphoric acid (zinc as a recommended system for liquid–liquid extraction: the effect of impurities on kinetics, *Hydrometallurgy*, 25, 375–386, 1990.

61. Danesi, P.R., Chiarizia, R., and Muhammed, M., Mass transfer in liquid anion exchange processes I: Kinetic of the two-phase acid-base reaction in the system trilaurylamine–toluene-HCl–Water, *J. Inorg. Nucl. Chem.*, 40, 1581–1589, 1978.

62. Czapla, C., *Zum Einfluss von Grenzflächenladungen auf den Reaktionsmechanismus organischer Säuren bei der Reaktivextraktion*, Shaker Verlag, Aachen, 2000.

63. Stern, O., The theory of the electric double-layer, Zeitschrift Elektrochemie und Angew. Physikal. Chemic, 30, 508–516 (1924).

64. Lyklema, J., De Coninck, J., and Rovillard, S., Electrokinetics. The properties of the stagnant layer unraveled, Langmuir, 14 (20), 5659–5663, 1998.

65. Czapla, C. and Bart, H.-J., Influence of the surface potential on interfacial kinetics, *Ind. Eng. Chem. Res.*, 40, 2525–2531, 2001.

66. Hüttinger, K.J. and Schegk, J.R., Chancen des Tensid-Einsatzes bei der Flüssig/Flüssig-Extraktion, *Chem.-Ing.-Tech.*, 56 (7), 554–555, 1984.

67. Raatz, S. and Härtel, G., in *DECHEMA — Monography Mass Transfer of Metals Surfactant-Loaden l/l-Interface*, Wiley - VCH, Weinheim, 2000, pp. 317–339.

68. Wildberger, A. and Bart, H.-J., Influencing the rate of mass transfer in reactive extraction using high voltage, in *Proceedings of the ISEC 2002 South African Institute of Mining and Metallurgy*, Marshalltown, 2002, 251–256.

69. Baier, G., Graham, M.D., and Lightfoot, E.N., Mass transport in a novel two-Fluid Taylor vortex extractor, *AIChE J.*, 46, 2395–2406, 2000.

70. Baier, G. and Graham, M.D., Two-fluid Taylor–Couette flow with countercurrent axial flow: linear theory for immiscible liquids between co-rotating cylinders, *Phys. Fluids*, 12 (2), 294–303, 2000.

71. Zhu, X., Campero, R.J., and Vigil, R.D., Axial mass transport in liquid–liquid Taylor–Couette–Poiseuille flow, *Chem. Eng. Sci.*, 55, 5079–5087, 2000.

72. Wildberger, A., Zum Stoffübergang bei der Flüssig-Flüssig Extraktion im elektrischen Hochspannungsfeld, Dissertation, TU Kaiserslautern, Kaiserslautern, 2004.

73. Gneist, G. and Bart, H.-J., Influence of high-frequency AC fields on mass transfer in solvent extraction, *J. Electrostatics*, 59, 73–86, 2003.

74. Gneist, G. and Bart, H.-J., Electrostatic drop formation in liquid–liquid systems, *Chem. Eng. Technol.*, 25 (9), 899–904, 2002.

75. Newman, A.B., The drying of porous solids: diffusion and surface emission equations, *AIChE J.*, 27, 203–220, 1931.

76. Kronig, R. and Brink, J., On the theory of extraction from falling droplets, *Appl. Sci. Res.*, A2, 142–154, 1950.

77. Yamaguchi, M., Electrically aided extraction and phase separation equipment, in *Liquid–Liquid Extraction Equipment*, Godfrey, J.C. and Slater, M.J. Eds., John Wiley and Sons, 1994, pp. 588–624, chap. 16.

78. Yamaguchi, M., Application of electric fields to solvent extraction, in *Electric field Applications in Chromatography Industrial and Chemical Processes*, Tsuda, T. Ed., Vol., VCH, 1995, pp. 185–203, chap. 10.

79. Gneist, G., Untersuchung des Einflusses hochfrequenter elektrischer Felder auf die Hydrodynamik und den Stoffaustausch in der Extraktion, Dissertation, TU Kaiserslautern, Kaiserslautern, 2003.

80. Marr, R. and Draxler, J., in *Membrane Handbook Applications*, Ho, W.S.W. and Sirkar, K.K. Eds., Van Nostrand Reinhold, New York, 1992, pp. 701–717.

81. Gneist, G. and Bart, H.-J., Droplet formation in liquid/liquid systems using high frequency AC fields, *Chem. Eng. Technol.*, 25 (2), 129–133, 2002.

82. Tsouris, C., DePaoli, D.W., Feng, J.Q., Basaran, O.A., and Scott, T.C., Electrostatic spraying of non-conductive fluids into conductive fluids, *AIChE J.*, 40 (11), 1920–1923, 1994.

83. Vignes, A., Hydrodynamique des dispersions, *Genie Chimique*, 93, 129–142, 1965.

84. Klee, A.J. and Treybal, R.E., Rate of rise or fall of liquid drops, *AIChE J.*, 2 (4), 444–447, 1956.

85. Henschke, M., *Auslegung pulsierter Siebboden-Extraktionskolonnen*, Habilitationsschrift, RWTH Aachen, Aachen, 2003.

Part IV

Hetero-Aggregate Finely
Dispersed Systems

Foams

19 Gas Bubbles within Electric Fields

Patrice Creux, Jean Lachaise, and Alain Graciaa
University of Pau, Pau, France

CONTENTS

I. INTRODUCTION

The study of electrokinetic phenomena has evolved considerably, since the first studies were carried out by Quincke, Helmholtz, and Lipman in the nineteenth century. In the last 30 years, at least five books have been devoted to the subject [1–5]. Among the parameters which can be deduced from electrokinetic measurements, ζ potential is profoundly related to colloid science because the "stability" of dispersions, emulsions, and foams depends on its value.

A foam is a dispersion of gas, usually air, in a liquid or solid medium [6]. It may be used to advantage a certain performance in a wide range of applications, but it may also form inadvertently and cause severe problems.

Solid foams have important applications as useful materials. The presence of gas bubbles results in a substantial reduction in the average density of the material. It also reduces the thermal conductivity, making solid foams very suitable for insulation materials. Solid foams find applications as lightweight material in mechanical components. In this case, the challenge is to achieve an effective compromise between weight and strength. In foods, such as bread, icecream, or marshmallows, solid foams help to give a pleasant texture.

By far, the largest technological application of liquid foams is in mineral froth flotation, and this process uses a substantial fraction of the world production of amphiphiles. Froth flotation is also used in other separation processes such as the deinking of recycled paper. Other uses of liquid foams include fire fighting, cleaning processes, drinks such as champagne or beer froth, foods such as whipped cream or egg white, and preparation of solid foams after solidification of the liquid continuous phase.

In the household, foams can also be a nuisance as, for example, when they form in boiling liquids, such as milk, or when, by mistake, a liquid detergent designed for cleaning by hand is used in a dishwasher or washing machine. The same type of phenomenon can readily occur in industrial processes but with more troublesome consequences. Thus, controlling foam "stability" is an issue of considerable practical importance.

Among the many factors which affect froth flotation or foam stability, the ζ potential of the bubbles play an important role. Although there is now abundant information on the ζ potential of solid particles, there has been a lack of information on the ζ potential of bubbles until recent years. This lack was due to the difficulty in measuring their electrophoretic mobility, which is often much lower than their rising mobility in the gravity field. However, for about 20 years, new methods have been developed to overcome this difficulty. This chapter is devoted to a review of the recent progress on the study of the behavior of gas bubbles within electric fields with the intention of providing information on the ζ potentials of gas bubbles. The first part of the chapter presents the experimental methods of measurement developed and the elementary precautions to take before their use. The second part presents the relevant results. The third part presents the paths followed to interpret the origin of the ζ potentials measured.

II. MEASUREMENT OF THE ζ POTENTIAL OF BUBBLES

The difficulties which are encountered during the study of the behavior of bubbles are numerous. The mean of bubbles is the high rising velocity of the bubbles. This velocity is given by the relation:

$$v_r = \frac{2}{9} \frac{\rho - \rho_0}{\eta} a^2 g \tag{19.1}$$

where ρ and η are the density and the dynamic viscosity of the immersing liquid, respectively, ρ_0 the density of the gas bubble, a the radius of the bubble, and g the acceleration due to gravity. For an 1 mm diameter bubble, the product $(\rho - \rho_0) a^2$ is about 100,000 higher than that for droplets of standard emulsions. Thus, the rising velocity of a bubble is much higher than the majority of the velocities of the colloidal domain.

The method which exploits the natural rising velocity of bubbles, without any external electric field, is the so-called flotation potential method [7–16]. Indeed, the charge-raising bubbles create an electric current, which develops a measurable electric potential difference, $\Delta\Phi$, between two sufficiently distant points (Dorn's effect). The relation between the ζ potential of the bubbles and $\Delta\Phi$ is

$$\zeta = \frac{\eta}{(\rho - \rho_0)\varepsilon g R} \frac{H}{\Delta H} \Delta\Phi \tag{19.2}$$

where ε, R, H are the dielectric permittivity, the electric resistance, and the height of the dispersion of bubbles in the liquid located between the two electrodes, which measure $\Delta\Phi$, respectively, and ΔH is the increment of H upon introduction of the gas bubbles.

Simple in its principle, this method is not easy to use. $\Delta\Phi$ is always small (at the very most, a few millivolts) and as the concentration of the bubbles increases, bubble–bubble interactions occur, so that Equation (19.2) is no longer valid.

All the methods of measurements that we present subsequently use an external electric field, which provides an electrophoretic velocity V_e to the bubbles. It is well known that the relation between the V_e, the ζ potential of the bubble, and the electric field E applied is:

$$V_e = \frac{\varepsilon f(\kappa a)}{\eta} \zeta E \qquad (19.3)$$

where ε the dielectric permittivity of the immersing liquid, κ the reciprocal thickness of the electrical double layer surrounding the bubble, and $f(\kappa a)$ the Henry function. As the size of the bubbles under consideration here is far larger than the thickness of the electric double layer ($\kappa a \gg 1$), $f(\kappa a)$ is equal to 1 and Equation (19.3) is reduced to the Smoluchowski equation.

$$V_e = \frac{\varepsilon}{\eta} \zeta E \qquad (19.4)$$

Then, the value of ζ could be deduced from measurements of E and V_e. However, difficulty comes from the fact that for bubbles, electrophoretic velocity is much lower than the rising velocity. So, all the methods that we report subsequently present solutions for overcoming or sparing this difficulty.

A. METHODS DEPENDENT ON GRAVITY

1. Method Using the Deviation by a Horizontal Electric Field

The method consists in the observation of the motion of a single rising bubble generated from electrodes. As a uniform horizontal electric field is imposed, the charged bubble deviates from its original vertical trajectory. The measured electrophoretic transverse deviation of the rising bubble in the stationary plane (where the electroosmotic velocity is 0) can be related to the bubble ζ deviation [17]. But, as the applied electric field generated convection currents, a so-called microelectrophoretic technique was developed by reducing the cell dimension to a millimeter order of magnitude [18]. Then, smaller bubbles were generated from dissolved air [19] and microbubble suspensions were introduced from the side of the cell [20].

Recently, the microelectrophoretic technique received two improvements [21]. First, electrodes were designed in such a manner that micrometer-sized bubbles are produced over the entire cross-section of the electrophoresis cell, which allows to select a bubble easily in the stationary plane. Second, a motorized vertical translation stage controlled by a computer is implemented. Thus, when bubbles rise, the electrophoresis cell mounted on the translation stage is made to move downward so that the bubbles can be kept in the field of view of the microscope. As a result, the movement of bubbles with diameters up to 80 μm can be readily followed and bubble trajectory can be traced for 4–8 sec.

2. Method Adding or Subtracting an Electric Force

This is a recent method which derives from the development of laser techniques [22,23]. It consists in the use of a double laser Doppler electrophoresis apparatus to determine bubble electrophoretic mobility by measurement of the difference in the bubble-rise rates, with and without an electric field. The latter is applied parallel to the bubble-rise vector, causing either a decrease or an increase in the natural rise rates of the bubbles according to its direction.

Electro-osmosis is neglected because the diameter of the cell is 100 times that of the bubble.

3. Method Minimizing the Gravity Effect

The gravity effect can be minimized using very small bubbles. Thus, Kim et al. [24] have generated only nanobubbles in the range 300–500 nm by ultrasonic action. These nanobubbles seem to be sufficiently stable in the gravity field to allow valuable measurements of the electrophoretic mobility for at least 1 h.

B. Methods Independent of Gravity

1. Method Using Weightless Conditions

The high mobility of the bubbles issued from the Archimedes' thrust would be eliminated if the experiments could be carried in weightless conditions in an orbital space station. In such circumstances, once a bubble had been formed, it would no longer escape the investigator; it could be subject to an electric field alone, which would impose only moderate movement. In the meantime, under these ideal conditions measurements independent of gravity can be performed by the so-called method of the spinning bubble.

2. Method of the Spinning Bubble

To counteract Archimedes' thrust, a long time ago, Quincke, Mc Taggart, and Alty [25–28] placed the bubble in a cylindrical tube filled with an aqueous phase and rotating about its symmetrical axis. The difference between the centrifugal forces acting on the bubble and on the aqueous phase maintains the bubble along the axis of rotation. This system was later abandoned, as the electric field applied to displace the bubble created an electro-osmotic flux within the tube, which biased the results [29].

Recently, Graciaa et al. eliminated the electro-osmosis by grafting an appropriate polymer onto the walls of the glass tube. This grafting neutralizes the surface charges of the glass and forms a highly viscous zone close to the walls of the tube, which prevents the liquid from moving [30]. The electrophoretic behavior of such a spinning bubble has been studied by Sherwood [31]. The two forces acting on the bubble (Figure 19.1) are evaluated by Sherwood as:

- The electric force (driving force):

$$F_{elec} = 19.2 \varepsilon a^2 \sqrt{\frac{\rho \Omega}{\eta}} \zeta E \qquad (19.5)$$

- The hydrodynamic force (resisting force):

$$F_{hydro} = \frac{16}{3} \rho a^3 \Omega V(t) \qquad (19.6)$$

where Ω is the tube spinning speed, and $V(t)$ is the instantaneous velocity of the bubble.

When the two forces are equilibrated, the so-called electrophoretic velocity V_e is reached and the relation between ζ, V_e, and E is

$$\zeta = \frac{\sqrt{\rho \eta \Omega}}{3.6 \varepsilon} a \frac{V_e}{E} \qquad (19.7)$$

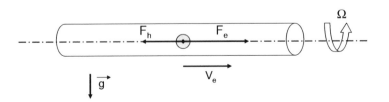

FIGURE 19.1 Motion of the bubbles in the horizontal spinning electrophorometer (F_e, electric force; F_h, hydrodynamic force; V_e, electrophoretic speed).

Obviously, this relation is different from that of the Smoluchovski because the motion of the bubble in the spinning electrophorometer is complex.

C. Indispensable Precautions

1. Purification of the Immersing Liquid

It is known that the surface of a bubble can easily adsorb alcohol molecules, surfactant molecules, ions, and, more generally, all sorts of impurities or particles present in the immersing liquid. This adsorption can drastically modify the electromobility of the bubbles. So, it is very important to purify the liquid carefully before introducing the bubble, to obtain valuable information on its intrinsic ζ potential.

These surface-active agents are often present in very low concentrations in the liquid. Their presence at the surface cannot be detected by any diversion of the interfacial tension. So, as a precaution, it is always useful to drain the liquid with masses of oxygen and hydrogen microbubbles generated directly in the liquid by electrolysis [22,23]. This drainage collects the undesirable molecules at the surface of the liquid; it is then easy to remove them. When coarse impurities are present as, for example, metal hydroxide precipitates, their elimination by ultracentrifugation is also possible [32].

As constant interfacial tension is no guarantee of absence of contamination, measurement of the upward velocity of a bubble in solution seems a much more sensitive criterion. Thus, the verification of the Hadamard–Rybczynski equation could be considered as a criterion, indicating that the solution has reached a purity beyond which there is no longer any contamination [22].

Furthermore, since the immersing liquid is most often an aqueous solution, attention must also be paid to water, which could dissolve CO_2 from the atmosphere. As it is known that at saturation, the pH can reach 5.6, the presence of a significant amount of carbonates and bicarbonates could affect the ζ potential of the bubbles. To avoid such problems, it is necessary to work under a neutral atmosphere or, at the very least, to drain the water with nitrogen bubbles systematically.

2. Control of Electro-osmosis

The phenomenon of electro-osmosis had introduced errors in the first mobility measurements carried out on bubbles in cell composed of a glass tube rotating around its own axis. The glass walls of the tube have electrical surface charges, and under the effect of the electric field, a liquid flow close to these glass walls is generated by electro-osmosis [29].

If the system is closed, an electro-osmotic reflux at the center of the cell takes place. This reflux can only be neglected when the cell is wide enough and its geometry is appropriate [22]. In all the other cases, the velocity of the bubble will be the result of the electrophoretic velocity and the reflux velocity. The electrophoretic velocity of the bubble can only be determined without correction on the so-called "stationary" levels of the cell, at which electro-osmotic flux and counterflux cancel each other [18,20,33–35].

However, if a feedback circuit is built in the electrophoretic cell, the velocity linked to electro-osmosis can be evaluated and then subtracted from the velocity measured to yield the electrophoretic velocity [36]. But, the operation is not easy.

Another possibility consists in grafting neutral and reticulated polymers onto the walls of the cell so as to reduce the surface electric charge drastically and to increase the viscosity of the interfacial layer for eliminating electro-osmosis. This type of processing has been the subject of extensive studies to determine the nature, the structure, and the molecular weight of the most effective polymers [37–39]. But, the maintenance of its efficiency must be evaluated according to the variations of the experimental conditions, because electro-osmosis can be restored by the adsorption on the polymer of an ionic compound of the immersing liquid. In particular, Creux [40] has shown that pH and hydrocarbon additives can affect the efficiency of hydropolymers but not that

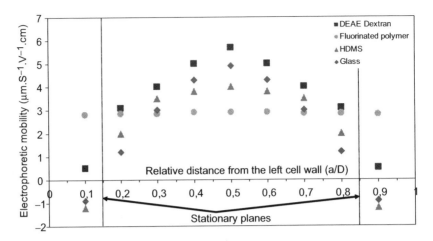

FIGURE 19.2 Profiles of the electrophoretic mobilities of polystyrene latex microspheres in the spinning cell for various polymers or surface treatments in the presence of a cationic surfactant (10 mM, TTAB) in the immersing liquid.

of fluoropolymers. As an example, profiles of the electrophoretic mobilities of polystyrene latex microspheres in the spinning cell are reported in Figure 19.2 for various polymers in the presence of a cationic surfactant tetradecyltrimethylammonium bromide (TTAB) as additive.

For the fluoropolymer, the profile is practically flat, indicating the absence of any measurable electro-osmosis. Better, for fluoropolymers, grafting is unnecessary because suppliers sell fluorined tubes of various lengths and diameters. These tubes present a good resistance to temperature and mechanical constraints and above all a good transparency, which is essential for a spinning bubble electrophoremeter. The extremely low dissolving of fluoride ions and metallic traces in the immersing liquid may contribute to the drastic decrease of electro-osmosis observed.

It was also proposed to use nonuniform electric fields, as the electro-osmotic phenomenon would take place much more slowly than the electrophoretic movement. Thus, mobility measurements could be made without any need to correct them for electro-osmosis [41]. This processing has been incorporated in devices, which are now proposed on the market [42].

Finally, a new electrode, the Brookhaven Uzgiris electrode, has been specially designed for removing electro-osmosis [24]. This electrode is entirely plunged within the cell and surrounded completely with liquids. Consequently, the electric field does not cut across the cell walls, in contrast to most of the commercial electrode systems.

III. EXPERIMENTAL RESULTS AND PROSPECTS

Each of the methods, the principle of which has been reminded briefly in the preceeding section, presents advantages and disadvantages. They have been used by numerous researchers under various conditions. We have already discussed their main results in recent reviews [43,44]. Let us recall the trends observed, report the new results obtained since our last review, and point out pioneering experiments which provide new prospects in the study of the electric properties of the gas–water interface.

A. Main Trends

Most of the investigators have found that the ζ potential of bubbles immersed in pure water is negative. But, their measurements are dispersed from some negative millivolt value to -100 mV. These variations can be attributed to the disparities of the methods of purification and to the difficulties of the measurements. However, the satisfying agreement of recent measurements may be noted for interfaces in which pure water is implied: -65 mV [30] and -60 mV [45,46]. The origin of that electronegativity is assigned to a preferential adsorption of hydroxide ions, probably under the influence of an orientation of the water dipoles near the interface with their positive poles directed toward the water bulk [23].

At the same time, the absolute value of the ζ potential of gas bubbles is found to decrease with pH decreases becoming zero for pH close to 3. At very low pH, it has even been claimed that the bubble could be positively charged, probably because of a preferential accumulation of H^+.

The magnitude of ζ potential of bubbles immersed in solutions of electrolytes are monotonously depressed by increasing the concentration of the mono- or divalent salts dissolved in the solutions. This behavior is attributed to the screening of the interactions between the applied electric field and the charged bubbles, induced by the salts. They also decrease as the acidity of the immersing solution increases. This decrease is all the more pronounced as the salinity is lower. For all the authors, the isoelectric point ranges (or would range by extrapolation) from pH 1.5 to 3.

In contrast to mono- or divalent salts, the trivalent salts can reverse the sign of the ζ potential at a certain concentration or pH. This behavior could be due to the precipitation of complexes or to a specific adsorption of the trivalent ions.

The use of amphiphilic additives as surfactants appears to be an extremely effective way to obtain bubbles with either extreme or zero ζ potential magnitude from the -60 mV of the bubbles in pure water. Anionic surfactants and, to a lesser extent, nonionic surfactants allow to reach ζ potential lower than -100 mV. Cationic surfactants allow to reach ζ potential higher than $+100$ mV after the reversal of the sign obtained at intermediate concentrations or pH.

Furthermore, it has been shown that the use of a binary mixture of surfactants gives information on the interfacial behavior of the two surfactants:

- Anionic and nonionic surfactants behave practically ideally relative to the composition of the mixture; the negative ζ potential of the bubble is included between the values for the two surfactants.
- Anionic–cationic surfactant mixtures make the ζ potential of the bubble close to zero over the wide mid-range composition, where the critical micelle concentration corresponds to the catanionic species.
- Fluorocarbon and hydrocarbon surfactants present a maximum of incompatibility near the equimolar composition of the mixture at which the electrostatic repulsion is enhanced by the lipophobic character of the fluorocarbon hydrophobic group. Since the ζ potential in the equimolecular case is almost half the value attained with the pure surfactants, it may be conjectured that the adsorption density is considerably reduced by the extra repulsion provided by the lipophobic effect of the fluorocarbon group.

When bubbles are generated by gas other than air (oxygen, hydrogen, nitrogen, chlorine, and carbon dioxide), the behaviors observed depending on pH, salinity, surfactants, and so on are similar to the behaviors of air bubbles, with perhaps an exception for gasses, such as carbon dioxide, which can dissolve easily in water.

B. NEW RESULTS

Since our last review, an interesting work has been published by Phianmongkhol and Varley on the measurement of ζ potential of air bubbles in solutions of various proteins, BSA, β-casein, and lysozyme [47]. They used a microelectrophoretic technique with a cylindrical cell coated according to the procedure described in our first paper on ζ potential measurement [30]. They found that the average value of the ζ potential of the bubbles immersed in phosphate buffer of pH 7.0 (ionic strength

0.005 M) was −63 mV. This compares well with our value of −65 mV reported for deionized water and of −60 mV reported both by Marinova et al. [45] and Bergeron [46] for similar bubbles. This agreement justified by the low ionic strength of the buffer and illustrates the progress carried out over the last few years in the difficult measurement of ζ potential.

For BSA and β-casein, Phiammongkhol and Varley [47] found that the magnitude of the negative ζ potential of the bubbles decreases with increasing protein concentration and ionic strength. They found that it is zero for lysozyme solutions. The work of Phianmongkhol and Varley shows the way for studies on the ζ potential of bubbles immersed in other protein or bioadditive solutions.

C. NEW PROSPECTS

The spinning bubble electrophorometer can maintain the bubble on the axis of rotation without any motion of translation as long as no electric field is applied. This is a significant advantage over the other techniques for studying the transient regime of the adsorption of surfactants at the surface of the bubble (Figure 19.3) and thus for obtaining information on their organization at this surface because the ζ potential is related to surfactant ion adsorption.

Such a study has been performed for fluorocarbon and hydrocarbon anionic surfactants [48]. It has been found that the fluorocarbon surfactant ions are more rapidly adsorbed than the hydrocarbon surfactant ions. The two surfactants, which exhibit an antipathy manifested by micellar demixing, compete for surface sites and their total adsorption is reduced when both are present. Furthermore, the variations of ζ potentials of the bubbles depending on compositions and concentrations of mixtures of the two surfactants in the solutions of which they are immersed are found to be closely correlated with the micellar phase diagram of these surfactants. These behaviors would deserve to be strengthened by measurements on other surfactant mixtures.

In the spinning bubble electrophorometer, the bubble moves along the horizontal axis of rotation of the cell under the influence of the electric field applied.

This motion can be stopped by tilting very slightly the axis of the cell. Schematically, at equilibrium, rising force, centrifugal force, and electric force annihilate themselves (Figure 19.4).

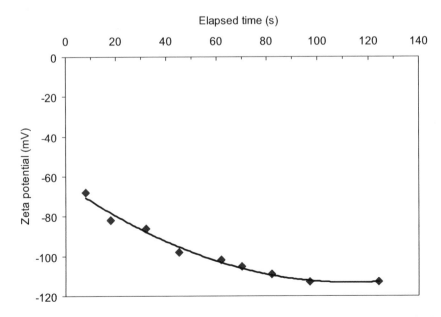

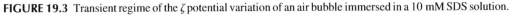

FIGURE 19.3 Transient regime of the ζ potential variation of an air bubble immersed in a 10 mM SDS solution.

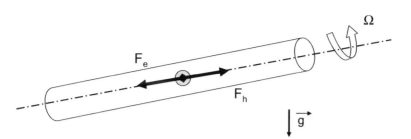

FIGURE 19.4 Equilibrium of the bubble in the tilted electrophorometer (F_e, electric force; F_h, global hydrodynamic forces).

Stopping of the bubble can be achieved either by varying the potential difference between the two electrodes for a given slope of the axis or by adjusting this slope for a given potential difference. As the rising force is high when compared with the electric force, the slope of the axis of rotation is always extremely small; it may be measured from the deviation at a long distance of a laser beam used as level. These very difficult measurements are in progress in our laboratory.

Recently, the ζ potentials of aqueous drops suspended in air were measured by the method used by Millikan for oil drops [49]. It was found that these potentials [50] have the same order of magnitude as the published ones for air bubbles generated in the same aqueous solutions and measured by electrophoresis. The same trends of variations were observed according to pH, valencies and concentrations of electrolytes, and ionic types of surfactants. Thus, measurements on aqueous drops seem to be a credible alternative to measurements on air bubbles for the knowledge of the electric properties of the air–aqueous solutions interfaces.

Finally, it would be interesting to study the ζ potential of bubbles immersed in polar liquids different from water or, why not, in organic liquids.

IV. ON THE ORIGIN OF THE ζ POTENTIAL OF BUBBLES

In the colloidal domain, most of the authors are getting on together for attributing the ζ potential of air bubbles immersed in water to specific adsorption of hydroxide ions at the air–water interface. An attempt to explain this specific adsorption is based on the fact that the static dielectric constant of the interfacial region between air and water is less than that of the adjacent aqueous phase [51]. Work is therefore required to transfer an ion from the aqueous phase into the interfacial region. This work is positive and thus unfavorable for ion adsorption; as it is greater for smaller ions than for larger ones, hydroxide ions would be adsorbed in preference to smaller hydrated protons.

Calculus of the numerical value of the ζ potential of a bubble is an extremely difficult problem. This potential is formally the electrostatic potential at the "plane of slip," but there is a high uncertainty about the location of this plane. As a first approximation, it is assumed to be located to one water molecular diameter away from the plane of charge or at a total distance of approximately 0.4 nm from the air–water interface. If we suppose the bubble surface is charged, ζ potential could be evaluated using the Poisson equation:

$$\frac{d^2 \psi}{dx^2} = -\frac{\rho(x)}{\varepsilon} \tag{19.8}$$

where $\psi(x)$ is the electric potential in a point located at the distance x from the interface, $\rho(x)$ the charge density at the same point, and ε the dielectric permittivitty of the medium.

At equilibrium, ions are distributed according to the Boltzmann relation so that $\rho(x)$ can be written as

$$\rho(x) = \sum_i z_i e c_{i\infty} \exp\left(\frac{-z_i e \psi(x)}{kT}\right) \tag{19.9}$$

where z_i is the valency of ions i, e the absolute value of the electron charge, $c_{i\infty}$ the concentration of ions i far from the interface, k the Boltzmann constant, and T the absolute temperature.

Report of relation (19.9) in relation (19.8) gives linear Poisson–Boltzmann differential equation:

$$\frac{d^2\psi}{dx^2} = -\frac{1}{\varepsilon}\sum_i z_i e c_{i\infty} \exp\left(-\frac{z_i e \psi(x)}{kT}\right) \tag{19.10}$$

In the case of electrode–electrolyte solution interfaces, the Poisson–Boltzmann equation has been modified for integrating many effects as, for example, finite ion size, concentration dependence of the solvent, ion polarizability, and so on. More often, this modification consists in the introduction of one or several supplementary terms to the energetic contribution in the distribution, which leads to modified Poisson–Boltzmann (MPB) nonlinear differential equations [52].

The situation is still more complex in the presence of surfactants. Recently, a self-consistent electrostatic theory has been presented to predict disjoining pressure isotherms of aqueous thin-liquid films, surface tension, and ζ potentials of air bubbles immersed in electrolyte solutions with nonionic surfactants [53]. The proposed model combines specific adsorption of hydroxide ions at the interface with image charge and dispersion forces on ions in the diffuse double layer. These two additional ion interaction free energies are incorporated into the Boltzmann equation, and a simple model for the specific adsorption of the hydroxide ions is used for achieving the description of the ion distribution. Then, by combining this distribution with the Poisson equation for the electrostatic potential, an MPB nonlinear differential equation appears.

The resulting MPB model is self-consistent. It contains only two adjustable parameters: the adsorption equilibrium constant and the maximum adsorption of the hydroxide ions. These parameters are optimized by matching the model calculations to measured disjoining pressure isotherms of nonionic surfactant thin films at three different ionic strengths. Without any further parameter adjustment, the proposed model predicts the disjoining pressure isotherms accurately over a range of pH, ionic strength, and surfactant concentration. It acceptably reflects the ζ potentials of the air bubbles considered, especially at low ionic strengths.

Improvement of this model would be probably obtained by considering the sizes of the ions at high ionic strengths, as this has just been performed for electrode–electrolyte interfaces [54]. Improvement would also be expected in taking the interfacial polarization gradient into account, as this has been recently proposed for solid–electrolyte interface [55,56]. But, extension of these improvements to the gas–liquid interface remains to be done.

V. CONCLUSIONS

Whatever the parameters considered (pH, salinity, nature of the salts, and nature of the additives), the trends observed for variations of the ζ potentials of gas bubbles immersed in aqueous solutions are globally consistent, even if there are still some disparities between authors with regard to the absolute values. These disparities probably originate from the diversity in the experimental systems and the relationships used to link mobilities with ζ potentials.

Thus, today, experimentation seems to have reached a satisfactory degree of reliability. It is possible to measure the ζ potential of such bubbles with sufficient accuracy to follow their evolution with respect to sensitive parameters. For example, the addition of surfactants makes these bubbles highly electronegative or highly electropositive according to the nature of the ionic surfactant added, which is of major importance for the stabilization of foams. It is also possible to neutralize these bubbles by adding a cationic surfactant or decreasing the pH drastically, which is also useful when, on the contrary, it is desired to destroy the foams.

Very recently, pioneer work has been performed on the ζ potentials of gas bubbles in protein solutions to obtain insight on formation and collapse of protein foams. As these foams play an important role in a number of processes, including foam formation in bioreactors, foam fractionation for protein recovery, and production of protein-based food and drinks, this work extends the study of the electric properties of the gas–liquid interfaces to biotechnology and food technology.

The negative potential of gas bubbles immersed in pure water is attributed to a specific adsorption of hydroxide ions in the first molecular layers of water at the interface. This preferential adsorption might be due to the positive work required to transfer an ion from the aqueous phase into the interfacial region. A MPB nonlinear differential equation has just been proposed to account quantitatively for the ζ potentials of gas bubbles immersed in electrolyte solutions with or without nonionic surfactant. Its resolution gives acceptable ζ potentials, especially at low ionic strengths. However, this interesting attempt of quantitative interpretation would deserve to be improved to account for the sizes of the ions at high ionic strengths and the part of the interfacial polarization gradients should be discussed. A hard theoretical challenge for the next years.

REFERENCES

1. Dukhin, S.S. and Deryaguin, B.V., Electrokinetic phenomena, in *Surface and Colloid Chemistry Series*, Matijevic, E., Ed., Vol. 7, John Wiley, New York, 1974.
2. Hunter, R.J., *Zeta Potential in Colloid Science, Principles and Applications*, Academic Press, London, 1981.
3. Kitahara, A. and Watanabe, A., Electrical phenomena at interfaces, in *Surfactant Science Series*, Schick, M.J. and Fowkes, F.M., Eds., Vol. 15, Marcel Dekker, New York, 1984.
4. Ohshima, H. and Furusawa, K., Electrical phenomena at interfaces, in *Surfactant Science Series*, Schick, M.J. Ed., Vol. 76, Marcel Dekker, New York, 1998.
5. Delgado, A.V. Ed., *Interfacial Electrokinetics and Electrophoresis*, Marcel Dekker, New York, 2001.
6. Evans, D.F. and Wennerström, H., *The Colloidal Domain*, Wiley-VCH, New York, 1999.
7. Dibbs, H.P., Sirois, L.L., and Bredin, R., Some electrical properties of bubbles and their role in the flotation of quartz, *Can. Metallurg. Q.*, 13 (2), 395–404, 1974.
8. Usui, S. and Sasaki, H., Zeta potential measurements of bubbles in aqueous surfactant solutions, *J. Colloid Interf. Sci.*, 65, 80–84, 1978.
9. Derjaguin, B.V. and Dukhin, S.S., *Proceedings of the Third International Congress on Surface Activity*, Universitatsdruckerei Mainz GmbH, Cologne, 1960, p. 324.
10. Dukhin, S.S., *Research in Surface Forces*, Derjaguin, B.V., Ed., Vol. 2, Consultants Bureau, New York, 1996, pp. 54–74.
11. Dukhin, S.S., Dorn's effect during strong retardation of bubble surface, *Kolloid Z.*, 45, 22–33, 1983.
12. Kuznetsova, L.A. and Kovarskii, N.Y., Evaluation of the charge of gas bubbles by measuring the bubble rise potential, *Colloid J.*, 57, 657–660, 1995.
13. Ozaki, M. and Sasaki, H., *Electrical Phenomena at Interfaces*, Ohshima, H. and Furusawa, K., Eds., Marcel Dekker, New York, 1998, pp. 245–252.
14. Ozaki, M. and Sasaki, H., *Interfacial Electrokinetics and Electrophoresis*, Delgado, A.V. Ed., Marcel Dekker, New York, 2002, pp. 481–492.
15. Brandon, N.P., Kelsall, G.H., Levine, S., and Smith, A.L., Interfacial electrical properties of electro-generated bubbles, *J. Appl. Electrochem.*, 15, 485–493, 1985.
16. Ozaki, M., Ando, T., and Mizuno, K., A new method for the measurement of the sedimentation potential: rotating column methods. *Colloids Surf. A, Physicochem. Eng. Aspects*, 159, 478–480, 1999.

17. Sirois, L.L. and Millar, G., Method to study surface electrical characteristics on a single bubble, *Can. Metallurg. Q.*, 12, 281–284, 1973.

18. Collins, G.L., Motarjemi, M., and Jameson, G.J., A method for measuring the charge on small bubbles, *J. Colloid Interf. Sci.*, 63, 69–75, 1978.

19. Kubota, K., Hayashi, S., and Inaoka, M., A convenient experimental method for measurement of zeta potentials generating on the bubble suspended in aqueous surfactant solutions, *J. Colloid Interf. Sci.*, 95, 362–369, 1983.

20. Yoon, R.H. and Yordan, J.L., Zeta potential measurements on microbubbles generated using various surfactants, *J. Colloid Interf. Sci.*, 113, 430–438, 1986.

21. Yang, C., Dabros, T., Li, D., Czarnecki, J., and Masliyah, J.H., Measurement on the zeta potential of the gas bubbles in aqueous solutions by microelectrophoresis method, *J. Colloid Interf. Sci.*, 243, 128–135, 2001.

22. Kelsall, G.H., Tang, S., Yurkadul, S., and Smith, A.L., Measurement of rise and electrophoretic velocities of gas bubbles, *J. Chem. Soc., Faraday Trans.*, 92, 3879–3885, 1996.

23. Kelsall, G.H., Tang, S., Yurkadul, S. and Smith, A.L., Electrophoretic behaviour of bubbles in aqueous electrolytes, *J. Chem. Soc., Faraday Trans.*, 92, 3887–3893, 1996.

24. Kim, J.Y., Song, M.G., and Kim, J.D., Zeta potential of nanobubbles generated by ultrasonification in aqueous alkyl polyglycoside solutions, *J. Colloid Interf. Sci.*, 223, 285–291, 2000.

25. Mc Taggart, H.A., The electrification at liquid gas surfaces, *Phil. Mag.*, 27, 297–314, 1914.

26. Mc Taggart, H.A., Electrification at liquid gas surfaces, *Phil. Mag.*, 28, 367–378, 1914.

27. Mc Taggart, H.A., On the electrification at the boundary between a liquid and a gas, *Phil. Mag.*, 44, 386–395, 1922.

28. Alty, T., The origin of the electrical charge on small particles in water, *Proc. Roy. Soc. A*, 112, 235–251, 1926.

29. Bach, N. and Gilman, A., Electrokinetic potential at gas–liquid interfaces, *Acta Physicochim. URSS*, 9, 1–18, 1938.

30. Graciaa, A., Morel, G., Saulnier, P., Lachaise, J., and Schechter, R.S., The zeta potential of gas bubbles, *J. Colloid Interf. Sci.*, 172, 131–136, 1995.

31. Sherwood, J.D., Electrophoresis of gas bubbles in a rotating fluid, *J. Fluid Mech.*, 162, 129–137, 1986.

32. Li, C. and Somasundaran, P., Reversal of bubble charge in multivalent inorganic salt solutions — effect of magnesium, *J. Colloid Interf. Sci.*, 146, 215–218, 1991.

33. Okada, K. and Akagi, Y., Method and apparatus to measure the ζ potential of bubbles, *J. Chem. Eng. Jpn.*, 20, 11–15, 1987.

34. Fukui, Y. and Yuu, S., Measurement of the charge on small gas bubbles, *AIChE*, 28, 866–868, 1982.

35. Kubota, K., Hayashi, S., and Inaoka, M., A convenient experimental method for measurement of zeta potentials generating on the bubble suspended in aqueous surfactant solutions, *J. Colloid Interf. Sci.*, 95, 362–369, 1983.

36. Hunter, R.J., *Zeta Potential in Colloid Science, Principles and Applications*, Academic Press, London, 1981, pp. 125–178.

37. Harris, J.M., Brooks, D.E., Boyce, J.F., Snyder, R.S., and van Alstine, J.M., Hydrophilic polymer coatings for control of electroosmosis and wetting, in *Materials Science in Space*, Spenser, Ed., New York, 1986, pp. 111–118.

38. Herren, B.J., Shafer, S.G., van Alstine, J., Harris, J.M., and Snyder, R.S., Control of electroosmosis in coated quartz capillaries, *J. Colloid Interf. Sci.*, 115, 46–55, 1987.

39. Knox, R.J., Burns, N.L., van Alstine, J.M., Harris, J.M., and Seaman, G.V.F., Automated particle electrophoresis: modelling and control of adverse chamber surface properties, *Anal. Chem.*, 70, 2268–2279, 1988.

40. Creux, P., Ph.D. thesis, Université de Pau, France, 2000.

41. Baygents, J.C. and Daville, S.A., Electrophoresis of drops and bubbles, *J. Chem. Soc. Faraday Trans.*, 87, 1883–1898, 1991.

42. Malvern Instruments Ed. (Orsay), France, Malvern News, 2, 6, 2000.

43. Graciaa, A., Creux, P., and Lachaise, J., Electrokinetics of gas bubbles, in *Interfacial Electrokinetics and Electrophoresis*, Delgado, A.V., Ed., Marcel Dekker, New York, 2001, pp. 825–836.

44. Graciaa, A., Creux, P., and Lachaise, J., Electrokinetics of bubbles, in *Encyclopedia of Surface and Colloid Science*, Marcel Dekker, New York, 2002, pp. 1876–1886.

45. Marinova, K.G., Alargova, R.G., Denkov, N.D., Velev, O.D., Petsev, D.N., Ivanov, I.B., and Borwankar, R.P., Charging of oil–water interfaces due to spontaneous adsorption of hydroxyl ions, *Langmuir*, 12 (8), 2045–2051, 1996.

46. Bergeron, V., Measurement of forces and structure between fluid interfaces, *Curr. Opin. Colloid Interf. Sci.*, 4, 249–255, 1999.

47. Phianmongkhol, A. and Varley, J., ζ potential measurement for air bubbles in protein solutions, *J. Colloid Interf. Sci.*, 260, 332–338, 2003.

48. Graciaa, A., Creux, P., Lachaise, J., and Schechter, R.S., Competitive adsorption of surfactants at air/water interfaces, *J. Colloid Interf. Sci.*, 261, 233–237, 2003.

49. Gu, Y. and Li, D., Measurements of the electric charge and surface potential on small aqueous drops in the air by applying the Millikan method. *Colloid and Surfaces A, Physicochem. Eng. Aspects*, 137, 205–215, 1998.

50. Paluch, M., Electrical properties of free surface of water and aqueous solutions, *Adv. Colloid Interf. Sci.*, 84, 27–45, 2000.

51. Schechter, R.S., Graciaa, A., and Lachaise, J., The electric state of a gas/water interface, *J. Colloid Interf. Sci.*, 204, 398–399, 1998.

52. Carnie, S.L. and Torrie, G.M., The statistical mechanics of the electrical double layer, *Adv. Chem. Phys.*, 56, 141–253, 1984.

53. Karraker, K.A. and Radke, C.J., Disjoining pressures, zeta potentials and surface tensions of aqueous non ionic surfactant electrolyte solutions: theory and comparison to experiment, *Adv. Colloid Interf. Sci.*, 96, 231–264, 2002.

54. Lamperski, S. and Outhwaite, C.W., A non primitive model for the electrode/electrolyte interface based on the Percus–Yevick theory, *J. Electroanal. Chem.*, 460, 135–143, 1999.

55. Ruckenstein, E. and Manciu, M., The coupling between the hydratation and the double layer interactions, *Langmuir*, 18, 7584–7593, 2002.

56. Huang, H., Manciu, M., and Ruckenstein, E., The effect of surface dipoles and of the field generated by a polarization gradient on the repulsive force, *J. Colloid Interf. Sci.*, 263, 156–161, 2003.

Fluosols

20 Structures and Substructures in Spray Pyrolysis Process: Nanodesigning

Vukoman Jokanovic
Institute of Technical Sciences, Serbian Academy of Sciences and Arts, Belgrade, Serbia, Serbia and Montenegro

CONTENTS

I. INTRODUCTION

Designing materials according to theoretically well-determined predictions has exceptional importance for more precise definition of materials structures, preparing conditions for organizing structures in a clear and predictable way. According to our research and theoretical models, only designing on all levels, that is, designing of the structures from the base, gives results that show exceptional accordance between theoretical expectations and experimental observations [1–4]. It is possible to create, in various systems, completely controlled structures on different levels of hierarchy (starting from the particles as a whole, down to their subelements or subparticles), in such way to be able to predict the development of these systems throughout all stages of the synthesis process.

Particles, whose sizes and population balances (the contribution of particle fractions in the whole particle system) can be predicted using theoretical models based on the principle of the harmonization of the physical field, give the possibility of predicting all relevant parameters of the structure and substructure to be designed, such as the spreading and the form of the pores, their sizes, number of their subelement contacts. Eventually, it also gives information about the type of structural organization one should obtain after designing.

This approach to designing makes available some specific synthesis to be organized in a fast, one-step manner. It also ascertains that some unique properties of the systems, very typical for the given system on the low level of structural organization (e.g., confined physical properties), to be retained in the process of enlargement of subelements into much bigger macroensembles (macroblocks), showing the same unique characteristics on the much higher level (macrolevel) of the organization [5–7]. In spite of expansion of subelements by their association, it is feasible to retain their individuality fully in the real organized structure by control of their contacts, on the level of dotted contacts, to get blocks of the so-called activated structures. Similar situation is also seen with catalytic and photoelectric characteristics of given specific structures in the process of organization transfer from the colloid-nano level to the submicron level.

Structure and substructures of obtained systems are always organized in agreement with theoretical model of harmonization of physical fields: (i) inside of system (physical field of the system itself) and (ii) external physical field (excitation physical field of given disturbance) [2–4,8–14].

These two fields determine the key (information or code), which can be associated with the system itself as its essential structural characteristic ("print" of the physical fields in which this system is produced).

This enables dualism in the screening. On one hand, the code enables reconstruction of the synthesis in its essential elements as such, and on the other, the code (complete spectra and combined code) enables organization of the system showing the self-organization of the system on the level of its structure and substructures, which are now given as the association of the codes which by its own geometry (code) can influence the organization of the material in the systems of greater organizational units [1–4,13].

This way of material organization shows that a system once organized possesses its own specific dynamics, giving the signal about itself, and its identity. If we were lucky enough to determine that code (combined code), use of such materials would have so exceptionally a great level of selection and would completely prove its precious value; consequently, such materials would have so selective a use in the fields, such as electronics, informatics, and medicine.

As the essential element from this follows multiplicity of existence of the given system (material) independent of its content, this could enable formation in the "same" material possessing different structures (conserved codes), which could reach multiple functional uses.

These ideas enable new strategies of the research in material sciences, for example in medicine, deposition of such structure on the well-chosen places in the organism may cause positive influence to the nearest environment, through spreading exactly controlled information created in the materials by corresponding physical fields during its synthesis.

This work is focused on SiO_2, TiO_2, $Ca_{10}(PO_4)_6(OH)_2$, and PW_8O_{26} as the models of the system. Owing to exceptionally large number of different applications of these systems, they are suitable to demonstrate that all these systems can function in a determined way and that harmonization of physical fields in which they are generated is exact, that determines the ways of their geometrical organization on the level of the structure and substructure.

II. GENERAL APPROACH TO THE SPRAY PYROLYSIS PROCESSES

The concept or the basis of spray pyrolysis method assumes that one droplet forms one product particle. To date, submicrometer- to micrometer-sized particles are typically formed in a spray pyrolysis process. A variety of atomization techniques have been used for solution aerosol formation, such as ultrasonic spray pyrolysis, electrospray pyrolysis, low pressure spray pyrolysis using a filter expansion aerosol generator (FEAG), salt-assisted spray pyrolysis, two-fluid pyrolysis method, etc. [15–18]. These atomization methods differ in droplet size, rate of atomization, and droplet velocity, which determine the residence time of the droplet in furnace chamber during spray pyrolysis. Two-fluid atomizers represent off-the-shelf technology capable of atomizing large quantities of liquid with minimum droplet size of only 10 μm.

One of the promising methods for producing satisfied quantities of a powder with narrow size distribution and nanometric mean diameter is electrospray pyrolysis method. In this method, a meniscus of a precursor (spray solution) at the end of capillary tube becomes conical when charged to a high voltage (several kilovolts) with respect to a counter electrode. The droplets are formed by continuous breakup of a jet extending from this liquid cone, known as "Taylor cone." Lenggaro, Xia, Okuyama, and Fernandez de la Mora, in their papers published from 2000 to 2003, described how this technique functions and how it is possible to measure online a size distribution of particles obtained from different types of precursor systems. For this purpose, they used differential mobility analyzer and a condensation nucleus/particle counter (CNC/CPC) [19–26].

The other very promising method is a two-step FEAG method developed by Kang and Park [27]. The FEAG consists a two-fluid nozzle for dispersing a liquid, a porous gas filter, and a vacuum pump. The liquid is sprayed through a pneumatic nozzle by a carrier gas onto a glass filter surface where it forms a thin liquid film [24]. This liquid film and carrier gas pass through the pores and expand into a low-pressure chamber. In a measurement using this method the mean droplet size was estimated to be around 2 μm [19–26].

Some advantages associated with each of these spray pyrolysis techniques for producing ceramic powders are given by the SAD method published by Xia et al., which focuses on a strategy of separation of nanocrystals contained in powder particles using some compounds (salts) which melt under maximal temperature of solid-state particle consolidation. The melted salts resemble eutectic mixture chlorides and nitrates of Li, Na, and K distributed on the nanocrystallite surfaces and they are prevented from agglomerating. After their removal, by washing the obtained powders, it is possible to get, finally, nanosized powders in different systems [19–26].

Finally, the ultrasonic atomization is shown as a very effective method for generating small droplets. The droplets produced by ultrasonic atomizer are 2–4 μm, but atomization rate is limited to <2 cm^3/min. Ultrasonic spray pyrolysis method is the most convenient method to obtain spherically shaped fine particles. The droplets are formed from feeding solution by means of an ultrasonic oscillator with frequency in the order of several megahertz or slightly less than 1 MHz [15,27].

A. Spray Pyrolysis Processes in Periodical Ultrasonic Fields

The possibility of generating a cloud of droplets by means of ultrasonic waves was first reported by Wood and Lomis [29]. Two different mechanisms have been reported to explain the ultrasonic atomization: capillary waves and cavitations. However, the interaction between these two approaches and limits in which one could predominate over the other depending on the different atomizing situation are challenging for immediate understanding.

The first study on stationary waves on the free surface of a liquid mass subjected to periodical ultrasonic field was reported by Faraday. In 1871, Kelvin finally derived the well-known equation for capillary waves [30,31]:

$$\lambda = \left(\frac{2\pi\sigma}{\rho f^2}\right)^{1/3} \tag{20.1}$$

where λ is the wavelength, σ the surface tension coefficient, ρ the liquid density, and f the frequency of the surface waves [31].

This work has been continued by Rayleigh, who modified Kelvin's equation and derived expression [32]:

$$\lambda = \left(\frac{8\pi\sigma}{\rho F^2}\right)^{1/3} \tag{20.2}$$

where F is the forcing sound frequency. From experimental measurements, it is shown that $f = F/2$. In 1917, Rayleigh derived the first mathematical model to explain the bubble collapse in incompressible liquids and that was the first attempt to explain the physical mechanisms involved in ultrasonic propagation in liquids. On the theoretical basis, Wood and Lomis made the first atomization of liquids with ultrasonic wave [29]. An explanation of this process based on cavitations produced under liquid meniscus was proposed by Solner [33]. Numerous papers (Bisa [34], Benjamin and Ursell [35], Sorokin [36], and Eisenmenger [37]) have pointed to unstable surface capillary waves as the origin of the droplet formation, mostly relying on simplified linear instability analysis. In 1962, Lang published his research with his well-known expression relating wavelength to droplet size through an empirical constant $k = 0.34$ [38].

A more elaborated theoretical model based on interfacial Taylor instability triggering the surface wave was developed by Peskin and Raco [39]. A thin layer of a liquid, wetting the surface of a solid resonator which vibrates to its plane, forms a chessboard-like pattern of stationary capillary waves. This phenomenon occurs when the vibration amplitude exceeds a threshold value. Further on, ligament breakup of the liquid occurs and droplets are hurled from the crests of the capillary waves. Together with the wavelength, they introduced wave amplitude and the sheet thickness as parameters to determine the droplet size [39].

In contrast, Eknadiosayants and Gersherson developed their theory based on cavitations (mentioned later in Barreras et al.). After these, there are numerous attempts to combine both theories. Some of them are made by Fogler and Timmerhaus [40], Boguslaskii and Eknadiosyants [41], Topp [42], Chiba [43], Basset and Bright [44], Miles and Sindayihebura and Bolle, and some experiments to value these approaches are given by Edwards and Fauve (1994) and, finally, Barreras et al. [45–50].

The model published in numerous papers is probably the most convenient approach for explaining all phenomena related to breakup mechanism of droplet formation [1,4]. This model quantitatively defines each line in the size droplets and size powder distribution spectrum. The mechanism of the droplet formation or particle genesis is fully determined by harmonization between the physical fields inherent to the system as the consequence of its physical characteristics: external, for example, ultrasonic, and internal, belonging to the system itself [1–4].

B. Precipitation Processes

The mechanism of droplet transformation into the particle, and degree reduction of the droplet through this process, depends on different parameters. The most important parameters are temperature gradient between the surface and the center of the droplet or particle, the viscoelastic properties of the droplet (character and level of its rigidity and droplet behavior during its collision with other aerosol droplets and coalescence), the thermodiffusion coefficient, and permeability of the shell formed on the droplet surface during its solidification [3].

The average size distribution of the final particle can be roughly determined from the size of atomized droplets and its concentration in the starting solution. The way of precipitation in a given system determines which kind of powders product is obtained. The full-density powders are obtained by volume precipitation and hollow-sphere powders by surface precipitation. If the solute concentration at the center of the droplet is greater or equal to the equilibrium saturation of the solute at the droplet temperature, then nuclei on the surface catalyze precipitation throughout the droplet, that is, volume precipitation. However, if the solute concentration at the center is less than the equilibrium saturation of the solute at the droplet temperature, then precipitation occurs only in that part of the droplet where concentration is higher than the equilibrium saturation, that is, surface precipitation occurs.

For volume precipitation to occur, it is necessary to have precursor-solute which should have the following properties [3,28,51]:

(i) A large difference between the solute critical supersaturation and equilibrium saturation should be used.

(ii) A high solubility and a positive temperature coefficient of solubility to satisfy the percoloration criteria.

(iii) Not to be thermoplastic or melt during the thermolysis stage of spray pyrolysis process.

(iv) Finally, it is recommended to use colloid sols and partially hydrolyzed alkoxide system, which very easily form three-dimensional networks when they are networked by gelling or polymerizing.

In contrast, for obtaining hollow spheres, all given earlier conditions are opposite. Precipitate layers of different thickness can be obtained depending on the concentration gradient at the onset of precipitation and depending on the shallow permeability. Also, if a salt precursor is melted before decomposing, it can inhibit removal of the entrapped solvent and cause formation of porous and irregular shaped particles.

In multicomponent systems, the aqueous precipitation is dependent on the differences between solubility of the solute and on the differences in the pH at which precipitation occur. If these differences are small and negligible, the precipitation is homogenous and the obtained phase distribution in this system is homogenous through its volume. If these differences are significant, the precipitation is heterogeneous and in the interstitial sites of first precipitated phase the second precipitated phase is placed [28,51–61].

1. Droplet Size Reduction

The droplet size reduction occurs through different stages of spray pyrolysis process: evaporation of the solvent from the surface of the droplet, diffusion of the solvent vapor away from the droplet in the gas phase, diffusion of solute toward the center of the droplet, shrinkage of the droplet, and change in droplet temperature and drying process, preferentially. The characteristic value for estimation of evaporation time needed for solvent evaporation and droplet precipitation is the evaporation rate given by [28,62]:

$$\frac{\mathrm{d}m}{\mathrm{d}t} = \frac{4\pi R D_\mathrm{v} M}{R_\mathrm{g}} \left(\frac{p_\infty}{T_\infty} - \frac{p_\mathrm{d}}{T_\mathrm{d}} \right) \tag{20.3}$$

where m is the droplet mass, t the evaporation time, p_∞ and T_∞ the ambient vapor pressure and temperature of the reactor, respectively, p_d and T_d the pressure and temperature at the surface of the droplet, respectively, M the molecular mass of the gas, R_g the gas constant, R the mean droplet radius, and D_v the diffusivity of solvent vapor in air.

To determine how the solute concentration changes as a function of drying conditions in the evaporation stage of spray pyrolysis, it is necessary to solve simultaneously the differential equations for solute diffusion and solvent evaporation as it was made using numerical, finite difference method by Lijn [63]. Once the concentration profile in a droplet is known in a function of the drying conditions, then the time at which the solute starts to precipitate can be found. In dealing with precipitation on the surface, the rise of temperature in the system deserves attention. If the rise in temperature is not uniformly reached throughout to all the droplets at the equilibrium saturation on the surface, the surface precipitation will occurr. According to this, the rate of solvent removal, dm/dt, in the presence of solids on the droplet surface is given by [63,64]:

$$\frac{dm}{dt} = \frac{4\pi R_c D_v}{1 + \left(\dfrac{D_v}{D_{cr}}\right)(\delta/R_c - \delta)}[\gamma_d - \gamma_\infty] \tag{20.4}$$

where R_c is the droplet radius at the time of precipitation, D_{cr} the diffusivity of vapor through the precipitate layer, δ the thickness of the crust, γ_∞ and γ_d the mass concentration of solvent vapor and bulk vapor, respectively, and D_v the diffusivity of solvent vapor in air. The vapor diffusivity through the pores in the precipitate layer is much slower than the diffusivity of solvent vapors in gas phase. When the solute starts to precipitate, the evaporation rate is significantly reduced and the droplet temperature rapidly rises to the ambient temperature — temperature of furnace tube. Before this, all processes have to be completed in the system, such as dehydroxylation, thermolysis, and phase transformations processes. Mostly, sintering process is excluded as a stage of spray pyrolysis process, because the residence of time droplet (and after solidification of the powder), in the furnace tube is very short. Therefore, mostly, the reduction of droplet size occurs through shrinkage process related to the evaporation and drying processes.

III. DESIGNING STRUCTURES IN SPRAY PYROLYSIS UNDER THE ACTION OF ULTRASONIC FIELDS

A. DESIGNING OF POWDER PARTICLE STRUCTURE

Force frequency of the ultrasound oscillator induces equivalent waves in the given liquid column, in direction perpendicular to direction of disturbance (transversal waves) and in direction parallel to direction disturbance (longitudinal waves). In general, the standing waves formed are ellipsoidal, basically determined by the relationship between the damping factors for transversal and longitudinal waves, depending very much on thickness of the sprayed solution liquid column. Besides, size distribution of the aerosol droplets is influenced by physical characteristics of that solution (surface tension and viscosity) and by geometry of the vessel containing the solution. For sufficiently thin liquid columns, damping factors of transversal and longitudinal waves differ only negligibly, that is why so obtained standing waves are spherical [1–3].

On the basis of 3D model of the capillary standing waves formed on the meniscus surface and analysis of harmonic function of generated disturbance the rate potential is given in the form [1–3,65–70]:

$$\rho\frac{\partial\varphi}{\partial t} + \rho g\xi - \frac{\sigma}{\rho}\left[\frac{\partial^2\xi}{\partial x^2} + \frac{\partial^2\xi}{\partial z^2}\right] = 0 \tag{20.5}$$

where φ is the rate potential of the capillary standing waves, σ the solution surface tension, g the force of inertia acting on the liquid in contact with ultrasound oscillator, ξ the amplitude of formed standing wave, x and z the coordinates equivalent to the given amplitude, and t the time, the final solution is obtained:

$$d = \left(\frac{\pi\sigma}{\rho f^2}\right)^{1/3} \tag{20.6}$$

where d is the mean diameter of the formed aerosol droplet and ρ is the solution density.

In the case of the ellipsoidal form, standing wave can be represented in the form of the Laplace equation, expressed in polar coordinates [1–3,65–70]:

$$\rho\frac{\partial^2\varphi}{\partial t^2}\bigg|_{r=R} - \frac{\sigma}{R^2}\left\{2\frac{\partial\varphi}{\partial r} + \frac{\partial}{\partial r}\left[\frac{1}{\sin\theta}\frac{\partial(\sin\theta\,\partial\varphi/\partial\theta)}{\partial\theta} + \frac{1}{\sin^2\theta}\frac{\partial^2\varphi}{\partial\varepsilon^2}\right]\right\} = 0 \tag{20.7}$$

where r is the radius of the aerosol droplet and ε and θ are the angles corresponding to equation transformed in polar coordinates. Solution of the given equation gives the set of radius dimensions equivalent to the given wave-damping factor:

$$d = \frac{1}{\pi}\left(\frac{2\sigma\pi}{\rho f^2}\right)^{1/3}[l(l-1)(l+2)]^{1/3} \tag{20.8}$$

where l is the integer, taking the values $l \geq 2$.

In contrast, the disturbance caused by the ultrasound oscillator induces characteristic oscillations within the given liquid column, corresponding to its geometry and its physical characteristics. Expression defining a number of possible values of the liquid column oscillation frequencies is obtained as a solution of the corresponding rate potential function for the standing wave formed on the meniscus surface, in the form [1–2]:

$$f = \frac{c}{2}\left[\frac{m^2}{a^2} + \frac{n^2}{b^2} + \frac{p^2}{h^2}\right]^{1/2} \tag{20.9}$$

where a, b, and h are dimensions of the liquid column and m, n, and p are integers, taking the values of degenerated wave function.

Distribution spectrum for the secondary particles at the thickness of the liquid column equal to h ($h \ll a$, b and $a \approx b$), connecting both degeneration factors: factor of force frequency degeneration caused by wave-damping factor and factor of characteristic frequency degeneration caused by change of the liquid column thickness, can be expressed in the following form [1–2]:

$$d_n = \left[\frac{\pi\sigma}{\rho p^2 c^2/4h^2}\right]\left[\frac{(2l(l-1)(l+2))^{1/3}}{\pi}\right] \tag{20.10}$$

Based on the obtained discrete values of the aerosol droplets size in size distribution spectrum, mean powder particle size can be calculated from the following equation [1–3]:

$$d_p = d_n\left(\frac{c_{pr}M_p}{\rho_p M_{pr}}\right)^{1/3} \tag{20.11}$$

where d_p is the powder particle diameter, ρ_p the powder density, M_p the powder molecular mass, c_{pr} the precursor concentration (solution used for spraying when forming powder particles), and M_{pr} the precursor molecular mass.

B. Designing of Powder Particle Substructure

After separation of the aerosol droplet from the meniscus surface, the droplet remains excited by the excitation transferred on it in the moment of its separation from the meniscus surface (its "birth"). Therefore, the droplet behaves as induced mechanical oscillator, whose characteristic frequency is determined by droplet geometry and frequency of excitation.

Diameter of nanoelements or nanodroplets and nanoparticles that constitute substructure of the given aerosol, that is, the secondary particle obtained by its solidification, can be determined by the wave equation of the centrally symmetric standing wave formed as the consequence of excitation of the ultrasound generator transferred onto the droplet, as it follows [1,4,68,69]:

$$\frac{\partial^2 \varphi}{\partial t^2} = c^2 \frac{1}{r^2} \frac{\partial}{\partial r}\left(r^2 \frac{\partial \varphi}{\partial r}\right) \qquad (20.12)$$

where c is the disturbance propagation rate and r is the radius of the aerosol droplet.

Solution of the given equation can be written in the following form [1,4]:

$$\tan(kr) = kr \qquad (20.13)$$

where k is wave number ($k = (2\pi f/c)$); numerical solution of this equation gives the series of terms [1,4]:

$$r = \frac{Nc}{f} \qquad (20.14)$$

where N is the value of numeric constant, varying for different solutions depending on substructural morphology of obtained particles. Values of number constant are given by following order: 4.49, 7.6, 10.8, 14, 24.0, etc.

Finally, using expression (20.7), it is possible to calculate diameters of subparticles formed as the consequence of such structural design.

C. COUPLING LEVEL AND PACKING FACTOR

Depending on coupling level, Equation (20.7) should be corrected for both structural and substructural design by the corresponding structure-packing factor, which is more influential if the system is more diluted and solidification temperature (the temperature inside reaction chamber — furnace temperature) is lower.

If the packing factor of the system (F_p) is defined as the volume ratio between the particle of theoretical density and actual mean-sized particle, it follows [1,71]:

$$F_P = \frac{V_T}{V_S} = \left(\frac{d_T}{d_S}\right)^3 \qquad (20.15)$$

where d_T and d_S are the values of theoretical and actual diameters of the mean-sized particle, respectively. From the determined value of the packing factor, it is possible to determine the number of subparticles (N_P) in the volume of secondary powder particle, according to expression:

$$N_P = \frac{6F_P}{\pi d^3} \qquad (20.16)$$

and a number of contacts per unit volume within the secondary particle, according to expression:

$$N_C = \frac{3F_P Z}{\pi d^3} \qquad (20.17)$$

where Z is the coordination number, which depends on the packing density, taking the values from 6 (cubic packing with the lowest coupling) to 12 (tetrahedral packing with the highest level of individual elements coupling). Packing density in randomly arranged spheres is most similar to the packing factor, which corresponds to orthorhombic arrangement (packing density is 60–64% of the theoretical density).

IV. EXPERIMENTAL RESULTS: STRUCTURES AND SUBSTRUCTURES IN DIFFERENT CERAMIC SYSTEMS

Experimental results obtained in designing structures, SiO_2, TiO_2, $Ca_{10}(PO_4)_6(OH)_2$, and PW_8O_{26}, are shown to enable comparison with results based on corresponding calculations, according to the already quoted theoretical model elaborated in Refs. [1–4,9,10,72].

The investigated systems include (i) solids; SiO_2, TiO_2, calciumhydroxyapatite and phosphorous-doped bronzes; (ii) colloids: SiO_2 sol and TiO_2 sol, $(NH_4)_2HPO_4$ and $Ca(NO_3)_2$; and (iii) solution: $(H_3PW_{12}O_{40} \cdot 29H_2O$ phosphorus-doped tungsten heteropolyacid) (Table 20.1 and Figure 20.1–Figure 20.4). HNO_3 and urea were used for pH adjustment and prolonging of the precipitation process.

Morphology of the obtained particles in all cases, was spherical or near to the spherical form, with average size from 363 nm (TiO_2) to 1690 nm (SiO_2) (Table 20.2). All Particles also show very characteristic substructure. Some of them, for instance TiO_2, were hollow spheres. Beside their diameters, the thickness of the ring fill-up with material (whole of the spheres) was experimentally determined. Its value was around 200 nm (Table 20.2).

Statistical treatment of the particle sizes and size distributions is performed using program ORIGIN 6, while the measurement of particles diameters has been carried out based on previous photographic magnification of $\times 100$, by which the error of measurement was decreased to ± 10 nm. Obtained results for all systems are shown in Table 20.1.

Powder S is obtained from 5.8 M solution of SiO_2 sol, powder B from 0.6 M solution of WPA-29 ($H_3PW_{12}O_{40} \times 29H_2O$), powder A from 0.065 M solution of hydroxyapatite precursor, and powder T from 10^{-2} M colloidal solution of TiO_2. It is obvious that obtained values for the mean size of powder particles and their size distribution show a discrepancy, and the main reason for this can be attributed to the differences in the concentration of initial precursors.

Size intervals of aerosol droplet diameters are concentrated mostly around the mean diameter value, from both sides, inside narrow area of distribution (for the system S, between 1500 and 2500 nm; for system T, between 280 and 480 nm; for system A, between 410 and 710 nm; and for system B, between 560 and 1220 nm).

Obviously, the precursor for TiO_2 had an exceptionally low concentration and preferred surface precipitation, while concentration of precursors for all other observed systems was much higher than in the case of TiO_2 and they preferred volume precipitation.

TABLE 20.1

Experimentally Determined Values of Particles Diameter, d_p (10 nm), and their participation I (%) for systems, SiO_2–S, TiO_2–T, $Ca_{10}(PO_4)_6(OH)_2$– A, and PW_8O_{26}–B (10 nm)

S		T		A		B	
d_{pS}	I_{pS}	d_{pT}	I_T	d_{pA}	I_A	d_{pB}	I_B
100	4	24	4	41	8	56	9
125	19	28	15	50	21	67	9
150	35	32	12	55	32	78	8
175	27	36	18	61	21	89	11
250	12	40	15	65	2	100	25
350	3	44	12	71	11	111	13
		48	13	81	2	122	13
		52	6	91	3	139	9
		56	5			167	3

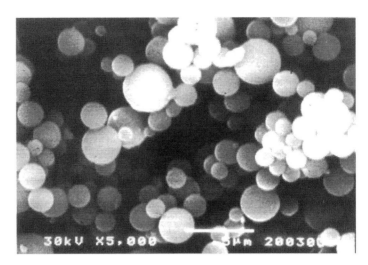

FIGURE 20.1 SEM micrograph of SiO_2 particles.

For the substructure of observed systems, measured data of subparticle diameter are given in Table 20.3.

From the presented data (Table 20.3), it can be seen that the particles of aerosol, which are of micrometer–submicrometer dimensions, are built from more fine particles — subparticles, that have sizes different from one system to the other. Diameters of the subparticles are conditioned primarily by the concentration of the precursors and by their factor of reduction during process of solidification, according to the Equation (20.11).

A. Calculations Related to the Structure and Substructure Based on Theoretical Model

Based on theoretical model, the spectrum of droplet diameters were calculated, together with values of compulsory frequencies of ultrasonic generator degenerated due to the damping factors associated

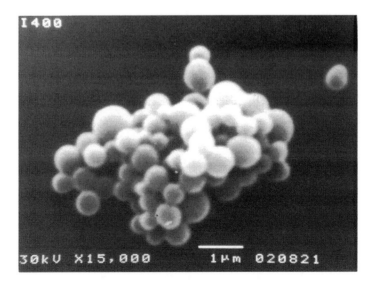

FIGURE 20.2 SEM micrograph of TiO_2 particles.

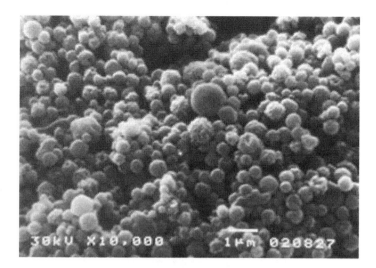

FIGURE 20.3 SEM micrograph of carbonated calciumhydroxyapatite particles.

with the thickness of liquid column. Also, calculation for the degree of appearance of discrete values of an aerosol droplet diameters and particle sizes obtained during the process of solidification is given by Equation 20.11 [1–4]. All these values are given in Table 20.4.

For the given values of particle diameters obtained by calculation, based on the structure model, the results are the following: (i) majority of discrete values of particle diameters approximately coincide (around 80% of all particles in all systems belong to the same interval) (ii) mean value of the particles determined by calculation and mean value of particle diameter determined experimentally are in close agreement (Table 20.5).

Differences of particles sizes determined experimentally and calculated theoretically (from 0.2% to 17.1%) show that the obtained values for particle diameters are close for all observed systems. Moreover, the agreement between these values is quite good even for TiO_2 particles, which precipitate only on the surface, in this way the full ring is formed (caused by their very low precursor

FIGURE 20.4 SEM micrograph of phosphor-doped tungsten bronze particles.

TABLE 20.2
Average Diameters of Aerosol Particles for the Systems, SiO_2–S, TiO_2–T, $Ca_{10}(PO_4)_6(OH)_2$–A, and PW_8O_{26}–B

Systems	S	T	A	B
d_a(nm)	1690	363	572	999

concentration), while the central part remain empty. In Ref. [1], it was shown in detail how it is possible to determine the thickness of the ring of the hallow particle and the way of packing of colloid 2.5 nm TiO_2 inside. With other observed systems (systems S, A, and B), similarity is even better (the error is minimized approximately to 5%).

Besides the theoretical estimate of size of the diameter of powder particles, the corresponding comparison with experimental values is presented in Table 20.6 and Figure 20.5–Figure 20.7.

Inaccuracies are larger than that in previous case. In agreement with the theoretical substructuring model, one of the main reasons is related to the chosen measured droplet and/or particle diameter values, because for each of them one order of discrete subparticle values can be found [1–4]. For estimation, for all members in the order, average droplet and particle diameter were chosen. In any case, given data demonstrate adequate accordance with experimentally obtained values, therefore the given model on the both levels of hierarchy can be used for the purpose of evaluating distribution spectrum of diameters of particles and subparticles of aerosol.

It confirms that in this type of process a discrete spectrum of powder distribution is always obtained. At the same time, it refers to the influence of a given physical field (physical field of synthesis) of the process of designing its structure and substructure.

V. GENERAL APPROACH TO THE DESIGNING OF STRUCTURES AND SUBSTRUCTURES IN THE PRESENCE OF OTHER PERIODICAL PHYSICAL FIELDS

In general, considering the physical fields to which a material is exposed during its synthesis, a strategy similar to that employed for designing materials is possible for synthesis. Therefore, the essential problem is to describe the process by a sufficiently exact model through essential mathematical and physical formalisms.

The systems given in this chapter, are rather diverse and distant by characteristics; some other systems have been investigated, and the results are available in Refs. [8–14,73–75]. In all cases, very good correlation between experimentally obtained results and theoretically estimated values for the diameter of particles and subparticles, as well as for their shares, has always existed.

The given structure can be observed profoundly, trying to find a "code", which carries the system in its structure, as the consequence of its genesis in the given physical field. Hence, it is possible to consider structural design as one specific characteristic of the material. Given material

TABLE 20.3
Average Values of Powder Subparticles Diameters for Systems, SiO_2–S, TiO_2–T, $Ca_{10}(PO_4)_6(OH)_2$–A, and PW_8O_{26}–B

Systems	S	T	A	B
d_{sp} (nm)	38	—	15	55

TABLE 20.4

Diameters of Aerosol Drops, d_d (nm), Degenerated Forced Frequencies of Ultrasonic Generator, f_d (MHz), shares of aerosol droplet diameters drop and particles and particles derived from them I (%) and diameters of powder particles, d_p (nm), for powders, SiO_2–S, TiO_2–T, $Ca_{10}(PO_4)_6(OH)_2$–A, and PW_8O_{26}–B

S	f_d	2.37	1.23	0.79				
	I	32	45	23				
	d_d	3600	5570	7500				
	d_p	1440	2230	3000				
T	f_d	1.5	1.8	2.25	2.62	3.0	3.38	3.75
	I	5.4	7.5	22	46	11	6	0.1
	d_d	1004	1118	1374	3278	3423	3577	4050
	d_p	119[a]	127[a]	159[a]	380[a]	422[a]	441[a]	469[a]
A	f_d	4.78	3.74	2.45	1.85	1.16	0.84	
	I	2	6	80	6	3	3	
	d_d	2100	2700	3280	4300	5400	6700	
	d_p	360	459	558	737	916	1144	
B	f_d	2.28	1.26	0.81	0.65			
	I	52	30	10	8			
	d_d	3460	5700	7690	9590			
	d_p	853	1320	1780	2220			

[a]To determine values for the size of powder particles TiO_2, the model of density packaging was applied, which is shown in Ref [1].

can be designed in such way so that it can show not only multiplicity of its structure, but a variety related to their specific characteristics as a "fingerprint" of physical field of its genesis. Importance of the structure designed in this manner could be exceptional and unexpected, because "this way modulated" structure behaves as the structure which remembers itself and creates possible conditions for the specific self-organization — self-assembling of material in its environment. This could be very significant for understanding behavior of structures in a real surroundings. The living systems, which themselves are behaving as "the modulated" specific structures, which in the long term, may produce significant impacts on surroundings (processes of bio-field "demodulation").

The importance of such research is in the fact that each structure appears always not as a unique, but fully harmonized spectrum of structures. This in whole describes the complexity of the physical fields in which these structures are generated.

TABLE 20.5

Experimental and Theoretical Average Particle Diameter for Systems, SiO_2–S, TiO_2–T, $Ca_{10}(PO_4)_6(OH)_2$–A, and PW_8O_{26}–B

Systems	S	T	A	B
$d_{pexp.}$ (nm)	1690	363	572	999
$d_{pteor.}$ (nm)	1800	301	573	1066
Difference (%)	4.5	17.1	0.2	6.7

TABLE 20.6
Experimental and Theoretical Average Subparticle Diameter for
Systems, SiO_2–S, TiO_2–T, $Ca_{10}(PO_4)_6(OH)_2$–A, and PW_8O_{26}–B

Systems	S	T	A	B
d_{spexp} (nm)	30	—	15	55
$d_{spteor.}$ (nm)	38	5	13	51
Difference (%)	21	—	13	7.3

A. FULL DENSITY PARTICLE SYSTEMS AND HOLLOW SPHERES

The powder systems S, A, and B, in the shape of full spheres are obtained, and the dominant mechanism of precipitation of these powders was volume precipitation. It is conditioned by relatively high concentration of given systems (expressed in percentage): with SiO_2 — 30%, with PW_8O_{26} — 20%, and with calciumhydroxysiapatite — 6.5%.

Only with TiO_2 system (concentration of the precursors was 0.96%) it was not able to obtain volume precipitation of the sample. To determine the moment of precipitation, and by that also the instance in which consolidation of the systems appears in the given volume, it is necessary to define concentration profile in the droplet, dependent on the drying conditions. A given profile is determined by the simultaneous solution of differential equations, that describe diffusion of the solution and evaporation of the dissolvent through diffusion in opposite direction, using the numerical method of finite differences given by Lee [63]. However, it is not possible to confirm, with certainty, that particles which are precipitated preferentially by volume precipitation are completely without cavities — voids (particularly in the case of the biggest particles). Generally, even for the systems with strictly preferred volume precipitation in some bigger particles such cases might be found.

This could be seen on some particles in the systems A and B. Some of the particles really demonstrate that inside of them vacancies exist, and that in these systems, although they preferably precipitate by volume, in some places, they do precipitate by surface.

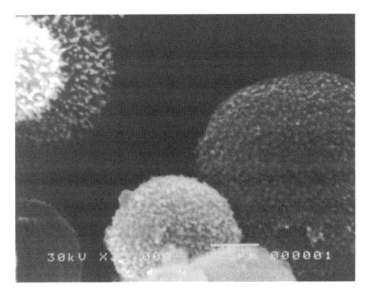

FIGURE 20.5 SEM micrograph of SiO_2 subparticles.

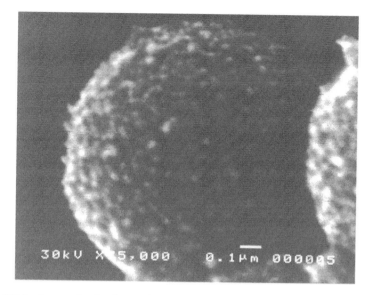

FIGURE 20.6 SEM micrograph of carbonated calciumhydroxyapatite subparticles.

With the colloid system that creates SiO_2 nanosols, mechanism of volume precipitation is present, almost completely due to its high concentration and the tendency of dense colloid systems to 3D linking inside complete volume of the sample.

B. Designing Porosity Inside of Particle Systems and Thin Films

Synthesis via colloid crystallization allows the pore size to be controlled in the range of nanometers to micrometers. Using the colloidal mixture of silica and polystyrene latex as a precursor, it is possible to obtain spherical shaped porous silica particles, that can be used as a catalyst in chromatography and as the material for controlling the release of drugs, as well as in microelectronics

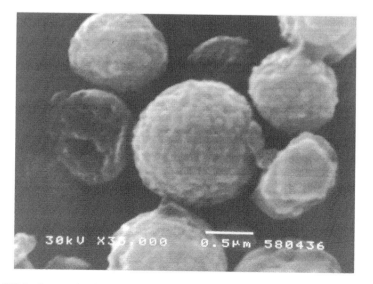

FIGURE 20.7 SEM micrograph of phosphor-doped tungsten bronze subparticles.

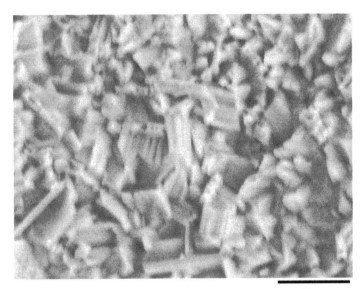

10μm

FIGURE 20.8 Micrograph of SiO$_2$ thin film.

and optoelectronics [76–82]. The obtained pores in silica depend on the size of latex particles. Micrographs of thin films silica, calciumhydroxyapatite, and phosphor-doped tungsten bronzes are shown in Figure 20.8–Figure 0.10.

It can be seen that the obtained films have very small roughness, especially the bronze film obtained on the silica substrate. The higher roughness of the calciumhydroxyapatite film is related to the more higher roughness of titanium substrate than that of silica substrate [83]. Obtained film of SiO$_2$ and phosphor-doped tungsten bronze has very small roughness and this method is very reliable for obtaining thin ceramic films on different kinds of substrates.

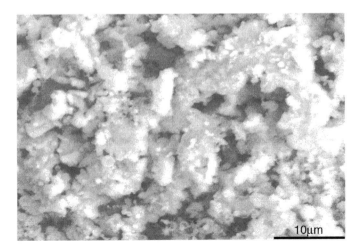

10μm

FIGURE 20.9 Micrograph of carbonated calciumhydoxyapatite thin film.

FIGURE 20.10 Micrograph of phosphor-doped tungsten bronzes.

VI. CONCLUSION

In this chapter, a summary of different techniques of spray pyrolysis with its advantages and disadvantages is presented. The process of spray pyrolysis with ultrasonic excitation is treated separately; for which, a theoretical model of structuring and substructuring of aerosol particles, SiO_2, TiO_2, carbonate calciumhydroxyapatite, and phosphor-doped wolfram bronze is developed.

All experimental results are associated with the average size of powder particles and subparticles and their distribution, and for all chosen systems; they show very good agreement with theoretically calculated values based on the model mentioned in this chapter. Numerous results connected to the research on other systems (Al_2O_3, spinel, mullite, and cordierite) show similar behavior.

From the given results, one may conclude that it is possible to consider physical fields as the main decisive factors in formation of the system structures, which carries "print of the field" as its completely defined code. This gives reference to the new possible strategies in the synthesis of the materials through the projection of the physical fields in which these syntheses take place. Also, it is possible from the specific structural information to elicit the other specific characteristics of the system and to establish complexity relevant to the potential function inside the system.

These differences could influence the process of self-assembling of the material in the environment of these structures, which possibly opens some new very selected and unexpected applications to the designing of materials in medicine.

REFERENCES

1. Jokanović, V., Spasić, A.M., and Uskoković, D., Designing of nanostructured hollow TiO_2 spheres obtained by ultrasonic spray pyrolysis, *J. Colloid Interf. Sci.*, 278 (2), 342–352, 2004.
2. Jokanović, V., Janaćković, Dj., and Uskoković, D., Influence of aerosol mechanism by ultrasonic field on particle size distribution of ceramic powders, *Ultrasonic Sonochem.*, 6 (3), 157–169, 1999.

3. Jokanović, V., Janaćković, Dj., Spasić, A.M., and Uskoković, D., Synthesis and formation mechanism of ultrafine spherical Al$_2$O$_3$ powders by ultrasonic spray pyrolysis, *Mater. Trans., JIM*, 37, 627–635, 1996.

4. Jokanović, V., Janaćković, Dj., Spasić, P., and Uskoković, D., Modeling of nanostructural design of ultrafine mullite powder particle obtained by ultrasonic spray pyrolysis, *Nanostruct. Mater.*, 12 (1–4), 349–352, 1999.

5. Saponjic, Z.V., Rakočević, Z., Dimitrijević, N.M., Nedeljković, J.M., and Uskoković, D.P., Tailor made synthesis of Q–TiO$_2$ powder by using quantum dots as building blocks, *Nanostruct. Mater.*, 10 (3), 333–339, 1998.

6. Anaćković, Dj., Jokanović, V., Kostić-Gvozdenović, Lj., and Uskoković, D., Synthesis of mullite nanostructured spherical powder by ultrasonic spray pyrolysis, *Nanostruct. Mater.*, 10 (3), 341–348, 1998.

7. Nedeljković, J.M., Saponjic, Z.V., Rakočević, Z., Jokanović, V., and Uskoković, D.P., Ultrasonic spray pyrolysis of TiO$_2$ nanoparticles, *Nanostruct. Mater.*, 9 (1–8), 125–128, 1997.

8. Janaćković, Dj., Jokanović, V., Kostić-Gvozdenović, Lj., Živković, Lj. and Uskoković, D., Synthesis, morphology and formation mechanism of mullite particles produced by ultrasonic spray pyrolysis, *J. Mater. Res.*, 11, 1706–1716, 1996.

9. Jokanović, V., Nikčević, I., Dačić, B., and Uskoković, D., Synthesis of nanostructured carbonated calciumhydroxyapatite by ultrasonic spray pyrolysis, *J. Ceram. Proc. Res.*, 5 (2), 157–162, 2004.

10. Jokanović, V., Jokanović, B., Nedeljković, J., and Milošević, O., Modeling of nanostructured hollow TiO$_2$ spheres obtained by ultrasonic spray pyrolysis, *Surface Colloid A*, 249, 111–113, 2004.

11. Janaćković, Dj., Jokanović, V., Kostić-Gvozdenović, Lj., Zec, S., and Uskoković, D., Synthesis and formation mechanism of submicrometer spherical cordierite powders by ultrasonic spray pyrolysis, *J. Mater. Sci.*, 32, 163–168, 1997.

12. Jokanović, V., Janaćković, Dj., Ćurčić, R., Živanović, P., and Uskoković, D., Synthesis of cordierite powder performed under the conditions of periodical ultrasonic fields activity, *Mat. Sci. Forum*, 282–283, 65–70, 1998.

13. Jokanović, V., Janaćković, Dj., and Uskoković, D., Drop size distribution and morphology of ultrafine spherical powders by ultrasonic spray pyrolysis, *Key Eng. Mater.*, 132–136, 197–200, 1997.

14. Janaćković, Dj., Kostić-Gvozdenović, Lj., Ćirjaković, R., Jokanović, V., Petrović-Prelević, J., and Uskoković, D., Synthesis of spinel powders by the spray pyrolysis method, *Key Eng. Mater.*, 132–136, 213–216, 1997.

15. Lefebvre, A.H., *Atomization and Sprays*, Hemisphere, New York, 1989, 95 pp.

16. Bayvel, L. and Orzechowski, Z., *Liquid Atomization*, Taylor & Francis, Washington, DC, 1993, 79 pp.

17. Iskandar, F., Langgero, I.W., Xia, B., and Okuyama, K., Functional nanostructured silica powders derived from colloidal suspensions by sol spraying, *J. Nanoparticle Res.*, 3, 263–270, 2001.

18. Mikrajuddin Iskandar, F. and Okyama, K., *In situ* production of spherical silica particles containing self-organized mesopores, *Nano Lett.*, 1, 231–234, 2001.

19. Okyama, K. and Lenggoro, I.W., Preparation of nanoparticles via spray route, *Chem. Eng. Sci.*, 58, 537–547, 2003.

20. Lengoro, I.W., Okuyama, K., de la Mora, J.F., and Tohge, N., Sizing of colloidal nanoparticles by electrospray and differential mobility analyzer methods, *Langmuir*, 18, 4584–4591, 2000.

21. Ehrig, R., Ofenloch, O., Schaber, K., and Deuflhard, P., Modelling and simulation of aerosol formation by heterogeneous nucleation in gas–liquid contact devices, *Chem. Eng. Sci.*, 57, 1151–1163, 2002.

22. Brenn, G., On the controlled production of sprays with discrete polydisperse drop size spectra, *Chem. Eng. Sci.*, 55, 5437–5444, 2000.

23. Mikrajuddin, Iskandar, F., and Okuyima, K., Single route for producing organized metallic domes, dots, and pores by colloidal templating and over sputtering, *Adv. Mater.*, 14, 930–933, 2002.

24. Xia, B., Lenggoro, I.W., and Okuyama, K., Novel route to nanoparticle synthesis by salt-assisted aerosol decomposition, *Adv. Mater.*, 13, 1579–1582, 2001.

25. Mikrajuddin, Iskandar, F., Okuyama, K., and Shi, G.F., Stable photoluminescence of zinc oxide quantum dots in silica nanoparticles matrix prepared by the combined sol–gel and spray drying method, *J. Appl. Phys.*, 89, 6431–6434, 2001.

26. Xia, B., Lenggoro, I.W., and Okuyama, K., Nanoparticle separation in salted droplet micro-reactors, *Chem. Mater.*, 14, 2623–2627, 2002.

27. Kang, Y.C. and Park B.S., A high-volume spray aerosol generator producing small droplets for low pressure applications, *J. Aerosol Sci.*, 26, 1131–1138, 1995.

28. Messing, G.L., Zhang, S.C., and Jayanthi, G.V., Ceramic powder synthesis by spray pyrolysis, *J. Am. Ceram. Soc.*, 76, 2707–2726, 1993.

29. Wood, W.R. and Lomis, A.L., The physical and biological effects of high frequency sound-waves of great intensity, *Phil. Mag.*, 4, 417–437, 1927.

30. Faraday, M., On the forms and states assumed by fluids in contact with vibrating elastic surfaces, *Phil. Trans. R. Soc. Lond.*, 52, 319–340, 1831.

31. Kelvin Lord (Thomson W.), Hydrokinetic solutions and observations, *Phil. Mag.*, 42, 362–377, 1871.

32. Rayleigh, L., On the pressure developed in the liquid during the collapse of a spherical cavity, *Phil. Mag. Ser.*, 6 (34), 94–98, 1917.

33. Solner, K., The mechanism of the formation of fogs by ultrasonic waves, *Trans. Faraday Soc.*, 32, 1532–1536, 1936.

34. Bisa, K., Dirnagl, K., and Eshe, E., Zerstaubung von Flussigkeiten mit Ultrashall (Ultrasonic spraying of liquids), *Siemens Z.*, 28, 341–344, 1954.

35. Benjamin, T.B. and Urssel, F., The stability of the plane free surface of a liquid in vertical periodic motion, *Proc. R. Soc. London A*, 225, 505–515, 1954.

36. Sorokin, V.I., The effect of fountain formation at the surface of a vertically oscillating liquid, *Soviet Phys. Acoust.*, 3, 281–291, 1957.

37. Eisenmenger, W., Dynamic properties of surface tension of water and aqueous solutions of surface active agents with standing capillary waves in frequency range from 10 kHz to 1.5 MHz, *Acoustica*, 9, 327–340, 1959.

38. Lang, R.J., Ultrasonic atomization of liquids, *J. Acoust. Soc. Am.*, 34, 6–8, 1962.

39. Peskin, R.L. and Raco, R.J., Ultrasonic atomization of liquids, *J. Acoust. Soc. Am.*, 35, 1378–1381, 1963.

40. Fogler, H.S. and Timmerhaus, K.D., Ultrasonic atomization studies, *J. Acoust. Soc. Am.*, 39, 515–518, 1965.

41. Boguslaskii, Y.Y. and Eknadiosyants, O.K., Physical mechanism of the acoustic atomization of a liquid, *Sov. Phys. Acoust.*, 15, 14–21, 1969.

42. Topp, M.N., Ultrasonic atomization — a photographic study of the mechaism of desintegration, *Aerosol Sci.*, 4, 17–25, 1973.

43. Chiba, C., Atomization of a liquid by immersed and convergent ultrasonic vibrator (case of distilled water), *Bull. J. Soc. Mech. Eng.*, 18, 376–382, 1975.

44. Bassett, J.D., Bright, W.W., Observation concerning the mechanism of atomization in an ultrasonic fountain, *J. Aerosol. Sci.*, 7, 47–51, 1976.

45. Edwards, W.S. and Fauve, S., Patterns and quasi-patterns in the Faraday experiments, *J. Fluid Mech.*, 278, 123–148, 1994.

46. Barreras, F., Amaveda, H., and Lozano, A., Transient high-frequency ultrasonic water atomization, *Exp. Fluids*, 33, 405–413, 2002.

47. Hinds, W.C., *Aerosol Technology*, Wiley, New York, 1982, pp. 124–153.

48. Fogler, H.S. and Timmerhaus, K.D., Ultrasonic atomization studies, *J. Acoust. Soc. Am.*, 39, 515–518, 1965.

49. Miles, J., Faraday waves, rolls versus squares, *J. Fluid Mech.*, 269, 353–371, 1994.

50. Sindayihebura, D. and Bolle, L., Ultrasonic atomization of liquid: stability analysis of the viscous liquid film free surface, *Atomiz. Spray*, 8, 217–233, 1998.

51. Jayanthi, G.V., Zhang, S.C., and Messing, G.L., Modeling of solid particle formation during solution aerosol thermolysis: the evaporation stage, *J. Aerosol Sci. Technol.*, 19, 478–490, 1993.

52. Moon, S., Chung, H.J., Woo, S.I., Hwang, C.S., Lee, M.Y., and Park, S.B., Effect of solution concentration on droplet size in ultrasonic aerosol generator, *J. Aerosol Sci.*, 28 (Suppl. I), 5525–5526, 1999.

53. Zhang, Z.Y., Meng, X.Q., Jin, P., Li, Ch.M., Qu, S.C., Xu, B., Ye, X.L., and Wang, Z.G., A novel application to quantum dot materials to the active region of superluminescent diodes, *J. Cryst. Growth*, 243, 25–29, 2002.

54. Li, Q. and Dong, P., Preparation of nearly monodisperse multiply coated submicrospheres with a high refractive index, *J. Colloid Interf. Sci.*, 261, 325–329, 2003.

55. Douglass, D.L. and Landuyt, J.V., The structure and morphology of oxide films during the initial stages of titanium oxidation, *Acta Metall.*, 14 (4), 491–503, 1966.

56. Ohmori, M. and Matijević, E., Preparation and properties of uniform coated particles: VII. Silica on hematite, *J. Colloid Interf. Sci.*, 150, 594–598, 1992.

57. Furlong, D.N. and Sing, K.S., The precipitation of silica on titaniumdioxide surfaces, I. Preparation of coated surfaces by electrophoresis, *J. Colloid Interf. Sci.*, 69, 409–419, 1979.

58. Mezza, P., Phalippon, J., and Sempere, R., Sol–gel derived porous silica films, *J. Non-Cryst. Solids*, 243, 75–79, 1999.

59. Caruso, F., Caruso, R.A., and Mohwald, H., Nanoengineering of inorganic and hybrid hollow spheres by colloidal templating, *Science*, 282, 1111–1114, 1998.

60. Zhang, F., Wang, W., Li, C., Wang, H., Chen, L., and Liu, X., Rutile type titanium oxide films synthesized by filtered arc deposition, *Surf. Coat. Tech.*, 110, 136–139, 1998.

61. Froba, M. and Reller, A., Synthesis and reactivity of functioned metal oxides in nanoscopic systems, *Prog. Solid. State Chem.* 27, 1–27, 1999.

62. Leong, K.H., Morphological control of particle generated from the evaporation of solution droplets: theoretical considerations, *J. Aerosol Sci.*, 18, 511–524, 1987.

63. van der Lijin, J., Simulation of heat and mass transfer in spray draying, *Agric. Res. Rep.*, No. 845, 1976.

64. Nesic, S. and Vodnik, J., Kinetics of droplet evaporation, *Chem. Eng. Sci.*, 46, 527–537, 1991.

65. Levič, G.V., *Fizičeskaja Hidrodinamika*, AN SSSR, Moscow, 1952, pp. 225–254.

66. McLachlan, N.W., *Theory and Applications of Mathieu Functions*, Oxford University Press, London, 1947, pp. 338–345.

67. Hercfeld, K.F. and Litovicz, T.V., *Absorption and Dispersion of Ultrasonic Waves*, Academic Press, New York, 1959, pp. 55–78.

68. Condou, E.U. and Odiskaw, H., *Handbook of Physics*, McGraw-Hill, New York, 1958, pp. 339–361.

69. Landau, L.D. and Lifšic, E.M., *Mehanika Neprekidnoe Sredii*, AN SSSR, Moscow, 1960, pp. 224–312.

70. Harmans, J.J., *Flow Properties of Disperse Systems*, North-Holland, Amsterdam, 1953, pp. 112–131.

71. Reed, J.M., *Introduction to the Principles of Ceramic Processing*, John Wiley & Sons, Inc., New York, 1988, p. 185.

72. Jokanović, V., Mioč, U.B., and Nedić, Z.P., Nanostructured phosphorous doped tungsten bronzes obtained by ultrasonic spray pyrolysis, *Sol. St. Ionics*, submitted for publication.

73. Spasić, A.M., Jokanović, V., and Krstić, D.N., A theory of electroviscoelasticity: a new approach for quantifying the behaviour of liquid–liquid interfaces under applied fields, *J. Colloid Interf. Sci.*, 186 (2), 434–446, 1997.

74. Spasić, A.M. and Jokanović, V., Stability of the secondary droplet–film structure in polydispersed systems, *J. Colloid Interf. Sci.*, 170 (1), 229–240, 1995.

75. Spasić, A.M., Electroviscoelasticity of liquid/liquid interfaces, in *Interfacial Electrokinetics and Electrophoresis*, Delgado, A.V., Ed., Marcel Dekker, New York, 2002.

76. Ahonen, P.P., Tapper, U., Kaupinnen, E.I., Joubert, J.C., and Deschanvres, J.L., Aerosol synthesis of Ti–O powders via in-droplet hydrolysis of titanium alkoxide, *Mat. Sci. Eng. A*, 315, 113–121, 2001.

77. Parak, W.J., Gerion, D., Pellegrino, T., Zanchet, D., Micheel, Ch., Williams, S.C., Boudreau, R., Le Gros, M.A., Larabell, C.A., and Alivastos, A.P., Biological applications of colloidal nanocrystals, *Nanotechnology*, 14, R15–R27, 2003.

78. Froba, M. and Reller, A., Synthesis and reactivity of functional metal oxides in nanoscopic systems, *Prog. Solid State Ch.*, 27, 1–27, 1999.

79. Fendler, J.N. Ed., *Nanoparticles and Nanostructured Films*, Wiley-VCH, Weinheim, 1998, pp. 5–25.

80. Iskandar, F., Lenggoro, I.W., Xia, B., and Okuyama, K., Functional nanostructured silica powders derived from colloidal suspensions by sol spraying, *J. Nanoparticle Res.*, 3, 263–270, 2001.

81. Kang, Y.C., Lim, M.A., Park, H.D., and Han, M., Ba^{2+} Co-doped Zn_2SiO_4:Mn phosphor particles prepared by ultrasonic spray pyrolysis, *J. Electrochem. Soc.*, 150 (1), H7–H11, 2003.
82. Kulak, A., Davis, S.A., Dujarin, E., and Mann, S., Controlled assembly of nanoparticle-containing gold and silica microspheres and silica/gold nanocomposite spheroids with complex form, *Chem. Mater.*, 15, 528–535, 2003.
83. Jokanović, V. and Uskoković, D., Calcium hydroxyapatite thin films on titanium substrate prepared by ultrasonic spray pyrolysis, *Mater. T. JIM*, 46, 228–235, 2005.

Polymer Membranes

21 Transfer Phenomena Through Polymer Membranes

Silvia Alexandrova, Dieudonné N. Amang,
François Garcia, and Véronique Rollet
University of Caen, Caen, France

Abdellah Saboni
University of Rouen, Rouen, France

CONTENTS

I. INTRODUCTION

The membrane separations represent a new unit operation whose principal advantage is the low energy requirements. The membrane separation processes can be classified according to the type of membrane used. There are three types of membranes, namely, porous (micro-filtration, ultrafiltration, and nanofiltration), nonporous or tight (dialysis, pervaporation, reverse osmosis, etc.), and liquid membranes (bulk, double emulsions, and supported). The second criterion, usually used to class these type of separation processes is the driving force of the separation: pressure gradient, concentration gradient, temperature gradient, (Table 21.1).

It is obvious that due to the diversity of membrane type and driving forces applied in each process, the mass transfer in membrane is governed by different mechanisms. For example, in a macroporous membrane the mass transfer does not rely on the chemical properties of the membrane, but on the contrary, in the case of tight membrane, the transfer and the separations are essentially due to the molecular interaction between the membrane and the solutes.

In this chapter, we will briefly present basic mass transfer mechanisms in polymer membranes and the mechanisms dominating in some membrane separations. In the first part of the chapter, we give a brief description of general transport mechanisms of solutes (charged or not) through a porous media (charged or not). In the second part, we discuss in detail some membranes processes and the transfer mechanisms, interesting from the theoretical point of view.

II. MECHANISMS OF MEMBRANE TRANSPORT

The polymer membrane (charged or not) can be considered as a selective barrier separating two solutions: feed and permeate.

The feed is introduced in the upstream side of the membrane and due to the membrane morphology and selectivity, some species of the feed selectively permeate through the polymer matrix and form the permeate stream (Q_p), and a part which does not pass through the membrane form the concentrate (Q_0). The concentrate contains the molecules or particles retained by the membrane. The flow fraction which crosses the membrane is the conversion rate Y defined by

TABLE 21.1
Membrane Separation Processes

Process	Type of membrane	Driving force
Reverse osmosis	Nonporous (tight)	Pressure gradient
Thermoosmosis	Nonporous (tight)	Temperature gradient
Vapor permeation	Nonporous (tight)	Vapor pressure
Pervaporation	Nonporous (tight)	Vapor pressure
Gas permeation	Nonporous (tight)	Pressure gradient
Dialysis	Nonporous (tight)	Concentration gradient
Electrodialysis	Nonporous (tight)	Gradient of electric potential
Liquid membranes	Liquid in a polymer matrix	Concentration gradient
Nanofiltration	Porous	Pressure gradient and solute–membrane interactions
Ultrafiltration	Porous	Pressure gradient
Microfiltration	Porous	Pressure gradient

$Y = (Q_p/Q_0)$. The tangential flow allows the accumulation and limitation of various species (particles, molecules, and ions) on the membrane.

In this part, we present the basic mechanisms which occur through the polymer film (membrane) caused by different driving forces.

A. DIFFUSION

The diffusional type of transfer consists in assuming that the solutes and the solvent are dissolved in the membrane and pass by diffusion due to the gradient of their chemical potential. This is the well-known "solution-diffusion" model developed by Lonsdale et al. [1] under three important presumptions: (i) the diffusion coefficient of the solute in the membrane is not dependent on its concentration; (ii) the solvent concentration in the membrane is constant; and (iii) the membrane properties are independent from the operating conditions. In this case, the flux of each species can be expressed as:

$$J_i = -\frac{\bar{D}_i \bar{C}_i}{RT} \left(\frac{\partial \mu_i}{\partial \bar{C}_i} \operatorname{grad} \bar{C}_i + \theta_i \operatorname{grad} P \right) \tag{21.1}$$

For water (solvent) when the concentration gradient is low, this expression becomes:

$$J_w = -\frac{\bar{D}_w \bar{C}_w \theta_w}{RTL} (\Delta P - \Delta \pi') \tag{21.2}$$

with

$$A = -(\bar{D}_w \bar{C}_w \theta_w / RTL) - \text{solvent permeability.} \tag{21.3}$$

For very selective membranes $\theta_i \operatorname{grad} P \ll (\partial \mu_i / \partial \bar{C}_i) \operatorname{grad} \bar{C}_i$ and the expression (21.1) for the solute flux density gives:

$$J_s = -\frac{\bar{D}_s}{L} \Delta \bar{C}_s = -\frac{\bar{D}_s}{L} \phi \Delta C_s \tag{21.4}$$

with

$$B = -\frac{\bar{D}_s \phi}{L} - \text{solute permeability} \tag{21.5}$$

and

$$\Delta C_s = C_m - C_p - \text{concentration gradient} \tag{21.6}$$

According to Equation (21.2), the flux increases as the pressure increases, while, according to Equation (21.4), the solute flux is independent of the pressure gradient. In fact, the solute flux increases somewhat with pressure increasing – this can be explained by some small leaks in the membrane.

This model, proposed by Lonsdale et al. [1], is only applied for the simple solutions (molecular solutes) and the noncharged membranes because it does not make it possible to take into account the coupling solute–membrane effects.

B. Osmosis

Osmosis is the term used for solvent diffusion through membrane. When the solvent diffusion is accompanied by the solute diffusion, the transfer is called dialysis. This phenomenon is observed when two solutions, concentrated and diluted, are separated by a membrane. In this system a diffusion of solvent from the diluted side to the concentrated side of the membrane appears due to a hydrostatic pressure existing between the two sides, called osmotic pressure.

The osmotic pressure for ideal solutions (or very diluted solutions) is related to the concentration of solute in the feed by the Van't Hoff law [2]:

$$\pi' = n_i C R T \tag{21.7}$$

The osmosis described by this equation is called normal osmosis and it is observed in dilute solutions and inert (noncharged) membranes. The osmosis of electrolytes solutions through charged ionic membranes is called anomalous osmosis because in this case the solute diffusion enhances (positive anomalous osmosis) or decreases (negative anomalous osmosis) the solvent transfer. The solvent transfer through a membrane under an electric current across the system is called electro-osmosis [3].

When a hydrostatic pressure, more important that the osmotic pressure, is applied to the concentrated side of the membrane, a diffusion of the solvent from the concentrated solution to the diluted solution is observed: this mass transfer phenomenon is called reverse osmosis [4].

C. Hydrodynamic Flow

When the membranes are macroporous and the fluid flow is laminar, a simple hydrodynamic theory can be applied to describe the transfer across the membrane. Three models are usually used in this case. The first is the Darcy's law developed for the flux through porous media [5]:

$$\frac{V}{St} = \frac{K' \Delta P}{\mu L} \tag{21.8}$$

In this expression the flux is proportional to the pressure gradient across the porous medium. The coefficient K' is named "permeability" and it depends on the properties of the medium. The viscosity μ reflects the frictions between the fluid and the porous medium: the drag force of solid and the viscous resistance.

The second hydrodynamic model is the Hagen–Poiseuille model. It is used to macroporous membranes with cylindrical straight pores of same diameter:

$$\frac{J_v}{S} = \frac{V}{St} = \frac{n' \pi r^4 \Delta P}{8 \mu L} = \frac{\varepsilon' r^2 \Delta P}{8 \mu L} \tag{21.9}$$

If the pore section is not circular, the Kozeny–Carman equation [6] can be used by introducing the hydraulic radius of pore:

$$\frac{J_v}{S} = \frac{V}{St} = \frac{\varepsilon^3 \Delta P}{\tau (1 - \varepsilon)^2 S'^2 \mu L} \tag{21.10}$$

where τ is the coefficient characterizing the pore.

The Kozeny–Carman model can be used for membrane considered as a compacted noncharged packed bed, the permeate flow through the filtration medium can be assumed laminar when the mean free path of fluid molecules is very small compared to the pore diameter.

D. DONNAN EFFECT

This phenomenon is observed in charged membranes. The fixed charges of the membrane absorb the solute counterions (counterions possess the opposite charge that the charged sites of the membrane) and reject the coions (coions have the same charge that the charged sites of the membrane) [7] (Figure 21.1).

Due to the existence of many charged species and fixed charges in the polymer, an electrical potential difference between the solution and the charged membrane appears. This electrical potential difference between the membrane and the solution is named Donnan potential (Φ_D) and it is introduced by:

$$\frac{\bar{a}_i}{a_i} = \frac{\bar{\gamma}_i \bar{C}_i}{\gamma_i C_i} = \exp\left(-\frac{Z_i \Im}{RT} \Phi_D\right) \tag{21.11}$$

under following assumptions: (i) the membrane charge is homogeneous; (ii) the membrane surface is not rough; and (iii) the membrane-swelling phenomenon is not taken into account.

The Donnan potential can be calculated from the condition of electro-neutrality of the system:

$$Z_m \bar{C}_m + \sum_i Z_i \bar{C}_i = Z_m \bar{C}_m + \sum_i Z_i C_i \frac{\gamma_i}{\bar{\gamma}_i} \exp\left(-\frac{Z_i \Im}{RT} \Phi_D\right) \tag{21.12}$$

This potential has a positive value for a cationic membrane and a negative value for an anionic membrane. The co-ions will be excluded by the membrane and the counterions will be absorbed in the membrane.

E. SORET EFFECT

The diffusion of solutes through a porous polymer membrane under the temperature gradient is named thermal diffusion or Soret effect. Few studies deal with this phenomenon. Its importance in the transfer through membranes is often neglected compared to the other coupled mechanisms of transfer.

The influence of this mechanism can be enough significant in the Knudsen regime [8], when the pressure potential is very low and the flux through the pore is due to the molecular diffusion. In this

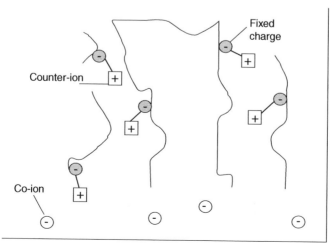

FIGURE 21.1 Charged cationic membrane.

case, it is important to take into account the variation of the diffusion coefficients in function of the temperature. The effusion can be calculated using the modified Knudsen equation for very thin membrane $(L \rightarrow 0)$.

$$J_{\mathrm{m}} = \frac{S''}{(2\pi RM)^{0.5}} \left(\frac{P_{\mathrm{a}}}{T_{\mathrm{a}}^{0.5}} - \frac{P_{\mathrm{b}}}{T_{\mathrm{b}}^{0.5}} \right) \tag{21.13}$$

III. MASS TRANSFER MODELING OF MEMBRANE PROCESS

The principal operating parameters influencing the transfer in a membrane separation processes are the hydrodynamic [9], pressure [10,11], solutes nature [12], pH [12], and of course, the nature, morphology, and thickness of the membrane. Usually, during the concentration increase of the compound interacting most strongly with polymer; the membrane swells, thus the diffusion coefficient increases [13]. Therefore the compounds' flow increases while the selectivity decreases [12].

The hydrodynamic conditions seem to have the greatest influence on the thickness of boundary layer (polarization of concentration and temperature) [14–16], usually existing at the solution–membrane interface.

When a selective transfer occurs through a membrane, there is an accumulation of the less transferred compound and an exhaustion of the more-transferred one in the boundary layer adjacent to the membrane. This phenomenon is known as "polarization of concentration." It can be also amplified by the composition of the feed and the selectivity of the membrane.

A schematic representation of the concentration profiles of the two involved species (1 and 2) is given on the Figure 21.2.

The "temperature polarization" principally occurs in pervaporation process. On the downstream (permeate) surface of the membrane, the permeate vaporization causes a cooling, which induces a difference in temperature on both sides of the membrane (Figure 21.3).

The temperature diminution in the membrane is much lower than that in the boundary layer [17]. This is due to the fact that a resistance to the heat transfer occurs in the boundary layer and the mass flow through the membrane is strongly dependent on the temperature.

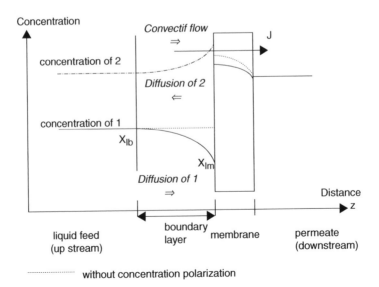

FIGURE 21.2 Concentration profile for the pervaporation on process (1: mainly sorbed compound, 2: fewer sorbed compound).

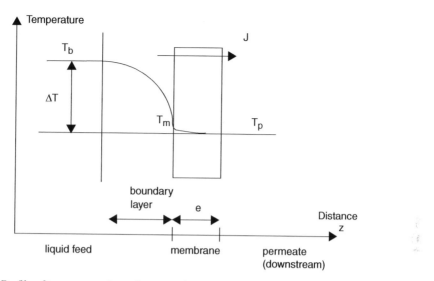

FIGURE 21.3 Profile of temperature in stationary regime by pervaporation.

For the practical evaluation of the membrane selectivity, a qualitative experimental parameter called the rejection rate, R_{obs}, can be used. This rate permits to have an idea of the selectivity of polymer according to easily measurable parameters: the permeate and the feed concentrations.

$$R_{obs} = \frac{c_0 - c_p}{c_0} = 1 - \frac{c_p}{c_0} \tag{21.14}$$

A. REVERSE OSMOSIS

1. Principle and Applications

Reverse osmosis is the transfer of solvent (water) through a semi-permeable membrane under the effect of a pressure gradient. As a flux is limited by the mass transfer in the membrane, the pressure gradient applied can be important. It is thus frequent to work with pressures of about 70 or 100 bar. The reverse osmosis process is usually used to obtain pure water from sea water, brackish water, or waste water.

2. Mass Transfer Modeling

Various models and mechanisms for the solvent and the solute transports through tight osmosis membrane have been developed [18–20]. The solution-diffusion model can be employed to calculate the pure water flux (Equation 21.2), j_w, and the solute flux (Equation 21.4), j_s [21].

The solvent flux depends on the hydraulic pressure applied across the membrane and on the difference in the osmotic pressure across the membrane [22]. Whereas the solute flux depends on the concentration gradient of the species (Equation 21.4).

From Equations (21.2), (21.4), and (21.14), it can be shown that R_{obs} is a function of pressure and concentration. An increase in pressure will increase solvent flux; consequently, with the solute rejection R_{obs} will increase.

Because of concentration polarization in the feed (upstream), the solute concentration at the membrane interface, c_m, will be higher than that in the bulk of the high-pressure side solution, c_B. The equation used for calculating c_m was derived from a consideration of the convective and diffusive mass transfer of the solute towards the membrane:

$$\phi = \frac{c_m - c_p}{c_B - c_p} = \exp\left(\frac{v'}{k}\right) \tag{21.15}$$

where the mass transfer coefficient, k, can be calculated from empirical equations [23]:

$$Sh = \frac{1.09}{\varepsilon} Re^{0.333} Sc^{0.333} \quad \text{for a hollow-fiber bundle} \tag{21.16}$$

$$Sh = 0.04 Re^{0.75} Sc^{0.333} \quad \text{for a spiral-wound membrane} \tag{21.17}$$

The permeation velocity, v', and the permeate concentration, c_p, can be calculated as follows:

$$v' = \frac{j_w + j_s}{\rho_p} \tag{21.18}$$

$$c_p = \frac{j_s}{v'} \tag{21.19}$$

and Q_P, Q_B, and c_B, can be calculated from the balance equations:

$$Q_p = v'S \tag{21.20}$$

$$Q_B = Q_0 - Q_p \tag{21.21}$$

$$c_B = \frac{Q_0 c_0 - Q_p c_p}{Q_B} \tag{21.22}$$

Another factor that must be considered in solving Equation (21.2) and Equation (21.4) is the pressure drop on the permeate-side of the membrane, P_p, and on the feed-side, P_B.

For a spiral-wound membrane, each of the permeate- and feed-side flows can be considered as a flow between two parallel plates with a length L', a width W, and a spacing ℓ. The pressure drop on the feed-side can then be calculated as follows:

$$\Delta P_B = \frac{12 Q_B x \mu}{W \ell^3} \tag{21.23}$$

For the hollow-fiber membranes, the flow on the feed-side is across a densely packed bed of fibers. The pressure drop in this membrane side can be calculated using Ergun's equation [24]:

$$\Delta P_B = \left[150 \frac{(1 - \varepsilon)^2}{\varepsilon^3} \mu \frac{u_s}{d_p^2} + 1.75 \frac{(1 - \varepsilon)}{\varepsilon^3} \rho \frac{u_s^2}{d_p} \right] (R_o - R_i) \tag{21.24}$$

where d_p is a specific diameter taken as 1.5 of the fiber diameter.

The permeate-side pressure drop can be calculated using the Hagen–Poiseuille's equation [24,25]:

$$\Delta P_p = \frac{16\mu}{r_i^4} r_o v' x^2 \tag{21.25}$$

So, Equation (21.16) and Equation (21.17) can be used to calculate the mass transfer coefficient, k. The concentration polarization factor is calculated from Equation (21.15). The pressure drops in the feed-side is calculated from Equation (21.23) and Equation (21.24). Then Equation (21.2) and Equation (21.4) are used to calculate the water and salt fluxes. An example of results obtained can be seen in Figure 21.4.

B. NANOFILTRATION

1. Principle and Applications

Nanofiltration is the most recent pressure driven membrane process. Many industrial applications are concerned in terms of separation, purification, or concentration of molecules: drinking water treatment, biotechnologies, food, textile, paper industries, effluent treatments, etc. Nanofiltration has replaced reverse osmosis in several applications due to lower energy consumption and higher permeate fluxes. Nanofiltration membranes are characterized by a porous structure with pore diameter less than 2 nm and generally possess a fixed charges. The cut-off of organic nanofiltration membranes is generally comprised between 100 and 1000 Da (g mol^{-1}). The operating pressures are generally comprised between 5 and 40 bar. The very high selectivity of the process is based on the size and the charge of the solutes [26,27]. Nevertheless, the mass transfer mechanisms concern also the fluids phase's hydrodynamic, chemical affinities, convection, diffusion, and at last electric effects when the solutes employed are charged. The nanofiltration mechanisms are consequently complex.

For uncharged solutes, the retention occurs by steric exclusion and consequently the solute retention is relative to the pore size of the membranes [26,28]. Thus, experiments are carried out with neutral solutes with different molecular weights (raffinose, maltose, galactose, glucose, etc.) in order to determine the mean pore size of the membrane.

Concerning charged solutes, the retention depends on the size, nature, concentration of the solutes [27], and the membrane properties. The nanofiltration membrane separation is generally fixed by the Donnan exclusion [26,27]. In order to determine the influence of the electrolyte nature on the retention, experiments are generally carried out with different salts, typically Na_2SO_4, NaCl and $CaCl_2$. When the membrane is negatively charged the retention sequence is $R_{obs}(Na_2SO_4) > R_{obs}(NaCl) > R_{obs}(CaCl_2)$, and when the membrane is positively charged the retention sequence is $R_{obs}(CaCl_2) > R_{obs}(NaCl) > R_{obs}(Na_2SO_4)$ [26,29–31]. Thus, the retention of bivalent coions is the highest and is more important than the retention of the monovalent co-ions caused by stronger electrostatic interactions. At last, the retention of bivalent counterions is the lowest [32,33]. Moreover, the retention decreases with increasing concentration according to the Donnan effect.

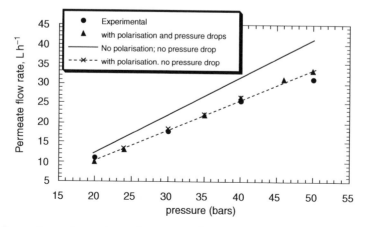

FIGURE 21.4 Comparison of experimental permeate flow rate for the spiral-wound membrane with theoretical values.

When the membrane pore size is relatively small, the retention is governed by steric effect and the retention sequence is in the same order that of hydrated ions size, which is close to pore size [26,28]. In addition to that, the retention sequence observed during nanofiltration experiment can be in the same order that of hydration energy, that is, the ion with the highest hydration energy is the most retained [34,35].

2. Mass Transfer Modeling

Concerning mass transfer modeling across nanofiltration membranes, the transport mechanisms occurring are convection, diffusion, and electromigration (when the solutes are charged). Taking into account the polarization concentration phenomenon, which can occur during the filtration operation (and thus the increase of the concentration at the membrane interface C_m), the solute real retention (R_{real}) can be calculated from the observed retention by:

$$R_{real} = \frac{R_{obs} \exp(J_v/k)}{1 - R_{obs}[1 - \exp(J_v/k)]} \tag{21.26}$$

wih

$$R_{real} = 1 - \frac{C_p}{C_m} \tag{21.27}$$

The mass transfer coefficient (k) is obtained experimentally or by different correlations depending of the geometry of the filtration cell.

a. Diffusion Model

Assuming that the solute transport in the membrane is principally diffusive, the solute flux can be obtained with analogy with the Fick's equation by the relation [29]:

$$J_s = -D^* \frac{d\bar{C}_s}{dx} \tag{21.28}$$

where D^* is an internal diffusion coefficient, which is specific to both solute and membrane.
Integration of Equation (21.28) on pore length L considering $J_s = J_v C_p$ leads to:

$$J_s = D^* \left[\frac{\bar{C}_s(x=0) - \bar{C}_s(x=L)}{L} \right] \tag{21.29}$$

Introducing an equilibrium partition coefficient (ϕ) for the solute between the membrane and the solution, and assuming identical conditions between the inlet ($x = 0$) and outlet ($x = L$) of a pore gives the expression:

$$\phi = \frac{\bar{C}_s(x=0)}{C_m} = \frac{\bar{C}_s(x=L)}{C_p} \tag{21.30}$$

Finally, the real retention obtained is:

$$R_{real} = 1 - \frac{1}{1 + J_v \beta} \tag{21.31}$$

with

$$\beta = \frac{L}{D^* \phi} \tag{21.32}$$

For uncharged solute, $\phi = \phi_s$ and the coefficient ϕ_s corresponds to the steric retention of the solute. When the solute is charged, $\phi = \phi_s \, \phi_E$ and the coefficient ϕ_E corresponds to retention due to electrostatic repulsions between charged solutes and membrane charged groups [36]. Experimental and calculated data for neutral and charged solutes are shown in Figure (21.5) and Figure (21.6) [29,30].

b. Irreversible Thermodynamics

The transport of solute in a membrane can also be described by irreversible thermodynamics considering the membrane as a black box. The solute flux is given by the Spiegler–Kedem equation [37]:

$$J_s = -\bar{P}_s \left(\frac{d\bar{C}_s}{dx} \right) + (1 - \sigma) J_v \bar{C}_s \tag{21.33}$$

where $\bar{P}_s$ is the locale solute permeability and σ the reflection coefficient.

The solute flux can be considered as the sum of diffusion and convection terms [38]. Integrating Equation (21.33) with the boundary conditions $\bar{C}_s = C_m$ at $x = 0$ and $\bar{C}_s = C_p$ at $x = L$ yield to the real retention expression:

$$R_{real} = \frac{\sigma(1 - F)}{1 - \sigma F} \tag{21.34}$$

with

$$F = \exp\left(-\frac{1 - \sigma}{P_s} J_v \right) \tag{21.35}$$

and

$$\lim_{J_v \to \infty} F = 0 \tag{21.36}$$

P_s is the overall solute permeability and is defined by:

$$P_s = \frac{\bar{P}_s}{L} \tag{21.37}$$

The reflection coefficient (σ) thus corresponds to the limit value of the salt retention at infinitely high water flux. Experimental and calculated data for neutral [29] and charged [30] solutes are shown, respectively, in Figure 21.5 and Figure 21.6.

c. Extended Nernst–Planck Equation

The extended Nernst-Plank equation is applied for charged solutes and can be written as [31,39,40]:

$$J_i = -P_i \left[\frac{d\bar{C}_i}{dx} + Z_i \bar{C}_i \frac{\Im}{RT} \frac{d\psi}{dx} \right] + J_v \bar{C}_i (1 - \sigma_i) \tag{21.38}$$

Electroneutrality condition for charged components is given by:

$$\sum_i Z_i \bar{C}_i = 0 \tag{21.39}$$

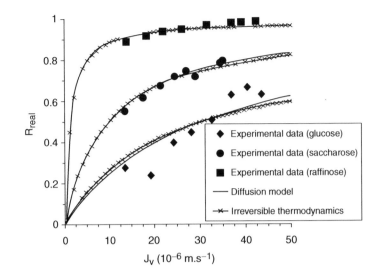

FIGURE 21.5 Experimental and calculated data for neutral solutes (10^{-1}mol l^{-1}) — Membrane BQ 01 (Osmonics) (Boucard, F., Contribution à la caractérisation des mécanismes de transport en nanofiltration: Expériences et Modèles, Ph.D. Thesis, University of Caen, France, 2000.).

Condition of zero electrical current through the membrane, if there is no external charge transport, gives the expression:

$$\sum_i Z_i J_i = 0 \tag{21.40}$$

Linearizing Equation (21.38) and considering that $C_i = C_{m,i}$ (for $x = 0$) [39,40] leads to:

$$J_i = -B_i \left[\Delta \bar{C}_i + Z_i C_{m,i} \frac{\Im}{RT} \Delta \psi \right] + J_v C_{m,i}(1 - \sigma_i) \tag{21.41}$$

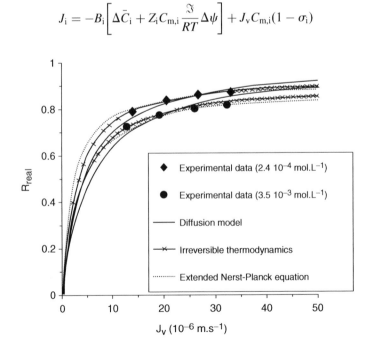

FIGURE 21.6 Experimental and calculated data for charged solutes ($NaNO_3$) — Membrane NF (FILMTEC) (From Garcia, F., Ciceron, D., Alexandrova, S., and Saboni, A., 16th International Congress of Chemical and Process Proceedings of the Engineering CHISA 2004, August, 22–26, 2004, Prague, Czech Republic).

with

$$B_i = \frac{P_i}{L} \tag{21.42}$$

The concentration difference across the membrane is given by:

$$\Delta \bar{C}_i = C_{p,i} - C_{m,i} \tag{21.43}$$

Finally, the real retention expression obtained for a binary system ($i = 1, 2$) is:

$$R_{\text{real}} = \frac{J_v R_{\infty,1}}{J_v + F_1} \tag{21.44}$$

with

$$R_{\infty,1} = \frac{\sigma_1 B_2 Z_2 - \sigma_2 B_1 Z_1}{B_2 Z_2 - B_1 Z_1} \tag{21.45}$$

and

$$F_1 = \frac{B_1 B_2 (Z_2 - Z_1)}{B_2 Z_2 - B_1 Z_1} \tag{21.46}$$

Experimental and calculated data for charged solutes are shown in Figure 21.6 [30].

C. DONNAN DIALYSIS

1. Principle and Applications

Donnan dialysis is an economical, simple technological, and energy-saving process; it is not applied mainly in industry because of its slow kinetic [41–45]. This process is applied to problems such as enrichment of trace levels of ions, metal separation, water softening [46–50], and in biomedical applications. In this last case, the most known application is haemodialysis (artificial kidneys) [51,52].

This process consists in exchanging counterions of solutions separated by a membrane. Thus the Donnan dialysis is presented as an ion counter-transport whose driving ion is that of which the concentrations difference between compartments is highest. The dialysis phenomenon is characterized by the solute (charged or uncharged) and the type of membrane used [charged (ion-exchange) or neutral].

When the dialysis phenomenon occurs in a neutral membrane, the membrane acts as a selective barrier [53]: it is the steric exclusion. In the case of charged membrane (ion-exchange membrane, IEM), the membrane selectivity is due to the fixed ionic groups. Three types of IEM can be distinguished [54]: cation-exchange membrane (acidic fixed ionic groups); anion-exchange membrane (basic fixed ionic groups); amphoteric membrane (uniform distribution within the membrane of fixed acidic groups intermingled with the fixed basic groups).

For IEM separating two electrolytes solutions, the electrical potential results in ions transfer through the membrane till electrolytes reach their equilibrium in the both side of the membrane.

2. Mass Transfer Modeling

The Donnan dialysis is based on Donnan equilibrium. This ionic transfer is generally described by Nernst–Planck's equation [49,55–57]:

$$J_i = -\frac{\bar{D}_i \bar{C}_i}{RT} \left(\frac{RT}{\bar{C}_i} \frac{d\bar{C}_i}{dx} + \theta_i \frac{dP}{dx} + \Im Z_i \frac{dE}{dx} \right) + \bar{C}_i u_c \tag{21.47}$$

The mass transfer modeling through an IEM also implies to know the properties of this membrane, namely, its electrochemical properties and its equilibrium state. Dialysis of model systems containing two salts ($NaNO_3$ and $NaCl$) is used to describe this process, although, mostly components can be encountered in real solutions to be treated.

a. Membrane Equilibrium

Ion-exchange equilibrium is characterized by the ion-exchange isotherm and some quantities such as the separation factor and the selectivity coefficient.

Ion-exchange isotherm: This isotherm shows the ionic composition of the ion-change membrane as a function of the experimental conditions (concentration, temperature, membrane, etc.).

According to the Donnan equilibrium [58], ions of the same sign as the fixed ionic groups (C_4) are excluded from the membrane. At equilibrium, the water distribution between the aqueous solution and the IEM is:

$$\Pi\theta_3 = -RT \ln \frac{\bar{a}_3}{a_3} \tag{21.48}$$

Moreover, the Donnan potential can then be written for a given ion as follows:

$$EZ_i \Im + \Pi\theta_i = RT \ln \frac{\bar{a}_i}{a_i} \tag{21.49}$$

The water ($\bar{C}_3$) and the counterions ($\bar{C}_i$) concentration in the membrane can be expressed as follows:

$$\bar{C}_3 = \frac{C_3 \gamma_3}{\bar{\gamma}_3} \exp\left(\frac{-\Pi\theta_3}{RT}\right) \tag{21.50}$$

$$\bar{C}_i = \frac{C_i \gamma_i}{\bar{\gamma}_i} \exp\left(\frac{-(\Pi\theta_i + EZ_i\Im)}{RT}\right) \tag{21.51}$$

At equilibrium, water concentration in ion-exchange membranes ($\bar{C}_3$) and swelling pressure (Π) are constant [59].

Separation factor and selectivity coefficient: The selectivity of the IEM for one counterion is often expressed by the separation factor K_1^2 or by the selectivity coefficient α_1^2.

$$K_1^2 = \frac{\bar{y}_2 y_1}{\bar{y}_1 y_2} \tag{21.52}$$

$$\alpha_1^2 = \frac{\bar{y}_2^{|Z_2|} y_1^{|Z_1|}}{\bar{y}_1^{|Z_1|} y_2^{|Z_2|}} \tag{21.53}$$

In the case of nitrate and chloride ions (species 1 and 2), $Z_1 = Z_2 = -1$, thus $K_1^2 = \alpha_1^2$. The Equation (21.53) becomes:

$$\bar{y}_1 = \frac{\alpha_1^2 y_1}{1 + (\alpha_1^2 - 1)y_1} \tag{21.54}$$

The selectivity coefficient can be obtained by fitting the experimental data with Equation (21.54) (Figure 21.7).

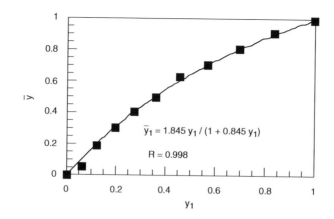

FIGURE 21.7 Nitrate equivalent ionic fraction in the membrane versus nitrate equivalent ionic fraction in the solution (ACS Tokuyama Soda membrane, electrolyte temperature 25°C). (From Nwal Amang, D., Alexandrova, S., and Schaetzel, P., *Desalination*, 159, 237–271, 2003. With permission.)

b. Electrical Conductivity

The electrical conductivity of an IEM is essentially due to the concentrations and motilities of fixed charges of the membrane. The electrical conductivity due to the counterion mobility [60] is directly connected with the diffusion coefficient [61]. During the IEM electrical conductivity experiments, the gradient of concentration and pressure are negligible [7], the Nernst–Planck's equation for two counterions (1 and 2) can be written as:

$$J_{1,2} = -\frac{Z_{1,2}\bar{D}_{1,2}\bar{C}_{1,2}\lambda\Im}{RT}\frac{dE}{dx}$$

(21.55)

Under the conditions of electroneutrality (Equation 21.39) and no current (Equation 21.40), the electrical conductivity λ is:

$$\lambda = \frac{\Im^2\bar{C}_4}{RT}\left[\bar{y}_1\left(\bar{D}_1 - \bar{D}_2\right) + \bar{D}_2\right]$$

(21.56)

The Equation (21.56) leads to a linear variation of the membrane electrical conductivity as a function of the equivalent ionic fraction of the nitrate ion in the membrane (Figure 21.8) [62].

c. Flux of Counterions

Ion-exchange membranes, while acting as a barrier for co-ion, are permeable for counterions. Thus an exchange (counter transport) of counterions between the solutions "a" and "b" is observed (Figure 21.9).

In this case, Nernst–Planck's equations for the two counterions can be written as follows:

$$J_1 = -\bar{D}_1\left(\frac{d\bar{C}_1}{dx} + \frac{\Im\bar{C}_1Z_1}{RT}\frac{dE}{dx}\right)$$

(21.57)

$$J_2 = -\bar{D}_2\left(\frac{d\bar{C}_2}{dx} + \frac{\Im\bar{C}_2Z_2}{RT}\frac{dE}{dx}\right)$$

(21.58)

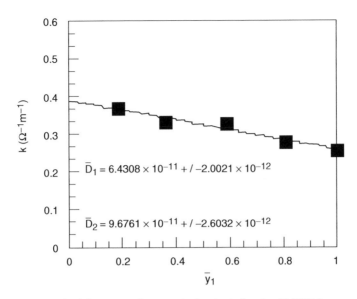

FIGURE 21.8 Electrical conductivity versus nitrate equivalent ionic fraction (ACS Tokuyama Soda membrane, electrolyte temperature 25°C). Diffusion coefficients are obtained using the electrical conductivity data fitting. (From Nwal Amang, D., Alexandrova, S., and Schaetzel, P., *Electrochim. Acta*, 48, 2563–2569, 2003. With permission.)

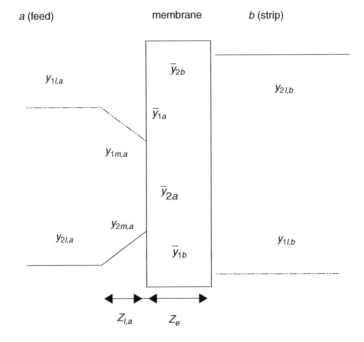

FIGURE 21.9 Concentration profiles in Donnan dialysis with concentration polarisation in the feed side of the membrane. (From Nwal Amang, D., Alexandrova, S., and Schaetzel, P., *Chem. Eng. J.*, 99, 69–76, 2004. With permission.)

Considering the case of monovalent counterions, the following assumptions are made [63]:

- Permselectivity of the membrane (exclusion of the co-ion Na^+);
- Electroneutrality in the membrane;
- Diffusion coefficients of counterions 1 and 2 in the solution are close [64];
- The equivalent ionic fraction of counterion 1 in the strip side of the membrane is close to zero ($y_{1,b} \approx 0$).

Under the assumptions, the Equation (21.57) and Equation (21.58) become:

$$J_1 = k_m \bar{C}_4 \ln\left[\frac{1 + \bar{y}_{1,a}(\bar{\beta} - 1)}{1 + \bar{y}_{1,b}(\bar{\beta} - 1)}\right] \tag{21.59}$$

with

$$\bar{y}_{1,a} = \frac{\alpha_1^2 \bar{y}_{1m,a}}{(\alpha_1^2 - 1)\bar{y}_{1m,a} + 1} [7] \tag{21.60}$$

and

$$k_m = \frac{\bar{D}_1}{L(\bar{\beta} - 1)} \tag{21.61}$$

In batch dialysis the exhaustion equation of the counterion 1 can be written as:

$$\frac{dy_{1,a}}{dt} = \frac{J_1 S}{V C_{5,a}} \tag{21.62}$$

Thus, the mass transfer coefficient can be calculated with parameter setting of the curves representing the equivalent ionic fraction of the counterion 1 in the feed according to time (Equation 21.63).

$$t = -\frac{V C_{5,a}}{S}\left[\left(\frac{1}{k_{1,a} C_{5,a}} + \frac{1}{\alpha_1^2(1 - \bar{\beta})k_m \bar{C}_4}\right)\ln y_{1,a} + \frac{\alpha_1^2 \bar{\beta} - 1}{\alpha_1^2(1 - \bar{\beta})k_m \bar{C}_4}(y_{1,a} - 1)\right] \tag{21.63}$$

The shape of the curve obtained using Equation (21.63) is represented in Figure 21.10 [65].

D. Electrodialysis

1. Principles and Applications

Although, electrodialysis was started as a modification of dialysis [54], the two phenomena are quite different. The driving force is an electrical potential due to an external electrical field [66–68]. Whereas in dialysis, the concentration gradient may diminish gradually as a result of the mass transport, in electrodialysis, the external electrical potential can easily be maintained until the desired degree of separation is achieved.

This process is widely used for the production of drinking and process water from rivers and sea water, and the treatment of industrial effluents which contain heavy metals [66].

In a typical electrodialysis cell, a series of anion- and cation-exchange membranes are arranged in an alternating pattern between an anode and a cathode to form individual cells. When an electrical potential is applied between the two electrodes, the positively charge cations move to the cathode, passing through the negatively charged cation-exchange membrane. On the other hand, the anions move to the anode, passing through the anion-exchange membrane and retained by

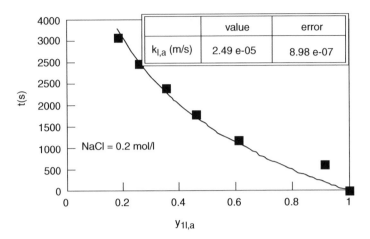

FIGURE 21.10 Experimental data fitting with Equation (21.63) for the calculation of the film mass transfer coefficient in the feed (From Nwal Amang, D., Ph.D. Thesis, University of Caen, France, 2000).

the cation-exchange membrane. At the end, ion concentration increases in alternate compartments with a simultaneous decrease of ions in other compartments [66].

2. Mass Transfer Modeling

The molar flux through the differential element of dilute solution compartment in a membrane cell pair in terms of mass balance and current density can be written as follows:

$$J_{\mathrm{m}} = -Q_{\mathrm{m}} \frac{\mathrm{d}C}{\mathrm{d}A_m} = \eta \frac{i}{\Im} = \frac{\eta}{\Im} \frac{\mathrm{d}I}{\mathrm{d}A_m} \tag{21.64}$$

With the assumption that η, J_{m}, and $Q_{\mathrm{m}} = Q/N'$ are constant in the cell compartment, we get:

$$\frac{\mathrm{d}I}{\mathrm{d}A_{\mathrm{m}}} = \frac{I}{A_{\mathrm{m}}}$$

Thus

$$V_{\mathrm{a}} \frac{\mathrm{d}C_{\mathrm{a}}}{\mathrm{d}t} = -\eta \frac{N'I}{\Im} \tag{21.65}$$

With

$$\eta = \xi Q^\alpha, \; I = \frac{E}{N'R} \; \text{and} \; R = \frac{p}{C} + q$$

By substituting η and I, the Equation (21.65) leads to:

$$-\frac{\Im V_{\mathrm{a}}}{E \xi Q^\alpha} \left(\frac{p}{C} + q \right) \mathrm{d}C = \mathrm{d}t \tag{21.66}$$

Integrating this equation leads to the following expression:

$$\frac{\Im V_{\mathrm{a}}}{EQ^\alpha} \left| v \ln \frac{C_0}{C} + v(C_0 - C) \right| = t \tag{21.67}$$

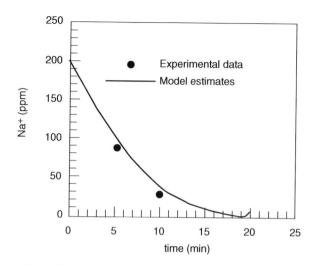

FIGURE 21.11 Fitness of model on concentration (ppm Na^+) versus time (min) (From Demircioglu, M., Kabay, N., Ersö, E., Kurucaovali, I., Safak, Ç., and Gizli, N., *Desalination*, 136, 317–323, 2001).

The shape of the curve (Figure 21.11) describing electrodialysis is given by Equation (21.67).

E. PERVAPORATION

1. Principles and Applications

Except membrane distillation, the pervaporation (term firstly used in 1958 by Binning et al. [69,70]) process is the only one with phase change: the permeated product is removed from liquid to vapor state.

Feed components permeate through the membrane and evaporate as a result of the partial pressure on the permeate-side being lower than saturation vapor pressure. This driving force is generally controlled by applying a vacuum in the downstream (permeate) side of the membrane (Figure 21.12). As an alternative, an inert carrier gas such as water vapor or air can probably be used to lower the partial pressure of the permeated components.

Pervaporation is an expensive process owing to the phase change, which is inextricably linked with the species transport across the membrane. However, because of its high selectivity, pervaporation is interesting when conventional separation processes (e.g., distillation) either fail or result in high specific energy consumption and high investment costs.

Some classical examples of application are the separation of azeotropique mixture [71], the isomeric mixture separation, the dehydration of organic mixtures, and fruit juice concentration, alcohol extraction from wines and beers, aroma extraction [72], etc.

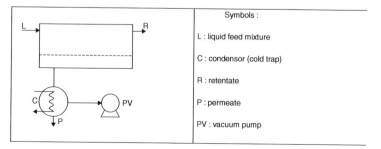

FIGURE 21.12 Usual pervaporation setup.

2. Mass Transfer Modeling

The commonly used mass transfer in pervaporation is the solution-diffusion model — a transfer occurs in three steps (Figure 21.13).

1. Sorption of the solvent molecules in polymer at the liquid–membrane interface. The membrane swells and behaves like a thin selective solvent layer. This step is generally fast. The sorption selectivity is mainly responsible of the efficiency of the membrane.
2. Molecular diffusion through the membrane due to chemical potential gradient. This step is generally the limiting one of the transfer.
3. Desorption on downstream-side of the membrane described by the traditional models of sorption. The downstream pressure is very low and this step is fast. As the pressure decreases abruptly in the membrane to become equal to that of the downstream face, the pressure gradient through the film can be neglected. Usually, the transport can be considered as an isobar and an isotherm.

According to the pervaporation implementation, various parameters can influence the desorption of the solutes having pervaporated through the membrane: the downstream pressure [73,74], the distance between the membrane and the cold trap [75], and the inert carrier gas flow rate [76].

The expression of molar flow is:

$$J_i = -D_i \frac{dC_i}{dx} - \frac{D_i C_i}{\gamma_i} \frac{d\gamma_i}{dx} \tag{21.68}$$

The first term corresponds to the Fick's first law, and the second describes the deviation from ideal behavior during transport.

The concentration polarization is a phenomenon, which influences the pervaporation efficiency. The boundary-layer resistance to the solute transport was reported by Psaume et al. [9] for the recovery of trichloroethylene in aqueous solutions. They showed that under certain operating conditions, transport through the membrane was determined by the hydrodynamic conditions in the feed-side, and thus the resistance of the membrane becomes relatively negligible. Colman et al. [77] showed that resistance to the transfer of the boundary layer is a limiting factor of the dehydration of isopropanol by pervaporation. Various others studies [16,78,79] highlighted the importance of the hydrodynamics on resistance to the solute mass transport in pervaporation.

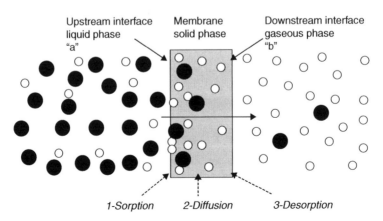

FIGURE 21.13 Mass transfer of two species in pervaporation process.

For a pervaporation cell used, the hydrodynamics can be controlled by respecting the proportions of a perfectly stirred reactor. The diameter of the cell with baffled double-envelope is of 0.07 m. The membrane is installed at the bottom of the cell. This assembly also allows the temperature control of the membrane. Thus, it is possible to suppose that the experiments are carried out without polarization of temperature.

In this case, to describe the influence of the hydrodynamics in the cell and of the operating conditions on the mass transfer through the membrane, the following correlation can be used [80]:

$$Sh = 0.003Re^{1.25}Sc^{1/3} \tag{21.69}$$

At the downstream interface, as the molecular diffusion is significant in a gaseous medium with very low pressure ($\sim$ a few mbar), the polarization of concentration is negligible (Figure 21.14) [80].

Generally, it is allowed that pervaporate transport is isothermal, but there is however a thermal effect due to evaporation on the permeate-side. The temperature is one of parameters influencing the transfer: acceleration of the diffusion in the film boundary layer and the membrane, modification of sorption on the membrane (generally, lowering), and improvement of desorption in the downstream interface.

The mass flow through the membrane is strongly dependent on the temperature. Thus, for a pure compound, flow can be described with Arrhenius-type law taking into account the temperature polarization:

$$j = j_{ref}\exp\left[\frac{-E_a}{R}\left(\frac{1}{T_m} - \frac{1}{T_{ref}}\right)\right] \tag{21.70}$$

The mass flow of species i is connected to the heat flux density Q_{proc}, which is a function of the heat transfer coefficient in film and of stirring rate N in the reactor (Figure 21.15).

To describe the influence of the hydrodynamics of the pervaporation cell and the operating conditions on the heat transfer through the membrane in a standard reactor, the following correlation can be used [80]:

$$Nu = 0.003Re^{1.25}Pr^{1/3} \tag{21.71}$$

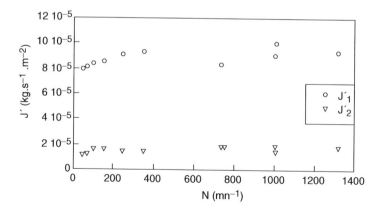

FIGURE 21.14 Ethanol and water flows according to the stirring rate [water–ethanol feed (19.5% mass of water) at 60°C, membrane PVA-PAA (90-10) of thickness of 25.2 μm], membrane temperature controlled at 60°C (Rollet, V., Transfert de matière en pervaporation, application à la déshydratation de mélanges complexes, Ph.D. Thesis, University of Technology of Compiègne, France, 2000.).

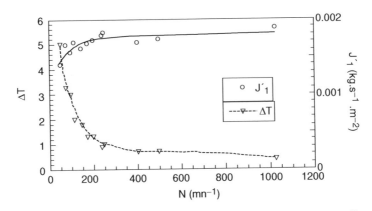

FIGURE 21.15 Water flow and difference in upstream–downstream temperatures, according to stirring speed N [feed of pure water at 60°C, membrane PVA-PAA (90-10) of thickness of 48 μm]. The water flow is predicted using Equation 21.88 with $E_a = 45.8 \times 10^6$ J mol^{-1} and $T_0 = 333.15$ K (From Rollet, V., Ph.D. Thesis, University of Technology of Compiègne, France, 2000.).

3. Simultaneous Temperature and Concentration Polarizations

The mass transfer through the membrane, during pervaporation with simultaneous polarizations of temperature and concentration, can be described using this system of three Equation (21.72)– Equation (21.74).

$$j = j_0 \exp\left[\frac{-E_a}{R}\left(\frac{1}{T_m} - \frac{1}{T_{ref}}\right)\right] = \frac{h_f}{\Delta H_{vap}}(T_b - T_m) \qquad (21.72)$$

$$j_1 = \rho_b k_f\left(x'_{1b} - x'_{1m}\right) \qquad (21.73)$$

$$j_1 = K_m x'_{1m} \exp\left[\frac{-E_a}{R}\left(\frac{1}{T_m} - \frac{1}{T_{ref}}\right)\right] \qquad (21.74)$$

In the case of water–ethanol mixture at 19.5% in mass of water at 60°C, membrane PVA-PAA (90-10) of thickness of 25.2 μm, the various coefficients obtained are [80]:

$$E_a = 45.8 \times 10^6 \text{ J mol}^{-1} \qquad T_{ref} = 333.15 \text{ K} \qquad h_f = 0.596\, Re^{0.93} \text{ W m}^{-2} \text{K}^{-1}$$
$$\Delta H_{vap} = 2356 \times 10^3 \text{ J kg}^{-1} \quad k_f = 1.05 \times 10^{-8}\, Re^{0.93} \quad K_m = 4.94 \times 10^{-4} \text{ kg s}^{-1} \text{ m}^{-2}$$

The appearance of boundary layers in pervaporation could be highlighted. These phenomena of polarizations of temperature and concentration influence the transfer during pervaporation. The amplitude of these two polarizations varies according to operational hydrodynamic conditions; the boundary layers of polarization decrease with the increase stirring rate.

The heat and the mass transfer at the bottom of a flat-bottomed cell, with standard dimensions of a stirred reactor, can be described, respectively, by a Nusselt- and Sherwood-like correlation.

IV. CONCLUSION

In this chapter, we have presented the more used theories to explain the mass transfer mechanism in a polymer solid (porous or tight, charged or not) membranes. It is clear that the membrane structure and the nature of the solution forms a system with complex interactions between the solutes, solvent, driving force, and the polymeric membrane. Often many basic mass transfer mechanisms intervene in certain membrane separations, and the mass transfer modeling is not very easy, because we must take into account any interactions between the components of the system.

NOTATIONS

a	solution activity in the aqueous solution (mol m^{-3})
$\bar{a}$	solution activity in the polymer (membrane) (mol m^{-3})
A	pure water (solvent) permeability constant (s m^{-1})
A_m	effective area of an exchange membrane (m^2)
B_i	overall permeability of component i (m s^{-1})
c	mass concentration (kg m^{-3})
C	molar concentration (mol m^{-3})
$\bar{C}$	molar concentration in the membrane (mol m^{-3})
C_m	molar concentration of solute at membrane interface (mol m^{-3})
D	diffusion coefficient in the solution (m^2 s^{-1})
D^*	specific (membrane or solute) diffusion coefficient (m^2 s^{-1})
$\bar{D}$	diffusion coefficient in the membrane (m^2 s^{-1})
d_p	specific diameter m
ΔH_{vap}	enthalpy of vaporization (J kg^{-1})
ΔP_B	pressure drop in the feed side Pa
E	electrical potential V
E_a	activation energy (J mol^{-1})
F	coefficient defined by Equation (21.35) and Equation (21.36) –
F_1	extended Nernst–Planck's equation parameter (m s^{-1})
$\Im$	Faraday constant (C mol^{-1})
G	pore geometric factor m
h	heat transfer coefficient (W m^{-2} K^{-1})
$(1/h)$	membrane area per unit volume of fiber bundle (m^{-1})
i	current density (A m^2)
I	current intensity A
j	mass flux (kg m^{-2} s^{-1})
J	molar flux (mol m^{-2} s^{-1})
J_v	volumic flux (m s^{-1})
k	mass transfer coefficient (m s^{-1})
K'	permeability m^2
K_m	global membrane mass transfer coefficient (kg s^{-1} m^{-2})
$k_{l,a}$	film mass transfer coefficient (m s^{-1})
k_m	membrane mass transfer coefficient (m s^{-1})
K_1^2	separation factor –
ℓ	spacing between to plates (Equation 21.29) m
L	membrane thickness or pore length m
L'	membrane length m
n	number of species –
n'	number of pore –
N'	number of cell section –
N	stirring rate (s^{-1})
Nu	Nusselt number –
p	constant in Equation (21.66) –
P	pressure Pa
P_i	permeability of component i (m^2 s^{-1})
Pr	Prandtl number –
$\bar{P}_s$	solute permeability (m^2 s^{-1})
P_s	overall solute permeability (m s^{-1})

P^*	overall ion permeability (m s^{-1})
q	constant in Equation (21.66) –
Q	volumetric flow rate (m^3 s^{-1})
r	fiber radius or pore radius m
R	universal gas constant – ideal gas constant (J mol^{-1} K^{-1})
R	fiber bundle radius m
Re	Reynold's number ($2r_0 u_s \rho / \mu$) –
R_{mass}	mass transfer resistance s m^{-1}
R_{obs}	solute observed retention (solute rejection rate) –
R_{real}	real retention of solute –
$R_{\infty,1}$	real retention when $J_w \rightarrow \infty$ defined by Equation 21.45 –
S	membrane area m^2
S'	membrane specific surface (m^2 m^{-3})
S''	surface area of pore m^2
Sc	Schmidt number ($\mu / \rho D$) –
Sh	Sherwood number ($2kr_0/D$) –
t	time s
T	temperature K
u_c	convective flow rate (m s^{-1})
u_s	superficial velocity for an empty bed (m s^{-1})
v	velocity (m s^{-1})
V	volume m^3
v'	permeation velocity (m s^{-1})
W	membrane width m
x	distance along the membrane module axis m
x	geometrical coordinate m
x'	mass fraction –
y	ionic fraction –
Y	conversion rate –
Z	valence of the ion –

Greek symbols

α	constant in Equation (21.66) –
α_1^2	selectivity coefficient –
β	diffusion model parameter (s m^{-1})
γ	activity coefficient –
$\bar{\gamma}$	activity coefficient in the membrane –
ε	void fraction – porosity –
ε'	pore area per unit membrane area –
η	current efficiency
θ	molar volume (m^3 mol^{-1})
λ	electrical conductivity (Ω^{-1} m^{-1})
μ	water viscosity (Pa s)
v	constant in Equation (21.67) –
ξ	constant in Equation (21.66) (s m^{-3})$^{\alpha}$
π'	osmotic pressure Pa
Π	swelling pressure of the polymer Pa
ρ	density (kg m^{-3})
σ	reflection coefficient –

τ	coefficient characterizing the pore –
υ	constant in Equation (21.67) –
ϕ	concentration polarization factor (Equation 21.15) –
ϕ	global partition coefficient –
ϕ_E	charges effect partition coefficient –
ϕ_F	steric partition coefficient –
Φ_D	Donnan potential V
ψ	electrostatic potential V

Subscripts

a	feed solution (retentate), upstream
b	strip solution (permeate)
B	bulk: remaining salt in the feed tank
f	feed, film
i	species i (component i); inside/inner
l	liquid
m	membrane, at the membrane interface
o	outside/outer
p	permeate, downstream
ref	at a temperature of reference
s	solute
w	water
0	initial, initially in the feed
1	counterions (e.g., nitrate ion)
2	counterions (e.g., chloride ion)
3	water
4	fixed ionic groups
5	coion

REFERENCES

1. Lonsdale, H.K., Merten, U., and Riley, R.L., TITRE, *J. Polym. Sci.*, 9, 1341, 1965.
2. Merten, U., *Desalination by Reverse Osmosis*, MIT Press Cambridge, 1966.
3. Hwang, S.-T. and Kammermeyer, K., *Membranes in Separations*, John Wiley & Sons, New York, 1975.
4. Sourirajan, S., *Reverse Osmosis*, Academic Press, New York, 1970.
5. Darcy, H.P.G., Les Fontainer Publiques de la ville de Dijon, Victor Dalmont, Paris, 1856.
6. Kozeny, J., *Sitzber. Acad. Wis. Wien. Math-Naturw.*, 136, 271, 1927.
7. Helfferich, F., *Ion Exchange*, McGraw-Hill, New York, 1962.
8. Knudsen, M., *Ann. Phys*, 28, 75, 1909.
9. Psaume, R., Aptel, Ph., Aurelle, Y., Mora, J.C., and Bersillon, J.L., Pervaporation: importance of concentration polarization in the extraction of trace organics from water, *J. Membr. Sci.*, 36, 373–384, 1988.
10. Greenlaw, F.W., Prince, W.D., Shelden, R.A., and Thompson, E.V., Dependance of diffusive permeation rates on upstream and downstream pressures. I. Single component permeant, *J. Membr. Sci.* 2, 141–151, 1977.
11. Greenlaw, F.W., Shelden, R.A., and Thompson, E.V., Dependence of diffusive permeation rates on upstream and downstream pressures. II. Two component permeant, *J. Membr. Sci.*, 3, 333–348, 1977.

12. Böddeker, K.W., Bengtson, G., and Böde, E., Pervaporation of low volatility aromatics from water, *J. Membr. Sci.*, 53, 143–158, 1990.
13. Long, R.B., Liquid permeation through plastic films, *Ind. Eng. Chem. Fundam.*, 4, 445–451, 1965.
14. Baker, R.W., Wijmans, J.G., Athayde, A.L., Daniels, R., Ly, J.H., and Le, M., The effect of concentration polarization on the separation of volatile organic compounds from water by pervaporation, *J. Membr. Sci.*, 137 (1–2), 159–172, 1996.
15. Wijmans, J.G., Athayde, A.L., Daniels, R., Ly, J.H., Kamaruddin, H.D., and Pinnau, I., The role of boundary layers in the removal of volatile organic compounds from water by pervaporation, *J. Membr. Sci.*, 109 (1), 135–146, 1996.
16. Mi, L. and Hwang, S., Correlation of concentration polarization and hydrodynamic parameters in hollow fiber modules, *J. Membr. Sci.*, 159 (1–2), 143–165, 1999.
17. Gooding, C.H., Modelling of some relatively simple concepts on the heat transfer aspects of pervaporation, in Bakisk, R., Ed., Proceeding of the First International Conference Pervaporation Processes in the Chemical Industry, Bakish Materials Corporation, Englewood, NJ, 1986, pp. 171–181.
18. Van Gauwbergen, D. and Baeyens, J., Modelling reverse osmosis by irreversible thermodynamics, *Separ. Purif. Technol.*, 13, 117–128, 1998.
19. Jonson, G., Overview of theories for water and solute transport in UF/RO membrane, *Desalination.* 35, 21–38, 1980.
20. Sourirajan, S. and Matsura, T., *RO/Ultrafiltration Principles*, NRC, Ottawa, Canada, 1985.
21. Al-Bastaki, N.M. and Abbas, A., Improving the permeate flux by unsteady operation of RO desalination unit, *Desalination*, 123, 173–176, 1999.
22. Aly, S.E., Combined RO/VC desalination system, *Desalination*, 58, 85–97, 1986.
23. Treybal, R.E., *Mass-transfer operations*, 3rd ed., McGraw-Hill, New York, 1980, p. 75.
24. Sekino, M., Precise analytical model of hollow fiber reverse osmosis modules, *J. Membr. Sci.*, 85, 241–252, 1993.
25. Hermans, J.J., Physical aspect governing the design of hollow fiber modules, *Desalination*, 26, 45–62, 1978.
26. Schaep, J., Van der Bruggen, B., Vandecasteele, C., and Wilms, D., Influence of ion size and charge in nanofiltration, *Separ. Purif. Technol.*, 14, 155–162, 1998.
27. Peteers, J.M.M., Boom, J.P., Mulder, M.H.V., and Strathmann, H., Retention measurements of nanofiltration membranes with electrolyte solution, *J. Membr. Sci.*, 145, 199–209, 1998.
28. Nyström, M., Kaipia, L., and Luque, S., Fouling and retention of nanofiltration membranes, *J. Membr. Sci.*, 98, 249–262, 1995.
29. Boucard, F., Contribution à la caractérisation des mécanismes de transport en nanofiltration: Expériences et Modèles, Ph.D. Thesis, University of Caen, France, 2000.
30. Garcia, F., Ciceron, D., Alexandrova, S., and Saboni, A., Mass transfer modelling in nanofiltration membrane: Comparison between two models, 16th International Congress of Chemical and Process Proceedings of the Engineering CHISA 2004, August, 22–26, 2004, Prague, Czech Republic.
31. Tsuru, T., Nakao, S., and Kimura, S., Calculation of ion rejection by extend Nernst–Planck equation with charged reverse osmosis membranes for single and mixed electrolyte solutions, *J. Chem. Eng. Jpn.*, 24, 511–517, 1991.
32. Schaep, J., Vandecasteele, C., Mohammad, A.W., and Bowen, W.R., Modelling the retention of ionic components for different nanofiltration membranes, *Separ. Purif. Technol.*, 22–23, 169–179, 2001.
33. Labbez, C., Fievet, P., Szymczyk, A., Vidonne, A., Foissy, A., and Pagetti, J., Retention of mineral salts by a polyamide nanofiltration membrane, *Separ. Purif. Technol.*, 30, 47–55, 2003.
34. Paugam, L., Taha, S., Dorange, G., Jaouen, P., and Quemeneur, F., Mechanism of nitrates ions transfer in nanofiltration depending on pressure, pH, concentration and medium composition, *J. Membr. Sci.*, 231, 37–46, 2004.
35. Pontie, M., Buisson, H., Diawara, C.K., and Essis-Tome, H., Studies of halide ions mass transfer in nanofiltration — application to selective defluorination of brackish drinking water, *Desalination*, 157, 127–134, 2003.
36. Schaep, J., Vandecasteele, C., Mohammad, A.W., and Bowen, W.R., Analysis of the salt retention of nanofiltration membranes using the Donnan-steric partitioning pore model, *Separ. Purif. Technol.*, 34 (15), 3009–3030, 1999.

37. Spiegler, K.S. and Kedem, O., Thermodynamic of hyperfiltration (reverse osmosis) criteria for efficient membranes, *Desalination*, 1, 311–326, 1966.
38. Afonso, M.D. and de Pinho, M.N., Transport of $MgSO_4$, $MgCl_2$, and Na_2SO_4 across an amphoteric nanofiltration membrane, *J. Membr. Sci.*, 179, 137–154, 2000.
39. Timmer, J.M.K., van der Horst, H.C., and Robbertsen, T., Transport of lactic acid through reverse osmosis and nanofiltration membranes, *J. Membr. Sci.*, 85, 205–216, 1993.
40. van der Host, H.C., Timmer, J.M.K., Robbertsen, T., and Leenders, J., Use of nanofiltration for concentration and demineralization in the dairy industry: model for mass transport, *J. Membr. Sci.*, 104, 205–218, 1995.
41. Wallace, R.M., Concentration and separation of ions by Donnan membrane equilibrium, *Ind. Eng. Chem. Process. Des. Dev.*, 6, 423–431, 1967.
42. Kim, B.M., Donnan dialysis for removal of chromates and cyanides, *AIChE Symp. Ser.*, 76, 184–192, 1980.
43. Wodzki, R., Sionkowski, G., and Pieta, T.H., Recovery of metal ions form electroplanting rinse solutions using the Donnan dialysis technique, *Pol. J. Environ. Stud.*, 5, 45–50, 1996.
44. Dieye, A., Larchet, C., Auclair B., and Diop, M., Elimination des fluorures par dialyse ionique croisé, *Eur. Polym. J.*, 34, 67–75, 1995.
45. Sionkowski, G. and Wodzki R., Recovery and concentration of metal ions, I. Donnan dialysis, *Separ. Sci. Technol.*, 30, 805–820, 1995.
46. Cengeloglu, Y., Kir, E., Ersoz, M., Buyukerkek, T., and Gezgin, K., Recovery and concentration of metals from red mud by Donnan dialysis, *Coll. Surf. A: Physicochem. Eng. Asp.*, 223, 95–101, 2003.
47. Palaty, Z. and Zakova, A., Transport of nitric acid through the anion-exchange membrane Neosepta-AFN, *Desalination*, 160, 51–66, 2004.
48. Velizarov, S., Reis, M.A., and Crespo, J.G., Removal of trace monovalent inorganic pollutants in an ion exchange membrane bioreactor: analysis of transport rate in a denitrification process, *J. Membr. Sci.*, 217, 269–284, 2003.
49. Ruiz, T., Persin, F., Hichour, M., and Sandeaux, J., Modelisation of fluoride removal in Donnan dialysis, *J. Memb. Sci.*, 212, 113–121, 2003.
50. Tongwen, X. and Weihua, Y., Industrial recovery of mixed acid ($HF + HNO_3$) from the titanium spent leaching solutions by diffusion dialysis with a new series of anion exchange membranes, *J. Membr. Sci.*, 220, 89–95, 2003.
51. Takagi, R. and Nakagaki, M., Ionic dialysis through amphoteric membranes, *Separ. Purif. Technol.*, 32, 65–71, 2003.
52. Raff, M., Weldch, M., Göhl, H., Hildwein, H., Storr, M., and Wittner, B., Advanced modelling of highflux hemodialysis, *J. Membr. Sci.*, 216, 1–11, 2003.
53. Donnan, F.G., Theory of membrane equilibria and membrane potentials in the presence of non-dialysing electrolytes. A contribution to physical-chemical physiology. *J. Membr. Sci.*, 100, 45–55, 1995.
54. Hwang, S.-T. and Kammermeyer, K., Chapter XIV: *Membrane and their Preparation. Membranes in Separations*, John Wiley & Sons, New York, pp. 421–456, 1975.
55. Miyoshi, H., Yamagami, M., and Kataoka, T., Characteristic coefficients for equilibrium between solution and Neosepta or Selemion cation exchange membranes. *J. Chem. Eng.*, 37, 120–124, 1992.
56. Miyoshi, H., Yamagami, M., and Kataoka, T., Characteristic coefficients of cation exchange membranes for bivalent cations in equilibrium between the membrane and solution. *J. Chem. Eng.*, 39, 595–598, 1994.
57. Dong-Syau, J., Chien-Cheng, H., and Fuan-Nan, T., Combined film and membrane diffusion controlled transport of ion through charge membrane. *J. Membr. Sci.*, 90, 109–115, 1994.
58. Donnan, F.G., The theory of membrane equilibria, *Chem. Rev.*, 1, 73–90, 1925.
59. Nwal Amang, D., Alexandrova, S., and Schaetzel, P., Bi-ionic chloride–nitrate equilibrium in a commercial ion-exchange membrane, *Desalination*, 159, 237–271, 2003.
60. Pourcelly, G., Psistat, P., Chapotot, A., Gavach, C., and Nikonenko, V., Self diffusion and conductivity in Nafion membranes in contact with $NaCl + CaCl_2$ solutions, *J. Membr. Sci.*, 110, 69–78, 1996.

61. Nikolaev, N.I., Diffusion in ion exchange membrane, Khimiya, Moscow, 1980.

62. Nwal Amang, D., Alexandrova, S., and Schaetzel, P., The determination of diffusion coefficients of counter ion in an ion exchange membrane using electrical conductivity measurement, *Electrochim. Acta*, 48, 2563–2569, 2003.

63. Nwal Amang, D., Alexandrova, S., and Schaetzel, P., Mass transfer characterization of Donnan dialysis in a bi-ionic chloride–nitrate system, *Chem. Eng. J.*, 99, 69–76, 2004.

64. Robinson, R.A. and Stokes, R.H., *Electrolyte solutions: The measurement and interpretation of conductance, chemical potential and diffusion in solutions of simple electrolytes*, 2nd ed. Butterworths, London, 1959.

65. Nwal Amang, D., Ph.D. Thesis, University of Caen, France, 2000.

66. Demircioglu, M., Kabay, N., Ersö, E., Kurucaovali, I., Safak, Ç., and Gizli, N., Cost comparison and efficiency modelling in the electrodialysis of brine, *Desalination*, 136, 317–323, 2001.

67. Kraaijeveld, G., Sumberova, V., Kuindersma, S., and Wesselingh, H., Modelling electrodialysis using the Maxwell–Stefan description, *Chem. Eng. J.*, 57, 163–176, 1995.

68. Almadani, H.M.N., Water desalination by solar powered electrodialysis process, *Renew. Energy*, 28, 1915–1924, 2003.

69. Binning, R.C. and James, F.E., New separate by membrane separation, *Pet. Ref.*, 37, 214–218, 1958.

70. Binning, R.C., Lee, R.J., Jennings, J.F., and Martin, E.C., Separation of liquid mixtures by permeation, *Ind. Eng. Chem.*, 53, 45–50, 1961.

71. Van Hoof, V., Van den Abeele, L., Buekenhoudt, A., Dotremont, C., and Leysen, R., Economic comparison between azeotropic distillation and different hybrid systems combining distillation with pervaporation fort the dehydration of isopropanol, *Separ. Purif. Technol.*, 37, 33–49, 2004.

72. Willemsen, J.H.A., Dijkink, B.H., and Togtema, A., Organophilic pervaporation for aroma isolation–industrial and commercial prospects, *Membr. Technol.*, February, 5–10, 2004.

73. Ten, P.K. and Field, R.W., Organophilic pervaporation: an engineering science analysis of component transport and the classification of behaviour with reference to the effect of permeate pressure, *Chem. Eng. Sci.*, 55 (8), 1425–1445, 2000.

74. Baudot, A., Souchon, I., and Main, M., Total permeate pressure influence on the selectivity of the pervaporation of aroma compounds, *J. Membr. Sci.*, 158 (1–2), 167–185, 1999.

75. Beaumelle, D. and Marin, M., Effect of transfer in the vapor phase on the extraction by pervaporation through organophilic membranes: experimental analysis on model solutions and theoretical extrapolation, *Chem. Eng. Process.*, 33 (6), 449–458, 1994.

76. Nguyen, Q.T. and Nobe, K., Extraction of organic contaminants in aqueous solutions by pervaporation, *J. Membr. Sci.*, 30, 11–22, 1987.

77. Colman, D.A., Naylor, T.D., Pearce, G.K., and Whitby, R.D., Application of membrane pervaporation to dehydration of alcohols in chemicals and pharmaceuticals processing, Green, A., Ed., Proceedings of the Intl. Tech. on Mem. Sep. Proc., Brighton, UK, May, 1989, Springer, Berlin, pp. 1–17.

78. Hwang, S.-T. and Raghunath, B., Concentration polarization in the boundary layer of phenol pervaporation, Li, N.N., Ed., Proceedings of the Intl. Cong. on Mem. and Mem. Tech., Vol. 1, Chicago, USA, North American Membrane Society, 1990, pp. 322–324.

79. Mi, L., Vane, F.R., Alvarez, and Giroux, E.L., Reduction of concentration polarization in pervaporation using vibrating membrane module, *J. Membr. Sci.*, 153 (2), 233–241, 1999.

80. Rollet, V., Transfert de matière en pervaporation, application à la déshydratation de mélanges complexes, Ph.D. Thesis, University of Technology of Compiègne, France, 2000.

Multiphase Dispersed Systems

22 Gas–Flowing Solids–Fixed Bed Contactors

Aleksandar P. Duduković and Nikola M. Nikačević
University of Belgrade, Belgrade, Serbia and Montenegro

CONTENTS

I. INTRODUCTION

The idea of contacting gas and fine solids particles (flowing solids) inside a packed bed of other solids was patented more than 50 years ago [1]. The first realization was in France for heat recovery, but many other applications were under consideration in recent years. They include various separation processes, as well as catalytic chemical reactors with separation *in situ* and integrated processes with heterogeneous chemical reaction and simultaneous heat exchange in a bed of catalyst.

In this type of equipment, fine solid particles are introduced at the top of the column containing a packed bed, while gas can be introduced either at the bottom for countercurrent contacting or at the top when cocurrent operation is desired. In the literature, such columns are named "gas–flowing solids–fixed bed contactors," or "raining packed bed contactors," or "solids trickle flow contactors," while the titles in the Russian literature were more descriptive.

Gas–flowing solids–fixed bed contactors can be considered as two-phase or three-phase systems. In the first case, gas and flowing solids are the contacting phases, while the packing serves only to enable better contact between the flowing phases. In such a case, there are no limitations in geometry and design of packing elements. In the other case, the packing elements could be considered as an additional, third phase, as in heterogeneous catalytic reactors. In these cases, the geometry of packing elements is limited by the process requirements, that is, by the shape of catalyst particles.

II. FLUID DYNAMICS

First studies of countercurrent flow of gas and fine solid particles inside the packed bed were carried by Kaveckii and Plankovskii [2]. Most of the later studies were devoted to experimental evaluation of two basic design parameters: pressure drop and flowing solids holdup [3–18].

A. Dynamic and Static Holdup

Flowing solids holdup is defined as the volume fraction occupied by the flowing solids inside the packing. It is a common practice to separate this holdup into two parts: static and dynamic holdup. Static holdup represents the fraction of fine solids that are at rest on the packing elements, while dynamic holdup represents the flowing solids moving through the packing. Both are usually experimentally determined by simultaneous closing of entering flowing solids and gas flows. From the amount of flowing solids that drain out from the bed, the dynamic holdup is determined, while the static holdup can be determined from the amount of flowing solids retained by the packing, the procedure as reviewed later.

B. Preloading and Loading

Similar to gas–liquid systems, in gas–flowing solids–fixed bed countercurrent contactors, three regimes were observed: preloading, loading, and flooding. At low gas flow rates (preloading), gas essentially does not affect the flow of solids, and consequently, flowing solids dynamic holdup is nearly independent of gas flow rate. At higher gas flow rates (loading regime), gas slows down the flow of solids, and there is a strong interaction between the gas flow and flowing solids. In the loading regime, dynamic holdup increases with gas flow rate, and the pressure drop increases faster than in the preloading regime. In Figure 22.1, a typical behavior of dynamic holdup as a function of gas flow rate is presented, while in Figure 22.2, typical results for pressure drop are shown.

C. Flooding and Segregation

If gas flow rate is further increased, the relative velocity between the gas and flowing solids eventually reaches a point where the solids terminal velocity is approached. The flowing solids phase starts to accumulate in the upper part of the packing, and these unstable conditions are called flooding.

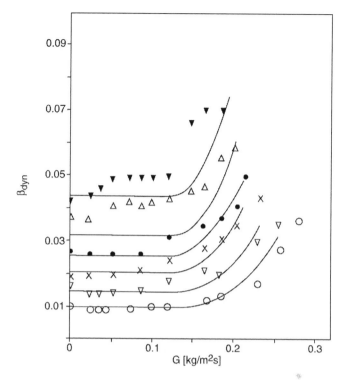

FIGURE 22.1 Typical results for dynamic holdup in gas–flowing solids–fixed bed contactors. Dynamic holdup as a function of superficial gas velocity for different values of flowing solids flow rates. Dynamic holdup increases with S; different symbols correspond to different flowing solids mass fluxes in the range 1.32–6.13 kg/m² sec. (From Roes, A.W.M. and van Swaaij, W.P.M., *Chem. Eng. J.*, 17, 81, 1979. With permission.)

Flooding is characterized with sudden, sharp increase in both pressure drop and dynamic holdup. This regime, which should be avoided, was registered in many of the laboratory and pilot-plant scale experiments. However, in some of industrial scale experiments, in columns of large diameter, there was no flooding, but a distinct segregation of gas and flowing solids flows occurred [8,18].

D. EXPERIMENTAL RESULTS

A number of experimental studies on determination of pressure drop and dynamic holdup were carried over the years, using columns of different constructions and dimensions and a variety of packing elements and flowing solids with a wide range of mean diameters and properties. In Table 22.1, the sources of experimental data are reviewed together with basic information on experimental conditions in their experiments. The range of obtained data is presented in Figure 22.3 for pressure drop and in Figure 22.4 for dynamic holdup.

Static holdup was not intensively studied, because it plays a minor role in the efficiency of the column [18] and because the technique for its determination is more complicated and time consuming. According to some authors [8], the static holdup is essentially independent of gas flow rate. However, other authors [12] found a slight increase of static holdup with gas flow rate, though an explanation for such phenomenon was not offered.

E. CORRELATIONS

Empirical correlations for prediction of pressure drop and dynamic holdup in countercurrent gas–flowing solids contactors were offered, based on the experimental data available in the literature.

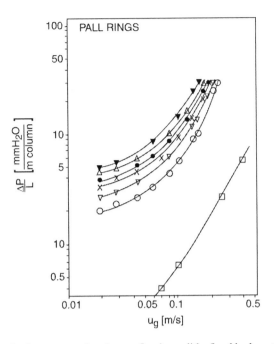

FIGURE 22.2 Typical results for pressure drop in gas–flowing solids–fixed bed contactors. Pressure drop as a function of superficial gas velocity for different values of flowing solids flow rates. Different symbols correspond to different flowing solids mass fluxes in the range 0–6.13 kg/m² sec. $\Delta p/L$ increases with S, square symbols are for $S = 0$. (From Roes, A.W.M. and van Swaaij, W.P.M., *Chem. Eng. J.*, 17, 81, 1979. With permission.)

For pressure drop along the packed bed, separate equations for preloading and loading regimes were offered [19], together with Equation (22.1) which defines the Reynolds number at the loading point, that is, at the transition from preloading to loading regimes:

$$Re_{\text{load}} = 0.1289 Ar^{0.48} \left(\frac{d_{\text{s}}}{d_{\text{eq}}} \right)^{-1.11} \left(\frac{G}{S} \right)^{0.23} \varepsilon^{0.85} \qquad (22.1)$$

where d_{s} is the diameter of flowing solids particles, d_{eq} the equivalent diameter of packing particles, G and S the gas and solids mass fluxes, respectively, ε the packed bed voidage, and Ar the dimensionless Archimedes number.

The pressure drop per unit bed height in the preloading regime can be found from Ref. [19]:

$$\frac{\Delta p}{L} = 3.057 Re^{0.44} Ar^{-0.19} Fr^{0.12} \left(\frac{d_{\text{e}}}{D} \right)^{-1.13} (1 - \varepsilon)^{0.51} \varepsilon^{-2.41} \qquad (22.2)$$

where Fr is dimensionless Froude number for flowing solids. For loading regime, the following equation was offered [19]:

$$\frac{\Delta p}{L} = 3.528 Re^{0.49} Ar^{-0.18} Fr^{0.1} \left(\frac{d_{\text{e}}}{D} \right)^{-1.35} (1 - \varepsilon)^{1.08} \varepsilon^{-2.45} \qquad (22.3)$$

TABLE 22.1
Studies of the Hydrodynamics of Gas–Flowing Solids–Fixed Bed Contactors

Ref.	Column Diameter (m)	Packing Type and Size (mm)	Void Fraction of Packing	Flowing Solids Phase: Type and Size (μm)	Solid Mass Flux (kg/m² sec)	Superficial Gas Velocity (m/sec)
Kaveckii and Planovskii [2]	0.127	Raschig rings, 15 × 15	0.64	Silicagel, 366	0.14–0.56	0.16–0.46
Claus et al. [3]	0.092	Cylindrical screens, 20 × 20	0.97	Sand, 235	9–42	0.02–1.7
Roes and van Swaaij [4]	0.076	Raschig rings, 10 × 10 × 1	0.80	FCC, 70	1.1–6.0	0.02–0.19
		Pall rings, 15 × 15 × 2	0.86		1.3–6.0	0.02–0.23
		Cylindrical screens, 10 × 10 × 0.5	0.97		1.3–6.0	0.02–0.17
Large et al. [5]	0.32	Pall rings, 15 × 15	0.94	Sand, 190	0.66–1.68	0.22–1.72
Saatdjian and Large [7]	0.125	Pall rings, 15 × 15	0.93	Sand, 205	0–3	0–4
Verver and van Swaaij [8]	0.10 × 0.10	Regulary stacked packing, 15 × 15		FCC, 70	0.03–0.8	0–0.2
				Sand, 255	0.1–2.4	0–0.9
				Sand, 425	0.2–2.2	0–1.4
				Steel shot, 310	0.3–4.8	0–2.8
				Steel shot, 880	0.3–4.0	0–5.4
Westerterp and Kuczynski [9]	0.025	Raschig rings, 7 × 7 × 3	0.45	FCC, 80	0.5–2.0	0.03–0.16
		Kerapak,	0.75			0.03–0.26
Kiel [10]	10 × 10	Regulary stacked packing, 3	0.61	Glass beads, 490	0.1–1	0.2–1
				Glass beads, 740	0.43–1	0.5–1
Predojević [14]	0.111	Raschig rings, 12 × 12 × 2.4	0.61	Sand, 253	0.16–2.5	0.06–0.46
		Raschig rings, 30 × 30 × 2.3	0.85			
		cer. beads, 19	0.47			
		Pall rings, 23 × 8 × 0.1	0.96			
		Raschig rings, 12 × 12 × 2.4	0.61	Propant, 642	0.14–2.5	
		Raschig rings, 30 × 30 × 2.3	0.85			
		cer. beads, 19	0.47			
		Pall rings, 23 × 8 × 0.1	0.96			

(Table continued)

TABLE 22.1 *Continued*

Ref.	Column Diameter (m)	Packing Type and Size (mm)	Void Fraction of Packing	Flowing Solids Phase: Type and Size (μm)		Solid Mass Flux (kg/m² sec)	Superficial Gas Velocity (m/sec)
Stanimirović [17]	0.111	Raschig rings, cer. beads,	0.61 0.47	Semolina,	348	0.07–1.09	0.12–0.46
		12 × 12 × 2.4 19					
		Raschig rings, cer. beads,	0.61 0.47	Sand,	253	0.18–2.6	
		12 × 12 × 2.4 19					
		Raschig rings, cer. beads,	0.61 0.47	Corundum,	331	0.23–3.46	
		12 × 12 × 2.4 19					
Pjanović [18]	0.0274	Glass beads, 5	0.41	Poliamid,	105	0.12–0.13	0.02–0.12
		Glass beads, 6	0.44			0.12–0.58	
		Raschig rings, 5 × 5	0.71			0.13–0.58	

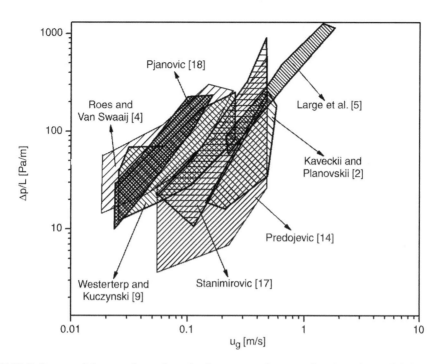

FIGURE 22.3 Survey of the experimental results for pressure drop as a function of superficial gas velocity.

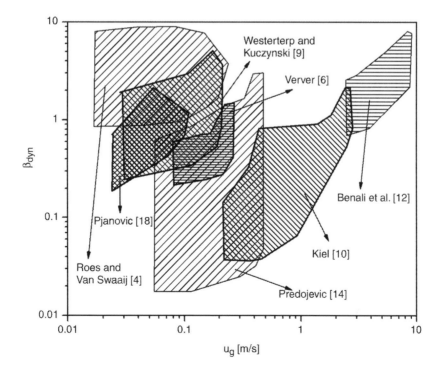

FIGURE 22.4 Survey of the experimental results for dynamic holdup as a function of superficial gas velocity.

The dynamic holdup (β_{dyn}) in the preloading regime can be calculated from the following empirical equation [20]:

$$\beta_{dyn} = 2196.2 Re_s^{1.21} Ar^{-0.88} \left(\frac{S^2}{\rho_s \cdot \rho_g \cdot u_g^2} \right)^{0.582} \left(\frac{d_s}{d_{eq}} \right)^{2.41} (1-\varepsilon)^{1.42} \varepsilon^{0.279} \qquad (22.4)$$

where Re_s is particle Reynolds number.

For the loading regime, it was reported as [20]:

$$\beta_{dyn} = 15570.7 Re_s^{1.57} Ar^{-1.24} \left(\frac{S^2}{\rho_s \cdot \rho_g \cdot u_g^2} \right)^{0.509} \left(\frac{d_s}{d_{eq}} \right)^{2.93} (1-\varepsilon)^{1.46} \varepsilon^{-1.45} \qquad (22.5)$$

Alternatively, a single equation could be used independent of the regime [20]:

$$\beta_{dyn} = 6560.7 Re_p^{1.29} Ar^{-1.01} \left(\frac{S^2}{\rho_s \cdot \rho_g \cdot u_g^2} \right)^{0.521} \left(\frac{d_s}{d_{eq}} \right)^{2.57} (1-\varepsilon)^{1.52} \varepsilon^{0.933} \qquad (22.6)$$

Equation (22.6) is convenient as it does not require prior determination of the regime, and it gives an average error of 26.9% for the loading regime, which is very close to the error of Equation (22.5), being 26.4% (based on 270 available loading data points). However, the difference in errors between predicted and experimental data for the preloading regime is higher for Equation (22.6), than for Equation (22.4) by about 8%.

F. Cocurrent Flow

Most of the fluid dynamic studies of gas–flowing solids–fixed bed contactors were devoted to countercurrent flow systems, because of the higher efficiency of countercurrent operations for most of the processes when compared with cocurrent operations. However, there is an upper limit for gas flow rate in countercurrent systems, due to flooding. Hence, the cocurrent operation system is an interesting alternative for higher gas flow rates, particularly for very small particles. Further, in some of the proposed applications, cocurrent contacting is a desirable flow pattern [22].

Only a very few experimental data on the cocurrent flow of gas and flowing solids through the packed bed have been reported. Kiel [10] studied the cocurrent gas–solids downflow over regularly stacked packing, while Barysheva et al. [21] used a packed bed of spheres in their experimental investigations.

For average dynamic holdup, a decrease with increasing gas flow rate was found for some types of regular packing, while for other types, a slight increase was detected [10]. The presence of specially designed packings increased the average flowing solids dynamic holdup by a factor of 10 in the range of operating conditions applied in this study. However, it induced a considerable increase in pressure drop [10].

For packed bed of spheres, Barysheva et al. [21] found that flowing particle mean residence time is essentially independent of gas flow rate. Mean solids velocity decreases with an increase of flowing particles diameter, this effect becoming more pronounced at higher solids flow rates. Further, flowing solids velocity decreases with an increase in flowing solids flow rate. This is a consequence of the interaction between individual flowing solids particles.

Pressure drop experimental data for different organized packings were reported by Kiel [10]. It was found that the total pressure drop is slightly positive, for very low gas flow rates, that is, the pressure at the bottom of the column is higher than at the top. Consequently, at very low gas flow rates, pressure drop due to gas–flowing solids interaction is larger than pressure drop due to friction between gas and packing. It implies that the solids flow faster than the gas. However, when gas flow rate starts to increase, pressure drop becomes more and more negative, and thus average solids velocity is lower than gas velocity. The collisions between flowing solids and packing elements increase the value of relative (slip) velocity between gas and flowing solids.

III. MATHEMATICAL MODELING

The development of a phenomenological model to describe the fluid dynamics of the flowing solids to enable prediction of important operating parameters and optimization of the packing configuration was the objective of a number of studies [4,6,8,22–24]. Owing to very complex interactions in these systems, some of the aspects of fluid dynamic behavior are still under investigation and some of the phenomena seem difficult to describe rigorously. Models proposed up to date are based on a number of simplifications and most of them contain empirical parameters.

Countercurrent flow of gas and flowing solids inside the packed bed has some analogies with gas–liquid systems, as preloading, loading, and flooding regimes. However, in many aspects, the nature of these phenomena are different and equations for gas–liquid systems cannot be applied to flowing solids systems.

A. AVERAGE SOLIDS AND SLIP VELOCITIES

The average flowing solids particle velocity can be defined as:

$$\bar{u}_s = \frac{S}{\beta_{dyn}\rho_s} \tag{22.7}$$

while the mean relative velocity between flowing solids particles and gas (slip velocity) is:

$$\bar{u}_R = \bar{u}_s + u'_g = \frac{S}{\beta_{dyn}\rho_s} + \frac{G}{(\varepsilon - \beta)\rho_g} \tag{22.8}$$

where $\beta = \beta_{dyn} + \beta_{st}$ is total flowing solids holdup.

For the preloading regime, Roes and van Swaaij [4] found that the average particle velocity is independent of gas and flowing solids flow rates. A constant value of average particle velocity in preloading zone was found to be 0.17 m/sec in their experiments (three types of packing: Pall rings, Raschig rings, and cylindrical screens). When the value of the average particle velocity is known, the value of the dynamic holdup for the preloading regime can be found from Equation (22.6). However, it is reasonable to assume that the mean particle velocity will depend on packing properties, and so $\bar{u}_s$ has to be determined from separate experiments for each type of packing.

Roes and van Swaaij [4] concluded from their experiments for the loading regime that under these conditions, the relative (slip) velocity is constant, having the value of 0.31 m/sec. The physical reasons for this constant value of slip velocity were not discussed. In fact, it was experimentally determined later [9] that relative velocity depends on the properties of both flowing phases and on the packing geometry.

B. TRICKLE FLOW MODEL

Westerterp and Kuczynski [9] considered the flow of solids as a flow of rivulets (trickles) rather than of individual particles. They assumed that in the preloading regime, the solids phase flows almost exclusively in the form of trickles. It was further assumed that some gas is dragged along by the trickles flowing downward. The trickles release the gas when they collide with the walls of fixed bed elements. The released gas flows upward together with the main stream, and in that way, local circulation of gas is produced. After collision, the stream of particles slides over the packing surface, and then again starts falling by gravity, voids in trickles are increasing and a new portion of gas is captured and dragged downward. The distance between collisions depends on the geometry of the packing, and the mean solids velocity is the average between the velocity of falling and sliding trickles over the packing surface. Accordingly, the increase in pressure drop is caused by trickles for two reasons: (1) the reduction of free cross-section available for gas flow, due to the presence of trickles and (2) the recirculation of gas, which increases the upward gas flow.

The void fraction occupied by gas in solids trickles was introduced and defined as:

$$\varepsilon_{tr} = \frac{\text{Gas volume in trickles}}{\text{Total trickle volume}}$$

which was assumed to be constant. The real gas velocity according to this approach is:

$$u_{g,corr} = \frac{G/\rho_g + S/\rho_s \left[\varepsilon_{tr}/(1 - \varepsilon_{tr})\right]}{\varepsilon - \beta_{st}(\rho_s/\rho_{s,PB}) - (S/\rho_s u_s (1 - \varepsilon_{tr}))} \tag{22.9}$$

The upward gas flow is increased by the second term in the numerator, while the free cross-sectional area available for flow is reduced by the static solids holdup and the trapped gas (second term in denominator) and by the area occupied by trickles (third term in denominator). The use of real gas velocity, defined by Equation (22.9), in the Ergun equation allows the determination of ε_{tr} by comparison between calculated and experimental results for pressure drop. Westerterp and Kuczynski [9] have found high values for ε_{tr}, around 0.975 for their experimental conditions.

The dynamic holdup in the preloading regime was successfully predicted using this approach [9], but the empirical parameter ε_{tr} had to be determined from the same experimental data. Westerterp and Kuczynski [9] did not derive a relation for reliable prediction of dynamic holdup in the loading range. Instead, for rough estimation, the use of Equation (22.7) and Equation (22.9) was suggested.

C. Particle Flow Model

In countercurrent gas–flowing solids systems, dynamic holdup (i.e., the volumetric solids concentration) usually has low values, as seen in Figure 22.4. Consequently, in first approximation, solids flow can be considered as a sum of single particle flows. In these systems, particles repeatedly collide with the packing, which retards their motion, causing a higher residence time of the flowing solids. The relative velocity between particles and gas does not usually reach the terminal velocity. The acceleration of the particles in the downward direction through the gas flowing upward is given by the momentum equation:

$$\rho_s \frac{du_s}{dt} = F_G - F_B - F_D \tag{22.10}$$

where F_G, F_B, and F_D are gravitational, buoyancy, and drag forces per unit solids volume, respectively and u_s is the local particle velocity in the downward direction:

$$u_s = \frac{dz}{dt} \tag{22.11}$$

Equation (22.10) in describing particle motion is applicable for the period between two successive collisions with packing elements.

Verver and van Swaaij [8] and Kiel and van Swaaij [23] introduced the concept of a "packing cell" for regularly stacked packing. In this approach, regularly stacked packing is considered as a two-dimensional array of identical packing cells and the fluid dynamic modeling is restricted to a single cell. Duduković et al. [24,25] considered flowing particles behavior in an "average void" as representative of a whole bed, for both random and regular packings.

The flowing solids–packing interaction was taken into account through the initial solids velocity in the vertical direction, at the entrance of the packing cell or the packing void. In some of the approaches [8,23], the initial solids velocity is treated as an empirical factor, which had to be determined from the experiments. The other solution [24,25] was to assume the initial velocity to be zero, that is, on average, the particles upon collision loose all of their velocity in downward direction. Analysis by Kiel [10] proved that the values of initial velocity are small, both positive and negative, and dependant on packing geometry and flowing solids flow rate.

The forces in the Equation (22.10) are given by:

$$F_G - F_B = (\rho_s - \rho_g)g \tag{22.12}$$

and

$$F_D = C_D \frac{3\rho_g}{4d_s}(u_s + u_g')^2 \tag{22.13}$$

where u_g' is the mean gas velocity in the voids:

$$u_g' = \frac{u_g}{\varepsilon - \beta} = \frac{G}{(\varepsilon - \beta)\rho_g} \tag{22.14}$$

This value of u_g' was additionally corrected by some authors [8,23] to account for the nonuniformity of packing porosity and for the wakes that packing elements produce in the gas stream, especially in organized packing with free space between the packing elements. Verver and van Swaaij [8] introduced an "effective packing porosity," while Kiel and van Swaaij [23] defined an "effectiveness factor." Both of these empirical parameters have to be determined from experimental data.

The drag coefficient in Equation (22.13) is a function of the particle Reynolds number based on relative velocity:

$$Re_s = \frac{\rho_g d_s(u_s + u_g')}{\mu} \tag{22.15}$$

For calculation of the drag coefficient, the equation proposed by Turton and Levenspiel [26] can be used:

$$C_D = \frac{24}{Re_s}\left(1 + 0.173\,Re_s^{0.6567}\right) + \frac{0.413}{1 + 16300\,Re_s^{-1.09}} \tag{22.16}$$

or some other similar equation [27].

1. Dynamic Holdup Predictions

The average dynamic flowing solids holdup can be found from Equation (22.6):

$$\beta_{\text{dyn}} = \frac{S}{\bar{u}_s \rho_s} \tag{22.17}$$

if the average solids velocity for the "average void" or "packing cell," $\bar{u}_s$, is known. It is assumed that it represents the average dynamic holdup for the whole bed.

Most of the proposed models [8,23–25] basically deal with the solution of the system of differential Equation (22.10) and Equation (22.11). Some of the specific assumptions were mentioned here, but are given in more detail in the original papers, together with numerical and computational methods. The empirical parameters involved were found by trial and error method, based on experimental results [8,23].

Verver and van Swaaij [8] found a good agreement between calculated and experimentally determined values of average particle velocity ($\bar{u}_s$) with optimal values of empirical parameters. However, both empirical parameters (the effective bed void and initial velocity) were found to have different values for each of the flowing solids materials tested. Kiel and van Swaaij [23] obtained a good agreement between computed and experimental results, for best values of empirical parameters found to be -0.05 m/sec for initial velocity and 1.35 for effectiveness factor.

Duduković et al. [24] introduced an empirical parameter (ψ) to take into account the acceleration of particles in the bed, instead of solving Equation (22.10) and Equation (22.11). The average velocity was given as:

$$\bar{u}_s = \psi u_{s,t} \tag{22.18}$$

where the terminal velocity is found from:

$$u_{s,t} = u_{R,t} - u'_g = \sqrt{\frac{4d_s g(\rho_s - \rho_g)}{3\rho_g C_D} - \frac{G}{\rho_g(\varepsilon - \beta)}} \tag{22.19}$$

The empirical parameter ψ in Equation (22.18) does not need to be experimentally determined for each case, but can be evaluated from a correlation given by Equation (22.20):

$$\psi = 0.4186 \sqrt{\frac{\rho}{\rho_s} \frac{d_V}{d_s}} + 0.0042 \tag{22.20}$$

where d_V is the equivalent void diameter, defined as:

$$d_V = \frac{2d_{\text{eq}}}{3(1 - \varepsilon)} \tag{22.21}$$

where d_{eq} being equivalent packing diameter. The agreement between predicted and experimental values was very good, considering that all the available data from the literature were taken into account, involving a very wide range of experimental conditions [24]. An average error between predicted and experimental data was 20.8% for 452 data points.

Models proposed by Verver and van Swaaij [8] and Kiel and van Swaaij [23] are semiempirical in nature, based exclusively upon experimental data taken in a single study and were not tested on data of other authors. In addition, the model proposed by Duduković et al. [24] is semiempirical, but presents a correlation, based on available data, for the empirical parameter ψ, which allows the prediction of dynamic holdup without any experiments in a system of interest. The model was tested on numerous data, but further testing on systems which were not part of the database for the correlation of Equation (22.20) is needed.

As an alternative to the earlier semiempirical approach, Duduković et al. [25] presented a model free of any empirical parameters, involving a number of simplifying assumptions. The model is based on solving simultaneously Equation (22.12)–Equation (22.16) to find the particle velocity as a function of vertical distance along the void. An average particle velocity can be found for the whole void, if the shape of the void is known and mathematically described. However, the geometry of the voids depends on the type of packing and differs in configuration and orientation even in the same bed. To obtain a solution which would be of general applicability, a simplifying approximation was introduced by describing an "average void" as a vertical double cone with diameter and height equal to the average void diameter (d_V), as presented in Figure 22.5. The average particle velocity was found as a mean integral value for a whole void, allowing calculation of dynamic holdup by the use of Equation (22.17). The comparison of predicted and experimental values, based on 564 data points from the literature, led to the average discrepancy of 31.3%. This could be considered as very good agreement, taking into account that there are no empirical parameters involved, and that the data cover a wide range of experimental conditions in both preloading and loading regimes.

2. Pressure Drop Predictions

Another important design parameter that can be predicting using the models described earlier is pressure drop. The usual [8,10,23,24] simplified assumption rests on the additivity of two pressure drop contributions given by:

$$\frac{\Delta p}{L} = \left(\frac{\Delta p}{L}\right)_{PB} + \left(\frac{\Delta p}{L}\right)_{FS}$$

(22.22)

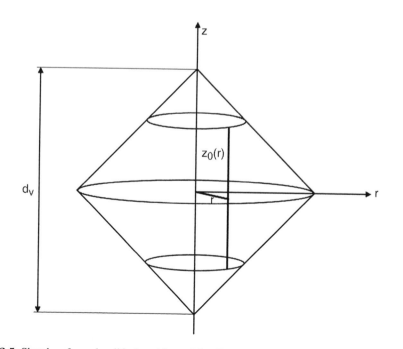

FIGURE 22.5 Sketch of a simplified void model. (From Duduković, A.P., Nikačević, N.M., and Kuzeljević, Ž.V., *Ind. Eng. Chem. Res.*, 43, 7445, 2004. With permission.)

where $(\Delta p/L)_{\mathrm{PB}}$ is the contribution of gas–packed bed interactions, while $(\Delta p/L)_{\mathrm{FS}}$ is the contribution of gas–flowing solids interaction in the voids. The first term can be found from the Ergun equation:

$$-\left(\frac{\Delta p}{L}\right)_{\mathrm{PB}} = \left(\frac{A}{Re_{\mathrm{PB}}} + B\right)\frac{u_{\mathrm{g}}\rho_{\mathrm{g}}}{d_{\mathrm{eq}}}\frac{1 - \varepsilon_{\mathrm{corr}}}{\varepsilon_{\mathrm{corr}}^3} \qquad (22.23)$$

where $\varepsilon_{\mathrm{corr}} = \varepsilon - \beta$ and Re_{PB} is the Reynolds number for gas flow through the packed bed:

$$Re_{\mathrm{PB}} = \frac{u_{\mathrm{g}}\rho_{\mathrm{g}}d_{\mathrm{eq}}}{\mu(1 - \varepsilon_{\mathrm{corr}})} \qquad (22.24)$$

It was recommended [9,24] that constants A and B in the Ergun equations should be determined for each packing separately, for single fluid flow.

Some authors [8,23] found it convenient to define a relative pressure drop, γ, as a ratio of the drag force exerted by the solids on the gas and the gravity force on the solids [23]:

$$\gamma = \frac{F_{\mathrm{D}}}{\beta\rho_{\mathrm{s}}g} \qquad (22.25)$$

The average relative pressure $\bar{\gamma}$ can be found by integration, taking into account that particle velocity, and consequently, the drag force are changing along the packing cell. The predicted values of the average relative pressure drop agreed well with experimental results of Kiel and van Swaaij [23] for the same values of empirical parameters determined for dynamic holdup. However, the error for pressure drop data was higher than for dynamic holdup, approximately by a factor of 2.

Starting from Equation (22.22), Duduković et al. [24] predicted pressure drop in flowing solids systems, using the value of dynamic holdup, as predicted previously by the same model. The second term in Equation (22.22) in their approach takes into account the mutual effects between particles. Their predictions were compared with 435 literature data points, giving an average error of 40.1%.

3. Cocurrent Flow

A very few studies on mathematical modeling of cocurrent gas–flowing solids–fixed bed contactors were published [10,22]. Kiel [10] assumed that a hydrodynamic phenomenon of cocurrent flowing solids systems allows a similar modeling approach as in the case of countercurrent flow. Therefore, the particle flow model originally developed for countercurrent gas–flowing solids flow [10,23] was applied. The authors started with analogous governing equations as described for countercurrent flow and introduced the same empirical parameters. The best values of empirical parameters (the hydrodynamic effectiveness factor and the solid velocity just after the collision of a flowing solid with a packing element) differed considerably from the corresponding values for the countercurrent system. The differences were attributed to a different gas flow pattern above the packing elements. With these values of empirical parameters, the predicted values for pressure drop gave a good agreement with experimental values. However, the model failed to predict the values for dynamic holdup, which would be in reasonable agreement with experimental results. The authors suspect that this is due to the specific gas flow pattern which should be studied further.

Barysheva et al. [22] developed a new model for flowing particles in the packed bed. The model is based on the statistical calculation of individual particle path in a packed bed column. Model gave good predictions for residence time distribution for low values of the solids flow rate. Differences between predictions and experimental values at higher gas flow rates were attributed to the interaction between the flowing solids particles, which was not included in the model.

D. Additional Considerations

Equation (22.10) and Equation (22.11) could be rearranged to give:

$$u_s \frac{du_s}{dz} = \frac{\rho_s - \rho_g}{\rho_s} g - \frac{3C'_D \rho_g}{4d_s \rho_s} (u_s + u'_g)^2 \tag{22.26}$$

In flowing solids systems, the voids are usually far too short for particles to reach the terminal velocity. Consequently, for low values of u'_g, the second term in Equation (22.26) can be neglected:

$$u_s \frac{du_s}{dz} = \frac{\rho_s - \rho_g}{\rho_s} g \tag{22.27}$$

and a well-known equation for free fall without drag effects is obtained:

$$u_s = \sqrt{2gz} \tag{22.28}$$

We could assume that this is the local solids velocity in the preloading regime, that is, that in the preloading regime, the drag forces on flowing solids can be neglected. To get the mean particle velocity in an average void, and consequently, in a column as a whole, the average value for the whole void needs to be found. This mean value depends on the length of free fall, that is:

$$u_{sL} = \frac{1}{L} \int_0^L \sqrt{2gz} \, dz = \frac{2}{3} \sqrt{2gL} \tag{22.29}$$

The length of fall in a void of equivalent diameter d_V will vary depending on position, from $L = 0$ to $L = d_V$. For a rough estimation, an average particle velocity in a void can be taken to be:

$$\bar{u}_s = \frac{1}{d_V} \int_0^{d_V} u_{sL} \, dL = \frac{4}{9} \sqrt{2g d_V} \tag{22.30}$$

The result of Equation (22.30) is, under this assumption, the constant value of average solids velocity in the preloading regime. Comparison of the values calculated using a very simple Equation (22.30) and the values obtained by Roes and van Swaaij [4] for their experimental systems yields excellent agreement. Detailed discussion of this issue will be the subject of a future publication.

If terminal velocity is reached in the voids, Equation (22.26) gives:

$$u_{Rt}^2 = (u_s + u'_g)^2 = \frac{\rho_s - \rho_g}{\rho_g} g \frac{4d_s}{3C_D} \tag{22.31}$$

Equation (22.31) leads to a constant local slip velocity. Evidently, for some packing geometries and operating conditions, an overall constant slip velocity is observed [4,9]. This is an indication that flowing particles experience essentially constant slip under such conditions.

When the gas velocity in the bed, u'_g, exceeds a critical value (and is higher than the relative particle terminal velocity, $u'_g \geq u_{Rt}$), the sign of the right side of Equation (22.26) changes as the particles enter the bed. The flowing particles are now decelerated and tend to accumulate at the top of the bed. Entry into the bed is made difficult, if not impossible, due to even higher gas velocity in the narrowest passages in the voids of the packing. This is the condition of incipient flooding, as will be discussed in more detail in a future publication.

IV. FLOW PATTERN AND CONTACTING

It is well known that plug flow is a desirable flow pattern in countercurrent separations. Real equipment always deviates from this ideal, but increasing deviation is disadvantageous because of the reduction in the driving force for mass transfer. Consequently, the residence time distribution in both flowing phases in gas–flowing solids–fixed bed contactors plays an important role in design for practical applications. It can be expected that the mutual interaction between gas and flowing solids will affect the flow pattern of both phases, but presently, it is not possible to predict it theoretically.

For the flowing solids phase, a complication in interpretation of the residence time distribution is caused by the nature of the solids holdup. The dynamic holdup is usually assumed to be the operating holdup, and static holdup is often treated as a "dead" part of flowing solids. However, if there is an exchange between static and dynamic holdup, this affects the residence time distribution of flowing solids, and consequently, the contactor performance.

A. EXCHANGE BETWEEN THE DYNAMIC AND STATIC FLOWING SOLIDS

Tracer technique was applied [28] to study the exchange between the fraction of flowing solids moving downward (characterized by dynamic holdup) and part of solids that are resting on packing elements (static holdup). A step signal in the inlet of the flowing solids was produced by switching from powder of one color to the other. Polyamide particles of 105 μm mean diameter were used, gray and red, all other characteristics being the same. At different time intervals after a step signal, flows of gas and flowing solids are closed simultaneously and flowing solids are drained out (dynamic part), leaving behind the stagnant portion of the solids particles. After emptying the column, the flowing solids particles were separated from the packing, weighted, and its content analyzed by scanning technique. Typical experimental results [28] are presented in Figure 22.6 as a fraction of tracer in the flowing solids, which remained in the bed (static fraction) as a function of time. The exchange between stagnant and dynamic parts of flowing solids is obvious from Figure 22.6. However, the fraction of tracer does not reach its maximum value of one after very

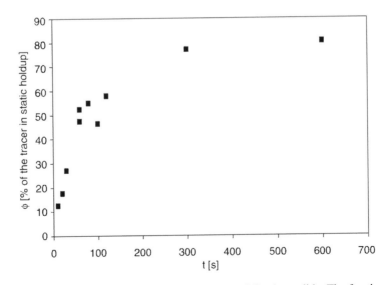

FIGURE 22.6 Exchange between dynamic and static portion of flowing solids. The fraction of exchanged static flowing solids as a function of time. (From Duduković, A.P., Nikačević, N.M., Pjanović, R.V., and Kuzeljević, Ž.V., *J. Serb. Chem. Soc.*, 70 (1), 137, 2005. With permission.)

long time, but there is a horizontal asymptote, implying that a part of stagnant zone does behave as a dead zone, free of any exchange with moving particles. From Figure 22.6, it could be estimated that for the experimental conditions employed, the dead zone occupied about 20% of the stagnant zone [28].

Defining the volumetric exchange rate for flowing solids per unit bed volume (f), the balance for the active part of the stagnant zone was given as:

$$(\beta_{st} - \beta_{dead})V_c \frac{d\varphi}{dt} = v_{T,in} - v_{T,out} = f V_c(1 - \varphi) \tag{22.32}$$

It was assumed that after the step signal, essentially all the particles in the flowing zone are tracer particles (because the front moves very fast when compared with the studied exchange phenomena). In contrast, the flow of particles from the stagnant zone contains both tracer and inert particles, that is, φ fraction. Fraction φ refers to the active part of the stagnant zone, while ϕ refers to the whole stagnant zone. If φ is replaced by ϕ, according to

$$\varphi(\beta_{st} - \beta_{dead})V = \phi \beta_{st}V \tag{22.33}$$

solving Equation (22.32) gives:

$$\phi = \left(1 - \frac{\beta_{dead}}{\beta_{st}}\right)\left(1 - e^{-(f/\beta_{st} - \beta_{dead})t}\right) \tag{22.34}$$

The logarithmic plot of Equation (22.34) gives a straight line, and the exchange rate (f) can be determined from the slope. For their experimental conditions, Duduković et al. [28] found

$$f = 0.274 \times 10^{-3} \, \text{m}^3/\text{m}^3 \, \text{sec}.$$

B. RESIDENCE TIME DISTRIBUTION

Roes and van Swaaij [29] used tracer technique to determine the residence time distribution and the extent of axial mixing in both flowing phases. The axially dispersed model was used to describe the degree of axial mixing.

1. Axial Dispersion in the Gas Phase

Imperfect pulse technique was applied [29,30] to study the residence time distribution in the gas phase. Helium was injected in the air stream at the entrance, and tracer concentration was determined at the exit stream. At zero solid mass flux (i.e., no flowing solids in the system), the Bodenstein number ($Bo_g = u_g d_s/D_g \varepsilon$) was found to be close to the value of 2, so that the height of one Pall rings layer corresponds to a single ideally mixed unit [29,30].

In countercurrent gas–flowing solids system, it was found that a very small solids flow rate drastically increases gas–phase dispersion. However, the Bodenstein number increases again with flowing solids rate, that is, the axial dispersion, D_g, decreases again. The Bodenstein number is also increasing with gas flow rate (Figure 22.7). For the conditions of practical importance, axial dispersion is still higher than that in single gas flow and layers of the two to five Pall rings height are equivalent to a single ideal mixing unit.

2. Axial Dispersion in the Flowing Solids Phase

The residence time distribution of the flowing solids phase was determined [29,30] by the tracer technique, using white and black particles. A special solids tracer injector was constructed and the concentration of black particles in the exit stream was measured by a reflection technique.

The Bodenstein number for the flowing solids phase ($Bo_s = U_s d_s/D_s \beta$) was found to be almost independent of the gas flow rate and to increase with solids flow rate. Near the flooding point, the

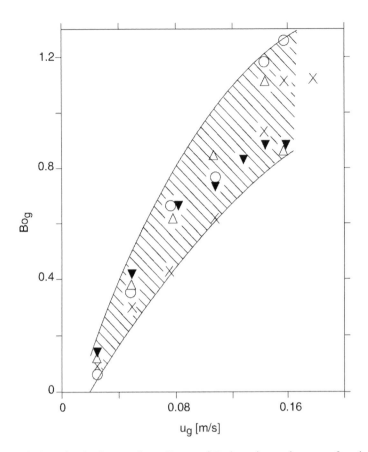

FIGURE 22.7 Axial dispersion in the gas phase. Range of Bodenstein numbers as a function of superficial gas velocity. Different symbols correspond to different values of flowing solids mass fluxes in the range 0.081–5.83 kg/m^3 sec. (From Roes, A.W.M. and van Swaaij, W.P.M., *Chem. Eng. J.*, 18, 13, 1979. With permission.)

increase of axial dispersion in the flowing solids phase was registered. In Figure 22.8, the range of Bodenstein number values obtained for different values of gas flow rate is presented as a function of the flowing solids mass flux. For practical conditions, 5–15 Pall ring layers correspond to a perfect mixing unit [29,30].

It should be noted that in this approach, the authors [29,30] used two differently defined flowing solids holdup parameters. Besides the total holdup, the corrected holdup was used, which did not take into account the fraction of flowing solids, named by the authors as "permanent fraction of static holdup." This fraction was experimentally determined by vibrating the column, after stopping the flowing streams and after the drainage of dynamic part of flowing solids occurred. After additional amount of solids particles drained, for the rest of flowing solids, it was assumed to behave like a part of the packing. Corrected holdup values gave much better agreement with the results of tracer analysis.

The "permanent fraction of static holdup" defined by Roes and van Swaaij [29] and the "dead part of the static zone" defined by Duduković et al. [28] describe the same phenomenon and are probably related, but may not be identical. Further investigations should check the possible effects of the characteristics of vibrations on the permanent fraction of flowing solids, as well as the dependence of the dead part of the static zone on flowing solids and gas flow rates.

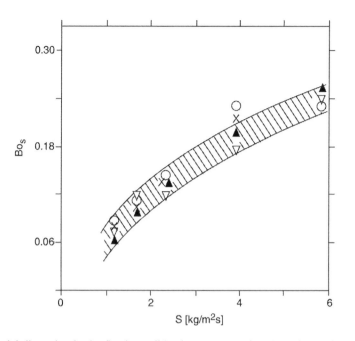

FIGURE 22.8 Axial dispersion in the flowing solids phase. Range of Bodenstein numbers as a function of flowing solids mass flux. Different symbols correspond to different values of superficial gas velocity in the range 0–0.130 m/sec. (From Roes, A.W.M. and van Swaaij, W.P.M., *Chem. Eng. J.*, 18, 13, 1979. With permission.)

V. HEAT TRANSFER

The high values of the heat transfer coefficients are one of the most favorable advantages of gas–flowing solids–fixed bed contactors. High heat transfer rates between fine solids and gas, along with low pressure drop in the system, make this type of equipment very attractive for applications. This has been established in industrial practice as raining packed bed exchangers [31].

The essential reasons for their significant thermal performance are good radial distribution of gas and flowing solids and large interfacial area available for heat transfer. Thermal properties have been studied experimentally [7,32–34], and theoretical models have been proposed [7,32]. The heat transfer properties of gas–flowing solids–fixed bed contactors not only depend on gas and solids flow rates and type of packing, but also on the portion of static holdup and axial dispersion.

A. Experimental Results

Experimental investigations were carried out only with sand as a flowing solid phase, but with the use of different packing constructions. Saatdjian and Large [7] investigated the thermal performance of a raining packed bed heat exchanger in a column of 0.32 m I.D. filled with 15 Pall rings. Experimental results were obtained with sand particles and the temperature of gas in the range between 200 and 350°C. Solids holdup, which was an important characteristic for interpretation of thermal behavior, was measured in a smaller column of 0.125 m I.D., but with the same solids and gas mass fluxes as used in the bigger one.

Verver and van Swaaij [32] obtained their results in $0.1 \times 0.1 \text{ m}^2$ columns packed with regularly stacked elements specially developed for "gas–solid trickle flow contactor." The authors reported significant heat transfer rates, with the number of transfer units from two to four for

0.5 m of packing, depending on gas and solids mass flow rates, as well as the packing construction used.

Boumehdi et al. [33] used several layers of horizontal tubes as a fixed bed. The heat transfer flows through the tubes which are perpendicular to the flow of the gas and flowing solids. Their experimental results demonstrated that the overall heat transfer coefficients are 20% higher than those predicted by the Colburn correlation and also two times higher than those obtained with gas alone.

B. MATHEMATICAL MODELS

Proposed models were based on a single particle flow behavior, not accounting for the effects of agglomeration and trickling. They were developed from energy balances for the gas and the flowing solids phase.

Saatdjian and Large [7] observed that the heat transfer rate of the gas–flowing solids–fixed bed exchanger is determined by the fraction of active flowing solids. They developed an expression for thermal efficiency (dimensionless temperature), which was derived from energy balances for both the solids and gas phases:

$$A_c S C p_s dT_s = -\alpha A_s \frac{S A_c dz}{\rho_s V_s u_s}(T_g - T_s) \tag{22.35}$$

$$\rho_g u_g \varepsilon C p_{,g} \frac{dT_g}{dz} + \frac{6\alpha S}{\rho_s u_s \alpha}(T_g - T_s) = 0 \tag{22.36}$$

Further, the theoretical thermal efficiency (η), gained from numerical solution of differential equations, was corrected so that it could take into account the fraction of fully suspended particles. This fraction was obtained from pressure drop measurements with or without the presence of flowing solids. In an approach similar to Westerterp and Kuczynski [9], the authors used the Ergun equation to determine the "effective packing porosity." They compared the calculated and experimental results for pressure drop. Figure 22.9 shows reasonably good agreement between theory and experimental results. The theoretical thermal efficiency increases as the gas velocity increases. It can be seen from Figure 22.9 that the real efficiency has a maximum. The reason for this is the segregation phenomenon which occurs at the higher gas velocities, when there is no full contact between phases.

Some discrepancies between the predicted and measured values could be caused by the fact that the data for holdup had been extrapolated from the smaller to the bigger column, neglecting important difference in the column diameter–packing dimension ratio. Energy losses to the surroundings were also neglected. The reason for disagreement between model and experiments could also lie in the use of the original coefficients in the Ergun equation (A, B), which are unsuitable for non-spherical packing elements.

Verver and van Swaaij [32] assumed that the thermal efficiency greatly depends on the initial solids distribution. The reason for lower values of the heat transfer rate, when packing elements, such as Pall rings or similar, are used, could be wall flow or channeling of solids, which is due to insufficient radial distribution of the packing. They developed two regularly stacked packing constructions, which produced a rapid radial solids distribution throughout the column. Furthermore, one of the advantages of regularly stacked packing is the reduction of the portion of stagnant flowing particles, as a result of which nearly all the particles are suspended in gas.

The authors [32] took into account the axial dispersion in both phases through the number of overall transfer units (NTUs), which consists both of the number of true transfer units determined by the interfacial transfer rate and the number of dispersion units. The column was divided into sections in which temperatures and thermal properties of both phases were approximately constant.

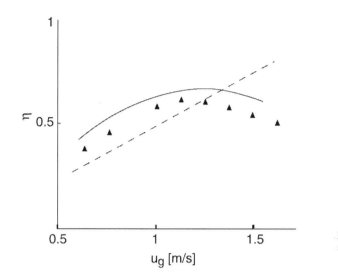

FIGURE 22.9 Comparison of the theoretical and real thermal efficiency in gas–flowing solids–fixed bed heat exchangers. (▲) experimental; (– – –) theoretical from energy balances; (—) corrected for "effective packing porosity." (From Saatdjian, E. and Large, J.F., *Chem. Eng. Sci.*, 40, 693, 1985. With permission.)

Differential equations were obtained from heat balance over each phase. From experimentally determined inlet and outlet temperatures, the heat transfer rate constant was derived by numerical calculation. The heat loss rate was assessed from temperature drop over the column, when solids flow was absent. The overall heat transfer rate constant is slightly affected by axial dispersion in the solids and gas phases (20% of overall performance at most). The proposed model [32] was based on single-particle flow. At higher solids mass fluxes, heat transfer is affected by agglomeration, which was not included in the model (Figure 22.10). From the Figure 22.10, it can be seen that the NTUs varies according to different packing types used and it increases with increasing gas and solids mass fluxes.

VI. MASS TRANSFER

Mass transfer rates in gas–flowing solids–fixed bed contactors are expected to be high, according to fluid dynamics and heat transfer behavior. Somewhat lower values of mass transfer coefficients than those expected were reported in the literature [6,35–37]. The reasons for that are the effects of segregation as well as strong influence of axial backmixing. Apart from this, mass transfer rates depend on size and structure (porosity) of flowing solids [36].

A. Experimental Results

Mass transfer behavior was examined through adsorption experiments in systems with different packing types used. Roes and van Swaaij [35] used experimental arrangement which consisted of two packed columns, an adsorber and a stripper interconnected by pneumatic transport lines. The column was filled with dumped Pall rings, the solid phase was a freely flowing catalyst carrier, and the gas phase was air at ambient conditions containing Freon-12 as adsorbing component.

Roes and van Swaaij [35] (Pall rings) and Verver and van Swaaij [6,37] (double-channel baffle column) experimentally obtained values of the mass transfer rate constant, which were much lower than values calculated from experimental solids holdup data and the well-know Ranz–Marshall correlation [38,39]. The low experimental values are to be attributed to particle-shielding phenomena due to the formation of less diluted suspensions or trickles.

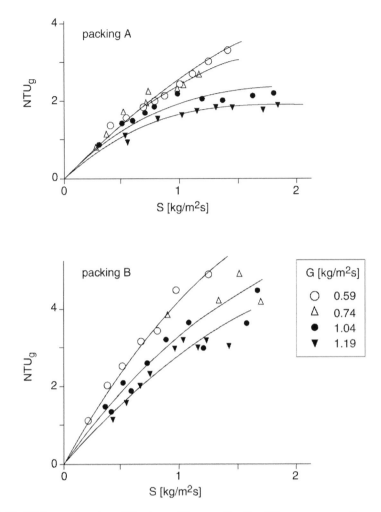

FIGURE 22.10 NTUs as a function of flowing solids mass flux for different gas mass fluxes. Comparison of experimental and predicted values for two different types of packing elements. (From Verver, A.B., van Swaaij, W.P.M., *Powder Technol.*, 45 (2), 119–132, 1986. With permission from Elsevier.)

Adsorption of water on 640 and 2200 m diameter molecular sieve, at ambient conditions, was used by Kiel et al. [36] to obtain mass transfer coefficients. They used a column with a cross-sectional area of 0.06×0.06 m^2 and a packing height of 0.27 and 0.53 m. The packing consisted of a bank of regularly stacked bars. Experiments resulted in an estimation of the actual gas–solids mass transfer coefficient, because the pore diffusion resistance could not be eliminated completely. In contrast, in these experiments, the agglomeration effects were avoided through the use of larger flowing solids particles. Mass transfer coefficients amount to 40–80% of the theoretical values, which were calculated by applying the Ranz–Marshall correlation for a single sphere in an undisturbed gas flow (Figure 22.11).

B. Mathematical Models

Simple theoretical models were proposed [35,36] to describe mass transfer phenomena in gas–flowing solids–fixed bed contactors. They included axial dispersion, without taking into account segregation effects and formation of trickles.

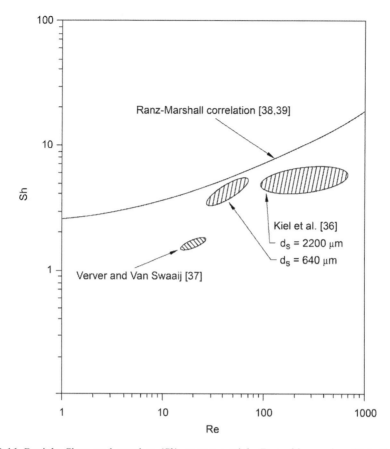

FIGURE 22.11 Particle Sherwood number (*Sh*) versus particle Reynolds number (*Re*). Comparison of results in gas–flowing solids–fixed bed contactor and Ranz–Marshall correlation for single sphere.

Roes and van Swaaij [35] interpreted their results with the number of overall transfer units:

$$N_{k,ov} = \frac{1}{1-F} \ln \frac{C_{g,in} - C_{s,out}/m}{C_{g,out} - C_{s,in}/m} \qquad (22.37)$$

where F is the extraction factor, which was evaluated from experiments. The number of overall transfer units decreases if the gas velocity increases, but it increases slightly if solid mass flux increases at the given gas velocity.

Roes and van Swaaij [35] described the adsorption process with the model of interphase mass transfer (the number of true transfer units, N_k) and axially dispersed plug flow:

$$\frac{dC_g}{d\xi} - \frac{1}{Pe_g} \frac{d^2 C_g}{d\xi^2} + N_k \left(C_g - \frac{C_s}{m} \right) = 0 \qquad (22.38)$$

$$\frac{d(C_s/m)}{d\xi} + \frac{1}{Pe_s} \frac{d^2 (C_s/m)}{d\xi^2} + N_k F \left(C_g - \frac{C_s}{m} \right) = 0 \qquad (22.39)$$

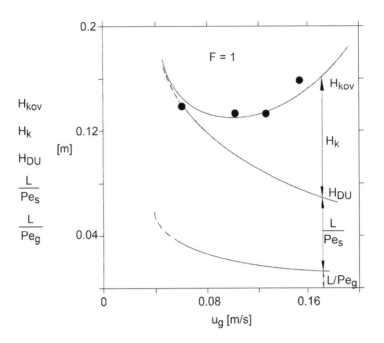

FIGURE 22.12 Different factors determining column performance as a function of the superficial gas velocity for $F = 1$. (From Roes, A.W.M. and van Swaaij, W.P.M., *Chem. Eng. J.*, 18, 29, 1979. With permission from Elsevier.)

The boundary conditions were represented by Danckwerts' equations [40]. The authors used the experimental results for axial dispersion from their previous investigations [29,30]. The differential equations were solved using several well-known approximations, which relate the number of overall transfer units with the number of true transfer units, as well as with the axial dispersion and with the extraction factor.

Figure 22.12 presents different factors which determine the column performance as a function of the superficial gas velocity for $F = 1$. The height of overall transfer unit can be expressed as:

$$H_{k,ov} = H_k + \frac{L}{Pe_g} + \frac{L}{Pe_s} \tag{22.40}$$

At lower gas velocities, the axial dispersion of gas, and especially that of the solid phase, is the limiting factor for column performance. At higher gas velocities, mass transfer limitations become important. For conditions of practical importance, the height of a true mass transfer unit corresponds to four to nine Pall ring layers.

Kiel et al. [36] used one-dimensional two-phase model to calculate gas–solids mass transfer coefficients. Basic assumptions for the model were (i) gas–solid mass transfer is the only resistance for adsorption and (ii) the effective area for mass transfer is equal to the external surface area of the spheres. Radial effects were neglected due to the good radial distribution properties of the regularly stacked packing. Axial dispersion in the gas phase was also estimated from the experimental results presented by Roes and van Swaaij [29,30].

The experiments with different inlet concentrations showed that the apparent gas–solids mass transfer coefficient value increases when the inlet concentration decreases. This indicated the existence of the pore diffusion. The pore diffusions resistance could not be eliminated completely, so the first assumption of the model is not reliable. Consequently, the mass transfer coefficients presented in the work are conservative estimates.

VII. APPLICATIONS

Gas–flowing solids–fixed bed contactor was patented as early as 1948 [1]. The first industrial realization occurred in France in 1965, named the raining packed bed exchanger [31] for heat recuperation process. Since then, it has been efficiently exploited. Potential applications are gas purification, adsorption processes, drying, etc. Another interesting concept would be the use of gas–flowing solids–fixed bed catalytic reactors for the equilibrium reactions with separation of products *in situ.*

A. HEAT RECUPERATION

French company Saint-Gobain (a glass manufacturer) and one of its subsidiaries TNEE (Tunzini-Nessi Enterprises d'Equipments) developed the gas–flowing solids–fixed bed heat exchanger, commercially named "Saturne" [31,34]. The TNEE application was used for the thermal reclaiming of foundry sands. Removal of the organic residues from the used sand may be done either mechanically or thermally. Thermal treatment is favorable because the total destruction of the organic residues can be obtained without affecting the particle characteristics. To avoid high energy consumption, two gas–flowing solids–fixed bed heat exchangers were integrated with a fluidized bed combustion unit (Figure 22.13).

The industrial operating unit has corroborated the experimental data obtained through both laboratory and pilot tests. This specially designed system operates with zero supplemental fuel consumption when used sand that contains more than 1% of organic compounds. Besides sand reclamation, gas–flowing solids–fixed bed concept may be used for the other applications such as heat recovery from ashes, from hot dirty gases, etc.

Boumehdi et al. [33] suggested gas–flowing solids–fixed bed heat recovery system with layers of horizontal tubes as a fixed bed. According to their concept, the heat transfer medium flows inside the tubes, which are perpendicular to the flow directions of gas and flowing solids. The authors presented three types of heat transfer processes: (i) transfer from a hot gas to solids and heat transfer

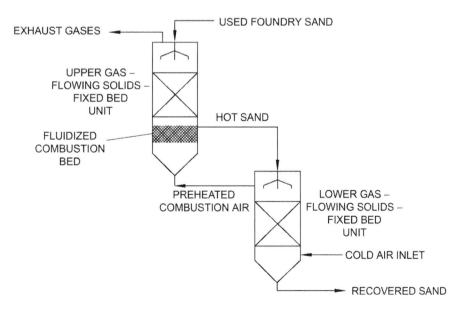

FIGURE 22.13 Foundry sand reclamation system consisting of two gas–flowing solids–fixed bed heat exchangers and a combustion fluidized bed. Conceptual design. (From Le Lan, A., Niogret, J., Large, J.F., and McBridge, J.A., *AIChe Symp. Ser.*, 80 (236), 110–115, 1984. With permission.)

medium, (ii) transfer from solids to a heat transfer medium (gas is employed to improve the contact between the phases), and (iii) transfer from a heat transfer medium to solid particles (e.g., desorption — where gas is also used to improve the contact and to carry away the products of desorption). Experimental investigations were performed on a pilot-plant level for two configurations: for multistage arrangement and for tubes forming a compact bundle. The latter was found to be much more suitable for industrial applications.

B. ADSORPTION

Gas–flowing solids–fixed bed contactors, according to their mass transfer and fluid dynamics properties, are suitable for various environmental applications, which involve adsorption of pollutants. Several investigations [6,10,36,37,41] have been carried out at pilot-plant level, and the results were promising for the design of industrial equipment. In the presented works [6,10,36,37,41], the adsorption process is followed by the chemical reaction on the flowing solids phase.

Kiel [10] obtained results from bench-scale "gas–solid trickle flow reactor" for simultaneous removal of SO_x ad NO_x from flue gas. Besides gas–flowing solids–fixed bed adsorber, the pilot plant contained a fluidized-bed regenerator and a pneumatic transport system for continuous sorbent circulation. The experimental program was limited to flue gas desulfurization (NO_x removal was not taken into account). The experiments were carried out with specially developed silica-supported CuO sorbent. Two different operation procedures have been considered: (i) the one with complete sorbent regeneration and subsequent introduction of reduced sorbent particles into the adsorber and (ii) the one without a sorbent regeneration, where oxidized (and partially sulfated) sorbent particles are introduced into the adsorber. Process operation with simultaneous oxidation of sorbent is clearly favorable compared with the operation with separate preoxidation. The simultaneous oxidation leads not only to large SO_2 removal degrees (up to 50%) during the period of relatively fast oxidation but also after this period. The SO_2 removal remains approximately five times larger than the one predicted by the model in the case of separate preoxidation. The research indicated that the gas–flowing solids–fixed bed contactor could be an efficient adsorber used in a full-size plant for dry regenerative flue gas desulfurization with supported CuO sorbent.

Kiel et al. [41] proposed a model, which predicts the performance of a full-scale gas–flowing solids–fixed bed absorber, for flue gas desulfurization. An one-dimensional, two-phase axially dispersed model has been applied. The model was derived from separate mass and heat balances for both gas and (porous) solid phases in the case of noncatalytic gas–solid reaction, which is first order in the gas phase.

This model resulted in the axial profiles for the four independent variables, namely, the gas-phase and the solid-phase temperatures and the concentrations of the gaseous and solid reactant. These axial profiles were calculated numerically for conditions based on the experimental findings from previous studies. Under isothermal conditions, the model equations can be solved analytically.

The SO_2 removal efficiencies over 95% can be achieved in a gas–flowing solids–fixed bed adsorber with a length of 15 m. Furthermore, the model [41] predicts a large temperature peak for both phases in the absorber if the heat capacity ratio (defined as the ratio of mass flux times specific heat capacity for both phases) is close to one. This temperature peak is a result of a combination between exothermic sulfation reaction and efficient countercurrent gas–solid heat exchange. The magnitude of the temperature peak decreases with a decreasing gas–solid heat-exchange efficiency, that is, with either a decrease of the gas–solids heat transfer coefficient or of the reactor length.

Verver and van Swaaij [6,37] studied the oxidation of H_2S by O_2 producing elemental sulfur in a gas–flowing solids–fixed bed reactor. In this reactor, one of the reaction products, that is, sulfur, had been removed continuously by a flowing catalyst or adsorbent. The arrangement of the reactor was similar to that of a single-channel zig-zag column. The NaX zeolite embedded in a porous, free-flowing catalyst carrier had been used as a catalyst and sulfur adsorbent. The mass transfer rate from the gas–solid suspension was evaluated experimentally, using the much faster reaction of H_2S with SO_2 and the same catalyst.

To describe trickle flow of particles in the reactor, Verver and van Swaaij [37] proposed a three-phase model, the respective phases being particle-free gas phase, gas–solids suspension, and catalyst. Within the gas–solids suspension phase, diffusion of the reactants is parallel to the reaction, which occurs on the catalyst. Mass transfer occurs in the zone between the gas phase and the outer surface of the suspension phase. According to the trickle-phase model, the overall rate constant can be defined as:

$$k_{ov} = \frac{1}{(1/k_g a') + (1/\beta k_r \eta_{tr})}$$

(22.41)

where a' is external surface area of trickle phase and η_{tr} is "effectiveness factor of trickle phase," which can be derived from the Theile modulus. This three-phase, trickling flow model can be the extension of a simple two-phase mass transfer models presented earlier [35,36], especially in the case of small particles where agglomeration effects are considerable.

From the experimental results obtained in the "trickle flow reactor" and model predictions, it could be concluded that for the H_2S—O_2 reaction, no mass transfer limitations are to be expected below about 200°C (Figure 22.14). However, at about 300°C, both external mass transfer and diffusion within the dense solids suspension are likely to offer substantial resistance to the reaction. Ultimately, at higher temperatures, the conversion rate can be determined by gas-phase mass transfer rate only, similar to the much faster reaction of H_2S with SO_2 (Figure 22.14).

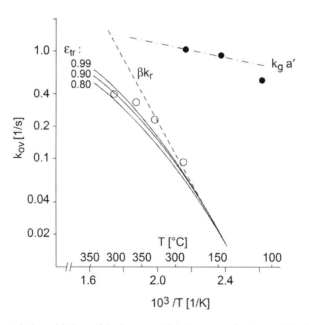

FIGURE 22.14 The catalytic oxidation of hydrogen sulfide in a gas–flowing solids–fixed bed reactor. The effect of temperature on the overall rate constant. Experimental results: (●) H_2S—SO_2 experiments (negligable chemical reaction limitations); (○) H_2S—O_2 experiments (overall rate constant); calculated values: —·—·—) mass transfer controls, empirical correlation by Verver; (– – –) no mass transfer limitations; (——) three-phase model for different values of void fraction in trickles. (From Verver, A.B., The Catalytic Oxidation of Hydrogen Sulphide to Sulphur in a Gas-Solid Trickle Flow Reactor, Ph.D. thesis, University of Twente, Enschede, The Netherlands, 1984. With permission.)

C. MULTIFUNCTIONAL REACTORS

1. Chemical Reactors with Separation *In Situ*

The classical processes for heterogeneous catalytic equilibrium reaction require high investment and operating costs. The ammonia and the methanol synthesis are the first-in-class examples among many other industrial processes, because of the extensive consumption of these chemicals. The main disadvantage of typical reactors comes from the equilibrium limitations. Owing to unfavorable chemical equilibrium, in these processes, only a fraction of the raw materials is converted, requiring high pressure. After the reactor, the nonconverted reactants must be separated from the products. The products are removed by condensation, and the reactants are recirculated to the reactor. The consecutive cooling and heating of the nonconverted gaseous reactants in and out of the condensing section are highly energy consuming. Owing to the recirculation, the flow through the reactor is highly increased, increasing the pressure drops over catalyst bed and condenser. In addition, reactor rates are reduced as equilibrium is approached, and consequently, longer catalyst bed is required.

To avoid the equilibrium restrictions, the idea was to use highly selective adsorbent to remove the product *in situ*, as soon as it is formed. This could be achieved introducing a third, flowing solids phase (selective adsorbent) in the catalytic reactor, countercurrently to the gas phase. Moving the equilibrium would lead to higher conversion, and consequently, all negative effects caused by recirculation would be reduced. If the product could be taken away completely, in principle, the reaction could go to completion.

Westerterp et al. [42,43] made a techno-economical evaluation of a methanol plant (of 1000 tpd) based on the gas–flowing solids–fixed bed principle. They compared economical performance of that system with a Lurgi methanol plant, which is designed in a classical manner, although it is highly optimized. The authors based their assessment on the research project, which included pilot-plant tests [44] and modeling of "gas–solid–solid trickle flow" reactor [45]. They predicted significant energy savings in the reactor section itself: 75% on circulation energy and 50% on cooling water. Furthermore, the raw materials consumption could be decreased by 5–10%, due to the fact that there would not be any material losses in purge gases. Large savings could also be expected in the synthesis of gas generation section, because it would no longer be necessary to produce a stoichiometric composition in the feed stream. Finally, the required amount of catalyst per unit of methanol produced could be reduced by 40%. However, the authors believed that the capital investments in the system would be equal to that in a Lurgi plant, because of the large cost of the solids handling system required. Nevertheless, large operating cost savings could make this concept attractive for realization.

The research team of the Twente University of Technology [44] designed a pilot-plant for the synthesis of methanol from CO and H_2, based on gas–flowing solids–fixed bed concept. The reactor consisted of three gas–flowing solids–fixed bed reactors with cooling sections in between. The catalyst was Cu on alumina and the selective adsorbent for methanol was a silica–alumina powder. From experimental results, Kuczynski et al. [44] concluded that the complete conversion is attainable in a single-pass operation. The initial high reaction rate could be maintained over the whole reactor length because of the effective *in situ* product removal. The authors also presented a possible way of operation at incomplete conversion, in which the nonconverted reactants, together with non-adsorbed products, leave the reactor at the top. The experimental results were compared with theoretical model developed by the authors [45], and the agreement is good over the whole range of conversions.

A model which Westerterp and Kuczynski [45] proposed was one-dimensional steady state. To derive a model, they assumed that the concentrations and the temperature would not change in radial direction. Furthermore, they assumed that the adsorption of the reaction product would be instantaneous, so that the adsorption equilibrium would exist over the entire reactor length. Therefore, the assumptions implied that the height of a reaction unit is much higher than the height of a mass-transfer unit, as well as the heights of axial mixing units for both solid and gas phase. The model was derived starting with the equations for: material balance for component A (CO), total material balance, and energy balance. These were all presented in dimensionless form.

The relevant dimensionless groups in the model were Damköhler number (Da) and adsorption number (E). Da represents dimensionless residence time in the reactor. The adsorption number E is a ratio of product that will be adsorbed and the amount that will leave the reactor on the top.

Numerically obtained results showed that at sufficiently high Da values complete conversion is attainable and that the driving force for the reaction remains high over the entire catalyst bed. Owing to the high solids flow, the methanol content of the gas phase remains low, so that the reaction rate keeps increasing, despite the temperature increase. At higher values of E, the methanol concentration decreases in the column as well as at the reactor outlet. Consequently, the driving force for the reaction increases. An increase in E also leads to a reduction of temperature rise in the reactor, which is a result of a higher heat adsorbing capacity of the increased solids stream.

The authors [45] showed that the improvements of the cooling system would lead to reduction of the solids recycle ratio and also to lowering of the temperature level. Therefore, they suggested an adiabatic reactor with intermediate cooling as the best solution for methanol synthesis operating on gas–flowing solids–fixed bed principle.

Figure 22.15 presents the comparison between gas–flowing solids–fixed bed reactor ($E = 10$) and the one without the adsorption of product ($E = 0$). Both reactors operate isothermally, but with two cases presented, one for 500 K and the other for 540 K. From Figure 22.15, it can be seen that at low conversions, there is no big distinction in performance of different reactors. In contrast, a great improvement was found for gas–flowing solids–fixed bed reactor in and above the zone of equilibrium conversions. A comparison of the gas–flowing solids–fixed bed reactors ($E = 10$) at two

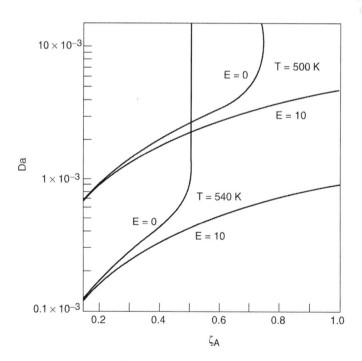

FIGURE 22.15 Methanol synthesis with ($E = 10$) and without ($E = 0$) selective product adsorption. Damköhler number (Da) versus the overall CO conversion (ζ_A) at isothermal conditions for two different temperatures. (From Westerterp, K.R. and Kuczynski, M., *Chem. Eng. Sci.*, 42, 1871, 1987. With permission from Elsevier.)

different temperature levels shows that at full conversion, the capacity of the reactor which operates at the higher temperature is much higher. However, at high temperatures, the adsorption capacity of the powder is low, thus demanding a high level of recycling ratio. This would hardly be economical, so in the practice, the gas–flowing solids–fixed bed reactor has to operate at relatively low temperature levels.

2. Chemical Reactors with Heat Supply or Removal *In Situ*

In highly exothermic or endothermic heterogeneous chemical reaction, input or output of heat in the reaction zone is needed. For chemical reactors with fixed bed of catalyst, it was suggested [21,46] that heat supply or removal can be provided by flowing solid particles. This concept was proposed [46] for heterogeneous catalytic dehydrogenation of hydrocarbons, but in principle, it could be used for any heterogeneous catalytic reaction with high value of reaction heat. The gaseous mixture of reactants is introduced at the top of the column, together with flowing solids at the same tempera-ture. Flowing solids act as a heat carrier, supplying the heat for endothermic or absorbing the heat for exothermic reactions. Owing to much higher heat capacity of solids than of the gaseous mixture, the temperature gradient along the reactor becomes very small. Moreover, moving particles have strong influence on the rate of heat and mass transfer rates in a catalyst bed.

VIII. CONCLUSIONS

Gas–flowing solids–fixed bed contactors can be used either as two-phase continuous contactors between flowing solids and gas where the fixed bed structure can be specially designed for distri-bution of both flowing phases or as three-phase contactors where the fixed bed has an active role (e.g., catalyst bed).

Most of the research up to date was primarily directed toward the determination and prediction of major fluid dynamic properties, including heat and mass transfer rates. Although there is still lack of data and prediction models for some of the parameters, many favorable features have been iden-tified: low pressure drop, high heat and mass transfer rates, and low axial dispersion of gas and flowing solids.

Numerous applications of this type of contactors were proposed for a number of separation processes (e.g., gas purification, adsorption, and drying), as well as for multifunctional reactors. Although the feasibility of some of these ideas was confirmed on the bench or pilot-plant scale, the only process that has been realized on industrial scale up to date was for heat recovery. This application of gas–flowing solids–fixed bed contactors has been efficiently exploited in France since 1965.

NOMENCLATURE

A, B	constants in the Ergun equation, Equation (22.23)
A_c	column cross-section area (m^2)
A_s	surface area of one solid particle (m^2)
Ar	Archimedes number for flowing solids particles $[= d_s^3 \cdot (\rho_s - \rho) \cdot \rho \cdot g/\mu^2]$
a	surface area of packing per unit bed volume (m^2/m^3)
a'	interfacial area of the trickle phase $[= (8\pi\beta/(1 - \varepsilon_{tr})A_c)^{1/2}]$ (m^2/m^3) reactor
a_{int}	internal surface area of solid per cubic meter of solid (m^2/m^3)
Bo_g	gas-phase Bodenstein number $(= u_g d_s/D_g \varepsilon)$
Bo_s	solids-phase Bodenstein number $(= U_s d_s/D_s \beta)$
C_D	drag coefficient

C_g	concentration in gas phase (kg/m^3)
C_s	surface concentration on flowing solids (kg/m^2)
Cp_g	gas heat capacity (J/kg K)
Cp_s	solids heat capacity (J/kg K)
D	column diameter (m)
Da	Damköhler number ($= R_{P,R}\rho_C A_c L / G_{m,in}$)
D_g	gas-phase axial dispersion coefficient (m^2/sec)
D_f	gas diffusion coefficient (m^2/sec)
D_s	solids-phase axial dispersion coefficient (m^2/sec)
d_{eq}	equivalent diameter of packing [$= 6(1-\varepsilon)/(a+4/D)$](m)
d_S	flowing solids particle diameter (m)
d_V	equivalent diameter of voids, Equation (22.21), (m)
E	adsorption number $= M \cdot m \cdot p/(\rho_g G_{m,in} y_{A,in} R_g T_R)$
F	extraction factor ($= u_g/ma_{int}u_s$) (m)
F_B	buoyancy force per unit solids volume (N/m^3)
F_D	drag force per unit solids volume (N/m^3)
F_G	gravity force per unit solids volume (N/m^3)
Fr	Froude number for flowing solids ($= S/\rho_s^2 g d_{eq}$)
f	rate of exchange between dynamic and static holdup per unit bed volume (m^3/m^3 sec)
G	gas mass flux (kg/m^3 sec)
G_m	molar gas flow rate (mol/sec)
g	gravity acceleration (m/sec^2)
H_{DU}	height of overall dispersion unit ($= L/Pe_g + L/Pe_s$) (m)
H_k	height of true transfer unit ($= u_g/k_g \cdot a$) (m)
$H_{k,ov}$	height of overall transfer unit, Equation (22.41), ($= L/N_{k,ov}$) (m)
k_g	gas-phase mass transfer coefficient (m/sec)
k_{ov}	overall rate constant (1/sec)
k_r	chemical reaction rate constant (1/sec)
L	fixed bed height (m)
M	solids mass flow rate (kg/sec)
m	adsorption equilibrium constant (m)
N_k	number of true transfer units ($= kaL/u_g$)
$N_{k,ov}$	number of overall transfer units, Equation (22.37)
p	pressure (Pa)
Pe_g	Pecklet number for gas phase ($= Bo_g L/d_{eq}$)
Pe_s	Pecklet number for solid phase ($= Bo_s L/d_{eq}$)
Re	particle Reynolds number ($= u_g d_{eq}\rho_g/\mu$)
Re_{load}	Reynolds number at preloading–loading transition point, Equation (22.1)
Re_{PB}	packed bed Reynolds number, Equation (22.24)
Re_s	particle Reynolds number based on relative velocity, Equation (22.15)
R_g	gas constant ($= 83,144$ J/mol K)
R_P	production rate (mol/kg sec)
$R_{P,R}$	production rate at reference temperature (mol/kg sec)
S	mass flux of flowing solids (kg/m^2 sec)
t	time (sec)
T	absolute temperature (K)
Th_{tr}	Thiele modulus for trickle phase [$= 1/\varepsilon_{tr}(Vk_r\beta/8\pi L\tau D_f)^{1/2}$]
T_g	temperature of gas (K)
T_R	reference temperature (K)
T_s	temperature of solids (K)
U_s	superficial solids velocity (m/sec)

u_g	superficial gas velocity (m/sec)
u_g'	effective superficial gas velocity $[= u_g/(\varepsilon - \beta)]$ (m/sec)
$u_{g,corr}$	corrected superficial gas velocity, Equation (22.9)
u_R	relative velocity between gas and flowing solids (m/sec)
$\bar{u}_R$	mean relative velocity (m/sec)
u_{Rt}	terminal relative velocity (m/sec)
u_s	particle velocity (m/sec)
$\bar{u}_s$	mean particle velocity (m/sec)
u_{sL}	local mean particle velocity, Equation (22.29), (m/sec)
u_{st}	terminal particle velocity (m/sec)
$v_{T,in}$	volumetric flow rate of tracer particles from flowing to static zone (m^3/sec)
$v_{T,out}$	volumetric flow rate of tracer particles from static to flowing zone (m^3/sec)
V	reactor volume (m^3)
V_s	solid particle volume (m^3)
z	axial coordinate along the packed bed column (m)
α	heat transfer coefficient between gas and particles (W/m^2 K)
β	flowing solids holdup
β_{dead}	dead part of static solids holdup
β_{dyn}	dynamic solids holdup
β_{st}	static solids holdup
ε	fixed bed void fraction
ε_{corr}	corrected fixed bed void fraction open to gas flow $(= \varepsilon - \beta)$
ε_{tr}	void fraction in solids trickles, Equation (22.8)
γ	relative pressure drop, Equation (22.25)
ϕ	fraction of tracer in static solids holdup
η	thermal efficiency $[\eta = (T_s(z=0) - T_s(z=L))/T_g(z=0) - T_s(z=L)]$
η_{tr}	effectiveness factor of trickle phase $(= \tan h(Th_{tr})/Th_{tr})$
φ	fraction of tracer in active part of static solids holdup
μ	gas dynamic viscosity (kg/(m sec))
ρ_C	bulk density of the catalyst (kg/m^3)
ρ_g	gas density (kg/m^3)
ρ_S	skeletal density of the flowing solids particles (kg/m^3)
τ	tortuosity factor
ξ	dimensionless length coordinate
ζ_A	conversion of reactant A
ψ	empirical parameter, Equation (22.18)

REFERENCES

1. De Directie Van De Staatsmijnen in Lumburg. Procede pour augmenter la concentration des particules solides dans un courant de milieu gazeuz, French Patent 978287, 1948.
2. Kaveckii, G.D. and Planovskii, A.N., Flow study of solids in up flowing gas in packed columns (Issledovanie techeniya tverdogo zernistogo materiala v voshodyaschem potoke gaza v kolonah s nasadkoi), *Khim. Tekhnol. Topl. Masel.*, 11, 8, 1962.
3. Claus, G., Vergnes, F., and Le Goff, P., Hydrodynamic study of gas and solid flow through a screen-packing, *Can. J. Chem. Eng.*, 54, 143, 1976.
4. Roes, A.W.M. and van Swaaij, W.P.M., Hydrodynamic behavior of a gas–solid counter-current packed column at trickle flow, *Chem. Eng. J.*, 17, 81, 1979.
5. Large, J.F., Naud, M., and Guigon, P., Hydrodynamics of the raining packed-bed gas–solids heat exchanger, *Chem. Eng. J.*, 22, 95, 1981.

6. Verver, A.B., The Catalytic Oxidation of Hydrogen Sulphide to Sulphur in a Gas-Solid Trickle Flow Reactor, Ph.D. thesis, University of Twente, Enschede, The Netherlands, 1984.

7. Saatdjian, E. and Large, J.F., Heat transfer in raining packed bed exchanger, *Chem. Eng. Sci.*, 40, 693, 1985.

8. Verver, A.B. and van Swaaij, W.P.M., The hydrodynamic behavior of a gas–solid trickle flow over a regularly stacked packing, *Powder Technol.*, 45, 119, 1986.

9. Westerterp, K.R. and Kuczynski, M., Gas-solid trickle flow hydrodynamics in a packed column, *Chem. Eng. Sci.*, 42, 1539, 1987.

10. Kiel, J.H.A., Removal of Sulphur Oxides and Nitrogen Oxides from Flue Gas in a Gas-Solid Trickle Flow Reactor, Ph.D. thesis, University of Twente, Enschede, The Netherlands, 1990.

11. Duduković, A.P., Predojević, Z.J., Petrović, D.Lj., and Pošarac, D., The influence of the type of reactor height on the flow characteristics of a gas-solid-solid reactor, *J. Serb. Che. Soc.*, 57 (5–6), 309–317, 1992.

12. Benali, M., Shakourzadeh-Bolouri, K., and Large, J.F., Hydrodynamic characterization of dilute suspensions in the gas-solid-solid packed contactors, *Proceedings of the Seventh Found. Conf. Fluid.*, 651–658, 1992.

13. Benali, M. and Shakourzadeh-Bolouri, K., The gas-solid-solid packed conctactor: hydrodinamic behavior of counter-current trickle of coarse and dense particles with a suspension of fine particles, *Int. J. Multiphase Flow*, 20, 161, 1994.

14. Predojević, Z.J., Fluid Dynamics of Countercurrent Gas–Solid–Packed Bed Contactor, Ph.D. thesis, Faculty of Technology, Novi Sad, Yugoslavia, 1997.

15. Predojević, Z.J., Petrović, D.Lj., and Duduković, A.P., Flowing solids dynamic holdup in the counter-current gas–flowing solids–fixed bed reactor, *Chem. Eng. Commun.*, 162, 1, 1997.

16. Predojević, Z.J., Petrović, D.Lj., Martinenko, V., and Duduković, A.P., Pressure drop in a gas–flowing solids–fixed bed contactor, *J. Serb. Chem. Soc.*, 63, 85, 1998.

17. Stanimirović, O.P., Prediction of Pressure Drop and Dynamic Holdup of Solid Phase in a Gas–Solid–Packed Bed Reactor, B.S. thesis, Faculty of Technology, Novi Sad, Yugoslavia, 1998.

18. Pjanović, R., The Basic Characteristics of Three-Phase Gas-Solid-Solid Contactor, M.S. thesis, Faculty of Technology and Metallurgy, Belgrade, Yugoslavia, 1998.

19. Predojević, Z.J., Petrović, D.Lj., and Duduković, A.P., Pressure drop in a countercurrent gas–flowing solids–packed bed contactor, *Ind. Eng. Chem. Res.*, 40, 6039, 2001.

20. Nikačević, N.M., Duduković, A.P., and Predojević, Z.J., Dynamic holdup in a countercurrent gas–flowing solids–packed bed contactors, *J. Serb. Chem. Soc.*, 69 (1), 77, 2004.

21. Barysheva, L.V., Borisova, E.S., Khanaev, V.M., Kuzmin, V.A., Zolotarskii, I.A., Pakhomov, N.A. and Noskov, A.S., Motion of particles through the fixed bed in a gas-solid-solid downflow reactor, *Chem. Eng. J.*, 91 (2–3), 219–225, 2003.

22. Guigon, P., Large, J.F., and Molodstof, Y., Hydrodinamics of raining packed bed heat exchangers, in *Encyclopedia of Fluid Mechanics*, Cheremissinoff, N., Ed., Gulf Publishing Co., Houston, 1986, pp. 1185–1214.

23. Kiel, J.H.A. and van Swaaij, W.P.M., A Theoretical Model for the Hydrodynamics of Gas–Solid Trickle Flow Over Regularly Stacked Packings, *AIChE Symp. Ser.*, 85 (270), 11–21, 1989.

24. Duduković, A.P., Nikačević, N.M., Petrović, D.Lj., and Predojević, Z.J., Solids holdup and pressure drop in gas–flowing solids–fixed bed contactors, *Ind. Eng. Chem. Res.*, 42, 2530, 2003.

25. Duduković, A.P., Nikačević, N.M., and Kuzeljević, Ž.V., Modelling and predictions of solids dynamic holdup in gas–flowing solids–fixed bed contactors, *Ind. Eng. Chem. Res.*, 43, 7445, 2004.

26. Turton, R. and Levenspiel, O.A., Short note on the drag correlation for spheres, *Powder Technol.*, 47, 83, 1986.

27. Končar-Djurdjević, S., Zdanski, F., and Duduković, A., Drag in single fluid flows, in *Encyclopedia of Fluid Mechanics*, Cheremisinoff, N.P., Ed., Gulf Publishing Co., Houston, 1986, Vol. 1, pp. 443–453.

28. Duduković, A.P., Nikačević, N.M., Pjanović, R.V., and Kuzeljević, Ž.V., Exchange between the stagnant and flowing zone in gas–flowing solids–fixed bed contactors, *J. Serb. Chem. Soc.*, 70 (1), 137, 2005.

29. Roes, A.W.M. and van Swaaij, W.P.M., Axial dispersion of gas and solid phases in a gas–solid packed column at trickle flow, *Chem. Eng. J.*, 18, 13, 1979.

30. Roes, A.W.M., The Behaviour of a Gas–Solid Packed Column at Trickle Flow, Ph.D. thesis, Twente University of Technology, The Netherlands, 1978.

31. Compagnie de Saint-Gobain, Produit intermediaire pour la fabrication du verre et autres silicates, et procede et appareillages pour sa fabrication, French Patent 1469109, 1965.

32. Verver, A.B., van Swaaij, W.P.M., The heat-transfer performance of gas–solid trickle flow over a regularly stacked packing, *Powder Technol.*, 45 (2), 119–132, 1986.

33. Boumehdi, P., Guigon, P., and Large, J.F., Heat recovery from solids in a raining bed exchanger, *Heat Recovery Syst.*, 5, 407–414, 1985.

34. Le Lan, A., Niogret, J., Large, J.F., and McBridge, J.A., Use of RPBE in an efficient foundry sand reclamation unit, *AIChe Symp. Ser.*, 80 (236), 110–115, 1984.

35. Roes, A.W.M. and van Swaaij, W.P.M., Mass transfer in a gas–solid packed column at trickle flow, *Chem. Eng. J.*, 18, 29, 1979.

36. Kiel, J.H.A., Prins, W., and van Swaaij, W.P.M., Mass transfer between gas and particles in gas–solid trickle flow reactor, *Chem. Eng. Sci.*, 48, 117, 1993.

37. Verver, A.B., van Swaaij, W.P.M., Modeling of a gas–solid trickle flow reactor for the catalytic oxidation of hydrogen sulfide to elementar sulfur, *Inst. Chem. Eng. Symp. Ser.*, 87, 177–184, 1984.

38. Ranz, W.E. and Marshall, W.R., Evaporation from drops, *Chem. Eng. Prog.*, 48 (3), 141–146, 1952.

39. Ranz, W.E. and Marshall, W.R., Evaporation from drops, *Chem. Eng. Prog.*, 48 (4), 173–180, 1952.

40. Danckwerts, P.V., Continuous — flow systems. Distribution of residence times, *Chem. Eng. Sci.*, 2, 1–13, 1953.

41. Kiel, J.H.A., Prins, W., and van Swaaij, W.P.M., Modelling of non-catalytic reactions in a gas–solid trickle flow reactor: dry, regenerative flue gas desulphurisation using a silica-supported copper oxide sorbent, *Chem. Eng. Sci.*, 47, 4271, 1992.

42. Westerterp, K.R. and Kuczynski, M.A., Retrofit methanol plants with this converter system. Hydrocarbon processing, *Int. Ed.*, 65 (11), 80–83, 1986.

43. Westerterp, K.R., Bodewes, T.N., Vrijiland, M.S., and Kuczynski, M.A., Two new methanol converters. Hydrocarbon processing, *Int. Ed.*, 67 (11), 69–73, 1988.

44. Kuczynski, M., Oyevaar, M.H., Pieters, R.T., and Westerterp, K.R., Methanol synthesis in a countercurrent gas-solid-solid trickle flow reactor. An experimental study, *Chem. Eng. Sci.*, 42, 1887, 1987.

45. Westerterp, K.R. and Kuczynski, M., A model for a counter-current gas-solid-solid trickle flow reactor for equilibrium reactions. The methanol synthesis, *Chem. Eng. Sci.*, 42, 1871, 1987.

46. Zolotarskii, I.A., Pakhomov, N.A., Barysheva, L.V., Kuzmin, V.A., Noskov, A.S., Zudilina, L.Yu., Lakhmostov, V.S., and Khanaev, V.M., Method of Catalytic Dehydrogenation of Hydrocarbons, Russian patent 2178399, 2002.

23 Reaction and Capillary Condensation in Dispersed Porous Particles

Nickolay M. Ostrovskii
Boreskov Institute of Catalysis, Omsk, Russia

Joseph Wood
University of Birmingham, Birmingham, U.K.

CONTENTS

I. INTRODUCTION

A. GENERAL

Solid heterogeneous catalysts are typical finely dispersed systems. Depending on the manufacturing method, the porous structure of catalyst grain is formed by numerous microparticles or nanoparticles bound together. The diameter of these particles varies from a few nanometers to hundred nanometer. For example, mixed hydroxide or carbonate catalysts are normally prepared by precipitation, leading to a final crystallite size of 3–15 nm in the precipitated catalyst [1]. The void space between particles in the catalyst represents the pore-structure, where active centers such as metal nanoparticles are located. Such a structure can be random (irregular), which is typical for most porous catalysts, or well-ordered, that is characteristic for zeolites.

During a catalytic reaction in porous structures, complicated mutual interactions of different processes takes place. These include adsorption–desorption, surface-catalyzed reaction, gas–liquid phase transition, and capillary phenomena. If the reactor feed consists of a single phase (gas or liquid), usually the same phase exists within the pores of the catalyst pellet. Some exceptions occur, notably trickle bed reactors, where both gas and liquid are fed to the reactor. However, even if the reactor feed is a single phase under inlet conditions, condensation of gas or vapor, and evaporation of liquid can lead to phase changes within the reactor. In this chapter, we concentrate in particular on capillary condensation in the catalyst particles, whereby the reactor feed is gas phase but condensation to liquid occurs in the finer pores of the catalyst as a result of capillary forces. It happens when the vapor pressure in small capillaries (pores) is less when than the dew point. Typical pore radii in which capillary condensation occurs are in the range 3–50 nm, which is typical of the pore sizes found in many industrial catalysts. As a result of the capillary condensation process, a configuration of pore filling as illustrated in Figure 23.1 occurs, whereby the smaller pores are filled with liquid and the larger pores with gas.

In this chapter, we consider the influence of capillary condensation upon diffusion/reaction processes taking place in porous catalysts and its role in catalyst deactivation. In reactions accompanied by capillary condensation, a two-phase transport/reaction process occurs in the catalyst pellets. We discuss the influence of pore structure upon the extent pore space filled by condensate and consider the transport processes that occur within the pores. The partial filling of catalyst pellets by liquid can lead to interesting effects such as different reaction rates in gas and liquid, changes in permeability, pore-blocking of gas-phase transport by liquid-filled pores, and modified rates of catalyst poisoning and coke deposition upon the catalyst surface. It is important to understand how capillary condensation can influence catalytic reactions (such as motor fuel hydrotreatment in fixed bed catalytic reactors), so that the process designer can allow for the aforesaid effects in selecting the operating temperature, pressure and catalyst pore size distribution to optimize the reaction conditions.

B. Background

The phenomenon of capillary condensation is widely used to investigate and measure the pore structure of catalysts because the extent of condensation or adsorption occurring at certain pressure depends upon the pore radius of the adsorbent [2]. Nitrogen sorption makes use of this effect for the

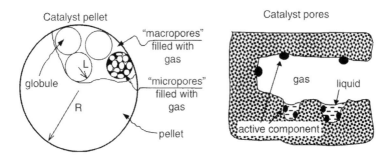

FIGURE 23.1 Bidisperse globular structure of pellet. R, pellet radius; L, globule radius. "Micropores," inside globule; "Macropores," between globules.

measurement of the pore structure of adsorbents and catalyst, and the results are interpreted using the Kelvin equation [3]. This important equation tells us the critical radius of pores below which capillary condensation occurs at a fixed pressure:

$$\ln\left(\frac{P_m}{P_S}\right) = -\frac{2\sigma V_m}{RT(r_K - \delta)} \tag{23.1}$$

Nevertheless, there are only a few publications devoted to the capillary condensation during catalytic reactions. The papers published by D.H. Kim and Y.G. Kim [4] were probably the first works concerning this subject. They experimentally observed several steady states of the reaction rate for cyclohexene hydrogenation over Pt/Al_2O_3, which were explained to be the result of capillary condensation of the reagents. These experiments were analyzed and interpreted in detail by Jaguste and Bhatia [5]. Using a theoretical analysis, Gurfein and Danyushevskaya [6] defined the limitations for the occurrence of capillary condensation in a catalyst particle and for its influence on the reaction.

Ostrovskii and Bukhavtsova published several experimental and theoretical works on capillary condensation in catalytic reactions. Capillary condensation was found to accompany some gasphase catalytic processes, in particular hydrotreating of jet fuel fractions [7]. The effects of gas–liquid interfacial surface, intra-particle diffusion, and of the ratio of gas to liquid reaction rates under conditions of capillary condensation were estimated [8]. The experimental study of p-xylene hydrogenation on Pt/SiO_2 (as a model reaction) was carried out in order to demonstrate the influence of capillary condensation on reaction kinetics and process dynamics, and corresponding model was proposed [9]. Finally, the poisoning of the catalyst under capillary condensation was also considered [10].

Experimental verification of the models has been carried out using equipment ranging from a thermogravimetric analyzer [11], a gradientless recycle reactor [9], to a single-pellet diffusion reactor [12,13].

Shapiro and Stenby [14] formulated a multicomponent version of the Kelvin equation, which is used to analyze the capillary condensation of nonideal mixtures at high pressure. Wood [13] has used this equation to calculate the critical pore radius at which capillary condensation occurs for a mixture of hydrocarbons and hydrogen. On this basis, Wood and Gladden [15] proposed an algorithm for calculating the critical radius below which capillary condensation occurs in combination with a pore network model of vapor diffusion and reaction. It was used to demonstrate the influence of network topology and porous structure upon the catalyst effectiveness factor in the presence of capillary condensation. Theoretical investigations of thiophene hydrodesulfurization were carried out in order to demonstrate the influence of capillary condensation on catalytic hydrotreating processes [15,16]. Experimental investigations of hydrotreating catalyst deactivation were also made [17].

C. Pore Networks

The models that may be used to represent the porous medium and transport processes within fall into two broad categories: continuum and discrete models. The former model is the simplest type, where the porous material is treated as a continuum. This type of approach is valid when the characteristic length for the variation in the macroscopic concentrations is much larger than the linear dimension of a statistically representative region of the pore space.

Pore network models are an example of a discrete model. The earlier pore network models consisted of parallel pores [18] and randomly oriented cross-linked pores [19]. Bethe lattice [20], and regular networks [21] have also been used to represent catalyst structures. Pore network models have been used to analyze the complicated interactions between diffusion and reaction that may occur in catalyst particles, for example Sharatt and Mann [21] used their cubic network to study diffusion and reaction in a porous catalyst. They showed that tortuosity varies with the Thiele modulus, the extent of this variation depending upon the pore-size distribution. Hollewand and Gladden [22] developed a random pore network model, which they applied to the problem of diffusion and reaction for the case of a single, isothermal, irreversible first order reaction. The random network is shown schematically in Figure 23.2a. The use of a random model attempts to provide a more realistic representation of a catalyst pellet, as in an industrial catalyst, sample pores are randomly oriented rather than fixed on a grid pattern. A comparison of tortuosities predicted from simulations on random and regular networks was made, and it was found that the random network model predicts significantly different tortuosities from the regular network simulation.

To represent catalyst pellets with a bimodal pore-size distribution using a pore network, some special considerations of the macro- and micro-structural features are necessary. Bidisperse catalyst pore structures consist of distinct regions of micro-voids connected together by a network of macro-voids. Hollewand and Gladden [23] modeled bidisperse pellets by setting up a network of macro-pores, then assigning micropores branching off the nodes of the macroporous network, as shown in Figure 23.2b.

Ramirez-Cuesta et al. [24] have developed a model which accounts for some further features of pore networks such as:

1. Heterogeneity — the basic entities of the pore space are not simply cylinders, but have size distributed bodies and throats.
2. Correlation — small pore throats are statistically more likely to be connected to small pore bodies and large throats to large bodies, not assigned at random.
3. Non-uniform connectivity — the number of throats connected to a pore is not constant throughout the network.

However, it must be remembered that when analyzing complex diffusion-reaction problems with percolation phenomena, it is desirable to use as large a network as possible. Therefore, in terms of computer memory, there is a trade-off between using a smaller network of more complex geometry and using a larger network of cylindrical pores.

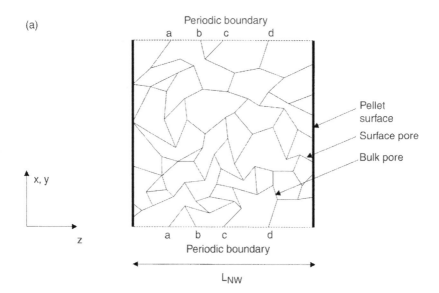

FIGURE 23.2 (a) Random network. (b) Bidisperse random network with correlated micro- and macro-porous regions (simulations are three-dimensional).

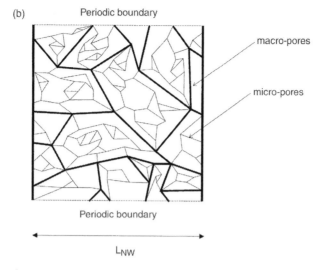

(b)

Periodic boundary

macro-pores

micro-pores

Periodic boundary

L_{NW}

FIGURE 23.2 *Continued.*

D. Capillary Condensation in Pores

The critical radius of pore below which capillary condensation occurs depends upon the properties of the fluid such as surface tension, fluid–solid interaction (such as the contact angle made by the liquid with the pore wall), and the relative pressure referred to the dew point. For pure fluids, it is well known that the critical radius of capillary condensation can be estimated from the Kelvin Equation (23.1). However, for industrial catalytic reactors, the Kelvin equation is of limited use in predicting the onset of capillary condensation because:

- The reactor feeds are usually multicomponent in nature
- The vapor phase is non-ideal
- The liquid phase may be compressible

In order to overcome these problems, Shapiro and Stenby [14] have proposed a multicomponent version of the Kelvin equation which is applicable to gas mixtures at high pressures:

$$\frac{P_c}{P_d} = \left(\frac{V_{VL}}{V_L}Z_{av} - 1\right)(\chi - 1) + \frac{V_{VL}}{V_L}Z_{av}(\chi - 1)^2 \tag{23.2}$$

In the multicomponent model, relative pressure is defined as $\chi = P/P_d$, where P_d is the dew point pressure. The average compressibility ratio Z_{av} is the ratio of the compressibility of the vapor at pressure P to the vapor compressibility at the dew point P_d. As the dew point is approached, Z_{av} approaches unity, and so in the neighborhood of the dew point may be taken as unity.

An important feature of the new generalized Kelvin equation is the presence of a thermodynamic parameter, the "mixed volume":

$$V_{VL} = \sum_{i=1}^{N_c} x_{Ld}^i V_V^i \tag{23.3}$$

Then the critical (Kelvin's) radius for capillary condensation can be obtained from the equation:

$$P_c = \frac{2\sigma \cos \theta}{r_K} \tag{23.4}$$

II. POSSIBLE FEATURES AND EFFECTS

When a catalyst pellet is partially liquid filled, some interesting and unusual features can occur, compared with the single phase. These effects can be summarized as:

- The pore volume and surface area available for gas-phase reaction is decreased.
- Reaction kinetics in the condensed liquid differ from the gas-phase kinetics.
- Catalyst poisoning occurs with different rates in the gas- and liquid-filled pores.
- Reaction rates are influenced by capillary vapor–liquid equilibria and by mass transfer.
- Regions of the catalyst pore structure become blocked off or disconnected from the bulk gas at the catalyst surface due to the presence of liquid-filled pores.
- The fraction and distribution of condensate-filled pores can have multiple steady states for a given temperature and pressure, depending on the history of the pellet.

These effects are considered in more detail in the following discussion. Ostrovskii et al. [7] have found that in their calculations of capillary condensation of jet fuel during hydrotreatment, condensate filled pores constituted 30% of the overall pore-volume and 50% of the surface area at a temperature of 300°C and pressure of 40 atm. Clearly, this will have a significant effect on the number of active sites of the catalyst that are accessible for gas-phase reaction. The extent to which the overall reaction rate is modified will depend on the relative reaction rates in the gas and liquid phase.

Such peculiarities are obvious especially in hydrocarbon processes that are carried out in the presence of hydrogen. Here the reaction conditions in large and small pores differ dramatically, mainly due to low solubility of gases in liquids (especially H_2 in hydrocarbons) shown in Table 23.1.

From studies in trickle bed reactors, Sedriks and Kenny [25] suggested that the overall rate of reaction in partially liquid-filled catalysts r_{tot}, depends on the rate in the gas phase r_G, the liquid phase r_L, and the wetted fraction of the active catalyst surface f, as:

$$r_{tot} = fr_L + (1-f)r_G \qquad (23.5)$$

The fraction of dry catalyst was found contribute significantly or even dominate the overall rate of reaction. The effect was attributed to the higher rates of mass transfer and lower rates of heat transfer in gas-filled pores compared with liquid-filled pores. Similarly, Wood and Gladden [15] have assumed that gas-phase reaction dominates the overall rate in their simulations of hydrodesulfurization of diethyl sulfide, by neglecting liquid-phase reaction. Ostrovskii et al. [7] have

TABLE 23.1
Conditions in Large and Small Pores for a Catalyst Close to the Dew Point

Parameter	Large Pores (gas)	Small Pores (liquid)
Total concentration, mol/l	1–3	5–10
H_2/hydrocarbons, mol/mol	3–5	10^{-3}
Diffusivity, cm^2/s	10^{-2}–10^{-3}	10^{-4}–10^{-6}
Adsorption equilibrium	$b_{H_2}P_{H_2} \approx \Sigma b_i P_i$	$b_{H_2}C_{H_2} \ll \Sigma b_i C_i$
Adsorption energy	Heat of adsorption	Heat of condensation
Reaction rate	Rate limiting step	Dissolution of H_2
Deactivation	Coking	Tar formation
Self-regeneration	By hydrogen	Tar solubility

Note: b_i are constants of adsorption equilibrium.

found that reaction kinetics of benzene hydrogenation over Pt/Al_2O_3 catalysts differ between the gas and liquid phase. The reaction was found to be third order with respect to hydrogen in the gas phase, though first order only in the liquid phase. In addition, the hydrogen solubility in the liquid phase influenced the rate of reaction. The rate of coke deposition in gas- and liquid-filled pores can also be differed if the rates of any side reactions that produce coke are phase dependent.

Regarding the rate of liquid-phase reaction, it is necessary to know the concentration of reactants within the capillary condensed phase in order to apply the appropriate rate equation. For the simple case, the liquid-phase composition can be estimated using Henry's law [8], or for multicomponent mixtures at high pressures a flash calculation can be carried out using an equation of state in combination with an appropriate algorithm, such as the stability test procedure of Michelsen [26,27]. The flash calculation determines the composition of liquid in equilibrium with a vapor-phase mixture in unconfined space. Although providing an estimate of liquid composition, it should be noted that in narrow capillaries the equilibrium diagram may differ from the bulk, and therefore the liquid composition in the capillary may not match the calculation exactly.

It was explained earlier that condensation in the catalyst pores leads to loss of surface area and volume available for vapor-phase reaction and diffusion. This effect may be greater than anticipated from the liquid-filled fraction alone if isolated areas of vapor-filled pores become blocked off from the bulk gas at the catalyst surface by regions of condensed liquid. Such problems can be solved using the techniques of percolation theory [28,29].

The percolation problem arises by assigning pores in a network as present (conducting) or absent with probabilities p and $q = (1 - p)$, respectively. Two sites are said to be "conducting" if there is a path between them consisting solely of occupied bonds. For an infinite lattice, there exists a critical conducting fraction, the percolation threshold p_c. In application to the capillary condensation problem, pores are considered to be "conducting" if filled with gas, and "nonconducting" if filled with liquid. Then the percolation threshold represents the case where there are just enough pores filled with condensate to allow diffusion of vapor between the two sides of the network, and at this point the pores connected to the vapor-filled path of pores are said to belong to the "backbone" of the network. Hoshen and Kopelman [30] have presented an algorithm for counting clusters of pores that belong to the backbone fraction of a percolating pore network. In catalytic reactions accompanied by capillary condensation, the algorithm has been useful in identifying which pores are connected to the pellet surface by pathways of other vapor-filled pores.

III. MASS TRANSFER IN PELLETS

A. Approximate Calculations of the Effect of Capillary Condensation

The porous structure of most catalysts is polydisperse. Therefore, capillary condensate fills only part of the pore-space — mostly small pores. The bidisperse globular structure (Figure 23.1) is convenient to consider as a model for rough estimations of the influence of external mass transfer and intraparticle diffusion on the total reaction rate. Such an analysis was made by Ostrovskii and Bukhavtsova [8]. According to this model, the only pores inside globules (micropores) will fill with liquid, and space between globules (macropores) fill with gas. Then the total porosity can be written as

$$\varepsilon_0 = \varepsilon_G + \varepsilon_L, \qquad \varepsilon_G = \varepsilon_0(1 - \varphi_L), \qquad \varepsilon_L = \varepsilon_0\varphi_L, \qquad \varepsilon = 1 - \varepsilon_0(1 - \varphi_L) \qquad (23.6)$$

It is convenient to compare the features of mass transfer in a trickle-bed (completely wetted pellet) and in catalyst pellets under condition of capillary condensation. In trickle-bed reactors, the interfacial gas–liquid surface (S_T) is slightly less (due to porosity) than the external surface of pellet S_R, $S_T \leq S_R\varepsilon_G$, and is proportional to $(1 - \varepsilon)\,\varepsilon_G/R$.

Pellets containing capillary condensate are surrounded by gas flow and the interfacial surface S_C is equal to the meniscus area in pores filled with liquid, which corresponds to the external surface of globules S_L, $S_C \approx S_L \varepsilon_L$, and is proportional to $(1 - \varepsilon)\varepsilon_L/L$.

Usually $R = 1$–5 mm and $L = 100$–1000 nm. Then the ratio of S_C and S_T is:

$$\frac{S_C}{S_T} = \frac{\varepsilon_L R}{\varepsilon_G L} \approx 10^3 - 10^5 \tag{23.7}$$

Thus, the gas–liquid interfacial area at capillary condensation is significantly higher than in trickle-bed and does not limit the mass transfer.

Under steady state, the reaction rate in "wetted" zone of pellet r_L is equal to the diffusion flux in the gas phase through the external surface of pellet F_R and is also equal to the diffusion flux of components into "wetted" globules F_L.

Assuming that (grad $C = 1$), F_R and F_L can be expressed as:

$$F_R \sim \frac{D_R S_R \varepsilon_G}{R}, \qquad D_R = 10^{-2} - 10^{-3}\,\text{cm}^2/\text{s}$$

$$F_L \sim \frac{D_L S_L \varepsilon_L}{L}, \qquad D_L = 10^{-4} - 10^{-5}\,\text{cm}^2/\text{s}$$

Because of $S_R \sim (1 - \varepsilon)/R$ and $S_L \sim (1 - \varepsilon)/L$, we can estimate

$$\frac{F_L}{F_R} \approx \frac{D_L R^2 \varepsilon_L}{D_R L^2 \varepsilon_G} = 10^5 - 10^6 \tag{23.8}$$

This value indicates that diffusion in capillary condensed liquid and the mass transfer through the gas–liquid interfacial film does not limit the overall rate of process under capillary condensation, unlike the situation in trickle-bed. Therefore, only diffusion in the gas phase (in large pores) or the catalytic reaction rate in the liquid (in small pores) is the rate limiting steps.

Taking into account this estimation, Ostrovskii and Bukhavtsova [8] have considered the simplified model of reaction/diffusion in catalyst particle under capillary condensation. According to Equation (23.7), we can suppose that diffusion limitation inside the globule (Figure 23.1) is negligible, even in the case where it is filled with liquid. Then the diffusion/reaction equation has the form

$$(1 - \varepsilon)D_R \left(\frac{S_R}{V_R}\right)^2 \frac{d^2 C_G}{d\rho^2} = k_G(1 - \varepsilon)C_G + \varepsilon k_L C_L \tag{23.9}$$

Using Henry's law $C_L = HC_G$, a simple equation, we can obtain:

$$\frac{d^2 C_G}{d\rho^2} = \psi^2 \beta C_G, \qquad \psi = \frac{V_R}{S_R}\sqrt{\frac{k_G}{D_R}}, \qquad \beta = 1 + \frac{\varepsilon}{1 - \varepsilon}\gamma, \qquad \gamma = \frac{H k_L}{k_G} \tag{23.10}$$

where ψ is Thiele parameter, k_G, k_L are reaction rate constants, and H is Henry's constant.

Let us express the effectiveness factor as the ratio of observed and kinetic reaction rates ($\eta = r_0/r_{kin}$):

$$r_{kin} = \beta k_G (1 - \varepsilon)C_G^o, \qquad r_0 = \beta k_G (1 - \varepsilon)\int_0^1 C_G(\rho)\,d\rho$$

In strong diffusion region, the following approximate expression holds

$$\eta \approx \frac{1}{\psi\sqrt{\beta}} \tag{23.11}$$

Also the relative rate of reaction ($\chi = r_{0L}/r_{0G}$) was used to characterize the processes occurring in the pellet. It equals the ratio of observed rate with (r_{0L}) and without (r_{0G}) capillary condensation. In the latter case $\gamma = 1$ and $\beta = 1/(1-\varepsilon)$. Therefore

$$r_{0G} = \frac{k_G C_G^o \sqrt{1-\varepsilon}}{\psi}, \qquad r_{0L} = \frac{k_G C_G^o \sqrt{\beta(1-\varepsilon)}}{\psi}$$

and consequently

$$\chi \approx \sqrt{\beta(1-\varepsilon)} = \sqrt{(1-\varepsilon)+\varepsilon\gamma} \tag{23.12}$$

The plots of dependencies (23.11) and (23.12) are represented in Figure 23.3 as $\eta\psi$ and χ functions of pellet "microporosity" ε and γ parameter.

At the absence of liquid phase, $\gamma = 1$ and $\chi = 1$, and effectiveness factor becomes

$$\eta_G = \frac{\sqrt{1-\varepsilon}}{\psi} \tag{23.13}$$

From r_{0L} and r_{0G}, it follows that $\eta_L = \eta_G/\chi$ or $\chi = \eta_G/\eta_L$. The value of η_G decreases with increasing value of ε because of decreasing of effective diffusivity $D_E \approx (1-\varepsilon)D_R$. This dependence for η is represented in Figure 23.3a by dotted line at $\gamma = 1$.

If capillary condensation leads to excluding the active centers in "micropores" from the reaction ($\gamma = 0$, $\beta = 1$), then the effectiveness factor is determined only by Thiele modulus ($\eta \approx 1/\psi$), and relative rate of such pellet ($\chi = \sqrt{1-\varepsilon}$) decreases with increasing ε according

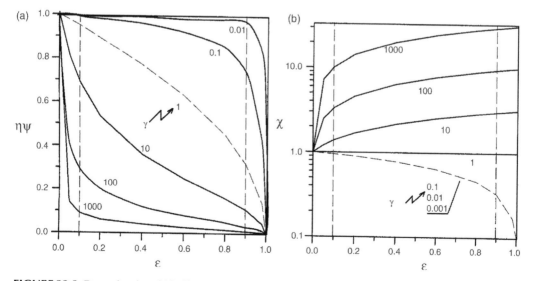

FIGURE 23.3 Dependencies of (a) effectiveness factor (η) and (b) relative rate (χ) on ε and γ parameters. (From Ostrovskii, N.M. and Bukhavtsova, N.M. *React. Kinet. Catal. Lett.*, 56, 391–399, 1995. With permission.)

to dotted line in Figure 23.3b. The estimates represented here are approximate and, obviously, limit range is $0.1 < \varepsilon < 0.9$ shown in Figure 23.3 by vertical lines.

Note that the relative rate χ does not depend on Thiele modulus ψ, because the diffusion resistance is concentrated in gas phase only.

It is clearly seen from the Figure 23.3, that if the reaction rate in liquid phase is less than those in gas phase ($\gamma \ll 1$), then χ and η slightly depend on ε and γ, respectively. It comes from a small contribution of the liquid-phase reaction in overall conversion rate. The process in pellet proceeds then in kinetic or in diffusion region depending on Thiele parameter value ψ.

In the opposite case, at $\gamma \gg 1$, the main contribution to the conversion occurs in the liquid phase. Therefore χ and η are determined mainly by $\gamma = Hk_L/k_G$ value and reaction is shifted to the strong diffusion region.

B. TRANSPORT PHENOMENA INSIDE CATALYST PARTICLE

1. Types of Diffusion

There are essentially three diffusion mechanisms that occur within a pore. They are [31]:

- *Bulk diffusion*: molecule–molecule collisions dominate over molecule–wall collisions. This type of diffusion is important at high pressures.
- *Knudsen diffusion*: molecule–wall collisions predominate over molecule–molecule collisions. This type of diffusion is very important in micro-pores.
- *Surface diffusion*: adsorbed species on the walls of the pores may be transported along the pore walls. This mechanism is important for strongly adsorbed species and in micro-pores.

In multicomponent systems, which prevail in catalytic processes, some curious effects occur that were predicted by Toor [32]:

- *Osmotic diffusion*: diffusion of a component takes place in spite of the absence of a driving force.
- *Reverse diffusion*: diffusion of a component takes place in a direction opposite to its driving force, which means that components flow in a direction opposite to their concentration gradients.
- *Diffusion barrier*: diffusion flux of a component is zero, despite a large driving force.

These three phenomena cannot be described by the traditional version of Fick's law, but are fully consistent with the modern multicomponent diffusion theory [33]. Multicomponent diffusion in porous media can be represented by the "dusty gas" model, which is described in the following section.

Bulk and Knudsen diffusion mechanisms occur in series and it is always wise to take both mechanisms into account, rather than assuming "controlling" only [34]. Surface diffusion occurs in parallel to the other mechanisms, and its contribution to the total species flux may be quite significant, especially at elevated pressures [34].

In order to calculate bulk and Knudsen diffusion fluxes simultaneously, the "dusty gas" model is often applied [35–37], where giant dust molecules distributed uniformly in space represent the pore walls.

2. Diffusion and Flow in Surface Adsorbed and Capillary Condensed Adsorbate

Apart from diffusion in continuum phase, the transport of surface-adsorbed molecules and capillary condensate takes place in meso- and macroporous media. In order to model transport of adsorbable vapor at elevated pressure, it is necessary to consider the type of adsorption occurring: monolayer adsorption, multilayer adsorption, or capillary condensation [38]. Models for surface diffusion have been proposed by Krishna [38], and various theoretical treatments for multilayer diffusion have been proposed [11,39]. It should be noted that at high pressure, multilayer adsorption, and capillary condensation can coexist within the pore structure of an adsorbent. Models taking account of this situation have been proposed by Lee and Hwang [40], Rajniak and Yang [20], Kainourgiakis et al. [41] and Kikkinides et al. [42]. They predicted that in membrane type processes, coexistence of multilayer adsorption, and capillary condensation can lead to improvements in permeability, as a result of capillary forces.

C. Estimation Using Fick's Law

Let us consider now the combination of models representing transport of gas and liquid in porous structures with kinetic expressions describing reaction at the active sites of the catalyst. This model considers the effect of the catalyst pore structure in more detail than the one presented in Section III.A. Wood and Gladden [15] used a network model to simulate the effects of diffusion/reaction accompanied by capillary condensation.

A random pore network model was constructed using an algorithm of Hollewand and Gladden [22]. As illustrated in Figure 23.2, the nodes of the network are randomly distributed in space, then connected to their nearest neighbors to achieve a desired connectivity. The cylindrical pores of the network connect two nodes together, such that each node is connected to the number of neighbors corresponding to the coordination number or connectivity of the network. At two faces of the network that constitute the pellet surfaces the pores are open ended, and at the other four faces periodic boundary conditions apply.

Wood and Gladden [15] simulated the following hydrodesulfurization reaction, as an example:

$$C_4H_{10}S \quad + \quad 2H_2 \quad \rightarrow \quad 2C_2H_6 \quad + \quad H_2S$$
$$\text{Diethyl Sulfide (A)} \quad \text{Hydrogen (B)} \quad \text{Ethane (C)} \quad \text{Hydrogen Sulfide (D)}$$

Although the kinetics of such reactions are often of Langmuir–Hinshelwood form [43], it was necessary to make some simplifying assumptions in order to solve the model. As hydrogen (B) is present in great excess in the hydroprocessing reactor, hydrogen concentration C_B, is effectively constant. Therefore, the diffusion process may be approximated as binary diffusion of diethyl sulfide (A) through hydrogen (B) within the porous medium, with pseudo first order reaction. For the case of the binary diffusion of a dilute component, Fick's law can be used to model the process.

The one-dimensional diffusion/reaction equation for a single pore is given by

$$D_{\text{eff}} \frac{d^2 C_A}{dz_i^2} - k_v C_A = 0 \tag{23.14}$$

The Thiele modulus for a catalyst pore network is given by

$$\Phi_{\text{NW}} = L_{\text{NW}} \sqrt{\frac{k_{vp}}{D}} \tag{23.15}$$

where k_{vp} is the reaction rate constant per unit void volume for the network and L_{NW} the length of the pore network.

The solution of the diffusion/reaction equation throughout the network is achieved by solving simultaneously the one-dimensional equation for each pore in the network. A dimensionless distance along the pore is defined as $Z_i = z_i/l_i$.

A dimensionless concentration is defined which relates C_{Ai}, the concentration at any point within the network to C_{AS}, the pellet surface concentration $C_{ri}(z_i) = C_{Ai}(z_i)/C_{AS}$.

The dimensionless diffusion/reaction equation for a pore then follows from Equation (23.14), and can be expressed in terms of the dimensionless variables as

$$\frac{d^2 C_{ri}}{dZ_i^2} - \phi_i^2 C_{ri} = 0 \qquad (23.16)$$

where ϕ_i is a local Thiele modulus for a single pore.

The solution of Equation (23.16) throughout the network was achieved by solving simultaneously the one-dimensional equation for each pore, assuming that at the nodes of the network no adsorption or reaction takes place. The node material balance can be expressed by an equation similar to Kirchoff's law as

$$\sum_{i=1}^{Z_c} \pi r_i^2 J_i = 0, \qquad (23.17)$$

The pore effectiveness factor is given by the ratio of the moles of reactant consumed in the presence of diffusion resistance, to the moles of reactant consumed in the absence of diffusion resistance. The single pore effectiveness is used to calculate the effectiveness factor for the network

$$\eta_{NW} = \sum_{i=1}^{N} \nu_i \eta_i \qquad (23.18)$$

where N is the number of pores in the network.

The pellet effectiveness factor is defined as the ratio of the calculated reaction rate to the rate that would be observed with no diffusional limitation or capillary condensation. The network effectiveness factor is equated to the pellet effectiveness by mapping the network solution onto the continuum solution. For a rectangular slab, the Thiele modulus and effectiveness factor are related by the expression

$$\eta = \eta_{NW} = \frac{\tanh \Phi}{\Phi} \qquad (23.19)$$

The effects of capillary condensation were included in the network model, by calculating the critical radius below which capillary condensation occurs based on the vapor composition in each pore using the multicomponent Kelvin Equation (23.2). Then the pore radius was compared with the calculated critical radius to determine whether the pore is liquid- or vapor-filled. As a significant fraction of pores become filled with capillary condensate, regions of vapor-filled pores may become locked off from the vapor at the network surface by condensate clusters. A Hoshen and Kopelman [30] algorithm is used to identify vapor-filled pores connected to the network surface, in which diffusion and reaction continue to take place after other parts of the network filled with liquid. It was assumed that, due to the low hydrogen solubility in the liquid, most of the reaction takes place in the gas-filled pores. The diffusion/reaction simulation is repeated, including only vapor-filled pores connected to the network surface by a pathway of other vapor-filled pores.

In order to investigate the effect of catalyst pore structure and reactor operating pressure upon the pellet effectiveness factor, simulations were carried out over the range of variables shown in Table 23.2. The PSD of the pore network was assigned from a Normal distribution of radii in a random fashion. The mean and standard deviation of the Normal distribution are given in Table 23.2. The random network was generated 10 times for each set of conditions, and the results of these simulations then averaged.

Figure 23.4 shows the effectiveness factor plotted against network Thiele modulus obtained for four simulation pressures in the range 27.5–27.8 bar, for which capillary condensation occurs.

TABLE 23.2
Values of Input Parameters to the Random Pore Network Simulations of Diethyl Sulphide Desulfurization in a Medium Meso-Pore Catalyst

Number of nodes	200–2000 (reference value 1000)
Connectivity	Random network 3–7 (reference value 6)
	Regular network 6
Grid dimensions	1 μm³
Mean pore size	128–184 Å (reference value 136 Å)
Standard deviation pore size	16–40 Å (reference value 28 Å)
Temperature	375 K
Pressure	27.5–27.8 bar (reference value 27.8 bar)
Hydrogen surface concentration	0.95 mole fraction
Diethyl sulfide surface concentration	0.05 mole fraction

Note: The values in brackets identify the parameters used in the reference lattice.

These results are compared with those of a simulation in which pore filling by capillary condensation is omitted from the algorithm such that the network contains only vapor.

As seen in Figure 23.4, at low Thiele moduli below 0.1, the strong reaction-controlled limit, capillary condensation causes a reduction in catalyst effectiveness factor by 4–7% depending

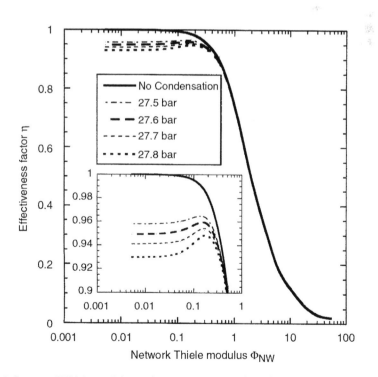

FIGURE 23.4 Influence of Thiele modulus and pressure upon catalyst effectiveness factor. Simulations were performed on a random network of 1000 nodes, defined by the reference input parameters given in Table 23.2, for pressures of 27.5, 27.6, 27.7, and 27.8 bar, at which capillary condensation occurs to varying extents. The inset of the figure shows an enlargement of the curves at Thiele moduli below 1.0. (From Wood J. and Gladden L.F., *Chem. Eng. Sci.*, 57, 3033–3045, 2002. With permission.)

upon the value of pressure. In this regime, pore diffusion resistance is weak, and reaction takes place throughout the vapor-filled part of the catalyst particle. Condensable species can penetrate deep into the pore network. Under these conditions, resistance to diffusion is low, and so reductions in the connectivity of vapor-filled pore space caused by pore blockage with liquid has little influence upon the effectiveness factor. The reduction in effectiveness is thought to be due to the loss of gas-filled pore volume that is displaced by the capillary condensate. At low Thiele moduli approximately 15% by number of the surface and bulk pores are filled with condensate, "surface" referring to those pores that are directly connected to the pellet surface, whilst "bulk" refers to pores indirectly connected to the network surface. Since the fractions of surface and bulk pores filled with condensate are approximately equal, condensate must be distributed quite evenly between the surface and interior of the pellet.

In the intermediate range 0.1–0.5 of Thiele moduli, resistance to diffusion in the pores is more significant. The presence of pores filled with condensate leads to a reduction in the connectivity of the vapor-filled pore space and so the tortuosity of the vapor-filled part of the pore structure increases. In this regime, the reduction in catalyst effectiveness factor due to capillary condensation is thought to result from both increased tortuosity of the vapor-filled pore space and reduced vapor phase pore volume. The inset in Figure 23.4 shows, on an enlarged scale, the effect of capillary condensation for $\Phi_{NW} < 1.0$; the effectiveness factor is observed to pass through a maximum at $\Phi_{NW} \sim 0.3$. The maximum occurs because on one hand, the effectiveness factor for the vapor-filled parts of the pellet increases as Thiele modulus decreases, but on the other hand the extent of capillary condensation is also higher at low values of Φ_{NW}, leading to blockage of some pores that would otherwise be available for reaction. In this regime, the percentage of surface pores filled with condensate is significantly higher than the percentage of bulk pores filled, for example at Φ_{NW}, 13% of surface pores are filled with liquid, whereas only 5% of bulk pores are liquid-filled. This effect occurs because vapor present in the surface pores of the network is richer in the condensable sulfur compound than vapor in the bulk pores.

At values of the Thiele modulus above 0.5, tending toward the strong diffusion-controlled limit, most of the reaction occurs near the surface of the catalyst pellet. It is seen from Figure 23.4 that the catalyst effectiveness is not significantly influenced by capillary condensation under these conditions. The percentage of both surface and bulk pores filled with condensate is less than 1% for Thiele moduli above 0.5, consistent with the mass flux of condensable reactant being confined to the entrance region of surface pores in the lattice. These results indicate that the influence of capillary condensation upon catalyst effectiveness is strongly dependent upon the Thiele modulus. At the strong reaction-controlled limit, capillary condensation reduces catalyst effectiveness considerably whereas at the strong diffusion-controlled limit, it has virtually no effect.

Figure 23.4 also illustrates the influence of pressure upon catalyst effectiveness under conditions of capillary condensation. At low Thiele moduli, where capillary condensation has an appreciable influence upon effectiveness factor, the effectiveness factor decreases as pressure increases. As the pressure increases, the dew point of the mixture is approached and the fraction of pores filled with condensate increases. The reduction in effectiveness factor with increasing pressure is explained by the larger fraction of condensate-filled pores that result as the dew point is approached.

Figure 23.5a shows effectiveness factor plotted against mean pore size for three different standard deviations of pore size. At a fixed value of standard deviation, the catalyst effectiveness factor increases with increasing pore size. This effect results as increasing the mean pore size of the PSD leads to a reduced possibility of the radius of a particular pore being below the critical radius for capillary condensation decreases, such that capillary condensation is less likely. Figure 23.5b shows catalyst effectiveness plotted against the standard deviation of the PSD for three mean pore sizes. At a constant mean pore size, the catalyst effectiveness factor decreases with increasing standard deviation. This effect occurs because a wide PSD incorporates more small pores which act as constrictions to diffusion and in which capillary condensation is more likely to occur. In this

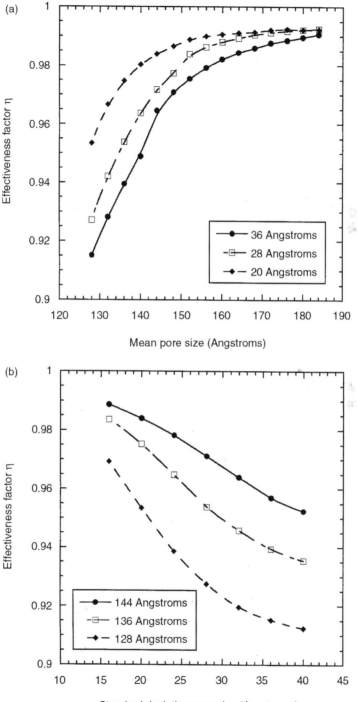

FIGURE 23.5 (a) Influence of mean pore size upon catalyst effectiveness factor. The standard deviation of the pore-size distribution is shown in the key of the figure. Simulations were performed on a random network. (b) Influence of standard deviation of pore-size distribution upon catalyst effectiveness factor. The mean of the pore-size distribution is shown in the key of the figure. The number of nodes, network connectivity and pressure are defined by the reference input parameters given in Table 23.2. (From Wood J. and Gladden, L.F., *Chem. Eng. Sci.*, 57, 3033–3045, 2002. With permission.)

figure, the influence of standard deviation is more pronounced for networks of smaller mean pore size, as condensation is more likely to occur in the smaller pores of the catalyst.

The results presented earlier illustrate that the catalyst effectiveness factor is sensitive to the properties of the PSD under conditions where capillary condensation occurs. The model could be used to optimize the pore structure and process conditions so as to control the extent of capillary condensation occurring and to increase the catalyst effectiveness factor.

D. MASS TRANSFER CALCULATIONS USING DUSTY GAS MODEL

The models mentioned so far are limited in their application as they represent only first order reaction kinetics with Fickian diffusion, therefore do not allow for multicomponent diffusion, surface diffusion or convection. Wood et al. [16] applied the algorithms developed by Rieckmann and Keil [12,44] to simulate diffusion using the dusty gas model, reaction with any general types of reaction rate expression such as Langmuir–Hinshelwood kinetics and simultaneous capillary condensation. The model describes the pore structure as a cubic network of cylindrical pores with a random distribution of pore radii. Transport in the single pores of the network was expressed according to the dusty gas model as

$$-RT\frac{r_i^2}{8\mu}\left(\mathbf{c}\frac{d^2c_{tot}}{dz^2} + \frac{d\mathbf{c}}{dz}\frac{dc_{tot}}{dz}\right) - \mathbf{D}^{-1}\frac{d^2\mathbf{c}}{dz^2} - \frac{d}{dz}(\mathbf{D}^{-1})\frac{d\mathbf{c}}{dz} - \frac{2}{r_i}\mathbf{vr}(\mathbf{c}, T) = 0 \qquad (23.20)$$

where **r** is the vector of reaction rates and **v** is a vector of stoichiometric coefficients.

The material balance condition must apply at the nodes of the network such that no adsorption or reaction occurs there, that is, no accumulation at the nodes. For each pore a boundary value problem has to be solved. Adopting the methods of Rieckmann and Keil, it was necessary to discretize each pore using a finite difference scheme. This led to a very large system of nonlinear equations, which were solved using the Schur complement method, described by Rieckmann and Keil [44].

A major advantage of the models of Rieckmann and Keil [44] and Wood et al. [16] was that, unlike earlier studies, any general reaction kinetics could be incorporated into the computer simulation. This means that reactions of industrial interest, which are typically represented by Langmuir–Hinshelwood kinetics, could be represented. Wood et al. [16] chose the hydrodesulfurization of thiophene as an example reaction, using the kinetic measurements of Van Parijs et al. [45] in their simulations. The kinetic model assumes that first thiophene is hydrogenated to butene, then butene is hydrogenated to butane:

Thiophene	+	Hydrogen	$\rightarrow$	Butene	+	Hydrogen Sulfide
C_4H_4S	+	H_2	$\rightarrow$	C_4H_8	+	H_2S
Butene	+	Hydrogen	$\rightarrow$	Butane		
C_4H_8	+	H_2	$\rightarrow$	C_4H_{10}		

The two steps of the hydrogenation were assumed to take place on two different types of active site: σ sites and τ sites. The rate of thiophene hydrogenation on σ sites is represented by the equation

$$r_{t\sigma} = \frac{k_{t\sigma}K_{t\sigma}K_{H_2\sigma}P_TP_{H_2}}{[1 + (K_{H_2}P_{H_2})^{1/2} + K_{T\sigma}P_T + K_{H_2S\sigma}P_{H_2S\sigma}/P_{H_2}]^3} \qquad (23.21)$$

and butene hydrogenation, which takes place on τ sites, is represented by

$$r_{A\tau} = \frac{k_{B\tau}K_{B\tau}K_{H_2\tau}P_BP_{H_2}}{[1 + (K_{H_2\tau}P_{H_2\tau})^{1/2} + K_{A\tau}P_A + K_{B\tau}P_B]^2} \tag{23.22}$$

where T represents thiophene, B butene and A butane.

The model representing diffusion/reaction involved solution of the transport equations for each single pore simultaneously to give concentration profile in the pore network. The calculations related to capillary condensation were performed in the same way as for the Fickian model, described in Section III.C.

The model parameters used in the pore network simulation of diffusion/reaction are:

- Pore size distribution and connectivity, which may be obtained from experimental nitrogen adsorption or mercury porosimetry measurements [46–48].
- Kinetic rate expressions, which may be obtained from experimental measurements in a Berty reactor or a similar type of reactor.
- Transport parameters such as binary diffusion coefficients and viscosities. For the purposes of the model described here, subroutines were coded to calculate diffusivity and viscosity. The binary diffusivity D_{ij} was estimated using the method of Fuller [49,50]. The viscosity was calculated according to the method of Lucas [51]. The necessary data and equations were obtained from Reid et al. [52].

Simulations were performed under the conditions shown in Table 23.3, using the kinetic parameters reported by Van Parijs et al. [45]. Figure 23.6 shows concentration profiles in a pellet for the reactants and products of the thiophene hydrogenation.

Figure 23.7a shows the conversion of thiophene at the pellet center plotted as a function of pressure. On increasing the pressure from 5 to 10 bar a slight increase in thiophene conversion is observed. This is due to a slight increase in the rate of reaction resulting from increased adsorption of the reacting species at the active sites of the catalyst. As the pressure is increased above 20 bar, the conversion of thiophene decreases very sharply, and at 21.5 bar the conversion of thiophene at the pellet center is negligible. Again the decreased conversion in networks containing capillary

TABLE 23.3
Input Parameters and Conditions for the Simulation of Multicomponent Diffusion/Reaction Accompanied by Capillary Condensation

Number of nodes	1000
Number of finite difference points per pore	6
Pellet dimension	12 mm
Mean pore size	151.2 Å
Standard deviation pore size	27.4 Å
Connectivity	4
Temperature	550.5–583 K
Pressure	5–21.5 bar
Thiophene concentration	0.2 mole fraction
Hydrogen concentration	0.6 mole fraction
Dodecane concentration	0.2 mole fraction

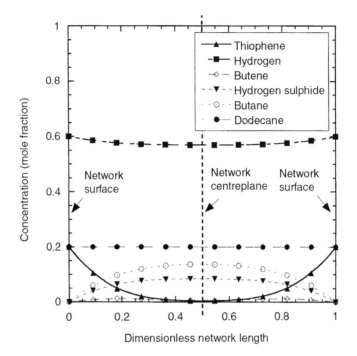

FIGURE 23.6 Concentration profiles in a pellet during thiophene hydrotreatment at a temperature of 573 K and a pressure of 20 bar. The feed composition is 60 mole% hydrogen, 20 mole% thiophene, 20 mole% dodecane. Dodecane is present as an inert carrier of thiophene. Data points represent the results of simulations and lines are smooth curves fitted to the data to guide the eye. (From Wood, J., Gladden, L.F., and Keil, F.J., *Chem. Eng. Sci.*, 57, 3047–3059, 2002. With permission.)

condensation is due to a reduction in the active surface area of the catalyst and reduced transport and reaction in dead end pores compared with "flowing" pores. As shown in Figure 23.7b, the fraction of pores on a percolating pathway decreases dramatically as the pressure approaches 21.5 bar, in the region of the dew point.

IV. REACTION KINETICS

In previous sections, we have considered some physical phenomena, which can complicate the processes occurring in a catalyst particle and influence the global (observable) reaction rate or effectiveness factor. Meanwhile, the presence of liquid condensate in some pores can also affect the intrinsic kinetics of the reaction due to the features discussed in Section II.

In order to elucidate this problem, Bukhavtsova and Ostrovskii [9] have studied experimentally the reaction kinetics on the catalysts with (and without) capillary condensation. The model catalysts Pt/SiO_2 with approximately the same characteristics except porous structure (Table 23.4) were used. Two modifications of support SiO_2 were used: KCK-1 with relatively large pores, and KCM-5 with small pores (Figure 23.8). Except pore size distribution, the platinum distribution among the pores with different size was measured by adsorption method [53].

The model reaction of *p*-xylene hydrogenation was chosen in order to provide the mild conditions of the experiments in both gas and capillary condensed phases, and to avoid the influence of side reaction and catalyst deactivation. The recycle type of gradientless reactor was used that provides uniform temperature and concentration profiles within all the catalyst packing. The catalyst particles (0.25–0.50 mm) provide a negligible intraparticle limitation of mass- and heat-transfer.

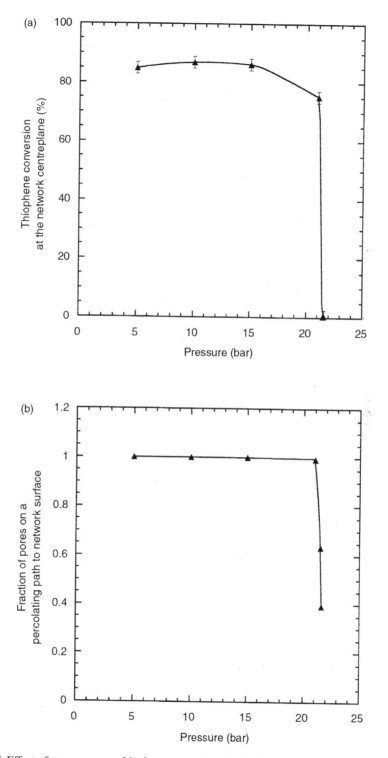

FIGURE 23.7 Effect of pressure upon thiophene conversion during hydrotreatment. Simulation temperature 553 K. (a) Thiophene conversion at network centerplane. (b) Fraction of pores on a percolating path to the pellet surface. The line is shown to guide the eye; error bars represent the spread in thiophene conversion values resulting from simulations performed upon 5 separately generated pore networks. (From Wood, J., Gladden, L.F., and Keil, F.J., *Chem. Eng. Sci.*, 57, 3047–3059, 2002. With permission.)

TABLE 23.4
Catalyst Characteristics

	Pt/KCK-1	Pt/KCM-5
[Pt] (%)	1.0	0.9
Average pore radius (nm)	10.0	3.0
Bulk density (g/cm^3)	0.40	0.55
Pt dispersion	0.51	0.70
Particle size (mm)	0.25–0.50	0.25–0.50
Pore volume (cm^3/g)	0.35	0.47
Specific area (m^2/g)	130	370

The reaction conditions are presented in Table 23.5. There r_K is the radius of pores, in which the capillary condensation is possible at initial partial pressure of p-xylene P_x^o. The values of r_K are calculated using Kelvin equation and take into account the thickness of adsorption film $\delta = 0.4$–1.0 nm.

Under conditions excluding the capillary condensation, the reaction kinetics follows a Langmuir–Hinshelwood type Equation (23.23). This is clearly seen from Figure 23.9a for Pt/KCK-1 at 60, 80, 100°C, and from Figure 23.9b for Pt/KCM-5 at 80°C.

$$k_G = k \frac{bP_x}{1 + bP_x} P_H^3 \tag{23.23}$$

However, the obvious deviation from this dependency was observed when the capillary condensation occurs, for example on Pt/KCM-5 at 60°C (Figure 23.9b).

The Pt fraction in pores filled with liquid at 60°C is represented by figures and dotted line in Figure 23.9b. This fraction of Pt (acting in liquid phase) increases with increasing P_x that results in a decrease of the reaction rate. However, the reaction rate reduction stops at $P_x = 0.045$ atm

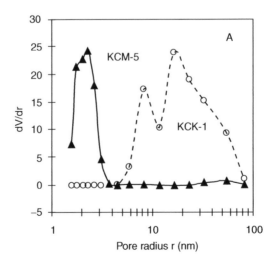

FIGURE 23.8 The distribution of differential (dV/dr) pore volume for KCK-1 (open symbol) and KCM-5 (filled symbol). (From Bukhavtsova, N.M. and Ostrovskii, N.M. *React. Kinet. Catal. Lett.*, 65, 322–329, 1998. With permission.)

TABLE 23.5
Experimental Conditions

T (°C)	P_x^o (atm)	r_K (P_x^o) (nm)	Fraction of pores with $r_p < r_K$ (%)	
			KCK-1	KCM-5
60	0.058	15.0	30	98
80	0.015	1.3	0	0
	0.058	2.5	0	70
	0.134	15.0	30	98
100	0.015	1.0	0	0
	0.058	1.4	0	0
	0.134	2.5	0	70

(corresponding to $r_K = 6$ nm) because of absence of the pores with radius $6 < r_p < 20$ nm in Pt/ KCM-5 (Figure 23.8). Further, the reaction concentrates in macropores, where approximately 40% Pt is located.

From these data, and also from dynamic experiments (Section V), the reaction rate in capillary condensed liquid was estimated. The kinetics may be represented by the equation

$$r_L = k_L P_H x \tag{23.24}$$

where $x = P_x/P_x^s$ is mole fraction of p-xylene in liquid according to Raoult–Dalton's law.

The total reaction rate, presented in Figure 23.9, was calculated as $r_{tot} = r_L \varphi_L + (1 - \varphi_L) r_G$. It is seen from Figure 23.9b that the reaction rate in liquid phase is lower than in gas phase. At 60°C and $P_x = 0.045$ atm, the ratio $r_G/r_L \approx 20$. Note that k_L includes Henry coefficient for hydrogen, the low solubility of which is responsible for the difference in reaction rates and in kinetic

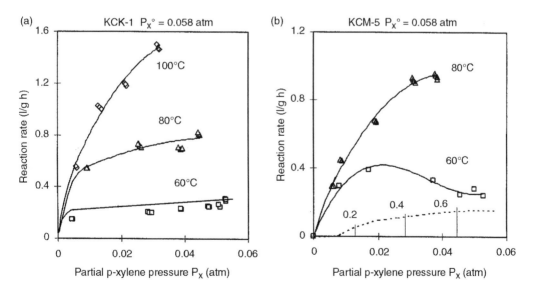

FIGURE 23.9 Steady-state kinetic curves over Pt/KCK-1 (a) and Pt/KCM-5 (b) Points — experiments, lines — model prediction. Fraction of Pt in pores filled with liquid at 60°C is shown by figures and dotted line. (From Bukhavtsova, N.M. and Ostrovskii, N.M., *React. Kinet. Catal. Lett.*, 65, 322–329, 1998. With permission.)

Equation (23.23) and Equation (23.24). The temperature dependency of the overall reaction rate r_{tot} is presented in Figure 23.10.

Similar kinetic experiments were carried out by Fatemi et al. [54] for the reaction of thiophene hydrogenolysis at the elevated temperature (170–190°C) and pressure (9.5 atm) that are close to conditions of industrial hydrotreating processes. They used n-heptane as a feed containing 3.24% of thiophene. Owing to the preferential condensation of n-heptane rather than thiophene, they did not observe a relationship between reaction rate and thiophene conversion, like the trend shown in Figure 23.9. So, the dependency of the reaction rate vs. temperature was used (Figure 23.11). To represent their experimental data, Fatemi et al. [54] have used a model similar to (Equation (23.23) and Equation (23.24)).

$$r_G = k_G y/(1 + Ky); \qquad r_L = k_L x \qquad (23.25)$$

where y and x are the mole fractions of thiophene in gas and liquid phases, respectively.

It is necessary to note that there are no special indicators for the presence of capillary condensation during catalytic reaction. Only careful examination of the experimental data provides a reliable interpretation. For example, Ostrovskii [55] has reported that the reaction of benzene hydrogenation demonstrates similar kinetic behaviors in wide range of temperature (Figure 23.12).

All the experiments in Figure 23.12, and other published experiments in the range 20–200°C, can be fitted using the same equation:

$$r = k \frac{bP_b}{1 + bP_b} P_H^3 \qquad (23.26)$$

where P_b is partial pressure of benzene, b is its adsorption coefficient.

Nevertheless, the temperature dependency of adsorption coefficient $b = b_o \exp(Q_b/RT)$ has a turning point at $\sim 70°C$. At $T > 70°C$, $Q_b \approx 12$–22 kcal/mol, but at $T < 70°C$, $Q_b \approx 7$–8 kcal/mol, that is equal to the heat of benzene condensation. A simple estimation shows that at $P_b = 0.02$–0.04 atm and $T = 50°C$, the capillary condensation of benzene is possible in pores with $r_p = 1.5$–20 nm. In this case, the Equation (23.26) may be used only at $T > 70°C$, but at $T < 70°C$ the combination of Equation (23.23) and Equation (23.24) is necessary. Certainly, it does not effect on the description of kinetic data, but is important for subsequent analysis of the process, especially in transient regimes.

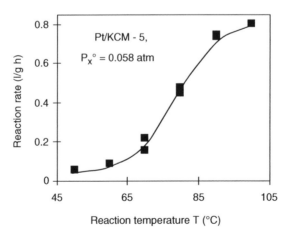

FIGURE 23.10 Reaction rate in steady-state regime vs. temperature. Points — experiments, line — model (23.23, 23.24). (From Bukhavtsova, N.M. and Ostrovskii, N.M., *React. Kinet. Catal. Lett.*, 65, 322–329, 1998. With permission.)

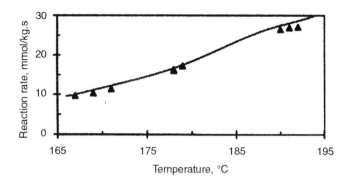

FIGURE 23.11 Reaction rate in steady-state regime vs. temperature. (From Fatemi, S., Moosavian, M.A., Abolhamd, G., Mortazavi, Y., and Hudgins, R.R., *Can. J. Chem. Eng.*, 80, 231–238, 2002. With permission.)

V. TRANSIENT REGIMES

The performance of a catalytic reactor in which capillary condensation occurs may not be uniquely defined for a certain temperature and pressure, but may also depend on the history of the operating conditions [4,56,57]. This effect was investigated by Kim and Kim [4] for the hydrogenation of cyclohexene, who reported transitions and hysteresis effects upon varying the bulk mole fraction of hydrogen in two similar catalysts, but of different activity. It was found that the effectiveness factor follows several possible "branches," depending on the mole fraction of hydrogen, which is related to the extent of capillary condensation occurring in a pellet. Bhatia attempted to explain the results based on considerations of hysteresis in the capillary vapor–liquid equilibrium. Capillary vapor–liquid equilibrium results since the shape of the liquid meniscus at the end of the pore changes depending on whether the liquid is condensing or evaporating: during condensation in a vapor-filled pore, the meniscus takes a cylindrical shape, and during evaporation in a liquid-filled pore it takes a hemispherical shape. Strictly, the Cohan equation is used to determine the conditions for condensation and the Kelvin equation is used to predict when evaporation occurs. Jaguste and Bhatia [5] obtained good agreement (Figure 23.13) between their model and the experimental data of Kim and Kim.

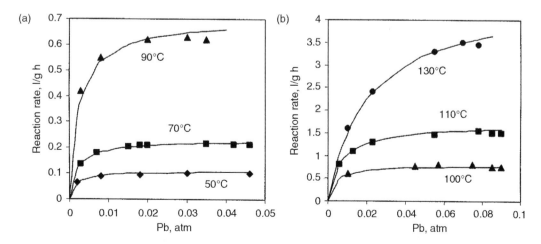

FIGURE 23.12 Benzene hydrogenation kinetics over Pt/Al_2O_3 at $P = 1$ atm. Points — experiments, lines — model (23.26). (From Ostrovskii, N.M., *Khimicheskaya Promyshlenost*, 1, 13–19, 1995. With premission.)

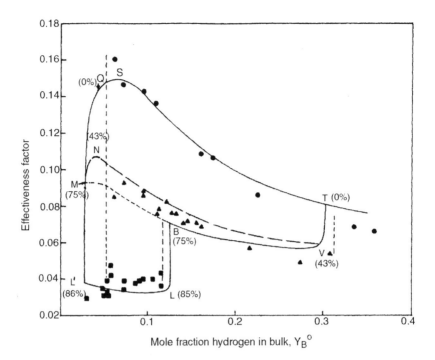

FIGURE 23.13 Catalyst effectiveness factor vs. mole fraction of hydrogen in the bulk for low activity catalyst at a temperature of 295 K. Lines represent the model predictions of Jaguste and Bhatia (1995), points represent experimental data of Kim and Kim [4]. Numbers in parentheses indicate the percent liquid filling of the pore structure. (From Jaguste, D.N. and Bhatia, S.K., *A.I.Ch.E.J.*, 37, 650–660, 1991. With permission.)

Bukhavtsova and Ostrovskii [9] have studied the effect of capillary condensation on the reaction rate of *p*-xylene hydrogenation in transitions by variations of the feedstock flow and temperature. In order to avoid the intraparticle diffusion limitation, the crushed catalyst particles were used (0.25–0.5 mm).

A. FLOW RATE VARIATION

In experiments on Pt/KCK-1, capillary condensation does not influence the reaction rate. Therefore, its relaxation to a steady-state value after the switching of flow rate proceeds rapidly. So, the kinetic function $r = f(P_x)$ in transitions practically coincides with the steady-state dependencies, represented in Figure 23.9a. This is due to the absence of the pores less than 5 nm in Pt/KCK-1. A similar result is observed on the catalyst Pt/KCM-5 at conditions where the capillary condensation is negligible.

In contrast, when the capillary condensation embraces a considerable part of pores (Pt/KCM-5, 60 and 80°C), then both steady-state kinetics change (as in Figure 23.9b) and also the reaction rate relaxation becomes slower. In Figure 23.14 the relaxation dependency for Pt/KCM-5 at 80°C and $P_x^o = 0.134$ atm are shown. The sequence of reaction rate measurements is marked by figures, and the direction of movement along the dynamic curve is shown by arrows.

The dynamic experiments start at high flow rate of feedstock. Therefore, the *p*-xylene conversion is small and its partial pressure is close to the initial value ($P_x \approx P_x^o = 0.134$ atm). The fraction of pores filled with liquid reaches 90–95% (Table 23.5), and the fraction of Pt in these pores accounts for approximately 60%.

After switching to the lower flow rate the conversion increases, P_x decreases and the reaction rate falls sharply. However, the vaporization of liquid begins because of phase equilibrium shift in

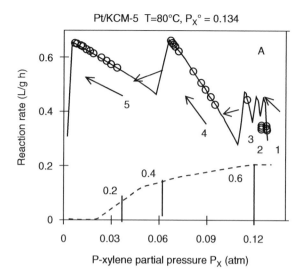

FIGURE 23.14 Reaction rate in dynamic mode of operation using flow rate variation: experimental (points) and predicted by model (solid lines). The fraction of Pt in pores filled with liquid is shown by figures and dotted line. Arrows are shown the sequence of experiments. (From Bukhavtsova, N.M. and Ostrovskii, N.M., *React. Kinet. Catal. Lett.*, 65, 322–329, 1998. With permission.)

pores. This leads to the emptying of some pores and the reaction rate increase slowly (0.5–1 h) to a new steady-state value. The products of reaction cannot replace the *p*-xylene in liquid because they have lower boiling temperatures.

B. TEMPERATURE VARIATION

The experiments were carried out at constant flow rate in two stages. At first, the steady-state dependency having S-form was obtained using slow operation of temperature (Figure 23.10 and v in Figure 23.15). From this curve, the activation energy and heat of *p*-xylene chemisorption were estimated. As the temperature is lowered to the range 50–55°C, filling of all pores with liquid occurs, which permits a more exact determination of the kinetic parameters in liquid.

At the second stage, rapid temperature variation was used (1–3°C/min). The results of dynamic experiments (○ in Figure 23.15) are reproduced many times when the cycles of heating and cooling are repeated. The initial and end points of the hysteresis coincide with steady-state values of the reaction rate, because the catalyst pores in these points are filled entirely with either liquid or gas.

In a heating period (Figure 23.15), there is not enough time to empty all the pores with radius $r_p > r_K$ at current temperature. Therefore, the reaction rate becomes lower than the steady-state value. In a cooling period, on the contrary, there is not enough time for the capillary condensation to fill the pores. As the fraction of pores filled with liquid is smaller than in steady-state, the reaction rate proves to be higher than the steady-state value.

C. MATHEMATICAL MODEL

For the interpretation of these experiments, Bukhavtsova and Ostrovski [9] have formulated a simplified model. It is important to remember that experimental conditions excluded the concentration and temperature gradients inside catalyst particle and bed. The reaction mixture was gaseous at input and output of the reactor. Therefore, the composition of liquid condensed in pores is adapted to gas-phase composition according to Henry's or Raoult–Dalton's laws $x^* = Hy$.

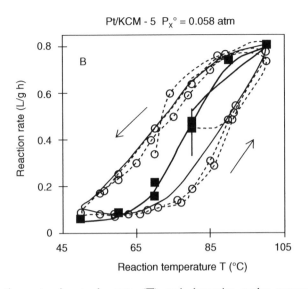

FIGURE 23.15 Reaction rates in steady state (■) and dynamics under temperature variation (○): experimental (points) and predicted by model (solid lines). Arrows are shown the sequence of experiments. (From Bukhavtsova, N.M. and Ostrovskii, N.M., *Can. J. Chem. Eng.*, 65, 322–329, 2002. With permission.)

The volume of liquid in the catalyst is equal to the volume of pores having radius below the critical value determined by Kelvin equation at given P_x and T. The rate of filling or emptying of pores is equal to the rate of interfacial mass transfer. In accordance with these statements, and taking into account that $P_H \approx$ const, the model equations are

reaction rates:

$$r_G = \frac{k_G by}{(1 + by)}; \quad r_L = k_L x \tag{23.27}$$

gas composition:

$$\varphi_G \frac{dy}{dt} = V_F(y_o - y) - (\varepsilon_o - \varphi_L)r_G - \beta a(x^* - x) \tag{23.28}$$

liquid composition:

$$\frac{d(x\varphi_L)}{dt} = \beta a \gamma(x^* - x) - \varphi_L r_L \tag{23.29}$$

liquid volume:

$$\frac{d\varphi_L}{dt} = \beta a \gamma[yP_0 - xP_x(r_K)] \tag{23.30}$$

Formally, the dynamic model has to be completed by adding an equation for the temperature T. However, the calculations show that the temperature relaxation time is less than that for concentration by two orders of magnitude. It permits us to exclude the equation for T from the model. The same calculation confirmed that without liquid condensation and evaporation, the reaction-rate hysteresis does not occur. Thus, the hysteresis is not attributed to the well-known multiplicity of steady-states in a gradientless reactor.

Most parameters of Equations (23.27)–(23.30) were determined from the steady-state experiments, where

$$\frac{dy}{dt} = \frac{dx}{dt} = \frac{d\varphi_L}{dt} = 0$$

Kinetic parameters for gas-phase reaction were estimated in the region without capillary condensation. In this case the reactor model is reduced to one equation:

$$V_F(y_0 - y) = \varepsilon_0 r_G(y) \tag{23.31}$$

The reaction rate constant in liquid phase was determined at the condition of complete filling of pores with liquid. In this case $\varphi_L = \varepsilon_0$, and the reactor model includes two equations:

$$V_F(y_0 - y) = \left(\frac{\varepsilon_0}{\gamma}\right) r_L(x) \tag{23.32}$$

$$\beta a(x^* - x) = \left(\frac{\varepsilon_0}{\gamma}\right) r_L(x) \tag{23.33}$$

Practically, the only parameter determined from dynamic experiments was the mass transfer coefficient (βa), because the sensitivity of steady-states to this parameter is low.

Using Equation (23.27)–Equation (23.33), the mathematical simulation of steady-state and dynamic experiments was carried out. The results of calculation are represented in Figures 23.9, 23.10, 23.14, and 23.15 (lines — calculation, points — experiment). It is easily seen that the model demonstrates a good agreement with experiments in both steady-state and dynamic regimes. This confirms the interpretation of the experiments as the influence of capillary condensation on kinetics and dynamics of catalytic reaction.

Fatemi et al. [54] also observed reaction rate hysteresis with temperature cycling by experiment. Thiophene hydrogenation was chosen as an example of a hydrotreating reaction and n-heptane was used as a solvent. Experiments were carried out for two Ni–Mo/γ-Al$_2$O$_3$ samples: first a narrow pore catalyst with BJH average pore radius of 26.9 Å and secondly a wide pore catalyst with BJH average pore radius of 70.3 Å. As shown in Figure 23.16, it was found that the narrow pore catalyst exhibits a more pronounced hysteresis loop than the wide pore catalyst. This difference could be explained since a larger volume of gas condenses and liquid evaporates from the catalyst with narrow pores, which are more susceptible to capillary condensation.

A model was developed to interpret the experimental results, which is similar to the model, Equation (23.27)–Equation (23.30). The Kelvin equation was used to calculate the critical radius of capillary condensation, and the thickness of the adsorbed layer on the walls of vapor-filled pores was calculated using the method of Gregg and Sing [2]. A Langmuir–Hinshelwood expression was assumed to represent the kinetics in vapor filled pores (Equation (23.25a)) and the reaction in liquid-filled pores was taken as first order (Equation 23.25b)).

VI. CATALYST DEACTIVATION

There are several reasons for catalyst deactivation during catalytic reaction:

- Poisoning of active centers by feed contaminants
- Coke deposition on active surface in the case of hydrocarbons conversion
- Sintering of active component and porous structure
- Phase transformation in catalyst under reaction condition.

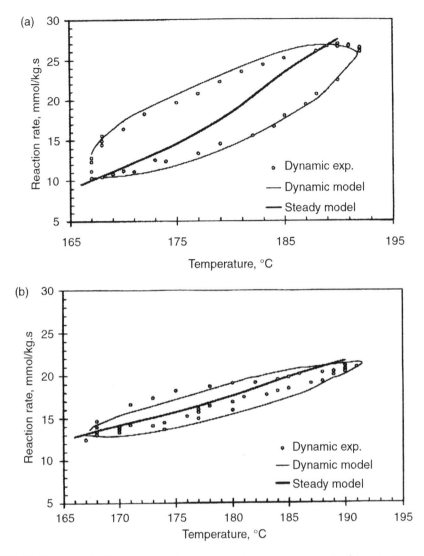

FIGURE 23.16 Hysteresis in the rate of reaction at constant pressure as a result of temperature cycling: (a) narrow pore catalyst; (b) wide pore catalyst. (From Fatemi, S., Moosavian, M.A., Abolhamd, G., Mortazavi, Y., and Hudgins, R.R., *Can. J. Chem. Eng.*, 80, 231–238, 2002. With permission.)

The reaction conditions (temperature, pressure, and concentration of reactants) significantly influence the mechanism and rate of deactivation. For the case of capillary condensation, the aforesaid considerations become more complex, because the deactivation mechanisms may differ in the gas and liquid phase.

Bukhavtsova and Ostrovskii [10] have studied the influence of capillary condensation on the poisoning of catalyst Pt/SiO_2 with sulfur (thiophene) in the reaction of p-xylene hydrogenation. The catalyst was poisoned with a p-xylene solution of thiophene (3.3%) in the pulse regime (by adding 1 μl portions). The most interesting results were observed under conditions, where the reaction first occurs in the gas phase and then moves to capillary condensation regime due to the poisoning. In other words, the p-xylene conversion is initially high, its partial pressure is low, and poisoning occurs in the gas phase. Upon the next thiophene pulse, a decrease in the p-xylene conversion causes an increase in its partial pressure, which becomes sufficient for capillary condensation to occur. Upon further thiophene pulses, poisoning occurs under the conditions of capillary condensation. The behavior of the dynamic curves (on the rate vs. time plots) does not change, but deactivation slows down (Figure 23.17).

The experiments were analyzed by the following deactivation equation [58]

$$X = k\tau(1 - X)a, \qquad \frac{da}{dt} = -\frac{k_P P_x^0 (a - a_S)}{1 - a_S} \tag{23.34}$$

Catalyst activity (a) can be expressed via conversion (X)

$$a = \frac{(1 - X_0)X}{(1 - X)/X_0}, \qquad a_S = \frac{(1 - X_0)X_S}{(1 - X_S)/X_0}$$

and after solving Equation (23.34), we can obtain

$$\ln\left(\frac{X}{1 - X} - \frac{X_S}{1 - X_S}\right) = \ln\left(\frac{X_0}{1 - X_0} - \frac{X_S}{1 - X_S}\right) - \frac{1 - X_S}{(1 - X_S)/X_0} k_P P_x^0 t \tag{23.35}$$

This equation provides the fitting of experiments by straight line and gives the possibility to evaluate the rate constant of poisoning k_P. For this purpose, each pulse of poisoning presented in Figure 23.17a was treated separately (see Figure 23.17b). As it clearly seen from Figure 23.17b, the capillary-condensed liquid caused a decrease in the slope of the curves characterizing the poisoning dynamics.

It was found that the rate of poisoning under capillary condensation is 2–6 times slower than in gas phase. The poisoning rate constant is decreased with increasing of liquid volume in catalyst pellet (Figure 23.18). Here φ_L is proportional to the fraction of Pt located in pores are filled with liquid, that was measured by adsorption method [53]. The retardation of poisoning is due to both thiophene diffusion in the liquid and a change in the thiophene/p-xylene ratio near the active sites when compared with the gas phase.

On the other hand, the poisoning of Pt under capillary condensation requires 1.5 times less thiophene than in gas phase (Figure 23.19). The complete poisoning of platinum in the gas-phase reaction and under the conditions of capillary condensation requires 0.38–0.40 and 0.26–0.28

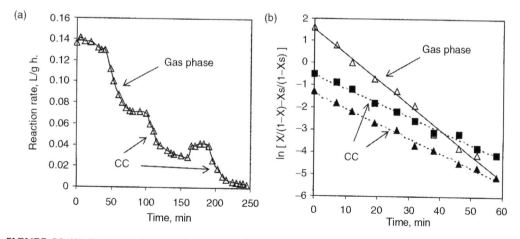

FIGURE 23.17 Catalyst poisoning dynamics. Pt/KCK, $T = 60°C$, $P_x^0 = 0.058$ atm. (a) Experiments; (b) fitting of experiments by Equation (23.35). (From Bukhavtsova, N.M. and Ostrovskii, N.M., *Kinet. Catal.*, 43, 81–88, 2002. With permission.)

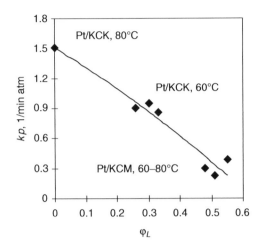

FIGURE 23.18 Poisoning rate constant vs. of liquid fraction in pellet (Taken from Bukhavtsova, N.M. and Ostrovskii, N.M., *Kinet. Catal.*, 43, 81–88, 2002. With permission.)

sulfur atoms per one platinum atom, respectively. This may be attributed to a dramatic decrease (by a factor of approximately 1000) in the H_2 concentration in the liquid as compared to the gas phase because of its low solubility (approximately 10^{-3} mol/mol).

Under deactivation of catalysts by coke deposition not only blocking of active centers take place, but also changes in catalyst pore structure occurs, according to several mechanisms [59]. These changes may involve pore narrowing, pore-mouth plugging and pore blockage [60]. As for the capillary condensation, it can initiate or retard these mechanisms. On the other hand, changes to the porous structure can increase the capillary condensation (in the case of pore narrowing) or reduce it (in the case of pore blockage). Additionally, the catalyst permeability and intrapartical diffusivity are usually changed. For example, Marafi and Stanislaus [61] found that during the initial few hours of gas–oil hydrotreating, there was significant coke deposition in the narrow pores (<50 Å) of the catalyst, leading to a 40% reduction in catalyst surface area during the first 3 h of operation. Loss of the catalyst internal surface area may result from the blockage

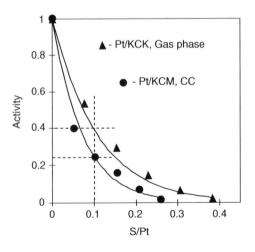

FIGURE 23.19 Complete poisoning of Pt by S in gas phase and under capillary condensation (From Bukhavtsova, N.M. and Ostrovskii, N.M., *Kinet. Catal.*, 43, 81–88, 2002. With permission.)

of pathways leading from the pellet surface to open pores, as well as by pore plugging or narrowing. The effective diffusivity of reactants diffusing into the catalyst has been shown to decrease with increasing coke content [62].

Wood [13,17] has studied the self-diffusivity of pentane and heptane imbibed in coked hydrotreatment catalyst pellets using pulsed-field gradient nuclear magnetic resonance (PFG-NMR). A set of aged industrial catalyst samples were used in the experiment, which had different levels of coke deposited on their surface, ranging from the fresh catalyst (0 wt. % coke) to an aged catalyst with 8 wt.% coke. Figure 23.20 shows a relationship between effective diffusion coefficient and coke content of the samples, which is approximately linear; effective diffusivity of both pentane and heptane decreasing with increasing coke deposition. It was estimated [17] that the reduction in effective diffusion coefficient could lead an increase in the Thiele modulus of up to 40% and to a reduction of approximately 10% in the catalyst effectiveness factor. However, the most influential consequence of coke deposition upon catalyst deactivation was considered to be the reduction in pore volume of the catalyst; pore volume being 48% lower in the coked catalyst compared with the fresh catalyst.

Coking of the catalyst pore structure can also lead to modifications of the tortuosity factor. Ren et al. [63] studied diffusion of heptane in coked alumina catalyst samples using NMR methods. They found that tortuosity of an alumina catalyst increased from 2.4 when fresh to 3.2 after 16 wt. % coke had been deposited on its surface. Wood and Gladden [17] also observed increases in tortuosity experienced by pentane and heptane probe molecules of 19 and 57%, respectively, in a coked hydroprocessing catalyst compared with a fresh catalyst sample.

VII. CAPILLARY CONDENSATION IN INDUSTRIAL PROCESSES

Capillary condensation is likely to occur in reactors where the operating temperature and pressure required for the reaction is close to the dew point of the reacting mixture. Both Ostrovskii et al. [7] and Wood [13] have shown that capillary condensation accompanies certain reactions such as motor fuel hydrotreatment. In order to predict the onset of capillary condensation, we must use the Kelvin equation, although for most reactions the feed will be a mixture of several reactants and therefore the multicomponent version of the Kelvin equation must be selected for the calculation.

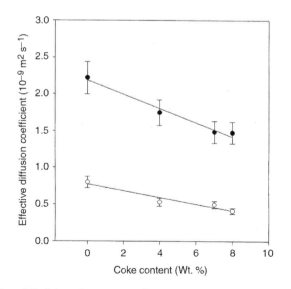

FIGURE 23.20 Effective diffusivity of pentane and heptane imbibed in coked alumina catalyst pellets measured using PFG-NMR, plotted against coke content of the catalyst determined by LECO combustion analysis. Solid symbols (pentane), open symbols (heptane). The lines are a linear fit to the data.

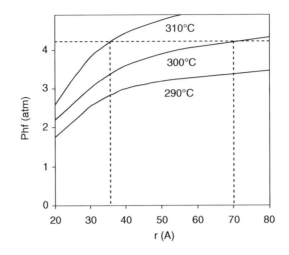

FIGURE 23.21 Capillary condensation in kerosine HDS-HDA process. P_{hf}, heavy component. (From Ostrovskii, N.M., Bukhavtsova, N.M., and Duplyakin, V.K., *React. Kinet. Catal. Lett.*, 53, 253–259, 1994. With permission.)

Ostrovskii et al. [7] have estimated, using Kelvin equation, the probability of capillary condensation in kerosine hydrodesulphurization and hydrodearomatization process (HDS and HDA) on the catalyst Ni–Mo/Al$_2$O$_3$, at $T = 290$–$310°C$, $P = 40$ atm. It was shown that at $300°C$, all the pores having radius less than 70 Å could be filled with liquid heavy fraction (Figure 23.21). This fraction (boiling temperature 220–240°C) attains 10% of feed and its partial pressure is $P_{hf} = 4.2$ atm. When the operation temperature increases to $310°C$, only the pores with $r_p < 35$ Å are filled with liquid. On the other hand, a temperature decrease to $290°C$ leads to blocking of all meso pores (<200 Å). The pores with $r_p < 35$ Å constitute approximately 30% of the overall pore volume and 50% of specific area (Figure 23.22).

Wood [13] has used the multicomponent Kelvin equation to calculate the critical pore radius at which capillary condensation occurs for a mixture of C$_5$–C$_{11}$ hydrocarbons and 50 mole % hydrogen. The composition is intended to represent a light feed for hydroprocessing. In order to calculate

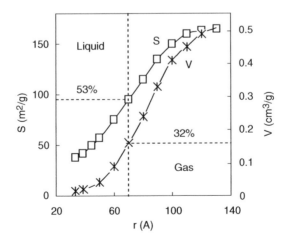

FIGURE 23.22 Specific area (S) and pore volume (V) distribution among the pores in HDS-HDA process. (From Ostrovskii, N.M., Bukhavtsova, N.M., and Duplyakin, V.K., *React. Kinet. Catal. Lett.*, 53, 253–259, 1994. With permission.)

the radius, it was necessary to perform a flash calculation, which was done using the algorithm of Michelsen [26,27]. The algorithm was found to be stable even for mixtures containing a significant amount of hydrogen, which has a much higher volatility than the hydrocarbons present.

Critical condensation radii for the hydrotreater mixture were calculated at a number of temperatures and over a range of pressures. Figure 23.23a shows capillary condensation radius plotted as a function of relative pressure, at temperatures 500, 520, and 540 K. The dew point of the mixture is approximately 32 bar. The calculated critical radii lie mostly in the meso-porous region between 20 and 500 Å for the temperatures 520 K and 540 K and in the micro- and meso-porous regions at 500 K, the transition from the micro-porous region below 20 Å occurring a relative pressure of 0.9. In the microporous region pore filling is likely to occur by means of multilayer adsorption rather than capillary condensation, hence it must be remembered that the Kelvin equation calculations are not physically meaningful for the micro-pores. It is noted that for all curves, there is an exponential rise in condensation radius close to the dew point, showing that in this region small changes in the pressure can lead to large changes in the fraction of pores filled with condensate. In order to determine the fraction of the pore volume in a particular catalyst that are filled with condensate, we must compare the critical diameter at which capillary condensation occurs with the pore-size distribution of the catalyst determined experimentally by nitrogen adsorption (Figure 23.23b). The area under the curve to the left of the critical radius represents condensate-filled pore volume.

Although pore network models are useful in simulating the behavior of catalyst pellets, the process design engineer is more interested in the performance of a catalytic reactor as a whole. Therefore, it is desirable to predict the effects of pellet-scale phenomena upon the behavior of the fixed bed, by using the solution of the single pellet models as inputs to the macroscopic mass and energy balances for the reactor. Sotirchos and Zarkanitis [64] used this approach in modeling the performance of a fixed bed coal gas desulfurization unit. Wood [13] used a

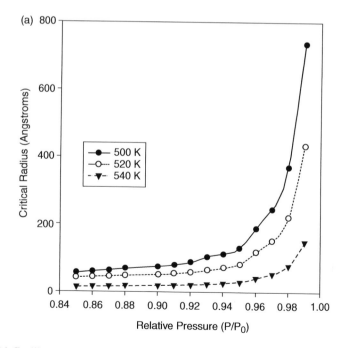

FIGURE 23.23 (a) Capillary condensation radius for typical hydrotreater light oil feed mixed with 50 mole % hydrogen, plotted as a function of relative pressure referred to the dew point. The dew point is in the region of 32 bar. (b) Pore size distribution for an alumina hydrotreatment catalyst support.

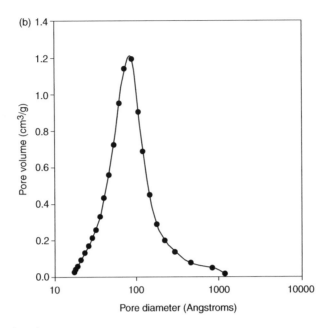

FIGURE 23.23 *Continued.*

similar approach to model the performance of a fixed bed hydrodesulfurization reactor under conditions of capillary condensation. The macroscopic fixed bed reactor model was based on the NINAF model of Tarhan [65].

The catalyst effectiveness factor η_1 was calculated from the pore network model of Wood and Gladden [15] under the conditions on which capillary condensation was expected. The pore network model was solved over a range of temperatures from 553 to 580 K and for several pressures in the interval 20–40 bar to create a database of effectiveness factors for input to the macroscopic reactor model. The hydrodesulfurization of 1 mole % diethyl sulfide in an inert dodecane carrier was considered, with a molar gas : oil ratio of 4. The catalyst was taken to have a connectivity of 6 and a normal distribution of pore sizes with a mean of 136 Å and standard deviation of 28 Å. By using the results of the pore network simulation as input to the macroscopic fixed bed reactor model, capillary condensation at the scale of the catalyst pellets was accounted for.

Figure 23.24 shows the concentration profiles of the sulfur compound through the fixed bed reactor for the pressures in the range 22.3–22.38 bar, using the pore network model to calculate the catalyst effectiveness factor. In addition, shown is the curve with effectiveness factor calculated from the theoretical relationship for a vapor-filled slab catalyst, $\eta_1 = (\tanh \phi)/\phi$. The curves for pressures 22.3 and 22.38 bar show a lag in the decrease of reactant concentration over the first 1 m of the bed, compared with the concentration profile calculated from the theoretical relationship. This effect occurs since the pore network model takes account of the reduction in effectiveness factor caused by partial wetting of the catalyst pellet, whereas the theoretical relationship for the flat plate does not allow for this effect. A reduction in the conversion of diethyl sulfide through the bed is observed for the models that allow for capillary condensation, compared with the vapor-filled slab catalyst. The effect of capillary condensation is particularly evident near the reactor inlet, where the feed is close to its dew point. Owing to exothermic reaction, the temperature increases through the bed length and as a result of pressure drop, the pressure decreases. Therefore, capillary condensation does not occur to a significant extent after the first 1 m of the bed. Consequently, the shape of the concentration profiles is quite similar after the first 1 m of the bed. However, it is noted that the reactant concentration at the bed exit increases with pressure because of the effects of capillary condensation near the bed entrance.

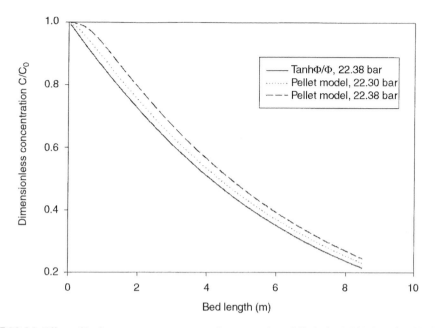

FIGURE 23.24 Effect of bed entrance pressure upon the conversion of diethyl sulphide in a fixed bed reactor. Lines represent fixed bed reactor simulations with catalyst effectiveness factor calculated from the random pore network model, including the influence of capillary condensation.

VIII. CONCLUDING REMARKS

In this chapter, the effect of capillary condensation upon catalytic reactions in porous media has been reviewed. It was shown that capillary condensation could have a strong influence upon catalytic reactions: on its kinetics, transient dynamics, and catalyst pellet effectiveness factor. The reaction rate in the liquid phase is usually slower than in the gas phase due to the difference in adsorption equilibrium, and due to low solubility of hydrogen in the liquid (in hydrotreatment processes).

The influence of capillary condensation upon catalyst effectiveness factor has been assessed both by approximate calculations and by pore network simulations. It was found that catalyst effectiveness could be affected by the presence of capillary condensation, depending on the ratio of reaction rates in the gas and liquid phases. The effectiveness factor under conditions of capillary condensation is sensitive to operating conditions of the reactor, such as pressure, and to properties of the catalyst pore structure like pore-size distribution and connectivity. Once the catalyst pellet contains some pores filled with liquid, the kinetics of the process become dependent upon the phase equilibria of the system. This can lead to multiple steady states in the reaction rate as a function of temperature or pressure, because the current state of the catalyst pellet depends on the history of temperature and pressure profiles to which it has been subjected.

Capillary condensation was also shown to influence catalyst deactivation, although sometimes deactivation was slower in the liquid-filled pores compared with the gas-filled pores. Coking of hydrotreatment catalysts was found to lead to significant losses in pore volume and increases in tortuosity.

Generally, the capillary phenomena in nanoscale porous structure of the catalyst can affect the macroscale performance of catalytic process in industrial reactor. It was shown that industrial hydrotreatment processes operate under conditions that are susceptible to capillary condensation, and this can affect the overall performance of the fixed bed reactor in the process. It is therefore important to carefully select the catalyst pore structure and operating temperature and pressure of the process to control the extent of capillary condensation occurring. Reliable methods of simulating and predicting the effect of capillary condensation therefore are a valuable tool to the process design engineer.

NOTATIONS

a	interfacial gas–liquid area (cm^2/cm^3)
a, a_S	relative activity
b	adsorption coefficient
$\mathbf{c}$	vector of concentrations in a pore (mol/m^3)
C_{tot}	total concentration (mol/m^3)
C_A	concentration of reactant A (mol/m^3)
C_G, C_L	concentration in gas and in liquid (mol/m^3)
C_{ri}	dimensionless concentration
$\boldsymbol{D}$	matrix of diffusivities (m^2/sec)
D	diffusion coefficient (m^2/sec)
D_{eff}	effective diffusion coefficient (m^2/sec)
D_R, D_L	diffusivity in gas and liquid phases (cm^2/sec)
f	fraction of catalyst surface wetted by liquid
F_R, F_L	flux of diffusion
J_i	flux of reactants entering pore junction
k_G, k_L	reaction rate constants in gas and in liquid phases
k_P	rate constant of poisoning
k_v	first order reaction rate constant per unit volume of catalyst (sec^{-1})
K	adsorption constant (bar^{-1})
l_i	length of pore i
L	radius of globule, cm; length of catalyst pellet (m)
L_{NW}	length of pore network representing catalyst pellet (m)
N	number of pores in the network
N_c	number of components
P	pressure (atm)
P_c	capillary pressure (atm)
P_d	dew point pressure (atm)
P_m	vapor pressure over liquid meniscus (atm)
P_S	pressure of saturated vapor (atm)
r_i	pore radius (m)
R	radius of pellet (cm); universal gas constant
r_c, r_p	radius of pores (cm)
r_G, r_L	reaction rates in pores filled with gas and liquid
r_K	radius of pores in Kelvin equation (cm)
r_{tot}	total rate of reaction
S_R, S_L	specific external surface of pellet and globule (cm^2/cm^3)
T	temperature (K)
t	time
v_i	fraction of total pore volume contained in pore i
$\boldsymbol{v}$	vector of stoichiometric coefficients
V_C	volume of catalyst in the reactor (cm^3)
V_F	feed space velocity (cm^3/sec)
V_m	molar volume of liquid (cm^3/mol)
V_{VL}	mixed volume (cm^3)
V_V^i	specific volume of the vapor phase
V_L	volume of liquid (cm^3)
V_R	pellet volume (cm^3)

X, X_0, X_S	conversion
x, y	molar fractions in liquid and gas
x_{Ld}^i	liquid phase mole fraction of component i
Z_i	dimensionless distance along a pore
z_i	distance along pore i (m)

Greek letters

β	parameter in Equation (23.10); mass transfer coefficient (cm/sec)
γ	parameter in Equation (23.10); ratio of liquid and gas molar volume
δ	thickness of adsorption film (cm)
ε	fraction of globules' volume in pellet
ε_o	total porosity
$\varepsilon_G, \varepsilon_L$	fractions of pellet being filled with gas and with liquid
ϕ_i	local Thiele modulus for a single pore
χ	relative pressure; relative rate of reaction
θ	contact angle
$\varphi_G = V_G/V_C$	gas and catalyst volume ratio in reactor
$\varphi_L = V_L/V_C$	fraction of liquid in catalyst pellet
μ	dynamic viscosity (N s m^{-2})
σ	surface tension
τ	contact time
ρ	dimensionless radius of pellet
ψ	modified Thiele parameter
η	effectiveness factor
η_t	effectiveness factor in pore i
η_{NW}	effectiveness factor for pore network
Φ	Thiele modulus for catalyst pellet
Φ_{NW}	network Thiele modulus

Subscripts (Equation (23.21)–Equation (23.22))

A	Butane
B	Butene
H_2	Hydrogen
H_2S	Hydrogen sulfide
T	Thiophene
σ	sigma sites
τ	tau sites

REFERENCES

1. Wijngaarden, R.J., Kronberg, A., and Westerterp, K.R., *Industrial Catalysis*, Wiley: Weinheim, Germany, 1998.
2. Gregg, S.J. and Sing, K.S.W., *Adsorption, Surface Area and Porosity*, 2nd ed., Academic Press, London, 1967.

3. Thompson, W.T., On the equilibrium of vapour at a curved surface of liquid, *Phil. Mag.*, 42, 448–452, 1871.

4. Kim, D.H. and Kim, Y.G., An experimental-study of multiple steady-states in a porous catalyst due to phase-transition, *J. Chem. Eng. Jap.*, 14, 311–317, 1981.

5. Jaguste, D.N. and Bhatia, S.K., Partial internal wetting of catalyst particles: hysteresis effects, *AIChE J.*, 37, 650–660, 1991.

6. Gurfein, N.S. and Danyushevskaya, N.M., *J. Phys. Chem.*, 63, 77–82, 1989 (in Russian).

7. Ostrovskii, N.M., Bukhavtsova, N.M., and Duplyakin, V.K., Catalytic reactions accompanied by capillary condensation. 1. Formulation of the problems, *React. Kinet. Catal. Lett.*, 53, 253–259, 1994.

8. Ostrovskii, N.M. and Bukhavtsova, N.M., Catalytic reactions accompanied by capillary condensation. 2. Mass transfer inside a pellet, *React. Kinet. Catal. Lett.*, 56, 391–399, 1995.

9. Bukhavtsova, N.M. and Ostrovskii, N.M., Catalytic reactions accompanied by capillary condensation. 3. Influence on reaction kinetics and dynamics, *React. Kinet. Catal. Lett.*, 65, 322–329, 1998.

10. Bukhavtsova, N.M. and Ostrovskii, N.M., Kinetics of catalyst poisoning during capillary condensation of reactants, *Kinet. Catal.*, 43, 81–88, 2002.

11. Jaguste, D.N. and Bhatia, S.K., Combined surface and viscous flow of condensable vapour in porous media, *Chem. Eng. Sci.*, 50, 167–182, 1995.

12. Rieckmann, C. and Keil, F.J., Simulation and experiment of multicomponent diffusion and reaction in three-dimensional networks, *Chem. Eng. Sci.*, 54, 3485–3493, 1999.

13. Wood, J., Two-phase transport in porous catalyst particles, Ph.D. thesis, University of Cambridge, Cambridge, 2001.

14. Shapiro, A.A. and Stenby, E.H., Kelvin equation for a non-ideal multicomponent mixture. *Fluid Phase Equilibria.*, 134, 87–101, 1997.

15. Wood, J. and Gladden, L.F., Modelling diffusion and reaction accompanied by capillary condensation using three-dimensional pore networks. Part 1. Fickian diffusion and pseudo-first-order reaction kinetics, *Chem. Eng. Sci.*, 57, 3033–3045, 2002.

16. Wood, J., Gladden, L.F., and Keil, F.J., Modelling diffusion and reaction accompanied by capillary condensation using three-dimensional pore networks. Part 2. Dusty gas model and general reaction kinetics, *Chem. Eng. Sci.*, 57, 3047–3059, 2002.

17. Wood, J. and Gladden, L.F., Effect of coke deposition upon pore structure and self-diffusion in deactivated industrial hydroprocessing catalysts, *Appl. Catal. A: Gen.*, 249, 241–253, 2003.

18. Wheeler, A., Reaction rates and selectivity in catalyst pores, *Adv. Catal.*, 3, 249–327, 1951.

19. Johnson, M.F.L. and Stewart, W.E., Pore structure and gaseous diffusion in solid catalysts, *J. Catal.*, 4, 248–252, 1965.

20. Rajniak, P. and Yang, R.T., Unified network model for diffusion of condensable vapours in porous media, *AIChE J.*, 42, 319–331, 1996.

21. Sharatt, P.N. and Mann, R., Some observations on the variation of tortuosity with Thiele modulus and PSD, *Chem. Eng. Sci.*, 42, 1565–1576, 1987.

22. Hollewand, M.P. and Gladden, L.F., Modelling of diffusion and reaction in porous catalysts using a three-dimensional network model, *Chem. Eng. Sci.*, 47, 1761–1770, 1992.

23. Hollewand, M.P. and Gladden, L.F., Representation of porous catalysts using random pore networks, *Chem. Eng. Sci.*, 47, 2757–2762, 1992.

24. Ramirez-Cuesta, A.J., Cordero, S., Rojas, F., Faccio, R.J., and Riccardo, J.L., On modelling, simulation and statistical properties of realistic three-dimensional porous networks, *J. Porous Mat.*, 8, 61–76, 2001.

25. Sedriks, W. and Kenney, C.N., Partial wetting in trickle bed reactors — the reduction of crotonaldehyde over a palladium catalyst, *Chem. Eng. Sci.*, 28, 559–568, 1973.

26. Michelsen, M.L., The isothermal flash problem. Part I. Stability, *Fluid Phase Equilib.*, 9, 1–19, 1982.

27. Michelsen, M.L., The isothermal flash problem. Part II. Phase split calculation, *Fluid Phase Equilib.*, 9, 21–40, 1982.

28. Broadbent, S.R. and Hammersley, J.M., Percolation processes I. Crystals and mazes. *Proc. Camb. Phil. Soc.*, 53, 629–641, 1957.

29. Stauffer, D. and Aharony, A., *Introduction to Percolation Theory*, Taylor and Francis, London, 1992.

30. Hoshen, J. and Kopelman, R., Percolation and cluster distribution. I. Cluster multiple labelling technique and critical concentration algorithm, *Phys. Rev. B.*, 14, 3438–3445, 1976.
31. Krishna, R., Problems and pitfalls in the use of the Fick formulation for intraparticle diffusion, *Chem. Eng. Sci.*, 48, 845–861, 1993.
32. Toor, H.L., Diffusion in three component gas mixtures, *AIChE J.*, 3, 198–207, 1957.
33. Keil, F.J., Diffusion and reaction in porous networks, *Catal. Today.*, 53, 245–258, 1999.
34. Feng, C. and Stewart, W.E., Practical models for isothermal diffusion and flow of gases in porous solids, *Ind. Eng. Chem. Fundam.*, 12, 143–147, 1973.
35. Mason, E.A. and Malinauskas, A.P., *Gas transport in porous media: the dusty gas model*, Elsevier, Amsterdam, 1983.
36. Wesselingh, J.A. and Krishna, R., *Mass transfer*, Ellis Horwood, Chichester, 1990.
37. Gilliland, E., Baddour, R.F., and Russel, J.L., Rates of flow through microporous solids, *AIChE J.*, 4, 90–96, 1958.
38. Krishna, R., Multicomponent surface diffusion of adsorbed species — a description based on the generalised Maxwell–Stefan equations, *Chem. Eng. Sci.*, 45, 1779–1791, 1990.
39. Flood, E.A. and Huber, M., Thermodynamic considerations of surface regions, adsorbate pressures, adsorbate mobility and surface tension, *Can. J. Chem.*, 33, 203–214, 1955.
40. Lee, K.H. and Hwang, S.T., The transport of condensable vapours through a microporous Vycor glass membrane, *J. Coll. Interf. Sci.*, 110, 544–555, 1986.
41. Kainourgiakis, M.E., Stubos, A.K., Konstantinou, N.D., Kanellopoulos, N.K., and Milisic, V., A network model for the permeability of condensable vapours through mesoporous media, *J. Membr. Sci.*, 114, 215–225, 1996.
42. Kikkinides, E.S., Tzevelekos, K.P., Stubos, A.K., Kainourgiakis, M.E., and Kanellopoulos, N.K., Application of effective medium approximation for the determination of the permeability of condensable vapours through mesoporous media, *Chem. Eng. Sci.*, 52, 2837–2844, 1997.
43. Girgis, M.J. and Gates, B.C., Reactivities, reaction networks, and kinetics in high-pressure catalytic hydroprocessing, *Ind. Eng. Chem. Res.*, 30, 2021–2058, 1991.
44. Rieckmann, C. and Keil, F.J., Multicomponent diffusion and reaction in three-dimensional networks, *Ind. Eng. Chem. Fundam.*, 36, 3275–3281, 1997.
45. Van Parijs, I.A., Hosten, L.H., and Froment, G.F., Kinetics of hydrodesulphurisation on a CoMo/γ-Al$_2$O$_3$ Catalyst. 1. Kinetics of the hydrodesulphurisation of thiophene, *Ind. Eng. Chem. Prod. Res. Dev.*, 25, 431–436, 1986.
46. Seaton, N.A., Determination of the connectivity of porous solids from nitrogen sorption measurements, *Chem. Eng. Sci.*, 46, 1895–1909, 1991.
47. Portsmouth, R.L. and Gladden, L.F., Determination of pore connectivity by mercury porosimetry, *Chem. Eng. Sci.*, 46, 3023–3036, 1991.
48. Liu, H., Zhang, L., and Seaton, N., Determination of the connectivity of porous solids from nitrogen sorption measurements — II. Generalisation, *Chem. Eng. Sci.*, 47, 4393–4404, 1992.
49. Fuller, E.N., Schettler, P.D., and Giddings, J.C., A new method for prediction of binary gas-phase diffusion coefficients, *Ind. Eng. Chem. Res.*, 58, 19–27, 1966.
50. Fuller, E.N., Diffusion of halogenated hydrocarbons in helium. The effect of structure on collision cross sections, *J. Phys. Chem.*, 75, 3679–3685, 1969.
51. Lucas, K., Pressure dependence of the viscosity of liquids — a simple estimate, *Chem. Ing. Tech.*, 53, 959–960, 1981.
52. Reid, R.C., Prausnitz, J.M., and Poling, B.E., *The Properties of Gases and Liquids*, 4th ed., McGraw Hill, New York, 1987.
53. Belyi, A.S., Smolikov, M.D., Fenelonov, V.B., Gavrilov, V.Yu., and Duplyakin, V.K., *Kinet. Catal.*, 27 (703), 1414, 1986 (in Russian).
54. Fatemi, S., Moosavian, M.A., Abolhamd, G., Mortazavi, Y., and Hudgins, R.R., Reaction rate hysteresis in the hydrotreating of thiophene in wide- and narrow-pore catalysts during temperature cycling, *Can. J. Chem. Eng.*, 80, 231–238, 2002.
55. Ostrovskii, N.M., Catalytic reactions, accompanied by capillary condensation, *Khimicheskaya Promyshlenost.*, 1, 13–19, 1995 (in Russian).
56. Bhatia, S.K., Steady-state multiplicity and partial internal wetting of catalyst particles, *AIChE J.*, 34, 969–979, 1988.

57. Bhatia, S.K., Partial internal wetting of catalyst particles with a distribution of pore size, *AIChE J.* 35, 1337–1345, 1989.
58. Ostrovskii, N.M., *Catalyst Deactivation Kinetics*, Nauka, Moskva, 2001 (in Russian).
59. Dadvar, M., Sohrabi, M., and Sahimi, M., Pore network model of deactivation of immobilized glucose isomerase in packed-bed reactors — I: two-dimensional simulations at the particle level, *Chem. Eng. Sci.*, 56, 2803–2819, 2001.
60. Froment, G.F., Modelling of catalyst deactivation, *Appl. Catal. A: Gen.*, 212, 117–128, 2001.
61. Marafi, M. and Stanislaus, A., Effect of initial coking on hydrotreating catalyst performance, *Appl. Catal. A: Gen.*, 159, 259–267, 1997.
62. Al-Bayaty, S., Acharya, D.R., and Hughes, R., Effect of coke deposition on the effective diffusivity of catalyst pellets, *Appl. Catal. A: Gen.*, 110, 109–119, 1994.
63. Ren, X.H., Bertmer, M., Stapf, S., Demco, D.E., Blümich, B., Kern, C., and Jess, A., Deactivation and regeneration of a naphtha reforming catalyst, *Appl. Catal. A: Gen.*, 228, 39–52, 2002.
64. Sotirchos, S.V. and Zarkanitis, S., Pellet model effects on simulation models for fixed-bed desulphurization reactors, *AIChE J.*, 35, 1137–1147, 1989.
65. Tarhan, M.O., *Catalytic Reactor Design*, McGraw Hill, New York, 1983.

24 Particle Production Using Supercritical Fluids

Dejan Skala and Aleksandar Orlovic
University of Belgrade, Belgrade, Serbia and Montenegro

CONTENTS

The unique characteristics of gases at supercritical conditions, which are between those for liquids (density) and gases (viscosity and diffusivity) enable that different processes could be realized as for example: the selective extraction of specific components from the complex mixtures, the particle micronization, the supercritical fractionation, the production of materials with unusual and specific structure, etc. Many similar processes, based on supercritical fluids use are today the part of intensive investigations, while others have already been patented and applied in the pilot-plant scale. Obviously, in the recent past there is not always the advantage for all investigated supercritical processes because of high financial investment necessary for their implementation in industry compared to other conventional methods. However, it is important to notice the substantial increase of scientific research published in many journals, which include the new potential use and application of supercritical fluids.

Manufacturing different substances from natural sources was realized at the industrial scale by using different supercritical fluids as well as supercritical fractionation or purification of the conventional plant extracts. There are large number of such examples: production of some active components from the plant materials, the elimination of solvent residue from the extract obtained by using common chlorinated organic solvent, production of synthetic drugs, the monomer extraction from implants produced by polymerization, and elimination of some toxic and hazardous substances from different sources and materials. Maximum interest for supercritical fluids application were obviously given for the development of specific drugs for controlled release of active substance in the human body and different human organs. Such an application is realized after solving the technology of producing nano and microparticles with uniform distribution of their sizes (microspheres, microcapsules etc.).

Treatment of specific materials with supercritical fluids seems to be a more promising field in the future in different industrial branches: the polymers purification, production of expanded polymers, and production of a highly porous materials (catalyst support) and its impregnation. Furthermore, development of supercritical fluid (SF) application for production of specific particles effective in color and coating, the aerogels synthesis and gel drying, which might be then used as insulating materials, production of different coatings and SF use for surface treatment and cleaning, the production of ceramics with specific characteristics, the production of carbon fibers, and other composite, etc., are future expansions of SF use.

I. PROPERTIES OF SUPERCRITICAL GASES

Supercritical fluid (SF) presents the state of some substance, which is at temperature above its critical temperature (T_c) and compressed above its critical pressure (P_c), above which no applied pressure can force the substance into its liquid state (Figure 24.1). SF is characterized with specific behavior: its density (from 0.1 to 0.9 g/cm^3 at 75–500 bar) is very similar as those of liquid; and it has a low viscosity, small diffusivity and surface tension which are the main characteristics of gases.

Only small changes in pressure and temperature in the vicinity of critical states for some substances can greatly influence the density of SF. This characteristic of SF is a parameter closely related to its solubilizing power and, therefore, by simple changing of operating parameters without any influence on the structure of SF, the equivalent of a series of different solvents can be easily provided.

The critical parameters of different compounds which are presented in Table 24.1, indicates that these values as well as dissolving characteristics (density) are very important parameters for choosing the right working fluid in the processes based on supercritical fluids. This evidence allows carbon dioxide, in the first place, as supercritical one in different processes, because of its relatively low critical temperature and pressure. Carbon dioxide is widely applied for producing heat-sensitive and attractive products such as flavors, pharmaceuticals, and labile lipids. Other advantages of CO_2 as working fluids are its nontoxicity, nonflammability, nonoxidizing agents which is possible to reduce and suppress many unwanted degradation processes and formation of artifacts.

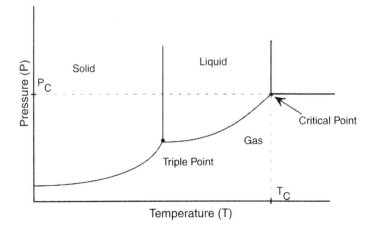

FIGURE 24.1 $P–T$ phase diagram for some substances.

TABLE 24.1
The Critical Parameters for Some Compounds

Compounds	T_c (°C)	P_c (bar)	ρ_c (g/cm³)
Methane	−82.5	46.4	0.16
Ethylene	10.0	51.2	1.22
CO_2	31.1	73.8	0.47
Ethane	32.4	48.8	0.20
Propylene	91.9	46.1	0.24
Propane	97.2	42.5	0.22
Ammonia	132.5	112.8	0.24
Pentane	187.1	33.7	0.23
i-Propanol	235.4	47.6	0.27
Methanol	240.6	79.9	0.27
Ethanol	243.5	63.8	0.28
i-Butanol	275.1	43.0	0.27
Benzene	289.0	48.9	0.30
Water	374.2	220.5	0.32

Carbon dioxide is also inexpensive and posses a relatively high dissolving power compared to other supercritical fluids.

The most applied operating parameters of different processes based on the use of SFs are: reduced pressure ($P_r = P/P_c$) in the range 1.01–1.05 and reduced temperature ($T_r = T/T_c$) in the range 1.01–1.1 [1]. Solubility of different substances in the supercritical fluids depends on many factors among them dependence on temperature and pressure are the most important (Figure 24.2). In the region of high pressure (E_2, Figure 24.2) the changes of density with temperature increase is less expressed but increases of the vapor pressure of solute leads to a higher solubility of solute. Decrease of solubility by temperature increase evidently exists at low pressure (S_2) and such anomaly is called retrograd solubility, and this region of working pressure and temperature is a retrograde. At these conditions, the increase of vapor pressure could not be enough to compensate decrease of solubility caused by decrease of supercritical fluid density (Figure 24.2 and Figure 24.3).

Solubility of some substances in the supercritical fluids, among different parameters, mostly depends on the vapor pressure, the substance polarity, and substance molar mass. Compounds of smaller molar mass and a higher vapor pressure at supercritical condition are more soluble in SFs, compared to other with lower vapor pressure and higher molar mass. An enhancement factor (dimensionless parameter) is defined as the ratio of solubility of substances in the SF (solvent) compared to its solubility in the ideal gas. Usually, this parameter has the common value between 10^4 and 10^6 [3]. Different mechanisms reported in the literature have been used for explaining the enhancement of solute solubility in supercritical fluids. They included the hydrogen bonding, the charge transfer complex formation, dipole–dipole noninduced and induced interactions, and solute–solvent with and without cosolvent interactions.

Supercritical carbon dioxide has a small dielectric constant (ε) and small degree of polarity in the unit of volume (α/V), so it possess a higher affinity to the nonpolar solutes. It is not a good solvent for polar compounds and compounds with high molar mass. Although, its polarity can be varied by the change of its density, the maximal range of dielectric constant is between 1 and 1.6, which is not enough for the substantial increase of solubility for many and special polar substances. Addition of some polar cosolvent (methanol, water, acetone) to the supercritical CO_2 (or so called modifier), from several to less than 10 mol%, has been used for polar substances

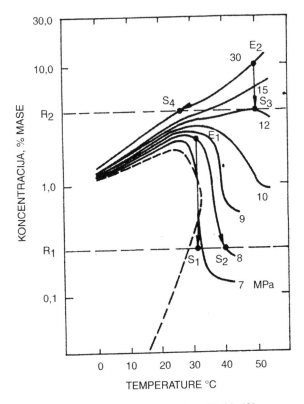

FIGURE 24.2 Solubility of naphthalene in supercritical carbon dioxide [2].

solubility increase and for substances with high molar mass [4]. Page et al. [5] have made a comprehensive review of commonly used modifiers.

It was shown that increased content of phospholipids might be obtained if supercritical extraction of corn germs with CO_2 and ethanol as a cosolvent (less than 10 mass%) were used [6]. The increased content of phospholipids in extract was detected when a large amount of cosolvent was

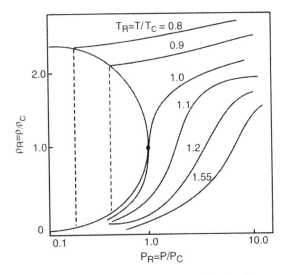

FIGURE 24.3 Dependence of reduced density in the vicinity of critical point.

used for supercritical extraction. Also, the proteins solubility was increased with a larger amount of alcohols (cosolvent) in supercritical CO_2 but there were no change in the content of fatty acids in extracted oils [6].

As already mentioned, the solubility of different substances depends on the dielectric constant of supercritical fluids. Water under subcritical and supercritical conditions is an example for possible and drastic changes of dielectric constant with pressure and temperature. Water at high temperature but very close (and below) to critical and under a high pressure is termed as subcritical fluid. Solubility of different substances in subcritical or supercritical state of water depends on dielectric constant of water. This characteristic of water has a value between 2 and 80 depending on pressure and temperature as shown in Figure 24.4. Such drastic change of dielectric constant indicates that water might be polar as well as nonpolar similar to some organic solvent (hydrocarbons).

That is why nonpolar organic substances, for example, polycyclic aromatic hydrocarbons (PAH) are practically insoluble in water under normal condition ($\varepsilon = 80$) when solubility might be determined only as few parts per billion (ppb). By increase of temperature and under mild pressure a drastic change in dielectric constant of water increases the solubility of pure soluble compounds to the level of several percents by mass. At subcritical condition of water ($250°C$ and 50 bar) when dielectric constant of water is three times lower ($\varepsilon = 27$), the quantitative extraction of PAH is possible.

The phase diagrams are often used to indicate the possible solubility level of different substances in a supercritical fluid over a wide range of temperature and pressure. The phase behavior in many cases is very complex in the vicinity of critical point. Its understanding and interpretation requires use of different thermodynamic models for explaining the complex interaction between the small molecules of nonpolar solvents, polar cosolvents, and very often large molecule of solutes. The equation of states has a wide application in modeling the phase behavior of complex supercritical mixture. Using the combination of different EOS and the models for prediction, the excess of Gibbs energy (e.g., UNIFAC model) necessary for determining behavior of real system, different and complex models were developed which can be used for phase behavior prediction.

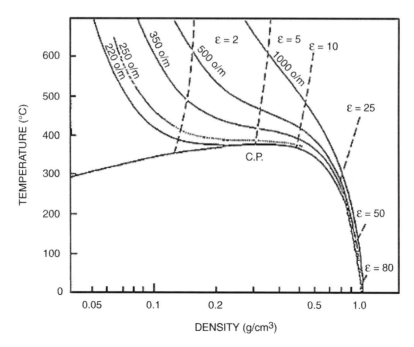

FIGURE 24.4 The water dielectric constant as a function on temperature [7].

Another very interesting behavior of supercritical fluids from the standpoint of their application in chemical reaction is relatively large solubility of hydrogen in supercritical CO_2 and in some other supercritical fluids. Determined concentration of hydrogen in the supercritical mixture containing hydrogen (85 bar partial pressure) and CO_2 (120 bar) at 50°C is 3.2 M (mol/dm^3), while hydrogen concentration in other conventional solvents (tetrahydrofurane) under the same pressure and temperature is only 0.4 M. This phenomena is used to realize some heterogeneous reaction as a homogeneous one thus enabling much faster reaction rate and the change of selectivity in the case of complex reaction of hydrogenation.

II. NANOSCALE PARTICLE PREPARATION

The particle design is a major development of supercritical fluids applications, with potential applications in the pharmaceutical, food, cosmetic, and some chemical industries. This section presents a concise survey of the published material classified according to the different processing concepts currently used to manufacture particles or other dispersed materials.

A. RAPID EXPANSION OF SUPERCRITICAL SOLUTIONS

Rapid expansion of supercritical solutions (RESS) is the method in which a pressurized solution is rapidly expanded through an adequate nozzle, causing an extremely fast nucleation of the solid solute in a micronized form. This process is attractive due to the absence of organic solvent during processing, but its application is restricted to products reasonably soluble in supercritical fluids (commonly used fluid is carbon dioxide). The basic concept of RESS was first described more than a 100 years ago [8], while modern development started after the pioneering works of Krukonis [2,3] and especially the Battelle Institute research team [9,10].

As presented in Figure 24.5, a supercritical fluid saturated with the substrates is expanded through a heated nozzle into a low-pressure chamber in order to cause an extremely rapid nucleation of the substrates in a micronized form.

The pure carbon dioxide is pumped to the desired pressure and preheated to extraction temperature using a heat exchanger. The supercritical fluid is then percolated through the extraction unit packed with one or more substrates. In the precipitation unit, the supercritical solution is expanded through a nozzle that needs to be heated in order to avoid plugging by substrates precipitation. Typically used expansion nozzles are capillaries of diameter < 100 μm or laser drilled nozzles of 20–60 μm in diameter. The process allows control over several parameters affecting

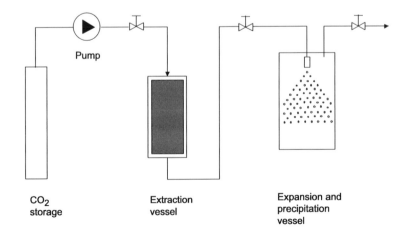

FIGURE 24.5 The RESS process.

the expansion: the solute concentration, the pre-expansion temperature and pressure, the nozzle dimensions, and background expansion conditions. Application of an expansion nozzle defines the path of the expanding solution, which allows manipulation of the solvent density at various stages of the expansion and permits some degree of control over the solute nucleation characteristics. The rate of nucleation is given by Equation (24.1) and Equation (24.2) [11].

$$J = \Theta \alpha_C V_{2,\text{mol}} v_2^2 \sqrt{\frac{2\sigma_G}{\pi m_{2,\text{mol}}}} \exp\left(-\frac{\Delta G_{\text{crif}}}{k_B T}\right) \qquad (24.1)$$

$$\frac{\Delta G_{\text{crif}}}{k_B T} = -16 \frac{\pi}{3} \left(\frac{\sigma_G V_{2,\text{mol}}^{2/3}}{k_B T}\right)^3 \frac{1}{\left[\ln S - V_{2,\text{mol}}(p_2 - p_{2,S})(1/k_B T)\right]^2} \qquad (24.2)$$

where Θ is the nonisothermal factor, α_C the condensation factor, $V_{2,\text{mol}}$ the solute molecular volume, $m_{2,\text{mol}}$ the solute molecular mass, σ_G the interfacial tension, v_2 the solute concentration in vapor phase, p_2 the partial pressure of the solid, $p_{2,S}$ the saturation vapor pressure of the solid, S the saturation ratio, k_B a Boltzmann's constant, and T the absolute temperature.

The morphology of the resulting solid particles has been observed to vary with changes in RESS processing conditions. Products obtained using RESS are uniform sub-micrometer scale particles, thin films, and fine fibers. If dissolved species of a low vapor pressure solid exist in sufficient concentration in the supercritical fluid prior to expansion, they nucleate and grow rapidly in the expansion jet as the fluid solvating capacity drops, which results in the generation of fine powders. Particle size, agglomeration, and other characteristics can be affected by the chemical properties of the solute, its concentration in the solution prior to expansion, temperature, pressure drop, distance of impact of the jet against the surface, dimensions of the expansion vessel, nozzle geometry, and the phase behavior of the solvent in the expansion jet [8,11,12]. For example, micronization of Ibuprofen by RESS using carbon dioxide indicated decrease in crystallinity of the obtained material when compared to the original one [13]. No clear dependence of the particle size on the extraction pressure was observed, while an increase in the spraying distance increased the particle size. However, an increase in the pre-expansion temperature, capillary length, and collision angle, were found to have reducing effect on the average size of Ibuprofen particles. It should be noticed that the initial investigations of RESS were focused on micronization of pure substrates in order to obtain very fine particles with narrow size distribution [12,14], while the recent publications are related to mixture processing in order to obtain microcapsules of a substrate inside a carrier [15] or to bioavailability of pharmaceutical agents [16].

The RESS process can be implemented in relatively simple equipment, it can produce very fine particles and it is solvent free, eventhough particle collection from the gaseous stream is difficult. Other disadvantages of the process are: high gas or substrate ratios due to the low solubility of the typical substrates, high pressures and temperatures, and large volumes of pressurized equipment.

B. SAS — SUPERCRITICAL ANTI-SOLVENT PRECIPITATION AND RELATED PROCESSES

Supercritical fluid anti-solvent processes have been recently proposed as alternatives to liquid anti-solvent processes commonly employed in the industry. The key advantage of the supercritical processes over liquid ones is the possibility to completely remove the anti-solvent by pressure reduction. This step of the process is problematic in case of liquid anti-solvents since it requires complex post-processing treatments for the complete elimination of liquid residues. Furthermore, the supercritical anti-solvent is characterized by diffusivity that can be up to two orders of magnitude higher than those of liquids. Therefore, its very fast diffusion into the liquid solvent produces the supersaturation of the solute and the precipitation in micronized particles with diameters that are not possible to obtain using liquid anti-solvents or other methods.

Supercritical anti-solvent micronization can be performed using different processing methods and equipment [17]. Different acronyms were used by the various authors to indicate the micronization process. It has been referred to as GAS (gas anti-solvent), PCA (precipitation by compressed anti-solvent), ASES (aerosol solvent extraction system), SEDS (solution enhanced dispersion by supercritical fluids), and SAS (supercritical anti-solvent) process [8,17]. Since the resulting solid material can be significantly influenced by the adopted process arrangement, a short description of the various methods is presented below.

In GAS or SAS, a batch of solution is expanded by mixing it with a supercritical fluid in a high-pressure vessel (Figure 24.6). Due to the dissolution of the compressed gas, the expanded solvent exhibits a decrease of the solvent power. The mixture becomes supersaturated and the solute precipitates in the form of microparticles. As shown in Figure 24.6, the precipitator is partially filled with the liquid solution of solid substance. The supercritical anti-solvent is then pumped up to desired pressure and introduced into the vessel, preferably from the bottom in order to achieve a better mixing of the solvent and anti-solvent. After a specified residence time, the expanded solution is drained under isobaric conditions in order to clean the precipitated particles. In this mode of operation, the rate of supercritical anti-solvent addition can be an important parameter in controlling the morphology and the size of solid particles.

Another method (ASES) involves spraying of the liquid solution through an atomization nozzle as fine droplets into a chamber filled with supercritical anti-solvent. The fast dissolution of the supercritical fluid into the liquid droplets is followed by a large volume expansion of droplets. This is followed by a reduction of the liquid solvent power, causing a sharp rise in the supersaturation within the liquid mixture. As a consequence, fine and uniform particles are being formed. The supercritical fluid is pumped to the top of the high pressure vessel by a high-pressure pump. Once the system reaches the steady state (temperature and pressure), the active substance solution is introduced into the high-pressure vessel through a nozzle. Particles are collected on a filter at the bottom of the vessel. The fluid mixture (supercritical fluid plus solvent) exits the vessel and flows to a depressurization tank where the conditions (temperature and pressure) allow gas–liquid separation. After stopping the spraying of solution, the pure supercritical fluid continues to flow through the vessel to remove residual solvent from the particles. As in the previous case (GAS, SAS, and PCA) the performed operation is not at steady state and it is difficult to analyze the effect of the process parameters on the final characteristics of the powders. Besides that, batch operation is usually not suitable for industrial application.

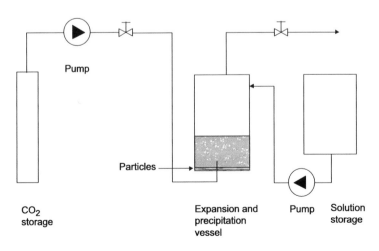

FIGURE 24.6 The GAS, PCA, or SAS process.

In order to overcome the limitations of batch processing, a continuous SAS process was developed and reported by Reverchon et al. [18]. In continuous operation, the liquid solution and the supercritical anti-solvent are continuously delivered to the precipitation chamber either in co-current or counter-current mode. This mode of operation provides the possibility to investigate the influence of different process variables on the evolution of micronized particles. The important process parameters, which can be studied at steady state operation are: the flow rates, the solution/anti-solvent ratio, and pressure. Beside those advantages, continuous steady state operation is also suitable for industrial application. A key role in the continuous operation and in the ASES process is played by the liquid solution injection device. The injector is designed to produce liquid jet break-up and to form small micronic droplets that expand in the precipitator. The solid solute is released when its local concentration exceeds the saturation limit. Various injection devices have been proposed in the literature. Some authors propose a nozzle of various diameters ranging from 5 to 50 μm [17]. Other authors have used small internal diameter capillaries [19], or vibrating orifices [20]. The latter one produces a spray by superimposing a high frequency vibration on the liquid jet that exits from an orifice. It is also possible to use a premixed injector in which the liquid solution is mixed into the flowing CO_2 before entering the precipitator, or to use stirred autoclaves. The type of injection device used can also strongly influence the precipitation process: the size of droplets, their coalescence, and mixing of the different fluids. Some authors view that the initial droplet size, formed at the nozzle, does not have an effect on final particle size [21]. They also suggest that the particle size is probably more determined by the mass transfer effects.

The SEDS method was developed by the Bradford University [22,23] in order to achieve smaller droplet size and intensify mixing of supercritical fluid and solution for increased transfer rates. The supercritical fluid is used both for its chemical properties and as "spray enhancer" by mechanical effect: a nozzle with two coaxial passages allow introducing the supercritical fluid and a solution of active substances into the particle formation vessel where pressure and temperature are controlled. The high velocity of the supercritical fluid (Reynolds number in the order of 10^5) allows breaking up the solution into very small droplets. Furthermore, the conditions are setup so that the supercritical fluid can extract the solvent from the solution at the same time as it meets and disperses the solution.

A key role in the precipitation by supercritical anti-solvent is played by the volumetric expansion of the liquid solvent. This phenomenon is the result of massive dissolution of the supercritical anti-solvent into the liquid phase. Two methods have been used to obtain volumetric expansion data. The first is the direct observation of the expansion of the liquid solvent (solution) through a view window. At a fixed temperature, a given quantity of solvent is charged in the vessel and pressure is progressively increased by adding the anti-solvent. The expansion of the liquid phase is monitored by measuring the increase of the liquid level inside the vessel. The second method on the other hand, measures the variation of the density of liquid solvent–supercritical anti-solvent mixture at increasing pressures. Using the known density of the liquid phase and its composition, the volume expansion is calculated by a modified Peng–Robinson equation of state [24]. As stated previously the liquid expansion reduces the solubility of the solute and precipitation takes place. The precipitation consists of initial nucleus formation and subsequent growth of the nucleus. The rate of initial nucleus formation is:

$$J = Z \exp\left(\frac{\Delta G_{max}}{RT}\right) \tag{24.3}$$

where J is the rate of nucleus formation, Z the collision frequency (calculable from the classical kinetic theory) and ΔG_{max} the Gibbs free energy.

At low volume expansion, particles are being formed by precipitation from a liquid phase at the bottom of the precipitator. The liquid phase saturated with anti-solvent is formed as the result of vapor–liquid equilibrium at operating pressure and temperature. Under these conditions the precipitation of samarium acetate connected nanoparticles was observed by Reverchon et al. [18]. Our investigations indicated a similar structure of aluminum oxi-chloride submicron particles, after precipitation under similar conditions (100 bar, 313 K). These particles shown in Figure 24.7 were obtained by the nonhydrolytic sol–gel synthesis from aluminum trichloride and diethyl ether as precursors dissolved in carbon tetrachloride. Supercritical anti-solvent precipitation was achieved from the solution of particles in carbon tetrachloride using supercritical carbon dioxide as an anti-solvent.

At intermediate expansion levels, expanded droplets (balloons) have been observed by Reverchon et al. [18]. Spherical balloons were also observed by Dixon et al. [25] during the processing of polystyrene. Different nucleation structures can appear and different kinds of balloons have been observed. For example, during the precipitation of samarium acetate, empty shells of solute with a continuous surface were observed [18]. This is explained by the solute precipitation, which starts at the supercritical fluid–liquid interface and then propagates inside the liquid attracting the solute toward the separation surface. This mechanism results in the formation of hollow spherical structures [17].

At very large expansion levels (asymptotic expansion) nanoparticles are produced by balloons disintegration. An example of yttrium acetate nanoparticles is reported in Figure 24.8. These particles are very small (100–200 nm) and have a very narrow particle size distribution [18].

Other additional growth mechanisms have also been observed and they can superimpose on the one described earlier. These mechanisms can produce more complex particle geometries. The first one is the coalescence of nanoparticles. The physical coalescence is characterized by a particle-to-particle interaction, for example, by impact during the precipitation process. These particles can be separated by sonication. Chemical coalescence mechanism is the result of interaction between particles and solvents, which can lead to the fusion of nanoparticles in groups where the single particle has no more distinct identity. It is possible to take the advantage of coalescence and to produce larger or smaller particle aggregates.

The further growth of the solid particles in the solution inside the precipitation chamber is another mechanism that can strongly modify the morphology of particles. Although slow, this mechanism is responsible for the formation of more complex morphologies. A pronounced effect of co-solvent can lead to specific morphologies as well. This effect is well known in supercritical extraction processing and has been used to improve the solubility of poorly soluble

FIGURE 24.7 SEM image of the submicron sol–gel obtained aluminum oxi-chloride particles.

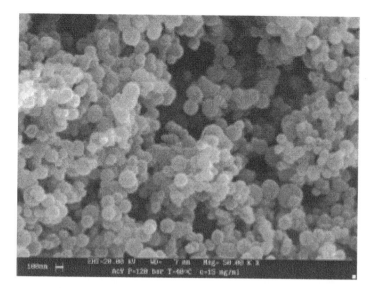

FIGURE 24.8 SEM image of yttrium acetate nanoparticles precipitated from DMSO (15 mg/ml) at 120 bar and 313 K, with mean diameters of about 150 nm. (Reprinted from Reverchon, E., *J. Supercrit. Fluids*, 15, 1–21, 1999. With permission from Elsevier.)

compounds. The formed complex (a result of solvent–solute interaction) is a loosely bound aggregate but this fact does not necessarily imply that it is destroyed when the anti-solvent is added. As a result of co-solvent action, no solute or only small quantities of solute are found (when compared with the injected quantity) in the precipitator since the solute is dissolved in the liquid–supercritical fluid mixture. Therefore, the solute is recovered in the liquid solvent collection chamber [26]. Another possible problem is the formation of a liquid phase at the bottom of the precipitation chamber even when asymptotic expansion of the liquid solvent has been obtained based on pure solvent curves. This problem can be the result of the modification of the solvent–anti-solvent phase equilibrium induced by the solute.

C. THE INFLUENCE OF DIFFERENT METHODS AND PROCESS PARAMETERS

The above described process modes can be divided into two large groups: precipitation from a liquid-rich phase and a supercritical fluid rich phase. The resulting particles obtained using these different process modes are usually quite different. Typically smaller and amorphous particles are obtained using the precipitation from the supercritical fluid rich phase. This fact could be explained by the faster expansion of the liquid phase (due to high anti-solvent or solvent ratio) and the larger liquid-to-supercritical fluid contact surface area produced by the formation of fine droplets. In liquid batch precipitation, the precipitation vessel has to be pressurized from atmospheric pressure up to the final pressure and this procedure results in the unsteady operation of the system and precipitating conditions. In general, when larger particles or crystalline materials are required, liquid batch processing should be adopted which can lead to more ordered crystal structures than the original material [27]; if smaller and amorphous particles are required, anti-solvent rich precipitation is the best choice.

Contradicting results have been obtained by different authors about the influence of pressure on the particle size during gas batch and continuous operation [17,19,20,28]. Concerning reduction of the precipitation pressure some authors found a particle size decrease [19,28], while others found the process insensitive to this parameter and some observed a particle size increase [17]. Similarly, the temperature reduction produced a particle size decrease according to certain authors [19], had no influence according to the others [17] and produced larger particles according to another group [28].

Particle size was relatively insensitive to solute concentration in the liquid according to some authors [17,19], while others observed that backing away from saturation conditions may be a better choice to prepare more uniform particles. A marked particle size increase and PSD enlargement with increasing concentration was observed by Reverchon et al. [18] in the SAS precipitation of yttrium, samarium, and neodimium acetates. This result is well illustrated in Figure 24.9a–c that are referred to samarium acetate precipitation from DMSO at the same pressure and temperature but at 10, 40 and 65 mg/ml concentration, respectively. From these SEM images the large increase of samarium acetate particle size and PSD is evident.

Very different behaviors can be obtained by changing the liquid solvent. For example, in the case of amoxicillin and tetracycline precipitated from DMSO and NMP, the SAS processing using the first solvent was completely unsuccessful since both antibiotics were extracted from the precipitation chamber, whereas, precipitation from NMP was successful and antibiotic nanoparticles were produced [17]. Similarly, the micronization of salbutamol was successful with dimethylsulfoxide (DMSO) while methanol and ethanol–water mixture were unsuccessful [29]. Hong et al. [30] have studied the influence of different solvents, the pressure, temperature, and the flow rate of solution on the SAS processing of organic pigments (Bronze red). Their results indicate that the size and morphology of the particles were influenced by the temperature, pressure, and flow rate when ethanol was used as the solvent. On the other hand, the application of acetone as solvent resulted in low sensitivity to temperature and pressure but the influence of the flow rate was significant. SEM micrographs indicate the influence of temperature on the particle morphology with ethanol as a solvent [30]. Chemical structure of the solute plays an important role, since compounds having a relatively simple molecular structure can form only primary particles, while compounds having a more complex chemical structure can exhibit further growth processes thus producing complex morphologies. Solubility, phase behavior of the system, and SAS processing of the solute are also influenced by the presence of another solute [31].

D. PGSS — Particles from Gas-Saturated Solutions/Suspensions

Since the solubilities of compressed gases or supercritical fluids in liquids and solids are usually high (significantly higher than the solubilities of liquids and solids in supercritical fluids), the main idea behind PGSS process consists in dissolving the supercritical fluid in melted or liquid-suspended substance. This leads to a so-called gas-saturated solution or suspension that is further expanded through a nozzle with the formation of solid particles or droplets (Figure 24.10). A specific modification of the process in which the compressed fluid behaves as a co-solvent in the initial solution phase, has been named the depressurization of an expanded liquid organic solution (DELOS) [32].

Typically, this process allows forming particles from a great variety of substances that need not be soluble in supercritical carbon dioxide, especially from some polymers that absorb a large concentration (10–40 wt.%) of CO_2 [8]. This process can also be performed with suspensions of active substrates in a polymer or other carrier substance leading to composite microspheres.

By dissolving the compressible media in a liquid, a so-called gas-saturated solution is formed. By expansion of such a solution in an expansion unit (e.g., a nozzle) the compressed medium is evaporated and the solution is cooled. Owing to the cooling caused by evaporation and the Joule–Thompson effect, the temperature of the two-phase flow after the expansion nozzle is lowered. At a certain point, the crystallization temperature of the substance to be solidified is reached, and solid particles are formed and cooled further.

The expansion phenomena (generation of low temperature by expansion) of compressible media is well described by the Joule–Thomson effect, which may be used as basic information on the applicability of a certain gas in the PGSS-process. If heat-transfer, changes in kinetic energy, and work-transfer for the flow process through an expansion valve are neglected, then the process may be considered to be isenthalpic [33]. The effect of change in temperature

FIGURE 24.9 SEM images showing the effect of concentration on particle size, and PSD of samarium acetate nanoparticles: (a) 10 g/ml DMSO; (b) 40 mg/ml DMSO; (c) 65 mg/ml DMSO. (Reprinted from Reverchon, E., *J. Supercrit. Fluids*, 15, 1–21, 1999. With permission from Elsevier.)

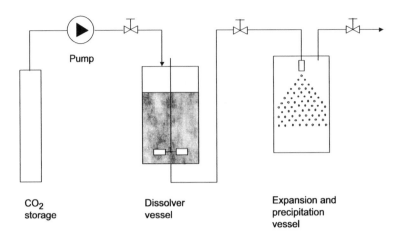

FIGURE 24.10 The PGSS process.

for an isenthalpic change in pressure is represented by the Joule–Thomson coefficient, μ_{jt}, defined by:

$$\mu_{jt} = \left(\frac{\partial T}{\partial P}\right)_h \tag{24.4}$$

The Joule–Thomson coefficient represents the slope of isenthalpic lines in the P–T projection. In the region where $\mu_{jt} < 0$, the expansion results in an increase of temperature, whereas in the region where $\mu_{jt} > 0$ the expansion results in a reduction of temperature. The latter area is recommended for the PGSS process application.

When the supercritical fluid has a relatively high solubility in the molten heavy component, the S–L–V curve can have a negative dP/dT slope. The second type of three-phase S–L–V curve shows a temperature minimum [34]. In the third type, where the S–L–V curve has a positive dP/dT slope, the supercritical fluid is only slightly soluble in the molten heavy component, and therefore the increase of hydrostatic pressure will raise the melting temperature and a new type of three-phase curve with a temperature minimum and maximum may occur [35]. In general, a system with a negative dP/dT slope and with a temperature minimum in the S–L–V curve could be processed by the PGSS. A thermodynamic model describing three-phase equilibrium in binary or ternary systems, which uses Peng–Robinson equation of state and one or two binary inter-action parameters, can be used satisfactorily to predict equilibrium data important for PGSS, RESS, and SAS processing [36].

The PGSS process has several advantages, which favor its application for large-scale process-ing. This process has advantages in the processing of low-melting point, highly viscous, waxy, and sticky compounds, even if the obtained particles are not of submicron size [37,38]. The appli-cation of the PGSS process has been investigated until now for the following products: polymers, waxes and resins, natural products (extracts from spices, phospholipids, menthol), fat derivatives, etc. The compressible fluids which have been used are: carbon dioxide, propane, butane, dimethyl ether, freon, ethanol, and nitrogen.

E. OTHER PROCESSES

Reverse microemulsions are systems in which fine aqueous droplets are uniformly dispersed in a continuous low-polarity fluid phase by the presence of surfactant shells [9]. Individual aqueous phase droplets are typically of 3–20 nm in diameter, and their size is strongly dependent on the molar water or surfactant ratio. Although, they are thermodynamically stable and able to exist in optically clear solutions for extended periods of time, these droplets are very dynamic at the molecular level. Surfactant molecules associated with aqueous droplets exchange rapidly with monomers dissolved in the continuous phase. Also, collisions between droplets lead to rapid restructuring and repartitioning of surfactant molecules and aqueous core contents on a short time scale. Reverse microemulsions offer a unique environment for chemical reactions. The dispersed aqueous droplets behave as miniature chemical reactors accommodating hydrophilic species in an otherwise hydrophobic environment. Low values of water or surfactant ratio result in highly ordered water molecules in the droplets due to association with the surfactant polar head groups. At higher values of water or surfactant ratio, droplet core usually resembles bulk water making it possible to dissolve various water-soluble compounds. Supercritical fluids offer specific characteristics which could be beneficial for reverse microemulsion systems. For example, diffusion of reacting species through the supercritical continuous phase is considerably improved, thereby improving the kinetics of diffusion limited reactions. Additionally, the stabilities of supercritical fluid microemulsions are highly pressure sensitive, with the single-phase microemulsions separating into two or more distinct phases at some pressure boundary. This characteristic offers a possibility to shift equilibrium of certain reactions by manipulating the pressure of the system. Silver nanoparticles obtained by the reduction of silver nitrate in bis(2-ethylhexyl)sulfosuccinate (AOT) reverse micelles in compressed propane [39]. The particle size distribution plot indicates narrow size distribution mostly from 2 to 5 nm and maximum around 3 nm. Salaniwal et al. [40] have studied reverse micelles in supercritical carbon dioxide using molecular dynamics simulation.

The supercritical fluids have received considerable attention as solvents for the synthesis of a number of ceramic, metal, and other materials. These new applications have been developed in order to improve the characteristics of the obtained powders, such as chemical homogeneity or unique structural properties. Unlike previously described processes (RESS, SAS, and PGSS) which are principally physical transformation, the chemical transformation of materials in supercritical fluid is the principle of different reactive processes.

Particles can be produced by thermal decomposition with a supercritical solvent where precursors are thermally decomposed in a supercritical media. At the end of the reaction, supercritical solvent is depressurized and the solvent, turned to gas phase, separates from the particles which remain in a highly divided state. The advantage of processing in a supercritical media is that high nucleation rate and low crystal-growth rate can be achieved, leading to the formation of very fine particles. Furthermore, the high density of the supercritical fluid avoids aggregation problems encountered in liquid–solid separation steps of conventional wet chemical processes [41]. When the thermal stability of the precursor does not allow dissolution in the supercritical fluid, a variant process can be used: a sol–gel reaction is conducted at high pressure and temperature, followed by supercritical drying. Fine powders (diameter: 50 nm) can be obtained using this process [8]. The hydrothermal processes typically designate processes conducted at relatively high temperatures and pressures in the presence of water. Under the supercritical hydrothermal conditions very fine crystalline oxide powders can be produced. Simple oxides and more complex compositions can also be generated. The produced powders are commonly pure, crystalline, and with highly uniform particle-size distributions.

The supercritical fluid can be used as a solvent and a reactant. The processes of this type described in the literature commonly use supercritical water. Hydrothermal synthesis has been used for microparticles or large crystals processing. In order to prepare oxide powders, low-cost precursors (oxides, hydroxides, or salts) are first dissolved in water and the solution is introduced into a reactor operated at supercritical conditions. This process has a potential to manipulate the direction of crystal growth, morphology, particle size, and size distribution, due to the controllability of thermodynamics and transport properties by pressure and temperature [42]. Moreover, high reaction rate leads to ultra-fine powder formation while crushing and calcination steps used in conventional processes are avoided. It is also possible to apply reducing or oxidizing atmosphere by the introduction of oxygen, hydrogen, or other gases. On the other hand, filtration, washing, and drying are still necessary, and corrosion problems associated with supercritical water are serious disadvantages of the process.

Specific physicochemical properties of the supercritical fluids offer flexible alternatives to established processes like chemical vapor deposition (CVD), which is used in the preparation of high-quality metal and semiconductor thin films on solid surfaces. Watkins et al. [43] reported a method named chemical fluid deposition (CFD) for the deposition of CVD-quality platinum metal films on silicon wafers and polymer substrates. The process proceeds through hydrogenolysis of dimethyl-(cyclooctadiene)platinum(II) at 353 K and 155 bar.

F. INORGANIC PARTICLES OBTAINED USING SUPERCRITICAL FLUIDS

The application of RESS in the processing of inorganic particles is relatively limited, since the solubility of inorganic solids in most of the supercritical fluids is very low. For example, solubility of SiO_2 in supercritical water at 773 K and 100 MPa is 2600 ppm and solubility of Al_2O_3 in supercritical water at the same conditions is only 1.8 ppm [9]. Therefore, RESS is considered for application only in the case of special inorganic materials. The literature data of processing of the inorganic material using RESS are shown in Table 24.2.

As it can be seen from Table 24.3, processing of inorganic material using SAS and related processes is dominantly applied to obtain organic compounds which serve as precursors for the synthesis of inorganic solids. An interesting route to synthesize ceramic material was proposed

TABLE 24.2
Inorganic Solids Obtained by the RESS

Substrate	SC Fluid	Results	Ref.
SiO_2	Water	>1.0 μm thick film and $0.1-0.5$ μm diameter spheres	[44]
GeO_2	Water	5 μm agglomerates or $0.3-0.5$ μm diameter spheres	[44]
AgI	Acetone	Ag film after reaction at 873 K	[45]
Ag triflate	Et_2O	Ag film after reaction at 873 K	[45]
$Al(hfa)_3$	Pentane	Al film after reaction at 953 K	[45]
$Al(hfa)_3$	N_2O	Al_2O_3 film after reaction at 373 K	[45]
$Cr(acac)_3$	Acetone	Cr film after reaction at 1073 K	[45]
$Cr(hfa)_3$	N_2O	Cr_2O_3 film after reaction at 373 K	[45]
$Cu(oleate)_2$	Pentane	Cu film after reaction at 1013 K	[45]
$Cu(thd)_2$	N_2O	Cu film after reaction at 973 K	[45]
$Cu(thd)_2$	N_2O	CuO film after reaction at 373 K	[45]
$In(acac)_3$	CO_2	In film after reaction at 873 K	[45]
$Ni(thd)_2$	Pentane	Ni film after reaction at 873 K	[45]
$Pd(tod)_2$	Pentane	Pd film after reaction at 873 K	[45,46]
$Si(OC_2H_5)_4$	N_2O	SiO_2 film after reaction at 373 K	[45]
SiO_2, KI	Water	20 μm agglomerates	[44]
$Y(thd)_3$	N_2O	Y film after reaction at 960 K	[45]
$ZrO(NO_3)_2$	Ethanol	Particles 0.1 μm	[44]
$Zr(tfa)_4$	Et_2O	Zr film after reaction at 873 K	[45]
$Pb(NO_3)_2$	NH_3, EtOH, MeOH	PbS nanoparticles 4 nm diameter	[47]

Abbreviations: MeOH, methyl alcohol; EtOH, ethyl alcohol; thd, *bis*(2,2,6,6-tetramethyl-3,5-heptanedionato); acac, *tris*(2,4-pentandionato); hfa, *tris*(1,1,1,5,5,5-hexafluoro-2,4-pentanedionato); tfa, tetrakis(1,1,1-trifluoro-2,4-heptanedionato); tod, *bis*(2,2,7-trimethyl-3,5-octanedionato).

TABLE 24.3
Inorganic Solids Obtained by the SAS and Related Processes

Process	Substrate or Solvent	SC Fluid	Results	Ref.
SAS	NH$_4$Cl/DMSO	CO$_2$	Particles 1–5 μm	[48]
SAS	BaCl$_2$/DMSO	CO$_2$	Particles 1–7 μm cubic or needle-like	[48]
SAS/ASES	SmAc/DMSO	CO$_2$	Particles 0.1–0.3 μm	[17,18]
SAS/ASES	YAc/DMSO	CO$_2$	Balloons; particles 0.08–2 μm	[17,18]
SAS/ASES	NdAc/DMSO	CO$_2$	Particles 0.1 μm	[17,18]
SAS/ASES	GdAc/DMSO	CO$_2$	Particles 0.2–0.4 μm	[49]
SAS/ASES	EuAc/DMSO	CO$_2$	Particles 0.2–0.4 μm	[49]
SAS/ASES	ZnAc/DMSO	CO$_2$	Particles 50–150 nm	[50]
SAS	VOHPO$_4 \cdot 0.5H_2O$/isopropanol	CO$_2$	Catalyst spheres 50–700 nm	[51]
SAS	SiO$_2$/acetone	CO$_2$	Spheres 1.2–2.2 μm, fibers	[52]

Abbreviations: DMSO, dimethylsulfoxide; TTIP, titanium tetraisopropoxide; Ac, acetate.

by Moner-Girona et al. [52]. Spherical silica aerogel particles were obtained by the hydrolytic sol–gel method in acetone followed by the particles precipitation in supercritical carbon dioxide.

Supercritical fluids have received considerable attention as solvents for the synthesis of ceramic or similar materials. One method applies thermal decomposition of precursors in a supercritical fluid, which is at the end of the reaction depressurized and removed from the system. After removal of the solvent, obtained inorganic material is typically in the form of micron or submicron particles. The second method uses supercritical fluid as a solvent and a reactant. In this case, the typical supercritical fluid is water, and the process is called hydrothermal synthesis. Inorganic solids obtained by the decomposition in the supercritical fluid or by the synthesis using supercritical reactant, are listed in Tables 24.4 and Table 24.5, respectively. As it can be seen from the presented data different morphologies can be obtained on the micrometer or nanometer scale.

G. Polymer Solid Particles Obtained Using Supercritical Fluid Processing

Polymer processing with supercritical fluids can result in the formation of very fine powders, fine fibers, microspheres, microballoons, and thin films. As indicated in Table 24.6 very fine particles with submicron diameters and ultra-thin films (6 nm) were obtained using RESS. Tepper and

TABLE 24.4
Inorganic Solids Obtained by the Decomposition in Supercritical Fluid

Substrate or Precursor	SC Fluid	Results	Ref.
MgAl$_2$O$_4$/MgAlTSB	Ethanol	Spheres 0.5–2 μm	[41]
MgO/MgChe, MgAc	CO$_2$ + ethanol	Particles 0.5–2 μm	[53]
TiO$_2$/TTIP	Ethanol	Spheres 0.5–2 μm	[41]
TiO$_2$/TTIP	CO$_2$ + isopropanol	—	[54]
TiO$_2$/TTIP	CO$_2$ + aq.surf	Spheres 0.1–2 μm	[55]
TiO$_2$/TTE	Ethanol + water + Et$_2$O	Particles 60–400 nm	[56]

Abbreviations: AlTSB, aluminum tri-sec butoxide; TTIP, titanium tetraisopropoxide; TTE, titanium tetraethoxide; Et$_2$O, diethyl ether; MgChe, magnesium chelate; MgAc, magnesium acetate; aq.surf, aqueous solution of surfactant.

TABLE 24.5
Inorganic Solids Obtained by the Synthesis with Supercritical Reactant

Substrate or Precursor	SC Fluid	Results	Ref.
$AlOOH/Al(NO_3)_3$	Water	Hexagonal, rhombic or needle like particles; size 100–600 nm	[57,58]
$Co_3O_4/Co(NO_3)_2$	Water	Octahedral particles 50 nm	[57,58]
$Ti(OH)_4/TTIP$	CO_2 or H_2O	Spheres 70–110 nm	[59]
$SiO_2/TEOS$	CO_2	Particles <1 μm	[52]
$Fe_2O_3/Fe(NO_3)_3$, $FeCl_2$, $Fe_2(SO_4)_3$	Water	Spheres 50 nm	[57,58]
$Fe_3O_4/Fe(NH_4)_2$	Water	Spheres 50 nm	[57,58]
$TiO_2/TiCl_4$	Water	Spheres 20 nm	[57,58]
$LiCoO_2/LiOH$, $Co(NO_3)_2$	Water + O_2	Particles 700 nm	[42]

Abbreviation: TTIP, titanium tetraisopropoxide.

Levit [65] obtained uniform PDMS microspheres with a typical diameter of about 2–3 μm as shown in Figure 24.11. Although, different polymers were processed with RESS as indicated in Table 24.6, its application is relatively limited due to the low solubility of polymers in supercritical solvents.

Different polymers (and applied solvents) micronized using SAS and related processes are shown in Table 24.7. As it can be seen different morphologies were obtained, including particles, fibers, and microspheres.

PGSS of polyethyleneglycols (PEGs) was investigated by Weidner et al. [75]. PEGs are water-soluble polymers, which are owing to their physiological acceptance widely used in pharmaceutical, cosmetic, and food industries. Depending on the molecular weight, PEGs can be either liquid ($M < 600$ kg/kmol) or solid ($M > 600$ kg/kmol). In case of molecular weights lower

TABLE 24.6
Polymers Obtained by the RESS

Substrate	SC Fluid	Results	Ref.
Polystyrene	Pentane	1 μm diameter fibers or 20 μm diameter spheres	[60]
Poly(carbosilane)	Pentane	Particles <0.1 μm diameter, or 1 μm diameter fibers	[44,60]
Poly(caprolactone)	Chlorodifluoromethane	Powder or fibers	[61]
L-PLA	Carbon dioxide	Particles 10–90 μm	[62]
PGA	Carbon dioxide	Particles 10–20 μm	[63]
D-PLA	Carbon dioxide	Particles 10–20 μm	[63]
PMMA	CCl_2F_2	Particles 200–600 nm	[64]
PEMA	CCl_2F_2	Particles 200–600 nm	[64]
PDMS	Carbon dioxide	Particles 2–3 μm	[65]
Teflon AF2400 (Du Pont)	Carbon dioxide	Thin films approx. 6 nm thick	[66]
PEHA	Carbon dioxide	Particles 270–340 nm	[67]

Abbreviations: L-PLA, poly-L-lactic acid; PGA, poly-glycolic acid; D-PLA, poly-D-lactic acid; PMMA, poly-methyl methacrylate; PEMA, poly-ethyl methacrylate; PDMS, poly(dimethyl-siloxane); PEHA, poly(2-ethylhexyl acrylate).

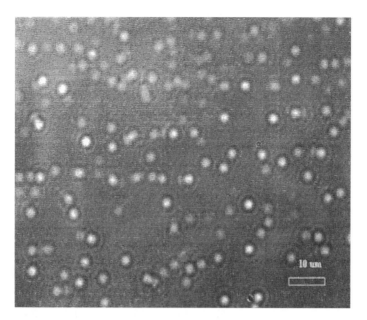

FIGURE 24.11 PDMS particles obtained by RESS. (Reprinted from Tepper, G. and Levit, N., *Ind. Eng. Chem. Res.*, 39, 4445–4449, 2000. With permission from American Chemical Society.)

than 2000 kg/kmol and higher than 10,000 kg/kmol, the solid polymer is either too greasy or too hard to be processed by milling. An alternative route for producing fine particles of PEGs is PGSS process, since the solubility of carbon dioxide in PEGs above a certain chain length is independent of the molecular weight of PEG and depends only on the temperature and pressure of the system. Fibers, spheres, and sponges of PEG were obtained using PGSS with carbon dioxide [75].

TABLE 24.7
Polymers Obtained by the SAS and Related Processes

Process	Substrate/Solvent	SC Fluid	Results	Ref.
ASES	L-PLA/acetone	CO_2	Particles 1–10 μm	[26]
ASES	Polystyrene/toluene	CO_2	Particles 0.1–20 μm	[20]
SAS	PLGA/CH_2Cl_2	CO_2	Particles 40–60 μm	[68]
SAS	L-PLA/CH_2Cl_2	CO_2	Particles 0.5–5 μm	[19]
SAS	Polyamide/DMSO	CO_2	Particles 2–10 μm, fibers	[69]
SAS	Polyamide/DMF	CO_2	Fibers	[69]
SAS	L-PLA/DMSO	CO_2	Particles 1.25–1.75 μm	[70]
SAS	Dextran/DMSO	CO_2	Particles 130–150 nm	[70]
SAS	HPMA/DMSO	CO_2	Particles 150 nm	[70]
SAS	L-PLA/chloroform	CO_2	Particles >200 nm	[71]
SAS	PC + SAN/THF	CO_2	Particles <10 μm	[72]
SAS	Polyamide/DMAc + LiCl	CO_2	Fibers	[73]
SAS	Polystyrene/toluene	CO_2	Spheres 1–20 μm	[25]
ASES	PLCG/methylenechloride	CO_2	Microparticles	[74]

Abbreviations: L-PLA, Poly-L-lactic acid; DMSO, dimethylsulfoxide; DMF, *N,N*-dimethylformamide; HPMA, poly-(hydroxypropylmethacrylamide); PC, polycarbonate; SAN, poly(styrene-co-acrylonitrile); THF, tetrahydrofuran; DMAc, *N,N*-dimethylacet amide; PLCG, poly lactide-co-glycolide.

The polymer industry is a major user and emitter of volatile organic compounds (VOCs) which are toxic and cause harm to the environment. As a result of growing pressure to reduce emission, there has been considerable effort devoted to find alternative nonpolluting solvents. DeSimone and others have shown the applicability of supercritical carbon dioxide as a viable medium for a number of polymerization reactions [76–78]. An example of application in processing of polymeric powders is continuous precipitation polymerization of vinylidene fluoride in supercritical carbon dioxide [79–81]. This polymer is produced commercially in batch reactors by either emulsion or suspension method, at monomer pressures from 10 to 200 bar and at temperatures from 283 to 403 K. In the emulsion process, the latex is coagulated, thoroughly washed, and then spray dried to form powder. The suspension method requires thorough washing and drying as well. In both processes, large quantities of waste water are generated and large quantities of energy are required to dry the polymer. Continuous precipitation polymerization in supercritical carbon dioxide eliminates generation of waste water, reduces energy consumption and results in the formation of micrometer-scale particles of poly(vinylidene fluoride) [80].

H. Particles with Specific Drugs

Poorly soluble drugs are the main problem in pharmaceutical drug formulation. The low bioavailability of such drugs makes the intravenous injection of such solution impossible [82]. Moreover, the poor solubility is usually associated with low dissolution rates and low drug absorption. For avoiding such problems in pharmaceutical drug preparation different supercritical fluid techniques were recently used showing many advantages.

Some examples of commercial active component production and production of substances with defined and uniform particle sizes (organic and inorganic materials) realized on pilot plant by using the RESS are given in Table 24.8. Other processes were also tested for synthesis of the particles with uniform size distribution as well as production of particles with specific structure (gas antisolvent recrystallization, GASR; precipitation with a compressed antisolvent, PCA; solution enhanced dispersion of solids, SEDS; particles from gas-saturated solutions, PGSS) as shown in Table 24.9. All these processes are of special interest in pharmaceutical industry and in the production of different polymers.

I. Encapsulation and Coating Using Supercritical Fluids

Among the prospective applications of supercritical fluids are also the coating and encapsulation of: inorganic compounds, metal micro and nanoparticles, active pharmaceutical ingredients, and others. Over the last decade, polymer microcapsules of inorganic nanoparticles have been attracting much attention in the catalytic, cosmetics, printing, and electronic industry [100]. Another potential field of application is in coating of metal powders with thin films [101], which are common components of the pyrotechnic and solid propellant compositions. Coating of metal powders reduces deterioration through corrosion and aggregation caused by moisture or other aggressive substances, and it also reduces flammability. The increasing trend in research and development of drug delivery systems will intensify in the pharmaceutical industry. The distribution of active substance directly to the target will enhance the treatment efficiency and reduce the doses and related side-effects, while many efficient drugs will have to be reformulated in order to allow control of delivery location and rate. This last criterion is of particular importance for long-term treatments like cancer or various chronic diseases. The microencapsulation could offer solutions by enhancing controlled delivery of active substances and material stability.

For microencapsulation of inorganic nanoparticles, preparation methods like *in situ* polymerization may be used [102]. However, these methods often require toxic organic solvents and surfactants. Furthermore, the removal of residual surfactants or solvents is needed, since they cause faults in the product. An environmentally benign solvent like supercritical carbon dioxide can be used to encapsulate inorganic nanoparticles. But its use is limited because of the low

TABLE 24.8
RESS[a] and GASR[b] Processes

Product	SCF	Pressure (bar)	Temperature (°C)	Particle Size (μm)	Ref.
Steroids[a]	CO_2	130–250	40–60	1–10	[83]
Theophyllinen[a]	CO_2	225	65	0.4	[84]
Salicylic acid[a]	CO_2	223	45	<4	[84]
Naproxen[a]	CO_2	190–210	60	1–20	[62]
L-PLA[a] (poly-L-lactic acid) + naproxen	CO_2	170–200	90–115	10–90	[62]
L-PLA[a]	CO_2	210	100–120	1–20	[62]
Flavone[a]	CO_2	250	35	10	[85]
PGA (poly-ethylene glycolic acid)[a]	CO_2	180–200	55	10–20	[63]
D-PLA[a] (poly-D-lactic acid)	CO_2	200	55	10–20	[63]
Lovastatin[a]	CO_2	125–400	55	0.04–0.3	[86]
β-Estradiol[a]	CO_2	345	55	—	[87]
β-Carotene[a]	C_2H_4+ toluene	300	70	20	[88]
Stigmasterol[a]	CO_2	100–150	100	0.05–2	[89]
Ibuprofen[a]	CO_2	130–190	35	<2	[90]
Acetaminofene[b]	Ethanol	71	20	—	[91]
Hidrokortizon[b]	DMSO	104	—	0.2–1	[68]
Indometacin[b]	CH_2Cl_2	200	—	8.2	[92]
Piroksikam[b]	CH_2Cl_2	90–200	—	6.0	[92]
Timopentin[b]	CH_2Cl_2	90–200	—	6.6	[92]
Hidrokortizon acetate (HCA)[b]	DMF	150	—	8.0	[93]
L-PLA 94000[b]	CH_2Cl_2	81.6	—	1–5	[85]

[a]REES process; [b]GASR process.

Abbreviations: DMSO, dimethylsulfoxide; DMF, dimethylformamide.

solubility of most of the polymers and inorganic solids in supercritical carbon dioxide. A novel method, rapid expansion of supercritical solution with nonsolvent (RESS-N), has been reported by Mishima et al. [103,104]. The process differs from conventional RESS as it applies organic cosolvents. Since polymers are usually insoluble in the supercritical carbon dioxide at temperatures and pressures of interest, several cosolvents are used to enhance solubility of the polymer in supercritical media. These solvents are nonsolvents for polymer at atmospheric pressure and they are only sparingly soluble in the polymer particles produced during expansion. As a result of RESS-N, the polymer precipitates and coats the inorganic nanoparticles [15]. RESS-N method was used to coat TiO_2 nanoparticles which are of interest as cosmetic, printing, and electronic materials. The obtained particles were in the order of 10–30 μm and not adhesive to each other.

Debenedetti et al. [105] obtained microparticles by the RESS co-precipitation of a drug (lovastatin) and a biodegradable polymer (poly(DL-lactic acid)). The co-precipitation of the polymer and the drug has led to a heterogeneous population of microparticles consisting of microspheres containing a single lovastatin needle, larger spheres containing several needles, microspheres without protruding needles, and needles without any polymer coating.

The protein–polymer microcapsules can be obtained by supercritical anti-solvent techniques [8]. Homogeneous protein–polymer mixtures were contacted with supercritical carbon dioxide in order to produce microspheres with diameter ranging from 1 to 5 μm and containing around 80% of protein. Production of PLA microparticles containing insulin, lysozyme, and chemotrypsin is claimed. SAS crystallization of a pharmaceutical (naproxen) and a biodegradable poly(L-lactic acid) was reported [8]. The results from SAS studies showed very small spherical particles

TABLE 24.9
Anti-solvent Experiments for Pharmaceutical Products

Material	Solvent	Process	Particle Size (μm)	Ref.
Insulin	DMSO	SAS	2–4	[94]
Indomethacin	CH_2Cl_2	PCA	8.2	[95]
Piroxicam	CH_2Cl_2	PCA	6.0	[95]
Thymopentine	CH_2Cl_2	PCA	6.6	[95]
Hydrocortisone	DMSO	PCA	0.2–1	[96]
MPA	THF	PCA	2.5–3	[89]
HCA	DMF	PCA	8.0	[89]
SX	Acetone	PCA	10	[97]
Insulin	DMSO	SAS	1–5	[98]
Lysozyme	DMSO	SAS	1–5	[98]
Trypsin	DMSO	SAS	1–5	[98]
CPM	CH_2Cl_2	PCA	1–5	[99]
Indomethacin	CH_2Cl_2	PCA	1–5	[99]
NaCrGlyc	MeOH	GAS	0.1–2	[99]

Abbreviations: MPA, methylprednisolone acetate; HCA, hydrocortisone acetate; SX, salmeterol xinafoate; CMP, chlorophenylamine maleate; NaCrGlyc, sodium cromoglycate; DMSO, dimethylsulfoxide; THF, tetrahydrofurane; DMF, dimethylformamide.

composed of a naproxen core surrounded by a polymer shell. The ASES technique was used for the coprecipitation of a model drug, parahydroxybenzoic acid (*p*-HBA) with the biodegradable polymers, poly(lactide-co-glycolide) (PLCG) and poly(L-lactic acid) (PLA) [8].

Supercritical fluid-soluble substrates can be easily impregnated inside porous media. Domingo et al. [8] used this process to prepare controlled drug delivery systems. In this work, zeolite, several amorphous–mesoporous inorganic matrices (silica gel, alumina, and florisil), and a polymeric matrix (amberlite) have been impregnated with various thermally unstable organic compounds (benzoic acid, salicylic acid, aspirin, triflusal, and ketoprofen) by diffusion from saturated supercritical carbon dioxide solutions. Copper and iron chelate complexes were impregnated into a polyarylate matrix from the solution in supercritical carbon dioxide by Said-Galiev et al. [106].

The precipitation methods previously described (RESS, SAS, PGSS, and thermal decomposition) can be used for particle coating, while those currently used in the industry are the Wurster process [107] and fluidized bed coating [108]. The latter was also investigated for application at supercritical conditions [109]. It was found that operating parameters have influenced the morphologies of coated particles. Kobe Steel Ltd. patented a process related to the RESS concept for microparticles formation and coating [8]. Primary microparticles are formed by rapid decompression of a supercritical solution of the core material. These particles are mixed with a supercritical solution of the coating agent and depressurized to form microcapsules. Coating with multiple layers can be achieved using this process. The method for coating polymeric thin films on particles involves a recirculation system that includes dissolution of the polymer into the supercritical solvent and coating the particles through a temperature swing operation in the fluidized bed [8]. This system has been tested with hydroxyl-terminated polybutadiene (HTTB) as coating polymer and particles of salt (30–500 μm). The film thickness was as low as 0.2 μm. In the method patented by Mainelab [8], the coating agent is solubilized into supercritical carbon dioxide. In a stirred vessel, particles of an active substance are dispersed into the supercritical solution. By changing the pressure and the temperature, the solubility of the coating agent in the supercritical fluid can be reduced so that it can precipitate on the particles. Microcapsules of active substance are collected after depressurization. This process allows having a good control of the structure (composition, thickness) of microcapsule without using organic solvents. This process was used to encapsulate substances such as: dyes, antibiotics, vitamins, and proteins. Coating agents that are soluble in supercritical carbon dioxide were used, such as waxes, glycerides, alcohols, fatty acids, and esters. Similar process reported by Glebov et al. [110] is used for coating of metal particles. In another method patented by Mainelab [8], a suspension of active substance in a solution of a slightly polar polymer (insoluble in liquid or supercritical carbon dioxide) in an organic solvent is used. This suspension is contacted with supercritical carbon dioxide so that the organic solvent is solubilized in such a manner that there is coacervation of the coating polymer onto the particles. Microcapsules are collected after decompression.

III. AEROGELS AND SOL–GEL PROCESSING

A. PREPARATION OF AEROGELS

Although, no official definition of aerogels exists, the term "aerogel" is conventionally applied to designate gels dried under supercritical conditions, which are highly porous and open texture materials. In most cases, the obtained dry solids exhibit amorphous structures when examined by X-ray diffraction and they are thermodynamically metastable. Due to their metastable character, aerogels can develop attractive physical and chemical properties which are of interest in a range of applications.

1. Sol–Gel Synthesis

First step in producing an aerogel is a sol–gel synthesis, depicted in Figure 24.12, with inorganic salts and metal alkoxides as common precursors.

Inorganic precursors, like metal salts of the transition metals, are solvated by water molecules according to:

$$M^{z+} + :OH_2 \longrightarrow M[(OH_2)]^{z+} \tag{24.5}$$

causing charge transfer from the filled bonding orbital of the water molecule to the empty d orbital of the transition metal [111]. This in turn causes the partial charge on the hydrogen to increase, thus increasing the water molecule acidity. Depending on the water acidity and the

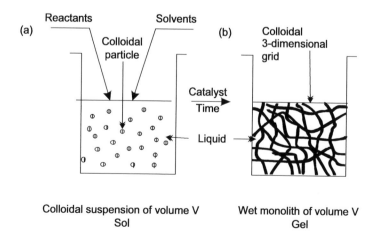

FIGURE 24.12 The sol–gel synthesis.

magnitude of the charge transfer, the following equilibrium has been established, which is defined as hydrolysis:

$$M[(OH_2)]^{z+} \Longleftrightarrow [M{-}OH]^{(z-1)+} + H^+ \Longleftrightarrow [M{=}O]^{(z-2)+} + 2H^+ \qquad (24.6)$$

The ligands present in noncomplexing aqueous media shown in the above equation are: aquo M-(OH_2), hydroxo M—(OH), and oxo M=O. The precise nature of the formed complex depends on the charge (z), coordination number and electronegativity of the metal atom, and the pH of the aqueous solution [111]. The partial charge model of Livage et al. [111,112], postulates that when two atoms are combined they acquire a partial positive (leaving group) or negative charge (nucleophile).

When a hydroxy bridge is formed between two metal centers by nucleophilic substitution reaction (where nucleophile is hydroxyl group and water is leaving group), then a condensation process is called as olation [113].

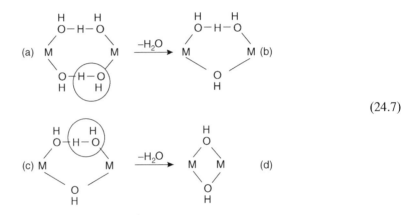

$$(24.7)$$

In case when an oxo bridge is formed between two metal atoms, the condensation process is called as oxolation. When the metal is coordinatively unsaturated, oxolation proceeds by nucleophilic addition with rapid kinetics [114,115], leading to edge- or face-shared polyhedra (Equation (24.8) and Equation (24.9)).

$$-M \quad + \quad M- \quad \longrightarrow \quad -M \quad M- \qquad {}_2(O)_2 \qquad (24.8)$$

$$-M \quad + \; O{-}M \quad \longrightarrow \quad -M{-}O{-}M- \qquad {}_2(O)_3 \qquad (24.9)$$

For coordinatively saturated metals, oxolation proceeds by two-step nucleophilic substitution reaction involving nucleophilic addition followed by water elimination.

$$M{-}OH + M{-}OH \longrightarrow M{-}O{-}M{-}OH \qquad (24.10)$$

$$M{-}O{-}M \longrightarrow M{-}O{-}M + H_2O \qquad (24.11)$$

Depending on the nucleophilic power of the starting precursor of the transition metal, solution pH, temperature, mixing speed, and condensation kinetics, precipitation or gelation can occur and different size oligomers can be formed [116]. The sol–gel parameters and primary particles sizes influence phase transformations, densification, and crystallization behavior during further processing of the gels [117,118]. It should be mentioned that most aqueous systems in addition to aquo, hydroxo, and oxo ligands, also contain anions introduced by the starting inorganic salts. These anions can compete for coordination to the metal centers and in many practical cases influence evolving particle morphology and stability [119,120]. The use of metallic salts as sol–gel precursors has recently seen a renewed interest with the use of organic solutions in which an organic slow "proton scavenger" is dissolved [121,122]. This procedure can lead to gel monoliths of Cr, Fe, Al, Zr, and other cations.

Metal alkoxides are the most widely used sol–gel precursors since they readily hydrolyze with water:

$$\equiv Si - OR + H_2O \quad \rightarrow \quad \equiv Si - OH + ROH \qquad (24.12)$$

Hydrolysis reaction is followed by alcohol or water condensation reactions and under typical conditions condensation commences before complete hydrolysis of the alkoxide.

$$\equiv Si - OR + HO - Si\equiv \quad \rightarrow \quad \equiv Si - O - Si\equiv + ROH \qquad (24.13)$$
$$\equiv Si - OH + HO - Si\equiv \quad \rightarrow \quad \equiv Si - O - Si\equiv + H_2O \qquad (24.14)$$

Numerous investigations have shown that variations in the synthesis conditions: water or alkoxide ratio, the catalyst type and concentration, the solvent, temperature and pressure, cause modifications in the structure and properties of the obtained product. Under acidic conditions (pH < 2.5) it is likely that an alkoxide group is being protonated in a rapid first step (Figure 24.13). Electron density is withdrawn from the silicon atom, thereby making it more susceptible to attack by water. The attacking water molecule acquires a partial positive charge, thereby reducing the positive charge of the protonated alkoxide and making alcohol a better leaving group. The transition state decays by displacement of alcohol [123]. Rate of hydrolysis is increased by the acid concentration and by substituents that reduce steric crowding around silicon.

Under basic conditions (pH > 2.5) it is likely that water dissociates to produce nucleophilic hydroxyl anions in a rapid first step. The hydroxyl anion then attacks the silicon atom. Iler [124] and Keefer [125] propose a mechanism in which hydroxyl anion displaces OR— with inversion of the silicon tetrahedron (Figure 24.14), while Pohl and Osterholz [126] favor a mechanism involving a stable five-coordinated intermediate which decays through second transition state in which any of the surrounding ligands can acquire a partial negative charge.

Since silicon acquires a formal negative charge in the transition state, both mechanisms are quite sensitive to inductive and steric effects. However, steric factors are more important since silicon acquires little charge in the transition state.

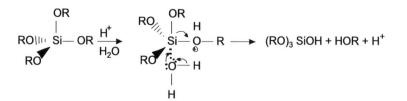

FIGURE 24.13 Acid catalyzed hydrolysis mechanism.

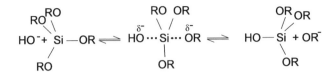

FIGURE 24.14 Base-catalyzed hydrolysis as proposed by Iler and Keefer.

It is generally believed that the acid-catalyzed condensation mechanism involves a protonated silanol species. Protonation of the silanol makes the silicon electrophilic and thus more susceptible to nucleophilic attack. The most basic silanol species are silanols contained in monomers or weakly branched oligomers, and these species are therefore most likely to be protonated. Condensation reaction therefore preferentially takes place between neutral species and silanols situated on monomers, chain end groups, etc. The most widely accepted mechanism for the condensation reaction under basic conditions involves the attack of a nucleophilic deprotonated silanol on a neutral silicate species. Silanols are deprotonated depending on their acidity, which depends on the other substituents on the silicon atom. When basic OR and OH are replaced with OSi, the reduced electron density on Si increases the acidity of the protons on the remaining silanols. This mechanism favors reactions between larger, more highly condensed species, which contain acidic silanols, and smaller, less weakly branched species. The condensation rate is maximized near neutral pH, where significant concentrations of protonated and deprotonated silanols exist. It is also believed that the base catalyzed mechanism involves penta- or hexacoordinated silicon intermediates. Investigations of Engelhardt et al. [127] on condensation of aqueous silicate at high pH using ^{29}Si NMR, indicate that a typical sequence of condensation products is monomer, dimer, linear trimer, cyclic trimer, cyclic tetramer, and higher order rings. These rings form the basic framework for the generation of discrete colloidal particles. Structural evolution of the gel is governed by the relative rates of hydrolysis, condensation, and reversible reactions (re-esterification, alkoholysis and, hydrolysis), which are dependent on the: solution pH, water or alkoxide ratio, solvent type, alkyl group structure, temperature, and pressure [128]. Compared to transition metals, silicon is generally less electropositive than metal atoms such as Al, Ti, and Zr, and therefore less susceptible to nucleophilic attack. These factors make the kinetics of hydrolysis and condensation of silicon considerably slower than observed in transition metal systems or group III systems.

An enormous number of possible multicomponent systems, makes it impossible to discuss them here in detail. However, it should be mentioned that there are two general approaches in the sol−gel synthesis of multicomponent systems: hydrolysis of mixed-alkoxide or metal organic precursors, and sequential addition of alkoxides to partially hydrolyzed precursors [129]. The first method invented by Dislich [130], is based on the idea to form complex through alcolation that contains all metals in proper stoichiometry. The second approach which is based on the sequential addition of alkoxides in the reverse order of their respective reactivities, was introduced by Yoldas [131,132]. The idea is that the newly added unhydrolyzed alkoxide molecules will preferentially condense with partially hydrolyzed sites on the polymeric species formed by the preceding hydrolysis and condensation, rather than reacting with themselves. The homogeneity of the product will depend on the size of the polymeric species to which the last component is added; and the most widely studied multicomponent systems are: $Al_2O_3-SiO_2$, $B_2O_3-SiO_2$ and TiO_2-SiO_2.

A relatively new nonhydrolytic sol−gel method, based on the reaction between alkoxides and halides of Al, Si, Ti, etc., or between metal halides and oxygen containing organic compounds (ethers, aldehydes, ketones, etc.), was developed by Vioux and coworkers [133–138]. Structural investigations of the obtained material using ^{27}Al NMR have revealed the presence of Al^{IV}, Al^{VI}, and metastable Al^{V} sites.

2. Supercritical Drying

Second phase of the sol–gel process is the removal of solvent and sol–gel synthesis byproducts from the solid gel matrix, or drying of the obtained gel. Drying at elevated temperature and under vacuum or atmospheric pressure evaporates liquid from the gel porous structure and produces xerogels. Partially emptied pores and existence of the pore size distribution in gels lead to cracking of the pore walls due to capillary pressure (Equation (24.15)) and result in uneven stresses. The most probable model to explain structural collapse of the gel during drying is [139]: severe local stresses generate point defects at the irregular drying front, which are then propagated by the macroscopic stresses resulting in cracks. As the result, obtained xerogels are materials with structure different than the original gel. They are typically less porous, with low pore volumes and surface areas.

$$\Delta P = 2\gamma(\cos\theta)/d \qquad (24.15)$$

Since capillary pressure is the main source of shrinkage and cracking, Kistler [140] reasoned that those problems could be avoided by removing the liquid from the pores above the critical point of the liquid. In this way, the existence of two fluid phases in the gel pores and liquid–vapor interface could be avoided, and hence the capillary pressure. Typically in the process of supercritical drying, the sol or wet gel is placed in an autoclave and heated until pressure and temperature are above solvent critical pressure or temperature. The pressure and temperature are increased in such a way that the interphase boundary is not crossed. This can be accomplished by two different procedures [141,142] (Figure 24.15). Line 1 represents the case in which an additional solvent volume is added to the extractor. Additional solvent evaporation during heating increases pressure inside the extractor above the critical point. Instead of solvent, another procedure (line 2) uses an inert gas which is applied from the beginning of heating. Once the critical point is passed, the solvent is vented out at a temperature higher than critical solvent temperature, thereby eliminating condensation and occurrence of two phases. The resulting gel, called an aerogel, has a volume similar to that of the original sol. This process makes it possible to produce large monolithic gels [143]. Monolithicity of the aerogels is influenced however by a number of sol–gel parameters: molar ratio of the reactants, type and concentration of the catalyst, gel ageing, and type of solvent [144,145].

Other sources of stress can generate during supercritical drying. During heating in the extractor, expansion of the pore liquid causes dilatation of the gel network, which compresses the pore liquid. Liquid flux towards the surface of the gel which offsets the increasing pressure is limited by the low gel permeability. At higher heating rates, the network stretching can produce significant stresses which can cause cracking. The most severe stresses occur when the gel is in contact with the extractor wall, so the radial expansion and radial flow are prevented [146].

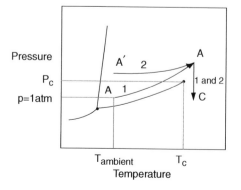

FIGURE 24.15 Different types of processing during supercritical drying.

Kistler was first to recognize that silica gel would dissolve if subjected to supercritical conditions when the gel pores contained water, so he exchanged the liquid by washing the gel with ethanol [147]. It is customary to exchange solvents, but even evacuation of organic solvents (alcohols, ethers, etc.) would sometimes result in gel dissolution. The main reason for this lies in the fact that typical sol–gel solvents have high values of critical parameters (P_c, T_c), which increases their solvent power. Critical parameters of certain typical sol–gel solvents (water, methanol, ethanol, and carbon dioxide) are shown in Table 24.1. A way to bypass this problem is to exchange sol–gel solvent with liquid carbon dioxide, which can be transformed into supercritical fluid at relatively moderate conditions [141,147].

Another possibility is to use supercritical carbon dioxide as an extracting fluid to remove solvent (or solvent mixture with sol–gel reaction byproducts) from the gel network. The procedure is conducted either by washing the gel with carbon dioxide following subsequent transformation of carbon dioxide to supercritical fluid and its removal from the extractor, or by reaching the desired supercritical condition in the extractor followed by continuous flow of supercritical carbon dioxide through the extractor [148–151]. In both cases, drying is principally extraction of the solvent (or mixture) with supercritical carbon dioxide. Since there is an extraction process taking place in a highly porous gel structure during drying process, the main resistance of the process is diffusion of the solvent through the highly developed gel porous network. In case of microporous gels even Knudsen diffusion can play an important role. This fact is the main disadvantage of the process, since drying of larger gel monoliths can require very long drying times [150–154]. In order to obtain large crack-free aerogels using supercritical carbon dioxide drying, it was found necessary to maintain drying pressure and temperature above the binary solvent–carbon dioxide critical curve [149].

Regardless of the supercritical drying procedure, aerogels as it has been shown over the past 15 years are branched polymeric materials having a fractal nature and structure different than corresponding xerogels [155]. They have: lower bulk densities, larger overall porosities, larger surface areas, different pore structures and pore size distributions, and different surface chemistry [156–159]. This could be illustrated by the comparison of mullite aerogel and xerogel obtained by the nonhydrolytic sol–gel method [158] (Table 24.10).

These structural differences are well observed on the micrographs recorded after heating at 1400°C (Figure 24.16). Similarly, investigations of textural properties of titania, zirconia, and niobia xerogels and aerogels by Suh and Park [159], have revealed substantially larger porosities and surface areas of aerogels compared to xerogels. Besides that, pore size distributions were quite different, with aerogels having medium pore sizes in the mesoporous region.

Structural features of the aerogels are influenced by the sol–gel process and physicochemical transformations occurring during supercritical drying. Pore size distributions of titania, zirconia, and niobia aerogels are affected by the acid-catalyst concentration in the sol–gel phase [159].

TABLE 24.10
Crystallite Size (in nm) of Mullite Xerogel and Aerogel Heated at Different Temperatures as Indicated by XRD

Gel Type	900°C	1000°C	1400°C
Xerogel	32.0	38.0	101.0
Aerogel	12.0	14.6	20.8

Reprinted from Janackovic et al., *Nanostructured Materials*, Pages 147–150, 1999. With permission from Elsevier.

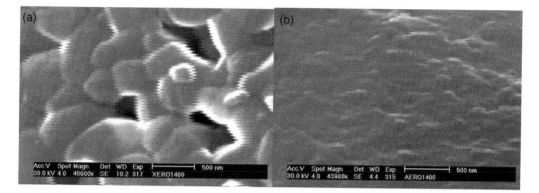

FIGURE 24.16 SEM micrographs of mullite (a) xerogel and (b) aerogel heated at 1400°C.

Results of Woignier et al. [160] indicate that physicochemical transformations during supercritical drying of silica aerogel can be associated with an acceleration of the ageing effect, resulting in the condensation of the dangling bonds of the network leading to the formation of new links and increasing the connectivity of the network. SAXS and TEM investigations of Himmel et al. [161] of silica, alumina, and mixed alumina or silica aerogels obtained by supercritical carbon dioxide drying, indicate that aerogels are characterized by a distribution of interconnected spherical particles whose size increases with alumina content (10–60 nm in diameter), which form branched clusters. The alumina aerogels were found to be composed of flat disc-like regions of approximately 20×25 nm lateral extension having an average thickness of 2 nm. These platelets were found to form particles larger than 150 nm. Similar investigation of alumina aerogel dried with supercritical carbon dioxide was performed by Keysar et al. [162]. The results obtained using SAXS and TEM indicate smallest aggregate in the aerogel with the size of about 2 nm, and the secondary cyclic structure of six aggregates surrounding the seventh one with diameter of about 8–10 nm.

Aging time of the gel can have a significant impact on the final aerogel structure. Increasing value of viscosity of the aging alumina gel, with asymptotic value reached after 1–2 weeks, results in the increased surface area of the aerogel [163]. As mentioned earlier, supercritical drying media influences the structure of obtained aerogel. Drying with supercritical alcohol media results in hydrophobic aerogels since the surface of the material is covered by corresponding alkyl groups [164]. Aerogels obtained by supercritical carbon dioxide drying are on the other hand dominantly hydrophilic. Size of the alcohol alkyl groups affects aerogel properties such as pore sizes, while alcohol supercritical drying can result in aerogels with excellent transparency. However, use of carbon dioxide as drying media is advantageous in case of organically modified silica aerogels, since mild conditions during supercritical drying allow incorporation of sensitive functional organic groups [165]. In case of Pd dispersed on alumina aerogel, it was found that Pd is in the form of oxide after supercritical drying with alcohol media, while drying with supercritical carbon dioxide yields Pd in a reduced form [166]. Supercritical drying with carbon dioxide resulted in the evolution of pentahedral-coordinated aluminum structure in alumina aerogel, while aerogel dried at ethanol supercritical conditions contained no such structure [167]. Additional important supercritical drying parameters are the temperature and pressure, and they can influence significantly the obtained aerogel structure. Investigations of Brodsky and Ko [168] aimed at determination of the influence of supercritical carbon dioxide drying temperature on the structural properties of zirconia, niobia, and titania–silica aerogels, revealed increasing pore volume with increasing drying temperature in case of all three oxides but with different impact on the pore size distribution. There was an increase in average pore diameter for zirconia and titania–silica but not in case of niobia (after calcination at 773 K). Further, a higher drying temperature facilitated crystallization of tetragonal zirconia and anatase titania, but a lower drying temperature facilitated the anatase to rutile transformation of titania, and crystallization of a low temperature modification (TT phase) for niobia. The structural evolution of niobia and titania–silica altered their acidic properties. Influence of carbon dioxide supercritical drying temperature and pressure on the aerogel structure is demonstrated further in the case of alumina or silica aerogel with zinc chloride [169]. It was established that high density carbon dioxide (high pressure and low temperature conditions) removed zinc chloride from the aerogel surface by extraction. Combination of carbon dioxide high pressure and temperature (moderate density) on the other hand, caused separation of the alumina or silica network into Al- and Si-rich phases and the decrease of catalytically favorable mixed Al—O—Si bonding.

B. Application of Aerogels

Due to their unique physical and chemical properties, aerogels represent a new class of solids with significant application potential [121,170–172]. Aerogels are among the best insulating materials known. Extremely fine porous structure of aerogels with porosities exceeding 90% and highly compartmentalized gaseous phase inside the pores, make possible a combination of excellent thermal insulating properties and optical transparency [173–176].

It was recognized relatively early by the Teichner group [177] that low refractive index silica aerogel constitutes a convenient medium for the Cherenkov counter. These devices are used in the identification of electrically charged particles in high-energy physics experiments [121,172,178–180]. Photoluminescent aerogels can be produced by encapsulation of various photoluminescent dopants in silica during the gelation process. Even simple cations introduced by a salt during sol–gel synthesis, as in the case of Al^{3+} in Si—O—Al structure, can provide increase of the luminescent intensity (in this case a green one) [121].

Essential to application of aerogels as electrode materials is to be able to reversibly intercalate ions in the aerogel network. Since aerogels are highly porous, a large amount of small ions can indeed be intercalated in the network. Moreover, since they are exceptional thermal insulators, aerogels can be applied in high-temperature batteries. The most widely studied aerogels for electrode applications are: vanadium oxide [181], manganese oxide [182], and molybdenum oxide [183] in which lithium ions are intercalated. A material like vanadium oxide aerogel can serve as high capacity (500–600 mAh/g), high-energy positive electrode in lithium batteries [184]. Besides Li ions, the polyvalent cations such as Mg^{2+}, Zn^{2+}, and Al^{3+} can also be reversibly intercalated [185]. However the kinetics and cycle performance of these positive electrodes need further improvement.

Application of aerogels as capacitor or supercapacitor electrodes, requires a reversible immobilization of a large quantity of electrical charge carriers in the aerogel structure. Moreover, the aerogel itself must conduct electricity, which explains the fact that carbon aerogels were mostly studied for this application [186,187]. Due to their high surface area and porosity, carbon aerogels can store more electrical energy than conventional capacitors. Specific capacitance of up to 77 F/cm^3 can be achieved with mixed CmRF-based carbon aerogel [186], while carbon aerogel button cell supercapacitor shows no significant degradation after 80,000 charging and discharging cycles [187]. The energy stored in aerogels can be released rapidly to provide an instant power of up to 7.5 kW/kg [121].

As a highly branched and porous structure, aerogels can serve as an encapsulation structure for various chemical entities. In pharmacy or agriculture where many applications require a progressive release of an active substance, aerogels could serve as a nanovessels to carry active drugs, fungicides, herbicides, or pesticides. Biomaterials can be encapsulated in aerogels as well. It was found that *Pseudomonas cepacia* lipase was the highly active biocatalyst when immobilized on silica, or alumina or silica aerogel [188]. Hydrophobic aerogels containing the $CF_3(CH_2)$— group (CF_3 aerogels) were found to absorb oil successfully [189]. The results have indicated that CF_3 aerogels have absorbed as much as 237 times their weight, regardless of $CF_3(CH_2)$— group concentration. High adsorption capacity of aerogels could be successfully used in numerous applications like: storage, filtration, and removal of different chemical compounds. Removal of dibenzothiophene from gas oil for fuel cells application is one of the examples [190]. Similarly, hydrophobic silica aerogel was investigated as modifier of granulated activated carbon for the removal of uranium from aqueous solutions [191] and as long life nuclear waste storage material [192]. Unique structure of aerogels can also facilitate their application as functional chemicals. Nanostructured aerogels of ferric oxide [193], titanium borate [194], copper borate [195], lanthanum borate [196], and magnesium borate [197] were investigated as oil lubricating additives.

A possibility to produce highly porous material like aerogel using process which provides opportunity to ideally mix starting precursors (sol–gel process), presents an advantage in catalyst design. This is equally important for the synthesis of catalytic mixed oxides and supported metal catalysts [198,199]. A good example is zirconia (zirconium oxide, ZrO_2) based catalysts which

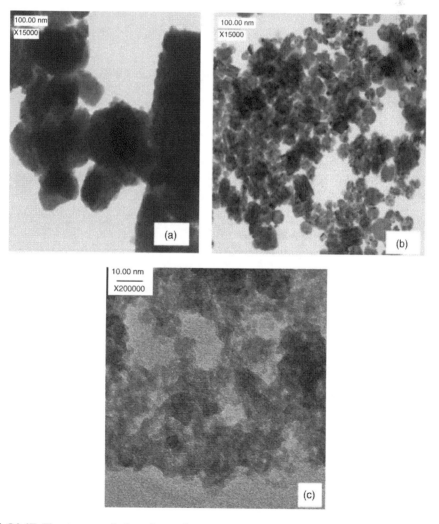

FIGURE 24.17 The texture of chromia catalysts: (a) thermal decomposition of the precursor (macro crystalline chromia), (b) evaporative drying of the gel (microcrystalline chromia), and (c) high temperature supercritical drying of the gel (nanocrystalline chromia). (Reprinted from Rotter, H., Landau, M.V., Carrera, M., Goldfarb, D., and Herskowitz, M., *Appl. Catal. B*, 47, 111–126, 2004. With permission from Elsevier.)

have attracted considerable research attention. Application of additives can modify the surface structure of zirconia and substitution of zirconium with dopant cations can result in a rise in anion vacancy concentrations and conductivity, which constitutes the basis of its potential catalytic application. With the application of sol–gel method and supercritical drying, the high surface area aerogels with high activity could be obtained. A particular application of zirconia which was studied extensively in recent years, is its application as a solid acid catalyst [200–203]. Another catalytic oxide which attracted considerable academic and commercial interest in the last ten years is titania [204–206]. Alumina, silica, and alumina or silica are also well-known and widely used catalysts or catalyst supports, while probably the best known example is the amorphous alumina or silica catalytic cracking catalyst [207]. These oxides are also widely used as catalyst supports for a range of catalytically active phases as in the case of hydrogenation catalysts [208,209]. The advantages of alumina or silica aerogel as catalytic material were demonstrated in the case of solid Friedel-Crafts alkylation catalysts [169,210]. The textural differences of conventional, xerogel and aerogel oxide catalyst are well demonstrated in the case of chromia catalyst (Figure 24.17) [211].

REFERENCES

1. Brenneck, J.F. and Eckert, C.A., Phase equilibria for supercritical fluid process design, *AIChe J.*, 35, 1409–1427, 1989.
2. McHugh, M.A. and Krukonis, V.J., *Supercritical Fluid Extraction*, Butterworth Publishers, Stoneham, 1986.
3. McHugh, M.A. and Krukonis, V.J., Eds., *Supercritical Fluid Extraction*, Butterworth Publishers, Boston, 1994.
4. Clifford, T., Eds., *Fundamentals of Supercritical*, Oxford Science Publication, 1999 (Chapter 2).
5. Page, S.H., Sumpter, S.R., and Lee, M.L., Fluid phase equilibria in supercritical fluid chromatography with CO_2 based mixed mobile phase — a review, *J. Microcolumn Separ.*, 4 (2), 91–122, 1992.
6. Snyder, J.M., Friedrich, J.P., and Christianson, D.D., *JAOCS*, 61, 1851–1856, 1994.
7. http://www.kobelco.co.jp/p108/p14/sfe03.htm.
8. Jung, J., Perrut, M., *J. Supercrit. Fluids*, 20, 179–219, 2001.
9. Matson, D.W. and Smith, R.D., *J. Am. Ceram. Soc.*, 72, 871–881, 1989.
10. Petersen, R.C., Matson, D.W., and Smith, R.D., *J. Am. Chem. Soc.*, 108, 2100–2102, 1986.
11. Helfgen, B., Türk, M., and Schaber, K., *J. Supercrit. Fluids*, 26, 225–242, 2003.
12. Mohamed, R.S., Debenedetti, P.G., and Prud'homme, R.K., *AIChE J.* 1989, 35, 325–328.
13. Kayrak, D., Akman, U., and Hortacsu, Ö., *J. Supercrit. Fluids*, 26, 17–31, 2003.
14. Cihlar, S., Türk, M., and Schaber, K., *J. Aerosol Sci.*, 30, S355–S366, 1999.
15. Matsuyama, K., Mishima, K., Hayashi, K., and Matsuyama, H., *J. Nanopart. Res.*, 5, 87–95, 2003.
16. Türk, M., Hils, P., Helfgen, B., Schaber, K., Martin, H.-J., and Wahl, M.A., *J. Supercrit. Fluids*, 22, 75–84, 2002.
17. Reverchon, E., *J. Supercrit. Fluids*, 15, 1–21, 1999.
18. Reverchon, E., Della Porta, G., Di Trolio, A., and Pace, S., *Ind. Eng. Chem. Res.*, 37, 952–958, 1998.
19. Randolph, T.W., Randolph, A.D., Mebes, M., and Yeung, S., *Biotechnol. Prog.*, 9, 429–435, 1993.
20. Dixon, D.J., Johnston, K.P., and Bodmeier, R.A., *AIChE J.*, 39, 127–139, 1993.
21. Rantakylä, M., Jäntii, M., Aaltonen, O., and Hurme, M. *J. Supercrit. Fluids*, 24, 251–263, 2002.
22. Shekunov, B.Yu., Hanna, M., and York, P., *J. Cryst. Growth*, 198/199, 1345–1351, 1999.
23. Bristow, S., Shekunov, T., Shekunov, B.Yu., and York, P. *J. Supercrit. Fluids*, 21, 257–271, 2001.
24. Kordikowski, A., Schenk, A.P., Van Nielen, R.M., and Peters, C.J., *J. Supercrit. Fluids*, 8, 205–216, 1995.
25. Dixon, D.J., Luna-Barcenas, G., and Johnston, K.P., *Polymer*, 35, 3998–4005, 1994.
26. Bleich, J., Muller, B.W., and Wassmus, W., *Int. J. Pharm.*, 97, 111–117, 1993.
27. Yeo, S.-D., Kim, M.-S., and Lee, J.-C., *J. Supercrit. Fluids*, 25, 143–154, 2003.
28. Gao, Y., Mulenda, T.K., Shi, Y.-F., and Yuan, W.-K. *J. Supercrit. Fluids*, 13, 369–374, 1998.
29. Reverchon, E., Della Porta, G., and Pallado, P., *Powder Technol.*, 114, 17–22, 2001.

30. Hong, L., Guo, J.Z., Gao, Y., and Yuan, W.-K., *Ind. Eng. Chem. Res.*, 39, 4882–4887, 2000.
31. Warwick, B., Dehghani, F., Foster, N.R., Biffin, J.R., and Regtop, H.L., *Ind. Eng. Chem. Res.*, 39, 4571–4579, 2000.
32. Ventosa, N., Sala, S., and Veciana, J., *J. Supercrit. Fluids*, 26, 33–45, 2003.
33. Prausnitz, J.M., Lichtenhaler, R.N., and de Azevedo, E.G., Molecular Thermodynamics of Fluid Phase Equilibria, Prentice Hall Inc.,, pp.371, 1986.
34. Tuminello, W.H., Dee, G.T., and McHugh, M.A., *Macromolecules*, 28, 1506–1510, 1995.
35. Weidner, E., Wiesmet, V., Knez, Ž., and Škerget, M., *J. Supercrit. Fluids*, 10, 139–147, 1997.
36. Kikic, I., Lora, M., and Bertucco, A., *Ind. Eng. Chem. Res.*, 36, 5507–5515, 1997.
37. Senčar-Božič, P., Srčič, S., Knez, Ž., and Kerč, J., *Int. J. Pharm.*, 148, 123–130, 1997.
38. Kerč, J., Srčič, S., Knez, Ž., and Senčar-Božič, P., *Int. J. Pharm.*, 182, 33–39, 1999.
39. Cason, J.P., Khambaswadkar, K., and Roberts, C.B., *Ind. Eng. Chem. Res.*, 39, 4749–4755, 2000.
40. Salaniwal, S., Cui, S., Cochran, H.D., and Cummings, P.T., *Ind. Eng. Chem. Res.*, 39, 4543–4554, 2000.
41. Chhor, K., Bocquet, J.F., and Pommier, C., *Mater. Chem. Phys.*, 32, 249–254, 1992.
42. Adschiri, T., Hakuta, Y., and Arai, K., *Ind. Eng. Chem. Res.*, 39, 4901–4907, 2000.
43. Watkins, J.J., Blackburn, J.M., and McCarthy, T.J., *Chem. Mater.*, 11, 213–215, 1999.
44. Matson, D.W., Fulton, J.L., Petersen, R.C., and Smith, R.D., *Ind. Eng. Chem. Res.*, 26, 2298–2306, 1987.
45. Hansen, B.N., Hybertson, B.M., Barkley, R.M., and Sievers, R.E., *Chem. Mater.*, 4, 749–752, 1992.
46. Hybertson, B.M., Hansen, B.N., Barkley, R.M., and Sievers, R.E., *Mater. Res. Bull.*, 26, 1127–1133, 1991.
47. Sun, Y.-P., Guduru, R., Lin, F., and Whiteside, T., *Ind. Eng. Chem. Res.*, 39, 4663–4669, 2000.
48. Yeo, S.-D., Choi, J.-H., and Lee, T.-J., *J. Supercrit. Fluids*, 16, 235–246, 2000.
49. Reverchon, E., De Marco, I., and Della Porta, G., *J. Supercrit. Fluids*, 23, 81–87, 2002.
50. Reverchon, E., Della Porta, G., Sannino, D., and Ciambelli, P., *Powder Technol.*, 102, 127–134, 1999.
51. Hutchings, G.J., Lopez-Sanchez, J.A., Bartley, J.K., Webster, J.M., Burrows, A., Kiely, J.C., Carley, A.F., Rhodes, C., Hävecker, M., Knop-Gericke, A., Mayer, R.W., Schlögl, R., Volta, J.C., and Poliakoff, M., *J. Catal.*, 208, 197–210, 2002.
52. Moner-Girona, M., Roig, A., Molins, E., and Llibre, L., *J. Sol–Gel Sci. Technol.*, 26, 645–649, 2003.
53. Chhor, K., Bocquet, J.F., and Pommier, C., *Mater. Chem. Phys.*, 40, 63–68, 1995.
54. Gourinchas-Courtecuisse, V., Bocquet, J.F., Chhor, K., and Pommier, C., *J. Supercrit. Fluids*, 9, 222–226, 1996.
55. Tadros, M.E., Adkins, C.L.J., Russick, E.M., and Youngman, M.P., *J. Supercrit. Fluids*, 9, 172–176, 1996.
56. Kim, T.-H., Lim, D.-Y., Yu, B.-S., Lee, J.-H., and Goto, M., *Ind. Eng. Chem. Res.*, 39, 4702–4706, 2000.
57. Adshiri, T., Kanazawa, K., and Arai, K., *J. Am. Ceram. Soc.*, 75, 1019–1022, 1992.
58. Adshiri, T., Kanazawa, K., and Arai, K., *J. Am. Ceram. Soc.*, 75, 2615–2618, 1992.
59. Reverchon, E., Caputo, G., Correra, S., and Cesti, P., *J. Supercrit. Fluids*, 26, 253–261, 2003.
60. Matson, D.W., Petersen, R.C., and Smith, R.D., *J. Mater. Sci.*, 22, 1919–1928, 1987.
61. Lele, A.K. and Shine, A.D., *AIChE J.*, 38, 742–752, 1992.
62. Kim, J.-H., Paxton, T.E., and Tomasko, D.L., *Biotechnol. Prog.*, 12, 650–661, 1996.
63. Tom, J.W. and Debenedetti, P.G., *Biotechnol. Prog.*, 7, 403–411, 1991.
64. Lele, A.K. and Shine, A.D., *Ind. Eng. Chem. Res.*, 33, 1476–1485, 1994.
65. Tepper, G. and Levit, N., *Ind. Eng. Chem. Res.*, 39, 4445–4449, 2000.
66. Gallyamov, M.O., Vinokur, R.A., Nikitin, L.N., Said-Galiyev, E.E., Khokhlov, A.R., Yaminsky, I.G., and Schaumburg, K., *Langmuir*, 18, 6928–6934, 2002.
67. Shim, J.-J., Yates, M.Z., and Johnston, K.P., *Ind. Eng. Chem. Res.*, 40, 536–543, 2001.
68. Subramaniam, B., Rajewski, R.A., and Snavely, K., *J. Pharm. Sci.*, 86, 885–890, 1996.
69. Yeo, S.-D., Debenedetti, P.G., Radosz, M., and Schmidt, H.-W., *Macromolecules*, 26, 6207–6210, 1993.
70. Reverchon, E., Della Porta, G., De Rosa, I., Subra, P., and Letourneur, D., *J. Supercrit. Fluids*, 18, 239–245, 2000.
71. Bothun, G.D., White, K.L., and Knutson, B.L., *Polymer*, 43, 4445–4452, 2002.

72. Mawson, S., Kanakia, S., and Johnston, K.P., *Polymer*, 38, 2957–2967, 1997.

73. Yeo, S.-D., Debenedetti, P.G., Radosz, M., Giesa, R., and Schmidt, H.W., *Macromolecules*, 28, 1316–1317, 1995.

74. Engwicht, A., Girreser, U., and Müller, B.W., *Biomaterials*, 21, 1587–1593, 2000.

75. Weidner, E., Knez, Ž., Steiner, R., third in Proceedings of the Int. Symp. on High Pressure Chem. Eng., Zürich, pp. 223–228, 1996.

76. Canelas, D.A. and DeSimone, J.M., *Adv. Polym. Sci.*, 133, 103–140, 1997.

77. Kendall, J.L., Canelas, D.A., Young, J.L., and DeSimone, J.M., *Chem. Rev.*, 543, 3–563, 1999.

78. Hagiwara, M., Mitsui, H., Machi, S., and Kagiya, T., *J. Polym. Sci. Part A-1*, 6, 603–608, 1968.

79. Charpentier, P.A., Kennedy, K.A., DeSimone, J.M., and Roberts, G.W., *Macromolecules*, 32, 5972–5975, 1999.

80. Charpentier, P.A., DeSimone, J.M., and Roberts, G.W., *Ind. Eng. Chem. Res.*, 39, 4588–4596, 2000.

81. Saraf, M.K., Gerard, S., Wojcinski II, L.M., Charpentier, P.A., DeSimone, J.M., and Roberts, G.W., *Macromolecules*, 35, 7976–7985, 2002.

82. Müller R.H., Jacobs C., and Kayser O., Nanosuspensions as particulate drug formulations in therapy rationale for development and what we can expect for the future, *Adv. Drug Delivery Rev.*, 47, 319, 2001.

83. Cathpole, O.J., Hockman, S., and Anderson, S.R., *J. High Pres. Chem. Eng.* 1, 309, 1996.

84. Subra, P. and Debenedetti, P.G. High Pressure Chemical Engineering, Rudolf von Rohr, Ph. and Trepp, Ch. Eds., Elsevier, Amsterdam, 49–54, 1996.

85. Mishima, K., Matsuyama, K., Uchiyama, H., Ide, M., Shim, J.-J., and Bae, H.-K., Fourth in *Proceedings of the International Symposium on Supercritical Fluids*, Sendai, Japan, pp. 267–270, 1997.

86. Kwauk, X. and Debenedetti, P.G., *J. Aerosol. Sci.*, 34, 445–469, 1993.

87. Krukonis, V., A.I.Ch.E. Meeting, San Francisco, November, pp.140f, 1984.

88. Chang, C.J. and Randolph, A.D., *AIChE J.*, 35, 1876–1882, 1989.

89. Ohgaki, K., Kobayashi, H., Katayama, T., and Hirokawa, N., *J. Supercrit. Fluids*, 3, 103–107, 1990.

90. Charoenchaitrakool, M., Dehghani, F., and Foster, N.R., in *Proceedings of the Fifth Conference on Supercritical Fluids and their Applications*, Garda (Verona), pp.485–492, 1999.

91. Wubbolts, F.E., Kersch, C., and van Rosmalen, G.M., in *Proceedings of the fifth Meeting on Supercritical Fluids*, Nice (France), pp. 249–256, 1998.

92. Bleich and Muller, B.W., *Microencapsulation*, 13, 131–139, 1996.

93. Schmitt, W.J., Salada, M.C., Shook, G.G., and Speaker III, S.M., *AIChE J.*, 41, 2476–2486, 1995.

94. Tavana, A. and Randolph, A.D., *AIChE J.*, 35, 1625, 1989.

95. Berends, E.M., Bruinsma, O.S.L., and van Rosmalen, G.M., *J. Cryst. Growth*, 128, 50, 1993.

96. Furuta, S., Rousseau, R.W., and Teja, A.S., in *Proceedings of the AIChE Annual Meeting*, USA, Miami, 1992.

97. Reverchon, E., Donsì, G., and Gorgoglione, D., *J. Supercrit. Fluids*, 6, 241, 1993.

98. Reverchon, E., Della Porta, G., Taddeo, R., Pallado, P., and Stassi, A., *Ind. Eng. Chem. Res.* 34, 4087, 1995.

99. Stahl, E., Quirin, K.W., and Gerard, D., Eds., *Dense Gases for Extraction and Refining*, Springer-Verlag, Berlin, 1987.

100. Benita, S., *Microencapsulation: Method and Industrial Applications*, Marcel Dekker Inc., New York, 1996.

101. Tom, J.W., Debenedetti, P.G., and Jerome, R., *J. Supercrit. Fluids*, 7, 9–29, 1994.

102. Hirai, T., Saito, T., and Komasawa, I., *J. Phys. Chem. B*, 105, 9711–9714, 2001.

103. Mishima, K., Matsuyama, K., Tanabe, D., Yamauchi, S., Young, T.J., and Johnston, K.P., *AIChE J.*, 46, 857–865, 2000

104. Matsuyama, K., Mishima, K., Umemoto, H., and Yamaguchi, S., *Environ. Sci. Technol.*, 35, 4149–4155, 2001.

105. Debenedetti, P., Tom, J.W., Yeo, S.D., and Lim, G.B., *J. Controlled Release*, 24, 27–44, 1993.

106. Said-Galiev, E., Nikitin, L., Vinokur, R., Gallyamov, M., Kurykin, M., Petrova, O., Lokshin, B., Volkov, I., Khokhlov, A., and Schaumburg, K., *Ind. Eng. Chem. Res.*, 39, 4891–4896, 2000.

107. Shelukar, S., Ho, J., Zega, J., Roland, E., Yeh, N., Quiram, D., Nole, A., Katdare, A., and Reynolds, S., *Powder Technol.*, 110, 29–36, 2000.

108. Sudsakorn, K. and Turton, R., *Powder Technol.*, 110, 37–43, 2000.

109. Schreiber, R., Vogt, C., Werther, J., and Brunner, G., *J. Supercrit. Fluids*, 24, 137–151, 2002.
110. Glebov, E.M., Yuan, L., Krishtopa, L.G., Usov, O.M., and Krasnoperov, L.N., *Ind. Eng. Chem. Res.*, 40, 4058–4068, 2001.
111. Livage, J., Henry, M., and Sanchez, C., *Prog. Solid State Chem.*, 18, 259–342, 1988.
112. Livage, J. and Henry, M., in *Ultrastructure Processing of Advanced Ceramics*, Mackenzie, J.D., Ulrich, D.R., Eds., Wiley, New York,, pp. 183, 1988.
113. Rollinson, C.L., in *The Chemistry of Coordination Compounds*, Bailar, J.C., Ed., Reinhold, New York,, pp. 448, 1956.
114. Freedman, M.L., *J. Am. Chem. Soc.*, 80, 2072–2077, 1958.
115. Schwarzenback, G. and Meier, J., *J. Inorg. Nucl. Chem.*, 8, 302–312, 1958.
116. Petrovic, R., Milonjic, S., Jokanovic, V., Kostic-Gvozdenovic, L.j., Petrovic-Prelevic, I., and Janaković, Dj., *Powder Techn.*, 133, 185–189, 2003.
117. Petrović, R., Janaćković, Dj., Božović, B., Zec, S., and Kostić-Gvozdenović, Lj., *J. Serb. Chem. Soc.*, 66, 335–343, 2001.
118. Petrović, R., Janaćković, Dj., Zec, S., Drmanić, S., and Kostić-Gvozdenović, Lj., *J. Mater. Res.*, 16, 451–458, 2001.
119. Matijević, E., Sapieszko, R.S., and Melville, J.B., *J. Coll. Interf. Sci.*, 50, 567–581, 1975.
120. Hamada, S. and Matijević, E., *J. Chem. Soc. Faraday Trans. I*, 78, 2147, 1982.
121. Pierre, A.C. and Pajonk, G.M., *Chem. Rev.*, 102, 4243–4265, 2002.
122. Gash, A.E., Tillotson, T.M., Satcher Jr., J.H., Hrubesh, L.W., and Simpson, R.L., *J. Non-cryst. Solids*, 285, 22–28, 2001.
123. Uhlmann, D.R., Zelinski, B.J., and Wnek, G.E., in *Better Ceramics Through Chemistry*, Brinker, C.J., Clark, and D.E., Ulrich, D.R., Eds., North-Holland, New York, pp. 59–70, 1984.
124. Iler, R.K., in *The Chemistry of Silica*, Wiley, New York, 1979.
125. Keefer, K.D., in *Better Ceramics through Chemistry*, Brinker, C.J., Clark, D.E., D.R. Ulrich, Eds., North-Holland, New York, pp. 15–24, 1984.
126. Pohl, E.R. and Osterholz, F.D., in *Molecular Characterization of Composite Interfaces*, Ishida, H., Kumar, G., Eds., Plenum, New York, pp. 157, 1985.
127. Engelhardt, V.G., Altenburg, W., Hoebbel, D., and Wieker, W.Z., *Anorg. Allg. Chem.*, 418, 43, 1977.
128. Brinker, C.J. and Sherer, G.W., in *Sol–Gel Science;* Academic Press Inc., San Diego, CA, pp. 212–215, 1990.
129. Petrović, R., Janaćković, Dj., Zec, S., Drmanić, S., and Kostić-Gvozdenović, Lj., *J. Sol–Gel Sci. Technol.*, 28, 111–118, 2003.
130. Dislich, H., *Angew. Chem. Int. Ed. Engl.*, 10, 363, 1971.
131. Yoldas, B.E., *J. Mater. Sci.*, 12, 1203–1208, 1977.
132. Yoldas, B.E., *J. Mater. Sci.*, 14, 1843–1849, 1979.
133. Acosta, S., Corriu, R.J.P., Leclercq, D., Lefevre, L., Mutin, P.H., and Vioux, A., *J. Non-cryst. Solids*, 170, 234–242, 1994.
134. Adrianainarivelo, M., Corriu, R., Leclercq, D., Mutin, P.H., and Vioux, A., *J. Mater. Chem.*, 6, 1665–1672, 1996.
135. Arnal, P., Corriu, R., Leclercq, D., Mutin, P.H., and Vioux, A., *J. Mater. Chem.*, 6, 1925–1932, 1996.
136. Corriu, R.J.P., Leclercq, D., Lefevre, L., Mutin, P.H., and Vioux, A., *J. Non-cryst. Solids*, 146, 301–303, 1992.
137. Acosta, S., Corriu, R., Leclercq, D., Mutin, P.H., and Vioux, A., *J. Sol–Gel Sci. Technol.*, 2, 25–28, 1994.
138. Acosta, S., Corriu, R., Leclercq, D., Mutin, P.H., and Vioux, A., *Mater. Res. Soc. Symp. Proc.*, 346, 345–350, 1994.
139. Brinker, C.J. and Sherer, G.W., in *Sol–Gel Science*, Academic Press Inc., San Diego, CA, pp. 493–498, 1990.
140. Kistler, S.S., *J. Phys. Chem.*, 36, 52–64, 1932.
141. Phalippou, J., Woignier, T., and Prassas, M., *J. Mater. Sci.*, 25, 3111–3117, 1990.
142. Rao, A.V., Pajonk, G.M., and Parvathy, N.N., *J. Mater. Sci.*, 29, 1807–1817, 1994.
143. Zarzycki, J., Prassas, M., and Phalippou, J., *J. Mater. Sci.*, 17, 3371–3379, 1982.
144. Rao, A.V. and Parvathy, N.N., *J. Mater. Sci.*, 28, 3021–3026, 1993.
145. Stolarski, M., Walendziewski, J., Steininger, M., and Pniak, B., *Appl. Catal. A: Gen.*, 177, 139–148, 1999.

146. Scherer, G.W., *J. Non-Cryst. Solids*, 145, 33–40, 1992.

147. Tewari, P.H., Hunt, A.J., and Lofftus, K.D., *Mater. Lett.*, 3, 363–367, 1985.

148. Moses, J.M. Willey, R.J., and Rouanet, S., *J. Non-Cryst. Solids*, 145, 41–43, 1992.

149. van Bommel, M.J. and de Haan, A.B., *J. Mater. Sci.*, 29, 943–948, 1994.

150. van Bommel, M.J. and de Haan, A.B., *J. Non-Cryst. Solids*, 186, 78–82, 1995.

151. Orlović, A., Petrović, S., Radivojević, D., and Skala, D., *Chem. Ind.*, 56, 244–248, 2001.

152. Orlović, A., Petrović, S., and Skala, D., *J. Serb. Chem. Soc.*, 70, 125–136, 2005.

153. Wawrzyniak, P., Rogacki, G., Pruba, J., and Bartczak, Z., *J. Non-Cryst. Solids*, 225, 86–90.

154. Wawrzyniak, P., Rogacki, G., Pruba, J., and Bartczak, Z., *J. Non-Cryst. Solids* 2001, 285, 50–56, 1998.

155. Fricke, J. and Emmerling, A., *J. Am. Ceram. Soc.*, 75, 2027–2036, 1992.

156. Sunol, S.G., Keskin, O., Guney, O., and Sunol, A.K., in *Innovations in Supercritical Fluids;* Hutchenson, K.W., Foster, N.R., Eds., ACS, Columbus, OH, pp. 258–268, 1995.

157. Vong, M.S.W., Sermon, P.A., and Sun, Y., *J. Chim. Phys.*, 93, 1016–1033, 1996.

158. Janackovic, Dj., Orlovic, A., Skala, D., Drmanic, S., Kostic-Gvozdenovic, Lj., Jokanovic, V., and Uskokovic, D., *NanoStruct. Mater.*, 12, 147–150, 1999.

159. Suh, D.J. and Park, T.-J., *Chem. Mater.*, 8, 509–513, 1996.

160. Woignier, T., Phalippou, J., Quinson, J.F., Pauthe, M., and Laveissiere, F., *J. Non-Cryst. Solids*, 145, 25–32, 1992.

161. Himmel, B., Gerber, Th., Bürger, H., Holzhüter, G., and Olbertz, A., *J. Non-Cryst. Solids*, 186, 149–158, 1995.

162. Keysar, S., De Hazan, Y., Cohen, Y., Aboud, T., and Grader, G.S., *J. Mater. Res.*, 12, 430–433, 1997.

163. Keysar, S., Cohen, Y., Shagal, S., Slobodiansky, S., and Grader, G.S., *J. Sol–Gel Sci. Technol.*, 14, 131–136, 1999.

164. Tajiri, K., Igarashi, K., and Nishio, T., *J. Non-Cryst. Solids*, 186, 83–87, 1995.

165. Huesing, N., Schwertfeger, F., Tappert, W., and Schubert, U., *J. Non-Cryst. Solids*, 186, 37–43, 1995.

166. Mizushima, Y. and Hori, M., *J. Mater. Res.*, 10, 1424–1428, 1995.

167. Mizushima, Y., Hori, M., and Sasaki, M., *J. Mater. Res.*, 8, 2109–2111, 1993.

168. Brodsky, C.J. and Ko, E.I., *J. Non-Cryst. Solids*, 186, 88–95, 1995.

169. Orlović, A., Janaćković, Dj., and Skala, D., *Catal. Comm.*, 3, 119–123, 2002.

170. Hrubesh, L.W., *J. Non-Cryst. Solids*, 225, 335–342, 1998.

171. Fricke, J. and Emmerling, A., *J. Sol–Gel Sci. Technol.*, 13, 299–303, 1998.

172. Akimov, Yu.K., *Instrum. Exp. Tech.*, 46, 287–299, 2003.

173. Herrmann, G., Iden, R., Mielke, M., Teich, F., and Ziegler, B., *J. Non-Cryst. Solids*, 186, 380–387, 1995.

174. Yoldas, B.E., Annen, M.J., and Bostaph, J., *Chem. Mater.*, 12, 2475–2484, 2000.

175. Kwon, Y.-G., Choi, S.-Y., Kang, E.-S., and Baek, S.-S., *J. Mater. Sci.*, 35, 6075–6079, 2000.

176. Forest, L., Gibiat, V., and Woignier, T., *J. Non-Cryst. Solids*, 225, 287–292, 1998.

177. Cantin, M., Casse, M., Koch, L., Jouan, R., Mestran, P., Roussel, D., Bonnin, F., Moutel, J., and Teichner, S.J., *Nucl. Intrum. Meth.*, 118, 177–182, 1974.

178. Akimov, Yu.K., Zrelov, V.P., Puzynin, A.I., Filin, S.V., Filippov, A.I., and Sheinkman, V.A., *Instrum. Exp. Tech.*, 45, 634–639, 2002.

179. Aschenauer, N. et al., *Nucl. Instrum. Meth. Phys. Res. Sect. A*, 440, 338–347, 2000.

180. Sakemi, Y., *Nucl. Instrum. Meth. Phys. Res. Sect. A*, 453, 284–288, 2000.

181. Giorgetti, M., Passerini, S., Smyrl, W.H., and Berrettoni, M., *Chem. Mater.*, 11, 2257–2264, 1999.

182. Passerini, S. Coustier, F. Giorgetti, M., and Smyrl, W.H., *Electrochem. Solid-State Lett.*, 2, 483, 1999.

183. Dong, W. and Dunn, B., *J. Non-Cryst. Solids*, 225, 135–140, 1998.

184. Owens, B.B., Passerini, S., and Smyrl, W.H., *Electrochim. Acta*, 45, 215–224, 1999.

185. Le, D.B., Passerini, S., Coustier, F., Guo, J., Soderstrom, T., Owens, B.B., and Smyrl, W.H., *Chem. Mater.*, 10, 682–684, 1998.

186. Li, W., Probstle, H., and Fricke, J., *J. Non-Cryst. Solids*, 325, 1–5, 2003.

187. Probstle, H., Schmitt, C., and Fricke, J., *J. Power Sources*, 105, 189–194, 2002.

188. Nuisson, P., Hernandez, C., Pierre, M., and Pierre, A.C., *J. Non-Cryst. Solids*, 285, 295–302, 2001.

189. Reynolds, J.G., Coronado, P.R., and Hrubesh, L.W., *J. Non-Cryst. Solids*, 292, 127–137, 2001.

190. Haji, S. and Erkey, C., *Ind. Eng. Chem. Res.*, 42, 6933–6937, 2003.
191. Coleman, S.J., Coronado, P.R., Maxwell, R.S., and Reynolds, J.G., *Environ. Sci. Technol.*, 37, 2286–2290, 2003.
192. Woignier, T., Reynes, J., Phalippou, J., and Dussossoy, J.L., *J. Sol–Gel Sci. Technol.*, 19, 833–837, 2000.
193. Hu, Z.S., Dong, J.X., and Chen, G.X., *Trib. Int.*, 31, 355–360, 1998.
194. Hu, Z.S. and Dong, J.X., *Wear*, 216, 87–91, 1998.
195. Hu, Z.S., Dong, J.X., Chen, G.X., and Lou, F., *Powder Technol.*, 102, 171–176, 1999.
196. Hu, Z.S., Dong, J.X., Chen, G.X., and He, J.Z., *Wear*, 243, 43–47, 2000.
197. Hu, Z.S., Lai, R., Lou, F., Wang, L.G., Chen, Z.L., Chen, G.X., and Dong, J.X., *Wear*, 252, 370–374, 2002.
198. Ward, D.A. and Ko E.I., *Ind. Eng. Chem. Res.*, 34, 421–433, 1995.
199. Kung, H.H. and Ko E.I., *Chem. Eng. J.*, 64, 203–214, 1996.
200. Ward, D.A. and Ko, E.I., *J. Catal.*, 150, 18–33, 1994.
201. Ward, D.A. and Ko, E.I., *J. Catal.*, 157, 321–333, 1995.
202. Boyse, R.A. and Ko, E.I., *Catal. Lett.*, 38, 225–230, 1996.
203. Miller, J.B. and Ko, E.I., *Chem. Eng. J.*, 64, 273–281, 1996.
204. Notari, B., *Catal. Today*, 18, 163–172, 1993.
205. Dutoit, D.C.M., Schneider, M., and Baiker, A., *J. Catal.*, 153, 165–176, 1995.
206. Hutter, R., Mallat, T., and Baiker, A., *J. Catal.*, 153, 177–189, 1995.
207. van Hoof, J.H.C., in *Chemistry and Chemical Engineering of Catalytic Processes;* Prins, R., Schuit, G.C.A., Eds., Sijthoff & Noordhoff; Alphen aan den Rijn, NL, pp. 161–181, 1980.
208. Skala, D.U., Šaban, M.D., Orlović, A.M., Meyn, V.W., Severin, D.K., Rahimian, I.G.-H., and Marjanović, M.V., *Ind. Eng. Chem.Res.*, 30, 2059–2065, 1991.
209. Jovanović, N.N., Stanković, M.V., Marjanović, M.V., and Skala, D.U., *J. Serb. Chem. Soc.*, 54, 145–154, 1989.
210. Orlović, A.M., Janaćković, Dj.T., Drmanić, S., Marinković, Z., and Skala, D.U., *J. Serb. Chem. Soc.*, 66, 685–695, 2001.
211. Rotter, H., Landau, M.V., Carrera, M., Goldfarb, D., and Herskowitz, M., *Appl. Catal. B: Environ.*, 47, 111–126, 2004.

Part V

Hetero–Aggregate Finely Dispersed Systems of Biological Interest

Biocolloids

25 Effects of Electrical Field on the Behavior of Biological Cells

Yung-Chih Kuo
National Chung Cheng University, Chia-Yi, Taiwan, R.O.C.

Jyh-Ping Hsu
National Taiwan University, Taipei, Taiwan, R.O.C.

CONTENTS

I. INTRODUCTION

Over the past century, the interdisciplinary knowledge about colloids and interfacial phenomena has gradually developed into a prosperous field in science and engineering. Both theoretical development and experimental study on macroscopic and microscopic systems have become a great significance [1]. Recently, the issues related to biological cells have attracted much attention for fundamental research and application. For the interaction between biological cells and bio-interfaces or among cells, deposition, adsorption, and flocculation are the most important phenomena. The knowledge about deposition of cells onto various bio-surfaces is crucial to understand numerous physiological disorders commonly encountered in practice, for example, the secondary tumor growth. Cancer cells are transported by blood circulation from the position of primary tumor to the vascular endothelium of other body tissues during the spread of malignant tumor. For the deposition of biological cells, two stages are in series, migration to the vicinity of a solid interface by external field followed by adhesion, are the widely adopted mechanisms [2]. Also, investigations on bio-adsorption behavior and flocculation among biological cells in an electrolyte solution are essential in understanding the biomedical performance of cells. Observations for bio-adsorption phenomena include adherence of platelets on a blood-vessel lumen to generate embolisms [3], adhesion of bacteria on a tooth wall to erode dental enamels [4], and adsorption of various proteins on an air–water interface to form bubbles [5]. In industrial process, immobilization technique of microbes on microcarriers in a bioreactor plays an important role. In addition, typical examples for bio-flocculation are: abnormal aggregation of dysfunctional platelets inside pathological blood-vessel, yielding fatal consequences like thrombogenesis or coronary thrombus formation and bleeding, and collagen self-assembly under physiological condition in connective tissue.

Often, the electrostatic contribution plays an important role in understanding the structure of a cellular suspension and the thermodynamics of ionic system [6–8]. Since the topic related to the so-called diffuse electrical double layer (DEDL), a hypothetical film of liquid solution containing mobile ions accumulated near a charged surface, was recognized as one of the great challenges to the basic understanding of colloidal behavior in dispersions, this subject has attracted a great number of research efforts. In addition, the wide-ranging applications of ionic suspensions in chemical, biochemical, and photoelectrical processes support the attention to study on the relevant DEDL issues. To estimate the electrical contribution in DEDL, the information about the electrostatic potential distribution, the most important property of a charged system, is required. Many of the concepts of DEDL have been improved owing to a large quantity of publications, started from the Göuy–Chapman theory (GCT) [9], where the equilibrium DEDL was adopted and the spatial ionic concentration can be typically estimated by solving the mean-field Poisson–Boltzmann equation (PBE) for a point-charge model (PMC) of electrolyte in continuum-solvent solution. The GCT and lots of the following theoretical achievements are based on using the PBE in a way equivalent to that later invoked by Debye and Hückel [10] in their ionic theory of strong electrolytes. Although, more complicated approaches [11–13], including molecular simulations, Percus–Yevick, hipernetted chain, mean spherical approximation, and optimized cluster expansions, have been proposed, the DEDL is often described by a mean-field level of the relatively simple PBE because it yields sufficiently reliable estimations [14–16]. On the other hand, the solution to the general PBE is, however, nontrivial except for a few limited cases because the natural characteristics of ionogenic species in dielectric medium and the sophisticated conditions on colloidal interfaces leads to nonlinear versions of the PBE.

Since pairwise interactions are dominant in dilute suspension, it becomes crucial to quantify the DEDL interaction between two colloids. In the mid-twentieth century, the Derjaguin–Landau–Verwey–Overbeek (DLVO) theory, a central canon in colloid and interfacial science, was constructed by considering particulate pairwise interactions in electrolyte solution and by incorporating two opposing contributions, the electrostatic repulsion of an overlapping DEDL and the London–van der Waals attraction between two identical lyophobic colloids [9], and considerable qualitative trends concluded from the experimental results have been adequately explained [17,18]. Subsequent advancement based on the original DLVO theory have been achieved through extending mathematical models in one of the three following approaches:

1. Consideration to arbitrary valences of electrolytic species in a dielectric solution [19]
2. Satisfaction of physiochemical descriptions for particulate boundaries [20]
3. Estimation for the influences of curvature effects of boundaries [21].

In liquid state physics of the prototypal GCT and the classic DLVO theory, all correlations for ionic characteristics and their interactions are entirely neglected and the only difference between cations and anions is in the valence of charge they bear. For real electrolyte solutions, there is nearly no reason to expect the vanishing of the finite sizes of ionic species. Historically, Stern first attempted to expand GCT to include the ionic sizes by a steric model of closest approach for ions next to a charged rigid surface. The Stern's model, normally referred to the modified Göuy–Chapman theory (MGCT) in literature, contains a particular repulsion zone in the vicinity of closest approach, possessing lower dielectric constant than the bulk liquid phase. Graham called the location of closest approach from a charged wall as the outer Helmholtz plane. In MGCT, the short-range interaction among electrolyte ions assume insignificant and the rigid cores of all ions are the same. Note that the ideas proposed later about the size effects on DEDL properties usually use a reasonable MGCT in a way of employing an accurate form, which manifests the excluded volumes of ions. Certainly, electrolyte ions are regarded as charged mobile particles in dielectric medium [22,23], and from then on, the influence of ionic rigid core on the electrical properties of DEDL became one of the focal points in theoretical development [24]. By invoking the concept of finite dimensions of ionic species, which may be composed of cations–solvent or anions–solvent structure near the vicinity of a charged surface, the asymmetric differential capacity on low-potential electrodes was demonstrated to be the result of an unequal effective sizes of anions and cations [25]. In a study of the effects of unequal ionic sizes and asymmetric electrolytes with valences 2:1 and 1:2 on the potential difference across a DEDL, incorporation of non-Coulombic effects such as specific adsorption for the interpretation of a nonzero electrostatic potential at a point of zero charge on a surface was concluded to arise naturally due to the effects of unequal sizes and asymmetric valences of the surrounded ions [26]. Moreover, a Monte Carlo simulation was adopted to study finite sizes of charged species on the distribution of ionic concentration in DEDL near an infinite planar surface immersed in a strong electrolyte solution [27]. In charged cavities, effects of ionic sizes have been simulated through grand canonical ensemble Monte Carlo with various ionic dimensions and surface charge densities [28,29].

Although, instances of applying the pristine frameworks of the GCT, MGCT, or DLVO theory are ample in literature, these successful theories, without doubt, show defects because colloidal rigidity is assumed. Biological entities, such as mammalian cells and collagen molecules, belong to one of the most important classes of soft biocolloids. For biological cells, the rigid surface model requires modification by introducing a membrane space on the outer cellular border. In fact, the peripheral zones of biological cells usually stretch out to form an ion-penetrable membrane, where fixed charge arises mainly from absorption of the ionic species in the nearby DEDL or are generated from dissociation of the ionogenic functional groups in the membrane layer. In other words, the fixed charge is spread over a finite volume in a three-dimensional space or the soft polar interface, rather than over a two-dimensional hard wall. This implies that, instead of the concept of surface charge scattered on a solid rigid surface, a membrane charge is fastened in and spread over a three-dimensional compartment. Typical example is an ion-penetrable charged glycoprotein layer, roughly 15 nm thick, stretched over the superficial margin of the external lipid phase on human erythrocyte [30]. Hence, it turns out to be unsatisfactory to the physical insights by regarding the biological membranes as rigid external boundaries, and the rigid surface model or the hard colloid model becomes inappropriate for describing the interaction among biological cells. Two decades ago, a mathematical model considering the permeable characteristics of planar membrane bearing fixed charge uniformly distributed over an exterior membrane layer was successfully developed to reflect more realistic feature and rational aspect for biological cells [31,32]. Furthermore, a theory, which includes regions of different permittivity, but assumes that a three-dimensional layer can be replaced by a monopole and a dipole layer, was proposed for evaluation of ionic transportation across soft polar interfaces [33,34]. Various regions in phospholipid bilayers and their influence on the permeability of species were also discussed [35–37]. Since the charged surface layer of biological cells usually has a capacitance of about 1 $\mu F/cm^2$ and a specific resistance of 10^9–10^{11} Ω cm, the cell membrane can be regarded as a dielectric. In a study of a system, comprising two dielectric membranes carrying no fixed charges in an electrolyte solution, it was concluded that the change in potential gradient within cell membrane could be on the order of 10^3–10^5 V/cm when the distance between two cells is about 10–20 Å [38]. A series of works on a liquid solution containing a symmetric electrolyte and two ion-penetrable membranes with uniform distribution of fixed charges was published [39–42]. Based on the theory, expressions for the interaction potential energy and the electrostatic force between an ion-penetrable membrane and a solid rigid surface under the assumptions of linearized PBE and uniformly distributed fixed charge in the membrane phase were derived. The analysis was also extended to various combinations of interaction pair, and expressions for critical coagulation concentration (CCC) were derived. By applying an ion-penetrable charged-membrane model, we have proposed calculation algorithm [43,44] and computed the essential quantities of an electrical double layer, such as the thermodynamic properties [45–49] and the net penetration charge [50], and to determine the crucial factors often encountered in a suspension of biological cells, namely, the interaction potential [51,52], CCC [53], and the stability ratio, W [54]. Here, perturbation method was employed to solve governing nonlinear PBE; approximate analytical expressions for electrostatic potential, stability ratio, and CCC of the counter ions were derived for a suspension of negatively charged biological cells in arbitrary $a{:}b$ electrolyte. Furthermore, the adsorption [55] and deposition [56,57] of biological cells, each covered with an ion-penetrable membrane, onto a charged surface was analyzed. The presence of a membrane phase was concluded to be advantageous to the rate of adsorption. A membrane version of MGCT by incorporating the size effects of freely mobile cation, anion, and fixed charge was proposed to estimate the electrostatic potential distribution [58]. It was concluded that the spatial variation in electrostatic potential was dramatically influenced by the sizes of charged species. Also, the validation of the model was performed by calculating the DEDL properties [59], and interaction force and energy barrier [60,61]. Even if the electrostatic potential is low, the existence of ionic size was found to have a significant effect on the apparent electrical field due to the presence of the membrane, and the net penetration charge was also dramatically influenced by the size of charged species. The applicability of the membrane MGCT was further demonstrated by the estimation of mobility [62], adsorption [63,64] of biological cells and critical coagulation concentration [65], and stability ratio [66,67] of a dispersion of biological cells. Recently, a theoretical model for a biological cell comprising a rigid core and a cation-absorptive membrane was developed for the estimation of bio-flocculation [68].

II. FIXED CHARGE

A. DISSOCIATION OF FUNCTIONAL GROUPS OR PROTON ABSORPTION

The origin of the fixed charge may arise out of the dissociation of functional groups, which the biological membrane bears or from the absorption of protons by those functional groups. Typical reactions in the membrane phase include:

$$R\text{—}COOH \Longleftrightarrow R\text{—}COO^- + H^+ \tag{25.1}$$

$$R'\text{—}NH_2 + H^+ \Longleftrightarrow R'\text{—}NH_3^+ \tag{25.2}$$

Equation (25.1) and Equation (25.2) represent the development of negative fixed charges from acidic groups and positive fixed charges from basic groups. If the size of the functional groups requires consideration, the two classes of fixed charges may be arranged in membrane so that the margin of the leftmost one is coincident with the interface of uncharged core and membrane, and the rightmost one is coincident with the interface of membrane and double layer.

B. Successive Dissociation

Suppose that both acidic and basic functional groups are present in the membrane layer, the dissociation of these functional groups can be generally expressed by

$$
\text{AH}^{(p-1)-}_{Z_a-(p-1)} \Longleftrightarrow \text{AH}^{p-}_{Z_a-p} + \text{H}^+, \quad K_{a,p}, \quad p = 1, 2, \ldots, Z_a \tag{25.3}
$$

$$
\text{BH}^{[Z_b-(p-1)]+}_{Z_b-(p-1)} \Longleftrightarrow \text{BH}^{(Z_b-p)+}_{Z_b-p} + \text{H}^+, \quad K_{b,p}, \quad p = 1, 2, \ldots, Z_b \tag{25.4}
$$

In (25.3) and (25.4), Z_a is the number of dissociable protons of the acidic group AH_{Z_a}, Z_b the number of absorbable protons of the basic group B, $K_{a,p}, p = 1, 2, \ldots, Z_a$, and $K_{b,p}, p = 1, 2, \ldots, Z_b$ are the corresponding equilibrium constants for those reactions. It can be demonstrated that the concentration of negative fixed charges, $N^{i=1}_-$, and that of positive fixed charges, $N^{i=1}_+$, are [69]

$$
N^{i=1}_- = \frac{N^{i=1}_a Q_a}{1 + P_a} \tag{25.5}
$$

$$
N^{i=1}_+ = N^{i=1}_b \left(Z_b - \frac{Q_b}{1 + P_b} \right) \tag{25.6}
$$

In these expressions, $N^{i=1}_a$ and $N^{i=1}_b$ are, respectively, the overall concentration of acidic groups and that of basic groups, and

$$
P_q = \sum_{s=1}^{Z_v} \left[\prod_{i=1}^{s} \frac{K_{q,i}}{C_{\text{H}^+}} \right], \quad q = a, b \tag{25.7}
$$

$$
Q_q = \sum_{p=1}^{Z_v} \left[p \prod_{i=1}^{p} \frac{K_{q,i}}{C_{\text{H}^+}} \right], \quad q = a, b \tag{25.8}
$$

where the concentration of protons, C_{H^+}, can be formulated by

$$
C_{\text{H}^+} = C^0_{\text{H}^+} \exp(-\varphi) \tag{25.9}
$$

In (25.9), $C^0_{\text{H}^+}$ and φ are, respectively, the bulk concentration of H^+ and the scaled electrostatic potential defined by

$$
\varphi = \frac{e\phi}{k_B T} \tag{25.10}
$$

where e and ϕ are, respectively, the elementary charge and electrical potential, and k_B and T denotes the Boltzmann constant and the absolute temperature, respectively.

A general expression for the scaled fixed charge distribution N arising from the functional group dissociation and proton absorption in the membrane can be evaluated by

$$
N = \frac{N_A(N^{i=1}_+ - N^{i=1}_-)}{an^0_a} \tag{25.11}
$$

Substituting (25.5) and (25.6) into (25.11) yields

$$N = \frac{N_A}{a n_a^0} \left[N_b^{i=1} \left(Z_b - \frac{Q_b}{1 + P_b} \right) - \frac{N_a^{i=1} Q_a}{1 + P_a} \right] \tag{25.12}$$

where a and n_a^0 are valance of cations and the number concentration of cations in the bulk liquid phase, respectively.

C. ABSORPTION OR CHELATION OF METALLIC IONS

The fixed groups in the membrane phase can also assume to receive charge by absorption or chelation of electrolyte cations penetrating from the diffuse electrical double layer. The absorption or chelation reaction can be described by the following stoichiometric equilibrium [68]:

$$(A_n M_m)^{z+} \Longleftrightarrow n A^{c-} + m M^{a+} \tag{25.13}$$

where A and M represent, respectively, the fixed functional group and the cation, n and m are, respectively, the number of fixed functional group and that of cation involved in absorption or chelation, and z and c denote, respectively, the valence of the absorbed or chelated membrane groups and that of the original fixed functional groups. Note that the charge balance leads to $z = ma - nc$. Three typical examples for (25.13) are presented subsequently.

$$R{-}COONa \Longleftrightarrow R{-}COO^- + Na^+ \tag{25.14}$$

$$(R{-}SO_3)_2 Ca^{2+} \Longleftrightarrow 2R{-}SO_3 + Ca^{2+} \tag{25.15}$$

$$(R{-}NH_2)_2 Al^{3+} \Longleftrightarrow 2R{-}NH_2 + Al^{3+} \tag{25.16}$$

Equation (25.14) stands for the absorption/desorption equilibrium, and A and M denote, respectively, R—COO and sodium (Na), $n = 1$, $m = 1$, $a = 1$, $c = 1$, and $z = 0$ in the case. Equation (25.15) symbolizes the absorption/desorption or chelation equilibrium, and A and M represent, respectively, R—SO₃, and calcium (Ca), $n = 2$, $m = 1$, $a = 2$, $c = 0$, and $z = 2$ in the case. And the Equation (25.16) corresponds to the absorption/desorption or chelation equilibrium, and A and M represent, respectively, R—NH₂, and aluminum (Al), $n = 2$, $m = 1$, $a = 3$, $c = 0$, and $z = 3$ in the case. Generally, $c \geq 0$. If K denotes the equilibrium constant of the stoichiometric equation for dissociation of cation–functional group complex in (25.13), we have

$$K = \frac{[A^{c-}]^n [M^{a+}]^m}{[(A_n M_m)^{z+}]} \tag{25.17}$$

where the species in brackets means its concentration. The conservation of the fixed functional groups in membrane phase leads to

$$N_0 = [A^{c-}] + n[(A_n M_m)^{z+}] \tag{25.18}$$

where N_0 is the total concentration of the fixed functional groups. The Boltzmann distribution of cations is described by

$$[M^{a+}] = \left(\frac{n_a^0}{N_A} \right) \exp(-a\varphi) \tag{25.19}$$

where N_A denotes the Avogadro's number. The definition of scaled concentration of fixed charge is

$$N = \frac{N_A\{z[(A_n M_m)^{z+}] - c[A^{c-}]\}}{a n_a^0} \tag{25.20}$$

From (25.17) to (25.19), $[A^{c-}]$ and $[(A_n M_m)^{z+}]$ can be obtained. Substituting these results into (25.20), N is evaluated. The number of fixed groups involved in the cationic absorption or chelation is generally 1 or 2, and the analytical expressions for N can be expressed by

$$N = \frac{N_0 N_A}{a n_a^0} \frac{z\left(\dfrac{n_a^0}{N_A}\right)^m e^{-am\varphi} - cK}{K + \left(\dfrac{n_a^0}{N_A}\right)^m e^{-am\varphi}}, \quad n = 1 \tag{25.21}$$

$$N = \frac{4zN_0 \left(\dfrac{n_a^0}{N_A}\right)^m e^{-am\varphi} - K(z+2c)\sqrt{1 + 8\dfrac{N_0}{K}\left(\dfrac{n_a^0}{N_A}\right)^m e^{-am\varphi}} + 2cK}{8a\left(\dfrac{n_a^0}{N_A}\right)^{m+1} e^{-am\varphi}}, \quad n = 2 \tag{25.22}$$

Note that for the case of $n = 1$, if

$$K > \frac{z}{c}\left(\frac{n_a^0}{N_A}\right)^m e^{-am\varphi} \tag{25.23}$$

the electricity of the membrane changes. For the case of $n = 2$, the critical K value for the change of membrane electricity can be determined by

$$K > \frac{4zN_0 \left(\dfrac{n_a^0}{N_A}\right)^m e^{-am\varphi}}{(z+2c)\sqrt{1 + 8\dfrac{N_0}{K}\left(\dfrac{n_a^0}{N_A}\right)^m e^{-am\varphi}} - 2c}, \quad n = 2 \tag{25.24}$$

D. Nonuniform Distribution of Functional Groups

1. Planar Membrane

The nonuniform distribution of functional groups in planar membrane layer of biological cells immersed in an $a{:}b$ electrolyte solution, N_j, can be formulated as [63]

$$N_j = \begin{cases} \dfrac{2ZN_A N_{0,a}[1 + \alpha(X - X_i)]}{a n_a^0(A+2)}, & j = 1 \\[3ex] \dfrac{ZN_A N_{0,a} A\{1 + \exp[\alpha(X - X_i)]\}}{a n_a^0[\exp(A) + A - 1]}, & j = 2 \end{cases} \tag{25.25}$$

where

$$A = \alpha(X_o - X_i) \tag{25.26}$$

$$X = \kappa r \tag{25.27}$$

$$X_i = \frac{\kappa \sigma_i}{2} \tag{25.28}$$

$$X_o = D - X_i \tag{25.29}$$

$$\kappa^2 = \frac{e^2 a(a+b)n_a^0}{\varepsilon_0 \varepsilon_r k_B T} \tag{25.30}$$

In these expressions, κ, σ_i and D are, respectively, the reciprocal Debye screening length, the effective radius of fixed functional groups in the cellular membrane and the scaled membrane thickness, X the scaled distance, r represents the distance measured from cellular core–membrane interface, e denotes the elementary charge, ε_0 and ε_r are the permittivity of a vacuum and the relative permittivity, respectively, α represents a nonuniform feature index, which is a parameter characterizing the radial nonuniform distribution of fixed functional groups in the membrane; N_j and j are, respectively, the scaled concentration of fixed functional groups and a distribution type index, which is a parameter characterizing the distribution of the fixed functional groups in membrane ($j = 1$, linear type; $j = 2$, exponential type); and $N_{0,a}$ and Z denote the average concentration of fixed functional groups and their valence in membrane, respectively. Note that X_i and X_o are, respectively, the location of the inner plane of fixed charge (IPFC) and that of the outer plane of fixed charge (OPFC). Hence, the fixed charges present in the zone of $X_i \le X \le X_o$. If the fixed functional groups are completely charged, $Z e N_A N_{0,a}$ represents the average density of the fixed charge in the membrane zone.

2. Curvilinear Membrane

For curvilinear geometry, N_j can be described by [64]

$$N_j = \begin{cases} \dfrac{Z N_A N_{0,a} \left(\dfrac{X_o^{\tau+1} - X_i^{\tau+1}}{\tau+1} \right) [1 + \alpha(X - X_i)]}{a n_a^0 \displaystyle\sum_{s=0}^{1} \dfrac{(1 - \alpha X_i)^{1-s} \alpha^s \left(X_o^{\tau+1+s} - X_i^{\tau+1+s} \right)}{\tau+1+s}}, & j = 1 \\[4ex] \dfrac{Z N_A N_{0,a} \left(\dfrac{X_o^{\tau+1} - X_i^{\tau+1}}{\tau+1} \right) \{1 + \exp[\alpha(X - X_i)]\}}{a n_a^0 \left\{ \dfrac{X_o^{\tau+1} - X_i^{\tau+1}}{\tau+1} + \displaystyle\sum_{s=0}^{\tau} \dfrac{(-1)^{\tau-s}\tau!}{\alpha^{\tau-s+1}s!} \left\{ X_o^s \exp[\alpha(X_o - X_i)] - X_i^s \right\} \right\}}, & j = 2 \end{cases} \tag{25.31}$$

In this expression, τ is a parameter characterizing the membrane geometry ($\tau = 1$, cylinder; $\tau = 2$, sphere). The relevant properties of N_j function are summarized subsequently. The distribution of the fixed functional groups described in (25.31) has a characteristic that assures the space-averaged concentration of fixed charges is constant despite α and j, that is,

$$
\begin{aligned}
2\pi L_c a e n_a^0 \int_{X_i}^{X_o} N_j X \, \mathrm{d}X &= N_{0,a} Z e N_A \cdot L_c \pi (X_o^2 - X_i^2), \quad \tau = 1 \\
4\pi a e n_a^0 \int_{X_i}^{X_o} N_j X^2 \, \mathrm{d}X &= N_{0,a} Z e N_A \cdot \frac{4}{3}\pi(X_o^3 - X_i^3), \quad \tau = 2
\end{aligned} \tag{25.32}
$$

where L_c is the scaled characteristic length of a cylindrical cell. It is worth to notice that as $\alpha = 0$, $N_j = ZN_AN_{0,a}/an_a^0$ for the cases of $j = 1, 2$ and $\tau = 1, 2$. This implies that as $\alpha = 0$, the fixed functional groups reduce to the typical uniform distributions with concentration $N_{0,a}$. In contrast, when α approaches infinity, the asymptotic behaviors of the N_j function for the cases $\tau = 1, 2$, are

$$N_1(X = X_i) = 0, \qquad N_1(X = X_o) = \frac{N_{0,a}ZN_A(\tau + 2)\sum_{s=0}^{\tau} X_o^{\tau-s}X_i^s}{an_a^0\sum_{s=0}^{\tau}(1 + \tau - s)X_o^{\tau-s}X_i^s} \tag{25.33}$$

$$N_2(X_i \leq X < X_o) = 0, \qquad N_2(X = X_o) \longrightarrow \infty \tag{25.34}$$

The distribution of fixed acidic and basic groups in membrane layer contributes critical influences on the variations in electrical properties of interacting double layers, rendering a variety of performances for the behavior related to biological phenomena. If the concentration of nonuniformly distributed functional groups in the membrane is defined by

$$P_j = \begin{cases} \dfrac{ZN_AN_{0,a}[1 + \alpha(X - X_i)]}{an_a^0\sum_{s=0}^{1}\dfrac{(1 - \alpha X_i)^{1-s}\alpha^s(X_o^{\tau+1+s} - X_i^{\tau+1+s})}{\tau + 1 + s}}, & j = 1 \\[2em] \dfrac{ZN_AN_{0,a}\{1 + \exp[\alpha(X - X_i)]\}}{an_a^0\left\{\dfrac{X_o^{\tau+1} - X_i^{\tau+1}}{\tau + 1} + \sum_{s=0}^{\tau}\dfrac{(-1)^{\tau-s}\tau!}{\alpha^{\tau-s+1}s!}\{X_o^s \exp[\alpha(X_o - X_i)] - X_i^s\}\right\}}, & j = 2 \end{cases} \tag{25.35}$$

we have

$$aen_a^0\int_{X_i}^{X_o} P_jX^\tau \, dX = N_0ZeN_A \tag{25.36}$$

Comparing (25.32) with (25.36), it can be figured out that P_j is a successful expression for the description of the total amount of fixed functional groups in a membrane since P_j demonstrates an aspect that the total amount of fixed charges is constant for any distributive style of fixed functional groups, α and j, and for any geometry of membranes, τ, X_i and X_o. However, substituting (25.35) into electrical governing equation, such as the Poisson–Boltzmann equation, necessitates tedious revisions for succeeding calculation because membrane-shape correlation in the formulation of P_j is overlooked.

III. ELECTROSTATIC POTENTIAL DISTRIBUTION

A. POISSON–BOLTZMANN EQUATION

Instead of the Nernst–Plank equation for the flux of ionic transportation, the equilibrium Poisson equation can be selected as the starting point for describing the electrical potential because the time required for cellular motion, electrophoresis, coagulation, and deposition is much longer than the diffusion equilibrium of ionic profile in interacting double layers [70–72]. The spatial variation in the electrostatic potential can be described by the Poisson equation

$$\nabla^2\phi = \frac{\rho_c}{\varepsilon} \tag{25.37}$$

where ∇^2 is the Laplace operator, ρ_c denotes the space charge density and ε represents the dielectric constant of the system which is defined by,

$$\varepsilon = \varepsilon_0\varepsilon_r \tag{25.38}$$

First, we may suppose the charge in the system, including $a{:}b$ electrolyte ions and fixed functional groups, can be treated as point charge. Based on the condition of electrothermal equilibrium and the differential Gauss law of an electric field, the scaled electrostatic potential for a planar membrane can be described by the following PBE

$$\frac{d^2\varphi}{dX^2} = \frac{-\exp(-a\varphi) + \exp(b\varphi) - wN_j}{a+b} \tag{25.39}$$

where w denotes a membrane index ($w = 1$ for membrane phase; $w = 0$ for liquid solution). N_j denotes the scaled fixed charge concentration in this section and the following related derivations. For the case of nonpoint charge, that is, the size of charge requires considering (25.39) can be modified to become the following form:

$$\frac{d^2\varphi}{dX^2} = \frac{-u\,\exp(-a\varphi) + v\,\exp(b\varphi) - wN_j}{a+b} \tag{25.40}$$

where u and v are the region index for the presence/absence of cations and anions, respectively. The spatial variation in the scaled electrostatic potential around a nonplanar cell can be governed by

$$\frac{d^2\varphi}{dX^2} + \frac{\tau}{X}\frac{d\varphi}{dX} = \frac{-u\,\exp(-a\varphi) + v\exp(b\varphi) - wN_j}{a+b} \tag{25.41}$$

If the sizes of fixed positive and negative charge are not identical (25.40) can be rewritten to become

$$\frac{d^2\varphi}{dX^2} = \frac{-u\,\exp(-a\varphi) + v\,\exp(b\varphi) - wN_{j,\mathrm{p}} + xN_{j,\mathrm{n}}}{a+b} \tag{25.42}$$

where w and x are, respectively, the index for the existence of fixed positive and negative charge, and $N_{j,\mathrm{p}}$ and $N_{j,\mathrm{n}}$ are the scaled concentration of fixed positive and negative charge, respectively. For a system containing binary cations of valence a_1 and a_2, the right-hand side of (25.37) requires revision, and (25.40) becomes

$$\frac{d^2\varphi}{dX^2} = \frac{v\,\exp(b\varphi) - u[(1-\chi)\,\exp(-a_1\varphi) + \chi\,\exp(-a_2\varphi)] - wN_j}{(a_1+b) + (a_2 - a_1)\chi} \tag{25.43}$$

where $\chi = a_2 n_{a_2}^0 / bn_b^0$ is the charge portion of cation with valence a_2 in bulk liquid phase, and n_b^0 and $n_{a_2}^0$ are, respectively, the number concentration of anion and cation with valence a_2 in bulk liquid phase. In this case, we have

$$bn_b^0 = a_1 n_{a_1}^0 + a_2 n_{a_2}^0 \tag{25.44}$$

where $n_{a_1}^0$ is the number concentration of cation with valence a_1 in bulk liquid phase. The effect of finite sizes of the charged species is usually significant in the case of low membrane potential. In this case (25.40) can be approximated by

$$\frac{d^2\varphi}{dX^2} = \frac{2}{ak}\left[(ua + vb)\varphi + v - u - wN_j\right] \tag{25.45}$$

where

$$k = 2\left(\frac{1+b}{a}\right)$$ (25.46)

B. BOUNDARY CONDITIONS

Figure 25.1 illustrates a schematic graph for the interaction between a planar membrane and a charged rigid surface. Here, L denotes the scaled distance between cell membrane and the charged hard wall. Let σ_{an} and σ_{ca} be the effective diameters of anions and cations, respectively. Without loss of generality, we assume that $\sigma_i > \sigma_{an} > \sigma_{ca}$. The fixed groups are arranged so that the margin of the leftmost one coincides with the core–membrane interface and that of the rightmost one coincides with the membrane–liquid interface. Referring to Figure 25.1, the system is divided into eight regions: (I) $0 < X < X_{ca}$, which is the inner uncharged membrane; (II) $X_{ca} < X < X_{an}$, which contains cations only in the cell membrane; (III) $X_{an} < X < X_i$, which contains both cations and anions in the cell membrane; (IV) $X_i < X < X_o$, which contains all the charged species; (V) $X_o < X < D$, which is the outer uncharged membrane containing both cations and anions; (VI) $D < X < X_{an'}$, which is the liquid phase of the electrical double layer containing both cations and anions; (VII) $X_{an'} < X < X_{ca'}$, which denotes the region containing cations only, close to the charged surface; and (VIII) $X_{ca'} < X < D + L$, which represents the charge-free region, close to the charged surface. The scaled symbols are defined by $X_{ca} = \kappa\sigma_{ca}/2$, $X_{an} = \kappa\sigma_{an}/2$, $X_{an'} = D + L - X_{an}$ and $X_{ca'} = D + L - X_{ca}$. Note that $-X_c < X < 0$ is the region of charge-free zone in the interior core of a biological cell, where $X_c = \kappa r_c$ is the scaled radius of cellular core, r_c being the linear radius of the cellular cores. When cationic absorption by fixed groups occurs, a space surrounded by side chains of a functional group or that among several side chains of functional groups allow the insertion of cations from nearby fluid. Hence, cations and ionogenic membrane groups become a new complex species whose size is approximately the same dimension of original fixed functional groups in membrane. For fundamental understanding concerning the effects of sizes of charged species, an inclusion of a region for the appearance of cation–functional group complex seems to be not crucial. For convenience, a region vector for the presence/absence of fixed charge, anion, and cation (w, v, u), can be defined by $(w, v, u) = (0, 0, 0)$, $(0, 0, 1)$, $(0, 1, 1)$, $(1, 1, 1)$, $(0, 1, 1)$, $(0, 1, 1)$, $(0, 0, 1)$, and $(0, 0, 0)$ represent regions I–VIII, respectively.

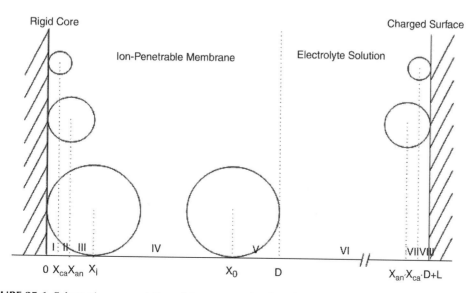

FIGURE 25.1 Schematic representation of the system related to charged biological cells.

The boundary conditions associated with (25.40) or (25.45) are assumed as

$$\varepsilon_{r,c}\frac{d\varphi}{dX} \to 0 \quad \text{as} \quad X \to 0 \tag{25.47}$$

$$\varepsilon_{r,m}\left(\frac{d\varphi}{dX}\right)_{X=X_{ca}^+} = \varepsilon_{r,m}\left(\frac{d\varphi}{dX}\right)_{X=X_{ca}^-} \quad \text{and} \quad \varphi(X_{ca}^+) = \varphi(X_{ca}^-) \quad \text{at} \quad X = X_{ca} \tag{25.48}$$

$$\varepsilon_{r,m}\left(\frac{d\varphi}{dX}\right)_{X=X_{an}^+} = \varepsilon_{r,m}\left(\frac{d\varphi}{dX}\right)_{X=X_{an}^-} \quad \text{and} \quad \varphi(X_{an}^+) = \varphi(X_{an}^-) \quad \text{at} \quad X = X_{an} \tag{25.49}$$

$$\varepsilon_{r,m}\left(\frac{d\varphi}{dX}\right)_{X=X_i^+} = \varepsilon_{r,m}\left(\frac{d\varphi}{dX}\right)_{X=X_i^-} \quad \text{and} \quad \varphi(X_i^+) = \varphi(X_i^-) \quad \text{at} \quad X = X_i \tag{25.50}$$

$$\varepsilon_{r,m}\left(\frac{d\varphi}{dX}\right)_{X=X_o^+} = \varepsilon_{r,m}\left(\frac{d\varphi}{dX}\right)_{X=X_o^-} \quad \text{and} \quad \varphi(X_o^+) = \varphi(X_o^-) \quad \text{at} \quad X = X_o \tag{25.51}$$

$$\varepsilon_{r,m}\left(\frac{d\varphi}{dX}\right)_{X=D^+} = \varepsilon_{r,m}\left(\frac{d\varphi}{dX}\right)_{X=D^-} \quad \text{and} \quad \varphi(D^+) = \varphi(D^-) \quad \text{at} \quad X = D \tag{25.52}$$

$$\varepsilon_{r,dl}\left(\frac{d\varphi}{dX}\right)_{X=X_{an'}^+} = \varepsilon_{r,dl}\left(\frac{d\varphi}{dX}\right)_{X=X_{an'}^-} \quad \text{and} \quad \varphi(X_{an'}^+) = \varphi(X_{an'}^-) \quad \text{at} \quad X = X_{an'} \tag{25.53}$$

$$\varepsilon_{r,dl}\left(\frac{d\varphi}{dX}\right)_{X=X_{ca'}^+} = \varepsilon_{r,dl}\left(\frac{d\varphi}{dX}\right)_{X=X_{ca'}^-} \quad \text{and} \quad \varphi(X_{ca'}^+) = \varphi(X_{ca'}^-) \quad \text{at} \quad X = X_{ca'} \tag{25.54}$$

$$\varphi = \varphi_L \quad \text{at} \quad X = X_L = D + L \tag{25.55}$$

In these expressions, φ_L is the scaled electrostatic potential on a charged rigid surface, X_L the scaled distance between the cellular core and the charged rigid surface, and $\varepsilon_{r,c}$, $\varepsilon_{r,m}$, and $\varepsilon_{r,dl}$ denote, respectively, the relative permitivities of the cellular core, the membrane layer, and the electrical double layer. Commonly, the dielectric properties in biological cells imply that $\varepsilon_{r,c} < \varepsilon_{r,m} < \varepsilon_{r,dl}$. We also consider the general case in which the membrane-layer thickness and the double-layer thickness can be comparable. This suggests that the potential at the interior core may not reach the Donnan potential. Note that the electrostatic potential in the interior core is not constant, nor is the electrostatic potential at the membrane–liquid interface; that is, both of them vary with L. Equation (25.47) and Equation (25.55) may imply that the biological system is under electroneutrality.

C. Several Cases

1. Low Potential

Equation (25.44) can be solved analytically for the case of uniformly distributed fixed charge, where N_j in (25.44) is replaced by a constant scaled charge distribution N, and low electrical potential. If the amount of fixed charge in a membrane is small, a considerable amount of counterions and coions may penetrate into the membrane. Hence, the effect of finite sizes of the charged species is significant in the case of low membrane potential. For an isolated cell, only regions I–VI need to be considered, and the boundary conditions are shown in (25.47)–(25.52) and

$$\varepsilon_{r,dl}\left(\frac{d\varphi}{dX}\right) \to 0 \quad \text{as} \quad X \to \infty \tag{25.56}$$

The analytical solution of electrostatic potential is presented below. Region I is free of charges, where $(w, v, u) = (0, 0, 0)$, and, therefore (25.45) reduces to

$$\frac{d^2\varphi}{dX^2} = 0 \tag{25.57}$$

Solving this equation subject to (25.47) yields

$$\varphi = \varphi_c, \text{ in region I} \tag{25.58}$$

where φ_c is the scaled electrical potential in the core of biological cell or the scaled membrane potential. Since region II contains cations only $(w, v, u) = (0, 0, 1)$, (25.45) becomes

$$\frac{d^2\varphi}{dX^2} = \frac{2}{k}\varphi - \frac{2}{ak} \tag{25.59}$$

Integrating (25.59) subject to (25.48) and (25.58) gives

$$\varphi = C_1 \sinh\left(\sqrt{\frac{2}{k}}X\right) + C_2 \cosh\left(\sqrt{\frac{2}{k}}X\right) + \frac{1}{a}, \text{ in region II} \tag{25.60}$$

where

$$C_1 = -\left(\varphi_c - \frac{1}{a}\right)\sinh\left(\sqrt{\frac{2}{k}}X_{ca}\right) \tag{25.61}$$

$$C_2 = \left(\varphi_c - \frac{1}{a}\right)\cosh\left(\sqrt{\frac{2}{k}}X_{ca}\right) \tag{25.62}$$

Both cations and anions are present in region III $(w, v, u) = (0, 1, 1)$, and (25.45) becomes

$$\frac{d^2\varphi}{dX^2} = \varphi \tag{25.63}$$

Integrating this expression subject to (25.49) and (25.60) yields

$$\varphi = C_3 \sinh(X) + C_4 \cosh(X), \text{ in region III} \tag{25.64}$$

where

$$C_3 = \left[\sqrt{\frac{2}{k}}\cosh(X_{an})\cosh\left(\sqrt{\frac{2}{k}}X_{an}\right) - \sinh(X_{an})\sinh\left(\sqrt{\frac{2}{k}}X_{an}\right)\right]C_1$$
$$+ \left[\sqrt{\frac{2}{k}}\cosh(X_{an})\sinh\left(\sqrt{\frac{2}{k}}X_{an}\right) - \sinh(X_{an})\cosh\left(\sqrt{\frac{2}{k}}X_{an}\right)\right]C_2 - \frac{1}{a}\sinh(X_{an})$$

$$\tag{25.65}$$

$$C_4 = \left[\cosh(X_{an})\sinh\left(\sqrt{\frac{2}{k}}X_{an}\right) - \sqrt{\frac{2}{k}}\sinh(X_{an})\cosh\left(\sqrt{\frac{2}{k}}X_{an}\right)\right]C_1$$
$$+ \left[\cosh(X_{an})\cosh\left(\sqrt{\frac{2}{k}}X_{an}\right) - \sqrt{\frac{2}{k}}\sinh(X_{an})\sinh\left(\sqrt{\frac{2}{k}}X_{an}\right)\right]C_2 + \frac{1}{a}\cosh(X_{an})$$

$$\tag{25.66}$$

In region IV, since all charged species are present $(w, v, u) = (1, 1, 1)$, (25.45) reduces to

$$\frac{d^2\varphi}{dX^2} = \varphi - \frac{2N}{ak} \tag{25.67}$$

Integrating this expression subject to (25.50) and (25.64) we obtain

$$\varphi = C_5 \sinh(X) + C_6 \cosh(X) + \frac{2N}{ak}, \text{ in region IV} \tag{25.68}$$

where

$$C_5 = C_3 + \frac{2N}{ak} \sinh(X_i) \tag{25.69}$$

$$C_6 = C_4 - \frac{2N}{ak} \cosh(X_i) \tag{25.70}$$

Regions V and VI contain electrolyte ions, but free of fixed charge $(w, v, u) = (0, 1, 1)$. Therefore (25.45) becomes

$$\frac{d^2\varphi}{dX^2} = \varphi \tag{25.71}$$

Integrating this expression subject to (25.51), (25.52), and (25.56) yields

$$\varphi = C_7 \exp(-X) + C_8 \exp(X), \text{ in regions V and VI} \tag{25.72}$$

where

$$C_7 = \frac{N}{ak}[\exp(X_o) - \exp(X_i)] - \frac{1}{2}(C_3 - C_4) \tag{25.73}$$

$$C_8 = 0 \tag{25.74}$$

It can be shown that N_0 and φ_c have the following implicit relation:

$$N = \frac{ak}{2} \frac{C_3 + C_4}{\exp(-X_i) - \exp(-X_o)} \tag{25.75}$$

The apparent scaled electric field due to the presence of a membrane, E_a, can be expressed by

$$E_a = \frac{\varphi_{ml} - \varphi_c}{D} \tag{25.76}$$

where

$$\varphi_{ml} = C_7 \exp[-(X_o + X_i)] \tag{25.77}$$

It can be shown that the net penetration charge per unit area of a membrane–liquid interface, Q_{sl}, is

$$Q_{sl} = -Q_t + E_a D + \varphi_c \tag{25.78}$$

where Q_t is defined by

$$Q_t = \frac{Ze^2 N_A N_0}{\varepsilon_0 \varepsilon_r k_B T \kappa^2} \int_{X_i}^{X_o} dX = \frac{ZN_A N_0 (X_o - X_i)}{a(a+b)n_a^0} \tag{25.79}$$

For interaction between two identical cells, only regions I–VI need to be considered, and the boundary conditions are (25.47)–(25.52) and

$$\left(\frac{d\varphi}{dX}\right) \rightarrow 0 \quad \text{as} \quad X \rightarrow X_m \tag{25.80}$$

where X_m is the location of mid-plane between two biological cells. Equation (25.47) and Equation (25.80) also imply that the system under consideration is at electroneutrality. Under the conditions of uniform fixed charge distribution and low potential, the analytical solution is the same as those given in (25.57)–(25.74). However, N value is regulated by X_m and φ_c as shown subsequently.

$$N = \frac{ak}{2C_9}[C_3 \cosh(X_m) + C_4 \sinh(X_m)] \tag{25.81}$$

where

$$C_9 = [\sinh(X_o) - \sinh(X_i)]\cosh(X_m) - [\cosh(X_o) - \cosh(X_i)]\sinh(X_m) \tag{25.82}$$

The electrostatic interaction force between two membranes F_R can be evaluated by

$$F_R = -\int_0^{\phi_m} (aen_a - ben_b)\, d\phi \tag{25.83}$$

where ϕ_m is the mid-plane electrostatic potential between two identical cellular membranes and n_a and n_b represent, respectively, the number concentrations of cation and anion, and n_a and n_b can be defined by

$$n_a = n_a^0 (1 - a_\varphi) \tag{25.84}$$
$$n_b = n_b^0 (1 + b_\varphi) \tag{25.85}$$

Substituting (25.84) and (25.85) into (25.83), we obtain, after integrating,

$$F_R = \frac{a^2 k}{4} n_a^0 k_B T \varphi_m^2 \tag{25.86}$$

where φ_m is the scaled electrostatic potential on the mid-plane between two identical cellular membranes. The electrostatic interaction energy, Φ_{el}, can be evaluated by

$$\Phi_{el} = \frac{2}{\kappa} \int_{X_m}^{\infty} F_R\, dX_m \tag{25.87}$$

Let us consider two idealized cases, which are used most often encountered in the relevant studies, that is, constant potential and constant charge density. In the former, the fixed charge density is regulated by separation distance between two cellular membranes. From (25.72), φ_m is expressed as, after rearrangement,

$$\varphi_m = \left(\frac{C_3 + C_4 C_{10}}{C_9}\right) \sinh(X_m) + \left(\frac{C_4 + C_3 C_{10}}{C_9}\right) \cosh(X_m) \qquad (25.88)$$

where

$$C_{10} = [\cosh(X_o) - \cosh(X_i)] \cosh(X_L) - [\sinh(X_o) - \sinh(X_i)] \sinh(X_L) \qquad (25.89)$$

Substituting (25.88) into (25.86) yields an explicit expression for the electrical interaction force. Substituting this resultant expression into (25.87) leads to, after integration

$$\Phi_{el} = \frac{a^2 k n_a^0 k_B T}{4\kappa} E_1^2 \sum_{i=1}^{\infty} (-D_1)^{i-1} \exp[-2i(X_m - X_c)] \qquad (25.90)$$

where

$$D_1 = \frac{\sinh(X_o) - \sinh(X_i) + \cosh(X_o) - \cosh(X_i)}{\sinh(X_o) - \sinh(X_i) - \cosh(X_o) + \cosh(X_i)} \qquad (25.91)$$

$$E_1 = (C_4 + C_3)D_1 + C_4 - C_3 \qquad (25.92)$$

For the case of constant charge density, the membrane potential is regulated by the separation distance between two membranes. Solving (25.81) for φ_c yields

$$\varphi_c = \frac{1}{a} + \frac{F_1 \cosh(X_m) + F_2 \sinh(X_m)}{G_1 \cosh(X_m) + G_2 \sinh(X_m)} \qquad (25.93)$$

where

$$F_1 = \frac{(2N/k)[\sinh(X_o) - \sinh(X_i)] + \sinh(X_{an})}{a} \qquad (25.94)$$

$$F_2 = -\frac{(2N/k)[\cosh(X_o) - \cosh(X_i)] + \cosh(X_{an})}{a} \qquad (25.95)$$

$$G_1 = A_2 \cosh\left[(2/k)^{1/2} X_{ca}\right] - A_1 \sinh\left[(2/k)^{1/2} X_{ca}\right] \qquad (25.96)$$

$$G_2 = B_2 \cosh\left[(2/k)^{1/2} X_{ca}\right] - B_1 \sinh\left[(2/k)^{1/2} X_{ca}\right] \qquad (25.97)$$

$$A_1 = (2/k)^{1/2} \cosh(X_{an}) \cosh\left[(2/k)^{1/2} X_{an}\right] - \sinh(X_{an}) \sinh\left[(2/k)^{1/2} X_{an}\right] \qquad (25.98)$$

$$A_2 = (2/k)^{1/2} \cosh(X_{an}) \sinh\left[(2/k)^{1/2} X_{an}\right] - \sinh(X_{an}) \cosh\left[(2/k)^{1/2} X_{an}\right] \qquad (25.99)$$

$$B_1 = \cosh(X_{an}) \sinh\left[(2/k)^{1/2} X_{an}\right] - (2/k)^{1/2} \sinh(X_{an}) \cosh\left[(2/k)^{1/2} X_{an}\right] \qquad (25.100)$$

$$B_2 = \cosh(X_{an}) \cosh\left[(2/k)^{1/2} X_{an}\right] - (2/k)^{1/2} \sinh(X_{an}) \sinh\left[(2/k)^{1/2} X_{an}\right] \qquad (25.101)$$

Equation (25.72) leads to

$$\varphi_m = [-F_1 + G_1(\varphi_c - 1/a)] \sinh(X_m) + [-F_2 + G_2(\varphi_c - 1/a)] \cosh(X_m) \qquad (25.102)$$

Substituting (25.93) and (25.102) into (25.83) yields an explicit expression for the electrical interaction force. Substituting this expression into (25.87) leads to, after integration

$$\Phi_{el} = \frac{a^2 k n_a^0 k_B T}{4\kappa} H_1^2 \sum_{j=1}^{\infty} (-K_1)^{j-1} \exp[-2j(X_m - X_c)] \qquad (25.103)$$

where

$$H_1 = (F_1 - F_2) - (F_1 + F_2)K_1 \tag{25.104}$$

$$K_1 = \frac{G_1 - G_2}{G_1 + G_2} \tag{25.105}$$

2. Negatively Charged Membranes

Consider the case of moderate-to-high electrical potential in which the membrane phase contains uniformly distributed negative charge ($Z < 0$). For instance, the dissociation of sulfonic acid yields the fixed charged group $-SO_3^-$. For an isolated cell, the potential distribution is governed by (25.40), and the associated boundary conditions are (25.47)–(25.52) and (25.56). Solving (25.40) subject to (25.47) gives

$$\varphi = \varphi_c, \text{ in region I} \tag{25.106}$$

Integrating (25.40) subject to (25.48) and (25.106) yields

$$X = X_c + k^{1/2} Y_c [\sin^{-1}(1) - \sin^{-1}(Y/Y_c)], \text{ in region II} \tag{25.107}$$

or

$$\varphi = \varphi_c + \frac{2}{a} \ln \left\{ \sin \left[\sin^{-1}(1) - \frac{X - X_c}{k^{1/2} Y_c} \right] \right\}, \text{ in region II} \tag{25.108}$$

where Y_c is the value of Y when $\varphi = \varphi_c$, and

$$Y = \exp\left(\frac{a\varphi}{2}\right) \tag{25.109}$$

Integrating (25.40) subject to (25.49) and (25.108) leads to

$$X = X_{an} + f_2 - f_2(Y_{an}), \text{ in region III} \tag{25.110}$$

where Y_{an} is the value of Y at $X = X_{an}$, and

$$f_2 = (- \text{ or } \mp) k^{1/2} \int f_1^{-1/2} \, dY \tag{25.111}$$

$$f_1 = 1 + aC_{11} Y^2 + \frac{a}{b} Y^k = \left\{ \sum_{s=1}^{\infty} \left[(-1)^s \frac{\prod_{i=1}^{s}(2i-1)}{\prod_{i=1}^{s}(2i)} \left(aC_{11} Y^2 + \frac{b}{a} Y^k \right)^s \right] \right\}^{-2} \tag{25.112}$$

$$C_{11} = - \left[\frac{\exp(-a\varphi_c)}{a} + \frac{\exp(b\varphi_{an})}{b} \right] \tag{25.113}$$

In (25.113), φ_{an} is the value of φ at $X = X_{an}$. The negative sign on the right-hand side of (25.111) is applicable to $K_2 \sigma_i < \sigma_{an}$ and $N_0 < N_{0,c}$, shown as case (A), and the $\mp$ sign is applicable to $K_2 \sigma_i \geq \sigma_{an}$ and $N_0 \geq N_{0,c}$, shown as case (B), K_2 and $N_{0,c}$ being a specific constant and the critical fixed charge density, respectively. For the present arrangement of charge, $K_2 < 1$. In case (B), $d\varphi/dX$ is negative at $X = X_{an}$ but positive at $X = X_i$. This implies that there is a plane in region III on which $d\varphi/dX$ vanishes, or the electrostatic potential possesses a local minimum. Because $d\varphi/dX = 0$ suggests that the net charges vanish, this plane is defined as the plane of zero charge

(PZC). If $X_{z,3}$ denotes the location of the PZC, the negative sign needs to be selected from the $\mp$ sign on the right-hand side of (25.111) for range $[X_{ca}, X_{z,3}]$, and a positive sign chosen for range $[X_{z,3}, X_i]$. Integrating (25.40) subject to (25.50) and (25.110) leads to

$$X = X_i + f_4 - f_4(Y_i), \text{ in region IV} \tag{25.114}$$

where

$$f_4 = (\mp \text{ or } +)k^{1/2} \int f_3^{-1/2} \, dY \tag{25.115}$$

$$f_3 = 1 + aC_{22}Y^2 + \frac{a}{b}Y^k - 2NY^2 \ln Y$$

$$= \left\{ \sum_{s=1}^{\infty} \left[(-1)^s \frac{\prod_{i=1}^{s}(2i-1)}{\prod_{i=1}^{s}(2i)} \left(aC_{22}Y^2 + \frac{b}{a}Y^k - 2NY^2 \ln Y \right)^s \right] \right\}^{-2} \tag{25.116}$$

$$C_{22} = C_{11} + N\varphi_i \tag{25.117}$$

Here, Y_i is the value of Y at $X = X_i$ and φ_i the value of φ at $X = X_i$. The $\mp$ sign on the right-hand side of (25.115) should be used for case (A) and the positive sign on the right-hand side of (25.115) for case (B). The former can be deduced from (25.50) and (25.51). Because $d\varphi/dX$ is negative at $X = X_i$ but positive at $X = X_o$, there is a plane in region IV on which the electrostatic potential possesses a local minimum. If $X_{z,4}$ denotes the location of the PZC in region IV, then the negative sign should be selected from the $\mp$ on the right-hand side of (25.115) for region $[X_i, X_{z,4}]$ and the positive sign for region $[X_{z,4}, X_o]$. Integrating (25.40) subject to (25.51), (25.52), and (25.56) gives

$$\varphi = \frac{2}{a} \ln \left(\frac{1 + D_{11}}{1 - D_{11}} \right), \text{ in regions V and VI} \tag{25.118}$$

where

$$D_{11} = \left[\tanh\left(\frac{a\varphi_o}{4} \right) \right] \exp[-k_3(X - X_o)] \tag{25.119}$$

$$k_3 = \begin{cases} \dfrac{(k-2)k_1 + 2k_2}{k}, & \text{if } k > 4 \\[3mm] \dfrac{2k_1 + (k-2)k_2}{k}, & \text{if } k < 4 \end{cases} \tag{25.120}$$

$$k_1 = \frac{2}{k^{1/2}} \left| \left(\frac{k}{2} \right)^{2/(k-2)} - 1 \right| \tag{25.121}$$

$$k_2 = \frac{2}{k^{1/2}} \tag{25.122}$$

In (25.119), φ_o is the value of φ at $X = X_o$.

Equation (25.51) provides the following implicit relation between N_0 and φ_c:

$$N = \frac{C_{11} + (1/a) + (1/b)}{\varphi_o - \varphi_i} \tag{25.123}$$

If $K_2\sigma_i = \sigma_{an}$, there can be a critical N_0, denoted by $N_{0,c}$, which induces a local minimum in the electrostatic potential located at X_i; that is, the PZC is located at the IPFC. In this case,

$f_1(Y_i) = f_2(Y_i) = 0$ and Y_i can be determined by (25.111) as

$$1 + aC_{11}Y_i^2 + \frac{a}{b}Y_i^k = 0 \tag{25.124}$$

From (25.110), we have

$$K_2 = \frac{X_{an}}{X_i} = 1 - \frac{f_2(Y_i) - f_2(Y_{an})}{X_i} \tag{25.125}$$

Along with the analysis presented earlier, there is only one PZC on which the electrostatic potential possesses a local minimum, and $X_{z,3}(\varphi_{z,3})$ and $X_{z,4}(\varphi_{z,4})$ cannot coexist. In case (B), the PZC is located in region III. From (25.110), the scaled electrical potential on this PZC, $\varphi_{z,3}$, can be calculated by

$$\varphi_{z,3} = -\frac{1}{a}\ln\left[\exp(-a\varphi_c) + \frac{a}{b}\exp(b\varphi_{an}) - \frac{a}{b}\exp(b\varphi_{z,3})\right] \tag{25.126}$$

On the other hand, based on (25.111), the value of Y at $X = X_{z,3}$, $Y_{z,3}$, satisfies

$$1 + aC_{11}Y_{z,3}^2 + \frac{a}{b}Y_{z,3}^k = 0 \tag{25.127}$$

Substituting the solution of (25.127) into (25.110), $X_{z,3}$ can be calculated. In the same way, the PZC in case (A) is located in region IV, and the scaled electrostatic potential on the PZC, $\varphi_{z,4}$, can be evaluated from (25.114) as

$$\varphi_{z,4} = \varphi_i - \left\{\frac{\exp(-a\varphi_c) - \exp(-a\varphi_{z,4})}{a} + \frac{\exp(b\varphi_{an}) - \exp(b\varphi_{z,4})}{b}\right\}\Big/N \tag{25.128}$$

On the other hand, based on (25.115), the value of Y at $X = X_{z,4}$, $Y_{z,4}$, satisfies

$$1 + aC_{22}Y_{z,4}^2 + \frac{a}{b}Y_{z,4}^k + 2NY_{z,4}^k \ln Y_{z,4} = 0 \tag{25.129}$$

Substituting the solution of (25.129) into (25.114), $X_{z,4}$ can be calculated. It is worth to note that since charge does not accumulate as $X \to \pm\infty$, it is not likely to have PZCs in region V in both cases (A) and (B).

3. Positively Charged Membranes

Let us consider a positively charged membrane $(Z > 0)$ under the same conditions described in the previous section. A typical instance is the ammonium group $-NH_3^+$. The electrostatic potentials in regions I and II are described respectively by (25.106) and by (25.107) or (25.108). The electrostatic potential in region III is described by (25.110) with $f_1^{-1/2}$ expressed as

$$f_1 = \left(\left(\frac{b}{a}\right)^{1/2}Y^{-k/2}\left\{1 + \sum_{s=1}^{\infty}\left[(-1)^s\frac{\prod_{i=1}^s(2i-1)}{\prod_{i=1}^s(2i)}\left(bC_{11}Y^{2-k} + \frac{b}{a}Y^{-k}\right)^s\right]\right\}\right)^{-2} \tag{25.130}$$

In region IV, the electrical potential is given by (25.114) with $f_3^{-1/2}$ expressed as

$$f_3 = \left(\left(\frac{b}{a}\right)^{1/2} Y^{-k/2}\left\{1 + \sum_{s=1}^{\infty}\left[(-1)^s \frac{\prod_{i=1}^{s}(2i-1)}{\prod_{i=1}^{s}(2i)}\right.\right.\right.$$
$$\left.\left.\left.\times\left(bC_{22}Y^{2-k} + \frac{b}{a}Y^{-k} + \frac{2b}{a}NY^{2-k}\ln Y\right)^s\right]\right\}\right)^{-2} \tag{25.131}$$

In addition, the signs $\mp$ or $+$ on the right-hand side of (25.115) need to be replaced by $-$ or $\pm$. The negative sign is used for case (A), and $\pm$ sign used for case (B). In the latter, there is a PZC in region IV on which φ possesses a local maximum. If $X_{z,4}$ denotes the location of the PZC, the positive sign needs to be selected from $\pm$ for region $[X_i, X_{z,4}]$ and the negative sign for region $[X_{z,4}, X_o]$. In region V, it can be demonstrated that

$$\varphi = -\frac{2}{b}\ln\left(\frac{1+D_{22}}{1-D_{22}}\right) \tag{25.132}$$

where

$$D_{22} = \left[\tanh\left(-\frac{b\varphi_0}{4}\right)\right]\exp[-k_3(X - X_0)] \tag{25.133}$$

In (25.133), k_3 is the same as that defined in (25.120) except that k is replaced by $k' = 2(1 + a/b)$. The relation between N_0 and φ_c is the same as (25.123). If $K_2\sigma_i = \sigma_{an}$, there is a critical N_0, denoted by $N_{0,c}$, in which both $X_{z,3}$ and $X_{z,4}$ approach X_i; that is, the PZC coincides with IPFC. In this case, Y_i and K can be determined by (25.124) and (25.125), respectively. Since the distribution of space charge is not continuous at $X = X_{ca}$, $X = X_{an}$, $X = X_i$, and $X = X_o$, $d^2\varphi/dX^2$ contains jumps on these planes. φ_i cannot be estimated by setting the right-hand side of (25.40) equal to zero, even if φ–X curve seems to have a saddle point on IPFC. In case (B), φ possesses a local minimum in region III and a local maximum in region IV. $\varphi_{z,3}$, $Y_{z,3}$, $X_{z,3}$, $\varphi_{z,4}$, $Y_{z,4}$, and $X_{z,4}$, can be calculated by (25.110), (25.126), (25.127), (25.128), (25.129), and (25.114), respectively. In case (A), the PZC does not exist unless the whole system remains at electroneutrality as described in (25.47) and (25.56). As X increases, φ monotonically decreases. Because charge does not accumulate as $X \to \pm\infty$, it is impossible to have PZC(s) in region V for both cases (A) and (B).

4. Numerical Scheme

Since the semi-analytical expressions are presented in the previous two sections, the solution to the potential distribution requires numerical analysis. We suggest the following numerical procedure to evaluate the electrostatic potential and to determine the critical condition under which the match for the PZC and the IPFC occurs. Consider first a negatively charged membrane. For a given φ_c, the steps below are taken: [i] φ_{an} is calculated from (25.108) and (25.49), and C_{11} is evaluated by (25.113). [ii] Based on (25.110), (25.111), (25.112), and (25.50), Y_i is calculated by an iterative method, which can be used to estimate φ_i. [iii] For a guessed value of N, denoted by N_g, $\varphi_{o,g}$ is calculated by (25.123), $Y_{o,c}$ evaluated by (25.114), (25.115), (25.116), and (25.51) through an iterative method, and $\varphi_{o,c}$ calculated. If the condition

$$\left|\varphi_{o,g} - \varphi_{o,c}\right| < h_1 \tag{25.134}$$

is satisfied, h_1 being a prespecified small value, $N_g = N$ and $\varphi_{o,g} = \varphi_o$. If (25.134) is not satisfied, return to step [iii] with a newly guessed N_g. These steps are repeated until the criterion (25.134) is satisfied. [iv] To determine the critical condition, Y_i must satisfy (25.124). If (25.124) is satisfied by the present Y_i, K_2 is calculated from (25.125) and $N_0 = N_{0,c}$. If not, return to step [i] with a newly guessed φ_c. This is repeated until (25.124) is satisfied. The PZC is determined by (25.126), (25.127), (25.110), (25.128), (25.129) and (25.114).

The numerical procedures for the estimation of the electrostatic potential and the critical condition for a positively charged membrane are the same as those used for a negatively charged membrane except that (25.112) and (25.116) are replaced, respectively, by (25.130) and (25.131). The procedure for determining the PZC is the same as that applied for a negatively charged membrane.

5. Other Cases

If anions are smaller than cations, that is, $\sigma_{an} < \sigma_{ca}$, the results for a negatively (positively) charged membrane are the same as those for a positively (negatively) charged membrane except that the values of a and b need to be changed with each other. If all the charged species can be viewed as point charges, i.e., σ_{ca}, σ_{an}, $\sigma_i \to 0$, regions I–III vanish and (25.48), (25.49), and (25.50) reduce to (25.47). The present model for ionic sizes reduces to the point charge model for a charged membrane. Since electrolyte ions are much smaller than biological cells, the curvature effect can be neglected near the rigid core of a particle. In this case, the electrical potentials in regions I–III are the same as those for a planar particle, and the analytical expressions for the electrical potential in regions IV, V, and VI reduce to the classic results.

IV. PHENOMENA RELATED TO BIOLOGICAL CELLS

A. CELLULAR MOVEMENT

Suppose that the Newton's second law of motion governs the movement or locomotion of charged biological cells. We have

$$m_c \left(\frac{dv_c}{dt} \right) = F_D + F_H = -\left(\frac{RT}{a_c} \right) \left(\frac{\partial \Phi_D}{\partial H_c} \right) - 6\pi\eta a_c \xi v_c \tag{25.135}$$

where

$$\xi = \frac{4}{3} [\sinh(\beta)] \sum_{i=0}^{\infty} \frac{i(i+1)}{(2i-1)(2i+3)}$$

$$\times \left\{ \frac{2\sinh[(2i+1)\beta] + (2i+1)\sinh(2\beta)}{4\sinh^2[(i+\frac{1}{2})\beta] - (2i+1)^2\sinh^2(\beta)} - 1 \right\} \tag{25.136}$$

$$\beta = \cosh^{-1}(1 + H_c) \tag{25.137}$$

$$H_c = \frac{H}{\kappa a_c} = \frac{X_L - D}{\kappa a_c} \tag{25.138}$$

$$a_c = \frac{X_0 + D}{\kappa} = r_0 + \frac{D}{\kappa} \tag{25.139}$$

$$\Phi_D = \Phi_{el} + \Phi_{vdw} = 2\pi\kappa^3 a_c^3 \int_{H_c}^{\infty} \int_{\ell}^{\infty} \frac{F_R(\ell')}{RT} d\ell' d\ell$$

$$+ \frac{A_{132}}{3RT} \left[\frac{1}{2} \ln\left(\frac{H_c + 2}{H_c} \right) - \frac{H_c + 1}{H_c(H_c + 2)} \right] \tag{25.140}$$

$$\frac{F_R}{an_a^0 k_B T} = \frac{1}{b}(e^{b\varphi} - 1) + \frac{1}{a}(e^{-a\varphi} - 1) - \left(\frac{a+b}{2X_c^2}\right)\left(\frac{d\varphi}{dX_r}\right)^2$$
$$+ w\int_\varphi^0 N_{j,p}\, d\varphi + s\int_\varphi^0 N_{j,n}\, d\varphi \tag{25.141}$$

$$X_r = \frac{r}{r_c} \tag{25.142}$$

In these expressions, m_c, v_c, F_D, F_H, X_r, a_c, H_c, H, t, η, ξ, ρ, R, Φ_D, Φ_{vdw}, and A_{132} are, respectively, the mass and velocity of the cells, the DLVO force, the hydrodynamic force, the scaled distance, the linear radius of the cells, the dimensionless closest half surface-to-surface distance between two cells, the scaled half separation distance between two cells, the time, the viscosity of liquid phase, the hydrodynamic retardation factor, the density of the cells, the gas constant, the scaled DLVO potential, the scaled van der Waals potential, and the Hamaker constant of the system, and i, ℓ, and ℓ' are dummy variables. Note that the fourth and fifth integral terms on the right-hand side of (25.141) represent, respectively, the contribution to the electrostatic repulsion force when the fixed positive and negative charge in the membrane phase of a cell appears. Equation (25.135) can be rewritten to become

$$\frac{d^2 H_c}{dt^2} + \frac{9\eta\xi}{2a_c^2\rho}\frac{dH_c}{dt} + \frac{3RT}{4\pi a_c^5\rho}\frac{\partial\Phi_D}{\partial H_c} = 0 \tag{25.143}$$

B. Electrophoresis

The electrophoresis of biological cells occurs when an electrical field was applied to the system where the cells exist. The fluid flow around the biological cells in the radial direction is governed by the following Navier–Stokes equation.

$$\eta\frac{d^2 u_f}{dr^2} - wfu_f + \rho_c(r)E = 0, \quad 0 \le r \le \infty \tag{25.144}$$

where u_f and f are, respectively, the velocity distribution of surrounding fluid and the friction factor of the cell membrane, and E the electrical strength of the applied external field. The third term on the left-hand side of (25.144) represents the electrical driving force for fluid flow, and ρ_c is related to the electrostatic potential distribution. Equation (25.144) can be rewritten to become

$$\frac{d^2 U}{dX^2} - w\lambda^2 U = \frac{\widehat{L}}{a+b}[-u\exp(-a\varphi) + v\exp(b\varphi)] \tag{25.145}$$

where

$$U = \frac{u_e}{U_0} \tag{25.146}$$

$$\lambda^2 = \frac{f}{\eta\kappa^2} \tag{25.147}$$

$$\widehat{L} = \frac{\varepsilon_0\varepsilon_{r,dl}k_B TE}{\eta e U_0} \tag{25.148}$$

where U_0 is the electrophoretic velocity of the cell. The boundary conditions associated with (25.145) are

$$U = 0 \quad \text{as} \quad X \longrightarrow 0 \tag{25.149}$$

$$\left(\frac{dU}{dX}\right)_{X=X_{ca}^-} = \left(\frac{dU}{dX}\right)_{X=X_{ca}^+} \quad \text{and} \quad U(X_{ca}^-) = U(X_{ca}^+) \tag{25.150}$$

$$\left(\frac{dU}{dX}\right)_{X=X_{an}^-} = \left(\frac{dU}{dX}\right)_{X=X_{an}^+} \quad \text{and} \quad U(X_{an}^-) = U(X_{an}^+) \tag{25.151}$$

$$\left(\frac{dU}{dX}\right)_{X=X_i^-} = \left(\frac{dU}{dX}\right)_{X=X_i^+} \quad \text{and} \quad U(X_i^-) = U(X_i^+) \tag{25.152}$$

$$\left(\frac{dU}{dX}\right)_{X=D^-} = \left(\frac{dU}{dX}\right)_{X=D^-} \quad \text{and} \quad U(D^-) = U(D^+) \tag{25.153}$$

$$U \longrightarrow -1 \quad \text{and} \quad \frac{dU}{dX} \longrightarrow 0 \quad \text{as} \quad X \longrightarrow \infty \tag{25.154}$$

Here, (25.149) and (25.154) are chosen for a flow field of a stationary cell. The definition of the electrophoretic mobility of biological cells, μ, is defined by

$$\mu = \frac{U_0}{E} \tag{25.155}$$

C. STABILITY OF CELL SUSPENSION

1. Stability Ratio

One of the most significant and fundamental quantities characterizing a cell suspension is the stability ratio that is closely related to the interactions between two biological cells and measures the effectiveness of potential energy barrier in preventing biological cells from coagulation. From the kinetic point of view, W is defined by (rate of rapid coagulation)/(rate of slow coagulation) or (frequency of collisions between cells)/(frequency of collisions leading to coagulation) [73]. The numerator of the above definition can be obtained by the Smoluchowski theory for Brownian coagulation in the absence of a potential barrier, while the expression for the denominator, which includes the term of interaction potential energy, can be calculated by the Fuchs theory for stability ratio of a suspension. The stability ratio of a cell suspension can be evaluated by

$$W = \int_1^\infty \frac{1}{X_s^2} \exp(\Phi_D) \, dX_s \tag{25.156}$$

where

$$X_s = X/2(X_c + D) \tag{25.157}$$

$$X_c = \frac{r}{a_c} \tag{25.158}$$

2. Critical Coagulation Concentration

CCC is one of the most significant characteristics of cell dispersion. The experimental observations for inorganic colloidal suspension reveal that the variation of CCC as a function of the valence of counterion follows roughly the inverse sixth power law, the classic Schulze–Hardy rule [74,75]. For negatively charged colloids, the CCC ratio of cations of valences 3, 2, and 1 is $3^{-6}:2^{-6}:1^{-6}$, or roughly 1:11:729. Based on the DLVO theory, which considered the electrical repulsive force and the van der Waals attractive force between two colloids, the Schulze–Hardy rule was successfully deduced. The original derivation of the Schulze–Hardy rule is based on planar hard colloidal model having a high constant surface potential and symmetric electrolyte solution. When biological cells are considered, the interaction force and potential energy can be modified as presented in (25.111) and (25.140). The condition for critical coagulation concentration of a cell suspension is described below.

$$\Phi_D = 0 \quad \text{and} \quad \frac{d\Phi_D}{dX_m} = 0 \tag{25.159}$$

Equation (25.159) can be used to evaluate the CCC of counterions.

D. Deposition

The gravitational sedimentation is usually considered as the most important contribution to the cellular transport from a stagnant solution. However, the interaction forces involved in cellular adhesion on biomaterial surfaces include the gravitational force, the van der Waals attraction force, the electrostatic repulsion force, the short-range repulsion force such as the Born repulsion, and others. The first three forces, however, offered sufficient and qualitative explanations for experimental results [76–78]. Therefore, it can assume that the scaled total energy, Φ_T, comprises the scaled double-layer energy, the scaled van der Waals energy, and the scaled gravitational energy, Φ_g. That is,

$$\Phi_T = \Phi_D + \Phi_g = \Phi_{el} + \Phi_{vdw} + \Phi_g \tag{25.160}$$

whre

$$\Phi_g = \frac{4}{3k_BT} \pi a_c^3 \Delta\rho g h \tag{25.161}$$

In (25.161), $\Delta\rho$ is the difference between the density of cell and that of the surrounding fluid, g the gravitational constant and h the minimum distance between the cell membrane and a collector surface.

Due to the sedimentation of biological cells from the bulk liquid phase and the deposition of biological cells to a charged surface, the number of cells at the secondary minimum, n_2, varies with time. The former can be viewed as the input of cells into the secondary minimum while the latter the output of cells from the secondary minimum. The temporal variation of n_2 can be described by

$$\frac{dn_2}{dt} = \frac{2a_c^2\Delta\rho g B}{9\eta}\left(1 - \frac{n_2}{n_{2m}}\right)^5 - \frac{dn_1}{dt} \tag{25.162}$$

where n_1 and n_{2m} are, respectively, the number concentration of cells on the charged surface, and that at the secondary minimum corresponding to the monolayer coverage, and B the initial number concentration of cells in the suspension. The rate of deposition is not only proportional to the number of cells at the secondary minimum but also depends on the probability for a cell possessing the sufficient energy to overcome the potential barrier. The temporal variation of n_1 can be described by

$$\frac{dn_1}{dt} = n_2 k_B T \left[\left(-\frac{d^2\Phi_T}{dh^2}\bigg|_{h_{max}} \frac{d^2\Phi_T}{dh^2}\bigg|_{h_{min}} \right)^{0.5} \frac{h_{max}}{12\eta\pi^2 a_c^2} \right] \exp(-\Delta\Phi_T) \tag{25.163}$$

where $\Delta\Phi_T = \Delta\Phi_{T,max} - \Delta\Phi_{T,min}$, h_{max}, h_{min}, $\Phi_{T,max}$, and $\Phi_{T,min}$ being, respectively, the distance from the collector surface to the position of the primary maximum, that to the position of the secondary minimum, the scaled total energy at the primary maximum, and that at the secondary minimum. The initial conditions associated with (25.162) and (25.163) are assumed as

$$n_1 = 0, \qquad t = 0 \tag{25.164}$$

$$n_2 = n_{2m} = 0.35 n_m, \qquad t = 0 \tag{25.165}$$

In these expressions, n_m is the number concentration of biological cells on the collector surface for compact monolayer coverage.

NOTATIONS

A	parameter for nonuniformly distributed functional groups $(-)$
A_1	parameter describing relation between φ_c and X_m $(-)$
A_2	parameter describing relation between φ_c and X_m $(-)$
A_{132}	Hamaker constant (kg m^2/s^2)
a	valence of cations $(-)$
a_1	valence of cations $(-)$
a_2	valence of cations $(-)$
a_c	linear radius of biological cells (m)
B	initial number concentration of cells in suspension (m^{-3})
B_1	parameter describing relation between φ_c and X_m $(-)$
B_2	parameter describing relation between φ_c and X_m $(-)$
b	valence of anions $(-)$
C_1	parameter describing low potential distribution in region II $(-)$
C_{10}	parameter describing relation between φ_m and X_m for low potential case $(-)$
C_{11}	parameter for expression of electrostatic potential in region III $(-)$
C_2	parameter describing low potential distribution in region II $(-)$
C_{22}	parameter for expression of electrostatic potential in region IV $(-)$
C_3	parameter describing low potential distribution in region III $(-)$
C_4	parameter describing low potential distribution in region III $(-)$
C_5	parameter describing low potential distribution in region IV $(-)$
C_6	parameter describing low potential distribution in region IV $(-)$
C_7	parameter describing low potential distribution in regions V and VI $(-)$
C_8	parameter describing low potential distribution in regions V and VI $(-)$
C_9	parameter describing relation between N and φ_c for low potential case $(-)$
C_H^+	concentration of protons (mol/m^3)
$C_{H^+}^0$	bulk concentration of protons (mol/m^3)
c	valence of original fixed groups $(-)$
D	scaled membrane thickness $(-)$
D_1	parameter for explicit expression of Φ_{el} for the case of constant potential $(-)$
D_{11}	parameter for expression of electrostatic potential in regions V and VI $(-)$
D_{22}	parameter for expression of electrostatic potential in regions V and VI $(-)$
E	electrical strength of the applied external field (V/m)
E_1	parameter for explicit expression of Φ_{el} for the case of constant potential $(-)$
E_a	apparent scaled electric field due to the presence of cell membrane (V/m)

e	elementary charge (C)
F_1	parameter describing relation between φ_c and X_m (−)
F_2	parameter describing relation between φ_c and X_m (−)
F_D	DLVO force (kg m/s^2)
F_H	hydrodynamic force (kg m/s^2)
F_R	scaled electrostatic repulsion force (−)
f	friction factor of the cell membrane (−)
f_1	parameter for expression of electrostatic potential in region III (−)
f_2	parameter for expression of electrostatic potential in region III (−)
f_3	parameter for expression of electrostatic potential in region IV (−)
f_4	parameter for expression of electrostatic potential in region IV (−)
G_1	parameter describing relation between φ_c and X_m (−)
G_2	parameter describing relation between φ_c and X_m (−)
g	gravitational constant (m/s^2)
H	scaled separation distance (−)
H_1	parameter for explicit expression of Φ_{el} for the case of constant charge (−)
H_c	scaled closest half distance between two cell membranes (−)
h	minimum distance between cell membrane and collector surface (m)
h_1	prespecified small value for convergence of solution (−)
h_{max}	distance from collector surface to the position of primary maximum (m)
h_{min}	distance from collector surface to the position of secondary minimum (m)
i	dummy variable for expression of hydrodynamic retardation factor, for equilibrium constants of successive dissociation of functional groups, for explicit expression of Φ_{el} or for electrostatic potential distribution in regions III and IV (−)
j	distribution type index of fixed functional groups
K	equilibrium constant for dissociation of cation-functional group complex (−)
K_1	parameter for explicit expression of Φ_{el} for the case of constant charge (−)
K_2	radius ratio of anion to fixed charge (−)
$K_{a,p}$	equilibrium constants for acidic group dissociation (mol/m^3)
$K_{b,p}$	equilibrium constants for basic group dissociation (mol/m^3)
k	valance parameter for $a:b$ asymmetric electrolyte (−)
k_1	valance parameter for electrostatic potential (−)
k_2	valance parameter for electrostatic potential (−)
k_3	combined valance parameter for the expression of electrostatic potential in regions V and VI (−)
k_B	Boltzmann constant (kg m^2/s^2 K)
L	scaled distance between cell membrane and charged hard wall (−)
$\tilde{L}$	scaled external electrical field (−)
L_c	scaled characteristic length of a cylindrical cell (−)
ℓ	dummy variable for electrostatic interaction force integration (−)
ℓ'	dummy variable for electrostatic interaction force integration (−)
m	number of cations involved in each absorption (−)
m_c	cell mass (kg)
N	scaled concentration of fixed charge (−)
N_0	total concentration of fixed functional groups (mol/m^3)
$N_{0,a}$	average concentration of fixed functional groups (mol/m^3)
$N_{0,c}$	critical fixed charge density (mol/m^3)

N_A	Avogadro's number $(-)$
$N_a^{i=1}$	overall concentration of acidic groups (mol/m^3)
$N_b^{i=1}$	overall concentration of basic groups (mol/m^3)
N_g	initially guessed N value for numerical calculation $(-)$
N_j	scaled concentration of fixed functional groups $(-)$
$N_{j,n}$	scaled concentration of fixed negative charge (mol/m^3)
$N_{j,p}$	scaled concentration of fixed positive charge (mol/m^3)
$N_+^{i=1}$	concentration of positive fixed charge (mol/m^3)
$N_-^{i=1}$	concentration of negative fixed charge (mol/m^3)
n	number of fixed groups involved in each absorption $(-)$
n_1	number concentration of cells on charged surface (m^{-3})
n_2	number concentration of cells at secondary minimum (m^{-3})
n_{2m}	number concentration of cells at secondary minimum corresponding to monolayer coverage (m^{-3})
n_a	number concentrations of cation (m^{-3})
n_a^0	number concentration of cations in the bulk liquid phase (m^{-3})
$n_{a_1}^0$	number concentration of cation with valence a_1 in bulk liquid phase (m^{-3})
$n_{a_2}^0$	number concentration of cation with valence a_2 in bulk liquid phase (m^{-3})
n_b	number concentrations of anion (m^{-3})
n_b^0	number concentration of anions in the bulk liquid phase (m^{-3})
n_m	number concentration of cells on the collector surface for compact monolayer coverage (m^{-3})
p	number of dissociable or absorbable protons in functional groups $(-)$
P_p	parameter for successive dissociation of fixed ionogenic groups $(-)$
P_j	alternative expression for the scaled concentration of fixed functional groups $(-)$
Q_p	parameter for successive dissociation of fixed ionogenic groups $(-)$
Q_{sl}	net penetration charge per unit area of membrane–liquid interface (C/m^2)
Q_t	total charge per unit area of membrane–liquid interface (C/m^2)
q	valence a or b
R	gas constant $(\text{kg m}^2/\text{s}^2\,\text{K})$
r	distance measured from the core-membrane interface (m)
r_c	linear radius of cellular core (m)
s	dummy variable for equilibrium constants of successive dissociation of functional groups, for distribution of functional groups in curvilinear membrane or for electrostatic potential distribution in regions III and IV $(-)$
T	absolute temperature (K)
t	time (s)
U	scaled fluid velocity $(-)$
U_0	scaled electrophoretic velocity $(-)$
u	region index for the presence/absence of cations $(-)$
u_f	velocity distribution of surrounding fluid (m/s)
v	region index for the presence/absence of anions $(-)$
v_c	cell velocity (m/s)
W	stability ratio of cell suspension $(-)$
w	region index for the presence/absence of fixed charge or fixed positive charge $(-)$
X	scaled distance measured from core-membrane interface $(-)$
X_{an}	scaled effective size of anion $(-)$
$X_{an'}$	starting position of charge-free region close to charged surface $(-)$
X_c	scaled radius of cellular core $(-)$
X_{ca}	scaled effective size of cation $(-)$
$X_{ca'}$	starting position close to charged surface for excluding anions $(-)$

X_i	scaled size of fixed group (location of IPFC) $(-)$
X_L	scaled distance between cellular core and charged rigid surface $(-)$
X_m	location of mid-plane between two biological cells $(-)$
X_o	location of OPFC $(-)$
X_r	scaled distance based on hard-core radius $(-)$
X_s	scaled distance $(-)$
$X_{Z,3}$	location of PZC in region III $(-)$
$X_{Z,4}$	location of PZC in region IV $(-)$
x	region index for the presence/absence of fixed positive charge $(-)$
Y	exponential form of scaled electrostatic potential $(-)$
Y_{an}	Y value at $X = X_{an}$ $(-)$
Y_c	Y value in cellular core $(-)$
Y_i	Y value on IPFC $(-)$
Y_o	Y value on OPFC $(-)$
$Y_{o,c}$	calculated Y_o value $(-)$
$Y_{Z,3}$	Y value at $X = X_{Z,3}$ $(-)$
$Y_{Z,4}$	Y value at $X = X_{Z,4}$ $(-)$
Z	valence of fixed charge in membrane $(-)$
Z_a	number of dissociable protons of acidic group $(-)$
Z_b	number of dissociable protons of basic group $(-)$
z	valence of cations-absorbed fixed groups $(-)$

Greek Symbols

α	nonuniform feature index for radial distribution of fixed functional groups $(-)$
β	parameter related to hydrodynamic retardation factor $(-)$
ε	permittivity of the system $(C/V\ m)$
ε_0	permittivity of a vacuum $(C/V\ m)$
ε_r	relative permittivity $(-)$
$\varepsilon_{r,c}$	relative permitivities of cellular hard core $(-)$
$\varepsilon_{r,dl}$	relative permitivities of liquid $(-)$
$\varepsilon_{r,m}$	relative permitivities of membrane $(-)$
η	viscosity of liquid solution (kg/ms)
κ	reciprocal of Debye screening length (m^{-1})
λ	scaled friction factor of the cell membrane $(-)$
μ	electrophoretic mobility (m^2/Vs)
ξ	hydrodynamic retardation factor $(-)$
ρ	cell density (kg/m^3)
ρ_c	space charge density (C/m^3)
σ_{an}	effective diameters of anion (m)
σ_{ca}	effective diameters of cation (m)
σ_i	effective diameters of fixed charge (m)
τ	curvature index $(-)$
Φ_D	scaled DLVO potential $(-)$
Φ_{el}	scaled double-layer potential $(-)$
Φ_g	scaled gravitational potential $(-)$
Φ_T	scaled total potential $(-)$
$\Phi_{T,max}$	scaled total energy at primary maximum $(-)$
$\Phi_{T,min}$	scaled total energy at secondary minimum $(-)$

Φ_{vdw}	scaled van der Waals potential $(-)$
ϕ	electrostatic potential (V)
ϕ_m	mid-plane electrostatic potential between two identical cellular membranes (V)
φ	scaled electrostatic potential distribution $(-)$
φ_{an}	scaled electrostatic potential at $X = X_{an}$ $(-)$
φ_c	scaled electrostatic potential in cellular core or the scaled membrane potential $(-)$
φ_i	scaled electrostatic potential on IPFC $(-)$
φ_L	scaled electrostatic potential at charged rigid surface $(-)$
φ_m	scaled electrostatic potential on the mid-plane between two identical cellular membranes $(-)$
φ_{ml}	scaled electrostatic potential at membrane−liquid interface $(-)$
φ_o	scaled electrostatic potential on OPFC $(-)$
$\varphi_{o,c}$	calculated φ_o value $(-)$
$\varphi_{o,g}$	initially guessed φ_o value for numerical calculation $(-)$
$\varphi_{Z,3}$	scaled electrostatic potential at $X = X_{Z,3}$ $(-)$
$\varphi_{Z,4}$	scaled electrostatic potential at $X = X_{Z,4}$ $(-)$
χ	charge portion of cation with valence a_2 in bulk liquid phase

Superscript

$i = 1$	membrane phase $(-)$

Subscript

$+$	positive charge $(-)$
$-$	negative charge $(-)$
a	acidic groups
b	basic groups

Abbreviations

CCC	critical coagulation concentration
DEDL	diffuse electrical double layer
DLVO	Derjaguin−Landau−Verwey−Overbeek
GCT	Göuy−Chapman theory
IPFC	inner plane of fixed charge
MGCT	modified Göuy−Chapman theory
OPFC	outer plane of fixed charge
PBE	Poisson−Boltzmann equation
PCM	point-charge model
PZC	plane of zero charge

REFERENCES

1. Hunter, R.J., *Foundations of Colloid Science*, vol. I, Oxford University Press, Oxford, 1989, pp. 1−5.
2. Hiemenz, P.C., *Principles of Colloid and Surface Chemistry*, 2nd ed., Marcel Dekker, New York, 1986, pp. 59−110.

3. Hanson, S.R. and Harker, L.A., Blood coagulation and blood-materials interactions, in *Biomaterials Science: An Introduction to Materials in Medicine*, Ratner, B.D., Hoffman, A.S., Schoen, F.J., and Lemons, J.E., Eds., Academic Press, San Diego, 1996, pp. 193–199.

4. Cranin, A.N., Sirakian, A., and Klein, M., Dental implantation, in *Biomaterials Science: An Introduction to Materials in Medicine*, Ratner, B.D., Hoffman, A.S., Schoen, F.J., and Lemons, J.E., Eds., Academic Press, San Diego, 1996, pp. 426–435.

5. Horbett, T.A., Proteins: structure, properties, and adsorption to surfaces, in *Biomaterials Science: An Introduction to Materials in Medicine*, Ratner, B.D., Hoffman, A.S., Schoen, F.J., and Lemons, J.E., Eds., Academic Press, San Diego, 1996, pp. 133–141.

6. van Aken, G.A., Lekerkerker, H.N.W., Overbeek, J.Th.G., and De Bruyn, P.L., Adsorption of monovalent ions in thin spherical and cylindrical diffuse electrical double layers, *J. Phys. Chem.*, 94 (22), 8468–8472, 1990.

7. Hsu, J.P. and Kuo, Y.C., Approximate analytical expressions for the properties of an electrical double layer with asymmetric electrolytes, *J. Chem. Soc. Faraday Trans.*, 89 (8), 1229–1233, 1993.

8. Kuo, Y.C. and Hsu, J.P., Electrical properties of a charged surface in a general electrolyte solution, *Chem. Phys.*, 236 (1), 1–14, 1998.

9. Shaw, D.J., *Introduction to Colloid and Surface Chemistry*, 4th ed., Butterworth-Heinemann, Oxford, UK, 1992, pp. 177–184.

10. Myers, D., *Surfaces, Interfaces, and Colloids: Principles and Applications*, 2nd ed., John Wiley & Sons, New York, 1999, 85–88.

11. Rasaiah, J.C. and Friedman, H.L., Integral equation methods in the computation of equilibrium properties of ionic solutions, *J. Chem. Phys.*, 48 (6), 2742–2752, 1968.

12. Stell, G. and Lebowitz, J.L., Equilibrium properties of a system of charged particles, *J. Chem. Phys.*, 49 (8), 3706–3717, 1968.

13. Lebowitz, J.L. and Percus, J.K., Mean spherical model for lattice gases with extended hard cores and continuum fluids, *Phys. Rev.*, 144 (1), 251–258, 1966.

14. Friedrichs, M., Zhou, R., Edinger, S.R., and Friesner, R.A., Poisson–Boltzmann analytical gradients for molecular modeling calculations, *J. Phys. Chem. B*, 103 (16), 3057–3061, 1999.

15. Hsu, J.P. and Kuo, Y.C., Approximate analytical expressions for the properties of an electrical double layer with asymmetric electrolytes: cylindrical and spherical geometries, *J. Coll. Interf. Sci.*, 167 (1), 35–46, 1994.

16. Kuo, Y.C. and Hsu, J.P., Electrical properties of charged cylindrical and spherical surfaces in a general electrolyte solution, *Langmuir*, 15 (19), 6244–6255, 1999.

17. Overbeek, J.Th.G., The rule of Schulze and Hardy, *Pure Appl. Chem.*, 52, 1151–1161, 1980.

18. Healy, T.W., Chan, D., and White, L.R., Colloidal behavior of materials with ionizable group surfaces, *Pure Appl. Chem.*, 52, 1207–1219, 1980.

19. Graham, D.C., Diffuse double layer theory for electrolytes of unsymmetrical valence types, *J. Chem. Phys.*, 21 (6), 1054–1060, 1953.

20. Ninham, B.W. and Parsegian, V.A., Electrostatic interaction between surfaces bearing ionizable groups in ionic equilibrium with physiologic saline solution, *J. Theor. Biol.*, 31, 405–428, 1971.

21. Glendinning, A.B. and Russel, W.B., The electrostatic repulsion between charged spheres from exact solution to the linearized Poisson–Boltzmann equation, *J. Coll. Interf. Sci.*, 93 (1), 95–104, 1983.

22. Blum, L., Theory of electrified interfaces, *J. Phys. Chem.*, 81 (2), 136–147, 1977.

23. Henderson, D. and Blum, L., Some exact results and the application of the mean spherical approximation to charged hard spheres near a charged hard wall, *J. Chem. Phys.*, 69 (12), 5441–5449, 1978.

24. Valleau, J.P., Ivkov, R., and Torrie, G.M., Colloid stability: the forces between charged surfaces in an electrolyte, *J. Chem. Phys.*, 95 (1), 520–532, 1991.

25. Valleau, J.P. and Torrie, G.M., The electrical double layer. III. Modified Göuy–Chapman theory with unequal ion sizes, *J. Chem. Phys.*, 76 (9), 4623–4630, 1982.

26. Bhuiyan, L.B., Blum, L., and Henderson, D., The application of the modified Göuy–Chapman theory to electrical double layer containing asymmetric ions, *J. Chem. Phys.*, 78 (1), 442–445, 1983.

27. Torrie, G.M. and Valleau, J.P., A Monte Carlo study of an electrical double layer, *Chem. Phys.*, 65 (2), 343–346, 1979.

28. Sloth, P. and Sorensen, T.S., Hard, charged spheres in spherical pores. grand canonical ensemble Monte Carlo calculations, *J. Chem. Phys.*, 96 (1), 548–554, 1992.

29. Sorensen, T.S. and Sloth, P., Ion and potential distribution in charged and non-charged primitive spherical pores in equilibrium with primitive electrolyte solution calculated by grand canonical ensemble Monte Carlo simulation, *J. Chem. Soc. Faraday Trans.*, 88 (4), 571–589, 1992.

30. Kawahata, S., Ohshima, H., Muramatsu, N., and Kondo, T., Charge distribution in the surface region of human erythrocytes as estimated from electrophrotic mobility data, *J. Coll. Interf. Sci.*, 138 (1), 182–186, 1990.

31. Ohshima, H. and Ohki, S., Donnan potential and surface potential of a charged membrane, *Biophys. J.*, 47 (1), 673–678, 1985.

32. Ohshima, H., Makino, K., and Kondo, T., Potential distribution across a membrane with surface charge layers: effects of nonuniform charge distribution, *J. Coll. Interf. Sci.*, 113 (2), 673–678, 1986.

33. Aguilella, V., Mafe, S., and Manzanareas, J., Double layer potential and degree of dissociation in charged lipid monolayers, *Chem. Phys. Lipids*, 105 (2), 225–229, 2000.

34. Aguilella, V., Belaya, M., and Levadny, V., Ion transport through membranes with soft interfaces. the influence of the polar zone thickness, *Thin Solid Film*, 272 (1), 10–14, 1996.

35. Martini, M.F. and Disalvo, E.A., Effect of polar head groups on the activity of aspartyl protease adsorbed to lipid membranes, *Chem. Phys. Lipid*, 122 (1–2), 177–183, 2003.

36. Martini, M.F. and Disalvo, E.A., Influence of electrostatic charges and non-electrostatic components on the adsorption of an aspartyl protease to lipid interfaces, *Coll. Surf. B*, 22 (3), 219–226, 2001.

37. Bakas, L.S., Saint-Pierre Chazalet, M., Bernik, D.L., and Disalvo, E.A., Interaction of an acid protease with positively charged phosphatidylcholine bilayers, *Coll. Surf. B*, 12 (2), 77–87, 1998.

38. Ohshima, H., A model for the electrostatic interaction of cells, *J. Theor. Biol.*, 65, 523–527, 1977.

39. Ohshima, H. and Kondo, T., Electrostatic repulsion of ion penetrable charged membranes: role of Donnan potential, *J. Theor. Biol.*, 128, 187–194, 1987.

40. Ohshima, H. and Kondo, T., Double-layer interaction regulated by the Donnan potential, *J. Coll. Interf. Sci.*, 123 (1), 136–1142, 1988.

41. Ohshima, H. and Kondo, T., pH dependence of electrostatic interaction between ion-penetrable membranes, *Biophys. Chem.*, 32 (3), 161–166, 1988.

42. Ohshima, H. and Kondo, T., Numerical data on the double-layer interaction between ion-penetrable membranes regulated by the Donnan potential, *J. Coll. Interf. Sci.*, 133 (2), 523–526, 1989.

43. Hsu, J.P. and Kuo, Y.C., An algorithm for the calculation of the electrostatic potential distribution of an ion-penetrable membrane carrying fixed charges, *J. Coll. Interf. Sci.*, 171 (2), 483–489, 1995.

44. Hsu, J.P. and Kuo, Y.C., An algorithm for the calculation of the electrostatic repulsion between surfaces coated with a charged membrane, *Coll. Polym. Sci.*, 273 (9), 881–885, 1995.

45. Hsu, J.P. and Kuo, Y.C., Approximate analytical expressions for the properties of a double layer with asymmetric electrolytes: ion-penetrable charged membranes, *J. Coll. Interf. Sci.*, 166 (1), 208–214, 1994.

46. Kuo, Y.C. and Hsu, J.P., Exact solution of the linearized Poisson–Boltzmann equation: ion-penetrable membrane bearing fixed charges, *J. Chem. Phys.*, 102 (4), 1806–1815, 1995.

47. Hsu, J.P. and Kuo, Y.C., Properties of a double layer with asymmetric electrolytes: cylindrical and spherical particles with an ion-penetrable membrane, *J. Coll. Interf. Sci.*, 171 (2), 331–339, 1995.

48. Hsu, J.P. and Kuo, Y.C., Solution to the Poisson–Boltzmann equation for particles covered by an ion-penetrable charged membrane, *J. Chem. Soc. Faraday Trans.*, 91 (8), 1223–1228, 1995.

49. Hsu, J.P. and Kuo, Y.C., Properties of a double layer with asymmetric electrolytes: ion-penetrable membrane carrying nonuniformly distributed fixed charges, *J. Membrane Sci.*, 108 (1–2), 107–119, 1995.

50. Hsu, J.P. and Kuo, Y.C., Net penetration charges of an ion-penetrable membrane in a general electrolyte solution, *J. Coll. Interf. Sci.*, 176 (1), 256–263, 1995.

51. Hsu, J.P. and Kuo, Y.C., Interactions between a charge-regulated particle and a charged surface, *J. Chem. Soc. Faraday Trans.*, 91 (22), 4093–4097, 1995.

52. Hsu, J.P. and Kuo, Y.C., Electrostatic interactions between particles with an ion-penetrable charged membrane, *J. Chem. Phys.*, 103 (1), 465–473, 1995.

53. Hsu, J.P. and Kuo, Y.C., Critical coagulation concentration of cations for negatively charged particles with an ion-penetrable surface layer, *J. Coll. Interf. Sci.*, 174 (1), 250–257, 1995.

54. Hsu, J.P. and Kuo, Y.C., The stability ratio for a dispersion of particles covered by an ion-penetrable charged membrane, *J. Coll. Interf. Sci.*, 183 (1), 184–193, 1996.

55. Hsu, J.P. and Kuo, Y.C., Adsorption of a charge-regulated cell to a charged surface, *Langmuir*, 13 (16), 4372–4376, 1997.

56. Kuo, Y.C. and Hsu, J.P., and Chen, D.F., Deposition of biocolloids on a charged collector surface: an ion-penetrable membrane model, *Langmuir*, 17 (11), 3466–3471, 2001.

57. Kuo, Y.C., Hsieh, M.Y., and Hsu, J.P., Deposition of charge-regulated biocolloids on a charged surface, *J. Phys. Chem. B*, 106 (16), 4255–4260, 2002.

58. Hsu, J.P. and Kuo, Y.C., Modified Göuy–Chapman theory for ion-penetrable charged membranes, *J. Chem. Phys.*, 111 (10), 4807–4816, 1999.

59. Kuo, Y.C. and Hsu, J.P., Double-layer properties of an ion-penetrable charged membrane: effect of sizes of charged species, *J. Phys. Chem. B*, 103 (44), 9743–9748, 1999.

60. Kuo, Y.C. and Hsu, J.P., Electrostatic interactions between ion-penetrable charged membranes: effect of sizes of charged species, *Langmuir*, 16 (15), 6233–6239, 2000.

61. Kuo, Y.C., Hsieh, M.Y., and Hsu, J.P., Interactions between a particle covered by an ion-penetrable charged membrane and a charged surface: a modified Göuy–Chapman theory, *Langmuir*, 18 (7), 2789–2794, 2002.

62. Huang, S.W., Hsu, J.P., Kuo, Y.C., and Tseng, S.J., Effect of ionic sizes on the electrophoretic mobility of a particle with a charge-regulated membrane in a general electrolyte solution, *J. Phys. Chem. B*, 106 (8), 2117–2122, 2002.

63. Kuo, Y.C., Adsorption of a biocolloid onto a rigid charged surface: characteristics of functional group, membrane and electrolytic solution, *J. Chin. Inst. Chem. Eng.*, 34 (2), 117–186, 2003.

64. Kuo, Y.C., Effects of particulate curvatures and sizes of charged species on the adsorption of biocolloids bearing bnonuniformly distributed fixed charges, *J. Chem. Phys.*, 118 (17), 8023–8032, 2003.

65. Hsu, J.P., Huang, S.W., Kuo, Y.C., and Tseng, S.J., Effect of ionic sizes on critical coagulation concentration: particle covered by a charge-regulated membrane, *J. Phys. Chem. B*, 106 (16), 4269–4275, 2002.

66. Hsu, J.P., Huang, S.W., Kuo, Y.C., and Tseng, S.J., Stability of a dispersion of particles covered by a charge-regulated membrane: effect of the sizes of charged species, *J. Coll. Interf. Sci.*, 262 (1), 73–80, 2003.

67. Kuo, Y.C., Application of a modified Göuy–Chapman theory to the stability of a biocolloidal suspension, *Langmuir*, 19 (14), 5942–5948, 2003.

68. Kuo, Y.C., Effects of sizes of charged species on the flocculation of biocolloids: absorption of cations in the membrane layer, *J. Chem. Phys.*, 118 (1), 398–406, 2003.

69. Hsu, J.P. and Kuo, Y.C., The electrostatic interaction force between a charge-regulated particle and a rigid surface, *J. Coll. Interf. Sci.*, 183 (1), 194–198, 1996.

70. Overbeek, J.Th.G., Double-layer interaction between spheres with unequal surface potential, *J. Chem. Soc. Faraday Trans.*, 1, 84 (9), 3079–3091, 1988.

71. Kuo, Y.C. and Hsu, J.P., An algorithm for the calculation of the electrostatic forces between identical-charged surfaces in electrolytes, *J. Coll. Interf. Sci.*, 156 (1), 250–252, 1993.

72. Hsu, J.P. and Kuo, Y.C., Approximate analytical expression for surface potential as a function of surface charge density, *J. Coll. Interf. Sci.*, 170 (1), 220–228, 1995.

73. Overbeek, J.Th.G., Recent developments in the understanding of colloid stability, *J. Coll. Interf. Sci.*, 58 (2), 408–422, 1977.

74. Hsu, J.P. and Kuo, Y.C., An extension of the Schulze–Hardy rule to asymmetric electrolytes, *J. Coll. Interf. Sci.*, 171 (1), 254–255, 1995.

75. Hsu, J.P. and Kuo, Y.C., The critical coagulation concentration of counterions: spherical particles in asymmetric electrolyte solutions, *J. Coll. Interf. Sci.*, 185 (2), 530–537, 1997.

76. Ruckenstein, E. and Prieve, D.C., Dynamics of cell deposition on surfaces, *J. Theor. Biol.*, 51, 429–438, 1975.

77. Ruckenstein E., Marmur, A., and Gill, W.N., Coverage dependent rate of cell deposition, *J. Theor. Biol.*, 58, 439–454, 1976.

78. Ruckenstein E. and Chen, J.H., Kinetically caused saturation in the deposition of cells — effects of saturation at the secondary minimum and excluded area, *J. Coll. Interf. Sci.*, 128 (2), 592–601, 1989.

26 Modeling Mesoscopic Fluids with Discrete-Particles — Methods, Algorithms, and Results

Witold Dzwinel, Krzysztof Boryczko
AGH University of Science and Technology, Kraków, Poland

David A. Yuen
University of Minnesota, Minneapolis, USA

CONTENTS

I. INTRODUCTION

Mesoscopic features embedded within macroscopic phenomena in colloids and suspensions, when coupled together with microstructural dynamics and boundary singularities, produce complex multiresolutional patterns, which are difficult to capture with the continuum model using partial differential equations, i.e., the Navier–Stokes equation and the Cahn–Hillard equation. The continuum model must be augmented with discretized microscopic models, such as molecular dynamics (MD), in order to provide an effective solver across the diverse scales with different physics. The high degree of spatial and temporal disparities of this approach makes it a computationally demanding task. In this survey we present the off-grid discrete-particles methods, which can be applied in modeling cross-scale properties of complex fluids. We can view the cross-scale endeavor characteristic of a multiresolutional homogeneous particle model, as a manifestation of the interactions present in the discrete particle model, which allow them to produce the microscopic and macroscopic modes in the mesoscopic scale. First, we describe a discrete-particle models in which the following spatio-temporal scales are obtained by subsequent coarse-graining of hierarchical systems consisting of atoms, molecules, fluid particles, and moving mesh nodes. We then show some examples of 2D and 3D modeling of the Rayleigh–Taylor mixing, phase separation, colloidal arrays, colloidal dynamics in the mesoscale, and blood flow in microscopic vessels. The modeled multiresolutional patterns look amazingly similar to those found in laboratory experiments and can mimic a single micelle, colloidal crystals, large-scale colloidal aggregates up to scales of hydrodynamic instabilities, and the macroscopic phenomenon involving the clustering of red blood cells in capillaries. We can summarize the computationally homogeneous discrete particle model in the following hierarchical scheme: nonequilibrium molecular dynamics (NEMD), dissipative particle dynamics (DPD), fluid particle model (FPM), smoothed particle hydrodynamics (SPH), and thermodynamically consistent DPD. An idea of a powerful toolkit over the GRID can be formed from these discrete particle schemes to model successfully multiple-scale phenomena such as biological vascular and mesoscopic porous-media systems.

A. MOTIVATIONS

Mesoscopic flows are important to understand because they hold the key to the interaction between the macroscopic flow and the microstructures. This is especially true in complex fluids, which involve colloidal mixtures, thermal fluctuations, and particle–particle interactions. Complex fluids take on a homogeneous appearance at macroscopic scales but appear very disordered and heterogeneous over atomistic scales. However, they possess an ordered structure over a mesoscopic length-scale. For example, in polymer solutions, the intermediate length-scale can be the size of a polymer chain. In a colloidal suspension, the mesoscopic length scale is the size involving colloid particles, i.e., 10–1000 nm. The properties of these systems are often determined by their mesoscopic structures, endowing a complex fluid with unique and interesting features. In many technological and physical processes, such as emulsification, formation of nanojets [1], water desalination, food production, gel filtration, and transport processes in sedimentary rocks or blood flow in tiny blood vessels, the microscopic or microstructural effects dominate over the hydrodynamics and take place in the realm of mesoscopic flows. For example, blood is a system rich in rheological properties, exhibiting shear-thinning, viscoelastic, and sedimenting behavior. But still, only qualitative correlations have ever been made between the observed microstructure and the reported rheology. Central to understanding these correlations is the determination of the intrinsic link between the rheology and the microstructure.

The successes enjoyed by nanosciences in many fields [2–10] have resulted in a need for adequate theory and large-scale numerical simulations in order to understand what the various roles are played by surface effects, edge effects, or bulk effects in nanomaterials. The dynamics of colloidal particle transport calls not only for passive transport, but also for additional processes such as agglomeration/dispersion, driven interfaces, adsorption to pore wall grains, and biofilm interactions [4,11–14]. In many cases, there is a dire need to investigate these multi-scale structures, ranging from nanometers to micrometers in complex geometries, such as in vascular and porous systems [4,15–17].

B. Diverse Spatial and Temporal Scales

Capillary network and porous material in the mesoscale can be viewed as a complex system consisting of two mutually interacting constituents, namely, the colloidal suspension and the wall material. The complexity of these components is connected mainly with the development of multiple spatio-temporal scales involved in a proper description of their physical, chemical, and geometrical properties. In the case of colloidal suspension, the multiple scales come from:

1. The spatial factor connected with the size of the colloidal bead differing by a few orders of magnitude from the sizes of solvent molecules.
2. Dynamic factor involved by chemical reactions and thermal fluctuations occurring over time scales, which are several orders of magnitude smaller than those associated with hydrodynamic flows.

The multi-scale hierarchy of the circulatory system can be an example of such a complex system. Depending on the spatial scale, both blood and capillary vessels behave in completely different ways. In the largest arteries and veins having diameters on the order of 10^{-2} m, where blood pressure is approximately 10 kPa, it can be regarded as a continuum fluid. The artery walls are elastic and controlled by sophisticated neural mechanism.

Macroscopic vessels represent only a small fraction of the circulatory system. Approximately 10^{10} blood vessels are the capillaries whose diameters are comparable with the dimensions of the red blood cells, i.e., 5–10 μm [18]. In the smallest capillaries with diameters 10^{-5} m, where the pressure drops to 1 kPa, blood flow represents a composite system with sharp interface between liquid and solid phases. In this spatial scale, the red blood cells (RBC) describe the phase volume with distinct elastic properties. In contrast to the macroscale, at the microscopic scale blood flow can be viewed at as the collective motion of an ensemble of microscopically interacting discrete particles. Unlike in large blood arteries, in the blood capillaries the wall consists of a layer of endothelial cells [18] responding to the shear flow.

Due to the flow of the colloidal suspension, namely blood, the phenomena occurring in the capillary network involve multiple spatio-temporal scales. Thus it is an extremely intricate physical system impossible to modeling within a single numerical paradigm. Physical phenomena developing in complex materials are usually described within hierarchical, multi-level numerical models. In these models, each level is responsible for different spatio-temporal behavior and passes out the averaged parameters to the level, which is next in the hierarchy. In realistic cross-scale simulations, communication between the levels must be bidirectional. Critical phenomena occurring due to hydrodynamic instabilities and mixing (e.g., combustion and cement hardening) [19–23] or fracture dynamics (crack propagation) [8,24–27] are well-documented examples of such a complex cross-scale behavior.

C. Discrete-Particles and Continuum Models

Neglecting the scales below the atomistic level, we can employ two principal physical paradigms and respective computational models for fluid simulation: the particle paradigm and the continuum approach (see Figure 26.1).

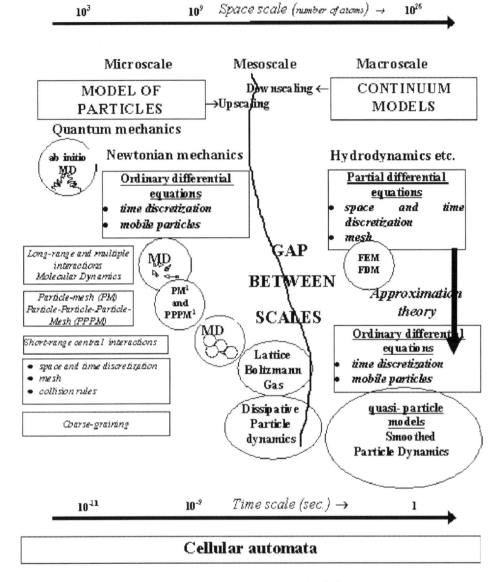

FIGURE 26.1 Particle and continuum models used in computer simulations.

The particle paradigm bases on interacting atoms, whose temporal evolution obeys the Newtonian laws of motion, and operates in the microscale. Molecular dynamics (MD) technique is the most prominent computational realization of this model [28–30]. The method consists in the solution of the set of ordinary differential equations in time for each particle by using numerical schemes. As a result, the positions and velocities of particles are updated during the simulation. The global physical parameters of the particle ensemble are computed by temporal and spatial averaging of the statistical functions of particles' positions and velocities, as well as by computing the correlation functions and higher moments.

The paradigms constructed on the basis of mass, momenta, and energy streams continuity are used to quantify the macroscopic properties of fluids. Their mathematical models are represented by a set of nonlinear partial differential equations. Computer implementation of these models is based on the space and time discretization on a fixed or reconfigurable mesh. This can be accomplished by using finite elements (FEM) or finite differences (FDM) techniques. The resulting set of linear equations is solved at each timestep and temporary values of density, momenta, and energy in the mesh nodes are computed. For more complex and nonlinear mathematical models, the set of nonlinear equations have to be solved at each timestep, which is difficult for continuum methods, especially for problems with steep gradients in physical properties.

As shown in Figure 26.1, the wide gap opens up between the particle and continuum paradigms. This gap cannot be spanned using statistical mechanical methods only. The existing theoretical models to be applied in the mesoscale are based on heuristics obtained via downscaling of macroscopic models and upscaling particle approach. Simplified theoretical models of complex fluid flows, e.g., flows in porous media, non-Newtonian fluid dynamics, thin film behavior, flows in presence of chemical reactions, and hydrodynamic instabilities formation, involve not only validation but should be supported by more accurate computational models as well. However, until now, there has not been any precisely defined computational model, which operates in the mesoscale, in the range from 10 Å to tens of microns.

The simplest way to conduct the simulations of mesoscopic phenomena, consists in downscaling of the continuum approach or upscaling (or coarse graining) the particle model. The limitation of the first approach is determined by the granular properties of matter, which are revealed for very small system. The behavior of granular fluid is different from that in the bulk. For example, very small systems may exhibit solid–liquid coexistence over a range of temperature different than that for large systems [31]. However, new approaches such as:

   (i) introducing, in a consistent way, fluctuations to the equations of fluid dynamics [32];
  (ii) using theory of approximation in the equations of hydrodynamics [33]; and
 (iii) employing direct numerical simulations (DNS) schemes for solving fluid dynamics
        problems with sharp interface between phases [34].

allow for simulating colloidal systems in scales of hundreds of nanometers.

The upscaling of the particle model can be realized by increasing the number of particles. It is the natural approach to the cross-scale computations. However, despite very efficient parallel implementations of nonequilibrium molecular dynamics (NEMD) method, only approximately 10^9 molecules (the sample 0.5 μm of size) in the time of nanoseconds can be currently simulated using the most powerful massively parallel systems [8,24,27,35]. The nanoscale MD simulations (e.g., [89,24,25,27,36]) reveal very interesting collective and complex behavior of the particle systems and their importance cannot be underestimated, especially, when they are applied for investigations of new materials, mixing phenomena, chemical reactions, interfacial phenomena, etc. Such atomistic simulations are valuable for obtaining constitutive relations, which can be used in the macroscale models. Some of subcontinuum fluid problems, which can be solved by using large-scale NEMD computations such as: rupture in solids or coalescence of liquid drops that occur on interfacial scales, nonphysical singularities resulting from such continuum approaches, spreading of wetting films on solid surfaces, and behavior of non-Newtonian fluid are discussed in Ref. [31].

In general, the upscaling from atomistic to mesoscopic scale realized by increasing the number of atoms is unrealistic. Therefore, fundamental simplifications have to be made in the model for simulating larger scales. The simplifications are:

1. Constructing of the coarse-graining procedures in which microscopic fluctuations are
   eliminated or replaced with simpler stochastic models.
2. Crude discretization of the particle system.

The last item concerns not only space and time discretization but also drastic simplification of the collision operator and particle motion rules as well, e.g., by employing cellular automata, lattice gas models [37,38], or stochastic models such as stochastic rotation dynamics [39].

The other way for simulating systems over wider spatio-temporal scales consists of employing heterogeneous hybridized cross-scale models.

D. Cross-Scale Models

In Figure 26.2, we show a schematic diagram depicting the spatio-temporal, hierarchical nature of hybrid numerical model, which can be used effectively in the cross-scale modeling of flows of colloidal suspensions in porous media. Despite its conceptual correctness, the methodological and computational disadvantage involved in this grandiose scheme is quite obvious. The hybridized model composed of heterogeneous mathematical and numerical concepts, such as MD and FEM or FDM techniques, involve different:

1. Discretized primitives: particles vs. finite elements.
2. Numerical approaches: summation of intermolecular forces and integration of ordinary differential equations opposed to solving multidimensional and nonlinear partial differential equations.

As shown in Figure 26.1, a large computational gap opens between these two approaches [40]. The size of a sample in which the microscopic effects can be averaged out by using continuum hydrodynamics is approximately hundreds of nanometers in the transition region between quantum mechanics and classical mechanics. For modeling such a system employing MD in three dimensions, more than 10^9 particles have to be simulated in 10^5–10^6 timesteps [27,35]. Matching continuum and MD models requires hundreds of iterations per timestep. This means a lot of floating point operations per timestep, at least 10^{12}.

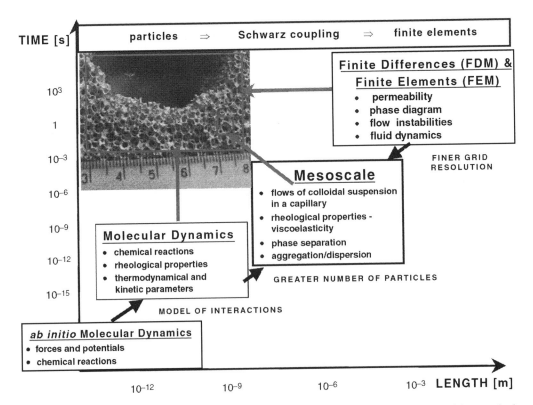

FIGURE 26.2 Schematic diagram showing the applicability of various kinds of numerical models to study the diverse spatial and temporal scales associated with the multitudinous phenomena spanning over multiple scales in the modeling of colloidal dynamics and flows in porous media.

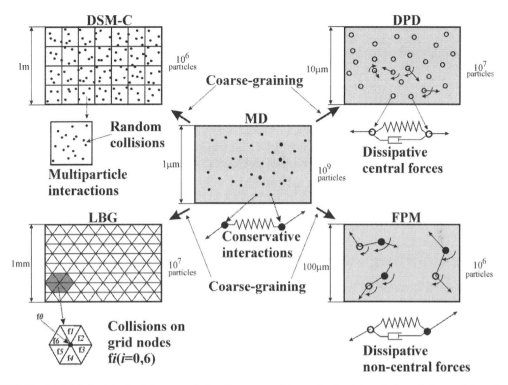

FIGURE 26.3 Methodological framework for discrete particle methods with a listing of the spatial scales capable to be captured today and number of particles simulated on reasonably large shared-memory computer systems [22].

As shown in Figure 26.3, both lattice Boltzmann gas (LBG) [37,38], direct simulation Monte-Carlo (DSM-C) [41], and off-grid particle methods such as DPD [42], and fluid particle method (FPM) [43] can be treated within a common methodological framework. This framework in the mesoscale consists the successive coarse-graining of the underlying molecular dynamics system. We can list its components as:

1. Common primitives, particles, defined by a set of attributes whose physical meaning is different depending which spatio-temporal scale is currently being considered.
2. Local collision operator Ω, which is defined as a sum of short-ranged additive forces or collisions between particles.
3. The rules of particle system evolution, which can be treated as a component of particle dynamics.

As shown in Figure 26.3, the methods can capture different scales from this general philosophy of successive coarse-graining. Because the direct simulation Monte-Carlo is used for modeling flows of rare gases and the systems defined by large Knudsen number, macroscopic scales can be reached by using relatively small number of particles. In the modeling of mesoscopic flows, great success has been enjoyed by lattice gas (LG) and LBG, which were used among others in modeling colloidal suspensions [44–49] and porous media (e.g., [50–55]). This success results from the computational simplicity of the method, which comes from coarse-grained discretization of both the space and time and drastic simplification of collision rules between particles. Today we can simulate by using LBG, the scales up to milimeters, and by using grids with $10^7 - 10^8$ collision nodes.

The off-grid particle methods, such as DPD and FPM, can capture easily mesoscopic scales of hundreds of micrometers employing up to 10^7 fluid particles currently, i.e., the scales in which temperature fluctuations and depletion forces interact with mesoscopic flows. Therefore, gridless particle methods can mimic the complex dynamics of fluid particles in the mesoscale more realistically than LBG. The FPMs also save computational time taken by molecular dynamics for calculating thermal noise. Instead, in DPD and FPM, we introduce the random Brownian force.

Here it will be recommended that we should use multi-level gridless particle approach for simulating mesoscopic phenomena, such as colloidal dynamics in complex geometries. These particles can be treated as fluid packets of the size reflecting with great fidelity, the spatial scale of simulation. The particles interact with each other and their dynamics is governed by the Newtonian equations of motion. We show that, by combining fluid particles with particles modeling solid colloidal beads, we can couple together macroscopic hydrodynamic modes with microstructures emerging due to interactions among the beads and between the beads and fluid particles. Thus, unlike in MD-FEM and LBG approaches, we can also control more precisely both the microstructural parameters and interactions between microscopic and macroscopic modes.

E. OUTLINE

In the first section, we present briefly the best candidates, which can be combined into a homogeneous, particle based, cross-scale numerical solver. First, we compare on-grid and off-grid methods. Then we describe the principles of MD, which supplies methodological basis for the family of off-grid discrete-particle methods. We discuss the equations describing interactions between atoms for different formulations of MD method and the issues for its coarse-graining representations spanning from DPD, FPM, multi-level schemes, smoothed particle dynamics (SPH), and thermodynamically consistent DPD (TC-DPD). In the following section, we present some computational aspects of discrete-particle simulations. The efficient algorithms for calculation of interparticle forces, implementation of various kinds of boundary conditions, and numerical schemes used for integrating Newtonian equations of motion are presented. Domain decomposition and load-balancing schemes, vital procedures for efficient parallelization of particle codes, are also briefly explained. We also show the role of the mutual nearest neighborhood clustering method in extracting microstructures spontaneously created due to flow. The results from high-performance NEMD, DPD, and FPM simulations of phenomena occurring in complex, mesoscopic fluids are demonstrated in the following section. Here, we give a brief overview of such phenomena as the Rayleigh–Taylor mixing, thin film evolution, phase separation for binary fluid, colloidal dynamics resulting in creation of micelles, colloidal arrays, colloidal aggregates, and dispersion of microstructures. Finally, we present the principal components of the particle model of red blood cells dynamics in capillary vessels and some results from modeling of blood clotting in choking point due to fibrinogen. In the last section, we discuss the possibility of using the discrete-particle methods proposed earlier as the components of the problem-solving environment (PSE) in GRID service network.

II. DISCRETE-PARTICLES: ALGORITHMS AND METHODS

From the standpoint of traditional fluid dynamics, a general problem in modeling of colloidal dispersion comes from a basic difficulty in defining physically consistent models, which can couple together continuum and discrete approaches. Continuum models such as the Navier–Stokes equations for fluid mechanics or the Cahn–Hillard equation for metallurgical phase separation (e.g., see [56]) and crystallization processes [57], which are usually based on simple conservation laws, represent "top-down" way of chemical–physical description of matter, i.e., from large to small length scales. This paradigm can be used successfully for simple Newtonian fluids and constrained crystals. For complex fluids with complicated physics such as two-phase flow and compaction (e.g., see [58,59]), however, equivalent phenomenological representations are usually unavailable. Therefore, it must be approximated by empirically derived constitutive relations and other empirical parameters, such as surface tension [58], obtained from computationally complex direct numerical simulations (DNS) [34] or heterogeneous models combining both continuum and discrete-particle models (e.g., fluid particle dynamics (FPD) by Ref. [60]).

Conversely, the modeling approach can be based on the microscopic description working from the smallest scales upward along the general lines of the program for statistical mechanics. Discrete-particles techniques represent such an approach.

A. Off-Grid and On-Grid Methods

As mentioned in the Section I.A, the upscaling procedure involving particles for covering larger spatio-temporal scales, consists in coarse-graining of particle system or in simplifications of collision operator and rules of particle motion. In the first case, we can assume that the particles do not represent the atoms but larger pieces of matter such as clusters of atoms, lamps of fluid or droplets. These clusters interact with each other with some deterministic force $F(r)$ (e.g., as it is in DPD) or stochastic (e.g., as it is in direct simulation Monte-Carlo) collision operator. By assuming the short-range interactions between particles within the predefined cut-off radius, the list of closest neighbors because each particle changes dynamically. As shown in Figure 26.4a, the particles are not constrained to the lattice and can move freely in the Cartesian space according to Newtonian laws of motion or stochastic dynamics rules. Each particle is described by its position and velocity.

Fundamental simplifications of particle–particle collision operator and particle-motion rules, result in the on-grid cellular automata models of fluid dynamics, i.e., LG and LBG schemes. LG hydrodynamics [37,38] describes the approach to fluid dynamics using a microworld constructed as an automaton universe, where the microscopic dynamics is not on the basis of a description of

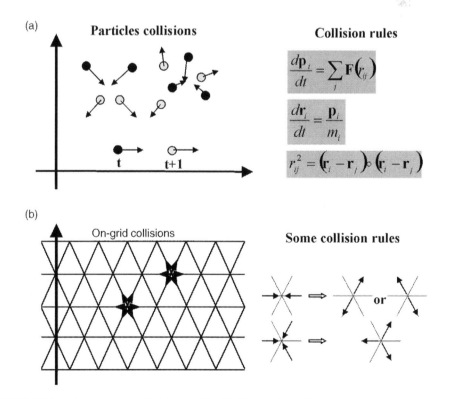

FIGURE 26.4 The schemes of on-grid (a) and off-grid (b) particle methods.

interacting particles, but on the basis of laws of symmetry and invariance of macroscopic physics. We imagine point-like particles residing on a regular lattice where they move from node to node and undergo collisions when their trajectories meet. This situation is shown in Figure 26.4b. If the collisions occur according to some simple logical rules, and if the lattice has a proper symmetry, then the automation shows global behavior very similar to that of real fluids. The LG methods are very efficient computationally because they are based on very simple particle dynamics and the neighbors for each particle are known or can be computed easily. The number of possible states for a given particle is finite (e.g., 12 for 6 directions and 1 or 2 lattice sites per timestep). Therefore, a particular lattice site can be represented with a finite number of bits. The movement and collision of particles in this system can be computed with integer arithmetic. Therefore, the number of calculations needed to change a temporal state of a single LG particle is much smaller than for molecular dynamics method. Thus, more LG particles can be simulated and larger physical systems can be explored. Mass, momentum, and energy are intrinsically conserved and the computation is unconditionally stable. However, due to large microscopic fluctuations, which should be averaged out, the huge lattices (more than 10^7) have to be simulated to cover mesoscopic length scales.

The advantage of the LG method over particle approach, however, is somewhat misleading. It may happen that the results of application of the LG methods are not more efficient than MD. This concerns the cases where the simplifications in the physical model cannot be compensated by the number of simulated particles. Moreover, unlike LG techniques, the particle approaches operate for strictly defined length and time scale. The drawback of the lattice approaches is that the dynamics are constrained by the configuration of the lattice and the system evolves in dimensionless time. This precipitates many problems, e.g., with 3D models, averaging, boundary conditions on shaped bodies, simulation of shock phenomena, and calculations of realistic fluid parameters. In Figure 26.5 we display the final snapshots from diffusion-limited aggregation DLA (e.g., see [61]) for off-grid and on-grid algorithms. Two cases were simulated: allowing smaller and larger thermal fluctuations. For off-grid system, larger thermal fluctuations produce more compact agglomerate, whereas the on-grid system yields similar clusters for both cases. We can conclude that off-grid method is more accurate, adapting to changing physical conditions and resolving multiresolutional situations such as greater concentration of particles in critical regions. For on-grid algorithms, unrealistically dense lattices or multi-grid and wavelet-based methods have to be applied to model multi-resolutional features [16,62,63]. Moreover, the on-grid LG techniques fit poorly to the cross-scale computations, though some interesting heterogeneous models were constructed [64]. The scales bridging problem would be here even more acute than for the MD-FEM heterogeneous cross-scale computations [65,66].

Unlike the LG schemes, the LBG method represents coarse-graining approach realized on the lattice. The integer particle population of LG is replaced with floating point numbers reflecting aggregated properties of clusters of particles. By this way, the microscopic degrees of freedom smaller than the mesoscopic are eliminated in the coarse-grained mesoscopic model. The number of bits required for storage of a single-lattice state is at least 32. The conservation of mass, momentum, and energy is now limited by the precision of floating point errors; such errors lead to stability problems. However, due to coarse-graining procedure, LBG systems are able to cover much larger temporal-scales than LG particle systems.

Similar schemes of coarse-graining can be realized by using off-grid discrete-particles models. In the following sections, some competitive ideas to LBG approach will be presented.

B. Molecular Dynamics

MD is the most fundamental computational technique used for a direct simulation of temporal evolution of interacting particle ensembles [28,29]. Many other particle-based techniques, such as the Brownian and Langevin dynamics [28], constitute in fact the simplification of the MD method. They were devised to cover larger length, and time scales than MD.

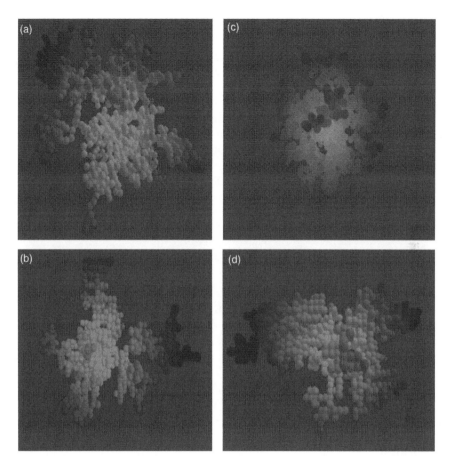

FIGURE 26.5 The results of diffusion-limited aggregation (DLA) for (a) off-lattice method with small fluctuations, (b) off-lattice method with large fluctuations, (c) on-lattice method with small fluctuations, and (d) on-lattice method with large fluctuations.

The system of MD particles is defined by a set of boundary and initial conditions and by inter-actions between particles represented by a force $\mathbf{F}(r_{ij})$ and the torques $\mathbf{N}(r_{ij})$. The particle system evolves according to the Newtonian equations of motion

$$\dot{\mathbf{v}}_i = \frac{1}{m_i} \sum_j \mathbf{F}(r_{ij}), \qquad \dot{\mathbf{r}}_i = \mathbf{v}_i \tag{26.1}$$

$$\dot{\boldsymbol{\omega}}_i = \frac{1}{I_i} \sum_j \mathbf{N}(r_{ij}), \qquad \mathbf{N}_{ij} = -\frac{1}{2} \mathbf{r}_{ij} \times \mathbf{F}_{ij} \tag{26.2}$$

$$r_{ij}^2 = \mathbf{r}_{ij} \circ \mathbf{r}_{ij}, \qquad \mathbf{r}_{ij} = \mathbf{r}_i - \mathbf{r}_j, \quad i = 1, \ldots, N, \quad j \in R(i) \tag{26.3}$$

which can be described (Equation 26.1–Equation 26.2) in a discrete form as

$$\Delta \mathbf{p}_i^n = \sum_j \mathbf{F}^n(r_{ij}) \cdot \mathbf{e}_{ij}^n \Delta t, \qquad \Delta \mathbf{r}_i^n = \frac{\mathbf{p}_i^n}{m_i} \Delta t, \tag{26.4}$$

where $\mathbf{r}_i^n$ is the position of particle i in timestep n; $\mathbf{v}_i$, ω_i, $\mathbf{p}_i$ are its translational velocity, rotational velocity and momentum, respectively, $R(i)$ is the set of indices of particles, which are in the interaction range of particle i, Δt is the integration timestep, n is the number of current timestep, while N is the number of particles. The torques vanishes assuming that the forces $\mathbf{F}()$ are central.

The type of forces depends on the MD model exploited. A simulation is realistic — that is, it mimics the behavior of the real system — only to the extent that interatomic forces are similar to those that real atoms (or, more exactly, nuclei) would experience when arranged in the same configuration. Forces are usually obtained as the gradient of a potential energy function $\Phi(\mathbf{r}_i)$, depending on the positions of the particles. The realism of the simulation therefore depends on the ability of the potential chosen to reproduce the behavior of the material under the conditions at which the simulation is run. The forces can be defined as follows:

$$\mathbf{F}(\mathbf{r}_i) = -\nabla\Phi(\mathbf{r}_i) + \mathbf{F}_i^{\text{extended}}(\mathbf{r}_i, \mathbf{v}_i, \omega_i) \qquad (26.5)$$

where $\mathbf{F}^{\text{extended}}(\ldots)$ is an extended force (e.g., velocity-based friction). Typically, the potential $\Phi(\mathbf{r})$ is given by

$$\Phi(\mathbf{r}) = \Phi^{\text{bonded}}(\mathbf{r}) + \Phi^{\text{nonbonded}}(\mathbf{r}) + \Phi^{\text{external}}(\mathbf{r})$$

$$\Phi^{\text{bonded}}(\mathbf{r}) = \Phi^{\text{bond}}(\mathbf{r}) + \Phi^{\text{angle}}(\mathbf{r}) + \Phi^{\text{dihedral}}(\mathbf{r}) + \Phi^{\text{improper}}(\mathbf{r}) \qquad (26.6)$$

$$\Phi^{\text{nonbonded}}(\mathbf{r}) = \Phi^{\text{electrostatic}}(\mathbf{r}) + \Phi^{\text{n-body}}(\mathbf{r})$$

For the long-range interactions such as gravitational and electrostatic the forces acting on each particle i can be computed by using Particle-Mesh (PM) approximation of the Gauss equation

$$\Delta\Phi^{\text{electrostatic}}(\mathbf{r}) = 4\pi\varepsilon_0\rho(\mathbf{r}) \qquad (26.7)$$

where $\rho(\mathbf{r})$ is the charge density in $\mathbf{r}$. The PM method [67] treats the force as a field quantity by approximating it on a mesh. Differential operators, such as the Laplacian, are replaced by finite difference approximations. Potentials and forces at particle positions are obtained by interpolating on the array of mesh-defined values. Mesh-defined densities are calculated by assigning particle attributes (e.g., "charge") to nearby mesh points in order to create the mesh-defined values (e.g., "charge density"). The PM method is basically unacceptable for studying close encounters between particles. This method is good for simulations where you want a "softening" of the inverse square law force. In general, the mesh spacing should be smaller than the wavelengths of importance in the physical system. Another disadvantage of using the mesh-based methods is that they have difficulties handling nonuniform particle distributions. This means that the PM methods offer limited resolution. To overcome the limited resolution, some researchers have developed PM algorithms which employ meshes of finer gridding in selected subregions of the system. These finer meshes permit a more accurate modeling of regions of higher density. If we wish to further refine the grid due to large dynamical changes in the system (e.g., due to shock waves), then we can apply moving grids or adaptive grids.

The P3M method solves the major shortcoming of the PM method — low resolution forces computed for particles near each other. This method supplements the interparticle forces with a direct sum over pairs separated by less than about $3 \cdot \Delta x$, where Δx is the grid spacing. The interparticle forces are split into a rapidly varying short-range part and a slowly varying long-range part. The PP method is used to find the total short-range contribution to the force on each particle and the PM method is used to find the total slowly-varying force contributions. The disadvantage of the P3M algorithm is that it becomes too easily dominated by the direct summation part and becomes unacceptably slow for larger systems. There are various adaptation methods explored by many researchers in the mid-1980s through the mid-1990s, to improve the PM and P3M methods such as: the Nested Grid Particle-Mesh and Tree-Codes methods [http://www.amara.com/papers/nbody.html; [68]]. However, the most successful algorithm

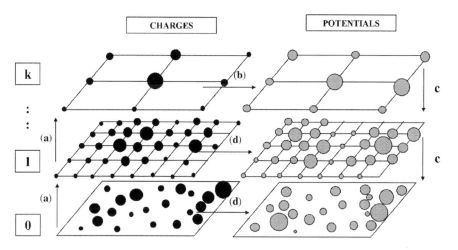

FIGURE 26.6 The multi-level scheme of the Multi-Grid model (a) Cluster to coarser grid, (b) Compute potential on the coarsest grid, (c) Interpolate potential values from coarser grid, and (d) Make local corrections.

solving the problem of the long-range interactions is based on the Multi-Grid separation of scales (see Figure 26.6).

The Fast Multipole Method (FMM) is a tree code that uses two representations of the potential field. As shown in Figure 26.7, the two representations are: far field (multipole) and local expansions. The two representations are referred to as the "duality principle." This method uses a very fast calculation of the potential field $\Phi^{electrostatic}(\mathbf{r})$. Therefore, the complexity of the method is $O(N)$. The strategy of the FMM is to compute a compact expression for the potential $\Phi^{electrostatic}(\mathbf{r})$, which can be easily evaluated along with its derivative, at any point. It achieves this by evaluating the potential as a "multipole expansion," a kind of Taylor expansion, which is accurate when R^2 (where R is the distance between particle and multipole) is large. The FMM method shares the quadtree and divide-and-conquer paradigm of Barnes-Hut [http://www.icsi.berkeley.edu/cs267/lecture27/lecture27.html; [69]].

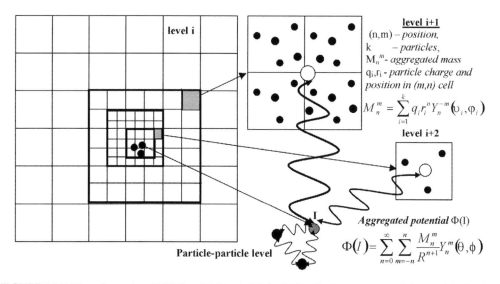

FIGURE 26.7 The schematic of MD Fast Multipole Method idea. In the lowest, particle–particle level, the interactions on particle "I" are computed from all the particles which are located in the same and neighboring cells, while the interactions from the larger cells are approximated.

Periodic boundary conditions (PBC) represent the most common trick to simulate the system of unlimited number of interacting particles by limited number of interacting lattices where each of them stands for a particle and its replicas. The Ewald summation was introduced in 1921 as a technique to sum the long-range interactions between particles and all their infinite periodic images efficiently. Ewald recast the potential energy of particle system, a single slowly and conditionally convergent series, into the sum of two rapidly converging series plus a constant term. The lattice sum with the Coulomb potential is usually expressed as

$$\Phi^{electrosttic}(\mathbf{r}_{ij}) = \frac{1}{4\pi\varepsilon_0} \frac{1}{2} \sum_{\vec{n}} \sum_{i=1}^{N} \sum_{j=1}^{N} \frac{q_i q_j}{|\mathbf{r}_{ij} + \vec{n}|} \tag{26.8}$$

where $\sum_{\vec{n}}$ is the sum over all lattice vectors $\vec{n} = (L_x n_x, L_y n_y, L_z n_z)$, where the values of $n_{x,y,z} \in \mathbf{N}$ and $L_{x,y,z}$, are the dimensions of the unit MD cell.

This sum is absolute convergent if the potential term

$$\Phi(\mathbf{r}) \leq A|\mathbf{r}|^{-3-c}$$

for large enough $|\mathbf{r}|$ and $A > 0$, $c > 0$.

To overcome the conditionally and insufficient convergence of Equation (26.8) for the Coulomb interactions, the sum is split into two parts by the following trivial identity:

$$\frac{1}{r} = \frac{f(r)}{r} + \frac{1-f(r)}{r}$$

The basic idea is to separate the fast variation part for small $r = |r|$ and the smooth part for large r. In particular, the first part should decay fast and be negligible beyond some cut-off r_{cut} distance, whereas the second part should be smooth for all r such that its Fourier transform can be represented by a few terms. Each point charge in the system is viewed being surrounded by a Gussian charge distribution $\rho(\mathbf{r})$ of equal magnitude and opposite sign, that is

$$\rho_i(\mathbf{r}) = q_i \alpha^e e^{(-\alpha^2 r^2)}/\sqrt{\pi^3} \tag{26.9}$$

The Ewald summation achieve an overall work complexity of $O(N^{2/3})$ for selected splitting parameter α and cut-off parameters, which limit the number of components in infinite real and reciprocal sums (see for more details [70]). The better computational efficiency can be achieved for mesh-based Ewald method. It approximates the reciprocal-space term of the classical Ewald summation by a discrete convolution on an interpolating grid using the discrete Fast Fourier Transform (FFT). By choosing an appropriate value of α, the computational complexity can be reduced to $O(N \log N)$. The accuracy and speed are additionally governed by the mesh size and interpolation scheme, which makes the choice of optimal parameters even more difficult (Figure 26.8).

The realistic modeling, simulating millions of atoms and molecules, involves more heuristic approach to define the interparticle forces. Instead of integrating the Gauss equation (which is very demanding computationally for large systems), the potential function $\Phi^{n-body}(\mathbf{r})$ is "guessed" in the form of an analytical formula or is derived experimentally. In particle–particle version of MD, it is assumed that the interactions can be approximated by using the central and n-body effective potentials satisfying the convergence condition Equation (26.9). In the past, the most potentials were constituted by pairwise interactions, but this is no longer the case. It has been recognized that the two-body approximation is very poor for many relevant systems, such as metals and semiconductors. Various kinds of many-body potentials are now of common use in condensed matter simulation, and are be briefly reviewed in [http://online.physics.uiuc.edu/courses/phys466/fall04/lnotes/pot.html#common]. However, the two-body approximation is still reasonable for many dense fluids and solids, where instead of pure electrostatic interactions

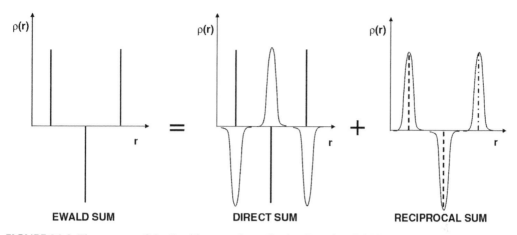

FIGURE 26.8 The concept of the Ewald summation reflecting Equation (26.10).

the screened and effective potentials can be used. Some simple nonbonded and bonded potential functions are presented in Table 26.1 and Table 26.2, respectively.

In this survey, we use only two-body short-ranged potentials. Many-body forces are still too demanding computationally to be used as a computational paradigm for cross-scale computations.

Assuming that the forces acting on every particles are defined, by integrating Equation (26.1)–Equation (26.3), the computer calculates particle trajectories in a 6N-dimensional phase space (3N positions and 3N momenta). However, such trajectories are usually not particularly relevant by ourselves. Molecular dynamics is a statistical mechanics method. Like Monte Carlo, it is a way to obtain a set of configurations distributed according to some statistical distribution function or statistical ensemble. According to statistical physics, physical quantities are represented by averages over configurations distributed according to a certain statistical ensemble. A trajectory obtained by molecular dynamics provides such a set of configurations. Therefore, a measurement of a physical quantity by simulation is simply obtained as an arithmetic average of the various instantaneous values assumed by that quantity during the MD run.

Statistical physics is the link between the microscopic behavior and thermodynamics. In the limit of very long simulation times, we could expect the phase space to be fully sampled, and in that limit this averaging process would yield the thermodynamic properties. In practice, the runs

TABLE 26.1
The Selected Nonbonded Potential Functions

Yukawa	$\Phi(r_{ij}) = kq_iq_j\dfrac{\exp(-kr_{ij})}{r_{ij}}$	(26.10)
Hard sphere	$\Phi(r_{ij}) = A_{ij}e^{-B_{ij}r_{ij}}$ or $\Phi(r_{ij}) = \dfrac{C_{ij}}{r_{ij}^{12}}$	(26.11)
Dispersion	$\Phi(r_{ij}) = -\dfrac{D_{ij}}{r_{ij}^{6}}$	(26.12)
Buckingham	$\Phi(r_{ij}) = A_{ij}e^{-B_{ij}r_{ij}} - \dfrac{D_{ij}}{r_{ij}^{6}}$	(26.13)
Lennard–Jones	$\Phi(r_{ij}) = 4\varepsilon_{ij}\left(\left(\dfrac{\sigma_{ij}}{r_{ij}}\right)^{12} - \left(\dfrac{\sigma_{ij}}{r_{ij}}\right)^{6}\right)$	(26.14)
Stillinger–Weber (3-body potential)	$\Phi(r_{ij,k}) = \phi(r_{ij}) + g(r_{ij})g(r_{ik})\left(\cos\theta_{ijk} + \dfrac{1}{3}\right)^{2}$	(26.15)

TABLE 26.2
The Selected Bonded Potential Functions

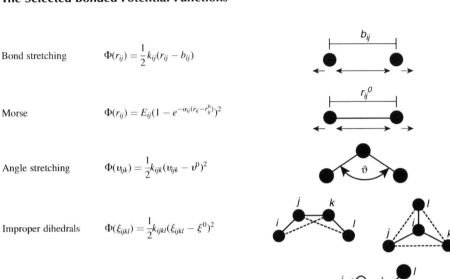

Bond stretching	$\Phi(r_{ij}) = \frac{1}{2}k_{ij}(r_{ij} - b_{ij})$	(26.16)
Morse	$\Phi(r_{ij}) = E_{ij}(1 - e^{-\alpha_{ij}(r_{ij}-r_{ij}^0)})^2$	(26.17)
Angle stretching	$\Phi(\upsilon_{ijk}) = \frac{1}{2}k_{ijk}(\upsilon_{ijk} - \upsilon^0)^2$	(26.18)
Improper dihedrals	$\Phi(\xi_{ijkl}) = \frac{1}{2}k_{ijkl}(\xi_{ijkl} - \xi^0)^2$	(26.19)
Proper dihedrals	$\Phi(\psi_{ijkl}) = k_\psi[1 + \cos(n\psi_{ijkl} - \psi_0)]$	(26.20)

are always of finite length, and we should exert caution to estimate when the sampling may be good ("system at equilibrium") or not. In this way, MD simulations can be used to measure thermodynamic properties and therefore evaluate, for example, the phase diagram of a specific material (see for more information in Ref. [29]).

Beyond this "traditional" use, MD is nowadays applied for other purposes also, such as studies of nonequilibrium processes NEMD; [25,31,71,72], granular dynamics [73], and as an efficient tool for optimization of structures overcoming local energy minima [74].

Because large-scale NEMD simulations (i.e., these involving 10^7–10^9 atoms) can bridge time scales dictated by fast modes of motion together with the slower modes, which determine the viscosity, it can capture the effects of varying molecular topology on fluid rheology resulting, e.g., from chemical reactions (e.g., see [75]) or mixing [76,77] with complicated velocity fields. However, in order to capture spatio-temporal scales occurring, e.g., in capillary or porous systems from Figure 26.2, filled with colloidal suspension and defined by the conditions of:

1. Colloidal beads of sizes larger than 10 nm.
2. Material structures with pores or capillary diameters ranging from one micron to millimeters,

we need billions or more of MD particles modeled in millions of timesteps [27].

C. Dissipative Particle Dynamics

Mesoscopic regimes involving scales of the porous system exceeding 1 ns and 1 μm entail the fast modes of motion to be eliminated in favor of a coarse-grain representation (see Figure 26.1). At this level, the particles will represent clusters of atoms or molecules, the so-called dissipative particles [45]. Flekkoy and Coveney have shown the feasibility to link and pass the averaged properties of molecular ensemble onto dissipative particles by using the bottom-up approach from molecular dynamics by means of a systematic coarse-graining procedure. The dissipative particles are represented by cells defined on the Voronoi lattice, [e.g., see 78,79] with variable masses and volumes (see Figure 26.9). The notion of the Voronoi cells allows for a very clear statement of the problem of coupling continuum equations and molecular dynamics. This is important when the continuum description breaks down due to complex molecular details in certain regions such as the contact line between two fluids and a solid, or the singularity of the tip in propagating fracture [26].

Entire representation of all the MD particles can be achieved in a general way by introducing a sampling function

$$f_k(\mathbf{r} - \mathbf{r}_k) = \frac{\theta(\mathbf{r} - \mathbf{r}_k)}{\sum_l \theta(\mathbf{r} - \mathbf{r}_l)} \qquad (26.21)$$

where the positions $\mathbf{r}_k$ and $\mathbf{r}_l$ define the centers of dissipative particles, $\mathbf{r}$ is an arbitrary position, and $\theta(\mathbf{r})$ is the Gaussian function. The mass, momentum, and internal energy E_k of the kth dissipative particle are then defined as

$$M_k = \sum_i f_k(\mathbf{r}_i), \qquad \mathbf{P}_k = \sum_i f_k(\mathbf{r}_i) m \mathbf{v}_i \qquad (26.22)$$

$$\frac{M_k U_k^2}{2} + E_k = \sum_i f_k(\mathbf{r}_i) \left(\frac{m\mathbf{v}_i^2}{2} + \frac{1}{2} \sum_{j \neq i} \Phi_{MD}(r_{ij}) \right) \equiv \sum_i f_k(\mathbf{r}_i) \varepsilon_i, \qquad (26.23)$$

where $\mathbf{v}_i$ is the velocity of ith MD particle having identical masses m, $\mathbf{P}_k$ is the momentum of the kth dissipative particle, and $\Phi_{MD}(r_{ij})$ is the potential energy of the MD particle pair i, j separated by a

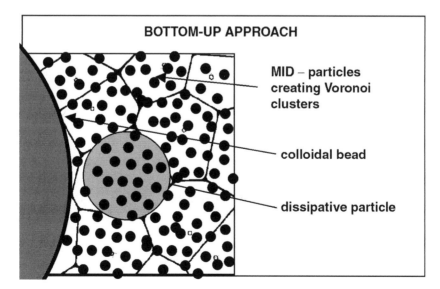

FIGURE 26.9 The schematics of atomistic bottom-up coarse-graining paradigm by Ref. [45]. The Voronoi cell contains particles, which are closest to the cell center. The centers of all the Voronoi cells correspond to the centers of coarse-grained representation of the system — dissipative particles — which can be approximated by spheres.

distance r_{ij}. The particle energy ε_i contains both the kinetic term and a potential term. In order to derive the equations of motion for dissipative particles, the time derivatives of Equation (26.23) must be resolved [45]. Finally, after averaging over the velocities, masses and interactions on the Voronoi lattice, we can obtain the following

$$
\frac{d\mathbf{P}_k}{dt} = M_k \mathbf{g} + \sum_l \langle \dot{M}_{kl} \rangle \frac{\mathbf{U}_k + \mathbf{U}_l}{2}
$$

$$
- \sum_l L_{kl} \left(\frac{p_{kl}}{2} \mathbf{e}_{kl} + \frac{\eta}{r_{kl}} [\mathbf{U}_{kl} + (\mathbf{U}_{kl} \cdot \mathbf{e}_{kl})\mathbf{e}_{kl}] \right) + \sum_l \tilde{\mathbf{F}}_{kl} \tag{26.24}
$$

where $\mathbf{U}_{kl}$ is the velocity of dissipative particles k, l; p_{kl}, a pressure term between k and l dissipative particles resulting from conservative MD interactions, L_{kl}, a parameter of the Voronoi lattice, η, the dynamic viscosity of the MD ensemble and the last summation symbolizes the fluctuations of the coarse-grained representation. This procedure links all the forces between the dissipative particles to a hydrodynamic description of the underlying molecular dynamics atoms. The method may be used to deal with situations in which several different dynamical length-scales are simultaneously present. To increase the computational efficiency, the Voronoi cells can be approximated by spheres (see Figure 26.9). Using additional simplifications, such as the unification of dissipative particle sizes, we can arrive at a model, which converges to the DPD.

In DPD [42], the two-body interactions Ω between two fluid particles i and j are assumed to be central and short-ranged. This can be defined as a sum of a conservative force $\mathbf{F}_C$, dissipative component $\mathbf{F}_D$, and the Brownian force $\mathbf{F}_B$. The Brownian factor represents the thermal fluctuations averaged out due to coarse-graining procedure. The following equations show the basic formula describing the interparticle forces.

$$
\mathbf{F}_C = \Pi \cdot \omega(r_{ij}) \cdot \mathbf{e}_{ij}, \qquad \mathbf{F}_D = \gamma \cdot M \cdot \omega^2(r_{ij}) \cdot (\mathbf{e}_{ij} \circ \mathbf{v}_{ij}) \cdot \mathbf{e}_{ij},
$$

$$
\mathbf{F}_B = \frac{\sigma \cdot \theta_{ij}}{\sqrt{\Delta t}} \cdot \omega(r_{ij}) \cdot \mathbf{e}_{ij} \qquad \omega(r_{ij}) = \frac{3}{n \cdot \pi r_{cut}^2} \left(1 - \frac{r_{ij}}{r_{cut}} \right)
$$

$\theta_{ij} \in (-1, 1)$ — random number, n–particle density

$\Omega(r_{ij}, \mathbf{p}_{ij}) = \mathbf{F}_C + \mathbf{F}_D + \mathbf{F}_B \quad$ for $r_{ij} < r_{cut}$

$$
\Omega(r_{ij}, \mathbf{p}_{ij}) = 0 \quad \text{for } r_{ij} > r_{cut} \tag{26.25}
$$

The cut-off radius r_{cut} is defined arbitrarily and reveals the range of interaction between the fluid particles. DPD model with longer cut-off radius reproduces better dynamical properties of realistic fluids expressed in terms of velocity correlation function [80]. Simultaneously, for a shorter cut-off radius, the efficiency of DPD codes increases as $O(1/r_{cut}^3)$, which allows for more precise computation of thermodynamic properties of the particle system from statistical mechanics point of view. A strong background drawn from statistical mechanics has been provided to DPD [43,80,81] from which explicit formulas for transport coefficients in terms of the particle interactions can be derived. The kinetic theory for standard hydrodynamic behavior in the DPD model was developed by Marsh et al. [81] for the low-friction (small value of γ in Equation (26.25)), low-density case and vanishing conservative interactions $\mathbf{F}_C$. In this weak scattering theory, the interactions between the dissipative particles produce only small deflections.

The strong scattering theory, where the friction between the DPD particles is large (but still for $\Pi \to 0$ in Equation 26.6), was considered by Masters and Warren [82]. In that paper, the Fokker–Planck–Boltzmann equation has been replaced by the Boltzmann equation with a finite scattering cross section. For large friction, in the collective regime, the dynamics is controlled by mode coupling effects by including an internal energy variable such that total energy becomes a conserved quantity.

A serious problem appears with matching the potentials to the realistic interactions in complex colloidal fluids. A few examples of successful matching of DPD method to the real complex fluid have appeared. Groot and Warren [83] devised the equation of state of a simple DPD fluid, which is used for mapping out the conservative part of DPD interactions onto a mean-field theory of polymer mixtures, i.e., the Flory–Huggins theory. An important theoretical work was also carried out by Español [43], whose fluid particle model represents the generalization of mesoscopic DPD [42]. Much less was done with bridging microscopic molecular dynamics with DPD, which could enable to construct a single, homogeneous, particle-based cross-scale model.

In our simulations published earlier [20,71,84], we map the DPD interactions onto the macroscopic parameters of fluid, for example, viscosity, compressibility and diffusion coefficient, by using continuum limit equations obtained from kinetic theory. For example, for a given density ρ_k and sound speed c_k in kth DPD fluid, we computed the scaling factor Π_k from the following continuum equations [81]

$$P_k = \frac{n \cdot \Pi_k \langle r \rangle}{2 \cdot D} \qquad P = \frac{1}{2} c_k^2 \cdot \rho_k \qquad (26.26)$$

where $\langle r \rangle = 1/2 r_{\text{cut}}$ and $D = 2$ for two-dimensional system (D is dimension of the system), but n is a mean number of particles in r_{cut} radius. The σ coefficient in the Brownian component of DPD collision operator (Equation 26.25) can be computed with the following formula [81]

$$\sigma_{kl}^2 = 2 \cdot \gamma_{kl} k_B T \cdot m_{kl} \qquad (26.27)$$

where T_0 is the temperature of the system. The value of γ — scaling factor in the dissipative component $\mathbf{F_D}$ (Equation (26.7)) — comes from [82]

$$v_{kl} = \frac{1}{2} \cdot \frac{\gamma_{kl} n \langle r^2 \rangle}{D(D+2)} \qquad (26.28)$$

where v is kinematic viscosity.

As shown in [81], under some conditions, this approximation yields realistic results but it does not resolve the problem of matching the spatio-temporal scales of DPD simulation to the actual scales. By increasing the spatio-temporal scale, the continuum limit approximation gets worse leading to unrealistic "freezing" of the particle system in whole [71]. This is due to increasing role of conservative component of DPD collision operator, which is usually omitted in transport coefficients obtained from kinetic theory [43,81,83]. Successful matching of DPD fluid properties to the real fluid in spatio-temporal scale under interest, involves more precise definition of the conservative DPD forces. This could be accomplished by means of cross-scale computations [40,45], where molecular dynamics can be used for calculation of interactions between clusters of MD atoms [45]. It may appear that the more sophisticated shape of the weight function $\omega(r)$ should be considered instead of the linear model usually assumed.

There exist several other methods, which generalize the dissipative particle technique, such as:

1. The fluid particle model [43].
2. The bottom-up atomistic approach — coarse-graining procedure replacing atoms interacting with conservative and central forces with clusters of atoms interacting with dissipative forces [85].
3. The top-down thermodynamically consistent approach — the fragments of continuum medium is approximated by fluid particles [32,86].

D. FLUID PARTICLE MODEL

FPM finds itself situated between classical DPD and new formulations of DPD: bottom-up approach from the tiny to the large length-scales, and the top-down method from the large to small scales.

One drawback of DPD is the absence of a drag force between the central particle and the second one orbiting about the first particle. To eliminate this side effect, a greater number of neighbors can be considered in calculating forces, which decreases the computational efficiency. Unlike the classical formulation of DPD method, fluid particles interact via additional noncentral forces, producing a drag between circumventing particles. This saves computational time spent for searching of the neighboring particles.

FPM represents a generalized SPH method for which, unlike in SPH, angular momentum is conserved exactly. The fluid particles are represented by their centers of mass, which posses several attributes, such as the mass m_i, position $\mathbf{r}_i$, translational and angular velocities and a type. These "droplets" interact with each other by two-body forces dependent on the type of particles. This type of interaction involves a sum of:

1. The repulsive conservative force $\mathbf{F}^C$.
2. The frictional forces $\mathbf{F}^T$ and $\mathbf{F}^R$ (translational and rotational), proportional to the relative velocities of the particles.
3. The Brownian forcere $\tilde{\mathbf{F}}$ presenting the microscopic degrees of freedom smaller than the mesoscopic scales, which have been previously eliminated in the coarse-grained mesoscopic model.

The equations for forces are becoming more complex than for the classical DPD technique and are as

$$\mathbf{F}_{ij} = \mathbf{F}_{ij}^C + \mathbf{F}_{ij}^T + \mathbf{F}_{ij}^R + \tilde{\mathbf{F}}_{ij}$$

$$\mathbf{F}_{ij}^C = -F(r_{ij}) \cdot \mathbf{e}_{ij}$$

$$\mathbf{F}_{ij}^T = -\gamma \cdot m \mathbf{T}_{ij} \cdot \mathbf{v}_{ij}$$

$$\mathbf{F}_{ij}^R = -\gamma \cdot m \mathbf{T}_{ij} \cdot \left(\frac{1}{2}\mathbf{r}_{ij} \times (\boldsymbol{\omega}_i + \boldsymbol{\omega}_j)\right) \qquad (26.29)$$

$$\tilde{\mathbf{F}}_{ij} dt = (2k_B T \gamma \cdot m)^{1/2} \left(\tilde{A}(r_{ij}) d\bar{\mathbf{W}}_{ij}^S + \tilde{B}(r_{ij}) \frac{1}{D} \mathrm{tr}[d\mathbf{W}]\mathbf{1} + \tilde{C}(r_{ij}) d\bar{\mathbf{W}}_{ij}^A\right) \cdot \mathbf{e}_{ij}$$

$$\mathbf{T}_{ij} = A(r_{ij})\mathbf{1} + B(r_{ij})\mathbf{e}_{ij}\mathbf{e}_{ij}$$

where $\boldsymbol{\omega}$ is the angular velocity and $\mathbf{W}$ are random matrices consisting of independent Wiener increments weighted by scalar weighting functions $F(r_{ij})$, $A(r_{ij})$, $B(r_{ij})$, $\tilde{A}(r_{ij})$, $\tilde{B}(r_{ij})$, $\tilde{C}(r_{ij})$ defined in Ref. [42].

As was shown in Ref [42], the single component FPM system yields the Gibbs distribution as the steady-state solution to the Fokker–Planck equation under the condition of detailed balance. Consequently, it obeys the fluctuation–dissipation theorem, which defines the relationship between the normalized weight functions, which are chosen such that

$$A(r) = \frac{1}{2}\left[\tilde{A}^2(r) + \tilde{C}^2(r)\right], \qquad B(r) = \frac{1}{D}\left[\tilde{B}^2(r) - \tilde{A}^2(r)\right] + \frac{1}{2}\left[\tilde{A}^2(r) - \tilde{C}^2(r)\right] \qquad (26.30)$$

For the DPD method $A(r) = 0$, consequently, $\tilde{A}(r)$, $\tilde{B}(r)$, $\tilde{C}(r) = 0$ and $F(r) \propto B(r)$, which means that all the DPD forces are central.

Owing to degree of a freedom allowed by the model in selecting the weight functions, we may assume that

$$\tilde{A}(r) = 0, \qquad A(r) = B(r) = \left(1 - \frac{r}{r_{cut}}\right)^2, \qquad F(r) = -\Pi_k \cdot \frac{3}{\pi \cdot r_{cut}^2 n} \left(1 - \frac{r}{r_{cut}}\right) \qquad (26.31)$$

where, r_{cut} is a cut-off radius, which defines the range of particle–particle interactions. For $r_{ij} > r_{cut}$, $F_{ij} = 0$.

The first assumption is recommended in Ref. [42]. We can postulate that the rest of weight functions are the same as in DPD [42,81]. Owing to an additional drag between particles caused by the noncentral interactions, we can also reduce the computational load assuming that the interaction range is shorter than for DPD fluid.

The FPM method can predict the transport properties of the fluid, thus allowing to adjust the model parameters by using the equations of continuum limit for the partial pressure P [43,81]:

$$P = \frac{n \cdot \int \mathrm{d}\mathbf{r} r F(r)}{2 \cdot D} \qquad (26.32)$$

and formulas in kinetic theory [43] for respectively bulk viscosity ν_b, shear viscosity ν_S and rotational viscosity ν_R

$$\nu_b = \gamma \cdot n \left[\frac{A_2}{2D} + \frac{D+2}{2D} B_2\right] + c^2 \frac{1}{\gamma \cdot Dn(A_0 + B_0)} \qquad (26.33)$$

$$\nu_S = \frac{1}{2}\gamma \cdot n \left[\frac{A_2}{2} + B_2\right] + c^2 \frac{1}{2\gamma n(A_0 + B_0)} \qquad (26.34)$$

$$\nu_R = \frac{\gamma n A_2}{2} \qquad c^2 = \frac{k_B T}{m} \qquad (26.35)$$

and

$$A_0 = \int \mathrm{d}\mathbf{r} A(r), \qquad B_0 = \frac{1}{D}\int \mathrm{d}\mathbf{r} B(r),$$

$$A_2 = \frac{1}{D}\int \mathrm{d}\mathbf{r} r^2 A(r), \qquad B_2 = \frac{1}{D(D+2)}\int \mathrm{d}\mathbf{r} r^2 B(r)$$

The kinetic theory for FPM has been developed for deriving transport coefficients by assuming that conservative forces are absent. For nonzero pressures, the transport coefficients computed from Equation (26.33)–Equation (26.35) can be used as the first approximation in an iterative procedure, which matches the coefficients of the FPM forces.

The results of our test runs [21,22,87] show that the FPM model satisfies the basic physical constraints:

1. Total angular and translational momenta are preserved.
2. The actual thermodynamical pressure computed from the virial theorem is constant. Its average is 5–10% larger than P assumed (for $M = 50.000$ particles).
3. The actual thermodynamical temperature computed from the average kinetic energy of the particle system is constant and deviates no more than 2–5% from the value of T.
4. The rotational kinetic energy is approximately one-third of the total kinetic energy.

Because the compressibility of the fluid is low (for a large value of π), there are some quantitative differences between the theory and the simulation. The kinetic theory formulas have been developed in the limit where no conservative forces are present. The actual transport coefficients computed from generalized Einstein and Green–Kubo formulas [88] are larger more than 50% than those predicted from the classical the kinetic theory [89]. Therefore, the transport coefficients computed from the theory can be used only as the first coarse approximation. The most precise matching can be done by using a new generic formulation of DPD model, which is a natural generalization of both DPD and FPM [31].

The temporal evolution of the FPM particle i is described by the Newtonian laws of motion given by Equation (26.1)–Equation (26.3). Because the particles have nonzero angular momenta, we can investigate the positive feedback effects of elastic interactions.

E. MULTILEVEL DISCRETE-PARTICLE MODEL

As shown earlier, the FPM model represents a generalization not only of DPD but also of the molecular dynamics (MD) technique. It can be used as DPD by setting the noncentral forces to zero or MD, by dropping the dissipative and Brownian components. The fluid particle model holds an advantage over DPD but only for larger scales where the fluid particles are adequately large and can interact only with their closest neighbors. In this situation, DPD is less efficient because many more particles than for FPM should be involved for creating a drag between circumvented DPD particles. DPD is computationally more efficient than FPM at smaller scales, for which the interaction range (r_{cut}) of the potential must be longer.

The three-level numerical model is designed for simulating complex fluids. Three types of particles are defined accordingly by:

Colloidal Particles with an interaction range $\geq 2.5 \times \lambda$, where λ is a characteristic length, equal to the average distance between particles. The CP–CP interactions can be simulated by a soft-sphere, energy-conserving potential with an attractive tail. The CP–CP forces conservative in nature and can be simulated by the two-body Lennard–Jones potential (see Equation (26.14)).

Dissipative particles, the "droplets of fluid" represented by solvent particles (SP) located in the closest neighborhood of the colloidal particles with an interaction range $\geq 2.5 \times \lambda$. The SP–SP and CP–SP forces represent only the two-body central forces given by Equation (26.25).

Fluid particles (FP), the "lumps of fluid" represented by the particles in the bulk solvent, with an interaction range $\leq 1.5 \times \lambda$. Noncentral forces are included within this framework (see Equation (26.29)).

Interactions among colloidal particles have been studied for more than 50 years. Colloids can be regarded as complex many-body systems described by highly approximate treatments drawn from classical statistical physics. As follows from the conventional (Derjaguin, Landau, Verwey, Overbeek) DLVO theory, the long-range electrostatic interaction between colloidal spheres can be modeled by a screened-Coulomb repulsion [90]. Additional interactions come from hydrodynamic [91] and depletion forces [92]. Some experimental findings [91,93] show that like-charged macro ions have been attracted to one another by short-ranged forces. This fact cannot be explained by the conventional theories. The simulation results described in Ref. [94] show that the fluctuation of the charge distribution by the small ions results in the attraction between micro ions. The mean force is a combination of hard-sphere and electrostatic force. As an approximation $\Phi(r_{ij})$ (see Equation (26.14)) of the mean force between colloidal beads, we use the sum of the Lennard–Jones (L–J) force and a very steep force with a soft core. In Figure 26.10 we depict that the approximation is very close to the mean force obtained from large-scale Monte–Carlo

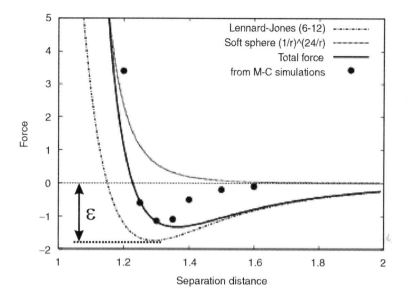

FIGURE 26.10 The model of the two-body conservative forces representing the CP–CP interactions [21].

calculations performed for a real colloidal mixture [94]. The force well depth is proportional to ε, the minimum of L–J potential, which is called here the cohesion factor. Unlike in our previous papers exploiting two-level model [20,84,95], we introduce an additional noncentral dissipative force between colloidal beads. This force is responsible for dissipation of energy due to bead collisions. The colloidal particles are larger and heavier than FPM "droplets" and their interactions are singular. Therefore, thermal fluctuations are transferred from the bulk of fluid and are partly dissipated inside the colloidal agglomerates ($\tilde{\mathbf{F}}_{ij} = 0$, if i and j are colloidal particles). There is also no drag between colloidal beads ($B(r_{ij}) = 0$) because the binder layer covering colloidal beads is assumed to be very thin.

The bead-"droplet" interactions are simulated by employing FPM forces. This is justifiable on the following grounds:

1. The bead-"droplet" forces cannot be singular.
2. Nonzero Brownian component ($\tilde{\mathbf{F}}_{ij} \neq 0$) comes from the fluid "droplet".
3. Viscous drag ($B(r_{ij}) \neq 0$) comes from the fluid "droplet" and the electrolyte binder.
4. The electrolyte concentration in electrolyte–solvent mixture surrounding the agglomerate is low, thus electrostatic bead-"droplet" interactions are negligibly small.

Because the colloidal bead contains a hard core, we have modified repulsive part of conservative $\mathbf{F}^C$ bead-"droplet" forces, thus making it steeper than for "droplet"–"droplet" interactions.

Solvent particles are represented by DPD particles and they are employed for simulations involving a larger cut-off radius. In the case of three-level computations in which the interaction range of CP particles is $\geq 2.5 \times \lambda$, DPD particles are defined only within the nearest neighborhood of CP particles. We employ the FPM model for simulating the rest of the fluid system. Owing to the comparable size for the free types of particles, the timestep for integrating the Newtonian equations of motion is uniform. Thus the three-level system consists of three different procedures, each representing a particular technique of interparticle interactions.

F. Smoothed Particle Hydrodynamics

In the macroscale, the particles can also be considered as moving mesh nodes, as it is so conceived in SPH [86,96] and moving particles semi-implicit method (MPS) [96]. Both of the algorithms belong to the wider class of approximation methods defined in Ref. [33]. The particle system is defined by mass m_j distribution in space, where fluid density in r is given, by the approximation formula

$$\rho_i(\mathbf{r}) = \sum_{j=1}^{N_i} m_j W_{ij}(\mathbf{r} - \mathbf{r}_j, h) \tag{26.36}$$

where $W(\mathbf{r}_{ij}, h)$ is a a smoothing (or interpolation) kernel function and h defines a cut-off radius for which if $|\mathbf{r}_{ij}| > h \Rightarrow W(\ ) = 0$. The role of this function is to transform the point representation of mass of particle to a spatially spread density. It is a bell-shaped function, which has a continuous and smooth first and usually second derivative. The simplest function, which meets these requirements is the Gaussian

$$W(r, h) = \frac{1}{h^\nu \pi^{\nu/2}} e^{-(r/h)^2}. \tag{26.37}$$

The simple Gaussian has the practical disadvantage that is not finite in extent. To overcome this problem, we can use the second-order accurate form of Monaghan and Lattanzio kernel, which is given by the formula

$$W(r, h) = \frac{1}{\pi h^3} \begin{cases} 1 - \frac{3}{2}\left(\frac{r}{h}\right)^2 + \frac{3}{4}\left(\frac{r}{h}\right)^3, & 0 \leq \frac{r}{h} < 1 \\ \frac{1}{4}\left[2 - \left(\frac{r}{h}\right)\right]^3, & 1 \leq \frac{r}{h} < 2 \\ 0, & \frac{r}{h} \geq 2 \end{cases} \tag{26.38}$$

From the principles of approximation [96] and statistical mechanics, we can show that for a continuum function $f(\ )$ the spatial average of this function and its gradient are defined respectively as

$$\langle f(\mathbf{r}) \rangle = \sum_{i=1}^{N} f(\mathbf{r}_i) \cdot W(\mathbf{r} - \mathbf{r}_i, h)\frac{m_i}{\rho_i}, \qquad \langle \nabla f(\mathbf{r}) \rangle = \sum_{i=1}^{N} f(\mathbf{r}_i) \cdot \nabla W(\mathbf{r} - \mathbf{r}_i, h)\frac{m_i}{\rho_i}. \tag{26.39}$$

This spatial averaging over different length-scales transforms the partial differential equations of continuum hydrodynamics into a set of nonlinear ordinary differential equations for the particle velocity $\mathbf{v}$ and internal energy ε

$$\frac{d\mathbf{v}_i}{dt} = -\sum_{j=1}^{N_i} m_j \left(\frac{P_i}{\rho_i^2} + \frac{P_j}{\rho_j^2}\right) \cdot \nabla W(\mathbf{r}_i - \mathbf{r}_j, h),$$

$$\frac{d\varepsilon_i}{dt} = -\frac{P_i}{\rho_i^2}\sum_{j=1}^{N_i} m_j(\mathbf{v}_i - \mathbf{v}_j) \circ \nabla W(\mathbf{r}_i - \mathbf{r}_j, h) \tag{26.40}$$

where N_i is the number of particles in $O(\mathbf{r}_i, h)$ sphere. The first equation can be interpreted as the equation of motion for a set of nodes i and j in which transient pressures and densities are P_i, P_j and ρ_i, ρ_j, respectively, and which "interactions" are derived in a canonical manner from the force laws of continuum hydrodynamics. In order to compute the local pressure, an equation of state $P_i(\rho_i, u_i)$

should be specified. For an ideal gas, the equation of state is

$$P_i = (\kappa - 1)\rho_i\varepsilon_i \qquad (26.41)$$

where κ is the ratio of specific heat capacities of the gas and ε_i is the internal energy per unit mass for particle i, also referred to as specific internal energy. We can also use the polytropic equation of state

$$P_i = K\rho_i^\kappa \qquad (26.42)$$

where K is a measure of the entropy of the gas, which can be globally constant or can vary both globally and in local terms. These two equations assume that the ratio of specific heats κ is unchanging throughout the simulation. The following equation

$$P_i = \frac{k\rho_i T_i}{\mu_i} \qquad (26.43)$$

has no such the restriction, as the temperature appears explicitly in the equation of state where k is Boltzmann's constant, T_i is the temperature, and μ_i is the mean molecular weight associated with particle i. In order to avoid large fluctuations in the SPH particle system, the artificial viscosity term in Equation (26.40) has to be introduced [86,96]. Similarly to MD, DPD, and FPM, the evolution of the smoothed particle system is then described by the Newtonian equations of motion Equation (26.1)–Equation (26.3).

The following features make SPH useful for simulation macroscopic flows:

- SPH will automatically follow complex flows and can easily maintain constant mass resolution. This can be achieved by using spatially variable smoothing lengths $h(r)$, and keeping a constant number of neighbors within radius $\sim h$, (typically one requires 30–70 neighbors within 2h.) This gives a large dynamic range in density, not an unusual situation for astrophysical problems. Typically, densities ranging dynamically over more than six orders of magnitude can be represented.
- SPH only performs calculations in the relevant regions, i.e., where the mass is located. No computational time will be spent in empty regions.
- It is relatively easy to include other physical processes in SPH. The equations retain their original mathematical form to a high degree. Many types of equations of states have been used, sometimes very complex.

The SPH has also disadvantages, which can be enumerated as follows:

- Simulating shock phenomenon is not the strongest part of SPH. SPH relies on the use of artificial viscosity, in order to resolve shocks. Typical shock resolution is then around three smoothing lengths. State-of-the-art finite difference schemes, like the Piecewise Parabolic Method (PPM) [98], can achieve far better (at comparable resolution). Typically, SPH needs numbers of particles in excess of 100,000, when shocks are present.
- SPH works best if the smoothing lengths are variable, but the resolution will then be highest in high density regions. Low-density regions may then be poorly resolved. The transition of sound waves into shocks, as they propagate into a rarefied medium, is an example where SPH would not be suitable.
- In generic SPH spherical kernel functions are used. The distribution of neighbors to a particle should then be roughly isotropic for the interpolation formulas to work properly. This will not be the case if thin (less than $\sim h$), sheets or disks forms. Although, this is mainly a question of

using sufficient resolution (enough many particles), three dimensional SPH does not correctly represent "two-dimensional" SPH when a computational region contracts to a sheet.

In order to bridge mesoscopic scales with the scales described by SPH technique, Serrano and Español [32] propose a new version of dissipative particle dynamics, the thermodynamically consistent DPD (TC-DPD).

G. THERMODYNAMICALLY CONSISTENT DPD

The thermodynamically consistent DPD method at present represents a superset for the classical DPD, FPM, and smoothed particle dynamics models. The main features of this new approach in comparison to the classical DPD are that:

1. Apart from position $\mathbf{r}_i$ and velocity $\mathbf{v}_i$, the volume v_i (or density ρ_i), temperature T_i and entropy S_i of the particle X_i are the relevant dynamical variable, thus $X_i \equiv (\mathbf{r}_i, \mathbf{v}_i, v_i, T_i, S_i)$ and
2.

$$\rho_i = \sum_j W(r_{ij}), \qquad v_i = \frac{1}{\rho_i}, \qquad \tilde{\mathbf{e}}_{ij} = W'(r_{ij})\mathbf{e}_{ij} \qquad (26.44)$$

where $W(\ )$ is a Gaussian weighting function.
3. The forces are given in terms of discrete versions of the gradient of the stress tensor.

This approach involves finite volume Lagrangian discretization of the continuum equations of hydrodynamics through the Voronoi tessellation technique shown in Figure 26.11.

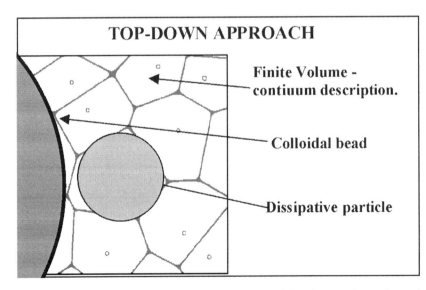

FIGURE 26.11 The schematics of coarse-graining idea realized by the top-down thermodynamically consistent DPD model. The Voronoi cell represents a fragment of continuum fluid. This fragment can be treated as a dissipative particle with variable mass, volume, temperature and entropy. The thermodynamically consistent DPD particles interact with forces dependent not only on their positions and velocities but on the current thermodynamic states of interacting particles as well.

Finally, we can describe the evolution of the fluid particle system by the reversible (SPH)

$$\dot{\mathbf{P}}_i = -\sum_j \tilde{\mathbf{e}}_{ij}\left(\frac{p_j}{\rho_j^2} + \frac{p_i}{\rho_i^2}\right) \tag{26.45}$$

and irreversible terms for momentum P* and entropy S*

$$\dot{\mathbf{P}}_i^* = -\sum_j \frac{1}{\hat{v}} \gamma_{ij}(T_i, T_j)(\mathbf{v}_{ij} \cdot \tilde{\mathbf{e}}_{ij})\tilde{\mathbf{e}}_{ij} + \sum_j \tilde{\mathbf{F}}_{ij},$$

$$T_i\frac{\mathrm{d}}{\mathrm{d}t}(S_i^* - \tilde{S}_i) = \sum_j f(T_i, T_j, \hat{v}, \mathbf{v}_{ij}, \tilde{\mathbf{e}}_{ij}) \tag{26.46}$$

where $\hat{v}$ is the average volume of i and j particles, γ_{ij}, the viscosity, $\tilde{\mathbf{F}}, \tilde{S}$, are the random factors for the force and entropy, respectively. The temperature T_i, the chemical potential and the pressure for each of fluid particles can be computed as corresponding partial derivatives of internal energy function $\varepsilon(S^*, N, V)$ which form can be approximated by the fundamental equation of ideal gas (as it is in Ref. [32]) or derived experimentally for a real fluid.

The formalism of thermodynamically consistent dissipative particle dynamics represents a consistent discrete model for the Lagrangian fluctuating hydrodynamics. Equation (26.45)– Equation (26.46) conserve the mass, momentum, energy and volume. The irreversible (produced) entropy S* is a strictly increasing function of time in the absence of fluctuations. Thermal fluctuations represented by $\tilde{\mathbf{F}}, \tilde{S}$ are consistently included, which lead to an increase of the entropy and to correct for the Einstein distribution function [31].

Viscous forces between a pair of fluid particles depend not only on the velocities of this pair but also on the velocities of the nearest neighbors. Therefore, we need to retain a great deal of information about the fluid state. Moreover, unlike in classical DPD, thermodynamically consistent DPD method adaptively controls the spatio-temporal scale of the model. The transport properties of thermodynamically consistent DPD fluid are known *a priori*. The size of the thermal fluctuations is given by the typical size of the volumes of the particles, scaling as the square root of this volume. The need to incorporate thermal fluctuations in a particular system will be determined by the external length-scales that need to be resolved. In the case of submicron colloidal particles, we require to resolve the size of this particle with fluid particles of size an order of magnitude or two smaller than the diameter of the colloidal particle. For these small volumes, fluctuations are important and lead to the Brownian motion. The grains, one order of magnitude greater, require fluid particles to be much larger, in which case thermal fluctuations are small or negligible. The thermodynamically consistent DPD represents a truly multiscale discrete-particle model, which shows its true ability by the effect of controlling the amplitude of thermal fluctuation depending on the spatio-temporal scale simulated. This, for example, allows for studying both the lamellar and disordered structures in binary fluids in multiple scales.

Summarizing, the hierarchy of gridless particle methods is presented in Figure 26.12. The methods establish a sound foundation for cross-scale computations, ranging from nanometers to centimeters. They can provide a framework to study the interactions between microstructures and large-scale flow, which are of value in blood flow [22,99–101] and other applications in polymeric and blood dynamics.

We should emphasize strongly here that molecular dynamics forms the centerpiece from which the other techniques are derived and are employed for attaining greater length-scales. Therefore, the numerical models of the particle methods having similar framework are very interesting for modeling multiscale phenomena. As shown in Refs. [40,82,99], by generalization of the well-known numerical MD models onto mesoscopic scales, we can solve many technical problems by correctly

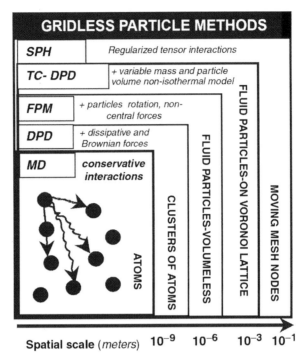

FIGURE 26.12 Schematic diagram illustrating the various hierarchies of the gridless particle methods and their resolved length-scales, spanning from nanometers to macroscopic dimensions.

implementing other discrete-particle methods. Some of these mainly algorithmic and implementation problems we demonstrate in the following section.

III. COMPUTATIONAL ASPECTS OF DISCRETE-PARTICLE SIMULATIONS

As we showed in the previous section, the discrete-particles model represents a multiresolutional approach for modeling complex phenomena from atomistic scales to the scales of meters. By changing the notions of the "particle," the "interactions" and by controlling the magnitude of fluctuations, we can use the same computational framework for simulating different scales. This is a real advantage over hybridized continuum-discrete models [24,65]. On the other hand, the high computational demand required by particle-based codes is the principal disadvantage of this approach. However, the efficiency of discrete-particle methods can be considerably increased by using modern algorithms, processor architectures and parallel systems.

The common computational framework of the discrete-particle approach bases on four types of algorithms:

1. The algorithms used for efficient computations of interparticle forces. The principal problem is how to avoid crude schemes in which all particle–particle forces are evaluated. Assuming that the number of particles is equal to N, for two-body forces the crude scheme is of order $O(N^2)$.
2. The numerical schemes used for integration of the Newtonian equations of motion. Stable, accurate and efficient schemes are in great demand. Unlike in the standard MD codes, in FPM and DPD due to the random Brownian force, the equations of motion are stochastic

differential equations (SDE). Therefore, new numerical schemes, different than those used for ordinary differential equations, are necessary for integration.

3. Complicated boundary conditions should be simulated.
4. Clustering procedures are required for extracting microstructures developed due to simulation.

To make calculations more efficient, the particle codes are usually run on multiprocessor systems in broad variety of computer architectures and parallel interfaces. In this section, we present basic issues exploited in construction of a typical discrete-particle code based on the FPM.

A. CALCULATION OF INTERPARTICLE FORCES

We consider here an isothermal three-dimensional system, which consists of M particles. The particle system is simulated within a rectangular box. The box can be closed or periodic. The particles of uniform or various types can be distributed randomly in the box, i.e., this multicomponent system can be perfectly mixed initially or separated by a sharp interface (stratified, circle, rectangular, and random shape). The particles defined by mass m_i, position $\mathbf{r}_i$, velocity $\mathbf{v}_i$ and angular velocity $\boldsymbol{\omega}_i$ interact with each other via a two-body, short-ranged forces given by Equation (26.25)–Equation (26.29). We assume also that, r_{cut} is a cut-off radius, which defines the range of FPM particle interactions. For $r_{ij} > r_{cut}$, $F_{ij} = 0$.

As shown in Figure 26.13, the box is divided into cubic cells of the edge size $l_C \sim r_{cut}$. For multicomponent fluid with different interaction ranges, we assume that $l_C \sim \max_k(r_{cut,k})$, where k means the kind of interaction. The forces are computed by using $O(M)$ order link-list scheme [67]. The force on a given particle includes contribution from all the particles that are closer than r_{cut} and which are located within the cell containing the given particle or within the adjacent cell (see Figure 26.13a). The link-list scheme can also be used for finding neighbors in the sphere of radius $r_{cut} + \delta$ (see Figure 26.13b), where $\delta \ll r_{cut}$. The lists of neighbors (so-called Verlet neighbor list [29]) calculated for each particle can be used during δ/v_{max} timesteps, where v_{max} is the maximal particle velocity in the system. Then the lists have to be updated by using the link-list scheme. On the one hand, such the hybrid algorithm combined with multiple-timesteps method can increase the computational efficiency more than twice.

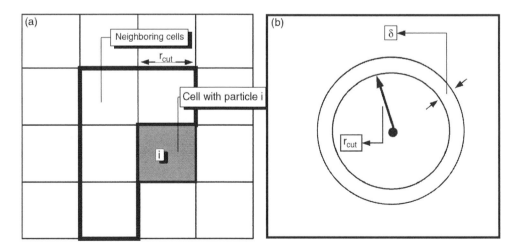

FIGURE 26.13 (a) The computational box divided into cells with size of the interaction range r_{cut}. (b) The Verlet list of neighbors is searched in the sphere with radius slightly larger than r_{cut}.

On the other hand, it involves high computational load and is usually inefficient in parallel environments.

Parallel computing requires decomposing the computation into subtasks and mapping them onto multiple processors. The total volume of the box is divided into P overlapping subsystems of equal volume, and each subsystem is assigned to a single processor in a P processors array. By using single program multiple data (SPMD) paradigm, commonly used for MD code parallelization, each processor follows an identical predetermined sequence to calculate the forces on the particles within assigned domain.

Among many parallel implementations of molecular dynamics code [8,24,30,35] two approaches for particles redistribution between processors are employed (see Figure 26.14).

1. The box is sliced along one coordinate and divided into identical sub-boxes (see Figure 26.14a).
2. The system is partitioned into a mesh of sub-boxes in x-, y-, and z-directions (see Figure 26.14b).

The particle positions and velocities from cells, which are situated on the boundaries between processor domains, are copied to the neighboring processor (see Figure 26.15). Thus the number of particles located in the boundary cells defines the communication overhead.

Let us assume that:

1. The system is confined in a box elongated in z-direction and the x, y cross-section of the box is a square of unit area.
2. The number of processors P, the system size and the length of the box $L_z \gg 1$ are constant.
3. The box is partitioned along z-axis onto processor domains.

For this case, the communication overhead t_{strips}, which is proportional to the area of the interface between processor domains, is constant and equal to 1. Let us assume that the box is partitioned additionally into n^2-mesh of identical sub-boxes on x, y-plane. The communication cost t_{box} for $n > 1$ is proportional to the area of walls (only half of them) of a single sub-box and is equal to $[2L_z/(P/n^2)] \cdot 1/n + 1/n^2$. For sufficiently long boxes with $L \gg 1$ the ratio of two overheads $\varphi = t_{box}/t_{strips} = [2L_z/(P/n^2)] \cdot 1/n + 1/n^2$ is greater than 1. This means that the communication overhead is lower, and consequently calculation communication ratio higher, for the first method

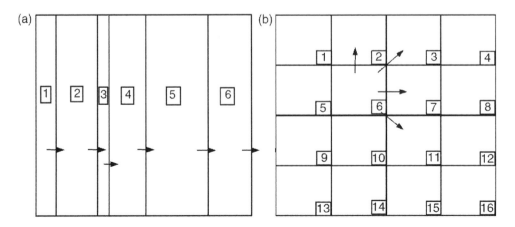

FIGURE 26.14 Two approaches for geometric parallelization of the computational box.

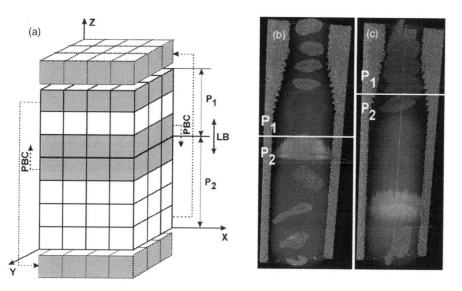

FIGURE 26.15 (a) The schematics of the geometrical decomposition of the computational box. The decomposition of the computational boxes for red blood cells flow in a capillary without (b) and with (c) load-balancing [102].

consisting in slicing the box onto strips along z co-ordinate. The value of φ is less than 1 for more regular computational boxes, e.g., a cubic computational box for which $P \sim n^3$ and $L_z = 1$. In this case, the second partition method dividing the box into cubic sub-domains is better.

Slicing the box along the z-axis considerably simplifies the routing of messages and enables sending them in unblocking way. Each processor sends the message only in one direction to its closest neighbor. The load balancing is easier and consists in shifting the boundaries of processor domains along one direction, whereas for the second method the load balancing schemes are very complex requiring irregular mesh.

Many parallel implementations of molecular dynamics codes employ neighbor tables for each particle for speeding-up the evaluation of forces. This increases considerably the memory requirements, communication overheads, and makes the code more complex. FPM has two-fourth times greater memory requirements than codes for molecular dynamics. Besides the positions and forces in highly optimized parallel codes for large-scale MD [8,35] (minimum 6 arrays), additional arrays must be allocated such as: the angular and translational velocities, torques and replicated arrays for velocities needed for integrating Newtonian equations of motion, that is, minimum 24 arrays. Moreover, the random number generator is invoked for computation of Brownian forces for each pair of interacting particles.

Therefore, the speed-up expected from application of neighbor tables can be compromised due to the effect of frequent cache misses resulting from its overload. The particles from boundary cells "cached" on the neighboring processors and those migrating from one processor to another must be updated every timestep (see Figure 26.15). Unlike in MD, the FPM forces (see Equation (26.29)) depend not only on the particle positions but also on translational and angular velocities. Moreover, besides reaction forces, the reaction torques must be updated. Thus, the communication overhead is almost three times greater for FPM than for MD. Because FPM fluid particle interacts only with their closest neighbors, the number of interactions per particles is smaller by factor of 4.5 than for a standard MD code. However, the number of arithmetic operations involved for evaluation of FPM interactions is greater, at least by the same factor, than for calculating the Lennard–Jones forces in MD code. Thus, we may expect that computational load per particle should be similar for these two cases. Summarizing, the high memory load in FPM will result in:

1. greater communication overhead,
2. more frequent occurrence of cache misses,

than in standard implementations of MD method in multiprocessor environment (e.g., see Ref. [35]).

B. Temporal Evolution of Fluid Particles

Integration of the Newtonian equations of motion in DPD and FPM is more complex than in MD. From Equation (26.25) and Equation (26.29), we note that the forces and torques depend not only on particle positions (as in MD) but translational and angular velocities as well. Moreover, due to the random Brownian force, the equation of motion are SDE. Numerical integration of SDE by using classical Verlet [29] and leap-frog schemes generate large numerical errors and artifacts [103], e.g., resulting in unacceptable temperature drift with simulation time. Therefore, very small timesteps should be used to obtain a reasonable approximation to the thermodynamical quantities. The Verlet and leap-frog schemes for DPD are following succeedingly.

1. Verlet

1. $\mathbf{v} \leftarrow \mathbf{v} + \dfrac{1}{2}\dfrac{1}{m}\left(\mathbf{F}^C \Delta t + \mathbf{F}^D \Delta t + \mathbf{F}^B \sqrt{\Delta t}\right)$
2. $\mathbf{r} \leftarrow \mathbf{r} + \mathbf{v}\Delta t$
3. Calculate $\mathbf{F}^C$, $\mathbf{F}^D(\mathbf{v})$, $\mathbf{F}^B$ $\hspace{4cm}$ (26.47)
4. $\mathbf{v} \leftarrow \mathbf{v} + \dfrac{1}{2}\dfrac{1}{m}\left(\mathbf{F}^C \Delta t + \mathbf{F}^D \Delta t + \mathbf{F}^B \sqrt{\Delta t}\right)$
5. Calculate $k_B T = \dfrac{m}{3N-3}\sum_{i=1}^{N} \mathbf{v}_i^2$

2. Leap-Frog

1. $\mathbf{v}_0 \leftarrow \mathbf{v}$
2. $\mathbf{v} \leftarrow \mathbf{v} + \dfrac{1}{2}\dfrac{1}{m}\left(\mathbf{F}^C \Delta t + \mathbf{F}^D \Delta t + \mathbf{F}^B \sqrt{\Delta t}\right)$
3. $\mathbf{v}_0 = \dfrac{1}{2}(\mathbf{v}_0 + \mathbf{v})$
4. $\mathbf{r} \leftarrow \mathbf{r} + \mathbf{v}\Delta t$ $\hspace{4cm}$ (26.48)
5. Calculate $\mathbf{F}^C$, $\mathbf{F}^D(\mathbf{v})$, $\mathbf{F}^B$
6. Calculate $k_B T = \dfrac{m}{3N-3}\sum_{i=1}^{N} \mathbf{v}_{0i}^2$

The predictor-corrector numerical schemes are both very time and memory consuming, which for high memory load for DPD and FPM will result in additional overheads. However, as shown in Figure 26.16, the two schemes such as DPD-VV [83] and especially the modern scheme described in Ref. [103], which applies thermostat, allow for using more than twice larger timesteps and greater accuracy than obtained for the Verlet and leap-frog algorithms. The schemes are following succeedingly.

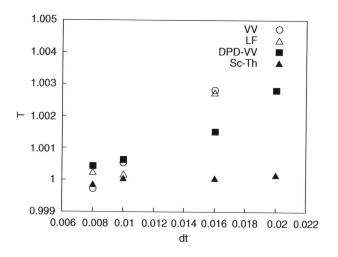

FIGURE 26.16 Increase of the error in calculated temperature T for the particle system ($N = 10^5$ particles) for DPD simulations, which use various timestep dt.

3. DPD-VV

1. $\mathbf{v} \leftarrow \mathbf{v} + \dfrac{1}{2}\dfrac{1}{m}\left(\mathbf{F}^C \Delta t + \mathbf{F}^D \Delta t + \mathbf{F}^B \sqrt{\Delta t}\right)$
2. $\mathbf{r} \leftarrow \mathbf{r} + \mathbf{v}\Delta t$
3. Calculate $\mathbf{F}^C$, $\mathbf{F}^D(\mathbf{v})$, $\mathbf{F}^B$
4. $\mathbf{v}_0 \leftarrow \mathbf{v}$
5. $\mathbf{v} \leftarrow \mathbf{v} + \dfrac{1}{2}\dfrac{1}{m}\left(\mathbf{F}^C \Delta t + \mathbf{F}^D \Delta t + \mathbf{F}^B \sqrt{\Delta t}\right)$
6. $\mathbf{v}_0 \leftarrow \dfrac{1}{\sqrt{2}}\sqrt{(\mathbf{v}_0^2 + \mathbf{v}^2)}$
7. Calculate $\mathbf{F}^D(\mathbf{v})$
8. Calculate $k_B T = \dfrac{m}{3N-3}\sum_{i=1}^{N}\mathbf{v}_{0i}^2$

$$(26.49)$$

4. SC-TH

1. $\dot{\eta} \leftarrow C(k_B T - k_B T_0)$
2. $\eta \leftarrow \eta + \dot{\eta}\Delta t$
3. $\gamma \leftarrow \dfrac{\sigma^2}{2 k_B T_0 m}(1 + \eta\Delta t)$
4. $\mathbf{v} \leftarrow \mathbf{v} + \dfrac{1}{2}\dfrac{1}{m}\left(\mathbf{F}^C \Delta t + \mathbf{F}^D \Delta t + \mathbf{F}^B \sqrt{\Delta t}\right)$
5. $\mathbf{r} \leftarrow \mathbf{r} + \mathbf{v}\Delta t$
6. Compute $\mathbf{F}^C$, $\mathbf{F}^D(\mathbf{v})$, $\mathbf{F}^B$
7. $\mathbf{v} \leftarrow \mathbf{v} + \dfrac{1}{2}\dfrac{1}{m}\left(\mathbf{F}^C \Delta t + \mathbf{F}^D \Delta t + \mathbf{F}^B \sqrt{\Delta t}\right)$
8. Compute $\mathbf{F}^D(\mathbf{v})$
9. Compute $k_B T = \dfrac{m}{3N-3}\sum_{i=1}^{N}\mathbf{v}^2$

$$(26.50)$$

The size of the timestep Δt should be estimated from the characteristic time scales for both rotational and translational motion. The mean collision time τ_{col} defines the time scale for the translational motion, which is given by:

$$\tau_{col} = \frac{\lambda}{\langle v_{rel} \rangle} \tag{26.51}$$

where $\langle v_{rel} \rangle$ is a relative velocity, λ, is the characteristic length-scale, which is equal to the average distance between particles.

Both the quality and numerical stability of the model can be estimated from the temporal behavior of the thermodynamic temperature Tth and dimensionless pressure $\delta = k_B \cdot T/(P/n)$. As shown in Ref. [95], the temperature Tth of the system, computed as the average kinetic energy of the FPM particle systems, fluctuates no more than 1.5%. Its average differs from the temperature T assumed (computed from detailed balance Equation (26.27)) on approximately 0.1%. For comparison, at the similar simulation conditions (but in 2D) and the same timestep, the equilibrium temperature Tth for DPD simulation of phase separation obtained in Ref. [104] is roughly twice its input value. The temperature drift (upward or downward, depending on the hardware and compiler used) caused by the round-off error, which is apparent for large number of timesteps, we have greatly reduced by using 64-bit compiler. The partial pressure Pth of FPM fluid computed from the viral theorem [29], can also be approximated accurately by the Equation (26.26) and Equation (26.32) [71,95].

C. Boundary Conditions

Periodic boundary conditions (PBC) simulate the system of unlimited number of interacting particles by limited number of interacting lattices where each of them stands for a particle and its replicas. When the distance between a particle and its nearest image is too short, long wavelength phenomena are cut and their energy is passed to the shorter waves, which go through the box generating numerical artifacts. Moreover, the commonly used computational box shape, such as rectangular prism, makes the system highly anisotropic. In Refs. [105,106], the minimum image convention is presented for noncubic boxes such as truncated octahedron, rhombic dedocahedron and hexagonal prism. In spite of the more symmetric geometry and savings in CPU time due to increase of the nearest image distance, the noncubic boxes are still not popular in particle simulations. There are at least two basic problems with noncubic boxes for simulating large particles ensembles.

1. Noncubic boxes involve noncubic cells in the linked-cells algorithm. This makes the code very clumsy (especially in 3-D) due to greater number of walls, edges and vortices in noncubic cells than for cubic ones, thus involving complicated nearest image convention schemes [107].
2. Domain decomposition is difficult for noncubic boxes.

In Ref. [107], a method for uniformization of the periodic box shape for small particle system was presented. As shown in Figure 26.17b, the periodic box can be divided into two, black and white, rectangles of the same size. Unlike for the periodic square, the box replicas are shifted creating checker board picture (see Figure 26.17b). For properly selected box sizes Lx, Ly, and Lz, we can reproduce different shapes. For example, the periodic hexagon can be simulated assuming that $Lx/Ly = 1/\sqrt{3}$ (see Figure 26.17c) while the box with Lx = 1, Ly = 1, and Lz = 2 (see Figure 26.17) corresponds to periodic rhombic dodecahedron [107].

The possibility of application of linked-lists method with cubic cells for noncubic periodic boxes is the great advantage of using the checker-board PBC. We present following the translation

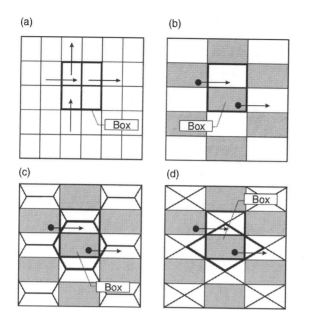

FIGURE 26.17 (a) Classical periodic boundary conditions, (b) checker-board periodic boundary conditions (CPBC), (c) and (d) various shapes of computational box simulated by CPBC.

scheme for renumbering the cell coordinates: Nx, Ny, and Nz, from the border of the computational box by replicating periodic rombic dodecahedron.

$$ix = INT ((float(Nx)/N) + 1) - 1$$
$$iy = INT ((float(Ny)/N) + 1) - 1$$
$$iz = INT ((float(Nz)/N) + 1) + INT (float (Nz)/N) - INT (float(Nz)/(2 * N)) - 2$$
$$iz = iz * [(abs(ix) + abs (iy) + abs(iz) - 1) \bmod 2]$$
$$Nx = Nx - N * ix$$
$$Ny = Ny - N * iy$$
$$Nz = Nz - N * iz$$

where N is here the number of cells in each direction of one half of computational box.

The parallel code for the checker-board periodic boundary conditions is relatively easy to implement by assuming that the box is decomposed by segmenting it along x or y coordinate (on Figure 26.15). For boxes elongated in z-direction, such a decomposition will increase communication time due to thin layers of domains and larger interface area between neighboring processor domains. Slicing the box along z-axis (see Figure 26.15) may generate even more serious problems with communication. The processors will communicate not only with their neighboring processors, but also with the distant processors. In this situation, communication time may depend strongly on the architecture and memory access time of the parallel system.

For simulating the flow in an elongated and periodic capillary, we have employed the hexagonal prism PBC. The checker-board PBC are realized only on x, y plane. This preserves more circular shape of the capillary section than for a periodic rectangular prism and allows us to employ the same strategy of domain decomposition as shown in Figure 26.14a and Figure 26.15b,c. We simulate the box with circular section in x, y plane with reflecting or dissipative boundaries in x- and y-directions by filling superfluous space with heavy or motionless particles. Applying this same method we can model more complex boundaries such as shown in Figure 26.18 and Figure 26.19. The most difficult problem is to simulate the flow using the same number of particles. Unfortunately, there is not any efficient and reasonable algorithm, which can replace classical

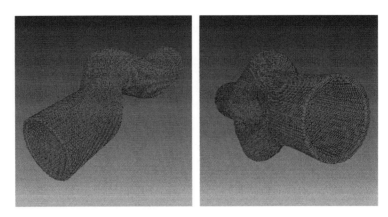

FIGURE 26.18 Modeled capillary vessels. The particles represent endothelial cells. They interact one with another via harmonic force to mimic the physics of elastic properties of the walls.

approach assuming the periodic boundary conditions in the direction of flow. To reduce inevitable errors resulting from this assumption, various techniques were implemented but any of them can be regarded as the best one.

D. CLUSTERING PROCEDURE

The patterns created in macrosopic flows, for which a homogeneous physical process dominates in multiple spatio-temporal scales, have typically self-similar fractal structures. In Refs. [21,87, 99,101], we show that the strong heterogeneities of the flow in the mesoscale co-produce complex multiresolutional patterns. The creation of micelles, colloidal arrays, colloidal agglomerates, and large-scale instabilities in fluid are the consequence of the competition between two coupled nonlinear processes: global motion of particle ensembles and local interactions between particles. These multiscale structures are complex due to the inflexibility of the description level with varying scale of observation. The detection of particle clusters for controlling their temporal behavior represents a very important aspect in visualizing and extracting the complex patterns. We have solved the problem for detecting clusters by using efficient $O(M)$ clustering algorithm (where M is the number of particles) inscribed in parallel structure of FPM code [99].

Clustering is a fundamental concept in pattern recognition and data mining (e.g., see [108,109]) but also has many applications in fields such as earthquake physics, astrophysics, and polymer fluid

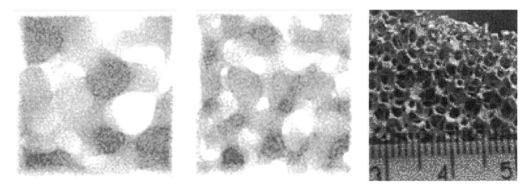

FIGURE 26.19 Two different models representing the structures of the same porosity but different average pore size, of the 3-D preform (here X–Y projections) generated by using droplets coalescence simulation [104]. The structures are compared to the ceramic preform used in an experiment.

dynamics. Clustering is used for classifying similar (or dissimilar) patterns represented by N-dimensional vectors. Depending on the data structures, different clustering schemes must be used. There are two principal approaches for classifying data. The first one consists in nonhierarchical extraction of clusters. This approach is used mainly for extracting compact clusters by using global knowledge of the data structure. The most known and simplest techniques base on K-means algorithm [108]. The main problem with K-means is that the final classification represents the local minimum of criterion function. Thus for multimodal functions, starting from different initial iterations, we can obtain different cluster structures. Because the global information about the data is required for the nonhierarchical schemes, the nonhierarchical algorithms also suffer from a high computational complexity and most of them require an *a priori* knowledge about the number of clusters. On the other hand, agglomerative clustering schemes [108] consist in the subsequent merging of smaller clusters into the larger clusters, based on a proximity criterion. Depending on the definition of the proximity measure, there exist many agglomerative schemes such as: average link, complete link, centroid, median, and minimum variance algorithm. The hierarchical schemes are very fast for extracting localized clusters with nonspherical shapes. All of them suffer from the problem of not having properly defined control parameters which can be matched for the data of interest and hence can be regarded as invariants for other similar data structures.

For finding the microstructures in colloidal suspensions, we have applied the agglomerative algorithm based on the mutual nearest neighborhood (MNN) concept [110], which is outlined as follows:

1. Find the list L_i of K nearest neighbors j of each particle i in R_{clust} radius and sort out the list in ascending order according to the distance between i and j particles. Thus $L_i(k) = j$ and k is the position of the particle j in the list. This procedure can be performed in parallel along with computation of forces in FPM code. To reduce the communication overhead, we use the parallel clustering algorithm off-line after simulation.
2. Assuming that $L_i(k) = j$ and $L_j(m) = i$, compute MNN(i, j) distances defined as: MNN$(i, j) = m + k$. The maximum MNN distance is less than $2K$.
3. Begin a classical agglomerative clustering algorithm (e.g., nearest linkage [103]) with the linked-lists concept, starting from the smallest MNN$(i, j) = 2$ value.
4. This is terminated upon reaching the greatest value of MNN.

The value of R_{clust} should be somewhat larger than the spacing between particles in aggregates ($R_{clust} \approx 0.2$–$0.3 \times r_{cut}$), and K value should be between 3–8. In Figure 26.20, we show the

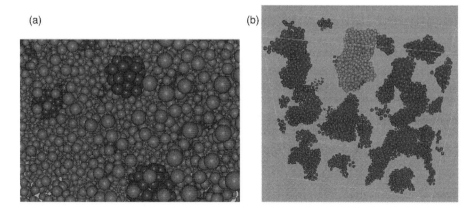

(a)   (b)

FIGURE 26.20 (a) Clusters of colloidal beads (dark gray) found by MNN algorithm for the two-level MD-SPH particle system. The gray particles represent a solvent. (b) Zoom-in of the system. Only colloidal clusters are shown. The largest one is colored in light gray.

cluster of colloidal beads obtained from aggregation process. These events are detected by using MNN algorithm [101,110].

The MNN clustering scheme usually applied for classifying N-dimensional patterns, is also truly promising tool for visualizing voluminous complex scientific data. Unlike the standard algorithms deployed in computer graphics, such as volume rendering, ray casting and ray tracing, the clustering schemes can be applied in extracting effortlessly valuable information from unstructured, irregular data, such as also found in finite-element calculations and wavelets besides the discrete-particle simulations treated here.

IV. COLLOIDAL DYNAMICS MODELED BY DISCRETE-PARTICLES

Here we present the results of high-performance simulation of mesoscopic fluids obtained by using discrete-particles methods described in the previous sections. Starting from the NEMD applied for simulating the Raleigh–Taylor mixing in the atomistic scale we show that by coarse-graining of scales we are able to model the same process involving mixing in the mesoscale, the droplet detachment process, phase separation and the thin-film evolution. We exploited multilevel particle approach for simulating micelles and colloidal crystals, creation of colloidal aggregates and their dispersion. Finally, we describe briefly the model of blood flow in capillary vessels, which is based completely on discrete-particle approach, and we discuss the results of modeling.

A. RAYLEIGH–TAYLOR INSTABILITY IN ATOMISTIC AND MESOSCOPIC FLUIDS

NEMD simulations in which collision operator Ω_{ij} is defined as a central, mostly short-ranged, conservative force have been used extensively in the past few years to study the rheology and instabilities in fluids represented by models of varying complexity [27,710pc102]. For example, for many years the problems of turbulent mixing have been thoroughly investigated theoretically, experimentally, and numerically [111–113]. The classical experiments have been carried with the temporal evolution of mixing involving two superimposed fluids, with different densities under a gravitational field (the Rayleigh–Taylor instability) or in shock waves (the Richtmyer–Meshkov instability). Numerous computer simulations, which are based on the Navier–Stokes equations, reveal the complex nature of the phenomenon, enable the validation of theoretical ideas concerning the initial stages of the process occurring in linear regime, and explain new mechanisms and properties of mixing in the nonlinear regimes.

In Ref. [20] we showed that the bubble-and-spikes regime of the Rayleigh–Taylor (R–T) instability (see Figure 26.21) is similar in both the molecular and in the macroscopic scales.

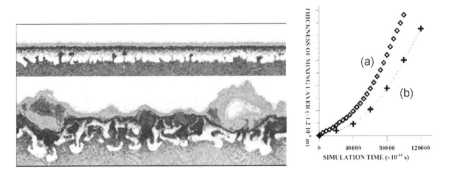

FIGURE 26.21 The snapshots from 2-D NEMD simulation of the Rayleigh–Taylor instability with free surface using one million Lennard-Jones particles [36,114,115]. The box length is about 0.5 μm long. The density contrast and microstructures such as bubbles and spikes are displayed. On the right a comparison of the mixing layer growth for the open (a) and closed (b) particle systems is displayed.

We simulated 2-D particle systems, which consisted of two particle layers of the same or different thickness. A layer consisting of heavy particles is placed upon the layer made of lighter particles. The layers represent the two Lennard–Jones fluids. The lighter one has the parameters of liquid argon while the heavier fluid X consists of particles whose parameters can be freely changed. Such an assumption can be justified, because we search for universal properties of mixing in molecular scale rather than for particular characteristics. Moreover, this assumption is very convenient, because the parameters of one fluid can be easily changed with respect to the second one. This approach allows us to explore more easily many different experimental conditions. We simulated particle ensembles consisting of $0.7 - 3 \times 10^6$ particles in $1.3 - 5 \times 10^5$ timesteps.

As shown in Ref. [113], the growth time of bubble layer h is given by the semiempirical relationship

$$h = A \cdot \alpha \cdot g \cdot t^2 \qquad \text{where } \alpha = \frac{\rho_1 - \rho_2}{\rho_1 + \rho_2} \qquad (26.52)$$

is the Atwood number, ρ_1 and ρ_2, are the densities of the heavy and light fluids respectively, A is the growth constant. The value of the growth constant $A \approx 0.07$ (in 3-D) and $A \approx 0.05$ (in 2-D) for microscopic mixing in 1 μm layer of Lennard–Jones fluid (see Figure 26.21), is approximately the same as that in the macroscale [113]. We also showed [20] that it remains stable for various physical conditions. The influence of fluid granularity on the speed of mixing can be observed only at a very early (exponential) stage of R–T instability. This start-up time is connected with spontaneous instabilities formation, which appears as the cumulative result of thermal fluctuations. We can conclude that the occurrence of the fluid instability in the microscale and its resemblance to the similar processes in the macroscale can expand the scope of discrete-particles applications for modeling greater spatio-temporal scales, especially mesoscopic phenomena where the thermal fluctuations cannot be neglected.

The similar problem of mixing and dispersion of heterogeneities in viscous and immiscible fluids is found universally in many fields in science and engineering. In spite of the recent advances in understanding the mixing of homogeneous fluids in the macroscale, realistic mixing problems of heterogeneous multicomponent fluids are inherently difficult due to the complex, multiple-scale nature of the flow fields, the intrinsic rheological complexity and the cross-scaling nature of the nonlinear physics, such as grain-boundary processes, chemical reactions, and geological processes with complex rheologies. For these reasons, mixing problems have traditionally been treated on a case-by-case basis. Classical modeling with partial-differential equations becomes intractable, if one wishes to examine all of the details simultaneously on a cross-scale basis. As shown in Refs. [116,117], current models incorporate two competing processes in mixing: breakup and coalescence. These two mechanisms involve one-sided interactions without any feedback from the microstructural dynamics back to the global flow structure. This problem of feedback of microstructural dynamics is very important in many applications, such as dispersion of solid nanoparticles through a polymer blend and dynamical mixing of crystals in magmatic flows.

Because of the many similarities encountered in NEMD simulation, we can justify the use of discrete-particle method over larger scales for simulating R–T instabilities. This extension to larger spatial scales can be realized by changing only the notion of the interparticle interaction potential by treating a large-sized particle as a cluster of computational molecules. This idea of up-scaling has already been adapted in the DPD method described in Section II. In Ref. [95], we show that in the turbulent R–T instability, the sustained acceleration causes the dominant spatial scales to evolve self-similarly and the initial scales are forgotten. The mixing layer depth h_T increases with time as $h_T \approx \alpha A_T G t^k$ as shown by many numerical (e.g., see [113]) and experimental results (e.g., see [112]). For the two end-members; two fluid layers of equal depths and very thin heavy fluid layer placed at the surface of lighter fluid, the values of $k = 2$ and $k = 1$ are assumed, respectively. As shown, our NEMD simulations for two superimposed microscopic Lennard–Jones fluids of equal depth yield the value of $k = 2$. Moreover, the proportionality

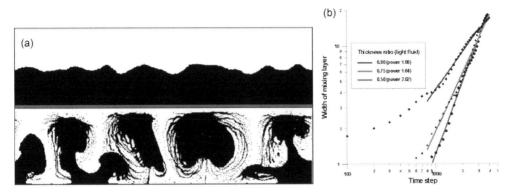

FIGURE 26.22 (a) The temporal evolution of interface length between heavy (white) and light (black) immiscible DPD fluids. The box length is about 5 μm long. (b) The mixing layer growth with time for 3-D DPD particle system for various fluid thickness ratios. The numbers on the left in the legend show the light fluid thickness in correspondence to the unit box height. The numbers in parentheses correspond to the fit of the time exponent k.

constant A_B from Equation 26.52 for bubble mixing layer contribution is approximately the same as these obtained in laboratory experiments and macroscopic simulations. A similar scenario can be found in DPD fluids, although, unlike in NEMD experiments, DPD fluids are immiscible producing slightly different mushroom microstructures (see Figure 26.22a) than NEMD simulations (see Figure 26.21). The growth of interface length with time for two DPD fluid layers of equal depth can be well approximated by $a + bt + ct^2$ polynomial in the period of time from the beginning of the mushroom structures formation to the breakup of stems. For the cases without mushrooms erosion, t^2 dependence can be conveyed to the speed of mixing layer. We can conclude from DPD simulation results depicted in Figure 26.22b that in the case of thinner region with heavy fluid, the time exponent k changes from 2 to 1. Thus the same behavior is found for both macroscopic and mesoscopic DPD particle system.

Unlike in NEMD models, the microstructures emerging due to competition between the breakup and coalescence processes can be studied by using DPD modeling. For example, in Figure 26.23, the four principal mechanisms, the same as those responsible for droplets breakup [118,119], can be observed in DPD simulation of the R−T instability. As shown in [116,119], moderately extended drops for capillary number close to a critical value, which is a function of dynamic viscosity ratio

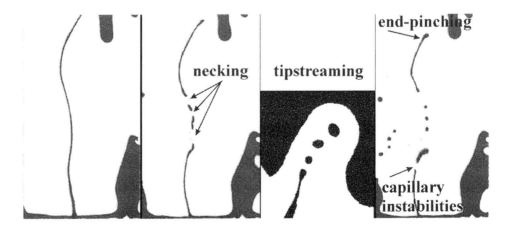

FIGURE 26.23 Four breakup microstructures observed on different portions of the same extended thread obtained from DPD simulation of turnover of thin heavy layer (black) in the bulk of lighter fluid (white). Both snapshots come from two-dimensional DPD simulations with 10^5 particles.

for the two superimposed fluids and the flow type, breakup by a necking mechanism. Owing to the elongation flow producing long threads (see Figure 26.23a) of the heavy fluid, according to Ref. [116], the necking mechanism causes the breakup in Figure 26.23b. After breakup, two daughter droplets are created. Another mechanism responsible for the stem breakup is shown in the Figure 26.23c. The bubble structure detaches thus producing from the stem tip a cluster of small droplets, which can overtake the large bubble. This process is similar to the tipstreaming mechanism, in which small drops break off from the tips of moderately extended pointed drop [119]. As displayed in Figure 26.23d supercritically extended drop in the presence of low shear rate and surface tension causes a breakup of the end-pinching type. When the drop is stretched to a highly extended thread, capillary instabilities are observed. The thread becomes unstable to small fluctuations and will eventually disintegrate into a number of drops consisting of secondary droplets between the larger primary drops.

In Figure 26.24 we display the snapshots from modeling of the R−T hydrodynamic instability using smoothed particle hydrodynamics in which 10^5 smoothed particles were modeled in about 3000 timesteps.

Although originally designed for astrophysical problems [86], as shown in [120], SPH can also be used for modeling polymers in the macroscale. However, smoothed particle hydrodynamics does not include thermal fluctuations in the form of a random stress tensor and heat flux as in the Landau and Lifshitz theory of hydrodynamic fluctuations. Therefore, the validity of SPH to the study of complex fluids is problematic at scales where thermal fluctuations are important.

B. Thin Film Evolution

Fluid flow with a thin layer over a solid surface is of considerable technological and scientific importance. Many industrial processes, ranging from spin coating of microchips, fast drying paint production, to the design of photographic films, are based on thin-film dynamics. When the films are subjected to the action of various mechanical, thermal or structural factors, they display interesting dynamical phenomena, such as wave propagation, wave steeping, fingering and the development of chaotic behavior.

The force driving the coating flow is usually gravity and/or an externally applied pressure. One boundary surface of the liquid layer is its interface with the supporting fluid, the other a fluid interface. If the ambient fluid is a dynamically passive gas, the film has a free surface as it flows down inclined planes. Coating flows are free-surface flows and as such are difficult be solved mathematically. The free surface is an integral part of the solution. In the solution scheme, it must be guessed

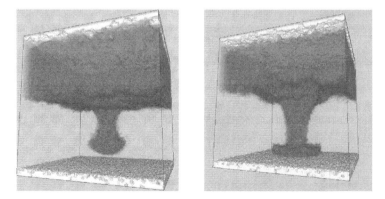

FIGURE 26.24 The snapshots from a simulation using smoothed particle hydrodynamics of the Rayleigh–Taylor instability for gases. The size of the box is approximately 1 cm. The dark grey system represents fluid which is 10 times heavier than the lighter one placed in the bottom (invisible in this figure). Approximately 10^5 SPH particles have been modeled.

and this guess must be iterated upon until it satisfies the flow and free surface conditions at all points. Such films can display additional rupture phenomena; creation of holes, spreading of fronts and the development of fingers. In principle, the film dynamics is governed by the set of Navier–Stokes (NS) partial differential equations (PDE) supplemented by appropriate moving boundary conditions. However, a full NS problem in extended spatial and time domains is difficult to solve, even with the most powerful modern computers. Therefore, a simpler, solvable description of the evolution, which is a sufficiently good approximation to the corresponding solution of the NS problem, is usually considered. The most popular case of such a simplification is known as the long-wave theory [121], in which the problem is reduced to a single PDE, which describes the evolution of the film thickness in one spatial variable. Lubrication equation is also commonly used to describe thin-film or liquid layer dynamics, driven by the large surface tension [122]. A simplified evolution equation (EE) is represented by the fourth-order nonlinear PDE. The EE solution exhibits steeping of the wavefronts, leading to wave-breaking in a finite time. A system of at least two coupled PDEs is needed for a good approximation of the long-wave regimes. However, the difficulty faced in attempting to solve them is correspondingly severe and can approach the level of difficulty in the original NS system. Moreover, for very thin films, molecular forces and surface tension varying with fluid thickness should also be taken into account. Therefore, for investigating its long-time evolution including droplet detachment and its fragmentation, the EE approximation may be insufficient.

In Ref. [123], we propose an entirely different numerical model of fluid film dynamics from those, which can be derived from the NS approach or its asymptotic expansions. The model is based on the DPD particle model and can be used for simulating thin-film dynamics in the mesoscale. Instead of changes of film thickness in nodal points in time according to the evolution equation discretized in both space and time, the temporal evolution of DPD particle system is governed by Newtonian laws of motion Equation (26.1)–Equation (26.4).

As shown in Figure 26.25, a thin film falling down inclined plane or a vertical wall produces fingers of different kind than these observed in the R–T mixing. In simulating falling sheet case, we begin with a dry wall and opens a gate at $x = 0$. This allows the viscous fluid of constant volume V to flow down the wall (along x-axis) with a straight contact line (parallel to z-axis) that moves according to the direction of the gravitational field. Some time after the release (the time depends on the fluid thickness, viscosity, physical properties of the wall surface etc.), a contact line spontaneously develops and produces a series of fingers of fairly constant wavelengths across the slope. Either long fingers develop with the sides parallel to the x-axis and with the roots

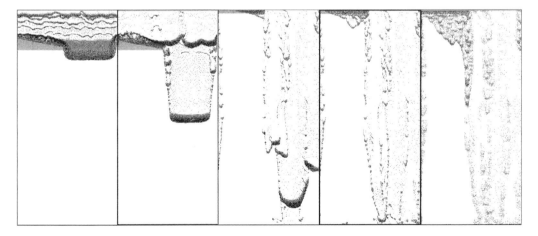

FIGURE 26.25 The snapshots from a thin film evolution simulated by using DPD. Complex microstructures create spontaneously due to flow.

fixed to the plate, or triangular fingers form, traveling downward with their roots moving downward. The morphology of the fingers depends on the contact angle. As the sheet moves downward, a bead or ridge forms behind the leading edge. This is formed due to recirculation flow down along the free surface toward the contact line and returning along the plate. This flow is caused by the presence of the contact line, which slows down the film drainage. High pressure near the contact line is responsible for ridge production [121].

From this scenario, we proposed in Ref. [123] a new 2-D numerical particle model of falling sheet evolution, which can be considered as a supplementary one to the EE theory. Let us consider a system in which the DPD particles are initially placed in the upper region of computational box. This region stands for a vertical wall covered by the particle fluid. The motionless obstacle made of particles is added, which depending on interparticle forces and their spacing can be both impermeable or permeable. The upper part of the box represents dry wall. Periodic and reflecting boundary conditions are assumed in z- and x-directions, respectively. There is not an additional supply of fluid to the system. Our system is two-dimensional. The y-dimension axis, which follows the fluid thickness, can be neglected, assuming that:

1. Fluid film is very thin. Therefore, there is a constraint on the distance between two particles $R_{ij}(x, y, z) \approx r_{ij}(x, 0, z)$ and $\Omega(R_{ij}) \approx \Omega(r_{ij})$ for $r_{ij} > h$ (where h is the fluid thickness, and $\Omega(\ )$ is the force defined by Equation (26.25)).
2. Interactions between particles are soft, for example, simulating a few particle layers placed one upon another, constant repelling force should be considered for $r_{ij}(x, 0, z) < \sigma$, then the approximation assumed earlier will also be valid for small r_{ij}.
3. The observed particle density represents a projection of the "real particle density" on x-z plane, therefore the large and small film thickness can be reflected by both high and low particle densities, respectively.

We showed in Ref. [123] that these theoretical results can be summarized as:

1. A ridge forms behind the leading edge.
2. The ridge has thicker regions of liquid advancing more rapidly than the thinner regions.
3. The contact-line resistance plays a "double role" not only in slowing-down but also by increasing the rate of spreading.

Let us assume further that a particle i undergoes large friction force, when the number of particles Neigh(i) in its vicinity (in cut-off radius r_{cut} sphere, see Equation (26.25)) is too small, i.e., when Neigh(i) < Neighmin. This particular procedure in the DPD algorithm is as follows:

$$\text{Damphi} = \frac{\lambda \Delta t}{2}$$

if(Neigh(i) < Neighmin)

    Damphi $\approx$ 1

else

    Damphi = small   (26.53)

endif

$$\mathbf{p}_i^{n+1/2} = \frac{(1 - \text{Damphi})}{1 + \text{Damphi}} \mathbf{p}_i^{n-1/2} + \frac{\Delta t}{1 + \text{Damphi}} \left(\sum_{j \in R_{cur}} \Omega_{ij}^n + m_i g \cdot \bar{\mathbf{e}}_Y \right)$$

The last equation in (26.53) represents the discretized and transformed Newtonian equation of motion (see Equation (26.1)–Equation (26.4)).

In order to demonstrate this algorithm, we study the particle system shown in Figure 26.15. As displayed in Figure 26.25, the U-shaped contact-line spreads very fast down a "dry" wall, due to larger liquid accumulation at the leading edge. The fingers, which appear later, are due to trailing end instability and rivulets created by the permeable obstacle. They are much slower than the U-shaped contact-line, despite they are falling down the "prewetted" wall. Such behavior comes from the intermolecular forces [124], which are comparable to the main driving force and seem to be powerful enough to exceed viscous dissipation in a ridge and hence overcome its accelerating tendency. Owing to the amplification of the accumulation effect by increasing the value of Neighmin, the initial flow becomes faster. The quasi-stationar bulges appear on the U-shaped edge of fluid (see Figure 26.25). As before, this structure is eventually destroyed by secondary fingers and rivulets.

The spontaneous emergence of avalanches, droplets and rivulets is very difficult to simulate with classical fluid dynamical models, due to the critical nature (self-organized criticality) and threshold character of these nonlinear phenomena. Therefore, the role of statistical fluctuations in thin-film dynamics cannot be underestimated, especially in the mesoscale. Unlike the classical approaches, we need not introduce any external and artificial perturbations. All phenomena occur spontaneously due to thermal noise inherent in the nonlinearly interacting particle dynamics.

The role of the thermal noise implemented as a Brownian force in DPD model is especially important factor in modeling of phase-separation process.

C. PHASE SEPARATION

Let us consider a mixture of DPD fluids each symbolized by an integer value $n = 1, 2, \ldots, M$. We define the value of $\Delta_{lk} = P_{kl} - P_k$ for $k \neq l$, which is equal to the difference between pressures (see Equation (26.26)) in "apparent" fluid 0 consisted of DPD particles for which $\Pi_0 = \Pi_{kl}$, (see Equation (26.25)) and fluid k. The value of Δ is responsible for fluid immiscibility. When $\Delta_{lk} \leq 0$ two fluids l and k are miscible, otherwise the two fluids will separate. The phase separation can occur in a different way. Two already mixed fluids may separate completely or may produce an emulsion [20,84]. For simplicity, let us focus on phase-separation problem for symmetric quenching in a binary liquid. We assumed that $\Delta = \Delta_{12} = \Delta_{21}$ and $P_1 = P_2$, in order to avoid system instability.

The growth kinetics of binary immiscible fluid and phase separation has been studied by using variety theoretical and computational tools. The time-dependent growth of average domain radius $R(t)$, which follows algebraic growth laws of the form

$$R(t) = t^b, \tag{26.54}$$

was investigated by using lattice gas automata, molecular dynamics, continuum model based on Langevin equations, LBG, and DPD [71,125]. In the absence of Brownian diffusion of interfaces the growth proceeds by the Lifshitz–Slyozov mechanism [104,126] and the power-low index b is set to one-third. They show that the scaling regime sets in at approximately the same domain size for various surface tensions assumed. For a very small surface tension, where the system prefers to order in a lamellar phase, a significantly different behavior is observed. After initial transients a region of logarithmic growth is detected, which corresponds to formation of lamellar microstructures (Figure 26.26a).

For isothermal DPD fluid model with intrinsic Brownian stochastic forces, both the Lifshitz–Slyozov mechanism and the lamellar regime cannot be observed [71]. The Brownian regime ($b = 1/2$) is not as stable as the Lifshitz–Slyozov regime. It persists for decreasing domain size with increasing surface tension and eventually disappears dominated by the inertial regime ($b \approx 2/3$) (see Figure 26.26b). However, as displayed in Figure 26.26c and d, the four regimes with $b = 1/3$, $1/2$, 1, and $2/3$ can be observed for phase-separation process modeled in 3D by

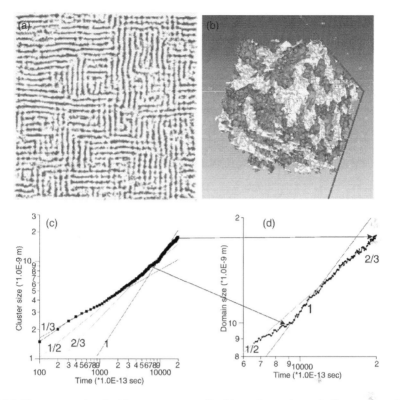

FIGURE 26.26 Phase separation for binary systems realized by a thermodynamically consistent DPD model with thermal fluctuations switched off (a) and on (b) (cross-section is shown to display lamellar structures). As shown in (c) and (d) using 3D TC-DPD simulation we can obtained all $b = 1/3, 1/2, 1$, and $2/3$ regimes in a single simulation [20,84,125].

using TC-DPD, which allows for self-adapting control of the amplitude of thermal fluctuations in various spatio-temporal regimes.

In the presence of only one phase, rarefied DPD gas with attractive tail in interparticle interaction forces, we can simulate condensation phenomenon. As shown in Figure 26.27, the microstructures appearing are different than those for binary fluids. The average cluster size $S(t) \sim R^2(t) \sim t^a$ increases much slower than in binary systems. Condensation patterns are more regular and resemble separate droplets rather than shapeless cluster structures. Therefore one can suppose that the mechanisms of growth in condensing gas must also be different than in separation of binary mixture.

Similarly as the processes of coagulation and break-up characterized for mixtures of fluids, the processes of micelles crystallization, colloidal agglomeration, and process of dispersion occurring in colloidal suspensions, can be simulated by using multilevel particle models described in Section II.E. This time, however, solid fraction, colloidal beads, has to be simulated by using different interaction paradigm.

D. COLLOIDAL ARRAYS, AGGREGATES, AND DISPERSION PROCESSES

As demonstrated already in Refs. [10,20,22,71,83,84,104,123,127] by changing just the nature of the conservative interactions between the fluid particles and by introducing also larger solid particles, we can easily model the different dynamics of colloids, micelles, colloidal crystals and aggregates. We consider two types of particles: solvent droplets and colloidal beads. We assume

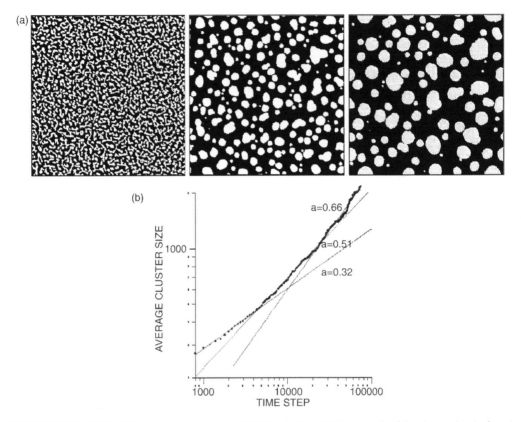

FIGURE 26.27 (a) The following snapshots from DPD simulation. (b) The growth of the cluster size (volume) in function of time for condensation of DPD droplets in vacuum. Three separate scaling regimes are depicted. The linear fits give a values in ascending order (0.32, 0.51, 0.66).

that the concentration of electrolyte in the solvent is low. In this case, we can neglect the long-range interactions and focus just on the short-range forces. Therefore, the electrolyte–solvent particles can be represented by DPD or FPM fluid particles. The colloidal agglomerates consist of primary particles, the colloidal beads. They can represent large charged ions, which are a few times larger than electrolyte–solvent droplets. We assume that the agglomerates are "wet," that is, colloidal beads are covered by electrolyte binder. This assumption allows us to use in the model mean forces similar to those obtained for charged macro ions in realistic colloidal mixtures (see section II.E).

We define the collision operator as

$$
\Omega_{ij}^n = \begin{cases} \pi_{kl} \cdot \omega_1(r_{ij}^n) - \gamma_{kl} M_i \cdot \omega_2(r_{ij}^n) \cdot (\mathbf{e}_{ij}^n \circ \tilde{\mathbf{v}}_{ij}^n) + \dfrac{\delta_{kl}\theta_{ij}}{\sqrt{\Delta t}}\omega_1(r_{ij}^n) & \text{if } k \text{ and } l \text{ solvent particles} \\[4mm] \dfrac{24\varepsilon_{kl}}{r_{ij}^n}\left\{ \left(\dfrac{\sigma_{kl}}{r_{ij}^n}\right)^6 - 2\left(\dfrac{\sigma_{kl}}{r_{ij}^n}\right)^{12} \right\} & \text{if } k \text{ or } l \text{ colloidal particle} \end{cases}
$$

$$(26.55)$$

In Figure 28, we show explicitly that initially perfectly mixed particles spontaneously create various micellar structures. Depending on the ratio χ of the depths of the potential wells $\varepsilon_{solvent-solvent}$ and $\varepsilon_{solvent-colloid}$, we can observe the emergence of lamellar, hydrophobic the hydrophilic colloidal arrays or coexistence of the two phases (Figure 26.28a). For other physical

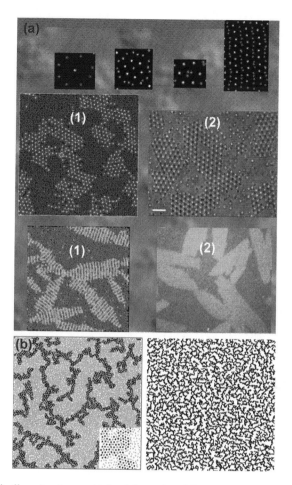

FIGURE 26.28 (a) Micellar structures obtained by using MD-DPD simulations (above) and (below) comparison of the microstructures obtained from simulations (1) to real colloidal arrays (2). (b) The colloidal agglomerates simulated in increasing spatial scales.

parameters, the micelles produce fractal-like colloidal aggregates with distinct multiresolutional structure (Figure 26.28b).

In larger scales, the colloidal beads do not "feel" discrete atomistic structure of the solvent and interact with "droplets" of fluid. Therefore, we assume that the bead-"droplet" interactions are simulated by employing FPM forces, having in mind that:

1. The bead-"droplet" forces cannot be singular.
2. Nonzero Brownian component ($\tilde{\mathbf{F}}_{ij} \neq 0$ in Equation (26.29), comes from the fluid "droplet."
3. Viscous drag ($B(r_{ij}) \neq 0$, Equation (26.29)–Equation (26.31)) comes from the fluid "droplet" and the electrolyte binder.
4. The electrolyte concentration in electrolyte–solvent mixture surrounding the agglomerate is low, thus electrostatic bead-"droplet" interactions are negligibly small.

Because the colloidal bead contains a hard core, we have modified repulsive part of conservative F^C bead-"droplet" forces, thus making it steeper than for "droplet"–"droplet" interactions.

By employing this version of two-level model, we have studied the agglomeration process in colloidal suspensions (see Figure 26.28c). For the cases of noncohesive systems, with a low concentration of colloidal beads, the asymptotic growth for $t \to \infty$ of the mean cluster size $S(t)$ is given by the power low $S(t) \propto t^\kappa$ where t is time and κ is the scaling-law index. In Ref. [84], we show that in dissipative solvent of high concentration of colloidal particles, the growth of mean cluster size can be described by the power law $S(t) \propto t^\kappa$. We have found the intermediate DLA (diffusion limited aggregation) regime, for which $\kappa = 1/2$. It spans for relatively long time. As shown in Ref. [84], the intermediate regime depends on physical properties of solvent as viscosity, temperature, and partial pressure. The character of cluster growth varies with time and the exponent κ shifts for longer times from $1/2$ to ≈ 1. This result agrees with the theoretical predictions for diffusion-limited cluster–cluster aggregation, which shows that for $t \to \infty$ the value of $\kappa = 1$ for a low colloidal particle concentration.

The reverse process to the agglomeration is a dispersion of a colloidal agglomerate due to flow. In Ref. [21], we study an accelerated flow of a colloidal slab in a periodic tube. Similar types of flows were studied earlier for solid–liquid mixing by assuming a constant shear rate [117]. By using an accelerated flow, we can investigate the granulation of colloidal agglomerate over a broad range of kinetic energies in the flow with a single simulation. The particles are confined within the rectangular box with periodic boundary conditions in y-direction and reflecting walls in x-direction. The aspect ratios of the slab to the box dimensions are $1:5$ and $1:10$ along x-and y-axes, respectively. As shown in Figure 26.29, the V-shaped profile of velocity field stabilizes after about 15,000 timesteps. Over the whole periodic box along both the x- and y-directions, we can observe strong correlations along the horizontal direction. The vertical correlations are weaker due to greater aspect ratio and vanishing velocity gradients in y-direction.

In Figure 26.30, we present the snapshots from 2-D and 3-D simulations of dispersion of colloidal agglomerate accelerated in a periodic box. We recognize several stages of granulation, which usually occur with some degree of overlap:

1. Imbibition, consisting in spreading off the liquid solvent into the colloidal cluster, and reducing the cohesive forces between the colloidal beads. This process corresponds to the wetting of a dry, porous solid by the liquid.
2. Fragmentation, consisting of shatter, producing a large number of smaller fragments in a single event, namely, rupture and breakage of a cluster into several fragments of comparable size, erosion, and gradual shearing off of small fragments of comparable size [117].
3. Aggregation, the process being the reverse of dispersion. Two traditional mechanisms can be recognized: nucleation, defined as the gluing together of primary particles due to the attractive forces; coalescence, is the process by which two larger agglomerates combine to form a granule.

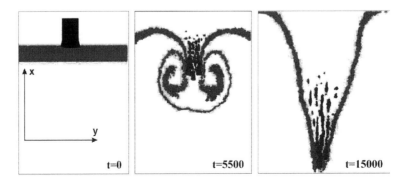

FIGURE 26.29 Velocity profiles with simulation time. The V-shaped profile becomes stable after about 15,000 timesteps.

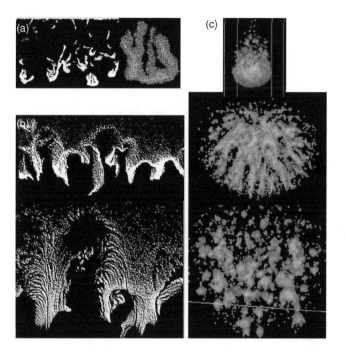

FIGURE 26.30 Snapshots from simulations of dispersion of colloidal slab made of solid particles in DPD and FPM fluids in 2-D (a,b) and 3-D (c). The gravitational acceleration is directed downward. We can observe crystallization process along the mixing front (a) for a properly defined interparticle force.

The appearance of similar multiscale structures can be observed in Figure 26.30a along the two-phase interface due to small-scale mixing. This nucleation results in rapid changes in fragmentation speed. As shown in Figure 26.30b,c and in Figure 26.31 in different type of solid–liquid flows (characterized with different kinds of particle–particle interactions ε_{SP-DP}), we can easily recognize the characteristic dispersion structures caused by the microstructural dynamics including phenomena such as rupture and erosion (Figure 26.30b and Figure 26.31), shatter (Figure 26.30c) and agglomeration [21]. The colloidal particles create long streaks, which shrink for less-vigorous flow. As shown in Figure 26.31, the characteristic streaks are formed also in 3-D particle systems due to accelerated shear flow. All of these dispersion phenomena would be very difficult to model, using partial differential equations such as the Cahn–Hillard equation.

We can model flows in the presence of larger, elastic microstructures besides systems consisting of solvent and colloidal particles of similar sizes. They can be built on many particles linked together with harmonic interactions. As shown in Refs [87,100], red blood cells (RBC) flowing in FPM fluid can be modeled in such a way. Owing to spatio-temporal scale, which can be captured by DPD, the modeling can be performed for a single capillary (see Figure 26.18) of different shapes with fixed, elastic and moving walls as we have employed in simulating blood flow in capillary vessels [87,100]. We can study many clotting factors in blood flow such as aggregation of RBC due to hydrodynamic and depletion forces and different geometry of capillary channels. Larger systems consisting of many capillaries can also be studied within the discrete-particle model in which the very notion that the particle must be redefined, as, it is defined in thermodynamically consistent DPD model [32]. In the following section, we present some basic assumptions and results of microscopic blood model.

E. DISCRETE-PARTICLE MODEL OF BLOOD

Blood is a physiological fluid, which consists of a suspension of polydisperse, flexible and, chemically and electrostatically active cells. These cells are suspended in an electrolytic fluid consisting

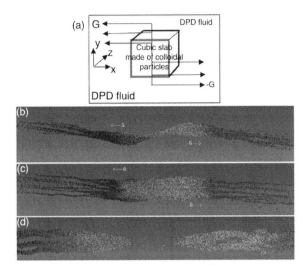

FIGURE 26.31 The particle system, which models the dispersion of a cubic colloidal slab positioned at the interface of counter flow sketched in (a). The flows are accelerated in both directions. The colloidal slab is made of colloidal particles (approximately 10^5 particles) and fluid consists of 10^6 DPD particles (invisible). The shade of gray of colloidal particle represents particle velocity. The dark gray and black colors indicate the largest velocity of particles. Figures (b) and (c) represent the projection of the colloidal particles after 2000 timesteps on x-y and x-z plains, respectively. In figure (d) the break-up instant (after 3000 timesteps) is displayed.

of numerous active proteins and organic substances. For many years, blood rheology has been researched experimentally, theoretically, and numerically.

A common mathematical description of the blood flow treats blood as a homogeneous fluid and uses the full three-dimensional, time-dependent, incompressible Navier–Stokes equations for non-Newtonian fluids (e.g., see [128–130]). Blood rheology can be described by an appropriate shear-thinning model or can be based on experimental viscosity data. Rheological measurements obtained when the dimensions of the measurement confinement are of at least two orders of magnitude larger than the microstructure in the fluid are termed bulk properties. Rheological measurements in smaller confinements will be different from these bulk measurements. For many applications, measurements in small confinements are required. Blood clotting in capillary vessels is such a case.

Macroscopic vessels represent only a small fraction of circulatory system, although the largest veins contain 50% of blood [18]. The vascular tissue is madeup of microscopic, capillary channels. There are approximately 10^{10} blood vessels whose diameters are comparable with the dimensions of the RBC. Therefore, the majority of defects in circulatory blood system occur in capillary vessels where blood flows less vigorously than in larger macroscopic arteries and arterioles. These defects can become very dangerous if they occur over a large area and in vital parts of the organism, for example, in the brain or eye. Bleeding over a vast capillary area is very feasible under extreme conditions, such as multiple g-acceleration exerted on jet pilots or extremely low viscosity of blood caused by drugs or alcohol.

Clustering of RBC is a very important and vital biological process (e.g., see [131]), which influences the rheological properties of blood and may lead to severe cardiovascular problems, such as anemia, ischemia, angina, thrombosis, and stroke. Clotting process is accomplished by the solidification of blood. It initiates three separate, but overlapping, hemostatic mechanisms:

(1) RBC clustering.
(2) The formation of a platelet plug.
(3) The production of a web of fibrin proteins around the platelet plug.

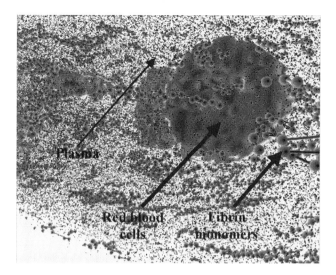

FIGURE 26.32 Three types of particles used for simulating the blood components. The smallest ones are the plasma particles. The fibrin monomers are larger and connected by sticks while the RBC are made of elastic grid of "solid" particles.

Unlike in the macroscale, where blood can be regarded as a continuous medium, blood flow in microscopic capillaries with diameters on the order of the red blood cell size, i.e., $5-8\mu m$, represents a system with sharp interface between liquid and solid phases (see Figure 26.32). In this spatial scale, the RBC represent the phase volume with distinct elastic properties. Contrary to macroscopic simulations, in the microscopic scales, the blood can be regarded as an ensemble of interacting discrete objects made of particles.

We model the blood flowing in capillary vessel of diameter $10-13\ \mu m$ and about $100\ \mu m$ length. We assume that the system modeled consists of a fragment of capillary and blood components such as plasma, RBC, and fibrinogen. As shown in Figure 26.32 and Figure 26.33, all of them can be constructed of discrete-particles. We defined two types of particles:

Solid particles: the "pieces of matter" placed in nodes of the elastic grid. They represent the elastic components of the vascular system, i.e., the endothelium wall and the red blood cells.

Fluid particles: used for modeling "portions" of the colloidal suspension.

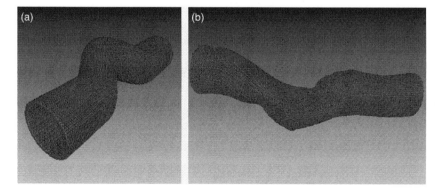

FIGURE 26.33 The snapshots of elastic capillary made of particles at the beginning of simulation (a) and after 3000 timesteps (b).

The particles, interactions between them, and the temporal evolution of the particle system are defined already by the discrete-particle model principles.

The definition of the collision operator is the principal factor, which distinguishes the particle types. The "solid particles" are the compounds of larger objects with recognized shape such as the red blood cells, capillary walls, and fibrinogen chains. These objects are made of particles, whose initial positions a_{ij} are specially arranged. For cohesion, the particles are coupled together by a bonded (see Table 26.2) particle-on-springs collision operator

$$\Omega(r_{ij}) = \chi \cdot (|r_{ij}| - a_{ij}) \tag{26.56}$$

We have assumed that in this case the particles interact only with an invariable list of neighbors. The capillary walls are physical constraints for the blood flow. We have assumed that:

1. The particle system is closed and the "infinite-like" system is modeled by using periodic boundary conditions.
2. The number of particles in the model remains constant. The same number of particles leaving the system must enter it from the other side.

The periodic boundary conditions produce the correlations between inlet and outlet streams which decrease in a larger system. The size of the capillary imposes the upper limit for the number of timesteps, which can be performed with negligible correlations. We model the quasi-elastic nature of capillary wall, assuming that the axial elasticity can be neglected. This model is considered to be physically correct because in the real blood vessel, the wall consists of a layer of endothelial cells [18], which can deform resisting the flow. As shown in Figure 26.33, the particles modeling the walls respond to the shearing blood flow. The wall particles interact with one another with forces similar to the solid particles in the red blood cells.

The RBC in a human vascular system are biconcave discs 2.5-μm thick at the edge and 1-mm thick at the center (see Figure 26.34). They can be envisaged as soft bags containing hemoglobin [18]. A membrane provides the cell with its shape, strength, and flexibility. They consist of lipid bi-layer supported by an extensive filament-like protein network, which is called the cytoskeleton. The mechanical properties of the individual blood cells produce various types of RBCs collective behavior, which considerably influence the entire blood system [18].

As shown in Figure 26.34, we assumed that the RBC is made of a mesh with particles-on-spring which models cytoskeleton network. The particle i interacts with particle j in their closest

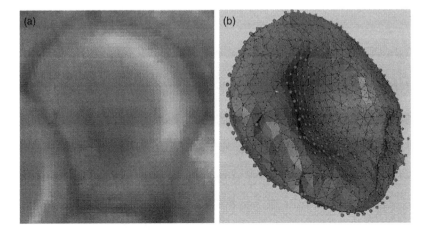

FIGURE 26.34 The (a) RBC and (b) biconcave disc modeled by using particles-on-springs paradigm.

neighborhood by conservative elastic force given by Equation (26.56), where χ defines the elasticity of the RBC and a_{ij} is the distance from a particle i to its nearest neighbor in equilibrium. The value of a_{ij} depends on the position of neighboring particle for the initial conditions. For the fcc mesh and for only the first layer of neighbors the value of $a_{ij} = 1$. Besides conservative force, the collision operator for solid particles includes an additional dissipative force (see Equation (26.29)). The dissipative term prevents RBC from breaking-up due to the collisions with the fast particles. Although the realistic RBC structure and its mesh model are different, the basic elastic properties of the model can be matched to the real blood cell by using mechanical principles [100].

The plasma particles are simulated by fluid particles. In contrast to the mesh of particles-on-spring, the ensembles of fluid particles form shapeless structures and have a variable number of neighbors within their interaction radius. In the lack of clotting factors, the collision operator can be defined as it is in the FPM described earlier (see section II).

Fibrinogen is a composite material consisting of six protein chains. Normally fibrinogen is dissolved in a blood plasma. It floats around, activating when a cut or injury causes bleeding. This involves a complex cascade of chemical reactions and path-signaling biological processes. The resulting protein (fibrin) has sticky patches exposed on its surface. The complementary shapes allow large numbers of fibrins to aggregate with each other. The long thread produced by the fibrin molecule cross-over each other and form fibrin gel that entraps the blood cells.

We assume that the fibrinogen has been activated already by the thrombin and all clotting factors are present. Because, the concentration of fibrinogen in blood is about 0.3% and its molecular weight is very large at 340,000 thus [132] we can expect about 10 fibrinogen molecules on the average for a single fluid particle. As we show in [102], due to the fibrinogen, the plasma particles can be defined as having a dual solid–liquid character with different properties depending on if the particles are bonded or not. The unlinked fluid particles interact with each other like the plasma particles (Equation (26.29)). They can merge each other with a given probability depending on their separation distance. The two particles, which create hydrated fibrin, interact one with another with the harmonic collision operator similar to Equation (26.56). The interactions with the other particles depend on the number of neighbors in the chain. For example, we have assumed that the probability for attaching more than two particles to a single monomer is equal to zero.

It is well known that human RBC can form clots whose formation depends on the presence of the proteins, fibrinogen, and globulin. The other components of the thrombotic process are aggregation and coagulation of RBC. The shear conditions can accelerate RBC clustering. The slower the blood flow, the smaller is the shear rate and the larger are RBC clusters. Necking of the microscopic vessels caused, e.g., by accumulation of cholesterol plagues in the vessel walls, slows down the flow. This stimulates the aggregation of larger and larger thrombi resulting in a positive feedback loop, which can eventually choke the flow.

In order to illustrate this fact, we have modeled the RBC flowing through the choking point of a periodic pipe. As shown in Figure 26.35a, the interactions of the blood cells with the walls and with the blood flow are responsible for RBC clustering. The simulation starts with the most convenient situation for clotting, i.e., all RBCs were placed in the perpendicular direction to the flow. In Figure 26.35a, the velocity field close to the strangulation has the lowest magnitude because of blockage by the blood cells.

However, the elastic discs are able to pass through, because of the positive feedback between disc elasticity and the hydrodynamic forces accompanied by the interactions between RBC and the wall. As shown in Figure 26.35b the fibrins accumulate mainly in the space filled by the RBC cluster where the density fluctuations is the largest. Because the RBC cluster is very tight due to the high acceleration, the sticky fibrins have length enough to glue the RBC up, producing the clot. As shown in Figure 26.36, this occurs even for a small probability p_0 of fibrin polymerization, even though the average fibrin length is small linking only five fluid particles in average. However, the longest fibrins, which are produced nearby and within the RBC cluster, are three times longer.

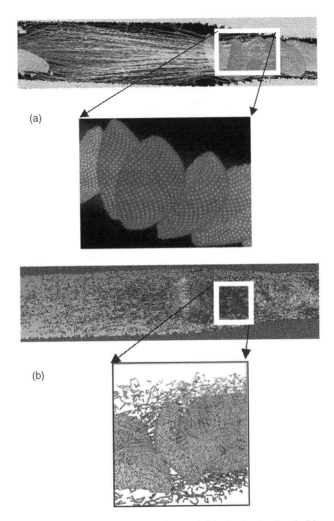

FIGURE 26.35 The snapshots of blood (a) aggregation and (b) clotting in the choking point.

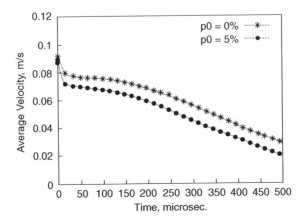

FIGURE 26.36 The average velocity of the blood with time for different probabilities p_0 of fibrin polymerization.

Our discrete-particle approach possesses the important properties of mesoscopic systems. It can model easily the heterogeneous nature of complex fluid suspension in the presence of fluctuations. This allows for simulating processes, which cannot be modeled by computational fluid dynamics codes. We showed that our microscopic blood model can be used for simulating microscopic, multi-component blood flow under extreme conditions in presence of high acceleration [100].

The computational problems involving multi-million particle ensembles, found in modeling mesoscopic phenomena, were considered only recently as the typical problems. Rapid increase of computational power of modern processors and growing popularity of coarse-grained discrete particle methods, such as dissipative particle dynamics, fluid particle model, smoothed particle hydrodynamics and LBG, allow for the modeling of complex problems by using smaller shared-memory systems [101].

However, the use of the discrete-particle method in simulating stagnant capillary flows in normal conditions (i.e., 2–3 mm/s) is still very demanding computationally. We estimated [101] that a single run will need approximately 6000 h of CPU time per single two core Power4/1300 chip. It may seem that employing high-performance computing and large shared-memory systems with greater number of parallel processors can be helpful overcoming this problem. Unfortunately, as we showed in Refs. [101,125], this issue is of limited use in the case of discrete-particle model. The main technical problem connected with efficient use of shared-memory machines in modeling of large ensembles of discrete-particles lies in highly irregular data structure they require. Many particle types, interaction forces, and irregular boundary conditions involve using nesting if statements and indirect addressing. This strongly limits an efficient use of new architectural issues of modern processors, which exploit explicit parallelism involving regularity of data structures and data independence. Greater computational efficiency can be gained by using different particle approaches, e.g., employing multiresolutional particle models [32] and more sophisticated timestepping schemes [103].

V. A CONCEPT OF PROBLEM-SOLVING ENVIRONMENT FOR DISCRETE-PARTICLE SIMULATIONS

On the one hand, extracting multiresolutional features from the results generated by continuum methods, involves modern feature-extraction techniques based on second-generation wavelets [133–135]. On the other hand, the raw data produced by particle codes comprise positions and velocities of particles. The detection and visualization of multiresolutional patterns created due to flow are crucial tasks for understanding complex flows in vascular and porous media. Fast algorithms and codes for the analysis and detection of microscopic coherent structures such as aggregates, clusters, droplets, etc. have to be constructed. In the case of out-of-core data mining, we propose to combine parallel clustering procedures, similar to those described in Refs. [100,136], with wavelet codes. The goal of cluster extraction is to collect statistical knowledge about micro-structural properties of complex systems at various spatial resolutions. This knowledge could also be used for bridging scales in subsequent coarse-graining procedures such as: NEMD–DPD–SPH. As shown in Figure 26.37, we have used modern visualization packages such as Amira [137] together with large display visualization systems (PowerWall) at the University of Minnesota (htpp://www.lcse.umn.edu).

The discrete-particle methods proposed earlier can be used as the components of the problem-solving environment (PSE) based on the conception of multiresolutional wavelets. As shown in Figure 26.38, the whole series of simulations can be performed over three different spatio-temporal levels similarly as it is for wavelets but here the various shapes of "wavelets" will depend on the model of particle (atom, DPD droplet, FPM drop, and SPH chunk of fluid) and consequently the interactions between particles. In fact, the shapes of short-ranged interaction can be treated as some sort of wavelets. The interactions are short-ranged with compact support and well localized

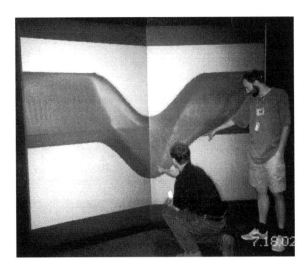

FIGURE 26.37 Blood flow in a curved capillary visualized on the PowerWall at the University of Minnesota (http://www.lcse.umn.edu).

in space. The final total forces acting on each particle are linear combinations of "wavelets" of various locations.

However, unlike wavelets we cannot get "details" for the whole macroscopic spatial domain but rather representative part of it. It does not matter for a homogeneous system but gets clumsily for a more interesting anisotropic system. Thus the global simulation should start from the coarsest SPH level ("approximation") and focalize on interesting areas in subsequent "details" (DPD and MD, respectively). This focalization procedure resembles thresholding of wavelets coefficients and setting them to 0 for all uninteresting parts of spatial domain [63]. From the coarse system, we remove the areas, which have to be simulated by using a more detailed model. We can find these regions self-adaptatively by using regular wavelets as shown in Ref. [135], exploiting clustering schemes (e.g., see [100,138]) or they can be extracted interactively by the user from visualized

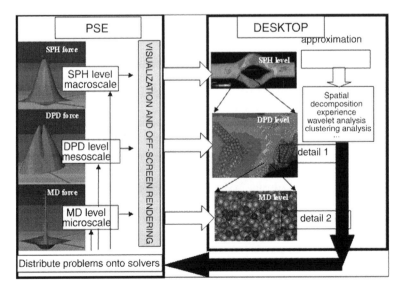

FIGURE 26.38 A conceptual scheme realizing the multiresolutional system within the framework of a problem solving environment (PSE) based on discrete-particles approach.

on-line snapshots from the simulations. Going down the scale axis we may also model specific areas and so on. In result we will obtain, such as it is in wavelets, multiresolutional approximation of the system. The user will define only the physical properties of the medium and will get "details" (MD and DPD) and "approximations" for each level.

Because of the prodigious amount of data produced by each model involving up to 10^8 particles and thus precipitating a few gigabytes at each timestep, they cannot be resent and preprocessed on the desktop computers. Extraction of features and microstructures from the simulation data involves specialized data-mining tools like clustering and classification schemes, which for huge data sets requires out-of-core computing. Visualization, which is crucial for such the line of global simulation allowing the user for control of the "thresholding" procedure, cannot be performed on-screen due to the lack of modern visualization software and adequate computational power for rendering the images of very high resolution. Their size exceeds many times the resolutions of the largest desktop displays. Therefore, we recommend using off-screen rendering systems such as presented in Ref. [139]. Summarizing, the data

- Must be stored on high storage devices.
- Redistributed adequately between disks on the disk array for further parallel visualization.
- Preprocessed by using data mining tool for extracting the microstructures.
- Rendered by using parallel system involving modern visualization software like Amira.
- The high-resolution images should be accessible, part by part, with the remote clients.

The global simulation on the GRID system (e.g., see [140–142]) shown in Figure 26.39 could be controlled by simple web tools remotely from the desktops and laptops, allowing the remote client to interrogate visually the images to extract the regions of interest (ROI) and, finally, to start the simulation over smaller spatial scales with finer timesteps. In this type of GRID system, all details in the implementation will be hidden and be self-adopted to the current load on the multiprocessing system.

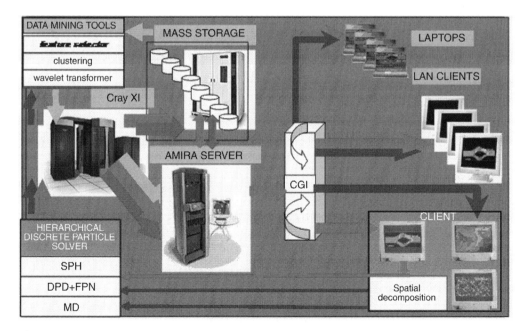

FIGURE 26.39 A proposed scheme for employing the GRID system for achieving a holistic approach in global cross-scaling modeling based on the discrete-particles approach. The Common Gateway Interface (CGI) is a standard for interfacing external applications with information server.

VI. CONCLUDING REMARKS

In recent years, new discrete-particle methods have been developed for modeling physical and chemical phenomena occurring in the mesoscale. The most popular are grid-type techniques such as cellular automata (CA), LG, LBG, diffusion- and reaction-limited aggregation [37] and stochastic gridless methods, e.g., DSMC used for modeling systems characterized by a large Knudsen number [41], and SRD [39]. Unlike DSMC, in SRD collisions are modeled by simultaneous stochastic rotation of the relative velocities of every particle in each cell.

Gridless discrete-particle methods have a few significant advantages over grid techniques. These advantages can be enumerated as follows:

1. In contrast to methods based on spatial discretization of partial differential operators, the dynamics of fluid particles develop over continuum space in real time, thus allowing for realistic visualization and statistical analysis, such as clustering.
2. Within the context of cross-scaling systems, they are homogeneous with both microscopic molecular dynamics and macroscopic smoothed particle hydrodynamics techniques. From a numerical standpoint, we do not need to switch over from particles to a static grid with control parameters, such as the Knudsen number.
3. The methods employing fluid particles are also homogeneous for solid–liquid simulations in which both solid and liquid can be represented by particles.
4. Many types of complex boundary conditions such as a free surface and complex porous structure can be easily implemented.
5. Particle methods are homogeneous from implementation point of view. Well-known sequential and parallel algorithms from MD and FPM simulations can be employed directly without any complications.

On the other hand, the LBG method can capture both mesoscopic and macroscopic scales even larger than those that can be modeled by discrete-particle methods. This advantage is due to computational simplicity of the method, which comes from coarse-grained discretization of both the space and time and drastic simplification of collision rules between particles. We can look at the validity of these simplifications by comparing them with more realistic discrete-particles simulation. We regard both DPD and LBG as being complementary computational tools for modeling the slow dynamics in porous media over wide spatio-temporal scales.

The aggregated DPD/LBG model has many other powerful features, which are lacking in the two models, when taken separately.

1. MD, DPD, and LBG methods together can capture both microscopic and macroscopic scales.
2. The common mesoscopic scale of confrontation of the two methods allows for more precise scales bridging and adjust more precisely the rheological parameters for both systems.
3. Both DPD and LBG methods are homogeneous within the context of solid–liquid simulations, i.e., both solid and liquid can be represented by particles.
4. They are also homogeneous in terms of implementing on a parallel computing, i.e., similar parallel algorithms based on geometrical decomposition and load-balancing schemes can be employed.

Large scale simulations must be carried out to model multiresolutional structures ranging from millimeters to micrometers emerging in vascular systems and porous media. To obtain satisfactory resolutions, we have to use at least 10 million DPD particles and huge LBG meshes with 10^8-10^9 sites. These large-scale modeling requires not only high-performance multiprocessor systems and fast parallel codes. Moreover, these calculations efforts several tetrabytes of data, which must be stored, co- and post-processed, analyzed, and visualized.

We show that to realize a complete cross-scale computational system, which combines large-scale computations, mass storage, data processing, and visualization, simultaneously making it user-friendly and remotely accessible, a new system–user interface and data-flow organization have to be implemented, such as in a GRID system (e.g., see [140–145]). Grid computing, and the software that enables it, is absolutely essential to building a cyber infrastructure that will help to improve the usefulness of mesoscopic modeling. Therefore, in the not-too-distant future with GRID computing coming to the fore, realistic large-scale and cross-scale simulations will be really affordable for performing intricate tasks from both conceptual and computational points of view. These two factors cannot be considered separately for hybridized cross-scale models. Conceptual structure of these models has to be easily mapped onto a modern computational environment, which is currently based on distributed and shared-memory computational resources and object-oriented and component way of programming. The discrete-particle approach meets all of these requirements under the aegis of GRID computing.

ACKNOWLEDGMENTS

We thank Dr. Gordon Erlebacher, Jim Rustad, and Geoffrey C. Fox concerning computational environments. Support for this work has come from the Polish Committee for Scientific Research (KBN) project 4T11F02022, the Complex Fluid Program of U.S. Department of Energy and from AGH Institute of Computer Science internal funds.

REFERENCES

1. Moseler, M., and Landman, U., Formation, stability, and breakup of nanojets, *Science*, 289, 1165–1169, 2000.
2. Banfield, J.F. and Navrotsky, A., Nanoparticles and the Environment, *Rev. Mineral. Geochem. 44*, Mineral. Soc. Am., Washington, D.C., 349, 2001.
3. Chelikowsky, J.R., The pseudopotential-density functional method (PDFM) applied to nanostructures, *J. Phys. D*, 33, R33–R50, 2000.
4. Davis, M.E., Ordered porous materials for emerging applications, *Nature*, 417, 813–821, 2002.
5. Goddard III, W.A., Computational materials chemistry at the nanoscale, *J. Nanoparticle Research*, 1, 51–69, 1999.
6. Groom, G.F. and Lockwood, D.J., Ordering and self organization in nanocrystalline silicon, *Nature*, 407, 358–361, 2000.
7. Hochella, M.F., Jr., Nanoscience and technology: the next revolution in the Earth sciences, *Earth Planetary Sci. Lett.*, 203, 593–605, 2002.
8. Nakano, A., Bachlechner, M., Campbell, T., Kalia, R., Omeltchenko, A., Tsuruta, K., Vashishta, P., Ogata, S., Ebbsjo, I., and Madhukar, A., Atomistic simulation of nanostructured materials, *IEEE Comput. Sci. Eng.*, 5 (4), 68–78, 1998.
9. Nakano, A., Bachlechner, M.E., Kalia, R.K., Lidorikis, E., and Vashishta, P., Multiscale simulation of nanosystems, *Comput. Sci. Eng.*, 3/4, 42–55, 2001.
10. Rustad, J.R., Dzwinel, W., and Yuen, D.A., Computational Approaches to Nanomineralogy, *Rev. Mineralogy Geochemistry*, 44, 191–216, 2001.
11. Albert, R., Barabasi, A-L, Carle, N., and Dougherty A., Driven interfaces in disordered media: determination of universality classes from experimental data, *Phys. Rev. Let.*, 81 (14), 2926–2929, 1998.
12. Benson, D.A., The Fractional Advection-Dispersion Equation: Development and Application. Ph.D Dissertation, U. Nevada-Reno, 1998.
13. Kechagia, P.E., Tsimpanogiannis, I., Yortsos, Y.C., and Lichtner, P.C., On the upscaling of reaction-transport processes in porous media with fast kinetics. *Chem. Eng. Sci.*, 57 (13), 2565–2577, 2002.

14. Knutson, C.E. and Travis B.J., A pore scale study of permeability reduction caused by biofilm growth, *Eos Trans. AGU*, 83 (47), Fall Meet Suppl., Abstract H71B-0797, 2002.

15. Davis, M.E. and Lobo, R.F., Zeolite and molecular sieve synthesis, *Chem. Mater.*, 4, 756–768, 1992.

16. Hou, T.Y. and Wu, X-H., A multiscale finite element method for elliptic problems in composite materials and porous media, *J. Comp. Phys.*, 134, 169–189, 1997.

17. Li, H., Eddaoudi, M., O'Keefe, M., and Yaghi, O.M., Design and synthesis of an exceptionally stable and highly porous metal-organic framework, *Nature*, 402, 276–279, 1999.

18. Fung, Y.C., Biomechanics, Springer-Verlag, New York-Berlin-Heidelberg, 1993, p. 515.

19. Devaney, J., Visualization of High Performance Concrete, http://math.nist.gov/mcsd/savg/vis/concrete/index.html, 2002.

20. Dzwinel, W. and Yuen, D.A., A multi-level discrete particle model in simulating ordered colloidal structures, *J. Coll. Interface Sci.*, 225, 179–190, 2000a.

21. Dzwinel, W. and Yuen, D.A., Mesoscopic dispersion of colloidal agglomerate in complex fluid modeled by a hybrid fluid particle model, *J. Coll. Inter. Sci.*, 217, 463–480, 2002.

22. Dzwinel, W., Yuen, D.A., and Boryczko, K., Mesoscopic dynamics of colloids simulated with dissipative particle dynamics and fluid particle model, *J. Molecul. Model.*, 8, 33–43, 2002.

23. Werner, A., Echtle, H., and Wierse, M., High Performance Simulation of Internal Combustion Engines, http://www.supercomp.org/sc98/TechPapers/sc98_FullAbstracts/Werner772/index.htm, 2001.

24. Abraham, F., Broughton, J.Q., Bernstein, N., and Kaxiras, E., Spanning the length scales in dynamic simulation, *Comp. Phys.*, 12 (6), 538–546, 1998.

25. Holian, B.L. and Ravelo, R., Fracture simulation using large-scale molecular dynamics, *Phys. Rev. B*, 51 (17), 11275–11285, 1995.

26. Mathur, K., Needleman, A., and Tvergaard V., Three dimensional analysis of dynamic ductile crack growth in a thin plate, *J. Mech. Phys. Solids*, 44, 439–464, 1996.

27. Vashishta, P. and Nakano, A., Dynamic fracture analysis, *Comput. Sci. Eng.*, 1 (5), 20–23, 1999.

28. Allen, M.P. and Tildesley, D.J., *Computer Simulation of Liquids*, Clarendon Press, Oxford, 1987.

29. Haile, P.M., *Molecular Dynamics Simulation*, Wiley&Sons, New York, 1992.

30. Rapaport, D.C., *The Art of Molecular Dynamics Simulation*, Cambridge University Press, Cambridge, UK, 1995.

31. Koplik, J. and Banavar, J.R., Physics of fluids at low Reynolds numbers — a molecular approach, *Comput. Phys.*, 12 (5), 424–431, 1998.

32. Serrano, M. and Español, P., Thermodynamically consistent mesoscopic fluid particle model, *Phys. Rev. E.*, 64, 046115, 2001.

33. Yserentant A., A new class of particle methods, *Numerische Mathemetik*, 76, 87–109, 1997.

34. Glowinski, R., Pan, T.-W., Hesla, T.I., Joseph, D.D., and Periaux, J. A fictitious domain approach to the direct numerical simulation of incompressible viscous flow past moving rigid bodies: application to particulate flow, *J. Comput. Phys.*, 169, 363–427, 2001.

35. Beazley, D.M., Lomdahl, P.S., Gronbech-Jansen, N., Giles, R., and Tomayo, P., Parallel algorithms for short range molecular dynamics, in World Scientific's Annual Reviews of Computational Physics III, World Scientific, 119–175, 1996.

36. Mościński J., Alda W., Bubak M., Dzwinel W., Kitowski J., Pogoda M., and Yuen D.A., Molecular dynamics simulations of Rayleigh–Taylor instability, *Ann. Rev. Comput. Phys.*, 5, 97–136, 1997.

37. Chopard, B. and Droz, M., Cellular Automata Modelling of Physical System, Cambridge University Press, Cambridge, 1998.

38. Rothman, D.H., Zaleski, S., Lattice gas models of phase separation: interfaces, phase transitions, and multiphase flow, *Rev. of Mod. Phys.*, 66 (4), 1417–1479, 1994.

39. Ihle, T. and Kroll, D.M., Stochastic Rotation Dynamics I: Formalism, Galilean invariance, and Green-Kubo relations, *Phys. Rev. E*, 67 (6), 066705, 2003.

40. Dzwinel, W., Alda, W., and Yuen, D.A., Cross-Scale numerical simulations using discrete particle models, *Molecul. Simul.*, 22, 397–418, 1999.

41. Bird, G.A., Molecular Dynamics and the Direct Simulation of Gas Flow, Oxford Science Publications, Oxford, 1994.

42. Hoogerbrugge, P.J. and Koelman, JMVA, Simulating microscopic hydrodynamic phenomena with dissipative particle dynamics, *Europhys. Lett.*, 19 (3), 155–160, 1992.

43. Español, P., Fluid particle model, *Phys. Rev. E*, 57 (3), 2930–2948, 1998.

44. Baudet, C., Hulin, J.P., Lallemand, P., and d'Humieres, D., Lattice-gas automata: a model for the simulations of dispersed phenomena, *Phys. Fluids A*, 1, 507–512, 1989.

45. Flekkøy, E.G., Lattice bhatnagar-Gross-Krook models for miscible fluids, *Phys. Rev. E.*, 52, 4952–4962, 1993.

46. Ladd, A.J.C. and Verberg, R., Lattice-Boltzmann Simulations of Particle-Fluid Suspensions, *J. Stat. Phys.*, 104, 1191–1251, 2001.

47. Stockman, H.W., Stockman, C.T., and Carrigan C.R., Modelling viscous segregation in immiscible fluids using lattice-gas automate, *Nature*, 348, 523–525, 1990.

48. Stockman, H.W., Li, C., and Wilson, J.L., A lattice-gas and lattice Boltzmann study of mixing at continuous fracture junctions: importance of boundary conditions, *Geophys. Res. Lett.*, 24 (12), 1515–1518, 1997.

49. Swift, M.R., Orlandini, E., Osbors, W.R., and Yeomans, J., Lattice Boltzmann simulations of liquid-gas and binary-fluid systems, *Phys. Rev. E.*, 54 (5), 5041–5052, 1996.

50. Chen, W. and Ortoleva, P., Reaction front fingering in carbonate-cemented sandstones, *Earth Sci. Rev.*, 29, 183–198, 1990.

51. Lutsko, J.F., Boon, J.P., and Somers, J.A., Lattice gas automata simulations of viscous fingering in a porous medium, in *T.M.M. Verheggen Numerical Methods for the Simulation of Multi-Phase and Complex Flow*, Springer-Verlag, Berlin, 124–135, 1992.

52. Manwart, C., Aaltosalmi, U., Koponen, A., Hilfer R., and Timonen, J., Lattice-Boltzmann and finite-difference simulations for the permeability for three-dimensional porous media, *Phys. Rev. E.*, 66, 016702, 2002.

53. Martys, N. and Chen, H., Simulation of multicomponent fluids in complex three-dimensional geometries by the lattice Boltzmann method, *Phys. Rev. E.*, 53, 743–750, 1996.

54. Rothman, D.H., Cellular automaton fluids: a model for flow in porous media, *Geophysics*, 53, 509–518, 1988.

55. Travis, B.J., Eggert, K., Grunau, D., Chen, S.Y., and Doolen, G., Lattice Boltzmann models for flow in porous media, in *Computing at the Leading Edge: Research in the Energy Sciences*, Mirin, A.A. and Van Dyke, P.T. Eds., National Energy Research Super-computer Center, Lawrence Livermore National Laboratory Rep. UCRL-TB-111084, 1993, pp. 42–47.

56. Pego, R.L., Front migration in the nonlinear Cahn-Hillard equation, *Proc. Roy. Soc. London, Ser. A*, 422, 261–278, 1989.

57. Kloucek, P. and Melara, L.A., The computational modelling of branching fine structures in constrained crystals, *J. Comput. Phys.*, 183, 623–651, 2002.

58. Bercovici, D., Ricard, Y., and G. Schubert, A two-phase model of compaction and damage. I. general theory, *J. Geophys. Res.*, 101 (B5) 8887–8906, 2001.

59. McKenzie, D.P., The generation and compaction of partial melts, *J. Petrol.*, 25, 713–765, 1984.

60. Tanaka, H. and Araki, T., Simulation method of colloidal suspensions with hydrodynamic interactions: fluid particle dynamics, *Phys. Rev. Lett.*, 85 (6), 1338–1341, 2000.

61. Meakin, P., *Fractals, Scaling and Growth far from Equilibrium*, Cambridge University Press, Cambridge, New York, 1998.

62. Moulton, J.D., Dendy, J.E. Jr., and Hyman J.M., The black box multigrid numerical homogenization algorithm, *J. Comp. Phys.*, 142, 80–108, 1998.

63. Vasilyev, O.V. and Paolucci, S., A dynamically adaptive multilevel wavelet collocation method for solving partial differential equations in a finite domain, *J. Comput. Phys.*, 125, 498–512, 1996.

64. Artoli, A.M.M., Hoekstra, A.G., and Sloot, P.M.A., 3D Pulsatile flow with the Lattice Boltzmann BGK Method, *Int. J. Mod. Phys. C*, 13 (8), 1119–1134, 2002.

65. Hadjiconstantinou, N.G. and Patera, A.T., Heterogeneous atomistic-continuum representation for dense fluid systems, *Int. J. Mod. Phys. C*, 8 (4), 967–976, 1997.

66. Xiantao Li, Weinan E., and Vanden-Eijnden, E., Some recent progress on multiscale modeling, Lecture Notes in *Comput. Sci. Eng.*, 39, 3–22, 2004.

67. Hockney, R.W., Eastwood, J.W., *Computer Simulation Using Particles*, McGraw-Hill Inc. 1981.

68. Pfalzner, S., and Gibbon, P., *Many-Body Tree Methods in Physics*, Cambridge University Press, New York, 1996, pp. 176.

69. Greengard, L., The numerical solution of the N-body problem, *Comput. Phys.*, 142–152, 1990.

70. Kittel, C., *Introduction to Solid State Physics*, Wiley, New York, 1986.

71. Dzwinel, W. and Yuen, D.A., Matching Macroscopic Properties of Binary Fluid to the Interactions of Dissipative Particle Dynamics, *Int. J. Modern Phys. C*, 11 (1), 1–25, 2000c.

72. Kröger, M., NEMD computer simulation of polymer melt rheology. *Rheol.*, 5, 66–71, 1995.

73. Herrmann, H.J., The importance of computer simulations of granular flow, *Comput. Sci. Eng.*, 1 (1), 72–73, 1999.

74. Dzwinel, W., Virtual particles and search for global minimum, *Future Generation Comput. Sys.*, 12, 371–389, 1997.

75. Alda, W., Yuen, D.A., Luthi, H.P., and Rustad, J.R., Exothermic and endothermic chemical reactions modeled with molecular dynamics, *Physica D*, 146, 261–274, 2000.

76. Pierrehumbert, R.T. and Yang, H., Global chaotic mixing on isentropic surfaces, *J. Atmos. Sci.*, 50 (15), 2462–2479, 1993.

77. Ten, A., Yuen, D.A., Podladchikov, Yu., and Larsen, T.B., Pachepsky, E., and Malvesky A.V., Fractal features in mixing of non-Newtonian and Newtonian mantle convection, *Earth Planet. Sci. Lett.*, 146, 401–414, 1997.

78. Augenbaum, J.M. and Peskin, C.S., On the construction of the Voronoi mesh on a sphere, *J. Comput. Phys.*, 14, 177–198, 1985.

79. Braun, J. and Sambridge M., A numerical method for solving partial differential equations on highly evolving grids, *Nature*, 376, 655–660, 1995.

80. Español, P. and Serrano, M., Dynamical regimes in DPD, *Phys. Rev. E.*, 59 (6), 6340–6347, 1999.

81. Marsh, C., Backx, G., Ernst, M.H., Static and dynamic properties of dissipative particle dynamics, *Phys. Rev. E.*, 56, 1976–1691, 1994.

82. Masters A.J. and Warren, P.B., Kinetic theory for dissipative particle dynamics. The importance of collisions, *Europhys. Lett.*, 48 (1), 1–7, 1999.

83. Groot, R.D. and Warren, P.B., Dissipative particle dynamics: bridging the gap between atomistic and mesoscopic simulation, *J. Chem. Phys.*, 107, 4423–4435, 1997.

84. Dzwinel, W. and Yuen, D.A., A two-level, discrete particle approach for large-scale simulation of colloidal aggregates, *Int. J. Modern Phys. C*, 11 (5), 1037–1061, 2000b.

85. Flekkøy, E.G. and Coveney, P.V., Foundations of dissipative particle dynamics, *Phys. Rev. Lett.*, 83, 1775–1778, 1999.

86. Monaghan, J.J., Smoothed Particle Hydrodynamics, *Annu. Rev. Astron. Astrophys.*, 30, 543, 1992.

87. Dzwinel, W., Boryczko, K., and Yuen D.A., A discrete-particle model of blood dynamics in capillary vessels, *J. Coll. Inter. Sci.*, 258 (1), 163–173, 2003.

88. Kubo, R., *Statistical Mechanics*, Wiley, New York, 1965.

89. Chapman, S. and Cowling, T.G., *The Mathematical Theory of Non-uniform Gases*, Cambridge University Press, Cambridge, 1990.

90. Daniel, J.C. and Audebert, R., Small volumes and large surfaces: the world of colloids, in *Soft Matter Physics*, Daoud, M. and Williams C.E., eds., Springer Verlag, 1999, p. 320.

91. Grier, D.G. and Behrens, S.H., Interactions in colloidal suspensions, in *Electrostatic effects in Soft Matter and Biphysics*, Holm C., Keikchoff P., and Podgornik R., Eds., Kluwer, 2001.

92. Yaman, K., Jeppesen, C., and Marques, C.M., Depletion forces between two spheres in a rod solution, *Europhys. Lett.*, 42, 221–226, 1998.

93. Larsen, A. and Grier, D.G., Like-charge attractions in metastable colloidal crystallities, *Nature*, 385, 230–233, 1997.

94. Prausnitz, J. and Wu, J., Resolving mysterious particle attraction in colloids, *En. Vision.*, 16/4, 18–19, 2000.

95. Dzwinel, W. and Yuen, D.A., Rayleigh–Taylor Instability in the Mesoscale Modelled by Dissipative Particle Dynamics, *Int. J. Modern Phys. C*, 12 (1), 91–118, 2001.

96. Libersky L.D., Petschek, A.G., Carney, T.C., Hipp, J.R., and Allahdadi, F.A., High Strain Lagrangian Hydrodynamics, *J. Comp. Phys.*, 109/1, 67–73, 1993.

97. Koshizuka, S. and Ikeda, H., MPS, moving particles semi-implicit method, http://www.tokai.t.u.tokyo.ac.jp/usr/rohonbu/ikeda/mps/mps.html, 1999.

98. Colella, P. and Woodward, P.R., The piecewise parabolic method (PPM) for gas-dynamical simulations, *J. Comp. Phys.*, 54, 174, 1984.

99. Boryczko, K., Dzwinel, W., and Yuen, D.A., Clustering revealed in high-resolution simulations and visualization of multi-resolution features in fluid-particle model, *Concurrency and Computation: Practice and Experience*, 15, 101–116, 2003.

100. Boryczko, K., Dzwinel, W., and Yuen D.A., Dynamical clustering of red blood cells in capillary vessels, *J. Molecul. Model.*, 9, 16–33, 2003.

101. Boryczko, K., Dzwinel, W., and Yuen D.A., Modeling heterogeneous mesoscopic fluids in irregular geometries using shared memory systems, *Molecul. Simul.*, 31 (1), 45–56, 2005.

102. Boryczko, K., Dzwinel, W., and Yuen, D.A., Modeling fibrin polymerization in blood flow with discrete-particles, *Comp. Models Prog. Biomed.*, 75, 181–194, 2004.

103. Vattulainen, I., Karttunen, M., Besold, G., and Polson, J.M., Integration schemes for dissipative particle dynamics: from softly interacting systems towards hybrid models, *J. Chem. Phys.*, 116 (10), 3967–3979, 2002.

104. Coveney, P.V. and Novik, K.E., Computer simulations of domain growth and phase separation in two-dimensional binary immiscible fluids using dissipative particle dynamics, *Phys. Rev. E.*, 54 (5), 5134–5141, 1996.

105. Adams D.J., Alternatives to the periodic cube in computer simulation: CCP5 Information Quarterly for Computer Simulation of Condensed Phases, 10:30, Informal Newsletter, Daresbury Laboratory, England, 1983.

106. Smith, W., The minimum image convention in non-cubic MD cell, CCP5 Information Quarterly for Computer Simulation of Condensed Phases, 30:35, Informal Newsletter, Daresbury Laboratory, England, 1989.

107. Dzwinel, W., Kitowski, J., and Moscinski, J., "Checker board" periodic boundary conditions in molecular dynamic codes. *Molecul. Simul.*, 7, 171–179, 1991.

108. Andenberg, M.R., *Clusters Analysis for Applications*, New York, Academic Press, 1973.

109. Theodoris, S. and Koutroumbas, K., *Pattern Recognition*, Academic Press, San Diego, London, Boston, 1998.

110. Gowda, C.K., and Krishna, G., Agglomerative clustering using the concept of nearest neighborhood, *Pattern Recogn.*, 10, 105, 1978.

111. Chandrasekhar, S., *Hydrodynamic and Magnetohydrodynamic Stability*, Clarendon Press, Oxford, 1961.

112. Read, K.I., Experimental investigation of turbulent mixing by Rayleigh-Taylor instability, *Physica D*, 12, 45–58, 1984.

113. Youngs, D.L., Numerical simulation of turbulent mixing by Rayleigh-Taylor instability, *Physica D*, 12, 32–44, 1984.

114. Dzwinel, W., Alda, W., Pogoda, M., and Yuen, D.A., Turbulent mixing in the microscale, *Physica D*, 137, 157–171, 2000u.

115. Alda, W., Dzwinel, W., Kitowski, J., Moscinski, J., Pogoda, M., and Yuen, D.A., Complex fluid-dynamical phenomena modeled by large-scale molecular dynamics simulations, *Comp. Phys.*, 12 (6), 595–600, 1998.

116. Ottino, J.M., Unity and diversity in mixing: stretching, diffusion, breakup, and aggregation in chaotic flows, *Phys. Fluids, A*, 3/5, 1417–1430, 1991.

117. de Roussel, P., Hansen, S., Khakhar, D.V. and Ottino, J.M., Mixing and dispersion of viscous fluids and powdered solids, *Advances Chem. Eng.* 25, 105–204, 2000.

118. Cohen, M.P., Brenner, J.E., and Nagel, S.R., Two fluid drop snap-off problem: experiments and theory, *Phys. Rev. Lett.*, 83, 1147–1150, 1999.

119. Stone, H.A., Dynamics of drop deformation and breakup in viscous fluids, *Ann. Revs. Fluid Mech.*, 26, 65–102, 1994.

120. Ellero, M., Kröger, M., and Hess, S., Viscoelastic flows studied by smoothed particle dynamics, *J. Non-Newtonian Fluid Mech.* 105, 35–51, 2002.

121. Oron, A., Davis, S.H., and Bankoff, S.G., Long-scale evolution of thin films, *Rev. Modern Phys.*, 69 (3), 931–980, 1997.

122. Myers, T.G., Thin films with high surface tension, *SIAM Rev.*, 40 (3), 441–462, 1998.

123. Dzwinel, W. and Yuen, D.A., Dissipative particle dynamics of the thin-film evolution in mesoscale, *Molecul. Simul.*, 22, 369–395, 1999.

124. Indeikina, A., Agarwal, A., Veretennikov, I., and Chang, H-C., Unexpected Phenomena in Contact-line Dynamics, 1998 Division of Fluid Dynamics Meeting, November 22–24, 1998, Philadelphia, USA.

125. Boryczko, K., Dzwinel, W., and Yuen D.A., Parallel Implementation of the Fluid Particle Model for Simulating Complex Fluids in the Mesoscale, *Concurrency and Computation: Practice and Experience*, 14, 137–161, 2002.

126. Gonnella, G., Orlandini, E., and Yeomans, J.M., Spinodal decomposition to a lamellar, phase: effects of hydrodynamic flow, *Phys. Rev. Lett.* 78, 9, 1695–1698, 1997.

127. Clark, A.T., Lal, M., Ruddock, J.N., and Warren, P.B., Mesoscopic simulation of drops in gravitational and shear fields, *Langmuir*, 16, 6342–6350, 2000.

128. Berger, S.A., Flow in Large Blood Vessels, Fluid Dynamics in Biology, Contemporary Math. Series, Eds. A.Y. Cheer and C.P. Van Dam, Amer. Math. Sec., Providence, 479–518, 1992.

129. Bitsch, L., Blood Flow in Microchannels, Master Thesis, Mikroelektronik Centret, MIC Dansk Polymercenter/Institut for Kemiteknik Technical University of Denmark, 2002, p. 93.

130. Quarteroni A., Veneziani A., and Zunino P., Mathematical and numerical modelling of solute dynamics in blood flow and arterial walls, *SIAM J. Numer. Anal.*, 39 (5), 1488–1511, 2002.

131. Diamond, S.L., Engineering design of optimal strategies for blood clot dissolution, *Annu. Rev. Biomed. Eng.*, 1, 427–461, 1999.

132. Bark, N., Foldes-Papp, Z., and Rigler, R., The incipient stage in thrombin-induced.brin polymerization detected by FCS at the single molecule level, *Biochem. Biophys. Res. Commun.*, 260, 35–41, 1999.

133. Erlebacher, G., Yuen, D.A., and Dubuffet, F., Current trends and demands in visualization in the geosciences, Electronic Geosciences, 6, 2001, http://link.springerny.com/link/service/journals/10069/free/technic/erlebach/index.htm.

134. Thompson D.S., Machiraju R., Dusi V.S., Jiang M., Nair J., Thampy S., and Craciun G., Physics-based Feature Mining of Computational Fluid Dynamics Data Sets, Special issue of *IEEE Comput. Sci. Eng.*, 22–30, 2002.

135. Yuen, D.A., Vincent, A.P., Kido, M., and Vecsey, L., Geophysical Applications of Multidimensional Filtering with Wavelets, *Pure. Appl. Geophys.*, 159, 2285–2309, 2002.

136. Faber, V., Clustering and the Continuous k-Means Algorithm, *Los Alamos Sci.*, 22, 138–149, 1994.

137. Amira, Advanced 3D Visualization and Volume Modeling, http://www.amiravis.com.

138. Jain, D. and Dubes R.C., Algorithms for Clustering Data, Prentice-Hall, Advanced Reference Series, 1988.

139. Yuen, D.A., Garbow Z.A., and Erlebacher, G., Remote Data Analysis, Visualization and Problem Solving Environment (PSE) Based on Wavelets in the Geosciences, submitted for publication in Visual Geosciences, 2004.

140. Foster, I., Kesselman, C., The grid: blueprint for a new computing infrastructure, Morgan Kaufmann, San Francisco, 1998 (www.mkp.com/grids).

141. Fox, G.C. and Furmanski, W., High-performance commodity computing, in *The GRID: Blueprint for a New Computing Infrastructure*, Foster, I. and Kesselman, C., Eds., Morgan-Kaufmann, San Francisco, 1999, Chapter 10, pp. 237–255.

142. Fox, G., Presentation on Web Services and Peer-to-Peer Technologies for the Grid ICCS Amsterdam April 24 2002 URL: (http://grids.ucs.indiana.edu/ptliupages/publications/presentations/iccsapril02.ppt).

143. Brezany, P., Bubak, M., Malawski, M., and Zając, K., Large-Scale Scientific Irregular Computing on Clusters and Grids, in *Proceedings of International Conference on Computational Science - ICCS*, Sloot, P.M.A. Kenneth Tan, C.J., Dongara, J.J., and Hoekstra, A.G., eds., , Amsterdam, April 21–24, Vol. I, Lecture Notes in Computer Science, no. 2330, Springer, 2002, 484–493.

144. Dutka, Ł., and Kitowski, J., Application of component-expert technology for selection of data-handlers in crossgrid, in *Kranzlmüller*, Kascuk, D., Dongarra, P., Volkert, J.J.,. Eds., Lecture Notes in Computer Science, Springer, 2002, pp. 25–32.

145. Fox G., Internal Presentation on Collaboration and Web Services April 4 2002 URL: (http://grids.ucs.indiana.edu/ptliupages/publications/presentations/collabwsncsaapril02.ppt).

27 Nonlinear Dynamics of DNA Chain – Peyrard–Bishop–Dauxois Model

Slobodan Zdravković
University of Priština, Serbia, Serbia and Montenegro

CONTENTS

I. INTRODUCTION

In this introductory chapter, we very briefly describe DNA molecule (Section A) and some basic mathematics (Section B and Section C), which is important to understand the rest of the text.

A. DNA Molecule

DNA molecule is the biggest known molecule. Its relative molar mass is of the order of 10^9 [1] and its density is approximately 1.7 g/cm^3 [2].

All macromolecules are chain-like polymers built up from smaller monomers. Usually only one family of subunits is used to construct each chain [3]. Amino acids are linked to other amino acids to form proteins, sugars are linked to form polysaccharides and nucleic acids are sequences of nucleotides.

Genetic role of DNA was established in 1943 [1] but its structure was not discovered until 1953 [4,5]. According to Watson–Crick model, DNA molecule is represented by two twisted strands.

FIGURE 27.1 A general scheme of DNA chain. (PHO stands for phosphate.)

Each strand is formed by the union of smaller molecules called nucleotides and each nucleotide is built up from phosphate, sugar, and either purine or pyrimidine basis.

The nucleotides are always linked together through a phosphate group (Figure 27.1) by covalent bonds [1]. However, different strands interact through basis by hydrogen bonds. Adenine (A) is always attached to thymine (T) by two hydrogen bonds, whereas guanine (G) and cytosine (C) are attached by three bonds. Covalent interactions are strong. For example, the energies of interaction are 348.6 kJ/mol and 336 kJ/mol for C—C and C—N bond, respectively [2]. The average energy of hydrogen bonds is around 16.5 kJ/mol, which is eight times the average energy of thermal motion of molecules at room temperature [1].

Much more information about this topic can be found in many books, one of them is Ref. [3].

B. SOLITONS

A traveling wave $\Phi_T(\xi)$ is a solution of wave equation which depends upon x and t only through $\xi \equiv x - ut$, where u is a constant. A solitary wave is a localized traveling wave [6]. An example is shown in Figure 27.2. This is a graphical representation of the function

$$\Phi(x - ut) = \frac{3u}{\alpha} \operatorname{sech}^2\left[\frac{\sqrt{u}}{2}(x - ut)\right] \tag{27.1}$$

which is a solution of Korteweg–de Vries (KdV) equation [6–9]

$$\Phi_t + \alpha\Phi\Phi_x + \Phi_{xxx} = 0 \tag{27.2}$$

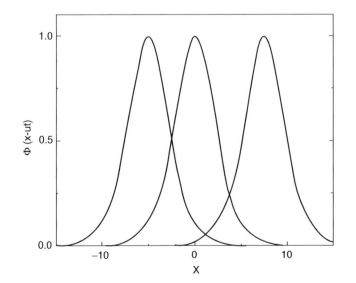

FIGURE 27.2 Bell type soliton $\Phi(x - 0.5t)$. (From left to right: $t = -10, 0, 15$.)

where indexes mean partial derivatives. Note that $u > 0$ and the soliton moving in the opposite direction would be given by the expression

$$\Phi(x + ut) = \frac{3u}{\alpha} \operatorname{sech}^2\left[\frac{\sqrt{u}}{2}(x + ut)\right] \tag{27.3}$$

By solitons, we usually assume solitary wave solutions of a wave equation which preserves its shape and velocity upon collision with other solitary waves [6]. In this chapter, we assume that solitons and solitonic waves are synonyms. The waves shown in Figure 27.2 are usually called bell-type solitons.

Solitary wave can also be defined as the traveling wave whose transition from one constant asymptotic state as $\xi \to -\infty$ to another as $\xi \to +\infty$ is localized in ξ [6]. An example would be a solution of sine-Gordon equation

$$\Phi_{xx} - \Phi_u = \sin \Phi \tag{27.4}$$

The solution is a so-called kink soliton

$$\Phi = 4 \arctan\left[\exp\frac{\pm(x - ut)}{\sqrt{1 - u^2}}\right] \tag{27.5}$$

where $+$ and $-$ refers to soliton and antisoliton, respectively. They are shown in Figure 27.3 and Figure 27.4.

C. Nonlinear Schrödinger Equation

In this chapter, we deal with another type of soliton called breather. Nonlinear Schrödinger equation (NSE) has a solution

$$i\Phi_t + \Phi_{xx} + k|\Phi|^2\Phi = 0 \tag{27.6}$$

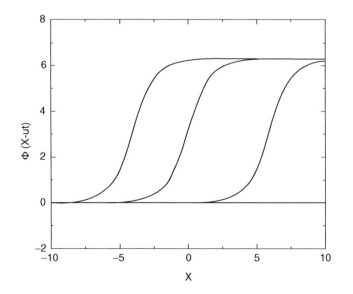

FIGURE 27.3 Kink type soliton $\Phi(x - 0.2t)$. (From left to right: $t = -20, 0, 30$.)

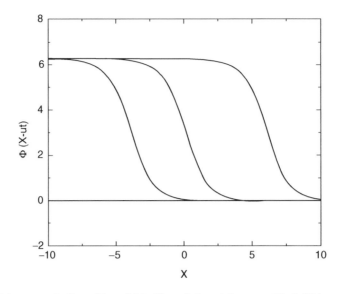

FIGURE 27.4 Kink type antisoliton $\Phi(x - 0.2t)$. (From left to right: $t = -20, 0, 30$.)

This solution can be written as [6]

$$\Phi = \Phi_0 \, \text{sech}\left[\Phi_0\sqrt{\frac{k}{2}}(x - u_e t)\right] \exp\left[\frac{iu_e}{2}(x - u_c t)\right] \tag{27.7}$$

where

$$\Phi_0 = \frac{u_e^2 - 2u_e u_c}{2k} \tag{27.8}$$

u_e and u_c are the envelope and the carrier velocities and

$$u_e > 2u_c \tag{27.9}$$

This is a modulated wave, that is, a modulated soliton, shown in Figure 27.5.

II. DNA DYNAMICS

Deoxyribonucleic acid (DNA) is certainly one of the most important biomolecules. Its double standard helical structure undergoes a very complex dynamics and the knowledge of that dynamics provides insights into various related biological phenomena such as transcription, translation, and mutation. The key problem in DNA biophysics is how to relate functional properties of DNA with its structural and physical dynamical characteristics. The possibility that nonlinear effects might focus the vibrational energy of DNA into localized soliton-like excitations was first contemplated by Englander et al. [10]. Although several authors [11–18] have suggested that either topological kink solitons or bell-shaped breathers would be good candidates to play a basic role in DNA nonlinear dynamics, there are still several unresolved questions about the issue. The hierarchy of the most important models for nonlinear DNA dynamics was presented by Yakushevich [19].

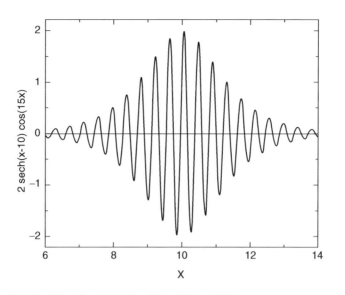

FIGURE 27.5 Modulated soliton (breather) $2 \operatorname{sech}(x - 10) \cos(15x)$.

A. PBD MODEL OF DNA MOLECULE

The B-form DNA in the Watson–Crick model is a double helix, which consists of the two strands, s_1 and s_2, Figure 27.6. Nucleotides of each strand are linked by the nearest-neighbor harmonic interactions along the chains. The strands are coupled to each other through hydrogen bonds, which are supposed to be responsible for transversal displacements of nucleotides.

According to Peyrard–Bishop (PB) model, one assumes a common mass m for all the nucleotides [16]. This is motivated by the fact that four different nucleotides differ in mass by at most 13% [20]. In addition, the same coupling constant k along each strand is assumed. These simplifications mean that we treat the DNA chain as a periodic one.

The PB model does not take helicosity into consideration (Figure 27.6). However, its extended version, developed by Dauxois [21,22], does. We refer to this model as Peyrard–Bishop–Dauxois (PBD) model. The helicoidal structure of the DNA chain implies that nucleotides from different strands become close enough so that they interact through water filaments. This means that a nucleotide at the site n of one strand interacts with both $(n + h)$th and $(n - h)$th nucleotides of

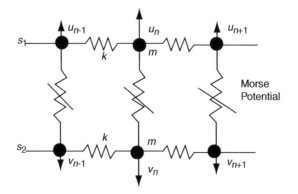

FIGURE 27.6 The simple model for DNA strands.

the other strand. In this chapter, we use $h = 4$ [21,22], although the value $h = 5$ might be better [23,24]. This comes from the fact that there are approximately ten nucleotides per one turn.

Introducing the transversal displacements u_n, v_n of the nucleotides from their equilibrium positions along the direction of the hydrogen bonds, the Hamiltonian for the DNA chain becomes [21,22]

$$H = \sum \left\{ \frac{m}{2} \left(\dot{u}_n^2 + \dot{v}_n^2 \right) + \frac{k}{2} \left[(u_n - u_{n-1})^2 + (v_n - v_{n-1})^2 \right] \right.$$
$$\left. + \frac{K}{2} \left[(u_n - v_{n+h})^2 + (u_n - v_{n-h})^2 \right] + D \left[e^{-a(u_n - v_n)} - 1 \right]^2 \right\} \quad (27.10)$$

Here k and K are the harmonic constants of the longitudinal and helicoidal springs, respectively. The last term in the Hamiltonian represents a Morse potential approximating the potential of the hydrogen bonds. Eventually, D and a are the depth and the inverse width of the Morse potential well, respectively.

It is more convenient to describe the motion of two strands by making a transformation to the center-of-mass co-ordinates representing the in-phase and out-of-phase transversal motions, namely

$$x_n = \frac{u_n + v_n}{\sqrt{2}}, \qquad y_n = \frac{u_n - v_n}{\sqrt{2}} \quad (27.11)$$

The dynamical equations, derived from the Hamiltonian (27.10), are

$$m\ddot{x}_n = k(x_{n+1} + x_{n-1} - 2x_n) + K(x_{n+h} + x_{n-h} - 2x_n) \quad (27.12)$$

$$m\ddot{y}_n = k(y_{n+1} + y_{n-1} - 2y_n) - K(y_{n+h} + y_{n-h} + 2y_n) + 2\sqrt{2}aD \left(e^{-a\sqrt{2}y_n} - 1 \right) e^{-a\sqrt{2}y_n} \quad (27.13)$$

Equation (27.12) describes usual linear waves (phonons) and Equation (27.13) describes non-linear waves (breathers). We restrict our attention on the nonlinear equation and assume that the oscillations of nucleotides are large enough to be anharmonic but still small enough so that the nucleotides oscillate around the bottom of the Morse potential. This safely allows the transformation

$$y = \varepsilon \Phi, \qquad \varepsilon \ll 1 \quad (27.14)$$

Equation (27.13) and Equation (27.14), and the expansion of exponential terms in Equation (27.13), yield

$$\ddot{\Phi}_n = \frac{k}{m}(\Phi_{n+1} + \Phi_{n-1} - 2\Phi_n) - \frac{K}{m}(\Phi_{n+h} + \Phi_{n-h} + 2\Phi_n)$$
$$- \omega_g^2 (\Phi_n + \varepsilon \alpha \Phi_n^2 + \varepsilon^2 \beta \Phi_n^3) \quad (27.15)$$

where

$$\omega_g^2 = \frac{4a^2 D}{m}, \qquad \alpha = \frac{-3a}{\sqrt{2}} \quad \text{and} \quad \beta = \frac{7a^2}{3} \quad (27.16)$$

To solve Equation (27.15), we use the semi-discrete approximation [25]. This means that we look for wave solutions of the form

$$\Phi_n(t) = F_1(\varepsilon nl, \varepsilon t)e^{i\theta_n} + \varepsilon \left[F_0(\varepsilon nl, \varepsilon t) + F_2(\varepsilon nl, \varepsilon t)e^{i2\theta_n} \right] + \text{cc} + O(\varepsilon^2) \quad (27.17)$$
$$\theta_n = nql - \omega t \quad (27.18)$$

where l is the distance between two neighboring nucleotides in the same strand, ω the optical frequency of the linear approximation of their vibrations, q the wave number (whose role will be discussed later), cc complex conjugate terms, and the function F_0 real.

Before we proceed in solving Equation (27.15), we give some explanations for Equation (27.17). If there were not the last term in Equation (27.15), the one with ω_g^2, which comes from the nonlinear term in Equation (27.13), we would expect the solution in the form $F_1 e^{i\theta_n} + cc$ instead of Equation (27.17). This would be a modulated wave with a carrier component $e^{i\theta_n}$ and an envelope F_1. We will see later that the modulation factor F_1 will be treated in a continuum limit whereas the carrier wave will not. In other words, the carrier component of the modulated wave includes the discreteness and the procedure is called semi-discrete approximation.

It was already pointed out that the function F_1 represents the envelope while the exponential term describes the carrier component of the wave. As if the frequency of the carrier wave is much higher than the frequency of the envelope, we need two time-scales, t and εt, for those two functions. Of course, the same holds for the coordinate scales.

As if there are terms with Φ_n^2 and Φ_n^3 in Equation (27.15), we cannot expect solution of this equation in the simple form $F_1 e^{i\theta_n} + cc$ and nonexponential term, as well as terms with $e^{i2\theta_n}$, should be incorporated into the expression for the solution. We do not worry about $e^{i3\theta_n}$ terms as if they would be multiplied by ε^3 which we neglect.

All the explanations given previously are rather intuitive and correspond to physics of the problem. However, the expression (27.17) has its mathematical basis. This is multiple-scale method or derivative-expansion method [9,26].

Now, we will solve Equation (27.15). It was already pointed out that the functions F_i would be treated in the continuum limit. So, taking this limit ($nl \to z$) and applying the transformations

$$Z = \varepsilon z, \qquad T = \varepsilon t \tag{27.19}$$

yields the following continuum approximation

$$F[\varepsilon(n \pm h)l, \varepsilon t] \longrightarrow F(Z, T) \pm F_Z(Z, T)\varepsilon l h + \frac{1}{2}F_{ZZ}(Z, T)\varepsilon^2 l^2 h^2 \tag{27.20}$$

where F_Z and F_{ZZ} mean corresponding derivatives with respect to Z. This leads to a new expression for the function $\Phi_n(t)$

$$\Phi_n(t) \longrightarrow F_1(Z, T)e^{i\theta} + \varepsilon\left[F_0(Z, T) + F_2(Z, T)e^{i2\theta}\right] + cc$$
$$= F_1 e^{i\theta} + \varepsilon\left[F_0 + F_2 e^{i2\theta}\right] + F_1^* e^{-i\theta} + \varepsilon F_2^* e^{-i2\theta} \tag{27.21}$$

where * stands for complex conjugate and $F_i \equiv F_i(Z, T)$.

From Equation (27.18)–Equation (27.21), we can straightforwardly obtain expressions for all the terms in Equation (27.15), some of them are

$$\dot{\Phi}_n = \varepsilon F_{1T}e^{i\theta} - i\omega F_1 e^{i\theta} + \varepsilon^2 F_{0T} + \varepsilon^2 F_{2T}e^{i2\theta} - 2i\varepsilon\omega F_2 e^{i2\theta} + cc \tag{27.22}$$

$$\Phi_{n+1} + \Phi_{n-1} - 2\Phi_n \longrightarrow 2F_1 e^{i\theta}[\cos(ql) - 1] + 2i\varepsilon l F_{1Z}e^{i\theta}\sin(ql)$$
$$+ \varepsilon^2 l^2 F_{1ZZ}e^{i\theta}\cos(ql) + \varepsilon^3 l^2 F_{0ZZ} + \varepsilon^3 l^2 F_{2ZZ}e^{i2\theta}\cos(2ql) + cc$$
$$+ 2\varepsilon F_2 e^{i2\theta}[\cos(2ql) - 1] + 2i\varepsilon^2 l F_{2Z}e^{i2\theta}\sin(2ql) \tag{27.23}$$

and

$$
\begin{aligned}
\Phi_n^3 \;\longrightarrow\; & 6\varepsilon F_0 |F_1|^2 + 3\varepsilon F_1^2 F_2^* + 3\varepsilon F_1^{*2} F_2 + e^{i\theta} 3|F_1|^2 F_1 + e^{-i\theta} 3|F_1|^2 F_1^* \\
& + e^{i2\theta}\big(3\varepsilon F_0 F_1^2 + 6\varepsilon |F_1|^2 F_2\big) + e^{-i2\theta}\big(3\varepsilon F_0 F_1^{*2} + 6\varepsilon |F_1|^2 F_2^*\big) \\
& + e^{i3\theta} F_1^3 + e^{-i3\theta} F_1^{*3} + e^{i4\theta} 3 F_1^2 F_2 + e^{-i4\theta} 3\varepsilon F_1^{*2} F_2^*
\end{aligned}
\tag{27.24}
$$

In the expressions for Φ_n^2 and Φ_n^3, only terms with ε^2 and ε should be included as if those two terms are multiplied by ε and ε^2 in Equation (27.15). So, the continuum version of Equation (27.15) becomes

$$
\begin{aligned}
& (\varepsilon^2 F_{1TT} - 2i\varepsilon\omega F_{1T} - \omega^2 F_1)e^{i\theta} - (4i\varepsilon^2\omega F_{2T} + 4\varepsilon\omega^2 F_2)e^{i2\theta} + cc \\
& = \frac{k}{m}\Big\{ 2F_1[\cos(ql) - 1]e^{i\theta} + 2i\varepsilon l F_{1Z}\sin(ql)e^{i\theta} + \varepsilon^2 l^2 F_{1ZZ}\cos(ql)e^{i\theta} \\
& \qquad\quad + 2\varepsilon F_2[\cos(2ql) - 1]e^{i2\theta} + 2i\varepsilon^2 l F_{2Z}\sin(2ql)e^{i2\theta} + cc \Big\} \\
& \quad - \frac{K}{m}\Big\{ 2F_1[\cos(qhl) + 1]e^{i\theta} + 2i\varepsilon hl F_{1Z}\sin(qhl)e^{i\theta} + \varepsilon^2 h^2 l^2 F_{1ZZ}\cos(qhl)e^{i\theta} \\
& \qquad\quad + 2\varepsilon F_2[\cos(2qhl) + 1]e^{i2\theta} + 2i\varepsilon^2 hl F_{2Z}\sin(2qhl)e^{i2\theta} + 4\varepsilon F_0 + cc \Big\} \\
& \quad - \omega_g^2\Big[F_1 e^{i\theta} + \varepsilon F_0 + \varepsilon F_2 e^{i2\theta} + 2\varepsilon\alpha|F_1|^2 + 2\varepsilon^2\alpha(F_0 F_1 + F_1^* F_2)e^{i\theta} \\
& \qquad\quad + \varepsilon\alpha F_1^2 e^{i2\theta} + 2\varepsilon^2\alpha F_1 F_2 e^{i3\theta} + 3\varepsilon^2\beta|F_1|^2 F_1 e^{i\theta} + \varepsilon^2\beta F_1^3 e^{i3\theta} + cc \Big]
\end{aligned}
\tag{27.25}
$$

From Equation (27.25), equating the coefficients for the various harmonics, we can get a set of important relations. For example, equating the coefficients for $e^{i\theta}$, we obtain

$$
\begin{aligned}
\varepsilon^2 F_{1TT} - 2i\varepsilon\omega F_{1T} - \omega^2 F_1 = & \; \frac{k}{m}\big\{ 2F_1[\cos(ql) - 1] + 2i\varepsilon l F_{1Z}\sin(ql) + \varepsilon^2 l^2 F_{1ZZ}\cos(ql) \big\} \\
& - \frac{K}{m}\big\{ 2F_1[\cos(qhl) + 1] + 2i\varepsilon hl F_{1Z}\sin(qhl) + \varepsilon^2 h^2 l^2 F_{1ZZ}\cos(qhl) \big\} \\
& - \omega_g^2\Big[F_1 + 2\varepsilon^2\alpha F_0 F_1 + 2\varepsilon^2\alpha F_1^* F_2 + 3\varepsilon^2\beta|F_1|^2 F_1 \Big]
\end{aligned}
\tag{27.26}
$$

Neglecting all the terms with ε and ε^2, we get a dispersion relation

$$
\omega^2 = \omega_g^2 + \frac{2k}{m}[1 - \cos(ql)] + \frac{2K}{m}[1 + \cos(qhl)]
\tag{27.27}
$$

This relation can be used to find the corresponding group velocity $d\omega/dq$ as

$$
V_g = \frac{l}{m\omega}[k\sin(ql) - Kh\sin(qhl)]
\tag{27.28}
$$

Similarly, equating the coefficients for $e^{i0} = 1$, we obtain

$$F_0 = \mu |F_1|^2 \tag{27.29}$$

where

$$\mu = -2\alpha \left[1 + \frac{4K}{m\omega_g^2} \right]^{-1} \tag{27.30}$$

However, for the next three harmonics ($e^{i2\theta}, e^{i3\theta}$ and $e^{i4\theta}$), the matter becomes more complicated. All of those harmonics give a relation

$$F_2 = \delta F_1^2 \tag{27.31}$$

but with different values for the parameter δ. For example, equating the coefficients for $e^{i2\theta}$ and neglecting all the terms with ε^2 and smaller, we get

$$\left\{ 4\omega^2 - \frac{2k}{m}[1 - \cos(2ql)] - \frac{2K}{m}[1 + \cos(2hql)] - \omega_g^2 \right\} F_2 = \omega_g^2 \alpha F_1^2 \tag{27.32}$$

which means that δ is not a constant but a function of ql. However, coefficients for $e^{i3\theta}$ and $e^{i4\theta}$ give constant values δ, $-\beta/2\alpha$ and $-3\beta/\alpha$, respectively.

In Ref. [27], it is given that the DNA dynamics using constant δ coming from the coefficients for $e^{i3\theta}$, whereas in Ref. [28] the parameter δ is given by Equation (27.31) and Equation (27.32). We will return to this point later.

Now we can derive a differential equation for F_1. To do this, we need new co-ordinates again defined as

$$S = Z - V_g T, \qquad \tau = \varepsilon T \tag{27.33}$$

which allows following transformations

$$F_Z = F_S, \quad F_{ZZ} = F_{SS}, \quad F_T = -V_g F_S + \varepsilon F_\tau, \quad F_{TT} = V_g^2 F_{SS} - 2\varepsilon V_g F_{\tau S} + \varepsilon^2 F_{\tau\tau} \tag{27.34}$$

According to Equation (27.26)–Equation (27.29), Equation (27.31) and Equation (27.34), we can finally obtain well-known NSE for the function F_1

$$iF_{1\tau} + PF_{1SS} + Q|F_1|^2 F_1 = 0 \tag{27.35}$$

where the dispersion coefficient P and the coefficient of nonlinearity Q are given by

$$P = \frac{1}{2\omega} \left\{ \frac{l^2}{m} [k\cos(ql) - Kh^2\cos(qhl)] - V_g^2 \right\} \tag{27.36}$$

and

$$Q = -\frac{\omega_g^2}{2\omega} [2\alpha(\mu + \delta) + 3\beta] \tag{27.37}$$

For $PQ > 0$, see Refs. [6,21,22], Equation (27.35) has an envelope-soliton solution

$$F_1(S, \tau) = A \, \text{sech}\left(\frac{S - u_e \tau}{L_e}\right) \exp\frac{iu_e(S - u_c\tau)}{2P} \tag{27.38}$$

The envelope amplitude A and its width L_e have the forms

$$A = \sqrt{\frac{u_e^2 - 2u_e u_c}{2PQ}} \tag{27.39}$$

$$L_e = \frac{2P}{\sqrt{u_e^2 - 2u_e u_c}} \tag{27.40}$$

and u_e and u_c are the velocities of the envelope and the carrier waves, respectively.
Finally, by setting

$$V_e = V_g + \varepsilon u_e \tag{27.41}$$

$$\Theta = q + \frac{\varepsilon u_e}{2P} \tag{27.42}$$

and

$$\Omega = \omega + \frac{\varepsilon u_e}{2P}(V_g + \varepsilon u_c) \tag{27.43}$$

Equations (27.17)–(27.19), Equation (27.29), Equation (27.31), Equation (27.33), and Eqaution (27.38) lead to

$$\Phi_n(t) = 2A \, \text{sech}\left[\frac{\varepsilon}{L_e}(nl - V_e t)\right] \left\{ \cos(\Theta nl - \Omega t) + \varepsilon A \, \text{sech}\left[\frac{\varepsilon}{L_e}(nl - V_e t)\right] \right.$$
$$\left. \times \left(\frac{\mu}{2} + \delta \cos[2(\Theta nl - \Omega t)]\right) \right\} + O(\varepsilon^2) \tag{27.44}$$

This is a sort of modulated solitonic wave. Such a soliton is called breather. To plot this function, we should determine a wavelength λ, defined as

$$q = \frac{2\pi}{\lambda} \tag{27.45}$$

where q appeared in Equation (27.18). Also, we need to know values for a couple of parameters present in Equation (27.44). This will be done in a next section.

A patient reader might ask why the parameter ε exists in the time scaling in Equation (27.33) but not in the space scaling. In other words, the question is why the co-ordinate τ has been defined as εT rather than $\tau = T$. It was already pointed out that the carrier components of the function (27.17) change faster than the envelope functions F_i. This means that the small parameter ε is present only in the envelope components F_i and this is why the scaling (27.19) was introduced. However, the definition (27.33) ensures that the time variation of the envelope function of (27.38), in units $1/\omega$, be smaller than the space variation of this function in units l [29]. Otherwise, there would be

$$-\frac{\partial f}{\partial t}\frac{1}{\omega} \gg \frac{\partial f}{\partial z}l \quad \text{for} \quad \tau = T$$

where f is the envelope component of Equation (27.38).

1. Determination of Wavelength λ

A procedure to determine the wavelength λ was explained in Ref. [27]. As if the parameter P is proportional to the width of the soliton, Equation (27.40), it was assumed to be positive, as well as group velocity V_g. To plot those functions versus ql, a following set of values, characterizing the DNA molecule, was chosen [21,22]:

$$\left.\begin{array}{ll} k = 3K = 24 \text{ N/m}, & a = 2 \times 10^{10} \text{ m}^{-1}, \quad m = 5.1 \times 10^{-25} \text{ kg} \\ l = 3.4 \times 10^{-10} \text{ m}, & D = 0.1 \text{ eV} \end{array}\right\} \quad (27.46)$$

as well as parameters characterizing a traveling wave [21,22]

$$u_e = 10^5 \text{ m/sec}, \quad u_c = 0, \quad \varepsilon = 0.007 \quad (27.47)$$

For $V_g > 0$ and $P > 0$, only a couple of intervals of values ql are allowed. The physical intuition leads towards a conclusion that the wavelength λ of a carrier wave ($q = 2\pi/\lambda$) should be an integer multiple of the spacing l. So, we obtain the following four possibilities [27]:

$$\begin{array}{ll} \lambda_1 = 6l & (ql = 1.05 \text{ rad}) \\ \lambda_2 = 7l & (ql = 0.9 \text{ rad}) \\ \lambda_3 = 8l & (ql = 0.78 \text{ rad}) \\ \lambda_4 = 9l & (ql = 0.7 \text{ rad}) \end{array}$$

All these four values lay in the first allowed interval for λ which reads $5.7l < \lambda < 9.5l$ or, equivalently, $0.66 \text{ rad} < ql < 1.11 \text{ rad}$, whereas other intervals do not contain any integer value for λ. For example, the second interval $2.3l < \lambda < 2.9l$ does not contain any integer multiple of l for the wavelength λ.

In Ref. [27], an argument to reject the value $ql = 1.05 \text{ rad}$ is given, and all the calculations were done for $ql = 0.78 \text{ rad}$. We will see later that the same value can be obtained using a different procedure.

It was already pointed out that a constant δ is used in Ref. [27]. Otherwise, the amplitude (27.39) would be imaginary. This means that we studied the DNA dynamics using δ coming from $e^{i3\theta}$ and did not use the lowest possible harmonic $e^{i2\theta}$. However, we should keep in mind two things concerning the issue we deal with. First, it is physically feasible that the lower harmonic plays a greater role in this perturbation approach so we prefer to analyze this case instead of the technically more comfortable constant value δ which has been used so far without clear justification. Second, it is apparent that the parameter K should be smaller than the harmonic constant of the longitudinal spring k. Consequently, it is expectable that the contributions of terms with k should be larger than those with K. However, in formulas (27.28) and (27.36), K is multiplied by either h or h^2 which might indicate that we cannot certain that $k = 3K$ is a good relationship for those two parameters.

So, in Ref. [28], we assumed the parameter δ to be nonconstant, but on the other hand, we tried to study the nonlinear dynamics of DNA considering the impact of different values for K. From Figure 27.7, we see that $Q > 0$ holds only for $K \leq 4.6 \text{ N/m}$ if $ql = 0.78 \text{ rad}$ is picked up. For

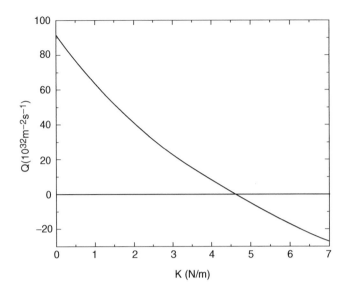

FIGURE 27.7 Nonlinear coefficient Q of the NSE as a function of K ($ql = 0.78$ rad).

$ql = 0.7$ rad the maximum value for K is 4.67 N/m. Of course, we can argue that those values for ql were obtained for $K = 8$ N/m in Ref. [27]. This is correct. However, using the same procedure for the possible ql values, explained earlier and in Ref. [27], we obtain $\lambda = 6l, 7l, 8l$ and $9l$, that is, $ql = 1.05, 0.9, 0.78$ and 0.7 rad, respectively, for both $K = 8$ N/m and $K = 6$ N/m. The value $K = 4$ N/m gives the same four values for ql and two more ($\lambda = 10l$ and $11l$, or $ql = 0.63$ rad and $ql = 0.57$ rad) in addition. For $K = 3$ N/m, we obtain $\lambda = 6l, 7l, \ldots, 13l$ while $K = 2$ N/m gives $\lambda = 6l, 7l, \ldots, 19l$. Hence, the value $ql = 0.78$ rad corresponds to both large and small values for the parameter K. This means that $ql = 0.78$ rad can be suitable choice for all acceptable values of K, but does not mean that this is the optimal value for our calculations. Thus, we need one more criterion for ql to be determined and we will show in Section 2 that this exists.

2. Density of Internal Oscillations of the Solitonic Waves in DNA

The breather-type solution, Equation (27.44), represents a sort of a modulated solitonic wave. From hyperbolic and cosine terms in Equation (27.44), we can recognize wave numbers of both the envelope and the carrier wave. In other words, we can see that the width of the envelope Λ and the wavelength of the carrier wave λ_c are

$$\Lambda = \frac{2\pi L_e}{\varepsilon} \tag{27.48}$$

and

$$\lambda_c = \frac{2\pi}{q + (\varepsilon u_e / 2P)} \tag{27.49}$$

This equation can be used to calculate the number of wavelengths of the carrier wave contained within the length of the envelope as

$$D_o = \frac{\Lambda}{\lambda_c} \tag{27.50}$$

We call D_o a density of internal oscillations (density of carrier wave oscillations) for short even though, strictly speaking, this is not density, that is, the number of wavelengths per unit length. From Equation (27.48)–Equation (27.50) and (Equation 27.40), we can easily obtain

$$D_o = \frac{u_e}{\sqrt{u_e^2 - 2u_e u_c}} \left(1 + \frac{2qlP}{\varepsilon l u_e} \right) \tag{27.51}$$

It should be noted that the dispersion coefficient P is defined by Equation (27.36) and is a function of ql.

Using Equation (27.46) and Equation (27.47), we can plot the function D_o versus ql and this is done in Figure 27.8 for three values of the parameter K [30]. From this figure, we see that the density of the internal oscillations of the soliton D_o reaches a maximum value for $ql \approx 0.77$ rad, which is extremely close to the value of 0.78 rad, which was earlier selected as the most favorable one. Should we accept this value, corresponding to the highest D_o, for our calculations? Certainly yes, since a higher D_o probably means a more efficient modulation. It is required for modulation, when both signals are cosine functions, that the frequency of the carrier component is much higher than the frequency of the envelope. In our case, however, there are no two frequencies, but we introduced D_o instead. Our intuition strongly suggests that nature wants modulation with D_o as high as possible, which we might call the most efficient modulation.

We should emphasize that the highest values of D_o do not have too profound meaning since those values simply depend on the choice of the somewhat arbitrary parameters ε and u_e. In fact, we are looking for the maximum of the product function $ql \cdot P(ql)$. However, we can make one more conclusion based on the analysis of how D_o depends on ql. For various K, we can obtain a set of maxima for the function D_o as follows:

$D_{om} = 7.74$ for $ql = 0.775$ rad if $K = 4.5$ N/m
$D_{om} = 7.08$ for $ql = 0.774$ rad if $K = 4$ N/m
$D_{om} = 6.43$ for $ql = 0.772$ rad if $K = 3.5$ N/m
$D_{om} = 5.77$ for $ql = 0.770$ rad if $K = 3$ N/m
$D_{om} = 5.11$ for $ql = 0.766$ rad if $K = 2.5$ N/m
$D_{om} = 4.46$ for $ql = 0.761$ rad if $K = 2$ N/m

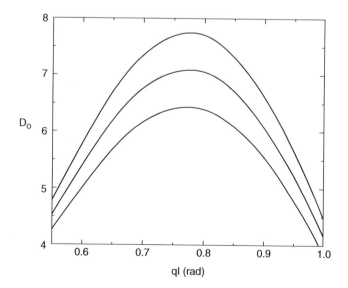

FIGURE 27.8 Density of internal oscillations (density of carrier wave oscillations) versus ql. (From top to bottom: $K = 4.5$ N/m, 4 N/m and 3.5 N/m.)

We can see that the highest values of the density of the internal oscillations D_o increase when the parameter K increases. Another important fact is that an optimal ql for any reasonable K does not differ significantly from our previously accepted value of $ql = 0.78$ rad. Otherwise, for $K = 0$ we obtain $D_{om} \approx 2$ for $ql = 0.53$ rad. This means that from the criterion of modulation, the PBD model [21,22] is better than the original PB model [16], which could be obtained from the former one by letting $K = 0$.

3. Modulated Solitonic Waves in DNA

In this section, we will plot our solitonic function $\Phi_n(t)$ given by Equation (27.44). However, before we do this, we want to study how the amplitude A depends on ql for various K. Note that the value A is the amplitude of the function $F_1(S, \tau)$ but is only about a half of the amplitude of the soliton $\Phi_n(t)$. Since the amplitude of the function y depends on the arbitrary constant ε, Equation (27.14), we concentrate on the value of A. This is why we plot the function $\Phi_n(t)$ rather than $y_n(t)$.

From Figure 27.9, we can see how A depends on ql for constant δ, that is, for $\delta = -\alpha/2\beta$, the value that was previously accepted for our calculations [27]. We selected $K = 4$ N/m, but for different K, both larger and smaller, the curves are almost the same. One can also see that $ql = 0.78$ rad is around the minimum of the amplitude $A(ql)$.

However, for δ given by Equation (27.31) and Equation (27.32), the situation looks completely different. Figure 27.10 shows how the amplitude A depends on ql for three values of K [30]. Instead of minimum (Figure 27.9), the function $A(ql)$ now has a maximum. The highest values are 6.50 nm at $ql = 0.76$ rad for $K = 4.5$ N/m, 3.61 nm at $ql = 0.76$ rad for $K = 4.3$ N/m, and 2.27 nm at $ql = 0.79$ rad for $K = 3.8$ N/m. We can see that these maxima decrease with smaller K. For $K = 4.58$ N/m, the maximum of the amplitude A is even 25 nm, which does not make physical sense, whereas for $K = 3$ N/m the function $A(ql)$ does not have any maximum at all.

Therefore, we can state three facts concerning the parameter K. First, there is a maximum value $K_m = 4.6$ N/m for $ql = 0.78$ rad. Second, higher K means larger D_o, that is, a better transmission of the signal through the DNA chain, as was explained earlier. Third, K should be small enough to ensure that the amplitude A is small enough so that the oscillating nucleotide does not exceed the depth of Morse potential well.

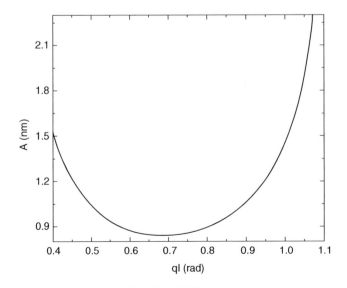

FIGURE 27.9 Amplitude A as a function of ql ($K = 4$ N/m, $\delta = $ const).

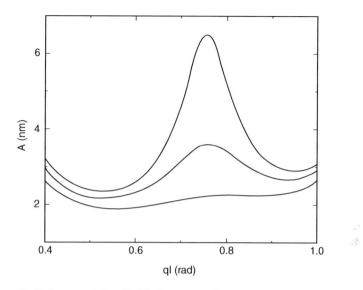

FIGURE 27.10 Amplitude A versus ql ($\delta = \delta(ql)$). (From top to bottom: $K = 4.5\,\text{N/m}$, $4.3\,\text{N/m}$ and $3.8\,\text{N/m}$).

Finally, we can plot our soliton $\Phi_n(t)$ given by Equation (27.44). In Figure 27.11–Figure 27.13, we show the elongation $\Phi_n(t)$ versus time for $K = 4.5\,\text{N/m}$, $4\,\text{N/m}$, and $2.5\,\text{N/m}$, respectively.

According to the shape of the curve in Figure 27.11 and the maximum of the function $\Phi_n(t)$, we can conclude that this value for K, that is, $K = 4.5\,\text{N/m}$, would not be acceptable. For $K = 4.6\,\text{N/m}$, $\Phi_n(t)$ has only positive values with the maximum reaching up to 500 nm which is totally unacceptable because the expansion of the Morse potential, Equation (27.15), presumes that the amplitude may not reach the Morse plateau! For $\Phi_n = 500\,\text{nm}$, from Equation (27.14) follows that $y_n = 4\,\text{nm}$ which is far beyond the plateau. However, Figure 27.12 and Figure 27.13 suggest that the values $K = 4\,\text{N/m}$ and $K = 2.5\,\text{N/m}$ might be acceptable. Of course, we should keep in mind that a higher K means more oscillations of the carrier wave in one envelope, that

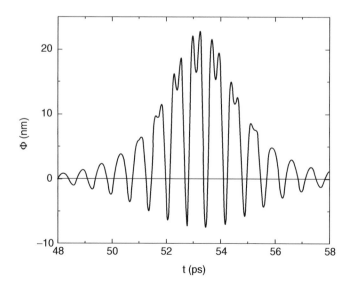

FIGURE 27.11 Elongation of the out-of-phase motion as a function of time ($n = 300$, $ql = 0.78\,\text{rad}$, $K = 4.5\,\text{N/m}$, $\delta = \delta(ql)$).

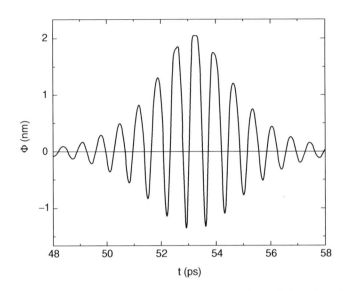

FIGURE 27.12 Elongation of the out-of-phase motion as a function of time ($n = 300$, $ql = 0.78$ rad, $K = 4$ N/m, $\delta = \delta(ql)$).

is, a higher D_o. For $K = 0$, only two internal oscillations can hardly be discerned which shows, as we stated earlier, the superiority of the PBD model [21,22] over the original PB version [16].

At this point, we can speak of a certain range for accepted values for the parameter K. In addition, as was pointed out earlier, K should be high enough to ensure a large D_o, but still not too large to bring about a very large amplitude.

It might be interesting to compare solitonic solutions $\Phi_n(t)$ given in Figure 27.11–Figure 27.13 with the same function obtained for constant δ [27]. In Figure 27.14 we show the solitonic wave $\Phi_n(t)$ for $K = 4$ N/m and constant δ, that is, $\delta = -\beta/2\alpha$. Comparing Figure 27.12 and Figure 27.14, both plotted with the same K, we see that the amplitude for the nonconstant δ (Figure 27.12) is about three times larger than that for the constant δ (Figure 27.14). In both

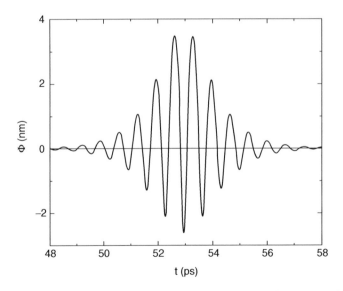

FIGURE 27.13 Elongation of the out-of-phase motion as a function of time ($n = 300$, $ql = 0.78$ rad, $K = 2.5$ N/m, $\delta = \delta(ql)$).

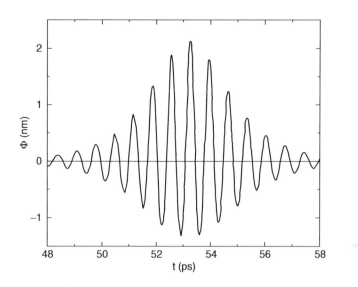

FIGURE 27.14 Elongation of the out-of-phase motion as a function of time ($n = 300$, $ql = 0.78$ rad, $K = 4$ N/m, $\delta = $ const).

cases positive amplitudes are higher than negative, which is a result of the term with the parameter δ in Equation (27.44). The shapes of both curves are almost the same. This is not surprising since D_o, the density of the carrier wave oscillations of the breather, does not depend on δ.

Finally, we need to point out a couple of advantageousness of this procedure (nonconstant δ) over the one when δ was constant. First, the amplitude $A(ql)$ has the maximum having very big values in a limit when the parameter K approaches its critical value, which suggests certain resonance behavior. This procedure solves the problem of a small amplitude discussed in Ref. [27].

Also, the existence of the upper limit of the parameter K, which was explained in Chapter 1, solves one more problem. Namely, in Ref. [21], optical and acoustical frequencies were compared. Optical frequency is defined by Equation (27.27). The same procedure for linear wave, Equation (27.12), instead of for that nonlinear, Equation (27.13), would lead to a so-called acoustical frequency [21,31]:

$$\omega_a^2 = \frac{4}{m}\left[k\sin^2\left(\frac{ql}{2}\right) + K\sin^2\left(\frac{qhl}{2}\right)\right] \tag{27.52}$$

Figure 27.15 was done for the values of the parameters given by Equation (27.46). We can see that there are four crossing points in the shown interval. The first two of them are at $q_1 = 0.184\ \text{Å}^{-1}$ ($ql = 0.624$ rad) and $q_2 = 0.278\ \text{Å}^{-1}$ ($ql = 0.946$ rad) [31]. It is interesting to point out that all the three accepted values for ql (Section 1 and Ref. [27]) belong to this interval, while the one that was rejected does not.

However, for $K < 6.4$ N/m, that is, for $K < a^2D$, those crossing points do not exist and $\omega_o > \omega_a$ in the whole zone [31]. This is shown in Figure 27.16. It might be interested to point out that those two curves touch each other if $K = a^2D$, and this resonance mode ($\omega_o = \omega_a$) occurs at $ql = 0.78$ rad/sec [31]. This is well known value first obtained in Ref. [27] and confirmed by maximization of the function $D_o(ql)$ [30].

B. THE INFLUENCE OF VISCOSITY

In this section, we study a more realistic DNA dynamics, taking surrounding of the DNA molecule into consideration. We discuss a possible interaction of DNA with its environment, water molecules for example, which may damp out vibrations of the nucleotides. This problem was studied in Ref. [27], where $K = 8$ N/m and constant $\delta = -\beta/2\alpha$ were used.

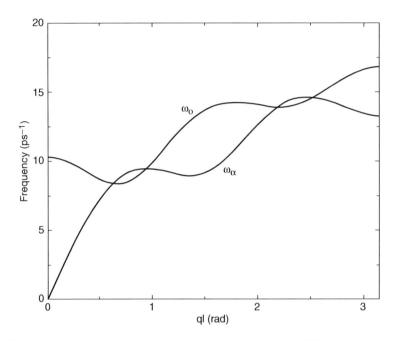

FIGURE 27.15 Optical and acoustical frequencies as a function of ql ($K = 8\,\text{N/m}$).

The impact of the medium can be taken into account by adding a viscous force [32] exerted on the nucleotide sequences

$$F_{\text{v}} = -\gamma \dot{y}_n \qquad (27.53)$$

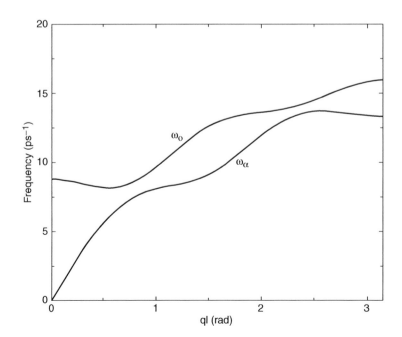

FIGURE 27.16 Optical and acoustical frequencies as a function of ql ($K = 4\,\text{N/m}$).

to the equation of motion, Equation (27.13). Here γ represents a damping coefficient and its value will be discussed later. This implies a new term in Equation (27.15) of the form

$$-\frac{\gamma}{m}\dot{\Phi}_n \qquad (27.54)$$

and, using the same procedure as in Section A, we obtain a new dispersion relation

$$\omega_\gamma^2 = \omega_g^2 + \frac{2k}{m}[1 - \cos(ql)] + \frac{2K}{m}[1 + \cos(qhl)] - i\frac{\gamma}{m}\omega_\gamma \qquad (27.55)$$

and the corresponding group velocity

$$V_\gamma \equiv V_{g\gamma} = \frac{l}{m}\frac{k\sin(ql) - Kh\sin(qhl)}{\omega_\gamma + i(\gamma/2m)} \qquad (27.56)$$

instead of the former expressions (27.27) and (27.28). The index γ refers to physical values with viscosity being taken into consideration and

$$\theta_\gamma \equiv \theta_{n\gamma} = q_\gamma nl - \omega_\gamma t, \quad q \equiv q_\gamma \qquad (27.57)$$

From Equation (27.27) and Equation (27.55), we obtain

$$\omega_\gamma^2 = \omega^2 - i\frac{\gamma}{m}\omega_\gamma \qquad (27.58)$$

It is clear that ω in Equation (27.58) is not exactly the same as ω in Equation (27.27) as if the former depends on $q_\gamma l$ and the latter on ql. However, we will see later that one can take the same value for $q_\gamma l$ for calculations as we did when viscosity was not taken into consideration, that is, $q_\gamma = q$. Equation (27.58) can be written as

$$\left[\omega_\gamma + i\frac{\gamma}{2m}\right]^2 = \omega^2 - \left(\frac{\gamma}{2m}\right)^2 \qquad (27.59)$$

and, defining the parameter b as the following ratio

$$b = \frac{\gamma}{2m\omega} \qquad (27.60)$$

we can easily obtain

$$\omega_\gamma + i\frac{\gamma}{2m} = \omega\sqrt{1 - b^2} \qquad (27.61)$$

and, from Equation (27.56),

$$V_\gamma = \frac{V_g}{\sqrt{1 - b^2}} \qquad (27.62)$$

Equation (27.29)–Equation (27.31) are not affected by the new term (27.54) added to Equation (27.15). New co-ordinates in Equation (27.33) will now be

$$S_\gamma \equiv S = Z - V_\gamma T, \quad \tau = \varepsilon T \qquad (27.63)$$

Instead of Equation (27.35) we now obtain a new cubic NSE as

$$\left(i - \frac{\gamma}{2m\omega_\gamma}\right)F_{1\tau} + P'F_{1SS} + Q'|F_1|^2 F_1 = 0 \tag{27.64}$$

where the adequate dispersion and nonlinearity parameters are given respectively

$$P' = \frac{1}{2\omega_\gamma}\left\{\frac{l^2}{m}[k\cos(ql) - Kh^2\cos(qhl)] - V_\gamma^2\right\} \tag{27.65}$$

$$Q' = -\frac{\omega_g^2}{2\omega_\gamma}[2\alpha(\mu + \delta) + 3\beta] \tag{27.66}$$

and, for the sake of simplicity, we use index S instead of S_γ to denote the derivative, as well as q instead of q_γ.

Using a very useful trick

$$i - \frac{\gamma}{2m\omega_\gamma} = \frac{i}{\omega_\gamma}\left(\omega_\gamma + i\frac{\gamma}{2m}\right) \tag{27.67}$$

and Equation (27.61), we finally obtain

$$iF_{1\tau} + P_\gamma F_{1SS} + Q_\gamma|F_1|^2 F_1 = 0 \tag{27.68}$$

$$P_\gamma = \frac{\omega_\gamma}{\omega}\frac{P'}{\sqrt{1-b^2}} \tag{27.69}$$

$$Q_\gamma = \frac{Q}{\sqrt{1-b^2}} \tag{27.70}$$

Therefore, the NSE of the same form as Equation (27.35) is derived. Of course, the coefficients P_γ and Q_γ differ from those in Equation (27.35). So, from Equation (27.38)–Equation (27.40), the solution of Equation (27.68) becomes

$$F_1(S, \tau) = A_\gamma \operatorname{sech}\left(\frac{S - u_e\tau}{L_\gamma}\right)\exp\frac{iu_e(S - u_c\tau)}{2P_\gamma} \tag{27.71}$$

where the amplitude and the width of the breather excitation are respectively

$$A_\gamma = \sqrt{\frac{u_e^2 - 2u_e u_c}{2P_\gamma Q_\gamma}} \tag{27.72}$$

$$L_\gamma = \frac{2P_\gamma}{\sqrt{u_e^2 - 2u_e u_c}} \tag{27.73}$$

Using the same procedure as in Section A, we finally obtain a complete expression for the damped soliton

$$\Phi_n(t) = 2A_\gamma \operatorname{sech}\left(\frac{\varepsilon(nl - V_{e\gamma}t)}{L_\gamma}\right) \times \left\{\cos\left(\Theta_\gamma nl - \Omega_\gamma t\right)\exp\left(-\frac{\gamma t}{2m}\right)\right.$$

$$\left. + \varepsilon A_\gamma \operatorname{sech}\left(\frac{\varepsilon(nl - V_{e\gamma}t)}{L_\gamma}\right) \times \left[\frac{\mu}{2} + \delta\cos(2(\Theta_\gamma nl - \Omega_\gamma t))\exp\left(-\frac{\gamma t}{m}\right)\right]\right\} + O(\varepsilon^2)$$

$$\tag{27.74}$$

where envelope velocity, phase and frequency are respectively given as follows

$$V_{e\gamma} = V_\gamma + \varepsilon u_e \tag{27.75}$$

$$\Theta_\gamma = q_\gamma + \frac{\varepsilon u_e}{2P_\gamma} \tag{27.76}$$

$$\Omega_\gamma = \omega\sqrt{1 - b^2} + \frac{\varepsilon u_e(V_\gamma + \varepsilon u_c)}{2P_\gamma} \tag{27.77}$$

The next step is to determine a convenient value for $q \equiv q_\gamma$. It was explained in Section 1 that the four values for ql were obtained according to the following requirements: $V_g > 0$, $P > 0$ and $\lambda \equiv \lambda_\gamma = Nl$ where N is an integer. Will something be changed if viscosity is taken into consideration? In other words, we should study the possibility that new values for λ might be obtained from the requirements explained earlier. The first requirement $V_\gamma > 0$ gives nothing new, as can be seen from Equation (27.62). For P_γ to be positive we obtain different values for $q_\gamma l$ for different g. Using the same procedure as in Section 1, we obtained the following intervals [27]:

$0.66 < ql < 1.11$ or $5.66l < \lambda < 9.52l$ for $g = 0$
$0.66 < ql < 1.08$ or $5.82l < \lambda < 9.52l$ for $g = 0.6$
$0.66 < ql < 0.99$ or $6.35l < \lambda < 9.52l$ for $g = 0.85$
$0.66 < ql < 0.86$ or $7.31l < \lambda < 9.52l$ for $g = 0.9$

Therefore, accepted values for λ are $6l$, $7l$, $8l$ and $9l$ for small viscosity ($g = 0$ and $g = 0.6$), $7l$, $8l$ and $9l$ for $g = 0.85$ while only $\lambda = 8l$ ($ql = 0.78$ rad) and $\lambda = 9l$ ($ql = 0.7$ rad) satisfy the requirements for very high g. So, $ql = 0.78$ rad will again be a good value for our estimations.

1. Determination of Damping Coefficient γ

The important issue is to estimate a value of the damping coefficient γ. To do this, in the context of dynamics of microtubules [32], we used simple fluid mechanics arguments. In that case the fluid is assumed to be water and its viscosity is temperature dependent. Taking physiological temperatures (approximately 300 K), we estimated the order of magnitude [32]

$$\gamma \sim 6 \times 10^{-11}\,\text{kg/sec} \tag{27.78}$$

Despite the significant difference in geometries of microtubule and nucleotide, as well as different surroundings, we assume that, for sequences of nucleotides, γ should be smaller but still match the same order of magnitude.

Thus, we will express the damping coefficient in units of 10^{-11} kg/sec, that is,

$$\gamma = g \times 10^{-11}\,\text{kg/sec} \tag{27.79}$$

Now we may estimate the values of the parameter g defined by Equation (27.60) and Equation (27.79) keeping in mind that the frequency ω involved in Equation (27.60) is a function of ql through the dispersion relation, Equation (27.27).

Therefore, the condition $b < 1$ implies that the inequality $\gamma < 2m\omega$ holds. In Figure 27.17, a maximum value of the constant g, corresponding to $\gamma_m = 2m\omega$, is shown.

For our accepted values for ql, we can calculate from the figure

$g < 0.90$ for $ql = 0.7$ rad
$g < 0.92$ for $ql = 0.78$ rad
$g < 0.99$ for $ql = 0.9$ rad
$g < 1.12$ for $ql = 1.05$ rad

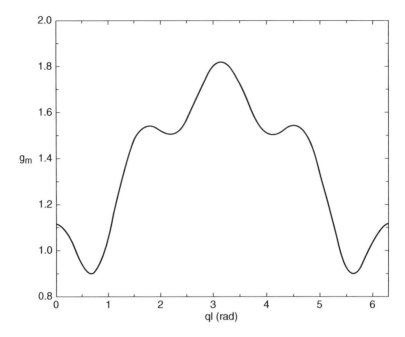

FIGURE 27.17 A maximum value for g versus ql.

A minimum of the function $2m\omega$ corresponds to $ql = 0.66$ rad and has the value 0.90.

One more interesting question deserves to be discussed. Is there an optimum value for the viscosity coefficient γ? Nature puts the DNA chain into solvent and we are looking for the appropriate viscosity, that is, for g required for the best functioning of the DNA molecule. According to the expression for the damped soliton, Equation (27.74), we can define the width of the soliton or "the wavelength of the envelope" as

$$\Lambda_\gamma = \frac{2\pi L_\gamma}{\varepsilon} \tag{27.80}$$

which, together with Equation (27.73), shows that the width Λ_γ of the soliton is proportional to the dispersion coefficient P_γ. So, the soliton will cover the most nucleotides for the value of g corresponding to the maximum of the function $P_\gamma(g)$. From Figure 27.18, we see how P_γ depends on viscosity, for $ql = 0.78$ rad. We calculated the maximum value of P_γ to be $P_{\gamma m} = 3.27 \times 10^{-6}\ \text{m}^2/\text{sec}$ for $g_m = 0.86\ (b_m = 0.93)$. It might be interesting to point out that for higher ql the maximum of the function $P_\gamma(g)$ becomes less sharp. For $ql = 1.05$ rad, for example, the function P_γ has no peak at all. On the other hand, for smaller ql, the peak is even sharper.

2. Demodulated Solitonic Waves in DNA

Before we plot our soliton, that is, the function $\Phi_n(t)$, we should examine the exponential terms in Equation (27.74). For $g = 0.8$, the first of them becomes $e^{-3.7t}$ where t is time in picoseconds. This means that it could make sense to keep those two terms in Equation (27.74) only for very small t, up to around 1 ps. However, even for so small values of t, the breather, like the one in Figure 27.12 or Figure 27.13, does not exist because the period of the carrier wave is longer than 2 ps. Therefore, only a "pure" soliton, that is, an envelope-type soliton, exists in the DNA chain and, neglecting

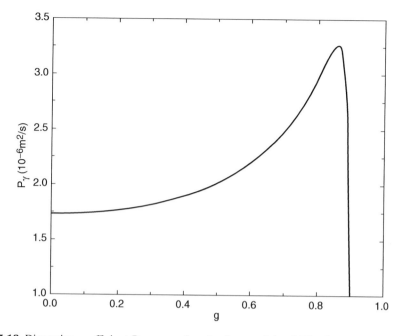

FIGURE 27.18 Dispersion coefficient P_γ versus viscosity factor g ($ql = 0.78$ rad).

exponential terms in Equation (27.74), we can obtain its expression as

$$\Phi_n(t) = \varepsilon\mu A_\gamma^2 \, \mathrm{sech}^2\left[\frac{\varepsilon}{L_\gamma}(nl - V_{e\gamma}t)\right] + O(\varepsilon^2) \qquad (27.81)$$

This function is shown in Figure 27.19 [27].

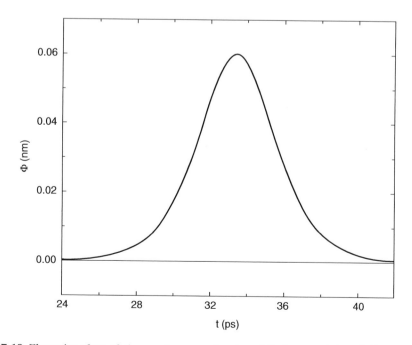

FIGURE 27.19 Elongation of out-of-phase motion versus time ($g = 0.8$, $\delta = $ const). ($\varepsilon = 0.007$, $u_e = 10^5$ m/s).

One can see that the amplitude of the soliton in Figure 27.19 is very small. However, the parameters characterizing a traveling wave solution (u_e, u_c, and ε) are very controversial. For Figure 27.19, we used the values given by Equation (27.47). Now, we change two of them, u_e and ε, to see their impact on the breather dynamics. For $u_e = 2.5 \times 10^5$ m/sec and $\varepsilon = 0.05$, function (27.81) is shown in Figure 27.20 [27]. In both cases, Figure 27.19 and Figure 27.20, the friction coefficient g is the same ($g = 0.8$).

We can easily estimate the speed and the width of the soliton shown in Figure 27.19. Those values, for $g = 0.85$ ($b = 0.92$), are [27]:

$$V_{e\gamma} \approx 3750 \text{ m/sec} \quad \Lambda \approx 58 \text{ nm} \tag{27.82}$$

This means that around 170 nucleotides in each strand are covered by the soliton.

The signals in Figure 27.19 and Figure 27.20 are demodulated solitons. There are no carrier components like waves in Figure 27.12–Figure 27.14. Therefore, instead of modulated signals (Equation (27.44), Figure 27.12–Figure 27.14), when viscosity is neglected, demodulated waves (Equation (27.81), Figure 27.19 and Figure 27.20) move through DNA molecule when viscosity is taken into consideration. So, there is analogy with engineering [33] where modulated signals are widely used because they can be transmitted at much longer distances than nonmodulated waves. The envelope is the wave that we want to transmit, for example music that we listen. Carrier waves, having much higher frequencies, can provide their transport to long distances. However, when this combined wave reaches its destination, that is, when we want to listen music, for example, we do not need the carrier wave any more. We get rid of it like passengers leave their train when get to the destination. So, the demodulation occurred when the signal is about to be accepted.

Let us get back to the DNA molecule. We already pointed out that viscosity provides demodulation. But, what are locations of demodulation and what are acceptors of this demodulated wave? It is well known that besides DNA molecule the other important molecules exist in cell nucleus. Especially important is messenger RNA molecule (m-RNA). To be more precise, m-RNA is formed in the nucleus from m-RNA polymerase. So, we expect that the piled material, from

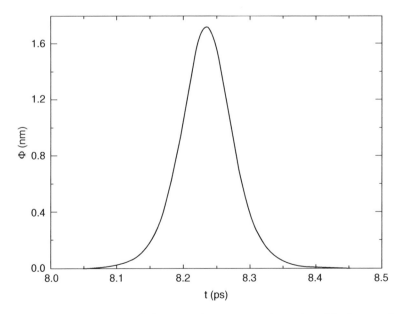

FIGURE 27.20 Elongation of out-of-phase motion versus time ($g = 0.8$, $\delta = $ const). ($\varepsilon = 0.05$, $u_e = 2 \times 10^5$ m/s).

which m-RNA is made, provides "big" viscosity, that is, demodulation agent for the carrier wave [33]. In other words, modulated wave moves through the DNA chain as long as this does not interact with the m-RNA polymerase molecules. When this wave reaches a segment where DNA molecule is surrounded by this material, that is, by m-RNA polymerase, the demodulation occurs. The lack of the carrier component in the solitonic wave means that the nucleotides of DNA molecule oscillate with small frequency. This provides a creative "key and lock" interaction with the m-RNA polymerase which yields the m-RNA formation [33]. Therefore, m-RNA polymerase molecules represent the big viscosity, which ensures the demodulation process. The demodulated signal, caring the useful information, is responsible for m-RNA transcription upon a unique strand of its DNA template.

3. Viscosity and Nonconstant δ

In Section A, the DNA dynamics was studied for both constant and nonconstant parameter δ. For example, the solitonic functions $\Phi_n(t)$ were shown in Figure 27.11–Figure 27.13 for nonconstant δ, and Figure 27.14 shows the soliton $\Phi_n(t)$ for $\delta = \text{const}$.

In Section B, however, we have dealt with constant δ only. For example, Figure 27.19 and Figure 27.20 were done for $\delta = -\beta/2\alpha$ [27]. The purpose of this section is to study the problem with nonconstant δ, when viscosity is taken into consideration. One should keep in mind that the parameter δ exists in the expression for Q only (Equation (27.37) and Equation (27.70)). This means that only the amplitude A_γ, in Equation (27.81), depends on this parameter. As if only A_γ is affected by the parameter δ, it was used in Ref. [34], for the estimations and figures, nonconstant δ neglecting viscosity. It was shown in Ref. [34] that the nonconstant δ solved the problem of small amplitudes mentioned earlier and in Ref. [27]. This can be seen from Figure 27.21, which was done for the same values of all the other parameters as Figure 27.19, except for K. If we compare Figure 27.21 and Figure 27.19, we see that the amplitude for nonconstant δ is higher (Figure 27.21) and that higher amplitude corresponds to a shorter wave, which is typical for the

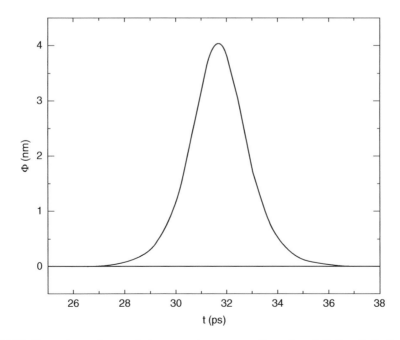

FIGURE 27.21 Elongation of out-of-phase motion versus time ($g = 0.8$, $\delta = \delta(ql)$, $K = 4.5\,\text{N/m}$). ($\varepsilon = 0.007$, $u_e = 10^5\,\text{m/s}$).

solitonic waves. However, as was already pointed out, δ used for Figure 27.21 is nonconstant but obtained using the procedure when viscosity was neglected. In what follows, more rigorous treatment will be done. In other words, a new nonconstant δ, when viscosity is taken into consideration, will be derived. It will be shown that this δ_γ is complex and brings about the complex parameter Q_γ in Equation (27.68). Such NSE, with the complex Q_γ, which is a special case of a complex Ginzburg–Landau equation to the best of author's knowledge, cannot be solved analytically.

It was already pointed out that viscosity was taken into consideration by adding the new term (27.53) into Equation (27.13), which brings about the term (27.54) in Equation (27.15). From (27.54) and (27.22), we can obtain a new term (NT) in Equation (27.25) to be

$$\text{NT} = -\frac{\gamma}{m}\left(\varepsilon F_{1T}e^{i\theta_\gamma} - i\omega F_1 e^{i\theta_\gamma} + \varepsilon^2 F_{0T} + \varepsilon^2 F_{2T}e^{i2\theta_\gamma} - 2i\varepsilon\omega F_2 e^{i2\theta_\gamma}\right) + \text{cc} \tag{27.83}$$

This NT does not affect expressions (27.29) and (27.30). How about Equation (27.31) and Equation (27.32). The expressions for δ were obtained by equating the coefficients for $e^{i2\theta}$ in Equation (27.25) and neglecting all the terms with ε^2. So, the only NT in Equation (27.32) is $i\frac{2\gamma}{m}\omega_\gamma F_2$, and Equation (27.32) becomes

$$\left\{4\omega_\gamma^2 - \frac{2k}{m}[1 - \cos(2q_\gamma l)] - \frac{2K}{m}[1 + \cos(2hq_\gamma l)] - \omega_g^2 + i\frac{2\gamma}{m}\omega_\gamma\right\}F_2 = \omega_g^2\alpha F_1^2 \tag{27.84}$$

Therefore, we obtain

$$F_2 = \delta_\gamma F_1^2 \tag{27.85}$$

$$\delta_\gamma = \omega_g^2\alpha\left[4\omega_\gamma^2 - \tau + i\frac{2\gamma}{m}\omega_\gamma\right]^{-1} \tag{27.86}$$

where

$$\tau \equiv \frac{2k}{m}[1 - \cos(2q_\gamma l)] + \frac{2K}{m}[1 + \cos(2hq_\gamma l)] + \omega_g^2 \tag{27.87}$$

Using Equation (27.58) and Equation (27.61), we finally obtain

$$\delta_\gamma = \omega_g^2\alpha\left[4\omega^2(1 - b^2) - \tau - i\frac{2\gamma}{m}\omega\sqrt{1 - b^2}\right]^{-1} \tag{27.88}$$

Note that the frequency ω in Equation (27.88) is not exactly the same as ω in Equation (27.27) as if q should be replaced by q_γ. However, it was already shown that we could assume the same value for both wave numbers (Section B).

So, the parameter δ_γ is complex and can be written as

$$\delta_\gamma = \delta_0 e^{i\varphi} \tag{27.89}$$

where

$$\delta_0 = \frac{\omega_g^2\alpha}{\sqrt{M^2 + N^2}} \tag{27.90}$$

$$\tan\varphi = \frac{N}{M} \tag{27.91}$$

$$M = 4\omega^2(1 - b^2) - \tau \tag{27.92}$$

and

$$N = \frac{2\gamma}{m}\omega\sqrt{1 - b^2} \tag{27.93}$$

Finally, this complex parameter δ_γ should be substituted into the expression for the coefficient of nonlinearity Q_γ, defined by Equation (27.37) and Equation (27.70). This means that NSE (27.68) with the complex Q_γ should be solved.

After some tedious mathematics, we can obtain the nonlinear parameter Q_γ as

$$Q_\gamma = Q_1 + iQ_2 \tag{27.94}$$

where

$$Q_1 = -4\alpha c\omega\omega_g^2\sqrt{1 - b^2} - \frac{\omega_g^2}{2\omega\sqrt{1 - b^2}}(2\alpha\mu - 2\alpha c\tau + 3\beta) \tag{27.95}$$

$$Q_2 = -2g \times 10^{-11}\frac{\omega_g^2\alpha c}{m} \tag{27.96}$$

and

$$c = \frac{\omega_g^2\alpha}{M^2 + N^2} \tag{27.97}$$

It is clear that for $g = 0$, the imaginary part Q_2 vanishes and Q_1 becomes Q given by Equation (27.37).

In Ref. [27], authors suggested that the most favorable viscosity g might be the value when the function $P(g)$ reaches maximum. This idea comes from the fact that the parameter P is proportional to the width of the soliton. From Equation (27.65) and Equation (27.69), and for the values of parameters used earlier, we can calculate this g_m. The ratio Q_2/Q_1 for this viscosity parameter g_m is approximately 0.3. This means that neither real nor imaginary part of the parameter Q_γ can be neglected.

However, in a limit $b \to 1$, N approaches zero and Q_1 becomes much higher than Q_2, as seen in Equation (27.95) and Equation (27.96). This certainly means that Q_2 can be neglected, but it is not likely that this limit has any physical meaning. Of course, the values of the parameters that were used for those calculations have not been precisely determined yet and this issue certainly requires further research.

4. "Small" Viscosity as an Alternative Approach

In Section B, we included the impact of viscous medium by adding the viscous force (27.53) to Equation (27.13). In that treatment, the viscous force was considered as competitive with other forces arising from Hamiltonian (27.1). The consequence of that approach was the outcome which showed the impact of viscosity being so strong that the soliton solution (27.74) decays almost instantaneously into its asymptotic form which is localized bell-shaped mode given by expression (27.81). However, there is an alternative approach where viscous force has features of small perturbation. We refer them as "big" and "small" viscosities [30].

So, we start from probably more realistic and favorable approach that viscous force has features of small perturbation. The viscous forces, exerted on the nucleotides n, are $-\varepsilon\gamma\dot{u}_n$ and $-\varepsilon\gamma\dot{v}_n$. The small ε, the same as in Equation (27.14), indicates that viscous force has the character of small perturbation. This leads to the effective damping force acting on the out-of-phase base pair motion as:

$$F_v = -\varepsilon\gamma\dot{y}_n \qquad (27.98)$$

Now starting from the perturbed equation of motion

$$m\ddot{y}_n = k(y_{n+1} + y_{n-1} - 2y_n) - K(y_{n+h} + y_{n-h} + 2y_n)$$
$$+ 2\sqrt{2}aD(e^{-a\sqrt{2}y_n} - 1)e^{-a\sqrt{2}y_n} - \varepsilon\gamma\dot{y}_n \qquad (27.99)$$

and performing the expansion explained, we find

$$iF_{1\tau} + PF_{1SS} + Q|F_1|^2F_1 = -\varepsilon\frac{\gamma}{2m\omega}F_{1\tau} \qquad (27.100)$$

This equation could be solved by the method of "slowly varying parameters" developed in [35]. The essence of the method is that the carrier wave number $\Theta = q + \varepsilon(u_e/2P)$ slowly changes with time through change of u_e. After some tedious calculations [30], we obtain the expression for the envelope velocity

$$V_e = V_g + \varepsilon u_{e0}\exp\left(-\frac{\gamma}{m}\varepsilon^2 t\right) \qquad (27.101)$$

On the basis of Equation (27.39) and Equation (27.40), it follows that the breather's amplitude decays with the same rate as envelope velocity, and the width of the breather spreads exponentially.

From the absolute viscosity of water ($T \approx 300\,\text{K}$), $\eta = 7 \times 10^{-4}\,\text{kg/msec}$, and considering a base as thin rigid rode, it could be roughly estimated $\gamma \sim 10^{-12}\,\text{kg/sec}$. With $\varepsilon = 0.007$, the breather's decay time is $t_d = m/\gamma\varepsilon^2 \approx 10^{-8}\,\text{sec}$ [30]. Starting with $V_e \approx 1.9 \times 10^3\,\text{m/sec}$ the path passed by the breather for $t_d = 10^{-8}\,\text{sec}$ reaches approximately $20 \times 10^{-6}\,\text{m}$ or 6×10^4 base pairs along DNA chain. This is quite favorable regarding expected role of breathers as long-range effects mediators in DNA [30].

In Ref. [33], we suggest that water molecules provide small and m-RNA polymerase big viscosity. Therefore, the procedure just explained in this section describes the impact of water surrounding to DNA. To describe the interaction of DNA with m-RNA polymerase, we need the idea of so-called big viscosity.

C. RESONANCE MODE IN DNA DYNAMICS

In Section A.3, optical and acoustical frequencies, given by Equation (27.27) and Equation (27.52), were compared. Figure 27.15 and Figure 27.16 were done for $K > a^2D$ and $K < a^2D$, respectively. The resonance mode was also defined. Namely, for $K = a^2D$, curves $\omega_o(ql)$ and $\omega_a(ql)$ have only one common point. In other words, those two curves touch each other at $ql = \pi/h$ and this is what we call the resonance mod.

Before we proceed, we should remind ourselves on the Morse potential, which is the last term in Equation (27.10). This function, shown in Figure 27.22, represents the potential of the hydrogen bonds between base pairs of different strands. Of course, a curve "a" corresponds to the larger force than a curve "b." As can be seen, smaller force means smaller value of the parameter a, as if a is the inverse width of the potential well. Of course, if the force between the nucleotides of different strands is larger than the amplitude of the oscillations will be smaller and vice versa.

Now, we are going to study the resonance mode, which means that $ql = \pi/h$ and $K = a^2D$ are assumed. We will study the resonance mode neglecting viscosity subsequently. However, it will be

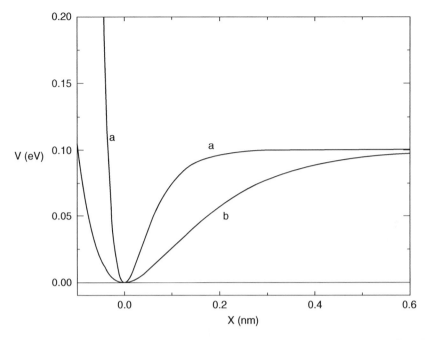

FIGURE 27.22 Morse potential (a: Large force, small amplitude, b: Small force, large amplitude).

seen that the wave moving through the DNA chain will be similar to the demodulated one, shown in Figure 27.19–Figure 27.21. This might mean that resonance includes demodulation effects.

For $a = \sqrt{K/D}$ one can obtain possible values of the parameter K according to the requirements that parameters P and Q be positive, which was explained in Section A.1. For example, Figure 27.23

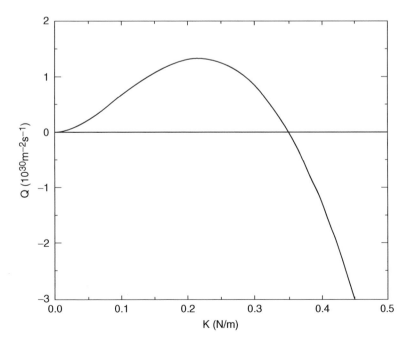

FIGURE 27.23 Nonlinear coefficient Q of the NSE as a function of K (Resonance mode).

shows how Q depends on K for $h = 5$, as well as the upper limit of the parameter K. So, for $D = 0.1\,\text{eV}$, and for the values of all the other parameters taken from Ref. [2], those intervals are:

$$0.07\,\text{N/m} < K < 0.35\,\text{N/m} \quad \text{for} \quad h = 5$$
$$0.16\,\text{N/m} < K < 0.82\,\text{N/m} \quad \text{for} \quad h = 4$$

We can argue that the value of the parameter a is very small if $K = a^2 D$, much smaller than what we find in references dealing with this issue. This is correct, but we should keep in mind that a from $K = a^2 D$ is the resonance a and is much smaller than the nonresonance a when $K < a^2 D$ and $\omega_o > \omega_a$ for any ql. In other words, Morse potential at DNA segments where resonance occurs is rather different from the potential at the rest of the chain (curves "b" and "a" in Figure 27.22).

The values of most parameters are still disputable and more detailed analysis should be done and the results will be published elsewhere. However, the purpose of this section is to show that the resonance mode can explain local opening of nucleotide pair during m-RNA formation and this will be shown subsequently.

It was explained in Section B.2 that the demodulated wave is responsible for m-RNA transcription. This transcription is, practically, the formation of m-RNA molecule from m-RNA polymerase molecules. Therefore, this occurs at the segments where the DNA chain is surrounded by m-RNA polymerase molecules.

For this to happen, the DNA chain should open locally. A possible explanation for this local opening of the DNA chain can be done using the idea of the resonance mode [36]. Namely, the m-RNA polymerase molecules interact with the DNA nucleotides. This interaction decreases the force between nucleotides of different strands, which increases the amplitude. The increase of the amplitude ensures that the parameter K goes down, as if the interaction between nth nucleotide of one strand, and $(n + h)$th and $(n - h)$th nucleotides of the other becomes smaller. It can be seen

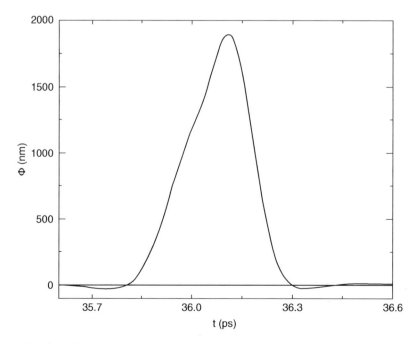

FIGURE 27.24 Elongation of the out-of-phase motion as a function of time (Resonance mode). ($n = 300$, $h = 5$, $K = 0.2\,\text{N/m}$, $\delta = \delta(ql)$).

from Figure 27.22 that the decrease of the force means that the parameter a decreases. So, both K and a go down, a^2 decreasing faster, and, finally, the resonance occurs ($K = a^2D$), which results in very big amplitude and, therefore, in the local opening of the DNA chain. One can argue that $K = a^2D$ is not a sufficient condition for resonance, as if $ql = \pi/h$ is required in addition. However, it has already been explained that $ql = \pi/h$ is probably the most favorable value for both, resonance and nonresonance mode.

The resonance solitonic wave $\Phi(t)$ is shown in Figure 27.24. This function is plotted according to Equation (27.44), but for $a = \sqrt{K/D}$. It is very interesting to realize that this soliton looks like demodulated signals, Figure 27.19–Figure 27.21, even though viscosity was neglected. Extremely big amplitude should not bother us as if viscosity was not taken into consideration.

III. CONCLUSION

The idea that solitons might play an important role in biopolymers comes from Davydov. In his article [37], he studied alpha-helical proteins and applied some achievements of nonlinear mathematics to biology. As for DNA, the nonlinear physics started in 1980 [10] when the first nonlinear Hamiltonian of DNA, as well as possibility of solitonic solution, was suggested. A crucial experimental research was explained in Ref. [38], representing victory of nonlinear over linear DNA physics.

In this chapter, one of the nonlinear DNA models is explained. This is extended PB or PBD model. This model represents a basis for research of DNA–protein interactions [28,39]. Also, the influence of ac field on the DNA dynamics was studied using the PBD model [20,28,40]. A few more examples can be found in Ref. [19]. Of course, some other models may also yield to solitonic excitations. As was pointed out at the beginning of Section II, the hierarchy of the most important models for nonlinear DNA dynamics was presented by Yakushevich [19].

One of the biggest problems we deal with is the fact that most parameters are still not accurately known. An example could be the values of the parameters characterizing Morse potential. The values (27.46) were taken from Refs. [21,22]. However, some authors have used different values, from $D = 0.33\,\mathrm{eV}$, $a = 1.8 \times 10^{10}\,\mathrm{m}^{-1}$ [16] to $D = 0.03\,\mathrm{eV}$, $a = 4.5 \times 10^{10}\,\mathrm{m}^{-1}$ [41]. We should keep in mind one important point present in a very basis of the procedure that we have used so far. Namely, Equation (27.15) was derived assuming $a\sqrt{2}y_n \ll 1$. Otherwise, the expansion of the exponential terms in Equation (27.13) would not be correct. This means that there should be $ay \leq 0.3$ for the error not to exceed approximately 10%. So, large a might not be compatible with the used theory.

Therefore, theorists can only estimate values of the parameters and use them for the calculations and analysis. Precise values, however, might come from experimental research. The last decade of the twentieth century will be remembered for the first successful experiments done on a single DNA molecule. This revolution in molecular biology started in 1996 [42,43]. Authors stretched the single DNA molecule and measured the applied force. In addition, the force that can unzip the strings of the DNA was measured [44–46]. Beside this progress, the values of the parameters are still unknown. Hopefully, the biophysical revolution is still in progress and keeps us in an optimistic mode.

ACKNOWLEDGEMENT

This project was financially supported by Ministry of Science and Environmental Protection of Serbia, entitled "Multiphase Dispersed Systems" 101822.

REFERENCES

1. Watson, J.D., *Molecular Biology of the Gene*, 3rd edn, W.A. Benjamin INC., Menlo Park, California, USA, 1976.
2. Volkenstein, M.V., *Biofizika*, Nauka, Moscow, 1981 (in Russian).
3. Alberts, B., Bray, D., Lewis, J., Raff, M., Roberts, K., and Watson, J.D., *Molecular Biology of the Cell*, 3rd edn, Garland Publishing, Inc., New York, London, 1994.
4. Watson, J.D. and Crick, F.H.C., Molecular structure of nucleic acid, *Nature*, 171 (4356), 737–738, 1953.
5. Watson, J.D. and Crick, F.H.C., Genetical implications of the structure of deoxyribonucleic acid, *Nature*, 171 (4361), 964–967, 1953.
6. Scot, A.C., Chu, F.Y.F., and McLaughlin, D.W., The soliton: a new concept in applied science, proceedings of the IEEE, 61 (10), 1443–1483, 1973.
7. Lamb, G.L., Jr., *Elements of Soliton Theory*, John Wiley & Sons, New York, 1980.
8. Das, A., *Integrable models*, World Scientific Lecture Notes in Physics, Vol. 30, World Scientific, Singapore, 1989.
9. Dodd, R.K., Eilbeck, J.C., Gibbon, J.D., and Morris, H.C., *Solitons and Nonlinear Wave Equations*, Academic Press, Inc., London, 1982.
10. Englander, S.W., Kalenbach, N.R., Heeger, A.J., Krumhansl, J.A., and Litwin, S., Nature of the open state in long polynucleotide double helices: possibility of soliton excitations, *Proc. Natl. Acad. Sci. USA*, 77, 7222–7226, 1980.
11. Yomosa, S., Solitary excitations in deoxyribonucleic acid (DNA) double helices, *Phys. Rev. A*, 30, 474–480, 1984.
12. Homma, S. and Takeno, S., A coupled base-rotator model for structure and dynamics of DNA, *Prog. Theor. Phys.*, 72, 679–693, 1984.
13. Zhang, Ch.-T., Soliton excitations in deoxyribonucleic acid (DNA) double helices, *Phys. Rev. A*, 35, 886–891, 1987.
14. Muto, V., Scott, A.C., and Christiansen, P.L., Thermally generated solitons in a Toda lattice model of DNA, *Phys. Lett. A*, 136, 33–36, 1989.
15. Volkov, S.N., Conformational transition. Dynamics and mechanism of long-range effects in DNA, *J. Theor. Biol.*, 143, 485–496, 1990.
16. Peyrard, M. and Bishop, A.R., Statistical mechanics of a nonlinear model for DNA denaturation, *Phys. Rev. Lett.*, 62 (23), 2755–2758, 1989.
17. Zhang, Ch.-T., Harmonic and subharmonic resonances of microwave absorption in DNA, *Phys. Rev. A*, 40, 2148–2153, 1989.
18. Yakushevich, L.V., Nonlinear DNA dynamics: a new model, *Phys. Lett. A*, 136, 413–417, 1989.
19. Yakushevich, L.V., *Nonlinear Physics of DNA*; Wiley Series in Nonlinear Science, John Wiley and Sons, Chichester, England, 1998.
20. Satarić, M.V., The influence of endogenous AC fields on the breather dynamics in DNA, *Physica D*, 126, 60–68, 1999.
21. Dauxois, T., Dynamics of breather modes in a nonlinear "helicoidal" model of DNA, *Phys. Lett. A*, 159, 390–395, 1991.
22. Dauxois, T. and Peyrard, M., Dynamics of breather modes in a nonlinear "Helicoidal" model of DNA, In *Lecture Notes in Physics 393*, Dijon, 1991, pp. 79–86.
23. Zoravković, S. and Satarić, M.V., Single molecule unzippering experiments on DNA and Peyrard–Bishop–Dauxois model, Submitted to *J. Mol. Biol.*
24. Gaeta, G., Crossing points and energy transfer in DNA nonlinear dynamics, *Phys. Lett. A*, 179 (3), 167–174, 1993.
25. Remoissenet, M., Low-amplitude breather and envelope solitons in quasi-one-dimensional physical models, *Phys. Rev. B*, 33 (4), 2386–2392, 1986.
26. Kawahara, T., The derivative-expansion method and nonlinear dispersive waves, *J. Phys. Soc. Japan*, 35 (5), 1537–1544, 1973.
27. Zdravković, S. and Satarić, M.V., The impact of viscosity on the DNA dynamics, *Physica Scripta*, 64, 612–619, 2001.
28. Zdravković, S., Satarić, M.V., and Tuszyński, J.A., Biophysical implications of the Peyrard–Bishop–Dauxois model of DNA dynamics, *J. Comput. Theor. Nanosci.*, **1**, 171–181, 2004.

29. Remoissenet, M. and Peyrard, M., Soliton dynamics in new models with parameterized periodic double-well and asymmetric substrate potentials, *Phys. Rev. B*, 29 (6), 3153–3166, 1984.

30. Zdravković, S., Tuszyński, J.A., and Satarić, M.V., Peyrard–Bishop–Dauxois model of DNA dynamics and impact of viscosity, Will be published in *J. Comput. Theor. Nanosci.*

31. Zdravković, S. and Satarić, M.V., Optical and acoustical frequencies in a nonlinear helicoidal model of DNA molecule, *Chin. Phys. Lett.*, 22 (4), 850–852, 2005.

32. Satarić, M.V., Tuszyński, J.A., and Žakula, R.B., Kinklike excitations as an energy-transfer mechanism in microtubules, *Phys. Rev. E*, 48 (1), 589–597, 1993.

33. Zdravković, S., Satarić, M.V., and Vukovic, D., Modulation and demodulation in DNA molecule, Submitted to *Int. J. Mod. Phys. B*.

34. Zdravković, S. and Satarić, M.V., DNA dynamics and big viscosity, *Int. J. of Mod. Phys. B*, 17 (31,32), 5911–5923, 2003.

35. Bogoliubov, N.N. and Mitropolskii, Y.A., *The Asymptotic Method in the Theory of Nonlinear Vibrations*, Nauka, Moscow, 1974 (in Russian).

36. Zdravković, S. and Satarić, M.V., Resonance and nonresonance modes in DNA dynamics, Submitted to *Phys. Rev. E*.

37. Davydov, A.S., Solitons in molecular systems, *Physica Scripta*, 20, 387–394, 1979.

38. Selvin, P.R., Cook, D.N., Pon, N.G., Bauer, W.R., Klein, M.P., and Hearst, J.E., Torsional rigidity of positively and negatively supercoiled DNA, *Science*, 255, 82–85, 1992.

39. Satarić, M.V. and Tuszyński, J.A., Impact of regulatory proteins on the nonlinear dynamics of DNA, *Phys. Rev. E*, 65, 051901, 2002.

40. Satarić, M.V. and Zdravković, S., Nonlinear dynamics of a DNA chain affected by endogenous AC fields, *Bioelectrochem. Bioenerg.*, 48, 325–328, 1999.

41. Peyrard, M., Nonlinear dynamics and statistical physics of DNA, *Nonlinearity*, 17, R1–R40, 2004.

42. Cluzel, P., Lebrun, A., Heller, C., Lavery, R., Viovy, J., Chatenay, D., and Caron, F., DNA: an extensible molecule, *Science*, 271, 792–794, 1996.

43. Smith, S.B., Cui, Y., and Bustamante, C., Overstretching B-DNA: the elastic response of individual double-stranded and single-stranded DNA molecules, *Science*, 271, 795–799, 1996.

44. Essevaz-Roulet, B., Bockelmann, U., and Heslot, F., Mechanical separation of the complementary strands of DNA, *Proc. Natl. Acad. Sci. USA*, 94, 11935–11940, 1997.

45. Cocco, S., Monasson, R., and Marko, J.F., Force and kinetic barriers to unzipping of the DNA double helix, *Proc. Natl. Acad. Sci. USA*, 98, 8608–8613, 2001.

46. Cocco, S., Monasson, R., and Marko, J.F., Force and kinetic barriers to initiation of DNA unzipping, *Phys. Rev. E*, 65, 041907, 2002.

28 Surface Modification of Dispersed Phases Designed for *In Vivo* Removal of Overdosed Toxins

Richard Partch
University of Florida, Gainesville, FL and Clarkson University, Potsdam, NY, USA

E. Powell
Clarkson University, Potsdam, NY, USA

Y-H. Lee
Kyungwon University, Sungnam City, Korea

M. Varshney
Hamdara University, New Delhi, India

D. Shah, R. Baney, D-W. Lee, D. Dennis, T. Morey, J. Flint
University of Florida, Gainesville, FL, USA

CONTENTS

I. SYNOPSIS

This chapter describes research by an interdisciplinary and international team focused on the preparation and evaluation of some dispersions of nanomaterials having potential application for injection into overdosed humans and to reverse cardiac toxicity by absorbing or binding the toxin. Specifically, the commonly overdosed therapeutics amiodarone, amitriptyline, and bupivacaine, and the illicit drug cocaine, are the deleterious chemicals under consideration for rapid reduction in concentration in blood.

The two types of injectable dispersed phases discussed are oil-in-water microemulsions capable of absorbing a toxin, and solid carrier nanoparticles functionalized with specific receptors for adsorbing such chemicals [1]. The ongoing effort has so-far established the efficiency of lowering the concentration of three of the listed toxins *in vitro* by both systems, and in the case of microemulsions, the *in vivo* reversal of toxicity of amitriptyline in rats.

II. INTRODUCTION

Death is inevitable but when it occurs due to exposure to an overdose of a natural or synthetic chemical for which there is no antidote, experts in engineering, medicine, and science can often join forces to develop a material designed to reverse the toxic effect and save lives. This has been a recurring phenomenon for centuries but is revitalized with the development of each new generation of chemicals that humans become exposed to either by intent or accident. Overdoses occur when patients or their physicians ignore dosage guidelines established for a therapeutic in or outside hospitals, when addicts succumb to increased use of drugs, and when human factions are in conflict (Figure 28.1).

Documented in medical statistics from the U.S. Center for Disease Control is data in the U.S. alone that there are over 100,000 hospitalizations annually due to poisoning, leading to over 20,000 deaths. Many of these are due to overexposure to the therapeutics: amiodarone (antiarrythmic), amitriptyline (antidepressant), and bupivacaine (local anesthetic) (Figure 28.2). These are extremely valuable prescription medicines but can cause cardiac arrest if overdosed. Bupivacaine is toxic to humans at approximately $12\ \mu M$ concentration. Amitriptyline has significant effect on the cardiovascular system at $0.3–0.8\ \mu M$ levels. Before the current research by the interdisciplinary collaborators, there were no antidotes for quickly reversing the concentration of these chemicals in blood. Therefore, they were a primary target for investigation. The design of experiments leading to the study included potential injectables composed of:

(1) Microgels [2] and microemulsions capable of absorbing the lipophilic molecules and deactivating them by phase transfer out of the blood medium.
(2) Encapsulated core-shell microemulsions having the absorbing property but greater physical integrity [3].

Drugs of Abuse	Therapeutic Drugs	Warfare Agents
● opiates	● anti-inflammatories	Soman, GD
● cocaine	● antibiotics	Sarin, GB
● cannabinol	● steroids	Tabun
● steroids	● anti-arrythmics	VX
	● antidepressants	
	● anesthetics	

FIGURE 28.1 Occurrence of death due to exposure to an overdose of chemicals.

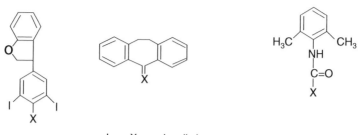

Amiodarone Amitriptyline Bupivacaine
Antiarrhythmic Antidepressant   Anesthetic

where X = aminoalkyl

FIGURE 28.2 Structure of medicines amiodarone (antiarrythmic), amitriptyline (antidepressant), and bupivacaine (local anesthetic).

(3) Solid tubular [4] or spherical nanoparticles having ligands bound to their surfaces for adsorbing and binding the therapeutics (Figure 28.3).

Development and evaluation of the microgels, the core-shell microemulsions, and the nanotubes with ligands for remediation of overdoses are referenced and outside the scope of this chapter. The current discussion focuses on the synthesis and physicochemical characterization of the microemulsion and spherical nanoparticle species, the proposed mechanisms by which each type of dispersed phase lowers the therapeutic or cocaine concentration in blood, analytical data on the efficiency of phase transfer of toxin out of normal saline, blood plasma, and whole blood, and efficacy of the microemulsions when employed *in vivo* in rats.

III. OIL-IN-WATER MICROEMULSION STUDIES

Emulsions and microemulsions abound in nature and in the synthetic chemical world. They are widely employed, for example, in food processing, paint manufacture and drug delivery.

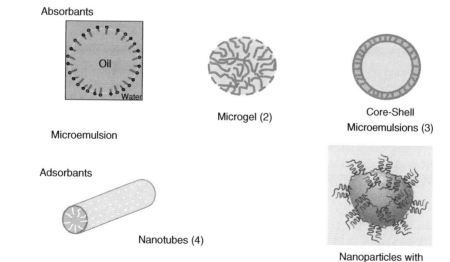

FIGURE 28.3 Composition of injectables used in the experimental design.
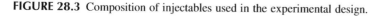

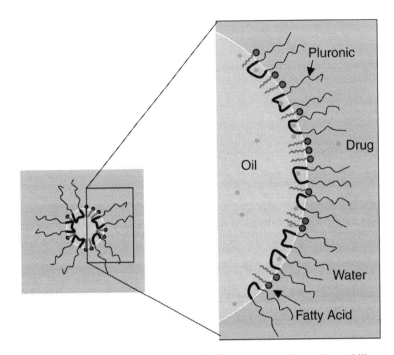

FIGURE 28.4 Addition of surfactants to the droplet of oil in water to enhance the stability.

Most are composed of droplets of an oil dispersed in water but the reverse is becoming more widely studied as new applications demand. In either case, the stability of the systems is enhanced by the addition of one or more surfactants, molecules of which have both hydrophobic and hydrophilic regions that transcend the interface between droplet and bulk medium (Figure 28.4).

An emulsion differs from a microemulsion by the size of the dispersed oil droplets, the latter being controlled by the type and ratio of the chemical components. Typically, oil droplets in an emulsion are more than 200 nm in diameter and diffract visible light, causing the dispersion tohave a milky appearance. A microemulsion having droplets less than 100 nm in diameter is anoptically isotropic, clear and is a thermodynamically stable multi-component system (Figure 28.5). In the present work, no distinction is made between a microemulsion and a swollen micelle.

Dinesh Shah [5–8] and others [9] have published extensively in the field of emulsions and microemulsions and is recognized internationally for his contributions to their application to drug delivery and other technologies. He and team member Varshney directed the microemulsion work reported in this chapter for removal of the types of therapeutics shown in Figure 28.2 from saline and biological fluids. Several combinations of nonionic and ionic surfactants, and oils, were used to obtain preliminary data on the preparation and stability of microemulsions and their possible employment as absorbants of toxins. Proof of concept has been achieved as described succeedingly using the relatively biocompatible components ethyl butyrate ($LD_{50} > 250$ mg/kg IV in dogs) as the oil, and AOT, pluronic [10–13] and fatty acid co-surfactants.

A. MICROEMULSION COMPOSITIONS AND DRUG EXTRACTION EFFICIENCIES

A typical procedure for the preparation of a microemulsion employed in the current study involves mixing, in 1% w/w proportions, ethyl butyrate with appropriate amounts of a pluronic, a sodium salt of a fatty acid in normal saline. An optimized composition for extraction of bupivacaine from normal saline at pH 7.4 contains 9 mM Pluronic F127, 36 mM sodium laurate and 155 mM ethyl

Intralipid Soybean   Microemulsion
Oil Emulsion

Particle Size :   ~ 400 nm   ~ 30 nm
Surface Area :   ~ 15 m²/ml   ~ 215 m²/ml

FIGURE 28.5 Difference between an emulsion and a microemulsion by the size of the dispersed oil droplets.

butyrate [14]. The slight turbitidy upon addition of the ester quickly dissipates during stirring. Dynamic light scattering and photon correlation spectroscopy was employed to determine oil droplet particle size in the microemulsions. The diameters depended on the amount and type of surfactants and ranged 11–40 nm with an average polydispersity index of 0.251 nm. These small sizes are compatible for recirculation in humans without disruption [15].

Pluronic is a generic name for a series of block copolymers composed of different ratios of more hydrophobic propylene oxide (PPO) monomers flanked by more hydrophilic ethylene oxide (PEO) monomers. Several commercially available analogs of Pluronics are shown in Figure 28.6 with the weight fraction of the hydrophilic PEO units ranging from 0.1 to 0.8.

These amphiphilic molecules aggregate and form micelles when mixed with water. Figure 28.4 shows Pluronic surfactant molecules at the interface between the oil core and water, with the hydrophilic PEO termini extending into the water phase. Figure 28.7 shows the HPLC results of bupivacaine extraction from saline or blood plasma by microemulsions using the same ratios of reagents and the same reaction conditions, but different Pluronic surfactants. Clearly, the Pluronic

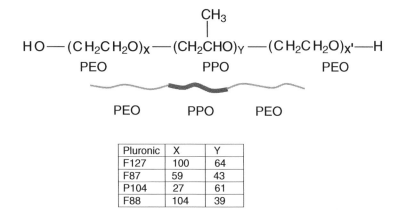

Pluronic	X	Y
F127	100	64
F87	59	43
P104	27	61
F88	104	39

FIGURE 28.6 Analogs of Pluronics with the weight fraction of the hydrophilic PEO.

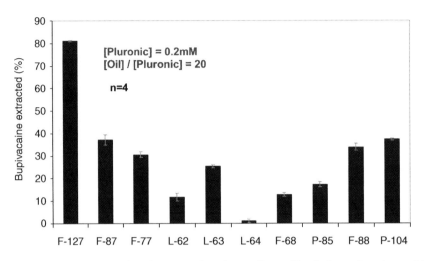

FIGURE 28.7 HPLC results of bupivacaine extraction from saline or blood plasma by micromulsions.

identified as F-127 is the most effective for the phase transfer of bupivacaine into the ethyl butyrate droplets. Experimental data suggests that 10 mM F127 may be optimum. The fundamental reason that F-127 Pluronic participates in better phase transfer of the drug into the microemulsion is not known but it is not due to stabilization of smaller or larger oil droplets because they average approximately 24 nm, which is in the middle of the range observed using several of the Pluronics.

As shown in Figure 28.7, fatty acid co-surfactants are also used in the formulation of the microemulsions. It has been determined that using a fatty acid co-surfactant having a chain length at least eight carbons long not only helps stabilize the dispersed phase in blood, but also enhances phase transfer of drug into the ethyl butyrate core (Figure 28.8). Also, extraction

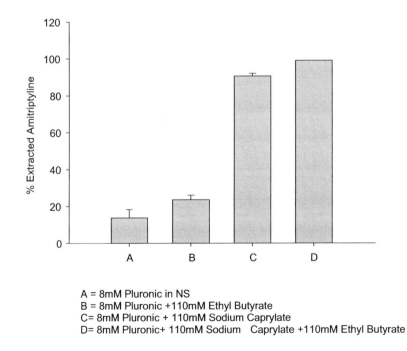

A = 8mM Pluronic in NS
B = 8mM Pluronic +110mM Ethyl Butyrate
C= 8mM Pluronic + 110mM Sodium Caprylate
D= 8mM Pluronic+ 110mM Sodium   Caprylate +110mM Ethyl Butyrate

FIGURE 28.8 Addition of fatty acid co-surfactant to stabilize and enhance the phase transfer.

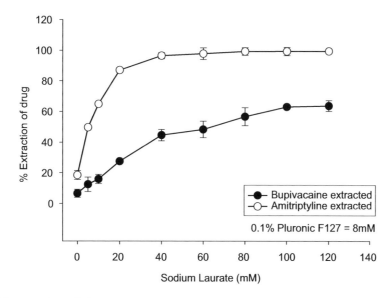

FIGURE 28.9 Extraction efficiency of amitriptyline and bupivacaine with the concentration of co-surfactant.

efficiency is increased by increasing the concentration of the ionic co-surfactant in microemulsions; and, depending on the chemical system described, amitriptyline extraction has a greater dependency than bupivacaine (Figure 28.9).

Extraction of bupivacaine from normal saline was also evaluated using microemulsions composed of Tween-80 and PEG surfactants, fatty acid salts, and ethylbutyrate. The HPLC data show that increasing Tween-80 from 0.07 to 0.20 M, while holding the PEG, sodium caprylate and ethyl butyrate concentrations constant, resulted in an increase in drug extracted from 7 to 22%. Using the same chemicals but increasing PEG concentration from 0.0 M to 0.5 M reduced by 60%, the amount of bupivacaine extracted.

Comparison studies on the efficiency of extraction of amitriptyline versus bupivacaine as a function of the type of ionic co-surfactant reveal that longer chain carboxylic acid salts greatly increase bupivacaine partitioning from normal saline. In addition, replacing such salts with the more organophilic dioctyl sulfosuccinate ester salt surfactant (AOT) enhances bupivacaine extraction even more (Figure 28.10).

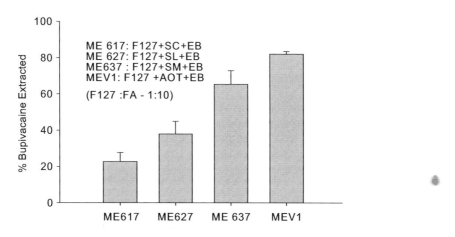

FIGURE 28.10 Results of bupivacaine extraction with organophilic dioctyl sulfosuccinate ester salt surfactants.

B. *In Vitro* and *In Vivo* Reversal of Drug-Induced Cardiac Arrest

Having established that a microemulsion can remove two of the therapeutic drugs from saline, blood plasma, and whole blood *in vitro*, attention was turned to determining if the potential antidote would be effective in restoring a toxin-induced decreased rate of heart beat, otherwise referred to as an increased QRS time interval, of an isolated rat heart previously exposed to toxin. In these experiments, carried out by Morey, Dennis, and Flint, the heartbeat was continually monitored while being infused first with saline first doped with bupivacaine and then with the microemulsion. The experiment was repeated using an intralipid macroemulsion, which, as shown in Figure 28.11, was much less effective in restoring the heartbeat to normal.

Thromboelastography and cell lysis experiments using whole blood verified that the microemulsions showing better drug removal properties did not cause excessive clot formation or rupture of cells.

Data suggesting that the overall program goal can be achieved came from the use of microemulsions 617 and 627 (see Figure 28.10 for compositions) to reverse *in vivo* cardiac arrest in rats induced by amitriptyline. When ME617 was administered 5 min prior to infusion of the drug, the increase in QRS interval was not as great as when infusing normal saline. Even more impressive was the fact that the increase in QRS interval due to prior drug application is lowered almost to normal within 30 min after ME627 is administered (Figure 28.12). This latter sequence of introduction of chemicals into the blood stream indicates that the microemulsions being developed may serve as antidotes for persons previously overdosed on amitrityline, bupivacaine, or related toxins.

C. Proposed Mechanism of Interaction of Drugs with Ethyl Butyrate Microemulsions

The three target drugs included in this study for removal from blood by microemulsions are all tertiary amino aromatics administered as therapeutics in the form of hydrochloride salts. Referring to Figure 28.2, it is noted that the aromatic portion of each drug exhibits hydrophobicity. Combining these salt and aromatic features, the molecules assume surfactant-like properties [16]. At physiological pH of 7.40, there is an equilibrium established between protonated and unprotonated amine functionalities, with the latter form making the drug molecules totally hydrophobic and more susceptible to partitioning into the oil core of microemulsion droplets. Amitriptyline has a

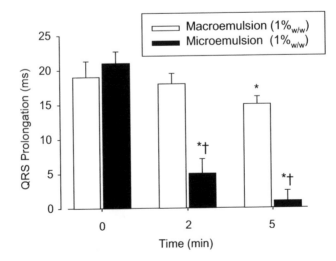

FIGURE 28.11 Effect of macroemulsions and microemulsions on QRS time interval.

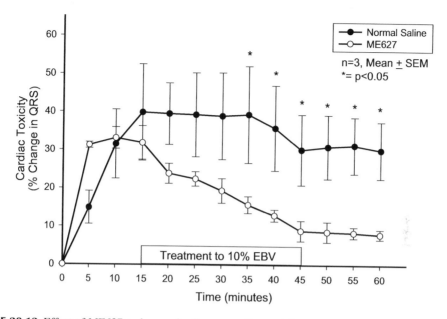

FIGURE 28.12 Effect of ME627 and normal saline on cradiac toxicity.

pK_a of 9.4 and is 99% protonated at pH 7.40; bupivacaine has a pK_a of 8.1 and is 83% protonated at the same pH [17]. These percentage differences play a major role in the amount of the two drugs that can be removed from normal saline by a microemulsion, and therefore presumably from whole blood at the same pH. HPLC results show that amitriptyline extraction declines from 100% at pH 6.5 to near zero percent when the saline is at pH 10; the amount of bupivacaine extracted increases as the pH of the saline increases. The data show that at physiological pH 99% amitriptyline but only 12% bupivacaine undergoes phase transfer into the ME617 microemulsion. In addition to pK_a differences between the drugs targeted for removal from blood, steric differences in molecular structures may also play a role in the phase transfer processes. Close examination of the molecular structures of amitriptyline and bupivacaine reveals that the aromatic and hydrophobic portion is sterically voluminous in the former drug and the amine portion is much less hindered. In bupivacaine, however, the amine nitrogen atom is positioned deep inside a cage-like hydrocarbon region, a suspect factor that controls the pK_a and ability of the drug to partition into oil (Figure 28.13).

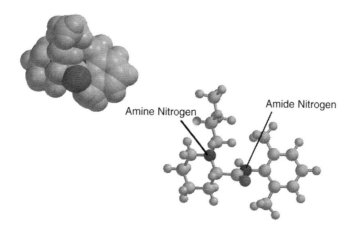

FIGURE 28.13 Molecular structures of amitriptyline and bupivacaine showing the positions of amine and amide nitrogen atoms.

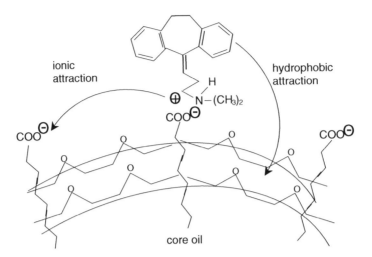

FIGURE 28.14 A hypothetical drawing of the approach of the protonated form of amitriptyline.

The previously described molecular features of amitriptyline and bupivacaine allow a prediction why amitriptyline undergoes partitioning into microemulsion oil droplets at or below physiological pH and why bupivacaine does so to a much lesser degree. Figure 28.14 is a hypothetical drawing of the approach of the protonated form of amitriptyline to the surface of a microemulsion droplet. Data in Figure 28.8 and Figure 28.9 show that fatty acid salts present in the co-surfactant shell around core ethyl butyrate droplets greatly enhances amitriptyline absorption. It is postulated that electrostatic attraction between the carboxylate groups protruding into the aqueous medium around a droplet, and the sterically unencumbered ammonium group, causes the drug to diffuse to the droplet surface, coincident with spacial tumbling of the hydrophobic aromatic portion of the molecule into the Pluronic, and subsequently the oil core.

The data in Figure 28.10 suggests that interfacial transport of bupivacaine into oil droplets is controlled more by hydrophobic than by coulombic interactions. Bupivacaine absorbs into oil better when the drug is in the unprotonated, neutral form, and when the co-surfactant with F127 Pluronic is more carbonatceous. Even when protonated, the amine group in bupivacaine is buried inside a hydrocarbon region of the molecule which may inhibit its near approach to the carboxylate groups and contribute to stronger hydrophobic interactions with the Pluronic and oil core components. This hypothesis may not be valid because bupivacaine extraction from saline should be improved by replacing Pluronic F127 with one having a larger number of hydrophobic propylene oxide units, Y in Figure 28.6, which is not this one. NMR spin-lattice relaxation time measurements are being undertaken to try to resolve these questions for both amitriptyline and bupivacaine interaction with microemulsion droplets.

IV. SURFACE MODIFIED PARTICLE TOXIN RECEPTOR STUDIES

A plethora of technical publications over the past several decades reveals that surface modification of solid core particles of several compositions, shapes, and sizes, using all types of chemical reagents and reaction techniques, is now well understood and applied to enhance the properties of many powders used in manufacturing and medicine [18–23]. In the context of the present work, the theme is to attach molecular units to the surface of carrier particles so that, when introduced into the blood stream, they will complex with and deactivate overdosed toxins (see Figure 28.3). In other words, to have controlled removal of drug/toxin molecules rather than controlled release. Team members Partch, Baney, Powell, Y.-H. Lee, and D-W. Lee postulated an approach employing the concept of charge-transfer complexes to capture the target drugs shown in Figure 28.2.

A wide range of inorganic and organic solid core particles exists to which receptor moieties can be attached. Both core and surface compositions must be biocompatible with blood and the receptors must be able to differentiate overdosed toxin molecules from natural ones having similar structures. If the modified carrier particles are less than 10 nm in diameter, they may be able to pass through the kidney, thus removing the toxin with them during urination. Alternatively, carrier particles too large to be removed by normal body function can be composed of biodegradable material of which smaller degradation units carrying toxin may be passed. To establish proof-of-concept that the objective could be achieved, silica particles and oligochitosan carrier molecules having covalently attached π-electron-deficient aromatic rings attached were synthesized and evaluated for efficiency of drug removal.

A. Silica Particle Synthesis and Characterization

The classic Stober method was employed for preparing several sizes of nearly monodisperse spherical silica particles [24,25]. Alternatively, higher surface area and more porous silica particles were prepared by the sol–gel method but incorporating pore templating molecules into the reaction mixture and subsequently removing them by solvolysis or thermal treatment [26–30]. An example experiment of the latter type and SEM photograph of the obtained nanoparticles is shown in Figure 28.15. 2,6-Dimethyl aniline and its N-acetylated derivative served as model compounds for bupivacaine (see Figure 28.2) during the synthesis and were removed from the silica by thermal treatment in air at 300°C. In the same fashion, bupivacaine was also used as a template, with the same result. No attempt was made to determine the difference in pore volume between the solid products.

B. Covalent Attachment of π Receptor Aromatic Rings to Silica Particles

The surface of silica particles has many exposed silanol —OH groups which are known to undergo facile reaction with silanes contained —Si(OR), —Si(Cl), and —Si(H) units, and with acyl or sulfonyl acid chloride functional groups [31,32]. Thus, preparation of the desired derivatized silica particles was easily accomplished by refluxing silica in a nonhydroxylic solvent with excess π receptor reagents such as 3,5- or 2,4-dinitrobenzoyl or dinitrobenzenesulfonyl chloride. This

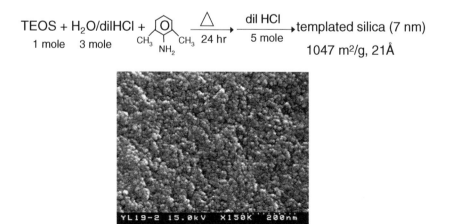

FIGURE 28.15 An experimental reaction to obtain silica particles and SEM photograph of the silica nanoparticles.

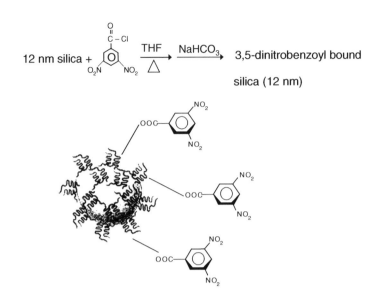

FIGURE 28.16 Reaction for obtaining 3,5-dinitrobenzoyl bound silica and the covalenty attached π receptor groups.

process yielded silica particles having numerous π receptor groups covalently attached to the silica (Figure 28.16), the number of which can be experimentally determined by titration of residual versus original —OH groups [33].

C. FUNCTIONALIZATION OF OLIGOCHITOSAN WITH π RECEPTOR AROMATIC RINGS

Attachment of π receptor groups to oligochitosan was achieved by team members, Lee and Baney employing the same chemistry as with silica [34]. Figure 28.17 reveals that the polymer has both hydroxyl and amino functionalities. The amino groups are more nucleophilic and when the polymer is reacted with an acyl or sulfonyl chloride, carboxamides or sulfonamides are formed, respectively. Specifically, oligochitosan having average molecular weight of 1150 Da and containing 8% water was dissolved in DMSO containing 2,4-dinitrobenzenesulfonyl chloride and stirred at 20°C for 48 h. Addition of ethanol caused the desired product to precipitate, which was collected by centrifugation, Soxhlet extracted with methanol, and dried in vacuum.

D. BACKGROUND INFORMATION ON AROMATIC CHARGE TRANSFER COMPLEXES

There is a long history on the subject of face-to-face functional group-induced interaction between the planar portions of two aromatic rings. The phenomenon has its genesis in the ability of some functional groups having positive Hammett sigma constants attached to benzene rings to reduce the density of the π electron cloud distributed over the six carbon atoms, and in the ability of some functional groups having negative Hammett sigma constants to increase the π cloud electron density

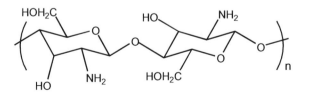

FIGURE 28.17 Polymer with both hydroxyl and amino functionalities.

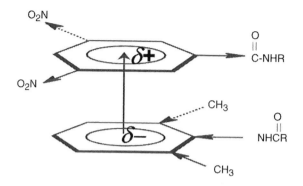

FIGURE 28.18 Effect of charge polarization on substituted benzene rings.

[35,36]. The former aromatic rings are called π acceptors and the latter π donors. Classic examples of π acceptors are trinitrobenzene and tetracyanoquinodimethane. Recently, $\pi-\pi$ charge transfer complexation has been employed for enantiomer separation [37], polymer stacking [38], assembly of microelectronic circuits [39], bimolecular analysis [40–42], and explanation of disulfide bond formation in biological quinhydrone systems [43]. Generally, $\pi-\pi$ complexation occurs with relatively small binding energies in the range of 8–20 kJ/mol. However, the phenomenon is identified in molecular recognition in biology [44–46].

The charge polarization in each type of substituted benzene ring causes the rings having opposite π cloud densities to align themselves as shown in Figure 28.18. In the upper ring, the two nitro and the carbonyl functionalities are powerful electron-withdrawing groups and induce that ring to exhibit positive charge character compared with an unsubstituted ring. The lower ring in the figure is oppositely charged due to the inductive and resonance input of electron density into the ring by the methyl and amido nitrogen substituents. In other words, the upper ring is a π acceptor and the lower ring is a π donor. It should be noted that the π donor ring has the same substitutent pattern as the benzene ring in bupivacaine (Figure 28.2).

The incentive to utilize for the first time $\pi-\pi$ complexation for selectively removing overdosed and toxic lipophilic aromatic compounds from blood originated from the work of Dust on binding dopamine derivatives to trinitrobenzene [47]. In that and other subsequent publications [42,48–50], spectroscopic methods are described for quantitative determination of complexation as well as how to calculate binding constants and activation energies.

When solutions of potential π donors are mixed with those having potential π acceptors, bathochromic shifts in UV/Vis spectra, and upfield shifts in NMR adsorbances serve as physical evidence that complexation has occurred. Figure 28.19 is representation of the UV/Vis spectral change that occurs when π acceptor N-methy-3,5-dinitrobenzamide (DNB) is complexed with ether bupivacaine hydrochloride or with its model compound 2,6-dimethylacetanilide.

E. NMR Evidence of $\pi-\pi$ Complexation

1. Solution Studies

As no previous information was available on $\pi-\pi$ complexation with the drugs targeted in this study, a series of experiments using model compounds of bupivacaine as well as the drug were carried out to establish the precedent [51]. 2,6-Dimethylacetanilide (DMAC) was selected as the appropriate model, and DNB as one of several π acceptors. Figure 28.20 shows the triplet and doublet proton resonances (CDCl$_3$) for DNB before and after addition of DMAC. The upfield shifts for these are 0.0412 and 0.1997 ppm, respectively. Magnetic anisotropy associated with a ring current shield effect causes the shift. The different magnitudes are due to the relative orientation of the substituents on the two stacked rings.

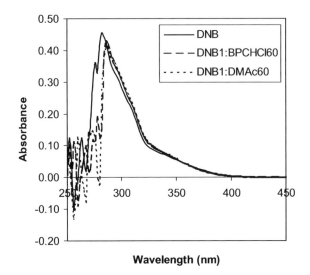

FIGURE 28.19 UV–VIS spectrophotograph representation of *N*-methyl-3,5-dinitrobenzamide with ether bupivacaine hydrochloride.

The magnitude of the upfield shifts in proton resonances when the free base of bupivacaine served as the π donor were 0.0662 and 0.2513 ppm, indicating that interaction between the drug and DNB was stronger than between DMAC and DNB. Measurements of the therapeutically administered hydrochloride salt of bupivacaine as π donor, mixed with DNB in 50/50 D_2O/CD_3CN,

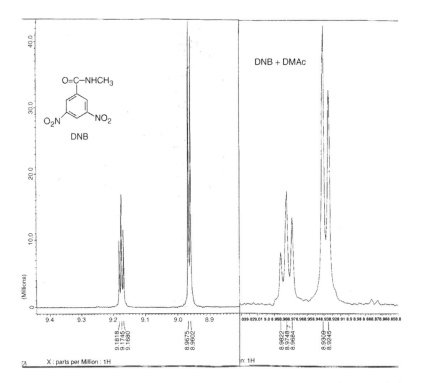

FIGURE 28.20 Spectrograph showing doublet and triplet proton resonances for DNB before and after addition of DMAC.

gave upfield shifts of 0.0275 and 0.0891 ppm. Hydrogen bonding of the salt with water molecules probably competes with the π acceptor for the donor.

Alternatives to DNB as π acceptor included N-ethyl-2,4-dinitrobenzenesulfonamide and N-methyl-pentafluorobenzamide. ^{19}F-NMR upfield shifts in the resonances of the three magnetically distinct fluorine atoms, when the latter π acceptor molecule was mixed with DMAC, are 0.3362, 1.6754 and 0.9570 ppm. This data is the first to be reported where fluorine resonance serves as a proof of complexation. When considering the goal of the current research, this may be useful since fluorine and trifluoromethyl substituted aromatics are tolerated by the human body more than nitro aromatics [52,53].

Proton NMR data was collected to obtain bupivacaine-binding constants and activation energies of π–π complexation using published procedures [42,47–50]. The magnitude of the upfield shift in π acceptor proton resonances reflects the strengths of complex formation. The data [51] yield enthalpies in the order: +1.75 kJ/mol for the DNB/2,6-dimethylaniline complex, +0.57 kJ/mol for the N-ethyl-2,4-dinitrobenzamide/dimethylaniline complex and −1.65 kJ/mol for the N-ethyl-2,4-dinitrobenzamide/bupivacaine complex. As stated earlier, the more hydrophobic carbon mass surrounding the amino nitrogen atom in bupivacaine may be the reason for the exotherm observed with the drug but not the model compounds.

2. NMR Studies with Oligochitosan

Most oligochitosan derivatives are water soluble [54–56]. The derivative formed by reaction with excess 2,4-dinitrobenzenesulonyl chloride was less but exhibited enough solubility to run NMR samples in D_2O, as well as in CD_3OD and CD_3Cl_3 (34). Using the oligochitosan derivatives as π acceptors and amitriptyline as donor in CD_3OD, the upfield shift in aryl proton resonance of the acceptor ranged from 0.015 ppm at 10 mM donor to 0.073 ppm at 200 mM. This data verifies that the large size of oligschitosan does not interfere with the desired π complexation reported earlier for lower-molecular weight acceptor molecules.

V. IN VITRO AND IN VIVO EFFECTIVENESS OF DISPERSED PHASE WITH π ACCEPTORS FOR REDUCING OVERDOSE TOXICITY

A. DERIVATIZED SILICA PARTICLES

The silica particles prepared as described in Figure 28.15, and then derivatized with dinitrobenzoyl π acceptor moieties as shown in Figure 28.16, proved effective in removing bupivacaine hydrochloride from normal saline. HPLC analysis showed that the core particles removed no drug from the liquid containing 1000–30,000 μM bupivacaine. However, 0.05% and 0.10% (w/v) derivatized nanoparticles exhibited plateau removal of the drug in amounts of 1800 and 4000 μM, respectively, when the original bupivacine was 10,000 μM or more. The proposed mechanism of removal is depicted in Figure 28.21 [1]. The preliminary results do not allow differentiation as to whether one- or two-point interaction of the drug with the particles occurs. The two-point interaction should be facilitated if the number of π acceptor units attached to a particle, and their freedom of movement on the end of the tethers, allows the basic nitrogen of a drug to diffuse close to and hydrogen bond with the residual silanol groups. Figure 28.13 reveals that such interaction is probably more reasonable for amitriptyline than for bupivacaine.

B. DERIVATIZED OLIGOCHITOSAN MACROMOLECULES

The QRS interval of excised rat heart or whole animal (in vivo) was not changed upon exposure to as much as 33 mg/kg of oligochitosan before or after derivatization with π acceptor units [34]. Neither were other detrimental physiological effects observed over a period of 4 weeks. The effective treatment of excised rat heart pre-exposed to sufficient amitriptyline to induce sodium channel

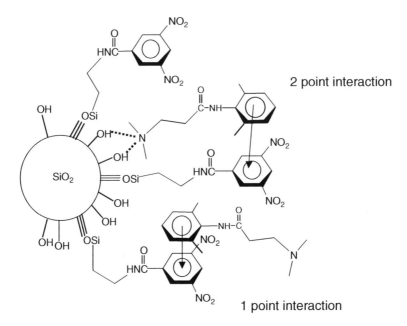

FIGURE 28.21 Mechanism for the removal of drugs.

blockage [57–59] proved the viable use of derivatized oligochitosan to reverse the effect of drug overdose. In these experiments, the percent QRS prolongation after treatment with the drug was 70% after 30 min. The result was essentially the same when oligochitosan was co-administered but when the derivatized oligochitosan was given, the prolongation was reduced to less than 30% in the same time interval.

VI. MICROEMULSIONS REVISITED

It is evident from the data presented in this chapter on the preparation of injectable nanodispersed species capable of lowering the concentration of the targeted drugs in blood, that the goal is achievable. However, the compositions of the microemulsions and surface-modified carrier particles must be optimized by the team before the antidotes can be used in humans.

Toward improving the microemulsions for removal of overdosed drugs, experiments have been carried out to make them more selective in absorbing different lipophilic molecules. One approach has proved successful in absorbing bupivacaine over cocaine.

A microemulsion containing Pluronic F127 and sodium caprylate co-surfactants has been prepared using the standard procedure in Section III.A. Ethyl butyrate as the oil was replaced by 1,3-bistrifluoromethylbenzene, the aromatic ring of which is a π receptor due to the electron withdrawing effect of the $-CF_3$ groups. The team conjectured that by using a π receptor as the oil core, the droplets would be more than only oil reservoirs and possibly be able to partition into them at different rates lipophilic drug molecules having different π donor capabilities. Figure 28.22 depicts the concept of a "smart" microemulsion oil core controlling the absorption of bupivacaine over that of cocaine. Both drugs have similar atomic and structural molecular compositions but the aromatic ring in cocaine is not as good a π donor [60].

Experimental proof that the concept indeed works is shown in Figure 28.23. This is the first reported data on incorporation of both hydrophobic and π acceptor properties into the oil core of a microemulsions. The microemulsion using ethyl butyrate does not differentially remove the

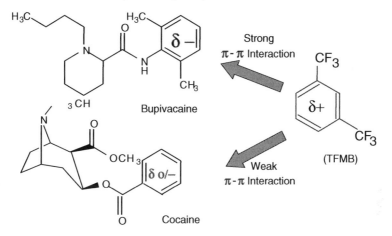

FIGURE 28.22 Concept of a "smart" microemulsion oil core controlling the absorption of bupivacaine over that of cocaine.

drugs from normal saline but the one using the π acceptor oil does. Having established this precedent the team anticipates making even further improvements.

VII. CONCLUSIONS AND PATH FORWARD

Summarized in this chapter is state-of-the-art data and their analyses for ongoing research in the formulation of three types of injectable and biocompatible dispersed phases showing promise as antidotes for a selection of commonly overdosed drugs. The oil-in-water microemulsions deplete

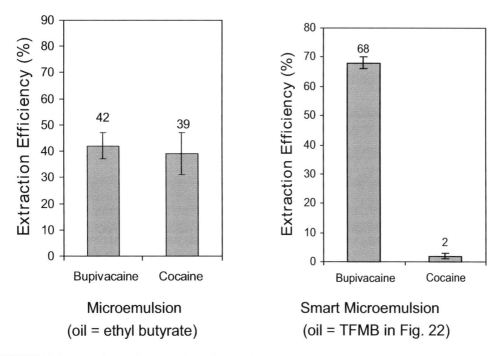

FIGURE 28.23 Experimental results using microemulsion and "smart" microemulsion in removing drugs.

amitriptyline and bupivacaine drugs through a combination of electrostatic and hydrophobic action. The oligochitosan molecule and solid silica nanoparticles, each derivatized with π acceptor dinitroaromatics, bind the same drugs as well as cocaine efficiently and rapidly, due to their π-donor characteristics.

Proof-of-concept that the research may lead to clinically viable remedies for some drug overdoses comes from results of *in vitro* and *in vivo* experiments. Preliminary experiments show that the cardiotoxicity of the drugs is successfully reversed when the microemulsion or oligochitosen phases are infused into excised rat hearts or whole animals.

The path forward by the interdisciplinary team will be to:

(1) Further improve the compositions of the microemulsions to make them more "smart" regarding drug adsorption.
(2) Adapt the π acceptor π donor concept for use with all three types of dispersed phases.
(3) Test the effectiveness of the phases to reverse the cardiotoxic effects of other than the three targeted therapeutics.
(4) Obtain fundamental information on drug interaction with the dispersed phases using NMR techniques and ^{13}C-enriched drugs [61].
(5) Enhance the biocompatibility of the optimized formulations.

ACKNOWLEDGMENTS

Financial and facility contributions are gratefully acknowledged from the National Science Foundation Engineering Research Center at the University of Florida (Grant EEC-94-02989), the New York State Center for Advanced Materials Processing at Clarkson University, the Department of Anesthesiology and Department of Materials Science and Engineering at the University of Florida, the Department of Chemistry at Clarkson University and the NIH for grant RO1 GM63679-01A1.

REFERENCES

1. A preliminary account has been published, Partch, R., Powell, E., Lee, Y-H., Varshney, M., Kim, S., Barnard, N., Shah, D., Dennis, D., and Morey, T., Injectable Nanoparticle Technology for In Vivo Remediation of Overdosed Toxins, in *Nanostructured Materials and Coatings for Biomedical and Sensor Application*, Gogotsi, Y., and Uvarova, I., Eds., Kluwer Publishers, Netherlands, 2003, pp. 27–40.
2. Somasundaran, P., Liu, F., and Gryte, C., Novel polyacrylamide nanogels for drug binding and drug deliver, *Polym. Mater. Sci. Eng.*, 89, 235–241, 2003.
3. Underhill, R., Jovanovic, A., Carino, S., Varshney, M., Shah, D., Dennis, D., Morey, T., and Duran, R., Oil-filled nanocapsules for lipophilic drug uptake: implications for drug detoxification therapy, *Chem. Mater.*, 14, 4919–4925, 2002.
4. Mitchel, D., Lee, S., Trofin, L., Li, N., Nevanen, T., Soderlund, H., and Martin, C., Smart nanotubes for bioseparations and biocatalysis, *J. Am. Chem. Soc.*, 124, 11864–11865, 2002.
5. Bagwe, R.P., Kanicky, J.R., Palla, B.J., Patanjali, P.K., and Shah, D.O., Improved drug delivery using microemulsions: rationale, recent progress, and new horizons, Critical Reviews Therap., *Drug Carrier Syst.*, 18, 77–140, 2001.
6. Palla, B.J., Shah, D.O., Garcia-Casillasa, P., and Matutes-Aquino, J., Preparation of nanoparticles of barium ferrite from precipitation in microemulsions, *J. Nanoparticle Research*, 1, 215–226, 1999.
7. Shah, D.O. Ed., *Micelles, Microemulsions, and Monolayers*, Marcel Dekker Inc., New York, 1998, 1–610.
8. Chhabra, V., Free, M.L., Kang, P.K., Truesdail, S.E., and Shah, D.O., Microemulsions as an emerging technology, *Tenside, Surfactants, Detergents*, 34, 156–168, 1997.
9. Sjoblom, J., Lindberg, R., and Friberg, S., Microemulsions-phase equilria characterization, structures, applications and chemical reactions, *Adv. Colloid Interface Sci.*, 65, 125–187, 1996.

10. Scottmann, T., Solubilization efficiency boosting by amphiphilic block copolymers in microemulsions, *Current Opin. Colloid Interface Sci.*, 7, 57–65, 2002.
11. Jakobs, B., Scottmann, T., Strey, R., Allgaier, J., Willner, L., and Richter, D., Amphiphilic block copolymers as efficiency boosters for microemulsions, *Langmuir*, 15, 6707, 1999.
12. Friberg, S., Mortensen, M., and Neogi, P., Hydrocarbon extraction into surfactant phase with non-ionic surfactants, *Sep. Sci. Technol.*, 20, 285, 1985.
13. Orringer, E., Casella, J., Ataga, K., Koshy, M., Adams-Graves, P., Luchtman-Jones, L., Wun, T., Watanabe, M., Shafer, F., Kutlar, A., Abboud, M., Steinberg, M., Adler, B., Swerdlow, P., Terregino, C., Saccente, S., Files, B., Ballas, S., Brown, R., Wojtowicz-Praga, S., and Grindel, J., Purified poloxamer 188 for treatment of acute vaso-occulusive crisis of sickle cell Di, A randomized controlled trial, *JAMA*, 286, 2099–3010, 2001.
14. Varshney, M., Morey, T., Shah, D., Flint, J., Moudgil, B., Seubert, C., and Dennis, D., Pluronic microemulsions as nanoreservoirs for extraction of bupivacaine from normal saline, *J. Am. Chem. Soc.*, 126, 5108–5112, 2004.
15. Allemann, E., Gurny, R., and Dekker, E., Drug-loaded nanoparticles: preparation method and drug-targeting issues, *Eur. J. Pharm. Biopharm.*, 39, 173, 1993.
16. Junquera, E., Romero, J., and Aicart, E., Behavior of tricyclic antidepressants in aqueous solution, *Langmuir*, 17, 1826–1832, 2001.
17. Strichartz, G., Sanchez, V., Arthur, G., Chafetz, R., and Martin, D., Fundamental properties of local anesthetics. II, *Anesth. Analg.*, 71, 158–170, 1990.
18. Keklikian, L. and Partch, R., Preparation of mixed phenylurethane/titania particles, *Colloids Surfaces*, 41, 327–337, 1989.
19. Nishida, Y., Iso, M., Matsuoka, M., and Partch, R., Study of polyimide particle surface modification, *Adv. Powdr. Technol.*, 15, 247–261, 2004.
20. Huang, C.-L., Partch, R., and Matijevic, E., Coating polyaniline on copper oxide, *J. Coll. Interface Sci.*, 170, 275–283, 1995.
21. Avella, M., Martuscelli, E., Raimo, M., Partch, R., Gangolli, S., and Pascucci, B., Polypropylene reinforced with silicon carbide whiskers, *J. Mat. Sci.*, 32, 2411–2416, 1997.
22. Dilsiz, N., Partch, R., Matijevic, E., and Sancaktar, E., Silver coating of magnetic particles for conductive adhesive applications, *J. Adhesion Sci. Technol.*, 11, 1105–1118, 1997.
23. Partch, R., in *Materials Synthesis and Characterization*, Perry, D. Ed., Plenum Press, New York, 1997, 1–17.
24. Iler, R., *The Chemistry of Silica*, Wiley, New York, 1979.
25. Brinker, J. and Schere, G., *Sol-Gel Science*, Academic Press, San Diego, 1989.
26. Dickey, F., Preparation of specific adsorbents, *Proc. Natl. Acad. Sci.*, 35, 227–229, 1949.
27. Raman, K., Anderson, M., and Brinker, C., Template-based approaches to the preparation of amorphous nanoporous silicas, *Chem. Mater.*, 8, 1682–1701, 1996.
28. Makote, R. and Collinson, M., Template recognition in inorganic-organic hybrid films prepared by the sol-gel process, *Chem. Mater.*, 10, 2440–2445, 1998.
29. Zimmermann, C., Partch, R., and Matijevic, E., Influence of plasma on surface properties of some inorganic oxide particles, *Colloids Surfaces*, 57, 177–185, 1991.
30. Katz, A. and Davis, M., Molecular imprinting of bulk, microporous silica, *Nature*, 403, 286–288, 2000.
31. Product Catalogue, Gelest, Inc., www.gelest.com.
32. Wang, H-Y., Ph.D. thesis, Clarkson University, 2000.
33. Peri, J. and Hensley, A., The surface structure of silica gel, *J. Phys. Chem.*, 72, 2926–2933, 1968.
34. Lee, D-W., Flint, J., Morey, T., Dennis, D., Partch, R., and Baney, R., Aromatic-aromatic interaction of amitriptyline: implications of overdosed drug detoxification, *J. Pharm. Sci.*, 94, 373, 2005.
35. Jaffe, H., A reexamination of the hammett equation, *Chem. Rev.*, 53, 191–261, 1953.
36. Hansen, O., Hammett series with biological activity, *Acta Chem. Scand.*, 16, 1593–1600, 1962.
37. Stinson, S., Chiral Cravings, *Chem. Eng. News*, 79, Dec. 10, 35–38, 2001.
38. Brunsveld, L., Folmer, B., and Meijer, E., Supramolecular polymers, *MRS Bulletin*, 25, 49–53, 2000.
39. Rouhi, A., Tinkertoy dreams, *Chem. Eng. News*, 79, July 30, 46–49, 2001.
40. Goodnow, T., Reddington, M., Stoddart, F., and Kaifer, A., Cyclobis(paraquat-p-phenylene), *J. Am. Chem. Soc.*, 113, 4335–4337, 1991.

41. Wasset, W., Ghobrial, N., and Agami, S., Spectroscopic studies on electron donor-electron acceptor interaction, *Spectrochim. Acta*, 47A, 623–627, 1991.
42. Fesik, S., Medek, A., Hajduk, P., and Mack, J., The use of differential chemical shifts for determining binding sites in proteins, *J. Am. Chem. Soc.*, 122, 1241, 2000.
43. Regeimbal, J., Gleiter, S., Trumpower, B., Yu, C-A., Diwakar, M., Ballou, D., and Bardwell, J., Disulfide bond formation involves a quinhydrone-type charge-transfer complex, *Proc. Nat. Acad. Sci.*, 100, 13779–13784, 2003.
44. Adams, H., Blanco, J., Chessari, G., Hunter, C., Low, C., Sanderson, J., and Vinter, J., *Chem. Eur. J.*, 7, 3494–3503, 2001.
45. Sinnokrot, M., Valeev, E., and Sherrill, C., Estimates of the Ab initio limit for $\pi-\pi$ interactions, *J. Am. Chem. Soc.*, 124, 10887–10893, 2002.
46. Hunter, C., The role of aromatic interactions in molecular reorganization, *Chem. Soc. Rev.*, 23, 101–109, 1994.
47. Dust, J., Charge-transfer and electron transfer processes in biologically significant systems, *Can. J. Chem.*, 70, 151–157, 1992.
48. Hanna, M. and Ashbaugh, A., Nuclear magnetic resonance study of aromatic molecular complexes, *J. Phy. Chem.*, 68, 811, 1964.
49. Foster, R. and Fyfe, C., NMR of Charge-transfer complexes, *Proc. Nucl. Magn. Reson. Spectrosc.*, 4, 1–9, 1969.
50. Neusser, H., and Krause, H., Binding energy and structure of Van der Waals complexes of benzene, *Chem. Rev.*, 94, 1829–1843, 1994.
51. Powell, E. and Partch, R., Remediation of overdosed of the local anesthetic bupivacaine by $\pi-\pi$ complex formation, 2004, submitted.
52. Yagupolskii, L., Bystrov, V., Stepanyants, A., and Fialkov, Y., Influence of substituents with trifluoromethyl group on the reactivity of aromatic compounds, *Z. Obshchei Khimii*, 34, 3682–3690, 1964.
53. Meyer, E., Costellano, R., and Diederich, F., Interactions with aromatic rings in chemical and biological recognition, *Angew. Chem. Int. Ed.*, 42, 1210–1250, 2003.
54. Domszy, J. and Roberts, G., Reactions of chitosan with dinitrofluorobenzene, *Int. J. Biol. Macromol.*, 7, 45–49, 1985.
55. Amiji, M., Surface modification of chitosan to improve blood compatability, *Recent Res. Devel. Polymer Science*, 3, 31–39, 1999.
56. Compare Lorenzo-Lamosa, M., Remunan-Lopez, C., Vila-Jato, J., and Alonso, M., Design of micro-encapsulated chitosan microspheres for colonic drug delivery, *J. Controlled Release*, 52, 109–118, 1998.
57. Pentel, P. and Benowitz, N., Tricyclic antidepressant poisoning: management of arrhythmias, *Med. Toxicol.*, 1, 101–121, 1986.
58. Pancrazio, J.J., Kamatchi, G.L., Roscoe, A.K., and Lynch, C., Inhibition of neuronal Na^{+} channels by antidepressant drugs, *J. Pharmacol. Exp. Ther.*, 284, 208–214, 1998.
59. Marshall, J.B. and Forker, A.D., Cardiovascular effects of tricyclic antidepressant drugs: therapeutic usages, overdose and management of complications, *Am. Heart J.*, 103, 401–414, 1982.
60. Henry, C., Cocaine Vaccine, in *Chem. Eng. News*, 82, June 28, 11, 2004.
61. Patist, A., Kanicky, J., Shukla, P., and Shah, D., Molecular packing in micelles and films, *J. Colloid Interface Sci.*, 245, 1–15, 2002.

29 Carbon Nanocapsules and Nuclear Applications

Enrique E. Pasqualini and Marisol López
Constituyentes Atomic Center, National Commission of Atomic Energy,
Buenos Aires, Argentina

CONTENTS

I. INTRODUCTION

Carbon nanocapsules [1–3] are polyhedrical shell structures that comprise several closed concentric graphene layers and a kernel. They have dimensions of the order of 50 nm and can have an interior gap between kernel and capsule. Filled nanocapsules were first identified in arc discharge experiments at very high temperatures by the simultaneous evaporation of graphite electrodes charged with rare earth elements. Several types of nanocapsules with different compounds inside have been already synthesized [4]. The polyhedrical capsules with up to 40 shells have interlayer spacings greater than in graphite and have chemical inertness similar as graphite. Rounded vertex or spherical capsules are formed when pairs of carbon atoms are eliminated from the shells of polyhedrical nanocapsules or multiwall nanotubes [5], and in this case interlayer spacings can be smaller than in graphite. Different synthetic routes and purification methods are being explored to overcome low production yields [6].

Characteristic physical properties can be enhanced depending on the material that is encapsulated, and additional physical, chemical, or biological characteristics can be tailored by coverage and funtionalization of the surface of the capsule. A wide spectrum of applications can be achieved by encapsulating magnetic compounds or radioactive materials. Feasibility studies in nuclear applications can be designed in the fields of radioactive sources and labeling, dosymetry, nuclear medicine, radioisotope production, nuclear fuels, and nuclear waste. Furthermore, nanocapsules can be functionalized or additionally covered for intermediate production steps or end purposes.

II. POLYHEDRICAL NANOCAPSULES

Typical nanocapsules [7] (Figure 29.1) have plane external faces as seen by high-resolution transmission electron microscopy (HRTEM), revealing the presence of several graphitic layers. At selected orientations, interlayer spacing of the carbon capsule can be measured, and strain and shell defects are identified (Figure 29.2).

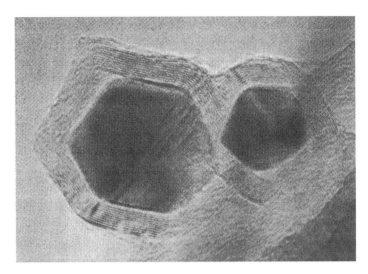

FIGURE 29.1 HRTEM of 30 and 40 nm size nanocapsules. (From Saito, Y., in *Fullerences: Recent Advaces in the Chemistry and Physics of Fullerences and Related Materials*, Kadish, K. and Ruoff, R., Eds., The Electrochemical Proceeding Series, Pennington NJ, 1994, pp. 1419–1447 (PV 94-24). With permission.)

X-ray diffraction (XRD) indicates that the interlayer spacing is slightly higher — in value and in dispersion — and has different stacking than in graphite [8] (Figure 29.3). Chemical inertness, similar to that of graphite, indicates that the kernel is totally surrounded. Thus, it can be said that nanocapsules are conformed by concentric polyhedra with plane faces of closed graphene sheets with a particular stacking order.

Graphene sheets are hexagonal arrays of carbon atoms as in the base plane of the crystallo-graphic graphite structure. Graphene sheets are formed by sp^2 hybridized carbon atoms connected to three neighbor atoms by σ bonds and bonded with other layers by van der Waals forces. Continuity of these sheets or faces can be resolved without great strain by bending bonds to conform the edges of the polyhedra. Structure continuity of the graphene sheet at the polyhedral angles is achieved by the extraction of one [9], two, or three 60° sectors, centered at any hexagon and with different possible orientations (Figure 29.4a). Joining the loose sides, polyhedral angles will be formed respectively with a five (Figure 29.4b), four (Figure 29.4c), and three (Figure 29.4d) atom polygon at the vertex. One way of joining these polyhedral angles to conform a polyhedron is with faces that are equilateral triangles or surfaces that can be decomposed in equilateral triangles. Elimination of more than three 60° sectors produces very acute angles that have not been observed in nanocapsules.

The general rule for perfect closure of polyhedra whose faces are polygons and vertices (atoms) are connected to three edges — coordination three — is satisfied by the condition upon the necess-ary relation between the quantity and type of polygons needed (Equation 29.1). This condition is derived from Euler's theorem, which relates the number of vertices (v), edges (e), and faces (f) in a polyhedron ($v + f = e + 2$) [10]. The number of polygons with i sides is identified by n_i ($i = 3$ is a triangle, $i = 4$ is a square, $i = 5$ is a pentagon, etc.):

$$12 = 3n_3 + 2n_4 + n_5 + 0n_6 - n_7 - 2n_8 - 3n_9 \cdots \tag{29.1}$$

The regular polyhedra that satisfy this condition are the tetrahedron, the cube, and the dodeca-hedron with triangular, square, and pentagonal faces, respectively. By conveniently truncating polyhedra at the vertices, new triangular or hexagonal faces are generated. Since any number of

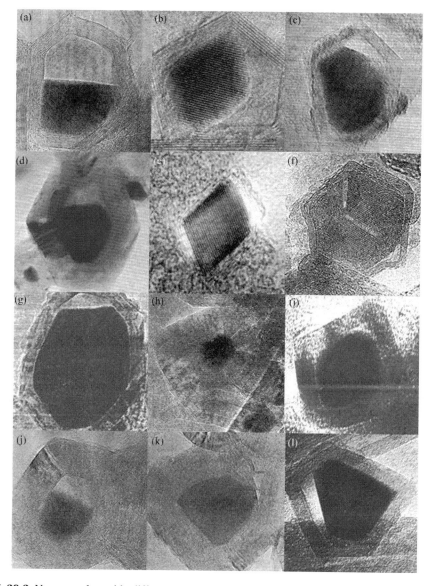

FIGURE 29.2 Nanocapsules with different kernels. For scale purposes, graphitic interlayer spacing is 0.34 nm. (a) LaC_2. (From Tomita, M., Saito, Y., and Hatashi, T., *Jpn. J. Appl. Phys.* 32 (2, 2B), L280–L282, 1993. With permission.) [2] (b) YC_2. (From Seraphin, S., et al., *Appl. Phys. Lett.*, 63 (15), 2074, 1993. With permission. © American Institute of Physics.) (c) NdC_2. (From Saito, Y., et al., *J. Phys. Chem. Solids.* 54 (12), 1850, 1993. With permission. © Elsevier Science Ltd.) (d) LuC_2. (From Saito, Y., Okuda, T., Kasuya, A., and Nishina, Y., *J. Phys. Chem.* 98 (27), 6627, 1994. With permission. © American Chemical Society.) [11] (e) UC_2. (From Pasqualini, E.E., Adelfang, P., and Nuñez Regueiro, M., *J. Nucl. Mater.*, 231, 176, 1996. With permission. © Elsevier Science Ltd.) [12] (f) B_4C. (From Zhou, D., et al., *Chem. Phys. Lett.*, 234, 236, 1995. With permission. © Elsevier Science Ltd.) (g) Mo_2C. (From Saito, Y., et al., *J. Crys Growth*, 172, 166, 1997. With permission. © Elsevier Science Ltd.) (h) ThC_2-WC-W_2C. Funasaka, H., Sugiyama, K., Yamamoto, K., and Takahashi, T., *J. Appl. Phys.*, 78 (9), 5321,1995. With permission. © American Institute of Physics.) [13] (i) ZrC. (From Bandow, S., et al., *Jpn. J. Appl. Phys.*, 32 (11B), L1678, 1993. With permission.) (j) La. (From Funasaka, H., et al., *Chem. Phys. Letts.*, 236, 280, 1995. With permission. © Elsevier Science Ltd.) (k) LaB_6 (From Funasaka, H., et al., *Chem. Phys. Lett.* 236, 282, 1995. With permission. © Elsevier Science Ltd.) (l) GdC_2 (From Subramoney, S., et al., *Carbon* 32 (3), 511, 1994. With permission. © Elsevier Science Ltd.)

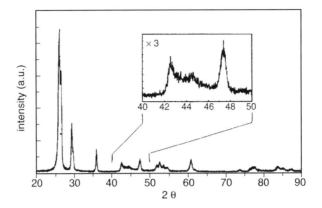

FIGURE 29.3 XRD pattern of partially purified UC_2 nanocapsules. The highest double peak corresponds to capsules and graphite interlayer spacings. The amplified zone between $42°$ and $46°$ evidences the layer stacking order different than in graphite.

hexagons can be used to fulfill Equation (29.1), triangular and hexagonal faces can be subdivided in hexagons, as if they where graphene sheets.

Triangles and squares surrounded by hexagons can be transformed by the elimination of one polygon edge in three pentagons (Figure 29.5a) and, two pentagons and two hexagons (Figure 29.5b), respectively. This edge elimination can be physically assimilated to the extraction of two contiguous carbon atoms that constitute the polyhedral shell. Rotation of edges can be associated with chemical bond rotations as in the case of a Stone–Wales transformation [14] in which a four-hexagons array will generate two pentagons and two heptagons (Figure 29.5c) lowering the atomic surface density. With these considerations, any polyhedra with coordination three can be constructed with hexagons, pentagons, and heptagons that are the most common configurations in carbon structures [15,16].

In the case of polyhedra with pentagons in the vertices and all hexagons at the faces, 12 pentagons are needed for obtaining a closed structure (Equation 29.1). With squares and triangles at the

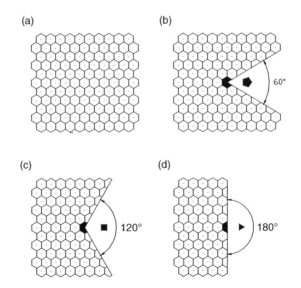

FIGURE 29.4 Construction of polyhedral angles from a graphene sheet (a) by the extraction of one (b), two (c), and three (d) $60°$ sectors.

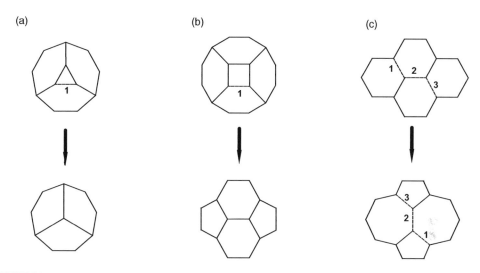

FIGURE 29.5 Local transformation of polygonal structures by edge elimination (a and b) or edge rotation (c).

vertex, the polyhedra will have respectively, six and four vertices. The polyhedra conformed are the truncated icosahedron, octahedron, and tetrahedron. The number of truncated triangular faces that can be decomposed with hexagons, for these polyhedra are twenty, eight, and four, respectively. Many other polyhedra can be constructed with different combinations of polyhedral angles [17] and sizes of the faces. In particular, since for example, single-wall nanotubes [18] also obey closure condition (Equation 29.1), the focus will be centered in regular polyhedra in which an inscribed sphere is tangent to all the polygonal faces that can be decomposed in hexagons.

By analyzing the cases of nested polyhedrical nanocapsules that have an inscribed tangent sphere, such as the truncated icosahedron, octahedron, tetrahedron, triangular bipyramid, and pentagonal bipyramid (Figure 29.6) further information can be obtained. Polyhedrical vertices can be constructed, as mentioned before (Figure 29.4), from a graphene sheet by the elimination of all the hexagons that are inscribed in an angle — centered in an hexagon — that is multiple of 60°. Choosing another hexagon of the folded sheet, the procedure can be repeated to form a second vertex. The segments joining the centers of hexagons at the vertices (Figure 29.7a) can be used as the basic edge dimension of truncated equilateral triangles that will generate the polyhedral faces. Repeating this procedure with other hexagon of the folded sheet at an equal distance as the edge dimension, a third vertex can be formed, and so on.

The folding of 20 truncated equilateral triangles by their sides and the replacement of the 12 vertices with pentagons, will produce a truncated icosahedron of equal truncated equilateral triangular faces (Figure 29.8). Similarly, a truncated octahedron can be formed with eight truncated equilateral triangular faces and eight squares in the truncated vertices. The tetrahedron is formed

FIGURE 29.6 Polyhedra that have an inscribed sphere tangent to all faces. From left to right: icosahedron, pentagonal bipyramid, octahedron, triangular bipyramid, and tetrahedron.

(a)   (b)

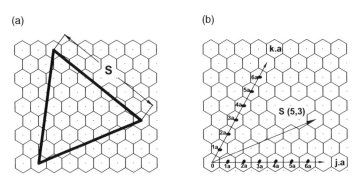

FIGURE 29.7 The edges of a basic triangular face of a truncated polyhedron (a) can be selected by integer numbers of the hexagonal coordinate base (b).

with four faces and four triangles at the vertices. The truncated triangular and pentagonal bipyramids will have two triangles and three squares at the vertices, and two pentagons and five squares, respectively. These polyhedra can be visualized by truncating vertices in Figure 29.6.

Many orientations of the graphene sheet are possible, and are defined by the primitive selection of the segment joining the centers of two hexagons (Figure 29.7a). The quantity of basic hexagons in the truncated equilateral triangular face will depend on the distance between the two hexagons first selected. If this distance is characterized by the position of the center of the hexagons in a coordinate system (Figure 29.7b) by vectors $a \cdot j$ and $a \cdot k$, where $a = 0.2461$ nm is the crystallographic base parameter of graphite and j and k have integer modulus, the distance from the origin to any other hexagon, that is, the side s of the equilateral triangles, will be:

$$s = a\sqrt{M}, \qquad M = j^2 + j \cdot k + k^2 \qquad (29.2)$$

where M is an integer variable.

Considering that the area of the equilateral triangle is $\sqrt{3}s^2/4$ and that of each hexagon is $\sqrt{3}a^2/2$, the number of hexagons in the equilateral triangle will be $s^2/(2a^2)$. Using Equation (29.2) and knowing that for each hexagon there are two carbon atoms (the six atoms conforming the hexagon are shared between three touching polygons), it is seen that M represents the number of atoms. The possible number of atoms in each face will therefore be: 1, 3, 4, 7, 9, 12, 13, 16, 19, 21, 25, 27, 28, 31, 36, 37, 39, 43, 48, 49, 52, ...

If concentric polyhedra are inscribed one inside the other forming shell structures, each shell will be characterized by the radius r of the corresponding inscribed sphere. The radius can be

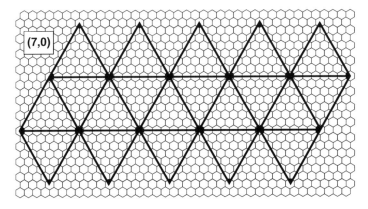

FIGURE 29.8 Construction of a truncated icosahedron from a graphene sheet.

TABLE 29.1

Quantity of atoms N in different polyhedra and constant c characterizing the radius of an inscribed sphere of radius r; $r = ca\sqrt{M}$

	Icosahedron	Pentagonal bipyramid	Octahedron	Triangular bipyramid	Tetrahedron
N	$20M$	$10M$	$8M$	$6M$	$4M$
c	0.75576	0.41777	0.40825	0.27217	0.20412

represented as $r = cs$, where c is a constant, different for each of the polyhedra — truncated icosahedron, octahedron, tetrahedron, triangular bipyramid, and pentagonal bipyramid — and that have respectively, 20, 10, 8, 6, and 4 faces (Table 29.1).

The M atoms in each face can be expressed as a function of the inscribed radius of the considered regular polyhedra characterized by the value of the parameter c (Table 29.1):

$$M = \left(\frac{r}{ca}\right)^2 \tag{29.3}$$

The difference in the number of atoms between two consecutive faces, $\Delta M_i = M_{i+1} - M_i$, with a separation between them of value d, can be calculated from the number of atoms in each face (M_i):

$$\Delta M_i = \frac{1}{(ca)^2}\left[(r+d)^2 - r^2\right] \tag{29.4}$$

Replacing the value of r, and rearranging terms, it is obtained that:

$$\Delta_i M = M_0\left(1 + 2\sqrt{\frac{M_i}{M_0}}\right), \qquad M_0 = \left(\frac{d}{ca}\right)^2 \tag{29.5}$$

where M_0 represents the minimum value that M can adopt once defined the interlayer spacing d.

Equation (29.5) is an integer equation where M_i and $\Delta_i M$ are integers and the following procedure can be used to find all possible solutions. M_0 (Equation (29.5)) is defined as the nearest possible integer number corresponding to the interlayer distance that is of interest. This value can be decomposed in such a way that $M_0 = p^2 M_0'$, where p and M_0' are positive integers and p^2 is the maximum square integer that can be obtained for the decomposition of M_0. General solutions will be obtained when $M_k = q^2 M_0'$, where q is a positive integer ($q = 1, 2, 3, \ldots$) and k a generic sub-index. Each of these general solutions M_k can be used as the first value M_1 of different sequence series with the same layer spacing. To obtain a complete sequence M_i, Equation (29.5) is used in a recursive form: from the first value M_1 it can be obtained $M_{i+1} = M_i + \Delta_i M$. From the q structures there will be p possible sequences that correspond to interlayer spacing d; others will only be different in the starting N_1. Particular solutions of Equation (29.5) for values M_0 noninteger are discarded because they do not generate integer sequences for the next layers. Shell structures with constant spacing can be expressed as:

$$M_i = M_0\left(i + \frac{q}{p} - 1\right)^2 \tag{29.6}$$

Constant interlayer spacing is possible when this equation is satisfied with $d = ca\sqrt{M_0}$.

The quantity of atoms in each face was only considered, and it was not necessary to use the total quantity of atoms N that is represented in Table 29.1 for the different polyhedra. From these values, the quantity of atoms in each shell can be known.

The constant interlayer spacings (Equation 29.5) that are nearest to the experimental XRD parameter of nanocapsules (0.342 nm) for the different truncated polyhedra — icosahedron, pentagonal bipyramid, octahedron, triangular bipyramid, and tetrahedron — correspond, respectively, to allowed M_0 in excess values of 4, 12, 12, 27, and 48. The interlayer spacings expressed in nanometers are, respectively, 0.3720, 0.3562, 0.3480, 0.3480, and 0.3480. None of these values match exactly to the observed interlayer spacing value and probably other polyhedra will also not match. The stacking of layers obtained from the — touching all faces inscribed sphere — polyhedra correspond to displaced layers (not rotated) and one of them, the icosahedron ($M_0 = 4$) has the same stacking as in graphite, that can be considered as the preferred orientation. During a detailed observation of HRTEM photographs, it is usual to find that at the vertices of nanocapsules, the interlayer spacing is increased. Generally, it is natural to think that steric effects will make match the number of atoms in each layer such that van der Waals energy is minimized. From these calculations, it can be inferred that interlayer spacing in nanocapsules is not strictly a constant as in regular crystallographic structures and that stacking corresponds to rotated layers.

III. ROUNDED CONCENTRIC STRUCTURES

Electron beam irradiation of multiwall nanotubes [19] and diamond [20], or heat treatment of carbon soot [21] can produce spherical, concentric wholly carbon shell structures. The onion like structures can have variable interlayer spacings [22], defects [23] or can probably be stable with constant interlayer spacing [24]. The formation mechanism of onion structures is a process in which atom extraction [25] forms vacancies at the external layers that migrate to the interior due to a stress-generating driving force [24]. The compression of the interior shells can transform them into diamond [22], produce a very defective particle, destroy the onion, or with controlled annealing conditions of tension relieving, a stable onion can reach an equilibrium condition. These various stages of mechanism controlled by the atom-extraction rate, diffusion of vacancies, and annealing conditions are the primary reasons why onions have not yet been obtained in high yields. High-current electron beam irradiation of carbon soot [26] and particular methods like arc discharge between two graphite electrodes submerged in deionized water [27] have been proposed as adaptable for onion mass production.

Similar conditions of carbon atoms extraction can be attained in the case of polyhedrical nanocapsules with other material inside [28] and the rounded vertex particles can have a stable structure [8,29] (Figure 29.9). In Figure 29.10 are shown the XRD spectra of uranium and gadolinium dicarbide nanocapsules in the region of (0,0,2) interlayer spacings. Deconvolution of peaks shows that there are three interlayer spacings that correspond to graphite, polyhedrical nanocapsules, and rounded vertex nanocapsules. The values of interlayer spacing for the latter are, respectively, 0.3374 ± 0.0002 and 0.3372 ± 0.0002 nm. Graphite has smaller interlayer spacing (0.3354 nm) and polyhedrical nanocapsules have the biggest (0.342 nm).

Calculations can be performed in spherical structures such as to simulate mean observed interlayer spacing (0.3373 nm) and infer the quantity of atoms in each shell of radius r. If the surface density ρ is constant [30], the quantity of atoms in each shell will be $N = 4\pi\rho r^2$. This equation is similar to Equation (29.3) and the same reasoning as before can be used to resolve the quantity of atoms N_i in consecutive layers with a constant d spacing:

$$\Delta_i N = N_0\left(1 + 2\sqrt{\frac{N_i}{N_0}}\right), \qquad N_0 = 4\pi\rho d^2 \tag{29.7}$$

The only condition imposed over Equation (29.7) is that closure conditions can be attained only with an even number of atoms [10]. The nearest even integer, N_0, matching the mean experimental value of interlayer spacing (0.3373 nm) and the atomic surface density of graphite planes

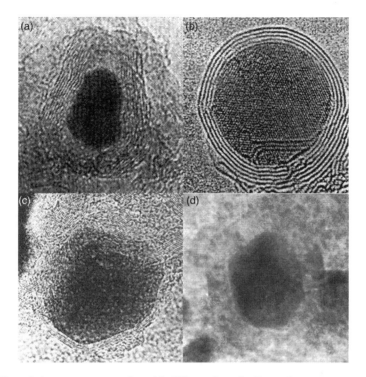

FIGURE 29.9 Rounded vertex nanocapsules with different kernels. For scale purposes, graphitic interlayer spacing is 0.34 nm. (a) Au. (From Ugarte, D., *Chem. Phys. Lett.* 209 (1–2), 701, 1993. With permission. © Elsevier Science Publishers B.V.) [28] (b) Co. (From Banhart, F., et al. *Chem. Phys. Lett.*, 292, 555, 1998. With permission. Elsevier Science B.V.) (c) Gd_2C_3. (From Majetich, S.A., Artman, J.O., Mc Hentry, M.E., Nuhfer, T., and Staley, S.W., *Phys. Rev. B* 48 (22), 16846, 1993. With permission. © The American Physical Society.) [31] (d) UC_2. (From Pasqualini, E.E., Unpublished.)

$(\rho_g = 4\sqrt{3}/(3a^2) = 38.13$ atoms/nm$^2)$ is 54. With this value of N_0, surface density is lower than in graphite and has to be 37.77 atoms/nm^2. Since N_0 can be expressed as $N_0 = p^2 N_0'$ with $p = 3$ and $N_0' = 6$, there will be three different sequences of the quantity of atoms per shell. In Table 29.2, are represented the quantity of atoms in the first five shells of these sequences obtained by resolving the equation $N_i = 54(i + q/3 - 1)^2$, similar to Equation (29.6).

The quantity of atoms in each shell does not imply anything about how they are arranged. If only 12 pentagons are used to satisfy Equation (29.1) the overall structure will be faceted [32]. Several attempts have been made to obtain local minimum energy structures [33], on fullerene analysis with holes in the zone of pentagons, or on rounding vertex with the elimination and rotation of pairs of carbon atoms producing pentagon–heptagon pairs [34]. It is possible to reduce the curvature along a pentagonal vertex with the introduction of five pentagon–heptagon pairs surrounding it (Figure 29.11). These transformations involve the elimination of ten atoms, produce a nearly planar structure, maintain the fivefold symmetry, still the structure obeys Euler's closure condition; from *ab initio* calculations it is demonstrated that there is practically no energy penalty since the binding energy per atom is very similar and strain is lower [29]. Pentagon–heptagon pairs can reduce overall strain energy and their presence is compatible with the lower surface density of shells in rounded structures compared with that of graphite. The presence of heptagons can facilitate the diffusion of some elements through this type of carbon layers.

As an overview of rounded concentric structures, it can be said that they have been detected as wholly carbon nanoparticles like onions, with and without a hole in the center, or a kernel of other material inside. They are the consequence of successive elimination of pair of carbon atoms from

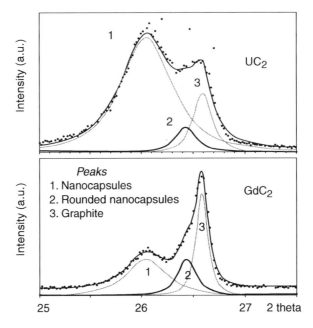

FIGURE 29.10 XRD of rounded vertex carbon nanocapsules of uranium and gadolinium dicarbides showing the deconvolution of the interlayer (0,0,2) peaks of (1) nanocapsules, (2) rounded vertex nanocapsules, and (3) graphite.

multiwall nanotubes or polyhedrical nanocapsules, and have a lower atomic surface density than in graphite layers, owing to the presence of pentagon–heptagon pairs.

IV. FORMATION KINETICS

Several materials such as pure metals, alloys, and carbides, have been nanoencapsulated in carbon (Figure 29.2). In the formation kinetics from a very high temperature, in which carbon and other elements are in the vapor phase, different situations can be present during the cooling process depending on the thermodynamic properties of the elements and their reactive capabilities. These situations are directly related to what kind of liquid phase is formed, how is the solidification process, and what are the transformations in condensed phases. For example, high-vapor pressure elements that do not form carbides have not been encapsulated in carbon nanocapsules [11,13,35,36]. Gold has been nanoencapsulated [28], but when heated by the electron beam of a HRTEM, diffuses through rounded carbon shells as a whole and is removed from the interior.

TABLE 29.2
Quantity of atoms N_i in consecutive shells of onion like structures with an interlayer spacing of 0.3373 nm and an atom surface density of 37.77 atoms/nm^2

i	$q = 1$	$q = 2$	$q = 3$
1	6	24	54
2	96	150	216
3	294	384	486
4	600	726	864
5	1176	1176	1350

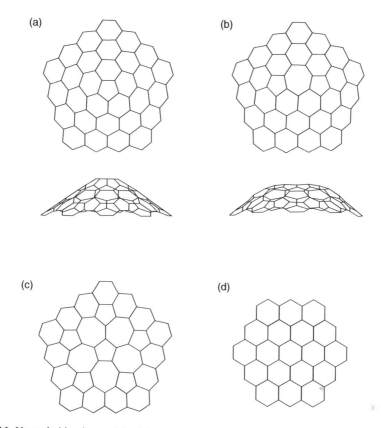

FIGURE 29.11 Up and side views of fivefold symmetry bowls (polyhedrical angles) with (a) one pentagon (80 atoms) and (b) one pentagon and a pair pentagon–heptagon (78 atoms). A fivefold planar structure is obtained in (c) with one pentagon and five pentagon–heptagon pairs (70 atoms). Structures where optimized using density functional theory (DFT) and structure (d) with all hexagons (54 atoms) was used as reference.

At least two different situations can be analyzed in the formation of carbon nanoencapsulation of materials. The first one is, as in the example of gold, where carbon and the other elements present do not form any compound or have very low carbon solubility. A possible scenario is that embryo clusters of an alloy or pure element are formed and nucleate as liquid droplets. Meanwhile, carbon atoms interact independently very fast in the usually present buffer inert gas. In such conditions, the evolution of carbon species will go through the formation of linear chains, ring structures, poly-aromatic carbon — graphene sheets — and even fullerenes [37]. One of the biggest energy barriers to be surpassed in this carbonaceous genesis is the cyclization of linear chains that can be catalyzed by the presence of the metallic elements. Polyaromatic carbons can have a dipolar moment with a perpendicular component when they are curved (bowl shape) and be paramagnetic when free radicals are present. Dipolar moment can favor the stacking of polyaromatic carbon, and para-magnetism can align them in a high local magnetic field as in the case of multiwall carbon nanotubes. At an appropriate temperature during the cooling stage, this graphene sheets can stick to the nanoscopic condensed phases, even stopping the growth of the metallic cluster size. If excess of carbon and adequate annealing conditions are present, graphene sheets will align them-selves forming graphite crystallites (Figure 29.10). When the thickness of the graphitic capsule is similar to the size of the crystallites, annealing can continue by the rotation of these crystallites con-necting the dangling bonds of their periphery. This process is very exothermic and could proceed very quickly so as to complete the bonding through all the graphitic coverage and reduce the stick-ing of more carbonaceous material. This mechanism can explain why polyhedrical nanocapsules does not have more than 20 or 30 layers, probably something similar as graphite crystallites have. This process of crystallites orientation and closure boundary conditions can be such that it leaves an excess volume inside the capsule, and can also be an explanation for the formation of the observed gap between kernel and capsule.

The other mechanism of nanocapsules formation is when carbon can be in solution in the presence of other liquid elements. The phase diagram [38] is such that as the solution solidifies, graphite is segregated towards the first cooling zone, that is, the external surface of the nanoscopic droplet. Carbon can be retained in solution or precipitated as a single carbide crystal inside the kernel. The segregated carbon can form graphite and as before, with appropriate annealing conditions, a complete coating of faceted closed layers.

Formation kinetics will depend upon the concentrations present in the vapor phase, properties of the encapsulated elements or compounds (such as carbon solubility and carbide formation), buffer gas conditions that controls cooling rates and the possibility of independent carbon condensation (polyaromatic carbon), and other process variables such as temperature and pressure. Catalytic effects can be present not only in the segregation of carbon in solution but also in the kinetic pathways of independent carbon species formed. Magnetic and electric fields can interact in these processes through the presence of paramagnetic (free radicals) and polar species.

During the formation process and especially if high temperatures are attained during an annealing process, carbon atoms can be eliminated from the condensed structures. This mechanism is a way in which filled multiwall nanotubes can surely be transformed into rounded vertices nanocapsules indicating an alternative pathway of formation. In the carbon pair elimination of shell structures, nanocapsules work as nanoscopic pressure vessels, and new allotropes or polymorphs with particular properties can be obtained.

Encapsulation conditions are not attained for elements that have high vapor pressure or compounds, like oxides that are reduced by carbon. For obtaining some control in the fabrication process, it is desirable to separate the different steps that can be present in the formation mechanisms involved and even reduce the temperatures needed for them. A possible strategy is to first produce the kernel nanoparticles, cover them with some poly aromatic hydrocarbon (PAH), pyrolize and anneal the overall structure. Pyrolisis temperatures are lower than 1000°C and hitherto many more solid materials with low melting points or low decomposition temperatures should be capable of being encapsulated. The hydrocarbon coating must not be greater than to form more than 50 shells, since bigger coatings do not produce continuous graphene layers. Experiments using catalysts manage similar conditions but they are intended for producing only carbon structures [39].

Nanocapsules have been extensively produced by the arc discharge method between graphite electrodes. The material to be encapsulated is charged in the anode by different methods that are usually described in each particular synthesis. Plasma torch, microwave sources [6], and water immersed arc discharge [40] — between others — are being used to increase yields and higher production capabilities. Technological efforts in producing nanocapsules are fundamentally important in the fields of magnetic [6] and nuclear materials [12]. The carbon capsule protects the interior material from environment and biological degradation. Since the layered structure is expected to have similar properties as highly oriented pyrolytic graphite (HOPG), transport properties are very different through the layers than parallel to them. For example, transport properties such as thermal conductivity and diffusion are more than 200 times higher in plane than through the layers [41,42]. Also, different techniques of nanoparticles self assembly [43] are of interest to obtain two-and three-dimensional structures with controlled organization.

V. PURIFICATION AND BIOLOGICAL COVERAGE

The produced material obtained from nanocapsules synthesis can contain fullerenes, empty and filled nanocapsules and nanotubes, nanoparticles embedded in amorphous carbon [44] or graphite crystallites [12], broken capsules, uncovered nanoparticles, and amorphous carbon. A general

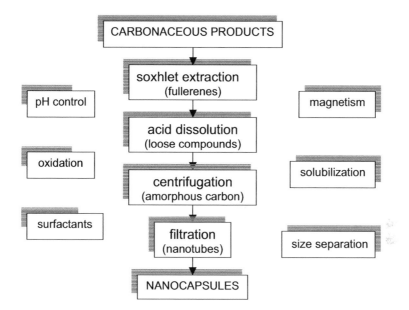

FIGURE 29.12 Nanocapsules purification flow chart and auxiliary techniques.

pathway can be mentioned to purify nanocapsules consisting of several steps (Figure 29.12). The first step usually is a soxhlet extraction of fullerenes in an appropriate solvent [45]. The next step is the dissolution of uncovered noncarbonaceous material by acid treatment [46], followed by a washing method. These two initial steps can reduce amorphous carbon present depending on solvents used, temperature, and pressure. A selective oxidation either in gas/liquid phase [47], or plasma etching [40] has as subject the dangling bonds of amorphous carbon, broken capsules, and graphite. The limit of this treatment is the attack of five-membered carbon rings — more reactive than graphite basal planes — that are present in nanocapsules [48]. Decantation, centrifugation, and different filtering techniques are being employed, if needed. Electrophoretic methods of separation could also be of great help.

The tendency of nonpolar nanocapsules dispersed in aqueous solutions is to further stick together by hydrophobic interactions. A stable colloidal suspension can be made for: size separation of nanocapsules and nanotubes, and eventually other carbonaceous material. Surface-active agents can modify the particles–suspending medium interface and prevent aggregation over long-time periods, thus overcoming the intrinsic instability of most liquid dispersions [49], which arises from van der Waals attractive forces between colloidal particles. They provide an additional repulsive force (electrostatic and steric), lower the surface energy, and change some rheological surface properties, which contribute to increase the stability of the colloidal suspension. In this case, semi-purified nanocapsules are dispersed by sonication in a surfactant, which is easily washed afterwards. The surfactant employed is sodium dodecyl sulfate (SDS) and better results are obtained when the SDS is slightly above the critical micellar concentration (CMC) [50] where molecules start to self-aggregate into micelles [51]. A colloidal suspension stabilized in water with SDS at a concentration of twice the CMC, takes part in the enhanced dispersion to perform a liquid-phase separation of carbonaceous material. The suspension is filtered in order to extract the larger particles, while leaving the smaller constituents in the filtrate. A typical laboratory suspension can be prepared from 500 ml distilled water, 2.5 g SDS, and 50 mg of semi-purified material strongly sonicated for 15 min. Sedimentation and centrifugation (at 5000 rpm for 10 min) removes particles larger than 500 nm from the dispersion. Best results are obtained with a two-step filtering through membranes (Millipore) of 0.6 and 0.2 μm porous size. More elaborated procedures can be designed by adapting methods using the advantages of column chromatography and vacuum filtration [52].

Solubilization of graphitic nanoparticles by functionalizing their surface [53], can surely help to separate them by size from nanotubes and uncovered noncarbonaceous particles. It can be used before the acidic treatment and its effectiveness will highly depend on the selective functionalization [54] of graphitic layers in comparison to amorphous carbon. Modification of some techniques used in the purification of single-wall nanotubes [55], by functionalization of their surface [56] and solubilization [57] can be adapted to nanocapsules. In the possibility of fluorinating nanocapsules as in the case of single-wall nanotubes, the damage introduced when fluorine atoms are driven off the graphitic surface [58] by heat treatment will probably not affect severely the multiwall material. Purification methods are strongly dependent on the compounds present in the carbonaceous material that are characteristic of different production methods. Physical properties of the nanoencapsulated material such as magnetism [31] or high density [8] can introduce shortcuts to purification pathways.

Biological species can be attached [59] directly to graphitic material [60,61], but a biological coverage that can simulate a cell membrane is a gateway for generating a much higher diversity for further biological functionalization. Protein immobilization and in particular antibodies attachment to this coverage can be used to reproduce biological functions such as cellular membrane trespassing or antigen–antibody coupling. The nanocapsules (nonpolar particles), separated by a modified method of size selection [62], and inserted in an aqueous solution can be covered with phospholipids (amphiphatic molecules). Tests were performed using 1,2 dipalmitoyl-*sn*-glycero-3 phosphocholine (DPPC) [63], because it is an integral part of biological membranes [64], and hitherto, nontoxic. When these molecules are placed in an aqueous dispersion of nanocapsules, they preferentially cover them with hydrophobic portions connected with the carbon surface. The phospholipid coverage works like a surfactant, avoiding the hydrophobic and van der Waals interactions that take place between the isolated carbon nanocapsules. The nanocapsules and the DPPC (10/1, wt/wt) in aqueous solution were sonicated for 10 min in a bath and then inserted in a rotary tube for 1 h in a thermostatic bath at 37°C. To observe this material by transmission electron microscopy (TEM), a drop of the dispersion was deposited in a carbon-coated grid. Other samples were prepared in the same way and immediately after the deposition of the dispersion in the grid, a drop of 2% ammonium molybdate solution was incorporated as a contrast agent that sticks to DPPC. In the samples without the contrast agent, nanocapsules were seen clearly. When observed with the molybdate, all nanocapsules were completely covered. DPPC concentration can control the quantity of phospholipids layers and minimum of empty vesicles.

Nanocapsules covered with phospholipids can be "functionalized" by attaching or inserting substances with particular properties, like monoclonal antibodies which can be used for targeting. Monoclonal antibodies produced in mammalian cell culture systems are becoming increasingly important as auxiliaries for the treatment of human diseases. Antibodies [65] attached to a phospholipid layer that cover radioactive nanocapsules can be transported by blood to a specific cell for binding to a tumor marker. Functionalization can also be used for self-assembly to obtain organized two- and three- dimensional structures [66].

VI. NUCLEAR APPLICATIONS

Nuclear applications of nanocapsules are related to the emitting physical properties of the encapsulated material. Emitted radiation can be electromagnetic of high energy (γ), electrons or positrons (β), alpha particles (^{4}He nucleus), or fission products [67]. These emitters can be in themselves radioactive or can be activated by a nuclear reaction, usually a neutron capture. The particular advantage of carbon nanocapsules in nuclear applications is related to the protective characteristics that the carbon capsule confers to the interior product. Experiments on irradiation of fullerenes have shown that knocked carbon atoms from one cage are found in another fullerene and even form dimers and trimers by a recoil-implantation mechanism [68]. The observed major damage of capsules in nanoencapsulated molybdenum irradiated in a nuclear reactor was produced by atoms' recoil and not by neutron bombardment [69]. Several isotopes can be activated [70], and nanoscopic sources obtained as in the case of nanoencapsulated cobalt transformed to ^{60}Co. Individual radioactive nanocapsules can be used as calibrated probes for diminishing contaminations (by interaction with other compounds).

Encapsulated radioactive emitters can be used in medical diagnosis when some particular benefits can be related to the encapsulation, as for example, preferential targeting. For example, best absorption in the lymphatic system seems to be achieved with particle size of 30–50 nm and medical application for studies and treatment involving the lymphatic system can be designed [71]. Preferential targeting can also be tailored by direct functionalization [72] of the capsule or functionalization in the surface of a biological cover. In order to use this covered nanocapsules as carriers for diagnosis or a treatment therapy, the biological material should not be recognized by the immunity system (B and T lymphocytes are the primary effectors of adaptive immune response). Nuclear nanocapsules can also be localized manually or, when combined with magnetic properties or other materials, localization can be enhanced by magnetic fields. Other properties of the encapsulated nuclear material, such as response to nuclear magnetic resonance (NMR), can also be used for biodistribution studies.

Boron neutron capture therapy (BNCT) is based on the nuclear reaction in which a thermal neutron and ^{10}B produces high-energy ^{7}Li and an α particle $-$ ^{10}B(n,α)^{7}Li. The requirements for using this nuclear reaction in therapeutic treatments involve selective localization and concentration of the radioactive isotope to maximize damage with a neutron flux innocuous to other tissue in the surroundings. Selectivity targeting can be enhanced with antibodies [73] attached to a biologically covered boron carbide nanoparticle. Trespassing a cellular membrane can be done with the help of other attached proteins and high boron concentrations can be localized near cell nucleus for neutron irradiation and ADN inhibalitation. High specificity and concentration of boron should be attained to maximize localized damage produced by the α-particle and ^{7}Li-reaction products whose penetration depth is approximately 10 μm.

Homogeneous reactors are being revisited for the production of medical radioisotopes [74–76]. The working principle is a solution of a fissile isotope that achieves criticality with the liquid volume control in a reactor and fission products are extracted from the circulating solution. Dispersed nanocapsules containing uranium dicarbide [77], using low enriched uranium (LEU, less than 20% ^{235}U), can have benefits in the separation process since the high energy fission products will pass through the nanocapsules and will be the only products in solution. Nanocapsules are extremely resistant to acidic media and benefits can be achieved in separation rates.

Size dimension of nanocapsules with fissile material makes them attractive for their usage as fuel for nuclear reactors. The high surface–volume ratio allows very quick extraction of heat from the interior to an out flowing gas, such as in a high temperature gas cooled reactor [78]. Fission products can be implanted in surrounding walls or nanocapsules could be sintered so as to retain fission products in the assembled structure (Figure 29.13). Extruded rods to form "hairs" or "needles" with a nucleus of 50 μm diameter and a cladding of 25 μm should not present problems in fabrication. Heat will be mainly transported parallel to graphitic layers that have extremely high thermal conductivity. An array of parallel pins can be assembled such that the cladding surface is more than ten times greater than that in conventional fuels. The main goal of this type of fuel is to reduce the fuel size of power reactors, especially for using them in ships, submarines, and rockets.

Proposals regarding nuclear waste nanoencapsulation are fundamentally oriented to incorporate an additional barrier for avoiding solubilization of radioactive materials in underwater streams in the case of leakage of the usual containers. Generally, the nuclear waste final deposition involves several materials, and nanoencapsulation at the time, is guided to work practically with unique compounds. Although, plutonium nanoencapsulation has not been reported, it is interesting to notice that if nanoencapsulation can be performed, chemical toxicity of this element will be highly diminished.

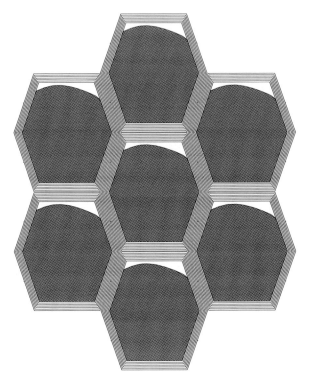

FIGURE 29.13 Sintered nanocapsules for nuclear fuel applications.

Nuclear applications of carbon nanocapsules with radioactive materials inside are in the very early steps of investigation. Because of time-consumming experiments, post-irradiation analysis, protocols, complex interdisciplinary work, and the usual requirement of big facilities to be employed, make advances very punctual. Feasibility studies are usually done to a deeper stage than in other areas of activity to balance research investment. Surely, results will be as promising as in other disciplines of nanotechnology.

REFERENCES

1. Ruoff, R.S., Lorents, D.C., Chan, B., Malhotra, R., and Subramoney, S., Single crystal metals encapsulated in carbon nanoparticles, *Science*, 259, 346–348, 1993.
2. Tomita, M., Saito, Y., and Hayashi, T., LaC$_2$ Encapsulated in graphite nano-particle, *Jpn. J. Appl. Phys.*, 32 (2, 2B), L280–L282, 1993.
3. Ruoff, R.S., Lorents, D.C., Malhotra, R., and Dyer, M.J., Carbon nanoencapsulates, US Patent 5,547,748 (August 20, 1996).
4. Saito, Y., Nanoparticles and filled nanocapsules, *Carbon*, 33, 979–988, 1995.
5. Zhou, O., Fleming, R.M., Murphy, D.W., Chen, C.H., Haddon, R.C., Ramirez, A.P., and Glarum, S.H., Defects in carbon nano-structures, *Science*, 263, 1744–1747, 1994.
6. McHenry, M.E. and Subramoney, S., Synthesis, structure, and properties of carbon encapsulated metal nanoparticles, in *Fullerenes; Chemistry, Physics and Technology*, Kadish, K.M. and Ruoff, R.S., Eds., Wiley-Interscience, New York, 2000 (Chapter 19), pp. 839–885.
7. Withers, J.C., Loufty, R.O., and Lowe, T.P., Fullerene commercial vision, *Full. Sci. Technol.*, 5 (1), 1–31, 1997.
8. Pasqualini, E.E., Nuclear nanocapsules and curved carbon structures, *Carbon*, 35, 783–789, 1997.
9. Fujita, M., Saito, R., Dresselhaus, G., and Dresselhaus, M.S., Formation of general fullerenes by their projection on a honeycombe lattice, *Phys. Rev. B*, 45 (23), 13834–13836, 1992-I.

10. Fowler, P.W. and Manolopoulos, D.E., *An Atlas of Fullerenes.* Clarendon Press, Oxford, 1995.

11. Saito, Y., Okuda, M., Yoshikawa, T., Kasuya, A., and Nishina, Y., Correlation between volatility of rare-earth and encapsulation of their carbides in carbon nanocapsules, *J. Phys. Chem.*, 98, 6696–6698, 1994.

12. Pasqualini, E.E., Adelfang, P., and Nuñez Regueiro, M., Carbon nanoencapsulation of uranium dicarbide, *J. Nucl. Mater.*, 231, 173–177, 1996.

13. Funasaka, H., Sugiyama, K., Yamamoto, K., and Takahashi, T., Synthesis of actinide carbides encapsulated within carbon nanoparticles, *J. Appl. Phys.*, 78 (9), 5320–5324, 1995.

14. Stone, A.J. and Wales, D.J., Theoretical studies of icosahedral C_{60} and some related species, *Chem. Phys. Lett.*, 128, 501, 1986.

15. Murry, R.L., Strou, D.L., Odom, G.K., and Scuseria, G.E., Role of sp^3 carbon and 7-membered rings in fullerene annealing and fragmentation, *Nature*, 366, 665–667, 1993.

16. Doyle, T.E., and Dennison, J.R., Vibrational dynamics and structure of graphitic amorphous carbon modeled using an embedded-ring approach, *Phys. Rev. B.*, 51 (1) 196–200, 1995-I.

17. Pasqualini, E.E., Mesaros, M., Navarro, G., and Saraceno, M., Nanocapsules: structure and formation kinetics, in *Fullerenes: Recent Advances in the Chemistry and Physics of Fullerenes and Related Materials*, Vol. 4, Ruoff, R. and Kadish, K. Eds., The Electrochemical Proceedings Series, Pennington, NJ, 1997, pp. 843–854, (PV 97-14).

18. Dresselhaus, M.S., Dresselhaus, G., and Eklund, P.C., *Science of Fullerenes and Carbon Nanotubes*, Academic Press, San Diego, 1996.

19. Ugarte, D., Curling and closure of graphitic networks under electron-beam irradiation, *Nature*, 359, 707–709, 1992.

20. Qin, L., and Iijima, S., Onion-like graphitic particles produced from diamond, *Chem. Phys. Lett.*, 262, 252–258, 1996.

21. de Heer, W.A., and Ugarte, D., Carbon onions produced by heat treatment of carbon soot and their relation to the 217.5 nm interstellar absorption feature, *Chem. Phys. Lett.*, 207, 480–486, 1993.

22. Banhart, F., and Ajayan, M., Carbon onions as nanoscopic pressure cells for diamond formation, *Nature*, 382, 433, 1996.

23. Zwanger, M.S., and Banhart, F., The structure of concentric-shell carbon onions as determined by high-resolution electron microscopy, *Phil. Mag. B*, 72 (1), 149–157, 1995.

24. Pasqualini, E.E., Concentric carbon structures, *Phys. Rev. B*, 56 (13), 7751–7754, 1997-I.

25. Banhart, F., Füller, T., Redlich, P., and Ajayan, P.M., The formation, annealing and self-compression of carbon onions under electron irradiation, *Chem. Phys. Lett.*, 269, 349–355, 1997.

26. Osawa, E., Japanese Patent, personal communication.

27. Sano, N., Wang, H., Chhowalla, M., Alexandrou, I., and Amaratunga, G.A., Nanotechnology: synthesis of carbon 'onions' in water, *Nature*, 414, 506–508, 2001.

28. Ugarte, D., How to fill or empty a graphitic onion, *Chem. Phys. Lett.*, 209, 99–103, 1993.

29. Pasqualini, E.E. and Mesaros, M., Onions: stresses, topology and proposed structures, in Proceedings of the 191th Meeting of the Electrochemical Society, Montreal, Canada, May 4–9, 1997.

30. Qiu, L.C. and Iijima, S., Spherical carbon clusters obtained from diamond, in *Fullerenes: Recent Advances in the Physics and Chemistry of Fullerenes and Related Materials*, Vol. 3, Ruoff, R. and Kadish, K., Eds., The Electrochemical Proceedings Series, Pennington, NJ, 1996, pp. 661–672, (PV 96-10).

31. Majetich, S.A., Artman, J.O., Mc Henry, M.E., Nuhfer, T., and Staley, S.W., Preparation and properties of carbon-coated magnetic nanocrystallites, *Phys. Rev. B*, 48 (22), 16845–16848, 1993-II.

32. Xu, C.H., and Scuseria, G.E., An O(N) tight-binding study of carbon clusters up to C_{8640}: the geometrical shape of the giant icosahedral fullerenes, *Chem. Phys. Lett.*, 262, 219–226, 1996.

33. Maiti, A., Brabec, C.J., and Bernholc, J., Zero and Finite Temperature Study of Single Fullerene Cages and Carbon "Onions" — Geometry and Shape, *Mod. Phys. Lett. B*, 7 (29 & 30), 1883–1895, 1993.

34. Terrones, M., and Terrones, H., The role of defects in graphitic structures, *Full. Sci. Technol.*, 4 (3), 517–533, 1996.

35. Saito, Y., Synthesis and characterization of carbon nanocapsules encaging metal and carbide crystallites, in *Fullerenes: Recent Advances in the Chemistry and Physics of Fullerenes and Related Materials*, Kadish, K. and Ruoff, R. Eds., The Electrochemical Proceedings Series, Pennington, NJ, 1994, pp. 1419–1447, (PV 94-24).

36. Majetich, S.A., Scott, J.H., Brunsman, E.M., Kirkpatrick, S., McHenry, M.E., and Winkler, D., Carbon arc nanoparticle growth and magnetism, in *Fullerenes: Recent Advances in the Chemistry and Physics of Fullerenes and Related Materials*, Ruoff, R. and Kadish, K. Eds., The Electrochemical Proceedings Series, Pennington, NJ, 1995, pp. 584–598, (PV 95-10).

37. Pasquali, E.E., and Lopez, M., Fullerene formation kinetics in carbon vapors, in *Fullerenes, The Exciting World of Nanocages and Nanotubes*, Vol. 12, Kamat, P.V., Guldi, D.M., and Kadish, K.M., Eds., The Electrochemical Proceedings Series, Pennington, NJ, 2002, pp. 815–820, (PV 2002-12).

38. Massalski, T. B., and Okamoto, H., *Binary Alloy Phase Diagrams*, 2nd ed, ASM, Materials Park, OH, 1990.

39. José-Yacamán, M., Terrones, H., Rendón, L., and Domínguez, J.M., Carbon structures grown from decomposition of a phenylacetylene and thiophene mixture on Ni nanoparticles, *Carbon*, 33 (5), 669–678, 1995.

40. Ang, K.H., Alexandrou, I., Mathur, N.D., Lacerda, R., Bu, I.Y. Y., Amaratunga, G.A.J., and Haq, S., Scalable production of carbon encapsulated Ni nanoparticles by water arc discharge: structural and magnetic properties, *J. Metastable Nanocrys. Mater.*, 23, 87–90, 2005 (http://www.scientific.net).

41. Pierson, H.O., Handbook of Carbon, Graphite, Diamond and Fullerenes, Noyes Publications, 1993.

42. Dayton, R.W., Oxley, J.H., and Townley, C.W., in Proceedings of the Symposium on Ceramic-Matrix Fuels Containing Coated Particles, Div. Tech. Info., USAEC, 1962.

43. Li, M., Schnablegger, H., and Mann, S., Coupled synthesis and self-assembly of nanoparticles to give structures with controlled organization, *Nature*, 402, 393–395, 1999.

44. Saito, Y., Nishikubo, K., Kawabata, K., and Matsumoto, T., Carbon nanocapsules and single-layered nanotubes produced with platinum-group metals (Ru, Rh, Pd, Os, Ir, Pt) by arc discharge, *J. Appl. Phys.*, 80 (5), 3062–3067, 1996.

45. Holmes Parker, D., Chatterjee, K., Wurz, P., Lykke, M.J., Pellin, M.J., and Stock, L.M., Fullerenes and giant fullerenes: synthesis, separation, and mass spectrometric characterization, in *The Fullerenes*, Kroto, H.W., Fischer, J.E., and Cox, D.E., Eds., Pergamon Press, 1993, pp. 32–41.

46. Lobach, A.S., Spitsina, N.G., Terekhov, S.V., and Obraztsova, E.D., Comparative study of purification methods for single walled carbon nanotubes, *Phys. Solid State*, 44 (3), 475, 2002.

47. Bonard, J.M., Forró, L., Ugarte, D., de Heer, W.A., and Châtelain, A., Physics and chemistry of carbon nanostructures, *Euro. Chem. Chron.*, 3, 9–16, 1998.

48. Colbert, D.T., Zhang, J., McClure, S.M., Nikolaev, P., Chen, Z., Hafner, J.H., Owens, D.W., Kotula, P.G., Carter, C.B., Weaver, J.H., Rinzler, A.G., and Smalley, R.E., Growth and sintering of fullerene nanotubes. *Science*, 266, 1218, 1994.

49. Dávalos-Orozco, L.A., del Castillo, L.F., Hydrodynamic Behavior of Suspensions of Polar Particles, in *Encyclopedia of Surface and Colloid Science*, Hubbard, A.T., Ed., Marcel Decker Inc., 2002, pp. 2375–2397.

50. Wanless, E.J., and Drucker, W.A., *J. Phys. Chem.*, 100, 3207, 1996.

51. Hunter, J. R., Introduction to Modern Colloid Science, Oxford University Press, Oxford, 1993.

52. Holzinger, M., Hirsh, A., Bernier, P., Duesberg, G.S., and Burghard, M., A new purification method for single-wall carbon nanotubes (SWNTs), *Appl. Phys. A*, 70, 599–602, 2000.

53. Unger, E., Graham, A., Kreupl, F., Liebau, M., and Hoenlein, W., Electrochemical functionalization of multi-walled carbon nanotubes for solvation and purification, *Curr. Appl. Phys.*, 2 (2), 107–111, 2002.

54. Liu, C., Fan, Y., Liu, M., Cong, H.T., Cheng, H.M., and Dresselhaus, M.S., Hydrogen storage in single-walled carbon nanotubes at room temperature, *Science*, 286, 1127–1129, 1999.

55. Georgakilas, V., Voulgaris, D., Vazquez, E., Prato, M., Guldi, D.M., Kukovecz, A., and Kuzmany, H., Purification of HiPCO carbon nanotubes via organic functionalization, *J. Am. Chem. Soc.*, 124 (48), 14318–14319, 2002.

56. Chen, Y., Haddon, R.C., Fang, S., Rao, A.M., Ekund, P.C., Lee, W.H., Dickey, E.C., Grulke, E. A., Pendergrass, J.C., Chavan, A., Haley, B.E., and Smalley, R.E., Chemical attachment of organic functional groups to single-walled carbon nanotube material, *J. Mater. Res.*, 1998, 13 (9), 2423–2431.

57. Zhao, B., Hu, H., Niyogi, S., Itkis, M.E., Hamon, M.A., Bhowmik, P., Meier, M.C., and Haddon, R.C., Chromatographic purification and properties of soluble single-walled carbon nanotubes. *J. Am. Chem. Soc.*, 123 (47), 11673–11677, 2001.

58. Gu, Z., Peng, H., Hauge, R.H., Smalley, R.E., and Margrave, J.L., Cutting single-wall carbon nanotubes through fluorination, *Nano Lett.*, 2 (9), 1009–1013, 2002.

59. Kaialstz, E., Shipway, A.N., and Willner, I., Chemically functionalized metal nanoparticles, in *Nanoscale Materials*, Liz-Marzán, L.M. and Kamat, P.V., Eds., Kluwer Academic Publishers, 2003, Chapter 2, pp. 5–78.

60. Davies, J.J., Green, M.L.H., Hill, H.A.O., Leung, Y.C., Sadler, P.J., Sloan, J.J., Xavier, A.V., and Tsung, S.C., The immobilization of proteins in carbon nanotubes, *Inorg. Chim. Acta*, 272, 261–266, 1998.

61. Jiang, K., Schadler, L.S., Siegel, R.W., Zhang, X., Zhang, H., and Terrones, M., Protein inmmobilization on carbon nanotubes via a two-step process of diimide-activated amidation, *J. Mater. Chem.*, 14, 37–39.

62. López, M., Manpower Fellowship — IAEA, ARG/01039. Clarkson University, Postdam, NY, USA, 2002.

63. López, M., Pasqualini, E.E., and Partch, R., Phospholipid coverage of carbon aggregates, in Proceedings of the 203rd Meeting of The Electrochemical Society, Paris, France, April 27–May 2, 2003.

64. Alberts, B., Johnson, A., Lewis, J., Raff, M., Roberts, K., and Walter, P., *Molecular Biology of the Cell*, 4th ed. 2002, Garland Science, NY, 2002, pp. 584–593.

65. Nolan, O. and O'Kennedy, R., Bifunctional antibodies: concept, production and applications, *Biochim. Biophys. Acta*, 1040 (1), 1–11, 1990.

66. Li, M., Schnablegger, H., and Mann, S., Coupled synthesis and self-assembly of nanoparticles to give structures with controlled organization, *Nature*, 402, 393–395, 1999.

67. Pfennig, G., Klewe-Nebenius, H., and Seelmann-Eggebert, W., Karlsruher Nuklidkarte. Druckhaus Haberbeck, D-32791 Lage, Lippe, 6th ed., 1995.

68. Ohtsuki, T., Masumoto, K., Kikuchi, K., and Sueki, K., Production of radioactive fullerene families using accelerators, *Mater. Sci. Eng.*, 217/218, 38–41, 1996.

69. Kasuya, A., Takahashi, H., Saito, Y., Mitsugashira, T., Shibayama, T., Shiokawa, Y., Satoh, I., Fukushima, M., and Nishina, Y., Neutron irradiation on carbon nanocapsules, *Mater. Sci. Eng.*, A217/A218, 50–53, 1996.

70. Braun, T., and Rausch, H., Radioactive endohedral metallofullerenes formed by prompt gamma-generated nuclear recoil implosion, *Chem. Phys. Lett.*, 288, 179–182, 1998.

71. Kolari, P.J., Maaranen, P., Kauppinen, E., Joutsensaari, J., Jauhiainen, K., Pelkonen, K., and Rannikko, S., Nanoparticles in biomedical applications: experiments with activated carbon, fullerene and ferrous oxide, *Med. Biol. Eng. Comput.*, 34 (1), 153–154, 1996.

72. Davis, J.J., Green, M.L.H., Hill, H.A.O., Leung, Y.C., Sadler, P.J., Sloan, J.J., Xavier, A.V., and Tsang, S.C., The immobilization of proteins in carbon nanotubes, *Inorg Chim. Acta.*, 272, 261–266, 1998.

73. Hawthorne, M.F., Biochemical applications of boron cluster chemistry, *Pure Appl. Chem.*, 1993, 63, 327.

74. Ball, R.M., Pavshook, V.A., and Khvostionov, V.Y., Present status of the use of LEU in aqueous reactors to produce Mo-99, in Proceedings of the Internatonal Meeting on Reduced Enrichment for Research and Testing Reactors, RERTR, 1998.

75. Ball, R.M., Characteristics of nuclear reactors used for the production of Mo-99. Production Technologies for Mo-99 and Tc-99m. IAEA -TECDOC-1065, February 1999, TECDOC 1065, IAEA, 1999.

76. Manzini, A.C., Progress on fission radioisotopes production in Argentina, in Proceedings of the International Meeting on Reduced Enrichment for Research and Testing Reactors, RERTR, Vienne, Austria, November, 7–12, 2004.

77. Pasqualini, E.E., Adelfang, P., and Nuñez Regueiro, M., Carbon nanoencapsulation of uranium dicarbide, *J. Nucl. Mater.*, 231, 173–177, 1996.

78. Proceedings of the Symposium on Ceramic-Matrix Fuels Containing Coated Particles, *Div. Tech. Info.*, USAEC, 1962.

30 Bioencapsulation in Polymer Micro- and Nanocarriers and Applications in Biomedical Fields

Elena Markvicheva
Shemyakin and Ovchinnikov Institute of Bioorganic Chemistry,
Moscow, Russia

CONTENTS

I. INTRODUCTION

Over the past several decades bioencapsulation has become one of the most promising techniques in various biomedical fields. Bioencapsulation is considered as special techniques for preparation of different polymer systems (hydrogel micro- and nanoparticles, as well as microparticles or micracapsules, etc.) with entrapped biomaterial, such as biologically active compounds (BAC) or alive cells. As BAC we can use proteins, including enzymes and high-molecular weight hormones, peptides, DNA and oligonucleotides, low-molecular weight antibiotics, and other drugs. Encapsulated cells could be of different origin: micro-organisms, plant or animal cells, but we would focus here only on animal cells. In this review, we would like to discuss several biomedical fields where various natural and synthetic polymers are widely employed to prepare polymer matrices with bioencapsulated biomaterial. In our opinion, a list of the most interesting and promising biomedical fields where nano- or microcarriers are used is as follows:

1. Controlled drug release delivery systems based on nano- or microparticles/microcapsules.
2. Development of new vaccines using nano- and microparticles/microcapsules.
3. Elaboration of novel DNA delivery nanocarriers to prepare genetically modified cells.
4. Somatic gene therapy using implanted microcapsules with entrapped genetically modified animal cells producing various therapeutic agents (proteins, peptides, etc.).

In this chapter, I would like to review polymer materials and bioencapsulation techniques focusing on the analysis of their merits and disadvantages, and to discuss some biomedical applications of nano- and microcarriers (both particles and capsules) with entrapped biomaterial, which have been recently reported in the literature.

II. PREPARATION OF MICRO- AND NANOCARRIERS WITH ENCAPSULATED ANIMAL CELLS OR BIOLOGICALLY ACTIVE COMPOUNDS

A. POLYMER MATERIALS FOR BIOENCAPSULATION OF BIOMATERIALS

Polymer nano- or microparticles/microcapsules used in biomedical fields are fabricated from bio-compatible biodegradable polymers. The list of these materials includes both natural materials, such as natural polysaccharides (alginate, cellulose, and chitosan as well as their many derivatives, etc.) and synthetic polymers. Natural polymers are attractive for cell immobilization due to their abundance and apparent biocompatibility. Synthetic polymers are used because of the high degree of control over the structure. Unlike natural materials, whose properties can vary from batch to batch, synthetic polymers provide highly consistent starting materials for any encapsulation technique.

As for animal cell entrapment in hydrogel microparticles or microcapsules, encapsulation procedure should proceed under physiological conditions within a short time (20–30 min), in order to provide cell viability, and to be as simple as possible because all manipulations are carried out under strictly sterile conditions. Taking into account all these requirements, it should be noted that the list of polymer materials and methods for animal cell encapsulation is rather limited. So-called alginate-based carriers (microparticles, micro- and nanocapsules) assure the favorable polymer systems for animal cell immobilization.

Alginic acid is an unbranched binary copolymer of 1–4 glycosidically linked α-L-guluronic acid (G) and its C-5 epimer β-D-mannuronic acid (M). The M- and G-residues are joined together in a block-wise fashion. Chemical and physical properties of alginate, entrapment of cells in alginate beads, and biocompatibility and selection of alginates for biomedical use are reviewed extensively by Melvik et al. [1]. The salts (and esters) of these natural polysaccharides are generally named alginates.

Alginate–calcium chloride systems are based on interfacial precipitation. Since 1980 when Lim and Sun proposed an original technique based on coating Ca-alginate beads with entrapped animal cells with poly-L-lysine (PLL) and subsequent dissolving Ca-alginate gel core with sodium citrate [2], many modifications of this technique have been studied. The list of these modifications includes coating of a polycation with an additional layer of alginate or another polyanion [3], formation of a double-membrane capsule [4] or a multiplayer capsule, as well as the replacement of PLL with other polycations [5].

(1-4)-linked 2 amino-2-deoxy-D-glucans (chitosans) are a family of natural biodegradable and biocompatible cationic polysaccharides commercially produced by the partial deacetylation of chitin from the reprocessing of seafood waste. Members of the chitosan family differ in terms of their molecular weight and deacetylation degree. Most chitosans are soluble only at acidic pH range which is the disadvantage, since encapsulating cells and tissues require physiological pH.

Chitosans of very low molecular weight (MM 3000–10000 Da), namely oligochitosans, are of great interest, since their usage leads to a shift in the pKa providing solubility and alginate–oligocitosan membrane formation at neutral pH [6]. A direct comparison showed that the toxic effects of PLL are significantly higher than that of chitosan [7]. PLL causes a necrotic cell death in higher concentrations, whereas in low concentrations it stimulates monocytes to tumor necrosis factor (TNF) production which has potent proinflammatory activities, increasing fibroblast proliferation, that is supposed to be the link to the observed fibrotic overgrowth of implanted microcapsules [8]. Coating a polycation with an outer polyanion layer or with a poly(ethylene glycol) (PEG) increases the biocompatibility of the capsules [9]. Calafiore et al. use poly-L-ornithine (PLO) instead of PLL. This polycation differs from PLL by a hydrophobic group (pendant group shorter by one CH$_2$ group), and in combination with a high Malginate (61%), it provides a tighter permeability and a more stable complex than the PLL-based one in alginate/PLL/alginate microcapsule [10].

Among synthetic polymers, we would like to mention polyamino acids, namely, PLL, PLO, poly(α-hydroxy esters), in particular poly(D,L-lactic acid) (PLA), poly(D,L-glycolic acid) (PGA) and their copolymers, namely, widely used poly(lactide-co-glycolide) (PLGA). Formed by a ring-opening polymerization, these polymers degrade via nonspecific hydrolysis of the ester linkage. The degradation products (lactic and glycolic acids) are naturally metabolized and easily cleared by the body. Microparticles based on poly(α-hydroxy esters), for instance PLA or PLGA are widely employed for encapsulation of peptides, proteins, DNA, and other compounds. The structure of poly-D,L-lactide-co-glycolide is shown in Figure 30.1.

As for polymer materials used for encapsulation of some low molecular weight drugs, such as cytostatics, antibiotics, etc., which are more stable compared with animal cells or biopolymers (DNA, proteins, peptides, etc.), the list of polymer materials used for their encapsulation is more extended. It includes a series of synthetic polymers as well as their copolymers with natural polymers. We could mention polyacrylates, for example, polyalkylcyanoacrylates [11], polyester–poly(ethylene glycol) [12], poly(epsilon-caprolactone) [13], polyethylenimine-dextran sulfate [14], dextran–HEMA (hydroxy-ethyl-methacrylate) [15].

B. Bioencapsulation Techniques

In order to keep alive cells or to avoid denaturation of proteins, DNA or other nonstable BAC, the methods for preparation of nano- or microparticle-based systems should be rather soft providing encapsulation procedure under physiological conditions (pH, temperature, etc.). We cannot describe in detail all methods presently used for bioencapsulation in nano- and microparticles (microcapsules) but would like to discuss some of them which are presently widely employed, or are rather promising to be used in the nearest future. All presently existing techniques for nano- and microcapsule production for biotechnological and biomedical applications are reviewed by Ian Marison et al. [16].

1. Bioencapsulation of Animal Cells in Alginate-Polycation Microcapsules

The scheme of one simple technique presently used for animal cell encapsulation into alginate-polycation (oligochitosan) microcapsules is shown in Figure 30.2 [17]. As well known, alginate and chitosan are oppositely charged polymers.

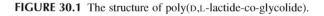

FIGURE 30.1 The structure of poly(D,L-lactide-co-glycolide).

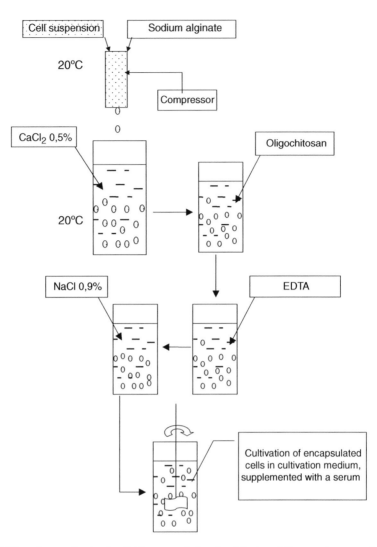

FIGURE 30.2 The scheme of animal cell bioencapsulation in alginate–chitosan microcapsules.

Cells are mixed with alginate solution and the mixture is dispersed dropwise into $CaCl_2$ solution to obtain insoluble hydrogel calcium alginate beads with encapsulated cells. After washing, the beads are transferred into chitosan solution, where an alginate–chitosan membrane on the surface of the beads is formed. After this a calcium alginate core can be dissolved using the chelating agent EDTA, and after washing with a physiological solution and then with a cultivation medium, one can get microencapsulated cells which can be cultivated *in vitro* or implanted into a patient. To provide a stable polyelectrolyte membrane (with the thickness of 30–100 μm) low molecular weight chitosan (oligochitosan, MM 3000–6000 Da) is used in this technique. Positively charged small oligochitosan molecules can easily diffuse into alginate matrix by cross-linking negatively charged alginate molecules and forming polyelectrolyte alginate–oligochitosan membrane. Instead of oligochitosan, PLL or PLO are used as polycations in this method. The obtained microcapsules can be coated with an outer alginate layer, in order to improve biocompatibility.

There are several nozzle extrusion methods for production of microbeads or microcapsules with entrapped cells, namely, simple dripping, concentric air jet, electrostatic droplet generation, rotating atomizers, and vibrating nozzle technique. All merits and disadvantages of these techniques were summarized and discussed earlier [18]. Among the different extrusion methods, an electrostatic extrusion technique that allows production in sterile conditions of beads less than 300 μm with a narrow size distribution is mentioned [19]. This technique is based on the use of electrostatic force to disrupt a liquid surface, in order to form a charged stream of fine uniform droplets. The technique was successfully used to encapsulate mammalian cells, such as hybridoma cells [20,21], islets of Langerhans [22], parathyroid cells [23], and human adenocarcinoma breast cells [17]. The growth of Chinese hamster ovary (CHO) cells within alginate–chitosan microcapsules is demonstrated in Figure 30.3.

2. Emulsion-based Techniques for Encapsulation of Low- and High-Molecular Weight Therapeutical Agents

An emulsion can be prepared by dispersing two immiscible liquids in the presence of a stabilizer or emulsifier. In the case of an aqueous core, it is called water-in-oil emulsion (w/o) and in a hydrophobic core phase is termed an oil-in-water emulsion (o/w). Emulsions are produced by adding a core phase to a vigorously stirred excess of a second-phase containing an emulsifier. The double emulsion technique, such as w/o/w is a modification of the basic method where by an emulsion is prepared using an aqueous solution in a hydrophobic wall polymer, for instance polyester. This emulsion is then poured using vigorous agitation into an aqueous solution containing a stabilizer [24]. The loading capacity of the hydrophobic core is limited by its solubility and diffusion to the stabilizer solution. This technique has found particular application for encapsulation of peptides, proteins, and hydrophilic pharmaceutical compounds [25]. The scheme of a standard microencapsulation procedure of peptides in PLGA microparticles using double emulsion-evaporation technique ($w_1/o/w_2$) is shown in Figure 30.4 [26].

To prepare a primary emulsion, a gelatin solution is mixed with a peptide solution, and the mixture is homogenized. This aqueous phase is dispersed in PLGA oil phase at 40°C. Then the emulsion is homogenized (Ultraturrax) and cooled to 15°C without stirring, and transferred into polyvinylalcohol (PVA) solution. After incubation of microparticles under stirring for 15 min, the temperature is raised up to 30°C to facilitate solvent elimination. Solid microparticles obtained are collected by filtration, washed with water, and dried overnight by lyophilization. Recently a Total Recirculation One-Machine System (TROMS) has been proposed for encapsulation of adenovirus in PLGA microparticles within the range of 7–10 μm [27]. The method is based on the turbulent injection of the phases, thus avoiding the use of aggressive homogenization techniques. In addition, the fabrication parameters can be easily adjusted to modulate the desired characteristics of microparticles obtained.

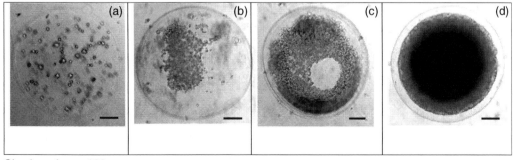

Size bar shows 100 μm.

FIGURE 30.3 *In vitro* cultivation of endostatin-transfected CHO cells in alginate–chitosan microcapsules (a) just after encapsulation, (b) day 5, (c) day 12, and (d) day 21.

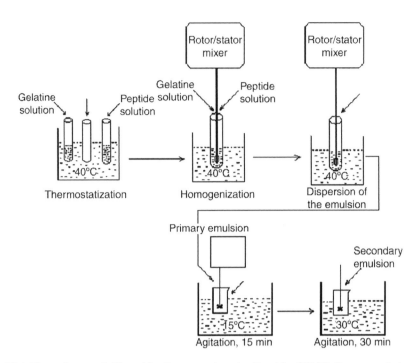

FIGURE 30.4 The scheme of Thrombin Receptor Agonist Peptide (TRAP-6) encapsulation in PLGA microparticles by double ($w_1/o/w_2$) emulsion-evaporation technique.

To enhance encapsulation efficiency, a spray-drying method could be proposed. Thus, this technique provides a significant improvement of betamethasone encapsulation efficacy into PLGA microparticles, namely, 90% encapsulation efficacy compared with 15% for $w_1/o/w_2$ double emulsion process [28]. An emulsification/spray-drying technique has also been proposed as an alternative to the double emulsion method for encapsulation of peptide drug, vancomycin into microspheres for the topical ocular delivery (encapsulation efficacy 84–99%) [29].

3. Layer-By-Layer Adsorption Technique for Encapsulation of Biomaterial

An original universal technique based on layer-by-layer (LbL) adsorption of oppositely charged macromolecules onto a surface of inorganic colloid particles has been recently elaborated [30,31]. Hollow nano- or microcapsules with entrapped biomaterial can be easily prepared by decomposing an inorganic core, although the microcapsule wall can provide desired release properties. The use of various polymer materials allows a proper shell design, in order to adjust required stability, biocompatibility, and affinity properties of the microcapsules.

To prepare microcapsules with entrapped protein (or DNA), $CaCO_3$ porous spherical microparticles (an inorganic core) with narrow bead-size distribution (mean size of 5 μm) have been elaborated (Figure 30.5).

These microparticles are incubated in protein (DNA) solution, and, after protein (DNA) adsorption, washed with water. Then freshly prepared PLL solution is added to them. To prevent aggregation, microparticles were treated with ultrasound, then after incubation on a shaker the microparticles were washed with 0.9% NaCl. In order to get the second polymer layer on $CaCO_3$ core surface, sodium alginate solution was added. After washing with water, the microcapsules are transferred to EDTA solution, and intensively mixed to dissolve $CaCO_3$ cores. The final protein (DNA) loaded microparticles contain 6–8 layers of polyelectrolytes (PLL-Alg-PLL-Alg...). Recently, the

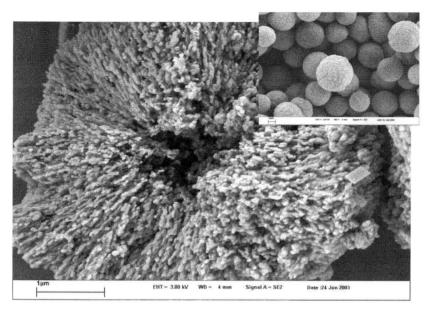

FIGURE 30.5 SEM images of macroporous $CaCO_3$ microparticles.

technique has been successfully used to encapsulate proteins (α-lactoalbumin, lysozyme, and horseradish peroxidase) [32] as well as some allergens, namely, protein Asp f2 and protein-encoding DNA (DNA-Asp f2) to develop a new vaccine against intracellular pathogens [33].

III. BIOMEDICAL APPLICATIONS OF POLYMER MICRO- AND NANOCARRIERS LOADED WITH BIOMATERIALS

A. CONTROLLED DRUG RELEASE SYSTEMS BASED ON NANO- AND MICROPARTICLES/MICROCAPSULES

Encapsulation of drugs provides the protection of a biologically active molecule by a polymer coating and release of this product following a defined profile. The properties of a drug delivery systems depend on the polymer materials used for bioencapsulation, the technique which has been chosen for preparation of nano- or microcarriers, and the type of bioencapsulated biomaterials (molecular weight, chemical structure, charge, etc.). As known, a drug-release rate can be controlled by several parameters, such as drug-loading degree in the matrix, hydrogel-hydratation degree, swelling kinetics of polymer matrix, etc. [34–36]. The methods for preparation of drug micro- or nanocarries can be classified into two main categories. First one gathers most of the methods based on polymerization reactions, and the second one deals with the use of preformed polymers (both natural or synthetic origin).

Chitosan-based nano- and microparticles are widely used for fabrication of controlled drug release systems. Numerous studies have demonstrated that chitosan and its derivatives (N-trimethyl chitosan, mono-N-carboxymethyl chitosan, etc.) are effective and safe for absorption enhance to improve mucosal (nasal, peroral) delivery of hydrophylic macromolecules, such as peptide and protein drugs as well as heparins [37,38]. This absorption enhancing effect of chitosan is caused by the opening of intercellular tight junctions, thereby favoring the paracellular transport of macromolecular drugs. Recently, a series of successful model chitosan-based polymer systems for mucosal drug delivery have been reported. Thus, Lim et al. [39] have proposed novel polymer microparticles based on combination of hyaluronic acid and chitosan hydroglutamate with entrapped gentamicin prepared by a solvent-evaporation method as an effective nasal delivery system. This formulation, as it has been demonstrated in a rabbit model, considerably enhanced the bioavailability of gentamicin (42.9%) compared with that of a gentamicin nasal solution (1.1%) or dry powder (2.1%). In addition, chitosan nanoparticles as doxorubicine (DOX) delivery systems are also mentioned [40]. In order to entrap a cationic, hydrophilic DOX molecule into positively charged chitosan nanoparticles, first positive charge of DOX was masked by complexing it with the polyanion, dextran sulfate. Confocal studies showed that DOX was not released in the cell-culture medium but entered the cells whereas remaining associated to the nanoparticles.

To accelerate healing of 2,4,6-trinitrobenzenesulphonic acid-induced colitis in rats, chitosan microcapsules loaded with the anti-inflammatory drug, 5-aminosalycylic acid (5-ASA), have been evaluated in Japan [41]. In conclusion, several research groups deal with development of chitosan-based microparticles with entrapped insulin as peroral delivery systems [42,43].

Microparticles used on another polysaccharide, hydroxypropyl methylcellulose were prepared by spray-drying technique and designed to provide the absorption of a high polar drug through nasal mucosa [44].

In order to study the kinetics of *in vitro* superoxide dismutase (SOD) release from microparticles of different origin, the drug was encapsulated by the spray-drying technique in three types of microcapsules, namely, in poly(ε-caprolactone) (PCL), in poly(D,L-lactide-co-glycolide) (PLGA) and in poly(D,L-lactide) (PLA) ones. The *in vitro* release profile showed that SOD was completely (100%) released from PLA in 48 h, and from PLGA and PCL within 72 h.

PLGA-based microparticles are being intensively studied as drug delivery systems. Thus, ibuprofen-loaded microspheres prepared by solvent-evaporation method based on o/w emulsion with an addition of a biodegradable oil Labrafil yielded the most suitable profile in case of intra-articular administration [45]. Forty milligram of the selected formulation was sufficient to provide *in vitro* "therapeutic" concentrations of ibuprofen (8 μg/ml) up to 8 days. Labrafil modulates the release rate of donor–acceptor substances, such as ibuprofen. Another example of promising PLGA microparticle-based system is long-term delivery of ivermectin to treat heartworm disease in dogs [46]. Analysis of pharmocokinetic data revealed sustained ivermectin release during at least 287 days. An interesting lidocaine delivery system based on PLA microparticles coated with chitosan has been reported [47]. The authors studied the effect of PLA microspheres coating with various chitosans on the initial burst and controlled drug release from the microspheres. Chitosan coating allowed to prolong the drug release up to 90 h compared with uncoated PLA microparticles, from which the drug has been released within 26 h.

Nanocapsules appear a promising approach as a drug system for topical application. The transport of chlorhexidine-loaded poly(ε-caprolactone) nanocapsules through full-thickness and stripped hairless rat skin was investigated [48]. The flexibility of nanocapsules assured a satisfying bioadhesion to the skin, whereas the rigidity of the carrier limited the molecular "spill" into the skin and controlled the drug delivery to the skin.

As for injectable nano- or microcapsules, they can contain various growth factors, for example, vascular endothelial growth factor (VEGF-165) and platelet-derived growth factor (PDFG) which accelerate a mature vascular network [49] or therapeutic agents for treating cerebral tumours [50].

In conclusion, the so-called *in situ* microparticle system (ISM-system) should be mentioned. This is a quickly developing approach to different biomedical fields, including cartilage and bone regeneration surgery. A novel drug delivery system forming biodegradable microparticles from liquid dispersed systems is proposed as an alternative to the currently used microencapsulation and implant methods. The micropaticles are formed *in situ* after injection of a dispersed system (e.g., an emulsion consisting of internal drug-containing polymer phase which is dispersed into an external phase) through the solidification of the internal polymer phase upon contacting with body fluids [51].

B. Development of New Vaccines Using Nano- and Microparticles/Microcapsules

Although current nonviral vaccine systems are less effective than viral vaccines, all research efforts are focused on the development of new nonviral vaccines, namely, the so-called DNA vaccines or genetic vaccines. A promising nonviral delivery system for genetic vaccines involves microencapsulation of antigen-encoding DNA. Injection of these formulations elicits a better immune response to plasmid-encoding antigens. The background and rationale for microparticle-encapsulated DNA vaccine delivery system has been reviewed earlier [52].

It should be noted that encapsulation approach could allow:

- To easily entrap in microcapsules specific ligands ensuring a targeted interaction with cell receptors.
- To encapsulate several DNA plasmids in one microcapsule.
- To protect DNA from serum nuclease cleavage.
- To mask DNA from pre-existing antibodies enabling effective priming of immune response.

As well known, there are several immunization routes. Among them mucosal (nasal, oral) vaccination has the striking advantage over others because it provides production of local antibodies at the sites where pathogens enter the body. More over, mucosal vaccination not only reduces costs and increases patient compliance, but also complicates the invasion of pathogens through mucosal sites [53]. Mucosal vaccination strategies which are aimed at gaining expression of transgens (using encapsulated DNA) within dendritic cells, in order to gain both T-helper and cytotoxic T cell (CTC) responses, are described by Howard and Alpar [54].

Thus, presently the emphasis is given to the development of mucosal (oral and nasal) vaccines in conjunction with microcapsulate systems.

To assure complete protection, many vaccines require repeated immunization. Over the past 20 years, several research groups have investigated the possibility of developing a single-administration vaccine (SAV). SAV formulations provide the repeated administration automatically. One approach is the development of injectable microsphere formulations containing the vaccine antigen that is released as a pulse 1–6 months after injection.

Because vaccines alone cannot be sufficiently taken up after mucosal administration, they need to be co-administered with penetration enhancers, adjuvants, or encapsulated in microparticles. Chitosan particles, powders, and solutions are promising candidates for mucosal vaccine delivery. The aspecific adjuvant activity of chitosans seems to be dependent on the deacetylation degree and the type of formulation.

Chitosan easily forms microparticles and nanoparticles which encapsulate large amounts of antigens, such as ovalbumin, diphtheria toxoid, or tetanus toxoid. It has been shown that ovalbumin-loaded chitosan microparticles are taken up by the Peyer's patches of the gut associated lymphoid tissue (GALT). This unique uptake demonstrates that chitosan particulate drug carrier systems are promising candidates for oral vaccination. In addition, after co-administering chitosan with antigens in nasal vaccination studies, a strong enhancement of both mucosal and systemic immune responses is observed [55].

Several studies in animal models have been carried out to prepare chitosan microparticles-based vaccines against influenza, pertussis, and diphtheria. The nasal chitosan vaccine delivery system has been tested for vaccination against influenza in humans [56]. The results showed that it was both effective and protective according to the Committee for Proprietary Medical Products (CPMP) requirements.

Present clinical findings have shown that immunization with PLGA microparticle-based plasmid formulations elicits T cell responses to viral antigens and suggest that this approach may be generally useful in immunotherapy. One of the examples of successful repeated immunization with plasmid DNA formulated in PLGA microparticles has been recently reported [57]. This formulation stimulated durable T cell responses to the weakly immunogenic, tumor-associated antigen cytochrome P450 1B1.

DNA delivery systems based on pH-sensitive poly-beta amino ester and poly(lactic-co-glycolic acid) can generate an increase of 3–5 orders of magnitude in transfection efficiency and are potent activators of dendritic cells *in vitro* [58].

PLA microparticles coated with polyethylene glycol (PEG-PLA) have been studied as carriers for nasal vaccine delivery. They are more stable compared with PLA microparticles and generate increasing and long-lasting immune response [59].

In conclusion, it is mentioned that a novel immunization strategy, so-called continuous antigenic stimulation system (CASS) which is based on microencapsulated cells producing a trangene [60]. Mice implanted with encapsulated C2C12 myoblasts secreting human factor IX (hFIX) elicited a strong humoral response detectable for 213 days after implantation. The mice had increasing IgG2a antibody titer, indicating a switch to Th1 profile immune response.

C. DNA Delivery Carriers for Preparation of Genetically Modified Cells

Transfection of eukaryotic cells to obtain genetically modified cells as producers of peptides and proteins is an effective tool for cell biotechnology, agriculture, and biomedicine. It should be noted that presently there are several approaches to fabrication of DNA delivery systems, for example, the use of adeno- and retroviruses, or cationic liposomes [61,62].

As for viral DNA carriers, they are nontoxic for cells and can provide rather good transfection efficacy. However, although most of the current medical protocols are based on viral DNA delivery systems, the use of these systems is limited for several reasons: in particular, because of possible biological peculiarities of viruses, such as oncogenicity. Among other disadvantages of these systems, it is mentioned that limited DNA quantities which can enter a cell as well as safety production issues.

Among liposome-based carriers, pH-sensitive liposomes are of great interest [63,64]. Cationic lipids of these liposomes contain amines with pK value within physiological range of 4.5–8.0. The liposome can interact with DNA molecule under acidic conditions and penetrate into the cell, and then, after changing pH and fusion with endosome membrane, to release DNA. However, the use of these liposomes is limited because they are rather expensive, are able to aggregate and can be immunogenic. Presently a lot of DNA vectors are commercially available (for instance lipofectin from Qiagen and some others).

DNA delivery devices based on polymer nanocapsules (200–800 μm) can be considered as a promising alternative approach to all DNA carriers mentioned earlier. There are at least several advantages of DNA delivery systems based on encapsulation over others:

- Ligands can be conjugated to the nanoparticle for targeting or stimulating receptor-mediated endocytosis.
- Lysosomalytic agents can be entrapped to reduce DNA degradation in the endosomal and lysosomal compartments.
- Other bioactive agents or multiple plasmids can be co-encapsulated.
- Bioavailability of DNA can be improved because of protection from serum nuclease degradation by the polymer matrix.
- Nanoparticles can be lyophilized for storage without loss of bioactivity.

Chitosan and its derivatives are considered as good candidates to create nonviral gene delivery systems due to high positive charges, low cytotoxicity, biodegradability, and their ability to improve the transport of macromolecules associated with chitosan across biological surfaces. Being positively charged, chitosan easily gives complexes with negatively charged

DNA molecules (polyplexes). Although there are a lot of reports on chitosan-based DNA nano-carriers varying in sizes (20–250 nm and higher), studies regarding effects and the chitosan-specific transfection mechanisms remain insufficient. It is shown that the level of transfection depends upon several factors, for instance chitosan molecular weight (from 22 kDa [65] to 400 kDa [66]), the ratio of chitosan nitrogen to DNA phosphate (N/P ratio), as well as serum concentration and pH of transfection medium, and finally on cell line [67].

Koping-Hoggard et al. [68] established the relationships between the structure and the properties of chitosan–pDNA polyplexes *in vitro* and *in vivo*. They compared polyplexes of ultrapure chitosan (UPC) of preferred molecular structure with those of polyethylenimine (PEI) polyplexes *in vitro* and after intratracheal administration to mice *in vivo*. UPC carriers were less cytotoxic than polyethylenimine (PEI) polyplexes and provided a better efficiency compared with that of commonly used cationic lipids. Low-molecular weight chitosan delivery systems were more efficient for cell transfection and less cytotoxic compared with PLL [65].

In order to enhance the efficacy of cell transfection, an introduction of a ligand into a polyplex macromolecule has been proposed. The ligand can interact only with cell receptor of the cells which should be transfected, and therefore a target DNA delivery can be provided. For example, for liver specificity lactobionic-acid-bearing galactose group was coupled with chitosan, and PEG was grafted to galactosylated chitosan (GC) for stability in water and enhanced cell permeability. The prepared GCP/DNA complexes were only transfected into hepatocytes Hep G2 having asialoglycoprotein receptors (ASGR), indicating specific interaction of ASGR on cells and galactose ligands on GCP [69].

However, being a natural polymer, chitosan has all disadvantages of natural materials mentioned earlier (see Section II.A). Presently, a lot of other polyplexes based on cationic synthetic polymers have been described in the literature. Among others, we could mention rather new DNA delivery nanoparticles based on poly(2-dimethylamino)ethyl methacrylate-co-poly(ethylene glycol) (PDMAEMA) [70,71] and PEGylated polyethylenimine (PEI/DNA) [72]. The last polyplexes have been encapsulated in PLGA microparticles and proposed for mucosal (oral) polyplex-based vaccination of Wistar rats.

The studies on preparation of new more efficient and cheaper polymer systems based on nanoparticles/nanocapsules are in progress.

D. Somatic Gene Therapy Using Implanted Microcapsules with Entrapped Genetically Modified Animal Cells Producing Therapeutic Agents

One can entrap into microcapsules not therapeutic agents but alive cells which can produce these compounds (recombinant proteins, including enzymes, hormones, peptide immunostimulators or immunomodulators, etc.). One of the most widely used techniques for animal cell encapsulation into alginate-polycation microcapsules has been described earlier (see Section III.A).

Recent advances in gene engineering have provided entirely new avenues for gene-based applications, especially in the area of health care. Genes are now widely used as templates and cells as mini-reactors to provide the end products. Entrapment of these genetically engineered cells in immunoprotective microcapsules would allow nonautologous cells to be implanted into any host, in order to deliver the desired gene product without triggering graft rejection. While "classical" gene therapy suggests genetic modification of host cells with following their perfusion to the same patient, so-called "somatic gene therapy" is based on design of universal nonautologous cell lines which can be used for any patient with the same disease. These nonautologous cell lines could be presented by mouse, rat, or other genetically modified cells. Thus, on one hand, microcapsules with semi-permeable membranes can provide the release of desired therapeutic product, and, on the other hand, they can protect the cell-producers from immunosystem of the host.

The list of diseases which can be possibly treated in the future using this approach is rather long. We would like to mention that there are several reports in the literature describing somatic gene therapy approaches to treat diabetes [73], anemia, hemophilia [74], lysosomal storage disease, for instance deficient of beta-glucuronidase [75], and diseases of the central nervous system [76], Parkinson's disease [77], and cardiovascular disease [78]. The same approach could be proposed to treat cancer or multiple sclerosis.

However, until now microencapsulation is not applied in clinical practice. The reasons for this is related to a lack of biocompatibility, limited immunoprotection properties, and hipoxia. As for microencapsulated pancreatic islet grafts, their survival could be supported by co-encapsulation with Sertoli's cells. Thus, grafts of co-microencapsulated Sertoli's cells with islets resulted in prolongation of the achieved normoglycemia in rats when compared with control animals (received only islets) [79]. Strategies that may well support encapsulated islet grafts include several issues, such as co-encapsulation of islets with Sertoli cells, the genetic modification of islet cells, the creation of an artificial implantation site, and the use of alternative donor sources [80].

Bioncapsulation in combination with one or more of these additional strategies may lead to a simple and safe transplantation therapy in the nearest future.

REFERENCES

1. Melvik, J. and Dornish, M., Alginate as a carrier for cell immobilization, in *Fundamentals of Cell Immobilisation Biotechnolog, Series Focus on Biotechnology*, Vol. 8A, Nedovic, V. and Willaert, R., Eds., Kluwer Academic Publishers, Dordrecht, 2004, pp. 33–52.
2. Lim, F. and Sun, A.M., Microcapsulated islet as a bioartificial endocrine pancreas, *Science*, 210, 908–910, 1980.
3. Thu, B., Bruheim, P., Espervik, T., Smidsrod, O., Soon-Shiong, P., and Skjak-Braek, G., Alginate-polycation microcapsules, II. Some Functional properties, *Biomaterials*, 17, 1069–1079, 1996.
4. Weber, C.J., Kapp, J.A., Hagler, M.K., Safley, S., Chryssohoos, J.T., and Chaikof, E.L., Long-term survival of poly-L-lysine-alginate microcapsulated xenografts in spontaneous diabetic NOD mice, in *Cell encapsulation technology and therapeutics*, Kuhtreiber, W.M., Lanza, R.P., and Chick, W.L., Eds., Birkhauser, N-Y-Boston, 1990, pp. 117–137.
5. Sawhney, A.S. and Hubbell, J.A., Poly(ethylene oxide)-graft-poly(L-lysine) copolymers to enhance the biocompatibility of poly(L-lysine)-alginate microcapsule membranes, *Biomaterials*, 13, 863–870, 1992.
6. Bartkowiak, A. and Hunkeler, D., Alginate-oligochitosan microcapsules: II. Control of mechanical resistance and permeability of the membrane, *Chem. Mater.*, 12, 206–212, 2000.
7. Carreno-Gomez, B. and Duncan, R., Evaluation of the biological properties of soluble chitosan and chitosan microspheres, *Int. J. Pharmac.*, 148, 231–240, 1997.
8. Strand, B., Ryan, L., Inet Veld, P., Kulseng, B., Rokstad, A.M., Skjak-Braek, G., and Espevik, T., Poly-L-lysine induces fibrosis on alginate microcapsules *via* the induction of cytokines, *Cell Transplantation*, 10, 263–275, 2001.
9. King, A., Strand, B.L., Rokstad, A.M., Kulseng, B., Andersson, A., Skjak-Braek, G., and Sandler, S., Improvement of the biocompatibility of alginate/poly-L-lysine/alginate microcapsules by the use of epimerized alginate as coating, *J. Biomed. Mater. Res.*, 64A (3), 533–539, 2003.
10. Calafiore, R., Basta, G., Luca, G., Boselli, C., Bufalari, A., Cassarani, M.P., Giustozzi, G.M., and Brunetti, P., Transplantation of minimal volume microcapsules in diabetic high mammalians, *Ann. NY Acad. Sci.*, 875, 219–232, 1999.
11. Miazaki, S., Takahashi, A., Kubo, W., Bachynski, J., and Loebenberg, R., Poly(N-butylcyanoacrylate (PNBCA) nanocapsules as a carrier for NSAIDs: *in vitro* and *in vivo* skin penetration, *J. Pharm. Phatm Sci.*, 6 (2), 238–245, 2003.
12. Ammeler, T., Marsaud, V., Legrand, P., Gref, R., Barratt, G., and Renoir, J.M., Polyester-poly(ethylene glycol) nanoparticles loaded with the pure antiesterogen RU 5: physico-chemical and opsonization properties, *Pharm. Res.*, 20 (7), 1063–1070, 2003.
13. Lboutounne, H., Charlet, J.F., Ploton, C., Flason, F., and Pirot, E., Sustained *ex vivo* skin antiseptic activity of chlorhexidine in poly(epsilon-caprolactone) nanocapsule encapsulated form and as a digluconate, *J. Control Release*, 82 (2–3), 319–334, 2002.

14. Tiyaboonchai, W., Woiszwillo, J., and Middaugh, C.R., Formulation and characterization of ampho-tericin B-polyethylenimine-dextran sulfate nanoparticles, *J. Pharm. Sci.*, 90 (7), 902–914, 2001.

15. Chung, J.T., Vlugt-Wensink, K.D.F., Hennink, W.E., and Zhang, Z., Mechanical characterization of biodegradable dex-HEMA microspheres for drug delivery, in *Proceedings of the XII International Workshop on Bioencapsulation*, Vitoria, Spain, September 24–26, 2004, pp. 93–96.

16. Marison, I., Peters, A., and Heinzen, Ch., Liquid core capsules for applications in biotechnology, in *Fundamentals of Cell Immobilisation Biotechnology, Series Focus on Biotechnology*, Vol. 8A, Nedovic, V., Willaert, R., Eds., Kluwer Academic Publishers, Dordrecht, 2004, pp. 185–204.

17. Markvicheva, E., Bezdetnaya, L., Bartkowiak, A., Marc, A., Goergen, J-L., Guillemin, F., and Poncelet, D., Encapsulated multicellular tumor spheroids as a novel model to study small size tumors, *Chem. Industry*, 57 (12), 585–588, 2003.

18. Heinzen, Ch., Berger, A., and Marison, I., Use of vibration technology for jet break-up for encapsula-tion of cells and liquids, in *Fundamentals of Cell Immobilisation Biotechnology, Series Focus on Biotechnology*, Vol. 8A, Nedovic, V., and Willaert, R., Eds., Kluwer Academic Publisher, Dordrecht-Boston-London, 2004, 257–275.

19. Bugarski, B., Obradovic, B., Nedovic, V., and Poncelet, D., Immobilization of cells and enzymes using electrostatic droplet generation, in *Fundamentals of Cell Immobilisation Biotechnology, Series Focus on Biotechnology*, Vol. 8A, Nedovic, V., and Willaert, R., Eds., Kluwer Academic Publishers, Dordrecht-Boston-London, 2004, 277–294.

20. Goosen, M.F., Mahmud, E.S.C., M-Ghafi, A.S., M-Hajri, H.A., Al-Sinani, Y.S., and Bugarski, M.B., Immobilization of cells using electrostatic droplet generator, in *Immobilization of Enzymes and Cells*, Bickerstaff, G.F., Ed., Humana Press, Totowa NJ, 1997, 167–174.

21. Bugarski, B., Vunjak, G., and Goosen, M.F.A., Principles of bioreactor design for encapsulated cells, in *Cell Encapsulation Technology and Therapeutics*, Kuhtreiber, W.M., Lanza, R.P., and Chick, W.L., Eds., Birkhauser, Boston, 1999, pp. 395–416.

22. Bugarski, M.B., Sajc, L., Plavsic, M., Goosen, M.F.A., and Jovanovic, G., Semipermeable alginate-PLO microcapsules as a bioartificial pancreas, in *Animal Cell Technology, Basic and Applied Aspects*, Vol. 8, Funatsu, K., Shirai, Y., and Matsushita, T., Eds., Kluwer Academic Publishers, Dordrecht, 1997, pp. 479–486.

23. Rosinski, S., Lewinska, D., Migaj, M., Wozniewicz, B., and Werynski, A., Electrostatic microencap-sulation of parathyroid cells as a tool for the investigation of cell activity after transplantation, *Landbauforschung Volkenrode,* SH 241, 47–50, 2002.

24. Benita, S., Drugs and the pharmaceutical sciences, in *Microencapsulation, Methods and Industrial Applications*, Benita, S., Ed., Vol. 73, Marcel Dekker Inc., New-York, 1996, pp. 587–632.

25. Frangione-Beebe, M., Rose, R.T., Kaumaya, P.T., and Schwendeman, S.P., Microencapsulation of a synthetic peptide epitope for HTLV-1 in biodegradable poly(D,L-lactide-co-glycolide) microspheres using a novel encapsulation technique, *J. Microencapsul.*, 18, 663–677, 2001.

26. Stashevskaya, K., Grandfils, Ch., Markvicheva, E., and Strukova, S., *In vitro* study of Thrombin Receptor Agonist Peptide release from PLGA microbeads, in *Proceedings of the XII International Workshop on Bioencapsulation*, Victoria, Spain, September 24–26, 2004, 295–298.

27. Garcia del Bario, G., Novo, F.J., and Irache, J.M., Loading of plasmid DNA into PLGA microparticles using TROMS (Total Recirculation One-Mashine System): evaluation of its integrity and controlled released properties, *J. Control Release*, 86, 123–130, 2003.

28. Chaw, C.S., Yang, Y.Y., Lim, I.J., and Phan, T.T., Water-soluble betamethasone-loaded poly(lactide-co-glycolide) hollow microcapsules as a sustained release dosage form, *J Microencapsul.*, 20 (3), 349–359, 2001.

29. Gavini, E., Chetoni, P., Cossu, M., Alvarez, M.G., Saettone, M.F., and Giunchedi, P., PLGA microspheres for the ocular delivery of peptide drug, vancomycin using emulsification/spray-drying as the preparation method: *in vitro/in vivo* studies, *Eur. J. Pharm Biopharm.*, 57 (2), 207–212, 2004.

30. Mohwald, H.D.E. and Sukhorukov, G.B., Smart Capsules, in *Multilayer Thin Films, Sequential Assembly of Nanocomposite Materials*, Decher, G. and Schlenoff, J.B., Eds., Wiley-VCH Verlag GmbH&Co KgaA, Weinheim, 2002, pp. 363–392.

31. Sukhorukov, G.B., Multilayer Hollow Microspheres, in *Dendrimers, Assemblies. Nanocomposites, MML Series*, Arshady, R. and Guyot, A., Eds., Citus Books, London, 2002, pp. 111–147.

32. Volodkin, D., Larionova, N., and Sukhorukov, G., Protein encapsulation via porous CaCO3 micro-particle templating, *Biomacromol.*, 5, 1962–1972, 2004.
33. Markvicheva, E., Svirshchevskaya, E., Grandfils, Ch., Bezdetnaya, L., Volodkin, D., Selina, O., Sukhorukov, G., Bartkowiak, A., Goergen, J-L., Guillemin, F., and Poncelet, D., Bioencapsulation of animal cells and biopolymers for biomedical applications, *J Chem. Technol. Biotechnol*, 2005.
34. Tanaka, T., Collapse of gels and the critical endpoint, *Phys. Rev. Lett.*, 40, 820–823, 1978.
35. Nirokawa, Y. and Tanaka, T., Volume phase transition in nonionic gel, *J. Chem. Phys.*, 81, 6379–6380, 1984.
36. Matsukata, M., Takei, Y., Aoki, T., Sanui, K., Ogata, N., Sakurai, Y., and Okano, T., Temperature-modulated solubility-activity alterations for poly(N-isopropyl-acrylamide)-lipase conjugates, *J. Biochem.*, (Tokyo), 116, 682–686, 1994.
37. van der Lubben, I.M., Verhoef, J.C., Borchard, G., and Junginger, H. E., Chitosan and its derivatives in mucosal drug and vaccine delivery, *Eur. J. Pharm Sci.*, 14 (3), 201–207, 2001.
38. Thanou, M., Verhoef, J.C., and Junginger, H.E., Oral drug absorption enhancement by chitosan and its derivatives, *Adv. Drug Deliv. Rev.*, 52 (2), 139–144, 2001.
39. Lim, S.T., Forbes, B., Berry, D.J., Martin, G.P., and Brown, M.B., *In vivo* evaluation of novel hyaluron/chitosan microparticulate delivery systems for the nasal delivery of gentamicin in rabbits, *Int. J. Pharm.* 23 (1), 73–82, 2002.
40. Janes, K.A., Fresneau, M.P., Marazuela, A., Fabra, A., and Alonso, M.J., Chitosan nanoparticles as delivery systems for doxorubicin, *J. Control Release*, 15 (73), 255–267, 2001.
41. Tozaki, H., Fujita, T., Terabe, A., Okabe, S., Muranishi, S., and Yamamoto, A., Validation of a pharmacokinetic model of colon-specific drug delivery and thetherapeutic effects of chitosan capsules containing 5-aminosalicilic acid on 2,4,6-trinitrobenzenesulphonic acid-induced colitis in rats, *J. Pharm. Pharmacol.*, 51 (10), 1107–1112, 1999.
42. Ramadas, M., Paul, W., Dileep, K.J., Anitha, Y., and Sharma, C.P., Lipoinsulin encapsulated alginate-chitosan capsules: intestinal delivery in diabetic rats, *J. Microencapsul.*, 17 (4), 405–410, 2000.
43. Tozaki, H., Komoike, J., Tada, C., Maruyama, T., Terabe, A., Suzuki, T., Yamamoto, A., and Muranishi, S., Chitosan capsules for colon-specific drug delivery: improvement of insulin absorption from the rat colon, *J. Pharm. Sci.*, 86 (9), 1016–1021, 1997.
44. Hascicek, C., Gonul, N., and Erk, N., Mucoadhesive microspheres containing gentamicin sulfate for nasal administration: preparation and *in vitro* characterization, *Il Farmaco*, 58 (1), 11–16, 2003.
45. Fernandez-Carballido, A., Herrero-Vanrell, R., Molina-Martinez, I.T., and Pastoriza, P., Biodegradable ibuprofen-loaded PLGA microspheres for intraarticular administration: effect of Labrafil addition on release *in vitro*, *Int. J. Pharm.*, 279 (1–2), 33–41, 2004.
46. Clark, S.L., Crowley, A.J., Schmidt, P.G., Donoghue, A.R., and Piche, C.A., Long-term delivery of ivermectin by use of poly(D,L-lactic-co-glycolic)acid microparticles in dogs, *Am. J. Vet. Res.*, 65 (6), 752–757, 2004.
47. Chiou, S.H., Wu, W.T., Huang, Y.Y., and Chung, T.W., Effects of the characteristics of chitosan on controlling drug release of chitosan coated PLLA microspheres, *J. Microencapsul.*, 18 (5), 613–625, 2001.
48. Lboutoounne, H., Faivre, V., Falson, F., and Pirot, F., Characterization of transport of chlorhexidine-loaded nanocapsules through hairless and wistar rat skin, *Skin Pharmacol. Physiol.*, 17 (4), 176–182, 2004.
49. Richardson, T.P., Petrs, M.C., Ennett, A.B., and Mooney, D.J., Polymeric system for dual growth factor delivery, *Nat. Biotechnol.*, 19, 1029–1034, 2001.
50. Chandy, T., Das, G.S., and Rao, G.H., 5-Fluorouracil-loaded chitosan coated polylactic acid microspheres as biodegradable drug carrier for cerebral tumors, *J. Microencapsul.*, 17 (5), 625–638, 2000.
51. Dashevsky, A. and Bodmeier, R., *In situ* forming biodegradable drug delivery systems, in *Macromolecular Approaches to Advanced Biomaterials Engineering Systems*, Proceedings of the NATO Advanced Research Workshop, Sofia, Bulgaria, November 8–11, 2003.
52. O'Hagan, D.T., Singh, M., and Ulmer, J.B., Microparticles for the delivery of DNA vaccines, *Immunol. Rev.*, 199, 191–200, 2004.
53. van der Lubben, I.M., Verhoef, J.C., Borchard, G., and Junginger, H.E., Chitosan for mucosal vaccination, *Adv. Drug. Rev.*, 52 (2), 139–144, 2001.
54. Howard, K.A. and Alpar, H.O., The development of polyplex-based DNA vaccines, *J. Drug Target.* 10, 143–151, 2002.

55. van der Lubben, I.M., Verhoef, J.C., Borchard, G., and Junginger, H.E., Chitosan and its derivatives in mucosal drug and vaccine delivery, *Eur. J. Pharm. Sci.*, 14 (3), 201–207, 2001.

56. Illum, L., Jabbal-Gill, I., Hinchcliffe, M., Fisher, A.N., and Davis, S.S., Chitosan as a novel nasal delivery system for vaccines, *Adv. Drug Deliv. Rev.*, 51 (1–3), 81–96, 2001.

57. Luby, T.M., Cole, G., Baker, L., Kornher, S., Ramstedt, U., and Hedley, M.L., Repeated immunization with plasmid DNA formulated in poly(lactide-co-glycolide) microparticles is well tolerated and stimulates durable T cell response to the tumor-associated antigen cytochrome P450 1B1, *Clin. Immunol.*, 112, 45–53, 2004.

58. Little, S.R., Lynn, D.M., Anderson, D.G., Puram, S.V., Chen, J., Eisen, H.N., and Langer, R., Poly-beta amino ester-containing microparticles enhance the activity of nonviral genetic vaccines, *Proc. Natl Sci. USA*, 101 (26), 9534–9539, 2004.

59. Vila, A., Evora, C., Soriano, I., Vila Jato, J.L., and Alonso, M.J., PEG-PLA nanoparticles as carriers for nasal vaccine delivery, *J. Aerosol. Med.*, 17 (2), 174–185, 2004.

60. Gomez-Vargas, A., Rosenthal, K.L., McDermott, M.R., and Hortelano, G., Continuous antigenic stimulation system (CASS) as a new immunization strategy, *Vaccine*, 22 (29–30), 3902–3910, 2004.

61. Oudrhiri, N., Vigneron, J.P., Peuchmaur, M., Leclerc, T., Lehn, J.M., and Lehn, P., Gene transfer by guanidinium-cholesterol cationic lipids into airway epithelial cells *in vitro* and *in vivo*, *Proc. Natl. Acad. Sci. USA*, 94, 1651–1656, 1997.

62. Kikuchi, A., Aoki, Y., Sugaya, S., Serikawa, T., Takakuwa, K., Tanaka, K., Suzuki, N., and Kikuchi, H., Development of novel cationic liposomes for efficient gene transfer into peritoneal disseminated tumor, *Hum. Gene Ther.*, 10, 947–955, 1999.

63. Vitiello, L., Bockhold, K., Joshi, P.B., and Worton, R.G., Transfection of cultured myoblasts in high serum concentration with DODAC:DOPE liposomes, *Gene Ther.*, 5, 1306–1313, 1998.

64. Zuidam, N.J. and Barenholz, Y., Characterization of DNA-lipid complexes commonly used for gene delivery, *Int. J. Pharm.*, 183, 43–46, 1999.

65. Lee, M., Hah, J.W., Kwon, Y., Koh, J.J., Ko, K.S., and Kim, S.W., Water-soluble and low molecular weight chitosan-based plasmid DNA delivery, *Pharm. Res.*, 18 (4), 427–431, 2001.

66. Wang, S.M.N., Gao, S.J., Yu, H., and Leong, K.W., Transgene expression in the brain stem effected by intramuscular injection of polyethylenimine/DNA complexes, *Mol. Ther.*, 3 (5), 658–664, 2001.

67. Ishii, T., Okahata, Y., and Sato, T., Mechanism of cell transfection with plasmid/chitosan complexes, *Biochim. Biophys. Acta*, 1514 (1), 51–64, 2001.

68. Koping-Hoggard, M., Tubulekas, I., Guan, H., Edwards, K., Nilsson, M., Varum, K.M., and Artursson, P., Chitosan as non-viral gene delivery system. Structure-property relationships and characteristics compared with polyethylenimine *in vitro* and after lung administration *in vivo*, *Gene Ther.*, 8 (14), 1108–1121, 2001.

69. Park, I.K., Kim, T.H., Park, Y.H., Shin, B.A., Choi, E.S., Chowdhury, E.H., Akaike, T., and Cho, C.S., Galactosylated chitosan-graft-poly(ethylene glycol) as hepatocyte-targeting DNA carrier, *J. Control Release*, 19 (76), 349–362, 2001.

70. Pantoustier, N., Moins, S., Wautier, M., Deg, P., and Dubois, P., Solvent-free synthesis and purification of poly(2-dimethylamino)ethyl methacrylate) by atom transfer radical polymerization, *Chem. Comm.*, 3, 340–341, 2003.

71. Grandfils, Ch. and Emonds, J.A., Optimization of poly(2-dimethylamino)ethyl methacrylate-co-poly (ethylene glycol)/DNA complexes designed for cell transfection, in *Proceedings of XII International Workshop on Bioencapsulation*, Vitoria, Spain, September 24–26, 2004, 81–84.

72. Howard, K.A., Li, X.W., Somavarapu, S., Singh, J., Green, N., Atuah, K., Ozsoy, Y., Seymour, L.W., and Alpar, H.O., Formulation of a microparticle carrier for one oral polyplex-based DNA vaccines, *Biochim. Biophys. Acta*, 1674 (2), 149–157, 2004.

73. Lanza, R.P., Ecker, D.M., Kuhtreiber, W.M., Marsh, J.P., Ringeling, J., and Click, W.L., Transplantation of islets using microencapsulation: studies in diabetic rodents and dogs, *J. Mol. Med.*, 77, 206–210, 1999.

74. Hortelano, G. and Chang, P., Gene therapy for hemophilia, *Art Cell Blood Subs Immob. Biotechnol.*, 28 (1), 1–24, 2000.

75. Ross, C.J., Bastedo, L., Maier, S.A., Sands, M.S., and Chang, P., Treatment of a lysosomal storage disease, mucopolysaccharidosis VII, with microencapsulated recombinant cells, *Hum. Gene Ther.*, 11 (15), 2117–2127, 2000.

76. Barsoum, S.C., Milgram, W., Mackay, W., Coblentz, C., Delaney, K.H., Kwiecien, J.M., and Chang, P.L., Delivery of recombinant gene product to canine brain with the use of microencapsulation, *J. Lab. Clin. Med.*, 142 (6), 399–413, 2003.

77. Kishima, H., Poyot, T., Bloch, J., Dauguet, J., Conde, F., Dolle, F., Hinnen, F., Pralong, W., Palfi, S., Deglon, N., Aebisher, P., and Hantraye, P., Encapsulated GDNF-producing C2C12 cells for Parkinson's disease: pre-clinical study in chronic MPTP-treated baboons, *Neurobiol. Disease*, 16 (2), 428–439, 2004.

78. Chen, L.G., Wang, Z.R., Wan, C.M., Xiao, J., Guo, L., Guo, H.L., Cornelissen, G., and Halberg, F., Circadian renal rhythms influenced by implanted encapsulated hANP-producing cells in Goldblatt hypertensive rats, *Gen. Ther.*, 11 (20), 1515–1522, 2004.

79. Luca, G., Calafiore, R., Basta, G., Ricci, M., Calvitti, M., Neri, L., Nastruzzi, C., Becchetti, E., Capitani, S., Brunelli, P., and Rossi, C., Improved function of rat islets upon co-microencapsulation with Sertoli's cells in alginate/poly-L-ornithine, *AAPS PharmSciTech.*, 2 (3), E 15, 2001.

80. de Groot, M., Schuurs, T.A., and van Schilfgaarde, R., Causes of limited survival of microencapsulated pancreatic islet grafts, *J. Surg. Res.*, 121 (1), 141–150, 2004.

31 Electrostatic Droplet Generation Technique for Cell Immobilization

Branko M. Bugarski and Bojana Obradovic
University of Belgrade, Belgrade, Serbia and Montenegro

Viktor A. Nedovic
Zemun, University of Belgrade, Belgrade, Serbia and Montenegro

Mattheus F. A. Goosen
University of Puerto Rico, Puerto Rico, USA

CONTENTS

I. INTRODUCTION

Electrostatic atomization and electrostatically assisted atomization have been employed in a variety of areas, including paint spraying [1], electrostatic printing [2], and cell immobilization [3]. The basic concept behind these applications involves electrostatic forces, which work to disrupt the liquid surface to form a charged stream of fine droplets. The effect of electrostatic forces on mechanically atomized liquid droplets was first studied in detail by Lord Rayleigh [4,5], who investigated hydrodynamic stability of a jet of liquid with and without applied electric field.

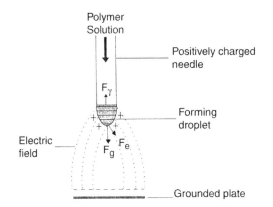

FIGURE 31.1 Schematic presentation of forces acting on a charged pendent droplet.

If a liquid is subjected to an electric field, a charge is induced on the liquid surface and mutual charge repulsion results in an outwardly directed force. Under suitable conditions, for example extrusion of the liquid through a needle, the electrostatic pressure at the surface forces the liquid drop to form a cone shape (Figure 31.1).

Surplus charge is released by emission of charged droplets from the tip of the needle. The emission process depends on several factors such as the needle diameter, distance from the collecting solution, and applied potential (i.e., strength of the electrostatic field) [6]. Under most circumstances, the electrical spraying process is random and irregular resulting in drops of varying size and charge generated at the capillary tip over a wide range of angles. However, if the configuration of the electrostatic generator is adjusted for liquid pressure, applied voltage, electrode spacing, and charge polarity, the spraying process can become quite regular and periodic.

A major concern in cell and bioactive agent immobilization has been the production of very small microbeads so as to minimize the mass transfer resistances associated with large beads (i.e., >1000 μm in diameter) [7]. Biocatalyst immobilization systems involved use of various gel-forming proteins (e.g., gelatin), polysaccharides (e.g., agar, calcium alginate, and κ-carrageenan), and synthetic polymers (e.g., polyacrylamide). Relatively large spherical carriers (i.e., 1000 μm in diameter) are normally formed by dispensing droplets into a hardening solution. A finer dispersion (100–300 μm) can be achieved by emulsifying the gel-forming solution in an oil phase and droplet hardening by internal gelling. On the other hand, Klien et al. [8] reported production of beads with diameters in the range 100–400 μm using compressed air to quickly force the gel solution through a nozzle. Surprisingly, few attempts have been made on the applications of electric fields to produce micron-size polymer beads for cell immobilization [9].

The objectives of this chapter are to review the studies on droplet formation mechanisms under the action of electrostatic forces and to determine key parameters critical for production of very small polymer microbeads (i.e., less than 100 μm in diameter). Specifically, attention is given to the effects of applied potential, needle size, polymer concentration, and electrode spacing, and geometry. An overview of theoretical models and experimental correlations for predictions of droplet diameter is presented.

II. FORCES ACTING ON DROPLET FORMATION

If gravity was the only force acting on the meniscus of a droplet, then large uniformly sized droplets of a radius r would be produced. Equating the capillary surface force to the gravitational force gives:

$$2\pi r_0 \gamma = \frac{4}{3}\pi \rho g r^3 \tag{31.1}$$

or

$$r = \left(\frac{3r_0\gamma}{2\rho g}\right)^{1/3} \tag{31.2}$$

where r_0 is the internal radius of the needle, γ the surface tension, ρ the liquid density, and g is the acceleration of the gravity. Equation (31.2) thus could be used to calculate droplet diameter if the liquid properties and the needle diameter are known. For example, in an usual setup with a 22 gauge needle ($r_0 = 203\ \mu m$) and alginate solution ($\rho = 1012\ kg/m^3$, $\gamma = 0.074\ N/m$), the diameter of the forming drop under influence of gravitational force should be 2.6 mm.

However, if electric field is applied to the needle and the liquid is flowing at a low rate, an electric charge is induced on the surface of the droplet forming at the needle tip, resulting in a mechanical force, directed normally outward from the surface (i.e., directly opposing the inward acting surface tension, [10–12], Figure 31.1). The electric field produces an additional term, F_e, in the force balance equation:

$$F_\gamma = F_g + F_e \tag{31.3}$$

In order for droplet to detach from the meniscus, the surface tension force (F_γ) must be overcome by the gravitational force (F_g) and the electrical force (F_e). Equation (31.3) can be also expressed as:

$$2\pi r_0\gamma = \frac{4}{3}\pi\rho g r^3 + qE \tag{31.4}$$

where q is the electric charge and E the strength of the electric field.

If sufficient charge is added to the liquid drop, it becomes unstable and disrupts, since the state of lowest energy will be the one in which the liquid is in the form of many small drops, rather than in one large drop. Rayleigh [4,5] showed that disruption occurs when the charge, q, on a droplet of radius, r, is given by:

$$q \geq 8\pi(\varepsilon_0\gamma r^3)^{0.5} \tag{31.5}$$

where ε_0 is the permittivity of air ($\varepsilon_0 = 8.85 \times 10^{-12}\ C^2/Nm^2$).

Electric field can be applied to liquid extrusion in several geometries (see Figure 31.2 for example). In the case of a parallel-plate setup where the needle is protruding through a charged plate (Figure 31.2a), the electrostatic force, F_e, exerted on the needle is defined by modifying the expression obtained by Taylor [13].

$$F_e = \frac{\varepsilon_0 V^2 L^2}{4H^2 \ln(2L/r) - 3/2} \tag{31.6}$$

where L is the length of the needle exposed in the electric field and H the distance of the electrode (Figure 31.2a).

In the case of a positively charged needle (Figure 31.2b), the stress produced by the external field at the needle tip is obtained using an expression developed by DeShon and Carlson [14]:

$$P_e = \frac{2\varepsilon_0 V^2}{r^2[\ln(4H/r)]^2} \tag{31.7}$$

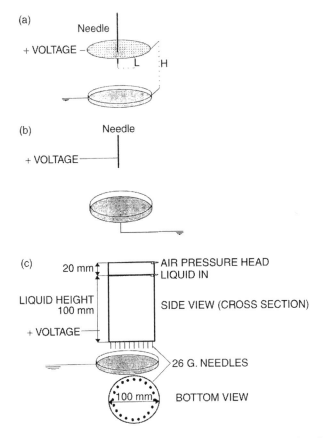

FIGURE 31.2 Electrode arrangements: (a) parallel-plate setup with plate charged positively; (b) positively charged needle setup; and (c) multi-needle parallel-plate device.

where P_e is the electrostatic pressure and H the spacing between tip and the grounded plate. In terms of the electrostatic force acting on the droplet hemisphere at the needle tip, Equation (31.7) can be written as:

$$F_e = \frac{4\pi\varepsilon_0 V^2}{[\ln(4H/r)]^2} \tag{31.8}$$

III. EXPERIMENTAL STUDIES

A. DROPLET FORMATION USING ELECTROSTATIC FIELD

A general experimental setup is presented in Figure 31.3 [15,16]. Sodium alginate was used as a model polymer. The polymer solution was extruded by a syringe pump (Razel, Scientific Instruments, Stanford, CT, USA) through a stainless steel, blunt edge needle (Chromatographic Specialties Inc., Brockville, Canada) and formed droplets were collected in a hardening solution (1.5% $CaCl_2$ solution, BDH, Toronto, Canada). The potential difference was controlled by a power supply unit (Model 30R, Bertan Associates, Inc., NY) with a maximum current of 0.4 mA and variable potential of 0–30 kV.

Three experimental setups were configured in order to investigate the effects of electrode arrangements on droplet formation and to test potentials for scale-up of the system (Figure 31.2, [15,17]).

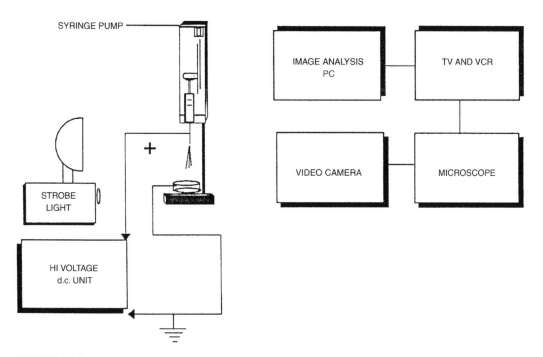

FIGURE 31.3 Schematic diagram of the experimental setup for electrostatic droplet generation. (From Bugarski, B., Amsden, B., Neufeld, R., Poncelet, D., and Goosen, M.F.A., *Can. J. Chem. Eng.*, 72, 517–522, 1994. With permission.)

In the first setup, a parallel-plate electrode system was employed to produce a uniform electric field in the same direction as that of the gravity (Figure 31.2a, [15]). The upper electrode was a positively charged plate with a $90°$ blunt needle protruding through a small aperture at the plate center. The other electrode was a grounded dish with the hardening solution and of the same dimensions as the upper charged plate.

In the second setup, the effect of a point-to-plane charge on droplet formation was examined. Namely, the electrodes were respectively the positively charged needle and the grounded dish with the hardening solution (Figure 31.2b, [15]).

Finally, for scale-up purposes, a cylindrical reservoir (15 cm height × 10 cm diameter) with 20 needles for the continuous production of polymer beads was designed (Figure 31.2c, [17]). The liquid flow rate was kept constant at 0.7 l/h (36 ml/h per needle) by adjusting the air pressure head above the polymer solution. A grounded collecting dish with the $CaCl_2$ solution was placed 2.5 cm below the needles. Twenty (22 gauge) stainless steel needles positioned 1.2 cm radially apart were attached to the cylindrical reservoir containing a liter of polymer solution and then connected to the power supply.

Several parameters affecting the droplet-formation mechanism and the microbead diameter were investigated, including the needle size, electrode spacing, sodium alginate concentration, and applied potential [18–20].

B. Image Analysis of Droplet Formation

Droplet formation under the influence of electrostatic forces was examined using stroboscopic light (Strobotac, GRC, MA, USA) at a defined frequency (50–400 per sec) along with a video or image analysis system [21]. Images were recorded when the frequencies of light and the forming beads were identical so that the object appeared as frozen. The image analysis system consisted of several components: a video camera (Panasonic Digital 5100), and a Sony CC camera (model AAVCD5), video adapter (Sony, CMAD5, Sony, Japan), video monitor (Sony Trinitron, PVM1342Q), VHS (NV-8950, Panasonic, Japan), PC, and software Java (Ver. 1.3) for the image analysis (Jandel Scientific, CA, USA). For close-up studies of droplet formation the video camera was connected to a microscope lens (Olympus SYH, Optical Co. Ltd, Japan) (Figure 31.3).

A sample containing 50–150 microbeads, was taken from each set of extrusion experiments and diameters of individual beads were determined from the images using the image analysis program. Average diameters of the beads and standard deviations with the maximum and minimum from each experimental set were computed automatically.

C. Determination of Microbead Size Distribution by Laser Light Scattering

Volumetric (volume of microbeads in each diameter class) and cumulative size distributions were determined by laser light scattering, with a 2602-LC particle analyzer (Malverin Instruments) according to the lognormal distribution model. The mean diameter and the arithmetic standard deviation were calculated from the cumulative distribution curve [22].

D. Extrusion of Animal Cell Suspensions Using Electrostatic Droplet Generator

Sensitivity of animal cells to a high potential (5–8 kV) was examined by extruding suspensions of insect cells (SF-9) [21] and murine bone marrow stromal cells (BMSC) [23] using the charged needle setup. The cell viability was assessed by staining the cells with trypan blue dye (only live cells can exclude the dye from the cell nucleus).

IV. RESULTS AND DISCUSSION

A. Parameters Affecting Microbead Size

In the case of the positively charged needle setup (Figure 31.2b), effects of electrode spacing on alginate bead size, produced with a 22-gauge needle, are shown in Figure 31.4a [19].

The electrode spacing was not found to be a significant parameter over the range investigated. For example, at an applied potential of 6 kV, the mean bead size decreased from 530 to 450 μm with a standard deviation of approximately 100 μm, as the electrode spacing was decreased from 4.8 to 2.5 cm. When the applied potential was increased to 12 kV, at a distance of 4.8 cm between the needle tip and collecting solution, the average bead diameter decreased to 340 μm. Keeping the applied potential constant at 12 kV, but reducing the electrode distance to 2.5 cm, resulted in only a slightly smaller bead size (300 μm).

Effects of alginate concentration on microbead diameter for two needle sizes are shown in Figure 31.4b and c [18,19]. When the concentration of alginate was decreased from 1.5% to 0.8% the average bead diameter decreased by 10–20%. Standard deviations also decreased at the lower polymer concentration, due to a more uniform bead size distribution. For example, at an applied potential of 5 kV, the mean bead diameter decreased from 440 ± 200 to 380 ± 80 μm for the beads prepared with alginate concentrations of 1.5% and 0.8%, respectively. However, in a similar study using a comparable procedure and alginate concentration of 2%, the obtained microbeads were almost double in size ranging from 600 to 1000 μm [3]. These results imply that the effects of alginate properties, such as concentration, viscosity, and surface tension on droplet formation under electrostatic field should be further investigated.

While the microbead diameter was not significantly affected by alginate concentration in the investigated range of 0.8–1.5%, it could be readily controlled by the needle size and applied potential (Figure 31.4b and c). The microbead size decreased as the needle size was reduced. For example at an applied potential of 4 kV, the mean bead diameter was reduced by a factor of three from approximately 1600 to 500 μm when the needle size was decreased from 22 to 26 gauge. On the

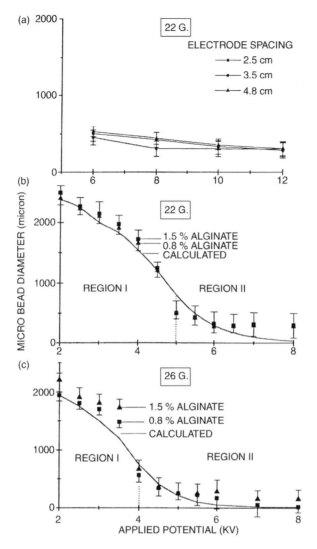

FIGURE 31.4 Microbead size as a function of applied potential in the positively charged needle setup; (a) effects of electrode spacing for a 22-gauge needle; (b) effects of alginate concentration for a 22-gauge needle; and (c) effects of alginate concentration for a 26-gauge needle. (From Poncelet, D., Neufield, R.J., Goosen, M.F.A., Bugarski, B., and Babak, V., *AIChE J.*, 45, 2018–2023, 1999. With permission.)

other hand, microbead size could be also decreased by increasing the applied potential up to a certain value above which there was no further decrease in bead diameter. Furthermore, in the case of the smaller needle, as the potential was increased above 6 kV, natural harmonic oscillations of the needle were observed resulting in a bimodal size distribution. Similar effects of needle size and applied potential were also found for extrusion of alginate suspension of yeast cells using electrostatic droplet generation technique [20].

B. Effects of Electrode Geometry on Microbead Size

In the case of the parallel-plate setup (Figure 31.2a), effects of electrode spacing and charge arrangement (i.e., different electric field and surface charge intensity) on polymer bead size are shown in Figure 31.5a [18]. As the potential between the electrodes in the parallel-plate setup, for example, increased from 6 to 12 kV at the electrode spacing of 4.8 cm, the average bead diameter decreased from 2300 to 700 μm. Reduction of electrode distance resulted in even smaller microbeads suggesting a strong influence of distance and potential on microbead size at this charge arrangement. For example, at 10 kV, reducing the distance from 4.8 to 2.5 cm resulted in a decrease in polymer bead diameter from 1500 to 350 μm. However, increase in the applied

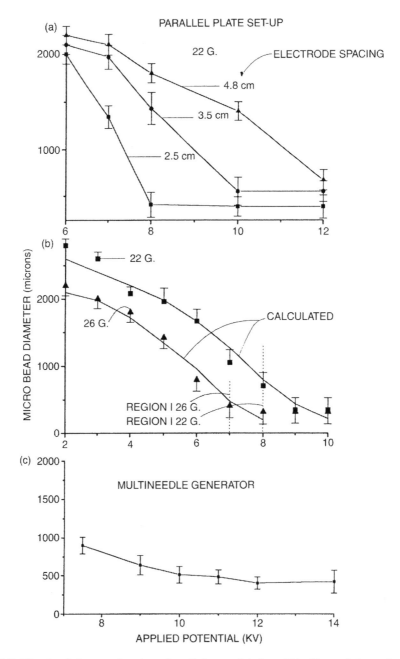

FIGURE 31.5 Microbead size as a function of applied potential; (a and b) effects of electrode spacing and needle size in the parallel-plate setup and (c) multi-needle device. (From Poncelet, D., Bugarski, B., Amsden, B., Zhu, J., Neufeld, R., and Goosen, M.F.A., *Appl. Microbiol. Biotechnol.*, 42 (2–3), 251–255, 1994. With permission.)

potential above 12 kV and decrease in electrode spacing did not further reduce the microbead size. This was probably due to a discharge between the plates accompanied by sparking as a result of air ionization in the space between the electrodes.

Similarly as found for the positively charged needle setup, needle size also significantly affected the microbead diameter in the parallel plate setup (Figure 31.5b, [18]). For a fixed electrode distance (2.5 cm), needle length (1.3 cm), and alginate concentration (1.5%) a decrease in bead diameter by a factor of two was observed for a wide range of applied potentials (2–8 kV) when the needle size was reduced from 22 to 26 gauge (Figure 31.5b).

The multiple needle device was essentially a scaled-up version of the parallel plate setup (Figure 31.2c, [17]). Results obtained were similar to those found for the single needle setup (Figure 31.2a). At an electrode distance of 2.5 cm, increase in the potential from 7 to 12 kV resulted in a decrease in bead diameter from 950 ± 100 to 400 ± 150 μm (Figure 31.5c). The device for continuous production of microbeads had a processing capacity of 0.7 l/h [17].

C. Mechanisms of Droplet Formation

In the absence of an electrostatic field with gravitational force acting alone, the mean bead diameter was 2400 ± 200 μm at a constant alginate flow rate of 36 ml/h and a 22-gauge needle. In this case, a droplet was produced every 1–2 sec. Each drop grew at the tip of the needle until its weight overcame the net vertical component of the surface tension force (Figure 31.6a, [21,24]).

Examination of the droplet formation under the influence of electrostatic forces revealed that an elongated cone is formed as the droplet meniscus advanced (Figure 31.6b, [21,24]). The forming droplet was drawn out into a long slender filament (potential of 4–5 kV, 2.5 cm electrode distance, 1.5% sodium alginate concentration, flow rate of 36 ml/h, 22-gauge needle). High charge density at the tip of the inverted cone reduced the surface tension of alginate solution [25] resulting in neck formation (Figure 31.6c). For more concentrated alginate (more than 1.5%), we observed that the neck elongated up to 1 mm before detachment (Figure 31.6d). While the main part of the liquid neck, persisted a new drop quickly formed, the long linking filament broke up into a large number of smaller droplets (Figure 31.6e, [21,24]). It was also observed that small (satellite) droplet formation usually occurred at higher potentials (above 6 kV), at which the elongation of the liquid neck prior to rupture was more pronounced. The largest of these satellite droplets was one half in diameter compared to the main drop, while the smallest was less than 20 μm in diameter.

When the alginate concentration was decreased from 1.5% to 0.8%, a difference was observed in droplet formation mechanism. For the low-viscosity alginate, elongation of the filament linking the new droplet and the meniscus at the tip of the needle was not as pronounced, resulting in more uniform bead sizes (Figure 31.7a–c, [18,26]).

The polymer droplet formation sequence is illustrated schematically in Figure 31.8. At the early stage (Figure 31.8a) shape of the liquid meniscus is almost spherical. When the electrostatic field is applied the meniscus is distorted into a conical shape as shown in Figure 31.8b. Consequently the alginate solution flows through this weak area at an increasing rate causing formation of a neck (Figure 31.8c and d). When the filament breaks away (Figure 31.8e), liquid meniscus on the needle tip is suddenly reduced for a short period of time until flow of the liquid causes the process to start again.

When the electrostatic potential was increased above 6 kV, harmonic natural needle oscillations were observed, only in the case of the thin and light 26-gauge needle [21,26]. High surface charge and the electric field on needle tip, produced a mechanical force causing needle vibration and resulting in an oscillating thread-like filament (Figure 31.9). Periodic oscillations of the electrically stressed meniscus at the needle tip caused intermittent formation and detachment of sinuously shaped filaments by a vigorous whipping action (Figure 31.9b). The long twisted filaments were formed as a result of a surface-energy component due to the surface charge [21]. In order to minimize the surface energy, molecular forces act to decrease the surface-to-volume

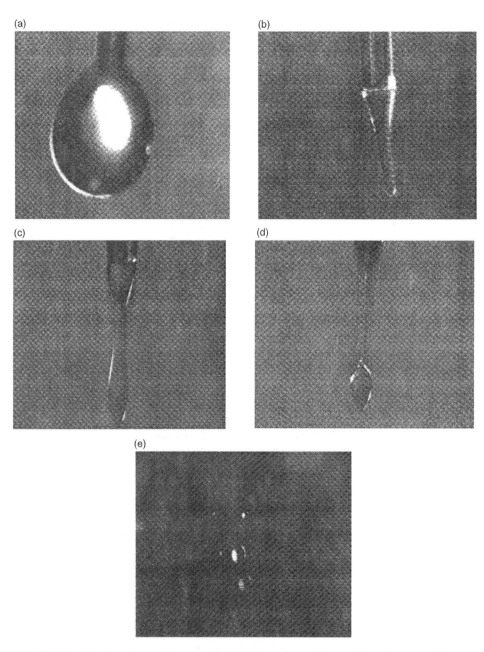

FIGURE 31.6 Droplet formation at the needle tip (1.5% sodium alginate): (a) without applied potential; (b–e) applied potential of 4–5 kV; (b) meniscus formation; (c) neck formation; (d) stretching of the liquid filament; and (e) break-up of the liquid filament. (From Nedovic, V.A., Obradovic, B., Poncelet, D., Goosen, M.F.A., Leskosek-Cukalovic, I., and Bugarski, B., *Landbauforschung Volkenrode*, SH 241, 11–18, 2002. With permission.)

ratio, while repulsive forces of the surface charge act to increase this ratio. Fragmentation of the filament results in a bimodal size distribution with peaks at 50 and 190 μm in diameter (Figure 31.11c). This phenomenon was not observed with a 22-gauge needle.

A comparative analysis of the two charge setups (Figure 2a and b) at the same applied potential and needle size, was carried out in order to get an insight into the droplet-formation mechanism [17].

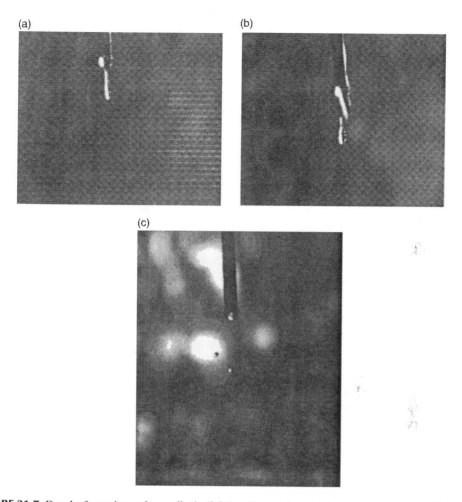

FIGURE 31.7 Droplet formation at the needle tip (0.8% sodium alginate); (a) meniscus formation; (b) filament detachment; and (c) dispersion of the liquid filament. (From Poncelet, D., Babak, V.G., Neufeld, R.J., Goosen, M.F.A., and Bugarski, B., *Adv. Coll. Int. Sci.*, 79 (2–3), 213–228, 1999. With permission.)

Examination of droplet-formation mechanisms at the needle tip using the image analysis or video system, revealed different formation modes for the two setups. At the same operating parameters (relatively high potential difference of 6 kV, 26-gauge needle, 2.5 cm electrode spacing, 36 ml/h flow rate) there was a noticeable difference in sizes of produced beads. In the parallel-plate charge setup, for example, we observed that at the given potential difference, the frequency of droplets leaving the needle tip was below that required to initiate spraying. The average mean diameter was found to be 1100 ± 200 μm (Figure 31.10a). In contrast, in the positively charged needle setup, a Taylor-cone like meniscus [13,27] was observed with a well-developed jet (80 μm diameter). This droplet formation mechanism resulted in microbeads of 170 ± 70 μm in diameter (Figure 31.10b), which is a decrease by a factor of seven as compared to the former charge setup.

D. MICROBEAD SIZE DISTRIBUTION

The mean bead size distribution curves obtained by plotting the relative frequencies versus bead diameter, typically resulted in a continuous function symmetrical about the mean value (Figure 31.11a;

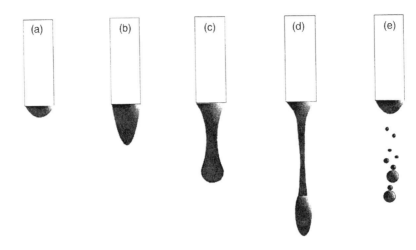

FIGURE 31.8 Schematic presentation of the droplet formation mechanism.

[16,18,19]). At the applied potentials of 4.5 and 6 kV, the mean bead size distribution was found to vary about the mean of 225 μm (Figure 31.11a) and 170 μm (Figure 31.11b), respectively. However, in the case of a naturally vibrating needle (applied potential of 7 kV), a bimodal bead size distribution was observed, with one peak at 50 μm and a second at 190 μm (Figure 31.11c, [16,18,19]).

E. FORCES ACTING ON DROPLET FORMATION

An optimization routine based on the force balance (Equation 31.4) was developed to assess the influence of applied potential on the bead diameter. The experimental data of bead radiuses, r, and applied potentials, V, in the region before the liquid jet appearance, were best fitted by the Equation (31.4) when the electric field, E, was approximated by:

$$E = \frac{V}{\sqrt{r}} \qquad (31.9)$$

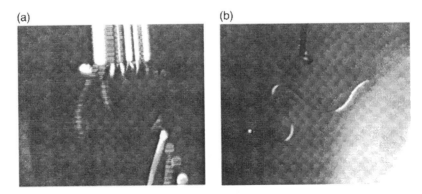

FIGURE 31.9 Needle oscillation (26-gauge, 7 kV applied potential): (a) needle vibration due to electrostatic forces, and (b) detachment of a thread-like liquid filament. (From Bugarski, B., Obradovic, B., Nedovic, V., and Poncelet, D., in *Fundamentals of Cell Immobilisation Biotechnology, Focus on Biotechnology*, Vol. 8a, Nedovic, V., Willaert, R.G., Eds., Kluwer Academic Publishers, Dordrecht, 2004, pp. 277–294. With permission).

In order to simplify the Equation (31.4), we have assumed that the internal needle radius, r_0, and the radius of the formed drop, r, were correlated by: $r_0 = k_i r$, where k_i is a constant dependent on the needle size. Substituting this expression and Equation (31.5) and Equation (31.9) into the Equation (31.4), a relationship between the applied potential and the droplet radius is obtained:

$$r = a\,(k_i - bV)^{0.5} \tag{31.10}$$

where a, b and k_i are the constants experimentally determined as $a = 3.3 \times 10^{-3}$ m, $b = 4.3 \times 10^{-5}$ V^{-1} and $k_{22} = 0.233$ for the 22-gauge needle and $k_{26} = 0.19$ for the 26-gauge needle [17].

The minimal potential required to initiate appearance of a continuous liquid jet, also referred as critical or minimal potential, can be found from positive roots of the Equation (31.10). This potential was calculated as 5.3 and 4.4 kV for 22- and 26-gauge needles, respectively, which was in good agreement with experimental observations by image analysis. At higher applied potentials bead diameters could not be estimated since roots of the Equation (31.10) become imaginary. Inserting the expression for the electric force exerted on the needle tip (Equation 31.8) in the Equation (31.3) results in:

$$V^2 = \frac{(\ln 4H/r)(c - dr^3)}{2\varepsilon_0} \tag{31.11}$$

with

$$c = r_0\gamma \quad \text{and} \quad d = \frac{2}{3}\rho g$$

where the values for the constant c were found to be 1.5×10^{-5} and 0.945×10^{-5} N for 22- and 26-gauge needles, respectively. A comparison was made between the measured and calculated diameters (Equation (31.11) and Figure 31.4; [18,19]). The general shape of calculated curves was similar to that of the experimental curves. Good agreement between experimental and calculated data was obtained in both regions for the 22-gauge needle (Figure 31.4a). For the smaller 26-gauge needle, predicted values in region I were underestimated, with an error of ± 10–25%, while calculated droplet diameters in the region II agreed well with the experimental data, with an error of $\pm 15\%$ (Figure 31.4b).

(a)   (b)

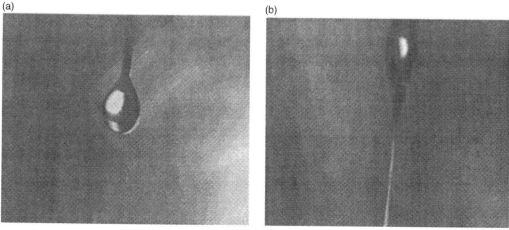

FIGURE 31.10 Comparative analysis of droplet formation mechanism (6 kV applied potential, 2.5 cm electrode distance): (a) parallel plate setup and (b) positively charged needle setup. Note the formation of the jet spray in (b). (From Poncelet, D., Bugarski, B., Amsden, B., Zhu, J., Neufeld, R., and Goosen, M.F.A., *Appl. Microbiol. Biotechnol.*, 42 (2–3), 251–255, 1994. With permission.)

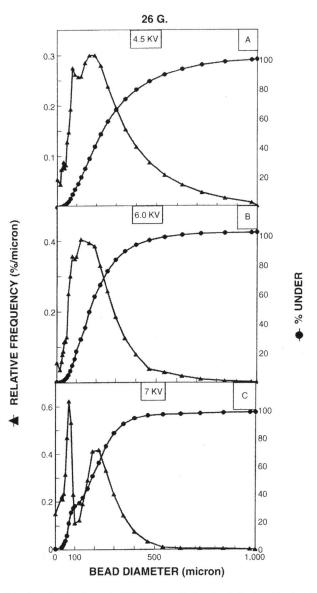

FIGURE 31.11 Droplet size distribution at different applied potentials (positively charged needle setup, 26-gauge, 2.5 cm electrode distance, 0.8% alginate concentration); (a) 4.5 kV; (b) 6 kV; (c) 7 kV. (From Poncelet, D., Babak, V.G., Neufeld, R.J., Goosen, M.F.A., and Bugarski, B., *Adv. Coll. Int. Sci.*, 79 (2–3), 213–228, 1999. With permission.)

In the case of the parallel-plate setup (Figure 31.2a), by combining Equation (31.3) and Equation (31.6), a relationship between the applied potential and the droplet radius could be obtained:

$$V^2 = \frac{(\ln 2L/r - 1.5)(c - dr^3)}{(L^2/4H^2)\varepsilon_0} \tag{31.12}$$

At a constant electrode distance ($H = 2.5$ cm) and needle length in the electric field ($L = 1.3$ cm), for both 22- and 26-gauge needles, droplet diameters predicted by the Equation (31.12) agreed well with the experimental values with an error of $\pm10\%$ (Figure 31.5b).

It is interesting to note that at the applied potential at which the liquid jet was first observed experimentally, the ratio of droplet radius to the internal needle radius (r/r_0) ranged from 2 to 4 for both needle sizes. As the potential was increased beyond the spraying potential, the ratio of droplet radius and the liquid jet radius was constant.

Let us examine different mechanisms of droplet formation in regions I and II for the charged needle setup (i.e., Figure 31.2b). The primary factor regulating droplet size in the region I is probably the intensity of the electric field in the vicinity of the forming droplet. Since the electric field increases as diameter decreases (Equation (31.9)), the intensity of the field is much higher at the meniscus tip. Detachment position is determined by the action of gravity and electrostatic field, and consequently, as the applied potential is increased the terminal droplet velocity increases. The result of increased velocity and reduced surface tension due to charge repulsion [28] is the increase in Reynolds and Weber numbers of the charged droplet. A transition occurs between low-frequency droplet generation (region I) and high-frequency droplet generation (region II). Further increase in the applied potential in the region II (above 8 kV) resulted in a slight increase in the mean bead diameter before discharge and air ionization occurred. This may have been due to a change in liquid surface tension as first proposed by Sample and Bollini [29]. Who showed that the dynamic surface tension is generally higher than the static surface tension for the same liquid. For example, the surface tension of water reached a dynamic value of 0.110 N/m, as compared to a static value of only 0.074 N/m. We can hypothesize that in the region I, charge repulsion is a dominant factor determining surface tension up to a certain value. However, when a sufficiently large charge is applied to the droplet, the resulting droplet velocity may be too high for droplet break-up to occur (i.e., there is no enough time for disintegration of the formed droplet).

F. Effects of Electrostatic Droplet Generation on Cell Viability

To assess the effects of the electrostatic field on animal cells' viability, suspensions of insect cells and murine BMSC were extruded using the electrostatic droplet generator. No detectable change in insect cell viability was observed after extrusion [21]. The initial cell density, 4×10^5 cell/ml, remained essentially unchanged at 3.85×10^5 and 3.8×10^5 cell/ml immediately after passing through the generator with an applied potential difference of 6 and 8 kV, respectively. Prolonged cultivation of these cells did not show any loss of cell density or viability (Table 31.1, [21]).

Application of the electrostatic potential of 4.6 kV to the suspension of murine BMSC in culture medium had a negligible effect on cell viability (97% versus 95% before and after the

TABLE 31.1
Effects of Electrostatic Droplet Generation on Insect Cell Viability[a] [21]

Time (days)	Cell Density (cells/ml) $\times 10^{-5}$	Cell Viability (%)
0	4	100
0[b]	3.8	93
2	7.5	90
4	10	94
6	13	92

[a]Growth in suspension after extrusion using the electrostatic voltage generator at applied potential of 8 kV with the positively charged needle.

[b]Immediately after application of high voltage.

extrusion experiment, respectively) [23]. Extrusion of cell suspension in 1.5% (w/w) sodium alginate (4×10^5 cells/ml, 24-gauge needle, 4.6 kV applied potential) resulted in formation of alginate microbeads with an average diameter of 680 ± 100 μm. The entrapped BMSC in alginate microbeads remained viable over 30 days of static cultivation [23,30].

V. CONCLUSIONS

In this chapter, an overview of the electrostatic droplet-generation method was presented with focus on droplet formation mechanisms and key parameters determining the size of resulting microbeads. An integrative approach was pursued in order to associate theoretical background of the phenomena with the qualitative and quantitative experimental analyses. The aim was to provide a resource for system optimization to achieve controlled production of very small uniform microbeads (i.e., less than 100 μm in diameter). Application of the charged needle setup resulted in significantly smaller microbeads as compared to the parallel-plate setup under the same operating conditions. Needle diameter and applied electrostatic potential were shown as main parameters affecting the microbead size in both setups. The greatest decrease in microbead size was observed when natural needle oscillations, caused by surface charge and high electric field, resulted in formation of whip-like liquid filaments breaking at the end of the needle. As a consequence, produced droplets had bimodal size distribution with a large fraction below 50 μm in diameter. Modification of the parallel-plate setup system in the form of a multi-needle device revealed possibilities for continuous production of uniform microbeads at high processing capacity. Finally, there was no detectable loss in viability after passing animal cells through the electrostatic droplet generator. This is a promising result, as it proves this technique amenable for cell immobilization.

ACKNOWLEDGMENT

This work has been funded by the Ministry of Science and Environmental Protection of the Republic of Serbia (Grant 1776).

NOTATIONS

E	electric field (V/m)
F_e	electrostatic force (N)
F_g	gravitational force (N)
F_γ	surface tension force (N)
g	acceleration of the gravity (m/s^2)
H	electrode distance (m)
k, a, b, c, d, e	constants in Equation (31.7) and Equation (31.8)
L	part of the needle length exposed to the electric field (m)
P_e	electrostatic stress (N/m^2)
q	electric charge (C)
r	radius of the forming droplet (m)
V	electrostatic potential (V)

Greek Symbols

γ	surface tension (N/m)
ε_0	permittivity of air (C^2/Nm^2)
ρ	density (kg/m^3)

REFERENCES

1. Balachandran, V. and Bailey, A.G., The Dispersion of Liquids using centrifugal and electrostatic forces, *IEEE Trans. Ind. App.*, IA20, 682–686, 1984.
2. Fillimore, G.L. and Van Lokeren, D.C., Multinozzle drop generator which produces uniform break up of continuous jets, In: *IEEE Meeting of the Industrial Application Society*, 1982, p. 991.
3. Keshavarz, T., Ramsden, G., Phillips, P., Mussenden, P., and Bucke, C., Application of electric field for production of immobilized biocatalysts, *Biotech. Tech.*, 6, 445–450, 1992.
4. Rayleigh, J.W.S., On the equilibrium of liquid conducting masses charged with electricity, *Phil. Mag.*, 14, 184–186, 1882.
5. Rayleigh, J.W.S., *The Theory of Sound*, Dover Publications, Inc., New York, 1945 (Reprint, Vol. II, 372).
6. Nawab, M.A. and Mason, S.G., The preparation of uniform emulsions by electrical dispersion, *J. Coll. Sci.*, 13, 179–187, 1958.
7. Fonseca, M.M., Black, G.M., and Webb, C., Reactor configuration for immobilized cells, in *Process Engineering Aspects of Immobilized Cell System*, The Institution of Chemical Engineers, Rugby, Warwickshire England, 1986, pp. 63–69.
8. Klein, J., Stock, J., and Vorlop, D.K., Pore size and properties of spherical calcium alginate bio-catalysts, *Eur. J. Appl. Microbiol. Biotechnol.*, 18, 86–91, 1983.
9. Goosen, M.F.A., O'Shea, G.M., Gharapetian, M.M., and Sun, A.M., Immobilization of living cells in biocompatible semipermeable microcapsules: biomedical and potential biochemical engineering applications, in *Polymers in Medicine*, Chiellini, E., Ed., Plenum Publ. Comp., New York, 1986, pp. 235–247.
10. Zeleny, G., Discharge from liquid points and method for measuring the electric field at their surfaces, *Phys. Rev.*, (Ser. 2) 3, 69, 1914.
11. Zeleny, G., Instability of electric liquid surfaces, *Phys. Rev.*, (Ser. 2) 10, 1–6, 1917.
12. Taylor, G.I., Disintegration of water drops in electric fields, *Proc. R. Soc. A*, 280, 383–397, 1964.
13. Taylor, G.I., The force exerted by an electric field on long cylindrical conductors, *Proc. R. Soc. A*, 291, 145–158, 1966.
14. DeShon, E.W. and Carlson, R., Electric field and model for electrical liquid spraying, *J. Coll. Sci.*, 28, 161–166, 1968.
15. Bugarski, B., Amsden, B., Neufeld, R., Poncelet, D., and Goosen, M.F.A., Effect of electrode geome-try and charge on the production of polymer microbeads by electrostatics, *Can. J. Chem. Eng.*, 72, 517–522, 1994.
16. Pjanovic, R., Goosen, M.F.A., Nedovic, V., and Bugarski, B., Immobilization/encapsulation of cells using electrostatic droplet generation, *Minerva Biotec.*, 12, 241–248, 2000.
17. Poncelet, D., Bugarski, B., Amsden, B., Zhu, J., Neufeld, R., and Goosen, M.F.A., Parallel-plate electrostatic droplet generator-parameters affecting microbead size, *Appl. Microbiol. Biotechnol.*, 42 (2–3), 251–255, 1994.
18. Poncelet, D., Babak, V.G., Neufeld, R.J., Goosen, M.F.A., and Bugarski, B., Theory of electrostatic dispersion of polymer solution in the production of microgel beds containing biocatalyst, *Adv. Coll. Int. Sci.*, 79 (2–3), 213–228, 1999.
19. Poncelet, D., Neufield, R.J., Goosen, M.F.A., Bugarski, B., and Babak, V., Formation of microgel beads by electrostatic dispersion of polymer solutions, *AIChE J.*, 45, 2018–2023, 1999.
20. Nedovic, V., Obradovic, B., Leskosek-Cukalovic, I., and Bugarski, B., Electrostatic generation of alginate microbeads loaded with brewing yeast, *Proc. Biochem.*, 37 (1), 17–22, 2001.
21. Bugarski, B., Smith, J., Wu, J., and Goosen, M.F.A., Methods for animal cell immobilization using electrostatic droplet generation, *Biotechnol. Tech.*, 7, 677–683, 1993.
22. Poncelet, B., Poncelet, D., and Neufeld, R. J., Control and mean diameter and size distribution during formulation of microcapsules with cellulose nitrate membranes, *Enzyme Microb. Technol.*, 11, 29–37, 1989.
23. Obradovic, B., Bugarski, D., Petakov, M., Jovcic, G., Stojanovic, N., Bugarski, B., and Vunjak-Novakovic, G., Cell support studies aimed for cartilage tissue engineering in perfused bioreactors, in *Progress in Advanced Materials Processes*, Mater. Sci. Forum, Vol. 453–454, Uskokovic, D.P., Milonjic, S.K., Rakovic D.I., Eds., Trans Tec Publications Ltd., Switzerland, 2004, pp. 549–555.

24. Nedovic, V.A., Obradovic, B., Poncelet, D., Goosen, M.F.A., Leskosek-Cukalovic, I., and Bugarski, B., Cell immobilisation by electrostatic droplet generation, *Landbauforschung Volkenrode*, SH 241, 11–18, 2002.

25. Hendricks, C.D., Jr., Charged droplet experiments, *J. Coll. Sci.*, 17, 249–259, 1962.

26. Bugarski, B., Obradovic, B., Nedovic, V., and Poncelet, D., Immobilization of cells and enzymes using electrostatic droplet generator, in *Fundamentals of Cell Immobilisation Biotechnology, Focus on Biotechnology*, Vol. 8a, Nedovic, V., Willaert, R.G., Eds., Kluwer Academic Publishers, Dordrecht, 2004, pp. 277–294.

27. Taylor, G.I. and Van Dyke, M.D., Electrically driven jets, *Proc. R. Soc. A*, 313, 453–475, 1969.

28. Landau, L.D., Lifshitz, E.M., and Pitaevskii, L.P., *Electrodynamics of Continuous Media*, Pergamon Press Ltd., Oxford, 1960, pp. 18–19.

29. Sample, S.B. and Bollini, R., Production of liquid aerosols by harmonic electrical spraying, *J. Coll. Sci.*, 41, 185–193, 1972.

30. Bugarski, D., Obradovic, B., Petakov, M., Jovcic, G., Stojanovic, N., and Bugarski, B., Alginate microbeads as potential support for cultivation of bone marrow stromal cells, Book of Abstracts, in *Proceedings of the Sixth Yugoslav Materials Research Society Conference YUCOMAT 2004*, Herceg Novi, Serbia and Montenegro, September 13–17, 2004, p. 112.

32 Micro-Biosensor Based on Immobilized Cells

Ljiljana Mojovic
University of Belgrade, Belgrade, Serbia and Montenegro

Goran N. Jovanovic
Oregon State University, Orlando, USA

CONTENTS

I. INTRODUCTION

Chromatophores are terminally differentiated, neuron-like cells containing pigmented granules that are responsible for the brilliant colors of fish, amphibians, reptiles, and cephalopods, and their dynamic color adaptations and adjustable camouflage capabilities [1,2]. Chromatophores are located in the skin and scale tissues of fish. They are biochemically related to nerve cells and share many of their important sensory properties and responses to biologically active chemical agents. Various biologically active and toxic substances can act on chromatophores as signaling molecules through the receptors placed on the cell surface. Chromatophore responses to these agents are mediated through a complex array of cell surface receptors, signal transduction pathways, and metabolic processes resulting in the movement of pigment granules along microtubules [3,4]. Figure 32.1 presents a

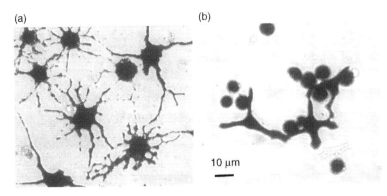

FIGURE 32.1 The morphological response of chromatophores isolated from *B. splendens*, the Siamese fighting fish, upon application of a biologically active agent. (a) Control cells, (b) cells treated with a chemical agent.

characteristic response of chromatophores to biologically active agents. Changes in pigment color and location within cells can be monitored microscopically. In addition to various environmental toxins such as heavy metals, organophosphate pesticides, polynuclear aromatic hydrocarbons [5], herbicides, fungicides, and some genotoxins [6], the cytosensor has the capability to detect a wide variety of potential toxicants from various classes of bacterial toxins to numerous cell-receptor agonists [7].

Furthermore, chromatophores have the potential to be used to detect unidentified and newly emerging food and water-borne microbial pathogens. Conformal assays (antibody- or DNA-based) offer good results in detecting the presence of a specific bacterium and its toxin. However, they cannot report on the biological activity of either the bacterium or its toxin, or can they be used to detect unidentified (genetically rearranged) or newly emerging (new toxin-producing variants) food and water borne microbial pathogens [8].

Because chromatophores are as complex and rich in physiological targets of drug and toxin action, they are very promising candidates for use as broad-ranging biosensors of utility in medical screening, pharmaceutical research, food toxicology, and environmental monitoring.

The primary objective of this work was to investigate the design of the micro-biosensor based on immobilized living chromatophores of Siamese fighting fish, *Betta splendens* and to develop enabling technologies like immobilization of chromatophores on an appropriate microcarrier, which would enable their movement through the biosensor device, whilst preserving their biosensing capabilities. For these purposes three different types of microcarriers glass, polystyrene, and gelatin beads were tested for their efficiency in binding fish chromatophores. The kinetics of cells attachment onto gelatin beads was also investigated, and the optimal conditions for cell binding were determined. In addition, the incorporation of ferromagnetic powder of iron (II, III oxide) into gelatin beads was studied and the functionality of the micro-biosensor was tested with the neurotoxin analog clonidine as a model toxin. A double-exponential model was proposed to describe the toxin-induced change of cell area covered with pigment.

II. DESIGN OF MICRO-BIOSENSOR

As shown in Figure 32.2, micro-biosensor prototype design is based on the microchannel architecture ($500 \times 700\ \mu m$ in cross-section, and length of 5 cm) and manufactured by using microlamination-machining technology [9]. The microscale system was chosen because it can provide precisely controlled microenvironments, faster response times, increased design flexibility, and has the potential for massively parallel operations. These attributes are uniquely suited for industrial, military, and research applications of high-throughput analysis, including testing

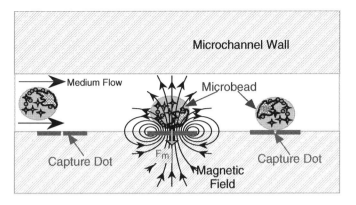

FIGURE 32.2 Microchannel type biosensor. Transport and capturing of microcapsules containing immobilized chromatophores within the magnetic field generated by "capture dot". (From Mojović, L. and Jovanović, G.N., *Chem. Ind.*, 57 (12), 605–610, 2003. With permission.)

devices for environmental sampling, medical screening, proteomics, combinatorial pharmaceutical development, and the detection of microbial pathogens.

Basic micro-biosensor element is a microbead or a microcapsule ($d = 250$ μm) containing two essential rudiments: (i) immobilized and live chromatophores isolated from Siamese fighting fish, *B. splendens* and (ii) inert ferromagnetic powder embedded in microcarrier matrix. Transport and positioning of microbeads within microchannel is facilitated by fluid flow, and by the interaction between the ferromagnetic material and a magnetic field. A specially designed "capture dot" device, presented in Figure 32.3, is a microsolenoid deposited in the microchannel walls [9–12]. Its main function is to generate the magnetic field needed for capturing of the microbeads.

III. IMMOBILIZATION OF CHROMATOPHORES

Immobilization of chromatophores on the surface of the microbeads represents a key step in bio-sensor development. Fish chromatophores are anchorage-dependent cells that require compatible surface for attachment, subsequent spreading, and growth [5]. Generally, immobilization on the surface of small beads or microcarriers is a suitable method for the cultivation of anchorage-dependent cells. Van Wezel [13] first reported immobilization of cells on mammalian cells on diethyla-minoethyl (DEAE)–Sephadex A50. Subsequently this type of carrier was improved by optimizing

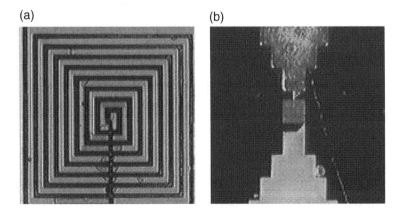

FIGURE 32.3 Operational capture-dot. (a) High magnification photomicrograph to show coil configuration, (b) lower magnification photomicrograph to show electrical leads.

the surface electrostatic charge allowing cells to grow to higher densities [14–17]. These findings led to the development of the first commercial microcarriers, which were based on derivatized Sephadex (Cytodex I, II, III, Pharmacia) and were utilized extensively for the commercial production of human vaccines [18]. Various other microcarriers, such as polyacrylamide [18,19], polyurethane foam [20], polystyrene [21,22], glass [22,23], etc., are reported to be suitable for various cell lines. However, a microcarrier of pure gelatin developed by Nilsson and Mosbach [24,25], and later commercialized as Cultispher (Percell, Biolytica, Sweden), has attracted a great deal of attention from researchers in cell culture area [26–30]. The main advantages of this type of carrier are efficient cell attachment due to biospecific binding, a high surface area, and good environmental protection that can be achieved [26]. In addition, cells can be recovered by dissolving the microcarriers with proteolytic enzyme.

Here we will present the results obtained by investigating three different types of microcarriers namely, glass, polystyrene, and gelatin beads for their efficiency in binding fish chromatophores. Various amounts of ferromagnetic powder of iron (II, III oxide) were incorporated in gelatin beads used for chromatophore immobilization.

IV. EXPERIMENTAL

A. ISOLATION OF PRIMARY CELL CULTURE

Fish chromatophores were isolated from the tails and fins of *B. splendens* fish according to the procedure described earlier [6,12]. Only red *B. splendens* fish which consisted of erythrophores (red-pigmented cells) were used in this study.

B. MICROCARRIERS

Three types of microcarriers were used in this study — glass (Sigma, $d = 150–212$ μm), polystyrene (BangsLabs, Inc., $d = 186$ μm), and macroporous ferromagnetic gelatin. Macroporous ferromagnetic gelatin beads containing various amounts of ferromagnetic powder (iron (II, III) oxide, powder $d_p < 5$ μm, Aldrich) were prepared according to the procedure described by Nilsson and Mosbach [19,25]. An appropriate amount of iron (II, III) oxide (mass fraction of 5, 10, 15, 20, and 25% on gelatin powder) was added to the water–gelatin solution (type I gelatin from porcine skin, Sigma) before further processing. At the end of processing, dry beads were sieved, and those with diameters between 180 and 300 μm were collected and cross-linked with glutaraldehyde (grade I, 50%, Sigma). All three types of microcarriers were hydrated in phosphate-buffered saline (PBS) without calcium and magnesium ions, washed extensively, and then resuspended in PBS at concentration of 5 g/l. Glass and gelatin microcarriers were autoclaved for 20 min at 121°C, while the polystyrene microcarriers were sterilized by incubating at 70°C for 2 h, as recommended by the manufacturer. After sterilization, microcarriers were kept in solution at room temperature.

In preparation of microcarriers, an appropriate amount of microcarrier stock suspension was transferred to a 50-ml sterile conical centrifuge tube and the beads were allowed to settle. After removing the supernatant, the microcarriers were washed twice with growth medium (L-15) and transferred to Erlenmeyer flasks, where the attachment of cells to beads was performed. When the effect of cell-attachment-promoting agents like fibronectin was studied, the appropriate amount of microcarrier was kept for 2 h prior to its use in PBS (20 ml) with 100 μl of fibronectin stock solution (Sigma), and then washed with growth medium before transfer into Erlenmeyer flasks. The number of microbeads per gram of beads was determined in order to optimize cell or bead ratio (λ). Beads were counted in a standard volume on a haemocytometer grid. An average value of 0.82×10^6 beads/g was found in repeated measurements for beads with 10% of ferromagnetic powder.

C. Cell Immobilization

The attachment of the cells to the beads was performed in siliconized (by Sigmacote, Sigma) Erlenmeyer flasks in L-15 media with very gentle stirring (30–50 rpm). L-15 media was enriched with 5% of fetal bovine serum (FBS, Hyclone, Lab). The rate of attachment of cells from inoculated microcarrier cultures was determined by counting the cells remaining in the culture. Culture samples (200 μl) were taken at 20-min intervals and allowed to settle for 1 min in an Ependorf tube. The microcarrier-free supernatant was introduced into a haemocytometer for cell counting.

Culture samples were examined microscopically to determine cell viability and toxin-sensitivity. Chromatophores that responded to addition of neurotoxin were considered alive and toxin-sensitive.

D. Analysis of Attachment Kinetics

The rate of disappearance of free cells was followed by an exponential decay curve:

$$C_t = C_0 \cdot 10^{-kt} \tag{32.1}$$

where C_t is the concentration of free cells (cells/ml) at time t, C_0 the original cell concentration (cells/ml), and k a rate constant.

This equation can be expressed logarithmically as:

$$\log C_t = \log C_0 - kt \tag{32.2}$$

Thus, a straight line would represent a first-order kinetics rate from a plot of $\log C_t$ versus time with a gradient (k). The value of k was interpreted at the specific attachment rate (min^{-1}).

E. Capturing of Microcapsules Using Magnetic Field

For these experiments, in order to protect the cells from environmental shear stress, an ~5 μm thin membrane made of alginate and poly-L-lysine was created around the immobilized micro-carrier [31]. The movement of microcapsules with immobilized chromatophores was observed microscopically in a glass microtube ($d = 700$ μm; $l = 5$ cm) with the magnetic field conduit embedded in the wall. The experimental setup is presented in Figure 32.4. Fluid (L-15 medium) velocities applied were in the range from 1.6 to 6.4 mm/s and corresponded to the predicted operational fluid velocities of the biosensor [11]. Fluid flow was provided by microsyringe pump (LabTronix, USA).

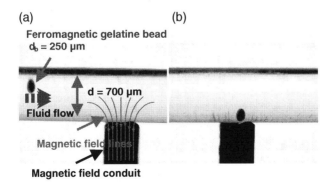

FIGURE 32.4 Capturing and releasing of gelatin bead with iron (II, III) oxide with magnetic field under fluid flow. Ferromagnetic gelatin bead in a micro-channel approaching magnetic field (a), and captured with magnetic field (b).

The magnetic field intensity needed to stop and capture microcapsules was determined for various fluid velocities and for different amounts of ferromagnetic material set in. Magnetic field intensity was measured by Gaussmeter, Model 410 (Lake Shore Cryotronics, Inc).

F. Testing of Immobilized Chromatophores with Clonidine

The response of the immobilized cells to different concentrations of a neurotoxin clonidine was monitored by a change in pigmented cell area induced by the neurotoxin as described earlier [32]. Image capture of immobilized chromatophores was performed at preset time intervals controlled by image-capturing software within 6 min with a digital Pulnix TMD-7DSP CCD camera connected to a Matrox computer. Captured images were analyzed to determine the change in pigmented cell area induced by the model toxin using proprietary software ("Cell cruncher") principally described elsewhere [5,33].

V. RESULTS AND DISCUSSION

A. Attachment of Cells to Microcarriers

Figure 32.5 shows the percent of attached cells to gelatin, glass, and polystyrene microcarriers, 3 h after cell inoculation. The best results were obtained with gelatin beads (95% of attached cells). Attachment to glass microcarrier resulted in significantly lower percent of cells attached (62%), while fish chromatophores showed the lowest affinity toward polystyrene microcarrier (17% of attached cells).

Gelatin beads have already been reported to be appropriate for various cell types, such as human fibroblast cells [26], pancreatic islet cells [28], Chinese hamster ovary (CHO) cells [27], green monkey kidney cells (Vero) [29], and human hepatocytes [30]. Cell attachments close to 100%, as well as high cell densities have been reported for gelatin microcarriers. These were attributed to a microcarrier chemical structure that enables biospecific binding of cells, as well as to a high surface area deriving from a porous structure. Thus, cells may also populate interior of the cavities and could withstand higher agitation rates than those on solid microcarriers [27].

Although, glass carriers are widely used in cell culture studies [22], the fish chromatophores showed only moderate affinity, lower than some other mammalian cell lines reported in the literature. Improvements in cell attachment might be achieved by either pretreatment with cell attachment-promoting agents, or by using special types of aluminum borosilicate glass with controlled pore size [23].

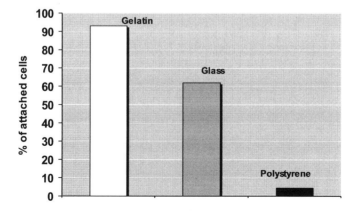

FIGURE 32.5 Attachment of fish chromatophores to gelatin, glass, and polystyrene microcarriers. Reaction conditions: L-15 media supplemented with 5% of FBS, pH = 7.4, time t = 3 h, stirring rate v = 40 rpm.

Lee et al. [34] immobilized BHK on a sulphonated polystyrene microcarrier (S80), and on surface-modified microporous polystyrene (Polyhipe). Performances of these carriers were reported to be equivalent to those of Cytodex (dextran microcarrier) and Cultispher (gelatin microcarrier). The study of Maroudas [35] has shown that sulphonated polystyrene with 2–5 negatively charged groups/nm^2 promote maximum cell attachment and spreading. Despite the fact that many cell types adhere better to polystyrene than to glass [22] fish chromatophores demonstrated poor attachment. This result may be a consequence of inappropriate surface charges present on the commercial support that was used. The gelatin beads gave the best results and were selected for further studies.

1. Kinetics of Attachment to Gelatin Beads

Attachment of fish chromatophores to gelatin beads containing various amounts [0–25% mass fraction of iron (II, III) oxide] was observed under conditions described earlier [10]. The kinetics of the attachment to gelatin beads containing 10% of ferromagnetic material is presented in Figure 32.6. To promote cell adhesion, gelatin beads were pre-treated with fibronectin.

After 140 min 95% of all cells present in the solution were attached on the microcarrier. Semi-logarithmic plots of unattached cell concentration as a function of time yielded straight lines, indicating first-order kinetics (Figure 32.6). The first-order attachment kinetics has been reported previously for immobilization of anchorage-dependent cells on DEAE-derivatized sephadex [15,17]. However, the authors used charged microcarriers and also reported an increase in the attachment rate with increasing exchange capacity of the microcarriers. Apparently, the kinetics of cell binding to charged microcarriers and the attachment rate constant are at least one order of magnitude higher than the one reported for the attachment to the biospecific macroporous gelatin carrier [27,29]. However, for both types of carriers the final attachment efficiency was reported as high as 90–100%.

Figure 32.6 shows that the attachment rate constant for fibronectin-pretreated beads ($k = 0.94 \times 10^{-2}\ min^{-1}$) is approximately 10% higher than for beads without pretreatment ($k = 0.85 \times 10^{-2}\ min^{-1}$). This result could be expected since proteins like fibronectin, vibronectin, laminin, and collagen make up the extracellular matrix between cells or between cells and substratum, and mediate cell attachment and spreading [22,23,36].

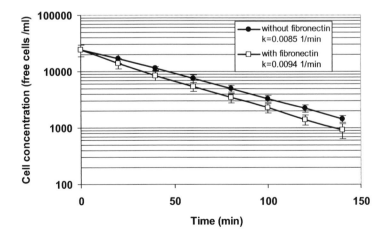

FIGURE 32.6 The kinetics of attachment of fish chromatophores on gelatin microcarriers with 10% of iron (II, III) oxide. Reaction conditions: L-15 media supplemented with 5% serum, pH = 7.4, time $t = 140$ min, stirring rate $\nu = 40$ rpm, cell/bead ratio $\lambda = 70$. Data presented are mean values of three experiments ± standard deviation. (From Mojović, L. and Jovanović, G.N., *Chem. Ind.*, 57 (12), 605–610, 2003. With permission.)

No significant effect of ferromagnetic material on the cell attachment rate constant was noticed in the range from 0 to 25% of iron (II, III) oxide concentrations used in this study (data not presented). Attachment rates for samples containing different amount of ferromagnetic material were found to be statistically indistinguishable from rates reported for gelatin beads without ferromagnetic material [10]. However, as shown in Figure 32.7, a significant effect of the concentration of ferromagnetic material on the cell viability was found. Cell viability was seriously affected on beads with 25% iron (II, III) oxide (Figure 32.7). The 10% mass fraction of iron (II, III) oxide may be considered appropriate for use in this biosensor study because it does not compromise cell viability and it supports complete cell functionality and toxin sensitivity.

2. Effect of Cell-to-Bead Ratio

The effect of the cell/bead ratio (λ = number of cells/bead) on the cell attachment rate constant and on the viability of immobilized cells was determined for gelatin beads with 10% of ferromagnetic material (Table 32.1). It is important to insure a large initial cell-to-bead ratio, which would not affect cell viability, and also to minimize the proportion of unoccupied beads during the immobilization process. Fish chromatophores are terminally differentiated cells and do not replicate in tissue culture, thus the initial cell/bead ratio will not increase with time, as reported for some other proliferating animal cells like Vero cells [10]. By microscopic examination, we observed that immobilized fish chromatophores stay functional, for example, responsive to clonidine for 2–4 weeks, although a small decrease in cell/bead ratio occurred due to apoptosis or cell death.

For each initial cell/bead ratio observed, viable immobilized cell/bead ratio was calculated from the total number of cells immobilized and measured after 24 h. Typical results are presented in Table 32.1. These results indicate that $\lambda = 70$ is an optimum value to use for immobilization of fish chromatophores. Higher initial cell/bead ratio causes lower cell viability, and therefore lower viable immobilized cell/bead ratio is achieved. Lower viability is most probably due to the shortage of living space and due to higher competition of cells for nutrients. At $\lambda = 70$, microscopic examination did not show the presence of unoccupied beads.

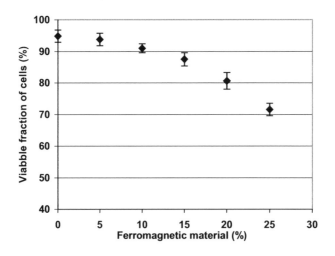

FIGURE 32.7 Effect of the amount of ferromagnetic material in gelatin beads on chromatophore viability. Reaction conditions: L-15 media supplemented with 5% serum, pH $= 7.4$, time $t = 140$ min, stirring rate $\nu = 40$ rpm, cell/bead ratio $\lambda = 70$. Data presented are mean values of three experiments $\pm$ standard deviation. (From Mojović, L. and Jovanović, G.N., *Chem. Ind.*, 57 (12), 605–610, 2003. With permission.)

TABLE 32.1
Effect of Cell/Bead Ratio on Fish Chromatophore Attachment Rate Constant k and on Cell Viability of Immobilized Chromatophores[a]

Initial Cell/Bead Ratio (λ)	Attachment Rate Constant $k \times 10^2$ min	Fraction of Viable Immobilized Cells (%)	Viable Immobilized Cell/Bead Ratio
15	0.74	92	12
30	0.78	91	25
50	0.84	88	41
70	0.85	83	55
80	0.85	70	52

[a]Reaction conditions: attachment was performed on gelatin microcarrier with 10% iron (II, III) oxide, $d = 250$ µm, in L-15 media supplemented with 5% serum at pH = 7.4 during $t = 140$ min with stirring rate $v = 40$ rpm. Viability of immobilized cells was measured after 24 h. The data presented are mean values of triplicate experiments.

B. CAPTURING OF MICROCAPSULES USING MAGNETIC FIELD

As shown in Figure 32.8, the magnetic field intensity needed to stop and capture gelatin microcapsules with immobilized chromatophores depends on the velocity of the fluid that moves microcapsules and on the amount of ferromagnetic material inside the capsules. However, the effect of ferromagnetic material in gelatin beads is more pronounced for the range of applied fluid velocities from 1.6 to 6.4 mm/s, which is the range of predicted operating velocities of fluid in the biosensor. As presented in Figure 32.8, higher magnetic field intensities are needed to capture the microbeads containing lower amounts of ferromagnetic material. These results are in agreement with the basic principles of magnetism [37,38], and also with the experimental results obtained with ferromagnetic particles fluidized in the magnetic field [39]. Generally, magnetic forces acting on ferromagnetic particles increase with the increase of the following parameters: magnetic field intensity, magnetic susceptibility (χ) of the ferromagnetic material, and volume fraction of ferromagnetic material contained in microbeads.

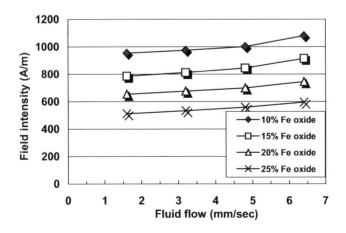

FIGURE 32.8 Effect of fluid flow velocity on the intensity of magnetic field needed to capture microcapsules with different content of ferromagnetic material.

Field intensities needed to capture microcapsules of immobilized chromatophores with 10% of ferromagnetic material that are considered to be the most appropriate for use in biosensor, were found to be in the range from 954 to 1092 A/m for the applied fluid velocities from 1.6 to 6.4 mm/s. These magnetic field intensities may be easily provided by solenoid or by some other type of magnetic field conduit.

C. Testing of Immobilized Chromatophores Using Clonidine

Figure 32.9 presents experimental video-image output at time intervals $t = 0, 50, 180, 360$ s after addition of toxin to fish chromatophores immobilized on gelatin microcarrier with 10% of ferromagnetic material. The aggregation of pigment granules induced by toxin in cells is obvious.

The response to clonidine is mediated by cell-surface receptors of the classic G-protein-linked type [1]. The mechanism of signal transduction is rather complex. Receptors that cause aggregation are linked to the G_i proteins, whose activation results in a decrease in cyclic adenosine monophosphate (cAMP) content in cells. On the other hand, receptors that cause dispersion are linked to G_s proteins, whose activation results in an increase in cAMP. This cAMP increase activates cAMP-dependent protein kinase (protein kinase A), which phosphorylates and activates other proteins, initiating a cascade of events resulting in pigment-granule dispersion. The long-range movements of pigment granules depend on polar microtubules and specific motor proteins bound to the pigment granules. There are two major families of motor proteins, kinesins that move their cargo outward, and dyneins that move the granules inward, toward the centrosome. The ability of these motor proteins to bind microtubules, and thus transport pigment granules, is regulated by the phosphorylations resulting from the signal cascade initiated by G-protein-linked receptor binding [1,3,40].

1. Fitting Experimental Data with a Double Exponential Model for Cell Area Decrease

The graph presented in Figure 32.10 shows an exponential decrease in cell area covered by pigment after addition of 50 nM clonidine.

A double exponential model presented with Equation (32.3) is proposed for this response. The model implies the existence of a heterogeneous population of cells, which can be roughly divided into two subpopulations, differing in the rate of cell reaction to the toxin exposure. It is interesting to note that the visual appearance of fast responding cells ("star-like" cells) and slow responding cells ("sheet-like" cells) can be easily discriminated under microscope and with the aid of the shape-recognition software.

$$A_t = A_0 - B(1 - e^{-Ct}) - D(1 - e^{-Et}) \qquad (32.3)$$

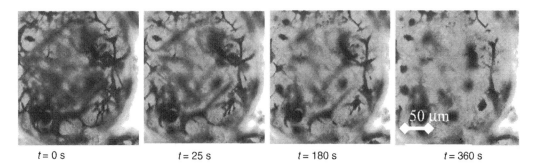

$t = 0$ s $\qquad$ $t = 25$ s $\qquad$ $t = 180$ s $\qquad$ $t = 360$ s

FIGURE 32.9 Video image frames of immobilized chromatophores beads with 10% of ferromagnetic powder at $t = 0, 25, 180$, and 360 s after exposure to clonidine ($c = 50$ nM). (From Mojović, L. and Jovanović, G.N., *Chem. Ind.*, 57 (12), 605–610, 2003. With permission.)

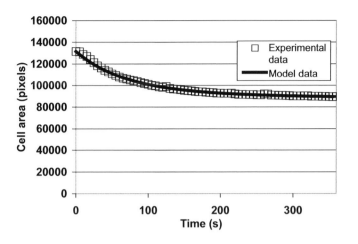

FIGURE 32.10 Decrease of the cell area of immobilized chromatophores with time, as a response to clonidine ($c = 50$ nM). Experimental data ($\square$), and model data (-), $\sigma = 0.009\%$.

At steady state ($t \rightarrow \infty$), the above equation reduces to:

$$A_\infty = A_0 - B - D \tag{32.4}$$

where A_0 is the initial area, A_t the cell area at time t, A_∞ the cell area at $t = \infty$, B the total area change of the first cell subpopulation (faster responding cells), C the rate constant for the first subpopulation, D the total area change of the second cell subpopulation (slower responding cells), E the rate constant for the second subpopulation.

Total percentage of area change (PAC) may be calculated as:

$$
\begin{aligned}
\frac{A_0 - A_\infty}{A_0 \cdot 100} &= \frac{A_0 - A_0 + B + D}{A_0 \cdot 100} \\
&= \frac{B + D}{A_0 \cdot 100}
\end{aligned}
\tag{32.5}
$$

Goodness-of-fit of the proposed model with experimental data is represented by the mean relative percentage deviation modulus σ (%) which is defined as:

$$\sigma = \frac{100}{N} \sum \frac{\varepsilon_i}{x_i} \tag{32.6}$$

where N is the number of trials, ε_i the experimental error, and x_i the model predicted value.

As shown in Figure 32.10, the proposed model and the experimental data are in very good agreement, with $\sigma = 0.009\%$.

Double exponential model equation that represents experimental data from the graph in Figure 32.10 is presented below:

$$A_t = 131565 - 26744.9(1 - e^{-0.0168t}) - 16368.9(1 - e^{-0.0078t}) \tag{32.7}$$

From the values of the coefficients B and D (26744.9 and 16368.9, respectively), it is obvious that the majority of cells are fast responding cells. The subpopulation of fast responding cells caused 20.3% of area change, while the population of slow responding cells caused 12.4%, of area change. Total PAC for applied clonidine concentration (50 nM) is 32.7% (Equation 32.5). Coefficients C and E (0.0168 and 0.0078, respectively) are rate constants, which are specific for the toxin used and prevailing experimental conditions. Higher rate constant C characterizes the subpopulation of faster responders, while the smaller constant E is a characteristic of the slower cell subpopulation.

Different reactive agents (environmental toxins, heavy metals, bacterial toxins, chlorinated hydrocarbons, drugs, etc.) may produce quite different cell reactions. The main differences could be in (i) mode of response: aggregation, dispersion, no reaction, or "freezing" of chromatophore; (ii) magnitude of response: partial aggregation, full aggregation, partial dispersion, full dispersion; (iii) kinetics of responses; and (iv) differences in responses of particular cell subtypes. These four main response features may be quantified with the parameters of the presented model. Toxins and agents could be classified into different categories according to chromatophore responses. It is already shown [5] that agents belonging to the same category show similar effect on chromatophore. For example, the category of adrenergic neurotoxins, is characterized by rapid and full aggregation, quick response time (<5 min), high sensitivity (>several nM) and high reproducibility. On the other hand, category of agents which elevate cAMP levels in cells causing dispersion, such as the melanocyte stimulating hormone (MSH), forskolin, and some bacterial toxins, exhibited slower response time (~20–80 min), but also high sensitivity and high reproducibility. However, the reaction to some chlorinated hydrocarbons is much slower and of lower sensitivity [6]. Figure 32.11 illustrates the dispersion of pigment granules caused by addition of MSH and adenylate cyclase from *Bordetella pertussis* [7].

An important issue that should be addressed is a testing and categorizing of substances or environmental factors that act as false positive agents. These agents should be identified and the cell responses should be analyzed and included into the biosensor database. The reduction of false positive and false negative responses is one of the important points in developing a reliable biosensor. Jovanović et al. [41] proposed an "orthogonal" approach as a way to improve the reliability of detecting potentially harmful agents by using two different cell-based sensing systems, for example, combining the fish chromatophores (*B. splendens*) with algal cells (*Mesotaenium caldariorum*). It was shown that the fish chromatophore system was able to classify 91% of the toxins into their actual group whereas the algal system classification efficiency was only 72%. The combination of the parameters for the two systems yielded a 100% correct toxin classification thus confirming the hypothesis that a combination of two orthogonal sensing systems facilitates better classification of the toxins by reducing the probability of false positive and false negative readouts.

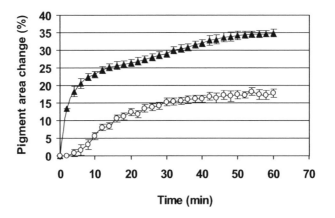

FIGURE 32.11 Change in pigment area (%) over time for adenylate cyclase from *B. pertussis* and MSH (-▲-) MSH 10 nM, (-o-) 5 μg/ml adenylate cyclase. Data are mean values of three measurements ± standard deviation. (From Dierksen, K., Mojovic, L., Caldwell, B., Preston, R., Upson, R., Lawrence, J., McFadden, P., and Trempy, J., *J. Appl. Toxicol.*, 24 (5), 363–369, 2004. With permission.)

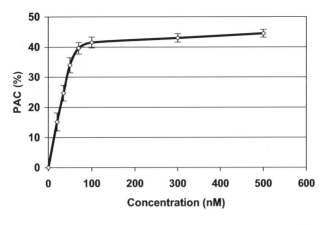

FIGURE 32.12 Effect of clonidine concentration on PAC of immobilized chromatophores. Data presented are mean values of three measurements ± standard deviation. (From Mojović, L. and Jovanović, G.N., *Chem. Ind.*, 57 (12), 605–610, 2003. With permission.)

Modeling of various toxin responses includes specifying and defining characteristic model parameters for different toxins. Linking these parameters to physiological and metabolic processes in cells may help to elucidate very complex mechanisms of cell reactions to toxins. Creation of a database of toxin responses may also assist in toxin identification. If, for example, two agents give similar exponential cell area decrease, specific parameters from Equation (32.3) are expected to be characteristic for the two toxins, thus providing a degree of discrimination. It is important to note that parameters A_0 and A_∞ may be evaluated independently from experimental observations, thus reducing the five-parameter model to only three degrees of freedom.

2. Effect of Toxin Concentration

Effect of toxin concentration on percentage of overall pigmented area change (PAC) from Equation (32.5) is presented in Figure 32.12. Percentage of area change increased for the clonidine concentration range from 20 to 100 nM, achieving a plateau value for higher toxin concentrations. As shown in Figure 32.12, concentration of 20 nM caused a significant response of fish chromatophores, 15.2% of area change. However, detection of lower concentrations is accompanied with higher error than detection of higher concentrations.

VI. CONCLUSION

Development of a micro-biosensor based on immobilized living chromatophores of Siamese fighting fish, *B. splendens*, for detection of microbial and environmental toxins was investigated in this study. Fish chromatophores were immobilized on ferromagnetic gelatin microbeads ($d = 250$ μm). Optimum conditions for the immobilization of fish chromatophores and optimum amount of ferromagnetic material incorporated in gelatin microbeads allowing good cell viability and toxin sensitivity was observed.

Movement of microcapsules with immobilized chromatophores within microchannels, and its capturing with magnetic field were studied in the range of predicted operational fluid velocities of biosensor from 1.6 to 6.4 mm/s. For these velocities, magnetic field needed for capturing and positioning of microcapsules is mostly dependent on the amount of ferromagnetic material, and was in the range from 512 to 1080 A/m.

The response of immobilized chromatophores to neurotoxin clonidine was monitored by measuring cell area covered by pigment. Percentage of area change is dose-dependent for this model toxin in a range of concentrations from 20 to 100 nM, achieving a plateau value for higher concentrations. The cell area decrease is shown to fit very well the proposed double exponential model, and the rate of cell reaction to various toxins may be described with model coefficients.

Currently, the testing of biosensors for a number of microbial toxins, biological agents, and pollutants is in progress, creating a large library of responses. Classification of agents according to specific features of cell responses, quantified by model parameters, is in progress. Further system development, miniaturization, and integration are expected.

ACKNOWLEDGMENTS

The authors thank all members of the Bio-MECS Team at Oregon State University for their contributions to the material cited herein. This work is supported by a grant from US National Science Foundation (BES-9905301).

REFERENCES

1. Fujii, R., The regulation of motile activity in fish chromatophores, *Pigment Cell Res.*, 13 (5), 300–319, 2000.
2. Fujii, R., Cytophysiology of fish chromatophores, *Internal. Rev. Cytol.*, 143 (2), 191–255, 1993.
3. Reese, E. and Haimo, L., Dynein, dynactin, and kinesin II's interaction with microtubules is regulated during bi-directional organelle transport, *J. Cell Biol.*, 151 (1), 155–166, 2000.
4. Danosky, T.R. and McFadden, P.N., Biosensors based on the chromatic activities of living, naturally pigmented cells, Biosens. *Bioelectron.*, 12 (9–10), 925–936, 1997.
5. Chaplen, F.W., Upson, R., McFadden, P.N., and Kolodziej, W.J., Fish chromatophores as cytosensors in a microscale device: detection of environmental toxins and microbial pathogens, *Pigment Cell Res.*, 15 (1), 19–26, 2002.
6. Mojovic, L., Dierksen, K., Upson, R., Caldwell, B., Lawrence, J., Trempy, J., and McFadden, P., Blind and naive classification of toxicity by fish chromatophores, *J. Appl. Toxicol.*, 24 (5), 355–361, 2004.
7. Dierksen, K., Mojovic, L., Caldwell, B., Preston, R., Upson, R., Lawrence, J., McFadden, P., and Trempy, J., Responses of fish chromatophore-based cytosensor to a broad range of biological agents, *J. Appl. Toxicol.*, 24 (5), 363–369, 2004.
8. Carlyle, C.A., Svejcar, M., Skinner, M.M., Preston, R.R., Jamerson, N.S., Fisher, C., Sellers, D., Robinson, D., Dierksen, K., Caldwell, B.A., Chaplen, F.W.R., McFadden, P.N., and Trempy, J.E., A cell based biosensor assay for toxin activity detection of bacterial origin, in *Proceeding of the 102nd ASM General Meeting*, Salt Lake City, UT, May 19–23, American Microbiology Society, 2002.
9. Paul, K.B. and Terhaar, T., Comparison of two passive microvalve designs for microlamination architectures, *J. Micromech. Microeng.*, 10 (1), 15–20, 2000.
10. Mojovic, L., Upson, R., Willard, C., Chaplen, F.W.R., and Jovanovic, G.N., Immobilization of fish chromatophores onto gelatin-based microcarriers, in *Proceedings of the sixth World Congress of Chemical Engineering*, Shollcross, D., Ed., Melbourne, Australia, September 23–27, 2001; (CD edition) ISBN 0 7340 2201 8; AICHE, Melbourne, 2001; 1–7.
11. Mojović, L., Upson, R., Chaplen, F.W.R., Peterson, C.A., Plank, T., and Jovanović, G.N., Development of a novel microreactor with a capture dot technology, in Abstract of Papers Part 1, 221th National Meeting of American Chemical Society, San Diego, CA, April 1–5, American Chemical Society, Washington, DC, 2001 (BIOT 119).
12. Mojović, L. and Jovanović, G.N., Immobilization of fish chromatophores for use as a micro-biosensor for biological toxins, *Chem. Ind.*, 57 (12), 605–610, 2003.
13. van Wezel, A.L., Growth of cell strains and primary cells on microcarriers in homogeneous culture, *Nature*, 216 (110), 64–65, 1967.

14. Levine, D.W., Wang, D.I.C., and Thilly, W.G., Optimization of growth surface parameters on micro-carrier cell culture, *Biotechnol. Bioeng.*, 21 (5), 821–845, 1979.

15. Hu, W.S., Meier, J., and Wang, D.I.C., A mechanistic analysis of the inoculums requirement for the cultivation of mammalian cells on microcarriers, *Biotechnol. Bioeng.*, 27 (5), 585–595, 1985.

16. Croughan, M.S., Hamel, J.P., and Wang, D., Effect of microcarrier concentration in animal cell culture. *Biotechnol. Bioeng.*, 32 (8), 975–982, 1988.

17. Himes, V.B. and Hu, W.S., Attachment and growth of mammalian cells on microcarriers with different ion exchange capacities, *Biotechnol. Bioeng.*, 29 (9), 1155–1163, 1987.

18. Fleischaker, R., Microcarrier cell culture, in *Large Scale Cell Culture Technology*, Lydersen, B.K., Ed., Hauser Publishers, Munich, 1987, pp. 59–81.

19. Nillson, K., Methods for immobilizing animal cells, *TIBTECH*, 5 (4), 73–79, 1987.

20. Matsushita, T., Ketayama, M., Kamihata, K., and Funatsu, K., Anchorage-dependent mammalian cell culture using polyurethane foam as a new substratum for cell attachment, *Appl. Microbiol. Biotechnol.*, 33 (3), 287–290, 1990.

21. Zuhlke, A., Roder, B., Widdecke, H., and Klein, J., Synthesis and application of new microcarriers for animal cell culture: design of polystyrene-based microcarriers, *J. Biomater. Sci. Polym. Ed.*, 5 (1–2), 65–78, 1993.

22. Koller, M.R. and Paputsakis, E.T., Cell adhesion in animal cell culture: physiological and fluid-mechanical implications, in *Cell Adhesion Fundamentals and Biotechnological Applications*, Hjortso, A.M. and Roos, W.J., Eds., Marcel Dekker, Inc., New York, 1995, pp. 61–111.

23. Panina, G.F., Monolayer growth systems: multiple processes, in *Animal Cell Biotechnology*, Vol. 1, Spier, R.E. and Griffiths, J.B., Eds., Academic Press, London, 1985, pp. 211.

24. Nilsson, K. and Mosbach, K., Preparation of immobilized animal cells, *FEBS Lett.*, 118 (1), 145–150, 1980.

25. Nilsson, K. and Mosbach, K., Macroporous particles for cell cultivation or chromatography, US Patent 5,015,576 (May 14, 1991).

26. Pettman, G.R. and Mannix, C.J., Efficient serial propagation of W138 cells on porous micro-carriers (Cultispher GL), in *Animal Cell Technology: Developments, Processes and Products*, Spier, R.E., Griffiths J.B., and MacDonald, C., Eds., Butterworth-Heinemann, Oxford, 1992, pp. 508–510.

27. Nikolai, T.J. and Hu, W.S., Cultivation of mammalian cells on microporous microcarriers, *Enzyme Microb. Technol.*, 14 (3), 203–208, 1992.

28. Malaisse, W.J., Olivares, E., Belcourt, A., and Nilsson, K., Immobilization of pancreatic islet cells with preserved secretory potential, *Appl. Microbiol. Biotechnol.*, 52 (5), 652–653, 1999.

29. Ng, Y.C., Berry, J.M., and Butler, M., Optimization of physical parameters for cell attachment and growth on macroporous microcarriers, *Biotechnol. Bioeng.*, 50 (6), 627–635, 1996.

30. Warner, A., Duvare, S., Muthing, J., Buntemeyer, H., Lunodorf, H., Straus, M., and Lehman, J., Cultivation of immortalized human hepatocytes HepZ on macroporous Cultispher G microcarriers, *Biotechnol. Bioeng.*, 68 (1), 59–70, 2000.

31. Bugarski, B., Jovanović, G., and Vunjak, G., Bioreactor systems based on microencapsulated animal cell cultures, in *Fundamentals of Animal Cell Encapsulation and Immobilization*, Goosen, M.F.A., Ed., CRC Press Inc., Boca Raton, FL, 1993, pp. 267–296.

32. Mojović, L. and Jovanović, G.N., Development of a microbiosensor based on fish chromatophores immobilized on ferromagnetic gelatin beads, *Food Technol. Biotechnol.* 43 (1), 1–7, 2005.

33. Pacut, A., Kolodziej, W., and Chaplen, F.W., Cytosensors for early detection of biological and chemi-cal threats-statistical approach, in *Neuronal Networks and Expert Systems in Medicine and Healthcare*, Proceedings of the fourth International Conference — NNESMED, Milos Island, Greece, June 20–22, Technological Education Institute of Crete, 2001, pp. 437–442.

34. Lee, D.W., Gregory, D., Haddow, D.J., Piret, M.J., and Kilburn, D.G., High density BHK culture using porous microcarriers, in *Animal Cell Technology: Developments, Processes and Products*, Spier, R.E., Griffiths, J.B., and MacDonald, C., Eds., Butterworth-Heinemann, Oxford, 1992, pp. 480–486.

35. Maroudas, N.G., Sulfonated polystyrene as an optimal substratum for the adhesion and spreading of mesenchymal cells in monovalent and divalent saline solution, *J. Cell Physiol.*, 90 (3), 511–519, 1977.

36. Ruoslahti, E. and Pierschdocher, M.D., New perspective in cell adhesion. *Science*, 238 (4), 491–495, 1987.

37. Craik, J., *Magnetism Principles and Applications*, John Wiley & Sons, New York, 1995, pp. 1–72.

38. Griffiths, D.J., *Introduction to Electrodynamics*, Prentice Hall Inc, New Jersey, 1999, pp. 202–242.

39. Conan, J.F., Stability of the liquid-fluidized magnetically fluidized bed, *AICHE J*, 42 (5), 1213–1219, 1996.

40. Nilsson, H., Melanosome and erythrosome positioning regulates cAMP-induced movement in chromatophores from spotted triplefin *Grahamina capito*, *J. Exp. Zool.*, 287 (3), 191–198, 2000.

41. Jovanović, G.N., Vissvesvaran G., and Chaplen, F.W.R., Biosensors: microscale design, testing and orthogonality, *Chem. Ind.*, 58 (6a), 92–95, 2004.

Index